GEOMECHANICS AND GEODYNAMICS OF ROCK MASSES

**United Nations
Educational, Scientific and
Cultural Organization**

**International
Competence Centre for
Mining-Engineering Education**

EUROCK2018:
Geomechanics and Geodynamics of Rock Masses

VOLUME 1

Editor

Vladimir Litvinenko

Rector, Saint-Petersburg Mining University, St. Petersburg, Russia
Chairperson of the EUROCK 2018 Organizing Committee

CRC Press
Taylor & Francis Group
Boca Raton London New York Leiden

CRC Press is an imprint of the
Taylor & Francis Group, an **informa** business

A BALKEMA BOOK

CRC Press/Balkema is an imprint of the Taylor & Francis Group, an informa business

© 2018 Taylor & Francis Group, London, UK

Typeset by V Publishing Solutions Pvt Ltd., Chennai, India
Printed and bound in Great Britain by CPI Group (UK) Ltd, Croydon, CR0 4YY

Published by: CRC Press/Balkema
              Schipholweg 107C, 2316 XC Leiden, The Netherlands
              e-mail: Pub.NL@taylorandfrancis.com
              www.crcpress.com – www.taylorandfrancis.com

ISBN: 978-1-138-61645-5 (set of 2 volumes)
ISBN: 978-1-138-61735-3 (Vol 1)
ISBN: 978-1-138-61736-0 (Vol 2)
ISBN: 978-0-429-46177-4 (eBook, Vol 1)
ISBN: 978-0-429-46176-7 (eBook, Vol 2)

Geomechanics and Geodynamics of Rock Masses – Litvinenko (Ed.)
© 2018 Taylor & Francis Group, London, ISBN 978-1-138-61645-5

Table of contents

Physical and mechanical properties of fractured rock

Geophysics in rock mechanics

Rock mass strength and failure

VOLUME 2

Mineral resources development: Methods and rock mechanics problems

Rock mechanics in petroleum engineering

Geodynamics and monitoring of rock mass behavior

Preface

Distinguished participants of the Symposium!

"Geomechanics and Geodynamics of Rock Masses" contains the contributions to the International European Rock Mechanics Symposium EUROCK-2018 (Saint Petersburg, Russia, 22–26 May 2018), and covers all high-tech solutions and advanced engineering developments in the field of geomechanics, from theory to application.

Taking into account the rapid growth of industry requirements, there is an obvious need to strengthen research activities maximally, to provide professional assessment and build science-based predictions in various industrial sectors to ensure the economy running and the state development.

At the same time there is a growing role of quality engineering service delivery to enterprises, which would be characterized by high scientific content and would provide the commercialization of research. It is of primary importance for scientific developments in geomechanics.

Today, it is impossible to imagine the design and operation of a mining enterprise without geodynamic classification of a deposit, detailed analysis of the rock mechanics, and geophysical analysis of the massifs geological aspects.

In conditions of global intersectoral and interdisciplinary integration, certain economic sectors (complexes) will be increasingly influenced by scientific and technological achievements as well as by economic and scientific changes in related industries.

Geoinformation technologies provide plentiful opportunities for the labour productivity improvement in prospecting, exploration and mining of mineral resources.

Based on information and communication infrastructure, the advances in aerial and space means of Earth remote sensing and the Global Positioning System enable to form integrated systems for managing the processes of prospecting, exploration, extraction and transportation of mineral resources.

Looking ahead, the development of biotechnologies may have revolutionary effect on the demand pattern in the mining industry:

– substitution of natural petroleum for biomass-based fuel;
– biological underground metal leaching instead of hazardous chemical leaching;
– ecosystem recovery through bioremediation of soils and grounds contaminated by hydrocarbons.

The technology of Earth sensing is being actively used in order to decrease the cost of labour force and minimize the influence on biosphere.

Science-based technologies are being intensively developed for the following geophysical studies:

– nuclear-geophysical and seismoacoustics;
– abyssal geoelectrics;
– geodetic gravimetry;
– geothermics;
– electroprospecting based on the Earth's natural electrical fields.

The major discovered deposits are located in hard-to-reach places. They are characterized by high resource and energy intensity of their exploration and the creation of necessary infrastructure.

Today's oil production is very costly. For instance, 30 years ago over a hundred barrels were produced from 1 barrel by means of energy, but nowadays, 10 barrels are averagely produced from 1 barrel and this figure is increasing sharply.

The ratio may be 1:1 in the coming 30–50 years. This business is not appropriate.

According to the US Geological Survey for 2014, undiscovered resources of conventional hydrocarbons in the Arctic are around 90 bln barrels (13 bln. tonnes) of oil, 44 bln. barrels (6.5 bil tonnes) of gas condensate and 1669 trn cubic feet (47 trn cubic meters) of natural gas. It equals to 13% of total undiscovered oil resources in the world, 30% of total undiscovered natural gas resources and 20% of the world gas condensate resources.

Grand-scale exploitation of mineral deposits in the Arctic zone implies the application of effective environmentally sound technologies and their development requires significant investment.

Serious risks are associated with climate changes capable to cause destruction of mining sites in the Arctic and ecological disasters.

Modern research in the field of mineral processing is aimed at increasing ecological and economic efficiency of technological processes.

Another important tendency is the transition to the maximal level of fineness (dispersiveness) of separation thus providing more enhanced extraction of useful component from the ore. In that regard, new systems for grinding to micro size will be developed.

Introduction of new technologies of processing will result in reducing water loss by 15–20%, electricity and reagents—by 30–50%.

Extracting of useful components from mining waste is currently becoming feasible.

Every year, 17.4 bln. tonnes of solid and liquid wastes are accumulated as a result of mining and mineral processing. Most of them are solid mining wastes.

In many countries technological wastes are considered to be an important segment of the economy resource base.

The increased use of renewable energy sources will call for their supplement with systems of energy accumulation, and "flexible" hydro- and gas generation to provide reserve duplication in the absence of suitable conditions of power production from renewable sources.

Work is ongoing to develop solutions for enhancing economic and technological efficiency of renewable generation, in particular to reduce demand in rare earth metals when manufacturing generating equipment.

For example, the project "Smart Mine" funded by the EU aims to develop a variety of technologies and know-how to create automated mines with minimal human presence and zero impact on the environment.

The key characteristics of a "smart mine" are:

- the highest standards of personnel safety;
- absence of harmful effect on landscapes;
- low specific hydrocarbon emissions;
- high rate of useful component extraction from ore;
- maximum use of telemechanics (remote control of processes) which makes it possible to deploy a considerable part of personnel outside production zones, but to big cities, which are comfortable for living;
- storage of waste rock within the mine (without hoisting), or production of building materials and other useful things from mine wastes;
- higher energy efficiency.

Major well-known technologies of enhancing oil recovery are characterized by high cost, energy and material consumption. As a rule, their implementation is effective at the oil cost of more than $70 a barrel.

In recent decade tertiary methods of advanced oil recovery have been used actively in the world including thermal, gas, chemical and microbiological ones.

Thermal methods include steam treatment, the initiation of interbedding combustion, the use of thermogas, displacement of oil and asphaltenes by hot water.

Chemical methods are based on displacement of oil by other substances: surfactant solutions, polymers, thickening agents, froth systems, alkaline solutions, acids and compositions of chemical reagents.

By enhancing oil recovery through tertiary methods, the additional 110–130 mln. tonnes of oil equivalent (2.5% of the world production) per year are produced in the world.

According to expert estimates, from 3 to 7% of recovered oil is wasted in oil fields, and the most part of pollutants (up to 75%) is emitted into the atmosphere, 20% – into water sources, 5% – into soils.

In the context of global warming the risk of oil pollution of the sea is growing due to a decline in ice sheet uniformity, better conditions for oil spills drift, the increased likelihood of coast line oil contamination.

High rates of growth are forecast for the waste management market. The most in-demand technologies will be those of less resource intensity, the complex use of raw materials, the prevention of negative impacts of pollutants on the environment.

Huge social responsibility falls on rock mechanics engineers when operating hydrotechnical structures, subways, mining deposits under cities and water bodies. The level of mining operations gets deeper with each passing year, increasing the impact on the enclosing rocks; the built-up density of mining and civil property is growing. This significantly complicates the extraction and leads to the need to develop new unconventional technical, technology-intensive solutions.

The areas of development of technologies and scientific-and-technological advance in the mineral and raw materials complex for the foreseeable future will be:

1. Creation of the equipment for the development and production of non-conventional hydrocarbons will provide conditions for the commercial development of new deposits. The use of these technologies will predetermine a multiple increase in the volume of recoverable reserves, expansion of extraction geography, introduction of resources alternative to traditional oil and natural gas to the hydrocarbon raw materials market (gas hydrates, shale gas, heavy oil and bituminous sands, high-gas-bearing coal methane, etc.).
2. The role of systems and methods for enhancing the oil recovery factor will increase, including controlled changes in the reservoir characteristics on depleted hydrocarbon fields and low-pressure fields. This is a combination of technological solutions, instruments and systems for chemical and physical impact on hydrocarbon-containing reservoirs as a whole and their individual components (hydrocarbon rocks, hydrocarbons themselves, water, etc.), contributing to enhanced oil recovery. New technologies will help improve the efficiency of hydrocarbon production in existing fields, reopen those that were previously abandoned, and start developing difficult-to-recover reserves. In the long term this will significantly prolonge life of already known deposits and will delay the moment of depletion of industrial stocks of traditional hydrocarbon raw materials for decades.
3. Introduction of integrated and deep processing of mineral raw materials to increase the extraction ratio of both major and associated components in the fields will ensure a significant increase in the efficiency of minerals processing and reduction in the volume of generation of production waste.
4. The discovery of new genetic types of deposits, as well as the expansion of the geography of prospecting and exploration of mineral deposits, will lead to a change in the geography of the countries—exporters and importers, and increased market competition.
5. Development of sustainable technologies for complex ore benefication can lead to reduction of the minimum commercial content, which will make it advantageous to involve unpayable ores into processing, and will make it possible to profitably use waste from concentrating mills. In addition, the introduction of new inventions will help to reduce the level of environmental pollution, including minimizing the areas for storing and dumping waste in industrial areas, thus eliminating the risk of highly toxic compounds entering ground, sewage waters and the atmosphere.
6. Introduction of roughing equipment used on the edge of an open-pit mine or in a mine and operating on different principles – gravitational, magnetic, electrical, flotation, pulsed,

radiation and radiation-thermal—will significantly reduce the primary cost of processing by lowering the cost of ore transportation to concentrating mills.

The ability to think systematically, at an interdisciplinary level is of great importance today. Geomechanics closely interacts with other branches of science in its scientific foundations, methods and means: mathematics, geophysics, thermodynamics, mine surveying, geology, geotechnology. Therefore, the challenge of our time is the need to ensure constant cooperation of experts of various areas in order to solve complex tasks.

I am confident that St. Petersburg Mining University, which is in the process of establishing the International Competence Center in Mining Education under the auspices of UNESCO, will be able to start the realization of its main goal this year—the integration of leading world scientists and mining specialists under its roof for the fruitful exchange of experience and the design of trends in the development of the industry.

Strengthening the international capacity to develop sound policies in the field of higher education and sectoral science will help the world community to address the challenges of equity, quality, inclusiveness and mobility.

Integration of scientific branch interaction into the system of technical and vocational education and training of specialists in the mineral and raw materials sector of the economy will contribute to the formation of optimal control systems for the processes of prospecting, exploration, extraction, transportation and processing of minerals.

The forecast of scientific and technological development of the world mineral and raw materials complex takes into account the need to create conditions for uniting the efforts of the world community in solving the following problems:

1. Exploration work, including new production areas, meeting economic and environmental requirements; development of geophysical methods of oil and gas exploration in non-conventional geological conditions; evaluation of oil reservoir productivity; methods of searching for zones of possible ore manifestation.
2. Methods of enhancing oil recovery, including controlled change in reservoir properties of reservoirs, which allows to increase the hydrocarbon extraction factor, including depleted fields and low-pressure gas fields.
3. Obtainment and use of non-conventional sources of raw materials, including hydrocarbon, including heavy oil, gas hydrates, shale gas, etc.
4. Physical, technical and physicochemical technologies for processing high-gas bearing coal seams with the prevention of coalmine methane emissions, including for the purposes of gaseous and liquid synthetic hydrocarbons production.
5. Technologies for efficient processing of solid minerals, including energy-saving complex processing of refractory natural and technogenic mineral ore with a high concentration of mineral complexes.
6. Use of waste from extraction and minerals processing on a commercial scale.
7. Ecologically safe marine exploration and extraction of various types of mineral resources in the extreme natural and climatic conditions of the World Ocean, the Arctic and the Antarctic.
8. Technologies of seismic exploration in ice-covered water areas.
9. Technologies for ensuring the integrated safety of operations on the continental shelf of the Russian Federation, in the Arctic and the Antarctic, including monitoring and forecasting of natural and man-made emergency situations.
10. Prevention and containment of oil spills, primarily in ice conditions, including technologies for detecting oil under the ice sheet.
11. Advanced technologies for seismic prospecting.
12. Advanced technologies of oil and gas production.
13. Advanced coal mining technologies.
14. New technologies for deep conversion of oil and gas condensate.
15. Effective technologies for the use of petroleum associated gas.
16. New technologies for deep conversion of natural gas with the production of liquid motor fuels and a wide range of chemical products.

17. Next-generation technologies for deep processing of solid fuels with integrated use of mineral parts.
18. New technologies for efficient transportation of natural gas.
19. New functional coatings for pipelines.
20. New membrane materials with a given pore size.
21. New types of catalysts.

I hope that experts and scientists of the International Society for Rock Mechanics (ISRM) will be able to meet and cooperate within the framework of the International Competence Center, enriching its activities with the in-demand skills and knowledge and making use of its advantages.

I would like to note that only together we can create a high-quality institutional potential in the field of natural and engineering sciences in the interests of ensuring sustainable development of society.

I am confident that **EUROCK-2018** will bring large benefits to the scientific and technical cooperation and the development of geomechanics as a separate field in the mining industry and will enable its participants to exchange their experience and valuable knowledge.

Vladimir Litvinenko
Rector of St. Petersburg Mining University
Chairman of the EUROCK 2018 Organizing Committee

Committees

ORGANIZING COMMITTEE

Vladimir Litvinenko	*Rector of St. Petersburg Mining University, Russia Chairperson of the EUROCK 2018 organizing committee*
Igor Sergeev	*Vice-Rector for Scientific Work, St. Petersburg Mining University, Russia*
Vladimir Trushko	*Head of the Department of Mechanics, St. Petersburg Mining University, Russia*
Arkady Shabarov	*Director of the Scientific Center for Geomechanics and Mining Problems, St. Petersburg Mining University, Russia*
Anatoly Protosenya	*Head of the Department of the Construction of Mining Enterprises and Underground Structures, St. Petersburg Mining University, Russia*
Victor Rechitsky	*Secretary of Russian Geomechanics Association of ISRM, Russia*
Valery Zakharov	*Director of Institute RAS IPKON, Russia*

INTERNATIONAL ADVISORY COMMITTEE

Walter Wittke	*ISRM Past President*
Doug Stead	*ISRM Vice President for North America*
Norikazu Shimizu	*ISRM Vice President at Large*
Stuart Read	*ISRM Vice President for Australasia*
Luís Lamas	*ISRM Secretary General*
Petr Konicek	*ISRM Vice President at Large*
William Joughin	*ISRM Vice President for Africa*
Seokwon Jeon	*ISRM Vice President for Asia*
Sergio A.B. Da Fontoura	*ISRM Vice President for South America*
Eda Freitas De Quadros	*ISRM President*
Manchao He	*ISRM Vice President at Large*
Charlie Chunlin Li	*ISRM Vice President for Europe*

INTERNATIONAL SCIENTIFIC COMMITTEE

Leandro R. Alejano	*Spain*
Alexander Barjakh	*Russia*
Giovanni Barla	*Italy*
Nick Barton	*UK*
Erast Gaziev	*Russia*
John Hadjigeorgiou	*Canada*
Michael Zhengmeng Hou	*Germany*
Erik Johansson	*Finland*

Iurii Kashnikov	*Russia*
Heinz Konietzky	*Germany*
Sergey Kornilkov	*Russia*
Anatoly Kozyrev	*Russia*
Srđan Kostić	*Serbia*
Charlie C. Li	*Norway*
Sergey Lukichev	*Russia*
Larisa Nazarova	*Russia*
Erling Nordlund	*Sweden*
Frederic Pellet	*France*
Igor Rasskazov	*Russia*
Anatoly Sashourin	*Russia*
Dick Stacey	*South Africa*
Manoj Verman	*India*
Christophe Vibert	*France*
Ivan Vrkljan	*Croatia*
Mikhail Zertsalov	*Russia*
Zhao Zhiye	*Singapore*
Yingxin Zhou	*Singapore*

Key lectures

Geomechanics and Geodynamics of Rock Masses – Litvinenko (Ed.)
© 2018 Taylor & Francis Group, London, ISBN 978-1-138-61645-5

Advancement of geomechanics and geodynamics at the mineral ore mining and underground space development

Vladimir Litvinenko

Rector, St. Petersburg Mining University, St. Petersburg, Russia

ABSTRACT: The article provides findings for topical geomechanical and geodynamic problems of mining companies, the problems arising from deposit and underground space mining.

The prediction of geomechanical processes and the justification of method parameters of mining by open-pit, subsurface and combined methods in complicated mining and geological conditions of tectonically stressed mass and under confined aquifers were carried out.

Geomechanics of deposits development in the Arctic area is studied.

The methodology of geomechanically safe underground space mining is stated for metropolises.

Keywords: geomechanics, depression, solid mass, stress pattern, mine, strength, model, numerical method, surface

1   INTRODUCTION

Geomechanical problems in deposit mining at great depth and in complicated hydrogeological and geodynamical conditions require experimental and theoretical methods development for stressed-deformed state of a rock mass, nonlinear geomechanics of saturated complex porous and fractured rock masses development, creating of the mass geomechanical model in space and methods for geodynamical zoning of rock masses.

Based on the recent geomechanics and geodynamics advances, the following problems were solved: geomechanically safe ore deposit mining in complicated mining and geological conditions and creation of calculation method for stressed-deformed state of a mass around mines in tectonic stress field affected by seismic waves from rock shocks and bulk blasts; creation of prediction methods for geomechanical processes at the mining of gas-saturated blanket deposits; monitoring of geomechanical processes and geodynamical zoning of rock masses, underground space mining for metropolises in complicated engineering-geological and city-planning conditions.

The Mining University possesses a complex of advanced equipment, which allows solving the entire range of aims related to the study of rock mass properties and modelling of conditions similar to real mass conditions and wide varieties of load modes. The processes of loading, gathering and processing the information are automated and enable to carry out compression, tension and bulk strength tests of hard and half rock with simultaneous recording of a number of indicators (deformation, load, acoustic emission, speed of elastic waves, pore pressure) with a process visualization of crack formation and development. Values of lateral and pore pressure may vary from 0 to 50 Mpa, axial load—up to 4600 kN, temperature—from −15 to +200°C.

Scientific studies are conducted with special software complexes based on a finite-difference approach for solution of continuum mechanical problems and discrete elements method which are designed to solve fractured, or block medium. Modelling of a blade shield moving through a mass is introduced into practice.

The more is depth, the more intense solid masses are, and the tendency to dynamic display of rock pressure. All the ore deposits mined at great depth are classified as bump hazardous. That is why one of the main problems of rock geomechanics is a prediction of stress-strain behavior of a mass and justification of safe ways and field mining methods at great depth conditioned with active dynamic phenomena. The further progress is needed in scientific works on recording properties of real rock mass of block, layered and complicated geology, having tectonic faults, and any type of destructions being under geostatic pressure and effect of tectonic forces and movements.

2 PREDICTION OF GEOMECHANICAL PROCESSES AND JUSTIFICATION OF SAFE PARAMETERS AND COMBINED METHODS OF DEPOSIT MINING IN A TECTONICALLY STRESSED MASS

·OAO Apatit is active in the business of underground, surface and combined mining of apatite-nepheline ores from six deposits.

Nearby surface and underground mines, quarries, which are close to limiting zone, and open pits, where adjacent stocks are extracted, critically complicate mining activities. The stability of mine outcroppings is negatively impacted by redistribution of gravity-tectonic stress field in adjacent mass [1,2] and the change in strength and deformation properties of a mass affected by blasting operations at a quarry [1,3]. Extraction practice of a quarry adjacent stock by underground method demonstrates that at mining activity the energy of dynamic phenomena also quantitatively increases in rock masses, and rock tectonic bumps and induced earthquakes are also possible to occur [4,5]. A number of fundamental papers [6,7,8] are devoted to the study of geomechanical process at combined ore mining, but not all problems have been solved yet. It is required to ensure stability of permanent and development mines [9] carried out in tectonically stressed adjacent mass (Table 1).

The stope is studied in the impact area of the Koashvinsky quarry (Figure 1) conditioned with operations at OAO Apatit Eastern mine, at extraction of which the topslicing and ore sublevel caving is applied. Height of sublevel is 20 m, distance between drill-haulage mines is 18 m. Room works are carried out from the mass to the quarry transversely to the stretch of ore body.

Table 1. Acting values of tectonic stress in mass.

Item No.	Depth from the surface, m	Value of stress along the stretch of ore body, MPa	Value of stress transversely to the stretch of ore body, MPa
1	5	5	2,5
2	50	45	20
3	250	54	28
4	450	60	35
5	650	66	40

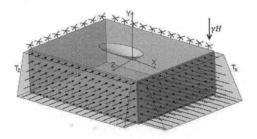

Design diagram of the finite element model

3D Model of the Koashvinsky quarry mass

Figure 1. Detail of finite-element model formed with account of the topographical relief and geological features of mass.

4

Finite element model is developed for a stope located at depth of 400 m from surface, on the distance from the open pit wall horizontally is 200 m (to the centre of block). At underground works the chosen block is the most complex in relation to the acting stress in solid mass (Figure 1).

All the rock formations in the model are given as heavy, homogeneous, isotropic, linearly deformed materials.

Boundary conditions are given as follows (Figure 1). On the edge of the model in direction of Ox and Oz axes, stresses are given obtained as a result of modelling of adjacent rock mass to the commencement of underground works.

The movements for opposite boundaries are prohibited along the corresponding directions. Boundary conditions along the top edge are given as distributed loads, values of which corresponds to the weight of the overlying rocks with the account of sub-extracted rock pressure formed by the moment of block extraction.

The algorithm for the problem solution includes five modelling stages: formation of stress and strain state of virgin mass, formed due to tectonic and gravity stress fields, with the effect of sub-developed rock pressure taken into account; development mines at all sublevels which are carried out by mass extraction within the boundaries of mines; fall of the first sublevel, which is implemented by assigning of decayed rock properties to the material within the boundaries of rock fall; fall of the second sublevel; fall of the third sublevel.

Thus, the model takes clearly into consideration the technology of consequential extraction of a sublevel with nominal idealization, i.e. a sequence of sublevel mining operations is not taken into account, in the same way as the stages of ore crushing, but ore fall is admitted along the area of the entire sublevel for a modelling stage.

The admitted idealization of the model allows analysing of stress and strain state of mass at the final stages of mines and fall of sublevels without analysis of stress development in the course of block caving.

Main maximal stresses are found in Oxy, Oxz, Oyz sectional planes cutting through the centre of the block and sublevel in order to ensure qualitative analysis of stress distribution within the considered model of the stope (Figure 2).

The analysis of stress distribution shows the stressed state of a mass has a critical cross-impact formed after putting the wall to the final position of gravity and tectonic stress field and bearing pressure from fallen ores and mass material.

Stresses are increasing with moving from the side wall to the foot wall and from ore mass to the caved mass.

In order to estimate quantitatively the stressed state of adjacent rock mass containing underground mines, changes in tangential stresses are recorded on the boundary.

The obtained data enable us to estimate quantitatively degree of impact from the open pit wall on stressed state in mass around the mines in adjacent rock mass. In roofs of the mines carried out perpendicular to the wall in the stress concentration area in the adjacent rock

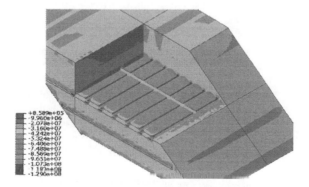

Figure 2. Distribution of main maximal stresses in a mass of the first sublevel with account of the quarry effect prior to room works.

mass, stresses are found to be 30–40% less than beyond the zone of effect, with compressive stress forming in mine walls within the zone of quarry effect, and tensile stress beyond the zone of effect.

Quite different, in terms of qualities, behaviour of stress distribution takes place for mines carried out parallel to the quarry. Increase of tangential stress concentrations is registered both on the boundaries: 1.5–1.6 more, and wall, as well as in the mine roof. At that, untypical change in stress values is registered per calculation stages, which may be explained by unsymmetrical formation of stress concentration zones on the mine boundary. Roofs and walls of mines, which are perpendicular to the open pit wall and mine beyond the zone of wall effect, are marked by 10–20% increase in stresses with the fall of overlying sublevels. Change in the stressed state in walls and roof is not regular in mines parallel to the open pit wall, which indicates the formation of stress concentration areas not in walls and roof but in points turned along the effect line of main stresses in adjacent rock mass.

3D model study of the stress and strain state of a mass around mines enables to reveal features of stress and strain state of a mass around mines in the zone of quarry effect, involving the enhanced concentration of compressive stresses in a roof of the mines perpendicular to the open pit wall and maximal component of tectonic stresses, and asymmetry of tangential stress distribution along the boundaries of mines parallel to the open pit wall and maximal component of tectonic stresses

Based on the calculation of a mass stress value, it is possible to conclude that the mines stability category decreases in the zone of quarry effect [10]. If, on the average, the second category of stability prevails for the block, then it becomes the third or the forth within the zone of quarry effect. The similar picture is observed in the areas of mine junctions and in the area of effect by room works. Therefore, in order to ensure safety of operations, type and parameters of supports for the mines within the zone of quarry effect shall be changed toward strengthening under III–IV stability categories.

3 CHOICE METHOD FOR GEOMECHANICALLY SAFE AND EFFICIENT PARAMETERS OF PILLARS

Current methods of parameter calculation for pillars do not regard the variety of interdependent factors related to the mine geological and engineering conditions of a deposit as well as technological features of applied mining methods. In particular, as it applies to apatitenepheline mines on the Kola Peninsula, it is needed to consider, among other factors, a nonuniform gravity and tectonic stress field, active in a mass, and for some cases – cross-effect of open and underground mining works.

To determine size of pillars, we suggest starting with the use of the existing experiential methods for finding parameters of chamber-and-pillar mining method based on the allowable chamber [4,5], and then calculating stressed state of pillars with the aid of 3D models with mining stages of ore body blocks. The method provides for defining the actual values of reserve strength ratios in pillars and, if their values are large—for finding [11] efficient parameters of pillars.

An option of the chamber-and-pillar mining method for Nyurpakkhskoye apatitenepheline deposit is considered for the depth of 400 m. Stress-strain behaviour of ores are admitted as follows: for ores: modulus of elasticity: $E = 61600$ MPa, Poisson ratio: $v = 0.28$, uniaxial compression strength for material: $R_{comp} = 150$ MPa; for enclosing rocks: $E = 64200$ MPa, Poisson ratio: $v = 0.25$, $R_{comp} = 200$ MPa. As failures in mass are typical for Nyurpakkhskoye deposit, this paperwork regards the medium fractured mass. Angle of incidence for the ore body is taken as 15°, its thickness – 40 m. The stressed state is defined in the context of pillar weight of overlying material and acting tectonic stresses and amounts to: vertical stress $\sigma_y = 10.59$ MPa; the maximal component of tectonic stresses, oriented along the stretch of ore body: $\sigma_{max} = 30.0$ MPa; the minimal component, oriented transversely to the stretch of ore body: $\sigma_{min} = 16.7$ MPa.

General view of the developed model is shown in Figure 3.

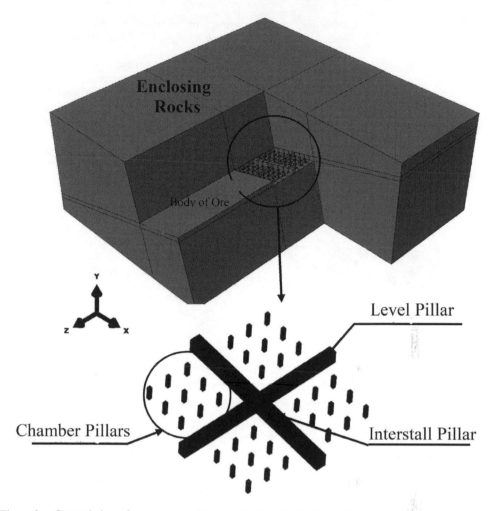

Figure 3.   General view of geometry and layout of pillars in the finite-element model.

Block mining modelling for the deposit is made in respect of main stages of its consequent extraction shown in Figure 4.

The relationship of stresses acting along the centre line of level and interstall pillars is determined at the first stage of modelling (Figures 5 and 6).

Analysis of dependences shows that in general only compressive stresses act in interstall and level pillars with demonstrating the asymmetric behaviour in the areas of intersection with pillars, perpendicular to them, which is determined by angle of incidence for ore body and the sequence of mining stages. A component, perpendicular to pillars, has the maximal stress values (for level pillars – σ_z; for interstall pillars – σ_x).

Growth in values of vertical stresses is typical for isolated interstall pillars with the opening of block chambers, at that growth is noted to attenuate with the removal of room works. Maximal vertical stress value amounts to 105.8 MPa at the final stage of the calculation model, which does not exceed the limit of strength for medium fractured mass equal to 150 MPa, and complies with the required actual reserve strength ratio for pillars.

In Table 2 actual reserve strength ratios are presented in pillars of various purposes for a base case estimated pursuant to allowable spans of mines [12, 13].

This calculation method of chamber-and-pillar mining parameters provides for the over-stated parameters of chamber-and-pillar mining and high strength reserve ratios.

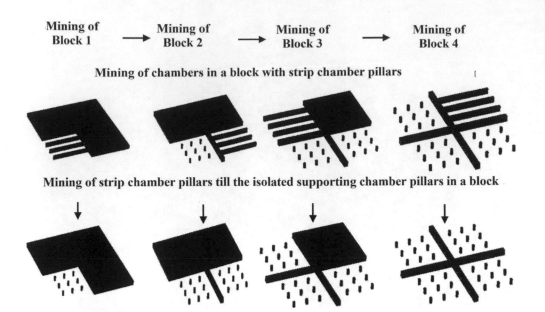

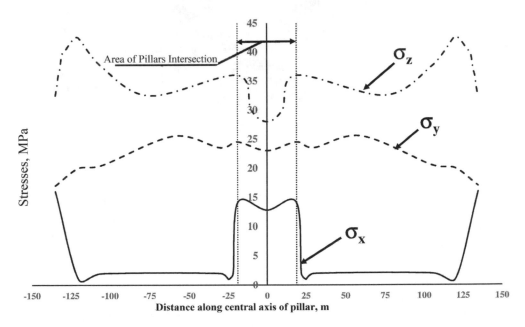

Figure 4. Stage sequence of block mining for ore body in finite-element model.

Figure 5. Relationship of stresses along central axes of a level pillar.

That is why let us consider 3 optimisation cases for the efficiency of pillar parameters: case 1 – decrease in width of the interstall pillar for 1 metre; case 2 – decrease in width of the level pillar; case 3 – decrease in width of interstall and level pillars for 1 metre. At that, the change of one of mining parameters results in the change of the stressed state of pillars in the whole system of pillars of various purposes.

The calculation demonstrates that in view of minimal reserve strength ratio equal to 2, it is appropriate to decrease the cross-section of insulated chamber pillars for 1 m and decrease the width either of a level, or an interstall pillar for 1 m.

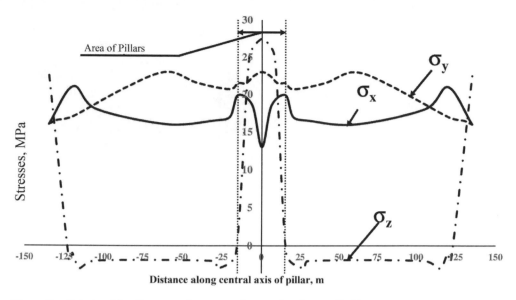

Figure 6.   Relationship of stresses along central axes of an interstall pillar.

Table 2.   Values of acting stresses and actual reserve strength ratios in pillars of various purposes.

Pillar	Direction of stresses	Compressive stress in pillar, MPa/Actual reserve strength ratio for a pillar			
		Base case	Case 1	Case 2	Case 3
Interstall	vertical	23.1/6.49	45.1/3.33	37.6/3.99	83.6/1.79
	along axis of a pillar σ_x	22.1/6.79	42.3/3.55	27.3/5.49	71.2/2.11
	transverse to a pillar axis σ_z	28.3/5.30	49.6/3.02	31.4/4.78	83.4/1.80
Level	vertical σ_y	25.9/5.79	36.8/4.08	56.2/2.67	94.5/1.59
	along axis of a pillar σ_z	42.1/3.56	61.2/2.45	71.2/2.11	137.4/1.09
	transverse to a pillar axis σ_x	16.0/9.38	28.1/5.34	36.1/4.16	72.4/2.07
Chamber	vertical σ_y	37.3/4.02	96.2/2.27	92.3/2.19	112.1/1.85

4   PREDICTION OF GEOMECHANICAL PROCESSES AT OPEN-PIT MINING

Automated algorithms are developed for analytic solutions of problems to define parameters of side slopes by various methods. The method of geometrical forces summation is developed which is based on introducing corrections to the method of algebraic forces summation. The engineering method is suggested for the calculation of slope stability in the 3D scene. The calculation findings, together with correlation of locations for more stressed surface of sliding obtained by the engineering method and those obtained by 3D numerical modelling show them to have a sufficient convergence (Figure 7).

The numerical modelling of an open pit walls and dumps is now becoming the norm at the analysis of their stability. The more complex geomechanical models applied for solving various problems allows for the more trusted findings.

Yet, backward calculations regarding the occurred deformations demonstrate that it is more reasonable to apply the finite element modelling method and Mohr-Coulomb criterion for the prediction of the slope deformation in the context of its stability when reserve ratios exceed the reference values. If the reserve ratio decreases further, physical modelling results should be used, see Figure 8. To extend the scope of application for the finite element

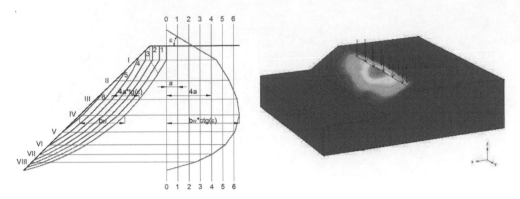

Figure 7. Instance of calculation of rectilinear slope stability by the ultimate equilibrium method and finite-element method.

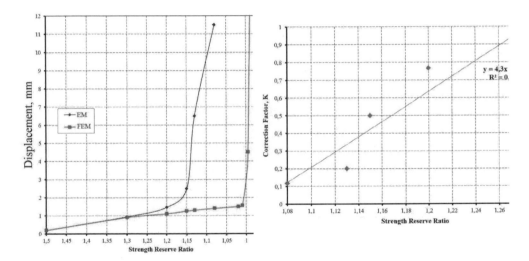

Figure 8. Development of deformations in a slope at various reserve strength ratios.

modelling it is advisable, depending on the slope stability degree, to introduce a correction into the modulus of deformation in accordance with the below proposed formula:

$$E_{\text{пр}} = E(4{,}3n{-}4{,}5),$$

where: n –strength reserve ratio;
   E – initial modulus of deformation.
   This formula is applicable at strength reserve ratio values from 1.08 to 1.3.
   In order to analyse impact from earthquakes on the stability of open pit walls and dump slopes, a method has been developed to enable taking into account not only intensity of seismic vibrations, but also magnitude at the same time (Figure 9).
   The application of physical modelling together with conventional calculation methods allows for the more accurate calculation of open pit wall parameters for the complicated engineering and geological conditions. Particularly, when stability of slopes is estimated with falling of mass in the range 65–90 degrees, existing reference documents suggest introducing a correction of about 10–15 degrees into the resulting angle of slope. However, at open-pit mining in areas with underdeveloped infrastructure the application of such a conventional solution may quite substantially affect the project profitability. The research, we carried out,

10

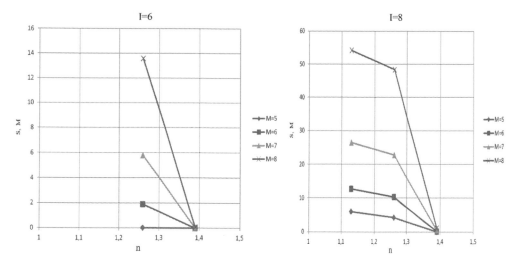

Figure 9.   Diagrams showing dependency of displacement for slopes with various initial reserve ratios at various intensity and magnitude of earthquakes.

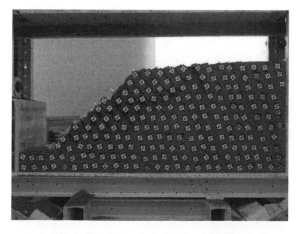

Figure 10.   Physical modelling of open pit wall, with layers falling towards mass within the range of 65–90 degrees.

shows such corrections should be introduced only at the very low strength along mine rock contacts and in severe hydrogeological conditions. At the same time, in permafrost rock conditions, the correction to the resulting angle amounts to around 5 degrees. The instance of physical modelling for the described instance is shown in Figure 10.

5   PREDICTION OF GEOMECHANICAL PROCESSES AT THE MINING IN THE COMPLICATED ENGINEERING, GEOLOGICAL AND HYDROGEOLOGICAL CONDITIONS UNDER CONFINED AQUIFERS

High degree of inundation is typical for Yakovlevskoye deposit. Ground water at the deposit is expanded in the sedimentary cover within which seven confined aquifers are distinguished. High-pressure Lower Carboniferous confined aquifer of 420–440 m head associated with limestone is located immediately above ore body, which deeper section displaces irregular clay layers. Yakovlevskoye deposit is outstanding regarding not only iron ore stock reaching

11

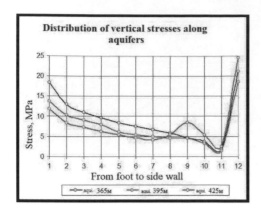

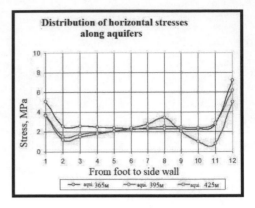

Figure 11. Diagram of vertical stress distribution.

Figure 12. Diagram of horizontal stress distribution.

65–70% iron, but also complexity of mining, geological and hydrogeological conditions. Tensile strength of high-grade iron ore in uniaxial compression does not exceed 1 MPa. A combination of high pressure of ground water and very low strength of ores greatly complicates the mining.

For the deposit development, a chamber mining is suggested with goaf stowing. Ore is extracted in layers from top to bottom. At the extraction under the Lower Carboniferous aquifer, a 65 m thick ore pillar is left in order to prevent an inrush of water from undrained confined aquifers overlying sedimentary.

Geomechanical bases [14, 15] are developed for friable iron ore mining in complicated conditions of mining and geology at the unique Yakovlevskoye deposit.

Figures 11 and 12 show the stress distribution in ore and crystalline mass prior to mining. Components of vertical and horizontal stresses are uniformly distributed and reach the largest values in contacts with enclosing rocks.

The untypical effect of ore body relaxation by harder and more solid enclosing rocks is disclosed.

Building of a concrete ceiling is suggested to ensure stability of the roofs for the first layer mining chambers. Loads on a protective ceiling, ore and stowing pillars are determined, with their parameters being found.

A regulation on complex hydrogeomechanical and hydrochemical monitoring is composed as a part of the system for monitoring, prediction and prevention of dangerous mining and geological processes development in underground mines.

5.1 *Geomechanics of deposits development in the Arctic area*

Prospects of deposits' development in the Arctic zone at the first stage require a systematic study of the properties of ice massifs. For these purposes ice physical and structural properties forming the Antarctic ice sheet in the area of Lake Vostok, which are characterized by evolutionary changes of the process of compaction and dynamometamorphism of ice rock layers have been studied. It is determined that variations of many structural characteristics for the depth are connected to climatic fluctuations which took place on ice sheet surface in the past [21].

The vertical ice profile (Fig. 13) shows that in depth intervals 20–3450 m grain size increased approximately 100 times: from 1 to 100 mm in diameter (estimated rock age at the depth of 3450 m) during 600 thousand years.

At the same time on the background of common tendency to the growth of ice crystal size with the increase of the depth fluctuations, which reveal clear correlation with the change of ice isotopic composition and concentration of aerosol impurities in it are observed.

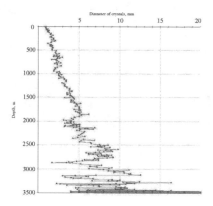

Figure 13.   Changing of ice grains size with depth.

The unknown earlier phenomenon that the cyclic variations on the surface of ice sheet, which are, in particular, caused by climatic factors, form the layered subparallel accumulation surface and differentiation of ice rheological properties has been discovered.

Ice sheet has the layered subhorizontal structure of the distribution of the material composition parameters, petrophysical, petrographic properties, as well as, respectively the velocity and direction of the ice masses flow [22].

Revealed dependences at the development of technology and drilling in the Antarctica ice sheet have been taken into account, this allowed to successfully penetrate in subglacial lake «Vostok» at the depth of 3769,3 m.

Achieved results were acknowledged by the world community as an outstanding scientific breakthrough and they can be used during the fields' development in the Arctic zone in future.

5.2   *Methodology of geomechanically safe underground development in metropolises*

The proposed methodology for geomechanically safe underground development in metropolises regards all stages of building and operation of an underground facility, including stages of the geological survey, designing of an underground facility, the construction of an underground facility, and operation of an underground facility.

Prediction methodology of geomechanical processes is based [16, 17, 18] on the following principles:

– representative and unbiased estimation of engineering, geological and hydrogeological conditions and features of soil mass,
– equations of state and laws of soil mass deformation, including those considering non-linear deformations of media and their creep,
– 3D geomechanical models of a mass [19], provided for the interaction of above-grade buildings and underground structures,
– stages of their construction,
– multi-variant numerical experiments allowing structural and technological modifications in construction facilities.

Mitigation of the impact caused by the underground construction on urban infrastructure, as well as building and structures located above the ground may be achieved by either the application of so-called low subsidence construction techniques for underground structures, or the application of the compensatory methods, or the reinforcement of the existing buildings and structures, or the combination of them all. The process of choosing the building protective measures may take considerable time and is of the iteration nature.

Let us consider the instance of the prediction of deformations of the land surface at the construction of complex 3D underground structure—transfer hub of two metro stations located in different altitudes, which includes 22 facilities (Figure 14).

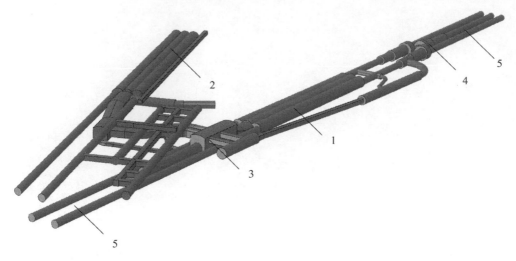

Figure 14.   Geometrical representation of space planning solution for the metro transfer hub: 1 – metro station No.1; 2 – metro station No.2; 3 – transfer hub; 4 – access track port; 5 – running tunnels.

The metro stations are initially divided in separate underground structures so that their cross-effect is maximally limited. After that, the sequence of construction for such underground structures is defined in accordance with a flowchart on the construction of a station. The local numerical models with fine detailed simulation of their construction method are developed for each structure. For long-stretching structures negligibly changing their configuration along their length, it is possible to consider the construction of part of the structure along the length only.

Then the numerical model is built for the entire metro station and ground mass (global model).

The global model represents the process of construction of the station in simplified form, with the most important stages of construction being distinguished, at that the predictions of ground surface deformations is carried out for the completion of each stage. Usually, the construction of one of the main station elements—station tunnel, port, auxiliary tunnel, connection area of tunnels and alike, is admitted as a stage of construction. Such an approach allows significant reducing time of calculation as the number of calculation stages does not usually exceed 15–20.

Pattern of land surface subsidence at different construction stages of the station is represented as diagrams at various enlarged stages of construction (Figure 15). Value of vertical displacements of land surface equaling to 1 mm is admitted as a border of the underground construction impact.

As is obvious from the represented results, pattern of trough of land surface subsidence extends with the construction of underground structure, affecting even greater area of land surface within the construction area of impact. The largest deformation of land surface forms over the areas with maximal concentration of underground structures, where displacement values reach 55 mm. The construction of large cross-section ports is to be given the special consideration, as those particular structures critically affect the pattern of land surface subsidence trough.

The obtained values of displacements for the underground structure rock boundaries and resulting displacements of land surface are calculated in the context of the accepted low subsidence construction techniques, all procedures of which are to be followed at actual works.

Based on obtained results of land surface deformation, the estimation of the obtained deformations permissibility is carried out, as well as development of mitigation measures of the underground construction impact on buildings located above the ground.

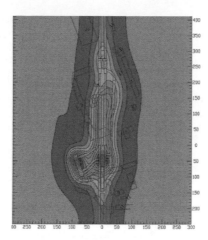

Figure 15.   Prediction of land surface subsidence (mm) after the construction completion of all underground structures.

This represented concept of the geomechanically safe underground development of metropolises has successfully approbated at 10 facilities of St. Petersburg and Moscow metro system. The obtained results demonstrate the represented method is able to provide the accident-free construction of complex underground structures having a complex space configuration, in complicated engineering and geological conditions.

REFERENCES

[1]   Kaplunov D.R., Kalmykov V.N., Kaplunov D.R., Rylnikova M.V. The Combined Geotechnology. Moscow: Publishing House «Ore and Metals», 2003, p. 560.

[2]   Enyutin A.N., Semenova I.E. On Stress-Strain State of Mass in Structural Elements in the Two-Stage Block Caving System at the Development of Pit Reserves. Challenges of Mineral Mining and Underground Development in North-West Russia, p. 3. – Apatity: Krasnoyarsk Research Centre, Russian Academy of Science Publishing, 2001. – pp. 49–55.

[3]   Rylnikova M.V., Kalmykov V.N., Meshcheryakov E.Yu. Practice of Adjacent Reserve Underground Mining of Uchalinsk copper-sulphide deposit. Mining Information-Analytical Bulletin – M.: MSMU, 1997. No. 3, p. 56–61.

[4]   Litvinenko V.S. Zubov V.P., Mikhaylenko O.V. Kholodninskoe Deposit: the Concept for Environmentally Safe Working, Main Technical Solutions, Prospects for Mining. Joural of Mining institute. 2012. Volume 196, p. 80–83.

[5]   Kozyrev A.A. Panin V.I., Maltsev V.A. System Approach to the Prediction and Prevention of Dynamic Phenomena in Mines. Mining Information-Analytical Bulletin – Moscow: MSMU, 2003. No. 12, p. 78–81.

[6]   Kozyrev A.A., Demidov Yu.V., Yenyutin A.N. and others. Geomechanical Support of Mining Operations Development Design for the Earlier Exhausted Space. Geomechanics at Mining Operations in Highly Stressed Masses – Apatity: Krasnoyarsk Research Centre, Russian Academy of Science Publishing, 1998, p. 25–37.

[7]   Litvinenko V.S., Boguslavsky E.I., Andreev M.N. Technology and Organization of Mining Works in Pit Reserves Extraction of Kimberlite Pipes of Yakutia in Complex Hydrogeological Conditions. Journal of the Mining institute. 2011. Volume 194, p. 79–83.

[8]   Kazikayev D.M. Geomechanical Processes at Ore Combined Mining and Remining. Moscow: Mineral Resources, 1989, p. 192.

[9]   Protosenya A.G., Kuranov A.D., Protosenya A.G. Prediction Method of Stress-Strain State of a Rock Mass at Combined Mining of Koashvinok Deposit Mining Journal No.1. Moscow: Publishing House «Ore and Metals», 2015, p. 17–20.

[10] Trushko V.L., Protosenya A.G., Matveyev P.F., Sovmen Kh.M. Geomechanics of Masses and Dynamics of Deep Ore Mines. Saint Petersburg: St. Petersburg State Mining Institute, 2000, p. 396.

[11] Protosenya A.G., Shokov A.N. Calculation of Pillar Parameters at Chamber-and Pillar Ore Mining with the Use of 3D models. News of Higher Education Institute. Mining Journal. 2015. No.11, p. 20–23.

[12] Methodology Guidelines on the Determining of Sizes of Chambers and Pillars at Chamber Non-Ferrous Metal Ore Mining. VNIMI, L.: 1972, p. 85.

[13] Petukhov I.M. and others. Calculation Methods in Mechanics of Rock Bumps and Outburst. Moscow, Mineral Resources, 1992, p. 256.

[14] Trushko V.L., Protosenya A.G., Plashchinsky V.F. Estimation of Stability of Outcrops and Calculation of Loads on Caving Supports at Yakovlevskoye Mine. Journal of the Mining institute. 2006. Volume 168, p. 115–122.

[15] Protosenya A.G., Trushko V.L. Prediction of Mine Stability in Low-Strength Iron Ores of Yakovlevskoye Deposit. Journal of Mining Science. 2013. No. 4, p. 49–61.

[16] Karasev M.A. Development of Nonlinear Elastic Transversely Isotropic model of Medium. Challenges of Geomechanics, Geotechnology, and Mine Survey. Journal of Mining institute. 2012. Volume 198, p. 202–206.

[17] Karasev M.A. Prediction of Land Surface Subsidence at Underground Construction of Deep-Laid Structures in urban conditions of St. Petersburg. Challenges of Geomechanics, Geotechnology, and Mine Survey//Journal of Mining institute. 2014. Volume 204, p. 248–252.

[18] Protosenya A.G., Karasev M.A., Verbilo P.E. The prediction of elastic-plastic state of the soil mass near the tunnel with taking into account its strength anisotropy. International Journal of Civil Engineering and Technology (IJCIET). Vol. 8, Issue 11, 2017, p. 682–694.

[19] Protosenya A.G., Ogorodnikov Yu.N., Demenkov P.A., Karasev M.A., Potemkin D.A., Kozin Ye.G. Mechanics of Underground Structures. 3D Models and Monitoring. Saint Petersburg: St. Petersburg Mining University-MANEB, 2011. – p. 355.

[20] Litvinenko V.S., Boguslavsky E.I., Korzhavykh P.V. Physical and Mathematical Modeling of Technological Parameters of the Horizone-Chamber Mining for Lower Horizon of the Gubkina Mine/Journal of Mining Institute. 2012. Volume 195, p. 115–119.

[21] Lipenkov V.Ya. Specific features of the Antarctic ice sheet structure in the area of station "Vostok" based on the results of petrostructural studies of ice core/V.Ya. Lipenkov, E.V. Polyakova, P. Dyuval, A.V. Preobrazhenskaya//Problems of the Arctic and Antarctic. 2007. Issue 76. p. 68–77.

[22] Vasilyev N.I., Dmitriev A.N., Lipenkov V.Ya. The results of drilling 5G borehole at the Russian station "Vostok" and the study of ice cores. – Journal of Mining institute. 2016. Volume 218, p. 161–171.

Geomechanics and Geodynamics of Rock Masses – Litvinenko (Ed.)
© *2018 Taylor & Francis Group, London, ISBN 978-1-138-61645-5*

Complete laboratory experimentation on hydraulic fracturing

Herbert H. Einstein & Omar Al-Dajani
Massachusetts Institute of Technology, USA

Bruno Gonçalves da Silva
*John A. Reif Jr., Department of Civil and Environmental Engineering, Newark College of Engineering,
New Jersey Institute of Technology, New Jersey*

G. Bing Li & Stephen Morgan
Massachusetts Institute of Technology, USA

ABSTRACT: Hydraulic fracturing is widely used to create new fractures or extend and open existing ones. However, what exactly happens in the field is not well understood because, in most cases, only indirect information in form of pumping records, microseisms and the in-situ stress field are known. The MIT Rock Mechanics Group has developed and used a unique test equipment, with which the hydraulic fracture propagation can be visually observed while acoustic emissions are simultaneously recorded. All this can be done under different far field (external) stresses and different hydraulic pressures and flow rates. Interestingly, it is also possible to observe how the hydraulic fluid moves in the fractures. This allows one to relate details of the fracturing process to the micro-seismic observations and the boundary conditions thus providing the complete information that the field applications cannot.

The testing equipment will be described first, followed by detailed descriptions of hydraulic fracturing experiments on granite and shale. These two rock types represent the typical usage of hydraulic fracturing: Granite for EGS (Engineered Geothermal Systems) and shale for hydrocarbon extraction.

1   INTRODUCTION

Hydraulic fracturing, often referred to as "fracking", is frequently discussed both in technical and general publications and this mostly in the context of petroleum engineering, specifically, hydrocarbon (oil and gas) extraction. It is discussed both because of its benefits by "unlocking" vast additional amounts of oil and gas, and because of its problematic environmental effects ranging from induced seismicity to groundwater contamination. Before getting into the details of why and how we want to better understand hydraulic fracturing, it is necessary to briefly review where hydraulic fracturing occurs or is used, which goes beyond hydrocarbon extraction.

Hydraulic fracturing can be a natural process, in which faults are (re)activated because of increasing pore pressures (Hubbard and Rubey, 1959; Zoback, 2010) or tensile fractures (joints) are caused by pore pressure changes (Secor, 1965; Pollard and Holzhausen, 1979; Engelder and Lacasette, 1990). Pore pressure changes, in turn, can be caused by tectonic processes or by temperature increase related to plutonic activity.

It appears that hydraulic fracturing in engineering has first been conceptualized and used in civil engineering in the 1920's in conjunction with grouting, where intentionally induced fractures filled with grout can be used to inhibit or reduce subsurface groundwater flow. Cambofort (1961) refers to this process as "claquage".

Hydrofracturing in the petroleum industry seems to have started in 1947 and is described in a paper by Clark (1947). At that time, the process involved injection of a high viscosity gel, which is then broken up to eventually allow oil to flow through the newly created fractures.

This is different from the process used today particularly in conjunction with unconventionals. These are very low (natural) permeability shales from which oil gas can only be extracted through an intense artificially created fracture network. Specifically, multiple fractures are induced from a horizontal borehole (well) using either the "plug and perf" or "sliding sleeve" method (see e.g. Shiozawa and McClure, 2014; Ahmed and Meehan, 2016; clearly a multitude of other detailed descriptions exist). The horizontal well is usually drilled in the direction parallel to the smallest principal natural horizontal stress; this facilitates the creation of hydraulic fractures perpendicular to the well and thus to the smallest horizontal stress. From this description, it is evident that one needs to know the in-situ stress field and much literature exists on how one can do this based on knowledge of the existing faulting regime (normal, reverse/thrust, strike-slip) (see e.g. Anderson, 1950; Zoback, 2010). Alternatively, or in addition, hydraulic fracturing often called mini-fracking is used to determine the stress field. This technique (see e.g. Fairhurst, 1964; Haimson, 1978) has been and is being also extensively used in civil and mining engineering.

Hydraulic fracturing or hydraulic stimulation of existing fractures is also used in EGS (Engineered Geothermal Systems). There the intention is to create a fracture system at depths greater than 5 to 6 km, circulate water or another fluid through the fractures to heat it to 180°C or greater and then transform the heat through a heat exchanger and power plant into electric energy (possibly using some of the heat also directly). The original version of EGS, called HDR (hot dry rock) to differentiate it from the well-established hydrothermal energy extraction, relied on the same hydraulic fracturing process as discussed for hydrocarbon extraction, i.e. creating artificial fractures in a largely unfractured rock mass (see e.g. Brown et al. 1999, 2012). EGS in contrast relies mostly on hydroshearing, i.e. the stimulation of existing fractures through pore pressure changes such that fractures displace in shear and produce greater apertures. It is interesting to note that this process is analogous to what has been mentioned above under natural geologic hydrofracturing processes, namely the activation of faults. The seismicity related to fault (re-)activation is one of the problems related to hydraulic fracturing and stimulation (see e.g. Cornet et al. 2007; Cornet, 2016; Jung, 2013; Mukuhira et al. 2013). See also NRC, 2013.

2 PROBLEM STATEMENT AND POSSIBLE SOLUTION

Several issues arise in the context of hydraulic fracturing. In the introduction above, the environmental effects (groundwater contamination, induced or triggered seismicity) have been mentioned. For this, one needs to know how hydraulic pressure produces new fractures or causes existing ones to displace in shear and/or open in tension. Hubbard and Willis (1957) produced path-breaking experimental and theoretical work to predict the creation of new fractures relative to existing stress fields. Many others have increased the understanding often using sophisticated experiments and models (Zoback et al.,1977; Rubin, 1983; Teufel and Clark, 1984; Cleary, 1988; Stoekhert et al., 2014; Stanchits et al., 2015; Lecampion et al., 2015). Similarly, fracture flow has received much attention including several studies by the National Research Council (NRC, 1996, 2015).

Somewhat related to the issue of environmental effects but particularly important for optimizing the gas/oil production from hydrocarbon reservoirs or heat in EGS is the creation of a specific fracture network in the ground. As mentioned earlier and discussed in the cited literature, one can theoretically relate fracture orientation to the in-situ stress field and use the observed injected volume of the fracking fluid to predict the overall extent of fractures. However, given the fact that one cannot directly see the created fractures the "observation" of fracturing is only indirect, namely, through measurement of the pressure-time behavior during the injection and observation of microseisms. Not surprisingly, researchers have tried to improve the knowledge by conducting increasingly sophisticated laboratory experiments and relate the results to reality through scaling laws (e.g. Bunger et al., 2005). Such laboratory experiments also contribute to fundamental understanding of the detailed mechanisms and thus form the basis of models.

Two types of pressurization devices were used. In Device 1 (shown in Figure 5) the entire face This approach to solving the problem is also what is presented below. We try to use known natural material, apply known external (far field) stresses, apply and measure hydraulic pressure in a variety of fracture geometries and, very importantly, observe the hydraulic fracturing process both visually and through acoustic emissions. This allows us to relate the visually observed fracturing process to the "indirect" observation of acoustic emission. While others, e.g. with CT Scanning (e.g. Kawakata et al., 1999) and transparent materials (e.g. Bunger et al., 2005) have attempted similar solutions our visual observations are unique.

3 EXPERIMENTAL SETUP

The principle of our experiments follows what was done in the past (Reyes and Einstein, 1991; Bobet and Einstein, 1998, 2002; Wong and Einstein 2009a, b) namely using prismatic blocks with pre-existing fractures, so called flaws, shown in Figure 1. The prismatic specimens are cut from cores or slabs of the particular rock, and the flaws are cut with saws (shale) or water jet (marble, granite). The specimen-preparation process is quite demanding and must be done with care (Al Dajani et al., 2017). Figure 2 shows the possible applied boundary conditions, namely uniaxial or biaxial external stresses without or with hydraulic pressure in the flaws. The equipment we use is shown in the photo of Figure 3 and the schematic of Figure 4. As mentioned above, what we do is unique with simultaneous visual and

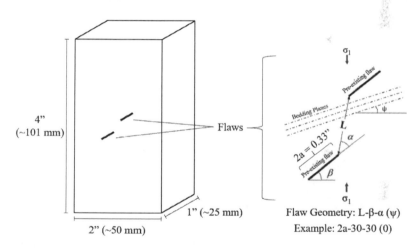

Figure 1. Typical prismatic specimen geometry with pre-existing flaws.—The flaw location and orientation can be varied.

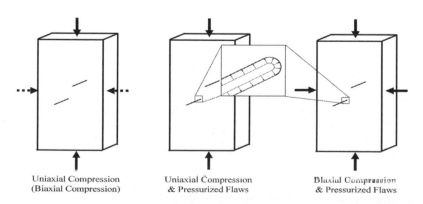

Figure 2. Possible loading conditions applied to prismatic specimen.

19

Figure 3. Hydraulic fracturing experimental setup: (a) Central data acquisition (b) Hydraulic fracture apparatus (PVA, LVDT, Pressure Transducer) (c) Lateral load (d) Axial load (e) High resolution images (f) High speed video (g) High resolution video (h) Load frame computer (i) High resolution camera computer (j) High speed video computer (k) Acoustic emissions system.

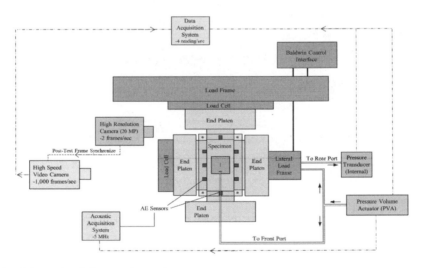

Figure 4. Schematic of the experimental setup.

AE observations. Specifically, the just mentioned boundary conditions are applied while we visually observe with high speed—and high resolution photography how new cracks emanate from the flaw(s), then propagate, and how this can lead to coalescence if there is more than one flaw. The photographic observations allow us to distinguish shear and tensile cracks and, if there is hydraulic fracturing, how the fluid flows in the newly created crack.

Two types of pressurization devices were used. In Device 1 (shown in Figure 5) the entire face is under water pressure while in Device 2 (Figure 6) only the flaws are pressurized under a local seal. In both cases the front is transparent to make visual observations possible, while the pressure is produced through a Pressure Volume Actuator. The placement of the AE sensors is somewhat different depending on the device, but a minimum of six sensors are placed as shown in Figure 7 while the schematic of the AE system is shown in Figure 8.

20

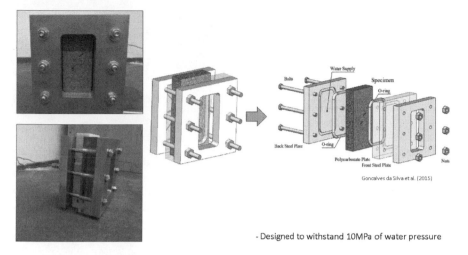

- Designed to withstand 10MPa of water pressure

Figure 5. Flaw pressurization device 1. Full-face pressurization; pressurizing fluid entering through the back.

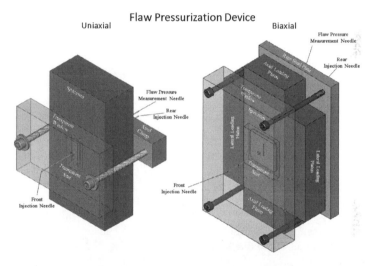

Figure 6. Flaw pressurization device 2. Local seal behind a transparent window; three "needles" are used to fill the flaw with the pressurization fluid and to measure the pressure in the flaw.
Note: The device is copyrighted by the authors.

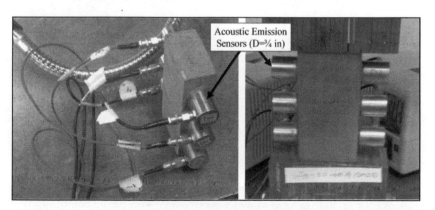

Figure 7. Placement of AE sensors.

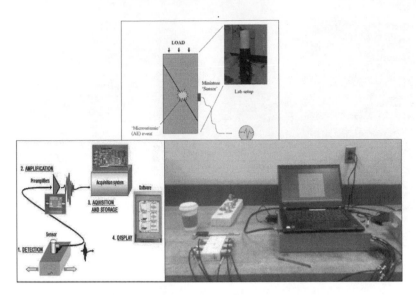

Figure 8. Schematic of AE recording system.

3 EXPERIMENTAL RESULTS

3.1 *Experimental evaluation process*

This is shown with the experiment on a single flaw inclined at 30° in granite subject of a constant vertical (uniaxial) stress of 5 MPa. Figure 9a, b show the time pressure/volume record. "Sketches" indicate when photos (either high resolution or high speed) were taken. They were then analyzed as in Figure 10, which also includes the evaluation process used in the analysis. Figure 10 shows the initial and final stages of the test the latter with a fully developed HF crack. Intermediate sketches such as Sketch 6 (Figure 11) show how the cracks develop and link up. Interesting is the fact that, as shown in Figure 11, so called "white patches" precede the development of visible cracks. White patches are the visual evidence of process zones consisting of microcracks as was described by Wong and Einstein (2009 a,b) and Morgan et al. (2013). The white patch and crack development can also be seen through acoustic emissions as shown in Figure 12, in which the events are categorized by amplitude. It is also possible to interpret the source mechanisms as will be discussed later. This example of HF propagation in an experiment shows that one can indeed relate visual and AE observations under specific external conditions.

3.2 *Granite*

Since hydraulic fracturing of granite (Barre Granite, see e.g. Morgan et al., 2013) was to some extent discussed above, the results presented in this section will be limited to summarizing the visually observed crack patterns and to aspects regarding acoustic emissions that have not been discussed before.

Gonçalves da Silva (2016) conducted an extensive series of hydraulic fracturing tests with Device 1 on double flaw geometries, in which the flaw inclination angle β (see Figure 1) was kept constant at 30° but with variation of the bridging angle α and applying either "0" external stresses or 5 MPa uniaxial stress. The observed cracking patterns are summarized in Figure 13 while the maximum water pressures are shown in Figure 14. Both for fundamental and practical purposes, it is interesting to know if hydraulic fractures can produce coalescence or not. As shown in Figure 13, coalescence does occur under all conditions if the flaws are significantly offset (bridging angle 60° or greater). For lower bridging angles, the occur-

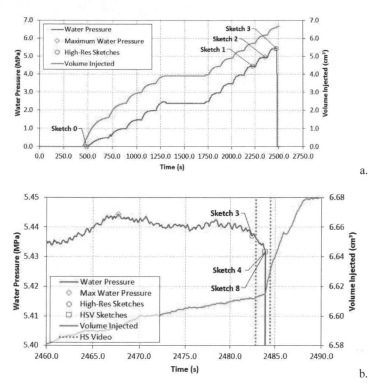

Figure 9. Typical pressure/volume—time behavior in hydraulic fracturing test. a. Entire experiment; b. Close-up of final stage. "Sketches" refers to images taken at that time (see Figs. 10 and 11).

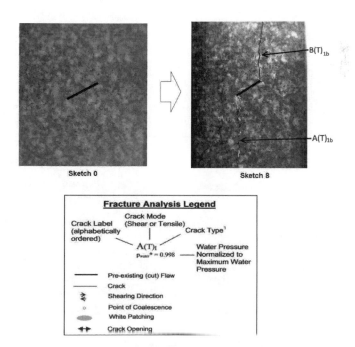

Figure 10. Images and superimposed sketches "0" and "8" of experiment shown in Fig. 9. The figure also shows the nomenclature used in the sketch.

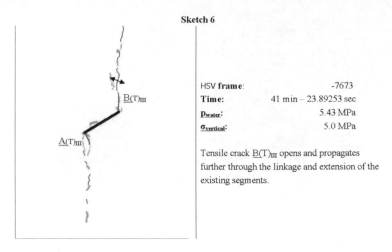

Sketch 6

HSV **frame**:	-7673
Time:	41 min – 23.89253 sec
P_water:	5.43 MPa
σ_vertical:	5.0 MPa

Tensile crack $\underline{B(T)}_{III}$ opens and propagates further through the linkage and extension of the existing segments.

Figure 11.   Detailed analysis of sketch "6" of experiment shown in Fig. 9. For Nomenclature, see Figure 10.

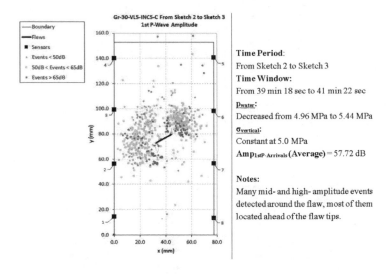

Figure 12.   Acoustic emissions of experiment shown in Fig 9.

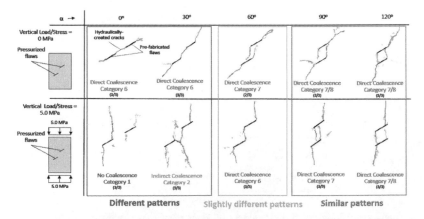

Figure 13.   Summary of hydraulic fracturing tests on granite—Induced cracking patterns.

24

rence and type of coalescence depend both on the flaw geometry and the external stresses. Given the scatter in Figure 14 (very low coefficients of determination) one can only say that the maximum water pressure increases if an external stress is applied.

The other item specifically related to granite is documented in Figure 15 with an interpretation of the acoustic events using moment tensor inversion for the same test as shown in

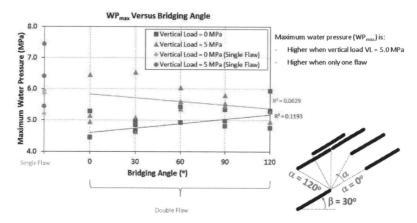

Figure 14.   Summary of hydraulic fracturing tests on granite—Maximum water pressure.

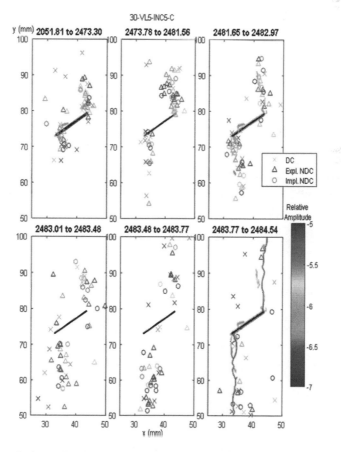

Figure 15.   Hydraulic fracturing tests on granite—interpretation of source mechanisms. DC = Double couple; NDC = Non- double couple explosion/implosion.

Fig 12. Interestingly, the source mechanisms obtained from AE show both tensile and shear mechanisms although the crack was visually identified as tensile ((Fig. 11). Investigations by Gonçalves da Silva (2016) show that the process zone (white patching) involves both mechanisms and some small-scale shearing in a crack that opens in tension.

3.3   *Shale*

Different types of shale, namely, Opalinus Chayshale from the Mt. Terri laboratory in Switzerland and Vaca Muerta shale from Argentina have been investigated. The Opalinus Clayshale comes from different locations in the Mt. Terri Lab and the facies therefore differ. Nevertheless, there is a reasonable consistency of the results (see Morgan, 2015). Opalinus Clayshale is not a source or reservoir rock. It was chosen because very carefully extracted cores were made available (courtesy Mt. Terri Laboratory and SwissTopo). Also, Al Dajani (2017) compared the behavior of Opalinus Clayshale to Vaca Muerta shale to make certain that the behavior of the two shales is comparable

The shale test series conducted so far differs from that on granite in that additional parameters were investigated, namely, the bedding plane orientations, application of biaxial external stresses and varied flow rates. All experiments were conducted with. pressurization Device 2 (see Figure 6). In tests with horizontal bedding planes under uniaxial external stress (Figure 16), it appears that when hydraulic fractures intersect bedding planes, flow occurs through some bedding planes before the hydraulic fracture continues, in other words, the hydraulic fracture is offset (Figure 17). The different types of HF crossings of bedding planes are shown in Figure 18. This behavior is reasonably well known (Fisher and Warpinski, 2011; Einstein, 1993) and is naturally very important since the crossing behavior contributes to the overall complexity of the induced fractures.

The effect of bedding plane orientation becomes more evident when conducting tests on specimens with differently oriented bedding planes (Figure 19) but even then, there is some stepping. Another observation is equally important: Figure 20 shows that the liquid (40 cp oil in this case) lags behind the crack tip. This behavior is also often cited in the literature (Christianovich et al., 1978; Daneshi, 1978; Feng and Gray 2017) but actually seeing it (and thus being able to measure the lag) in a natural material is unique.

When one varies the flow rate as shown in Figure 21a, the resulting HF crack pattern become increasingly complex with increasing flow rate. Simple energy relations explain this but additional tests need to be run to obtain accurate details. These flow rate experiments also involved AE readings which are shown in Figure 21b, and this shows that the just mentioned increasing crack complexity goes hand in hand with increasing number of events.

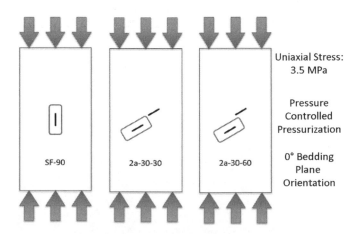

Figure 16.   Schematic of Hydraulic Fracturing Tests on Opalinus Clayshale with Horizontal Bedding Planes under Uniaxial Stress. Seal (blue outline) Indicates Pressurized Flaw.

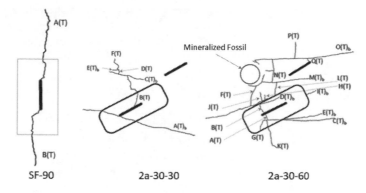

Figure 17. Cracking patterns induced by hydraulic fracturing in the tests of Fig. 16.

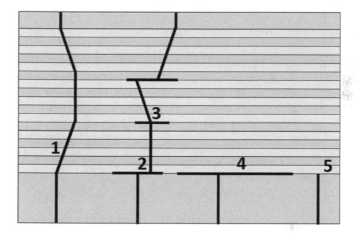

Figure 18. Schematic of hydraulic fracture crossing bedding planes.
1: direct crossing, 2: offset crossing with flow into bedding planes, 3: direct crossing with flow into bedding planes, 4: arrest with flow into bedding planes, 5: arrest.

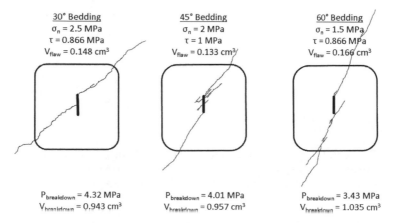

Figure 19. Cracking Patterns Induced by Hydraulic Fracturing in Tests on Opalinus Clayshale with a Single Flaw and Differently Oriented Bedding Planes.

Figure 20. Observation of Fluid Lag.

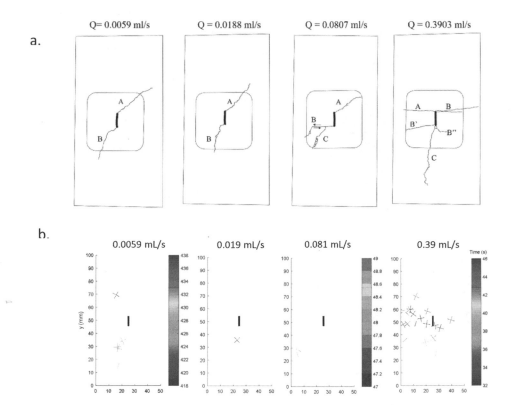

Figure 21. Hydraulic Fracturing in Tests on Opalinus Clayshale with Different Flow Rates; a. Cracking patterns; b. Source mechanisms (crosses = shear event; line = tensile event).

4 SUMMARY AND CONCLUSIONS

A unique experimental setup that allows one to observe the hydraulic fracture process both visually and with AE (acoustic emission) was developed at MIT. This setup was briefly described together with the parameters that can be controlled and the typical visual and AE observations obtained. The paper then shows that experiments on granite and shale produce results that relate the visual and AE observations; in other words, the underlying mechanisms can be explained. Such results not only provide fundamental understanding but also provide the basis for models that can eventually be upscaled to represent field conditions. Given that both granite and shale have been investigated, the applications will encompass both hydrocarbon extraction and Engineered Geothermal Systems.

ACKNOWLEDGEMENTS

The research on granite and the initial equipment development was sponsored by the US Dept. of Energy (Project Recovery Act: Decision Analysis for Enhanced Geothermal Systems) and TOTAL (Project MSGC—Multiscale Gasshale Collaboratory), while the shale work was sponsored by TOTAL (Project MSGC) and ARAMCO (fellowship). As mentioned in the text the Opalinus Clayshale was made available by the Mt Terri Lab of SwissTopo. The authors would like to express their gratitude for all this support.

REFERENCES

Ahmed, U., & Meehan, D.N. (Eds.). (2016). Unconventional Oil and Gas Resources: Exploitation and Development. CRC Press.

AlDajani, O.A. (2017). Fracture and Hydraulic Fracture Initiation, Propagation and Coalescence in Shale (Master's Dissertation, Massachusetts Institute of Technology).

AlDajani, O.A; Morgan, S.P; Germaine, J.T; Einstein, H.H. (2017 b). Vaca Muerta Shale—Basic Properties, Specimen Preparation, and Fracture Processes. 51st US Rock Mechanics/Geomechanics Symposium.

Anderson, E.M. (1951). The Dynamics of Faulting and Dyke Formation with Applications to Britain. Hafner Pub. Co.

Bobet, A.; Einstein, H.H. (1998). Fracture Coalescence in Rock-Type Materials under Uniaxial and Biaxial Compression. International Journal of Rock Mechanics and Mining Sciences, 35(7), 863–888.

Bobet, A.; Einstein, H.H. (2004). Crack Coalescence in Brittle Materials—Overview of Current Knowledge. Rock Engineering—Theory and Practice. Proceedings of the ISRM Symposium. W. Schubert (ed.): EUROCK 2004 and the 53rd Geomechanics Colloquy, Salzburg, Austria, pp. 475–478.

Brown, D.; DuTeaux, R.; Kruger, P.; Swenson, D.; Yamaguchi, T. (1999). Fluid Circulation and Extraction from Engineered Geothermal Reservoirs. Geothermics, 28(4), 553–572.

Brown, D.W.; Duchane, D.V.; Heiken, G.; Hriscu, V.T. (2012). Mining the Earth's Heat: Hot Dry Rock Geothermal Energy. Springer Science & Business Media.

Bunger, A.P.; Jeffrey, R.G.; Detournay, E. (2005). Application of Scaling Laws to Laboratory-Scale Hydrocarbon Fractures. Proceedings U.S. Rock Mechanics Symposium, Anchorage.

Cambefort H., (1961). L'Injection et ses problèmes, Bulletin Technique de la Suisse Romande, Vol. 87.

Christianovich, S.A.; Zheltov, Y.P. (1955). Formation of Vertical Fractures by Means of a Highly Viscous Fluid. In Proc. 4th World Petroleum Congress (Vol. 2, pp. 579–586).

Clark J.B. (1947). A Hydraulic Process for Increasing the Productivity of Wells, AIME Petroleum Transactions.

Cleary, M.P. (1988). The Engineering of Hydraulic Fractures State-of-the-Art and Technology of the Future. Jrnl. of Petroleum Tech. pp. 13–21.

Cornet, F.H. (2016). Seismic and Aseismic Motions Generated by Fluid Injections. Geomechanics for Energy and the Environment, 5, 42–54.

Cornet, F.H.; Bérard, T.; Bourouis, S. (2007). How Close to Failure is a Granite Rock Depth?. International Journal of Rock Mechanics and Mining Sciences, 44(1), 47–66.

Daneshy, A.A. (1978, February 1). Hydraulic Fracture Propagation in Layered Formations. Society of Petroleum Engineers.

Einstein, H.H. (1993). Modern Developments in Discontinuity Analysis—The Persistence-Connectivity Problem. In Comprehensive Rock Engineering. J. Hudson, ed. Pergamon Press.

Fairhurst C. (1964). Measurement of In-Situ Rock Stresses with Particular Reference to Hydraulic Fracturing Rock Mechanics Vol. 2 pp. 129–147.

Feng, Y.; Gray, K.E. (2017). Discussion on Field Injectivity Tests During Drilling. Rock Mechanics and Rock Engineering, 50(2), 493–498.

Fisher M.K.; Warpinski N.R., (2011). Hydraulic Fracture Height Growth Real Data, SPE Annual Techhnical Conference and Exhibition.

Gonçalves da Silva, B. (2016). Fracturing Processes and Induced Seismicity Due to the Hydraulic Fracturing of Rocks. MIT. Ph.D. Thesis.

Haimson, B.C. (1974), A Simple Method for Estimating In-Situ Stresses at Great Depths. Field Testing and Instrumentation of Rock ASTM STP 554, pp. 156–182.

Hubbard M.K., Willis D.G. (1957). Mechanics of Hydraulic Fracturing AIME Petroleum Transactions

Hubbard M.K.; Rubey, W.W. (1959). Role of Fluid Pressure in Mechanics of Overthrust Faulting, Bulletin of the Geological Society of America Vol. 70 pp. 115–166.

Jung, R. (2013). EGS—Goodbye or Back to the Future 95. In Effective and Sustainable Hydraulic Fracturing. InTech.

Kawakata, H.; Cho, A.; Kiyama, T.; Yanagidani, T.; Kusunose, K.; Shimada, M. (1999). Three-Dimensional Observations of Faulting Process in Westerly Granite under Uniaxial and Triaxial Conditions by X-ray CT Scan. Tectonophysics, 313(3), 293–305.

Lecampion, B.; Desroches, J.; Jeffrey, R.G.; Bunger, A.P.; Burghardt, J. (2015). Initiation versus Breakdown Pressure of Transverse Radial Hydraulic Fracturing, Theory and Experiments. Proc. Int'l. Congress of the ISRM.

Morgan, S.P.; Einstein, H.H. (2014). The Effect of Bedding Plane Orientation on Crack Propagation and Coalescence in Shale. In: Proceedings of the 48th U.S. Rock Mechanics/Geomechanics Symposium, Minneapolis, Minnesota.

Morgan, S.P.; Johnson, C.A.; Einstein, H.H. (2013). Cracking Processes in Barre Granite: Fracture Process Zones and Crack Coalescence. International Journal of Fracture, 180 (2), 177–204.

Mukuhira, Y.; Asanuma, H.; Niitsuma, H.; Häring, M.O. (2013). Characteristics of Large-Magnitude Microseismic Events Recorded During and After Stimulation of a Geothermal Reservoir at Basel, Switzerland. Geothermics, 45, 1–17.

National Research Council (2015) Fracture Characterization at Depth.

National Research Council. (1996). Rock Fractures and Flow. National Academies Press.

National Research Council. (2013). Induced Seismicity Potential in Energy Technologies. National Academies Press.

Pollard D.D.; Aydin A. (1988). Progress in Understanding Jointing over the Past Century. Bulletin of the Geological Society of America Vol. 100 pp. 1181–1204.

Pollard, D.D.; Holzhaisenn G. (1979). On the Mechanical Interaction between Fluid-Filled Fracture and the Earth's Surface Tectonophysics Vol. 53 pp. 27–57.

Reyes, O.; Einstein, H.H. (1991). Failure Mechanisms of Fractured Rock—a Fracture Coalescence Model. In: Proceedings of the 7th ISRM Congress, Aachen, Germany.

Rubin, M.B. (1983). Experimental Study of Hydraulic Fracturing in an Impermeable Material. Jrnl. of Energy Resources Tech. Vol. 105.

Sarmadivaleh, M.; Rasouli, V. (2015). Test Design and Sample Preparation Procedure for Experimental Investigation of Hydraulic Fracturing Interaction Modes. Rock Mech. and Rock Eng. Vol. 48.

Secor, D.T. (1965). Role of Fluid in Pressure Jointing, American Journal of Science Vol. 263, pp. 633–646.

Shiozawa, S.; McClure, M. (2014). EGS Designs with Horizontal Wells, Multiple Stages, and Proppant. In Proceedings of the 39th Workshop on Geothermal Reservoir Engineering, Stanford.

Stanchits, S.; Burghardt, J.; Surdi, A. (2015). Hydraulic Fracturing of Heterogeneous Rock Monitored by Acoustic Emission. Rock Mech. Rock Eng. Vol. 48.

Stoeckhert, F.; Brenne, S.; Molenda, M.; Alber, M. (2014). Hydraulic Fracturing of a Devonian Slate under Confining Pressure—with Emphasis on Cleavage Inclination. Rock Mech. and Rock Eng.

Teufel, L.W.; Clark, J.A. (1984). Hydraulic Fracture Propagation in Layered Rock: Experimental Studies of Fracture Containment. Jrnl. of Society of Petroleum Engineers. pp. 19–32.

Wong, L.N.Y.; Einstein, H.H. (2009a). Crack Coalescence in Molded Gypsum and Carrara Marble: Part 1. Macroscopic Observations and Interpretation. Rock Mechanics and Rock Engineering, 42 (3), 475–511.

Wong, L.N.Y.; Einstein, H.H. (2009b). Crack Coalescence in Molded Gypsum and Carrara Marble: Part 2. Microscopic Observations and Interpretation. Rock Mechanics and Rock Engineering, 42 (3), 513–545.

Zoback M. (2010), Reservoir Geomechanics, Cambridge University Press.

Zoback, M.D.; Rummel, F.; Jungi, R.; Raleigh, C.B. (1977). Laboratory Hydraulic Fracturing Experiments in Intact and Pre-Fractured Rock. Intl. Jrnl. Rock Mech. and Mining Sciences Vol. 14 (2).

Geomechanics and Geodynamics of Rock Masses – Litvinenko (Ed.)
© *2018 Taylor & Francis Group, London, ISBN 978-1-138-61645-5*

Strength and deformability of brittle polycrystalline materials in multiaxial stress-strain state: Rupture energy evaluation for brittle materials

Erast G. Gaziev
Center of Geodynamical Researches, Hydroproject Institute, Moscow, Russia

ABSTRACT: A phenomenological strength criterion for brittle polycrystalline materials (including rock) in multi-axial stress states is proposed. This criterion allows for the strength of materials to be assessed practically with any combination of principal stresses. It is based on the maximum distortion energy theory. It does not contain parameters which could require additional evaluation (excluding uniaxial compressive and tensile strength data). The results well agree with the available experimental data. This criterion can be used for assessment of strength of brittle polycrystalline materials, such as rocks and concrete, for strength analysis of concrete structures and their foundations by numerical methods.

Proposed criterion can also be used to assess the strength of the riprap. This conclusion is confirmed by numerous experiments conducted for a number of rockfill dams in Mexico, as well as to evaluate the strength of rockfill materials under uniaxial loading, which can be used as a classification parameter.

The process of material destruction is always associated with the release of the energy expended for loading.

The failure process of a material is always connected with the energy consumption. Work performed by external forces leads to the accumulation of energy by the constituting elements of the material structure; failure of the material suddenly releases part of the cumulative energy.

The researches carried out indicate that there is a close relationship between rupture energy and intensity of the stress applied at the moment of failure. This relationship is valid for both the uniaxial and triaxial strength tests confirming that during the failure process there is a simultaneous mobilization of the tensile and shear strengths whereas the rupture energy is the result of energy distortion.

Keywords: Principal stresses, strength criterion, intensity of stresses and deformations, stress-strain diagrams, rupture energy, strength of rockfill materials

1 INTRODUCTION

Determination of the strength of brittle polycrystalline materials, including rock, in a complicated stressed state is one of the fundamental problems of rock mechanics, which has not yet found a satisfactory practical solution. The destruction of such materials during combined loading, and especially brittle fracture, has been the subject of numerous theoretical and experimental studies in recent decades.

It was proposed many empirical strength criteria, each of which describes the strength within a certain range of loading the materials studied that points to the necessity of further studies in order to identify the nature of the destruction of such materials, including the study of micro- and macro-fracturing and deformation processes in the entire range of loading both in the pre-failure and post-failure zones.

The main feature of brittle rock behavior under triaxial loading is that in natural conditions, as a rule, it is deprived of the possibility of free lateral expansion and, therefore, with increasing one of acting principal stresses two others also increase. Moreover, this increase of the "lateral" principal stresses continues after "fracturing" of the material as a result of the increase of the coefficient of lateral expansion and tight conditions existing in the rock mass. Even a slight increase in the lateral compression causes the increases of the load-bearing capacity of the rock mass creating the so-called "hardening" effect.

Fig. 1 shows a typical stress-strain curve obtained by continuous recording of the stress and strain in uniaxial loading and breaking gypsum sample. It can be noted that the process of micro-destruction of the sample began almost from the very beginning of its loading, reaching the maximum value with the load reaching approximately 80% of the maximum compressive strength R_c, after which the macro-destruction has started.

The combination of the principal stresses that generate a sharp increase in the deformation of the sample with insignificant increase in the stress tensor shall be considered as the strength of the polycrystalline material subject to a triaxial stress state.

The state of stress at the moment of failure can be described by three principal stresses σ_1, σ_2, and σ_3, representing a point in the coordinate system of principal stresses.

If all the points, corresponding to the failure of the material are connected, we will obtain the strength surface in the coordinates of principal stresses:

$$f(\sigma_1, \sigma_2, \sigma_3) = 0 \tag{1}$$

It can be assumed that the strength surface will not exist under hydrostatic compression conditions ($\sigma_1 = \sigma_2 = \sigma_3$) when destruction is considered impossible. In addition, the condition for the hierarchy of principal stresses $\sigma_1 \geq \sigma_2 \geq \sigma_3$ requires consideration of the existence of a strength surface only where this condition is satisfied.

Numerous experimental studies of rock samples, cement and gypsum stones in 3-D stress state showed that the brittle materials under these conditions acquire the plastic properties and at the same time their strength increases significantly. At the same time there is an alignment of the lateral (second and third) principal stresses.

To "automatically" reproduce this process of increasing "lateral" stresses with increasing the first main stress in the Laboratory of Rock Mechanics of the Institute "Hydroproject", in 1982 a special installation was designed and manufactured (Fig. 2).

The sample was placed in a cage of steel plates connected in pairs with steel rods simulating the stiffness of the surrounding rock mass. The test was conducted with increasing a vertical compressive load imparted by a press (σ_1) with detent the lateral stresses (σ_2 and σ_3), and the acoustic emission in the sample body. At the same time the vertical and lateral deformations were measured.

The main advantage of this method of research is that, after reaching the limit state by the stress tensor, it continued to "slide" across the strength surface, thus making it possible to obtain not just a single point of the strength surface in one experiment as usually obtained in conventional tests with constant lateral stresses, but a series of consecutive points, i.e. the whole fragment of it (Fig. 3) representing a stress-strain diagram of the concrete specimen obtained with this equipment (Gaziev et al. 1984).

The moment when the stress tensor reaches the "strength surface" can be determined from the diagram of the sample deformation.

Fig. 4 shows photographs of the side faces of the sample as a result of the first destruction, and then the test has been continued with the fractured sample.

The studies have shown that after reaching the "strength surface" by the sample, followed by unloading and new loading, the sample deformation diagram returns to the same "strength surface" and continues to slide along it. This effect is explained by the fact that during the fracturing the sample lateral stresses σ_2 and σ_3 start increasing due to a sharp increase of the coefficient of dilatation (Poisson's ratio), as seen in Fig. 5, which shows the growth of lateral stresses of a sample of cement stone after the beginning of its destruction at 55 MPa.

2 RUPTURE CRITERION EVALUATION

As early as in 1855 E. Beltrami proposed to consider the amount of energy required for deformation as a criterion of strength of the material. However, this proposal has not been confirmed, since it is a well-known fact that during triaxial hydrostatic compression of a specimen, the material is capable of accumulating a huge amount of energy without evidencing visible indications of failure. Therefore, not all energy spent for deformation becomes determinant but only the component necessary for distorting the specimen.

This idea was advanced by J.C. Maxwell in his letter to W. Thomson of 1856: "I have strong reasons for believing that when [the strain energy of distortion] reaches a certain limit then the element will begin to give way". We see that Maxwell already had the theory of yielding which we now call the maximum distortion energy theory. But he never came back again to this subject and his ideas became known only after publication of Maxwell's letters (Timoshenko, 1953). It took many years until first M.T. Huber in 1904 and independently of him in 1913 von Mises (R. von Mises) came to the same idea and proposed a theory called the theory of maximum distortion energy of Huber-Mises (Timoshenko, 1953).

This theory that was proposed to describe the beginning of a plastic-type behavior of soft steel alloys is based on the fact that the limiting state starts when the so-called specific distortion energy becomes equal to a certain value. Several trials were made subsequently to develop a theory based on the distortion energy. One of such criteria describing the strength of a polycrystalline material subjected to a multiaxial stress state based on the same theory with a satisfactory accuracy of the results obtained was proposed by the author of this paper (Gaziev et al., 1984; Gaziev, 1996; Gaziev & Levtchouk, 1999).

Taking into account the considerable difficulties in obtaining a purely theoretical criterion of description of the failure process for such materials, the only reliable solution of this problem for developing the practicable criterion can be the creation of a phenomenological criterion that considers the most important factors involved in the material failure. The condition $\sigma_1 \geq \sigma_2 \geq \sigma_3$ must be satisfied.

For working out this phenomenological failure criterion it was assumed that the strength of brittle material is mainly conditioned by distortion energy of a sample. Then, the main parameters determining the strength of the rock under multiaxial stress state are the following:

– the first invariant of stresses:

$$J_1 = \sigma_1 + \sigma_2 + \sigma_3$$

– the second invariant of stress deviator:

$$J_2 = (\sigma_1 - \sigma_2)^2 + (\sigma_2 - \sigma_3)^2 + (\sigma_1 - \sigma_3)^2$$

– compressive strength (uniaxial compression) R_c;
– tensile strength (uniaxial tension) R_t (R_t as the strength parameter is always positive).

Further in this analysis, it will be assumed that the compression is positive and $\sigma_1 > \sigma_2 > \sigma_3$.

The following strength criterion for brittle materials under multiaxial complex stress state was proposed (Gaziev, 1996):

$$\frac{\sigma_* + m}{1 + m} = \left(\frac{\tau_* - m}{1 - m}\right)^n, \tag{2}$$

where all participating parameters are dimensionless:

$$\sigma_* = \frac{\sigma_1 + \sigma_2 + \sigma_3}{R_c}, \tag{3}$$

33

$$\tau_* = \sqrt{\frac{(\sigma_1 - \sigma_2)^2 + (\sigma_2 - \sigma_3)^2 + (\sigma_1 - \sigma_3)^2}{2R_c^2}} \qquad (4)$$

$$m = R_t/R_c. \qquad (5)$$

The expression for "n" was derived from experimental studies:

$$1.15 \leq n \leq 1.3 \qquad (6)$$

The proposed criterial relationship (2) describes the "strength surface" in the principal stress coordinates (Fig. 6).

The results of thorough investigations of Dr. M. Takahashi and H. Koide (1989), kindly sent at our disposal by Dr. Takahashi (studies of Shirahama and Izumi sandstones, Westerly granite, Yuubari shale and Yamaguchi marble), as well as the experimental results of Z.T. Bieniawski (1971), N.S. Parate (1969), G. Vouille and D. Laurent (1969) were used for verification of the criterion for different magnitudes of all three principal stresses (with $n = 1.3$)

To justify the proposed criterion for different combinations of principal stresses at failure, the linear dependence between the left and right parts of the equation (2) were used:

$$X = \frac{\sigma_* + m}{1 + m}$$

$$Y = \left(\frac{\tau_* - m}{1 - m}\right)^n.$$

These experimental results are presented in Fig. 7.

Good agreement of the criterion with experimental data is observed over a fairly wide spectrum of the principal stresses.

In all cases when there is no reliable tensile strength data, the suggested criterion allows for evaluating m value as a parameter of criterion equation (2), by processing the experimental data on triaxial stress state. Any simple mathematical method can be used.

Fig. 8 shows a diagram of strength of the siltstone enclosing underground structures of the Rogún hydropower station (Tajikistan). The strength was calculated in the same coordinates of equation (2) with $n = 1.15$ based on the results of triaxial tests performed in Turin (Dr. Barla) and Tehran (Samonian) (Rogún, 2015). Despite the fact that the studies were carried out in different laboratories in different series of siltstone samples, we can say that they are very well matched to the proposed criterion.

Later on with participation of the author in the Engineering Institute of the National Autonomous University of Mexico City a new installation has been created for testing the cubic samples measuring $15 \times 15 \times 15$ cm with independent application of the three principal stresses (Gaziev & Levtchouk, 1997). Photo of the installation is shown in Fig. 9.

For recording and analyzing the obtained experimental results, such as: the value of the applied load, the magnitude of deformation and acoustic emission occurring within the sample during its cracking, the program allows for simultaneous recording 32 analog signals at a rate of 10,000 samples per second (up to 100,000 samples per second when using only one channel). Measurements of the frequency and amplitude of acoustic emission was made by small-size microphone mounted directly on the load plate.

The program allowed for plotting the relationship between the measured parameters on the computer screen, as well as the dependence of these parameters on time producing the accumulation of information on the computer hard disk in "real time" for reconstructing the experiment and pre-processing the information (Gaziev & Levtchouk, 1997).

To obtain a "complete diagram" of deformation of the sample prior to and during the rupture a high-speed recording of the experimental results in real time was used. Such high-speed deformation recording made it possible to obtain the diagram shown in Fig. 1, as well as all diagrams $\varepsilon = f(\sigma)$ below.

Fig. 10 is a diagram of the concrete sample deformation (c.21) (Gaziev & Levtchouk, 1997) with initial values of confining stresses $\sigma_2 = \sigma_3 = 2.16$ MPa during 3 cycles of loading and unloading obtained in Mexico triaxial loading installation (Fig. 9).

The peak stress τ_i at the beginning of rupture can be determined from the criterion proposed, i.e.:

$$(\tau_i)_{cr} = R_t + (R_c - R_t)\left(\frac{\sigma_1 + \sigma_2 + \sigma_3 + R_t}{R_c + R_t}\right)^{\frac{1}{n}} \qquad (7)$$

To determine the moment when failure starts, the acting stress τ_i can be divided into its predetermined limiting value.

At the moment when $\tau_i/(\tau_i)_{cr} = 1$ the strength surface is reached and if the value of τ_i continues rising, the representative point "slides" for a certain time along this surface (the failure process in a three-dimensional state of stress) (Fig. 10).

Table 1 and Fig. 11 depict the data of a triaxial test performed on a cement stone specimen (c3). With the criterion proposed it is possible to determine the moment of rupture that occurred when the principal stresses reached the values of $(\sigma_1)_{cr} = 46.08$ MPa, $(\sigma_2)_{cr} = (\sigma_3)_{cr} = 0.98$ MPa. The corresponding strains became equal to $(\varepsilon_1)_{cr} = 0.00871$ and $(\varepsilon_2)_{cr} = (\varepsilon_3)_{cr} = -0.00345$. The magnitude of the critical stress was equal to $(\tau)_{cr} = 45.1$ MPa and that of the critical strain became $(\varepsilon)_{cr} = 0.012161$.

It can also be noted that the application of lateral pressure to the sample, which amounted approximately 2.5% of the uniaxial strength of the sample (0.98 MPa from 39.4 MPa) led to an increase in its strength by 17% (46.1 MPa instead of 39.4 MPa).

3 RUPTURE ENERGY EVALUATION

For analyzing the work performed by external forces it is convenient to operate with the so-called stress intensity (Bezukhov, 1961) that is determined from the following equation:

$$\tau_i = \frac{3}{\sqrt{2}}\,\tau_{oct} = \frac{1}{\sqrt{2}}\sqrt{(\sigma_1 - \sigma_2)^2 + (\sigma_2 - \sigma_3)^2 + (\sigma_1 - \sigma_3)^2} \qquad (8)$$

as well as from the deformation intensity that is in turn obtained from expression:

$$\varepsilon_i = \frac{1}{\sqrt{2}}\sqrt{(\varepsilon_1 - \varepsilon_2)^2 + (\varepsilon_2 - \varepsilon_3)^2 + (\varepsilon_1 - \varepsilon_3)^2} \qquad (9)$$

These two parameters are directly proportional to the square root of the second invariant of the deviatoric stress and strain tensor.

At the moment of failure the stress intensity $\tau_i = \tau_{cr}$. In the case of the uniaxial test τ_{cr} becomes equal to the unconfined compressive strength, $\tau_{cr} = R_c$, whereas for the «conventional» triaxial test (when $\sigma_2 = \sigma_3$) it assumes the value of the peak shear strength, $\tau_{cr} = (\sigma_1 - \sigma_3)$, just at the time when failure starts.

The work of the external distortion forces when failure starts or the rupture energy for a unit volume of the specimen can be calculated from,

$$E_{cr} = \int_0^{\varepsilon_{cr}} \tau_i(\varepsilon_i)\,d\varepsilon_i \qquad (10)$$

For a triaxial test, the moment when failure starts is determined by the phenomenological criterion (2).

Table 1 and Fig. 11 depict the data of the triaxial test performed on a cement specimen (c3).

Table 1. Concrete specimen (c3) triaxial test data.

σ_1 MPa	$\sigma_1 = \sigma_3$ MPa	ε_1	$\varepsilon_2 = \varepsilon_3$	ε_i	τ_i MPa	$(\tau_i)_{cr}$ MPa	$\tau_i/(\tau_i)_{cr}$
0.98	0.006	0.00011	−1.5E-05	0.000125	0.974	8.348	0.117
1.96	0.012	0.00022	−0.00003	0.00025	1.949	9.482	0.205
2.94	0.018	0.00033	−4.5E-05	0.000375	2.923	10.562	0.277
3.92	0.024	0.00044	−0.00006	0.0005	3.897	11.599	0.336
4.90	0.030	0.00055	−7.5E-05	0.000625	4.872	12.601	0.387
5.88	0.036	0.00066	−0.00009	0.00075	5.846	13.573	0.431
6.86	0.043	0.00077	−0.00011	0.000875	6.820	14.520	0.470
7.84	0.049	0.00088	−0.00012	0.001	7.795	15.445	0.505
8.82	0.055	0.00099	−0.00014	0.001125	8.769	16.349	0.536
9.80	0.061	0.0011	−0.00015	0.00125	9.743	17.236	0.565
14.71	0.092	0.00165	−0.00023	0.001875	14.614	21.459	0.681
19.61	0.123	0.0022	−0.0003	0.0025	19.485	25.409	0.767
22.55	0.144	0.00253	−0.00035	0.002875	22.405	27.687	0.809
23.53	0.147	0.002642	−0.00036	0.003002	23.382	28.426	0.823
24.51	0.157	0.002757	−0.00038	0.003133	24.353	29.168	0.835
25.49	0.167	0.002876	−0.00039	0.003269	25.324	29.905	0.847
26.47	0.186	0.002998	−0.00041	0.00341	26.284	30.650	0.858
27.45	0.196	0.003126	−0.00044	0.003561	27.255	31.374	0.869
28.43	0.206	0.003259	−0.00046	0.003722	28.225	32.093	0.879
29.41	0.225	0.003399	−0.0005	0.003897	29.186	32.821	0.889
30.39	0.235	0.003547	−0.00054	0.004088	30.157	33.530	0.899
31.37	0.255	0.003705	−0.00059	0.004299	31.118	34.248	0.909
32.35	0.275	0.003873	−0.00067	0.004539	32.078	34.961	0.917
33.33	0.294	0.004053	−0.00073	0.004785	33.039	35.669	0.926
34.31	0.314	0.004246	−0.00082	0.005066	34.000	36.373	0.935
35.29	0.343	0.004454	−0.00093	0.005379	34.951	37.085	0.942
36.27	0.373	0.00468	−0.00105	0.005725	35.902	37.794	0.950
37.25	0.392	0.004923	−0.00119	0.006108	36.863	38.484	0.958
38.24	0.431	0.005183	−0.00134	0.006527	37.804	39.197	0.964
39.22	0.461	0.005469	−0.00153	0.006999	38.755	39.893	0.971
40.20	0.500	0.005785	−0.00174	0.007524	39.696	40.598	0.978
41.18	0.559	0.006133	−0.00197	0.008102	40.618	41.325	0.983
42.16	0.608	0.006525	−0.00222	0.008747	41.549	42.035	0.988
43.14	0.686	0.00696	−0.0025	0.009457	42.451	42.780	0.992
44.12	0.775	0.00746	−0.00279	0.010253	43.343	43.533	0.996
45.10	0.882	0.00804	−0.00311	0.011151	44.216	44.307	0.998
46.08	*0.980*	*0.00871*	*−0.00345*	*0.012161*	*45.098*	*45.065*	*1.000*
46.64	1.049	0.0094	−0.00374	0.01314	45.588	45.512	1.002
46.62	1.765	0.0094	−0.00372	0.01313	45.441	45.662	0.996

The magnitude of the critical stress was equal to $\tau_{cr} = 45.1$ MPa and that of the critical strain became $\varepsilon_{cr} = 0.012161$.

The corresponding rupture energy resulted equal to $E_{cr} = 340$ kJ/m^3 (area of the hatched zone in the diagram in Fig. 11).

The results of several uniaxial tests carried out by various authors are presented in Table 2, whereas Table 3 contains the results of some triaxial tests performed at the Engineering Institute of the National Autonomous University of Mexico.

The diagram that relates to the rupture energy E_{cr} with the stress intensity τ_{cr} (or the unconfined compression strength in the case of uniaxial tests, R_c) is depicted in Fig. 12. A very reasonable relationship can be observed.

Table 2. Uniaxial strength tests.

The icon on the diagram in Fig. 12	Rock type	R_c, MPa	$(\varepsilon_i)_{cr}$	E_{cr} kJ/m³	Source
1	Diabase breccias	130–153	0.0143–0.0175	849–1139	A
2	Diabase and quartz	87–128	0.0121–0.0141	474–788	
3	Diabase	60–76	0.0102–0.0123	336–432	
4	Metadiabases breccias	90–95	0.0108–0.0121	416–528	
5	Metadiabases with quartz	53–58	0.0084–0.0105	217–265	
6	Metadiabase	30–35	0.0070–0.0094	157–187	
7	Quartz schist	40–45	0.0038–0.0089	191–282	
8	Chlorite schist	17–24	0.0071–0.0089	89–183	
9	Shale breccia	18–28	0.0115–0.0133	59–153	
S	Sandstone	143	0.0084	601	B
T	Tuff	44	0.0074	165	
DC	Diabase (Coggins)	341	0.00738	1531	C
B	Basalt (Lower Granite)	223	0.00761	1092	
c21	Concrete	51.7	0.0087	390	D
g	Gypsum	12	0.0029	19	

A. Pininska, J., Lukaszewski, P., 1991. The relationships between post-failure state and compression strength of Sudetic fractured rocks. Bulletin of the International Association of Engineering Geology, (43), 81–86.

B. Kawamoto, T., Saito, T., 1991. The behavior of rock-like materials in some controlled strain states. 7th International Congress on Rock Mechanics, Aachen (Germany), vol. 1, 161–166.

C. Miller, R.P., 1965. Engineering classification and index properties for intact rock. Thesis doctoral, University of Illinois, Urbana.

D. Gaziev, E., 2001. Rupture energy evaluation for brittle materials, International Journal of Solids and Structures, v. 38, pp. 7681–7690.

Table 3. Multiaxial strength testing.

The icon on the diagram in Fig. 12	Material	Stresses at rupture	τ_{cr} MPa	ε_{icr}	E_{cr} kJ/m³
c2	Concrete $R_c = 45$ MPa $R_t = 2.9$ MPa	$\sigma_{1cr} = 63.3$ MPa $\sigma_{2cr} = \sigma_{3cr} = 2.9$ MPa	43.35	0.00612	180
c3	Concrete $R_c = 39.4$ MPa $R_t = 2.7$ MPa	$\sigma_{1cr} = 46.1$ MPa $\sigma_{2cr} = \sigma_{3cr} = 1.0$ MPa	45.1	0.01212	340
c4	Concrete $R_c = 45$ MPa $R_t = 3.1$ MPa	$\sigma_{1cr} = 63.44$ MPa $\sigma_{2cr} = \sigma_{3cr} = 2.97$ MPa	60.47	0.00942	415
c5	Cement stone $R_c = 39.4$ MPa $R_t = 2.7$ MPa	$\sigma_{1cr} = 61$ MPa $\sigma_{2cr} = 4.65$ MPa $\sigma_{3cr} = 3.13$ MPa	57.14	0.01230	380

The fact that the rupture energy for both uniaxial and triaxial tests is described in terms of the same relationship based on the distortion energy evidences that the failure process of rock materials is induced by the joint action of normal and shear stresses. The normal tensile stresses develop the conditions necessary for the failure to occur (at a macroscopic level) under the action of shear stresses. The experimental work on the failure mechanism during shear performed with specimens on the models and in-situ showed that this process starts with development of tension-induced micro-cracks in the zone where the shear stresses will occur (Vouille & Laurent, 1969; Fishman & Gaziev, 1974; Pininska & Lukaszewski, 1991; Kawamoto & Saito, 1991). The same conclusion was derived by Martin and Chandler (1994) according to which tensile and shear strengths develop simultaneously.

4   ROCKFILL STRENGTH

For the case of triaxial compression tests of rockfill materials, when $R_t = 0$, equation (2) can be written in dimensionless form as

$$\frac{\sigma_1 + 2\sigma_3}{R_c} = \left(\frac{\sigma_1 - \sigma_3}{R_c}\right)^n. \tag{11}$$

The proposed expression for the rockfill strength evaluation well corresponds to the experimental results, and it is well supported by experimental data obtained for different types of rockfill materials. It is necessary to note, that R_c here is a virtual compressive strength of the rockfill, which cannot be determined directly from the experiment, however, can be used as a classification parameter for the rockfill material.

Consequently and based on the experimental data from two triaxial compression tests executed on the same material but with different confining stresses σ_3, the experimental values of R_c and of n can be calculated by means of the following expressions

$$n = \frac{\log \dfrac{(\sigma_1 + 2\sigma_3)_i}{(\sigma_1 + 2\sigma_3)_{i+1}}}{\log \dfrac{(\sigma_1 - \sigma_3)_i}{(\sigma_1 - \sigma_3)_{i+1}}} \tag{12}$$

$$\log R_c = \frac{1}{1-n}\left[\log(\sigma_1 + 2\sigma_3)_i - n\log(\sigma_1 - \sigma_3)_i\right] \tag{13}$$

in which the subscripts (i) and ($i+1$) correspond to tests (i) and ($i+1$).

On applying these expressions to the set of data on the triaxial compression tests carried out by R.J. Marsal (1977, 1980), A.A. Vega Pinto (1983), D. Marachi, C.K. Chan, & H.B. Seed (1972) it was found that the value of n can be estimated equal to the average value of 1.15.

Taking the magnitude of n as a constant value of 1.15, then

$$\frac{\sigma_1 + 2\sigma_3}{R_c} = \left(\frac{\sigma_1 - \sigma_3}{R_c}\right)^{1.15} \tag{14}$$

Presented in Fig. 13 comparison of this expression with the experimental results demonstrated its good accordance with these data.

5   CONCLUSIONS

1. The main feature of the behavior of brittle rock materials under triaxial loading is the limitation of its volume expansion. With an increase of one principal stress increases two others. Moreover, this increase of the "lateral" principal stresses continues during fracturing of the material as a result of increasing the coefficient of lateral dilatation and the current tightness in the rock mass. Even a slight increase in lateral compression increases the carrying capacity of the rock, so-called "hardening".

2. The strength of the polycrystalline material submitted to a triaxial state of stress can be considered as the result of the combination of the principal stresses that generate a sharp increase in the deformation of the sample.
3. The proposed phenomenological strength criterion allows for the strength evaluation practically for any combination of principal stresses in triaxial, as well as in biaxial stress states. It does not contain parameters which could require additional evaluation (excluding uniaxial compressive and tensile strength data).
4. The distortion energy spent by the rupture mechanism represents a parameter that is in good correlation with the strength and deformability characteristics of material. There exists a close relationship between the rupture energy and the intensity of the stress applied at the moment of failure. This relationship is valid for both the uniaxial and triaxial tests, therefore, confirming that the tensile and shear strengths are simultaneously mobilized and the rupture energy is determined from the distortion energy.
5. It has been proposed to use for analyzing the rupture energy of brittle polycrystalline materials the stress and deformation intensities expressed by equations (8) and (9).
6. The analytical expression (14) describing the strength of rockfill materials in triaxial stress state is well supported by experimental data obtained for different types of rockfills. Given the multiplicity of factors influencing the strength of rockfills, it might be preferable to use parameter R_c as a global index parameter of the rockfill material.

ACKNOWLEDGEMENTS

The author gratefully acknowledges the support for this work received from the Institute Hydroproject (Moscow, Russia), and the practical assistance in the experimental studies received from the collaborators of the Rock Mechanics laboratory of the same Institute.

The author expresses his deep gratitude to Professor Manabu Takahashi (Geological Survey of Japan) for his help in the noble providing the results of his experimental researches that played an important role in development of the presented strength criterion.

This research could not have been carried out without the involvement, although posthumously, of Professor Raúl J. Marsal (CFE, Mexico) who executed with immense talent and zeal the experimental work and who has inherited his data base and his enlightening publications. Analysis of the results of these studies as well as works of other authors formed the basis set in strength evaluation of rockfill materials.

In the analysis of the results of the triaxial tests of rockfill materials, an honorary researcher of the Engineering Institute of the National Autonomous University of Mexico, my friend and colleague Jesús Alberto Aramburu took part, whose bright memory I dedicate this report.

REFERENCES

Bezukhov, N.I., 1961. The bases of the theory of elasticity, plasticity and creep (in Russian). Editorial "Vysshaya Shkola" ("High School"), Moscow.

Bieniawski, Z.T., 1971. Deformational behaviour of fractured rock under multiaxial compression. In M. Te'eni (Ed), Structure, Solid Mechanics and Engineering Design, London, Wiley-Interscience, Part 1, 589–598.

Fishman, Yu.A., Gaziev, E.G., 1974. In situ and model studies of rock foundation failure in concrete block shear tests. 3rd International Congress of the ISRM, Denver (USA), vol. II-B, 879–883.

Gaziev, E., Morozov, A., Chaganian, V., 1984. Comportement expérimental des roches sous contraintes et deformations triaxiales. Revue Française de Géotechnique, Paris, (29), 43–48.

Gaziev, E., 1996. Criterio de resistencia para rocas y materiales frágiles policristalinos. Segunda Conferencia Magistral "Raúl J. Marsal", Sociedad Mexicana de Mecánica de Rocas, Mexico, 47 pp.

Gaziev, E., Levchouk, V. 1997. Study of the behavior of brittle polycrystalline materials in the post-failure stress-strain state (in Russian), XI Russian Conference on Rock Mechanics, St. Petersburg, p. 103–114.

Gaziev, E., Levtchouk, V., 1999. Strength characterization for rock under multiaxial stress states. 9th International Congress on Rock Mechanics, Paris (France), 601–604.

Gaziev, E., 2001. Rupture energy evaluation for brittle materials, International Journal of Solids and Structures, v. 38, pp. 7681–7690.

Gaziev, E., 2005. Rock foundations of concrete dams (in Russian). Editorial ASV, Moscow, 280 pp.

Kawamoto, T., Saito, T., 1991. The behavior of rock-like materials in some controlled strain states. 7th International Congress on Rock Mechanics, Aachen (Germany), vol.1, 161–166.

Marachi, D., Chan, C.K., & Seed, H.B., 1972. Evaluation of properties of rockfill materials, Journal of Soil Mechanics and Foundation Engineering, ASCE, Vol. 98, SM1, 95–114.

Marsal, R.J., 1965. Discussion, Proceedings of the 6th International Conference on Soil Mechanics and Foundation Engineering, Montreal (Canada), Vol. 3, 310–316.

Marsal, R.J., 1977. Research on granular materials. Rockfills and soil-gravel mixtures, Instituto de Ingeniería, Universidad Nacional Autónoma de México, Publicación E-25.

Marsal, R.J., 1980. Contribuciones a la mecánica de medios granulares, Comisión Federal de Electricidad, México.

Martin, C.D., Chandler, N.A., 1994. The progressive fracture of Lac du Bonnet granite. Int. Journal Rock Mech. Min. Sciences & Geomechanical Abstracts, 31(6), 643–659.

Miller, R.P., 1965. Engineering classification and index properties for intact rock. Thesis doctoral, University of Illinois, Urbana.

Parate, N.S., 1969. Critère de rupture des roches fragiles. Annales de l'Institut Technique du Batiment et des Travaux Publiques, Paris, (253), 149–160.

Pininska, J., Lukaszewski, P., 1991. The relationships between post-failure state and compression strength of Sudeten fractured rocks. Bulletin of the International Association of Engineering Geology, (43), 81–86.

Rogún HPP Construction Project. Powerhouse cavern complex., 2015. Siltstone rock characterization from laboratory tests, Appendices. Coyne et Bellier – Electroconsult Consortium.

Takahashi, M., Koide, H., 1989. Effect of the intermediate principal stress on strength and deformation behavior of sedimentary rocks at the depth shallower than 2000 m. Proceedings of the ISRM-SPE International Symposium "Rock at Great Depth", Pau (France), 19–26.

Timoshenko, S.P., 1953. History of strength of materials. McGraw-Hill Book Company, New York, Toronto, London, 368–371.

Vega Pinto, A.A., 1983. Previsâo do comportamento estructural de barragens de enrocamento, Laboratorio Nacional de Engenharia Civil, Lisboa.

Vouille, G., Laurent, D., 1969. Etude de la courbe intrinsèque de quelques granites. Revue de l'Industrie Minérale, Paris, Numero spécial, 15 juillet 1969, 25–28.

Geomechanics and Geodynamics of Rock Masses – Litvinenko (Ed.)
© *2018 Taylor & Francis Group, London, ISBN 978-1-138-61645-5*

Anisotropic and nonlinear properties of rock including fluid under pressure

Ian Gray, Xiaoli Xhao & Lucy Liu
Sigra Pty Ltd, Acacia Ridge, Queensland, Australia

ABSTRACT: This paper presents the mathematics and procedures to determine elastic rock properties from testing cylindrical core based upon orthotropic elastic theory. It also examines the extremely non-linear, but elastic, stress strain characteristics of some sandstones and what controls their Young's moduli and Poisson's ratios. The effects of fluid pressure changes within the rock are also considered. Two different modes of fluid behaviour are considered. The first is associated with poroelastic behaviour, while the second is associated with the effect of fluid within fractures. The use of these parameters leads to stress distributions and deformations that vary from those arrived at using conventional, but incorrect, assumptions of rock behaviour.

Keywords: rock, effective stress, Young's moduli, fluid pressure, rock, poroelastic

1   INTRODUCTION

Fluids may be thought to act mechanically within rock in two different ways. The first is by a poroelastic response to fluid pressure which affects the deformation of the rock mass. The second is by the direct action of the fluid within fractures. In each case the fluid pressure changes what may be considered to be the effective stress within the rock, but in different ways.

If we examine the general equation for effective stress within a rock mass it may be thought to follow the form of Equation 1 (Gray, 2017).

$$\sigma'_{ij} = \sigma_{ij} - \delta_{ij}\alpha_i P \qquad (1)$$

where: σ'_{ij} is the effective stress on a plane perpendicular to the vector i in the direction j.

σ_{ij} is the total stress on a plane perpendicular to the vector i in the direction j.

δ_{ij} is the Kronecker delta. If $i \neq j$ then $\delta_{ij} = 0$, while if $i = j$ then $\delta_{ij} = 1$.

α_i is a poroelastic coefficient affecting the plane perpendicular to the vector i. Its value lies between 0 and 1.

P is the fluid pressure in pores and fractures within the rock.

The Kronecker delta term is used because a static fluid cannot transmit shear.

The directional subscript indicating direction in the poroelastic coefficient is not usual practice where, for measurement reasons, only a scalar value is obtained.

If the choice of coordinates aligns with that of an open joint, then the poroelastic coefficient orthogonal to the joint is unity. More generally in a porous rock mass it lies somewhere between zero and unity. In a volcanic glass the values of the poroelastic coefficients are zero.

2 DEFORMATION AND THE DETERMINATION OF THE STIFFNESS MATRIX FROM ROCK CORE

The deformation of a rock mass is dependent upon the stresses it is subject to and its stiffness. If a rock is subject to a complex stress with six components of direct and shear stress it will deform to produce six strains. The correlation between stress and strain is a compliance matrix of 36 components. Determining all of these is practically impossible. The common simplification where the rock is treated as being isotropic, having a single value of Young's modulus and Poisson's ratio, is however frequently in error.

If we make a simplifying assumption that the rock mass is orthotropic then the compliance matrix has twelve components. These are shown in Equation 2 below.

$$
\begin{Bmatrix} \varepsilon_{11} \\ \varepsilon_{22} \\ \varepsilon_{33} \\ \gamma_{23} \\ \gamma_{31} \\ \gamma_{12} \end{Bmatrix} =
\begin{bmatrix}
\frac{1}{E_1} & -\frac{v_{21}}{E_2} & -\frac{v_{31}}{E_3} & 0 & 0 & 0 \\
-\frac{v_{12}}{E_1} & \frac{1}{E_2} & -\frac{v_{32}}{E_3} & 0 & 0 & 0 \\
-\frac{v_{13}}{E_1} & -\frac{v_{23}}{E_2} & \frac{1}{E_3} & 0 & 0 & 0 \\
0 & 0 & 0 & \frac{1}{G_{23}} & 0 & 0 \\
0 & 0 & 0 & 0 & \frac{1}{G_{31}} & 0 \\
0 & 0 & 0 & 0 & 0 & \frac{1}{G_{12}}
\end{bmatrix}
\begin{Bmatrix} \sigma_{11} \\ \sigma_{22} \\ \sigma_{33} \\ \tau_{23} \\ \tau_{31} \\ \tau_{12} \end{Bmatrix}
\tag{2}
$$

If we take a sample that is aligned perpendicularly to some plane of obvious symmetry then we have reduced the unknowns. An example of this is a core drilled perpendicularly to the bedding planes of a sedimentary rock. If we subject that core to testing in a triaxial loading rig which can apply a stress along the axis of symmetry of a round core sample and a confining stress perpendicular to that axis, we can measure the resulting axial and perpendicular strains with each stress increment. If there are axial strain gauges and at least three tangential strain gauges it is possible to calculate the major and minor tangential strains in addition to the axial strain. We can therefore derive three strains which can be assumed to represent the orthogonal cases.

Because we are dealing with principal stresses and strains the compliance matrix to be solved for the rock behaviour has therefore been reduced to nine components. It may be reduced still further to six unknowns because the matrix is symmetrical. This means that the off diagonal components are equal as shown in Equation 3. This symmetry provides a link between the values of Young's moduli and Poisson's ratios.

$$
\frac{v_{ij}}{E_i} = \frac{v_{ji}}{E_j}
\tag{3}
$$

However three principal strains and two loading cases does not provide an adequate basis to determine the six unknowns. To determine the three Young's moduli and associated Poisson's ratios it is necessary to assume something. The assumption that we have made to determine the values of Young's moduli and Poisson's ratios is shown in Equation 4. Here the geometric mean Poisson's ratio, v_a, is assumed to have the same value for the three combinations of v_{ji}.

$$
\sqrt{\left(v_{ij}v_{ji}\right)} = v_a
\tag{4}
$$

In a triaxial situation the axial load, in the 1 axis, may be changed and the associated strain changes measured. From this loading and strain measurement the axial Young's modulus, E_1, and the two Poisson's ratios, v_{12} and v_{13}, may be directly determined. However when the confining stress is changed, the situation is more complex to analyse because the 2 and 3 axis loadings are the same and applied simultaneously. Equation 5 describes strain in an elastic solid under three varying stresses.

$$\Delta\varepsilon_i = \frac{1}{E_i}\Delta\sigma_i - \frac{v_{ji}}{E_j}\Delta\sigma_j - \frac{v_{ki}}{E_k}\Delta\sigma_k \qquad (5)$$

Using the relationship of Equation 3, Equation 5 may be re-written as Equation 6.

$$\Delta\varepsilon_i = \frac{1}{E_i}\left(\Delta\sigma_i - v_{ij}\Delta\sigma_j - v_{ik}\Delta\sigma_k\right) \qquad (6)$$

Equation 6 can in turn be re-written using the relation of Equation 4 as Equation 7.

$$E_i = \frac{1}{\Delta\varepsilon_i}\left(\Delta\sigma_i - \sqrt{\frac{E_i}{E_j}}v_a\Delta\sigma_j - \sqrt{\frac{E_i}{E_k}}v_a\Delta\sigma_k\right) \qquad (7)$$

Three nonlinear equations of the form of Equation 7 may be derived for each of the principal Young's moduli. These may be solved simultaneously using some value of the geometric mean Poisson's ratio, v_a. A specific value of v_a will solve Equation 7 to provide the same value of E_1 as derived from a purely axial stress change at the same stress range. This value of v_a provides a basis for the determination of the other values of Young's moduli and Poisson's ratios. A series of step changes in axial and confining loading can thus be used to determine the orthotropic moduli of a core sample.

3   POROELASTIC EFFECTS

If we now consider the case where the rock contains fluid in its internal pore space Equation 5 may be rewritten in terms of effective stress as causing deformation simply by replacing $\Delta\sigma_i$ with $\Delta\sigma_i'$. By substituting Equation (1) into Equation (5) for effective stress, we may arrive at Equation (8) which describes deformation in terms of the three principal total stresses, fluid pressure and the poroelastic coefficients applying to these directions.

$$\Delta\varepsilon_i = \frac{1}{E_i}\Delta\sigma_i - \frac{v_{ji}}{E_j}\Delta\sigma_j - \frac{v_{ki}}{E_k}\Delta\sigma_k - \Delta P\left(\frac{1}{E_i}\alpha_i - \frac{v_{ji}}{E_j}\alpha_j - \frac{v_{ki}}{E_k}\alpha_k\right) \qquad (8)$$

As described previously it is possible to determine the orthotropic Young's moduli and Poisson's ratio for a core sample by a stepwise testing process involving changes in axial and confining stress. If we also incorporate cycles of fluid pressure variation into the test routine on the same strain gauged sample, it is possible to determine the values of the poroelastic coefficients via simultaneous solution of the three strain equations based on Equation 8 (Gray, 2017). The poroelastic coefficients so determined are in the direction of the axes of the principal orthogonal stiffness determined by the test procedure previously described. It is possible that the principal directions of the poroelastic coefficients are in fact different from the elastic ones. Unlike the work by Biot and Wills (1957) the poroelastic coefficients derived are a tensor. Their results are a function of the volumetric determination of poroelastic behaviour.

4 EXPERIMENTAL RESULTS

Let us examine the results of testing and analysis of Hawkesbury sandstones from the Sydney area of New South Wales, Australia. These samples have been core drilled approximately perpendicular to their bedding planes. The core diameter was 61 mm.

The first sample is of a porous medium grained sandstone. Figure 1 shows its Young's modulus perpendicular to the bedding plane plotted against axial (perpendicular to the bedding plane) and confining stress. Figure 2 shows the Young's modulus in a direction parallel to the

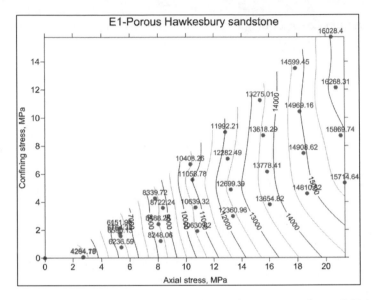

Figure 1. The Young's modulus perpendicular to bedding of a porous sandstone (MPa).

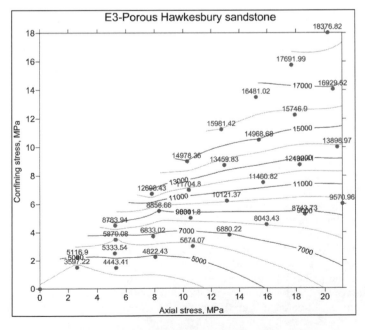

Figure 2. The Young's modulus parallel to bedding of a porous sandstone (MPa).

bedding plane. In each case Young's modulus increases with stress and is primarily a function of the stress in the same direction.

Figure 3 shows the Poisson's ratio associated with axial (cross bedding) stress and deformation parallel to the bedding plane. It characteristically shows an increase in Poisson's ratio with shear stress. Figure 4 shows the poroelastic coefficient in the direction of the axis of the sample. It also increases with shear stress. Both the increase in Poisson's ratio and poroelastic coefficient may be dependent on some level of dilation.

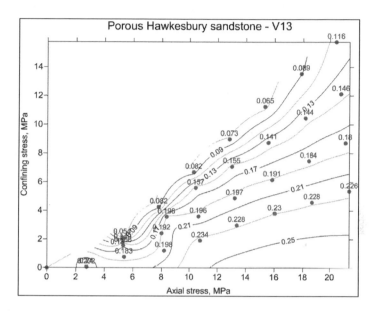

Figure 3.   The Poisson's ratio associated with cross bedding stress and deformation parallel to the bedding for a porous sandstone.

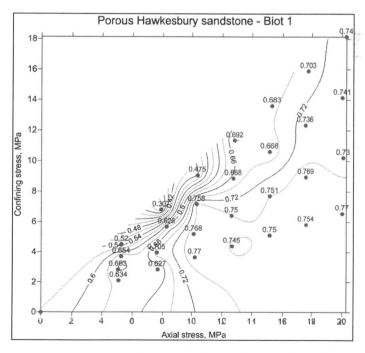

Figure 4.   The poroelastic coefficient referred to the cross bedding axis of a porous sandstone.

Testing of a fine to medium grained low porosity sample of Hawkesbury sandstone containing some silty and clay components provided some quite different characteristics. The value of the Young's modulus across the bedding plane (E_l) was dependent on both the axial and confining stresses and the poroelastic coefficients were very small.

Testing on granite from the Snowy Mountains area of New South Wales has shown generally isotropic behaviour with a virtually constant Young's modulus and Poisson's ratio and a negligible poroelastic coefficient. The poroelastic coefficient appears sensible given the crystalline nature of the rock, but is at variance with values reported by Detournay and Cheng, 1993.

5 FRACTURED ROCK BEHAVIOUR

The previous sections deal with poroelastic behaviour. The effective stress associated with this is an apparent effective stress dependent on the action of fluid within the pores, and micro fractures of the rock causing strain within the rock matrix. If the rock contains clear fractures then the term α_i describes a ratio of fracture area to total area over which fluid acts. This case has no relation to the poroelastic behaviour of the rock matrix.

6 CONCLUSIONS

This paper describes the process and mathematics involved in obtaining orthotropic elastic parameters including poroelastic behaviour from the triaxial testing of core. To do this requires the assumptions that the rock behaves in an orthotropic manner, that a likely axis of symmetry is common with the core axis and that a unique value of a geometric mean Poisson's ratio exists for each stress state.

Using the procedures and mathematics outlined the results of testing Hawkesbury sandstones show that the Young's moduli are highly dependent on the state of stress, varying some four to five fold from zero to 20 MPa axial and confining load. In a porous sample the Young's moduli seem to be dependent on the state of stress in the direction being measured. In the case of a low porosity sample the Young's moduli were dependent on both the axial and confining stress. In this sample the poroelastic coefficients were close to zero. Tests on some siltstones have shown less variation of modulus with stress, however the general trend of increasing modulus with stress exists. Some coals tested show an increase in stiffness of an order of magnitude. Generally we have found that sedimentary rocks we have tested show an anisotropy of less than 1.5:1 but with some exceptions which are nearly 5:1.

Most rock mechanics design is based on linear elastic models until strength based failure is reached. In addition the effects of fluid pressure are generally ignored. This paper shows that these assumptions are, in the case of the sedimentary rocks tested, quite incorrect. The elastic but very nonlinear behaviour is of particular importance. The consequences are that predicted deformations and stresses using linear elastic assumptions will be, in some cases, quite significantly in error.

REFERENCES

Biot, M A, & Wills, D G, 1957. The elastic coefficients of the theory of consolidation. ASME Journal of Applied Mechanics, 24:594–601.
Detournay, E & Cheng, AH-D, 1993. 'Fundamentals of poroelasticity', in C Fairhurst (ed), *Comprehensive Rock Engineering: Principles, Practice and Projects, Volume 2: Analysis and Design Methods*, Pergamon Press, pp. 113–171.
Gray, I, 2017. Effective Stress In Rock. Deep Mining 2017: Eighth International Conference on Deep and High Stress Mining – J Wesseloo (ed.) © 2017 Australian Centre for Geomechanics, Perth, ISBN 978-0-9924810-6-3.

Geomechanics and Geodynamics of Rock Masses – Litvinenko (Ed.)
© *2018 Taylor & Francis Group, London, ISBN 978-1-138-61645-5*

Dynamic rock support in burst-prone rock masses

Charlie C. Li
Norwegian University of Science and Technology (NTNU), Trondheim, Norway

ABSTRACT: Rockburst occurs in hard and strong rock after excavation when the in situ rock stresses are high. The "driving force" for rockburst is the energy released from the rock mass. The released strain and seismic energy is transformed to the kinetic energy of rock ejection during rockburst. To prevent out-of-controlled rock ejection, the released energy has to be absorbed by the rock support system when a rockburst event occurs. Therefore, it is required in dynamic rock support design that the support elements must be capable of absorbing a good amount of energy in burst-prone rock conditions. On the other hand, it is required that the displacement of the tunnel wall must be neither larger than the ultimate displacement capacity of the support elements nor the maximum allowable operational displacement. The types of rockburst, the dynamic loading conditions, the design principles of dynamic rock support, and typical yield rockbolts used for combating rockburst issues are presented in the paper.

Keywords: rockburst, dynamic loading, dynamic rock support, yield rockbolt, energy-absorbing rockbolt

1   INTRODUCTION

Serious rockburst events could occur in hard rock excavations when the depth is beyond 1000 m. The rock support system must be able to adapt to the dynamic loading condition. Attentions are paid to the strength of support elements in the traditional support design. Under dynamic loading conditions, however, it is required that the support elements must not only be strong but also deformable in order to avoid premature failure of the support elements. In other words, support elements must be energy-absorbent. The concept of yield support elements, such as yield rockbolts, was first proposed by Cook and Ortlepp in 1968 in the Bulletin of the Chamber of Mines of South Africa Research Organisation. One year later, Ortlepp (1969) published his field tests on yield rockbolts in a symposium in Oslo, Norway. Ortlepp was probably the first one who carried out dynamic field tests of rock support. His tests were summarized by himself (Ortlepp 1992) and others, for instance Stacey later (2012). Ortlepp carried out the tests on two rockbolt-mesh support systems installed in a tunnel, one with fully grouted conventional rockbolts and the other one with the yield rockbolts that he developed. The rockburst load was simulated by blasting, with blastholes 430 mm apart drilled parallel to the tunnel axis about 600 mm outside the tunnel perimeter. The first test failed because of the high blast intensity. The second one with reduced blast intensity proved that the support system, consisting of the yield rockbolts and a double layer of wire mesh, could contain the energy of the blast. The sketch of the tunnel profile after the test, shown in Fig. 1, clearly illustrates the effectiveness of the yield support system on the right hand side of the tunnel, and the ineffectiveness of the conventional support system on the left hand side. The field tests of Ortlepp demonstrated that use of a support system incorporating yielding rockbolts could successfully contain rock damage. It has become a common practice to combat rockburst with yield rockbolts in deep metal mines in the recent decade.

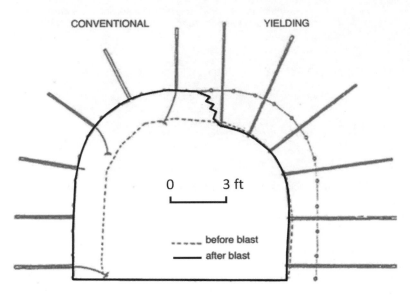

CONVENTIONAL   YIELDING

0   3 ft

- - - - before blast
———— after blast

Figure 1.   The tunnel profile after the blasting test (Ortlepp 1969, Stacey 2012).

In this paper, the types of rockburst and the relationships of the ejection velocity with the strain energy in the rock and the fault-slip seismic energy are talked about first. The design requirements and the factor of safety for dynamic rock support are presented afterward. Finally, some typical yield rockbolts for dynamic rock support are introduced.

2   TYPES OF ROCKBURST

Rockburst events can be classified to three types based on their triggering mechanisms and energy sources. In hard and massive rock, the tangential stress in the contour rock is so significantly eleveated that the intact rock in the tunnel wall simply explodes after excavation, Fig. 2. This is the so-called *strain burst* (Type 1). The burst energy of a strain burst event is simply contributed by the potential strain energy stored in the ejected rock. There is no seismic activities before the event with this type of rockburst (Fig. 2a). Seismicity with a limited intensity is generated after strain burst events (Fig. 2b). A strain burst event is often characteristized by thin slices of rock in the rock pile (Fig. 2c).

Underground excavation changes the stress state in the rock mass in such a manner that the tangential stress around the opening is elevated but the radial stress is reduced. The reduction in the radial stress would lead to a decrease in the normal stress on some pre-existing faults nearby and in turn the shear resistance on the faults is reduced. Slippage, therefore, may occur along some faults. Such fault slippage will generate stress waves that propagate spherically outward from the epicenter of the slippage. This is called mine seismicity in the mining industry.

In some cases, the tangential stress in the contour rock is significantly elevated after rock excavation, but they are not high enough to break the rock. The arrival of fault-slip seismic waves could trigger a rockburst event in the highly stress rock. This is the so-called *fault-slip strain burst* (Type 2), Fig. 3. The magnitude of this type of rockburst could be stronger than the strain burst of Type 1 because the burst energy is the sum of the strain energy in the burst rock and a portion of the fault-slip seismic energy. Seismic activities exist both before and after the event with this type of rockburst. It is only the fault-slip seismicity before the burst event (Fig. 3a), but burst seisimicity is also generated after the event (Fig. 3b).

In a relatively weak or fractured rock mass, the contour rock could become fractrued immediately after excavation, resulting in a fracture zone around the the tunnel. A portion

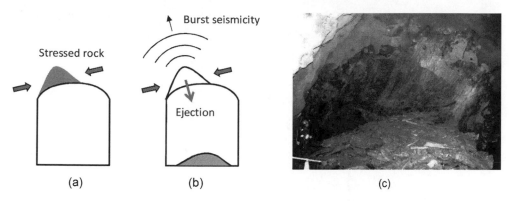

Figure 2. Type 1—Strain burst. Sketches illustrating (a) the stress concentration in the rock before rock ejection and (b) the seismicity after rock ejection, and (c) a strain burst event in a metal mine.

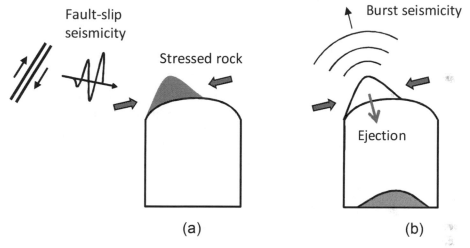

Figure 3. Type 2: Fault-slip strain burst. Sketches illustrating (a) the stress concentration and the fault-slip seismicity before rock ejection and (b) the rock ejection and the burst seismicity.

of the fractured rock may be ejected by fault-slip seismic waves when a fault-slip movement occurs in the rock mass, Fig. 4. This is the so-called *fault-slip rockburst* (Type 3). The ejection energy of this type of rockburst is mainly contributed by the fault-slip seismic waves. Some fault-slip events release a significant amount of energy. A fault-slip rockburst could be more violent than a strain burst and thus cause more serious damage to underground infrastructures than a strain burst does. It was once registered a fault-slip rockburst of up to 3.8 Mn in a deep metal mine in Canada (Counter 2014). Rock debris from a fault-slip burst is composed of rock pieces of different sizes, ranging from finely fragmented debris to large blocks. Fig. 4c shows the rock pile after the fault-slip rockburst event of 3.8 Mn in the deep metal mine in Canada. That event was triggered by a fault-slippage located approximately 100 m behind the rockburst position.

Hoek (2006) defined rockburst as "explosive failures of rock which occur when very high stress concentrations are induced around underground openings". Obviously Hoek referred to the rockburst of Type 1, that is, the strain burst. Kaiser et al. (1995) defined rockburst as "damage to an excavation that occurs in a sudden or violent manner and is associated with a seismic event". They meant the rockburst of Type 3, that is, the fault-slip rockburst. The essential characteristic of a rockburst event is its dynamic feature. We call it rockburst if the

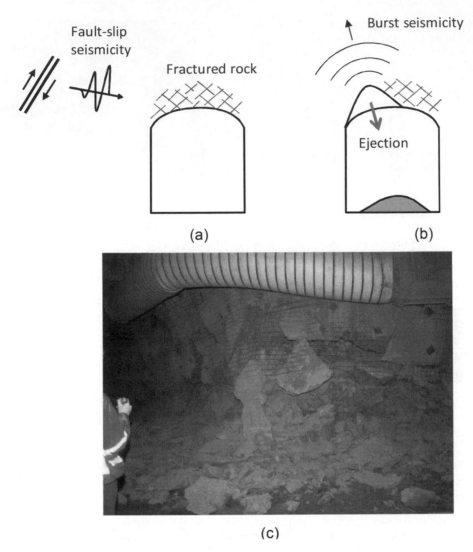

Fault-slip seismicity

Fractured rock

(a)

Burst seismicity

Ejection

(b)

(c)

Figure 4.   Type 3: Fault-slip rockburst. Sketches illustrating (a) the fractured rock before rock ejection and (b) the seismicity during rock ejection, and (c) the fragmented rock after a fault-slip rockburst event of 3.8 Mn in a deep metal mine (Counter 2014).

rock is ejected with a certain velocity. A proper definition for rockburst, therefore, may be simply *"damage to an excavation that occurs in a sudden or violent manner"*. The cause for a rockburst event can be either stress concentration or fault-slip or both.

3   DYNAMIC LOADING CONDITIONS

In a rockburst event, the failed rock is ejected in a certain ejection velocity that is associated with the strain energy in the rock as well as the fault-slip seismic energy.

3.1   *Ejection velocity in a strain burst (Type 1)*

Strain burst (Type 1) is caused directly by the stress concentration in the rock surrounding the underground opening. Strain burst does not involve any seismic activity prior to the rockburst event. It is simply owing to the energy release after rock failure. In other words, the strain energy stored in the failed rock is transformed to kinetic energy when the rockburst event occurs. The elastic strain energy (i.e. the potential energy) stored in the ejected rock party is expressed by:

$$\text{Potential energy} = \frac{m}{2\rho E}\sum \sigma_i^2 \qquad (1)$$

where m is the mass of the ejected rock, ρ the density of the rock, E is the Young's modulus of the rock and σ_i the average principal stresses in the rock party (i = 1, 2 and 3). The kinetic energy of the ejected rock is expressed by

$$\text{Kinetic energy} = \frac{1}{2}mv_1^2 \qquad (2)$$

where v_1 is the ejection velocity of the rock. The right sides of the above two expressions should be equal according to the law of energy conservation. Thus, the ejection velocity of a strain burst event is obtained as:

$$v_1 = \sqrt{\frac{1}{\rho E}\sum \sigma_i^2}. \qquad (3)$$

Assume that the density of a massive rock mass is 2700 kg/m³, the Young's modulus of the rock is 60 GPa, and the secondary rock stresses in the contour rock after excavation are σ_1 = 60 MPa, σ_2 = 20 MPa, σ_3 = 0 MPa with σ_1 and σ_2 parallel with the tunnel wall and σ_2 perpendicular to the wall. The ejection velocity of the rock when a rock burst event occurs is obtained from Eq. (3) as v_1 = 5 m/s which is a quite reasonable ejection velocity for a strain burst event in hard rock.

3.2   Ejection velocity in fault-slip bursts (Type 2 and Type 3)

With rockburst of Type 2 and Type 3, fault-slip seismicity is involved in the burst event. The kinetic energy of the ejected rock is equal to the sum of the elastic strain energy in the ejected rock and a portion of the seismic wave energy. The seismic wave energy is derived below. Assume that a fault-slip event generates a sinusoidal seismic wave (Fig. 5) that is expressed by:

$$u_x = A\sin(\omega t - kx) \qquad (4)$$

where u_x = particle displacement at position x, A = the displacement amplitude, ω = angular frequency, $\omega = 2\pi f$, f = frequency, t = time, k = wave number, k = ω/C, C = wave velocity and x = position.

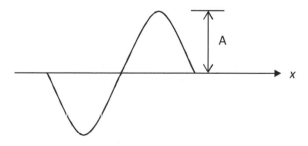

Figure 5.   A sinusoidal seismic wave pulse.

The seismic wave induces particle vibrations in the rock it passes through. The vibrations bring about a strain and stress in the rock so that a static strain energy density, w_s, is thus induced in the rock by the seismic wave. On the other hand, the wave propagation means that a kinetic energy component, w_k, is also induced by the seismic wave. Therefore, the total energy induced by the seismic wave, w, is the sum of the two components w_s and w_k, that is, $w = w_s + w_k$.

In the case of a longitudinal wave (i.e. the P wave), the normal strain induced by the wave is $\varepsilon_x = \partial u_x / \partial x$ and the normal stress is $\sigma_x = E\varepsilon_x$. The static wave strain energy density is expressed as $w_s = \sigma_x \varepsilon_x / 2$. The average static strain energy density of the seismic wave is then obtained as

$$\bar{w}_s = \frac{1}{4}\rho (PPV)^2 \tag{5}$$

where ρ is the density of the rock and PPV represents the Peak Particle Velocity, $PPV = A\omega$. The particle velocity is expressed by $\dot{u}_x = \partial u_x / \partial t$. The kinetic energy density is calculated as $w_k = \rho \dot{u}_x^2 / 2$. The average kinetic energy density is then obtained as

$$\bar{w}_k = \frac{1}{4}\rho (PPV)^2. \tag{6}$$

The total energy density in the rock, which is caused by the seismic wave, is thus

$$w = \bar{w}_s + \bar{w}_k = \frac{1}{2}\rho (PPV)^2. \tag{7}$$

Let v_2 represent the wave-induced velocity of the ejected rock. The following equilibrium must exist:

$$\frac{1}{2}mv_2^2 = wV \tag{8}$$

where V is the volume of the ejected rock, $V = m/\rho$. The ejection velocity v_2 is then obtained as:

$$v_2 = PPV. \tag{9}$$

The total ejection velocity is then obtained as

$$v = \sqrt{v_1^2 + v_2^2} = \sqrt{\frac{1}{\rho E}\sum \sigma_i^2 + PPV^2}. \tag{10}$$

The study by Yi and Kaiser (1993) showed that it is reasonable to assume the rock ejection velocity is equal to the peak particle velocity (PPV) under typical mining and seismicity conditions. The theoretical solution of Eq. (9) agrees with their conclusion if we only talk about the ejection velocity induced by seismicity. In the case of a fault-slip triggered rockburst event, the ejection velocity is a vector superposition of two components: the velocity due to the release of the strain energy in the rock, v_1, and the velocity due to the seismic wave, v_2, as expressed by Eq. (10). The near-field PPV of a fault-slip seismic event of Nuttli magnitude 3–4 is approximately 3 m/s according to Kaiser et al. (1995). The ejection velocity caused by the seismicity will be thus $v_2 = PPV = 3$ m/s according to Eq. (9). Assume that the rock has been subjected to the stresses as given in the example above, that is, $\sigma_1 = 60$ MPa, $\sigma_2 = 20$ MPa, $\sigma_3 = 0$ MPa, which would result an ejection velocity $v_1 = 5$ m/s. The total ejection velocity is thus obtained, according to Eq. (10), to be 5.8 m/s. It seems that the seismicity mainly plays a role of trigger in rockburst events of such magnitudes. The ejection velocity is mainly dependent on the prevailing stress state in the rock prior to the burst event.

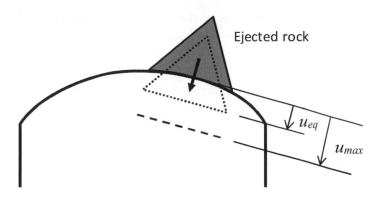

Figure 6. The equilibrium displacement u_{eq} and the maximum allowable displacement u_{max} related to a rockburst event.

4 DESIGN REQUIREMENTS

The basic requirement for rock support elements is their energy absorption capacity in burst-prone rock. In a rockburst event, a part of the released energy is converted to kinetic energy to eject the rock. It is required that the rock support elements in a support system must be able to absorb the kinetic energy in order to prevent the rock from being ejected. Let E_{ab} represent the total energy absorption of the support elements in a support system and E_{ej} is the kinetic energy of the ejected rock, which is expressed by:

$$E_{ej} = \frac{1}{2}mv^2 \tag{11}$$

The ratio of E_{ab} to E_{ej} has to be larger than 1 in order to avoid rock ejection, that is,

$$\frac{E_{ab}}{E_{ej}} > 1. \tag{12}$$

With a competent support system, the ejected rock will stop moving after a displacement u_{eq} (Fig. 6). it is required that the displacement u_{eq} must be smaller than the ultimate displacement u_{ult} of the support system in order to avoid failure of the system, that is, $(u_{ult}/u_{eq}) > 1$. In engineering practice, there usually exists a maximum allowable displacement, denoted as u_{max}, from the point of view of operation. For example, the radial displacement of a TBM tunnel usually is not allowed to be larger than 150 mm in order to avoid clogging of the TBM cutter head. In other words, the ratio of the u_{max} to the displacement at equilibrium, u_{eq}, must be larger than 1, that is, $(u_{max}/u_{eq}) > 1$. The value of the factor of safety in the burst-prone rock condition is the minimum one among the three ratios above, that is,

$$FS = \min\left(\frac{E_{ab}}{E_{ej}}, \frac{u_{max}}{u_{eq}}, \frac{u_{ult}}{u_{eq}}\right). \tag{13}$$

5 TYPICAL YIELD ROCKBOLTS

Yield (or energy-absorbing) rockbolts are the most powerful support elements for dynamic rock support so far. The first commercial yield rockbolt, the so-called cone bolt, was invented in the Southern Africa in the beginning of the 1990s (Jager 1992, Ortlepp 1992), but it was not widely accepted until the 2000s. In the past decade, a number of other yield rockbolts have appeared in the market. The typical yield rockbolts used for dynamic rock support are introduced below.

5.1 The cone bolt

The cone bolt consists of a smooth steel bar and a flattened conical flaring forged at the far end of the bolt shank. It has two versions, the original one for cement grout (Fig. 7a) and a modified one for resin grout (Fig. 7b). A blade is added at the end of the modified cone bolt for the purpose of resin mixing. The cone bolt is fully grouted in a borehole. Rock dilation will induce a load on the face plate, which then transfers the load to the bolt shank and the cone at the bolt end. The grout facing the conical side of the cone is crushed when the pull load is high enough and the cone then ploughs in the grout to do work.

A cone bolt can displace for a considerable distance if it ploughs as desired. A series of pull tests were once carried out on cement-grouted cone bolts in an underground mine in Sweden. The bolts displaced up to 900 mm at a pull load of approximately 170 kN. In order to achieve the desired ploughing mechanism, the shape of the cone and the crushing strength of the grout must have a satisfactory match. In reality, the strength of the grout varies because of variations in the type of grout material and the water–cement ratio or the mixing quality in the case of using resin grout. This leads to significant variations in the yield load of the cone bolts. Many static pull and dynamic drop tests have been carried out both in the field and in laboratories over the past decades. Fig. 8 shows representative results of such tests. The static yield load varies from approximately 60 kN to 150 kN (Fig. 8a). The drop test results shown in Fig. 8b are for 22 mm cone bolts tested with a kinetic energy input of 33 kJ. The dynamic yield load of those bolts is 150–175 kN for the 40 MPa resin grout, but drops to approximately 100 kN for the 20 MPa resin grout. It was observed in field tests that the cone of the bolts might plough little or not at all in some cases, so that the displacement was purely

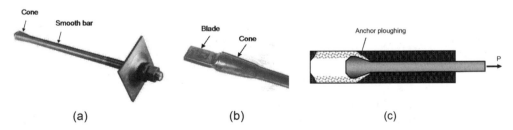

Figure 7.   The cone bolt. (a) The original version for cement grout, (b) the modified version for resin grout (Simser, 2001), (c) the work principle.

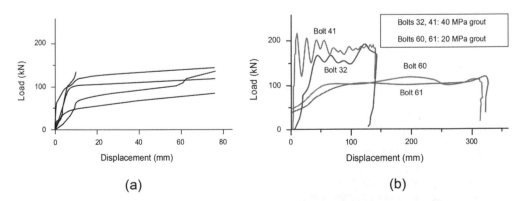

Figure 8.   Static and dynamic test results of modified cone bolts with resin grouts. (a) Static pull tests, redrawn after Simser et al. (2006), (b) dynamic drop tests, redrawn after Varden et al. (2008).

coming from the stretching of the bolt shank (Simser *et al.* 2006). In those cases, the yield load of the bolts was equal to the yield limit of the bolt steels. Such a large spread in the load capacity could cause uncertainty for rock support design.

The cone bolt has been used for rock support in many metal mines in Canada and Australia in the past decade.

5.2 The D-Bolt

The D-Bolt, invented in Norway, comprises a smooth steel bar and a number of integrated anchors along the bolt length (Fig. 9) (Li, 2010). The bolt is either cement or resin encapsulated in a borehole. The short anchors are firmly fixed in the grout, while the long smooth bar sections between the anchors elongate upon rock dilation. The bolt absorbs energy through full mobilization of the strength and deformation capacity of the bolt steel.

Static and dynamic test results for D-Bolts are presented in Fig. 10. The bolt sections tested are 22 mm in diameter and 1.5 m in length between the anchors. The ultimate static load and displacement are 260 kN and 165 mm, respectively (Fig. 10a), and the ultimate dynamic load and displacement are 285 kN and 220 mm, respectively (Fig. 10b). The bolt section absorbs approximately 60 kJ of energy prior to failure under dynamic loading. Every section of the bolt works independently; the failure of one section does not result in the loss of the entire bolt, with the remaining sections continuing to provide rock reinforcement. In general, the ultimate load of the D-Bolt is equal to the tensile strength of the steel and the ultimate displacement is approximately 15% of the bolt length. D-Bolts have been used in metal mines for dynamic rock support in Sweden, Canada, USA, Chile and Australia.

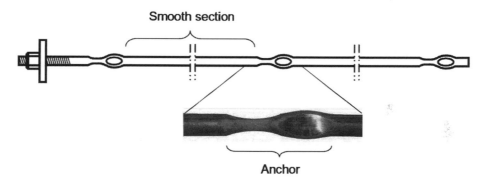

Figure 9.   The D-Bolt.

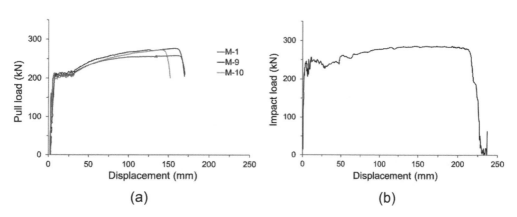

Figure 10.   Test results of the D-Bolt sections of 22 mm × 1.5 m (Li, 2012; Li & Doucet, 2012). (a) Static pull test results, (b) dynamic drop test result (drop mass 2897 kg, drop height 1.97 m and input energy 56 kJ).

5.3 The Yield-Lok

The Yield-Lok bolt consists of a round steel bar of 17.2 mm in diameter (Fig. 11). The anchor, or Upset, of the bolt is encapsulated in an engineered polymer coating. The bolt is encapsulated in the borehole with resin grout. The Upset ploughs within the polymer coating when the pull load exceeds the predefined load limit. The mechanics of the Yield-Lok is similar to the cone bolt. Static and dynamic test results of the bolt are shown in Fig. 12. The dynamic load is in general lower than the static load. Yield-Lok bolts are used in some metal mines for dynamic rock support in Canada.

5.4 The Garford solid bolt

The Garford solid bolt, invented in Australia, consists of a smooth solid steel bar, an anchor and a coarse-threaded sleeve at the far end (Fig. 13). This bolt is characterised by its engineered anchor, the inner diameter of which is smaller than the diameter of the solid bolt bar. The bolt is spun into the borehole, which is filled with resin cartridges. The resin is mixed by the threaded sleeve at the bolt end. The anchor is resin encapsulated in the borehole after installation. When the rock dilates, the pull load in the solid bar forces the solid bar to be extruded through the hole of the anchor. The yield load is determined by the difference in the diameters of the anchor hole and the solid bar. The ultimate displacement of the bolt is determined by the length of the bar tail contained within the threaded sleeve. Fig. 14 shows the dynamic test results of two 20 mm bolts, which were loaded with a kinetic energy input

Figure 11.   The Yield-Lok (Wu and Oldsen 2010).

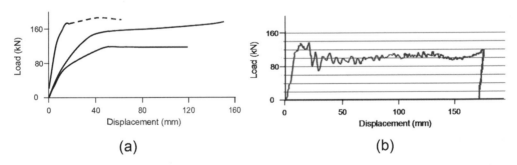

Figure 12.   Test results of the Yield-Lok bolts. (a) Static pull test results, (b) dynamic test result (drop mass 1115 kg, drop height 1.5 m and input energy 16.4 kJ). Redrawn from Wu & Oldsen (2010).

Figure 13.   The Garford solid bolt.

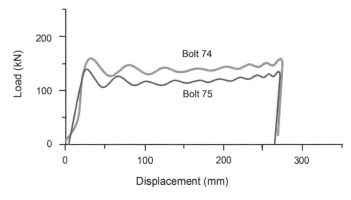

Figure 14.   Dynamic test results of 20 mm Garford bolts with impact input of 33 kJ. Redrawn after Varden *et al.* (2008).

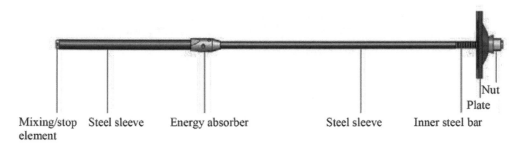

Figure 15.   The Roofex rockbolt.

of 33 kJ (Varden *et al.*, 2008). The Garford bolts are used in some Australian metal mines for dynamic rock support.

5.5   *The Roofex*

Roofex is not on the market anymore, but several publications on the bolt exist so that it is introduced here. Roofex is composed of an engineered anchor and a smooth bar (Fig. 15) (Charette & Plouffe, 2007; Galler *et al.*, 2011). Its work principle is similar to the Garford solid bolt, that is, the smooth solid steel bar is extruded through the hole of the anchor at the designed load level. The bolt is spun into a borehole that is filled with resin cartridges and the resin mixer at the end of the bolt mixes the resin. It is required that the anchor must be fully encapsulated in the resin grout. The smooth steel bar slips through the anchor to dissipate the energy and accommodate the rock displacement. The mechanics of the Roofex is similar to that of the Garford bolt, as described by Eq. (14), but the factor k would have a different value for Roofex because of the different shape of its anchor. Fig. 16 shows static pull and dynamic drop test results for Roofex rockbolts. The results indicate that the dynamic load of the Roofex bolt is much smaller than its static load.

5.6   *Durabar*

Durabar is another yielding rockbolt invented in South Africa (Ortlepp *et al.*, 2001). The bolt is composed of a smooth bar and a sinusoidally waved portion as well as the face plate and the nut (Fig. 17). The bolt has a smooth tail in the far end, which determines the maximum displacement capacity of the bolt. The bolt is fully grouted into a hole in the rock. It is

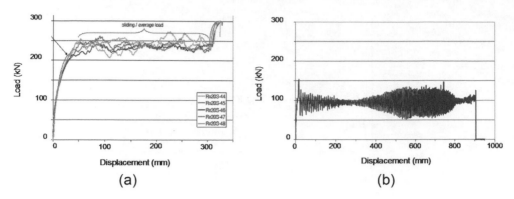

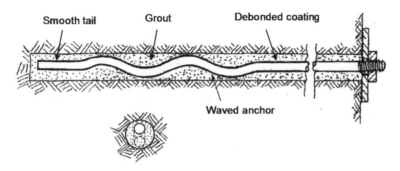

(a)   (b)

Figure 16.　Laboratory test results of Roofex R × 20 rockbolts (Galler *et al.*, 2011). (a) Static pull test results, (b) dynamic test result.

Figure 17.　Durabar (Ortlepp *et al.*, 2001).

required that the shape of the waved portion must be carefully configured and the full length of the bar is de-bonded from the grout in order that the desired performance of the bolt is achieved. The Durabar is able to yield by sliding through the hardened cement grout. As the load on the Durabar reaches the designed yield load, it starts to slip through the wave path created in the grout by the waved shape of the Durabar. The friction in the wave path is the main energy dissipation mechanism. The ultimate load of the Durabar is related to the slope angle of the wave portion and the frictional coefficient between the bolt shank and the grout. A normal load is induced on the wave portion as a pull load is applied to the bolt. The magnitude of the normal load is associated with the slope angle i that is represented by the ratio of the amplitude of the wave to the half-wave length.

The performance of Durabar is mainly dependent on the properties of the steel and requires that the strength of the grout exceeds a minimum value of about 25 MPa. A 16 mm 2.2 m long Durabar can dissipate 45 kJ over a 500 mm displacement under a static pull loading according to Ortlepp *et al.* (2001). The static load of the 16 mm Durabar stabilizes at a level of 80 kN, while its dynamic load drops to approximately 60 kN for a drop velocity of 3 m/s (Fig. 18).

5.7　The He bolt

The He bolt, invented in China, is composed of a solid bar, a cone-shaped piston, a sleeve, a face plate and a nut (Fig. 19). The far end of the solid bar is groove threaded, which is the anchor of the bolt. The diameter of the cone-shaped piston is slightly smaller than the inner diameter of the sleeve so that it is tightly assembled in the sleeve, which is attached to the face plate and the nut. The bolt is fully grouted in a borehole with either cement mortar or resin. Both the threaded anchor of the bolt at the far end and the sleeve are encapsulated in the

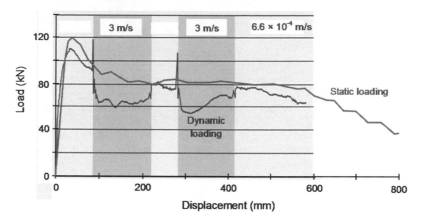

Figure 18.   Static and dynamic performances of Durabar. Redrawn from the Durabar brochure.

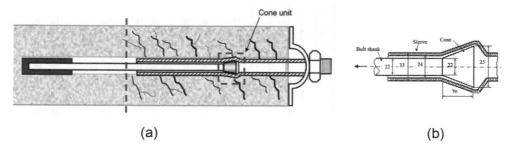

(a)   (b)

Figure 19.   The He bolt (He *et al.*, 2014). (a) The work principle, (b) a close-up sketch of the match between the cone and the sleeve.

grout according to the authors (He *et al.*, 2014). The cone slips in the sleeve to accommodate the rock dilation, Fig. 19a. In accordance with the drawing illustrating the match between the cone and the sleeve during slippage (Fig. 19b), it seems that the sleeve is radially expanded when the conical piston slips. Therefore, the pull resistance load of the bolt is not simply due to the friction between the cone and the sleeve wall, but also to the plastic deformation of the sleeve.

The bolt was both statically and dynamically tested in the laboratory. The diameter of the solid bars of the bolt specimens was 22 mm and the outer and inner diameters of the sleeves were 33 and 24 mm, respectively. The diameter of the conical piston was 0.7–1 mm larger than the inner diameter of the sleeve, that is, varying from 24.7 to 25 mm. Fig. 20a shows the static test result of the bolt specimen with a cone diameter of 24.9 mm. The stick–slip phenomenon occurred during slippage of the cone in the sleeve, with a load oscillation between 140 and 180 kN. The mean values of the static load were 108, 125 and 106 kN, corresponding to cone diameters of 24.7, 24.8 and 24.9 mm, respectively. It seems that the load capacity of the bolt is sensitive to the cone diameter.

Three He bolts were drop tested with a drop mass of 1000 kg and a drop height of 0.5, 0.7 or 1 m (He *et al.*, 2014). The average load of the bolts varied from 67 to 88 kN, which is smaller than the static load. Fig. 20b shows the dynamic test result of a He bolt tested with a drop mass of 1000 kg and a drop height of 0.7 m. It is seen that the amplitude of the load oscillations in the dynamic test is much larger than that in the static tests. It is noticed that all of the rockbolts were tested in open air instead of in real or simulated boreholes. It is not clear how the encapsulation affects the behaviour of the bolt when the bolt is installed in a borehole.

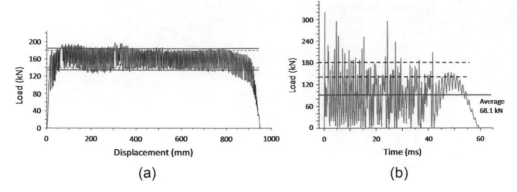

Figure 20.   Laboratory test results of the He bolts (He *et al.*, 2014). (a) The static pull test result of the bolt with a cone diameter 24.9 mm, (b) the drop test result of the bolt (drop mass 1000 kg, drop height 0.7 m).

6   CONCLUDING REMARKS

Rockburst events classified to three types on the basis of their triggering mechanisms and energy sources. Type 1 is called *strain burst* that is purely caused by the stress concentration. With a strain burst event, the intact rock explodes and the strain energy in the rock is transformed to fracture energy and burst energy. Type 2 is called *fault-slip strain burst*. With such a burst event, the feature of the burst is the same as Type 1, that is, the intact rock explodes, but the trigger is the fault-slip seismic waves. The burst energy in a fault-slip strain burst event are contributed both by the strain energy in the rock and the seismic waves. Type 3 is called *fault-slip rockburst*. With a burst event of Type 3, the fault-slip seismic waves both trigger the event and contribute the burst energy.

In burst-prone rock masses, it is required that the rock support system must be able to absorb the kinetic energy of the ejected rock in order to prevent out-of-controlled dynamic rock falls. All support elements in a support system must be energy absorbent, that is, not only strong but also deformable. The practice has proven that use of energy-absorbing (or yield) rockbolts is the most efficent means to combat rockburst issues in underground rock excavation. There are a number of energy-absorbing rockbolts avi31able on the market at present. Examples of them are the cone bolt, the D-Bolt and the Yield-Lok. An energy-absorbing bolt is characterised by its high load and displacement capacities, and it can dissipate a large amount of energy prior to failure. The existing energy-absorbing rockbolts absorb energy either through material stretching (the D-Bolt) or through friction or ploughing in the grout (all the other yield bolts).

The factor of safety of a rock support element, such as a rockbolt, cannot be calculated with the strength of the element and the load on it since the load on the support element is dependent on the deformation in underground excavation. The energy absorption and the ultimate displacement of the support elements, the amount of the released energy and the tunnel wall displacement have to be taken into account for the calculation of the factor of the safety. Under dynamic loading conditions like rockburst, the factor of safety of a support element is the minimum one of the following three ratios: the ratio of the energy absorption of the support element to the kinetic energy of the ejected rock, the ratio of the ultimate displacement of the element to the equilibrium displacement and the ratio of the maximum allowable displacement of the tunnel to the equilibrium displacement.

REFERENCES

Charette F, Plouffe M. Roofex—results of laboratory testing of a new concept of yieldable tendon. In: Potvin, Y. (ed.) *Deep Mining 07—Proceeding of the 4th International Seminar on Deep and High Stress Mining*. Australian Centre for Geomechanics. pp. 395–404.

Cook, N.G.W. & Ortlepp, W.D. (1968) A yileding rockbolt. *Chamber of Mines of South Africa Research Organisation Bulletin,* No. 14., 6–8.

Counter D. 2014. Kidd mine—dealing with the issues of deep and high stress mining—past, present and future. *DeepMining 2014—Proc. Of the 7th Int. Conf. on Deep and High Stress Mining,* 16–18 Sept. 2014, Sudbury, Canada. Australian Centre for Geomechanics. 3–22.

Galler, R., Gschwandtner, G.G. & Doucet, C. (2011) Roofex bolt and its application in tunnelling by dealing with high stress ground conditions. In: *ITA-AITES World Tunnel Congress,* Helsinki, Finland. 11p.

He, M., Gong, W., Wang, J., Qi, P., Tao, Z., Du, S. and Peng Y. 2014. Development of a novel energy-absorbing bolt with extraordinarily large elongation and constant resistance. *Int J Rock Mech Min Sci,* 67: 29–42.

Hoek E, 2006. *Practical Rock Engineering.* North Vancouver, British Columbia: Evert Hoek Consulting Engineer Inc.

Jager AJ, 1992. Two new support units for the control of rockburst damage. In: Kaiser, P.K. and McCreath, D.R. (eds.) *Proc Int Symp on Rock Support.* Rotterdam: Balkema. pp. 621–631.

Kaiser PK, McCreath DR and Tannant DD, 1995. Volume 2: *Rockburst Support.* In: C. Graham, ed. Canadian Rockburst Research Program 1990–1995. Ontario: CAMIRO Mining Division.

Li CC and Doucet C, 2012. Performance of D-bolts under dynamic loading conditions. *Rock Mech & Rock Engng,* 45(2), 193–204.

Li CC, 2010. A new energy-absorbing bolt for rock support in high stress rock masses. *Int J Rock Mech Min Sci,* 47(3), 396–404.

Li CC, 2012. Performance of D-bolts under static loading conditions. *Rock Mech & Rock Engng;* 45(2), 183–192.

Ortlepp WD, 1969. An empirical determination of the effectiveness of rockbolt support under impulse loading. *Proc Int Symp on Large Permanent Underground Openings,* Oslo, Sept 1969. Brekke, T.L. and Jorstad, F.A. (eds.). Universitats-forlaget. 197–205.

Ortlepp WD, 1992. The design of support for the containment of rockburst damage in tunnels—an engineering approach. In: *Rock Support in Mining and Underground Construction.* Rotterdam: Balkema, 593–609.

Ortlepp WD, Bornman JJ and Erasmus N, 2001. The Durabar—a yieldable support tendon—design rationale and laboratory results. In: *Rockbursts and Seismicity in Mines—RaSiM5.* South African Institute of Mining and Metallurgy. pp. 263–264.

Simser B, 2001. Geotechnical Review of the July 29th, 2001. West Ore Zone Mass Blast and the Performance of the Brunswick/NTC Rockburst Support System. Technical report, 46p.

Simser B, Andrieux P, Langevin F, Parrott T and Turcotte P, 2006. Field Behaviour and Failure Modes of Modified Conebolts at the Craig, LaRonde and Brunswick Mines in Canada. In: *Deep and High Stress Mining.* Quebec. 13p.

Stacey TR, 2012. A philosophical view on the testing of rock support for rockburst conditions. *The Journal of The Southern African Institute of Mining and Metallurgy,* 112, 703–710.

Varden R, Lachenicht R, Player J, Thompson A and Villaescusa E, 2008. Development and implementation of the Garford Dynamic Bolt at the Kanowna Belle Mine. In: *10th Underground Operators' Conference.* Launceston. 19p.

Wu YK and Oldsen J, 2010. Development of a New Yielding Rock Bolt – Yield-Lok Bolt. In: *Proc. Of the 44th US Rock Mechanics Symposium,* Salt Lake City, USA. Paper ARMA 10–197, 6p.

Yi X and Kaiser PK, 1993. Impact testing of rockbolt for design in rockburst conditions. *Int J Rock Mech Min Sci & Geomech Absstr,* 31, 671–685.

Deep mining rock mechanics in China—the 3rd mining technology revolution

Manchao He
State Key Laboratory for Geomechanics and Deep Underground Engineering,
China University of Mining and Technology, Beijing, China

ABSTRACT: With the increasing of mining depth, more and more mining activities are being conducted in deep strata. A series of hazards such as rockburst, large deformation, and collapse occur frequently in deep mines. Researches show that the main reason causing these deep mine hazards are related to the high stress level environment. Therefore, innovating the mining technology is one of the important ways to mitigate and to reduce mine hazards. In this paper, a study about the longwall mining method is presented in order to permit solving these high stress problems. The traditional longwall mining method was introduced at the beginning of 18th century the so-called as 121 mining method (1 working face, 2 excavation gateways and 1 remaining coal pillar). The method was widely used in China. According to the statistics in coal mines, more than 92% of the accidents occurred during the excavation of gateways. To reduce mining accidents associated with longwall mining method and to increase the productivity of the mining explorations, the author proposed a new longwall mining method in 2008, called the 110 mining method (1 working face, 1 excavation gateway and no coal pillar), including directional pre-splitting cutting and the use of large deformation bolt/anchor supporting systems. Using the new method, 50% of the gateways are no longer needed to be excavated; instead, they are formed by a controlled roof collapse. By reducing the need for gateway excavation, mining accidents and consequently costs could be significantly decreased. In this sense, the third mining technology revolution is underway in China.

Keywords: Mining innovation; Deep mining; Longwall mining method; Non-pillar mining; 110 mining method

1 INTRODUCTION

The traditional longwall mining method was introduced at the beginning of 18th century the so-called as 121 mining method (1 working face, 2 excavation gateways and 1 remaining coal pillar). This method was widely used in China until recently. In 2016, 3.41 billion ton coal were outputted and 13,000 km gateways were excavated by using the 121 mining method. According to the statistics, more than 91.6% of accidents happened in the gateways. And a series of hazards such as rockburst, large deformation, collapse became more challenging with the increasing of mining depth (He, 2004; 2005; He et al., 2014). Due to the high stress conditions, it was basically considered that the traditional 121 mining method was considered not suitable for deep mining purposes for safety and economical reasons (Zhai and Zhou, 1999; Li, 2000; Liu and Shi, 2007; Fei, 2008).

In 2008, the theory of "Cutting Cantilever Beam Theory" (CCBT) was first put forward. In this theory it can be noted that the ground pressure was used for the purpose of advanced roof caving by precutting to form a cantilever beam above the gob-side gateway. When the precutting was performed on the roof of gateway, the transmission of overburden pressure was cut off, which mitigated the periodic pressure when using the 121 mining method, and part of roof rock mass was driven down, forming one side of the gateway for the next stope mining cycle. The CCBT provides a new basis for the non-pillar mining, under which the "Longwall Mining 110 method" was developed (He et al., 2007; Zhang et al., 2011; Liu and Zhang, 2013; Song and Xie; 2012; Wang and Wang; 2012; Sun et al., 2014). The method 110 means that there is one working face, after the first mining cycle, and only needs one advanced gateway excavation, while the other one is automatically formed during the last mining cycle with no coal pillars left in the mining area by using this mining technology. The core idea of 110 mining method is that, firstly, the natural ground pressure is used to help human drive down part of the roof rock, instead of fully resisting it by an artificial supporting system and by a coal pillar; secondly, the gob roof rock is used to form one side wall of the gob-side gateway; and thirdly, the characteristic of broken expand for gob roof rock is used in gob to reduce the surface subsidence. This mining method will reduce 50% of gateway excavation in the stope and fulfill 100% coal pillar recovery, which achieves a significant reduction in mining costs and more important it will reduce the accidents in the stope. In this paper, China's mining associated theories, particularly the 121 mining method and 110 mining method will be discussed. The key technologies and features will be introduced, and also the numerical simulation methods that have been used to analyze the mining-induced stress distributions in the application of 110 mining method. The CCBT and 110 mining method will be considered to be the basis for China's next-generation mining industry development.

Up to now, 110 mining method has been successfully applied in many underground coal mines in some giant coal mining groups across China, e.g. China National Coal Group Corporation, Shen Hua Group, Sichuan Coal Industry Group Limited Liability Company, etc. The total length of the gateway tunnel created by the 110 method is more than 19,000 m. In the coal mines where the 110 method are employed, engineering disasters caused by the gateway excavation and coal pillars along the goaf are almost completely eliminated, such as roof accidents, rockburst, coal and gas outburst, and the else potential dynamical-events. At the same time, great economic benefits and noticeable social benefits were achieved due to the significant reduction of the gateway tunnel excavation and elimination of the coal pillars, as well as the safety production.

2 THE CCBT AND 110 MINING METHOD

Due to the limits of the traditional 121 mining method, the CCBT and 110 mining method were proposed in order to address the problems in longwall mining. The CCBT was verified in field by using advanced roof caving, and was first applied in 2010 to No. 2442 working face in Baijiao coal mine, Sichuan. In the project, a non-pillar mining technique was used in the gateway near the goaf formed automatically by advanced pressure relief and roof caving (Zhang et al., 2011). The CCBT was established on the basis of interactions of stress fields, supports, and surrounding rocks during the process of advanced pressure release and roof caving. One of the key technologies was the orientation cutting in the goaf side roof, which alters the roof

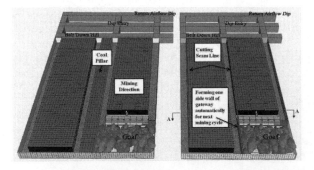

Figure 1.   3D-view schematic diagram of the long wall mining 121 mining method (left) & 110 mining method (right).

connection status and prevents stress propagation from the gob roof to entry roof. Other key technologies are involved to achieve CCBT. This is the case of a new supporting bolt or anchor with constant resistance and large deformation (CRLD), that is employed in the gob-side gateway roof supporting to keep the gateway stable during the advanced caving.

A three-dimensional schematic drawing of a coal panel mined by the 121 and 110 mining methods along the working face direction is illustrated in Fig. 1. It is seen in the figure, that the 121 mining method, requires the excavating of two gateways and retaining one coal pillar. When the coal is mined out, the two gateways will be abandoned and destroyed by the periodic pressure as a result of the mechanized for the stoping mining. In contrast, the mining of a panel using the 110 mining method requires two gateways that will be excavated during mining the first panel; in the subsequent stoping process, forming a gateway by cutting a tunnel-length slit beside the next mining panel. Therefore, the coal pillar between the two mined panels will be cancelled, i.e. only one gateway is enough.

3   KEY TECHNOLOGIES IN 110 MINING METHOD

For the fulfillment of 110 mining method, several key technologies are employed, as illustrated in Fig. 2, in a 3D view. The method includes the following steps: 1) cutting the roof with directional pre-splitting; 2) supporting the roof using CRLD bolt/cables; and 3) blocking gangue by hydraulic props. Thus, the 110 mining system with non-pillar mining and automatic formation of gob-side gateway for the next mining cycle by precutting and advanced roof caving is established.

For a more detailed description of these key technologies, another 3D schematic drawing of 110 mining methods is illustrated in Fig. 3. Firstly, cutting the roof with directional pre-splitting is performed to cut down the transmission passage of ground pressure in part of overlying rock strata, and the gob-side pressure is used to drive part of gob roof rock down, instead of totally resisting it. And the roof rock is used to form one side of the gateway wall, and the gob-side gateway is reserved for the next mining cycle.

After the working face advance, the upper strata will collapse, although the connection within upper strata is partially separated, but outside the slit depth, both part, part up the gateway and up the gob-side, are still connected and interact. When the upper strata in gob-side breaks down, between the two parts will have a relatively large shear force at this moment. The traditional support material in response to this instant impact will be destroyed, and the integrity of gateway cannot be guaranteed. Thus, for this situation, it was developed a new kind NPR (Negative Poisson Ratio) cable/anchor, the designated CRLD bolt/cable, which has a constant resistance and large deformation characteristics. When the gateway-side roof is affected by the drop-down shear force, the NPR cable would provide high resistance and at same time a large deformation. As shown in Fig. 3, the cables are applied to close the slit, forming a fulcrum, thus ensuring the integrity of the gateway.

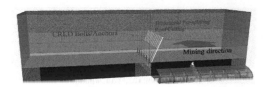

Figure 2.   3D-view schematic diagram of the 110 mining method.

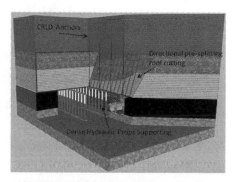

Figure 3.   Another 3D-view schematic diagram of the 110 mining method.

Finally, after the working face mined back, part of gateway roof will be driven down by the ground pressure, and gangues will be blocked by using hydraulic props and barbed wire which closed to the gob-side. So not only isolate the goaf and ensure the integrity of the gateway, and this gateway will be used by the next working face.

The following sub-sections mainly describe the mechanical properties of two key technologies.

3.1 *Directional pre-splitting roof cutting technology*

For the design of a pre-splitting roof technology the characteristics of high rock compressive strength and low tensile strength were comprehensively considered, and it was developed an appropriate blasting device to achieve the two-directional blasting in order to generate concentrated tensile stresses. The blasting device is employed with normal explosives, and the depth of boreholes is judged by the coal seam depth, gateway height and other conditions in the field. The depth of the boreholes varies from 1.5 to 5 m, or more. The explosive charge follows the general blasting design, normally from 2 to 8 packages of explosives with directional blasting device for different engineering conditions. The top plate is set in accordance with the direction of the formation of pre-splitting tensile fracture surfaces (Fig. 4). Field application results (Fig. 5) show that this technology can achieve good directional roof

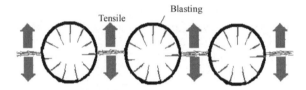

Figure 4.   Mechanism of directional pre-splitting roof cutting technology.

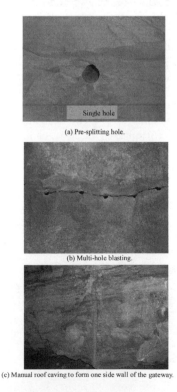

(a) Pre-splitting hole.

(b) Multi-hole blasting.

(c) Manual roof caving to form one side wall of the gateway.

Figure 5.   Photographs for field application of the pre-splitting cutting technology.

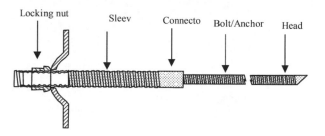

Figure 6.   The CRLD bolt/anchor.

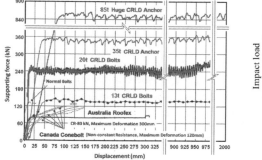

Figure 7.   Curves of different mechanical properties of CRLD bolts.

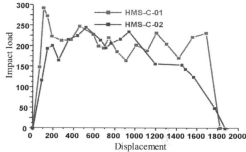

Figure 8.   Impact dynamics features of the CRLD support material.

pre-splitting according to the design at exact positions, and reach the designed depth along the roof with actively advanced pre-splitting roof cutting but will not destroy the gateway roof.

3.2   CRLD supporting system

The problems of mining pressure transfer are one of the key issues during advanced pre-splitting cutting and roof caving. Part of the roof in the existing gateway needs to be reserved. The traditional support system, with normal mesh, bolts and anchors, can be easily broken when surrounding rocks have large deformations. In this case, the manual roof caving will produce large tensile force to the gateway roof, although the precutting has been performed to reduce the force transition. For this reason, a new supporting material, the CRLD bolt, is used to control the gateway deformation and reserve the roof, as shown in Fig. 6. A large number of experiments have been conducted on this material. Testing results show that its mechanical properties are quite unique and can keep the designed constant resistance during elongation. As shown in Fig. 8, the CRLD bolt is able to adapt to the dynamic pressures generated by the gateway roof caving and effectively control part of the reserved roof. The CRLD bolt can also withstand various dynamic impacts, and high impact energy absorbing abilities are observed in both laboratory and field tests. Therefore, CRLD bolt can achieve high impact resistance and deformation energy released during roof caving, which can effectively guarantee the overall stability of gateway safety (He, 2014; He et al., 2017).

4   SHORT-ARM BEAM STRUCTURE IN 110 MINING METHOD

The CCBT was established on the basis of interactions of stress fields, supports, and surrounding rocks during the process of advanced pressure release and roof caving. One of the key technologies is the orientation cutting in the goaf side roof, which transfers the overburden pressure on the roof to the gob area. Before roof cutting, the gateway roof is one part

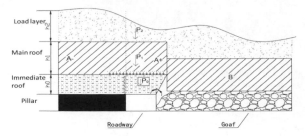

Figure 9. Structure model of short-arm beam.

of the long hanging roof structure, and their movements are intimately associated, just as in 121 mining method. After roof splitting, the gob roof strata fracture and cave under the action of roof weighting. The caved gangues expand to support and control the deformation of the upper main roof. Nevertheless, the entry roof remains stable under the entry-in support. By this way, roof above the gateway is completely protected. The pressure in the main roof pass itself through the lower portion strata, which is the part of the delivery pressure of rock can be considered to consist four boundaries: 1) Artificial boundary along the seam face. 2) The gateway roof free surface boundary. 3) The horizontal interface between disconnection and connection parts, since the deep end of seam face inside the rock formed a natural geometric parting, the plane parallel to the gateway roof and through the seam end can be considered as a stress boundary, which is passed down uniform loading. 4) At the coal wall side of the gateway forming a natural support surface, which can be considered to form a fixed end. As shown in the blue line circled on the roof rock strata in Fig. 3. Along the working face direction, the structure model of short-arm beam is shown in Fig. 9.

In Figure 9, rock formation B is the main roof above the goaf, which is driven by overburden pressure. Rock formations A− and A+ are the main roof above the pillar and gateway. Since part of main roof was cut off, rock B is complete separated from rock A+. Regarded the rock formation above the main roof as the upper load, the area of short-arm beam is between the dotted line and rock B, and under rock A+. The load on the short-arm beam consist of three parts: P_0 is the gravity of immediate roof (kN), P_1 is the gravity of main roof (kN), P_2 is the gravity of upper load (kN), h_0 is the depth of roof precutting (m), h_1 is the depth of main roof, h_2 is the depth of rock formation above the main roof. The essence of 110 mining method is the active control of the entry roof and effective utilization of the bulking characteristics of the gob roof rock. As the short-arm beam is controlled, the stability of the gateway can be well guaranteed.

5 NUMERICAL ANALYSIS OF THE MINING-INDUCED STRESSES

Based on the mining design and geological conditions of No. 1105 working face in Hecaogou 2# coal mine, Shanxi, China, a numerical model was developed including two working faces: No. 1105 and No. 1103, as green area shown in Fig. 10. The rock stress distribution was analyzed by numerical simulation. The results provide a guideline for future mining design and entry supports.

The rock mechanics properties used in the simulation are shown at Table 1. The roof strata are in ascending order sandy mudstone, gritstone. While the floor strata are in decline order argillaceous siltstone and fine sandstone. In the numerical simulation, there were seven types of strata considered, from the top to the bottom as follows: loess, sandstone, gritstone, sandy mudstone, coal, argillaceous siltstone and fine sandstone.

5.1 Boundary condition of the model

Considering the boundary effects of the model on the entry, the model size is selected to be $310 \times 55 \times 100$ m³, as shown in Fig. 11. In this case, the FLAC³D code was used to simulate the stress distribution around the gateroad (Itasca). The gravity stress is imposed to the body of the model. The displacement of X and Y directions are limited to the horizontal of the

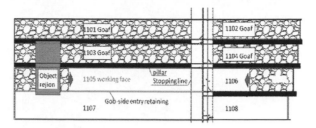

Figure 10. Panel location where the object region of numerical analysis.

Table 1. Rock mechanics properties of the surrounding rocks.

Rock formation	Bulk modulus (GPa)	Shear modulus (GPa)	Tensile strength (MPa)	Cohesion (MPa)	Internal friction angle (°)	Density (kg/m³)
Loess	1.02	0.58	0.35	0.21	21	2335
Sandstone	3.9	2.3	1.12	0.36	24	2567
Gritstone	4.12	2.45	1.31	0.49	26	2678
Sandy Mudstone	2.12	1.45	0.63	0.35	23	2587
Coal	0.85	0.48	0.32	0.18	21	1447
Argillaceous Siltstone	2.68	1.34	0.75	0.38	22	2543
Fine Sandstone	4.3	2.61	1.22	0.43	25	2693

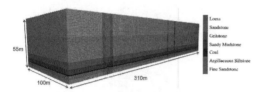

Figure 11. 3D numerical model.　　　　　　Figure 12. Boundary condition of the model.

model, and Z direction is fixed of the bottom boundary, as shown in Fig. 12. To highlight the effects of roof splitting, 110 mining method and 121 mining method were simulated in the same model, but different gateways. The roof of the air return way was cut, while the roof of the haulage gate was in its intact state (see Fig. 12).

5.2 Analysis of obtained results

During the simulation process, the variation of the vertical stresses in the front of working face seam were recorded. The simulation results were then imported into MATLAB to process. The 3D vertical stress distribution around the gateway is presented in Fig. 13.

After mining 30 m, the vertical peak stress of the seam appears in the certain range before the working face, and the stress level is gradually reduced to the original level outside the scope. Compared with the vertical stress in the face ends, 121 method side rises a vertical stress peak zone, about 20 m wide and a maximum of about 4 MPa.

With the deepening to the central region of working face, the vertical stress gradually reduced to 2.5 MPa. The other side 110 method also had a vertical stress peak zone, but only about 8 m wide and a maximum of about 3.2 MPa and trend was same with the deepening to the central region of working face.

The cycle pressure of the working face and shear stress on the coal seam were also obtained, as shown in Figs. 14 and 15. It is obvious that the cycle pressure and shear stress in

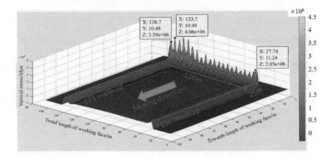

Figure 13.   Vertical stresses in the seam after mining 30 m (Unit: Pa).

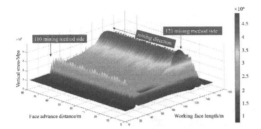

Figure 14.   The cycle pressure in the coal seam (Unit: Pa).

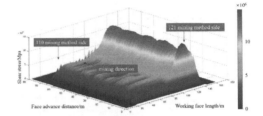

Figure 15.   The shear stress distribution of ahead regional of mining face after advancing 70 m (Unit: Pa).

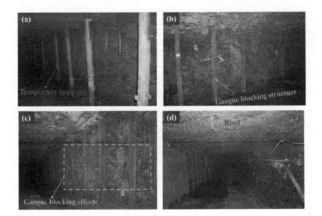

Figure 16.   Entry retaining effects using 110 mining method.

110 mining method was much smaller than those in 121 mining method. The maximum cycle pressure and shear stress in 121 mining method were 4.8 MPa and 14.3 MPa, respectively, while the maximum cycle pressure and shear stress in 110 mining method were only 3.6 MPa and 5.2 MPa, appropriately 25% and 63% decrease than those in 110 mining method. All these results indicate that the retained entry was in a low stress environment after adopting 110 mining method.

6   FIELD TEST OF 110 MINING METHOD

To verify the reliability of 110 mining method, field tests have been performed at many mines in China (He et al., 2017; Gao et al., 2017; He et al., 2015). Here, No. 1105 working face in Hecaogou 2# coal mine was taken as an example. Figure 16 shows the final entry retaining photos after adopting the new method. The gob roof collapsed into gangues along the splitting line. The caved gangues became another gateroad rib. During caving and compaction of the gangues, gangue prevention structures and metal nets were used to prevent the gangues from extending out into the gateway. It is clear that the cross-section of the retained entry could fully meet requirements of the next mining panel, revealing that 110 mining method is feasible and effective.

7 CONCLUSIONS

The paper presents major technological changes in China's mining science and technology in terms of its representative theories. Longwall 121 and 110 mining methods were introduced based on the theoretical basis, numerical analysis and field test. The main conclusions are drawn as follows:

1. The traditional longwall mining 121 method made important contributions to the development of China's mining science and technology.
2. With the increasing of mining depth, large deformation of surrounding rocks in deep tunnel becomes a challenging issue. Then the CCBT method using advanced roof caving is put forward. With the use of directional pre-splitting roof cutting, periodic pressures can be reduced or eliminated. The CCBT provides a basis for non-pillar mining and automatic tunneling technology, under which the longwall 110 mining method was established.
3. Special emphases were played on the numerical simulation of the geostress distribution found in the mining panel using the 110 method. At the same time, the stress distribution on the "short beam" left by the roof cutting when performing the 110 mining method was also investigated using numerical simulation.
4. The 110 mining methodology has been applied already with great success in many underground coal mines in China. Large benefits were introduced in terms of reducing substantially engineering disasters and in the involved costs in mining operations in tunnel excavations and by eliminating coal pillars. Therefore, a mining technology revolution in coal mines is underway in China.

REFERENCES

Fei X. The status-quo of support technology on gob-side entry retaining laneway and existing problem discussion. China Science and Technology Information, 2008, (3): 31–32 (in Chinese).

Gao Y., Liu D., Zhang X., He M. Analysis and optimization of entry etability in underground longwall mining. Sustainability, 2017, 9(11): 2079.

He M., Gao Y., Yang J., Gong W. An innovative approach for gob-side entry retaining in thick coal seam longwall mining. Energies, 2017, 10(11): 1785.

He M., Gong W., Wang J., Qi P., Tao Z., Du S., Peng Y. Development of a novel energy-absorbing bolt with extraordinarily large elongation and constant resistance. International Journal of Rock Mechanics & Mining Sciences, 2014, 67: 29–42.

He M., Li C., Gong W., Sousa L.R., Li S. Dynamic tests for a conrant-resistance-large-deformation bolt using a modified SHTB system. Tunnelling and Underground Space Technology, 2017, 64: 103–116.

He M., Zhang G., Qi G., Li Q., Jia Q., Zhou J. Stability control of surrounding rocks in deep entry of Jiahe coal mine. Journal of Mining & Safety Engineering, 2007, 24(1): 27–31 (in Chinese).

He M., Zhu G., Guo Z. Longwall mining "cutting cantilever beam theory" and 110 mining method in China-The third mining science innovation. Journal of Rock Mechanics and Geotechnical Engineering, 2015, 5: 483–492.

He M. Conception system and evaluation indexes for deep engineering. Chinese Journal of Rock Mechanics and Engineering, 2005, 24(16): 2854–2859 (in Chinese).

He M. Latest progress of soft rock mechanics and engineering in China. Journal of Rock Mechanics and Geotechnical Engineering, 2014, 6 (3). 165–179.

He M. Present situation and prospect of rock mechanics in deep mining engineering. In: Proceedings of the 8th Conference of Chinese Rock Mechanics and Engineering. Beijing: Science Press, 2004: 88–94 (in Chinese).

Itasca, Consulting Group Inc. Fast language analysis of continua in 3dimensions (version 5.0). User's Guide.

Li H. Roof strata control design for gob-side gateway. Chinese Journal of Rock Mechanics and Engineering, 2000, 19(5): 651–654 (in Chinese).

Liu X., Zhang G. Technology of roof cutting pressure relief gob-side entry retaining in soft fractured stratum. Coal Science and Technology, 2013, (S2):133–134 (in Chinese).

Liu Y., Shi P. Existing problem on long wall remaining coal pillars support mining. Journal of China Coal Society, 2007, 32(6): 565–569 (in Chinese).

Song R., Xie J. The application of pre-splitting roof cutting and pressure releasing technology at working face and gob-side gateway maintaining. Coal Science & Technology magazine, 2012, (3): 52–54 (in Chinese).

Sun X., Liu X., Liang G. Key parameters of gob-side entry retaining formed by roof cut and pressure releasing in thin coal seams. Chinese Journal of Rock Mechanics and Engineering, 2014, 33 (7):1449–1456 (in Chinese).

Wang J., Wang G. Discussion on gateway retained along goaf technology with roof breaking and pressure releasing. Coal Engineering, 2012, (1): 24–26 (in Chinese).

Zhai X., Zhou Y. Research on the filling body for gob-side gateway and its interaction with roof strata. Coal Mine Design, 1999, (8): 6–8 (in Chinese).

Zhang G., He M., Yu X., Huang Z. Research on the technique of no-pillar mining with gob-side entry formed by advanced roof caving in the protective seam in Baijiao coal mine. Journal of Mining & Safety Engineering, 2011, 28 (4): 511–516 (in Chinese).

Geomechanics and Geodynamics of Rock Masses – Litvinenko (Ed.)
© 2018 Taylor & Francis Group, London, ISBN 978-1-138-61645-5

Diagnostics and prediction of geomechanical objects state based on back analysis

Larisa Nazarova

Chinakal Institute of Mining of the Siberian Branch of the RAS, Novosibirsk, Russia

ABSTRACT: The formulations and general approaches to the back analysis problems in verification of geomechanical models and in estimation of the parameters that govern the state and properties of mine-technical objects of any scale are considered. For the illustrative purposes, the authors propose the methods: canister test data interpretation which allows quantitative evaluation of gas-kinetic characteristic od coal (gas content, coefficients of diffusion and mass exchange) by the value of pressure in the sealed vessel containing slurry coal; determination of rheological characteristics of petroliferous strata with high content of organic matter by the thermobaric test results; prediction of weak seismicity level in the Baikal rift zone, induced by variation of water-level in the Lake Baikal; diagnosis of state of antiseepage screen at liquid effluent pond dam by the piezometric measurements in terms of thermohydrodynamic model of thawed and frozen rock mass.

Keywords: non-linear geomechanical model, rock mass, coal, inverse problem, verification

1 INTRODUCTION

Geomechanical evaluation of hard mineral mining technologies, estimation of stability of structures in underground and surface mines, geodynamic zoning of areas, analysis of hydrocarbon exposure and recovery scenarios—this is a far from complete list of problems solution of which needs data on deformation-strength characteristics of rocks and reservoir properties, as well as on external natural and induced fields. The implementation approaches are:

1. traditional method (Jaeger et al., 2007) when properties of rocks (parameters of state equations) are determined at a laboratory scale, while boundary conditions are set based on direct (*in situ* stress measurements) (Zang & Stephansson, 2010) or indirect (for instance geodetic) data (Hudson, 1995; Nazarova, 1999);
2. back analysis (Sakurai, 2017) when mathematical modeling includes integrally laboratory or in situ data by minimizing objective function.

The second approach is more broad-based as it allows using data of different physical nature (electromagnetic, temperature, hydrodynamic) by means of introduction of relevant objective functions:

$$\Psi(\alpha_1,...,\alpha_m) = \sum_s \gamma_s \psi_s(\alpha_1,...,\alpha_m), \tag{1}$$

where arguments are free parameters of a selected mathematical model, and a minimum point $(\alpha_1^*,...,\alpha_m^*)$ ensures the best match of measurement data and simulation results. A proper set of non-negative weight numbers γ_s ($\mathring{a}\gamma_s = 1$) makes it possible to take into account quality of input data (relative accuracy of different measurement methods) and, sometimes, ensures unimodality of Ψ.

There are two basis types of the function ψ_s in (1):

$$\psi_s(\alpha_1,...,\alpha_m) = I_s^{-1} \sum_{i=1}^{I_s} [1 - w_s(p_i,\alpha_1,...,\alpha_m)/W_s(p_i)]^2; \qquad (2)$$

$$\psi_s(\alpha_1,...,\alpha_m) = \sqrt{I_s \sum_{i=1}^{I_s} [W_s(p_i) - w_s(p_i,\alpha_1,...,\alpha_m)]^2} \bigg/ \sum_{i=1}^{I_s} W_s(p_i), \qquad (3)$$

where $W_s(p_i)$ – input data of a s-th type recorded at an i-th point of space, and/or at an i-th moment of time; $w_s(p_i,\alpha_1,...,\alpha_m)$ – the conformable values calculated using the selected model at some values of free parameters. The functions (2) are used if W_s are obtained from measurements based on different physical principles.

The implementation of the back analysis in geomechanics has some peculiarities.

1. Most of geomechanical processes connected with mineral mining and stability of natural and anthropogenic objects are quasi-stationary processes; for this reason, the amount of input data (governing, in particular, the number m of arguments in Ψ), obtainable in a short time, is comparatively small.
2. The cost of "quantum" of information is, as a rule, much higher during in situ experimentation than in the seismic or electrical exploration.
3. For the analysis of stress state of large geological objects (lithospheric plates, cratons, terrains), the point measurements are useless; thus, it is advisable to use input data from satellite (SAR, InSAR, GPS) and seismology (trajectories of seismotectonic deformations) observations.

2 BACK ANALYSIS AND INVERSE PROBLEMS. SELECTION OF FREE PARAMETERS

Majority of continuum models describing deformation, heat and mass exchange processes in rocks are physically (more seldom, geometrically) nonlinear (except for linearly elastic models for infinitesimal strains, perhaps). There is no strict proof of existence and uniqueness of solution to the respective initial boundary value problems. For this reason, although often used in solving inverse problems (Tarantola, 2004), the gradient methods of minimum search of function of several variables (Avriel, 2003) do not much good. Therefore it is required to carry out brute force search of arguments of Ψ in *a priori* preset ranges. This approach is the ordinary practice of back analysis.

Evidently, amount and quality of input data govern resolvability of the problem

$$\Psi(\alpha_1,...,\alpha_m) \to \min \qquad (4)$$

for the selected number m of free parameters. In particular, in case that

$$\left| \alpha_q \Psi_{,\alpha_q} \right| << 1 \quad (1 \le q \le m),$$

it is impossible to evaluate α_q from the condition (4), and α_q should be withdrawn from the arguments of the objective function. Consequently, the parametric analysis of Ψ should precede solving of (4). To this effect, it is expedient to use synthetic input data

$$W_s(p_i) = (1 + \delta \xi_{is}) w_s(p_i,\alpha_1^0,...,\alpha_m^0), \qquad (5)$$

where δ – noise level, ξ_{is} – random values uniformly distributed over the interval $[-1,1]$, and $w_s(p_i,\alpha_1^0,...,\alpha_m^0)$ -exact solution of direct problem at some values of the model parameters marked with the superscript "0".

The paper presents the case studies of back analysis carried out for different scale geomechanical objects using synthetic input data, laboratory test results and *in situ* observations.

3 INTERPRETATION OF CANISTER TEST DATA: THEORY AND EXPERIMENT

3.1 *Preface*

Pre-mine degassing is an inherent element of coal mining. It allows mitigation of induced accidents in mines (coal and gas outbursts, rock bursts) (Seidle, 2011) and also furnishes an alternative and ecology-friendly energy source-methane (Nambo, 2001). Choosing an optimized pattern of degassing boreholes, as well as estimating degassing time and amount of gas recovered requires data on gas-kinetics indices of coal beds out of which the key parameter is a gas content S. The existent methods of *in situ* estimation of S are based on the lab experiments with coal samples, or borehole pressure measurements (Diamond & Schatzel, 1998, Seidle, 2011). The results are interpreted, as a rule, using integral approaches and simple models while useful information on the properties of coal and country rocks gets "lost."

3.2 *Model of gas emission in a sealed vessel*

Coal bed gas content is widely estimated using "canister test" (Standards Association of Australia, 1999) offered apparently in (Bertard et al., 1970). At a time $t = 0$ a borehole is drilled, a sample (coal slack, volume V_s) is taken out and at $t = t_0$ is placed in a sealed vessel (volume V_c, Fig. 1a), where pressure $P(t)$ is measured. Let the sample grain-size composition be known as $\{l_k, f_k\}$ $(k = 1,...,K)$, where f_k is the abundance of particles with the linear size $l_k = 2R_k$. (Fig. 1b) We replace each particle with a ball of the same volume and find the number of particles in fraction k as $n_k = 6 f_k V_s / \pi l_k^3$.

Suppose that methane instantaneously desorbs during sampling; in this case, gas emission from a spherical particle in a k-th fraction can be described using by equation (Nazarova et al., 2012)

$$C_{,t}^k = D(C_{,RR}^k + 2C_{,R}^k / R) \qquad (6)$$

with the initial condition $C^k(R_k,0) = S$ and the boundary conditions:

$$C_{,R}^k(0,t) = 0; \; C^k(R_k,t) = S(1 - t / t_0), \text{ if } t < t_0, Q^k(R_k,t) = -\beta G H(G) \text{ if } t \geq t_0. \qquad (7)$$

where $C^k = C^k(R, t)$ is the gas concentration in the particle of the k-th fraction; $G(t) = B(t) - C^k(R_k,t); B(t) = m(t) / V$; S is initial gas content; D and β are the coefficients of diffusion and mass transfer; $Q^k = -DC_{,R}^k$ is the gas flux (Fick law); H – Heaviside step function, $V = V_c - V_s$;

$$m(t) = 4\pi \sum_{k=1}^{K} n_k R_k^2 \int_{t_0}^{t} Q^k(R_k,t)dt \qquad (8)$$

is the mass of gas emitted from the sample.

Being the function of t, S, D and β the pressure P in the vessel is found from the Boyle–Mariotte law $P(t)V = p_0(V + m(t) / \rho_0)$ (ρ_0 is the gas density under the atmospheric pressure p_0)

$$P(t) = p_0(1 + B(t) / \rho_0). \qquad (9)$$

Figure 1.   Disassembled vessel (a); model of vessel interior with coal slack (b).

The system (6)–(8) was solved using the finite-difference scheme "cross" (Samarskii, 2001), with the time step 1 s and space step $0.01R_k$. The following values of the model parameters: $S = 12$ kg/m³, $V_c = 600$ cm³, $V_s = 300$ cm³, $\rho_0 = 0.72$ kg/m³ (CH_4), and $t_0 = 2$ min were chosen for simulation. The considered grading of the sample are given in the Table 1.

We considered the following measurement mode: at the times t_1 and t_2 the pressure was instantaneously dropped down to p_0, and the vessel was sealed again. Such test scheme is conditioned by the linear relationship between P and S: it is required to change the range of the pressure, otherwise the inverse problem discussed below will have non-unique solution.

The plots $P = P(t)$ at $D = 10^{-10}$ m²/s, $\beta = 10^{-8}$ m/s and various t_0 are shown in Fig. 2: with the increase of t_0 the magnitude of P decreases due to gas loss in the time interval $[0,t_0]$.

3.3 Inverse problem for finding gas-kinetic parameters by pressure measurement in a sealed vessel

Is it possible to determine gas-kinetic parameters — S, D and β — from measured pressure P(t)?

For the *in situ* experiments, five vessels were manufactured with the volume $V_c = 673$ cm³ (Fig. 1a), equipped with the sensors of pressure and temperature (absolute accuracy 10 Pa and 0.01°C, respectively). Coal was sampled from the same borehole at different t_0 in a heading face of the Berezovskaya Mine (Kuznetsk Coal Basin). Vessels with the samples were placed in a thermostat at a temperature 10°C. The temperature in the vessels varied from 8.8 to 10.3°C, therefore, on data interpretation the process was assumed isothermal.

Within 10–12 days of the experiment, pressure was dropped twice at $t_1 = 42$ hours and $t_2 = 111$ hours. The pressure $P_*(t)$ measured in the vessel No 5 is shown by the blue line in Fig. 3a. Upon the test completion, mesh analysis as well as V_s determination were performed, its results are given in Table 2.

Let us take a function $A(t,D,\beta) = [P(t,S,D,\beta) - p_0]/[P(t+t_1,S,D,\beta) - p_0]$, that, as follows from (9), is independent of S, and introduce an objective function

$$\Psi(D,\beta) = \sqrt{(T_2 - T_1)\int_{T_1}^{T_2}[A(t,D,\beta) - A_*(t)]^2\,dt}\ \Big/ \int_{T_1}^{T_2} A_*(t)\,dt$$

Table 1. Grain-size composition of coal sample used for simulation.

k	1	2	3	4	5	6
l_k, mm	0.40	1.0	2.0	3.0	5.0	8.0
f_k	0.25	0.15	0.15	0.25	0.10	0.10

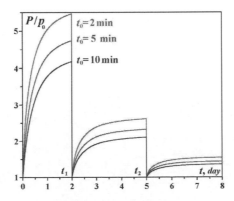

Figure 2. Pressure variation in the vessel at different times t_0.

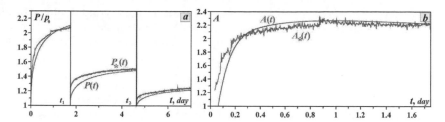

Figure 3. The pressure $P_*(t)$ in the vessel No 5 and the inverse problem solution $P(t)$ (a); the input data $A_*(t)$ and the related theoretical curve $A(t)$ (b).

Table 2. Fractional makeup of real coal samples.

Number of vessel	Average diameter of particle l_k, mm						t_0, s	V_s, cm^3
	0.13	0.40	0.78	1.5	2.5	4.0		
1	0.30	0.16	0.18	0.26	0.08	0.03	60	205
2	0.34	0.18	0.19	0.23	0.04	0.02	70	243
3	0.38	0.18	0.19	0.20	0.03	0.02	85	211
4	0.26	0.14	0.16	0.28	0.11	0.05	90	257
5	**0.28**	**0.12**	**0.14**	**0.23**	**0.13**	**0.10**	**125**	**252**

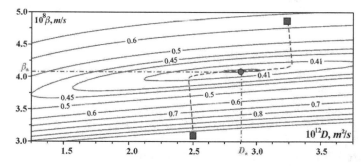

Figure 4. The isolines of the objective function Ψ, square—initial approximation of minimum search procedure, dash line—its trajectory, circle—inverse problem solution.

where $t_0 < T_1 < T_2 < t_1$, $T_2 + t_1 < t_2$ and

$$A_*(t) = (P_*(t) - p_0) / (P_*(t+t_1) - p_0) \qquad (10)$$

Figure 4 shows the level lines of Ψ at $T_1 = 7.75$ h, $T_2 = t_1$. According to the numerical experiments, Ψ is unimodal; its minimum (D_*, β_*) was searched using the modified conjugate gradient method (Nazarov et al., 2013). The blue line in Fig. 3b illustrates $A_*(t)$ calculated from (3.5) using the measured pressure $P_*(t)$, the red line in the same figure shows the theoretical curve calculated at $D_* = 2.81 \cdot 10^{-12}$ m^2/s and $\beta_* = 4.1 \cdot 10^{-8}$ m/s. With the known D_* and β_*, we can find the gas content S_* from the condition

$$\int_{t_0}^{t_1} [P(t, S, D_*, \beta_*) - P_*(t)]^2 dt \to \min.$$

The value of S_* appeared to be equal to 2.85 kg/m^3 (at a temperature of 9–10°C). The red line in Fig. 3 a plots the theoretical pressure $P(t)$ in the vessel: although the determination of

S_* used the data at $t>t_1$, there is a good conformity between P_* and P at $t>t_1$ also. The difference between the theoretical and experimental curves, in the author's opinion, ensues from insufficiently detailed determination of fractional makeup of the coal sample.

3.4 Resume

Method of interpreting the "canister test" data for the quantitative estimation of gas-kinetic parameters of coal bed (initial gas content S, coefficients of diffusion D and mass transfer β) based on the developed model of gas emission from coal sample placed in a sealed vessel. The model accounts for grain-size composition of the sample. The inverse coefficient problem is formulated to find S, D and β from the data of pressure measurements in the sealed vessel. It is shown that the inverse problem is uniquely solvable if the pressure is dropped during the experiment.

4 DETERMINATION OF RHEOLOGICAL CHARACTERISTICS OF OIL-BEARING ROCK (IN TERMS OF BAZHENITE)

4.1 Preface

The widely applied method of high-viscous oil recovery is thermal treatment of reservoirs (Baikov & Garushev, 1981), which has both positive (viscosity reduction) and negative effect, namely, drop in permeability: the increase in the temperature induces the decrease in the elasticity of rocks, which, given constant stresses, results in the higher deformation and in the narrowing of filtration channels. In order to estimate the change in the permeability and porosity of reservoirs under thermal treatment, it is required to know rheological properties of rocks. Let us discuss a simple method of finding Young modulus η and effective viscosity η of bazhenite by the data of thermobaric testing.

4.2 Interpretation of test data

During the tests, cylindrical specimens of bituminized argillite (Bazhenov Formation, Salym Field, height $H = 38$ mm, diameter $D = 28$ mm) were placed in a tube furnace. A load having the known weight created vertical stress σ. In the time $t \leq t_0$, a specimen was heated up to the temperature 150°C later on maintained for about five days (green line in Fig. 5a). At the moments t_m ($m = 1, \frac{1}{4}, 34$) the current specimen height h_m was recorded to evaluate axial strain $\varepsilon_m = 1 - h_m / H$ (circles in Fig. 5a). The process of viscoelastic deformation was described by the Voigt model (Bland, 1960):

$$E\varepsilon + \eta\dot{\varepsilon} = \sigma \qquad (11)$$

with the initial condition $\varepsilon(t_0) = 0$, where $\varepsilon(t) = 1 - h(t)/H$ - relative change in the height of the specimen. The solution (4.1) is given by

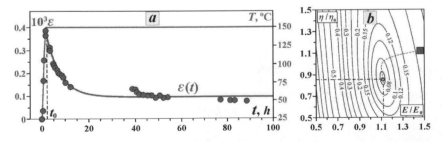

Figure 5. Test data and theoretical strain $\varepsilon(t)$ (a); the level lines of Ψ, square—initial approximation of minimum search procedure, dash line—trajectory, circle—inverse problem solution (b).

$$\varepsilon(t) = \sigma\left[1 - \exp(-E(t-t_0)/\eta\right]/E. \tag{12}$$

We introduce an objective function

$$\Psi(E,\eta) = \sqrt{30\sum_{m=5}^{34}\left[\varepsilon(t_m,E,\eta)-\varepsilon_m\right]^2}\Big/\sum_{m=5}^{34}\varepsilon_m$$

which isolines are shown in Fig. 5b. It turned out that Ψ has a single minimum $E_* = 1.29$ MPa, $\eta_* = 2.78\cdot10^{10}$ Pa·s (Fig. 5b, where $E_0 = 1.16$ MPa, $\eta_0 = 3.21\cdot10^{10}$ Pa·s); accordingly, the inverse problem of finding Young modulus and viscosity by the measured height h_m is uniquely resolvable. The blue line in Fig. 5a shows the strains $\varepsilon(t)$ calculated from (12) at $E = E_*$ and $\eta = \eta_*$.

4.3 Resume

The obtained Young modulus E_* appeared lower than in the in-place conditions by three orders of magnitude (Nemova, 2011), which is explained by the use of acoustic methods for the determination of in situ properties of rocks and by the strong dependence of rheological properties on bitumen content of bazhenite. The calculated viscosity η_* has the comparative order of magnitude with the difficult-to-weld plastic at the same temperature (Zaitsev, 1984).

5 WEAK SEISMICITY IN THE BAIKAL RIFT ZONE AND VARIATION OF WATER-LEVEL OF LAKE BAIKAL

5.1 Preface

Artificial reservoir water impoundment is known to increase seismic activity of an area, and seasonal variation of water level scales up the number and magnitude of induced dynamic events (Liu et al., 2011). Similar regularities hold true for the natural objects (Dyadkov, 2003), in particular, large water bodies (Ellsworth, 2013). Various mechanisms are proposed for the qualitative explanation of cause-and-effect relationship between seismicity and seasonal variation of physical fields (temperature, pore pressure etc.) (Nazarov et al., 2011, Huang et al., 2017). The difficulty of relating quantitatively, for instance, number and energy of earthquakes in a certain area and variation of stress field parameters lies in the its stochastic nature and, occasionally, in the lack of information required for estimating and/or modeling the latter.

Let us carry out back analysis of the spatial-temporal distribution of natural seismicity and variation of stress state in the Baikal Rift Zone (BRZ).

5.2 Geological and geophysical substantiation of geomechanical model

The selected model area of BRZ includes the Baikal Rift with the southwestern and northeastern wings. General structure of the Baikal Rift are governed, foremost, by its location close to the junction of two major tectonic elements in East Siberia - Siberian Craton and Central Asian Mobile Belt (Logachev, 2003).

Features of the modern structure and gapping of the zone may be governed by both the passive rifting ('push-off' of the Amur Plate from the Siberian Craton under the action of India-Eurasia collision) (Molnar&Tapponier, 1975) and the active rifting under the influence of bottom-up mantle flow (Dyadkov et al., 1997). The implemented geomechanical modeling of stress state in the lithosphere of Central Asia (Dyadkov et al., 1999) shows that it is most likely that both of these sources of rift genesis are active in BRZ (Dyadkov et al., 2004). Either in the first or the second case, extension takes place across the strike of the Rift according to GPS observations (Lukhnev et al., 2010), The passive rift genesis also presup poses northeastward compressing forces along the strike of the Rift.

The faulted structure of this area is examined in (Lunina et al., 2012). It is known that faults occur mostly near-vertically in the continental rift zones, including BRZ. This information as well as the data on physical properties of rocks, discontinuities (Dyadkov et al., 1999) and spatial distribution of hypocenters of quakes with the magnitude $M \geq 2.7$ (www. seismology.harvard.edu/data), which mostly take place near faulting areas, have been used in numerical modeling of BRZ.

The computational domain D was a parallelepiped 800×1800×150 km long the corresponding axis of the Cartesian coordinate system (x,y,z), BRZ was in the center of the domain, the ordinate axis was oriented to the North-East (Fig. 6a), and the z-axis — vertically downward. Deformation of blocks was described using an elastic model, deformation of faults—in accordance with the approach (Barton, 1986). The aim of this study is to relate weak seismicity of BRZ and seasonal variation in the water level of the Lake Baikal; for this reason, the side boundaries of D was set free from stresses, zero vertical displacements were set at the bottom boundary while the daylight surface of the lakescape was subjected to the vertical stresses $\sigma_{zz} = \rho g h$ (ρ-density of water, g-acceleration of gravity, $h = 1$ m-maximum water level difference in the Lake as per Federal Law 94-FZ as of May 1, 1999). The geomechanical model was implemented using FEM, the domain D was divided into 87425 tetrahedrons. Figure 6c presents the fragment of the daylight surface mesh, the faults are tinted.

Figure 7 demonstrates the numerical modeling example in terms of the distribution of the maximum shear stress τ in horizontal sections of D (spatial smoothing, span of 25 km). It is seen that the stress field is extremely nonuniform (as a consequence of the complex tectonic structure of BRZ) though the expended tendency for a reduction in τ is traceable with depth.

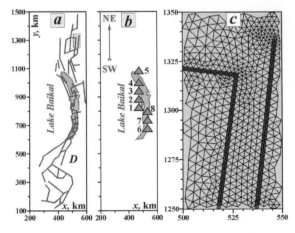

Figure 6.   BRZ faulted structure (a); arrangement of blocks under analysis (b); fragment of finite element mesh (c).

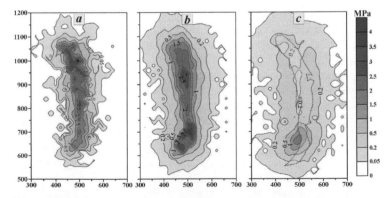

Figure 7.   Maximum shear stress τ (MPa) under seasonal variation of water level in the Lake Baikal at different horizontal sections: (a) $z = 10$ km; (b) $z = 20$ km; (c) $z = 50$ km.

Table 3. Spatial-temporal analysis results on seismic activity and stresses in the BRZ.

Number in Fig. 6b	Block	X_b, km	Y_b, km	$N_{ij}(2017)$	K_{ij}	δ, %	τ, MPa
1	$B_{9\,16}$	475	825	200	0.333	9.90	1.008
2	$B_{9\,17}$	475	875	250	0.398	8.81	1.282
3	$B_{9\,18}$	475	925	228	0.316	14.59	1.158
4	$B_{9\,21}$	475	975	358	0.569	5.22	1.789
5	$B_{9\,16}$	475	1075	262	0.450	9.07	1.299
6	$B_{10\,13}$	525	675	183	0.296	7.20	1.427
7	$B_{10\,14}$	525	725	211	0.324	4.81	1.509
8	$B_{10\,15}$	525	775	235	0.381	5.43	1.593

Comment: X_b and Y_b– coordinates of the center of a block.

Table 4. Coefficients in the empirical relation (5.2).

Blocks	A_1	A_2	δ, %
$B_{9\,16}$–$B_{9\,21}$	0.330	–0.018	7.13
$B_{10\,13}$–$B_{10\,15}$	0.513	–0.440	1.98

5.3 Analysis of weak seismicity

On evidence of (www.seismology.harvard.edu/data), over a period of 1963–2017, more than 15 thousand earthquakes with the magnitude M from 2.7 to 7.8 took place in the area under study, the depth of the hypocenters is guessed. For spatial-temporal analysis of weak seismicity, the domain D was split into parallelepipeds B_{ij} ($i = 1,…,16; j = 1,…,36$) 50×50 km in area size, and the cumulative number of earthquakes $N_{ij}(t)$ with $2.7 \leq M \leq 7.8$ between 1963 and 2017 was determined in each block (temporal sampling t — one month) with the linear approximation later on:

$$N_{ij}(t) \approx K_{ij}\,t. \tag{13}$$

Table 3 complies the calculation results for all blocks B_{ij} in the area of the Lake Baikal, where the relative error δ of the approximation (5.1) is not higher than 15% and $N_{ij}(t) > 150$. Expectedly, the events are arranged along the fault structures bordering the Lake (Fig. 6b).

The last column in Table 3 shows the average values of τ per blocks. It turns out that the number of events $N_{ij}(t)$ and coefficients K_{ij} grows with the increase in τ and can be fitted by linear functions, for example:

$$K(\tau) \approx A_1\tau + A_2 \tag{14}$$

where the coefficients A_1 and A_2 are given in Table 4.

5.4 Resume

The relations (5.1) and (5.2) correlate quantitatively the weak seismicity in BRZ and the change in stress state parameters due to seasonal variation of water level in the Baikal Lake. These relations, valid and true in the neighborhood of the Lake, are associated with the fault structures and can be used to predict level of seismic activity in the vicinity of faults.

6 IMPERVIOUS SCREEN DIAGNOSTIC OF DIKE DAM IN PERMAFROST AREA BASED ON PIEZOMETRIC DATA

6.1 Preface

Erection of various hydraulic engineering works (embankment, powerplants etc.) in permafrost zones disturbs natural heat balance, rises temperature of frozen rocks and causes new migration paths for fluid (Yershov&Williams, 2004). This results in the instability of a waterworks and in pollution of environment with production waste. Various aspects of modeling of heat-mass exchange in terms of objects in permafrost areas are discussed, for instance, in (Riseborough et al, 2008). Substantial research deals with analyzing behavior of large natural and man-made objects were carried out in (Chzhan, 2002, Velicogna et al., 2012; Buiskikh & Zamoshch, 2010).

Kumtor gold deposit was discovered in 1978 and put into operation in 1990. The deposit lies in the south-east Kyrgyzstan at a height of 4 km, in the permafrost zone (www.kumtor.kg). The winter and summer temperatures are, respectively, −35°C and +20°C (Nepomnyashchaya, 2007).

The mine uses open-pit workings method, gold is recovered by cyanidation, liquid waste goes to a tailing pond. The dam is 36 m high, 280 m wide and its extension is about 3 km (Osmonbetova, 2011). To prevent infiltration, an impermeable screen has been created: the dam surface is covered with 1.5 mm thick films that are sealed along edging. The weld seals are the weakest place of the protection cover. We proposed the diagnostic technique of screen integrity based on back analysis.

The thermohydrodynamic processes in the vicinity of the Kumtor mine dike dam and the diagnostics method for the state of the impervious screen are the subjects for study (Nazarova et al. 2015).

6.2 Direct problem

Figure 8a shows configuration of the object under analysis. The tailing pond is non-freezing over the entire service life, the fluid temperature T_w varies from 4 to 8°C. The state of the dike dam and underlying ground is monitored by means of measuring pressure and temperature at different depth in observation boreholes. As the extension of the object is much more than its lateral dimensions the analysis involves a 2D problem in the Cartesian coordinates (x, z) (Fig. 8b). The current dam state is of the prime importance, so we simplify the situation by forgetting the history of the change in the configuration of the dam in time.

The lower boundary of the computational domain (Fig. 8b) lies in the zone of permafrost at the permanent negative temperature T_i that is accepted to equal the temperature of −2°C of the neutral layer in the discussed geographical region (RF Construction norms, 1996). Let the dam and the underlying ground had temperature $T = T_i$ at the initial time moment $t = 0$, and later on the tailing pond is filled with the fluid with the temperature T_w, which is assumed constant in view of the continuity of the production process.

Thawing is much slower than the heat exchange "fluid-ground", therefore evolution of the temperature $T(t, x, z)$ is described by the equation of thermal conduction, considering convection (Bejan, 2013)

$$T_{,t} + \vec{V} \cdot \nabla T = \nabla \cdot (k_T \nabla T), \tag{15}$$

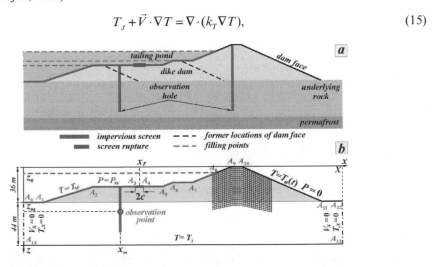

Figure 8. Scheme of object (a), boundary conditions and finite element mesh fragment (b).

where k_T is the temperature conductivity coefficient; $\vec{V}$ is the fluid velocity defined by the Darcy law:

$$\vec{V} = -k_F \nabla P / p_0,$$

k_F is the filtration coefficient; p_0 is atmospheric pressure. Filtration takes place only in the thawing zone ($T{>}0$) where the additional pressure $P(t,x,z)$ obeys the equation:

$$P_{,t} = \nabla \cdot (k_F \nabla P). \tag{16}$$

The boundary conditions are summarized in the Table 5.

The system (6.1), (6.2) with the boundary conditions (Table 5) and the initial data $T(0, x, z) = T_i$, $P(0,x,z) = 0$ was solved using FEM and the original codes (Nazarova et al., 2015). The linear size of an element was 2 m, the time step was 6 h. The calculation involved the following model parameters (Fig. 8b): $X = 600$ m, $z_0 = 2$ m, fluid density $\rho = 1020$ kg/m^3. The values of the coefficients of filtration k_F and temperature conductivity k_T corresponding to weakly permeable clays: thawed rock $k_F = 2.0 \cdot 10^6$ m/s, $k_T = 1.2 \cdot 10^6$ m^2/s; frozen rock $k_F = 0$, $k_T = 2.0 \cdot 10^6$ m^2/s (Goncharov, 2002).

6.3 *Search the location and timing of impervious screen rupture based on piezometric data*

To monitor the state of the dam, vertical observation holes are drilled (Fig. 8), temperature and pressure are measured at various depth in these holes. Let us consider how the pressure field responds the rupture of the impervious screen. Suppose that at the time $t = t_r$ at the point $x = x_r$, due to low-quality sealing (Ishchenko, 2007), a rupture $2c$ in length arises in the screen. In this case, in the section $x_r - c \le x \le x_r + c$ the hydrostatic pressure $\rho g(z_r - z_0)$ is set (Fig. 8b, Table 5), where z_r is the depth of rupture.

Evidently, the rupture is an additional source of fluid, and the infiltration toward the daylight surface of the dam intensifies. The colored lines in Fig. 9 show time variation of the

Table 5. Boundary conditions (Refer to Fig. 8b).

Section	Temperature	Pressure	Scholium
$A_0 \dots A_9$	$T = T_w$	$\vec{V} \cdot \vec{n} = 0$	intact screen
$A_3 A_4$	$T = T_w$	$P = \rho g(z_r - z_0)$	rupture of screen
$A_9 \dots A_{12}$	$T = T_a$	$P = 0$	daylight surface
$A_0 A_{14}, A_{12} A_{13}$	$T_{,x} = 0$	$V_x = 0$	symmetry
$A_{13} A_{14}$	$T = T_i$	–	neutral layer

Comment. $T_a = T_a(t)$ is the air temperature, $\vec{n}$ is the external normal to $A_0 \dots A_9$.

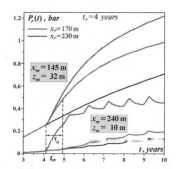

Figure 9. Pressure at observation points for different location x_r of rupture in the impervious screen.

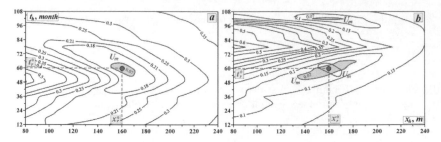

Figure 11.   Level lines of Ψ for: $x_m = 150$ m, $z_m = 32$ m (a); $x_m = 210$ m, $z_m = 30$ m (b).

pressure $P_r(t) = P(t, x_m, z_m)$ at $c = 1$ m; the coordinates x_m and z_m of the observation points as well as parameters t_r and x_r of hole are specified in the insets. For the comparison, the black lines show the pressure at the same points for the undamaged impervious screen. It turns out that the pressure sensors (accuracy 10 kPa) perceive the rupture of the impervious screen within the first months depending on the distance from the rupture position. The slope α of the curve $P_r(t)$ grows with closer distance to the hole in the screen. The behavior α is the indication of the integrity of the impervious screen.

Now, we perform the back analysis: search the time t_r and rupture point x_r in the impervious screen by the data of piezometric measurements. Let introduce an objective function:

$$\Psi(t_h, x_h) = \sqrt{t_* \int\limits_{t_m}^{t_m + t_*} [P(t, x_m, z_m, t_r, x_r) - P_*(t)]^2 \, dt} \bigg/ \int\limits_{t_m}^{t_m + t_*} P_*(t) dt, \qquad (17)$$

where $P_*(t)$ – pressure measured at the observation point (x_m, z_m), $P(t, x_m, z_m, t_r, x_r)$ – theoretical pressure at the same point, additional arguments of the function P imply that this is the direct problem (6.1), (6.2) solution at certain values of t_r and x_r. In (6.3) t_m is the start time of the pressure growth at the point (x_m, z_m); t_* is the time interval set empirically. The numerical experiments showed that t_* grows as k_S reduces; at the chosen model parameters, $t_* = 4$–6 months.

In order to examine structure of function Ψ, we synthesize input data similar to (1.4):

$$P_*(t) = [1 + \delta \xi(t)] P_0(t, x_m, z_m, t_r^0, x_r^0),$$

where δ is the relative amplitude of superimposed multiplicative noise. Figure 10 shows the isolines of Ψ for $\delta = 0.3$, $t_* = 6$ months, $x_r^0 = 160$ m, $t_r^0 = 4$ years for different x_m and z_m.

It turned out that Ψ is not always unimodal (Fig. 11b); for this reason, the objective function is to be considered for various observation points. The equivalence domain U_m (yellow-colored in Fig. 11) is a variation range of the objective function arguments such that Ψ is not higher than a certain small value ε (in our case, 0.07). The dimensions of U_m, which condition the accuracy ε of the back analysis, depend on the noise level δ in the input data.

Based on the results of the numerical experiments, it is impossible to use gradient methods to search the minimum of Ψ if the measurement points are close to the frozen zone since coefficients k_F and k_T are discontinuous functions in the vicinity of the zero isotherm and the objective function Ψ becomes nondifferentiable. Therefore the brute force method was deployed: the desired solution belongs to the intersection of the equivalence domains U_m of the objective function constructed for a number of points where the pressure is measured. Then, with the moderate noise in the input data, the back analysis problem is one-valued solvable.

6.4   Resume

In the framework of thermohydrodynamic model of dike dam in permafrost area the back analysis technique permits to find the location and onset time of impervious screen rupture by the piezometric data. Numerical experiments revealed that the formulated problem is resolvable if input data were recorded at a number of observation points and given the measurement error is moderate.

7 CONCLUSION

Verification is inherent element of geomechanical modeling of natural and man-made objects as well as deformation and failure processes in rock mass. Lab testing, full-scale experiments and field observation are the tools for acquisition of direct and indirect data for this purpose. At the same time the obtained information is only a set of numbers (sometimes noisy or stochastic which stipulates various uncertainties) unless they are adequately interpreted based on suitable model. Back analysis and inverse problems are the universal tool for accounting of information of different physical nature on the one hand and the large body of data processing technique providing the proper determination of model key parameters on the other hand. The paper exemplified various procedures for back analysis implementation in terms of small, medium and large-scale geomechanical objects.

ACKNOWLEDGEMENTS

The author is grateful for the partial support provided by the Russian Foundation for Basic Researches (Project No. 16-05-00573). Author wishes to express her appreciation and thanks to colleagues for the contribution they made in the experimental work.

REFERENCES

Avriel, M. (2003). *Nonlinear Programming: Analysis and Methods*. Dover Publishing, 532 p.
Baikov, N.K., Garushev, A.R. (1981). *Thermal Methods of Oil Recovery*. Nedra, Moscow, 286 p.
Barton, N.R. (1986). Deformation phenomena in jointed rock. *Geotechnique*, 36(2), 147–167.
Bejan, A. (2013). *Convection Heat Transfer*. Wiley, 4th Edition, 696 p.
Bertard, C., Bruyet, B., Gunther, J. (1970). Determination of desorbable gas concentration of coal (direct method). *Int. J. of Rock Mechanics and Mining Science,* 7(1), 43–65.
Bland, D.R. (1960). *The Theory of Linear Viscoelasticity*. New York. Pergamon Press. 125 p.
Buiskikh, A.A., Zamoshch, M.N. (2010). Prediction of thermal regime within a tailing dump under permafrost. *J. of Mining Science*, 46(1), 28–33.
Chzhan, R.V. (2002). Temperature Condition and Stability of Low-Head Hydraulic Power Systems and Ground Channels in Permafrost Zone. Yakutsk. Institute of Permafrostology SB RAS. 207 p.
Diamond, W.P., Schatzel, S.J. (1998). Measuring the Gas Content of Coal: A Review. *Int. J. of Coal Geology*, 35(1–4), 311–331.
Dyadkov, P.G., Nazarov, L.A., Nazarova, L.A. (1997). Numerical modeling of stress state of the Earth's crust and conditions of dynamic instability of seismically active faults during rift genesis. *Geologia i Geofizika*, 38(12), 2001–2010.
Dyadkov, P.G., Nazarov, L.A., Nazarova, L.A. (2004). 3D viscoelastic model of lithosphere in Central Asia: construction methodology and numerical experiment, *Fizich. Mezomekhanika*, 7(1), 91–101.
Dyadkov, P.G., Nazarov, L.A., Nazarova, L.A. et al. (1999). Seismic and tectonic activation of the Baikal Region in 1989–1995: experimental observations and numerical modeling of stress state changes. *Geologia i Geofizika*, 40(3), 373–386.
Dyadkov P.G. (2003). Features of induced seismicity at the Baikal Lake: relationship with the loading rate and regional activation of seismo-geodynamic process. *Fizich. Mezomekhanika*, 6(1), 55–61.
Ellsworth, W.L. (2013). Injection-induced earthquakes. *Science*, 341, 1225942.
Goncharov, S.A. (2002). *Thermodynamics*. Moscow. Publishing House of Moscow State Mining University. 440 p.
Huang, Y., Ellsworth, W.L., Beroza, G.C. (2017). Stress drops of induced and tectonic earthquakes in the central United States are indistinguishable. *Science Advances*. August 3(8), e1700772.
Hudson, J.A. (1995). Rock Testing and Site Characterization. Principles, Practice and Projects. Pergamon, 1001 p.
Ishchenko, A.V. (2007). *Infiltration Protection and Impervious Efficiency of Waterworks*. Rostov-on-Don: North Caucasian Research Center of Higher School. 256 p.
Jaeger, J.C., Cook, N.G.W., Zimmerman, R. (2007). *Fundamentals of Rock Mechanics*, 4th Edition. Wiley-Blackwell. 488 p.
Liu, S., Xu, L., Talwani, P. (2011). Reservoir-induced seismicity in the Danjiangkou Reservoir: a quantitative analysis. *Geophys. J. Int*. 185, 514–528.

Logachev N.A. (2003). Baikal Rift: history and geodynamics. *Geologia i Geofizika*, 44(5), 391–406.

Lukhnev, A.V., Sankov, V.A., Miroshnichenko, A.I. et al. (2010). Rotations and deformations of the ground surface in the Baikal–Mongolia region by GPS data. *Geologia i Geofizika*, 51(7), 1006–1017.

Lunina, O.V., Gladkov, A.S., Gladkov, A.A. (2012). Systematization of active faults for estimating seismic hazard. *Tikhookean. Geologia*, 31(1), 49–60.

Molnar, P., Tapponier, P. (1975). Cenozoic tectonics of continental collision. *Science*, 189(4201), 419–426.

Nambo, H. (2001). *The Abandoned Coal Mine Gas Project in Northern Japan*, presented at the 1st Annual Coalbed and Coal Mine Methane Conference, Denver, Colorado.

Nazarov, L.A., Nazarova, L.A., Karchevsky, A.L., Miroshnichenko, N.A. (2013). Pressure distribution in a hydrocarbon-bearing formation based on the daylight surface movement measurements. *J. of Mining Science*, 49(6), 854-861.

Nazarov, L.A., Nazarova, L.A., Yaroslavtsev, A.F. et al. (2011). Evolution of stress state and induced seismicity in operating mines. *J. of Mining Science*, 47(6), 707–713.

Nazarova, L.A. (1999). Estimating the stress and strain fields of the Earth's crust on the basis of seismotectonic data. *J. of Mining Science*, 35(1), 26–35.

Nazarova, L.A., Nazarov, L.A. Polevshchikov, G.Ya., Rodin, R.I. (2012). Inverse problem solution for estimating gas content and gas diffusion coefficient of coal. *J. of Mining Science*, 48(5), 781–788.

Nazarova, L.A., Nazarov, L.A., Dzhamanbaev, M.D., et al. (2015). Evolution of thermohydrodynamic fields at tailings dam at Kumtor mine (Kyrgyz Republic). *J. of Mining Science,* 51(1), 17–22.

Nemova, V.D. (2011). Bazhenov formation: structure, properties, lab test procedures. *SPE, Moscow section 11*.

Nepomnyashchaya, T.A. (2007). Macro-zoobenthos of the Kumtor river under anthropogenic impact. *KRSU Bulletin*, 7(1), 111–113.

Osmonbetova, D.K. (2011). Modern ecology in the Naryn river upstream zone and predictive estimates in view of the gold deposit development. *KRSU Bulletin*, 11(3), 50–53.

RF Construction norms and regulations 2.01.01-82 (1996). Construction climatology and geophysics, Ministry of Construction of Russia. 140 p. (in Russian)

Riseborough, D., Shiklomanov, N., Etzelmuller, B. et al. (2008). Recent advances in permafrost modeling. *Permafrost and Periglacial Processes*, 19, 137–156.

Sakurai, S. (2017). *Back Analysis in Rock Engineering*. London, CRC Press. 240 p.

Samarskii, A.A. (2001). *The Theory of Difference Schemes*. Marcel Dekker Inc. 762 p.

Seidle, J. (2011). *Foundations of coalbed methane reservoir engineering*. PennWell Corporation, Tulsa, Oklahoma, USA. 401 p.

Standards Association of Australia (1999). Australian Standard AS. 3980-1999: Guide to the Determination of Gas Content of Coal Seams. Direct Desorption Method. North Sydney, NSW.

Tarantola, A. (2004). *Inverse Problem Theory and Methods for Model Parameter Estimation*. Paris: Society for Industrial and Applied Mathematics. 352 p.

Velicogna, I., Tong, J., Zhang, T. Kimball, J.S. (2012). Increasing subsurface water storage in discontinuous permafrost areas of the Lena River basin, Eurasia, detected from GRACE. *Geophys. Res. Lett.*, 39, L09403.

Yershov, E.D. and Williams, P.J. (2004). *General Geocryology*. Cambridge University Press, 608 p.

Zaitsev, K.I. (1984). Welding of Plastic in Oil and Gas Industry Construction. Nedra, Moscow, 224 p.

Zang, A., Stephansson, O. (2010). *Stress Field of the Earth's Crust*. Springer Netherlands, 322 p.

Geomechanics and Geodynamics of Rock Masses – Litvinenko (Ed.)
© *2018 Taylor & Francis Group, London, ISBN 978-1-138-61645-5*

Rock mechanics and environmental engineering for energy and geo-resources

Frederic L. Pellet
MINES ParisTech—PSL Research University, France

ABSTRACT: Environmental issues become more and more pressing as the human population and its demands and needs drastically increase. In this paper, the issues raised by energy and geo-resource (both production and waste) management will be discussed with insights in rock mechanics and rock engineering. This paper will demonstrate that in most cases, a comprehensive multi-physics approach is needed to provide acceptable solutions to the assessment of encountered risks.

1 INTRODUCTION

As a new era (the 21st century) begins, the fields of rock mechanics and environmental engineering face tremendous challenges caused by the unprecedented increase in the human population and the resulting increase in its demands and associated needs. In fact, rock mechanics and rock engineering are relevant to many environmental issues, including the consequences of urban expansion, the development of transportation facilities, and the mitigation of natural and man-made hazards.

In the current paper, we will focus only on issues related to energy and geo-resource (both production and waste) management in the context of greenhouse gas limitations. It will be shown that among all these issues, many similarities exist related to the so called multi- physics couplings. Indeed, in many cases, temperature, pore pressure and changes in mechanical stresses must be linked (Selvadurai and Suvorov, 2017). Moreover, electro- chemical processes are also very often involved.

2 ENERGY AND GEO-RESOURCES

2.1 *Unconventional oil and gas*

In the last few decades, energy supply issues have led to the conclusion that the conventional production of resources (oil and gas) cannot meet the demand. Subsequent increases in the prices of resources have encouraged the development of unconventional resource extraction methods.

Unconventional resource extraction means drilling deeper (up to several kilometers) and using deviated wells that may end up in the horizontal direction. Greater depths also mean higher pressures (over 100 MPa) and high temperatures (up to 180°C). In such severe environments, new approaches involving Thermo-Hydro-Mechanical couplings were developed to analyze well bore stability and to predict pore pressure in the reservoirs (Elyasi et al., 2016).

At the same time, new geological formations were explored. All of them were tight rocks with very low permeability (less than 1 μDarcy or 10^{-18} m²). It was obvious that to recover most of the resource products (oil or gas), the reservoir permeability had to be significantly enhanced. The best way to achieve that goal was hydraulic fracking, which allows one to create paths to move the product from the reservoir to the well., However, in the case of tight rocks with reactive minerals (like the clays in shale), electro-chemical desorption is also necessary. Figure 1 shows a typical tight gas reservoir.

It should be specified that most of the geological formations under consideration are sedimentary, such as sandstone. Consequently, carbonate dissolution is also a possibility to increase the rock's porosity and permeability. This is achieved by the injection of a chemical product, such as an acidic solution (e.g., hydrochloric or hydrofluoric acid), which increases the porosity and leads to a multiple porosity medium (Gelet et al., 2012).

For coalbed methane production, CO_2 injection helps to desorb the methane molecule.

However, such methods raise some important environmental issues such as water table contamination. Therefore, the environmentally safe production of shale gas or oil requires that advanced rock mechanics investigations be performed. Hydraulic fracturing needs to be mastered, and Thermo-Hydro-Mechanical modeling, including fracture propagation prediction, must be run to assess the potential risks.

2.2 Geo-energy

Green energy production, such as the production of deep geothermal energy, also requires clear insights into rock mechanics. Indeed, some disasters, such as fault reactivation and its associated induced seismicity, can be triggered if the rock properties and geomechanical conditions are ignored (Pellet, 2017). Unlike reservoirs for unconventional resources or CO_2 sequestration, deep geothermal systems are located in igneous or plutonic crystalline geological formations. This means that a more-or-less dense network of discrete fractures pre-exists.

The efficiency of an Enhanced Geothermal System (EGS) depends on the magnitude of the exchange surface between the rock mass and the fluid which circulates from the injector well to the pumping well. To create or to increase this exchange surface, the rock is hydraulically fractured. Hence, a good understanding of the pore pressure generation and fluid flow through the fractured rock mass is required to model the fracture generation processes. In this context, dynamic fracture mechanics is a suitable tool to analyze the fractures' onset and propagation.

Today, numerical models such as the eXtended Finite Element Method (XFEM) allow one to perform such numerical modeling in simple conditions (Ngo et al. 2017). Other approaches such as the Finite Discrete Element Method (FDEM) are also available (Abuaisha et al. 2017). In any case, to perform realistic simulations, it is mandatory to account for Thermo-Hydro- Mechanical couplings (Stephansson et al., 2004). Figure 2 illustrates the concept of Enhanced Geothermal System.

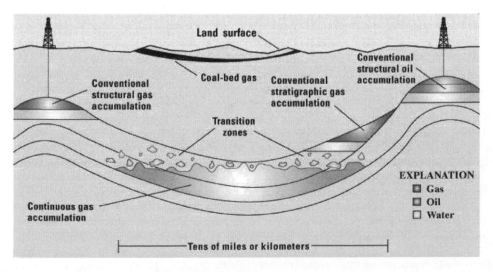

Figure 1.   Typical tight gas reservoir (source Total).

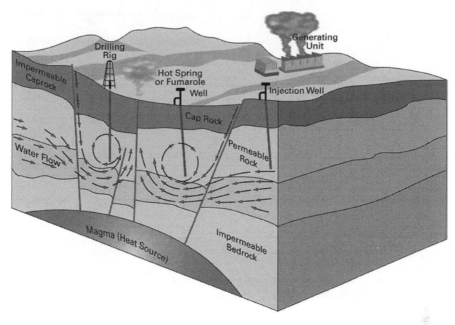

Figure 2.   Schematic of a conceptual EGS (British Geological Survey).

2.3 *Mining*

Historically, the mining industry has benefited a lot from scientific advances in rock mechanics. Mining requires work in hostile and heterogeneous surroundings that are never perfectly known. This environment constitutes the first source of dangers, such as the falling blocks in the galleries and large-scale instabilities of the rock mass, for the staff. Above all, personnel safety issues have led the mining industry to take an interest in rock mechanics and to support its development. In the past, many technologies such as rock support designs, efficient rock blasting or rock burst prediction have benefited from rock mechanics advances.

Today, the mining industry faces new challenges related to the optimization of mining exploitations and new technologies. For example, in situ leaching or "In Situ Recovery" (ISR), which is an alternative operating technique, brings new environmental concerns. ISR (Figure 3) consists of dissolving the element that one seeks to recover directly in the deposit by circulating a leaching solution between the injection wells and producers. While this technology reduces hazards (accidents, dust, radiation, etc.) for the staff and is low cost because there is no need for large uranium mill tailings deposits, groundwater restoration must be achieved after the termination of an in situ leaching operation. More specifically, the aquifer must be decontaminated from harmful chemical species. This process also requires numerical modeling accounting for hydro mechanical couplings, a good understanding of the involved chemical phenomena and reactive transport modeling.

3   UNDERGROUND STORAGE

3.1 *Carbon dioxide sequestration*

When considering energy production, it is also necessary to take into account the issue of managing any potentially harmful side effects. In this context, the reduction of carbon dioxide (CO_2) emissions to the atmosphere could be tackled by the CO_2 sequestration in geological traps such as deep reservoirs. Once again, rock mechanics is an essential tool to ensure that the storage conditions are environmentally safe. The major risk of CO_2 sequestration is groundwater table contamination.

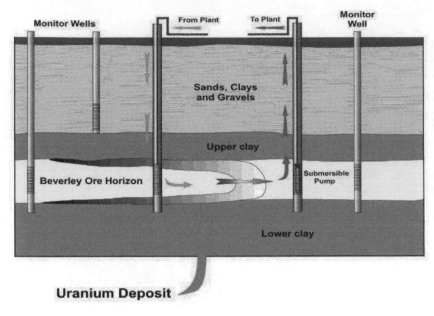

Figure 3.   Concept of in situ recovery (World Nuclear Association).

CO_2 is injected in super critical conditions that are conducive to three main possible sources of CO_2 leakages. The first one is related to the integrity of the sealing of the casing of the injector well (Lecampion et al., 2011). Indeed, the casing must be properly sealed with cement to the host rock to ensure long-term resistance. The cement annulus between the casing or the rock could be damaged by excessive cement shrinkage. Additionally, the bond between the cement and the casing or between the rock and the cement could be weakened and therefore act as a possible path for CO_2 leakage.

The second issue involves the cap rock integrity of the geological trap. The trap layers are made of argillaceous rock. Excess pore pressure and chemical reactions could contribute to a breach of the cap rock and therefore allow the gas molecules to migrate toward the biosphere (Selvadurai, 2013). Another side effect is the heave of the ground surface as explained by Siriwardane et al. (2016).

The last concern is related to fault reactivation and possible CO_2 leakage through existing faults. To prevent such failures, the geomechanical model must be properly defined. This means that the three principal stresses must be accurately determined and the excess pore pressure has to be limited to prevent fault slips (Rinaldi et al., 2014). Figure 4 presents a schematic view of CO_2 injection into a saline aquifer.

3.2   *Radioactive waste disposal*

Another concern related to energy production is the management of High Level radioactive Wastes (HLW). In many countries, one of the solutions being studied is the storage of radioactive waste in deep underground facilities. For this particular problem, time is a crucial parameter as rock mass behavior must be predicted over a very long period (several hundreds of thousands of years).

Many countries have studied this problem in different geological environments. Northern countries (Canada, Finland, Sweden) prefer crystalline rocks (e.g., granite, diorite, etc.) while others (Belgium, France, Switzerland) choose argillaceous rocks. Rock salt formations have also been under consideration in some countries (Germany, USA). In all cases, one of the major issues is the prediction of the extension of the Excavation Damage Zone (EDZ) around the

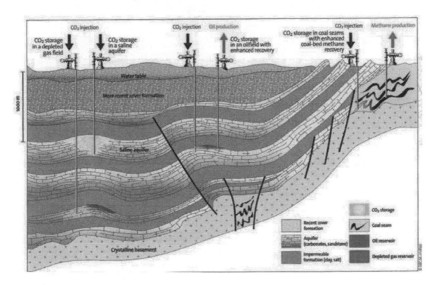

Figure 4. Schematic view of CO_2 injection different rock formations (source Total).

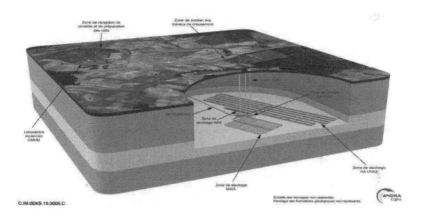

Figure 5. View of the geological disposal facility for the French project Cigéo (Armand et al. 2017).

underground openings. In fact, if cracks and fractures develop, radionuclide can spread in the geological environment and eventually contaminate the groundwater table and the biosphere.

In this context, the prediction of the EDZ extension requires numerical modeling based on a time-dependent constitutive equation (Pellet et al., 2009)., In addition, mechanically induced damage due to stress redistribution and thermo-hydro-mechanical processes have to be accounted for, as it was explained in a recent paper by Armand et al. (2017) that summarized the main outcomes of a large experimental program carried out in the Callovo-Oxfordian geological layer for the French project Cigéo (Figure 5).

3.3 *Underground storage of energy*

Currently, because of greenhouse gas regulations associated with energy demand increases, most countries are engaged in an energy transition process. The transition process is also focused on limiting the quantity of greenhouse gases released to the atmosphere. Alternative energy sources such as renewable electricity generation (e.g., solar and wind sources) are intermittent; therefore, there is a need to store energy.

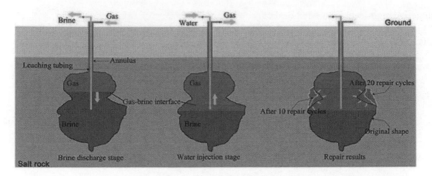

Figure 6. Different stages during the life of a cavern storage in rock salt (Li et al., 2017).

Different storage technologies are currently in use or in development. In most cases, the most suitable geological formation is rock salt because the specific properties of salt make it an ideal host rock for the location of stable, tight caverns suitable for storing gas (e.g. helium, hydrogen and methane) as well as compressed air. Salt caverns are mostly solution-mined in bedded salt or salt dome formations with sufficient thicknesses. The main feature of rock salt mechanical behavior is its propensity to creep (Bérest, 2013).

Under operation, the cycle of pumping and injecting fluid leads to drastic variations in pressure and temperature in the cavern. Modeling this cycle necessitates a thermodynamic approach including phase changes from liquid to gas and vice versa (Li et al. 2017). Recently, Rouabhi et al (2017) proposed a multiphase multi-component modeling approach to tackle this problem.

4 CONCLUSION

Environmental problems resulting from energy production and mineral resource exploitation can be largely solved through rock mechanics. One common point of all these issues is the need to take mutiphysics couplings into account.

Hydromechanical coupling associated with chemical transformations are fundamental components of the rock dissolution that occurs during CO_2 sequestration, unconventional production of oil and gas, in situ recovery of minerals and the storage of energy in rock salt formations. Hydromechanical couplings are also linked to all the problems caused by fault reactivation encountered in GeoEnergy's operation and in the storage of CO_2.

To address many of these problems, fracture propagation must be properly modeled. This is the case for radioactive waste disposal that involves a large time scale. Fracture propagation can also connect to existing faults and produce seismic waves, such as in the case of a geothermal system where dynamic fracture mechanics is useful.

For all these problems, many researchers have developed suitable constitutive equations. These constitutive models, which are rigorous on the theoretical level, are mostly calibrated against lab experiments. Few studies calibrated against in situ measurements are available. Therefore, some efforts must be made to back-analyze case studies using the inverse analysis technique. Inclusion of a probabilistic or stochastic approach would also help to more reliably assess the associated risks and would facilitate decision-making by stakeholders.

REFERENCES

AbuAisha M. (2017), Hydro-mechanically coupled FDEM framework to investigate near- wellbore fracturing in homogeneous and fractured rock formations, Journal of Petroleum Science and Engineering, 154:100–113.

Armand, G., Bumbieler, F., Conil, N., de la Vaissière, R., Bosgiraud, J.M., Vu, M.N. (2017) Main outcomes from in situ thermo-hydro-mechanical experiments programme to demonstrate feasibility of radioactive high-level waste disposal in the Callovo-Oxfordian claystone, Journal of Rock Mechanics and Geotechnical Engineering, **9**, 3: 415–427.

Bérest P., 2013: "The mechanical behavior of salt and salt caverns", Key Note Lecture. Proc. Eurock 2013, Balkema, 17–30.

Elyasi, A., Goshtasbi, K., Hashemolhosseini, H., (2016) A coupled thermo- hydro-mechanical simulation of reservoir CO_2 enhanced oil recovery, Energy & Environment, 27(5): 524–541.

Gelet, R., Loret, B., Khalili, N., (2012) Borehole stability analysis in a thermo-poro-elastic dual-porosity medium, International Journal of Rock Mechanics and Mining Sciences, vol. 50: 65–76.

Jinlong Li, Xilin Shi, Chunhe Yang, YinpingLi, Tongtao Wang, Hongling Ma, Hui Shi, Jianjun Li, Jiqin Liu (2017) Repair of irregularly shaped salt cavern gas storage by re-leaching under gas blanket, Journal of Natural Gas Science and Engineering, vol. 45: 848–859.

Lecampion, B., Quesada, D., Loizzo, M., Bunger, A., Kear, J., Deremble, L., Desroches, J. (2011) Interface debonding as a controlling mechanism for loss of well integrity: Importance for CO_2 injector wells, Energy Procedia 4: 5219–5226.

Ngo, D.T., Pellet, F.L., Bruel, D. (2017), Modeling of dynamic crack propagation under quasi-static loading, Proc. 15th International Conference of the International Association of Computer Methods and Advances in Geomechanics, IACMAG, Wuhan, China.

Pellet, F.L, Roosefid, M., Deleruyelle, F. (2009), On the 3D numerical modelling of the time- dependent development of the Damage Zone around underground galleries during and after excavation, Tunnelling and Underground Space Technology, vol 24, no 6, pp. 665–674.

Pellet, F.L. (2015), Micro-structural analysis of time-dependent cracking in shale, Environmental Geotechnics, vol. 2, n°2, pp. 78–86.

Pellet, F.L., (2016), Rock creep mechanics, Chapter 24, In Rock Mechanics and Engineering: Vol. 1 Principles, ISRM Book series, CRC Press/Balkema – Taylor & Francis Group, Leiden, pp. 745–770.

Pellet, F.L., Selvadurai, A.P.S. (2016), Rock damage mechanics, Chapter 3, In Rock Mechanics and Engineering: Vol. 1 Principles, ISRM Book series, CRC Press/Balkema – Taylor & Francis Group, Leiden, pp. 65–107.

Pellet F.L. (2017), Rock mechanics is meeting the challenge of geo-energies, Procedia Engineering, Symposium of the International Society for Rock mechanics, Eurock 2017, vol. 191, pp. 1104–1107.

Rinaldi, A.P., Rutqvist, J., Cappa, F., (2014), Geomechanical effects on CO_2 leakage through fault zones during large-scale underground injection., International Journal of Greenhouse Gas Control, 20: 117–131.

Rouabhi, A., Hévin, G., Soubeyran, A., Labaune, P., Louvet, F. (2017), A multiphase multicomponent modeling approach of underground salt cavern storage, Geomechanics for Energy and the Environment, **12:** 21–35.

Satter,A., Iqbal, G.M., Buchwalter, J.L., (2008) Practical Enhanced Reservoir Engineering: Assisted with Simulation Software, PennWell Corp., Tulsa, Oklahoma, 661 p.

Selvadurai, A.P.S. (2013), Caprock breach: A potential threat to secure geologic sequestration, Chapter 5, Geomechanics of CO_2 storage Facilities (G. Pijaudier- Cabot and J.-M. Pereira Eds.), John Wiley, Hoboken, NJ: 75–93.

Selvadurai, A.P.S., S, uvorov A.P., (2017), Thermo-Poroelasticity and Geomechanics, Cambridge University Press, Cambridge, 250 p.

Siriwardane, H.J., Gondle, R.K., Varre, S.B., Bromhal, G.S., Wilson, T.H. (2016), Geomechanical response of overburden caused by CO2 injection into a depleted oil reservoir, Journal of Rock Mechanics and Geotechnical Engineering, vol. 8, Issue 6: 860–872.

Stephansson, O., Hudson, J.A., Jing, L. (2004), Coupled Thermo-Hydro-Mechanical- Chemical Processes in Geo-Systems, Fundamentals, Modelling, Experiments and Applications, Elsevier Geo-Engineering Book series, vol. 2, Amsterdam, The Netherlands: Elsevier Scientific.

The development of geomechanical engineering in mining

V.L. Trushko
St. Petersburg Mining University, Russian Geomechanics Association, St. Petersburg, Russia

I.B. Sergeev
St. Petersburg Mining University, St. Petersburg, Russia

A.N. Shabarov
Research in Geomechanics and Mining, St. Petersburg Mining University, St. Petersburg, Russia

ABSTRACT: The article discusses new ways of setting up and developing engineering centres. It describes tasks which such centres need to solve due to the transition of the global economy to a new technological paradigm. It gives the example of the Centre for Research in Geomechanics and Mining at St. Petersburg Mining University as a successful research centre and suggests promising ways of developing engineering services for effective and safe mineral resources management.

Keywords: geomechanical engineering, centre, goals and objectives, development prospects, research and development, monitoring

1   INTRODUCTION

As the world economy is becoming more global, mining companies are facing new challenges due to the increased competition in the market, the influence of digitalization on the entire production cycle, the ever-increasing complexity of geological conditions, and other factors.

Macroeconomic trends in the industry, such as the volatility of prices for most mineral and energy resources, the growing role of alternative energy sources, the dependence of the company's capitalization on its mineral reserves, environmental policy, and compliance with social requirements (in particular that for the maximum level of safety) make it more essential than ever to optimize technical and economic performance.

Companies have to continuously carry out research and development aimed at increasing the profitability of operations, developing new deposits with complex geological structures, improving the techniques of mining and mineral processing, implementing measures for mitigating the negative impact on the environment, producing high value added products, etc.

Decades of experience gathered by Russian mining companies show that changes in the legal and regulatory framework do not take into account the impact of increasingly complex mining and geological conditions on conducting mining operations [1–3].

The development of Smart Fields technology and the use of automated miners require a major breakthrough in the engineering services sector and the development of new techniques for safe mining.

Engineering companies provide expert and consulting services to mining companies on an ongoing basis, aiming to solve specific production tasks. However, it should be noted that engineering services in their original understanding are gradually becoming a thing of the past due to the fact that it is becoming essential to introduce a new component – a scientific one – into them.

The world has moved into a post-industrial age, which is the age of information. In this new environment, a knowledge-based capital, the rapid adoption of new technology and effective management become the driving force for businesses. More and more often it proves impossible to introduce technological changes without the extensive use of scientific data and methods, such as analysis, synthesis, modelling, and the integration of different ideas and concepts with a view to designing an optimal solution [4].

As it is essential to take into consideration all design constraints and requirements simultaneously, industrial engineering should deal with the development of methodological, financial, technological, managerial, IT and other solutions, turning into a multifaceted system.

If there is no scientific component, engineering will be unable to perform its fundamental function – the research function – which involves the use of general scientific methods, concepts, experiments and logical tools for studying problems, discovering new principles and predicting processes with the aim of designing innovative solutions and products. Scientific support is often necessary in other processes based on engineering, such as design, construction, production management and production itself.

In other words, a modern engineering company should know how to combine the ability to create innovations and turn them into commercial proposals which will meet the needs of manufacturing companies.

This becomes particularly important for mining companies in light of the ever-increasing complexity of exploration and mining operations. Mining is going deeper and deeper, geological conditions are becoming more complex, mining equipment capacity is growing, new mining techniques are being introduced, and the negative environmental impact of mining is increasing. Simultaneously, geomechanical support is starting to play a more important role at all stages of mining, including design, construction, exploitation, reconstruction, and closure, as it strongly influences the effectiveness and safety of mining operations as well as the environment.

Setting up engineering centres that conduct a lot of research and are aimed at dealing with the most challenging issues in geomechanics can be considered as a step towards solving this problem.

2 ADVANTAGES OF ENGINEERING RESEARCH CENTRES

Currently, a large research centre that is part of a top university has certain advantages over private companies in terms of designing and providing engineering services:

1. the opportunity to use public money as an extra source for financing fundamental and applied research;
2. the opportunity to carry out research both for the government and private companies, meaning that private investments will stimulate the research centre to constantly analyze its performance as well as the quality of its designs;
3. the availability of complex and integrated laboratory facilities and the opportunity to set up research teams which will consist of professionals who are able to solve the entire set of scientific and production issues with the aim of conducting mining operations effectively and safely and producing marketable products;
4. a wide range of intellectual assets represented by postgraduate students and academic staff who drive and develop different branches of science;
5. the opportunity to organize international events using an established communications network that will serve as a venue for exchanging knowledge;
6. the opportunity to carry out research and development using not only standard equipment and software, but also customized solutions targeted at conducting specific research;
7. the opportunity to develop and get approval from the government for issuing new regulations aimed at ensuring safe and efficient use of mineral resources.

These advantages enable research centres to provide effective scientific and engineering support characterized by a number of synergistic effects. This is not characteristic of private engineering companies, whose range of services is substantially limited [5, 6].

In other words, today's universities are the place where innovations appear, which is due to the development of research. Universities can provide an extremely wide range of services in the mining industry, conduct interdisciplinary studies integrating engineering knowledge and the knowledge of economics, teach future professionals and help them develop both theoretical knowledge and practical skills, and serve as a link between Russian businesses and top technical universities all over the world.

Enhancing cooperation between research centres and mining companies can reveal existing technology issues and result in the development of new solutions for solving these issues. As a result, the following effects are produced:

– commercialization and implementation of new solutions, which makes the process of production more technologically developed;
– an increase in the skill level of research and production staff.

Such interaction makes it possible to develop mechanisms for assessing the effectiveness of applied research and outline promising directions for the development of new technologies, simultaneously training professionals who will be able to make use of them.

3   GOALS AND OBJECTIVES OF ENGINEERING CENTRES

The strategic goal of such centres is to carry out innovative fundamental and applied research in the growth areas of geotechnology. Based on comprehensive studies of geodynamic, geomechanical, hydrogeological and geophysical processes in undisturbed and disturbed rock masses, such engineering centres should:

– fulfil government orders for developing mineral deposits in a safe and efficient manner and solving issues in mineral resources management;
– fulfil orders coming from the mining, energy, and construction industries as well as regulatory bodies for solving issues in mining and ensuring geodynamic and environmental safety in industrially developed regions and production facilities;
– develop regulations aimed at ensuring safe and efficient use of mineral resources;
– provide scientific support to companies engaged in the design, construction, operation, conservation and closure of mining enterprises, underground oil and gas storage facilities, and other underground structures;
– conduct a pre-project analysis (analysing regulations and justifying investments) and prepare project documentation (feasibility studies and project descriptions) for new construction, reconstruction and upgrading of mining enterprises and facilities;
– participate in the assessment of geodynamically and environmentally hazardous facilities and the investigation of emergencies and industrial accidents resulting from the use of mineral resources;
– examine hazardous production facilities for industrial safety.

4   THE EXPERIENCE OF THE RESEARCH CENTRE AT ST. PETERSBURG MINING UNIVERSITY

The Centre for Research in Geomechanics and Mining was opened at St. Petersburg Mining University in 2007.

More than 350 studies on geodynamic, geomechanical, hydrogeological and surveying maintenance of mining operations have been carried out.

The centre boasts modern research facilities for studying physical and mechanical properties of rocks which can fully satisfy the needs of mining enterprises in terms of obtaining basic engineering data for solving geomechanical tasks at all stages of mining [7, 8].

Figs. 1 and 2 show equipment for analyzing acoustic emission in the process of testing rock samples.

Figure 1.   Equipment for analyzing acoustic emission in rock samples.

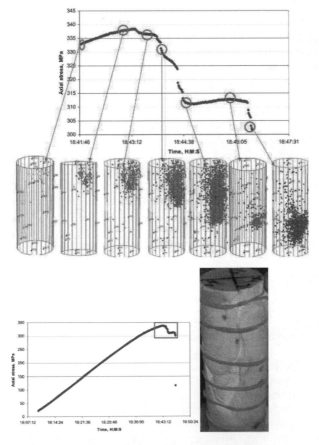

Figure 2.   Example acoustic emission research results when rock testing.

Properties of rocks are studied according to both standard methods (GOST, ISRM, ASTM) and customized methods which take into account specific geological features. Among the factors taken into account and influencing the process of making technology decisions are the stress pattern, rock temperature, groundwater conditions, methods of driving tunnels, as well as other factors.

Examples of services provided by the centre include geomechanical support of the companies which develop potash deposits in the Russian Federation (Verkhnekamskoye, Gremyachinskoye, Nivenskoye), the Republic of Kazakhstan (Zhilyanskoye), the Republic of Belarus (Starobinskoye), Turkmenia (Garlyk), the Kingdom of Thailand (Bamnet Narong), Russian gold and copper deposits, and diamond deposits [9–11].

The centre collaborates with Thyssen Schachtbau GmbH, SRK Consulting Ltd, and Weatherford in studying rock properties, which means that it has to constantly work on aligning regulatory policies existing in Russia and other countries.

A required aspect of deposit development geomechanical support, enabling effective management of processing, storage and interpretation of the results of various geomechanical engineering works, is drawing geomechanical models. The examples of making geomechanical models which contain data on physical and mechanical properties of rocks of host massif, its structural disturbance and stability assessment are given in Fig. 3 [12,13]. To obtain these data, it is necessary to perform the workload to determine physical and mechanical properties, to do

a)

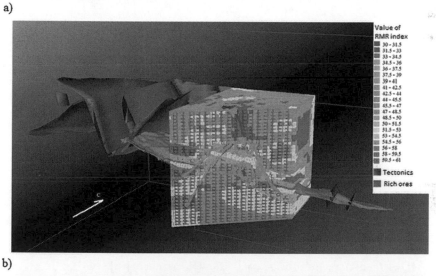

b)

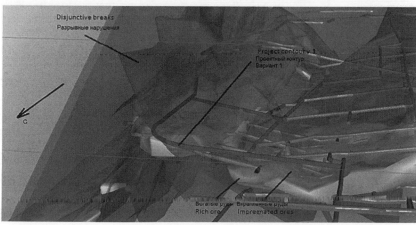

Figure 3.   Example of wireframe (a) and block (b) models of geotechnical factors visualization (RMR).

the fracturing mapping in the field, empirical and analytical calculations, mathematical modeling The Scientific Center of Geomechanics has developed, updated, added data or audited such models for a number of deposits: Goltsovoye (ZAO "Serebro Magadana), Maiskoye (OOO Gold Mining Company "Maiskoye"), Kukisvumchorr, Yukspor, Apatite Circus, Rasvumchorr Plato, Koashvinskoye (AO KF "Apatit"), Yakovlevskoye (OOO "Metalgroup") [14–16].

Such approach has allowed the researchers to study working conditions on the site of the "Big Gorst" Taimyr mine of PJSC MMC Norilsk Nickel in details. In this area, characterized by extremely difficult working conditions, the whole complex of geomechanical engineering surveys was carried out: the fracturing mapping was performed, rocks samples were selected and tested, a numerical forecasting model of stress state was built. The results enabled formulating recommendations for the works development on the site.

In general, the Center has completed a number of works associated with unique conditions of underground mining operations: (OOO "EMC Mining", OOO "SPb-Giproshakht", OOO "Himgorproekt", AO "Polymetal", OOO "Gipronickel Institute", OOO "Mining Chemical Engineering"), (PAO Gaisky GOK, AO Elkonsky GMK, OAO South Verkhoyansk Mining Company, PAO Priargunskoe Production Mining and Chemical Association, OAO "Combinat Magnezit"), AO KF "Apatite", OOO "Metalgroup" [17, 18].

A new level of geomechanical engineering for the Scientific Center is the implementation of the Comprehensive Programme of Instrumental-Methodological Works for ensuring safe operating of apatite nephelinic ores of AO Apatit. The works are being carried out in the following directions:

– defining parameters of the structure and morphology of the rock massif of the AO KF "Apatit" deposits;
– defining parameters and forecast of the rock massive stress state;
– analysing, processing and applying the region seismic activity data;
– creating and developing regional and local seismoacoustic and deformation methods of the massif control;
– creating databases, performing their comprehensive analysis and training artificial neural networks;
– hydrogeology works;
– Open mining maintenance studies.

Concerning the scientific and engineering support of open mining, both traditional and modern methods, tools and approaches are applied. In particular, the fracturing mapping is performed by photogrammetry and further processing of results by Sirovision program, Datamine software, the slopes stability is assessed in two-dimensional and three-dimensional formulation by the methods of limit equilibrium and finite elements using both own developments (MOFIS) and foreign software complexes (Rocscience) [19].

Complex geodynamic monitoring systems have been developed to forecast and prevent hazards while mining.

While carrying out researches, in addition to solving practical problems, the important scientific regularities are determined, among which is the dependence of the change in the displacements of the earth surface as it moves away from the dewatering wells. The obtained results made it possible to develop recommendations on improving the monitoring system for the slope stability and optimizing the technology of laying new starting points at a distance where the effect of drainage measures on the stability of their elevation marks is minimal.

5 DEVELOPMENT PERSPECTIVES

The promising direction in engineering service is the development of express methods of property studying and methods of the rock massif stress assessment using the "memory" effects of rocks. Express methods can simultaneously determine the complex of the rock properties (compressive resistance, tensile strength, deformation characteristics, strength characteristics) when testing samples of free shape. Based on the results of laboratory tests of the core of geological wells the stress estimation methods using the "memory" effects allows to determine the quantitative values of the main stresses of the massif at the time of core extraction.

The development of shaft-free and unmanned mining processes and their geomechanical monitoring, as well as the introduction of cloud technologies and satellite monitoring for geomechanical analysis of mining facilities is gaining the importance.

The directions of long-term development must be realized on the principle of the transition of quantity to quality. The accumulated results of geomechanical engineering should be used in the development of the regulatory and methodological framework, which should unify the approaches both in carrying out engineering work, and in developing methodological guidelines for determining the parameters of safe underworking constructions, determining the parameters of excavation support, size of pillars, stable outcrops for enterprises.

As the world experience shows, the best and reliable methods for determining the parameters of development systems are the methods obtained by statistical processing of data of real mining situations: rock bumps, deformation of timbering, pillars, loss of stability, or, on the contrary, retaining the stability of rock outcrops. In this regard, the development of both a regulatory framework, founded on engineering methods, and fundamental researches should accumulate the data received from mining enterprises. It should be noted that the processing of such data with modern analysis tools, such as artificial neural networks, provides means for obtaining the most reliable methods to forecast dangerous geomechanical and geodynamic phenomena.

REFERENCES

[1] Ensuring regulations for slope stability on coal mines/SPb: VNIMI, 1999. – 208 p.

[2] Methodological guidelines for determining the dimensions of chambers and pillars for underground mining of non-ferrous metal ores/Chita, 1988. – 126 p. (Ministry of Nonferrous Metallurgy of the USSR. All-Union Scientific Research, Design and Design Institute of Mining of Nonferrous Metallurgy VNIPIgortsvetmet, Chita branch).

[3] The first meeting on the organization of an all-Russian project for the preparation of a regulatory document on the justification of the stability of the of the side stability of quarries and tailings/ Mining Industry – No. 4, 2017. – p. 16–17.

[4] Sergeev I.B., Ponomarenko T.V. (Saint-Petersburg Mining University), Evolving and evaluation synergy effect in the integration of mining companies.—«Mining Information and Analytical Bulletin (scientific and technical journal)», M: «Limited Liability Company «Gornaya kniga» – 2013. No. 6. p. 316–323.

[5] Sergeev I.B., Lebedeva O.Y. (Saint-Petersburg Mining University), Universities and mining companies: cooperation for sustainable development.—«The Eurasian Scientific Journal», M: «Limited liability company «Publishing company «World of science», LLC – 2015. V. 7. No. 2 (27). p. 65.

[6] Sergeev I.B., Mineeva A.S. (Saint-Petersburg Mining University), State incentives for increasing the energy efficiency of industrial production: analysis of international experience.—«Mining Information and Analytical Bulletin (scientific and technical journal)», M: «Limited Liability Company «Gornaya kniga» – 2015. No. 40. p. 61–74.

[7] Ilinov M.D., Karmanskii A.T., Korshunov V.A., Gizatulina I.N. (Saint-Petersburg Mining University), Laboratory tests of the strength and filtration-capacitive parameters of coals under complex stress conditions regarding preliminary degassing of coal seams.—«GAS Industry», M.: «Limited Liability Company «Camelot Publishing». – No.: S672 (672) 2012 68–71.

[8] Repko A.A., Mezhuev V.G., Ilinov M.D., (Saint-Petersburg Mining University), Samorodov B.N., Marysyuk V.P., Babkin E.A. (ZF JSC "MMC "Norilsk Nickel".) A new method for studying the geomechanics of rock massifs. – Mining Journal. M: Ore & Metals Publishing house. – No. 12, 2004, p. 71–75.

[9] Shabarov A.N., Tcirel S.V., Morozov K.V., Rasskazov I.Y. The concept of complex geodynamic monitoring in underground mining/Mining Journal, Ore & Metals Publishing house – No. 9, 2017. – p. 59–64.

[10] Shabarov A.N., Tcirel S.V (Saint-Petersburg Mining University). Provision of geodynamic safety in underground mining/Mining Journal, M., Ore & Metals Publishing house – No. 9, 2017, – p. 65–70.

[11] Tcirel S.V., Shabarov A.N., Prosvetova A.A. (Saint-Petersburg Mining University), Forecasting risk assessment of geodynamic hazards in the design of mining operations. – « Mining Information and Analytical Bulletin (scientific and technical journal)», M: «Gornaya kniga», – No. 4, 2015, – p. 323–326.

[12] Kosuhin N.I., Shabarov A.N., Sidorov D.V. (Saint-Petersburg Mining University), Assessment of the stress-strain and shock-hazard state of the rock massif in the development of the Talnakh and Oktyabrskoye deposits in the zones of influence of large-amplitude tectonic disturbances. – «Marksheiyderskiiy vestnik», M «Scientific Research and Project Design Institute of Mining and Non-ferrous Metallurgy», 2015. No. 6. p. 39–42.

[13] Shabarov A.N., Zubkov V.V., Krotov N.V., (Saint-Petersburg Mining University) Podosenov A.A., Marshak N.A. (SPb-Hyproshaht, Saint-Petersburg), Design selection exploration of deposits taking into account the results of geodynamic zoning and geometrization of hazardous zones. – «Journal of Mining Institute», SPb «Saint-Petersburg Mining University».

[14] Tcirel S.V., Pavlovich A.A., Budilova V.V. Application of various methods and approaches for geomechanical justification of quarry side parameters/Mine surveying and subsoil, JLC Geomar Nedra – No. 6 (80), 2015. – p. 41–45.

[15] Trushko V.L., Protosenya A. G., Ochkurov V.I. (Saint-Petersburg Mining University), Prediction of the Geomechanically Safe Parameters of the Stopes during the Rich iron Ores Development under the Complex Mining and Geological Conditions. – International Journal of Applied Engineering Research, Vol. 11, No. 22, 2016.

[16] Tcirel S.V., Noskov V.A., Korchak P.A., Zhukova S.A. Estimation of economic efficiency of forecasting and management of the geodynamic state of the massif/Mining Journal, Ore & Metals Publishing house – No. 3, 2017. – p. 26–31.

[17] Trushko V.L., Trushko O.V., Potemkin D.A. (Saint-Petersburg Mining University), Provision of Stability of Development Mining at Yakovlevo Iron Ore Deposit. – International Journal of Applied Engineering Research, Vol. 11, No. 18, 2016.

[18] Trushko V.L., Protosenya A.G., Dashko R.E. (Saint-Petersburg Mining University), Geomechanical and hydrogeological problems of development of the Yakovlevsky deposit. – ««Journal of Mining Institute», SPb « Saint-Petersburg Mining University » – Vol. 185, SPb, 2010.

[19] Guidelines for open pit slope design/Editors John Read, Peter Stacey. – CRC Press/Balkema, 2009. – 509 p.

Review papers

Geomechanics and Geodynamics of Rock Masses – Litvinenko (Ed.)
© *2018 Taylor & Francis Group, London, ISBN 978-1-138-61645-5*

Geodynamic safety of subsurface management

Arkady N. Shabarov & Sergey V. Tsirel
*Research Center of Geomechanics and Mining Industry Issues, Saint-Petersburg Mining University,
St Petersburg, Russia*

ABSTRACT: Geodynamic risks presuppose breakdown risks in technological processes and engineering objects mining working, oil wells, pipelines and other communications, buildings and constructions, impact on people's and animals' health related to dynamic processes in the upper layer of the earth crust and often caused by natural phenomena as well as interaction of both natural and anthropogenic factors. The most challenging problems with geodynamical safeguarding occur in the old mining regions where the natural geodynamical balance of rock mass is disturbed.

The study evidences that active faults and phenomena related to them (geomechanical, hydrogeological, geoelectrochemical, etc.) contribute the most to geodynamical hazards. The hypocenters of most earthquakes and rock-tectonic bursts, more than two third of rock bursts and sudden gas coal outbursts in mines, up to half trunk pipelines breakages, more than half accidents in city underground communications, not less than a quarter of development of karst and gullying events, landslides, etc. account for the fault zones, comprising the first percentage of the Earth surface total area.

Manifestations of faults activity and fault zone properties, affecting mining, hydrocarbons extraction and other industry-related objects are considered in detail. Different fault zones are classified by risk type and level. It is shown that both mining and hydrocarbons extraction is affected by the most compressed spans of faults where they are closed joints in rock mass, notable for extreme strength and burst-hazard. Another relevant type of fault zones are split and tension zones, affecting the risks of above-ground and buried communications breakdowns; breakouts, rock falls and excessive water inflows occur when mining. Geological factors to identify fault zones of different types are described. It is proved that although mining alters the stress pattern of rock mass, most of hazards account for the faults already existing in the rock mass with abnormal tectonic stresses.

Three stages of hazardous dynamical events prevention are specified: geodynamical zoning and risk assessment, geodynamical survey, measures to reduce hazards. Geodynamical zoning technique which allows to identify potentially hazardous zones at early stages is considered in detail. The technique involves active faults identifying, block structure reconstruction, calculation of stress pattern with the help of mathematical and physical modeling, direct and indirect estimates of rock tension; exploring mechanisms of dangerous geodynamical events both above ground (pipelines breakages) and underground (rock and tectonic bursts). Techniques of comprehensive geodynamical survey in mines and shafts are considered, including different types of survey—seismic, deformational, acoustic, electro-magnetic, etc., and techniques of prognosis and prevention of dangerous events as well.

By way of illustration rock bursts prognosis in mines, open-pit slopes and benches stability analysis, mineral deposits locating and oil production stimulation and potentially breakage zones in trunk pipelines and underground communications are considered in the study.

Keywords: geodynamical hazards, faults, stress, deformation, rock bumps, mines, open pits, pipelines, underground communications

1 INTRODUCTION

Geodynamic risks presuppose breakdown risks in technological processes and engineering objects, mining working, oil wells, pipelines and other communications, buildings and constructions, impact on people's and animals' health related to dynamic processes in the upper layer of the earth crust and often caused by natural phenomena as well as interaction of natural and anthropogenic factors.

Notwithstanding development of technologies and the safeguards, the level of various geodynamic hazards in mining and other kinds of management subsurface resources keeps being high for reasons:

- growing underground space developing intensity; qualitative and quantitative forms of subsurface resources management;
- development of deeper, more complex deposits in difficult mining and geological conditions;
- general growth of geodynamic and seismic activity on the Earth starting in the 90 s and the 2000 decades;
- intensive mineral resources extraction technologies and underground and buried constructions building give the rock mass less time for stress relief and «getting accustomed» to new geomechanical conditions;
- a number of factors in old mining regions, firstly, large amount of extracted and moved rock mass, noticeable changes in hydro-geological modes cause an unstable state of regional geodynamical conditions, new faults propagation, seismic activity increase.

Though engineers in many fields encounter the above-mentioned geodynamic hazards, they are usually considered separately, in spite of their common character. In certain areas of engineering, for instance, when buried pipelines and various underground communications constructing, these hazards are given insufficient attention.

According to the analysis of different forms of mineral resources' use, the majority of various accidents occur in proximity of disjunctive breaks. Summarizing different studies [4–8 and others] we can conclude that the fault zones comprising the first percentage of total size of the Earth surface account for the following:

- epicenters and hypocenters of majority of earthquakes and mining-tectonic pulses (anthropogenic earthquakes);
- more than two-thirds of mining pulses and sudden bumps of coal and gas in coalmines;
- two-thirds of substructure damages;
- one third of track structure damages;
- up to a half of accidents on trunk pipelines;
- half of accidents on underground city communications of different uses;
- a quarter of cases of development of karst, mountain slip, arroyo cutting.

There exists statistically justified research which show that zones of influence of certain faults somehow cause anomaly of spreading different diseases.

Faults divide rock mass into geodynamic blocks. Geodynamic blocks have various stress patterns, different reservoir properties, fluid content, and, accordingly, its impacts on mining works, underground and buried facilities are different.

2 CONSIDERING THE ABOVE-MENTIONED INFORMATION, WE ARGUE THAT ACTIVE FAULTS, RECONSTRUCTION OF MASS BLOCK SYSTEM AND ROCK MASS STATE CONTROL IS THE BASIS OF GEODYNAMICAL SAFEGUARD IN MINING

For this reason, a technology including the three basic stages has been developed:

- geodynamic regional patterns, identification of potentially dangerous zones;
- geodynamic monitoring;
- measures to reduce hazards.

For the first time the technology has been developed for underground mining [4,5,9], then it was used for open cast mining, deposits' exploration and hydrocarbons' extraction, underground construction, pipelines, engineering communications, railways, etc. safeguarding.

At the first stage hazard zones are identified and hazards based on static 2D and 3D geological and structure models and stress-pattern fields are evaluated; besides, initial image of virgin rock is created by the means of geodynamic zoning (see below); further models consider geometric changes, properties and stress pattern of the mass, caused by mining and other kinds of mineral resources extraction.

However, static models do not take into account either all initial properties of rock mass or processes developed gradually including creep and formation of joints. Hence, geodynamic survey of the mass state in proximity of protected objects (mainly mining roadways) in real-time mode is needed. It can identify current hazard zones which need taking anti-burst measures and the ones to lower hazards.

2.1 *Geodynamic zoning of subsurface resources*

In geology essential means of locating faults are structure patterns based on drilling data and areal geophysical research.

However, application of these measures to identify potentially dangerous zones related to faulting is not sufficient. On the one hand, quite a few of known faults with wide amplitude of shifting can be recovered now, are inactive and not dangerous. On the other hand, newly-appearing disjunctive breaks with small amplitude of shifting can be more dangerous for above-ground constructions as well as for mining roadways. Insufficient density of wells does not allow to divide forming disjunctive and plicative breakages as well as to define accurately the position of a fault on the space between wells.

Consequently, in order to locate active faults and relevant block structure, other methods such as geophysical, geochemical and geomorphological, etc. should be used. Implementation of geodynamic zoning showed that two approaches are most effective—decryption of images of aerospace and cosmic space as well as analysis of forms of relief (morphometric approach). For decryption of cosmic-space images two methods are used: manual working of images or the one with a computer using appropriate software. Significance of morphometry is defined by the fact that for every known structure we can make more or less the image of it on relief and hydrographic network as a whole system of geomorphological system (exact task of morphometry). In turn, it can give opportunity to point out unknown structures using such properties ('reverse task') [9, 10].

Combination of methods of fault identification, reconstruction of current block structure, evaluation of its stress pattern and geodynamic hazards for various industrial objects in Russia is called **geodynamic zoning of mineral resources** [4]. While carrying out geodynamic zoning it is admitted that earth crust and its surface are hierarchical fractal structure in which one can point out up to 10 measuring levels (ranks).

The study by M.A. Sadovskiy, L.G. Bolhovitinov, V.F. Pisarenko [11] shows that geometric progression of ranks is not only mathematical abstraction that is convenient for analysis of tectonic structure as well as genuine structures of geological environment. According to [11] there are some isolated sizes of geological structures with transition coefficients (common ratio of geometric progression) which are inserted in each other. They range from 2,5 to 5 (3,5 on the average). Other researchers, first of all V.S. Kuksenko [12], consider transition coefficient which, according to different estimations, ranges from 2,7 to 5–7 to be an approximately constant quantity. The paper [13] presents a compromising opinion stating that there are isolated sizes caused by different geological factors, which can provide 1–2 levels down and up on the scale of sizes with transition coefficients ranging from 2,5–3 to 5–6 due to the processes of fragmenting and consolidation. At the same time connection of segments of various geometric progressions in a single row is often a formal process. Thus, notwithstanding the presence of formal constructions in the hierarchy of measuring levels of block structure, main measuring levels show a genuine structure of massif, its division into blocks with different strain pattern and other properties.

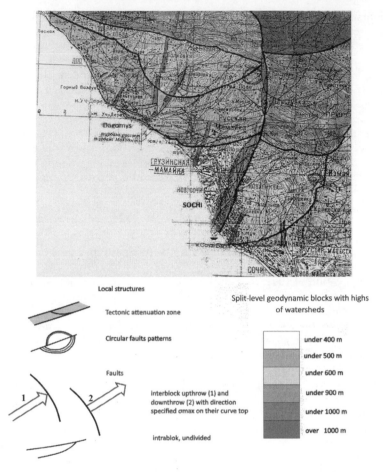

Figure 1. Example of geodynamic zoning with isolation of different types of faults, geodynamic blocks, tectonic attenuation zones and deformations (according to Y.G. Kobilyanskiy, coast of Northern Caucasus).

Mathematical modeling of stress pattern of rock mass with the help of various methods (finite element method, boundary element method, boundary integral equation method, etc.) is carried out on a reconstructed block structure of a deposit. Based on the structural and geological model of estimated stress as well as combination of other factors characterizing geodynamic activity (intensity of current tectonic movements, seismic activity, water inflow, methane abundance) final geodynamic model is designed. In order to justify the role of every factor, experience of mining works in dangerous deposits is used. It is analyzed with the help of different logical and statistical methods, including neural networks. Based on geodynamic model, potentially dangerous zones (zones of tectonic strain) and zones of tectonic weakening are identified [12].

2.2 *Geodynamic hazards in different blocks*

Estimation of potential geodynamic hazard includes:

- calculation of stress pattern with the help of mathematical modeling for initial state and in the process of building and exploitation of mines as well as other buried and underground objects;
- evaluation of rock-bump hazards, their aptitude to accumulate elastic energy;
- evaluation of other hazards connected with content and movement of fluids (underground waters, methane, etc.);

– evaluation of tectonic deformation of blocks.

Let us describe tectonic deformation in detail [14], it characterizes change of rock mass properties as well as current strain state. Figure 2 shows the method of evaluation of tectonic shearing deformations.

Figure 3 shows differences in number of dynamic conditions in blocks of various level of deformation. Significant differences also occur at other deposits, first of all, the coal ones (Kuzbass, Donbass, Vorkutskoye deposit and others).

$$D = dd_1 / ad$$

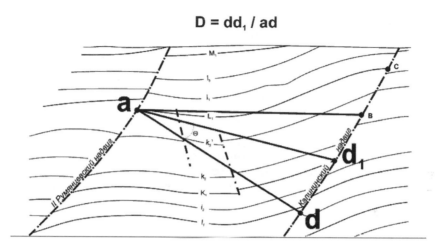

Figure 2.   Method of evaluation of tectonic shearing deformation of D for geodynamic blocks.

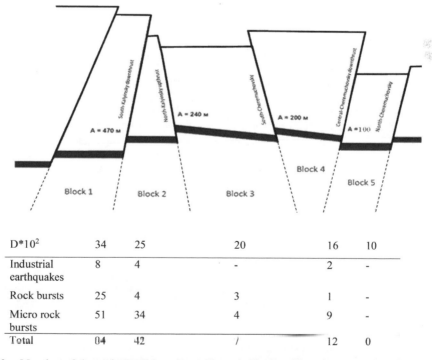

D*10²	34	25	20	16	10
Industrial earthquakes	8	4	-	2	-
Rock bursts	25	4	3	1	-
Micro rock bursts	51	34	4	9	-
Total	84	42	7	12	0

Figure 3.   Number of dynamic conditions in geodynamic blocks with various tectonic deformation of D at Severouralskoye bauxite deposit.

2.3 Specificities of fault zones

Though number of accidents and dynamic conditions differ significantly in blocks with different level of strain and deformation, majority of them takes place at border faults or intra-block ones. What are the reasons of such strong impact of faults on extraction of mineral resources, different forms of subsurface resources exploitation and even other aspects of people's life?

Upon first consideration it is possible to identify the following specificities of fault zones and forms of influence on subsurface resources exploitation related to them.

1. Variations of geomechanical characteristics of geological environment. Geodynamic activity of faults is primarily occurrence of anomalies of tectonic movements, both direct and oscillating, which take place currently, as well as stress state of rock mass.
2. Another specificity of fault zones, which is connected with the first one, is accumulated changes of the state and structure of rock mass. The most typical changes are increased rock jointing, revealed in the growing number of joints' systems of different orientation as well as in reducing the distance between them. In addition, not only does the average number of joints per unit length change but also distribution of distance between joints (size of separate ones). If distribution of distance between joints inside blocks is more or less symmetrical and can be approximated normal distribution, in the fault zones distribution becomes noticeably asymmetrical and is characterized as logarithmically normal distribution [16] (according to other data—exponential distribution or distribution of Rozin-Rammler [17, 18]).
   At the same time an increase of jointing, formation of fragmenting zones is not the only option of changing properties and structure of massif in fault zones. Under the influence of high voltage reduction of openness, closing of joints, increase of pore-water and gas pressure, solidification of the structure of rock mass and increase of its capacity to accumulate elastic energy can take place [5, 15].
3. Faults are basic ways of fluid movement, liquid and gas, which can be both indicator of activity and risk factor. In fault zones metamorphism and metasomatism resulting in essential changes of physical and mechanical and chemical properties of rock mass pass intensively. Attenuated zones of increased jointing around faults and especially their crossing may accumulate various fluids. Consequence of these processes is formation of veined deposits as well as concentration of coal methane which is both a risk factor while coal extraction and a mineral resource.
4. In the fault zones various anomalies of physical fields take place: gravitational, magnetic, electrical, thermal. Specific properties of fault zones, first of all concentration of tectonic mobility, make them more sensitive to external impacts. Movements caused by tidal impacts of the Sun and the Moon varying from semi-diurnal to long-period cycles take place primarily on border faults between tectonic blocks of different ranks [19,20].
   Also, mobile fault zones can respond to various oscillating processes including natural (meteorological, microseisms, seismic waves) as well as anthropogenic (industrial explosions, transportation vibrations). There is a supposition that small-amplitude reversed alternate motions of rock mass and subsoil in fault zones make a significant negative impact on strength of construction materials (fatigue failure) as well as supporting strength of subsoil.

2.4 Basic geodynamic hazards in subsurface resources extraction

Experience in mining activity shows that the greatest danger in subsurface resources extraction is connected with stress pattern and properties of rocks near faults, while in coal deposits it is related to collecting properties that lead to some variations of gas pressure and methane accumulation. Besides, as studies have shown [15, 21], intersection of two types of zones that are connected with disjunctive breaks influences mining works to a great extent. The most intensive tectonic strained zones are confined to closure of discontinuities, their bends, junctions with branch faults and all sites, where commissure of dislocation is a closely connected joint (zones of the 1st type, pic. 1). However, those parts where discontinuity is the zone of fragmenting are working less intensively, while zones of increased stresses are located on the

Figure 4. Typical forms of tectonic discontinuities of types I and II and related systems of joints.

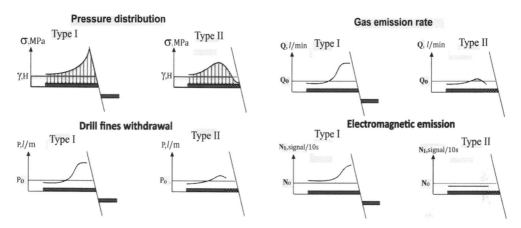

Figure 5. Typical allocation of characteristics for fault zones of types I and II.

far distance from discontinuity and have significantly less intensity (zones of the 2nd type, pic. 4). Graphs of changing stress state and properties of rock massif in fault zones of two types are shown using an example of coal stratum on pic. 5.

Fault zones of parts of tectonic dislocations (types I and II), have typical geological properties, firstly properties of jointing as well as collecting properties of rock masses. Intensive tectonic strained zones are connected not only with singular discontinuities but also with discontinuity systems. The most dangerous structures that form intensive tectonic strained zones are echelon system of discontinuities, convergence of discontinuities, fraying out at an acute angle (10°–15°), tapered structures.

Dangerous geological faulting characterized by intensively high or intensively low voltage [14, 22] is often not disjunctive break which is fully formed or in the process of forming. Primarily, these are monoclines and sharp changes of reservoir thickness (thickness of bank) of subsurface resources and layers of surrounding rocks. These disjunctive breaks do not have obvious geologic features and they can be skipped even during rigorous geological analysis of drill sample and initial development workings; however, later they can noticeably appear during extraction works. For example, there are such zones on the deposit of Talnakh ore cluster (Norilsk). These are preore faults intermined by intrusion. Before conducting mining works, high-strength intrusion blocks and links two parts of fault in one whole, yet when mining works are carried out nearby, the zone of increased pressure is formed, where formation of dislocations starts to take place. It can be accompanied by dangerous dynamic effects and rock breaking.

Though conducting of mining works change massif stress pattern significantly, majority of dangerous situations and geodynamic occurrences (primarily seismic occurrences and bursts) during stoping takes place in the tectonic strained zones which have already been identified in massif. It is in the area of tectonic strained zone that industrial impact causes formation of extremely strained **geodynamic dangerous zones** while stopping (support pressure).

Taking into account experience of developing stratified deposit in the zone of extraction activity, pressure increases by 30–80% (on the average—about 50%) because of industrial impact. At the same time the process of creating such zones, their quantity, size ad level of danger depend on strain of tectonic blocks and type of discontinuities.

The most powerful and dangerous dynamic occurrences in mines and ore mines are rock bursts (industrial earthquakes) which are related to active faults and strained zones. As in majority of naturally-happening earthquakes, mechanism of bursts presupposes elastic come-back after opening the blocking on a fault. Yet the reason of opening is that conducting mining works close to seismogenic fault from one or from both sides reduces powers which press sides of a fault. At the same time seismic waves from movement along the fault are triggers for secondary dynamic occurrences which cause additional dislocations in dangerous zones of mine roadways. Figure 6 shows an example of rock burst with wave energy of $10^{7-7.5}$ J at mine 'Cheremukhovskaya' (North Urals Bauxite Mine), which took place with an obvious movement of up to 10 cm along the fault of tectonic dislocation, which is a surrounding dislocation of Central Cheremukhovskiy upcast.

As studies have shown, preparation of rock bursts comprises two stages. During the first stage which lasts from half a year to 2 years strain axis in active wing is sub-parallel to a fault and a slow movement of wings relative to each other can be observed. During the second stage which lasts about 3 times less than the first one movements cease, pressure near fault is growing.

First of all, one can observe powers compressing fault in both wings, and then powers holding wings from movement are decreasing a bit; at the same time moving power in an active wing is increasing and overcoming linking of wings.

When fulfilling mining works in zones of II type, the situation is different. First of all, a slight increase of pressure is noticed, then it is reduced abruptly. Entrance in fault zone is not safe. When carrying out numerical modeling, *SRF* (*Strength Reduction Factor*) is used as a criterion of pressure [23, 24]. According to [24], level of danger in tectonic attenuated zones can be compared to level of danger in tectonic strained zones in proportion of σ_1 and break-down point for simple compression within the limit of 0,2–0,3. In fact, it is revealed in reducing stability of roofing and escarpment (including working face), inrush, rood break, increased gas generation, etc.

Peculiar danger in such zones relate to increased water inflow which can cause flood of working face and hazard of losing expensive mining equipment.

Experience of developing apatite-nepheline deposits on the Kola Peninsula showed that in conditions of high horizontal tectonic stress many firm rock masses keep their rock-bump hazard even in the zone of increased jointing that is revealed in heavy industrial load.

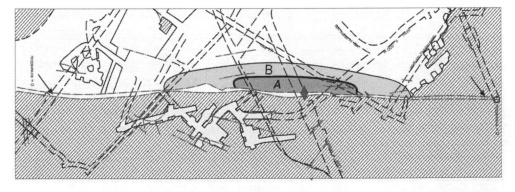

Figure 6.   Rock burst with shear in a break of continuity along the fault (North Urals Bauxite Mine 'Cheremukhovskaya'): A – area of shear of 8–10 cm; B – area of shear of ≥ 3 cm; arrows denote borders of areas of shear along the fault, diamond shows hypocenter of earthquake.

In order to ensure security in mines and pits a special approach to mine working conditions in those areas is required as well as constant complex survey (time-lapse seismic data, deformation, acoustic, electro-magnetic, hydrogeological, gaseous, etc.). The survey is primarily required in geodynamically active zones, where the dynamic occurrences are most likely to take place. However, rock mass together with workings inside it form an integral dynamic system. Condition of each of its parts depends on each other and change over time as mining activity continues and irreversible alterations accumulate. So the survey must comprehend the entire mine and take the regional geodynamical activity into consideration.

Nowadays geodynamics is surveyed mostly at two large-scale levels. The first one is the general-mine (regional) survey including the data of geophysical and geodetic networks, surveyor mine survey and, which is the most important, seismic survey by the mine seismic station. However, seismic survey, as well as the earthquake prediction system, can only distinguish the zones of seismic activity but it isn't able to make short-term prediction. First of all, it is most difficult to say what is actually occurring: energy-storage or, on the contrary, energy release. The prognostic value of even the most physically sound method grounded in the correlation of seismic energy and seismic moment and other related values is not very high due to the complexity and variety of geodynamic processes in the mining zones.

Therefore, the local survey is used to localize the most hazardous zones with the help of geomechanical and geophysical methods which indicates the stress ratio on the specific site of the mine working and localizes, in the first approximation, the hazard zone. However, it cannot indicate how near the destruction point is and what is happening at the rock mass deep. And abnormal hazard near the working face can imply both moderate local and extensive strain rise which can cause a heavy rock burst. That's why the hazard level and the number required of anti-burst measures is not clear as well.

For the accurate situation analysis and selecting anti-burst measures a more detailed appreciation of the rock mass state is required; above all, the deformation process occurring in it and local traits of the stress condition and fissuring processes. Two approaches are mostly used for the reconstruction of deflected mode—mathematical simulation and complex seismographic calculations [25, 26]. But both of these methods are based on the considerable idealization of the rock mass and they don't into consideration many of complex nonlinear traits of its behavior. Besides, mathematical simulation shows the static situation at a certain moment but not the processes development in dynamics.

The approach evolved consists of getting extra empirical data and its systematical processing to estimate the state of the rock mass, the geodynamical hazard and select the anti-burst measures without rejecting mathematical simulation of the stress fields and other approaches that are widely used.

The density of sensors arrangement and even the means of survey can change depending on the deflected mode traits of the rock mass site, the correlation between the recorded stresses and strains, the nature of the geodynamical hazards. In the first place the following zones are distinguished:

- zones of elastic deformations where the stress is near-critical but it doesn't reach the critical value; those zones are subject to microburst and dynamical cutter breaks forming hazards;
- zones of high tensions where the stress pattern is near hydrostatical so there is seismic quiescence. However, mining in proximity can cause drastic changes in the stress pattern, an increase of internal shears, fissuring and the hazard of unexpected dynamic processes including heavy rock bursts;
- zones of slight stress and large deformations (the areas of high jointing, unloaded faults, hazardous roof, worked out rock foliation, etc.). Prevailing hazards are roof and walls collapse, high water inflows, gassing. However, since not all the jointy rocks lose rock-burst hazard, with stress increase the hazard of rock burst may occur.

The described approach is being implemented within the framework of the Comprehensive program of safety working protection it the mines of JSC «Apatit» involving its employees and the executive staff of the «Fosagro» holding.

To fulfil the set task the design described above is supplement with the following points [27].

1. Detailed zoning to the microfaults about the first dozens meters length. To carry it out both visual observations and jointing measuring and electromagnetic sounding of walls of working are used.
2. Detailed investigation of physical properties of not only the basic rock but their difference and especially rock junction.
3. Sonic logging and investigation of process in the mass, amplitude and frequency analysis of signals coming from the mass their coherence and cluster generation of emission sources [28].
4. An essential aspect of the program is measurement both in deep rock and in border zones of working by means of three-component strain gages and operational strain measurement of the boreholes walls and workings themselves.
5. Stress measurements not only by means of classical de-stress mining techniques, but with partial de-stressing which allows to measure the stress pattern more promptly and in more gage points although with some sacrifice in accuracy.
6. Stationary seismo-acoustic survey in the most hazardous zones of rock mass. Operating band is 0,5 to tens of kilohertz [29]. It shows dynamic events energy with less accuracy, but it detects minor events which allows to identify the inception of dynamic processes development.

Table 1. Examples of describing different geodynamic situations by means of the comprehensive survey.

General regional survey	Intermediate (zonal) types of survey		Local survey	
Seismic survey	Strain survey	Stationary seismo-acoustic survey	Visual observations Sample drilling Strain analysis of the borehole walls Local geophysical (acoustic and electromagnetic methods)	Rock-burst hazardous state
+, ++↑, ↑↑	−, ±↓, ↓↑	+, ++↑, ↑↑	+, ++↑, ↑↑	Extremely hazardous state, immediate hazard of rock burst
+, ++↑, ↑↑	−, ±↓, ↓↑	+, ++↑, ↑↑	±, + ↑, ↓↑	Increased hazard of rock burst
+, ++↑, ↑↑	±, + ↑, ↑↑	+, ++ ↑, ↑↑	±, + ↓↑, ↓	Imminence of a seismic event in the deep of the rock mass
+, ++↑, ↑↑	±, + ↑, ↑↑	−, ±↓↑, ↓	−, ±↓↑, ↓	Stress relieving in the rock mass
−, ±↓↑	+, ++ ↑, ↑↑	+ ↑, ↑↑	−, ±	Imminence of sloughing
++ ↓	− ↓	+	−, ±,+	Seismic quiescence before a dynamic event in the deep of rock mass

Legend:
+ high level, ++ very high level, ± moderate level, − low level, ↑increase, ↑↑intensive increase, ↓↑ retention, ↓ decrease.

Depth strain gages and seismo-acoustic survey form an intermediate (zonal) survey level between the general-shaft and the local one. Besides, survey of the zonal level may incorporate driftmeters to measure block mobility, different measurement techniques of electromagnetic fields and electrical current attenuation when passing through rock mass, etc. The three-level system of comprehensive geodynamics survey allows to implement the described above approach to rock mass state estimate and selecting methods to prevent dynamic events or, on the contrary, trigger them off at a desired moment.

Several different approaches are tested for the joint analysis of the array of data obtained—complex tests construction, based on weighing of contribution of each factor, using neuronets and logical and physical diagrams of dynamic events preparation. An example of generalized logical diagram of geodynamic situation in different zones of rock mass is illustrated in the Table 1.

3 EXAMPLES OF USING GEODYNAMIC INFORMATION IN DIFFERENT WAYS OF SUBSURFACE RESOURCES EMPLOYMENT

3.1 *Coal and ore deposits mining*

Let's take using of information obtained with geodynamical zoning and regularly updated with geodynamical survey in the coal mines of Vorkuta coal-basin [30, 31]. The coalbeds of Vorkuta deposit are mined in several mines independently. The mine fields are situated in the adjacent areas, which causes intercoupling of geodynamical processes, provoked by mining. For that reason the role of regional stress fields anticipating and the geomechanical state of the rock mass at the entire deposit scale as means of geodynamical safeguarding of mining is increasing. At the stage of regional rock mass state control in the mines of Vorkuta coal-basin the optimal, providing the maximum protection effect, order of beds-take working was developed with additional measures of gas-draining in de-stressed zones.

As the second example the working of one of the areas of Kirov mine of JSC «Apatit» is considered. Its terrain surface is rugged with large horizontal stress tenser components, exceeding the vertical (hydrostatic) component by many times and has high rock-burst hazard of mined rocks. In the deflection mode calculations the rock strength parameters and deformation behavior of the rock mass obtained in the laboratories of the Scientific center of Saint-Petersburg mining university and Kola scientific center of Russian Academy of Sciences were used. The recording of structure attenuation was performed by means of the Bieniawski Rock Mass Rating (RMR) and Hoek and Brown Geological Strength Index (GSI) method [32, 33]. The figure 7 presents the prognostic hazard zones of dynamical events for an area in the Kukisvumchorr mine of the JSC «Apatit» before working and after working to the level +170 meters.

3.2 *Slopes and benches of open pits stability estimate*

Open pit slopes and benches stability depends on several factors—rock strength parameters and deformation behavior of the rock mass and its junctions, joints density and their directionality bedding nature, water-bearing nature, seismic impact of near and distant earthquakes and blasting workings as well. Among those factors critical role is played by splits alongside with rock strength, jointing and stress state alterations of the rock mass spans. Splits can burst open, grow, channel water inflow, cause strains, faults and slope spans landslides.

Therefore, the proper designing of open pit slopes, selecting slope gradient of working and idle benches and slope spans, blasting methods requires using of geodynamical zoning (Figure 8). As can be inferred from the figure, the splits, detected with geodynamical zoning and the ones discovered when excavating the pit perfectly match.

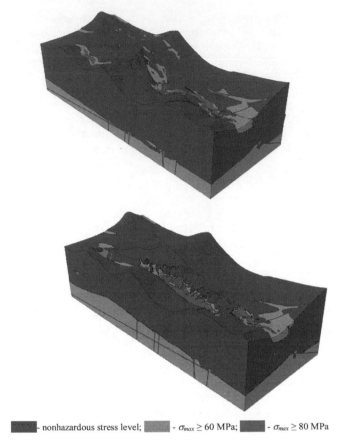

- nonhazardous stress level; $-\sigma_{max} \geq 60$ MPa; $-\sigma_{max} \geq 80$ MPa

Figure 7.   Prognostic zones hazardous of rock burst in natural state and after working to +170 m (after N. Beljakov and D. Kouranov).

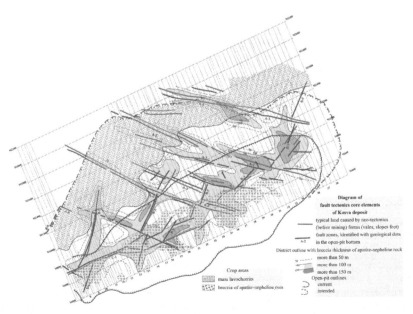

Figure 8.   «Koashva» open-pit of JSC «Apatit»: the faults, detected before and during the open-pit exploitation and their influence on ores bedding (after E. Melnikov).

116

3.3 Hydrocarbon reservoirs search

Geodynamic zoning can be used to locate areas that are expected to be oil and gas fields. Figure 9a shows the basic diagram of the fault zones of compressive stress and tensile stress formation where points A and B are «fixed» fault-closures and M to M^1 span is the peak displacement amplitude [38]. Figure 9b shows the most expected secondary breaks and tension zones where traps can be found filled with hydrocarbons. Number 1 designates the faults occurred due to the initial stress, forming the master fault AB. Numbers 2 and 3 – the faults occurred in different sides of the fault under the influence of the stresses altered by growing of the AB fault.

In fact, as a rule traps develop only in the lower of the two tension stress zones (correspondingly, in the hanging wall or in the foot wall). From another trap which evolves up the section and is usually synclinal, petroleum and gas dissipate in the oil-bearing strata.

Deposit boundaries, as a rule, do not expand beyond a single geodynamical block and mostly match it. That allows to not only detect the most expectable deposit locations, but anticipate their possible size.

3.4 Geodynamical hazards of trunk pipelines operation

The first thing to note is that the most hazardous for ground-surface and shallow-buried structures and pipelines are unloaded zones of faults.

According to [35–37] root accident causes in densely populated areas are actions of intruders, the external corrosion is in the second place. In less populated areas the external corrosion often becomes the root cause. For example, according to [38] in Poland the external corrosion contribution is up to 90%, yet faults are not mentioned among the factors boosting external corrosion. Meanwhile, direct evidences of accidents frequency increasing in areas of intersection with faults exist. For instance, in the Fedorovskiy oilfield site in Western Siberia recurring trunk, infield, gasoil pipelines and water conducts accidents are related with faults. Another clear-cut example of such correlation is the detailed corrosion damages propagation analysis performed by a public corporation «North West Trunk Pipelines» (Fig. 10).

Breakage mechanism analysis was performed in Krasnoturinsk gas pipeline run (Northern Urals) notable in «causeless», at first sight, frequent accidents points [39]. The run of only 37 km long has 44 out of 53 (83%) grouped in 11 «clusters», combining 2 to 5-10 accidents with total length of 4 km which is 11% of overall pipelines length.

The research has shown that the high accident rate of the run is caused by considerable mineral saturation of the ground waters while all the high-density-accident spans of pipelines lay next to unloaded faults intersections with high-voltage electric power lines. The statistical analysis shown that electro-corrosion is most likely responsible. The strong evidence is that in the dry grounds where the electrolytes coming through the faults are less dissolved by surface waters, the accidents rate if much higher than in the watered areas.

Another, simpler mechanism evidences itself in permafrost zones. The most hazardous for the pipelines are the intense soil heaving zones, taliks and buried ice sites as well. These

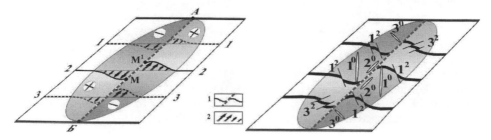

Figure 9. The basic diagram of basic diagram of the fault zones of «compression» (+) and «tension» (–) in vertical shear walls (the figure shows the horizontal section of the fault plane) and actual amount and location of secondary faults and open faults in the «tension» zones of the left shear (after N. Mishin). 1 – horizons crossed by the plane fault, 2 – differential amplitude of shearing.

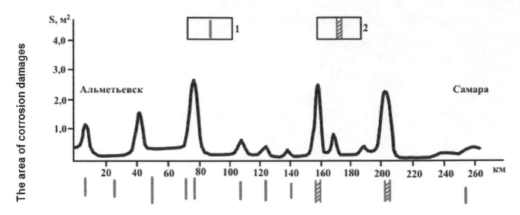

Figure 10. The corrosion losses distribution in the pipe of Almetyevsk-Samara pipeline diagram obtained by flaw detection. 1 – Paleozoic-Mesozoic laying faults; Archaean-Proterozoic laying deep faults.

Table 2. Stages of using geodynamical data and geodynamical hazard management in mining operation.

Stages of using geodynamical information
Appraisal survey identifying of ore-conductive and ore-distributing faults; oil and gas reservoirs search; selecting sites to extract coal bed methane, etc.
Exploration: setting the required prospecting borehole network density and openings; setting the resource category with regard to the faulting
Draft project making: deposit contouring, geo-structural modeling of the resources amount; deposit areas complexity estimation, losses and dilution prognosis, etc.
Project making: selecting mining system and the main rock pressure control method; selecting the mine-working order; mine layout; selecting district measures of geodynamical rock mass state management, etc.
Current planning: exploitation conditions adjusting; setting up an operation schedule; allowable walls advance rate estimate; selecting local gas and geodynamical events counter measures
Mining operation: Geodynamical state survey and current management, walls advance rate adjusting, etc.
Conservation: selecting the conservation method; shift through final parameters estimate; hydrogeological condition alterations prognosis, etc.
Tailings and debris storing: estimate of tailing pits, dumps and hydraulic waste disposals; keeping them available to be used as technogenic deposits

are exactly the sites where the heaviest pipeline deformations occur which can cause integrity damage. The research we performed in Taymyr showed that there is a clear-cut relation between taliks and buried ice expansion with fault blocks properties.

3.5 *Other ground-surface and buried communications*

As our research showed, railroad and highways are mostly affected by the most active faults, separating geodynamical blocks with the biggest difference in elevation. According to our

118

data obtained in the St. Petersburg-Moscow railroad and other North West cities of Russia, 85–90% of damages of substructure and about 35% of the track structure damages (earth bank breaks, rail surface bending) are related to a limited number of faults. Obviously, the key role is played by the big difference in thickness of deposits in various fault blocks and hydrogeological anomalies causing internal friction angle lowering and ground subsidence increase under vibrational impacts due to train movement.

The research of the breakdown rate in plumbing and sewerage systems in Saint-Petersburg has shown that more than half breakdowns account at fault zones. In winter with soil freezing cast-iron pipes and in summer and fall—steel ones are subject to breakage due to corrosion; polymeric pipes are the least subject to breakages. The research performed in other departments of the University, show that deepest and active faults affect negatively the subway tunnels state.

4   CONCLUSION

The research performed has shown that geodynamical processes, above all, in the active faults zones of influence both evolved and the ones yet in progress form ore deposits and oil-and-gas reservoirs on the one hand, and are fraught with different hazard for various kinds of subsurface resources exploitation. The most severe hazard of faults influence occurs when mining deep rock-burst-hazardous coal and ore deposits.

At the same time being aware of the faults location and their properties makes it possible to not only lower hazards but outline deposits, estimate geological reserves, select reasonable ways of mining, conservation of noneconomic reserves, etc. (Table 2).

In order to prevent geodynamical hazards when mining we recommend using the technology we develop, consisting of geodynamical zoning, potential hazard zones identifying (TDL and TRL in the first place), comprehensive geodynamical monitoring, planning and carrying out measures of hazard prevention.

REFERENCES

[1] Methods and systems of seismo-deformational survey of induced earthquakes and rock burst. Volume 1. Novosibirsk: SB RAS Press, 2009., 304 p.; Volume 2 Novosibirsk: SB RAS Press, 2010., 261 p. (In Russian).
[2] Averin A.P., Zaharov V.N., Filippov Y.A. Development of Geoinformation Systems of Monitoring Anthropogenic Geo-Dynamic and Gas-Dynamic Processes in Mineral Resources Development// Open education. 2010. №5. P. 4–12. (In Russian).
[3] Lazarevich T.I., Polyakov A.N. Mining survey of Kuzbass seismic and geodynamic safety//The «Mine Surveying Bulletin» magazine. 2010. № 01. P. 16–22. (In Russian).
[4] Petuhov I.M., Batugina I.M. Geodynamics of subsurface resources. – M.: «Nedra Communications LTD», 1999. – 383 p. (In Russian).
[5] Prognosis and prevention of rock bursts in the mines/ed. by I.M. Petuhov, A.M. Ilyin, K.N.Trubetskoy. – M.: AGN, 1997. – 376 p. (In Russian)
[6] Batugin A.S. To the estimate of geodynamical hazard//GIAB—Miner's Week – 2016. – workshop № 11. – P. 44–52. (In Russian).
[7] Kuzmin Y.O. Modern geodynamics and geodynamical hazard estimate when using subsurface resources. M.: Economy News Agency, 1999. – 220 p. (In Russian).
[8] Rudnik V.A., Melnikov E.K. Geo-active zones (GAZ) and geochemical specialized complexes (GSC) – dominant factor of life environment//Issues of ecological minerology and geochemistry, SPb., 1997. – P. 92–98. (In Russian).
[9] Geodynamical zoning of subsurface resources.//Methodical guidelines ed. by I.M. Petuhov, I.M. Batugina. L.: VNIMI, 1990. – 127 p. (In Russian).
[10] Filossofov V.P. Morphometric method of searching tectonic structures compendium. Saratov state university. 1960, 94 p. (In Russian).
[11] Sadovsky M.A., Bolhovitinov L.C., Pisarenko V.F. Geophysical environment deformation and seismic process. – M.: Nauka, 1987. – 1000 p. (In Russian).
[12] Jurkov S.N., Kuksenko V.S., Petrov V.A. Anticipating mechanical rupture basis // AS USSR – 1981. – V.259. – № 6. – P.1350–1353. (In Russian).

[13] Tsirel S.V. About the possible mechanism of hierarchical structure of geophysical environment// Academic conference thesis «New ideas in geological science», M., 2000. (In Russian).

[14] Practical applications of mining geo-mechanics. Mining mechanics engineering and surveying in the III millennium. SPb.: VNIMI press/Shabarov A.N., Tsirel S.V., Gusseva N.V., Dupak Y.N., Kobylyansky Y.G., Olovanny A.G., 2004 – P. 137–162. (In Russian).

[15] Shabarov A.N., Dupak Y.N., Batugin A.S. Tectonic stressed and de-stressed zones in mountain range//Coal. – 1994. – № 7. – P. 28–30. (In Russian).

[16] Chernyshev S.N. Rock mass joints. M.: Nauka, 1983. – 240 p. (In Russian).

[17] Hudson J.A., Priest S.D. Discontinuities and Rock Mass Geometry//International Journal or Hock Mechanics und Mining Sciences and Geomechanical Abstracts.–1979.–V.16., № 6. –p.339–362.

[18] Faddeenkov N.N. Analytical description of granulometric composition of blasted rock mass with regard of prior jointing//Journal of Mining Science. (In Russian).

[19] Melchior Paul J. The Tides of the Planet Earth. Oxford; New York.Pergamon Press, 1978, 609 p.

[20] Shapiro V.A., Korokina T.P. The results of geomagnetic survey and releveling data correlation within the bounds of Ural geomagnetic site. – Present-day crustal motion. M.: Nauka, 1980, p. 118–123. (In Russian).

[21] Tsirel S.V., Shabarov A.N., Prosvetova A.A. Predictive estimate geodynamic hazard associated with design of mining. – 2015. – № 4. – P. 323–326. (In Russian).

[22] Shabarov A.N., Shadrin M.A. Faults influence on rock-burst hazard of bauxitic deposits. Mining journal, 1992, № 11. – P. 56–58. (In Russian).

[23] Bieniawski Z.T., Engineering rock mass classifications. New York, John Wiley & Sons, 1989. – 251 p.

[24] Villaescusa E. Geotechnical Design for Sublevel Open Stoping. NY., SRC Press,2014.- 541 p.

[25] Xu, N.W., Tang, C.A., Sha, C., Liang, Z.Z., Yang, J.Y,Zou, Y.Y, 2010. Microseismic monitoring system establishment and its engineering applications to left bank slope of Jinping I Hydropower Station. Chinese Journal of Rock Mechanics and Engineering, 2010, V. 29, pp. 915–925.

[26] Seismicity in mining. Joint authors: Kozyrev A.A. ant others. – Apatites, 2002. 325 p. (In Russian).

[27] Shabarov A.N., Tsirel S.V., Morozov K.V., Rasskazov I.Y., Concept of integrated geodynamic monitoring in underground mining/Mining journal, # 9, V, 2017. P. 59–64. (In Russian).

[28] Rasskazov I.Y., Tsirel S.V, Rosanov A.O., Tereshkin A.A., Gladyr A.V. Using of seismo-acoustic survey to identify the nature of fracture nucleus in the rock mass//Journal of Mining Science, 2017. № 2 –P. 29–37. (In Russian).

[29] Gladyr A.V., Migunov D.S., Miroshnikov V.I., Lugovoy V.A. Design of the system of geoacoustical monitoring of the new generation//Mining Informational and Analytical Bulletin. 2010. № 9. P. 101–108. (In Russian).

[30] Shabarov A.N., Krotov N.V., Sidorov D.V., Tsirel S.V. Modern Methods and Means for Solving Forecast Issues and Prevention of Geodynamic Phenomena in Collieries, 21st World Mining Congress & Expo 2008, 7–12 September 2008-Poland. Krakow. 2008, p.137–142. (In Russian).

[31] Zubkov, V.V., Zubkova I.A., Kvatovskaya, E.E., Building zones of high rock pressure in analysis of suite of coal seams working projects//The «Mine Surveying Bulletin» magazine. – 2013. – № 1. – P. 60–63. (In Russian).

[32] Bieniawski, Z.T. Evaluation of the deformation modulus of rock masses using RMR. Comparison with dilatometer tests. Workshop: Underground Works under Special Conditions, 2007. –251 p.

[33] Hoek E, Carranza-Torres CT, Corkum B.: Hoek–Brown failure criterion – 2002 edition. Proceedings of the 5th North American Rock Mechanics Symposium, Toronto, Canada, vol. 1, 2002, pp. 267–273.

[34] Mishin N.I., Stepina Z.A., Panfilov A.L. Structural organization of ore fields. SPb.:«Avtor», 2007. – 232 p. (In Russian).

[35] Eiber, R.J., Jones, D.J.: An Analysis of Reportable Incidents for Natural Gas Transmission and Gathering Lines June 1984 through 1990, NG-18 Report No. 200. Battelle, August 1992.

[36] Gas pipeline incidents: 1970–1992. A report of the European Gas Pipeline Incident Data Group. Pipes&Pipelines International, July-August 1995. p. 9–12.

[37] Kharinovsky, V.V.: Operating integrity problem of gas pipeline constructions. Pipeline Technology. Ed.: DENYS, R. Elsevier, Amsterdam—Lausanne—New York—Oxford—Shannon—Tokyo, 1995. Vol. I, p. 35–43.

[38] Dietrich, A.: Technical Report: Task 4.1: Definition of different damage processes in pipelines. LIMATOG project document classification code: TRP_T4.1_(BB), First Draft. p. 1–4.

[39] Kobylyansky Y.G., Tsirel S.V. Mechanisms of influence of faults on pipeline accidents//Academic periodical «Georesources, geoenergetics, geopolitics». Issue 2(2), 2010. Available at: http://oilgas-journal.ru/index.html (In Russian).

Geomechanics and Geodynamics of Rock Masses – Litvinenko (Ed.)
© *2018 Taylor & Francis Group, London, ISBN 978-1-138-61645-5*

Methods and approaches to geomechanical ensuring of mining safety at potash mines*

A.A. Baryakh & V.A. Asanov
The Mining Institute of the Ural Branch of the Russian Academy of Sciences, Perm, Russia

ABSTRACT: Potash and salt deposit development is always associated with the risk of inrushes of fresh waters into stopes and mine flooding. This article considers experimental and theoretical approaches to mining safety and mine protection from flooding that are based on an informative interpretation using mathematical modelling results of instrumental and geophysical studies regarding mechanical properties and stress-strain state of under-mined rock mass. A comprehensive implementation of these methods provides adequate mathematical modelling results and allows increased validity of predictive estimates related to stability of structural elements and integrity of water-blocking strata when it comes to the room-and-pillar method.

Keywords: saliferous rocks, rib pillars, physical-mechanical properties, mathematical modelling, mass stress-strain state, stability

1   INTRODUCTION

An increase in demand for potash-magnesium ores that are used in many industries requires a considerable increase in potash salt extraction. This is achieved by both an increase in operating mine capacity and the construction of new enterprises (the development of the Ust-Yaivinsky, Polovodovsky, Talitsky sites at the Upper Kama potash deposit (the Upper Kama potash-magnesium salt deposit, UKPMSD), mines in Belarus, Uzbekistan and Tajikistan). The developed sites are characterized by difficult mining-geological and hydrogeological conditions (the presence of water-bearing and unstable rocks), which increases the risk of technogenic accidents. The main risk associated with the UKPMSD mines is an inrush of surface waters into stopes. The overlying rocks separating the water-bearing horizons from the mined-out space must preserve their continuity throughout the whole service life of a mine, thereby performing functions of a waterproof pillar – of the water-blocking strata (WBS). This is achieved by the use of the room-and-pillar method with continuous ore excavation and maintenance of all the overlying strata on the chain pillars that must preserve their bearing capacity throughout the whole service life of a mine [1]. The exploitation of the deposit at the mines of PJSC Uralkali has resulted in large mined-out spaces of tens of square kilometers that are still increasing. As long as mining operations are developed, the risk of discontinuity of the water-blocking strata and consequently that of mine flooding is constantly increasing.

The deposit is used to develop sylvinite beds KpII and АБ separated from each other by the technological interbed of 1.5–6.0 m in thickness, and the width of the rooms of overlying bed АБ is 3.2 m, that of KpII ranges from 5.2 to 6.1 m depending on a combine to be used. The width of the pillars of KpII bed ranges from 2.9 to 10.0 m, that of АБ from 5.8 to 12.9 m. The workable beds' roof (especially in case of KpII) is notable for the interlayers of argillaceous and saliferous rocks (cakes) that have weak stability and are usually subject to

---

*This work was financially supported by the Russian Science Foundation (Project No. 16-17-00101).

collapse when developed. In order to provide a stable state of the stopes' roof, KpII bed is developed either with a cake undercutting or with a protective bench left in the roof of the bed with 0.6 m thickness.

The state analysis of the mined-out spaces at the UKPMSD mines shows that certain sites, when the superimposed sylvinite beds are developed together with the "rigid" pillars intended for "infinite" service life, are notable for their considerable destructions 20–30 years later. This situation may be aggravated by a decrease in strength properties in the zones of the saliferous strata's abnormal structure. All this leads to a case when the rib pillars are gradually destructed and displacement of the undermined rocks is intensified. In recent years, this problem has become especially severe for the mines where the reserves near the cities of Berezniki and Solikamsk are being developed, as the displacement of the earth surface may result in cracks, deformed and destructed industrial and residential buildings. Many affecting factors make it rather difficult to determine the rib pillars' service life.

Mining operations and practices show that even if all the regulatory requirements are met, the pillar selvedges are destructed due to difficult mining-geological conditions, which in some cases causes critical deformation of the WBS, their discontinuity and an accidental inrush of fresh waters into the mined-out space. In the last 30 years the two mines, i.e. Berezniki Potash Deposit-3 and Berezniki Potash Deposit-1 (BPD-3, BPD-1) have been flooded at the deposit, and the Solikamsk Potash Deposit-2 (SPD-2) functions in an emergency mode.

The task of mining safety in the course of development of sylvinite beds can be divided into two stages. The first stage is to assess the degree of danger (risk) of discontinuity of the water-blocking strata (WBS) and to reveal potentially dangerous sites as far as emergency situations are concerned.

The second stage is to comprehensively monitor states of the load-bearing elements of the mining method and water-blocking strata at the distinguished dangerous sites of a mine take. This monitoring includes the experimental deformation and destruction studies with regard to elements of underground structures, the geophysical and surveying control of undermined rock mass' state and the theoretical interpretation of their results, which allows a quick consideration of local changes in mining-geological and mining-engineering conditions when managerial decisions aimed at preventing emergency situations are made.

2 METHODS AIMED AT DETECTING POTENTIALLY DANGEROUS SITES AND ASSESSING THE RISK OF THE WBS DISCONTINUITY

A detailed analysis of events starting or initial for emergency situations connected to the technology of mining operations, engineering design mistakes, rock mass reactions to implemented parameters of mining operations is taken as a basis for the method guaranteeing the numerical evaluation of a risk of water-blocking strata's discontinuity at potash mines. Besides, the method includes an analysis of uncertainties connected to incomplete knowledge of the water-blocking strata's structure, mechanical properties of rocks, etc. The identification of potentially dangerous sites at the developed areas is based on a comprehensive analysis, including the assessment of both indicators specified in regulatory documents and parameters characterizing a certain mining-engineering and geomechanical situation. Potential risks of mine take sites are analyzed using a "reason tree" that allows revealing combinations of mistakes when mining operations are designed and conducted, uncertainties connected to the rock mass structure that may potentially lead to an emergency situation (Fig. 1). Discontinuity risks related to the water-blocking strata are determined according to the following four main factors: natural development conditions, technogenic impacts on water-blocking strata, reactions to implemented development parameters, breaches of requirements of regulatory documents, organizational and administrative risks.

Discontinuity risks related to the water-blocking series of strata are determined by risks and factors included in this group in accordance with their weights on the basis of the hierarchy analysis method [2]. Weights of certain groups of factors are calculated according to results of statistical analyses of polled data taken from the representative sampling by experts.

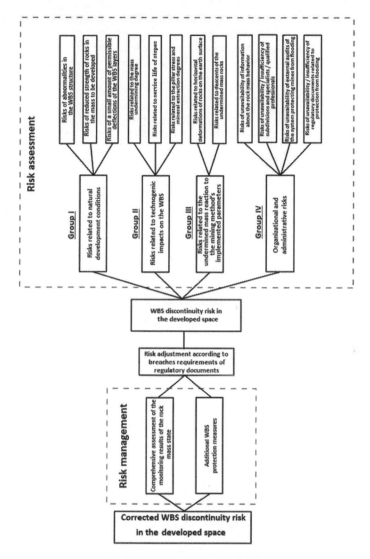

Figure 1.   General chart. Discontinuity risk assessment and management.

This stage of studies allows zoning of mine takes according to potential WBS discontinuity risks and identifying top-priority monitoring objects.

The long-term studies of structures, physical and mechanical properties, stress-strain states and destructions of saliferous rocks in the course of developing sylvinite beds at the UKPMSD mines allowed developing the principles intended to experimentally and theoretically assess the rib pillar stability and WBS state. This method consists in an informative interpretation of field study results with the use of mathematical modelling methods. Geological and surveying documentation obtained in the course of mining exploration and development of workable beds, inspection reports on stopes and mine openings, information on earth surface sedimentations, results from geophysical and seismological studies serve as initial information when assessing states of undermined rock mass at potentially dangerous sites. These sites are used to carry out studies of physical and mechanical properties of main lithologic types of saliferous rocks, assess stress-strain states and destructions of rib pillars and interbeds over time [3]. The experimental study results specify the models of deformation and displacement of the rock mass and earth surface when developing the workable beds.

123

These models consider the rib pillars' form, structural and textural features of the saliferous rock mass. Multivariant calculations of how the stress-strain state of the saliferous mass evolves in the course of mining operations due to certain mining-geological conditions allow assessing the stability degree of load-bearing elements during the room-and-pillar method and WBS discontinuity risk [4].

3 STUDIES OF PHYSICAL AND MECHANICAL PROPERTIES AND DEFORMATIONS OF QUASI-PLASTIC SALIFEROUS ROCKS

The mine takes of the UKPMSD are used to study mechanical properties of saliferous rocks considering the core of the underground mining exploration wells. The saliferous rocks of the productive strata are subject to section sampling.

The analysis of the study results shows that, despite the uniformity of main types of saliferous rocks, the lateral variability of their mechanical indicators changes within wide ranges and may differ by 2–3 times compared to average values. Such parameters' dispersion is actually caused by the geological structure of the mass, i.e. variability of its composition, structure, texture and degree of defectiveness peculiar to the saliferous rocks according to the nature of their formation.

The analysis of the large-scale test results shows that only 34% of the total number of experimental data correspond to the specified values of strength of red sylvinite equal to 23.0 MPa ± 10%. At the same time, 31% of the measurements shows that the strength is less than 20 MPa, of which up to 50% of the samples have reduced strength due to the fact that the samples have structural and textural defects, and 25% has defects of a technogenic origin.

It has been estimated that the reduction of rocks' strength composing a pillar by more than 10% leads to an increased degree of stress in rib pillars by 15% at average, and the reduction of strength by 25% leads to an increase by 45%. In this case, the rib pillars will be inherently deformed in a "soft" mode, with their limited service life. This leads to an increased technogenic load on the WBS and accelerated deformations of the earth surface.

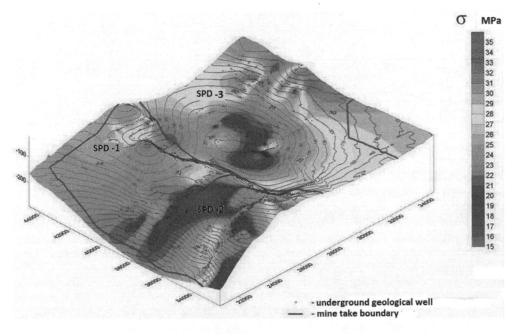

Figure 2. Distribution of the strength limit of red sylvinite of KpII bed along the area of the mine takes at the SPD-1, 2, 3.

In order to specify the models of saliferous mass deformation that are used when explaining safe mining conditions, volumetric loading experiments for samples of various scale levels are carried out, strength and deformation properties of contacts of the weakest link specifying the stability of exposures are investigated, rheological characteristics of saliferous rocks at long stresses are determined. The studies are conducted using the "rigid" test equipment units that allow varying deformation speeds and stress states within a wide range. Out-of-limit deformation of samples is controlled by both optical and acoustic-emission methods. The energy capacity of crack formation is assessed by the grain-size analysis of destructed material's' particles.

At present we have collected a large amount of data (physical and mechanical properties of saliferous rocks from more than 500 underground and surface wells) that allow revealing main regularities of changes in mechanical properties of saliferous rocks depending on their structure and bedding conditions. The forecast maps of distribution of strength and deformation properties along the area of the workable beds at the UKPMSD mines where the variability of mechanical properties of rocks according to both saliferous strata sections and deposit areas (Fig. 2) is observed have been created for the operating mines. As the sampling points have spatial references, the parameters of stoping works are determined on the basis of reliable information about properties of saliferous rocks at certain sites, which provides a more reliable explanation of the parameters of sylvinite beds' development.

4   EXPERIMENTAL STUDIES OF STRESS-STRAIN STATE OF SALIFEROUS MASS

The experimental studies show that the study of the saliferous mass stress states requires methods that avoid the necessity for model transfers from measured deformations to stresses, with measurement database considerably exceeding the typical degree of heterogeneity of rocks. These requirements are generally met by compensation methods with the use of different loading devices (hydraulic jacks, hydraulic cushions, etc.) that restore an initial stress state in the previously unstressed rock mass [5].

A large scope of experimental studies of stress states of saliferous rocks allows stating that the Upper Kama potash deposit is characterized by a rather difficult distribution of the stress field of the unbroken mass. Vertical stresses are usually determined by weights of overlying rocks (γH); and horizontal components at certain sites connected with abnormal structures of saliferous masses vary from 0.7 to 1.3 γH. With their depth, horizontal stresses increase according to a linear law.

The Goodman hydraulic jack that represents a logging probe is used to measure deformations of well walls on the basis of 200 mm under the influence of applied stresses and to study stress-strain states of rocks in the elements of the room-and-pillar method. The stress measurement method is based on the use of the stress compensation method with the registration of acoustic emission rates (AE) in the course of restoration of stresses near wells. The amount of the stresses existed in the mass is determined by spasmodic changes in acoustic emissions (memory effect) at the time when the wall pressure exceeds the initial stress [6, 7].

The comparative studies with the use of the slot de-stressing methods according to the scheme of the Mining Institute of the Ural Branch of the Russian Academy of Sciences and stress compensation with the use of a flat hydraulic cushion have been conducted in order to assess the reliability of stress measurement results obtained with the use of memory effects. The analysis of stress values obtained through different methods shows good convergence of their results. The average value dispersion does not exceed 15%.

The experimental stress studies conducted in the process of developing sylvinite beds show that the maximum of support pressure in the "fresh" rib pillars is at the depth of 1.0–1.3 m and can reach 1.8–2.0 γH. The central part of the pillar has the level of the vertical stresses higher than the weight of the overlying rocks by 25–40%. The horizontal stresses with the distance from the pillar's contour increase and are about 60–70% compared to the vertical ones. The middle part of the ceiling has the horizontal stresses ranging from 5.0 to 7.5 MPa when developing a series of sylvinite beds. Stresses close to the roof of KpII bed and in the floor of АБ bed decrease up to 1.5 MPa, which is confirmed by the inspection of destruction of the stope roof.

Ground pressure and moisture contained in air decrease strength properties of rocks and the peripheral mass gradually collapses, therefore there is a redistribution of existing stresses, and the zone of support pressure moves deep into the pillar. Eventually, the stress existing in the middle part of the pillar may exceed an admissible rate of stress of pillars ($With \geq 0.4$) [1]. In this case, the pillar passes from a "rigid" deformation mode to a "soft" one, which finally affects the amount of displacement of the whole overlying strata.

Long-term observations with the earth surface displacement and the deformation of the pillars and stope roofs are carried out in order to identify the interrelation of deformation of the rock mass during various periods of development of workable beds. The process of deformation and destruction of rocks, roof, floor and stope walls is studied at gauge stations equipped with contour and deep reference points. The earth surface displacement is controlled according to ground reference points of profile lines put on the surface over the sites to be developed.

The analysis of the study results shows that the sites with a high content of clay materials in the roof of the sylvinite beds to be developed (a southeast part of the mine take of the BPD-2) are notable for an intensive stratification of rocks of the roof of underlying KpII bed with the subsequent collapse of clay cakes within the first 1.0–2.5 months after treatment, at the same time all the layers of KpII – АБ interbed are involved in deformation. When stoping works approach the door of a gauge station, the processes of roof's stratification and floor heaving are become extremely activated in the chambers. During the further development of the beds the speed of deformation of the rocks in the floor of the chambers of АБ bed decreases up to the previous level.

In the northeast part of the mine take of the BPD-2 where a content of clay is much less, the main share of deformations is associated with the peripheral layer of the roof of KpII bed, and the values of deformations are of an order of magnitude less.

The instrumental measurement results are initial information for a model setting when calculating a stable state of the load-bearing elements of the room-and-pillar method through mathematical modelling.

5 THEORETICAL STUDIES OF STRESS-STRAIN STATE OF SALIFEROUS MASS DURING STOPING

The theoretical studies of evolution in the stress-strain state of saliferous mass as a result of mining operations are conducted in two stages. The first stage is to make detailed local mathematical models describing the process of deformation and destruction of the bearing elements of the room-and-pillar method (pillars, technological interbeds, rooms' roof and floor). The second stage includes the large-scale mathematical modelling of the WBS stress state with the subsequent criterion assessment of its long-term stability. The mathematical modelling of destruction of the bearing elements of the room-and-pillar method is usually carried out in the setting of a plane strain state. The state of the room block is described by an ideal elastic-plastic medium for which the connection between deformations and stresses at a prelimit stage is determined according to Hooke's law. Ultimate stresses in the area of compression are calculated according to Coulomb's and Moore's law. In the tensile area the ultimate stresses are reduced to a tensile strength limit. When modelling, the destruction of clay contacts between layers [8] as the "weakest" elements in the structure of the saliferous mass determining its stratification and the possibility of gravitational collapse is considered.

The mathematical model is calibrated on the basis of the backward analysis of field observations over deformations of the room block elements (Fig. 3). It is preferable to control results according to the connected field study results, for example, according to measured stresses (Fig. 4). Such approach provides adequate forecast estimates regarding the term of preserving the bearing ability of structural elements during the room-and-pillar method.

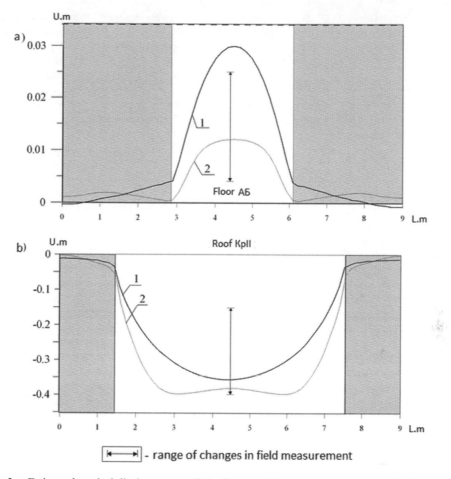

Floor АБ

Roof KpII

⟵⟶ - range of changes in field measurement

Figure 3. Estimated vertical displacements of the floor of АБ bed (a) and the roof of KpII bed (b) at stope excavation when a protective bench (1) is left and with cakes cutting (2).

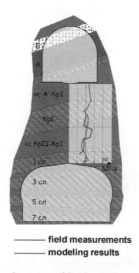

——— field measurements
——— modeling results

Figure 4. Comparing the estimated and measured horizontal stresses in the technological interbed.

The geomechanical analysis of the water-blocking strata stability is generally based on the same phenomenological approaches that have been stated above. However, changes in modelling scales impose certain requirements to the parametrical ensuring of geomechanical calculations. In particular, the mathematical modelling of changes in the stress-strain state of the undermined saliferous mass over time is based on a mathematical description of the schedules of increases in the earth surface's descents by hereditary viscoelasticity equations. The WBS mathematical model considers the features of their geological structure and state established according to the ground and mine seismic observation results. The geomechanical assessment of safe conditions of the WBS undermining is carried out in the two-dimensional or three-dimensional elastic-plastic setting by the finite element method with the use of its own software products and computational resources of the supercomputer of the Institute of Mathematics and Mechanics of the Ural Branch of the Russian Academy of Sciences through a dedicated channel.

The mined-out space is modelled by the medium which properties are decreased compared to rocks of a corresponding bed. The degree of reduction of mechanical properties is determined by actual parameters of the room-and-pillar method and, if necessary, may be rectified according to the mathematical modelling results of the local objects' state.

When carrying out geomechanical calculations, the natural anomalies distinguished according to the results of the geophysical works are considered as areas with decreased strength and deformation properties of saliferous rocks. The computation scheme aimed at determining the stress-strain state of the mass over time is implemented on the basis of modifying the variable elasticity modulus method specifying the deformation of not only all the elements of a geological section, but also of the developed beds [9]. This approach is advantageous due to the fact that it allows to separately describe the time nature of deformation of all the developed beds, to easily consider the difference in terms of their development and assess the state of the undermined mass at an arbitrary point of time. The risk of the water-blocking strata's discontinuity as a result of mining operations is analyzed according to the results of plastic deformations' localization that are physically treated as zones of shear cracks' development (in the compression area) or tensile cracks (in the tensile area) [10].

The adequacy of the mathematical modelling results is determined according to the results of comparing the estimated descents of the earth surface with the surveying observations over the development of the displacement process (Fig. 5).

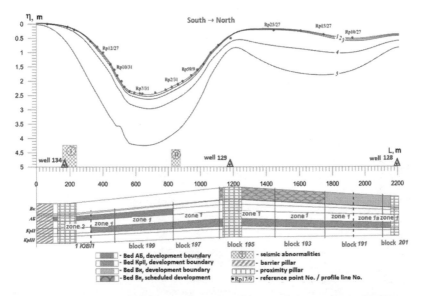

Figure 5.   Changes in the sedimentation of the earth surface in 2017 (1), 2020 (2), 2025 (3), 2050 (4) and the end of the displacement process (5).

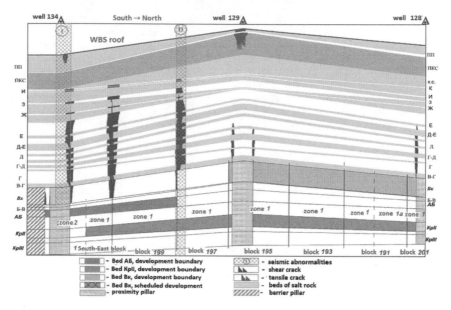

Figure 6. Technogenic discontinuity of the WBS.

Multivariant geomechanical calculations show that the zones potentially dangerous due to the water-blocking strata's discontinuity are associated with the selvedges of the undermined mass and with the sites of gradient turns of the shift trough. The destruction of the rock salt beds developed in the WBS begins with the formation of shear cracks in their lower part (Fig. 6). As long as the deformations increase, there is a rather slow development of the shift jointing zone up the WBS section. Separation cracks appear in the upper part only in case of the earth surface's considerable descents. Their main development in the WBS capacity takes place from top down. After the formation of tensile cracks the WBS is destructed rather quickly. As for shallow shift troughs, the rock salt beds are destructed only in case of the earth surface's very considerable descents. In this case the shear cracks over the selvedge of the mined-out space and the tensile in the WBS roof take place practically at the same time. As long as descents increase, the inversion mechanism of their development is observed: shear cracks grow from bottom up, tensile cracks from top down. When a certain amount of descents in the development of the upper fracturing zone is achieved, the formation of tensile cracks begins to dominate.

A part of potentially dangerous zones can be created in the distinguished areas of geological anomalies. In this regard, in the course of mining operations the degree of decrease in mechanical properties of rocks of salt strata is gradually specified according to the results of seismic works and, in accordance with the obtained information, expected geomechanical estimates of long-term stability of the WBS are rectified.

On the basis of the revealed regularities, the engineering criteria of ensuring safe conditions of the WBS undermining based on the regulation of the total capacity of the unbroken rock salt beds, the maximum descent and the prevention of formation of tensile cracks are suggested.

6 CONCLUSION

The developed principles aimed at experimental and theoretical evaluation of the state of the elements during the room-and-pillar method, as well as the state of the WBS allow increasing the reliability of geomechanical ensuring of mining safety at potash mines and protecting them from flooding.

REFERENCES

[1] Instructions on protection of mines from flooding and protection of undermined facilities in Verkhnekamsk Potash Deposits. – SPb. Perm, 2014. – 130 p.

[2] Saati T. Decision-making. The method of analyzing hierarchies. – M.: Radio and Communications, 1993. – 278 p.

[3] Baryakh A.A., Asanov V.A., Toksarov V.N., Evseev A.V. Methodology of geomechanical evaluation of long-term stability of inter-chamber pillars // Scientific and Technical Support of Mining: Collection of scientific papers Mining Institute n.a. D.A. Kunaev. – Vol. 87. The city of Almaty. – 2015. pp. 91–95.

[4] Baryakh A.A., Asanov V.A., Samodelkina N.A., Pankov I.L., Telegina E.A. Geomechanical protection of mines from the flooding of rocks/Mining Journal. – 2013. – Nr. 6. p. 30–34.

[5] Asanov V.A., Toksarov V.N., Evseev A.V., Beltyukov N.L., Anikin V.V. Instrumental method to control the stress-strain state of the near-boundary salt rock massif/Mining Journal. – 2013. – Nr. 6. – p. 40–44.

[6] Lavrov A.V., Shkuratnik VL, Filimonov Yu.L. Acoustoemission memory effect in rocks. – Moscow: Publishing House of Moscow State Mining University, 2004.

[7] Asanov V.A., Toksarov V.N., Evseev A.V., Beltyukov N.L. The Mining Information Analytical Bulletin. – 2010. – № 10. – Moscow: Publishing House of Moscow State Mining University.

[8] Baryakh A.A. Interaction of the layers in a salt massif. Communication 1. Mechanical properties of contacts/A.A. Baryakh, I.N. Dudyrev, V.A. Asanov, I.L. Pankov//Physico-technical problems of mining mineral deposits. – 1992. –Nr. 2. –p. 7–11.

[9] Baryakh A.A. Samodelkin N.A. About one approach to rheological analysis of geomechanical processes. Physical and technical problems of developing mineral resources. – 2005. – № 6.

[10] Baryakh A.A., Samodelkina N.A. Destruction of waterproof strata in large-scale mining operations. Part II//Physical and technical problems of developing mineral resources. – 2012. Nr. 6. pp. 12–20.

Geomechanics and Geodynamics of Rock Masses – Litvinenko (Ed.)
© 2018 Taylor & Francis Group, London, ISBN 978-1-138-61645-5

The Ural scientific school of geomechanics: Fundamental and applied research

S.V. Kornilkov, A.D. Sashourin & A.A. Panzhin
The Institute of Mining of the Ural Branch of RAS, Ekaterinburg, Russia

ABSTRACT: The paper presents the information about the formation of the scientific school of geomechanics in the Institute of Mining, Ural Branch of RAS. The School has held the path of scientific perception of the processes and phenomena in rock mass accompanying the development of mineral deposits. The paper is dedicated to fundamental problems of geomechanics and shows the demand for the results of research at the mining enterprises. The paper describes the main results of the fundamental and applied research performed by the Ural scientists-geomechanics within the period from 2012 to 2017. Discovery of the relationship between natural and technogenic catastrophes with the recent geodynamic movements expands the practical application of research results from the mining to other areas of the mining looking for the prognosis and prevention of natural and technogenic catastrophes.

Keywords: geomechanics, mining, stress-deformed state, geodynamic movements, stress fields, satellite geodesy, natural and technogenic catastrophes, mining areas, geological environment.

1 INTRODUCTION

In the tragic statistics of accidents and catastrophes of the last time the special place belongs to the sphere of the exploration and exploitation of georesources, including mineral and raw materials, energy and construction complexes. In this sphere of activity in Russia, the past three decades revealed more than twenty-five major accidents and catastrophes which could be combined by common real sources and causes. Errors in survey and designing, violations of the technological regulations in construction and operation, wear and other sides of the human factor encounter always to a greater or lesser extent, however their exaggeration in the practice of the accident cause investigation is not conducive to the disclosure of the real root causes of catastrophic phenomena. Meanwhile the results of fundamental research in the Geosciences testify that the processes and phenomena in the rock mass and above ground could cause serious failures of the underground objects of different purposes. Among them the leading role belongs to the deformation processes caused by the recent geodynamic movements of the earth's crust, as well as by the technogenic mining activities and by the creation of new mining objects. The study of regularities in development and manifestations of the rock mass deformation processes and using them to ensure the efficiency and safety of the mining constitute the main task of geomechanics. Given that modern society satisfy the main part of the needs in raw materials at the expense of mining, rapidly expanding the scope and depth of the extraction, the role of the geomechanics in the development of mining is difficult to overestimate.

2 ESTABLISHMENT AND DEVELOPMENT OF THE URAL SCIENTIFIC SCHOOL OF GEOMECHANICS

The intensive development of the mining industry and the high demand for the results of scientific research on the geomechanics have contributed to the development of research in this direction, and the Ural school of geomechanics formed, developed and runs on this basis [1]. In 1959 the Mining and Geological Institute of the Ural branch of the USSR Academy of Sciences had founded the laboratory of mining pressure and stability of open-pit sides headed by Professor, Doctor of Engineering Sciences M.L. Rudakov. Increase in the open-cut and underground mining output, involvement in the development of deposits under complex conditions of occurrence at great depths, severization of the requirements to the mining safety and efficiency—all this has advanced geomechanical production support to the leading role in mining and mining sciences, and the young scientists engaged in the 1960s and 1970s showed and saved their devotion to the geomechanics problems and form today the staffing basis of the scientific school.

The methods to study the stress-deformed state (SDS) of the rock mass developed in the 1970s and 1980s were widely used already in the initial period in a number of institutions and organizations of the USSR. In particular, during 1972–1981, the employees of the Institute conducted studies at the invitation of the Institute Hydroproject to determine the stress changes in the rock foundation of the world's highest arched dam of the HPS Inguri (Western Georgia).

With the transition of the underground mining to deeper horizons the prevention of the rock bumps became the serious problem that the Ural School of geomechanics started to solve since 1975 on all the iron ore mines in the Urals and Siberia. The timely prediction and prevention of the rock-bump hazard developed on the basis of geomechanics have allowed to minimize damage of accidents in the mines. One of the important problems for mines at the end of the 20th century was packing the voids when mining the large ore bodies. The natural controlled cover caving method developed by the Institute of Mining of the Ministry of Iron and Steel Industry of the USSR allowed the mines to put away the forced packing of mine goafs and provided the mining safety and produced the great economic effect.

The expansion of the research methodology of the rock pressure to the rock movement processes in the open-cut and underground mining in combination with large amounts of instrumental measurement of the ore deposit movements has allowed to identify the relationship between the parameters of the rock mass deformation in the movement trough and in the open-pit sides with the initial stress state. The identified patterns were the basis of essentially new concepts and theories of the movement process whose practical use ensured the widespread introduction of advanced solutions to protect the structures and to optimize the parameters of the open-pit sides at the mines of the Ural region and Kazakhstan, to make efficient use of the mines and carriers at the expense of reactivation of the part of the ore reserves of the safety pillars and to locate the overburden in the caving areas.

The proven theoretical provisions are reflected in the Rules for the protection of structures and natural objects against the harmful effects of the underground developments at the ferrous metal ore deposits in the Urals and Kazakhstan, these Rules are acting as a normative document until now [2].

New approach to the creation of the geomechanical model of the developed rock mass area has required new means of measurement and techniques to evaluate the stress state. Instead of small bases of stress measurement ranging from a few centimeters to the first tens of meters, the new method has been developed for stress measurement on the large bases reaching 1.5–2.0 kilometers. This method has been used to measure the stresses at 6 mines, and work is underway on the interpretation of the experimental data at 25–30 deposits that makes a serious contribution to knowledge of the rock mass SDS of the Ural region and the whole upper part of the earth's crust.

The research performed under the supervision of V.G. Zoteev has had the following results: the theoretical justification has been given for the deformation mechanism for the rocky slopes of deep and ultra-deep quarries; the analytic solution has been obtained for the three-dimensional problem on the stability of the slope with two and three surfaces of the weakening; the calculation method has been developed for the parameters of the distressing zone in the rock mass surrounding the borrow cut; the regularities have been established for deforming the dumping horizons in time; the method has been developed for selecting the technology and modes of stockpile filling.

The results of the research of the Ural school of geomechanics have been put into practice of many enterprises in our country and near abroad: JSC "Korshunovsky MPS", JSC

"Kachkanarsky MPS "Vanadium", JSC "Orenburg minerals", JSC "Uralasbest", JSC "Olenegorsky MPS", JSC "Kovdorsky MPS", JSC "Sokolovsko-Sarbaiskoe MPO" etc.

The scientific achievements of the school of geomechanics depended primarily on the contribution of individual scientists. Among them are N.P. Vlokh, A.D. Sashourin, A.V. Zubkov, V.G. Zoteev, V.E. Bolikov, Yu.P. Shupletsov, Ya.I. Lipin, O.V. Zoteev, V.P. Lelikov, S.M. Ushakov, Yu.G. Feklistov, V.I. Doroshenko, etc. who have passed all the stages of the school formation in the 1960s.

Years of the perestroika-era and reforms have had a negative impact on the state of the collective of the Institute of Mining of the Ural branch of Russian Academy of Sciences, including the Ural school of geomechanics. The Ministry of Iron and Steel Industry of the USSR had ceased to exist, and the mining enterprises being in need of scientific monitoring by the geomechanics were in a critical condition. In these circumstances and in order to avoid degradation the staff of the Institute began on the initiative of the geomechanics to act for the return to the structure of the Russian Academy of Sciences, and in 1994 the Institute had re-entered the Ural branch of Russian Academy of Sciences. In the academic status the staff of the Institute of Mining of UrB of RAS had found the "second breath", and in this period the scientific school has been established completely.

The combination of the fundamental research with the solving of the applied problems for the mining enterprises as they began to come out of the crisis has resulted in the further development of the scientific school of geomechanics. The first international conference "Geomechanics in Mining" had been held in 1994 and became traditional arousing great interest among the participants. Its topics, along with the traditional problems of SDS, included the problem of technogenic earthquakes as a logical result of the tectonic rock bumps at large-scale impact of mining works on the top of the lithosphere.

In recent years in connection with the expansion of economic activity the mining and other spheres encounter the serious problems in prediction and prevention of natural and technogenic catastrophes. The destruction of the underground and aboveground industrial and social objects, the firing breaks of the main oil and gas pipelines, the methane gas explosions in mines, the inundation of salt mines and other catastrophes with tragic consequences and material damage—all this required immediate disclosure of the nature, origin and mechanism of their manifestations for the development and implementation of measures to prevent and minimize the severity of the consequences [3].

The previous experience of studies showed that a wide range of catastrophes in the mining sphere is associated with the recent geodynamic movements taking place in the rock mass and on the surface of the earth. Their study required to conduct the large-scale integrated basic research with application of modern technologies of satellite geodesy and geophysical methods. The financial support of the RFBR has helped the Institute of Mining of UrB of RAS to create the Center of collective use of the unique equipment whose organization has allowed to raise the instrumentation and methodical support to the world level and to organize the unprecedented experimental work.

The application of modern technologies in the research practice has experimentally confirmed the previously developed concept of large-scale impact of mining on the SDS of the upper part of the lithosphere, as well as the possibility of technogenic earthquakes in areas of intensive mining. High-precision measurement of the earth's surface shifts with the use of the satellite GPS geodesy complexes have revealed two types of the recent geodynamic movements: the trend movements preserving the relatively stable speed and direction in the long period and the cyclic movements with alternating directions and wide frequency range [4]. Their discovery has allowed us to make the fundamental conclusion about the state and the properties of the rock mass in the upper part of the lithosphere. The geological environment, particularly in the areas of tectonic disturbances, appears to be in its natural continuous movement [5].

Identifying the continuity of movement of the geological environment has completed the building of a holistic view of the origins and regularities of the natural rock mass SDS. Its original quasistatic components are formed by the gravitational stress field and tectonic processes with the recent trend movements; and, finally, they are superimposed by the variable SDS formed by the recent cyclic geodynamic movements.

This concept about the nature of the formation of the natural rock mass SDS has served as a key to understanding the sources of the broad class of the natural and technogenic catastrophes in the mining objects and is a subject for the detailed studies of the Ural scientific school of geomechanics in the present period [6].

3 RESEARCH RESULTS FOR PERIOD 2012–2017

As noted above, the experimental works of the Ural school of geomechanics for the exploration of the recent geodynamic movements, by means of continuous and discrete observations using satellite geodesy, have revealed two types of the geodynamic movements: the trend movements of relatively constant movement speeds and directions; the cyclic movements of polyharmonic nature, including the duration of cycles from the first seconds to several hours and days [7]. The experimental determination of the parameters for the trend and cyclic recent geodynamic movements has been so far completed at more than 25 mining objects covering the territory of Russia and Kazakhstan from the Central region to Yakutia, and on their basis the database of recent geodynamic movements has been created [8]. It follows that both types of the recent geodynamic movements are taking place in all regions, regardless of their seismic or aseismic category [9].

The established data on the recent geodynamic movements coupled with the transition to the large base of measurements have made another significant step in the knowledge of the formation of the rock mass stress-deformed state, in particular, it has been found that it is a variable in time. Further refining and detailing of the structure of the stress and deformation fields have showed that the rock mass, having a priori hierarchical block structure under the conditions of constant mobility and variable stress-deformed state, is being subjected to secondary structuring [10]. In this case the recent geodynamic movements concentrate on the borders of the secondary structural blocks, and the structure of the stress-deformed state acquires the discrete nature defined by the secondary structural blocks [11].

Thus the formation of the rock mass SDS is determined by the following main factors: the hierarchical block structure; the constant mobility; the secondary structuring; the concentration of the recent geodynamic movements on the borders of the secondary structural blocks. Under their influence the mosaic and relatively homogeneous in its average integral parameters SDS forms in the real rock mass. With its relative homogeneity it would seem that it is difficult to expect the formation of the anomalous zones with dramatically different parameters [12] which could be considered as focal areas for development of catastrophic events. The further deepening of the heterogeneity in the structure of the rock mass SDS is due to the fractal nature of the borders between the neighboring secondary structural blocks and the above noted concentration of the recent geodynamic movements on these blocks. The reciprocal movements of the neighboring structural blocks represent the secondary movements. With fractality of their borders the ledges and troughs interact. The fragment of the borders of the neighboring blocks and the layout of their interaction are shown schematically in Fig. 1.

The frontal planes of the ledges on the neighboring blocks move over each other under the influence of the geodynamic movements, creating a zone of concentration of compressive stresses. At the same time their back planes diverge, forming the zones for relaxing the compressive stresses and deformations, the depression zones, in contrast to the zones of concentration, up to the full relaxing of the compressive stresses.

It is this chain of events is the basis of the formation of the structure and parameters of the rock mass SDS where the clusters of catastrophic events appear. The zone of concentration of the compressive stresses arising at the frontal planes represent a potential danger on the possibility of induced seismicity in the form of the technogenic earthquakes [13]. The depression zones manifest themselves in the form of subsidence troughs of the relaxed destructured mass up to the formation of the caving areas.

Further, on the basis of the study of deformation fields of the earth's surface in the area of various mining objects the complex technology has been developed for predicting the areas where deformation processes have emergency nature. The technology includes the stage of

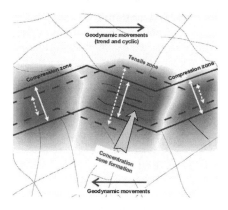

Figure 1. Diagram of concentration zone formation under influence of recent geodynamic movements in border blocks.

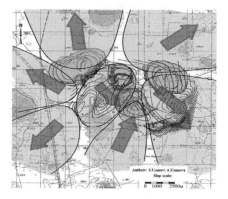

Figure 2. Deformation field of Kacharsky deposit.

the complex instrumental studies of the structural deformation parameters of the rock mass in the area of influence of the mining works, the stage of the geoinformation modeling of the geomechanical processes and the stage of the prediction of the areas of development of the emergency deformations on the basis of the regularities of the formation of the deformation fields (Fig. 2). The deformation field of the earth's surface in the mining object is characterized by multiple dynamic values which are visualized in the geoinformation model [14]: with altitude change isolines, offset vectors, isolines of the first invariant of the major deformations [15]. The prediction of emergency deformation processes has allowed to improve industrial safety of the mining enterprises, reduce the damage from the deformation processes by planning the construction of buildings and structures at safe territories.

The results of the field measurements of the geodynamic movements and the deformation of the rock block masses at different spatial scales under various mining and geological conditions, and the regularities were identified for forming the SDS system "underground construction—host rock mass" [16] taking into account the hierarchical block structure of the host rock mass, and on this basis it has been developed the method for stage-to-stage building of the geomechanical state model of this system. It has been justified that the boundary conditions of the system are defined by the superposition of the stresses and deformations obtained in the static statement of the elastic problem for three spatial and temporal scales [17] defined by the recent geodynamic movements of three hierarchies of structural elements of the rock mass; the large-scale lithospheric blocks determining the region's SDS, the structural blocks of the ore field determining the deposit sites' SDS, the structural blocks of the peripheral rock mass determining the SDS of the geotechnical system "underground construction—host rock mass." Performing the mathematical modeling takes into account the peculiarities of the transition between its successive stages determined by the scale levels in terms of the conformity of the types of boundary value problems, the detailing of the structural configuration of the reviewed rock mass and the dispersion of its physical properties (Fig. 3).

The most important task for the geomechanical behavior monitoring is timely identification of changes in the state of the open-pit sides and benches, including changes in its structural configuration for the safe mining works [18]. The identification of changes is impossible without accurate mappings of the structural anomalies of the investigated territory which requires the creation of a digital volumetric model of the open-pit space to ensure the validity of the geo-data mapping followed by analysis of the ongoing changes. For a number of deposits there are designed the large digital models of the open pits, allowing to track the real-time state changes of the open-pit sides and the mobility of the individual blocks and to monitor the changes in their structure at its followed filling with the derived geophysic data (Fig. 4). The developed algorithm of the geomechanical behavior monitoring includes the modern geodetic methods—aerial photography using the quadcopter in conjunction with the GPS technologies of spatial and altitude point positioning used in the 3D modeling and geophysical study methods of the structure of the near-edge rock mass which allows to perform the diagnostics of the structure and the state of the rock mass in the near-edge rock mass and the engineering facility foundations [19].

The stability and durability of the operation of the facilities in the mining sphere, as well as the efficiency and safety of the mining operations should be based on knowledge of the host rock SDS and their physico-mechanical properties. For long time operation of the facilities, such as the permanent mine openings, the concrete dams, the bridges etc. it is also required to know the load changes in the rock mass in time [20]. One of the reliable methods for monitoring the stress state changes of the objects in time is the measurement of the characteristics of the interference pattern in the inclusions—in the photoelastic sensors installed in the shot holes in the rock mass or in the holes of the structural elements of the objects. The deformation of the walls create the stress in the sensor wherein appears the time-dependent interference band pattern. The pattern is caused by the fact that the light beam released from the polariscope's polarizer and reflected in the mirror of the sensor is decomposed into two constituent planes wherein the light fluctuations coincide with the direction of the principal normal stresses in the representative points. Further the light passes through the analyzer where it is again brought to the same plane, and owing to varying speed through the sensor one beam part advances the other beam part, and the bringing to one plane in the analyzer occurs with a shift of phases referred to as the path-length difference which causes the phenomenon of interference. The Institute of Mining of the Ural branch of RAS has developed and manufactured in small series the compact mining polariscope PShK-S (Fig. 5) whereby the measurements of the stress state in the rock mass in the mines and at the capital underground construction objects were performed [21].

On the basis of the long-term experimental definitions of the rock mass SDS changes the hypothesis has been developed [22] whereby the stress state of the rock mass includes the following components: gravitational, semi-constant tectonic and time-variant. In the underground con-

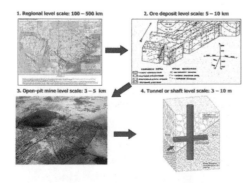

Figure 3. Large-scale assessment levels of SDS system "underground construction—host rock mass" illustrated by Donskoy MPS.

Figure 4. 3D modeling of structural configuration of open-pit side.

Figure 5.

structions the first two components result in their compression, and the third component results in increase of the compression or in its reduction with the cycle 11 years. As the mining works deepen, the first component causes it to expand, the second causes the compression, and the third component results in increase or reduction of the compression up to the extension. On this basis, the scientific justification has been performed for the geomechanical stabilization of the constructive elements of the mining systems of the underground and open-pit geotechnologies for the large steep deposits at depths more than 500 m at a high level of stresses.

Recently the Ural school of geomechanics believes that one of the most important tasks is the extension of the research results obtained in the field of geomechanics to the practical mining sphere, and its scientists regularly draw the attention of the Rostekhnadzor and other oversight bodies to the outdated level of the ideas about the geomechanical processes and phenomena in the normative documents regulating the security of the mining objects, including the acting regulatory documents on surveying for the construction, in the practice of the assessment of the operated mining object safety, in the practice of the examination of industrial safety of the projects, technical solutions, exploited hazardous objects, and in the oversight activities. Today the level of fundamental and applied research in the field of geomechanics allows us to solve specific tasks in increase of the mining production efficiency and safety, the problems of identifying the causes of natural and technogenic disasters and catastrophes in the mining sphere, to develop the predictive assessment of their manifestations and the technologies for reduction of risk and accident severity.

4 CONCLUSION

The long-term activities of the school of geomechanics is dedicated to the mining safety and efficiency, and the fundamental and applied studies cover the wide range of issues for interaction of the natural and technogenic systems of the mining enterprises and other mineral resources and ensure the development of the methods for the efficient exploitation of the deep-seated deposits, taking into account the geodynamic features, the structural configurations and the stress-deformed state of the minerable rock mass.

The Ural school of geomechanics is notable for the fundamental nature of the performed studies and their relevance in the applied sphere for solving the practical tasks of the mining enterprises with a vast geography—from the Kola Peninsula to Yakutia. The interesting scientific problems directly affecting the lives of people, the modern research technologies conforming to the world level, the demand for the scientific results in the enterprises—all this as a whole helped to engage the scientists and to stabilize the youth staff.

Analyzing the history of the Ural school of geomechanics, it can be concluded that the modern stage of its development which began with the return of the industry institute to the Russian Academy of Sciences runs with the rise and accumulation of the scientific capacity inspiring with hope for further fruitful development of the scientific school in discovering new phenomena and solving the problems in the field of human interaction with the Earth.

REFERENCES

[1] Sashourin A.D., Bolikov V.E., Balek A.E., Panzhina N.A. Modern geodynamics and problems of geomechanics in mining. *Gorny zhurnal* [Mining magazine]. 2005. No. 12. pp. 102–107. (in Russian).

[2] Rules for the protection of structures and natural objects against the harmful effects of the underground developments at the ferrous metal ore deposits in the Urals and Kazakhstan. Sverdlovsk: IM MISI USSR. 1990. (in Russian).

[3] Sashourin A.D. Geodynamic origins of major natural and geotechnic catastrophes. *GIAB* [Mining informational and analytical bulletin]. 2011. No. 11. pp. 225–236. (in Russian).

[4] Panzhin A.A. The study of harmonics of quasi-periodic recent deformations of the rock mass at large spatial and temporal databases. *GIAB* [Mining informational and analytical bulletin]. 2010. No. 9. pp. 313–321. (in Russian).

[5] Ruchkin V.I. Monitoring of the geodynamic activity of the rock mass. *GIAB* [Mining informational and analytical bulletin]. 2010. No. 9. pp. 354–360. (in Russian).

[6] Sashourin A.D. Formation of stress-deformed state of the hierarchical block of the rock mass. *Problemy nedropolzovaniya* [Mining problems]. 2015. No. 1 (4). pp. 38–44. (in Russian).

[7] Panzhin A.A., Makarov A.B. Modern methods of geodynamic monitoring with subsoil use. *Chyornaya metallurgiya* [Ferrous metallurgy]. 2014. No. 4 (1372). pp. 16–22. (in Russian).

[8] Sashourin A.D., Melnik V.V., Panzhin A.A. and others. Base of experimental data on characteristics of recent geodynamic movements. Certificate of state registration of databases of 08.08.2012 No. 2014620345. (in Russian).

[9] Kuzmin Yu. O. Modern geodynamics of dangerous fractures. *Fizika Zemli* [Physics of the Earth]. 2016. No. 5. pp. 87–101. (in Russian).

[10] Destruction of the earth's crust and processes of self-organization in the field of heavy technogenic influence. V.N. Oparin and others. Novosibirsk. Publ. SB RAS. 2012. 632 p. (in Russian)

[11] Ruchkin V.I., Konovalova Yu.P. Changing of stress-deformed state of geological environment under the influence of complex of natural and geotechnic geodynamic factors in mining enterprises. *Problemy nedropolzovaniya* [Mining problems]. 2015. No. 1 (4). pp. 32–37. (in Russian).

[12] Lovchikov A.B., Gorbatsevich F.F. About distribution of tectonic stresses in the near-surface layers of the earth's crust in vertical. *GIAB* [Mining informational and analytical bulletin]. 2015. Special edition No. 56. pp. 157–163. (in Russian).

[13] Ruchkin V.I., Zheltysheva O.D. Influence of technogenic loads on dynamics of stress-deformed state of rock mass. *Problemy nedropolzovaniya* [Mining problems]. 2015. No. 1 (4). pp. 26–31. (in Russian).

[14] Usanov S.V.. Monitoring technology for nonlinear deformations of buildings and constructions. S.V. Usanov, V.I. Ruchkin, O.D. Zheltysheva. *FTPRPI* [Physics and engineering problems of mining]. 2014. No. 6. (in Russian).

[15] Mazurov V.T., Panzhin A.A., Silaeva A.A. Structural modeling of movements obtained by geodetic data via visualization. *Geodeziya i kartograpiya* [Geodesy and cartography]. 2016. No. 3. pp. 35–40. (in Russian).

[16] Balek A.E., Sashourin A.D. Problem of assessing natural rock mass SDS at mountain massif by exploitation of mineral resources. *GIAB* [Mining informational and analytical bulletin]. 2016. No. 21 (special edition). pp. 9–23. (in Russian).

[17] Balek A.E., Sashourin A.D. Improvement of methods of in situ measurements of stress-deformed state of large areas of rock mass. *Vestnik of PNIPU* [Herald of Perm National Research Polytechnic University]. Geology. Oil-and-gas engineering and mining. 2014. No. 11. pp. 105–120. (in Russian).

[18] Sashourin A.D., Panzhin A.A., Melnik V.V. Sustainability of open-pit sides in order to protect the potential dangerous areas of transport berms. *Vestnik MGTU im. G.I. Nosova* [Herald of Nosov Magnitogorsk State Technical University]. 2016. vol. 14, No. 3. pp. 5–12. (in Russian).

[19] Zheltysheva O.D., Efremov E.Yu. Modern technologies for monitoring sustainability of open-pit sides. *Marksheideriya and nedropolzovanie* [Mine surveying and mining]. 2014. No. 5(73). pp. 63–66. (in Russian).

[20] Zubkov A.V., Biryuchyev I.V., Krinitsyn R.V. Study of rock mass stress-deformed state changes. *Gorny zhurnal* [Mining magazine]. 2012. No. 1. pp. 44–47. (in Russian).

[21] Invention patent No. 2587101 Russian Federation. Compact mining polariscope / A.V. Zublov, Yu. G. Feklistov. Pending 22.05.2014, publ. 10.06.2016., Bulletin No. 16.

[22] Zubkov F.V. Principle of formation of natural stress state of the earth's crust. *Litosfera* [Lithosphere]. 2016. No. 5. pp. 146–151. (in Russian).

Geomechanics and Geodynamics of Rock Masses – Litvinenko (Ed.)
© 2018 Taylor & Francis Group, London, ISBN 978-1-138-61645-5

Geomechanical substantiation of mining in rockburst-hazardous deposits

A.A. Kozyrev, V.I. Panin & I.E. Semenova
Mining Institute of the Kola Science Centre RAS, Russia

ABSTRACT: The article presents the research results of the Mining Institute of the Kola Science Centre RAS in the field of geomechanical support of mining operations at the rockburst-hazardous deposits of the Kola Peninsula for recent years. The methods developed help minimizing geodynamic risks when conducting large-scale underground and open mining operations. The authors consider examples of justification of the order of carrying out stoping operations with the use of complex full-scale and numerical methods. The main directions have been shown of development of geomechanical support of operations, variants of technological solutions leading to unloading of the rock mass sections, as well as local measures to reduce the level of acting stresses in the boundary zone of mine excavations. A brief analysis is given of powerful seismic events, tectonic rockbursts and mining-induced earthquakes.

The results of a prognostic study of geomechanical processes associated with a change in the stress-strain state during the mining of the adjacent deposits at the Khibiny apatite arc are presented. The authors outline a methodical approach to the development of a complex of multiscale numerical models of a volumetric stress-strain state (SSS) of a rock massif with the possibility of differentiated or complex accounting of the main geological and mining factors on the example of creating a geomechanical model of the Khibiny ore mining district. A geomechanical model has been created that allows considering the development of the complex of the adjacent Khibiny deposits and determining the order and direction of mining operations in rockburst-hazardous conditions, taking into account the enclosed ring structure of the rock massif, the effects of tectonic stresses, major radial faults, the relief of the surface, and parameters of ore bodies. Specific features of the stress-strain state in the Khibiny rock massif with successive excavation of the ore apatite-nepheline deposits confirm the hypothesis of transformation of the SSS types with depth and reorientation of the principal stresses in the blocks of the rock mass between the radial faults. The absolute altitude marks $-1000 \div -1200$ m have been defined, below which the vertical stresses prevail over the horizontal ones. The obtained results indicate that all the proven reserves of the Khibiny apatite arc will be worked out in conditions of evident tectonic compression of the rocks. It is possible to use the results obtained in making decisions on the development of other adjacent deposits in complex geomechanical conditions.

Keywords: stress-strain state, adjacent deposits, large-scale mining, management of geodynamic risks, numerical modelling, tectonically stressed rock massifs, ring structures, radial faults

1 INTRODUCTION

At present, about 200 billion tons of rocks are extracted from the earth's interior every year, and the scale of the minerals extraction continues to grow. The movement of such a volume of rock masses leads to changes affecting virtually all elements of the biosphere: water and air basins, the earth's surface, subsoil, flora and fauna. The mining-induced impact of mining operations also has a significant effect on the stress-strain state of large areas of the earth's crust and can cause geodynamic phenomena with catastrophic consequences [1–5].

The problem of rockbursts and mining-induced seismicity is now important for many developed mining regions of the world, including for most of the Kola Peninsula mines. Therefore, since the creation of the Mining Institute of the Kola Science Centre of the Russian Academy of Sciences, great attention has been paid to the development of the theoretical and practical geomechanical research.

The basis for geomechanical support and justification of principles and procedures for mining operations is the availability of:

– data on physical and mining-technological properties in relation to the problems of rock pressure and destruction of rocks;
– knowledge about the geodynamic structure of the rock massif and systemic fracturing;
– information on the initial stress field in ore bodies and deposits.

The Institute has performed studies in all three areas with the support of mining enterprises from the 60 s of the 20th century. The base of data and knowledge has been accumulated, which is constantly replenished.

Geomechanical support of mining operations at the stage of deposit exploitation can be divided into two parts. The first part assumes a long-term forecast of SSS when developing regulations for mining horizons, cut-off blocks, and block-pillars. The second one is the implementation of current assessments of the geomechanical situation involving a regional and local forecast based on the results of:

– monitoring of geological environment (seismic, deformation, tomographic, acoustic);
– visual observation of mining excavations;
– numerical modelling of SSS with accounting actual mining-engineering situation.

This approach makes it possible to predict the geomechanical situation as mining progresses, to support them and prevent dangerous dynamic phenomena at the regional mining enterprises. Let's consider the main provisions, achievements and problems in supporting mining operations on the example of the Khibiny ore mining district.

2 OBJECT OF RESEARCH. MODERN IDEAS ABOUT THE STRESS-STRAIN STATE OF THE KHIBINY APATITE ARC

The Khibiny rock massif is the largest raw material base for the production of mineral fertilizers, which has no analogues in the world and is represented by ten deposits: six deposits are operated by JSC Apatit, two are owned by Northwest Phosphorous Company (JSC NWPC), and two are in reserve. Over the deposits development period, more than 1.6 billion tons of ore were extracted (about 1/3 of all reserves) or more than 4.5 billion tons of rock. The deposit reserves are worked out with open and underground methods. The underground mining uses the variants of the caving system of ore and enclosing rocks. Extraction and transfer of large volumes of rock masses has resulted in a change in the relief and regional geodynamic regime.

An important factor from the point of view of the formation of stress fields in the Khibiny massif is that the rock massif belongs to tectonically-strained massifs. By definition, this is a complex of rocks of one genesis, occurred in areas of the earth's crust uplift separated by geological and structural boundaries from surrounding rocks. Characteristic features of tectonically stressed massifs are specific manifestations of rock pressure in the excavations, disk core, azimuth curvature of the borehole shafts, which is caused by the horizontal component of the stress tensor, significantly exceeding the vertical component. The results of in-situ determinations carried out in 1965–1995showed mainly the direction of the action of horizontal stresses, close to the occurrence of ring structures of the Khibiny massif (Fig. 1). It is believed that the Kola Peninsula region is located in the zone of action of sublatitudinal tectonic forces [6–8]. In subsequent years, the parameters of the stress-strain state in individual deposits were clarified. To date, the most complete picture has developed for the

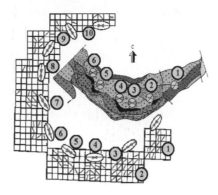

Figure 1.   Direction of the action of tectonic stresses in an intact rock massif by the results of in-situ determinations during 1965–1995. Mined and prospective deposits are shown with figures.

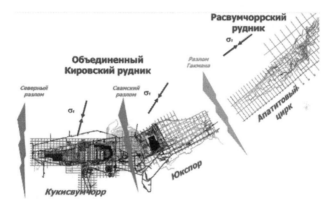

Figure 2.   Direction of the action of tectonic stresses in the deposits of JSC Apatit, mined by underground methods. Number 4 ÷ 6 in Figure 1.

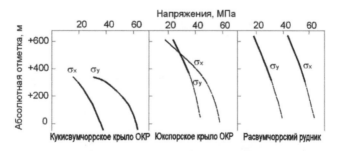

Figure 3.   Values of tectonic stresses in the deposits of JSC Apatit, mined by underground methods.

Khibiny apatite arc deposits, which are mined by the underground methods, where stress measurements by the doorstop method were carried out within the mine field on several horizons. Figures 2 and 3 show the effect of tectonic stresses and their magnitude at different depths at the Kukisvumchorr and Yukspor deposits and the Apatitovy circus deposit.

Data on the parameters of the deposits' stress field are compulsorily entered in the guidance documents of mining enterprises that regulate mining operations at the Khibiny deposits, prone to and hazardous for rockbursts [9, 10].

That is, excavation of the Khibiny minerals is carried out under rockburst-hazardous conditions associated with high tectonic stresses in the rock mass [11, 12]. In this case, the inevitable deepening of mining operations and their intensification leads to the transformation of the initial stress-strain state with the formation of both concentration zones and unloading of individual sections of the massif, as well as a change in the direction of action of the stresses. In general, the planned-to-mining levels are predicted to increase the background stress level with an aerial increase in potentially hazardous areas.

3   GEOMECHANICAL MODEL OF THE KHIBINY ROCK MASSIF. RESULTS ANALYSIS

In order to study the geomechanical processes associated with the change in the stress-strain state at the rock massif during the large-scale mining in high-stress rock massifs, a 3D finite element model of the Khibiny massif was developed, taking into account the main geological factors: embedded ring structure of the massif, radial faults, surface relief, parameters of ore bodies, and mining-technical factors: the geometry of open and underground stoping areas [13].

In connection with the elastic deformation of most of the rocks composing the Khibiny rock massif, up to the destruction and rocks destruction character in the vicinity of the excavations at the mined deposits, an elastic model has been chosen. The problem was solved both with the action of the dead weight of the rocks, and with the specification of additional boundary conditions in the form of horizontal loading in the sublatitudinal direction. The values of the horizontal stresses correspond to the data of long-term in-situ measurements, both in absolute values at different depths and in the direction. That is, a tectonic stress field approximating the real was simulated. The model dimensions are in the plan of 90 km × 75 km, the height is about 20 km (Fig. 4).

First of all, the parameters of the stress field for the intact rock massif have been determined. The obtained data were compared with the stress field without taking into account tectonics that is, taking into account only the dead weight of rocks and lateral repulsion. The layering of the rock massif and the fault structures were simulated in the same way. Taking into account only the dead weight of the rocks, the main factor affecting the stress distribution σ_{max} is the surface relief and ring structure of the massif. In the case of additional account of the existing tectonic forces, the influence of fault structures becomes more significant, the gradient of stress reduction increases (Fig. 5a).

The main part of the apatite ore reserves is located in the southern part of the Khibiny massif within the so-called Khibiny apatite arc, where excavation of minerals is carried out both underground and open-pit. When analysing the results of the vector field σ_{max}, the hypothesis of a reorientation of stresses in the vicinity of large fault structures, including the faults of the Saamsky and Gakman faults, was confirmed, with which the features of stress distribution at the Kukisvumchorr, Yukspor and Apatitovy circus are related. Direction of the action of the maximum stress component in some sections of the deposits is at an angle of more than 45° and even across the strike of the ore body with the established sublatitudinal direction of regional compression (Fig. 5b). At that, the largest concentrations

Figure 4.   Finite element model of the Khibiny rock massif.

142

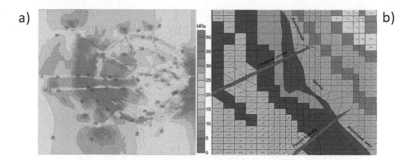

a)   b)

Figure 5.   Distribution of σ_{max} in the horizontal section at the zero point in the version taking into account the tectonics a) isolines for the area including the entire Khibiny massif; b) vector distribution for the site, which includes deposits mined by JSC Apatit by underground methods.

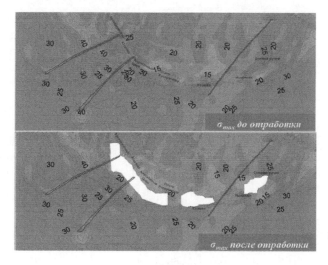

Figure 6.   Distribution of σ_{max} in horizontal section at the depth about 50 m. a) intact rock massif; b) complete extraction of proved reserves.

of compressive stresses occur at the ends of the fault structures and to the boundaries of the various-module media. In the faults themselves, the maximum stresses are oriented perpendicular to their strike and remain subhorizontal even at great depths.

In simulating successive mining of the apatite-nepheline arc ore deposits, up to the complete excavation of all explored reserves, a significant decrease in the compressive stresses σ_{max} was noted with the appearance of tensile stress zones σ_{min} and an increase in the area of these zones, as well as an increase in the absolute values of the tensions from the hanging wall of the ore deposit. Thus, with the complete excavation of all the reserves of the zone and imitations of the caving of the undermined rocks, the size of the tension zone in the plan reaches the dimensions of the extraction itself across the strike (Figure 7) with an increase in the absolute values of the stresses up to critical stresses comparable to the tensile strength of rocks. That is, in the rocks of the hanging wall there will be an active fracturing, which in the future will result in the processes of destruction and weathering to a significant flattening of the mountainous terrain in the Khibiny apatite arc area.

As the obtained data analysis result, the hypothesis has been confirmed of the transformation of the stress-strain state types with depth. The prevalence of vertical stresses above horizontal begins with absolute marks $-1000 \div -1200$ m. Since according to the geological prospecting data the ore bodies of the Khibiny apatite arc are wedged to an absolute elevation of -500 m, we can assume that all the reserves that have been discovered to date will be

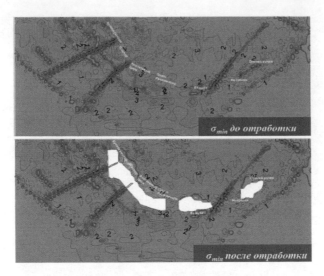

Figure 7. Distribution of σ_{min} in horizontal section at the depth about 50 m. a) intact rock massif; b) complete extraction of proved reserves.

mined under the conditions of a gravity-tectonic stress field with the prevalence of horizontal stresses over vertical ones. It should also be noted that as the reserves of apatite ore are extracted, the σ_{max} vectors at great depths are flattened, that is, the stress field is of even more expressed tectonic character. According to the forecasted data, the background level of the maximum compressive stresses at the mining horizons after the extraction of reserves to the zero mark is more than 60 MPa, whereas now the background values of σ_{max} do not exceed 50 MPa. At this level of stress, the vast majority of excavations will have a high category of rockburst hazard, which will require additional costs for fastening and will influence the increase in the cost of mining.

Thus, it was created a geomechanical model of the Khibiny massif. The model allows to consider the mining of the contiguous deposits complex and to determine the optimal order and direction for the mining development in rockburst-hazardous conditions. Also the model is the basis for the choice of the boundary conditions in the stress-strain state simulation of individual sections of Apatite arc, deposits and blocks.

4   FEATURES OF SEISMIC PROCESSES IN MINING OPERATIONS

To form powerful seismic events, including mining-induced earthquakes during mining operations, a combination of a number of conditions is necessary. First of all, a high level of horizontal tectonic stresses, determined by the corresponding tectonic-physical situation, that is, the presence of zones with large velocity gradients. In addition, the presence of the corresponding brittle high-strength rocks with tectonic irregularities within the development zone, favourable geomorphological conditions (mountainous terrain), the presence of large-scale mining, explosive impact during excavation and ore breakage, and seasonal water inflows into mining and much more.

The basic scheme for the preparation and implementation of large seismic events in the rock massif with the natural development of processes and under the influence of mining-induced impacts is presented in Figure 4 [14]. As follows from the data given, with the natural development of the process for a long time, the following three stages of preparation of the rupture are distinguished (Fig. 4a):

1. Accumulation of tectonic stresses with slowly developing deformation, thus small portions of energy are allocated in the form of weak shocks or earthquakes (a seismic background);

2. increase in stresses and the activation of deformations, accompanied by the appearance of foreshocks and the enlargement of the discontinuities; this stage is completed by the formation of a large rupture, which causes the main shock of earthquakes;
3. subsequent redistribution of stresses, the growth of the rupture, the manifestation of aftershocks.

Under the influence of mining (Figure 8b), stresses are concentrated in the vicinity of the stoping space and reach the limit values much earlier and are more often realized in the form of weaker shocks. With powerful explosive influences, the stress fields are added together, and the destruction is realized even earlier. Therefore, information on the stressed state of the intact rock massif, possible deformations and additional stresses due to large-scale excavation of rock masses, and the conditions for the release of accumulated energy are important for predicting tectonic rockbursts and mining-induced earthquakes.

In the zone of influence of mining operations, the rate of deformation of rocks increases by an order of magnitude or more in comparison with the natural one. Together with the dynamic impact of bulk blasts, this leads to processes of formation of a focal source; premature initiation of a series of weaker shocks due to a breakdown on the contacts of blocks in the presence of free space in the rock massif; the breaking of the barrier between the adjacent fissures and faults, between the faults and the stoping space, or between two adjacent stoping areas. In addition, with the use of mining systems with caving, a large part of the large seismic events occur during the formation of fracture fissures in the rocks of the hanging wall of the undermined massif [15].

The last large mining-induced event with a magnitude of about 3.5 occurred on October 21, 2010 at 12 hours 10 minutes Moscow time. The hypocentre of the event was located in the Kukisvumchorr wing of the United Kirovsky mine approximately on the level of +16 m (360 m from the surface) at some distance from the mine excavations. The released energy was of the order of $7 \cdot 10^9$ J. Strong shocks and horizontal vibrations lasting several seconds were recorded in the Kukisvumchorr village (0.5–1.5 km from the epicentre), in the Botanical Garden of the Kola Science Centre RAS (5 km) and Kirovsk town (7–12 km). In Apatity

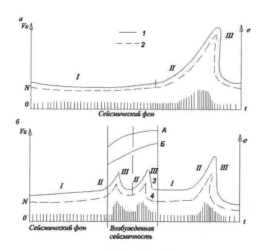

Figure 8. Character of changes in time t, of deformation rates V, tectonic stresses σ and seismic energy release N. a) during natural development of processes for a long time t; b) under mining operations influence;
1 – change of stresses σ; 2 – change of deformation rate V; 3, 4 – change of stresses and deformation rates when breaking certain rock mass volume;
I – stage of fault occurrence; II – stage of foreshocks manifestation; III – main earthquake shock, stresses redistribution and aftershocks manifestation;
A – change of extracted rock volume; B – change of stress concentration due to rock mass extraction.

Figure 9.   Destruction in the excavations of the haulage level + 90 m, United Kirovsky mine, as a result of mining-induced earthquake (photo A. Panteleev).

town (25–30 km), vibrations were noted selectively. As a result of this event, in a number of underground mine excavations there occurred cracking, peeling of concrete and spray-concrete support and collapse of the rock (Figure 9). No significant earthquake effects on the surface have been identified.

In work [16], the growing rupture was estimated, which amounted to about 700 m, while the size of the earthquake preparation zone is several kilometres, that is, it includes the entire industrial mine zone.

5   ROCKBURST FORECAST

The solution of problems of geomechanical substantiation and forecasting at a modern level is impossible without applying the full-scale and numerical methods, as well as an integrated monitoring system for the rock massif and includes the solution of several major tasks:

- study and constant refinement of the parameters of the initial stress field and identification of geodynamically active structures;
- rock mass monitoring during excavation operation using a set of regional and local methods (seismic, deformation, seismic, ultrasonic, visual inspection of excavations, etc.);
- current and long-term forecast of redistribution of the stress field as mining progresses with the allocation of zones of increased rockburst hazard and the search of an optimal variant for the development of mining operations and mining system parameters;
- substantiation of preventive measures, type and parameters of excavations fastening.

However, even if all of the above tasks are performed at the present level in conjunction with the relevant services of the mining enterprise, the probability of occurrence of critical situations is not zero. If such a situation takes place, then the need to promptly address the problems of localizing negative consequences and development of measures to take steps out of the situation jointly with the workers of the mining enterprise will come to the fore. The measures can be as follows: the revision of the plan of mining operations, additional measures for unloading or fastening excavations.

The Mining Institute KSC RAS is working on the development of regional and local forecast methods. The expert system was developed and is improved on the basis of the software complex Sigma GT (Certificate of state registration of the computer program No. 2012613935 dated April 27, 2012), which implements the finite element method in volumetric formulation and allows modelling the stress and strain field taking into account the main geological and mining factors [17, 18]. A series of finite-element models of the Khibiny apatite arc deposits of various details has also been formed. The software is used in mining enterprises such as Forecast and Prevention of Rockbursts Services JSC Apatit, JSC "PGGHO" (is being implemented at JSC NWPC) for rapid assessment and forecast of the geomechanical situation for annual and long-term planning, as well as for choosing technical solutions to ensure the safety of mining works. Calculations of SSS and categories of excavations are carried out at several scale levels, which allows retrospective, current and long-term forecasts of changes in the stress-strain state of complex natural-engineering systems. The results of calculations are the basis for choosing the safest mining system and its parameters that are safest in terms of geomechanical conditions and are used in planning the safe order of mineral extraction and anti-rockburst measures (Figure 10).

An important area is also the processing and analysis of data from automated control system for rock mass state. The complex estimation technique of seismic activity in the rock mass has been developed with the simultaneous analysis of recorded events by several prognostic criteria, differing by combining different in terms of physical sense and complementing each other individual criteria by bringing their values to a comparable type and isolating them by the values of a complex estimation of four zones according to the degree of seismic activity (damped, stable, pulsating, growing). The last two zones are the most dangerous. To share criteria that reflect different parameters of the seismic emission stream, the values of the complex estimation in the cell are calculated:

$$X'' = \frac{1}{N} \sum_{j=1}^{N} X'_j,$$

where N – number of individual prognostic criteria, X'_j – a value of a prognostic criterion.

Thus, a comprehensive assessment of a number of individual criteria allows taking into account the behaviour of each of them. The application of the integrated assessment allows taking into account the behaviour of all the individual criteria and more accurately estimating the prognostic parameters of the seismic regime in the spatial cell than when using one criterion. The geodynamic state of the massif is evaluated on the basis of cluster analysis of seismic activity (taking into account the parameters of seismic events: coordinates, energy, seismic moment), using the Kukisvumchorr and Yukspor deposits as an example. The assessment allows detailed analysis of the interrelation between clusters of seismic events and

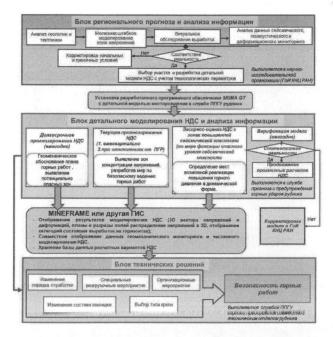

Figure 10. The integration scheme of the SSS forecast module into the geodynamic safety system of the mining enterprise.

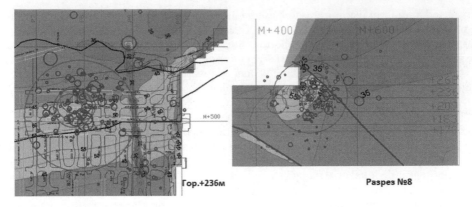

Figure 11.   Comparison of seismicity and stress distribution in a rockburst-hazardous block.

mining and geological factors on the basis of identification of their occurrence in a certain area of the rock massif (horizon, excavations) [19, 20].

The integration of seismic data and the results of numerical modelling of the rock massif's SSS allows for more accurate assessment of the geomechanical state of the rock massif, determining the cause of the occurrence of the dynamic phenomena, and also improving the quality of the assessment of potentially hazardous areas in the planning of mining operations. The coincidence of zones that are dangerous both in terms of seismic level and in terms of the level of acting stresses increases the reliability of assigning such sites to those that are rockburst-dangerous. This increases the safety of mining operations, when developing appropriate measures (Figure 11).

In recent years, the main task in the analysis and forecast of the geomechanical situation in the mining of rockburst-hazardous deposits is the integration of geological, mining and geomechanical data in a united information space. Almost all mining enterprises use geoinformation systems (GIS) today, on the basis of which it is logical to carry out such data integration. In the Mining Institute KSC RAS, a complex model of the Khibiny mining area was created on the basis of the MINEFRAME software, including terrain relief, geological environment objects (ore bodies, most significant tectonic disturbances), full-scale and predicted data on the stress field parameters, complexes of underground excavations and structural elements of open pits, dumps, tailing dumps and industrial infrastructure [21, 22].

The main difficulty in the joint analysis of seismic data and the design parameters of the stress field is the necessity of combining continuous and discrete data. In general, the results of the stress distribution are analysed for a certain configuration of stoping excavations with the distribution of seismic events at a certain point in time. At the same time, the same distribution of stress-strain state is associated with seismic activity zones that can migrate not only due to changes in the mining front [23]. The experience of complex analysis shows its relevance, practical importance and the need for further development with the development of forecast criteria for the localization of spatial and temporal parameters of hazardous zones.

6   DANGEROUS ZONES. PROPOSED TECHNOLOGIES AND ARRANGEMENTS

Analysis of seismic activity, areas of maximum damage, and results of numerical forecasts of the stress-strain state allowed determining the main zones of increased rockburst hazard in the mining of reserves from the Khibiny apatite arc deposits. The areas of the massif between the approaching large-scale mining operations, in fact, are block-pillars with an increased concentration of compressive stresses. On the other hand, tensile stress zones along the minimal component are observed in these sections. This combination of the joint action of compressive and tensile stresses is a factor that adversely affects the stability of the rock massif as a whole and of individual excavations in particular. Another zone with an increased level of rockburst hazard is the support pressure zone, where in most cases the concentration of compressive stresses is maximal. The solution to the problem here is largely due to the management and forecast of caving of the undermined rocks.

Normative documents provide for systematic extraction of reserves in rockburst-hazardous conditions, if possible without the formation of pillars, sharp corners and protrusions of the stoping operation front. Optimum in geomechanical conditions is the development of mining operations from the centre of the ore deposit to the flanks, or from one flank to the other. However, in reality, it is practically impossible to avoid the situation when it becomes necessary to mine a section between stoping spaces (underground and open). It is necessary to regulate the extraction of the block-pillar reserves, both from the point of view of technology and from the point of view of ensuring the safety of mining operations. Such blocks-pillars are formed in the following cases:

- when extracting reserves of adjacent deposits with a conventional boundary;
- when excavating reserves of deposits in a combined mining method;
- in the case of several cuts in one deposit, when excavating reserves in an underground method.

The problem is that at the initial design stage, open and underground mining operations at a deposit and even more so the mining operations at the adjacent deposits are regarded as independent. Therefore, at a certain point, the mining company faces the need to mine an adjacent area. In the case of rockburst-hazardous deposits, this situation requires the solution of a whole range of problems. Many of them can be solved on the basis of multivariate predictive modelling of the stress-strain state [24]. The mining technology with creation of protective zones (as a rule, it is a leading unloading zone in the hanging wall of the ore deposit) is proposed, which allows reducing the level of maximum compressive stresses in a large part of the rock massif by more than 2 times. An example of redistribution of stresses in the case of the implementation of this technology at the Yukspor deposit is shown in Figure 12.

A technique has been developed for predicting the parameters of the caving of overlapping rocks in rock massifs, taking into account the actual volumetric stress-strain state and using the destruction criteria, both in absolute values of the tensile deformations and in the

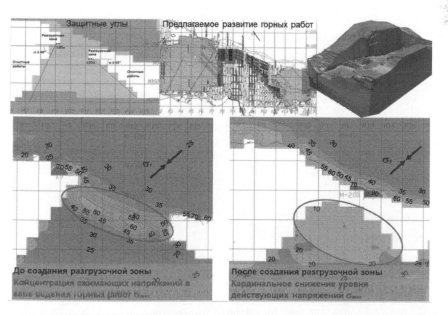

Figure 12.   Development of regional arrangements to reduce acting stresses in a zone of mining operations on an example of the Yukspor deposit, JSC Apatit.

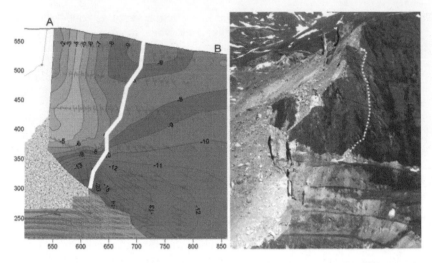

Figure 13.   Forecast of collapse of undermined rocks up to day surface, the Kukisvumchorr deposit, JSC Apatit.

orientation of their sites, accounting the disturbance of the rocks. The technology allows taking the necessary preventive measures on the basis of numerical modelling. An idea was formed of the staginess of caving of the rocks at the hanging wall up to the day surface under the tectonic forces: the development and accumulation of fissures in zones occurred in the boundaries of the completion of the undermined rocks and directly on the surface; merging of fissures and formation of a main fault; the formation of a single plane of collapse to the surface during the accumulation by the rock mass of energy potential, necessary to overcome the zone with reduced values of tensile deformations. An example of the prediction of the germination of a fracture in an undermined massif and its actual position on the surface is shown in Figure 13 [25].

The effectiveness of these measures is confirmed by the low level of individual geodynamic risk on Apatite mines [26].

7   CONCLUSIONS

Study and forecast of geomechanical processes in the geological environment of mining systems allow today to conduct mining operations in tectonically stressed rocks mass with minimal risk. However the subsoil of the Khibiny rock massif contains still more than two billion tons of ore. However, these reserves are characterized by a lower content of apatite and more complex mining conditions. In this regard, it is especially urgent to develop a strategy for the long-term, efficient and safe use of a unique ore base, the priority of involving the reserve deposits in the mining and the integrated use of mined ore. The solution of this problem requires the creation of a comprehensive self-learning system for the rockburst hazard forecast, which includes all the factors affecting the geodynamic state of the rock massif with the possibility of analysing them in a united information space and ranking by impact strength for specific mining conditions.

REFERENCES

[1]   Kozyrev A.A., Panin V.I., Maltsev V.A., Akkuratov M.V., Kozhin V.S. Change in the geodynamic regime and the manifestation of mining-induced seismicity during large-scale mining operations at the Khibiny apatite mines. // Problems of geodynamic safety.: VNIMI. 1997, pp. 66–71.

[2] Kozyrev A.A., Panin V.I., Svinin V.S. Geodynamic safety during ore deposits mining in highly-stressed rock massifs // Eurasian Journal, 2010. – № 9, – pp. 40–43.

[3] Sashurin A.D., Panzhin A.A. Actual problems of geomechanical support for the effective and safe mining of solid mineral deposits in the northern and northeastern regions of Russia//Mining informational and analytical bulletin (scientific and technical journal). 2015. No. S30. pp. 62–70.

[4] Maab T.H., Tanga C.A., Tangc L.X., Zhanga W.D., Wanga L., Rockburst characteristics and microseismic monitoring of deep-buried tunnels for Jinping II Hydropower Station/Tunnelling and Underground Space Technology. Vol. 49, June 2015, pp. 345–368

[5] Meifeng Cai, Prediction and prevention of rockburst in metal mines—A case study of Sanshandao gold mine/Journal of Rock Mechanics and Geotechnical Engineering. Vol. 8, Issue 2, April 2016, pp. 204–211

[6] Tectonic stresses in the earth's crust and stability of mine excavations/Turchaninov I.A., Markov G.A., Ivanov V.I., Kozyrev A.A. – L.: Nauka, 1978. – 256 p.

[7] Tectonic stresses and rock pressure in the mines of the Khibiny massif/Markov G.A. – L.: Nauka, 1977. – 213 p.

[8] V.I. Pozhilenko, B.V. Gavrilenko, D.V. Zhirov, S.V. Zhabin. Geology of ore regions of the Murmansk region. – Apatity: Edition of Kola Science Centre of the Russian Academy of Sciences, 2002. – 359 p.

[9] Guidelines for the safe management of mining operations in deposits prone and dangerous for rockbursts (Oleny Ruchey deposit) Kozyrev A.A., Semenova I.E., Rybin V.V., Zemtsovsky A.V., Fedotova Yu.V., Konstantinov K.N., Lobanov E.A., Suslov I.V., Volkov D.S., Panteleev A.V. Mining Institute of the Kola Science Centre of the Russian Academy of Sciences, JSC NWPC. Apatity, 2015.101 p.

[10] Guidelines for the safe management of mining operations in deposits prone and dangerous for rockbursts (Khibiny apatite-nepheline deposits) Kozyrev A.A., Semenova I.E., Rybin V.V., Panin V.I., Fedotova Yu.V., Konstantinov K.N. ., Salnikov I.V., Gadyuchko A.V., Belousov V.V., Korchak P.A., Streshnev A.A. Mining Institute of the Kola Science Centre of the Russian Academy of Sciences, JSC Apatit. Apatity, 2016.112 p.

[11] Kozyrev A.A., Panin V.I., Semenova I.E., Fedotova Yu.V., Rybin V.V. Geomechanical support of technical solutions for mining operations in highly-stressed rock massifs//Physical-technical problems of mining. 2012. № 2. pp. 46–55.

[12] Kozyrev A.A., Panin V.I. Evolution of geomechanical studies at the Mining Institute of the Kola Science Centre of the Russian Academy of Sciences in the period 1961–2011//Bulletin of the Kola Science Center of the Russian Academy of Sciences. 2012. No. 4 (11). pp. 81–93.

[13] Semenova I.E. Investigation of the transformation of the stress-strain state of the Khibiny apatite arc during large-scale extraction of minerals//Mining informational and analytical bulletin (scientific and technical journal). 2016. № 4. pp. 300–313.

[14] Kozyrev A.A., Panin V.I., Savchenko S.N., Maltsev V.A. et al. Sei8 smicity in mining—Apatity, KSC RAS, 2002, p. 325.

[15] A.A.Kozyrev. Geomechanical substantiation of safe mining of deposits in rockburst-hazardous conditions//Problems and tendencies of rational and safe development of geo-resources: Proceedings. Apatity-St. Petersburg, 2011. – pp. 251–264.

[16] Kozyrev A.A., Kagan M.M., Zhirov D.V., Konstantinov K.N. Deformational precursors of mining-induced earthquake at the United Kirovsk mine, JSC Apatit//Geodynamics and the stressed state of the Earth's interior: Proceedings—Novosibirsk: IM SB RAS, 2011.-T 2. – pp. 228–234.

[17] Kozyrev A.A., Panin V.I., Semenova I.E. Experience in applying expert assessing systems for the stress-strain state of the rock massif to select safe mining methods. Notes of the Mining Institute 2012. T.198. pp. 16–23.

[18] Semenova I.E. Creation of a 3D geomechanical model of the Khibiny ore mining district//Mining informational and analytical bulletin (scientific and technical journal). 2011. № S6. pp. 263–277.

[19] Kozyrev A.A., Fedotova Yu.V., Zhuravleva O.G. Complex forecast of changes in the seismic regime at the Khibiny apatite mines.//Modern methods of processing and interpretation of seismological data, Proceedings of the Sixth International Seismological School: - Obninsk: MS RAS, 2011. – pp. 170–174.

[20] Kozyrev A.A., Zhuravleva O.G., Fedotova Yu.V. On the clustering of seismic events during the mining of rockburst-hazardous deposits of the Khibiny massif. // Fundamental and applied problems of mining sciences Novosibirsk IM SB RAS after N.A. Chinakal, 2015. – № 2. – pp. 108–113.

[21] Lukichev S.V., Nagovitsyn O.V. Information support of development and exploitation of deposits from the Khibiny mining region. Physical and technical problems of mining. 2012. No. 6. pp. 98–105.

[22] Kozyrev A.A., Lukichev S.V., Nagovitsyn O.V., Semenova I.E. Geomechanical and mining-engineering modeling as a means of increasing the safety in mining of solid mineral deposits. Mining informational and analytical bulletin (scientific and technical journal). 2015. № 4. pp. 73–83.

[23] Kozyrev A.A., Zhuravleva O.G., Semenova I.E. Complexing of seismic data and numerical modeling results of the stress-strain state in the rock massif for the assessment of potentially rock-burst-hazardous zones during mining operations//The Fourth Tectonic-physical Conference at the Schmidt Institute of Physics of the Earth of the Russian Academy of Sciences "Tectonophysics and Current Issues in Earth Sciences". Moscow, IPE RAS. 2016. T. 2. pp. 77–81.

[24] Kozyrev A.A., Semenova I.E., Avetisian I.M., Zemtsovskii A.V. Methodological approaches and realization of joining zones mining in the rockburst hazardous conditions//International Multidisciplinary Scientific GeoConference SGEM. 2016. Book 1. V. II.P. 565–572.

[25] Kozyrev A.A., Semenova I.E., Avetisian I.M. Investigation of changes in the character of collapses of undermined thickness with increasing depth of mining at the Kukisvumchorr deposit, JSC Apatit. MIAB No. 5, 2011, pp. 11–20.

[26] Fedotova Y.V., Panin V.I. Geodynamic risk assessment experience at mining in high stressed deposits. INTERNATIONAL MULTIDISCIPLINARY SCIENTIFIC GEOCONFERENCE SGEM. 2016. P. 419–426.

Geomechanics and Geodynamics of Rock Masses – Litvinenko (Ed.)
© 2018 Taylor & Francis Group, London, ISBN 978-1-138-61645-5

The researches of burst-hazard on mines in Russian Far East

I.Ju. Rasskazov, B.G. Saksin, M.I. Potapchuk & P.A. Anikin
Mining Institute, Far Eastern branch, Russian Academy of Sciences, Khabarovsk, Russia

ABSTRACT: Features of manifestation of burst hazard on underground mines of the Far East and Transbaikalia are considered. All range of dynamic manifestations of rock pressure and the technogenic seismicity is registered on Far Eastern and Transbaikalia underground mines. The main approaches to a solution of the problem of the forecast and prevention of rock bursts are specified. The possibility of the assessment of the deflected mode of the rock massif of at various stages of development of ore bodies and substantiation of efficient preventive actions is shown on the example of vein type deposits. Basic principles of creation of system of complex geomechanical monitoring and results of application of instrumental methods and technical means of monitoring of burst hazard in the Far East mines are given in the article.

Keywords: Rock bursts, technogenic seismicity, geodynamic safety, deflected mode, numerical modelling, monitoring, forecast

1 INTRODUCTION

The geomechanics is an important constituent of mining science. Now the majority of technological processes at mining and mineral processing in a varying degree is based on the data obtained as a result of geomechanical researches. These researches are necessary for the determination of properties and a condition of rocks and ores in the course of technogenic transformation of a subsoil at various scale levels (Trubetskoy et al., 2010). Most current geomechanical problems in the development of mineral deposits are: ensuring of stability pit edge and pit pile (when conducting open mining operations). Displacement of rocks, protection and maintaining of excavations, stability of structural elements of systems of development, the problem of the forecast and prevention of dynamic manifestations of rock pressure and technogenic seismicity (Turchaninov et al., 1989; Makarov, 2006; Oparin, Malovichko, Kozyrev et al., 2009; Melnikov et al., 2015) is actual for underground mining of deposits. Geomechanical researches are of particular importance at a combined method of deposit development (Kaplunov, Rylnikova, 2012; Kazikayev, 2008). Increase of mining depth, deterioration of the mining-geological conditions of development result in number of new problems concerning geomechanics. The solution of these problems demands new scientific approaches and, in some cases, revision of traditional views, including theoretical basis of rock behavior under deflected mode (Adushkin, Oparin, 2012). All this demands further perfecting of methods and means of studying of the geological environment in the conditions of intensive technogenic influence, receiving and reconsideration of the new experimental data obtained in areas with various geodynamic mode. The problem of rock bursts is one of the most composite in geomechanics. In this connection the research, executed in the last 50 years in the Far East region are of a special interest. Large number of burst hazard various mineral are developed here.

2 PROBLEM OF GEODYNAMIC SAFETY DURING MINING OPERATIONS IN THE COMPOSITE MINING-AND-GEOLOGICAL AND THE BURST HAZARD CONDITIONS

First geomechanical studies in the Far East of Russia was caused by need of fight against the frequent cases of sudden emissions of coal, rock and gas, and rock bursts, as a result. It happened in the middle of the last century on coal mines of Primorsky Krai (generally Partizansky basin) and Sakhalin. Results of these researches are reflected in research by Bich and Muratov (Bich, Muratov, 1990) and other publications.

The following stage of intensive geomechanical researches was fulfilled in the middle of the 70th years of the 20th century. It was connected with appearance of dynamic manifestations of rock pressure in the tin and polymetallic underground mines (Hrustalnensky and Hingansky ore-dressing and processing enterprises, production association "Dalpolimetall", etc.) (Rasskazov, 2008). Studying of physical-mechanical properties of rocks and ores, assessment of parameters of fields of tension in the developed massifs, influence of mining operations to the initial deflected mode was executed within the next 10–15 years (Rasskazov, 2008; Baryshnikov et al., 1982; Freydin et al., 1992). It was established that burst hazard rock massifs usually consists of hard, weak and high-elastic rocks inclined to destruction in a dynamic form. Unequal-component compression stress exists in rock massif. Greatest components are focused in North-East and South-East direction. They exceed a gravitational component of rock mass weight by 1,5–3 times (Table 1). Dominance of horizontal compressive strength (caused by the tectonic forces) over vertical (gravitational) is defined by their geodynamic position within tectonic active Amur plate. This plate is characterized by high structural heterogeneity, tectonic dissociation and existence of areas of the increased tension (Rasskazov, Saksin, Petrov et al., 2014). Majority of burst hazard mineral deposits of the Far East region are located within such areas (Fig. 1).

The high tension of the rock massif of mineral deposit of the Far East region is observed in some cases in rather small mining depths. The high tension is one of the main reasons of hazard dynamic manifestations of rock pressure. First strong rock burst was registered at 120–300 m depth in Juzhnoe polymetallic deposit, located in East Primorye (Freydin et al., 1992). The thematic example is rock burst at 170 m mining depth (block 4–208), ore shrinkage is applied here. The ore body is presented by quartz-sulphide vein (0.3–3.5 m, an angle of incidence 50–60°); host rock is fine-grained sandstones. When ore pillar decreased up to 5 m, rock burst happened. It accompanied by 40 m^3 rock mass and top failure of a ventilating drift. At the same the soil heave on 0,7 m height was observed in a drift 205. The ascending order of the block development, decreasing pillar and influence of a tectonic zone of Eldorado breaks was the reason of rock burst.

Burst hazard of a big group of mineral deposits increased with increase of mining depths to 400–500 m. Growth of depth of mining operations resulted in the increase of intensity of dynamic manifestations. In particular, the first manifestations of rock pressure in a dynamic

Table 1. Results of in-situ measurements of stress in rock massifs of some burst hazard mineral deposits of the Far East region.

Deposit	Mining depth, m	Maximum stress, MPa	Ratio of principal stresses (σ_1:σ_2:σ_3)
Nikolaevsk	860	70–150	2,5:1,5:1
Juzhnoe	500	95	1,8:1:1
Vostok-2	500	50–70	2,5:1,3:1
Perevalnoe	600	50	2:1:1
Hingansk	550	50–70	3:1:1
Antej	700	50–100	2,5:1,4:1
Darasun	700	27	1,7:1,3:1
Irokindinsk	200	11	1,3:1,2:1

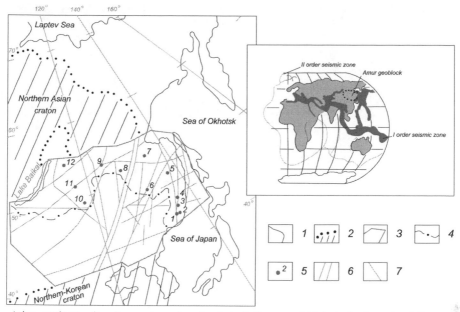

1 – Asian continent edge; 2 – craton edges (ancient platforms – rigid tectonic elements of the territory); 3 – Amur tectonic plate; 4 – Russian state border; 5 – burst hazard ore deposits: 1 – Nikolaev; 2 – Juzhnoe; 3 – Zabytoe; 4 – Vostok-2; 5 – Perevalnoye; 6 – Hingansk; 7 – Malomyr; 8 – Pioneer; 9 – Berizitovoye; 10 – Antej; 11 – Darasun; 12 – Irokindin; 6 – boundary deep break dividing the Amur plate into first order blocks; 7 – other, mainly interblock, neotectonic breaks within the Amur plate

Figure 1.   General position of the Amur tectonic plate and position of the ore burst hazard deposits.

form were noted at a construction stage in the process of bore construction and prospecting and preparatory developments on the Antej uranium deposit (East of Transbaikalia) (Rasskazov, 2008). In a consequence in the process of mining—preparatory and excavation works the dynamic ore chipping and a shooting were observed on the VII horizon (400 m depth). Development of excavation works especially in deeper horizons the intensity of dynamic manifestations of rock pressure sharply increased. More than 100 dynamic events were registered on the Antej deposit in the last decade in certain years. Rock pressure was shown in the form of ore shooting, ore chipping dynamic, roof failure, clicks and pushes inside the massif, sometimes accompanied by the destruction of backfilling and the cave soil, sometimes raising and deformation of railway lines took place.

The power of the dynamic phenomena has grown more than by 3 times since 2010. More than 20 strong dynamic events followed by a sharp strong sound, dust formation, massif quake, roof failure (1–1,5 m³), destruction of a backfilling are registered for the last several years on Antej deposit. The rock burst happened on January 29, 2011 in a block 6a-1110. As a result, more than 10 m³ of rock mass were thrown out of the right board and the soil of a pressure-relief drift No. 1. Support was destroyed for 80 m long; excavation soil a rose on 5–15 cm. By results of investigation of accident, it was established that the decrease of pillar thickness (block 6a-1110) led to cracks of contacts and movement in the vertical direction of the wedge-shaped ore block. This block is limited by 13 and 160 tectonic cracks. Rock burst accompanied by the energy emission, rock failure and timbering destruction of pressure-relief drift, seismic fluctuations including surface quakes.

The high level of geodynamic risk has remained in recent years on mines of JSC Dalpolimetall. Numerous cases of technogenic seismicity are registered here. They were followed by sound manifestations, clicks and a crash in the massif, chipping failure on various sites of the miner field. Consequences of rock burst on the Nikolaev mine (drift No. 1, – 420 m horizon) are shown in the Figure 2. Epicenter was under the bottom of the block 40 of the ore zone

Figure 2. Destructions of excavations as a result of rock burst (06.04.2016, Nikolaev polymetallic deposit).

East-1 and was followed by the strong sound effect, massif quake on all mine field, dusting of excavations. More than 10 m³ of rock mass were thrown out from the left-hand board and a roof of a drift as a result of rock burst. Mine caving were partially destroyed on −375, −390, −406 m horizons. By results of researches it is established (Rasskazov, Saksin et al., 2016) that manifestation of an burst hazard on the Nikolaev mine is bound with the tectonic structure of the rock massif. Mining excavation (5 million m³) along borders of the tectonic block resulted in activation of geodynamic processes. Shifts and motions (up to 5–10 cm) were observed on the block edges.

In general, the analysis of burst hazard on underground mines of the Far East region indicates the complication of a mining situation and increase in geodynamic risk when conducting mining operations. Complication of a mining situation caused by growth of the caved spaces and mining depths. Activation of the geodynamic processes proceeding in the form of reorganization and self-organization of the block massif of rocks in the natural and technogenic field of tension is observed. These processes are followed by shifts and motions along tectonic cracks of various scale level, emission of the considerable elastic energy and manifestation of technogenic seismicity.

In-depth complex geodynamic studies are necessary. Assessment of a geodynamic, seismic situation and the deflected mode of rock massifs; studying of regularities of geodynamic fields and processes in the field of technogenic influence of mining operations are actual. Application of seismic, seismoacoustic, geodetic, seismic-and-straining and other methods are essential for the collection of new information regarding deflected mode of the geological environment and precursors of the dynamic phenomena of a different energy level in the massif.

3 RESEARCH OF FEATURES OF FORMATION OF NATURAL AND TECHNOGENIC TENSION FIELDS INSIDE MASSIFS OF THE BURST HAZARD ORE DEPOSITS

For a solution of the problem of burst hazard and decrease in geodynamic risk by underground mining of the field it is necessary to estimate on an incipient state of its development change of the deflected mode of the rock massif under the influence of natural and technogenic factors. Having determined consistent patterns of redistribution of initial tension in structural elements of system of development, there is a possibility of management of rock pressure and jsubstantiation of a complex of actions for decrease in geodynamic risk.

In practice of geomechanical researches numerical modeling of the deflected mode by a finite element method is most widely applied to studying of natural and technogenic fields of tension (Fadeyev, 1987; Zienkiewicz, 2004). The most reliable results of researches provide

3D models considering technology of development of deposit, a geological structure of the massif and feature of its tectonic structure including features of interaction on borders of tectonic blocks. The known criteria of brittle and shift fracture and the power index of burst hazard, considering the module of recession of ore on the incredible chart of deformation, are used for the assessment of degree of a potential burst hazard of certain sites of the massif (Petukhov et al., 1992).

Results of in-situ measurements of physical-mechanical properties of rocks and parameters of natural fields of tension are used as boundary conditions at problems definition of numerical modeling of the deflected mode. At the initial stage of mining in the absence of the underground excavations excluding a possibility of direct measurements of tension at larger depths, valuable information can be obtained on the basis of complex interpretation of geologic-geophysical and the morphological-structural data, GPS observation data, applications of methods of the Earth remote sensing (Rasskazov, Saksin, Petrov et al., 2014; Saksin, Rasskazov, Shevchenko, 2015).

For more than 30-year period of researches the numerous circle of tasks of a research of features of formation of the natural and technogenic field of tension in massifs of the deposits which are characterized by the most various mining-and-geological conditions and developed with use of various technologies (Rasskazov, 2008; Rasskazov et al., 2010; Potapchuk, Kursakin, Sidlyar, 2014; Rasskazov, Saksin, Potapchuk, Usikov, 2014) was solved. The most intense sites of the miner field were established and prognosis cards of a stressed state at development of mining operations were designed at various stages of the development.

Ore deposits of non-ferrous and noble metals are developed and being prepared for the development in the present time Far East region. Some of them is referred to hazard category or to inclined to rock bursts. Vein type deposits with a steep dip angle of ore body make up a special group. They are Yujnoe, Berezovsky, Zabytoe, Southern Hingan and some other. By results of geodynamic division into districts is established that the horizontal compressive stresses focused from the subwidth to northeast near the fields (generally in a cross of an extension of ore bodies) and exceed a gravitational component by 1.8–2.5 times. The system of development with ore shrinkage and the system of development by subfloor drifts are mainly recommended for the development of this type of deposit.

Yujnoe deposit is the most burst hazard. About 100 and more dynamic manifestations of rock pressure is registered with the growth of mining depth. Quakes in the depth of the massif (seismic events) dominance; their share exceeds 40% of total number of the dynamic phenomena annually. Additional impact on burst hazard growth was exerted also by the following factors: significant increase of volume of excavated rock mass without backfilling, change of mining-and-geological conditions of development, change of parameters and elements of laying of ore bodies (thickness, angle of incidence, morphology of veins, etc.), changes of mechanical characteristics of a lithologic complex.

Results of laboratory and in-situ researches are used for 3D modeling of the deflected mode of the massif of the Yujnoe deposit. Bearing strata is characterized by average strength (≤ 118 MPa) and high values of an elastic modulus ($E \leq 104$ MPa); sulphide ores are close on elastic properties to bearing strata ($E = 26.6$–94.3 GPa), but are less strong (50–55 MPa). Ores are capable to accumulate the considerable potential energy and to faiture in a dynamic form, being potentially burst hazard.

By results of modeling it is established that mining depth more than 400 m leads to formation of zones of the increased tension in safety pillars and in edge of the massif, mainly in two top subfloors.

Growth of normalized stress (middle pressure) happens in direct ratio with increase of excavated length (L), reaching a maximum at the full block development (100 MPa and more, Fig. 3). The situation is complicated in case of the complete excavation of the close clearing block. In this case tension in the pillar increases in 25–30%. At the same time there is a risk of destruction of pillar of the top subfloor in a dynamic form.

Zabytoe tungsten deposit and Southern Hingansk manganese ore deposit have similar mining-and-geological and geomechanical conditions with Jyjnoe deposit. They are referred to category "Inclined to rock bursts". Their development is planned to be conducted by

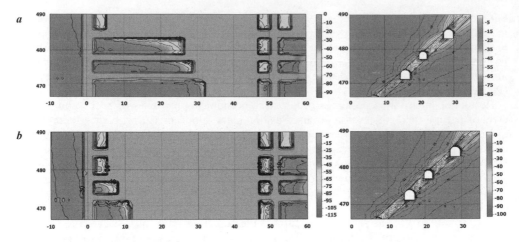

Figure 3. Distribution of average stress (σ_{av}) in structural elements of system of development on the Jyjnoe deposit (in a projection to the sloping plane, cross section to the ore body extension, pillar between excavating chambers, m = 3 m; $\beta = 40°$) length of the developed space: a – 20 m; b – 40 m.

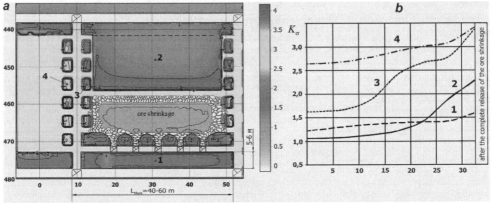

1–4 – characteristic point of the massif: in sill pillar (1), decreasing the temporary ore pillar (2) and safety pillars (3, 4)

Figure 4. Change of coefficient of concentration (K_σ) inside elements of system of development with ore shrinkage at various stages of block development: a – a projection to the sloping plane of an ore body; b – characteristic points of elements of system of development.

system with ore shrinkage. Rock pressure management assumes leaving the sill pillar and the under-drift of the pillar which are the main stress concentrators. Maximum stress is accumulated in in the sill pillar in mining depth more than 350 m in condition of full of excavation and destruction of upper pillars (average pressure of σ_{av} reaching 100 MPa, and intensity of tangents τ_{int} – 70 MPa).

High stress concentration (Fig. 4) is also observed in safety pillars and also in edge of the ore block, at the intermediate stage of development of the excavation block (with a 20 m shrinkage height). Increase in height of ore shrinkage leads to the growth of the main and tangential stresses in the pillar in 20–25%. Safety pillar from the waste clearing block are the most compressed. Especially difficult and potentially burst hazard situation forms at the complete development of the block (complete ore chute of the shrinkage out of the excavation block). Values of the first principal stresses approach to ultimate compressive strength and tangential- to ultimate shear strength.

Results of modeling and researches of regularities of formation of technogenic fields of tension in the developed rock massifs of the researched deposits and the common practice of burst hazard deposit development showed: 1) change of parameters of structural elements of systems of development and an order of development does not provide the guaranteed protection against hazard dynamic manifestations of rock pressure; 2) decrease of stress inside ore pillars by realization of preventive protective measures is necessary. The protective measures include creation of protective zones by drilling of pressure-relief wells, camouflet blasting or shaking blasting and a combination of these measures.

Different variants of pressure-relief wells orientation relatively the direction of action of the maximal stress were considered. Change of dip angle of ore body of the Jyjnoe deposit was considered. Jyjnoe deposit is developed by the system of subfloor drifts.

It is established that after development of a half of the clearing block (at 25 m length) the most efficient way of pressure-relief of safety pillar is well-drilling in a roof and the soil of mine caving of the top subfloor perpendicular to the direction of action of the maximal tension. These measures provide decrease of stress in the pillar in 20–25% (Table 2).

Effectiveness of creation of a protective zone by camouflet blasting of blasthole charges on pressure-relief (compensation) wells was estimated on Zabytoe and Southern Hingansk deposits. Decrease in degree of burst hazard is realized as a result of these measures. Main stresses are formed between horizons in safety pillars and in the subdrift pillar (5 m height, 3 m vein thickness, length of the backfilled excavating block—up to 60 m).

The choice of rational parameters of this preventive measures were carried out with the use of a method of numerical modeling. Change of length of unloading wells from 3 to 10 m, change of slope angle, relatively effective maximal stresses, from 0 to 75°. Diameter of wells was constant – 105 mm. The relation of the initial tension attached to sides of model was $\sigma_x = 2{,}5\sigma_z$. The generalized calculated scheme of the studied of model and results of model operation are provided on the Figure 5.

By results of modeling it is established: well-drilling at 30 and 60° angle relatively the ore bodies does not provide sufficient decrease in hazard stress concentrations; 45° angle leads

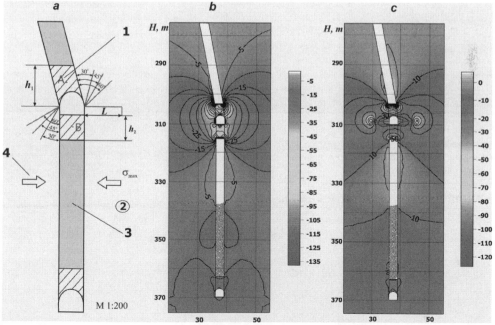

1 – ore pillar; 2– bearing strata (granite); 3 – waste area; 4 – direction of action of main stress σ_{max}; A, B – characteristic points in rock massif for analyst SSS; 30, 45... – angle of incline of pressure-relief wells relatively ore body deposition, l_h – well length (3–10 m); h_1, h_2 – height of sill pillar and underling pillar

Figure 5.   The calculated scheme (a) and a stress distribution (σ_{med}) in the rock massif: b – before preventive measures; c – after drilling of horizontal 10 m long pressure-relief wells.

to pressure-relief of boards of excavation; increase in well length leads to decrease in stress in the center of safety pillar in 15%. Significant increase of stress in the upper part of safety pillar reached 155 MPa (Fig. 5).

Camouflet blasting of blasthole charge in pressure-relief wells, drilled in the direction of action of the maximal tension. Pressure-relief wells length was 5 to 10 m through all the length of excavated block. These measures allow to decrease burst hazard degree safety pillars and the subdrift pillars. They also create the favorable conditions for the protection and maintaining of an excavation.

Structure and parameters of rational burst-hazard preventive actions for the development of the deep horizons of Yujnoe, Zabytoe and Southern Hingansk deposits are presented in Table 2. Expected decrease of coefficient of a stress concentration (K_σ) in 20–80% is also shown. This decrease is necessary and sufficient in the considered conditions.

4 INSTRUMENTAL MONITORING OF BURST HAZARD AND TECHNOGENIC SEISMICITY ON THE FAR EASTERN AND TRANSBAIKALIA MINES

An important role is played by the methods and technical means of assessment and monitoring of a geomechanical condition of the massif of rocks for the solution of the problem of the forecast of rock bursts and decrease in geodynamic risk. These technical means are used immediately in underground excavations (Turchaninov, Iofis, Kasparyan, 1989; Makarov, 2006; Rasskazov, 2008). On fields of the Far East region various measuring complexes are

Table 2. Efficient measures for the decrease of degree of burst hazard in structural elements of systems of development of vein fields.

Mining depth	450–500		300–350	
System of deposit development	sublevel drifts		ore shrinkage	
Azimuth of strike of ore body, degrees	55		345–360	
Burst hazard mining elements	safety pillar (two upper stories)		sill pillar	under drift pillar
Pillar parameters	6–7 m height, 3 m length		height and length up to 5 m	5 m height, up to 40 m length
Sort of control	drilling of pressure-relief wells		camouflet blasting of blasthole charge for pressure-relief wells	
Angle of the ore body incline β, degree	$45 \leq \beta < 60$	$75 \leq \beta < 90$	75	90
K_σ before the control (safety measures)	$1,8 \leq K_\sigma < 3,6$	$4,5 \leq K_\sigma < 4,1$	5	3,5

Optimal parameters of pressure-relief safety measures

Stage of application	excavation of half of the length (½ L) of the developed block		excavation of half of the height ½ h of the developed block	
Well length, m	6–8		5–10	
Well diameter, mm	105		105	
Drill spacing, m	no more than 0,25		0,4–0,5	
Drilling angle of well (relatively σ_{max} direction), degree	90	45	0	
Drilling angle (relatively ore body incline), degree	$45 \leq \alpha_2 \leq 60$	$15 \leq \alpha_2 \leq 45$	75	90
Reduction of stress concentration factor K_σ,%	20–80	30–40	25,2	55,8

applied: seismoacoustic, microseismic and straining methods and tools. Hardware from local portable registrars to the multi-channel systems of monitoring providing regional monitoring of all miner field or its sites (Rasskazov, 2008; Rasskazov, Lugovoy, Kalinov et al, 2013; Aksenov, Ozhiganov, 2014; Rasskazov et al., 2015) are most widely applied. They are necessary both for obtaining the initial information on properties and a condition of the rock massif, for modeling of the deflected mode and for verification of results of theoretical researches and the subsequent monitoring of geomechanical processes during mining.

The most reliable prediction of burst hazard can be received as a result of application of various instrumental methods united in the uniform system of complex geomechanical monitoring. Similar complex is created in uranium ores deposits of the East Transbaikalia (Rasskazov, Gladyr, Anikin, 2013). The system of monitoring unites multi-channel seismoacoustic and microseismic measuring complexes providing filing of seismoacoustic events with energy from 10 to 10^6 J in the frequencies range from 10 to 12000 Hz. System of monitoring also includes straining stations, with the laser deformograph capable to record crust shifts with 0.1 nm accuracy in the frequencies range from 0 to 1000 Hz (Dolgih, Privalov, 2009; Rasskazov, Dolgikh, Petrov et al., 2016). The technology of temporary synchronization giving the possibility to support uniform time in all measuring tools with a divergence no more than 10 ms is realized in system. All recorded data are integrated into uniform informational network that provides quick collateral interpretation of results of observations (Gladyr A.V., 2017). "MineFrame" software is used as the main integration platform complex (Lukichev S.V., Nagovitsyn O.V., 2010). It provides representation and the collateral analysis of results of geophysical monitoring. At the same time results of geomechanical monitoring are fixed in databases, are processed and represented in the form of cards of seismoacoustic and straining activity, schedules and charts of change of power, spectral and other parameters of geophysical fields.

By results of complex geomechanical monitoring on mines of the Far East it is established that important feature of behavior of potentially burst hazard massif is formation in it the active regions (the potential centers of the geodynamic phenomena). Clustering

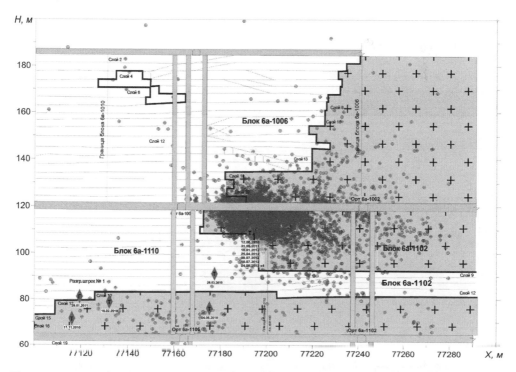

Figure 6.   A spatial distribution of the centers of seismoacoustic events according to monitoring on the Antej deposit (in a projection to the vertical plane).

process of sources of microdestructions confirms the formation of the potential centers of the geodynamic phenomena. Distribution of the centers of the seismoacoustic events in the massif of the field Antej reflecting process of the centers of rock bursts and technogenic seismicity is shown in the Figure 6.

For selection of the potential centers of large dynamic manifestations of rock pressure the technique of data interpretation is used (Guzev et al., 2018). This technique is based on methods of the cluster analysis and theory of casual counts and experimentally determined consistent pattern. According the said theory, distance (r_i) between all centers of microde-structions ($S_1...S_p$) a Poisson stream of intensity λ in the area U does not exceed 10 m, and their majority (93%) is distributed in the two first intervals on 2 m. The average minimum distance between points in a Poisson stream of M_r is defined from expression:

$$M_r = \int_0^\infty \exp\left(\frac{-4\pi\lambda R^3}{3}\right) dR = \frac{4\pi\lambda R^3}{3} = x^3 = \Gamma\frac{4}{3} \cdot dx \left(\frac{4\pi\lambda R^3}{3}\right)^{-1/3}.$$

Having created A matrix from paired average distances between the selected points, transition from a casual Poisson stream of points in the sphere of U to the casual count is formed:

$$\left\| \sqrt{(x_i - x_j)^2 + (y_i - y_j)^2 + (z_i - z_j)^2} \right\|_{i,j=1}^\rho$$

The solution of a matrix of A_t gives the chance to allocate the maximal ("huge") component of a connectivity of points from area U. The allocated centers of microdestructions from a row $S_1...S_l$ on the basis of the maximal component form themselves a focal zone of SAE-events and have a compendency (a possibility of join in a cluster) than differ from other events in selection (Fig. 7).

According to long-term mine observations it is established (Rasskazov, 2008) that the main signs of preparation of rock bursts are: growth of number of SAE-events twice and higher (N_{AE}); decrease of a mean-square distance (R_{cpp}) from a source to the center of the formed focal zone; growth of summary energy (E_{AE}) more than in 80%; reduction of a time interval between SAE-events (t_{cpp}) and reduction in the rate of migration (υ_{OZ}) epicenter of a focal zone up to 8 m/days. On this basis the complex indicator of burst hazard (K_{bh}) which is defined as the relation of the normalized (in comparison with data in 5 last days) values of summary energy and number of AE-events in the active region to the work of the normalized

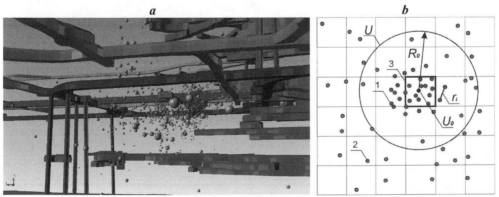

1 – connected among themselves AE-events are marked out with red; 2 – unconnected events (grey color); 3 – cluster with the greatest number of the centers of AE-events; U_0 – epicenter of SAE-events; R_0 – U sphere radius; r_i – distance between the centers of AE-events

Figure 7. Selection of potentially burst hazard of sites in the rock massif according seismoacoustic monitoring data: a) area of the forming center of the rock burst; b) the calculated scheme for selection of focal zones.

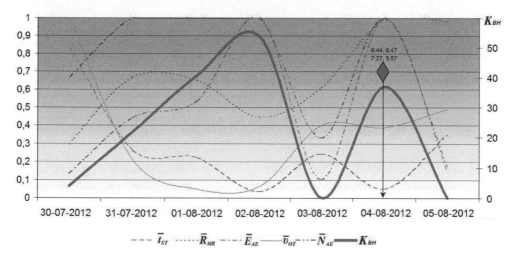

Figure 8. The change of the normalized parameters of seismoacoustic activity reflecting process of preparation of rock burst (04-08-2012, Antej deposit).

values of distance, time between the subsequent events and speeds of migration of the center of a zone of acoustic activity is developed:

$$K_{BH} = \frac{\bar{E}_{AE} \cdot \bar{N}_{AE}}{\bar{R}_{MR} \cdot \bar{t}_{ST} \cdot \bar{v}_{OZ}},$$

where E_{AZ} – summary energy of SAE-events in the focal zone (connected events), J; N_{AE} – number of SAE-events in the focal zone; R_{MR} – distance between subsequent SAE-events, m; t_{ST} – time between the subsequent AE-events, c; v_{oz} – speed of migration of the center of a focal zone, m/days.

Change of parameters of seismoacoustic activity and K_{bh} index before 4 seismic events around 6a-1102 block, Antej is shown in the Figure 8. Seismic events were followed by destruction of the backfilling massif and quake of buildings on the surface of the mine.

"Dangerous" category is given on the basis of the quantitative values of an indicator. Indicator is calculated by methods of the discrete analysis and mathematical statistics with the application of interval algorithms of a pattern recognition. For the Antej deposit the critical size of an index of an burst hazard was 5.11 ($K_{bh} > 5.11$). At the same time reliability of forecasts of dynamic manifestations for the last 5 years was rather high and was 84,4%.

In the course of geomechanical monitoring the interrelation of dynamic manifestations of rock pressure and seismic waves from natural (earthquakes) and technogenic (technological explosions) sources is also revealed. According to high-precision straining monitoring on the geodynamic ground near the Streltsov ore field it is established that seismic waves from the strong earthquakes exert noticeable impact on the straining field that is reflected in increase in seismoacoustic activity by 2–3 times in comparison with the average level. Seismic waves act as the trigger and initiate destruction in a dynamic form of sites of the massif of the rocks which are in extremely stressed state.

Field instruments for the express diagnostics of a condition of regional sites of the massif of rocks are applied to specification of borders the burst hazard of the zones revealed according to regional monitoring on the Far East mines. geoacoustic device for local assessment of burst hazard of "Prognoz L" (Fig. 9) demonstrated enough effectiveness. Complex geomechanical monitoring are verified and sites of excavations and elements of the rock designs demanding application of pressure-relief measures to be transfered to a not burst hazard state are defined by the by means of "Prognoz L".

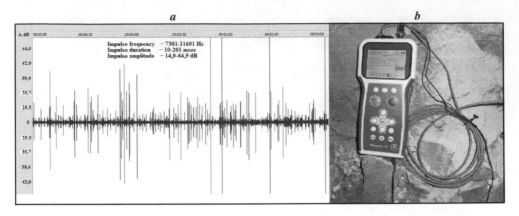

Figure 9. Signalogram (a) reflecting burst hazard of the rock massif, received by results of express assessment by the Prognoz L device (b).

5 CONCLUSIONS

Now the geomechanical situation on a number of underground mines of the Far East and Transbaikalia is illustrated by existence of a wide range of dynamic manifestations of rock pressure or classification of sites of the massif as inclined to rock bursts. The technogenic seismicity increases with the growth of volumes of the excavated underground spaces and depths of mining operations on mineral deposits. Technogenic seismicity is often followed by destructions of excavations. In-depth studies of burst hazard with application of a complex of the modern methods and technical means are required in modern conditions.

Wide range of geomechanical tasks, including identification of regularities of formation of natural and technogenic fields of tension in the burst hazard rock massif is solved by methods of numerical modeling of the deflected mode. It is established that the most burst hazard elements of systems of development of ore bodies are various type of a pillar with the 2–3 times higher tension inside in comparison with initial. A complex of preventive measures for the decrease of tension in ore pillar is applied. Effective parameters of these measures are accepted by results of theoretical and pilot studies.

For assessment of a state and properties of rocks, verifications of results of numerical modeling and geomechanical monitoring of the massif in the process of mine development on the Far East mines seismoacoustic, microseismic and straining methods and tools are applied. Hardware include local portable registrars, multi-channel measuring complexes. Join of measurement hardware in the uniform system of complex geomechanical monitoring provides the best effect. The revealed regularities of behavior of the burst hazard rock massif are the basis for the developed techniques of assessment of burst hazard which application provides rather high reliability of the forecast of the hazard geodynamic phenomena.

ACKNOWLEDGEMENTS

This research was supported by Russian Science Foundation grant 17-77-10071. Authors are grateful for the following researchers of Mining Institute FEB RAS who took part in the study: Kursakin G.A., Gladyr A.V., Kalinov G.A., Migunov D.S., Rasskazov M.I., Sidljar A.S., Tereshkin A.A.

REFERENCES

Adushkin V.V., Oparin V.N. From the phenomenon of sign-variable reaction of rocks to dynamic influences—to waves of oscillating type in strained the geo-medium//Journal of mining science: P. I, 2012. – No. 2; P. II, 2013. – No. 2; P. III, 2014. – No. 4; P. IV, 2016. – No. 1.

Aksenov A.A., Ozhiganov I.A. Application of a method of an acoustic emission for the forecast of burst hazard of the massif of rocks//Mining journal. – 2014. – No. 9. – pp. 82–84.

Baryshnikov V.D., Kurlenya M.V., Leontyev A.V. et al. To the issue of the deflected mode of the Nikolaev field//Physics and technology problems of development of minerals. – 1982. – No. 2. – pp. 3–11.

Bich Ya.A., Muratov N.A. Preventive measures for rock bursts. – Vladivostok: Dalnevost publishing house, 1990. – 248 p.

Dolgih G.I., Privalov V.E. Lasers. Laser systems. – Vladivostok: Dalnauka, 2009.

Gladyr A.V. System of integration of microseismic and geoacoustic data of geomechanical monitoring// Mining informational and analytical bulletin. – 2017. – No. 6. – pp. 220–234.

Guzev M.A., Rasskazov I. Yu., Tsitsiashvili G. Sh. Algorithm of potentially burst-hazard zones dynamics representation in massif of rocks by results of seismic-acoustic monitoring//Procedia Engineering. – 2017. – No. 191. – pp. 36–42.

Kaplunov D.R., Rylnikova M.V. The combined development of ore fields. M.: Mountain book, 2012. 344 p.

Kazikayev D.M. The combined development of ore fields. M.: Mining book, 2008. 360 p.

Lukichev S.V., Nagovitsyn O.V. Computer technology of engineering support of mining operations at development of fields of solid minerals//Mining journal. –2010. – No. 9. – pp. 11–15.

Makarov A.B. Practical geomechanics. M.: Mining book, 2006. 391 p.

Melnikov N.N., Kozyrev A.A., Panin V.I. Technogenic seismicity—a hazard anthropogenic factor when conducting mining operations in high-intense massifs//Geoecology, engineering geology, hydrogeology, geocryology. – 2015. – No. 5. – Page 425–433.

Oparin V.N., Malovichko A.A., Kozyrev A.A., et al. Methods and systems of seismostraining monitoring of technogenic earthquakes and rock bursts. – Novosibirsk: Publishing house Siberian Branch of the Russian Academy of Science, V. 1, 2009; V. 2, 2010.

Petukhov I.M. et al. Calculation methods in mechanics of rock bursts and emissions: handbook—M.: Subsoil, 1992. 256 p.

Potapchuk M.I., Kursakin G.A., Sidlyar A.V. Improvement of safety of development of bump hazardous vein deposits of Eastern Primorye//Eurasian mining. – 2014. – No. 1(21). – pp. 18–22.

Rasskazov I. Yu. Monitoring and management of rock pressure on mines of the Far East region. – M.: Mining book, 2008. – 329 p.

Rasskazov I.Yu., Potapchuk M.I., Osadchiy S.P., Potapchuk G.M. Geomechanical assessment of the applied technologies of development the burst hazard of fields of JSC Dalpolimetall//Mining Informational and Analytical Bulletin. M.: Publishing House World of Mining book, 2010. No. 7. pp. 137–145.

Rasskazov I.Yu., Lugovoy V.A., Kalinov G.A., et al. Development of measuring complexes for the assessment and control of burst–hazard during mining, Proceedings of the 8-th International symposium on rockbursts and seismicity in mines (Russia, Saint-Petersburg—Moscow. 1–7 September 2013), Obninsk—Perm, 2013. – P. 121–124.

Rasskazov I.Yu., Gladyr A.V., Anikin P.A., Svyatetsky V.S., Prosekin B.A. Development and modernization of the control system of dynamic manifestations of mountain pressure on mines of JSC PPGHO//Mining journal. – 2013. – No. 8 (2). – pp. 9–14.

Rasskazov I.Yu., Saksin B.G., Petrov V.A., et al. Present Day Stress Strain State in the Upper Crust of the Amur Lithosphere Plate // Izvestiya, Physics of the Solid Earth. 2014. Vol. 50. No. 3. pp. 444–452.

Rasskazov I.Yu., Saksin B.G., Potapchuk M.I., Usikov V.I. Geomechanical assessment of mining conditions in the khingansk manganese ore body//Journal of Mining Science. 2014. T. 50, No. 1. Page 10–17.

Rasskazov I.Yu., Migunov D.S., Anikin P.A., et al. New-generation portable geoacoustic instrument for rockburst hazard assessment//Journal of Mining Science. – 2015. – Vol. 51, No. 3. – P. 614–623.

Rasskazov I.Yu., Saksin B.G., Usikov V.I., Potapchuk M.I. A geodynamic condition of the ore massif of Nikolaev polymetallic field and peculiarities of burst-hazard phenomenon in the process of its development//Mining journal. – 2016. – No. 12. – pp. 13–19.

Rasskazov I.Yu., Dolgikh G.I., Petrov V.A., et al. Application of a laser deformograph in the system of complex geodynamic monitoring near the Streltsovsky ore field//Journal of Mining Science. – 2016. – No. 6. – pp. 29–37.

Saksin B.G., Rasskazov I.Yu., Shevchenko B.F. Principles of integrated analysis of modern deflected mode in the outer crust of the Amur plate//Journal of Mining Science. 2015. V. 51, No. 2. pp. 243–252.

Trubetskoy K.N. Chanturia V.A., Kaplunov D.R., Rylnikova M.V. Complex development of fields and deep processing of mineral raw materials. M.: Science, 2010. 438 p.

Turchaninov I.A., Iofis M.A., Kasparyan E.V. Fundamentals of mechanics of rocks. – Leningrad: Subsoil, 1989. – 488 p.

Fadeyev A.B. A finite element method in geomechanics. M.: Subsoil, 1987. 221 p.

Freydin A.M., Shalaurov V.A., Eremenko A.A. et al. Increase in effectiveness of underground mining of ore fields of Siberia and the Far East Novosibirsk: Science, SIF, 1992. – 177 p.

Zienkiewicz O.C. The birth of the finite element method and of computational mechanics//Int. J. Numer. Meth. Eng. 2004. N 60. pp. 3–10.

Geomechanics and Geodynamics of Rock Masses – Litvinenko (Ed.)
© *2018 Taylor & Francis Group, London, ISBN 978-1-138-61645-5*

Modeling geomechanical and geodynamic behavior of mining-altered rock mass with justifying mechanisms of initiation and growth of failure zones

Valerii Nikolaevich Zakharov & Olga Nikolaevna Malinnikova
Institute of Comprehensive Exploitation of Mineral Resources Russian Academy of Sciences, Moscow, Russia

ABSTRACT: The paper presents the basic results of physical and mathematical modeling of mining-induced processes causing failure in rock mass. These results are useful for the applied and basic research. The implemented studies have enabled updating mechanisms of failure in coal saturated with gases and determining effects preceding failure. Moreover, 3D geo-models have been constructed to describe sagging and lamination of roof rocks, spontaneous sloughing of coal, clustering of induced damages in rocks, propagation of blasting-caused wave of displacements and stresses in nonuniform rock mass, landsliding etc.

Keywords: rock mass, coal bed, failure, mining-induced effect, coal mine, ore mine, outburst, rock burst, landslide

1 INTRODUCTION

The actual practice of mining at deeper levels, under aggravating geological conditions and at the increased rate and intensity induced active geodynamic behavior and elevates risk of hazardous dynamic events and catastrophes in rock masses. In this connection, new demands are imposed on the research into geomechanical behavior and geodynamic response of rocks to the induced effect towards the enhancement of operations safety in modern highly productive mines.

The investigations carried out by the Institute of Integrated Mineral Resources Development—IPKON of the Russian Academy of Sciences aim to develop a new concept of mining geomechanics based on transition from justification of mining technologies to revealing of geomechanical laws of development in geosystems in order to handle global problems connected with subsoil management and preservation of mineral wealth. The expanded scope of mining geomechanics encompasses research methodology, new physical laws, methods and means of rock mechanics survey, geomedium monitoring, geomechanical and geodynamic control and other issues.

The exploration trends in the areas of basic and applied research actually undertaken by IPKON include: mining-induced subsidence and displacement in rock masses, geodynamic zoning, deformation and failure of fractured coal seams with high methane content, blasting, interaction between geomechanical and hydrogeological processes, geomechanics in resource-saving geotechnologies and environmental measures, methods and means of geodynamic control in rock mass damaged by mining. The research methods are laboratory tests, in-situ experimentation, mathematical modeling, statistical processing of field data and engineering approaches to recommendations on geomechanical control. The research findings are the subject of discussions at international and Russian conferences, and are published in reputable scientific periodicals.

Failure of gassy coal seams is an important scientific problem addressed in numerous researches both of the applied and basic nature. For coal mines, it is essential to reveal conditions of failure in a dynamic form (outbursts, rock bursts etc).

Based on the analysis and generalization of the earlier experimental data on the laboratory nonequicomponent triaxial compression tests by the Karman scheme, under varied lateral stresses and gas pressure to simulate environment of a production face, conditions of self-sustaining avalanche failure typical for gas-dynamic events in coal seams have been revealed [1–4].

The analysis of the axial stress σ_1—strain ε_1 curves shows that the increase in the lateral compression σ_3 sharply enhances the strength $\sigma_{1\,max}$ of coal promotes high concentration of stresses and governs the fracture mode under post-limit deformation. When $\sigma_3 > 13$–18 MPa, the post-limit deformation curves become horizontal (drop modulus $M \approx 0$), $\sigma_{1\,max}$ jumps and coincides with the residual strength σ_0, i.e. failure mode is mostly shearing (ductile failure of coal) (Figure 1).

Under the relatively low lateral compression $\sigma_3 < 5$–7 MPa, the strength reduces together with the energy density of failure to A'' or A''' (depending on σ_3), and failure takes place as brittle fracture. The post-limit deformation diagrams have a pronounced branch of drop in the stress σ_1, and the ultimate stress limit $\sigma_{1\,max}$ essentially exceeds the residual strength σ_0.

The experimentally determined criterion of failure in coal is given by $\sigma_{1\,max} \geq 3\,\sigma_3$ or $\beta = \sigma_3/\sigma_1 \leq 0.33$. As σ_3 drops from 13 to 2 MPa, the energy density of failure determined using the $\sigma_1 - \varepsilon_1$ curves for edium-strength coal sharply decreases from $A' = 15 \cdot 10^5$ J/m^3 to $A'' = 0.4 \cdot 10^5$ J/m^3, i.e. by 35–40 times.

In full-scale conditions, upon the delay in straining in the line of σ_3 from the face side (a typical situation of small amplitude dislocations), the strength of coal grows together with the stresses and energy density. Under an instantaneous advance of heading, for instance, by blasting, σ_3 releases in rock mass at a velocity near to sound speed, and the strength of adjacent coal block in the working face drops to a level governed by the change in σ_3. Given that $\beta \leq 0.33$ rock mass is oversaturated with energy. This is a characteristic point of bifurcation that conditions a further failure mode—normal fracturing and sloughing, or self-sustaining avalanche failure typical for gas-dynamic events when plastic strains decelerate and the excessive energy of elastic compression releases in the form of brittle fracture [5].

A feature of failure in tectonically damaged coal having high gas content in the working face is its ability to follow the mode of both shear and rupture depending on the ratio of the actual principal stresses σ_3 and σ_1 in the triaxial compression field, on the value of coalbed gas

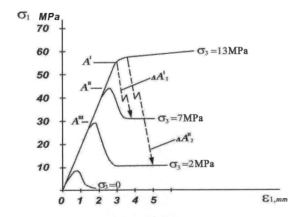

Figure 1.   Stress–strain curves for coal under triaxial compression.
A', A'', A''' – energy density of failure at different values of the stress σ_3,
$\Delta A'$ and $\Delta A''$ – release of excessive elastic energy under the drop in σ_3.

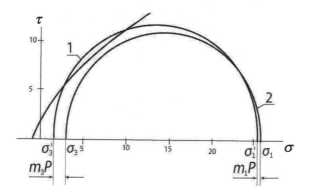

Figure 2. Influence of the pore pressure of gas on the actual stresses and strength of coal.
σ_1—axial stress that initiates failure; σ_3—lateral stress; 1, 2—Mohr's circles for coal saturated with gas at $P = 1.5$ MPa (1) and for drained coal (2); m_1 and m_3—indexes of dual porosity determined by the orientation of the stresses σ_1 and σ_3.

pressure P and on the dual porosity (voids and fractures) index m, which are variable in space and time.

Based on the data obtained in a series of triaxial compression tests, Mohr's circles and the envelope have been plotted (Figure 2). The values of the index m in deep rock mass subjected to high actual stresses depend only porosity of coal and cannot exceed $m_1 = 0.1$. In the working face, when the drop in σ_3 is reduced by excavations and genetic cleavage fractures open, the index m_3 in the orthogonal plane to the orientation of σ_3 jumps up to $m_3 = 1$. This results in the change in the actual stress $\sigma_3' = mP$, and Mohr's circle moves to position 1 (σ_1', σ_3'), i.e. it approaches the envelope the intersection of which means failure (Figure 1). The reduction in the strength of coal saturated with gas becomes particularly appreciable in the range of low values of σ_3, i.e. near coal exposure. When $\sigma_3 \leq P$ failure begins under the action of tensile stresses induced only by gas pressure.

The experimental criterion of strength and failure of coal under triaxial compression is given by: $\sigma_{1max} = \sigma_C + \chi(\sigma_3 - mP)$, where σ_{1max}—peak stress of failure initiation; σ_C—uniaxial compression strength; $\chi = 0.33$–empirical coefficient. Coal failure at the free surface (face) takes place upon satisfaction of the condition $mP \geq \sigma_r + \sigma_3$, where σ_r—rupture strength of coal.

Emission conditions of micron and submicron particles from the surface of different types of rocks and building materials subjected to quasi-static uniaxial compression are examined in the tests of specimens with a through cylindrical hole. In such tests, failure is recorded at an early loading stage owing to nonuniform distribution of stresses on the hole walls [6]. The test results prove emission of particles when loading is increased; the expansive growth in the number of particle is the indication of pre-microfailure of a specimen.

The experimental research into stress memory in rocks subjected to cyclic uniaxial compression, the nature and mechanism of which are related by many researchers with the defects that initiate and grow at different scales, have proved [7] that with an increase in the number of cycles, cycle loading (uniaxial compression) at the rate of 0.6 of the uniaxial compression strength results in the attenuation of particle emission while loading at the rate of 0.8–0.9 intensifies formation of particles as a consequence of accumulation of defects and, thus, weakening of a specimen. A jump in the particle emission is a pre-cursor of specimen failure.

The investigations of the dynamic effect exerted on rocks with the simultaneous recording of particles detached from the hole walls in a specimen give evidence of the interaction between the energy of the effect and the number of particles depending on the type, jointing, homogeneity, isotropy and other properties of rocks, which quantitatively and qualitatively differ in various types of rocks.

Currently the research staff of IPKON actively develop a fundamentally new method aimed to monitor stresses and strains in rock mass by recording submicron particles for the purpose of predicting geodynamic events in the form of rock bursts and other induced disastrous processes [8], which is a high-priority problem in deep-level mining in seismically active regions.

Analysis of structural characteristics of rocks. Failure processes connected with different scale fracturing are in many ways governed by structure which is characterized by shape and size of grains, porosity and microdefects. Structure along with the chemical composition conditions physical properties of materials, including strength. At IPKON the structure and physical properties of rocks are analyzed using various methods, including electron microscopy, X-ray microtomography, nano- and micro-indentation.

It is important to estimate parameters of microcracks as a structural element. To this effect, the tests on dynamic effect exerted on rock specimens placed in special preservation capsules were carried out. The specimens were subjected to pulsed loading until microcracking and failure. According to electron microscopy and analysis of microcracks in specimens taken out of preservation capsules, the minimum and maximum opening of microcracks ranged from 0.1 to 10.0 μm depending on the conditions and value of loading [9, 10].

The elastic and strength characteristics of rocks and minerals have been studied jointly with the researchers of the Tambov State University using the methods of nano- and micro-indentation [11]. The values of Young's modulus, hardness and fracture toughness are determined for rocks and some minerals. For instance, for quartz, it is typical that hardness is high at comparatively low toughness of fracture for separate quartz grains and grain interfaces with magnetite and hematite; furthermore, it appears that values of Young's modulus and hardness vary greatly in some rocks. The quantitative regularities are revealed in the change of Young's modulus and hardness in mineral components of ferruginous quartzite at different structural scales [11, 12]. The research findings are useful for developing an insight into the processes of deformation and failure of rocks, and make a framework for the further investigations of their structural and physical characteristics at different scales.

Mathematical modeling of mining-induced processes in rock mass allows a detailed preliminary analysis of interaction between a mine and enclosing rock mass, and enables studying geomechanical scenarios under man-made impact during opening-up of a mineral deposit, first working and actual mining, which is obligatory for safe and efficient mine performance. Earlier, the geomechanics research staff of IPKON used mostly the analytical approach, in particular, the theory of complex potentials. A significant advance was achieved, but many problems important for theory and applications yet remained unsolved because of the limitation of the approach. Numerical modeling spurred a new research programs in geomechanics. Recently, 3D mathematical modeling of state and behavior of rocks and mineral bodies enjoys a much wider application both in mine planning and design and in actual mining, with the focus on the detection of different damages in rocks.

Bedded deposit. The ample research is devoted to the description of state and behavior of rocks in the course of bedded deposit mining [13]. The most hazardous for the safety and serviceability of a mine is roof subsidence which takes place cyclically as a working face is advanced. Large spans of overlying strata are left unsupported and sag under gravity.

Sagging is determined by rock mass bedding and lamination over surfaces of weakened mechanical strength. In case that contact strength is low and contacts fall in the zone of tension of vertical stresses, gaping fractures appear in such areas. Reaching certain sag, the main roof caves during stoping. Prediction of the primary and secondary sagging of roof is necessarily included in the design of methods and means aimed at stability and safety of coal mines.

In Figure 3 for a quarter roof, the calculated geometry of gaping fractures and distribution of vertical displacements above mined-out void is illustrated. The vertical displacement in meters is imaged by the color for the purpose of better visualization (color scale is given at the bottom of the figure).

In the areas of maximal horizontal tension at the bottom surface of the roof, transverse fractures can appear. Such fractures initiate at a certain face advance in the center of each side of the rectangle and then propagate over the major portion of it. This process is depicted in Figures 4 and 5 in plan view, where 1 – coal seam; 2 – working face and direction of its advance; 3 – fractures above the boundary of mined-out area; 4 – fractures on the bottom surface of the main roof; 5 – mined-out area.

Figures 6a and 6b show the contour lines and spatial representation of maximum deformation at the top surface of the roof. Expectedly, at the points B and D (see Figure 5), the deformation reaches actually infinite values and failure of coal seam can initiate at these points. Along the segment BD (dashed line in Figure 5), deformation values are also considerably high, which is reflective of probable failure of the coal seam along this line or nearby it. Thus, we have an oval wedge of failure. When such fractures grasp the major portion of the roof, it will collapse [14].

When mining is carried out in coal seams under extended longwalls, sloughing and weakening of coal occurs over the whole face. The sloughing zone may develop unstably when the face is advanced in coal having sufficiently high shear strength and gradually approaches contact with low strength coal. In this case, sloughing may happen suddenly at coal edges and the sloughing zone expands greatly to 10–15 m and more within a short time of the order of 0.1 s.

Figure 7 demonstrates the scheme of a plane problem where $x = 0$ and $z = 0$ are the axes of symmetry. The mining depth $H = 500$ m and the coal seam thickness $2h = 2$ m, the initial length of the mined-out area is $L = 80$ м.

The roof rocks are assumed elastic with certain deformation characteristics: Young's modulus $E = 3.10^{10}$ Pa, Poisson's ratio $v = 0.25$. The coal seam is also elastic with $E = 2.10^9$ Pa, $v = 0.28$, and density $\rho = 1300$ kg/m³. At the coal and wall rock interface, force interaction

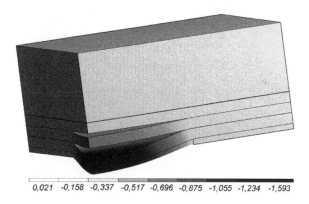

0,021 -0,158 -0,337 -0,517 -0,696 -0,875 -1,055 -1,234 -1,593

Figure 3.   Lamination in coal seam roof.

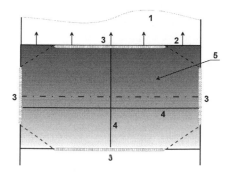

Figure 4.   Schematic drawing of mined-out cavity roof.

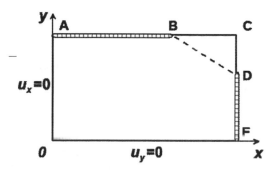

Figure 5.   Analytical model of mined-out cavity roof.

171

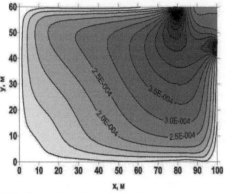 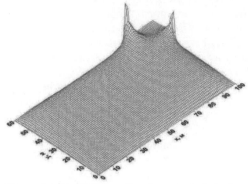

(a) isolines of major principal strains   (b) spatial distributions of major principal strains

Figure 6.   Major principal strains in the model.

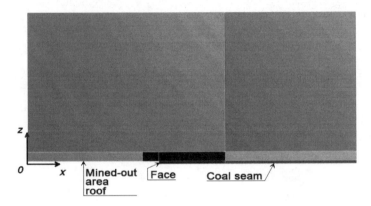

Figure 7.   Computational domain.

takes place in a certain mode. In a roof area 40 m <*x*<60 m the coefficient of friction is set, and in an area 60 m <*x*<100 m the condition of adhesive bonding is assigned, which means that mutual displacement of edges is prohibited though absolute displacement is allowed. A relaxation wave is generated in the coal seam and propagates with the gradually blurring front, and PPV drops in the coal seam; i.e. the distortional wave fades out as it propagates along the coal seam (Figure 8).

Vibrations in the rear of the wave have higher amplitudes than at the wave front. In this case, the relation $v=\sqrt{E/\rho(1-2v)}$ applicable for unlimited media, at the accepted parameters, yields v = 1900 m/s while the wave velocity in wall rocks is 4900 m/s.

In the model of plastic deformation of a coal seam, abutment pressure reaches maximum at a certain distance from the face, in 12–13 m at an estimated time limit. Plastic deformation and coal failure, in a varying degree, take place inside this zone.

Stress state of pillars in a bedded deposit was assessed with a view to revealing mechanisms of loading and deformation of pillars and probability of a natural equilibrium arch between safety pillars (Figure 9) [15,16].

The studies show that no apparent zones of vertical tensile stresses appear in rock mass as against mined-out area if unsupported by pillars. As a consequence, it is impossible that lamination takes place in the roof with the formation of natural equilibrium arch as a mine-technical structure within which rock mass is free from the external influence.

The author interprets this result as a "fictitious" arch of equilibrium in rock mass and pillars are only loaded inside this arch by a load γH_n. The value of H_n/H, i.e. the actual shape of the "arch" in percentage terms of H, is shown for each n-th pillar in Figure 10.

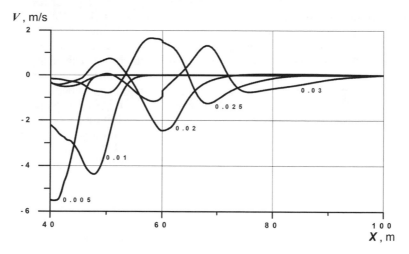

Figure 8.   PPV in the coal seam at different time of interaction between the seam and wall rocks with the friction factor k = 0.3.

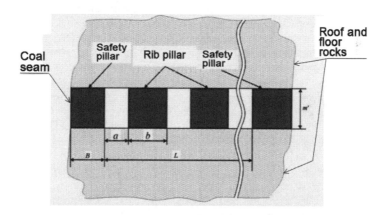

Figure 9.   Layout view of bedded deposit mining.

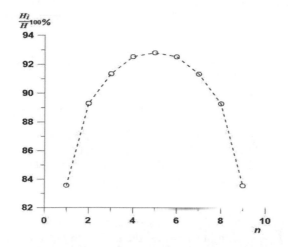

Figure 10.   Load applied to pillars in percentage terms of the maximum load.

173

Probable failure can essentially alter stress state in rock mass [16]. With the increase in ductibility of pillars due to damage and deformation, the load on them reduces and the geomechanical situation becomes similar to the case with the absence of pillars in mined-out area discussed above. The pillars are damaged under the roof subsidence the maximum value of which is given by $v|_{y=0} = -\frac{2(1-v^2)\gamma H}{E}L$, ($E$, v—deformation characteristics of roof rocks; H and L—depth of occurrence of coal seam and length of mined-out area, respectively; γ—average density of rocks). Pillars can compress by this value and prevent generation of tensile stresses in the roof. Under complete ductile failure of pillars, the bearing reaction becomes zero and this case corresponds to the situation when pillars are absent and natural equilibrium arch formation is possible. This means that the natural equilibrium arch within which overlying strata exert load on pillars can form as a consequence of failure of the pillars.

Such rock-made 'bridge' reduces load on pillars. The thickness of the 'bridge' is lesser significant than its distance from the mined-out void roof. At a certain critical distance, the 'bridge' loses its influence on the load of pillars. This means that the load on pillars is mostly formed by rock mass adjacent to mined-out area as a result of elastic recovery.

Modeling clusters of mining-induced damage in rock mass. Another research trend followed by IPKON in the area of geomechanics is calculation methodology for geomechanical behavior of rock mass with respect to its jointing. The recent studies focus on technological effect on jointing, which agrees with the concept of mining-altered geomedium as a new subject of the research in mining sciences developed at IPKON.

Among the induced factors in geomechanical behavior governing efficient and safe subsoil development, of great importance are deformation and subsidence of rock mass under the impact of underground excavations to be determined based on the laws founded on in-situ measurement results.

When stress state of rock mass varies, contacts of fractures irreversibly change (dislocate)—joints open and rock blocks move along them. The research was undertaken with the purpose to model random junction of dislocated fracture contacts as possible channels for free water flow based on the calculations of nonuniform stress state of rock mass, probabilistic assessment of rock mass strength and provisions of the theory of percolation.

In the first large-scale approximation, a rock mass was assumed a continuum. In the second approximation, the studied domain was divided into equal small sub-domains where the calculated stresses were averaged and assumed uniform. Each sub-domain had a block structure. The block and the joints were assumed damaged if the averaged Mises stress in the block exceeded the nominal probabilistic strength of the rock mass. The scope of the studies embraced probabilities and regularities of formation of chains (clusters) of damages in the disordered media or structures. Crucial influence on the wholesomeness of the studied structure is exerted by the formation of a cluster extending from one boundary of the structure to the other.

We modeled a safety (crown) pillar between an open pit bottom and an underground mined-out stope. In sub-domains, the equivalent Mises stress and strength of rocks were assumed normally distributed random values. Statistical analysis of clusters of damages was performed in each sub-domain. The clusters were connected using a special computer program in order to find possibility of an integral cluster in the whole domain of the safety pillar.

Figure 11 illustrates a potential induced damage at the left of the safety pillar. The damaged blocks are black-colored, and the largest assemblages of the damage blocks are grey. Figure 11a depicts the case when there is no cluster connecting the top and the bottom of the pillar, while Figure 11b shows the connecting cluster (grey color).

All in all, a series of 30 independent experiments was carried out (Monte-Carlo method), and the results were used to determine the mathematical expectation of probability of a connecting cluster P_{cl} and the confidence interval. At the confidence of 0.95, the wanted probability is $0.53 < P_{cl} < 0.55$. In other words, the probability of a connecting cluster in the analyzed geological conditions is approximately 0.5.

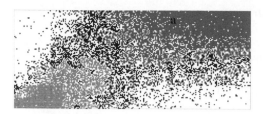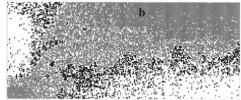

Figure 11. Probable clusters of induced damages at the left of the crown pillar: a—no connecting cluster; b—connecting cluster extends from the top boundary (bottom of an open pit) to the bottom boundary of the pillar (mined-out stope).

The static modeling displays two distinct facts—probability of a connecting cluster and potential configuration of the cluster. In the theory of percolation, the latter fact is of no importance while it is significant in actual practice. The induced damage cluster in jointed rock mass governs the structure of the generated channels of water flow [17].

In this context, the main inference of the modeling states that a rock mass in a near-critical state and subjected to even weak geomechanical disturbance can spasmodically change its integral permeability as a consequence of connection of separate damage clusters into a large end-to-end cluster, which may bring disastrous effects associated with water.

One geomechanics research trend connected with the induced fracturing is *modeling fracture propagation* with regard to dynamic action of natural water (*model of hydraulic fracturing*). This research includes the analysis of events of water inrush in mines and damage of confining layer between water-bearing beds. Initiation and growth of the main feeder fracture is considered as the case of natural hydraulic fracturing [18].

The developed model of hydraulic fracturing includes two forces—natural water pressure and rock mass stress. Under induced redistribution of stresses, the zone of low compressive stresses (and even tensile stresses) can form in rock mass. If a fracture initiated in this zone is filled with water, the axial pressure of water may exceed the compressive stress of rocks. In this case, hydraulic fracturing is possible.

The mathematical model of hydraulic fracturing uses two fracturing criteria—critical tension of rocks at the fracture tip and openness of the fracture along its whole length. Increment in the length of the fracture at the fracture tip is conditioned by the Griffiths—Irwin fracture theory. The direction of the fracture length increment is determined from the Ioffe hypothesis that a fracture grows in the direction perpendicular to the orientation of the major tensile tangential stresses at the fracture tip, σ_θ, when σ_θ at a certain distance from the fracture tip reaches a limit value.

Fracture propagation is a nonlinear process and the current conditions of the fracture growth depend on the current fracture geometry and the variable stress state of rocks. For this reason, the fracture development is calculated in small steps in consecutive order which describes the change in geometry of mined-out area, extension of the fracture and the other external effects.

Based on the discussion in [18], it may be concluded that the developed model of natural hydraulic fracturing is adequate. The model can be used to predict hazardous geological situations in mineral mining. The examples discussed in [18] also show that formation of main feeder fractures is connected with the development of anomalous displacements in rock mass.

Modeling displacements and dynamic stresses in real objects carried out at IPKON has resulted in a series of 3D computer geo-models. The modeling objectives include: stress distribution in rock mass around a roadway (in terms of Kuobass coal mine) which intersects a low-amplitude geological fault; analysis of variation in stresses and strains in an ore body and host rock mass under different variants of stoping and backfilling in Komsomol and Taimyr Mines; change in the geomechanical behavior of rock mass and ground surface before hazardous geomechanical processes in the influence zone of open pit mining in Tabor field; stability of pillars in different mining scenarios etc. [19, 20].

Modeling induced vibrations and oscillations in adjacent rock mass of coal face based on the analysis of experimental data on drilling-and-blasting designs and operation of tunneling and cutting-and-loading equipment has shown that vibrational energy mainly concentrates in the coal bed in the zone of extra load and in the zone of influence exerted by low-amplitude dislocations in coal.

The most severe exposure—explosion during shattering or blasting—generates influx of extra vibrational energy in coal and rock mass, which is comparable with the energy of elastic recovery of coal bed and with the gas energy. The determined values of displacements (1–2 mm) near the blasting site allow expecting initiation of failure at inter-molecular scale in coal with the breakage of the longest bonds in fibers and with the formation of active radicals (including methyl groups) typical for coal and gas outbursts, which can result in extra methane generation. In this manner, a source of a high-energy dynamic event with the increased methane release appears in face zone of coal bed [21, 22].

The process of development of deformation in time has been estimated using computer technologies [23, 24]. It has been proposed to estimate instability of natural and man-made slopes based on the generalized Mohr–Coulomb criterion with regard to stress–strain state of rocks. The most probable sliding surface is determined by means of step-by-step iteration by the scheme of the method of local variations. It is shown that the state and behavior of a landslide body should be estimated in the course of dynamic movement to find its terminal position. An example of such estimation is given in Figure 12.

The calculations with the varied determinants of the process, namely, coefficients of friction over sliding surfaces, strength and deformation characteristics of landslide body etc, arrive to a conclusion that the dynamic development of landslide is only possible at small angles of internal friction (< 4–5°) on potential sliding surfaces. The required value of the internal friction angle is conditioned by the average slope of sliding surfaces in sections of landslide. In some sections, where the angle of internal friction is higher than the slope, the landslide is in the steady-state condition; in the sections where the internal friction is lower than the sliding surface slope, sliding is possible both in static and dynamic modes.

Given the dynamic mode, landslide body acquires a considerable kinetic energy. In this case, the forces of friction blocking landsliding are opposed by the force of gravity and inertial forces of the moving body, while the coefficient of friction depends on the slide velocity and reduces as the velocity grows.

The examples from [23] show that assessment of landsliding probability needs estimating slope stability and calculation of instability consequences. It becomes crucially important to determine exactly the strength and deformation characteristics of contacts between layers composing a slope depending on external effects.

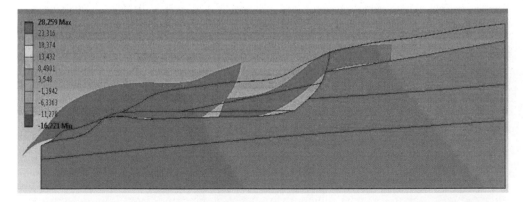

Figure 12.   Slide velocity 45 s after the initiation.

The research into geomechanics, geodynamics, gas-dynamics and hydrodynamics of rock mass altered by mineral mining, development of underground space and agglomeration of industries implemented by the scientific team of IPKON makes a fundamental framework for understanding physical laws of change in the lithosphere of the Earth under essentially intensified anthropogenic impact in the recent decades.

The theoretical and experimental result are used in the applied research, technological and design projects and in specific real-world problem handling at mining companies, such as selection of the most efficient mining technology; enhancement of mineral mining safety; preparation of regulatory documents; investigation of accidents; execution of examinations etc.

REFERENCES

[1] Feit G.N., Malinnikova O.N. Alteration of stress state initiating gas-dynamic events as face is advanced, Deformation and Failure of Materials with Defects and Dynamic Events in Rock Mass and Mines: XXIII Academician Khristianovich International School Proceedings, Simferopol, 2013, pp. 267–271. (in Russian).

[2] Zakharov V.N., Malinnikova O.N., Feit G.N. Geodynamics of initiation and assessment of risk of gas-dynamic events in mining-altered coal and rock mass, Problems and Prospects of Integrated Development and Preservation of Mineral Wealth: Academician Trubetskoy International School Proceedings, Moscow: IPKON RAS, 2014, pp. 87–90. (in Russian).

[3] Zakharov V.N., Malinnikova O.N., Feit G.N. Features of change in stress state and failure in coal face zone, Deformation and Failure of Materials with Defects and Dynamic Events in Rock Mass and Mines: XXIII Academician Khristianovich International School Proceedings, Simferopol, 2014, pp. 71–74. (in Russian).

[4] Zakharov V.N., Malinnikova O.N. Features of failure in coal face zone, Trigger Effects in Geosystems: III All-Russian Conference Proceedings, V.V. Adushkin, G.G. Kocharyan (Eds.), 2015, pp. 224–235. (in Russian).

[5] Zakharov V.N., Malinnikova O.N., Feit G.N. Development of the theory of self-sustaining avalanche failure in coal face zone, Subsoil Management Problems in the 21st Century in the Eyes of the Youth: Proceedings XI Int. School of the Young Scientists and Specialists, Moscow: IPKON RAS, 2014, pp. 5–7. (in Russian).

[6] Viktorov S.D., Kochanov A.N., Odintsev V.N., Osokin A.A. Emission of submicron particles during deformation of rocks, Izv. RAN, Ser. Fiz., 2012, Vol. 76, No. 3, p. 388.

[7] Osokin A.A., Analysis of uniaxial compression memory effects in rock specimens, XXV Academician Khristianovich International School Proceedings, Simferopol, 2015, pp. 2138–140. (in Russian).

[8] Trubetskoy K.N., Viktorov S.D., Osokin A.A., Shlyapin A.V. Rockburst prediction based on submicron particle emission in rocks under deformation and failure, Gorny Zhurnal, 2017, No. 6, pp. 16–20.

[9] Viktorov S.D., Kochanov A.N., Growth of micro-fractures in rocks under dynamic failure, Izv. RAN, Ser. Fiz., 2015, Vol. 79, No. 6, pp. 829–831.

[10] Kochanov A.N., Micro-fractures in solids in terms of rocks, GIAB, 2015, No. 7.

[11] Viktorov S.D. Golovin Yu.I., Kochanov A.N., Tyurin A.I., Shuklinov A.V., Shuvarin I.A., Pirozhkova T.S. Micro- and nano-indentation approach to strength and deformation characteristics of minerals, Journal of Mining Science, 2014, Vol. 50, No. 4, pp. 652–659.

[12] Golovin Yu.I., Tyurin A.I., Viktorov S.D., Kochanov A.N., Samodurov A.A., Pirozhkova T.S. Physical properties and micromechanisms of local deformation of thin subsurface layers in complex multi-phase materials, Izv. RAN, Ser. Fiz., 2017, Vol. 81, No. 3, pp. 390–394.

[13] Zakharov V.N., Malinnikova O.N., Trofimov V.A., Filippov Yu.A. Modeling in of geomechanical processes in rock minerals deposits, Proceedings of XXIV World Mining Congress: Mining in a World of Innovation, Rio de Janeiro/RJ, Brazil, 2016, pp. 338–347.

[14] Trofimov V.A., Malinnikova O.N., Filippov Yu.A. Stratification assessment of roof rocks in coal mining, GIAB, 2016, No. 2, pp. 119–126.

[15] Malinnikova O.N., Trofimov V.A., Filippov Yu.A. Calculation of loads on yielding pillars, Problems and Prospects of Integrated Development and Preservation of Mineral Wealth: Academician Trubetskoy International School Proceedings, Moscow: IPKON RAS, 2016, pp. 90–93. (in Russian).

[16] Trofimov V.A., Filippov Yu.A. Load on pillars in bedded deposit mining, Bezop. Truda Prom., 2017, No. 4, pp. 46–53.

[17] Militenko I.V., Militenko N.A., Odintsev V.N. Modeling induced dislocation in host rocks around excavations, Journal of Mining Science, 2013, Vol. 49, No. 6, pp. 847–853.

[18] Odintsev V.N., Militenko N.A. Water inrush in mines as a consequence of spontaneous hydrofracture, Journal of Mining Science, 2015, Vol. 51, No. 3, pp. 423–434.

[19] Zakharov V.N., Malinnikova O.N., Filippov Yu.A., Arshavsky V.V. Modeling mining-induced variation in stress state of rock mass, Proc. Conf. Geodynamics and Modern Rockburst-Hazardous Mining Technologies, Polar Division of Norilsk Nickel, Norilsk: 2012, pp. 77–83. (in Russian).

[20] Malinnikova O.N., Filippov Yu.A. Mining-induced change in the stress state of rock mass at tabor deposit, Problems and Prospects of Integrated Development and Preservation of Mineral Wealth: Academician Trubetskoy International School Proceedings, Moscow: IPKON RAS, 2014, pp. 122–125. (in Russian).

[21] Zakharov V.N., Malinnikova O.N., Trofimov V.A., Filippov Yu.A. Evaluation of concentrated energy efficiency in the vicinity of underground excavations and geological dislocations, Deformation and Failure of Materials with Defects and Dynamic Events in Rock Mass and Mines: XXIII Academician Khristianovich International School Proceedings, Simferopol, 2016, pp. 94–99. (in Russian).

[22] Zakharov V.N., Malinnikova O.N., Averin A.P. Modeling mining-induced vibration in production face area in coal–rock mass, Gorny Zhurnal, 2016, No. 12, pp. 11–22.

[23] Zakharov V.N., Malinnikova O.N., Trofimov V.A., Filippov Yu.A. Stability and creeping of landslide slope, Journal of Mining Science, 2014, Vo. 50, No., 6, pp. 1007–1016.

[24] Filippov Yu.A., Malinnikova O.N. Stability calculation of landslide slope in the Tuapse–Adler section of the North Caucasus Railway with ANSYS, Subsoil Management Problems in the 21st Century in the Eyes of the Youth: Proc. X Int. School of the Young Scientists and Specialists, Moscow: IPKON RAS, 2013, pp. 96–98. (in Russian).

Physical and mechanical properties of fractured rock

Geomechanics and Geodynamics of Rock Masses – Litvinenko (Ed.)
© *2018 Taylor & Francis Group, London, ISBN 978-1-138-61645-5*

Determination of the errors arising from apparatus and operator during applying the point load index

Deniz Akbay & Rasit Altindag

Department of Mining Engineering, Suleyman Demirel University, Isparta, Turkey

ABSTRACT: Point load index is used in the classification by the strength of the rocks, strength parameter of the rock materials in some rock mass classification systems, also to determine indirectly the other strength parameters such as uniaxial compressive strength and tensile strength. However, at the present time it is not recommended using of point load index to determine indirectly the uniaxial compressive strength and tensile strength because of the drawbacks of the test method. The point load index test is based on the principle that rock sample is broken between two conical platens. Failure load and specimen dimensions are used to calculate the point load strength index.

In this study, the point load index tests were applied on 7 different rock specimens which has prismatic on 15 different point load index test apparatuses (PLITA) which are based in different locations by the same operator. The errors of classic PLITA will be indicated and underlined and numerically supported by using the test results obtained. And then a new PLITA manufactured which eliminate the determined errors (the effect (experience, gender, fatigue, physiological/psychological status etc.) of the person doing the experiment) of the test method. Thus, more precise results could be evaluated.

Keywords: Point load index, rock mechanics, physical and mechanical properties, test apparatus design

1 INTRODUCTION

The point load index test is intended as an index test for the strength classification of rock materials, but it may also be widely used to predict other material strength parameters with which is correlated. It is an attractive alternative method, because it can provide similar data at a lower cost its ease of testing, simplicity of sample preparation, and possible field application. Many research works had been conducted to acknowledge with regard to point load index test and has resulted in widely used point load index and other parameters. However, more experimental works helps to substantiate the existing correlation. In order to estimate uniaxial compressive strength indirectly, index-to-strength conversion factors are constructed (ISRM, 1985; 2007; Zacoeb and Ishibashi, 2009).

The point load test has been considered as a cheap and useful testing method to estimate the strengths of rocks due to its ease of testing, simplicity of specimen preparation, and possible field application (Broch and Franklin, 1972; Bieniawski, 1975; Kahraman and Gunaydin, 2009; Basu and Kamran, 2010). It has been often reported as an indirect measure of the compressive or tensile strength of rocks (Fener et al., 2005; Cobanoglu and Celik, 2008; Heidari et al., 2012; Chau and Wong, 1996). Samples can be of various shapes, including cut cylindrical cores tested axially or diametrically, cut blocks, or irregular lumps (Brook, 1985; International Society for Rock Mechanics (ISRM), 1985; American Society for Test ing and Materials (ASTM), 2008). The point load test can be used to test both strong and weak rocks (Hardy, 1997; Heidari et al., 2012; Tsiambaos and Sabatakakis, 2004; Kahraman and Gunaydin, 2009). Establishing a proper correlation between the uniaxial compressive strength (UCS) and point load strength index ($I_{S(50)}$) is one of the most critical concerns in applying the point load test on various rock types. However, numerous experimental studies have shown that the conversion factors may vary with rock types and rock classes (igneous, metamorphic, and sedimentary rocks) (Tsiambaos and Sabatakakis, 2004; Fener et al., 2005; Kahraman and Gunaydin, 2009; Kahraman et al., 2005). Research is still actively on-going for certain rocks at some specific locations (Heidari et al., 2012; Singh et al., 2012). The relationship between the Brazilian tensile strength (BTS) and $I_{S(50)}$ is also often considered and assessed (Heidari et al., 2012; Li and Wong, 2013).

The point load test method was standardized by the ISRM in 1985. In this test, rock specimens (cylindrical, prismatic or irregular) loaded between two conical platens (of stipulated geometry and hardness) fail by the development of one or more extensional planes containing the line of loading and these failure modes are referred as valid failure modes whereas deviation from these failure patterns is indicated as a failure in invalid mode by the standards like ISRM and ASTM (ISRM, 2001, 2007; ASTM, 1995, 2008). The standards stipulate that the point load strength determined from a point load test that has lead to invalid specimen failure mode should be rejected. This is applicable for isotropic rocks as invalid failure in this case implies failure of the specimen governed by some preferential cracks or fabric that is not representative of the actual rock material. Following this stipulation, point load test has been successful in estimating UCS of isotropic rocks or rocks devoid of mesoscopic foliations. In case of anisotropic rocks, foliations/weakness planes are characteristics of the rocks and it is expected that mechanical behaviors of these rocks will be controlled by these planes under a certain state of stress. However, neither standards have a different stipulation about validity of failure modes for anisotropic rocks nor this subject has been explored comprehensively by researchers. Keeping the merit of the point load test and its success and wide use for isotropic rocks in mind, following two issues emerge to be absolutely important in rock engineering: (Brook, 1985) to check feasibility of utilizing point load test (with weakness planes at an angle with loading direction) for strength evaluation of anisotropic rocks and (Cargill and Shakoor, 1990) to evaluate predictability of UCS by point load strength for these rocks. The present study investigates both issues sequentially for schistose rocks (Basu and Kamran, 2010).

Failure load and specimen dimensions are used to calculate the point load strength index and this index is used in engineering geological studies. In this study, the problems related to the point load testing apparatus and the common mistakes of the operators are determined. Based on the experience of the researcher and the literature review, loading speed, failure time, no regular maintenance, error of indication of a measuring instrument, geometry of conical platens, axis shift of conical platens, length of loading arm, different opeators. For this reason, a new test apparatus has been designed to eliminate these disadvantages and uncertainties in this study and the 7 different rock specimens of the same size were tested uniformly at constant loading speed and the point load indexes were determined. Differences between manual loading and automatic loading have been demonstrated.

2 METHOD

The rocks obtained from various marble processing plants were brought to the SDU Mining Engineering Natural Stone Technology and Excavation Mechanics Laboratory and the test samples were prepared in accordance with the standards recommended by ISRM (1985, 2007), taking into account their planar positions and having certain shapes and volumes. In order to be able to represent rocks with different strength values, the samples used in the study were tried to be determined in such a way that a low strength to high strength scale would be formed. For this purpose, 7 different rocks were selected from different regions of Turkey (Table 1). Prismatic specimens of $30 \times 50 \times 50$ mm were prepared from rocks selected from various natural stone processing plants. For each test device, 10 samples are prepared from each sample. The prismatic specimens prepared later are drawn from the diagonal so that the operator can perform the loading operation from the exact midpoint of the specimen (Figure 1).

Table1. Geographical and geological origins of the rocks used in the study.

Kayaç Adı	Köken	Bölge
Lymra	Sedimentary	Antalya
Limestone-1	Sedimentary	Isparta
Limestone-2	Sedimentary	Isparta
Marble	Metamorphic	Muğla
Andesite	Igneous	Isparta
Granite	Igneous	Aksaray
Diabase	Igneous	Kayseri

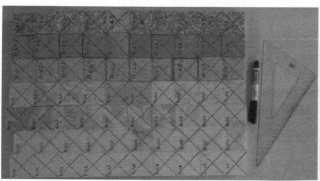

Figure 1.   Samples prepared for point loading strength tests.

At the beginning of study, an experimental preliminary study was done, and some observations were made during the it. The literature and the observations are combined with the problems and difficulties that have been encountered so far so that some measurements are predicted before the experiments. A form is created for this and this information is processed in this form.

2.1   Load indicator value measurement (calibration control)

A 1 ton-capacity load cell was purchased to provide control of the calibrations of the devices. As a first step in the tested devices, the values read on the display of the device by the load cell and the values read on the indicator of the load cell were recorded (Figure 2). Then, the video is displayed in the form of a slow shot and the values read on the indicator of the device and the values read on the indicator of the load cell are simultaneously noted. These values are then transferred to the Microsoft Office Excel program. With the help of the program, the values read in the indicator of the device and the values read in the indicator of the load cell are plotted and the relationship graphs are plotted and it is decided whether any correction coefficient is determined by looking at the relation between them and whether the read values are corrected.

2.2   Conical platen hardness value measurement

The standard proposed by ASTM (1995; 2008) suggests that conical platens are manufactured from tungsten carbide or hardened steel (Rockwell hardness value 58 HRc). Before the tests, the hardness values of the conical platens were measured with the hardness tester (Figure 3)

According to ISRM (1985; 2007), conical platens should have an angle of 60° and the radius of the end must be 5 mm (Figure 4). In order to control this, a simple mechanism was designed in the laboratory and the conical platens were put on this apparatus and pictures

Figure 2.   Load indicator value measurement.

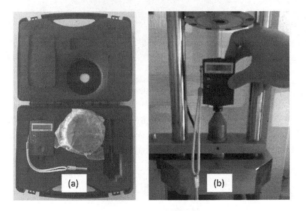

Figure 3.   (a) Hardness tester, (b) conical platen hardness measurement.

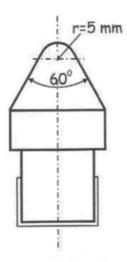

Figure 4.   The standard conical platens proposed by ISRM (1985; 2007).

Figure 5.   Apparatus for measuring conical platens.

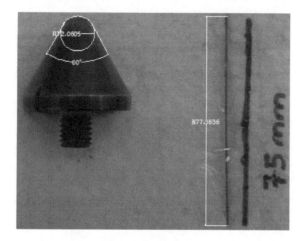

Figure 6.   Drawing of radius and radius of conical platen with software.

Figure 7.   The places where test equipments are located (symbolic).

185

were taken (Figure 5). These pictures are then drawn in the AUTOCAD program, with the radius and the radius of the ends of the conical platens drawn. Then, the actual radius values are calculated by using the reference length in the figure (Figure 6).

2.3 Point load index test apparatus used in the study

In the study, experiments were carried out on 15 different point load index apparatus in 12 different laboratories in 8 different universities (Figure 7) by the same operator in accordance with the standards specified in the ISRM (2007). The apparatus belong to 8 different brands, 6 have mechanical indicators and 9 have digital indicators. When the conical platens used in the apparatus are examined, it is determined that there are conical heads with 5 different geometries.

3 RESULTS

Before the point load index tests were carried out, measurements were made for the calibration of test apparatus and the necessary information was noted on measurement template. The experiments were carried out on 15 different point load index in accordance with the standards specified in the ISRM (2007). The information about used test apparatus were given in Table 2. The point load index test results were given in Table 3 and Figure 8.

It is seen when the Table 3 is examined that different values have been obtained for same samples for different test apparatus, although all experiments were made by the same the operator. It appears that standard deviations are also within acceptable limits. Referring to Figure 8, it is realized that each apparatus has its own characteristic. The $I_{S(50)}$ values of the samples were higher or lower in the same devices. The reasons for these differences in $I_{S(50)}$ values and also similarities in increases and decreases of $I_{S(50)}$ values, can be explained by the informations (Hrc hardness of conical platens, angle of conical platens, rounding radius of conical platens, length of loading arm) in the Table 2.

If the experiments were carried out by different operators, in addition to the parameters we have mentioned above, changes due to operator differences (age, gender, physical fitness, stress level, nutrition, ergonomics, health situation, fatigue etc.) would also be relevant. Although the experiments were done by the same person, different times and different times of the day may have caused changes in $I_{S(50)}$ values. Therefore, all the

Table 2. The general information of test apparatus.

Apparatus code	The arrival date of the lab (year)	Hrc hardness of conical platens	Angle of conical platens (°)	Rounding radius of conical platens (mm)	Length of loading arm (cm)
1	2012	62.63	59.5	5.8	51.0
2	1995	63.59	60.0	6.1	45.0
3	1995	59.27	60.5	5.2	45.0
4	2007	53.37	59.5	2.2	39.5
5	1999	59.14	60.5	6.2	50.5
6	2011	55.95	63.0	4.4	44.5
7	2000	51.44	59.5	5.9	35.0
8	1995	63.74	59.5	5.6	45.0
9	2014	56.9	62.5	4.0	47.0
10	2013	60.64	59.5	5.6	40.0
11	1995	59.55	60.5	5.6	45.0
12	2016	58.45	60.0	5.0	43.5
13	2008	54.45	60.0	3.9	39.5
14	2016	63.25	59.5	3.9	36.5
15	1995	62.02	60.5	5.4	44.5

Table 3. The point load index test results.

Apparatus code	Lymra $I_{s(50)}$ (MPa)	sd	Limestone-1 $I_{s(50)}$ (MPa)	sd	Limestone-2 $I_{s(50)}$ (MPa)	sd	Marble $I_{s(50)}$ (MPa)	sd	Andesite $I_{s(50)}$ (MPa)	sd	Granite $I_{s(50)}$ (MPa)	sd	Diabase $I_{s(50)}$ (MPa)	sd
1	4.03	0.38	3.11	0.42	4.39	1.15	2.94	1.05	6.76	0.84	6.82	0.40	8.32	0.71
2	3.42	0.28	2.78	0.09	3.73	0.84	3.38	1.61	5.48	0.86	6.92	0.52	8.51	0.66
3	2.87	0.10	2.75	0.09	3.11	0.39	2.38	0.39	4.67	0.77	5.86	0.44	6.96	0.71
4	3.05	0.14	2.81	0.19	3.15	0.95	2.57	1.22	4.64	0.30	5.51	0.65	6.74	0.45
5	5.54	0.74	4.59	0.45	6.05	1.58	5.89	1.84	8.00	0.57	9.04	0.50	10.37	1.09
6	3.94	0.31	2.91	0.46	3.73	0.97	3.59	1.80	5.85	0.73	6.52	0.79	7.38	0.51
7	4.49	0.79	3.54	0.48	4.11	0.95	2.94	1.75	7.20	0.75	7.80	0.62	8.98	0.98
8	4.13	0.11	2.30	0.04	4.00	0.16	3.94	0.12	6.32	0.64	7.45	0.54	8.53	1.10
9	3.71	0.34	2.13	0.39	2.91	1.24	1.95	1.37	4.89	1.12	5.28	0.70	6.87	1.33
10	3.95	0.47	3.37	0.30	4.86	0.96	3.59	1.02	6.51	1.19	7.09	0.56	8.82	1.13
11	4.53	0.53	3.61	0.54	4.83	0.66	4.66	1.73	7.15	1.09	7.62	0.65	8.85	1.13
12	3.18	0.41	3.00	0.46	3.47	0.54	3.89	1.29	4.47	0.76	5.90	1.11	5.97	1.33
13	3.99	0.76	3.33	0.27	4.23	0.47	2.63	1.01	6.63	0.54	6.64	1.48	8.72	0.75
14	3.72	0.33	2.90	0.28	4.19	1.09	3.30	1.52	5.97	0.65	6.18	0.41	6.80	2.78
15	4.19	0.34	2.55	0.14	3.86	0.95	2.84	1.48	4.76	1.03	6.07	0.73	7.85	1.42

$I_{s(50)}$: Point load strength; sd: Standart deviation

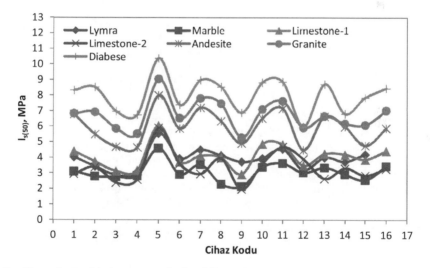

Figure 8. The point load index test results for different test apparatus.

factors originating from both the operator and the device must be evaluated together. So the reason for the decline in a $I_{s(50)}$ value for a sample cannot be explained only by the rounding radius of conical platens.

3.1 *Observed errors*

- The angle and rounding radius of conical platens
- Display errors
- Fracture time
- The length of loading arm
- Maintenance of the apparatus

187

4 CONCLUSION

In this study, various problems are observed in the application of point load index experiment which is widely used in engineering projects. There are errors both from the apparatus and from the operator. In this study, it was determined that blinding of conical platens, display errors, lack of periodical maintenance, and lack of a standard length of loading arm were determined as test apparatus errors. The operator must be absolutely trained on the experiment before applying it.

The point load index is widely used in the estimation of the uniaxial compressive strength and the tensile strength of rocks. In the past a large number of coefficients have been proposed in the literature to predict uniaxial compressive strength and the tensile strength. After identifying these errors, the reliability of these coefficients should be monitored. More accurate values will be obtained if the following 3 items are provided at least.

- The new apparatus must be electrical operated to remove the effect of the operator
- Rounding radius of conical platens must be 5 mm
- Regular maintenance of the apparatus should be done periodically

REFERENCES

ASTM, 1995. Standard Test Method for Determination of the Point Load Strength Index of Rock. Am. Soc. Test. Mater. 22, 1–9.
ASTM, 2008. Standard Test Method for Determination of the Point Load Strength Index of Rock and Application to Rock Strength Classifications. ss. 1–11.
Basu, A., Kamran, M., 2010. Point load test on schistose rocks and its applicability in predicting uniaxial compressive strength. Int. J. Rock Mech. Min. Sci. 47, 823–828.
Basu A, Kamran M (2010) Point load test on schistose rocks and its applicability in predicting uniaxial compressive strength. Int J Rock Mech Min Sci 47(5):823–828.
Bieniawski ZT (1975) The point-load test in geotechnical practice. Eng Geol 9(1):1–11.
Broch E, Franklin JA (1972) The point-load strength test. Int J Rock Mech Min Sci Geomech Abstr 9(6):669–676.
Chau KT, Wong RHC (1996) Uniaxial compressive strength and point load strength of rocks. Int J Rock Mech Min Sci Geomech Abst 33(2):183–188.
C obanoglu I, Celik SB (2008) Estimation of uniaxial compressive strength from point load strength, Schmidt hardness and P-wave velocity. Bull Eng Geol Environ 67(4):491–498.
Fener M, Kahraman S, Bilgil A, Gunaydin O (2005) A comparative evaluation of indirect methods to estimate the compressive strength of rocks. Rock Mech Rock Eng 38(4):329–343.
Hardy JS (1997) The point load test for weak rock in dredging applications. Int J Rock Mech Min Sci 34(3–4): 295.e1–e13.
Heidari M, Khanlari G, Torabi Kaveh M, Kargarian S (2012) Predicting the uniaxial compressive and tensile strengths of gypsum rock by point load testing. Rock Mech Rock Eng 45(2):265–273.
ISRM, 1985. Suggested Method for Determining Point Load Strength, içinde: ISRM. ss. 53–60.
ISRM, 2007. Suggested Method for Determining Point Load Strength, içinde: ISRM. ss. 125–132.
Kahraman S, Gunaydin O (2009) The effect of rock classes on the relation between uniaxial compressive strength and point load index. Bull Eng Geol Environ 68(3):345–353.
Kahraman S, Gunaydin O, Fener M (2005) The effect of porosity on the relation between uniaxial compressive strength and point load index. Int J Rock Mech Min Sci 42(4):584–589.
Li, D., Wong, L.N.Y., 2013. Point load test on meta-sedimentary rocks and correlation to UCS and BTS. Rock Mech. Rock Eng. 46, 889–896.
Singh TN, Kainthola A, Venkatesh A (2012) Correlation between point load index and uniaxial compressive strength for different rock types. Rock Mech Rock Eng 45(2):259–264.
Tsiambaos G, Sabatakakis N (2004) Considerations on strength of intact sedimentary rocks. Eng Geol 72(3–4):261–273.
Zacoeb, A, Ishibashi, K., 2009. Point Load Test Application for Estimating Compressive Strength of Concrete Structures From Small Core. ARPN J. Eng. Appl. Sci. 4, 46–57.

Geomechanics and Geodynamics of Rock Masses – Litvinenko (Ed.)
© 2018 Taylor & Francis Group, London, ISBN 978-1-138-61645-5

Determination of rock deformability using the coastal sections of concrete dams as a large-scale stamp

E.S. Argal & V.M. Korolev
Gidrospetsproekt Ltd., Moscow, Russia

ABSTRACT: Many high-pressure concrete dams are erected by leaving column cutting joints. After curing of the concrete the cutting joints are filled with cement grout under a certain amount of pressure. After hardening of the grout the dam becomes monolithic.

Before joints grouting hydraulic tests are conducted to determine the joints widening. When grouting the joints, which are separated from the sides of the canyon just by one section of the dam, discharge pressure is passed through this section on the rock massif. By cyclic loading with water pressure, (i.e., using it as a large-scale stamp), and measuring the additional and residual disclosures of the joint, deformation characteristics of the rocky massif in contact with the concrete dam can be determined. Formula for calculating the deformation modulus of the rocks foundation is given in the paper. Such studies, accomplished at the erection of Krasnoyarsk and Chirkej dams (Russia), have shown the ability to not only solve the above-mentioned task, but compact the superficial layer of the rocky massif when a sufficiently large discharge water pressure is applied.

1 INTRODUCTION

To determine the deformation characteristics of the bedrock, the stamps loading method is sometimes used. However, in the construction of large dams this method has a drawback: the size of the load-applying stamp is disproportionately small compared to the area of the dam foot. In addition, such experiments require a significant amount of work, time and money. But sometimes we can do the work with a new and inexpensive method.

2 THE NEW METHOD

Many high concrete dams are erected by leaving temporary column cutting (heat-shrink) joints that are filled afterwards by pumping cement grout via embedded tubes in the concrete. Grouting joints in arch dams are radial, i.e. intersections. For gravity dams intersection joints near the head surface are grouted sometimes. Before grouting the joints are hydraulically tested under a certain amount of water pressure. Knowing the dimensions of the loading platform, the amount of load applied and the opening displacement of the joint under the water pressure, the deformability of the system, including both the concrete and the rocky massif, can be determined.

The coastal section in the lower part usually is thin. Therefore, the deformation characteristics of the system are mainly determined by the rock properties. The grouting-check area is considerable large (up to 250 m^2). This method is quite simple and requires minimal additional costs. With the help of this method it's possible to identify the deformation features of the bedrock in the entire height of the canyon.

The deformation modulus of the rocks foundation can be determined by the following formula [Tsitovich, 1963].

$$E_d = Pb\omega \left(1 - \mu^2\right)/U_{av},$$

where E_d – the modulus of deformation (when calculating the elastic modulus E_e you need to subtract the residual displacement from the total displacement U), P – pressure, MPa; b – the width of the loading platforms (grouting check), m; U_{av} – the average settlement of the loaded square (the additional opening of the joint), m; ω – a coefficient, dependent on the correlation sides of the loading platform; μ – the Poisson's ratio.

3 EXPERIMENTAL DATA

The following table shows the data, obtained in the experimental grouting checks (Fig. 1) of the heat-shrink joints in the Krasnoyarsk dam, as well as in the abutment and plug in the Chirkej dam [Argal, Korolev, 1974]. The additional opening of the joint (Fig. 2) is determined by watch indicators that are embedded in the concrete and in the middle height of the check (in check 62-III the indicator was installed at the top of the check). In the same table it is shown the calculation results of the deformation characteristics of a two-tiered system of concrete-rock, using the above formula. Coefficient ω is adopted in calculations for flexible stamps, and the Poisson's ratio is equal to 0,2.

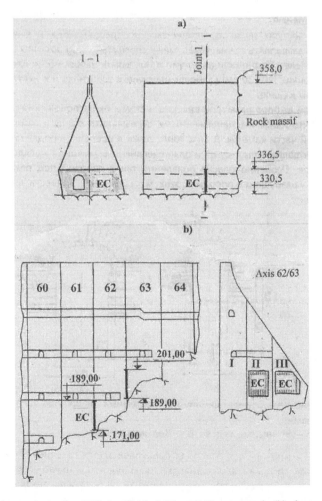

Figure 1. Experimental checks (EC) in Chirkej (a) and Krasnoyarsk (b) dams: 62, 63 ... – dam sections; I, II, III – columns.

190

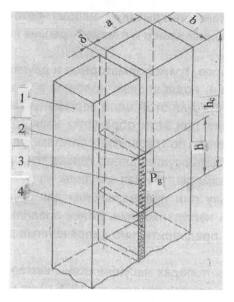

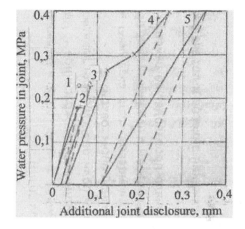

Figure 2. Transmission diagram on walls grouting joint: 1 – dam column; 2 – contour waterstop; 3 – grouting check; 4 – grouted check; h – cheks height; a, b, h_c – column sizes; δ – joint disclosure before grouting.

Figure 3. Water pressure in joint vs additional joint disclosure: 1, 2, ... – loading cycles.

Preliminary calculations, conducted for the abutment checks as for beams on the elastic foundation (with the bed coefficient 50 MPa), showed, that the additional opening of the joint due to the related sections is approximately half of the total. Therefore, the calculation data on the settlement load pad for this check were halved (the table enclosed in brackets). The ratio of the elastic modulus to the deformation modulus in this case is taken equal to the ratio of the total deformation to the elastic part.

Monitoring in other checks was carried out after grouting the contiguous joints (deformation lacking),

Fig. 3 shows a typical diagram for the additional opening joints on cyclic loading concrete dam parts by the pressure of injected water in the hydrostatic testing in abutment Chirkej dam. From the diagram you can see, that injection pressure and additional opening joints have a mixed relationship. The modulus of deformation is reduced with the increase in the workload. In this case, the deformation is influenced by the contiguous section.

The nature deformation of the two-layer system in general is similar to the nature deformation of concrete dam parts, but the percentage of the residual deformations of the system, especially during the first load cycles, as a rule, are higher, which is obviously,- connected with the partial closure of the cracks of the rock massif, that are oriented normal to the direction of the load, and with the compaction of their filler.

However, the data of the other checks, where adjacent sections almost not distorted, show that the deformation module of the two-layer system of concrete-rock decreases with the increase in the workload and the number of loading cycles, but at the same time the ratio of the elastic modulus typically increases. For loads under the previously applied, the system works absolutely elastically – the residual deformation does not increase.

The table shows that the strain of the rocky massif at the location of the abutment is significantly lower than in the plug boards. That is confirmed by other research data.

In view of the complexity of the process of deformation of the considered two-layer system and the interpretation of the results currently, they can only be used for the comparison with other data. To obtain the more precise definition of the deformation characteristics we

Table 1.

Dam		Chirkej						Krasnoyarsk			
Check		In the abutment		II-III-1-4				II-IV-1-4	62-III		
The size of the check, m	height	6,0		7,5				7,5	9,5		
	width	11,0		35,0				35,0	10,0		
The discharge pressure, MPa		0,41		0,55 0,6 0,7 0,9				4,0	6,0 1,0	1,3	0,9
The joint opening, total $M \cdot 10^{-3}$		0,37 0,4 0,39 (0,18) (0,2) (0,2)		0,1 0,12 0,14 0,2				0,08 0,23 $\frac{0,25}{0,35}$	0,6	0,72	0,62
	elastic	0,2 0,23 0,32		0,05 0,07 0,08				0	0,13 0,5	0,58	0,43
E_d, $MPa \cdot 10^3$		16,6 15,4 15,8		69 63 63 57				63	$\frac{30,2}{21,6}$ 15,6	15,3	12,3
								22			
E_e, $MPa \cdot 10^3$				138		126 142		0	51 18,7	19,0	17,6
E_e/E_d		1,85 1,75 1,2		2		2		0	1,92 1,2	1,25	1,44

(The denominator shows the additional opening joint after a few cycles of loading).

need to clarify the methodology of the testing, in particular, a more detailed examination of the effect of the magnitude and the duration of the load on the rock massif. For the ability to create the significant pressure in a joint and exclusion of the water ingress into the overlying checks it is appropriate to set the contour seals in the experienced checks for the increasing of the hermeticity (for example, double), and to drain the neighbour checks. To reduce the effect of the deformation of the concrete to the additional opening of the joints, we need to strive to reduce the thickness of the coastal sections, and to grout the experimental checks only after grouting the adjacent joints.

4 ADDITION

When the tube system is laid for grouting the concrete surface area with the rocky ground, the determination of the deformation characteristics of the bedrock becomes more specific nature: when we pump water in this zone the load is transferred directly to the base and thus simplifies the task. In such cases it is advisable to set some gauges in the contact zone for direct determination of deformations and stresses in the rock foundation.

Cyclic loading in the joints of contact zone by water pumping with high pressure can result in compaction of the upper rocks and an increase in its deformation modulus, because an increase in the number of loading cycles will make the share of the residual deformations decrease. After grouting the joints and subsequently turning into a stone grout «cans» deformation occurred in the bedrock layers adjacent to the concrete.

5 CONCLUSION

The method described above can be used only in the cases that a concrete dam is erected but it's joints, separated from the sides of the canyon just by one section of the dam, are not grouted yet. With the help of this method we can determine the deformation characteristics of the rock massif and even compact it.

REFERENCES

Argal E.S, Korolev V.M. Opredelenie deformativnyh kharakteristik skal'nogo osnovaniya plotin pri tsiklicheskom nagruzhenii ego krupnomasshtabnymy shtampami//Trudy koordinatsionnyh sovesh'aniy po gidrotekhnike, vyp. 91, L., 1974, c. 125–128.
Tsitovich N.A. Mekhanika gruntov. Gosstroyizdat, 1963.

Characterization of hydromechanical damage of claystones using X-ray tomography

C. Auvray
GeoRessources Laboratory, UMR 7359 CNRS/UL/CREGU, Vandœuvre-lès-Nancy Cedex, France

R. Giot
IC2MP Laboratory, UMR 7285, ENSI Poitiers, Poitiers Cedex, France

ABSTRACT: During the excavation of underground galleries in a rock mass, reorganisation of the stress field around the site results in damage to the surrounding material. In the framework of research undertaken at the ANDRA underground laboratory in France, galleries have been excavated in a Callovo-Oxfordian argillite formation. The damage has manifest as a number of fracture systems in the adjacent rock. Observations have shown that fractures generated during excavation and storage exploitation are able to self-seal during re-saturation of the site in the post-closure stage. For the present study, we performed a range of self-sealing experiments using a newly-developed triaxial compression cell placed within a nano-CT scanner with a 10 µm resolution. The samples contained an artificial fracture that had been generated in the laboratory, into which a fluid was injected. Two facies of claystone were tested. In the first facies ($CaCO_3$ content of 29.7%), an important but not total reduction in volume after 34 days of injection was observed, as well as a high decrease in permeability, even if safe claystone values were not recovered. In the second facies ($CaCO_3$ content of 59.4%), a low volume reduction and decrease in permeability were observed after 15 days of injection.

1 INTRODUCTION

During excavation of galleries at the ANDRA Meuse/Haute-Marne underground research laboratory, reorganisation of the stress field around the site has resulted in damage to the surrounding material. In the Callovo-Oxfordian argillites, this damage has manifest as a network of fractures organized into two zones: a metric-scale zone, in which the fracture system is dense and highly connected, and (ii) a multimetric zone, characterised by isolated fractures of poor hydraulic conductivity.

In the context of radioactive waste storage, the future of the damaged zone is of vital importance with regards to the transfer of water, gas and solutes along the galleries. Observations have shown that fractures generated during excavation and storage exploitation are able to self-seal during resaturation of the site in the post-closure phase. This leads to a significant reduction in water permeability in the damaged zone. Self-sealing is a result of a rearrangement of minerals and porosity in the fracture region. The main processes involved are (i) swelling between sheets of smectite phases, (ii) inter-particle swelling due to osmotic effects, and (iii) plugging or blockage of the fractures by particle aggregation. These processes lead to structural rearrangement within the self-sealed zone, which in turn determines the hydromechanical and transfer properties of the sealed fracture zone.

Even though the ability of argillite fractures to self-seal has already been demonstrated in experimental tests and on in situ samples (Bastiaens et al. 2007; Rothfuchs et al. 2007; Van Geet et al. 2008; Zhang et al. 2008; Zhang 2009; 2011; 2013), little information is available regarding the effectiveness of self-sealing under different mechanical, hydrodynamic or chemical conditions (e.g., as a function of fracture-opening size, applied stress, rate of water flow in the fracture, or the composition of the circulating fluids, for example, water from the site or from the cement).

The main objective of the present study was to demonstrate that the self-sealing process depends principally on the mineralogy of the argillites. For this, we considered the carbonate content of the argillite. Using tomographic imaging, measurements of the volume of residual space and permeability were calculated throughout the duration of the tests.

2 EQUIPMENT AND SAMPLES

The Multi-Scale HydroGeoMechanics team at the GeoRessources Laboratory has developed a triaxial compression cell that is transparent to X-ray and allows observation of rock samples placed under mechanical stress. The triaxial compression cell (Figure 1) is manufactured entirely from a thermoplastic of sufficient resistance that, moreover, is X-ray transparent (Auvray & Morlot 2017). The design of the compression cell allows cylindrical samples of 20 mm diameter and 40 mm height to be tested. The tomograph used in the experiments is a GE Phoenix Nanotom S CT scanner.

Samples were taken from two facies selected on the basis of their respective carbonate contents. The first, identified as UA, has a $CaCO_3$ content of 29.7%. The second, USC, has a

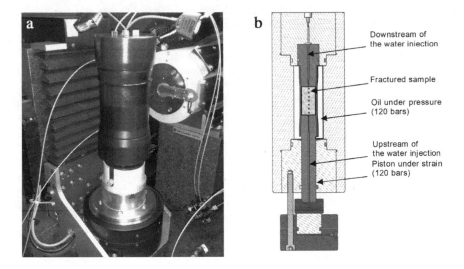

Figure 1. Transparent triaxial compression cell (a: placed with in the GE Phoenix Nanotom S CT scanner chamber; b: section).

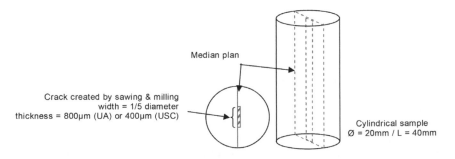

Figure 2. Sample and artificial crack (produced by coring, sawing and milling).

194

CaCO$_3$ content of 59.4%. Rock samples were artificially cracked, and sawn in two according to a plane containing the axis of the cylinder. One of the faces was machined so as to obtain a slit across 1/3 of the diameter, the opening of which was controlled precisely to be 400 μm for USC sample and 800 μm for UA sample (Figure 2). Water was then injected into the crack.

The water used in the framework of these experiments was a synthetic water equivalent to that found in situ at the ANDRA research laboratory, which has a chemical composition of NaCl = 1.755 g.l^{-1}, CaCO$_3$ = 0.0142 g.l^{-1} and Si/Al = 0.0010 g.l^{-1}.

3 EXPERIMENTAL PROTOCOL

The samples, isolated from the confining fluid by a waterproof Viton© membrane, were placed in the triaxial compression cell and a hydrostatic stress of 12 MPa was applied. This 12 MPa pressure is similar to the in situ stress conditions. Once the pressure had stabilised, fluid was injected into the end of the sample at a constant rate of 0.05 ml/min and a maximum injection pressure of 1 MPa. Once the maximum injection pressure had been reached, the volume injected was recorded.

Before placing the sample under isotropic strain, 3D tomography was performed at 24 voxel precision in order to visualise and calculate the volume of the artificial crack. During the tests, a series of tomographic images were acquired in order to monitor the self-sealing process. At the end of the test, and before disassembling the triaxial cell, a final 3D tomographic image was acquired in order to visualise and calculate the volume of the residual space after the fluid circulation phase.

During each phase of fluid circulation, the pressures and volumes of the injections were recorded as a function of time.

4 RESULTS

In the UA facies, a significant but not total reduction in volume after 34 days of injection was observed (Figure 3), as well as a high decrease in permeability, even if safe claystone values were not recovered (Figure 4). A two-phase kinematic developed: an initial rapid phase of several hours' duration and a subsequent slower phase lasting several weeks.

In the USC facies, a low reduction in volume and decrease in permeability were observed after 25 days of injection, and the kinetics were monotonous (Figures 5 and 6).

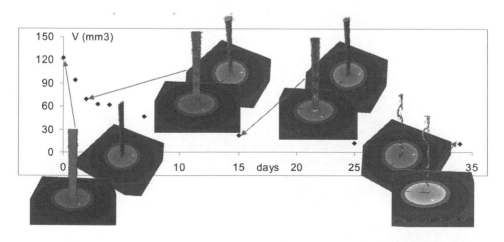

Figure 3. UA sample—3D reconstruction of the fracture before, during and after 34 days of water injection.

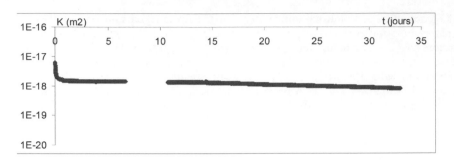

Figure 4. UA sample—permeability vs time.

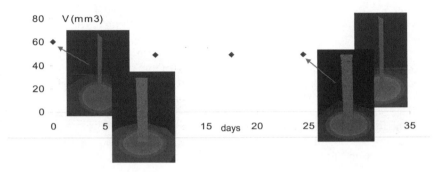

Figure 5. USC sample—3D reconstruction of the fracture before and after 25 days of water injection.

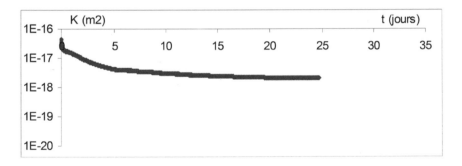

Figure 6. USC sample—permeability vs time.

5 DISCUSSION—PERSPECTIVES

The main driver of self-sealing in argillite is the swelling of clay minerals. A study of self-sealing, or of self-healing, is therefore closely linked to the study of clay shrinkage/swelling. Zhang et al. (2013) describe clay swelling as the sum of crystalline and osmotic swelling, the latter dominating over the former in conditions of high relative humidity (RH).

In an attempt to simulate these phenomena, a model of wet swelling under partially-saturated conditions might be considered on the basis of these observations. At high RH, such a model would be supported by one that allows the geochemical processes that act on the clay shrinkage/swelling properties to be taken into account, thus necessitating a finite-element type code and a reactive transport code. For a model of crystalline swelling, at low RH, various pre-existing models, such as the elasto-plastic models (Gens & Alonso,

1992, Thomas & He, 1998), the Barcelona Basic Model (Alonso et al, 1990) or the Barcelona Expensive Model (Alonso et al., 1999), might be considered as a basis. At high RH, the approach of Zhang et al. (2013) might be considered.

6 CONCLUSION

The development of a new compression cell coupled to a nanotomograph has allowed visualisation of a particular physical phenomenon—self-sealing—that until now could only be measured indirectly. We were able to monitor the self-sealing of a fracture while at the same time record measurements of the residual permeability (8.10^{-19} m^2), a value that approaches the permeability of the non-fractured rock (10^{-20} m^2). It remains necessary to validate the inferred physical phenomena using a behavioural model that integrates hydro-chemico-mechanical coupling.

The tests also allowed us to demonstrate that a certain amount of clay is required for self-sealing to occur. In effect, a significant reduction in the volume of residual space was not associated with a facies with a higher carbonate content, even if we did observe a slight reduction in permeability (3.10^{-18} m^2). A possible explanation for this is that during the passage of a fluid, the fluid becomes loaded with particles that have detached themselves from the lips of the fracture. This would lead to an increase in the viscosity of the fluid and, in turn, a progressive reduction in permeability.

A series of new experiments are currently being undertaken in order to determine a clay threshold or, conversely, a maximum carbonate content, for the occurrence of self-sealing. Analyses of the water exiting the system have also been envisaged in order to determine the presence of particles that could explain the reduction in permeability.

REFERENCES

Alonso, E.E., Gens A. and Josa A. (1990). A constitutive model for partially satured soil. *Géotechnique, 40, 3, 405–430.*

Alonso, E.E., Vaunat J. Gens A. (1999). Modelling the mechanical behavior of expansive clay. *Eng. Geol. 54, 1173–183.*

Auvray, C., Morlot, C., Giot, R.,Talandier, J. and Conil, N. (2017). Self-sealing of claystone under X-ray nanotomography: a newly-developped triaxial compression cell., *ICTMS-3rd International Conference on Tomography of Materials and Structures, Lund, Sweden.*

Bastiaens, W., Bernier, F. and Xiang, L.L. (2007). SELFRAC: experiments and conclusions on fracturing, self-healing and self-sealing processes in clays. *Physics and Chemistry of the Earth, 32, 600–615.*

Gens, A. and Alonso, E.E. (1992). A framework for the behavior of unsaturated expansive clay. *Can. Geotech. J. 29, 1013–1032.*

Rothfuchs, T., Jockwer, N. and Zhang, C-L. (2007). Self-sealing barriers of clay/mineral mixtures—The SB project at the Mont Terri Rock Laboratory. Physics and Chemistry of the earth, 32, 180–115.

Thomas, H.R. and He, Y. Modelling the behaviour of unsaturated soil using an elastoplastic constitutive model. *Géotechnique, 48, 5, 1998, 589–603.*

Van Geet, M., Bastiaens, W., and Ortiz, L. (2008). Self-sealing capacity of argillaceous rocks: review of laboratory results obtained from the SELFRAC project. *Physics and Chemistry of the earth, 33, S396-S406.*

Zhang, C-L. and Rothfuchs, T. (2008). Damage and sealing of clay rocks detected by measurements of gas permeability. *Physics and Chemistry of the earth, 33, S363–S373.*

Zhang, C-L. (2011). Experimental evidence for self-sealing of fractures in claystones. *Physics and Chemistry of the earth, 36, 1972–1980.*

Zhang, C-L. (2013). Sealing of fractures in claystone. Journal of Rock *Mechanics and Geotechnical Engineering, 5, 214–220.*

Zhang, C-L. (2009). Self sealing of fractures in argillites under repository conditions. *International conference and workshop in th framework of the EC TIMODAZ project and THERESA.*

Geomechanics and Geodynamics of Rock Masses – Litvinenko (Ed.)
© *2018 Taylor & Francis Group, London, ISBN 978-1-138-61645-5*

Creation of regional database of physical-and-mechanical characteristics of man-induced dispersed soils

Svetlana P. Bakhaeva & Dmitriy V. Guryev

T.F. Gorbachev Kuzbass State Technical University, Kemerovo, Russia

ABSTRACT: Herein, there are the results of generalization and analysis of physical-and-mechanical characteristics of man-induced dispersed soil composing the dam body—surface liquid-waste impoundment dike. There is the algorithm of formation of the regional table of physical-and-mechanical properties of soils.

Keywords: Physical-and-mechanical characteristics; man-induced dispersed soils; the regional table

1 INTRODUCTION

The increasing rate of extraction of mineral resources leads to intensive formation and accumulation of liquid wastes by mining enterprises. In order to protect environment, the wastes are provided to store in the liquid-waste impoundment dikes, the capacity of which is formed by construction of enclosure structures—earth dikes.

Operation of the earth dikes is accompanied by a risk of collapse due to their loss of stability, as a result there is a possibility of contamination of surface and underground waters, and soils by toxic substances, damage by a breakthrough wave of buildings and structures located in downstream entrance of the impoundment.

Safety criteria of the earth dikes are the stability coefficient that depends on the physical-and-mechanical characteristics of the soil dam body and base. The variation of these characteristics within a single structure may exceed 50%, consequently the issue regarding the choice of the calculated values is solved ambiguously (Bakhaeva S.P., 2016).

During development of the project documentation, the regulatory and design values of strength and deformation characteristics of soils are determined as per the tables of Annex B "The Code of Rules" (SP 22.13330.2011, 2011) or Annex G "The Code of Rules" (SP 11-105-97, 2000). The drawback of this method is in the fact that these tables were compiled for the entire territory of the former Soviet Union; a large part of soil bases and structures on the territory of Russia is not included in the above mentioned tables; a possible influence of the soil genesis is not taken into consideration; indicators for anthropogenic soils are lacked; the original data had been obtained by the old standards and processed in the 'pre-computer era', that proves their low accuracy; the use of the 'input' to the table for clayey soils as a porosity ratio, while their physical-and-mechanical characteristics are mostly determined by plasticity, humidity and water saturation ratio (Chernyak E.R., 2011).

Foreign scientists also bring up an issue of variability of characteristics (Xiaolei Liu, 2017; Breysse D., 2005) and evaluation of their influence on slope stability (Sivakumar G.L., 2004; Cho S.E., 1992; Edigenov M.B., 2014), and on the behavior of tunnels in soft rocks (Huang H.W., 2017).

Thus, a necessity of generalization of the available survey materials by the man-made soils, statistical methods and study of the dependence of the strength characteristics of soils from the physical correlation methods has been roused.

2 TARGET OF RESEARCH

The man-induced dispersed incoherent clay fill barrages—liquid-waste impoundment dikes from industrial enterprises.

3 INITIAL DATA

There are reports on engineering and geological surveys carried out on soil dikes – liquid-waste impoundment dikes from mining enterprises of Kuzbass. The results of the surveys were pre-processed, during which samples of soil have been extracted by the same type of drilling rigs (URB-2A-2); tested on the same devices by the scheme of unconsolidated – not-drained shift; by specific geological-genetic complex; located above (in natural state) and below (in saturated condition) of the depression curve (Guryev D.V., 2015).

4 RESEARCH METHODS

Mathematical statistics: verification of distribution law, comparison of the average values of sampling populations; single-factor analysis of variance; multiple comparisons.

5 RESULTS AND REVISING

In the process of the dike stability assessment, the issues of the choice of the strength characteristics for soils are solved ambiguously, as the coefficient of variation even within a single structure may reach 70%.

For example, for anthropogenic soils of the object N (Figure 1); the coefficient of variation of the density is 5%, angle of internal friction - 11%, adhesion is up to 60% (Table 1).

The analysis of engineering and geological survey data of the individual object and combination of objects of Kuzbass has found out [S. P. Bakhaeva, 2016] that variability of soil characteristics, defined by the coefficient of variation, has a small (up to 4%) date scattering in density ($\rho = 1,61$ of 2.23 g/cm³), slightly larger (up to 20%) on the angle of internal friction ($\varphi = 11°–35°$) and very significant (up to 70%) variation of adhesion (C = 5–140 kPa).

Reviewing the surveys', a hypothesis regarding a possibility of definition of the standard values of soils of the specific geological-and-genetic complex (loam) by method of generalization of experimental data, the procedure of which is shown in Figure 2, has been made.

By the given algorithm the physical-and-mechanical properties of the man-induced dispersed soil in Kuzbass has been processed. To insure the experiment, the initial data array

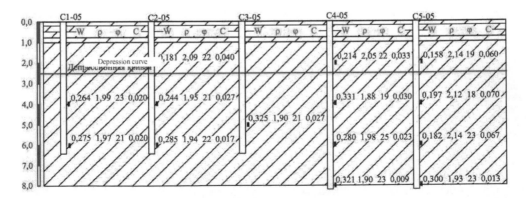

Figure 1.   Geological cross-section along the dike of the object N.

Table 1. Variability of physical-and-mechanical characteristics of soils for the object N.

# of well	Characteristics		
	φ, degree	C, kPa	ρ, g/sm³
1–05 (2)	21–23 (6%)	20 (0%)	1,97–1,99 (1%)
2–05 (3)	21–22 (3%)	17–40 (41%)	1,94–2,09 (4%)
4–05 (4)	19–25 (11%)	9–33 (45%)	1,88–2,05 (4%)
5–05 (4)	18–23 (11%)	13–70 (51%)	1,93–2,14 (5%)
		...	
Dike (25)	18–25 (10%)	5–83 (60%)	1,88–2,14 (4%)

Column 1: First digit is the ordinal number of the well, second digit—number of taken samples; columns 2–4: range of variability for individual wells, in parentheses—the coefficient of variation.

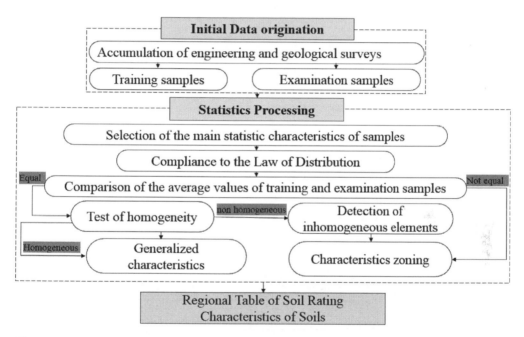

Figure 2. Logical structural scheme of formation of the regional table of the physical-and-mechanical.

was divided into two samples: (training and examination) presented as matrices (Guideline for the Regional Tables Drawing, 1981):

$$
\begin{pmatrix}
x_{11}; x_{21}; \ldots; x_{i1}; \ldots; x_{n1} \\
x_{12}; x_{22}; \ldots; x_{i2}; \ldots; x_{n2} \\
x_{1j}; x_{2j}; \ldots; x_{ij}; \ldots; x_{nj} \\
\cdot \quad \cdot \quad \cdot \quad \cdot \quad \cdot \quad \cdot \\
x_{1m}; x_{2m}; \quad ; x_{im}; \ldots; x_{nm}
\end{pmatrix}, \tag{1}
$$

where x_{ij} – the result of determination of the characteristics for i-sample ($i = 1,2,\ldots,n$) j-object ($j = 1,2,\ldots,m$).

The reviewing samplings of the soil characteristics have been converted into interval series. The visual analysis of histograms of the empirical distribution of the soil characteristics showed the closeness to the normal law. Based on the convergence estimate of the empirical distribution to the theoretical one (as per Pearson K. criterion χ^2) the obedience of the values of density, and angle of internal friction of soil (below the depression curve) to a normal distribution, the cohesion—to the lognormal distribution (Figure 3) has been established.

Insignificance in differences of the average values of the training and examination samples, determined by Student's t-test, has allowed to determine the generalized characteristics of soil for conditions of Kuzbass: for the angle of internal friction and density—average values of cohesion—modal (Table 2, column 2).

The presence of cohesion and density of soils located above the depression curve of heterogeneous elements has been found out by means of single-factor analysis of variance in samples. These elements and relevant objects have been revealed by multiple comparisons by Duncan test [Müller P., 1982]:

$$D_\phi = \frac{\left| \overline{x}_i - \overline{x}_j \right|}{\sqrt{\frac{S_{M\Gamma}}{2} \left(\frac{1}{n_i} + \frac{1}{n_j} \right)}}, \tag{2}$$

where $S_{M\Gamma}$ – variance within groups; $\overline{x}_i, \overline{x}$ – average values of the compared groups; n_i, n_j – sample sizes of the compared groups.

Based on the statistical processing of analyzed arrays the regional table of physical- and mechanical characteristics of cohesive particulate dispersive loamy soils for conditions of Kuzbass have been created (Table 2) [Guryev D. V., 2017].

In the process of research and statistical processing of experimental data on physical-and-mechanical properties of man-induced loamy soils the following have been figured out:

– the range of actual values of the characteristics is greater than the ones recommended by "The Code of Rules" SP 11-105-97 part III, table J.1 (SP 11-105-97, 2000) for cohesion – in 5.2 times; angle of internal friction – in 3.5 times; density – in 1.8 times;

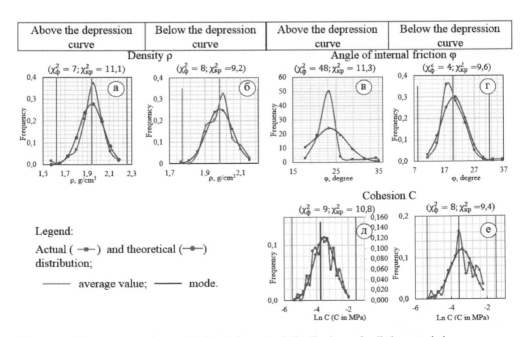

Figure 3.   Histograms of the empirical and theoretical distributions of soil characteristics.

202

Table 2. Regional database of physical-and-mechanical characteristics of man-induced loamy soils.

Index	Generalized data on Kuzbass	Region						The code of rules 11-105-97	
		Kemerovskiy	Leninsk-Kuznetskiy	Novokuznetskiy	Prokopyevskiy	Guryevskiy		Freshly formed	Packed
Soil located above the depression curve									
φ, degree	−[1] / 14−35	− / 14−32	− / 14−29	− / 18−33	− / 17−25	− / 15−35		15−20	18−26
C, kPa	24[2] / 5−140	21 / 5−90	27 / 5−140	28 / 13−78	33 / 24−41	27 / 10−50		30−50	39−65
ρ, g/sm³	1.95 / 1.61−2.23	1.93 / 1.76−2.11	1.95 / 1.62−2.11	1.98 / 1.61−2.14	1.90 / 1.83−2.99	1.99 / 1.63−2.19		1.50−1.80	1.95−2.30
Soil located below the depression curve									
φ, degree	20 / 11−32	20 / 14−32	18 / 16−24	21 / 18−26	20 / 18−26	22 / 19−23		−	−
C, kPa	25 / 5−117	26 / 8−83	24 / 10−57	29 / 5−83	23 / 5−117	28 / 12−67		−	−
ρ, g/sm³	1.98 / 1.77−2.14	1.97 / 1.77−2.08	1.95 / 1.89−2.99	1.98 / 1.84−2.14	2.12 / 1.85−2.13	1.99 / 1.93−2.04		−	−

[1]Characteristics of the angle of internal friction for the soils located above the depression the curve does not obey the normal law of distribution, so the average value of the indicator was not determined;

[2]There is a range in reciprocally, and the average value of soil characteristics—in the numerator.

– the differences in characteristics of soils located above and below the depression of the curve are within the error of the calculations (angle of internal friction $-1°$, cohesion – 1 kPa, density – 0.03 g/cm^3).

Therefore, at design stage of the earth fill structures, and in absence of engineering and geological surveys, one can be guided by generalized values of physical-and-mechanical characteristics of soils given in the Table 2.

The order of the survey data processing is presented in details in the Guidelines (Bakhaeva S.P., 2016).

REFERENCES

[1] Bakhaeva, S.P., Guriev, D.V., 2016. Analytical Prediction of Stability of Earth-Fill Dam. Proceedings of the 8th Russian-Chinese Symposium "Coal in the 21st Century: Mining, Processing, Safety": 188–192 (in China).

[2] Sivakumar, G.L., Mukesh, M.D., 2004. Effect of soil variability on reliability of soil slopes. Géotechnique, 5: 335–337 (in Great Britain).

[3] Cho, S.E., 1992. Effects of spatial variability of soil properties on slope stability. Engineering Geology, 92: 97–109 (in Netherlands).

[4] Huang, H.W., Xiao, L., Zhang, D.M., Zhang, J., 2017. Influence of spatial variability of soil Young's modulus on tunnel convergence in soft soils. Engineering Geology, 228: 357–370 (in Netherlands).

[5] Xiaolei, L., Maosheng, Z., Hong, Z., Yonggang, J., Chaoqi, Z., Hongxian, S., 2017. Physical and mechanical properties of loess discharged from the Yellow River into the Bohai Sea, China. Engineering Geology, 227; 4–11 (in Netherlands).

[6] Breysse, D., Niandou, H., Elachachi, S., Houy, L., 2005. A generic approach to soil–structure interaction considering the effects of soil heterogeneity. Géotechnique, 55 (in Great Britain).

[7] Akhlyustin, O.E. Regularities of variability of physical-and-mechanical properties of collapsible soils in Anapa district of Krasnodar region: abstract from the thesis for PhD scientific degree of (25.00.08)/Akhlyustin Oleg Evgenyevich; – Yekaterinburg, 2013. – 23 p.

[8] Guryev, D.V., 2015. Generalization of characteristics of dispersive soils of man-induced arrays on the example of Kuzbass. Bulletin KuzSTU, 3: 31–36 (in Russian).

[9] Guryev, D.V. Estimate of stability of the earth fill barrage considering spatial variability of the strength properties of man-induced loamy soils [Text]: abstract from the thesis for PhD scientific degree (25.00.16)/Guryev Dmitriy Vitalyevich; Publishing Center CIP KuzSTU. – Kemerovo, 2017 – 22 p.

[10] Edigenov, M.B., 2014. Variability of soil properties at the open-pit edges at Varvarinskoye Deposit, Kostanay region, Kazakhstan. News of the national Academy of Sciences of the Kyrgyz Republic, 2: 30–35 (in Kyrgyz).

[11] Generalization of physical-and-mechanical characteristics of the man-induced clayey soils: Guideline/Bakhaeva, S.P., Mikhaylova, T.V., Guryev, D.V. [and others]/KuzSTU, JSC "Kuzbassgiproshaht", "Geotechnical Engineering". – Kemerovo, 2016. – 45 p.

[12] Guidance for compilation of regional tables of normative and estimated values of soil properties/ PNIIIS Gosstroy of the USRR. – Moscow: Stroyizdat, 1981.

[13] SP 11-105-97. Engineering and geological surveys for construction. Part III: Regulations for operations in the regions with specific soils. – Moscow: PNIIIS Gostroy of Russia, 2000.

[14] SP 22.13330.2011. Foundations of buildings and structures – Moscow: Institute OJSC SIC "Construction", 2011.

[15] Chernyak, E.R., 2011. The future for regional tables of regulatory and estimated values of mechanical properties of soils. Engineering surveys, 9: 4–8.

Geomechanics and Geodynamics of Rock Masses – Litvinenko (Ed.)
© 2018 Taylor & Francis Group, London, ISBN 978-1-138-61645-5

Analysis of crack initiation and crack damage of metamorphic rocks with emphasis on acoustic emission measurements

Kirsten Bartmann & Michael Alber
Ruhr-University Bochum, Bochum, Germany

ABSTRACT: The process of compressive rock failure can be subdivided into 4 stages: Crack Closure (CC), Crack Initiation (CI), Crack Damage (CD) and peak strength (UCS). It is assumed that CI defines the lower limit of the long-term strength of rocks and is therefore an important stage in the complete failure process. Nevertheless, failure mechanisms and especially the determination of CI are not yet completely understood. This study aims to present various methods to detect CI and CD and to compare their results for a quartzite and a gneiss. For the latter, special focus was set on the foliated structure. Triaxial tests at minor confining pressures as well as uniaxial compression tests were conducted. In case of the triaxial tests one acoustic sensor was installed on the loading plate. In addition to passive recordings of acoustic emissions, we repeatedly transmitted acoustic pulses through the samples during uniaxial experiments. The arrival times and amplitudes of the recorded waves were evaluated and especially the latter leads to good results to determine CI. Orientated thin sections before and after a strength test were prepared to examine cracks and to identify the mineralogical composition by dint of a polarization and a scanning electron microscope.

Keywords: triaxial tests, crack initiation, crack damage, acoustic emission, metamorphic rocks, scanning electron microscope

1 INTRODUCTION

Engineering and geotechnical issues, for example tunnel constructions, often deal with compressive strength of rocks. For economic and safety reasons it is important to determine the rock strength accurately. The difficulty here is to define the in-situ rock spalling strength because it is significantly lower than the one detected in the laboratory. The formation of thin spall slabs near excavation boundaries in brittle rocks are referred to as spalling. It occurs under low confinement and might be violent (Diederichs et al. 2010). At higher confinement the stress envelope transitions to the CD defined envelope (Diederichs & Martin, 2010). Researchers like Martin and Christiansson (2009) focused on this issue and suggested that CI predicted from laboratory experiments coincides with the in-situ rock spalling strength. They ascertained that CI is 0.4 ± 0.1 of the unconfined compressive strength.

To date different methods have been suggested to determine CI and CD in laboratory experiments. The aim of this study is to compare a selection of these methods for metamorphic rocks. We put special emphasis on AE measurements and in particular on the arrival times and on the recorded waves of acoustic pulses. Moreover, we investigated the influence of confining pressure up to 2.5 MPa on the UCS, CI and CD.

2 ROCKS INVESTIGATED

We tested two different metamorphic rocks, a quartzite and a gneiss. To examine the influence of the foliation of the gneiss, the specimens were prepared in two ways: mica layers parallel and mica layers perpendicular to the axial loading. Hereafter, the abbreviation for the quartzite is Qz and for the gneiss CG.

Table 1 summarizes the basic geomechanical properties of the investigated rocks based on uniaxial and triaxial tests. For safety reasons only uniaxial compressive strength tests of Qz were performed. In total 25 triaxial tests on CG samples including 5 saturated specimens were conducted with varying confining pressures between 0.1 and 2.5 MPa. Furthermore, 8 uniaxial compression tests of CG and 14 of Qz were performed including 4 and 6 saturated samples, respectively.

The quartzite consists mainly of quartz as its name suggest. The other components are feldspar, a few clay minerals and muscovite (Fig. 1). Quartz often shows non-uniform extinction and its average grain size is 0.1–0.2 mm. Alteration processes as well as the Carlsbad and pericline twin laws help to identify feldspar (MacKenzie & Guilford, 1981). Its grain size is comparable with the one of quartz. Clay minerals cannot be further investigated via a polarization microscope. Their grain sizes vary greatly. Probably, the grade of metamorphosis is low because of the occurrence of clay minerals. Within higher grades of metamorphosis, these minerals change into mica minerals (Maresch et al., 2014). Table 2 gives an overview of the mineral's content.

Feldspar, biotite and quartz are the main components of CG. Furthermore, we observed pyroxene and amphibole (Fig. 2). The relief helps to distinguish quartz from feldspar. Their average grain size is 0.6 mm, respectively. Rarely, quartz shows non-uniform extinction and feldspar sometimes indicates pericline twinning. Biotite has a mottled texture at extinction, an average grain size of 0.5 mm and a brown intrinsic color (Barker, 2014). The mineral's content is given in Table 2.

Table 1.   Geotechnical properties of Qz and CG.

Rock type	ρ_{geom}· [g/cm^3]	v_p [km/s]	UCS [MPa]	Young's Modulus [GPa]	v [–]	σ_1–$\sigma_{3\ at\ failure}$ [MPa]
Qz	2.56 ± 0.005	5.20 ± 0.12	248.5 ± 27.6	30.1 ± 2.2	0.21 ± 0.07	–
CG ||	2.71 ± 0.01	3.86 ± 0.59	113.4 ± 19.9	19.0 ± 1.4	0.14 ± 0.04	149.0 ± 23.5
CG ⊥	2.74 ± 0.01	2.02 ± 0.14	164.4 ± 35.9	19.4 ± 1.3	0.08 ± 0.02	182.9 ± 35.2

Total sample number n = 47, mean values ± SD.

Figure 1.   Overview of Qz's mineralogical composition; PPL (left) and XPL (right).

Table 2.   Mineral modal content of Qz and CG.

Rock type	Feldspar [Vol.-%]	Quartz [Vol.-%]	Clay mineral [Vol.-%]	Muscovite [Vol.-%]	Biotite [Vol.-%]	Pyroxene [Vol.-%]	Amphibole [Vol.-%]
Qz	6	90	2	2	–	–	–
CG	53	27	–	–	16	3.7	0.3

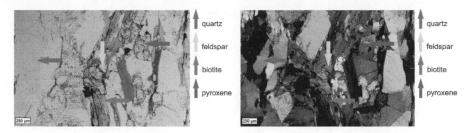

Figure 2.   Overview of CG's mineralogical composition; PPL (left) and XPL (right).

Table 3.   Methods to determine CI and CD.

Method	Reference	CI	CD
crack volumetric strain	Martin & Chandler, 1994	√	
volumetric strain	Cai, 2010	√	√
Instantaneous Poisson's Ratio	Diederichs et al., 2010	√	√
Average Young's Modulus	Eberhardt et al., 1998	√	√
AE	Slatalla & Alber	√	√

3   MATERIALS AND METHODS

The triaxial and uniaxial compression experiments were conducted servo-controlled with a MTS Teststar IIm using a stiff loading frame of capacity 4000 kN. The frequency of data acquisition was 10 Hz. 3 Linear Variable Differential Transformers (LVDT) with a resolution of 10^{-7} m recorded the displacements and a axial strain rate of 10^{-5} mm/mm/s was chosen to control the experiments. 2 LVDTs registered axial displacements and at mid of the sample the third LVDT measured circumferential displacements with the aid of a chain. According to ISRM the specimen's length to diameter ratio is approximately 2:1 and in this case 80 mm to 40 mm. After drilling and sawing cylindrical cores, the end planes were polished to insure an equally load on the whole sample surface.

The aluminum triaxial cell consists of a base plate welded with the circular centerpiece and a cover plate, where the wiring of the internal LVDT runs through. The rock specimen was surrounded by protective covers to ensure that no oil reached the specimen.

In addition to common uniaxial compression test, uniaxial compressive strength tests with AE monitoring were conducted. Therefore, 4 AE sensors were installed around the specimen in 90 degree steps. The experiments were stopped before the specimen's failure to protect the sensitive sensors. During the test one AE-sensor after another pulsed and the other 3 sensors detected the arrival time of the pulse signal. This was repeated every 10 MPa stress. Monitoring acoustic wave velocity changes is an indirect method to detect CC, CI and CD. We also analyzed the change of the amplitudes of the transmitted signals.

AE monitoring also provides information on the onset of CI and CD. The amplitudes per second of the recorded acoustic emissions show an increase with the start of CI and CD, respectively (Slatalla & Alber, 2008). Table 3 summarizes the methods for determining CI and CD including the according references.

4   RESULTS

Figure 3 depicts the results of uniaxial and triaxial tests, constituting confining pressure on the x-axis and differential stress on the y axis. The blue symbols represent saturated samples. As mentioned before, we conducted only uniaxial compression tests of Qz. The squares represent CG samples with mica layers parallel and the circles with mica layers perpendicular to the maximum principal stress.

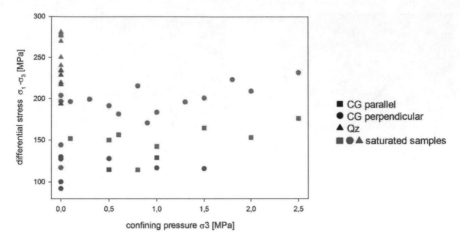

Figure 3.   Differential stress as a function of confining pressure for Qz and CG.

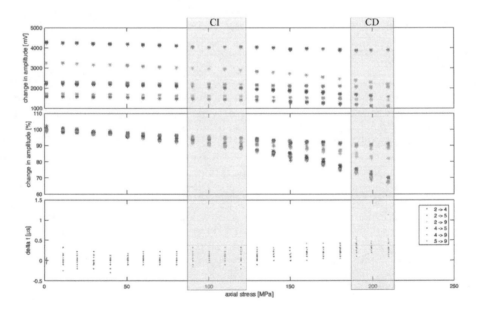

Figure 4.   Change in amplitude and arrival times of AE measurements of a Qz sample.

The diagrams below show exemplary the change of the amplitudes of the transmitted signals during the uniaxial test of Qz (Fig. 4). The upper subplot depicts the actual change in amplitude in [mV]. The second subplot presents the change in amplitude in [%] normalized to the first detected amplitude of the unloaded specimen. The last subplot shows the change of the arrival times in [μs] of the pulse signals reaching the recording sensors standardized to the first arrival time of the unloaded sample. The various colors indicate the several sensor pairs, whereby the number on the left side donates the pulsing and the number on the right side the recording sensor.

To visualize the induced cracks during a compression test a scanning electron microscope was used. Therefore, thin sections of an unstressed and of loaded samples up to CI or CD, respectively, were prepared. The photos below show an example of an unloaded sample of CG (Fig. 5, left) and of a specimen loaded up to CD, specifically $\sigma_1 = 160$ MPa and $\sigma_3 = 0.6$ MPa (Fig. 5, right). The stresses were kept for 15 minutes and afterwards the specimen

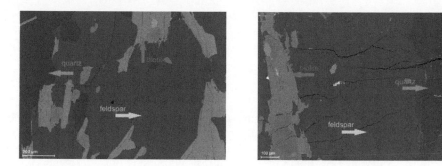

Figure 5. Scanning electron microscope picture of an unloaded sample (left) and a loaded sample of CG (right).

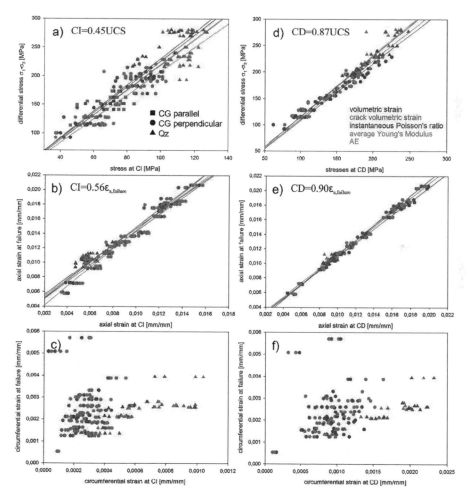

Figure 6. Stresses as well as axial and circumferential strains at CI (a-c) and CD (d-f) for different methods.

was removed. The minerals appear in different grey tones as the interaction of their atoms with the electron beam of the scanning electron microscope differs from each other.

The following diagrams summarize the uniaxial and triaxial test results of CG and Qz including saturated and dry specimens with regard to CI and CD (Fig. 6). Therefore, the determined stresses or strains at CI/CD are shown on the abscissae and the y-axis is either differential stress or the strain at failure. All strains are taken positive for convenience. The several colors delineate the different methods and the symbols typify the rocks in each diagram.

5 DISCUSSION AND CONCLUSION

As we might suspected the data set of Figure 3 shows that CG specimens with mica layers perpendicular to the maximum principal stress have higher compressive strengths than CG samples with parallel layers. Moreover, we observed a dependence of the compressive strength and the confining pressure as the strength rises with increasing confinement. From 1 MPa confinement this dependency is more obvious as the strengths fluctuate more with less than 1 MPa confining pressure.

Figure 4 illustrates that the amplitudes of the transmitted signals change during the test as the closing of cracks cause an increase of the amplitude and scattering at initiated cracks attenuates the wave and thus decreases the amplitude. The same applies to the arrival times of the pulse signal as the forming of cracks lead to slow the waves. This phenomenon is quite clear for a homogenous rock, in this case a quartzite. In case of an anisotropic rock (CG) this observation is more indistinct. Probably, the mica layers of CG affect the wave dispersion and their amplitudes.

The scanning electron microscope pictures present the difference of an unloaded and a loaded sample with regard to the cracks (Fig. 5). Even the unloaded CG sample shows cracks, especially at grain boundaries, although the crack density and opening of the loaded specimen is higher.

To plot stresses and strains at failure against the ones at CI and CD shows that the different methods lead to different results (Fig. 6). The difference for the axial stresses and strains at CD is relatively low. Regarding the lateral strains at CI and CD their data points are arranged in a point cloud.

To conclude, this study presents different ways to determine CI and CD. AE measurements and scanning electron microscope observations enhance the understanding of cracks forming during compression.

REFERENCES

Barker, A.J. (2014): *A key for identification of rock-forming minerals in thin-section.* Taylor & Francis Group, London, UK.

Cai, M. (2010): *Practical estimates of tensile strength and Hoek-Brown strength parameter mi of brittle rocks.* Rock Mech. Rock Eng. 43, pp. 167–184.

Diederichs, M.S., Carter, T., Martin, D. (2010): *Practical rock spall prediction in tunnels.* in: Proceedings of ITA World tunnel congress, Vancouver, pp. 1–8.

Diederichs, M.S., Martin, C.D. (2010): *Measurement of spalling parameters from laboratory testing.* In: Taylor & Francis Group (Eds.) Rock mechanics in civil and environmental engineering, London, pp. 323–326.

Eberhardt, E., Stead, D., Stimpson, B., Read, R.S. (1998): *Identifying crack initiation and propagation thresholds in brittle rock.* Can. Geotech. J. 35, pp. 222–233.

MacKenzie, W.S., Guilford, C. (1981): *Atlas gesteinsbildender Minerale.* Ferdinand Enke Verlag, Stuttgart.

Maresch, W., Schertl, H.-P., Medenbach, O. (2014): *Gesteine: Systematik, Bestimmung, Entstehung.* E. Schweizerbart'sche Verlagsbuchhandlung, Stuttgart.

Martin, C.D., Chandler, N.A. (1994): *The progressive fracture of Lac du Bonnet granite.* Int. J. Rock Mech. Min. Sci. & Geomech. Abstr. 31(6), pp. 643–659.

Martin, C.D., Christiansson, R. (2009): *Estimating the potential for spalling around a deep nuclear waste repository in crystalline rock.* Int. J Rock Mech. Min. Sci. 46, pp. 219–228.

Slatalla, N., Alber, M. (2008): *Characteristic acoustic emission sequences of rock samples under uniaxial loading.* 27th Int. Conference on Ground Control in Mining, pp. 107–112.

Geomechanics and Geodynamics of Rock Masses – Litvinenko (Ed.)
© *2018 Taylor & Francis Group, London, ISBN 978-1-138-61645-5*

Comparison between conventional and multi-sensor geotechnical core logging methods

Mahadi Bhuiyan & Kamran Esmaieli
Lassonde Institute of Mining, University of Toronto, Toronto, Canada

ABSTRACT: Collecting high-quality geotechnical data is essential for increasing efficiency in mine planning and design. Core logging is a fundamental practice used to obtain geological and geotechnical data. Conventional manual logging can be subjective and inconsistent and have low auditability. Moreover, manual logging does not exploit collection of other potential information at the time of logging. Multi-sensor digital logging can be beneficial for reducing manual logging inconsistency and providing greater resolution in physical rock properties. This study presents a comparison between manual and multi-sensor geotechnical core logging for samples obtained from an open-pit mine. Oriented core samples were logged using a multi-sensor core logging facility and results were compared to manually logged data. Discrepancy in discontinuity orientation and good agreement for RQD are found between the two methods. Multi-sensor digital data show good correlation between lithology, and ultrasound velocity and hardness. Effect of weathering on discontinuity wall strength is quantified using hardness test on discontinuity surface. In addition, multi-element data from pXRF measurements are shown to be a proxy for alteration mineralogy at discontinuity surface.

1 INTRODUCTION

The quality and quantity of field data can significantly influence the reliability of a geomechanical mine design. One of the fundamental methods that is commonly used in the mining industry to obtain geological and geotechnical data is core logging. The conventional manual core logging method is very subjective, time-consuming and inconsistent with low auditability. Over the last decade, some efforts have been made to improve geotechnical core logging and analysis using image analysis techniques (Olson et al. 2013, Orpen 2014) and multi-sensor core logging facilities, e.g., Geotek's MSCL. None of the developed imaging methods can take the 360° circumferences of the core samples, which can be essential for accurate estimation of joint orientation in core logging process. In addition, most of the developed multi-sensor core logging laboratories are focused on geochemical and mineralogical studies. These labs are very large and it is generally difficult to take them to the field.

Using a portable multi-sensor core logging facility can modernize field data collection for geomechanical rock mass characterization. It can decrease the subjectivity and inconsistency in the geotechnical core logging through measurement rather than assessment of rock properties. It makes the logging process much faster and efficient through better management of the collected data, reducing the potential errors in data transfer (digitization) and increasing the precision of the collected data. The method allows for maximizing the information that can be derived from rock cores as our limited sampling sources. This paper compares manual vs. multi-sensor geotechnical core logging data for geomechanical mine design.

2 METHODS

2.1 Core sample

For this study, ~38 m of oriented core sample was obtained from six different geotechnical borehole intervals drilled in an open-pit mine. Four lithologies are present in the core. Grano-diorite intrusives (GDI) consists of either: meta-intrusive diorite with a biotite foliation, or mafic metavolcanic basalt. Felsite (FVC) is a micro-crystalline quartz-feldspathic rock that preserves a strong phyllosilicate foliation in some sub-units. Banded iron magnetite (BIM) and metamorphosed greywacke and siltstone (SVC) are locally interbedded in the sample. The core is generally intact, i.e., un-weathered, and has high RQD and recovery.

2.2 Field and lab manual core logging

Field logging was conducted by qualified personnel in two different drilling programs. Original paper logs were not available, so the most primitive data in a digital database was obtained. The geomechanical rock mass characterization parameters logged are: RQD; discontinuity orientation and spacing; discontinuity roughness index (J_r) and alteration index (J_a); and intact rock strength index (ISRM, 1981).

The field data is limited, inconsistent, and has relatively low resolution. The J_r and J_a values are not available for ~34.5 m of the core. An erroneous J_a value of 1.15, and a single J_r value of 4 (discontinuous joints) is recorded for the remaining interval. A total of seven discontinuities were recorded for all sample intervals. The RQD measurements are reported at intervals greater than 1 m. The strength index is qualitative and does not report variability in strength at intervals less than 3 m. Given the issues with field data and to better compare manual logging with multi-sensor logging, the core samples were also logged in the lab. 74 discontinuities were identified over the entire core sample, a marked increase from the seven recorded during original logging, and characterized for joint orientation and surface condition. The surfaces of 28 discontinuities display staining or alteration. RQD was calculated per meter, and averaged over each sample's interval.

2.3 Multi-sensor core logging

The portable multi-sensor core logging facility at the University of Toronto was used for this study. Fig. 1 shows the main components of the facility. The stages of multi-sensor logging are listed:

1. A 360° core image scan was taken at a resolution of 10 pixel/mm. The images were processed in a digital logging software to create 3D virtual core. Depth and orientation information were imported from borehole logs and assigned to the virtual core.
2. Discontinuities were digitally logged. Orientation is assigned by approximating a sinusoidal plane to the discontinuity in an "unrolled" 2D core image (see Fig. 3 left). The plane's geometry is used to calculate its 3D orientation. The RQD is calculated by the length of

Figure 1.   Multi-sensor logging components. Left to right: Core scanner; Equotip hardness tester; UPV testing unit; and pXRF scanner.

core images and occurrence of discontinuities over a specified interval (one meter). Core loss is accounted for by gaps between images.

3. Hardness tests were performed every cm on along the reference line of the core, and orthogonal to discontinuity surfaces using an Equotip LEEB hardness testing unit. A mean of ten tests was taken as representative per measurement location (ASTM International, 2017).

4. An ultrasound pulse velocity (UPV) unit was used for measurements of P and S wave velocities along the core samples. UPVs were recorded at one cm and two cm spacing for P and S wave velocities, respectively. Measurements were taken both in-plane (non-rotated core) and normal (rotated core) to the reference line to study anisotropy (Barton, 2007; Aydin, 2014).

5. A portable X-Ray fluorescent analyzer (pXRF) was used for measurements on altered discontinuity surfaces, and at two cm spacing intervals on the core. The pXRF records the multi-element signature of the test surface, and is capable to detecting light elements up to magnesium. Four filters were used which cover a range of relevant rock-forming elements, e.g., Mg, Fe, K, Al.

3   RESULTS AND DISCUSSION

3.1   *Comparison of logging methods*

3.1.1   *Discontinuity orientation*
Table 1 presents a summary of differences between digital and manual discontinuity orientation measurements for all intervals.

Systematic bias in digital logging can arise from the approximation of 2D sinusoidal curves to true discontinuity geometry and the adequacy of user's selection of these planes. However, the removal of subjectivity and human error in measurement of alpha and beta angle means that precision is greater and random bias is reduced in digital logging. If digital measurements are considered as a benchmark, then a measure of variability in manual measurements is possible. The mean difference in both dip and dip direction is greater than 10°, which is significant. For example, manual measurements of a joint set dipping at 45° would have a mean error of +/− 30% (100% *13.5/45). Variability in the dip direction can appreciably influence kinematic analyses. The high maximum differences, 56° dip and 119° dip direction, occur in one interval, and may be attributed to two human errors: incorrect marking of the reference line and subjectivity in measurement of shallow-dipping discontinuities. Absolute difference in dip and dip direction is reduced considerably, after modifying the reference line in this interval from 180° to 0°. This suggests that the reference line was either not recorded at 180° or incorrectly drawn. Precision of manual measurement is more likely to reduce for the shallower discontinuities, since the dip vector becomes difficult to discern. The lab orientations of three shallow discontinuities in this interval are recorded as 10°/085°, 15°/303°, 29°/215°, whereas the digital measurements are 25°/203°, 26°/262°, and 28°/213°, respectively. The digital orientations suggest that these are part of a discontinuity set, but the low precision of lab orientations make this judgement improbable. The results suggest that the large error is due to the erroneous manual measurement.

Table 1.   Absolute difference in discontinuity orientations between digital and manual logging.

Orientation (°)	Digital vs. Lab (n = 74)			Digital vs. Field (n = 7)		
	Mean	Min	Max	Mean	Min	Max
Dip	13	0	56	14	2	23
Dip Direction	19	0	119	14	0	21

3.1.2 RQD

The mean and maximum absolute difference in RQD are 0.8%, and 1.7%, respectively, between manual and digital measurements. The result demonstrates that quantification of RQD using core image analysis is accurate. Digital logging is advantageous because it provides greater spatial resolution than the field data, which reports RQD at intervals greater than one meter. Digital logging can also determine RQD over variable intervals, which is useful when reporting for core runs.

3.2 Multi-sensor data analysis

3.2.1 Lithological variation using hardness and UPV

Multi-sensor log data allows quantification of physical variability within and amongst lithologies. Fig. 2 presents boxplots of hardness and rotated P wave velocity (normal to the core reference line) for different rock units. An appreciable variability in hardness amongst units is evident. In addition, the greatest variability within a unit occur for foliated units, i.e., BIM, foliated FVC, GDI, and SVC/BIM. Hardness of foliated FVC shows a greater spread than the non-foliated FVC, which is due to compositional banding. The largest spread occurs for the strongly foliated BIM. Hardness may be a better representation of intact rock strength (Verwaal and Mulder, 1993; Aoki and Matsukura, 2008) compared to the strength index test (ISRM, 1981) since it is a quantitative measure at a significantly greater spatial resolution. For example, in field logging, both GDI and FVC are assigned strength indices of R5 and R6, but the mean and range of GDI hardness is generally lower than both FVC sub-units. The UPV summary also highlight a variation within and amongst lithologies (Fig. 2 right). Variability is greatest for BIM, GDI, and foliated FVC, which suggests persistent foliation as a source of anisotropy. A substantial difference in mean and spread is observed between the FVC subunits. Characteristic ranges of hardness and UPV may be established with further statistical analyses, and can be used for discriminating between lithological zones for geomechanical design (Yasar and Erdogan, 2004).

A multi-sensor log for an interbedded SVC-BIM interval is presented in Fig. 3. Change in the hardness and UPV profiles can be correlated to the lithological transition from SVC to BIM. A relative decrease and increase, in hardness and P wave velocity, respectively, occur at the BIM unit. The SVC show a relatively constant hardness and UPV profile. A variability in hardness is observed for the BIM, which is due to the foliated layers of harder silicates and softer magnetite-grunerite.

3.2.2 Quantification of discontinuity wall strength using hardness test

Hardness tests on the discontinuity surface can quantify discontinuity wall strength (ISRM, 1978) and offer an improved understanding of joint alteration indices, i.e. J_a, and J_{con}, in rock

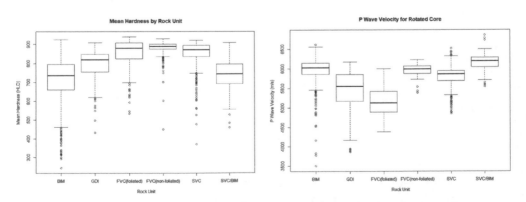

Figure 2.   Boxplot of hardness (left) and rotated P wave velocity (right). SVC/BIM represents interbedded zones.

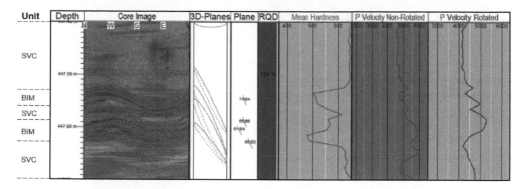

Figure 3. Multi-sensor log of interbedded SVC and BIM. SVC is the grey unit above and below the foliated green BIM.

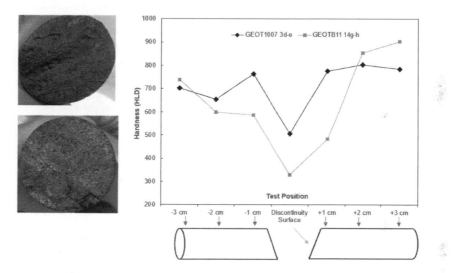

Figure 4. Left: Discontinuity surface 3d-e (top) and 14g-h (bottom). Right: Variation of hardness near and at discontinuity surfaces. Schematic of core shows hardness test position (not to scale).

mass classification systems. A comparison of hardness of two discontinuity surfaces, 3de and 14gh, is presented (Fig. 4, left). Both surfaces are generally planar, slightly rough, and weathered due to iron-oxide alteration. Surface 14gh has a greater degree of alteration. In manual logging, the joint surface characteristics correspond to a J_a of 1 and 2 for 3de and 14gh, respectively, and a J_r of 1.5 for the two joint surfaces.

Hardness profiles of the core near and at the discontinuity surface (Fig. 4, right) show that surface 14gh has a greater difference in hardness from the adjacent intact rock than 3de (109 and 135 HLD from average hardness above and below surface, respectively). This relative variation indicates that a greater degree of surface weathering can cause a reduction in hardness, which could imply a degradation in wall strength. The greater difference also quantifies the higher J_a value assigned to the more weathered 14gh surface.

3.2.3 *Multi-element signature as a proxy for discontinuity surface alteration*
Classification of alteration mineralogy on discontinuity surfaces is important in rock mass characterization. Wall strength of altered discontinuities is generally dictated by the type and amount of alteration minerals. Subjectivity in mineralogical classification may occur during manual logging, which would cause mischaracterization of the rock mass and lead to design ramifications. Thus, a quantitative method of characterizing discontinuity surface

215

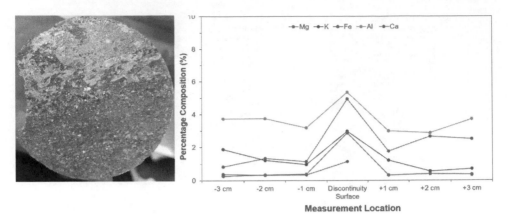

Figure 5. Left: Mica (muscovite) on discontinuity surface 13cd. Right: Plot of multi-element variation near and at 13cd.

alteration is desirable. Multi-element signatures can be used as a proxy for mineralogy given the geochemistry (Rowe et al., 2012). Fig. 5 highlights an example of the use of pXRF multi-element data as a proxy for alteration geochemistry. Discontinuity surface 13cd contains visible alteration of muscovite, a potassium-aluminum phyllosilicate. The pXRF measurement on the surface yields a greater elemental percentage of potassium and aluminum than measurements on the core above and below the discontinuity. If the multi-element signature of the intact rock is taken as a baseline, then this suggests that the increase in K and Al is due to the presence of muscovite at the surface. This information could be further analyzed using bulk mineralogy libraries and reconstruction of whole-rock geochemistry.

4 CONCLUSIONS

The study demonstrates that subjectivity and inconsistency of manual core logging can produce results that lead to considerable uncertainty in geomechanical design. The number of characterized discontinuities vary significantly between field and lab logging, and highlights the subjectivity of manual logging measurements. Bias and reduced precision in manual logging causes a notable discrepancy in discontinuity orientations between lab and digital measurements. This discrepancy is meaningful for design applications such as kinematic analyses in slope stability. Multi-sensor logging has considerable benefits for geomechanical design purposes. RQD determination with digital logging are in close agreement with manual measurements, which demonstrates the accuracy of core image analysis. Multi-sensor log data, i.e., hardness and UPV, can be used to quantify variability within and amongst lithological units. UPV profiles can be used to study anisotropy, which has important implications for rock mass characterization. This information is captured at a high spatial resolution and can be used to model small-scale variation in physical properties. Multi-sensor logging can also quantify subjectivity in qualitative rock mass classification parameters, such as intact rock strength index. Discontinuity wall strength is quantified with hardness testing, and the surface degradation is captured by the variation of surface hardness from the hardness of intact rock core. pXRF measurements provide multi-element data that could be as a proxy for alteration mineralogy of discontinuity surface.

ACKNOWLEDGEMENTS

This study was enabled by the support of Kinross Gold and the National Science and Engineering Research Council of Canada. The authors are grateful to David Eden, Stefano

Girardo and Christopher Bai for their integral roles. Lab photos were taken by Stefano Girardo.

REFERENCES

Aoki, H., & Matsukura, Y. (2008). Estimating the unconfined compressive strength of intact rocks from Equotip hardness. *Bull. Eng. Geol. and Environ., 67*(1), 23–29.

ASTM International. (2017). *ASTM A956/A956M-17 Standard test method for LEEB hardness testing of steel products.* doi:https://doi.org/10.1520/A0956_A0956M-17.

Aydin, A. (2014). Upgraded ISRM suggested method for determining sound velocity by ultrasonic pulse transmission technique. *Rock Mech. and Rock. Eng., 47*, 255–259.

Barton, N. (2007). *Rock quality, seismic velocity, attenuation, and anisotropy.* London; New York: Taylor & Francis.

ISRM. (1978). Suggested methods for the quantitative description of discontinuities in rock masses. *Int. J. Rock Mech. Min. Sci. & Geomech. Abstr., 15*, 319–368.

ISRM. (1981). *Rock characterization, testing & monitoring: ISRM suggested methods.* Pergamon Press.

Olson, L., Samson, C., & McKinnon, SD., (2013). The 2D and 3D imaging of core for fracture mapping, In Proc. 46th American Rock Mech. Symp., San Francisco, CA.

Orpen, J.L. (2014). Best practice and new technology in core drilling, logging and fracture analysis, 1st International Conference on Discrete Fracture Network Engineering, October 19–22, 2014, Vancouver, Canada.

Rowe, H., Hughes, N., & Robinson, K. (2012). The quantification and application of handheld energy-dispersive x-ray fluorescence in mudrock chemostratigraphy and geochemistry. *Chemical Geology, 122–131*, 324–325.

Verwaal, W., & Mulder, A. (1993). Estimating rock strength with the equotip hardness tester. Technical note. *Int. J. Rock Mech. Min. Sci. & Geomech. Abstr., 30*(6), 659 662.

Yasar, E., & Erdogan, Y. (2004). Correlating sound velocity with the density, compressive strength, and Young's modulus of carbonate rocks. Technical note. *Int. J. Rock Mech. & Min. Sci., 41*, 871–875.

Geomechanics and Geodynamics of Rock Masses – Litvinenko (Ed.)
© *2018 Taylor & Francis Group, London, ISBN 978-1-138-61645-5*

Graphical evaluation of 3D rock surface roughness: Its demonstration through direct shear strength tests on Bátaapáti Granites and Mont Terri Opalinus Claystones

I. Buocz
Project Partners Ltd—Consulting Engineers, Lausanne, Switzerland
Department of Engineering Geology and Geotechnics, Budapest University of Technology and
Economics, Budapest, Hungary

N. Rozgonyi-Boissinot & Á. Török
Department of Engineering Geology and Geotechnics, Budapest University of Technology and
Economics, Budapest, Hungary

ABSTRACT: The shear strength of rocks along discontinuities is one of the key parameters for the determination of rock slope stability, the stability of rocks during underground space development and tunneling. Its value is influenced by numerous factors including surface roughness, which is one of the most widely investigated discontinuity property. The present paper introduces a simple graphical methodology for the classification of the surface roughness of rocks, based on the example of two different rock types, Bátaapáti Granites (Hungary), and Mont Terri Opalinus Claystones (Switzerland). The 3D surface of 24 rock samples was digitized using a photogrammetric surface detection method with the help of the ShapeMetrix3D software. The plane of each rock surface was defined by fitting a linear regression plane to the surface data. The distance between the data points of the surface roughness model and the regression plane was measured, and cumulative frequency diagrams of the measured distance values were constructed. This procedure allowed to define three surface roughness categories. The methodology proposed represents a promising new approach to surface roughness quantification, which could improve shear strength estimation.

1 INTRODUCTION

Shear strength of rocks is one of the key input parameters for stability analyses of rock masses. The design of appropriate supporting systems (type and strength) used to ensure the ideal degree of safety for people interacting with any engineered rock surface depends on the results of these analyses. However, the value of the shear strength is influenced by several factors, such as the mechanical properties of the intact rock and the discontinuities, as well as the laboratory testing methods and testing machines (Barton, 1973, 2013; Grasselli 2001; Buocz, 2016, 2017a; Dzugala et al., 2017). Therefore, the exact calculation of this parameter is very challenging. Among others, surface roughness is one of the most widely investigated discontinuity property with significant influence on the direct shear strength of rocks. Forty years ago, Barton presented for the first time 10 typical 2D surface roughness profiles, which defined as many value intervals for the Joint Roughness Coefficient (JRC). Once implemented into his rock shear strength model, these values helped to provide an appropriate estimate of the shear strength (Barton and Choubey, 1977). With the fast development of technology and the increasing precision of methodologies for surface detection, i.e., laser scanning or photogrammetry (Gaich et al. 2006), the 3D analysis of surface roughness gained again a central attention (Ge et al. 2015). Different theories were elaborated for the quantification of surface roughness and the determination of values for JRC in three dimensions (Zhao, 1997; Grasselli 2001, Bae et al., 2011). However, due to its simplicity, the well-accepted 2D surface roughness determination still remains in use in practice.

In this article, two types of rocks are investigated, i.e., Bátaapáti Granites and Mont Terri Opalinus Claystones. Both types are characterised by natural discontinuity surfaces and were studied by performing direct shear strength tests. Prior to the experiments, the specimen surfaces were digitized by a 3D photogrammetric method, and their surface roughness was analysed according to a newly developed graphical technique. The results obtained from the presented methodology were then compared to the results from the shear strength tests.

2 ROCK MATERIALS, 3D SURFACE DETECTION, DIRECT SHEAR STRENGTH TEST

Bátaapáti Granites (Hungary) and Mont Terri Opalinus Claystones (Switzerland) have very different rock mechanical properties. Bátaapáti Granites are hard crystalline rocks, with a mean compressive strength of 117 MPa (Kovács et al., 2012; Barsi et al. 2012), whereas Mont Terri Opalinus Claystones are soft sedimentary rocks with bedding planes and a mean compressive strength of 10.5 MPa parallel to the bedding and 25.6 MPa normal to the bedding (Nussbaum and Bossart 2008). However, these rock types share the peculiar feature of being capable of storing radioactive waste; both lithotypes have been studied in detail, since the Hungarian granite is a host rock for low and medium level radioactive waste (Balla et al. 2006, Buocz et al. 2014a, Buocz 2016), while Opalinus Clay is a potential candidate for hosting radioactive waste in Switzerland (Nussbaum and Bossart 2008).

Samples of 50×50 mm were cut from the rock materials (6 sample pairs of Bátaapáti Granites and 18 sample pairs of Mont Terri Opalinus Claystones), each of them containing a natural open shear surface. Prior to the shear strength tests, the rock samples were fixed into the sample holder boxes, i.e. the Mont Terri Opalinus Claystone samples were embedded in high strength mortar, the Bátaapáti Granites in high strength concrete. The sample surfaces were digitized by means of a photogrammetric method (ShapeMetriX3D). For each pair of rock samples, relying on the hypothesis that the joints were fresh and perfectly matching, only one half of the pair was analysed by surface detection (Buocz 2016).

One part of the Bátaapáti Granites was sheared under 10 MPa normal stress, the other part under 5 MPa normal stress. The Mont Terri Opalinus Claystones were sheared under 1 MPa normal stress. The two types of rocks were tested by two different shear machines: one gave precise results up to 2 MPa, the other for higher normal stresses. Shear tests on both shear machines were carried out with 0.8 mm/min shear displacement velocity. From the 24 pairs of specimens, only 14 produced a peak shear strength value, 5 Bátaapáti Granites and 9 Mont Terri Opalinus Claystones. In what follows, the evaluation, was based on these shear strength values.

3 3D SURFACE QUANTIFICATION AND GRAPHICAL REPRESENTATION

As a result of the photogrammetric surface detection, a point cloud of ~30'000–60'000 points was generated for each surface and a linear regression plane was fitted to them by using the R code (R_Core_Team, 2011). For the quantification of the surface roughness, the distances between the points belonging to the surface and the regression plane were calculated. In order to obtain a reliable set of data, prior to the distance calculation, a raster of 1×1 mm was placed over the regression plane and, for each cell, one single surface point was calculated from the average of the points belonging to the corresponding cell. Frequency diagrams were calculated from the statistical frequency of these distance values, with class intervals of 0.1 mm. In order not to have results influenced by possible outliers, values below the 5th percentile and above the 95th percentile of the frequency distributions were neglected from the following calculations (Fig. 1).

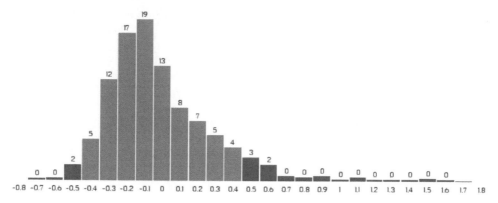

Figure 1.   Frequency diagram of the distance between regression plane and sample surface [mm]. In red: values under the 5th and above the 95th percentiles (Buocz, 2016).

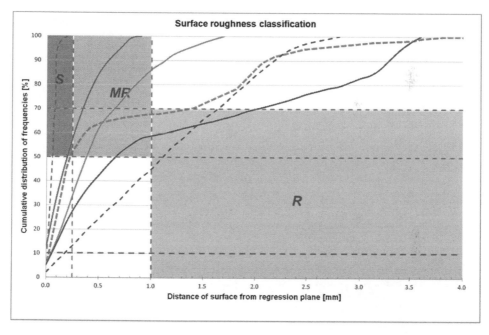

Figure 2.   Sample classification into smooth (red line), moderately rough (green line) and rough (blue line) sample surfaces, based on the computed average distances of the surface points from the regression plane and the cumulative distribution of the frequencies of these distances (Buocz, 2016).

Afterwards, from the frequency diagrams, cumulative frequency diagrams were calculated, in which the values of the frequencies were summed up regardless whether the corresponding point representing the surface was found to be located below or above the regression plane. The cumulative frequencies were then represented on a diagram, as a function of the distance of the sample surface from the regression plane. An empirical method was developed to graphically classify a rock surface in one of the following categories: samples with i) smooth surface, ii) moderately rough surface, iii) rough surface (Buocz 2016). The diagram contains three predefined areas. The paths of the cumulative frequency curves passing through these areas define which of the three above mentioned surface roughness categories the investigated samples belong to (Fig. 2). The red area, marked by "S", is the most decisive for the smooth surfaces; it corresponds to cumulative frequency values greater than the median, but with a distance smaller than 0.25 mm from the regression plane. The green area marked by "MR" is the most decisive for the moderately rough surfaces; it corresponds to cumulative frequency values greater than the median, just like the area "S", but with a distance of between 0.25–1.00 mm from the regression plane. The third area is denoted by "R" and is the most decisive for the rough surfaces; it corresponds to cumulative frequency values lower than 70%, with a minimum of 1.00 mm distance from the regression plane. Taking into consideration the three predefined areas and the path of the curves defining the surface roughness, a sample is considered to have:

- *Smooth surface*, if the curve passes through the "S" area only, or it falls both in "S" and "MR" at the same time. For these samples at least 10% of the sample surface points should lie directly on the linear regression plane;
- *Moderately rough surface*, if the curve passes through the "MR" area only, or it falls in the area of "S", "MR" and "R" at the same time.
- *Rough surface*, if the curve passes uniquely through the area "R", or "MR" and "R" at the same time.

4 EVALUATION OF DATA AND DISCUSSION

The cumulative frequency curves were represented in a diagram for the 14 rock samples, which produced a peak shear strength value. Based on the methodology described above, 4 Mont Terri Opalinus Claystone samples had a smooth surface and 5a moderately rough surface. Among the Bátaapáti Granites, 3 samples had a moderately rough surface, and 2a rough surface (Fig. 3).

The peak shear strength values were calculated partially based on the suggestions of the International Society for Rock Mechanics (ISRM, 1974, 2015), and partially by taking into consideration the effect of downslope/upslope shearing due to the deviation of the plane of

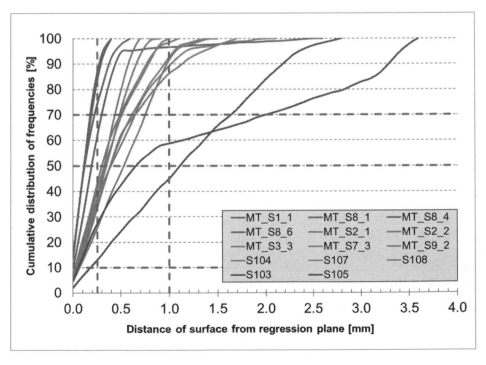

Figure 3. Mont Terri Opalinus Claystones and Bátaapáti Granites categorised according to surface roughness groups based on the new graphical approach proposed (red: smooth surface, green: moderately rough surface, blue: rough surface).

Table 1. Surface roughness categories of Bátaapáti Granite and Mont Terri Opalinus Claystone samples, values of normal stresses used for the direct shear strength test and corresponding peak shear strength values.

Rock type	Sample	Surface	Normal stress [MPa]	Peak shear strength [MPa]	Min [MPa]	Max [MPa]	Average [MPa]
Mont Terri Opalinus Claystone	MT_S1_1	smooth	1	0.439			
	MT_S8_1	smooth	1	0.393	0.393	0.480	**0.431**
	MT_S8_4	smooth	1	0.480			
	MT_S8_6	smooth	1	0.413			
	MR_S2_1	m. rough	1	0.512			
	MR_S2_2	m. rough	1	0.537			
	MR_S3_3	m. rough	1	0.516	0.364	0.598	**0.505**
	MR_S7_3	m. rough	1	0.598			
	MR_S9_2	m. rough	1	0.364			
Bátaapáti Granite	S103	rough	10	9.05	9.05	9.05	**9.05**
	S104	m. rough	10	8.15	8.15	8.15	**8.15**
	S105	rough	5	4.95	4.95	4.95	**4.95**
	S107	m. rough	5	3.91	2.44	3.91	**3.18**
	S108	m. rough	5	2.44			

the sample surface from the shear plane, caused by inaccurate encapsulation (Buocz et al. 2017b). For each surface roughness category of each rock type, and for each corresponding normal stress used throughout the shear test, the average values of shear strength were calculated (Table 1).

When the peak shear strength values are calculated from the direct shear test, several problems may arise (Barton 2013). In this study, peak shear strength values calculated from the direct shear tests are associated with the surface roughness of each sample; it is evident that the surface roughness influences the value of the peak shear strength. The average shear strength values of each surface roughness group for both rock types were compared, taking into consideration the different normal stress conditions. For Mont Terri Opalinus Claystones, samples with moderately rough surfaces showed peak shear strength values approximately 15% higher than those with smooth surfaces. For Bátaapáti Granites, the number of samples for which a peak shear strength value occurred was limited. For shear tests under 10 MPa normal stress, only one specimen was available for each of the two surface roughness categories. The one with rough surface presented however a peak shear strength value approximately 10% higher than the one with moderately rough surface. For shear tests under 5 MPa normal stress, one sample belonged to the surface roughness group "rough", and two to the "moderately rough". The former was characterised by a value approximately 35% higher. Although the relationship between the surface roughness and the peak shear strength confirmed the tendency expected prior to the study, further tests are needed to establish a more accurate relationship between these two parameters.

5 CONCLUSION

In this paper a new quantified graphical approach was introduced for determining the surface roughness of rock discontinuities, based on 3D surface detection results of natural discontinuities. The methodology is based on the calculation of frequency diagrams of the sample surfaces (distance between the surface roughness points and the linear regression plane fitted to a sample surface), from which cumulative frequency curves were determined. When these curves are represented on a diagram as a function of the distance of the surface from the regression plane, the sample surfaces can be classified into smooth, moderately rough and rough surfaces. Two rock types were investigated in this study: Mont Terri Opalinus Clay-stones and Bátaapáti Granites. The classification of the studied sample surfaces is in good agreement with the test results obtained from direct shear strength tests, i.e., the samples with rough surfaces had higher shear strength values than the ones with moderately rough surfaces, or the samples with moderately rough surfaces had higher shear strength values than the ones with smooth surfaces.

The methodology proposed represents a promising new approach to surface roughness quantification, which could contribute to improve methods for determining the shear strength.

ACKNOWLEDGEMENTS

The authors are thankful for the co-operation and the support of the following societies/institutions and their collaborators: Project Partners Ltd. – Consulting Engineers, 3GSM GmbH, Scientific Exchange Program (Sciex), Nagra, Swisstopo, RHK Kft., Mecsekérc Zrt., Budapest University of Technology and Economics—Department of Engineering Geology and Geotechnics, École Polytechnique Fédérale de Lausanne (EPFL) – LMR, International Society for Rock Mechanics (ISRM).

REFERENCES

Bae D., Kim K., Koh Y., Kim J. (2011): Characterization of Joint Roughness in Granite by Applying the Scan Circle Technique to Images from a Borehole Televiewer. Rock Mech Rock Eng, technical Note, Vol. 44; pp. 497–504.

Balla Z., Császár G., Gulácsi Z., Gyalog L., Kaiser M., Király E., Koloszár L., Koroknai B., Magyari Á., Maros Gy., Marsi I., Molnár P., Rotárné Szalkai Á., Tóth Gy. (2009): Explanatory Notes to the Geological Map – series of North-eastern Part of the Mórágy Block (1:10 000), (in Hungarian). Magyar Állami Földtani Intézet, pp. 15–17.

Barton N. (1973): Review of a new shear-strength criterion for rock joints. Engineering Geology, Vol. 7; pp. 287–332.

Barton N., Choubey V. (1977): The shear strength of rock joints in theory and practice. Rock Mech, Vol. 10, No. 1–2; pp. 1–54.

Barton N. (2013): Shear strength criteria for rock, rock joints, rockfill and rock masses: Problems and some solutions. Journal of Rock Mechanics and Geotechnical Engineering, Vol. 5, No. 4, pp. 249–261.

Barsi I., Görög P., Török, Á. (2012): Statistical analysis of test data deriving from rock mechanical tests carried out on Bátaapáti granitic rock samples, (in Hungarian). In: Török Á., Görög P. (eds.): Kőzetmechanika és kőzetkörnyezet szerepe a radioaktív hulladéklerakók kialakításánál, Budapest, Hungary, TERC Kereskedelmi és Szolgáltató Kft; pp. 113–122.

Buocz I., Rozgonyi-Boissinot N., Török Á., Görög P. (2014a): Direct shear strength test on rocks along discontinuities, under laboratory conditions. Pollack Periodica, Vol. 9, No. 3, pp. 139–150.

Buocz I. (2016): Parameters influencing rock shear strength along discontinuities: a quantitative assessment for granite and claystone rock masses of underground radioactive waste repositories. Doctoral dissertation, Budapest University of Technology and Economics, Civil Engineering, Department of Engineering Geology and Geotechnics; 206 p.

Buocz I., Rozgonyi-Boissinot N., Török Á. (2017a): Influence of Discontinuity Inclination on the Shear Strength of Mont Terri Opalinus Claystone. Periodica Polytechnica, Vol. 61, No. 3; pp. 447–453.

Buocz I., Rozgonyi-Boissinot N., Török Á. (2017b): The angle between the sample surface and the shear plane; its influence on the direct shear strength of jointed granitic rocks and Opalinus Claystone. Procedia Engineering, Vol 191., pp. 1008–1014.

Dzugala M., Sirkiä J., Uotinen L., Rinne M. (2017): Pull Experiment to Validate Photogrammetrically Predicted Friction Angle of Rock Discontinuities. Procedia Engineering, Vol 191., pp. 378–385.

Gaich A., Pötsch M., Schubert W. (2006): Basics and application of 3D imaging systems with conventional and high-resolution cameras. In: Tonon F., Kottenstette J. (eds): ARMA Workshop on Laser and Photogrammetric Methods for Rock Face Characterization, Golden Colorado, June 17–18, 2006.

Ge, Y., Tang, H., Ez Eldin, M.A.M., Chen, P., Wang, L., Wang, J. (2015): A Description for Rock Joint Roughness Based on Terrestrial Laser Scanner and Image Analysis. Scientific Reports. Vol. 5, 16999. pp. 1–10.

Grasselli G. (2001): Shear Strength of Rock Joints based on Quantified Surface Description. EPF-Lausanne, Switzerland; 124 p.

ISRM (1974); Franklin J.A., Kanji M.A., Herget G. and Ladanyi B., Drozd K. and Dvorak A., Egger P., Kutter H. and Rummel F., Rengers N., Nose M., Thiel K., Peres Rodrigues F. and Serafim J.L., Bieniawski Z.T. and Stacey T.R., Muzas F., Gibson R.E. and Hobbs N.B., Coulson J.H., Deere D.U., Dodds R.K., Dutro H.B., Kuhn A.K. and Underwood L.B. (1974): Suggested Methods for Determining Shear Strength. International Society for Rock Mechanics Commission on Testing Methods, In: Ulusay R., Hudson J.A. (eds.): The Complete ISRM Suggested Methods for Rock Characterization, Testing and Monitoring: 1974–2006, ISRM, Ankara, Turkey (2007); pp. 165–176.

ISRM (2015); Muralha J., Grasselli G., Tatone B., Blümel M., Chryssanthakis P., Yujing J.: ISRM Suggested Method for Laboratory Determination of the Shear Strength of Rock Joints: Revised Version. In: Ulusay R. (ed.): The ISRM Suggested Methods for Rock Characterization, Testing and Monitoring: 2007–2014. Springer; pp. 131–142.

Kovács L., Deák F., Somodi G., Mészáros E., Máté K., Jakab A., Vásárhelyi B., Geiger J., Dankó Gy., Korpai F., Mező Gy., Darvas K., Ván P., Fülöp T., Asszonyi Cs.. (2012): The revision and expansion of the Geotechnical Explanatory Report (GÉJ). Manuscript – Kőmérő Kft., Pécs, RHK-K-032/12; 312 p.

Nussbaum C., Bossart P. (2008): Geology. In: Bossart and Thury (eds.): Mont Terri Rock Laboratory Project, Programme 1996 to 2007 and Results, Rep. Swiss Geol. Surv., No. 3; pp. 29–38.

R_Core_Team (2011): R: A Language and Environment for Statistical Computing. R Foundation for Statistical Computing, Vienna, Austria.

Zhao, J. (1997): Joint surface matching and shear strength part A: joint matching coefficient (JMC). International Journal of Rock Mechanics and Mining Sciences. Vol. 34. No. 2; pp. 173–178.

Geomechanics and Geodynamics of Rock Masses – Litvinenko (Ed.)
© 2018 Taylor & Francis Group, London, ISBN 978-1-138-61645-5

Feasibility study on the influence of discontinuities on anisotropic rock masses regarding the stiffness

A. Buyer
Research and Teaching Associate, Graz University of Technology, Graz, Austria

L. Gottsbacher & W. Schubert
Graz University of Technology, Graz, Austria

ABSTRACT: For the design of rock engineering projects, rock mass parameters are usually determined by the use of classification systems (Q, GSI, RMR, etc.). A homogeneous rock mass with reduced parameters via empirical relationships is generated, and the influence of the anisotropy is lost during the process.

The influence of discrete discontinuities on the global rock mass stiffness is investigated in this feasibility study and the results compared to empirical solutions based on the GSI.

The results show that with one joint set oriented perpendicular to the direction of loading, the block volume is rather insignificant for the resulting E_{rm}; whereas the joint normal spacing parallel to the direction of loading and the block shape are the most influencing factors. In the cases, with the block volume and joint condition as input parameters, the E_{rm} is significantly underestimated, while the E_{rm}, based on RQD and $JCond_{89}$ still fits in the range of the numerical results. However, the numerical investigations are so far limited on an orthogonal oriented joint network under uniaxial compression and further investigations are necessary.

Keywords: Young's modulus, jointed rock mass, block volume, 3DEC

1 INTRODUCTION

When establishing a model for the design of rock engineering projects, simplifications are frequently required. One of those simplifications is not to model each and any discontinuity discretely, but consider their effect by at least partially "smearing" them into a continuum, and thus arrive at reduced rock mass parameters. For converting intact rock and discontinuity parameters into rock mass parameters, generally a kind of damage index (Q, RMR, GSI, etc.) and empirical correlations are used [1].

The shortcomings of such systems are among other aspects that the determination of the damage index is strongly biased, and the anisotropy is lost in the process. The results from different evaluation methods scatter in a wide range, as has been recently shown by [1].

Using for example, the block volume as an input for the determination of the rock mass properties, the stiffness and strength of the rock mass would be the same for different block shapes with identical block volume. Furthermore, all provided approaches so far consider the rock mass in two dimensions only and neglect the influence of joint spacing in the third dimension.

As the quality of the result of any analysis depends on the quality of the input data, improvements in the determination of rock mass parameters are urgently needed. To arrive at realistic rock mass parameters for the stiffness, a feasibility study was conducted, using numerical modelling on a rock mass with different joint set numbers, and block shapes. The results are compared to the Young's moduli (E_{rm}) according to [2] determined with the block volume dependant GSI(V_b, J_C) according to [3] and the RQD-dependant GSI(JCond$_{89}$, RQD) according to [4–6].

2 METHODOLOGY

The numerical model consists of a cubic block, intersected by a varying number of joint sets and an edge length of 10 meters. In general, the model resembles a discontinuous and blocky rock mass under a displacement controlled uniaxial compression (unrestrained lateral displacements). For the research, the number of joint sets and the joint normal spacing have been varied. The calculation was done in 17 steps, including four loading and three unloading cycles. The load steps with their assigned deformation are visualized in Figure 1.

Every load step was cycled 1000 times, allowing to reach equilibrium within reasonable computation time. The number was derived empirically; an increased number of computation cycles did not affect the resulting E_{rm} significantly. The numerical E_{rm} was calculated as the mean deformation modulus during the three unloading steps (step 4, 7 and 8, see Figure 1).

For the determination of the influence of the block shape and volume on the E_{rm}, five different block shapes have been predefined according to their ratios of the joint spacing in x-, y- or z-direction (Figure 2 and Table 1).

The material and joint parameters for the numerical model are listed in Table 2. An elastic-plastic constitutive model with a Mohr-Coulomb failure criterion was used.

The investigated cases cover an unjointed block model (Figure 3a), models with either a vertical or a horizontal joint set (Figure 3c and d), models consisting of two joint sets forming either columns or beams (Figure 3e and f) and a model with cubic shaped blocks generated by the intersection of three joint sets (Figure 3b). For each joint set, the spacing was set to 1 m.

As a comparison, the E_{rm} was calculated with the empirical solution according to Hoek & Diederichs [2] based on the anterior determination of the GSI according to Cai et al. [3] and Hoek et al. [4]. The first approach is based on the joint condition factor (J_C) in combination with the block volume (V_b). The second solution uses the RQD, to describe the joint network and the joint condition rating (JCond$_{89}$) according to [5] in order to describe the state of

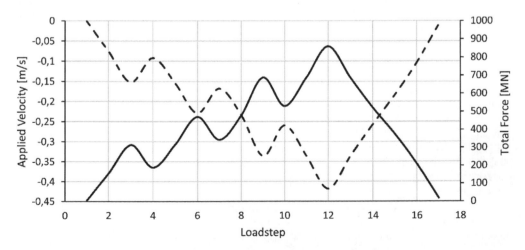

Figure 1.   Applied deformation velocities (dashed line) and the resulting total forces (solid line) on the upper and lower block boundaries according to their loading and unloading steps.

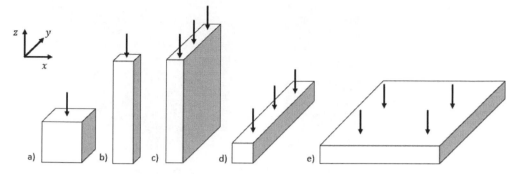

Figure 2.   Distinguished block shape classes: a) cubic, b) columnar, c) platy, d) beam and e) slab.

Table 1.   Shape classes according to the ratios of
the joint spacings in x-, y- and z-direction.

Shape class	x/z	y/z	x/y
cube	= 1	= 1	= 1
column	< 1	< 1	≤ 1
plate	< 1	> 1	< 1
slab	> 1	> 1	≤ 1
beam	≤ 1	> 1	< 1

Table 2.   Material and joint properties used in the 3DEC model.

	Parameter		Value
Material properties	density	[kg/m³]	2700
	Poisson's ratio	[–]	0.25
	Young's modulus	[MPa]	50,000
	cohesion	[MPa]	15
	dilatation angle	[°]	0
	friction angle	[°]	30
	tensile strength	[MPa]	0.7
	bulk modulus	[MPa]	33,333.3
	shear modulus	[MPa]	20,000
Joint properties	joint normal stiffness	[MPa/m]	500,000
	joint shear stiffness	[MPa/m]	5,000
	joint friction angle	[°]	27
	joint cohesion	[MPa]	0

weathering as well as roughness and opening. The input values of the analytical solutions are listed in Table 3.

According to [4, 5], the discontinuities can be described as slightly rough surfaces with an opening less than one millimetre and a state of moderate weathering. Likewise, the joints according to [3] can be described as slightly stepped joints with smooth to polished surfaces. The joints are filled with sand and clay (E_s ~ 25 MPa). There is only partial contact between the intact rock walls.

The chosen Young's modulus for intact rock E_i is again set to 50 GPa. The disturbance factor D was set to 0. With a minimum joint spacing of 100 cm, the RQD is 100% for all calculations. The joint network is arranged orthogonally and all joints are persistent.

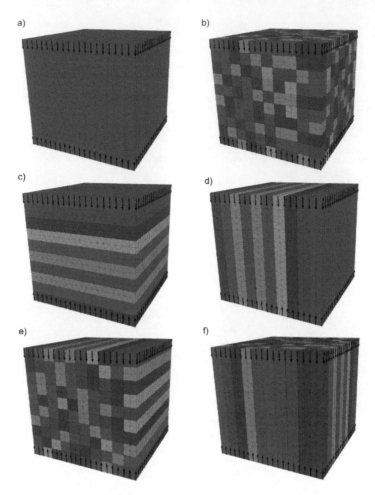

Figure 3. Modelled block shapes including a) a model without discontinuities, b) a model with three discontinuities ($s_{1,2,3}$ = 1 m), c) a model with slab shaped blocks (s_1 = 1 m), d) a model with plate shaped blocks (s_2 = 1 m), e) a model with beam shaped blocks ($s_{1,2}$ = 1 m) and f) a model with column shaped blocks ($s_{2,3}$ = 1 m); the black arrows indicate the direction of the applied velocities resp. load.

Table 3. Joint properties to calculate GSI(V_b, J_C) and GSI (RQD,JCond$_{89}$) according to [3–5].

Factor	Value
Joint waviness (J_W)	2.15
Small Scale Roughness (J_S)	0.9
Joint Alteration (J_A)	4.75
Joint Condition (J_C)	0.407
Joint Condition Rating (JCond$_{89}$)	22

3 RESULTS

In the following section, the results of the numerical modelling are directly listed with the results from the analytical solution.

The model set-ups of the block models with cubic shaped blocks either without or with three equally spaced discontinuity sets are shown in Figure 3 (a and b). In the case of an unjointed rock mass, the analytical E_{rm} equals per definition E_i which also equals the numerical E_{rm} (50 GPa).

Table 4. Results of the numerical and analytical calculation of the Young's modulus for the block model with three equally spaced discontinuity sets ($s_{1,2,3} = 1$ m).

	GSI		3DEC
Case	V_b, J_C	RQD, $JCond_{89}$	3 joint sets
V_b [cm³]	1.00E+06	1.00E+06	1.00E+06
GSI	48.72	83	–
E_{rm} [GPa]	14.20	45.50	44.53

Table 5. Results of the numerical calculation of the Young's modulus for both the slab and plate shaped block models with one discontinuity set in comparison to the analytically derived E_{rm}.

	GSI		3DEC	
Case	V_b, J_C	RQD, $JCond_{89}$	slab	plate
V_b [cm³]	1.00E+08	1.00E+08	1.00E+08	1.00E+08
GSI	67.60	83	–	–
E_{rm} [GPa]	34.31	45.50	45.97	50.0

Table 6. Results of the numerical and analytical calculation of E_{rm} for the column and beam shaped blocks.

	GSI		3DEC	
Case	V_b, J_C	RQD, $JCond_{89}$	column	beam
V_b [cm³]	1.00E+07	1.00E+07	1.00E+07	1.00E+07
GSI	57.21	83	–	–
E_{rm} [GPa]	22.85	45.50	50.0	45.17

Table 4 lists the results of the numerical and analytical calculations of E_{rm} for the numerical model with three joint sets and a block volume of $1 \cdot 10^6$ cm³.

The models with slab and plate shaped blocks are shown in Figure 3 (c and d). The calculated E_{rm} are listed in Table 5.

The results for the beam and columnar shaped blocks are listed in Table 6. The block models are shown in Figure 3 (e and f).

4 DISCUSSION

First, Figure 4 shows that the numerical E_{rm} ranges from 44.53 and 45.97 GPa and since the GSI(RQD, $JCond_{89}$) is constant, the analytical E_{rm} is also constant (45.50 GPa). Both results are significantly higher, than the block volume dependant E_{rm}, which varies from 14.20 to 34.31 GPa. The lowest stiffness is reached in the case of the smallest V_b (Table 4), which highlights the significant impact of the parameter on the results. However, this strong decline of the stiffness is not shown by the reference values, although the numerical solution also shows smallest stiffness at the smallest V_b (44.53 GPa).

Second, Table 5 shows a reduced E_{rm} for slab shaped blocks in the numerical solutions and no reductive effect of the joints on E_{rm} in the plate shaped model (both cases with an equal V_b). This differentiation is not reflected by the analytical solutions. Same counts for the cases of two joint sets (Table 6) with columnar and beam shaped blocks. Columnar shaped blocks

231

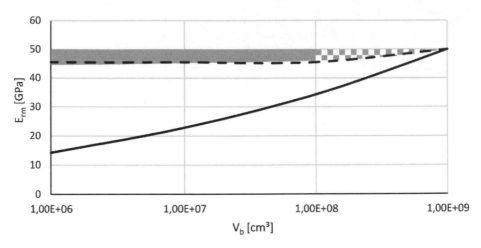

Figure 4. Block volume dependant E_{rm} (solid black line), E_{rm}, depending on $JCond_{89}$ and RQD (dashed black line) and the range of the numerical results is displayed as a gray field; the dotted gray field represents an estimated course of the numerical results.

do not seemingly influence the global deformability behaviour of the modelled rock mass, whereas beam shaped blocks lead to a reduction of E_{rm} to 90.34% of E_i. Due to a V_b of 10 m³, the block volume dependant E_{rm} is reduced to 45.8% E_i.

Third, in the modelled cases of an unjointed rock mass as well as plate and columnar shaped blocks, E_{rm} is not influenced by the discontinuities at all. All three cases have the same joint normal spacing s_3 (10 m), which is also the direction of loading, in common. Similarly, the E_{rm} of the models with three joint sets as well as the beam resp. slab shaped blocks, where s_3 is 1 m, varies between 45.19 ± 0,55 GPa.

The results from the numerical simulations lead to the conclusion that the joint normal spacing parallel to the direction of loading is rather significant for the determination of E_{rm}, more so than the actual block volume. This hypothesis is valid for different block shapes (cubic, columnar and plate shaped resp. cubic, slab and beam shaped). Especially in blocky rock masses with a reasonable low block volume, the analytical solution according to [3] might underestimate the actual rock mass stiffness, as long as compression is the only type of deformation and no reversible shearing occurs along the discontinuities. Still, the block volume seems to influence the resulting E_{rm}, however not on such a large extent as proposed by Cai et al. [3].

This underestimation and the reduction of the rock mass properties solely based on V_b and J_C was also criticised by Hoek et al. [4] and it seems that the proposed RQD-dependant determination of the GSI provides better results. Although, this approach still does not consider the apparent influence of the block shape or elongation according to the direction of loading.

However, further investigation on the influence of joints on the rock mass stiffness is necessary to validate the presented results. In detail, the rock mass behaviour under triaxial loading conditions has to be investigated, which might approximate "true" rock mass stiffness better. So far, the rock mass stiffness cannot be approximated with any proposed method satisfactorily. The used input parameters for the joint normal and shear stiffness and their influence on the resulting E_{rm} need also a detailed review and sensitivity analysis. Additionally, only an orthogonally aligned joint network was modelled, which did not consider irregular shaped blocks—another parameter which has to be considered.

5 CONCLUSION

The feasibility study presented here could highlight the potential underestimation of the rock mass stiffness by calculating GSI based block volume or the degree of fracturing, without considering the block shape and the orientation of elongated blocks in relation to the main direction of loading at constant joint conditions. It could also be shown that in general, the analytical solution for the block volume dependant E_{rm} overestimates the deformability of a jointed rock mass. However, with the updated GSI chart and its classification, basing on the RQD and the $JCond_{89}$, the E_{rm} was comparable to the results from the numerical modelling, which confirms the assumptions and suggestions of Hoek et al. [4]. Nonetheless, neither the block form nor the orientation of the blocks in relation to the direction of the main load is considered.

But still, the presented results are very basic and more research as well as more numerical models with higher order constitutive models should be run, in order to validate the preliminary results.

REFERENCES

[1] Zhang, L.: Evaluation of rock mass deformability using empirical methods – A review. Underground Space 2017; 2(1): p. 1–15. DOI: https://doi.org/10.1016/j.undsp.2017.03.003.
[2] Hoek E. & Diederichs, MS.: Empirical estimation of rock mass modulus. Int. J. of Rock Mechanics & Mining Sci. 2006; 43(2): p. 203–215. DOI: 10.1016/j.ijrmms.2005.06.005.
[3] Cai M., Kaiser PK., Uno H., Tasaka Y. & Minami M.: Estimation of rock mass deformation modulus and strength of jointed hard rock masses using the GSI system. Int. J. of Rock Mech. and Mining Sci. 2004; 41(1): p. 3–19. DOI: 10.1016/S1365-1609(03)00025-X.
[4] Hoek E., Carter TG. & Diederichs MS.: Quantification of the Geological Strength Index chart. 47th US Rock Mechanics/Geomechanics Symposium (ARMA) 2013. 9 p.
[5] Bieniawski, Z.T. 1989. Engineering rock mass classification. New York: Wiley Interscience. 272 p.
[6] Deere, DU.: Technical description of rock cores for engineering purposes. Felsmechanik und Ingenieurgeologie (Rock Mechanics and Engineering Geology) 1963; 1 (1): p. 16–22.

Geomechanics and Geodynamics of Rock Masses – Litvinenko (Ed.)
© 2018 Taylor & Francis Group, London, ISBN 978-1-138-61645-5

The effect of mining layout, regional pillars and backfill support on delaying expected shaft deformation at Bambanani mine

P.M. Couto
Middindi Consulting, Applied Science—Mining Engineering

ABSTRACT: The extraction of Bambanani Mine Shaft Pillar commenced in 2011. Mining of the shaft pillar commenced prior to the establishment of a decline system extending from 60 to 75 level. The decline would ultimately replace the Main shaft between 60 and 75 levels.

To justify stoping of the shaft pillar prior to completion of the takeover decline, initial numerical modelling of expected stress and strains in the main shaft was conducted using MinSim. The initial modelling was conservative, and modelling was conducted without the consideration of regional pillars and backfill support. The stress and strains where modelled to obtain an understanding of the expected progressive deformation during different phases of the pillar extraction.

The results indicated that some 20% of the pillar could be extracted before deformation in the shaft steelwork, brattice wall and concrete lining was expected.

In October 2017 the extraction of the pillar had reached 49% and during routine shaft examinations the expected deformation was less than predicted through the initial modelling. The completed extraction was then re-modelled utilising IMS Vantage with the inclusion of the correct mining layout, regional pillars and backfill support.

Results from the modelling and observations made during shaft examinations have indicated that the mining layout, regional pillars and the inclusion of backfill have had a positive impact on delaying the expected deformation to the shaft barrel.

1 INTRODUCTION

Bambanani east shaft pillar extends 800 m along dip and 700 m along strike with a total area of 560 000 m². The main shaft is situated in the middle of the square shaped pillar together with various other life of mine excavations.

The feasibility study conducted in 2011 identified two options for securing the shaft to ensure that the pillar was safely and economically extracted.

The options were:

1. Conventional shaft pillar extraction with modified steelwork consisting of a floating steelwork section with mining proceeding from an inner cut around the barrel towards the edge of the pillar in both mining directions along strike.
2. A decline option where the shaft is locked below a certain level, no modifications to the steelwork is done and a decline is developed outside the shaft pillar in de-stressed ground from where all mining operations will be commence and mine along strike in the one direction towards the shaft.

The second option was chosen as several important service excavations forming part of the sub-shaft infrastructure were located on reef or near reef. By commencing the extraction from the one side of the shaft, mining on the lower sub-shaft levels could continue. *(Report on the Bambanani Shaft, Shaft pillar extraction, BBE-11-70, K.R Brentley, Lucas & Associates Pty Ltd).*

2 LOCALITY

Bambanani Mine is situated in the Free State portion of the Witwatersrand Basin, between Welkom and Virginia in the Republic of South Africa. The mines boundaries include the President Steyn 1 and 2 shafts and borders against Harmony Gold Mine's Unisel shaft towards the south. To the east, the mine borders on the De Bron fault, which displaces the reef up towards Harmony Gold Mine's Saaiplaas #3 Shaft. The area underlain by accessible gold bearing reef in the Free State is about 50 km long and 11 km wide stretching from north of Allanridge in the north to Virginia in the south.

Bambanani is a seismically active mining operation with a approximately 10 000 to 12 000 events per calendar month.

3 SHAFT BARREL

Bambanani east shaft is concrete lined, circular in shape with a diameter of 11.0 m, and is split by means of a brattice wall. The brattice wall channels the down casting ventilation. The shaft comprises.of eight compartments, with four being personnel and material and the remaining four are used for rock hoisting. The shaft ascends from surface to a depth of 2548 mbs, with the reef intersection lying at 2079 mbs.

4 DESIGN SCENARIOS

Damage within the shaft barrel, in terms of steelwork buckling, concrete unravelling and brattice wall failure are because of displacement generated whilst mining. This displacement can be estimated by means of induced strain within the shaft.

Literature studies have indicated that the following induced strain levels will result in the following damage as indicated in Table 1. (Stacey, TR, December 2001).

Input parameters for elastic modelling conducted for all design options are as follows:

- Young Modulus – 50 GPa
- Poisson's Ratio – 0.20
- Vertical Virgin Stress – 0.027 MPa/m
- K Ratio – 0.5
- Stoping width – 2.0 m
- Fill width – 1.8 m
- Host Rock Material Strength – 200 MPa
- Grid Size – 20 m
- Downgraded (60% fill) hyperbolic type backfill formula (et) – 0.01 and (b) – 0.4.

The two elastic models namely MinSim and IMS Vantage were utilised in conjunction with actual field observations made in the shaft barrel which determined the input parameters.

Table 1. Damage criteria for shafts (STACEY, TR, Best Practice Rock Engineering Handbook for "other mines, SRK Consulting, OTH 602, December 2001).

Type of damage	Criterion	Reference
Steelwork damage and increased shaft maintenance	$\varepsilon_z^i < 0.2 \times 10^{-3}$ $\varepsilon_z^i < -0.4 \times 10^{-3}$	Wilson (1971) Van Emmenis and More O'Ferrall (1971)
Tensile fracture of concrete lining	$\varepsilon_z^i < 0.4 \times 10^{-3}$ $\varepsilon_z^i < -0.51 \times 10^{-3}$ to -0.3×10^{-3} $\varepsilon_z^i < -0.2 \times 10^{-3}$	Esterhuizen (1980) Kratzsch (1983) Budavari (1986)
Compressive fracture of concrete lining	$\varepsilon_z^i < 0.7 \times 10^{-3}$	Jager and Ryder (1999)

Two models were used purely to evaluate if the initial design was accurate without considering regional pillars and backfill in a full mini longwall mining setup. IMS Vantage was used as a calibration model for both the MinSim model and the field observations.

Model A

The initial model was conducted using the calibrated input parameters, mentioned above. The model was constructed without backfill and without any regional stability pillars. Figure 1 indicates the mining steps in a mini—longwall effect with a total extraction.

Model B

The regional pillar model utilised the same calibrated design parameters, however an actual representation of the current mined out area was digitised with similar extraction volumes when compared to the initial design. This was modelled like this to ensure that each mining step could be compared equally throughout the three different scenarios. Refer to Figure 2.

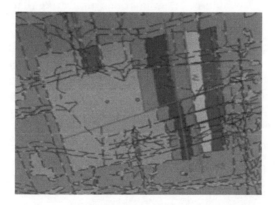

Figure 1.   Plan view of the mining steps utilised for model A.

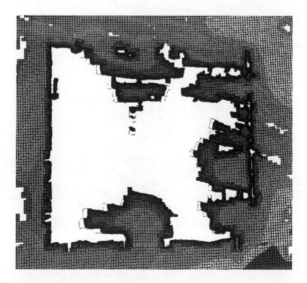

Figure 2.   Plan view of the actual mining used in model B and C.

Model C

All the standard input parameters where utilised in this model as previously done in model A and B. The mining steps in relation to the extraction were modelled the same throughout all three models. The hyperbolic formula for Backfill was introduced in this model. Refer to Figure 2.

5 ANALYSIS AND RESULTS

The results for model A are indicated in Figure 3 below. The mining steps have been allocated a percentage extraction based on the initial model generated in 2011 utilising MinSim. Reasoning for this was based on confirming that both MinSim and Vantage produced comparable results with the same calibrated input parameters. A red line has been placed at 0.3 millistrains in Figure 3 as the baseline where deformation within the shaft is expected according to Table 1. According to the model's visible deformation would have occurred at 19% extraction as initialled design for in 2011.

The induced strain in the shaft for model A, B and C at 49% extraction is plotted in Figure 4. If the mining layout was followed as per model A then the expected induced vertical strain in the shaft would be 1.8 millistrains, indicating that the shaft would have experienced severe damage. Model B is represented by the green line in Figure 4 which indicates that if the current layout with regional pillars but no backfill as its primary support, the current shaft would experience some 0.4 millistrains. Model C represents the current situation which includes backfill and estimated induced vertical strains of 0.36 millistrains. Observations during a shaft exam confirm these results. Steelwork damage and increased shaft maintenance is the current condition as represented by model C. The modelling confirms the field observations and allows for a more accurate modelling in the future of the shaft extraction. A summary of the modelling results is represented in Table 2.

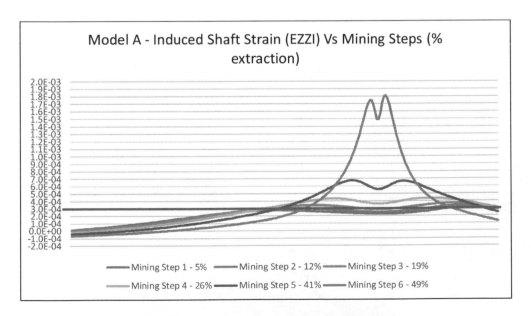

Figure 3.   Induced strain per mining step/percentage extraction with the shaft.

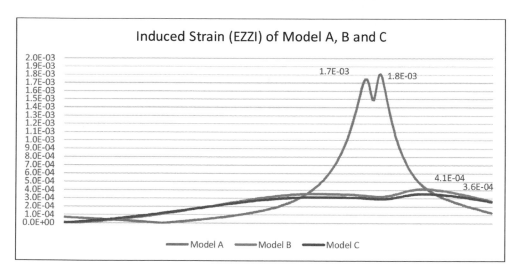

Figure 4. Comparison of all three models and their current peak induced strain.

Table 2. Improvement of mining layouts, regional pillars and backfill when mining a shaft pillar.

Model	Mining layout	Peak EZZI (millistrains)	Extraction	Variance	Reduction of deformation	Backfill benefit
A	Longwall	1.80	19%	0%	0%	N/A
B	Mini-longwall with regional pillars	0.41	49%	30%	77%	N/A
C	Mini-longwall with regional pillars and backfill.	0.36	49%	30%	80%	12.2%

6 CONCLUSION

Field observations and continuous modelling to identify accurate input parameters is important to obtain accurate modelling results irrespective of which elastic modelling software is utilised. Calibration and constant modelling is essential to maintain more accurate rock engineering results. The modelling indicated that there is a significant reduction in shaft deformation when different mining layouts, regional pillars and backfill are utilised in conjunction.

This system might have other limitations in terms of mining practicality to an extent, however the overall stability benefit is significant.

In conclusion the delay of displacements and subsequent shaft deformations had a positive spin off in that the main shaft was used for a longer period than expected, because of operational constraints and changes during the shaft pillar extraction, which were not originally catered for in the initial design. This allowed for increased time to complete the decline system which due to various reasons was behind schedule.

ACKNOWLEDGMENTS

Jacques Gerber—Institute of Mine Seismology—Manager Seismological Services.
Kevin Brentley—Brentley, Lucas and Associates—Rock Engineering Consultant.

REFERENCES

Du G., T. Judeel, AngloGold, Study of the effects of shaft pillar extraction on a vertical shaft located in highly stratified and poor quality rock masses.

JD Applegate, Rock Mechanics Aspects of Sequential Grid Mining, Faculty of Engineering, University of Witwatersrand, Johannesburg, as partial fulfilment towards a Master of Science in Engineering.

Malan, D.F. Guidelines for measuring and analysing continuous stope closure behaviour in deep tabular excavations, The Safety in Mines Research Advisory Committee (SIMRAC), 2003, page 1–67.

Ryder, J.A. and Jager, A.J. (1999). A Handbook on Rock Engineering Practice for Tabular Hard Rock Mines, SIMRAC.

Ryder, J.A. and Jager, A.J. (2002) A Textbook on Rock Mechanics for Tabular Hard Rock Mines, SIMRAC.

Stacey, T.R. Best Practice Rock Engineering Handbook for "other mines, SRK Consulting, OTH 602, December 2001.

Stradling, A.W. Backfill in South Africa: Developments to classification systems for plant residues, Minerals Engineering Volume 1, Issue 1, 1988, Pages 31–40.

Geomechanics and Geodynamics of Rock Masses – Litvinenko (Ed.)
© *2018 Taylor & Francis Group, London, ISBN 978-1-138-61645-5*

Fractured clay rocks as a surrounding medium of underground structures: The features of geotechnical and hydrogeological assessment

R.E. Dashko & P.V. Kotiukov
Department of Hydrogeology and Engineering Geology, Saint Petersburg Mining University, Russia

ABSTRACT: The article presents the results of studying the fracturing of clay rocks (claystone) which are the surrounding medium of underground structures located on the territory of St. Petersburg and the Leningrad Region. The presence of systems and zones of increased fracturing having various orientation, intensity and nature that complicate the structure of the clay massif was established. The influence of the fracture intensity on strength and deformation ability of such rocks has been studied experimentally. Particular attention is paid to the assessment of fissured clay permeability in connection with the possibility of upward or downward flow of groundwater which is able to contact with clay blocks and underground structures. Data on the investigation of the chemical and physico-chemical interactions of clay rocks with groundwater of different compositions are presented. The influence of the groundwater upward flow on the change in the stress-strain state of the surrounding medium and the development of tunnel displacements is shown. The role of groundwater in the corrosive destruction of constructional and hydroisolating materials is studied. It has been experimentally proved that, during the process of groundwater flow through fissured clay rocks, biofilms form on the walls of fractures, and not inside the blocks.

1 INTRODUCTION

In the design and construction of underground structures in clay rocks, the problem of estimating their fracturing takes one of the central places. The intensity of fracturing has a strong influence on the strength, deformability, water and gas permeability of clay rocks as well as on the possibility of rock spelling, groundwater breakthrough, development of gas-dynamic phenomena and other dangerous natural and man-made processes that require special methods of tunnel driving and ensure the stability of workings for the whole period of their operation. As construction practice shows, fracturing can be revealed not only in magmatic or metamorphic rocks, which are crystalline formations, but also in most sedimentary deposits having a high degree of lithification among which are the Upper Vendian (also called «Proterozoic») and Lower Cambrian clay rocks. Widely occurring in the territory of St. Petersburg and the Leningrad region, they are the surrounding medium for various underground structures such as tunnels and metro stations, sewage collectors, gas storages, etc. These circumstances determine the practical significance of the studies.

2 FRACTURING OF UPPER VENDIAN AND LOWER CAMBRIAN CLAY ROCKS

The Upper Vendian and Lower Cambrian clay rocks had been formed more than 500 million years ago and, for a long geological time, had been under relatively high pressures (up to 6–10 MPa), temperatures and the influence of secondary processes that led to their consolidation, the formation of cementing structural bonds and, as a consequence, to an increase in strength and decrease in their physical and chemical activity as well as the ability to develop plastic deformations. Such clay rocks have to be considered as brittle or brittle-plastic medium depending on the level of the volumetric stress state, the action of water and aqueous solutions. The formation of fracturing in such rocks occurs under the influence of tectonic, glaciotectonic forces, temperature differential and unloading of rocks. The degree of fracture opening is depending on their origin. The permeability of the fissured clay massif is indicated by the secondary formations of iron hydroxides, carbonate and gypsum deposits on the fracture walls.

The fracturing of clay rocks has a various genesis. On the background of lithogenetic fissures, typical for sedimentary formations, it was revealed the tectonic fracture systems dividing the massif into separate blocks. According to the results of studies, conducted by prof. R.E. Dashko, there are zones of weathering fissures, which are usually superimposed on tectonic fracturing, in the upper part of the section to depths of about 20–25 m in Lower Cambrian clays and up to 40 m in Upper Vendian clays. Outside the areas adjacent to the tectonic faults, the intensity of rock disintegration normally decreases with depth and the size of blocks gradually increases. These regularities of clay rock disintegration should be taken into account when assigning the depth of the routes of transport structures and other underground excavations (Dashko, 2001).

3 INFLUENCE OF FRACTURING ON THE MECHANICAL PROPERTIES OF CLAY ROCKS

The presence of fracturing is of decisive importance in the study of the strength, stability, deformability of clay rocks, which determine the features of their mechanical behavior during interaction with underground structures, as well as their water and gas permeability (Dashko, 2016). Tables 1 and 2 show the results of studying the claystone shear strength parameters in different depth zones taking into account their fracturing intensity. It is important to emphasize that in the analysis and evaluation of the strength of lithiated clay rocks, their microfissuring plays an important role, which is often not fixed visually, but its effect can be studied even in laboratory conditions when investigating the scale effect.

Evaluation of the fractured clay massif strength parameters is performed using the structural weakening coefficient of rocks. For consolidated clay, this value varies from 0.3 to 0.7. Lower values are usually used to determine the strength of a fractured massif consisting of stiff clay rocks.

The presence of micro fractures in clays leads to the appearance of a scale effect, which appears in dependence of the experimental results on the sample size (Table 3). It is most clearly observed when testing samples of clays selected from tectonic fault zones for which the maximum degree of disintegration is characteristic.

Table 1. Comparison of strength and deformability parameters of low cambrian clay rocks according to triaxial tests (Dashko, 2001).

| Specimen location | Depth from clay roof, m | Strength parameters | | Modulus of deformation E_o, MPa |
		Cohesion C, МПа	Angle of internal friction φ, degrees	
Outside of tectonic fault zones	0.0–3.0	0.035–0.05	0	15–20
	3.0–8.0	0.075–0.17	0–2	19–24
	8.0–17.0	0.220–0.34	6–8	20–25
Within the limits of fault zones	0.0–3.0	0.027–0.04	0	1.5–2.2
	3.0–5.0	0.034–0.078	0–4	3.0–6.6
	5.0–8.0	0.15–0.19	0–6	6.2–10.5

Note: Specimen square $F = 25$–26 sm^2.

Table 2. Shear strength parameters of vendian clay rocks in different depth zones with considering the fissuring intensity (Dashko, 2001).

Zone	Depth, m	Specimen (via triaxial tests)				Massive			
		C, MPa		φ, degrees		C, MPa		φ, degrees	
I	0–20.0	<u>0.18</u>		<u>5</u>		<u>0.13</u>		<u>5</u>	
		0.09		2		0.05		2	
	10.0–20.0	<u>0.86*</u>	<u>0.3**</u>	3*	24**	<u>0.43*</u>	<u>0.12**</u>	3*	24**
		0.45	0.05	18		0.22	0.02	18	
	20.0–30.0	<u>1.4</u>		<u>22</u>		<u>0.60</u>		<u>22</u>	
		0.82		11		0.33		11	
	30.0–40.0	<u>2.04</u>		<u>23</u>		<u>0.82</u>		<u>23</u>	
		1.10		18		0.56		18	
II	40.0–60.0	<u>2.80</u>		<u>23</u>		<u>1.12</u>		<u>23</u>	
		1.9		19		0.62		19	

Note[1]: the values of C and φ for clay rocks outside the fault zones are given in the numerator and inside the zones—in denominator.
Note[2]: *– quasiplastic clays; ** – non-plastic clays with different values of C and φ.

Table 3. Impact of scale effect on the strength and deformability of clay rocks.

Specimen square, sm²	20–26.5	40–48	98
Unconfined compressive strength, MPa	0.70–0.92*	0.24–0.50	0.105–0.140
	0.81/6**	0.34/7	0.12/5
Modulus of deformation, MPa	14–20	8–10	3–6
	16/6	8.5/7	4/5

Note: * – range of the parameter value; ** – mean value of the parameter/number of tests.

The investigations made it possible to derive an empirical formula (1) for determining the unconfined compressive strength taking into account the scale effect:

$$R_{cæc} = R_{max} \cdot \exp[k \cdot (\frac{F_{min}}{F_{max}} - 1)],$$ (1)

where R_{max} is the unconfined compressive strength of specimen with maximal square, k is empirical coefficient, F_{min} и F_{max} is minimal and maximal square of specimen, respectively.

The deformation ability of clay rocks as well as their strength depends on the occurrence depth and the position in cross-section relative to the fault zones. Outside such zones, the value of clay deformation modulus tends to increase with depth. In fault zones, it varies in depth irregularly and is characterized by a relatively wide range of values, which changes from 20 to 180 MPa and more (laboratory tests), that indicates a high degree of the clay massif disintegration.

4 IMPACT OF GROUNDWATER FLOW THROUGH FRACTURED CLAY ROCKS ON UNDERGROUND STRUCTURES

Considering the interaction of underground structures with lithiated clays, the scale effect should also be introduced when analyzing their permeability and using in calculations the values of the hydraulic conductivity obtained from the results of field tests. This will avoid errors associated with an incorrect assessment of the fractured aquitard permeability. The fractured-block structure of the clay rock massif leads to the fact that the value of their hydraulic conductivity rises by 2–3 orders of magnitude (up to 10^{-1}–10^{-2} m/day) in comparison with clays, which are fine-porous media, and in the fault zones can increase up to 1.0–1.5 m/day.

In the design practice, when driving tunnels in clayey massif, they are usually analyzed as an aquiclude that provides the absence of groundwater flow whose corrosive aggressiveness towards the tunnel lining is totally ignored. At the same time, the results of the survey of tunnels driven in Upper Vendian clays confirm that the clayey stratum is permeable, first of all, for the confined aquifers being traced below. This is evidenced by numerous seepages which are observed in the exploited tunnels of St. Petersburg Metro (Dashko, 2015). The number of seepages in the sections of the tunnels driven in the zones of tectonic faults (within the buried valleys) amounts to 79–99% of the total number of seepages (Fig. 1).

The study of the interaction between tunnels underground structures and the surrounding medium shows that, depending on the hydrodynamic conditions in the underground environment, both upward and downward flow is possible, accounting for which is of fundamental importance and can be expressed in the following sentences: 1) the chemical and physico-chemical interaction of individual clay blocks with groundwater of different composition; 2) the direct effect of groundwater on the stress-strain state of the tunnel; 3) an activation of the underground microbiota vital activity and associated natural and man-made processes (biochemical gas generation, biocorrosion, etc.).

The downward flow through the massif of fractured clays is associated with quaternary aquifers among which the most important are unconfined groundwater. They are characterized by a high degree of contamination due to inorganic and organic compounds and, as a result, have increased mineralization (more than 1 g/dm^3), a significant content of chlorides, sulfates, ammonium, organic components, and aggressive carbon dioxide. During the interaction of clay rocks with groundwater polluted by sewage, their ability to swell rises steadily, as a result of which these clays gradually become less dense and lose their bearing capacity (Table 4, Fig. 2).

The possibility of upward flow is due to the presence of the Vendian complex aquifer system, underlying the Upper Vendian clays, whose piezometric level is above the water table. The groundwater of this complex has increased mineralization (about 4 g/dm^3) and chloride-

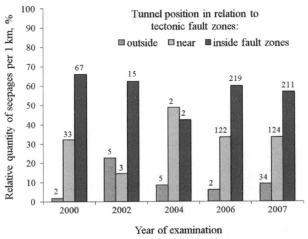

Note: absolute number of seepages per 1 km of the tunnel is shown above histograms

Figure 1. Number of seepages in tunnels depending on their location relative to the tectonic fault zones (according to: 2000–2006 – tunnel survey service, 2007 – Mining University).

244

Table 4. The effect of sewage on the parameters of clay rock swelling.

Initial water content, %	Free swell index (%)/ Time of swelling (days)		Water content after swelling,%		Swelling pressure, MPa	
	water	liquid waste	water	liquid waste	water	liquid waste
24	18/3	24/32	38	43	0.08	0.27
23	6.4/4	13.5/8	34	36	–	–
20	7.9/3	13.2/26	32	37	0.19	0.37
16	18.5	24/41	37	43	0.35	>0.5

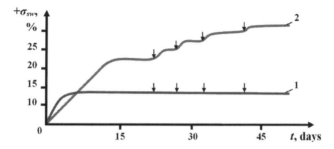

Note: arrows indicate addition of sewage and distilled water

Figure 2. Dynamics of clay rock swelling in water (1) and in sewage (2).

sodium composition. Moving along the fractures, they react (ion exchange) with cations of the clay particle diffuse layer. An increase of the potassium (Na^+) content in the clay particle diffuse layer leads to the dispergation of clays, as a result of that, their hydrophilicity increases. Such physicochemical processes (in the absence of swelling) lead to a decrease in the shear strength parameters of the clay almost by half (Fig. 3). In this case, they turn into a quasiplastic state, in which the shear strength value is almost independent of pressure increase ($\varphi = 0$, c $\neq$ 0).

The upward groundwater flow from the Vendian complex aquifer system affects the change in the stress-strain state of the clay massif which is the surrounding medium for deep tunnels that results in the development of tunnel uneven displacements. In this case, the lifting of tunnels accompanied by the opening of joints and fractures of lining is especially dangerous. It should be added that the mineralized chloride sodium groundwater of the Vendian complex aquifer are aggressive towards materials which are concrete, reinforced concrete and metals. Groundwater are capable of causing their progressive destruction with the produce of the secondary formations (stalactites, efflorescence, etc.) on their surface. In the chemical composition of aqueous extracts, prepared from samples of destroyed materials and secondary formations, anomalously high concentrations of chlorides (up to 112 g/dm³) and potassium ions (up to 84 g/dm³), confirming the impact of chloride sodium groundwater from Vendian complex aquifer system, were found.

The additional effect of aqueous solutions on clays with rigid structural bonds causes not only physicochemical processes, but also activation of the underground microbiota activity. In the groundwater flow, there is often a transfer of microorganisms that are sorbed, adhere to the surface of fractures in rock blocks with the formation of biofilms. Experimental microbial study confirmed the tendency to microorganism penetration into the clay layer along the fractures. The content of microbial mass in terms of the total protein was 10.3 µg/g in block and 72.0 µg/g on the surface of fractures that indicates the transfer of microorganisms and their metabolic products through a system of fissures.

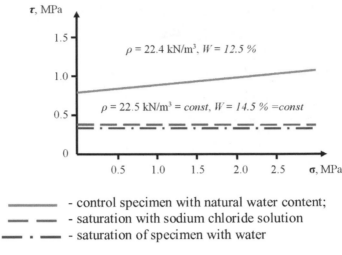

- control specimen with natural water content;
- saturation with sodium chloride solution
- saturation of specimen with water

Figure 3. Shear strength diagram of clay rocks with initial water content and with addition of water and sodium chloride solution.

Microorganisms are able to produce active metabolites such as mineral and organic acids, soluble and slightly soluble gases, as well as enzymes (proteins), acting as a catalyst for biochemical processes in the underground environment. As a result, a negative transformation of the composition, physical and mechanical parameters of clay rocks occur. When slightly soluble biochemical gases accumulate, a significant gas-dynamic pressure is formed (up to 3 atm, rarely more) that causes the transformation of the surrounding medium stress-strain state and leads to a decrease in the settlement intensity or to the lifting of underground structures. When gas dissipates, significant settlements are observed. The development of alternating deformations of tunnels contributes to the appearance of fractures in lining and to the failure of their integrity and watertightness. The presence of water-soluble gases (carbon dioxide, hydrogen sulfide, ammonia) contributes to an increase in the aggressiveness of groundwater towards materials of underground structures. Oxidation of methane leads to the accumulation of CO_2 in water, which causes the development of carbon dioxide aggressiveness.

5   CONCLUSION

1. Analysis and geotechnical assessment of clay rocks as a surrounding massif for underground structures should be based on their consideration as a fractured-block medium that implies the need to take into account the micro- and macrofracturing in the calculations of long-term stability and safe operation of the designed facilities.
2. The presence of fracturing in the clay rock massif defines its permeability the value of which is determined by the nature and intensity of the rock disintegration, and also by the acting pressure gradients. Change in hydrodynamic pressure leads to alternating displacements of tunnels depending on the pressure vector.
3. The interaction of flowing groundwaters having different chemical composition and biochemical specificity with clay rocks causes the development of physico-chemical processes (ion exchange, swelling, dispersion) in the clayey layer leading to negative changes in their strength, deformability and transition to a quasiplastic state.
4. When designing underground structures in such geotechnical conditions, it is necessary to consider the development of corrosion processes of tunnel lining, including biocorrosion, which is confirmed by studying the nature of the destruction of concrete, reinforced concrete and cast iron, as well as the backfilling materials from the tunnels of the St. Petersburg Metro driven in fissured clay rocks.

ACKNOWLEDGEMENTS

This publication was supported by a grant of the Russian Science Foundation under Contract 16-17-00117.

REFERENCES

[1] Dashko, R.E. & Eremeeva, A.A. 2001. Engineering geological features of the bedding clays of St. Petersburg as a medium for underground structures, Proceeding, International Symposium "Engineering geological problems of urban areas", Vol. 2, pp. 675–681.
[2] Dashko, R. & Karpova, Y., Engineering geology and geotechnics of fractured clays as building base and surrounding medium (by the example as clayey bedrocks in Saint-Petersburg), Proceedings, International Multidisciplinary Scientific GeoConference Surveying Geology and Mining Ecology Management, SGEM, Vol. 3, 2016, pp. 85–92.
[3] Dashko, R.E. & Kotiukov P.V., Geotechnical analysis of long-term stability of Saint Petersburg Metro tunnels in Upper Vendian Clay, Proceedings, International Multidisciplinary Scientific GeoConference Surveying Geology and Mining Ecology Management, SGEM, Vol. 2, 2015. pp. 353–360.

Geomechanics and Geodynamics of Rock Masses – Litvinenko (Ed.)
© *2018 Taylor & Francis Group, London, ISBN 978-1-138-61645-5*

New methods to fit a Hœk Brown failure criterion to data sets from multiaxial laboratory tests

Angéline Defay & Siegfried Maïolino
Département Laboratoire de Lyon, Cerema, Direction Centre-Est, Lyon, France

ABSTRACT: In this study, a fitting method for the Hoek-Brown parameters is presented, where the measurement of error is based on the distance between the yield surface and the experimental points in the stress vector space. The distance between experimental data and the 3d curve is defined as the minimum distance obtained through closest point projection. A least squares method based on this distance is used with a grid search algorithm. The algorithm is tested on various data sets that can be found in the literature. The results are compared with the values obtained by other authors, and both a standard Hoek Brown failure criterion and a modified smooth version are fitted.

1 NORM IN STRESS SPACE

Many rocks present a non linear Mohr's envelope. In order to take into account this characteristic many criteria have be developed since the 1960s; one of the best known is the Hœk-Brown criterion [1], which has been developed so it can be used for a wide range of rocks. The Hœk-Brown criterion can be written as follow, with σ_{ci} the Uniaxial Compressive Strength of the intact rock, m_b the value the Hœk-Brown constant for intact rock (value between 4 and 33).[1]:

$$f(\underline{\underline{\sigma}}) = (\sigma_I - \sigma_{III}) - \sigma_{ci}\sqrt{1 - m_b\frac{\sigma_I}{\sigma_{ci}}} \tag{1}$$

Our aim is to fit Hœk-Brown's criterion parameters (m_b and σ_{ci}) using a last square method. In order to minimize square of the distances between experimental data and yield surface, we choose to use the distance in the stress space, associated to the following scalar product: for two symmetric second order tensors, $\underline{\underline{T_1}}$ and $\underline{\underline{T_2}}$:

$$\underline{\underline{T_1}} \cdot \underline{\underline{T_2}} = \underline{\underline{T_1}} : \underline{\underline{T_2}} = \mathrm{Tr}\underline{\underline{T_1}}\underline{\underline{T_2}} \tag{2}$$

Hence defining the following norm (Frobenius norm) for a symetric second order tensor $\underline{\underline{T}}$:

$$\|\underline{\underline{T}}\| = \sqrt{\underline{\underline{T}} : \underline{\underline{T}}} \tag{3}$$

Using this norm, the norm of the stress tensor, can be written as a function of σ_m and J_2:

$$\|\underline{\underline{\sigma}}\|^2 = 3\sigma_m^2 + 2J_2 \tag{4}$$

---

1. Traction stresses are positive, and the principal stresses ordered as follow : $\sigma_I \geq \sigma_{II} \geq \sigma_{III}$.

2 REPRESENTATION OF CRITERIA IN THE DEVIATORIC PLANE

When the mean stress σ_m is constant, a yield surface can be reduced to its representation in the deviatoric plane (π-plane) as a yield surface admits an equivalent polar expression [2]:

$$\sqrt{J_2} = \sigma^+ g_p(\theta) \tag{5}$$

- θ being the Lode angle: $\theta = \frac{1}{3} \arcsin\left(\frac{-3\sqrt{3} J_3}{2\sqrt{J_2}^3} \right)$
- $\sigma^+(\sigma_m) = \sqrt{J_2}_{\,/\theta=\frac{\pi}{6}}$, the deviatoric radius, gives the yield function in the meridional plane $(\sigma_m, \sqrt{J_2})$, for $\theta = \frac{\pi}{6}$. This value of the Lode angle corresponds to the condition of a classical triaxial test, or compressive triaxial test ($\sigma_I = \sigma_{II} > \sigma_{III}$);
- the function $g_p(\theta)$ is the shape function of the yield function in the deviatoric plane. We have $(g_p(\frac{\pi}{6}) = 1)$. It gives directly the value of the extension ratio $g_p(-\frac{\pi}{6}) = L_S$.

For Hœk-Brown criterion, we will use two shape functions, both using the scaled internal pressure [3]: $P_i = \frac{s}{m_b^2} - \frac{\sigma_m}{m_b \sigma_{ci}}$ whose value is strictly positive.

1. The shape function of the original Hœk-Brown criterion [4]:

$$g_p(\theta) = \frac{-\sin\left(\theta + \frac{2\pi}{3}\right) + \sqrt{\sin^2\left(\theta + \frac{2\pi}{3}\right) + 12 P_i \cos^2 \theta}}{2 \dfrac{-1 + \sqrt{1 + 36 P_i}}{3} \cos^2 \theta} \tag{6}$$

2. The shape function associated to the following yield function, a smooth Hœk-Brown criterion[5]:

$$f(\underline{\underline{\sigma}}) = \frac{3\sqrt{3}}{2}(1 - L_s)J_3 + (L_s^2 + 1 - L_s)\sigma^+ J_2 - \sigma^{+3} L_s^2 \tag{7}$$

With $L_s = 2\frac{-1 + \sqrt{1 + 9 P_i}}{-1 + \sqrt{1 + 36 P_i}}$, the extension ratio of the original Hœk-Browncriterion.

For the latter case (7), solving a third degree equation leads to the explicit expression of g_p:

$$g_p(\theta) = \begin{cases} 2\sqrt[6]{r} \cos\left(\dfrac{1}{3} \arccos \dfrac{-d}{\sqrt{r}} \right) - \dfrac{1 - L_s + L_s^2}{3(-1 + L_s)\sin(3\theta)} & \text{si } \theta \in \left[\dfrac{-\pi}{6}; 0 \right[\\[4mm] \dfrac{L_s}{\sqrt{1 - L_s + L_s^2}} & \text{si } \theta = 0 \\[4mm] 2\sqrt[6]{r} \cos\left(\dfrac{-2\pi}{3} + \dfrac{1}{3}\arccos \dfrac{-d}{\sqrt{r}} \right) - \dfrac{1 - L_s + L_s^2}{3(-1 + L_s)\sin(3\theta)} & \text{si } \theta \in \left] 0; \dfrac{\pi}{6} \right] \end{cases} \tag{8}$$

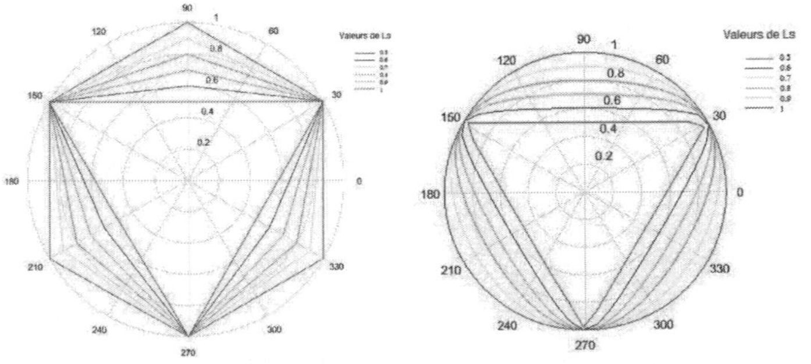

(a) Original criterion   (b) Smooth criterion

Figure 1.   Representation of g_p as a function of L_S.

with $d = \frac{2(1-L_s+L_s^2)^3}{27(-1+L_s)^3 \sin^3(3\theta)} - \frac{L_s^2}{(-1+L_s)\sin(3\theta)}$ and $r = \frac{(1-L_s+L_s^2)^6}{729(-1+L_s)^6 \sin^6(3\theta)}$.

Figure 1 presents the two forms of g_p for various values of the extension ratio.

3 CLOSEST POINT PROJECTION OF EXPERIMENTAL RESULTS ON THE YIELD SURFACE

3.1 Projection in the deviatoric plane

For the sharp criterion (original), calculation of the distance between experimental point and closest point projection, is straightforward. For the smooth criterion, the problem of projection in the deviatoric plane is represented at Figure 2.

We have $\sigma^+ g_p(\theta) = \sqrt{J_2}$ and we define $\rho = \frac{\sqrt{J_{2,exp}}}{\sigma^+}$, so we can write [6]:

$$\| \overrightarrow{\sigma\sigma}_{exp} \|^2 = \| \underline{s} \|^2 + \| \underline{s}_{exp} \|^2 - 2 \| \underline{s} \| \cdot \| \underline{s}_{exp} \| \cos\left(\theta - \theta_{exp}\right) \tag{9}$$

as we have $\| \underline{s} \| = \sqrt{2J_2} = 2\sigma^+ g_p(\theta)$ and $\| \underline{s}_{exp} \| = \sqrt{2J_{2,exp}} = \sqrt{2}\rho$, we can define $\| \overrightarrow{\sigma\sigma}_{exp} \|^2 = 2d^2$ so that:

$$d(\theta)^2 = \left(\sigma^+ g_p(\theta)\right)^2 + \rho^2 - 2\sigma^+ g_p(\theta)\rho \cos\left(\theta - \theta_{exp}\right) \tag{10}$$

The θ value of the projected point, is the value that minimizes $d^2(\theta)$. Deriving equation (10) with respect to θ, we can equivalently find the value of θ that is the root of the derivative:

$$2\sigma^{+2} g_p(\theta)g_p'(\theta) - 2\rho\sigma^+ \left[g_p'(\theta)\cos\left(\theta - \theta_{exp}\right) - g_p(\theta)\sin\left(\theta - \theta_{exp}\right)\right] = 0 \tag{11}$$

3.2 Determination of the projection point

We use a grid search method to determine the projected point of a given experimental point, i.e. examining among points on the yield surface the one that satisfies the condition of projection. We are prensenting the principle of the tools and implementation of this method.

For a given point of the yield surface, the tensor normal to the yield surface $\frac{\partial f}{\partial \sigma}$ is known. Hence if $\frac{\partial f}{\partial \sigma}$ tensor $\overrightarrow{\sigma\sigma}_{exp}$ are colinear, hence $\underline{\sigma}$ is the closest projection of $\underline{\sigma}_{exp}$ on the yield surface.

3.2.1 Expression of tensors

We introduce the orthoradial tensor, $\underline{\upsilon}$ [4]:

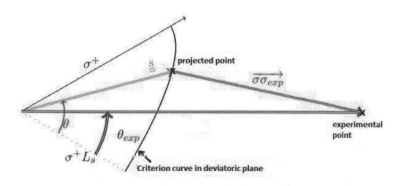

Figure 2. Closest point projection in the deviatoric plane for J_3 dependent criterion[6].

251

$$v = 3\frac{\sqrt{3}}{2}\frac{1}{J_2}\underset{=}{s^2} - \sqrt{3}\underset{=}{1} - \frac{9\sqrt{3}J_3}{4J_2^2}\underset{=}{s} \tag{12}$$

Hence those three tensors, $\mathcal{B} = \left\{ \underset{=}{\mathbb{I}}, \underset{=}{s}, \underset{=}{v} \right\}$ constitute an orthogonal basis, of the space of symmetric second order tensors. One can notice, that only $\underset{=}{\mathbb{I}}$ is independent from σ, and $\underset{=}{s}$ and $\underset{=}{v}$ define a polar base of π planes. Hence the gradient of the yield surface can be orthogonally decomposed in this base.

$$\frac{\partial f}{\partial \underset{=}{\sigma}} = g_u \underset{=}{\mathbb{I}} + g_s \underset{=}{s} + g_v \underset{=}{v} \tag{13}$$

With [4]:

$$g_u = \frac{3L_s^2 \sigma^{+2} - (1 - L_s + L_s^2)J_2}{\sqrt{3}\sqrt{1 + 36P_i}} + 5(1 - L_s)\frac{2L_s \sigma^{+3} + (1 - 2L_s)\sigma^+ J_2 + \dfrac{3\sqrt{3}J_3}{2}}{24m_b \sigma_{ci}\sqrt{P_i}} \tag{14}$$

$$g_s = \frac{9\sqrt{3}(1 - L_s)J_3}{4J_2} + (1 - L_s + L_s^2)\sigma^+ \tag{15}$$

$$g_v = J_2(1 - L_s) \tag{16}$$

Similarly, we can orthogonally decompose $\overrightarrow{\underset{=}{\sigma\sigma}}_{exp}$, in a first time:

$$\overrightarrow{\underset{=}{\sigma\sigma}}_{exp} = h_u \underset{=}{\mathbb{I}}_3 + h_s \underset{=}{s} + h_v \underset{=}{v} \tag{17}$$

Determination of the hydrostatic part is straightforward, as:

$$h_u = \sigma_{m,exp} - \sigma_m \tag{18}$$

We can observe that the deviatoric part of $\overrightarrow{\underset{=}{\sigma\sigma}}_{exp}$, can be split in two orthogonal components, a radial $h_s \underset{=}{s}$, and a orthoradial, $h_v \underset{=}{v}$; but those components cannot be expressed easily, as this polar coordinates are linked to the projection on the yield surface, and not to the experimental stress $\underset{=}{\sigma}_{exp}$. So it will be more efficient to study the norm of $h_s \underset{=}{s} + h_v \underset{=}{v}$, as this deviatoric part of $\overrightarrow{\underset{=}{\sigma\sigma}}_{exp}$ can be expressed as: $\left\| h_s \underset{=}{s} + h_v \underset{=}{v} \right\| = \sqrt{2}d$, with d is the minimum of 10.

3.2.2 Minimization of distance

AS the hydrostatic and deviatoric parts are orthogonal, we have

$$\left\| \overrightarrow{\underset{=}{\sigma\sigma}}_{exp} \right\|^2 = 3h_u^2 + 2d^2 = 3\left(\sigma_{m,exp} - \sigma_m \right)^2 + 2d^2 \tag{19}$$

So we will minimize, with respect to σ_m the following function:

$$D^2 = 3\left(\sigma_{m,exp} - \sigma_m \right)^2 + 2d^2 \tag{20}$$

However, d^2 is a function of σ_m and moreover its derivative cannot be expressed easily. So, we used a grid search method to compute the value of σ_m minimizing (20). Time is reasonable for original Hœk-Brown, but computation time is more important for the smooth criterion.

3.2.3 Determination of the projection for the smooth criterion

This method was developed for the smooth criterion. We will find value of σ for witch $\frac{\partial f}{\partial \underset{=}{\sigma}}$ and $\overrightarrow{\underset{=}{\sigma\sigma}}_{exp}$ are collinear. For that we will use the definition of the dilatancy angle, i.e. ratio between the hydrostatic and deviatoric part of $\frac{\partial f}{\partial \underline{\sigma}}$ and consider the ratio between the hydrostatic and deviatoric parts of $\overrightarrow{\underline{\sigma\sigma}}_{exp}$.

First we will compare hydrostatic and deviatoric part. Comparison of the hydrostatic part is straightforward and we can evaluate $\frac{h_u}{g_u}$.

The deviatoric part $g_s\underline{s}+g_v\underline{v}$ and $h_s\underline{s}+h_v\underline{v}$ are generally not collinear, but when $\underline{\sigma}$ is closest point projection of $\underline{\sigma}_{exp}$ in the deviatoric plane. In the latter case, we have $\left\|h_s\underline{s}+h_v\underline{v}\right\|=\sqrt{2}d$ with d defined by (10).

Finally, sufficient condition of collinearity between $deriv f\underline{\sigma}$ and $\overrightarrow{\underline{\sigma\sigma}}_{exp}$ is that: $\frac{|\sigma_{m,exp}-\sigma_m|}{|g_u|}=\frac{\sqrt{2}d}{\|g_s\underline{s}+g_v\underline{v}\|}$ En utilisant $\|g_u\underline{1}\|=\sqrt{3}g_u$ we finally get:

$$\sqrt{3}\,|\,\sigma_{m,exp}-\sigma_m\,|=\sqrt{2}d\tan\delta \qquad (21)$$

Where δ is dilatancy angle defined by $\tan\delta=\frac{g_u\|\underline{\mathbb{L}}_3\|}{g_s\|\underline{s}\|+g_v\|\underline{v}\|}$

Equations (14), (15) and (16) are used to calculated $\tan\delta$, then the roots of (21) can be computed. Considering if $\sigma_{m,exp}$ is inside of outside the yield surface, we can determine is σ_m is greater or lower than $\sigma_{m,exp}$, and then know which root of (21) is the value of σ_m that minimizes the distance.

4 RESULTS

Implementation was made in Matlab and grid search is used fo find values m_b and σ_{ci}. Table 1 present results find on triaxial test datas and compared to values identified by Colmenares et Zoback [7] and Figure 3 present fitted surfaces and experimental results in the stress space.

For Dunham Dolomite, we can notice that Colmenares et Zoback find great variations upon the values of σ_{ci} when fitting different criteria: 235 MPa for Mohr-Coulomb, 380 for Lade and 340 for Wiebols and Cook, leading to a mean value of 318 MPa. In this paper, we found values of σ_{ci} in this range, and results are close of those calculated using only triaxial tests.

Table 1. Results.

Data set	Original HB [7]		Original Hœk-Brown			Smooth Hœk-Brown		
	m_b	σ_{ci}	m_b	σ_{ci}	d^2	m_b	σ_{ci}	d^2
Dunham Dolomite	8	400	10	304	6.1165e + 03	7	328	$2.572 \cdot 10^3$
Solenhofen Limestone	4.6	370	7	291	1.8263e + 03	5	302	$1.7805 \cdot 10^3$
Shirahama Sandstone	18.2	65	39	24	654.6821	21	39	118.3444
Yuubari Shale	6.5	100	13	54	373.3447	9	64	$235.5316 \cdot 10^3$

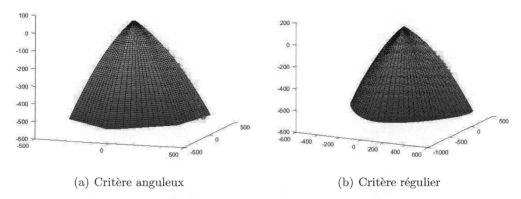

(a) Critère anguleux   (b) Critère régulier

Figure 3. Experimental results and fitted yield surfaces for Dunham Dolomite.

One can notice that values with the two different criteria are generally closer between themselves than from those of Colmenares et Zoback. So one we can assume, if when fitting and comparing criteria, that choice of fitting method can lead to discrepancies greater than the differences between criteria.

REFERENCES

[1] E. Hoek and E.T. Brown. Empirical strength criterion for rock masses. *Journal of the geotechnical engineering division*, September 1980.
[2] O.C. Zienkiewicz and G.N. Pande. Some useful forms of isotropic yeld surfaces for soil and rock mechanics. In *Numerical methods in soil and rock mechanics, Karlsruhe*, pages 3–16, September 1975.
[3] Carlos Carranza-Torres and C. Fairhurst. The elasto-plastic response of underground excavations in rock masses that satisfy the Hoek-Brown failure criterion. 36 :777–809, 1999.
[4] Siegfried Maïolino. *Fonction de charge générale en géomécanique : application aux travaux sou-terrains*. PhD thesis, École Polytechnique, 2006.
[5] Siegfried Maïolino. Proposition of a general yield function in geomechanics. *C. R. Mécanique*, 333(3) :279–284, 2005.
[6] Siegfried Maïolino. Alternative to return-mapping algorithm for computing plastic strain. application to dilatant materials. 2016.
[7] M.D. Zoback L.B. Colmenares. A statistical evaluation of intact rock failure criteria constrained by polyaxial test data for five different rocks. *International Journal of Rock Mechanics and Mining Sciences*, (39) :695–729, 2002.

Geomechanics and Geodynamics of Rock Masses – Litvinenko (Ed.)
© *2018 Taylor & Francis Group, London, ISBN 978-1-138-61645-5*

Application of a new clustering method for automatic identification of rock joint sets

Sayedalireza Fereshtenejad & Jae-Joon Song
Department of Energy Systems Engineering, Seoul National University, Seoul, South Korea

ABSTRACT: The analysis of data collected on rock discontinuities often requires that the data be separated into joint sets or groups. A statistical tool that facilitates the automatic identification of groups of clusters of observations in a data set is always required. For this purpose several algorithms were formerly proposed which are mostly based on the Fuzzy K-means method. This is a widely used partitioning technique, where optimization involves minimizing the sum of squared distances between objects and the means of clusters (centroids). However, empirical studies have shown some drawbacks of this popular clustering method, such as the tendency to produce spherical clusters and the scale dependency. More importantly, it has been found that this method tends to generate equal sized clusters, often referred to as the equal-size problem, which is not adequate in distinguishing clusters with different sizes. One explanation for such a tendency is that a partition based on the minimization of the trace of dispersion matrix is equivalent to the maximum likelihood partition when the data is assumed to come from multivariate normal mixture distributions with equal covariance matrices. This paper introduces the application of a clustering method which applies an optimization criterion based on the estimated covariance matrix to overcome the so-called equal-size problem associated with the K-means method, while maintaining computational simplicity. Advantages of the proposed method are demonstrated using artificial data with low dimensionality.

Keywords: Joint sets delineation, Clustering analysis, K-means algorithm, Equal-size problem

1 INTRODUCTION

Rock discontinuities are the most important features to control the stability of rock slopes and underground openings. The first step in determining the mechanical and hydraulic properties of rock masses is to characterize the number of joint sets and their representative orientation (Cacas et al., 1990; Long et al., 1985; Piggott and Elsworth, 1989; Zhang and Einstein, 2000; Song et al., 2001; Fereshtenejad et al., 2016). The dip and dip direction are the most common geometrical properties which are considered in joint clustering. The first method of tackling the problem of subjectivity in clustering discontinuities was proposed by Shanley and Mahtab (1976), and was then further improved by Mahtab and Yegulalp (1982). Their method utilizes a rejection scheme based on a randomness test for the objective partition of discontinuity data. The angle of search cone which is constructed to find out dense discontinuity normals has a significant effect on the number and size of the sets delineated during their proposed clustering process. Harisson (1992) was the first who applied pattern recognition algorithms based on fuzzy objective functions for clustering discontinuities considering their orientations. Sirat and Talbot (2001) proposed a method based on the use of artificial neural networks for classifying discontinuity sets. Rodriguez and Sitar (2006) demonstrated the use of a spectral clustering algorithm for the identification of rock discontinuity sets based on discontinuity orientations. They claimed that the convergence is fast, and the algorithm can be easily and efficiently implemented using popular software packages for numerical analysis. Dershowitz et al. (1996) proposed an algorithm for definition of fracture sets based on a probabilistic, geological approach to fracturing. The algorithm defines the fracture set as group of fractures with statistically homogeneous properties including not only orientation but also geological information such as termination modes, displacement, and mineralization. Zhou et al. (2002) demonstrated a methodology that clusters discontinuities into subsets based on multiple attributers, so that discontinuities within the same subset will have similar geometric properties. Tokhmechi et al. (2011) applied the coefficient of variation and principal component analysis (PCA) of joint properties to improve clustering process when a large number of joint properties are considered.

Despite the fact that various clustering algorithms are available, K-means clustering approach is still widely applied to the clustering of joint data. However, there are several drawbacks associated with conventional K-means algorithm which should be considered and modified to avoid arising any possible deficiency during clustering process. The so-called equal-size problem is one of the controversial issues of K-means clustering method which has not been sufficiently studied and needs to be addressed. The clustering method tends to generate equal sized clusters which is not simply adequate in distinguishing clusters with different sizes. However, no treatments or alternatives have been reported yet to remove this limitation in rock mechanics area. This paper introduces the application of a clustering method which applies an optimization criterion based on the estimated covariance matrix to overcome the so-called equal-size problem associated with the K-means method, while maintaining computational simplicity.

2 EQUAL-SIZE PROBLEM ASSOCIATED WITH APPLICATION OF K-MEANS CLUSTERING

The application of K-means is narrowed down to well separable, equal density and round-shaped (convex) clusters. Furthermore, this algorithm tends to generate clusters containing a nearly equal number of objects, which is referred to as the "equal-size" problem. In its quest to minimize the within-cluster sum of squares, the K-means algorithm gives more weight to larger clusters. Therefore, K-means algorithm is not capable to find unbalanced clusters of varying sizes, and densities. The main goal of K-means algorithm is to partition dataset into k clusters based on the minimization of a criterion over all k clusters. The most widely used clustering criterion is the sum of the squared Euclidean distances between each data point and the centroid (cluster center) of the subset which contains the data point. The criterion which is also called clustering error could be expressed by trace of dispersion matrix. For clustering n objects into k clusters the dispersion matrix W_m for each cluster is defined as

$$W_m = \sum_{l=1}^{n_m} \left(x_{ml} - \bar{x}_m \right) \left(x_{ml} - \bar{x}_m \right)'$$

(1)

where, n_m, $m = 1, ..., k$, is the number of objects in k^{th} cluster, x_{ml}, $l = 1, ..., n_m$, is l^{th} object of kth cluster, and $\bar{x}_m$ is the centroid of each cluster. The optimization criterion of the K-means method which is minimization of the sum of squared Euclidean distance between objects and its cluster mean, $d^2(x_{ml}, \bar{x}_m)$, is equivalent to minimization of $tr(W)$.

$$tr(W) = \sum_{m=1}^{k} \sum_{l=1}^{n_m} d^2 \left(x_{ml}, \bar{x}_m \right) = \sum_{m=1}^{k} \sum_{l=1}^{n_m} \left| x_{ml} - \bar{x}_m \right|^2.$$

(2)

Symons (1981) mentioned that when the cluster sizes are nearly equal, the method employing the determinant of the within-groups sum-of-squares matrix performs very well; however, when the group sizes are disparate, the $|W|$ criterion tends to favor partitions of equal size. To explain the reason of this deficiency, Yan (2005) applied the technique which had been proposed by Marriott (1982) for investigating the properties of several clustering optimization criteria. The technique detects the changes in the value of K-means criterion when a single object is added to the current data set. By adding a point x to cluster k, new dispersion matrix for the cluster is derived by

$$W_m^* = W_m + d_m d_m'$$ (3)

where, $d_m = (x - \bar{x}_m)\left(\dfrac{n_m}{n_m+1}\right)^{\frac{1}{2}}$. Therefore, $tr(W^*)$ is derived by

$$tr(W^*) = tr(W) + d_m d_m' = tr(W) + \left(1 - \dfrac{1}{n_m+1}\right)\|x - \bar{x}_m\|^2.$$ (4)

Suppose the distances between the new data x and two distinct cluster centroids are equal. $tr(W^*)$ becomes smaller provided that the new point is assigned to the cluster containing less data points. Hence, there is a higher possibility for a data point being assigned to clusters possessing fewer data points. Therefore, the optimization criterion of K-means clustering method is not appropriate for partitioning of clusters which are not equal in size.

3 NEW CLUSTERING ALGORITHM

Equal-size problem could be avoided by selecting an appropriate data processing method for clustering discontinuities. Yan (2005) proposed a method to eliminate the so called "equal-size" problem associated with the application of K-means clustering. The optimization criterion in the proposed method is based on the estimated covariance matrix of all the observations in each cluster $(\hat{Q}_m)$.

$$\hat{Q}_m = \dfrac{1}{n_m-1}\sum_{l=1}^{n_m}(x_{lm} - \bar{x}_m)(x_{lm} - \bar{x}_m)'$$ (5)

Given a fixed number of clusters, g, the proposed partitioning criterion minimizes the summation of $tr(\hat{Q}_m)$. over the number of clusters.

$$\sum_{m=1}^{g} tr(\bar{Q}_m) = \sum_{m=1}^{g} \dfrac{1}{n_m-1} tr(W_m) \equiv Q(g)$$ (6)

In this method, hill-climbing algorithm is designed to search for the best partition of data which optimizes the applied clustering criterion. This is a generic term covering many algorithms trying to reach an optimum by determining the optimum along successive directions (Besset, 2015). In fact, Hill-climbing is an iterative algorithm that starts with an arbitrary solution to a problem, then attempts to find a better solution by incremental change in a single element of the solution. The following steps elaborate how the algorithm minimizes the proposed optimization criterion.

According to the given number of clusters, g points are randomly selected as primary centroids of g clusters, $C_1, \ldots, C_g$. Then, the objects of data set are assigned to the clusters based on their distances to initial cluster centroids. Each of them is allocated to the cluster whose centroid is nearest to. The centroids of clusters are recalculated considering the mean of objects which are previously assigned to each cluster. Thereupon, the partitions are updated by changing a single object's membership which brings the largest reduction in $Q(g)$. When x_i which is currently assigned to C_k is moved to C_l ($k \neq l$), the value of the criterion changes from $Q(g)$ to $Q^*(g)$. Let the resulting estimated covariance matrices be $\hat{Q}_{k(i-)}$ and $\hat{Q}_{l(i+)}$ for C_k and C_l, respectively. Thus, the difference between $Q(g)$ and $Q^*(g)$, $\Delta(k \rightarrow l, i)$ is derived as below.

$$\begin{aligned}\Lambda(k \rightarrow l, i) &= Q(g) - Q^*(g) - \left[tr(\hat{Q}_k) + tr(\hat{Q}_l)\right] - \left[tr(\hat{Q}_{k(i-)}) + tr(\hat{Q}_{l(i+)})\right]\\ &= \dfrac{n_k d(i,C_k) - tr(W_k)}{(n_k-1)(n_k-2)} + \dfrac{tr(W_l)}{(n_l-1)n_l} - \dfrac{d(i,C_l)}{n_l+1}\end{aligned}$$ (7)

257

where $d(i, C_k)$ and $d(i, C_l)$ are defined as below.

$$d(i,C_k) = (x_i - \bar{x}_k)'(x_i - \bar{x}_k) \tag{8}$$

$$d(i,C_l) = (x_i - \bar{x}_l)'(x_i - \bar{x}_l) \tag{9}$$

The largest reduction in $Q(g)$ which is denoted by $\Delta(i)$ determines whether x_i will be reassigned to another cluster and $c(i)$ determines the destination cluster if $\Delta(i)$ gives the movement permission.

$$\Delta(i) = \max_{\substack{1 \le l \le g \\ l \ne k}} \Delta(k \to l, i) \tag{10}$$

$$c(i) = arg \max_{\substack{1 \le l \le g \\ l \ne k}} \Delta(k \to l, i) \tag{11}$$

If $\Delta(i)$ is positive, moving the x_i from C_k to another cluster will reduce $Q(g)$. The value of $\Delta(i)$ should be computed for all objects and destination cluster will be eventually decided based on that value. Provided that the membership of an object is changed, the centroids of the object's previous and current clusters should be updated. The calculation of $\Delta(i)$ should be repeated for all existing objects. The algorithm is stopped when no movement of a single object is observed during a complete pass through all objects. This algorithm finally end up with a local optimum of the criterion.

4 CASE STUDY

To investigate the applicability of the proposed algorithm, the method is applied to an artificial datasets. It was previously mentioned that the detrimental effects of the equal-size problem associated with K-means algorithm could be substantially significant when joint sets' centroids are closer and the number of sampled joints of each set is much different. This example is mentioned to find out the capability of the proposed clustering method to partition the sets which are not linearly separable and clearly distinct. The Fisher distribution which is the most commonly used probability density function for discontinuity orientations was considered (Fisher, 1953). MCSDO (Zheng et al., 2014), a program developed in Microsoft Excel environment for generating discontinuity orientations based on the Fisher distribution, was applied to generate two unequal-sized artificial joint sets whose borders are very close to each other. Table 1 lists the orientation characteristics which were originally considered to generate 390 discontinuities for joint set 1 and 60 discontinuities for joint set 2. Fisher constant (K) is a measure of the degree of clustering, or preferred orientation, within the population (Priest, 1993). For large K, the mass is concentrated on a small portion of the sphere around the mean orientation and for $K = 0$, the mass is uniformly distributed (Mardia, 1972). Figure 1 illustrates the lower hemisphere stereographic projection of joints normals whose orientations were generated following Fisher distribution function. For comparison, both K-means algorithm and the proposed clustering method were applied to partition whole dataset into two joint sets via a developed MATLAB script. Figure 2 shows the results of both clustering methods. As seen, none of the methods were able to classify all the data based on their predefined membership. This was quite predictable since a few members of two clusters are so close to each other

Table 1. The orientation which was originally used for generation of two joint sets.

Joint set	Dip (°)	Dip direction (°)	Fisher constant (K)	Number of joints
1	20	200	30	390
2	30	70	40	60

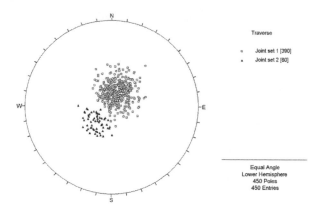

Figure 1. Stereographic projection of joints normals whose orientations were generated obeying Fisher distribution.

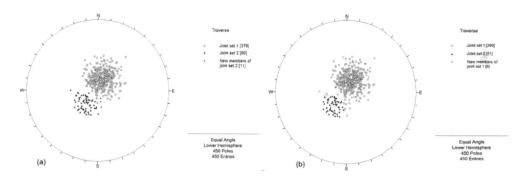

(a) (b)

Figure 2. (a) Red filled triangles are the joint normals which were wrongly assigned to joint set 2 after applying *K*-means clustering, (b) Red unfilled rectangles are the joint normals which were incorrectly assigned to joint set 1 after applying the proposed clustering method.

Table 2. Clustering results obtained applying different methods for generated data set.

	Joint set 1			Joint set 2		
	Before clustering	*K*-means method	Proposed method	Before clustering	*K*-means method	Proposed method
Dip (°)	24.1	24.36	24.2	30.17	27.74	30.45
Dip direction (°)	199.07	200.25	196.6	52.21	68.7	45.7
Number of joints	390	379	399	60	71	51

and the sets are not linearly separable. After applying *K*-means clustering, 11 joints originally assigned to joint set 1 were allocated to another set. Figure 2a depicts the poles of those joints with red filled triangles. After applying the proposed algorithm for clustering purpose, 9 joints originally assigned to joint set 2 were allocated to joint set 1. Figure 2b illustrates the poles of those joints with red unfilled rectangles. Table 2 lists the mean dip direction and mean dip of both joint sets before clustering and after clustering applying *K*-means and proposed algorithm. The comparison of the results obtained from two clustering approaches shows the superiority of the proposed method over *K*-means clustering method. As seen in Table 2, the mean dip and dip direction of joint set 1 obtained as a result of applying both applied clustering methods are quite close to those of the predefined joint set 1. Applying K-means algorithm leads to slightly better result for dip direction of joint set 1.

However, a significant discrepancy is observed between the original orientation of joint set 2 and the orientation obtained by using K-means clustering method. The resulted discrepancies obtained for dip and dip direction are 2.43° and 16.49°, respectively. 16.49° is not a negligible error and will have a great influence on the further analysis. On the other hand, the mean dip resulted after applying the proposed method for joint set 2 deviates less than 0.3° from that of predefined joint set 2. More importantly, the mentioned large discrepancy obtained after using K-means method for the dip direction of joint set 2 was reduced to 6.51° by applying the proposed method. Furthermore, the adjusted Rand index (Hubert and Arabie, 1985) was applied to assess the degree of agreement between predefined and clustered partitions. Basically, this index quantifies the similarity between two clusters. Its maximum value is 1 and its expected value in the case of random clusters is 0 (Yeung and Ruzzo, 2000). A larger adjusted Rand index means a higher agreement between two partitions and indicates that the clustering algorithm recover most of the classes. The adjusted Rand index for comparing the predefined clusters and those obtained by using K-means and the proposed clustering methods were computed 0.8724 and 0.8844, respectively. The calculated indices demonstrates the advantages of proposed algorithm compare to commonly used K-means clustering algorithm in this case.

5 CONCLUSIONS

K-means algorithm tends to generate clusters containing a nearly equal number of objects, which is referred to as the "equal size" problem. In its quest to minimize the within-cluster sum of squares, the K-means algorithm gives more weight to larger clusters. Therefore, the algorithm is not capable to find unbalanced clusters of varying sizes, and densities.

Equal-size problem could be avoided by selecting an appropriate data processing method for clustering discontinuities. This paper introduces the application of a clustering method which applies an optimization criterion based on the estimated covariance matrix to overcome the so-called equal-size problem associated with the K-means method, while maintaining computational simplicity. The main advantage of the introduced method is that its implementation is very convenient: it is defined without assuming any underlying distribution of the data; the designed algorithm for finding the optimal partition is time-efficient.

The application of proposed clustering method suffers from a number of limitations which previously mentioned as disadvantages of K-means clustering method. An important step in cluster analysis is to determine the best estimate of the number of clusters in a data set, which has a deterministic effect on the clustering results. The proposed method as well as conventional K-means algorithm requires pre-determination of the number of clusters to proceed with the computation. If the number of clusters is overestimated, however, the proposed method is capable of forming true clusters while the redundant clusters contain a few data within a very small region around their centers, whereas the K-means method tends to separate data into equal clusters of smaller sizes. Moreover, the application of the introduced clustering method is narrowed down to well separable and round-shaped (convex) clusters. The shortcomings of proposed method can be partially modified by using the same processes employed to improve K-means algorithm.

ACKNOWLEDGEMENT

This research was supported by Basic Science Research Program through the National Research Foundation of Korea (NRF) funded by the Ministry of Education, Science and Technology (NRF-2016R1D1 A1B03936033).

REFERENCES

Besset, D.H., 2015. Object-oriented implementation of numerical methods; An introduction with Small-talk, https://github.com/SquareBracketAssociates/NumericalMethods.

Cacas, M.C., Ledoux, E., de Marsily, G., Tillie, B., Barbreau, A., Durand, E., Feuga, B., Peaudecerf, P., 1990. Modeling fracture flow with a stochastic discrete fracture network: calibration and validation: 1. The flow model. Water Resour. Res. 26: 479–489. http://dx.doi.org/10.1029/WR026i003p00479.

Dershowitz, W., Busse, R., Geier, J., Uchida, M., 1996. A stochastic approach for fracture set definition. In: Aubertin M, Hassani F, Mitri H (eds) Proceedings of the 2nd North American rock mechanics symposium (NARMS '96), rock mechanics: tools and techniques, Montreal, Quebec, June 1996, pp. 1809–1813.

Fereshtenejad, S., Afshari, M.K., Yarahmadi Bafghi, A., Laderian, A., Safaei, H., Song, J.J., 2016. A discrete fracture network model for geometrical modeling of cylindrical folded rock layers. Engineering Geology. 215:81–90. https://doi.org/10.1016/j.enggeo.2016.11.004.

Fisher, S.R., 1953. Dispersion on a sphere. Proc. R. Soc. Lond. A, 217: 295–305. DOI: 10.1098/rspa.1953.0064.

Hubert, L., Arabie, P., 1985. Comparing partitions. Journal of Classification, 2:193–218. DOI:10.1007/BF01908075.

Long, J.C.S., Gilmour, P.,Witherspoon, P.A., 1985. A model for steady fluid flow in random three-dimensional networks of disc-shaped fractures. Water Resour. Res. 21: 1105–1115. http://dx.doi.org/10.1029/WR021i008p01105.

Mahtab, M.A., Yegulalp, T.M., 1982. A rejection criterion for definition of clusters in orientation data. American Rock Mechanics Association.

Mardia, K.V., 1972. Statistics of directional data. London and New York: Academic Press.

Marriott, F.H.C., 1982. Optimization methods of cluster analysis. Biometrika, 69:417–421. DOI: https://doi.org/10.1093/biomet/69.2.417.

Piggott, A.R., Elsworth, D., 1989. Physical and numerical studies of a fracture system model. Water Resour. Res. 25:457–462. http://dx.doi.org/10.1029/WR025i003p00457.

Priest, S.D., 1993. Discontinuity Analysis for Rock Engineering, 1st Edn. Chapman & Hall, London.

Shanley, R.J., Mahtab, M.A., 1976. Delineation and analysis of clusters in orientation data. J Math Geol. 8(1):9–23. doi:10.1007/BF01039681.

Sirat, M., Talbot, C.G., 2001. Application of artificial neural networks to fracture analysis at the Aspo HRL, Sweden: Fracture Sets Classification. Int. J. Rock Mech. Min. Sci. 38, 621–639. https://doi.org/10.1016/S1365–1609(01)00030–2.

Song, J.J., Lee, C.I., Seto, M., 2001. Stability analysis of rock blocks around a tunnel using a statistical joint modeling technique. Tunn. Undergr. Space Technol. 16:341–351. http://dx.doi.org/10.1016/S0886-7798(01)00063-3.

Symons, M.J., 1981. Clustering criteria and multivariate normal mixtures. Biometrics, 37:35–43. DOI: 10.2307/2530520.

Tokhmechi, B., Memarian, H., Moshiri, B., Rasouli, V., Ahmadi Noubari, H., 2011. Investigating the validity of conventional joint set clustering methods. Eng Geol.; 118:75–81. https://doi.org/10.1016/j.enggeo.2011.01.002.

Yan, M., 2005. Methods of determining the number of clusters in a data set and a new clustering criterion (Ph.D.thesis), Virginia Polytechnic Institute and State University.

Yeung, K.Y., Ruzzo, W.L., 2000. An Empirical Study on Principal Component Analysis for Clustering Gene Expression Data, Technical Report UW-CSE-2000-11-03, Dept. of Computer Science & Eng., Univ. of Washington.

Zhang, L., Einstein, H.H., 2000. Estimating the intensity of rock discontinuities. Int. J. Rock Mech. Min. Sci. 37:819–837. http://dx.doi.org/10.1016/S1365-1609(00)00022-8.

Zheng, j., Deng, J., Yang, X., Wei, J., Zheng, H., Cui, Y., 2014. An improved Monte Carlo simulation method for discontinuity orientations based on Fisher distribution and its program implementation, Computers and Geotechnics, 61: 266–276, https://doi.org/10.1016/j.compgeo.2014.06.006.

Zhou, W., Maerz, N.H., 2002. Implementation of multivariate clustering methods for characterizing discontinuities data from scanlines and oriented boreholes. Comput. Geosci. 28 (7), 827–839. http://dx.doi.org/10.1016/S0098 3004(01)00111-X.

Geomechanics and Geodynamics of Rock Masses – Litvinenko (Ed.)
© 2018 Taylor & Francis Group, London, ISBN 978-1-138-61645-5

Experimental evidences of thermo-mechanical induced effects on jointed rock masses through infrared thermography and stress-strain monitoring

Matteo Fiorucci, Gian Marco Marmoni & Salvatore Martino
Department of "Sapienza", Research Center for Geological Risk (CERI) and Earth Sciences, University of Rome, Rome, Italy

Antonella Paciello
Italian National Agency for New Technologies, Energy and Sustainable Economic Development, (ENEA)—Casaccia Research Center, Rome, Italy

ABSTRACT: The Mediterranean area is subjected to considerable daily and seasonal thermal variations due to intense solar radiation. This effect influences the long-term behaviour of jointed rock masses, operating as thermal fatigue process and acting as a preparatory factor for rock failures. In order to quantify intensity and influence of thermo-mechanical effects on rock slope stability, a multi-parametric monitoring is operative in an abandoned limestone quarry at Acuto (central Italy) by remote (i.e. IR Thermography) and direct (i.e. thermocouple and strain sensors) sensing techniques. The monitoring system has been focused on an intensely jointed rock block. Several joints bound a prismatic-shaped volume, isolate it from the rock wall behind and cause its protrusion respect to the quarry wall so increasing proneness to fall and toppling. Preliminary data analyses highlight the cyclical deformative response of block to thermal forcing, revealing the effect of sun radiation and exposure because of the heating of the rock surface.

1 INTRODUCTION

Short-time transient phenomena such as rainfalls, earthquakes, or excavations are the most common factors responsible for stress-field variations, which lead to slope failure. In specific meteorological conditions, rock masses can also be sensitive to thermomechanical forcing related to cyclic expansion and contraction that can operate over long time, inducing cumulative deformations. Cyclic thermal stresses are, indeed, regarded to operate as fatigue processes (Hall, 1999) acting along joints of a fractured rock mass (Pasten et al., 2015) and extending in depth even below the thermal active layer (Gischig et al., 2011). These effects, ascribable to preparatory factors, can induce irreversible deformations along joints and favour slope instabilities (Gunzburger et al., 2005 and reference therein). In order to quantify the effect of intense solar radiation and evaluate the predisposition of the rock to thermally-induced deformations, a thermal monitoring was carried out through integrated direct and indirect (i.e. thermographic) sensing techniques by the coupling use of a rock thermometer and thermal camera (Mineo et al., 2015; Pappalardo et al., 2016). IR-imaging represents a reliable tool that allow to quickly derive temperature distribution at the rock-air interface, quantifying the effect of solar radiation and measuring the heat propagation within a discontinuous medium.

2 ACUTO TEST-SITE

The Acuto test-site is located in central Italy on the Mt. Ernici ridge, where Mesozoic limestones widely outcrop (Fantini et al., 2016). An abandoned quarry was selected to host a multi-parametric monitoring system devoted to analyse the thermo-mechanical behaviour of rock masses at rock-block scale. The N20E trending quarry wall is exposed to intense solar radiation and predisposed to gravitational instabilities by rock fall, sliding or toppling mechanisms. The rock mass is intensely jointed due to persistent low-angle stratification and high-angle N80 W conjugate joint sets that isolates a prismatic rock block (Fig. 1), so providing kinematic arrangement for planar and wedge slides. Laboratory tests were performed to define the main physical properties of Mesozoic limestone. More in particular, values of weight per unit of volume were evaluated (γ_{sat}: 26.45 kN/m³; γ_n: 26.44 kN/m³; porosity: 1.10%). The average Uniaxial Compression Strength (UCS) of the rock, by Point Load tests, resulted of 130 MPa. According to the Deere & Miller (1966) chart, this value corresponds to a Young modulus of about $6*10^4$ MPa. The rock mass properties were defined through the RMR classification, which assigned to Mesozoic limestone a score of 68.

A jointed rock block about 20 m³ sized was selected as main target for the installation of a multi-parametric monitoring system, consisting of a rock thermometer (Type-K Thermocouple) and geotechnical devices, including six HBM® strain gauges and four extensometers, devoted to monitoring the strain on micro-fractures and open joints, respectively. This block is isolated by the rock wall behind and protruded respect to the quarry face by opened fractures (Fig. 1) and it is exposed to sun radiation along two sides: a front-side exposed to the East and a back-side exposed to the South.

The installed multi-sensor monitoring system also includes a weather station for monitoring local weather conditions equipped with rain gauge, air thermometer, hygrometer and anemometer. The monitoring system has been implemented with remote survey analysis via Infrared Thermography (IRT) to evaluate the role of solar radiation and rock-block exposure on rock temperature in the thermal active layer and on surveyed joints in inducing slope instabilities.

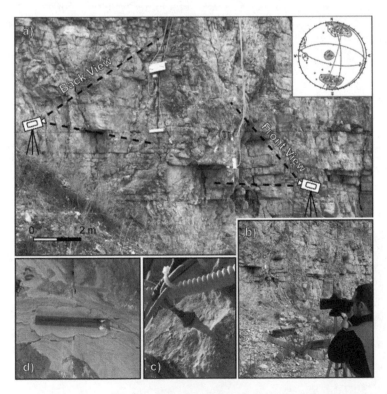

Figure 1.   a) Point of view of IR Thermographic survey respect to the monitored rock-block. b) Indirect surveys performed in the quarry face by handle thermal camera Testo-885-2 based on tripod. Details of rock thermometer c) and strain gages d) are also shown.

3 SURVEY TECHNIQUES AND METHODS

Infrared Thermography (IRT) technique allows to determine the temperature of rock mass by capturing its emitted infrared radiation (wavelength = 0.8–14 μm) and consequently converting it to temperatures by means of a proportionality constant represented by emissivity. The outputs result, represented in a false colour optical image, consist in a matrix of pixels that represent absolute temperatures in the interested zone (Pappalardo et al., 2016). IRT surveys were experimented in sunny days, to evaluate daily amplitudes and trend of rock temperature at the rock-air interface, using a Testo® 885-1 thermal camera (measuring range between −30 and 100°C; accuracy of measurements ± 2°C; FOV 30° × 24°; image resolution 320*240 pixel; maximum geometric resolution of 1.7 mrad). Thermal indirect monitoring was performed by acquisition of standardised thermal images over the diurnal hours with a regular time interval of 30 minutes, from two different points of view located 5 m apart from both front- and back-side of the monitored rock-block (Fig. 1a). This distance is small enough to guarantee a spatial resolution of about 8.5 mm and allow the distinction of the joints.

A quantitative analysis of thermal images was carried out by extraction of thermal time-series from some rock mass zones selected on both rock-block sides. At this aim, thermal images were converted to thermal matrixes from .csv to. ASCII file format by using IRSoft® open-source software and a Phyton compiled script. Matrixes were then imported in GIS environment where geometric distortions were corrected by relative georeferencing processes. Thermal time-series were extracted by selected control points on the georeferenced images for different zones of the rock mass. To guarantee a statistical significance of the recorded thermal values, the so derived temperatures were averaged by measuring on nine neighbour control points.

In order to evaluate the influence of temperature variations in the mechanical behaviour of rock mass and constrain its role in the stability of the quarry face, a direct thermal and mechanical monitoring system was implemented by installing a thermocouple (measuring range between −30 and 100°C; resolution of 0.1°C) on the front-face of the rock block at a depth of 8 cm. Strain and temperature time-series were derived for all the strain gauges and extensometers installed on micro-fractures and open joints with 1-minute sampling step. All the data were stored on local Campbell Scientific datalogger and remotely managed by a Cloud-based database.

4 RESULTS

The recorded measurements of air temperature, rock temperature and rock strain, have been derived by direct and remote sensing techniques, thanks to the devices directly installed on the rock-block. Every day, a heating ramp and a cooling ramp have been identified for temperature and strain values, highlighting a cyclical thermal behaviour for the rock-block. The heating of the rock block results in daily cyclic effective deformation strains. In fact, after each daily thermal cycle, the rock mass accumulates a small effective deformation. This deformation was derived for each daily cycle through a customised Unix script, which allowed to compute the value of the effective daily cumulative strain, i.e. resulting from a complete daily thermal cycle (i.e. between two thermal minimum – Fig. 2). These cumulative values computed over long-time make it possible to output inelastic deformations related to thermo-mechanical effects suffered by the rock-block.

The recorded time series of temperature values show relevant differences between rock—and air—temperature, since the first one always exceeds the air temperature due to direct solar radiation over the rock block. The maximum temperature every day recorded into the rock-block occurred around 10:00, i.e. anticipating the air temperature peak that occurs around 13:00. The heating and cooling rates derived by the rock-thermometer measurements, reveal different average trends with a fast heating phase (experienced during the hours of direct solar radiation) significantly shorter than the cooling phase. More in particular, the last one

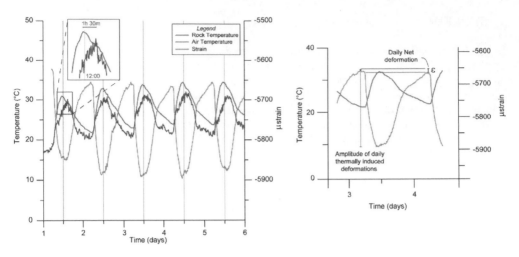

Figure 2. Example of daily strain cycles within the rock mass due to cyclic thermal input (left); computation of the daily effective deformation (right).

can be divided in two sub-phases: a first sub-phase, during which the rock cools quickly, and a second sub-phase, during which the rock cools slowly.

The coupling of IRT techniques allowed to derive the heating and cooling phases of the surficial temperature at the rock-air interface for both sides of the rock block, from 8:00 to 17:00 (solar times) for each day. The obtained temperature time-series evidenced the role of solar radiation in the heating of the rock mass, showing a clear dependency of thermal input on both sun paths and rock-block exposure. The front-side of the rock block is irradiated since the dawn to the midday, while the back-side becomes lights up since the first hours of the afternoon to the sunset. For example in a sunny day during Summer 2016 (maximum air temperature 34°C, minimum air temperature 22°C, average air humidity 39%, average wind speed 1.98 m/s), the rock-block in its front-side experienced a maximum temperature up to 41 +− 0.5°C. This peak of temperature was reached at 10:30, when a maximum is achieved even at the rock thermometer located on the same block-side. On the back-side, shadowed at 10:00 (inset in Fig. 3), the maximum value is reached around 14:00, according to the attitude of the rock-block faces. The maximum temperature value recorded by IR-camera on the front-side is about 8°C higher than the ones detected within the rock by thermocouple at depth of 8 cm. This difference reaches more than 10°C during the second part of the day, when air temperature decreases, and sunrays light up the back-side of the block (Fig. 3a). Moreover, the measured temperature difference reaches the maximum value during the daily hours corresponding to an intense solar radiation, decreasing asymptotically down to a minimum value after the midday, when the shadow involves the overall block (around 15:00). These temperature values, derived by IRT, are in good agreement with the ones that would be obtained scaling the temperature of the rock matrix to the surface at rock-air interface, assuming a 2D conductive propagation of heat flow in a semi-infinite half space (Carslaw and Jaeger, 1959).

The IRT survey was also applied to monitor temperature across the main open joint, which isolates the rock block from the quarry wall. During the heating phase, the air temperature within the joint is almost 6°C lower than the surrounding rock temperature, because of different thermal conductivity and specific heat capacity between the two media. During the cooling phase, the air temperature within the joint and the rock temperature gradually decrease towards the same value, so reaching isothermal conditions when the shadow involves the rock-block. The environmental air temperature is consistently lower than the one within the joint over all the monitoring period (half day cycle), revealing that the rock temperature influences the air temperature within the main joint (Fig. 3b).

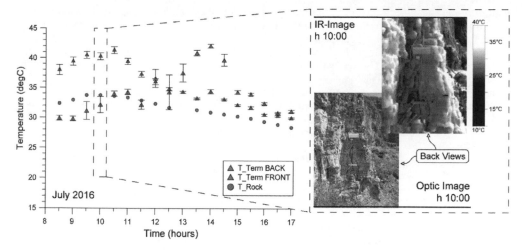

Figure 3. a) Plot of temperature vs. time derived by IRT survey that shows time temperature values of the rock-block at the rock-air interface for both rock-block side; the temperature of the rock at 8 cm depth on front-side and surveyed by thermocouple is also shown; b) Plot of temperature vs. time derived by IRT survey that shows time temperature values across the main open fracture behind the rock block.

5 CONCLUSIONS AND PERSPECTIVE

The multi-parametric monitoring system installed on a rock-block exposed to intense solar radiation, allowed to reconstruct the temperature and strain variations in the rock mass, highlighting the link between rock deformation and the daily thermal input. This monitoring was integrated by remote IRT, which allow to discern the significant incremental contribution of heat radiation on the rock-air interface, revealing a suitable technique for the study of thermalisation of discontinuous rock masses.

Preliminary IRT surveys, carried out over several days, will be extended over longer period in order to evaluate seasonal thermo-mechanical effects on the isolated rock-block or the rock mass behind. These differential heating conditions can influence the rock block stability, causing inelastic strains along joints and leading to possible rock failures (i.e. fall or toppling). Future experiments will aim to perform stress-strain numerical analysis to identify cause-to-effect relations between thermal forcing and induced strains, with the purpose of assessing preparatory conditions for slope instability over longer time-scales, i.e. resulting after daily and seasonal cyclical stress changes.

ACKNOWLEDGMENTS

This research is framed in the project "Influenza delle proprietà termo-meccaniche di ammassi rocciosi fratturati ai fini della stabilità e della mitigazione del rischio da frana" (Year 2016, prot. P.I. Matteo Fiorucci) funded by Sapienza University of Rome. The authors wish to acknowledge P. Mazzanti and NHAZCA S.r.l. for technical support to the thermographic monitoring. We are also grateful to G. Grechi, A. Serafini for supporting data acquisition and L.M. Giannini for suggestions provided in the analysis of the IR-images dataset.

REFERENCES

Carslaw H.S. and Jaeger J.C., 1959. Conduction of heat in solids – 2nd ed., Oxford: Clarendon Press.
Deere D.U. and Miller R.P., 1966. Engineering classification and index properties of rock. Tech. Report Air Force Weapons Lab., New Mexico, 65–116.

Fantini A., Fiorucci M., Martino S, Marino L., Napoli G., Prestininzi A., Salvetti O., Sarandrea P., Stedile L., 2016). Multisensor system designed for monitoring rock fall: the experimental test-site of Acuto (Italy). Rend. Online Soc. Geol. It., 41, pp. 147–150.

Gischig V.S., Moore J.R., Evans K.F., Amann F., Loew S., 2011. Thermomechanical forcing of deep rock slope deformation: Conceptual study of a simplified slope – Journal of Geophysical Research, 116. doi:10.1029/2011JF002006.

Gunzburger Y., Merrien-Soukatchoff V., Guglielmi Y., 2005. Influence of daily surface temperature fluctuations on rock slop stability: case study of the Rochers de Valabres slope (France) – Int. J. Rock Mech. Min. Sci., 42: 331–349.

Hall K., 1999. The role of thermal stress fatigue in the breakdown of rock in cold regions. Geomorphology 31:47–63.

Mineo S., Pappalardo G., Rapisarda F., Cubito A., Di Maria G., 2015. Integrated geostructural, seismic and infrared thermography surveys for the study of an unstable rock slope in the Peloritani Chain (NE Sicily). Eng Geol 195:225–235. doi: 10.1016/j.enggeo.2015.06.010.

Pappalardo G., Mineo S., Perriello Zampelli S., Cubito A., Calcaterra D., 2016. InfraRed Thermography proposed for the estimation of the Cooling Rate Index in the remote survey of rock masses. Int J Rock Mech Min Sci 83:182–196. doi: 10.1016/j.ijrmms.2016.01.010.

Pasten C., Garcia M., Cortes D.D., 2015. Physical and numerical modelling of the thermally induced wedging mechanism. Géotechnique Letters 5, 186–190, http://dx.doi.org/10.1680/geolett.15.00072.

Geomechanics and Geodynamics of Rock Masses – Litvinenko (Ed.)
© *2018 Taylor & Francis Group, London, ISBN 978-1-138-61645-5*

A correlation between thermal conductivity and P-wave velocity of damaged granite

P.K. Gautam
Department of Earth Science, Indian Institute of Technology, Powai, Mumbai, India

A.K. Verma & P. Sharma
Department of Mining Engineering, Indian Institute of Technology (Indian School of Mines), Dhanbad, Jharkhand, India

T.N. Singh
Department of Earth Science, Indian Institute of Technology, Powai, Mumbai, India

ABSTRACT: Knowledge of thermal conductivity will require proper understanding of the thermophysical behaviour of the host rocks i.e. granites. The purpose this study is to understand the correlation between thermal conductivity and the physical properties of damaged granite exposed to high temperature (up to 600°C). The result showed that: (1) the mass loss rate increases exponentially with the increase of temperature and reaches to 8.46% at 600°C; (2) thermal conductivity decreases linearly with the increase of temperature; (2) the damage threshold temperature of granite specimens is found to be 300°C and the limit temperature is 600°C; (4) there is a good linear relationship between the thermal conductivity and P-wave veolocity, with the correlation coefficient (R^2) of the fitting line is 0.989; (5) Microscopic observations of rock specimens revealed mainly intergranular microcracks in quartz, the opening of cleavage planes and the deformation in mica.

Keywords: Thermal Conductivity, Granite, Temperature, Intergranular crack, Intragranular crack

1 INTRODUCTION

Damaged rock caused by heat needs to be considered in dealing with many rock engineering applications along with temperature is one of the most significant ones. The thermal conductivity of rock depends not only on its mineral composition and texture but also on its degree of crystallisation, which affects the physico-mechanical properties. With increasing geotechnical engineering structures concerning with high-temperature problems such as building materials and rock mass in underground works, geothermal energy recovery system or extreme situations such as tunnel fires, fires in excavations in the vicinity of nuclear waste repositories and deep wells for injection of carbon dioxides, etc. it is important to study the influence of heat conduction on the physical and mechanical properties of rocks. While reliable formation temperature estimates can be obtained locally from borehole measurements no existing tool allows for large-scale estimation of rock formations temperature away from available boreholes. Therefore, if good correlations are established between thermal conductivity and the P-wave velocity properties of rocks, this would be helpful for geothermal energy and radioactive waste storage application. An important empirical relationship between thermal conductivity and physical properties was proposed by different researchers (Incropera & Witt (1985); ASTM, 1990; ASTM C,1990; Salmon, 2001; Xamán et al., 2009). The hot plate method (BS874 1986) is typical of the steady-state method of determining the thermal conductivity. The thermal conductivity is determined by the measurement of temperature gradients in the rock and heat input (ASTM C 1045-90 1990) (Yasar & Erdogan 2004; Chaki et al., 2008; Sharma & Singh 2008; Singh et al., 2007; Kahraman, 2002; Khandelwal & Ranjith 2010; Khandelwal & Singh 2009; Khandelwal, 2013; Jha et al., 2016; Verma et al., 2016). In this work, the main focus lies on the variation of mass loss rate and thermal conductivity after thermal treatment. Based on data sets from different temperature, the aim of this study is to set up an empirical relationship between thermal conductivity and physical properties after thermal treatment at ambient temperature.

2 AREA OF STUDY

The rock samples were collected from Jalore district, Rajasthan, India. It was investigated for its mechanical, thermal, and thermomechanical behaviour. This district is situated between Latitude 24°37′ & 25°49′ and Longitude 71°11′ & 73°05′ and is bounded by Banner district in NW, Sirohi district in SE, Pali district in NE and Banaskantha district of Gujarat state in SW. The average dry density of granite specimens is about 2.62 g/cm^3 and the natural content of water is about 0.11%.

3 LABORATORY INVESTIGATION

Specimens were prepared as per the ISRM (1981) standards (Ø54 × 32 mm^2 and Ø54 × 129 mm^2 cylinders). The end faces of the specimens were made parallel with an accuracy of 0.5 mm and the tangent planes of specimens of both sides were parallel with an accuracy of 0.2 mm. Twenty-one specimens were made, and they were divided into seven groups heated to different temperatures and 3 specimens were assigned to each group to ensure data accuracy. Before testing, the specimens were air dried at 102°C for 24 h to remove any moisture. Heating treatment was conducted using programmable high-temperature furnace at the heating rate of 10°C/min to avoid thermal shock. Specimens were categorised into 7 groups based on heating temperature room temperature, 100°C, 200°C, 300°C, 400°C, 500°C and 600°C. The programmed temperature, once reached, is kept constant for 12 h. When the heating process was over, we kept specimens at designated temperature for 30 min in order to ensure specimens heated evenly, and then let them cool naturally.

3.1 Determination of thermal conductivity

Thermal conductivity (k) is defined as the quantity of heat (Q) transmitted through a unit thickness (ΔL) in a direction normal to a surface of unit area (A) due to a unit temperature gradient (ΔT) under steady-state conditions when the heat transfer is dependent only on the temperature gradient using Eq. (1).

$$k = \frac{Q \times \Delta L}{A \times \Delta T} \qquad (1)$$

Thermal conductivity was measured using Guarded Hot Plate method and is one of the most common methods for measuring the thermal conductivity of rock specimens. The setup is usually vertical with the hot brass plate at the top, the specimens in between and then the cold brass plate at the bottom. The specimens of 54 mm diameter and 32 mm thickness were used for the experiments.

3.2 Determination of longitudinal wave velocity

In this system, whose transducer frequency is 54 kHz, the sampling interval is 1–200 μs, emission voltage is 500 or 1000 V and frequency bandwidth of 200 Hz to 200 kHz. We used silicone gel as a coupling agent between granite specimens and ultrasonic transducers in order to transmit the ultrasonic energy to the specimen and also do not affect the measurement of others viscous coupling. The distance that the pulse travelled in the material must be measured to calculate the velocity, using the formula and results of longitudinal wave velocity, pulse from the emitter to receiver end transducer in the rock specimens using Eq. (2) as per ISRM (1978a) standard. The specimens of 54 mm diameter and 32 mm thickness were used for the experiments.

$$\text{Pulse elocity} = \frac{\text{Path length}}{\text{Transit time}} \text{ m/s} \qquad (2)$$

4 RESULTS AND DISCUSSION

4.1 *Variation of thermal cracks and mineralogy of granite*

Thermal cracks in granite specimens were observed after high temperatures treatment as shown in Fig. 1. When the temperature is lower than 300°C, thermal cracks only develop at the edges of the surface. When the temperature is above 300°C the minerals in rock expand differently and generate thermal stress. When the thermal stress is greater than the adhesive force between mineral particles, cracks will develop. These thermal stress changes eventually influence the physical response of the rocks. Increasing the temperature causes variation in the microstructural characteristics of the rock such as grain size, grain shape, mineralogy, etc. (Fig. 2). It is observed that a large number of fractures developed on the idiomorphic are feldspars with straight grain boundaries while the quartz shows straight and interlobate grain boundaries. Microscopic images of thin sections of granites are shown along with the grain size variation and found to be mainly in the range of 0.6 mm–2.7 mm (Fig. 3 and Table 1). The amount of intergranular or intragranular cracking is dependent on the stress state, rock type and mineralogy of the rock mass. Intergranular cracks occur along grain boundaries, preexisting faults and micro-cracks of the rock, while intragranular cracks occur through weaker mineralogical constituents.

4.2 *Variations of mass loss rate and thermal conductivity after different temperatures*

Measurement of mass loss has been carried at different temperatures by taking the mass of each rock specimens before and after the heating at designated temperature. It can be observed that up to 300°C, the mass loss rate is almost constant (Fig. 4). This is because free water

Figure 1. Thermal cracks on the surface of granite after high-temperature treatment.

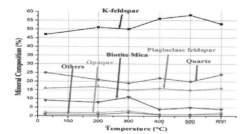

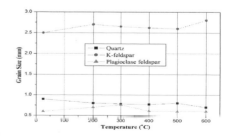

Figure 2. Effect of temperature on the mineral composition of granite.

Figure 3. Effect of temperature on the grain size of granite.

271

Table 1. Variation of thermal conductivity and physical properties in Jalore granite at elevated temperatures.

Temperature (°C)	Statistical parameters	Thermal conductivity (W/mK)	P-Wave velocity (m/s)	Damage coefficient
25°C	Max.	3.12	5710	0
	Min.	3.11	5161	0
	Avg.	3.12	5492	0
	Std. dev.	0.0072	221.43	0
200°C	Max.	2.94	4891	0.18
	Min.	2.93	4562	0.09
	Avg.	2.94	4761	0.13
	Std. dev.	0.0051	135.77	0.0392
300°C	Max.	2.49	4571	0.3
	Min.	2.48	3719	0.2
	Avg.	2.48	4146	0.25
	Std. dev.	0.0043	389.48	0.0436
400°C	Max.	2.39	3411	0.46
	Min.	2.38	2769	0.39
	Avg.	2.39	3169	0.42
	Std. dev.	0.0008	312.42	0.0318
500°C	Max.	2.01	2183	0.62
	Min.	1.99	1972	0.59
	Avg.	2.00	2079	0.62
	Std. dev.	0.0014	98.84	0.0181
600°C	Max.	1.88	1451	0.84
	Min.	1.86	839	0.74
	Avg.	1.87	1196	0.78
	Std. dev.	0.0073	275.81	0.0424

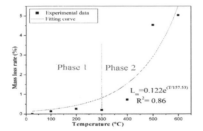

Figure 4. Variation of mass loss rate after high-temperature.

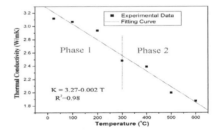

Figure 5. Variation of thermal conductivity after high temperature.

would escape at around 100°C and the bound water escapes between 200–300°C. Cracks are mainly intergranular taking advantage of mica boundaries as shown in Fig. 4. Intergranular cracks are more frequent in quartz-feldspar (both alkaline and plagioclases) borders than in quartz–mica or feldspar–mica. Fig. 1 shows that when the heating temperature is higher than 300°C, the colour of granite is changed significantly (Sun et al., 2017). This change resulted from the loss of water of a different kind. Above 300°C, the loss of crystal water and structural water would lead to the loss of constitutive water and dehydroxification, which results in the exponential mass loss rate, damage of mineral crystal lattice skeleton and increase in the number of defects of granite. Intragranular and intergranular cracks may have formed during heating or may be the result of the growth of preexisting cracks. The most remarkable reaction is the inversion of quartz from α-to-β phase at 573°C. The high value of R^2 (0.86) indicates the good correlation between mass loss rate and temperature.

The thermal conductivity of granite after heat treatment at different temperature has been measured using the steady-state method as shown in Fig. 5. It can be observed that the

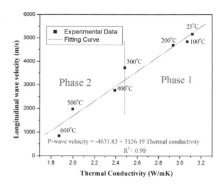

Figure 6. Correlation between thermal conductivity and longitudinal wave velocity.

thermal conductivity decreases with increase in temperature. Variation of thermal conductivity can be divided into two phases. Phase 1, represents the temperature range from room temperature to 300°C and phase 2 from 300°C to 600°C inter-granular fracture and widening of micro-cracks along with the formation of new cracks are key factors which affect thermal conductivity of rock specimens. In phase 2, thermal conductivity falls largely due to more physical damage induced because of high temperature. Above temperature from 600°C environment causes intergranular fracture and intragranular fracture along with other physical phenomena such as propagation and widening of microcracks, joining of cracks and formation of relatively larger cracks. Correlation of thermal conductivity and temperature shows a linear trend with a high value of correlation coefficient (R2 = 0.98).

4.3 Correlation of thermal conductivity with P-wave velocity

In phase 2, longitudinal wave velocity, and the thermal conductivity, both decreases rapidly (Fig. 6). The heating of rock specimens above 300°C causes the change in microstructure by different mechanisms such as the development of new cracks, coalescence of preexisting and new cracks, propagation of cracks to adjacent locations and widening of cracks. Additionally, when the temperature is lower, cracks only develop at the edges of grains. When the temperature reaches a threshold, the temperature may either be caused by thermal expansion and altered by microcrack network or driven by the structural damage of rocks (Fig. 2). In phase 2 the result of these mechanisms shows an increase of the crack density and size of cracks which may be called as a cavity. Such major changes in the physical condition of specimens cause the change in thermal conductivity and longitudinal wave velocity. Therefore, it can be inferred that the longitudinal wave velocity and thermal conductivity of granite are related to each other. To find out the relation between P-wave velocity and thermal conductivity, a linear equation has been proposed and a fitting curve has been plotted. It shows the goodness of the proposed equation with a value of R^2 as high as 0.98, indicating a very good correlation of thermal conductivity and P-wave velocity.

5 CONCLUSION

The following observations and conclusions were made:

1. A consistent colour change with increasing temperature was observed for all specimens where the colour had changed from pink brownish to reddish from the preheated temperature of 25°C to 600°C, irrespective of the cooling condition.
2. The mass loss rate, porosity and the damage coefficient increases with temperature. This may be caused by the gradual loss of structural water and crystalline water. Besides, the rapid development of thermal cracks (caused by the different expansion of minerals) after 300°C may lead to this result. Consequently, the conclusion can be drawn that 300°C should be the damage threshold temperature of the Jalore granite.

273

3. The longitudinal wave velocity, elastic modulus, peak stress and tensile strength were found to decrease however, the failure strain always showed an increasing trend with heat treatment. The crack propagation process in any rock mass is largely influenced by the mineralogical composition because cracks normally propagate through weaker planes of the rock mass. From 300°C–600°C cracking modes identified are intergranular and intragranular. At higher temperature, the thermal damage accumulates continuously with the increase in heating temperature.

ACKNOWLEDGEMENTS

The authors greatly appreciate and acknowledge the scholarship provided under the SERB-NPDF Project of the Department of Science and technology, India.

REFERENCES

ASTM (1990). Standard Test Method for practice for the calculation of thermal transmission properties from steady state heat flux measurements. American Society for Testing and Materials, Pennsylvania ASTM C 1045-90.

ASTM, C. (1990). Test method for thermal conductivity of refractories by hot wire, Platinum Resistance Thermometer Technique. American Society for Testing and Materials, Philadelphia ASTM C 1113-90.

Chaki, S., Takarli, M. & Agbodjan, W.P. (2008). Influence of thermal damage on physical properties of a granite rock: porosity, permeability and ultrasonic wave evolutions. CBM, 22(7), 1456–1461.

Gautam, P.K., Verma, A.K., Jha, M.K., Sarkar, K., Singh, T.N. & Bajpai, R.K. (2016). Study of strain rate and thermal damage of Dholpur sandstone at elevated temperature. Rock Mechanics and Rock Engineering, 49(9), 3805–3815.

Gautam, P.K., Verma, A.K., Maheshwar, S. & Singh, T.N. (2016). Thermomechanical analysis of different types of sandstone at elevated temperature. RMRE, 49(5), 1985–1993.

Incropera, F.P. & De Witt, D.P. (1985). Fundamentals of heat and mass transfer.

ISRM (1978b). Suggested methods for determining tensile strength of rock materials. Int. J. Rock Mech. Min. Sci. Geomech. Abstr 15, 101–103.

Jha, M.K., Verma, A.K., Maheshwar, S. & Chauhan, A. (2016). Study of temperature effect on thermal conductivity of Jhiri shale from Upper Vindhyan, India. BOEG, 75(4), 1657–1668.

Kahraman, S. (2002). Estimating the direct P-wave velocity value of intact rock from indirect laboratory measurements. IJRMMS, 39(1), 101–104.

Khandelwal, M. (2013). Correlating P-wave velocity with the physico-mechanical properties of different rocks. PAAG, 170(4), 507–514.

Khandelwal, M. & Ranjith, P.G. (2010). Correlating index properties of rocks with P-wave measurements. Journal of Applied Geophysics, 71(1), 1–5.

Khandelwal, M. & Singh, T.N. (2009). Correlating static properties of coal measures rocks with P-wave velocity. International Journal of Coal Geology, 79(1), 55–60.

Rummel, F. & Vanheerden, W. (1978). Suggested Methods For Determining Sound-Velocity. International Journal Of Rock Mechanics And Mining Sciences, 15(2), 53–58.

Salmon, D. (2001). Thermal conductivity of insulations using guarded hot plates, including recent developments and sources of reference materials. MST, 12(12), R89.

Sharma, P.K. & Singh, T.N. (2008). A correlation between P-wave velocity, impact strength index, slake durability index and uniaxial compressive strength. BOEG, 67(1), 17–22.

Singh, T.N., Sinha, S. & Singh, V.K. (2007). Prediction of thermal conductivity of rock through physico-mechanical properties. Building and Environment, 42(1), 146–155.

Sun, H., Sun, Q., Deng, W., Zhang, W. & Lü, C. (2017). Temperature effect on microstructure and P-wave propagation in Linyi sandstone. Applied Thermal Engineering.

Verma, A.K., Jha, M.K., Maheshwar, S., Singh, T.N. & Bajpai, R.K. (2016). Temperature-dependent thermophysical properties of Ganurgarh shales from Bhander group, India. Environmental Earth Sciences, 75(4), 1–11.

Xamán, J., Lira, L. & Arce, J. (2009). Analysis of the temperature distribution in a guarded hot plate apparatus for measuring thermal conductivity. ATE, 29(4), 617–623.

Yasar, E. & Erdogan, Y. (2004). Correlating sound velocity with the density, compressive strength and Young's modulus of carbonate rocks. International Journal of Rock Mechanics and Mining Sciences, 41(5), 871–875.

Geomechanics and Geodynamics of Rock Masses – Litvinenko (Ed.)
© *2018 Taylor & Francis Group, London, ISBN 978-1-138-61645-5*

Characterization of shear behavior of discontinuities of Brioverian schist

Simon Guiheneuf & Damien Rangeard
Laboratoire de Génie Civil et Génie Mécanique, INSA de Rennes, Rennes, France

Véronique Merrien-Soukatchoff
CNAM, Equipe Géotechnique, Département, ICENER, Laboratoire Géodésie Géomatique Géosciences Aménagement et Droit Foncier (GeF), Paris, France

Marie-Pierre Dabard
Géosciences-Rennes, Université de Rennes 1, CNRS UMR 6118, Campus de Beaulieu, Rennes, France

ABSTRACT: In order to better characterize brioverian schist of Rennes Bassin (Brittanny) the mechanical behavior of the discontinuities (schistosity and/or stratification planes) extracted from different excavation sites was tested. A Casagrande shear box, rather adapted to characterize soils, was used which was a challenge due to the brittleness and the unmoldability of the material. Yet, the results demonstrate that this type of equipment, much more economical than a classic rock shear test installation, is adapted to characterize Hard Soil or Soft Rock (HSSR) such as brioverian schist.

The experiments highlight the influence of the petrography of the specimen on the friction angle. Moreover, tests realized under two different water condition (dry and after immersing the joint) show the differentiated influence of the presence of water according to the nature of the rock.

Keywords: schist, shear test, hardening rock-soft soil

1 INTRODUCTION

Brioverian schists are located in the western part of France (Figure 1, Debelmas 1974) and concomitant with the same deposits in other European countries. The deposition era of the Brioverian formation is still uncertain; first estimates placed it between the Archean and the Cambrian period but more recent work improved this dating and distinguished the Brioverian massif of Brittany and the Brioverian massif of Normandy do not date from the same time period (Dabard, 1990).

The formation of the Brioverian schists of Brittany (at least the deposit of the sedimentary layers) happened between 750 million years and 540 million years ago.

The cycles of progradation and retrogradation (variation of see level) induced variation in lithology and of the granulometry of the sediments. During a progradation period, the sediments were deposited in thick layers (up to several meters) from very fine siltone to medium sandstone. The sediments deposited during retrogradation period consist of alternation of thin-bedded siltstones and fine-grained sandstones surmounted by a thin layer of graphitic shale. Those original sediments were then subjected to deformations (schistosity and folds) and to a greenschist-facies metamorphism at a maximum temperature of 500°C and maximum pressure of 1000 MPa during the hercynian orogenesis. This summary of geological history is significant to understand mechanical behaviour of the rock mass.

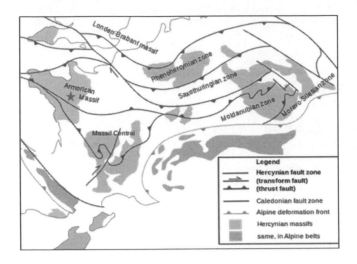

Figure 1. The Brioverian massif in Europe (modified from Franke, 1992) and Rennes city (red star).

In consequence of the geological history, the brioverian rocks are affected by different discontinuities (stratification, schistosity, fracturation) which are the main direction of potential failure. The characterisation of the shear behaviour of this plane is a key parameter for investigating the behaviour of retaining structure in this type of material. This paper is focused on the characterization of the friction angle of the discontinuities regarding to the material lithology and water content.

2 MATERIALS AND METHODS

The shear behavior of the discontinuities is investigate using a direct shear tests realised using an equipment mainly used to characterize soils. Maximal vertical and horizontal loads are limited to 5 kN. Shear speed was maintained constant during the test.

The main advantage of this equipment is to be easy to use and less expensive than a classical rock shear equipment whereas the main drawback is that the vertical load cannot be adjusted to preserve the vertical stress constant during the test and though it varies with the shear surface. However, the range of normal loads used for the test was low enough to limit the impact of the material (Le Cor et al. 2014).

The tests are realised on open discontinuities extracted from different excavation sites in Rennes, in different lithologies. The adopted methodology for preparing samples and realised the tests is those described by Le Cor et al. (2014). According to this procedure, the lower and the upper part of the tested samples were embedded in mortar, and then stored at 20°C and relative humidity of 50%.

The direct shear tests are then performed under constant normal load, at a constant shear speed. The normal and shear stress are then corrected regarding to the variation of the shearing surface during the test.

Considering the low number of samples available by lithology, we carried out 4 to 5 cyclic shear tests on each sample. After each cycle, the normal load was increased before starting the next one. This technique can lead to a decrease of the mechanical strength of the joint tested due to the degradation of the shear surface (Lee et al., 2001; Jafari et al., 2013; Pellet et al., 2013). The impact of the cyclic shear tests was studied on previous sample (Le Cor, 2014) by comparing the results (friction angle) on a single cycle and those with the same vertical load but obtained after the specimen has undergone several cycles (up to 5) with lower vertical loads. The comparison revealed that the influence of the cycles was limited to 10% (decrease of the friction angle for specimens having undergone several cycles).

Table 1. Granulometry and type of tested discontinuities by site.

Site	1	2	3	4	5	6	7	8
Silstone			FSi S1	FSi S1		FSi S1	FSi S1	FSi S1
Sandstone	FSa S1		FSa S1		FSa F, S1			MSa S1
Graphitic shale	VFSi S1	VFSi S1		VFSi S1				

(VFSi: Very fine siltstone: 2 μm < X < 4 μm; FSi: Fine siltstone: 4 μm < X < 8 μm; FSa: Fine sandstone: 125 μm < X < 250 μm; Medium sandstone: 250 μm < X < 500 μm).

Each sample was tested at a constant shear speed of 0.5 mm/min until a horizontal displacement equal to 3 to 4 mm. This length was sufficient to reach the residual shear strength of the majority of the samples.

The tested materials are extracted from 8 sites. Three main types of lithology were identified: grey siltstone (25 samples from 8 sites), sandstone (13 samples from 4 sites), and graphitic shale (7 samples from 3 sites). The discontinuities tested are generally schistosity planes S1 that sometimes coincide with the stratification S0. In sandstone materials, few fractures different from schistosity from site 5 were also sampled and tested (Table 1). The granulometry was determined from thin-section analysis, and the adopted classification reported in Table 1 is those proposed by Blott and Pye (2001).

Because graphitic shale were deposit in very thin layers, they are rarely intercepted by excavation works and we have very few samples of this lithology.

For each material, samples are tested under two conditions, dry and wet. In dry condition, the shear test is realized at 20°C and 50% of relative humidity, whereas in wet condition, in order to have a wet shear surface, the sample is immersed 10 minutes in water before testing.

3   RESULTS

Typical shear stress versus shear displacement curves obtain from one sample are displayed on Figure 2. The normal stress used ranged from 74 kPa to 600 kPa. No shear stress peek was reached during the test, but a constant value of the shear stress was observed after a small displacement of 0.2 to 0.5 mm. As expected, the shear stress increase with the normal stress and, during shearing back, the shear strength is reduced compared to the opposite direction. These tendencies were observed on most of the more rough samples. This can be explained by the degradation of the shear surface during the first part of the test, asperities are destroyed and only an increase of the normal load can mobilized higher shear stress. This phenomenon was observed by other authors (Krahn & Morgenstern, 1979; Jing et al., 1992).

The average shear stress along the plateau for each normal stress and for each sample (one lithology on one site) allows to derive the Mohr-Coulomb parameters (Figure 2b). Bellow, we focused only on the friction angle. Indeed, the value of the cohesion is linked to the relative orientation of the shear plane and the loading direction.

The friction angle in dry conditions are presented Figure 3a, 3b and 3c for discontinuities in graphitic shale, grey fine siltstone and fine sandstone.

For discontinuities from graphitic shale, the friction angle is very small (always lower than 20°). The average value is similar for all the tested sites. The low shear-strength of those discontinuities is linked to the composition of this graphitic shale, which has a high clay content, and very fine grains. Deformation and metamorphism have led to schistosity planes in which the Iso-orientation of phyllosilicates and graphite gives a glossy surface (Figure 4).

For discontinuities collected in fine grey siltstone, variable friction angles are encountered depending of the site. However, the values are rather similar with an average value around 21°

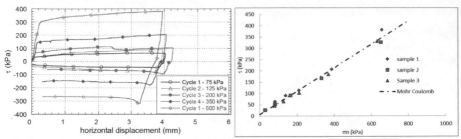

a – shear –displacement curve (sample 1)   b– stress path in Mohr Coulomb diagram (all samples)

Figure 2.   Shear tests results for fine siltstone—site 3.

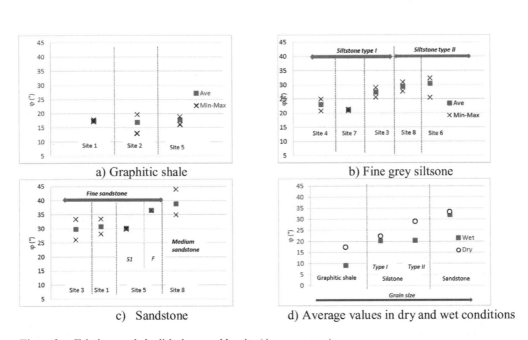

a) Graphitic shale

b) Fine grey siltsone

c) Sandstone

d) Average values in dry and wet conditions

Figure 3.   Friction angle by lithology and by site (Ave = average).

for two sites when the average value is higher than 27° for the three other sites. This difference can be partly attributed to the mineralogy of the clay content of the material. Indeed, while the grain size is similar for all sites, those for which the friction angle is lower (sites 4 and 7) contained swelling clay, when the siltstones from the other sites (3, 8 and 6) only contained chlorite and illite. The first group, sample with swelling clay, is noted "type I", and the second group, sample without swelling clay, is noted "type II" on Figure 3b.

Friction angle of the discontinuities from sandstones materials, is higher than those in materials with finer grain size (graphitic shale and fine siltstone), related to more rough rock wall. The friction angle increases with the granulometry of the materials containing the tested discontinuities

The friction angle measure in medium sandstone (site 8) is higher than those measured in fine sandstone (site 3,1 and 5). However, the grain size of the material is not the only param-eter influencing the friction angle. For site 5, the friction angle of crack planes is higher than those of schistose planes from the same material (fine sandstone).

The synthesis of all the results presented in Figure 3d highlight the effect of lithology on the shear behavior of the discontinuities. When tested in wet condition, the discontinu-ity exhibit a smaller friction angle. Similar observations on the decrease of the mechanical

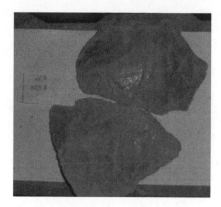

Figure 4.   Open discontinuity in graphitic shale (schistosity plane, site 3).

parameters with water content have been made by several authors on clay rocks discontinuities (Hoang, 2010; Pellet et al., 2013, Barros et al. 2014). However, the effect of water is more noticeable for the materials with higher clay content and the clay mineralogy also impacts the shear result: the material containing swelling clay have the smaller friction angle. Let us notice that for graphitic shale, the effect of water is especially important: the friction angle moving from 17° in dry condition to 9° in wet condition.

4   CONCLUSIONS

The use of Casagrande box to characterize open rock discontinuities is an interesting way to measure shear strength parameters on rock discontinuities of soft rocks. Compared to classical rock mechanics equipment it is a less expensive investment that can be used for a characterization under low normal stress that is representative of the stress state of rock mass close to the surface.

Shear tests in brioverian schists are quite rare and our tests contribute to a better knowledge of this material in terms of measured friction angles. The tests carried out on discontinuities on brioverian breton schists from Rennes city area demonstrated otherwise the importance of the grain size of the material containing the discontinuities on the friction angle and the role of water content, especially for the graphitic shale.

The presence of thin layers of this poor quality material have important consequences on the behavior of works during excavation and on the support capacity of the retaining wall when it could be difficult to identify, even more to sample these layers in vertical drilling.

REFERENCES

Barros, R.S., Oliveira, D.V., Varum, H., Alves, C.A.S., Camoes, A. 2014. Experimental characterization of physical and mechanical properties of schist from Portugal. Construction and Building Materials, 50, 617–630. doi:10.1016/j.conbuildmat.2013.10.008.

Blott S.J., Pye K. 2001. – Technical communication gradistat: a grain size distribution and statistics package for the analysis of unconsolidated sediments. Earth Surf. Process. and Landforms, 26, pp. 1237–1248.

Dabard, M.P. 1990. Lower Brioverian formations (Upper Proterozoic) of the Armorican Massif (France): geodynamic evolution of source areas revealed by sandstone petrography and geochemistry, Sedimentary Geology, 69, 45–58.

Debelmas, J. 1974. – Géologie de la France, 1: vieux massifs et grands bassins sédimentaires, In: French Doin éditeurs, Paris.

Franke, W. 1992. Phanerozoic structures and events in central Europe In D.J. Blundell, R. Freeman & S. Mueller (Eds.) A continent revealed – The European Geotraverse, 297 pp., Cambridge University Press.

Hoang, T.T.N. 2010. – Etude du comportement d'un milieu rocheux fracturé: Application à la réalisation du tunnel de St Béat. Doctoral dissertation from l'école des ponts Paristech. 279 pages.

Jafari, M., Amini, Hosseini, K., Pellet, F., Boulon, M., Buzzi, O. 2003. Evaluation of shear strength of rock joints subjected to cyclic loading. Soil Dynamics and Earthquake Engineering, 23, (7), pp. 619–630. doi:10.1016/S0267-7261(03)00063-0.

Jing, L., Nordlund, E., Stephansson, O. 1992. – An experimental study on the anisotropy and stress-dependency of the strength and deformability of rock joints. International Journal of Rock Mechanics and Mining Sciences & Geomechanical Abstracts, 29 (6), pp. 535–542.

Krahn, J., Morgenstern, N.R. 1979. – The ultimate frictional resistance of rock discontinuities. International Journal of Rock Mechanics and Mining Sciences & Geomechanical Abstracts, vol. 16, pp. 127–133.

Le Cor, T., Rangeard, D., Merrien-Soukatchoff, V., Simon, J. 2014. Mechanical characterization of Weathered Schists. Engineering Geology for Society and Territory, Vol. 6, pp. 809–812. (Proceedings of 12th International IAEG Congress, Torino, Italy, 15–19 Sept. 2014).

Le Cor, T. 2014. Etude du comportement de terrains anisotropes lors de travaux de génie civil. Phd Thesis, INSA de Rennes, France.

Lee, H., Park, Y., Cho, T., You, K. 2001. – Influence of asperity degradation on the mechanical behavior of rough rock joints under cyclic shear loading. International Journal of Rock Mechanics and Mining Sciences, 38(7), pp. 967–980. doi:10.1016/S1365-1609(01)00060-0.

Nagy, E.A., Samson, S.D., D'Lemos, R.S. 2002. – U-Pb geochronological constraints on the timing of Brioverian sedimentation and regional deformation in the St. Brieuc region of the Neoproterozoic Cadomian orogen, northern France. Precambrian Res., 116, 1–17.

Pellet, F.L., Keshavarz, M., Boulon, M.M. 2013. Influence of humidity conditions on shear strength of clay rock disconti-nuities. Engineering Geology, 157, 33–38. doi:10.1016/j.enggeo.2013.02.002.

Geomechanics and Geodynamics of Rock Masses – Litvinenko (Ed.)
© 2018 Taylor & Francis Group, London, ISBN 978-1-138-61645-5

Effects of stress, anisotropy and brittle-to-ductile transition on fracturing and fluid flow in shales

Marte Gutierrez

Department of Civil and Environmental Engineering, Colorado School of Mines, Golden, CO, USA

ABSTRACT: This paper aims at providing a better understanding of the fracturing behavior of shales due to shearing in relation to stress level, anisotropy and brittle-ductile transition, and the effects of fracturing to the changes in permeability and seismic velocities in shales. Triaxial compression tests were performed using a High-Pressure and High-Temperature (HPHT) triaxial cell on samples of Mancos shale cored horizontally parallel to the bedding plane. During triaxial shearing, the continuous changes in sample permeability, and the P and S wave velocities were simultaneously monitored during loading. The formation of fractures in the test samples during shearing was characterized. Structural anisotropy of shale is evaluated in terms of the shear wave splitting parameter. The results show complicated interactions between initial heterogeneity, brittle-to-ductile transition and stress level on the fracturing and the ensuing permeability changes during shearing of shales. An interesting observation was that tensile fracturing can occur in a compressional regime in shales due to shearing along pre-existing undulating bedding plane laminations. Notwithstanding these complex interactions, it was found that the shear wave splitting parameter can be a useful index for monitoring fracturing and permeability changes of shales during to shearing.

1 INTRODUCTION

Understanding the hydro-mechanical behavior of shale formations is becoming increasingly important as shales play important roles in the production of unconventional oil and gas resources as well as in geological carbon sequestration where shales typically form the cap rock above CO_2 storage reservoirs. Shales typically have very low permeability ranging from pico to micro Darcy orders (e.g., Brace, 1984; Neuzil, 1994) due to their very tight matrices that are mainly composed of kerogen, other organic matters, clay minerals and silica. Shale permeability is highly stress sensitive (Spencer 1989; Gutierrez et al. 2000). The wide-ranging variation of matrix permeability of shales can be attributed to their microstructure, which is often strongly anisotropic because of fissility and laminations along the bedding plane. The stress sensitivity of shale permeability becomes more pronounced when shale is fractured (Gutierrez et al., 2000).

Fracturing is often the only way to provide high permeability pathways for fluid flow and to extract fluid from or to inject fluids into shales. Fracturing and the resulting hydraulic properties of the fractured shale are strongly dependent on the tensile and shear strength properties but also on brittleness or ductility, which in turn depends on the stress level during fracturing, the diagenetic history of the shale and pore fluid composition. The mechanical response particularly failure and fracturing of shales are also strongly influenced by microstructure. The stress sensitivity and anisotropy of fracturing behavior of shales is significant due to their laminated microstructure. However, the prominent layered microstructure of shale has not been fully considered even though it can result in strong hydro-mechanical anisotropy that can interact with the stress dependence and brittle-to-ductile transition in shale fracturing behavior.

The main goal of this paper is to improve the understanding of the hydro-mechanical characteristics of shale and their dependencies on stress level and microstructure. This improved understanding is crucial to reliably predicting the fracturing and fluid flow and leakage behavior of shale in may geoengineering applications. Three major topics are investigated in this paper. First is the effect of intrinsic structural anisotropy and stress anisotropies on fracturing behavior including induced fracture morphology. Second is the effect of fracturing on the changes in permeability during shearing. Third is the use of seismic wave velocities, particularly shear wave splitting, on characterizing the natural and fracturing-induced hydro-mechanical anisotropy of shales.

2   TEST SAMPLE AND TRIAXIAL TESTING

Triaxial compression tests were carried out on Mancos shale plug samples, which were sheared at constant confining pressure conditions in a high pressure-high temperature (HPHT) triaxial cell equipped with 0.5 MHz-piezoelectric transducers. Two test samples were horizontally cored with 38 mm diameter from a block of Mancos shale retrieved from Douglas Creek Arch located between the Uinta and Piceance basins in Colorado, USA. Mancos shale is an organic rich shale of Cretaceous age and has a pronounced layered microstructure. The grain density is 2.68 g/cm^3, and the porosity values of the tested samples were 5.8 and 6.6%. The samples were dried at 40°C prior to triaxial testing until the weight of the sample became constant. The horizontally-cored sample is enclosed in a rubber sleeve and placed in a triaxial cell that is capable of applying 69 MPa of cell pressure at a temperature of up to 150°C. Both ends of core sample are confined with axial pistons through grooved metal plates to distribute compressed air. The cell pressures P_c of the two samples were maintained at 2 and 12 MPa using a precision ISCO 260D syringe pump. The axial load is generated by pushing an axial piston with another syringe pump (ISCO, 100DX) at a constant strain rate of 8.1×10^{-3}%/min near failure. The piston displacement is measured by using a LVDT whose accuracy is ±0.25% of 1 mm at full scale. The permeability changes of the test samples during shearing were characterized by injecting dry air from one end of core sample. Air permeability was converted to liquid permeability using the correction from Klinkenberg (1941). The air flow rate is measured by using a mass flow meter whose capacity is 200 ml/min and turndown ratio is 1:100. The air pressure gradient was monitored by using a differential pressure transducer whose accuracy is ±0.25% of 860 kPa of full scale capacity.

During shearing, permeability, and compressional and shear wave velocities were measured simultaneously with the stress-strain and permeability response. The caps confining the both ends of core plug are equipped with piezoelectric transducers to measure the P and S wave velocities of the core sample along core axis. The seismic waves are generated by piezoelectric transducers at one end and registered by a second set of piezoelectric transducers at the other end. The natural frequency of the transducers is 500 kHz. Fig. 1 illustrates the directions of oscillation of transducers. Two different triads of shear wave transducers (indicated as SH an SV) shake the cap in orthogonally different directions. The bedding plane

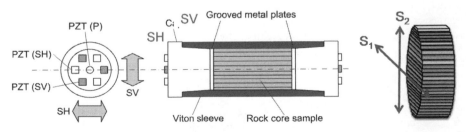

Figure 1.   Details of the P and S wave velocity measurements showing relationship between oscillation directions of shear wave velocities relative to bedding plane of shale.

of core sample is aligned parallel to the oscillation direction of SH wave. The other wave (SV) oscillates in the direction perpendicular to the bedding plane. The extent of the intrinsic anisotropy of the shale samples used in the study is quantified based on the degree of shear wave splitting that emerges when a shear wave encounters a discontinuity. Shear waves entering in an anisotropic rock split into two different components oscillating in the respective preferred directions and propagating at different speeds. The degree of intrinsic anisotropy is quantified by means of the shear wave splitting parameter defined as the difference between the split shear wave velocities normalized by the faster shear wave velocity (Thomsen, 1986). The shear wave splitting parameter is also defined as $(V_{SH} - V_{SV})/V_{SH}$, where V_{SH} and V_{SV} is the shear velocities parallel and perpendicular to the bedding plane, respectively.

3 STRESS-DEPENDENT SHEAR AND PERMEABILITY BEHAVIOR OF MANCOS SHALE

Fig. 2a compares the shear stress-strain curve and the change of permeability of shale sample during shearing at $P_c = 2$ MPa. The deviatoric stress is given by $q = \sigma_a - P_c$, where σ_a is axial stress. The sample reaches the yield point at 0.7% of axial strain. In the pre-yield stage, the stress-strain response is almost linear. The permeability gradually decreased from 30 to 12 µD with increasing axial stress in this stage. The decrease of permeability can be attributed to closure of microcracks in the sample. Upon reaching the yield shear stress at around 64 MPa, the sample suddenly lost its stiffness and strain softening ensued. At the same time, the permeability started to abruptly increase from 15 µD. After 1.2% of axial strain, where the permeability reached 380 µD, the permeability hardly changed. The shearing behavior of another core tested at $P_c = 12$ MPa is shown in Fig. 2b. The stress-strain response of this shale shows brittle behavior with a well define peak and strain softening. The yield point occured at 1.44% of axial strain and 115 MPa of shear stress. The overall trend of permeability change is similar with that observed at $P_c = 2$ MPa. However, the change of permeability became more gradual compared to the previous test. In the pre-yielding stage, the permeability decreased from 83 to 15 µD until the stress reached the yield point. The amount of permeability decrease in pre-yield stage is higher than that at $P_c = 2$ MPa. There is no sign of permeability increase until the sample reached the yield point, whereupon the permeability immediately started to increase. The permeability increased from 15 to 50 µD observed in the very early part of the post-yielding stage and occurred within a small change of axial strain from 1.44 to 2.05%. However, 50 µD of post-yield permeability is still lower than the initial permeability value which is 83 µD. Subsequently, the permeability gradually and continuously increased up to 150 µD until the sample reached 11% of axial strain.

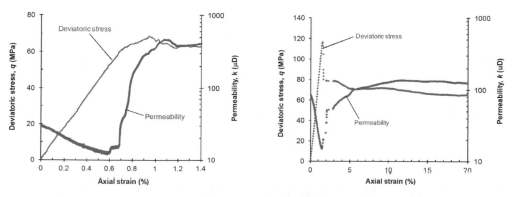

Figure 2. Changes in stress-strain response and permeability of Mancos shale at confining stress of 2 MPa (left), and 12 MPa (right).

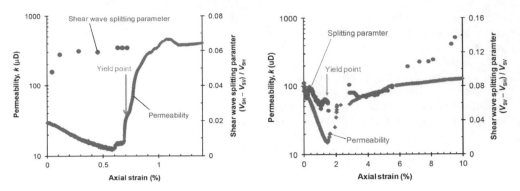

Figure 3. Changes in shear wave splitting parameter and permeability of Mancos shale during shearing at confining stress of 2 MPa (left), and 12 MPa (right).

4 STRUCTURAL ANISOTROPY EVOLUTION OF SHALE DURING SHEAR FRACTURING

Evaluation of the degree of structural anisotropy of Mancos shale sample was attempted through the shear wave splitting parameter $(V_{SH} - V_{SV})/V_{SH}$. Fig. 3 shows the shear wave splitting parameter changes together with the permeability measurements during shearing of the two tests. For the test at 2 MPa confining stress, the value of splitting parameter slightly increased from 0.048 to 0.062 when the axial stress increased up to the yield stress. But this change of splitting parameter is negligibly small, because the increment of splitting parameter after 0.11% of axial strain is only 0.004. The shear wave splitting parameter tested at $P_c = 12$ MPa intermittently decreased from 0.084 to 0.051 before reaching the yield point. At this period, the permeability decreased consistently. Immediately after reaching the yield point, the splitting parameter value started to increase. The increase of splitting parameter occurred gradually and continuously up to 10% of axial strain. The change of shear wave splitting parameter, which is initially declining and subsequently increasing, correlates well with the permeability change.

5 RELATIONSHIP BETWEEN FRACTURE MORPHOLOGY AND HYDRO-MECHANICAL BEHAVIOR OF SHALE

After the triaxial tests, the shale core samples were carefully retrieved from the triaxial cell to visually characterize the induced fracture morphologies. Pictures of the samples after shearing with highlighted induced fractures are shown in Fig. 4. Note that the samples were fully intact prior to shearing. For the shale sample tested at $P_c = 2$ MPa, three distinct narrow vertical fractures are observed after shearing. These fractures propagate mainly along the bedding plane of shale sample. For the sample sheared at a confining pressure of $P_c = 12$ MPa, some cracks are observed in the radial direction of core sample after shearing. More significantly, it can be seen that the two conjugated main shear fractures were created. The other fractures running in the radial direction are not considered to develop in an earlier stage of shearing, because these fractures are located on the principal plane of stress on which no shear stress acts.

The observed formation of shear-induced fractures shown in Fig. 4 is consistent with brittle-to-ductile transition behavior of soft rocks showing stress dependency of fracturing behavior as discussed in Nygaard et al. (2006). At low confining stress, brittle vertical fractures were developed along the bedding plane, while conjugated shear fractures formed at high confining stress due to less brittle response. A notable observation is that the vertical fractures, which are typically observed only under tensile stresses or extensional lateral strains,

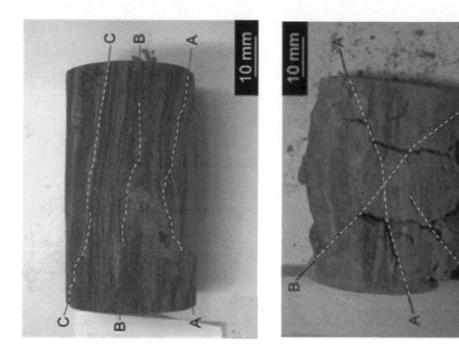

Figure 4. Mancos shale samples after shearing at confining stresses of 2 MPa (left) and 12 MPa (right). Dotted lines indicate fractures induced after shearing.

formed for the sample loaded at the confining stress of 2 MPa, which is still a relatively high level of compressive load. These vertical fractures are attributed to local shear stresses generated along the length of the bedding plane due to the undulations. In turn, the local tensile stresses potentially occurred along the bedding planes at each side of plane to ride up the bedding plane undulations. The outcome is the activation and opening up of the bedding planes. The results conclusively show that vertical tensile/extensional can form even in compressional stress regime.

From the results presented in Fig. 3 above, it is clear that the evolution of permeability corresponds to the fracturing process. For the test at P_c = 2 MPa, the shear wave splitting parameter value hardly changed before yielding. Correspondingly, the permeability decreases but the amount of change is negligibly small. Significant fracturing does not occur in the pre-yielding stage. The sudden increase of permeability observed immediately after yielding can be attributed to development of multiple fractures which suddenly opened along the bedding planes of shale sample within the narrow range of axial strain. At P_c = 12 MPa, the gradual and continuous increases of the shear wave splitting parameter and permeability in the post-yield stage suggest that the observed shear fractures were developed gradually with increasing axial strain. The initial decline of permeability observed before yield point is considered to be caused by the closure of microcracks and compression of pore space, because the degree of structural anisotropy evaluated with shear wave splitting parameter simultaneously decreases.

The observed changes of permeability due to closing microcracks and opening fractures during triaxial test can be organized in a simple relationship between the permeability and shear wave splitting parameter as shown in Fig. 5. In the pre-yield stages, the permeability decreases with decreasing shear wave splitting parameter. Fracture evolution after yielding increases both the permeability and shear wave splitting parameter. Interestingly, the two test results obtained at the different confining stress conditions are consistent in this relationship. The shear wave splitting parameter appears to be a good index parameter that can be used as indicator of permeability changes in shale subjected to shearing.

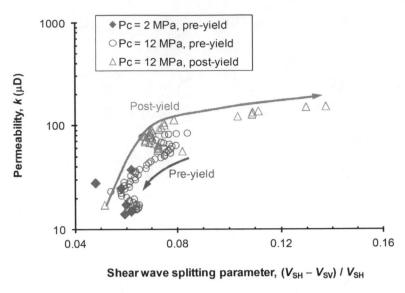

Figure 5.   Relationship between permeability and shear wave splitting parameter.

6   CONCLUSIONS

Triaxial compression tests were carried out on dry core plugs of organic-rich Mancos shale at 2 and 12 MPa of confining stresses to relate the change of hydro-mechanical and seismic properties of shale with development of structural anisotropy due to shear fracture development. The air permeability and P and S wave velocities of shale samples were measured simultaneously with the stress-strain response during the triaxial tests. The following are the main conclusions from the study: 1) Permeability parallel to bedding plane decreases pre-peak (due to compaction), but increases significantly post-peak due to fracturing; 2) The shear wave splitting parameter decreases and increases correspondingly with permeability in the pre- and post-failure stages; 3) Induced fracture morphology changes from tensile to shearing fractures with increasing confining pressure; 4) Initial heterogeneity appears to enhance the brittleness of shale, and to affect the failure mode (i.e., tensile fracturing along the bedding plane at low confinement, and branching of shear fractures along bedding planes at high confinement); 5) Tensile fracturing can occur in a compressional regime in shales during shearing along pre-existing undulating bedding plane joints; and 6) The shear wave splitting parameter can be a useful index for monitoring permeability changes and fracturing of shales during to shearing.

REFERENCES

Brace W.F. (1984). Permeability of crystalline rocks—New in situ measurements. *J. Geophys. Res.*, vol. 89, no. B6, pp. 4327–4330.

Gutierrez M., L.E. Øino and R. Nygaard. (2000). Stress-dependent permeability of a de-mineralised fracture in shale. *Mar. Petrol. Geol.*, vol. 17, no. 8, pp. 895–907.

Klinkenberg, L.J. (1941). The permeability of porous media to liquids and gases. *Drilling and Production Practice*, American Petroleum Inst., pp. 200–213.

Neuzil, C.E. (1994). How permeable are clays and shales? *Water Res. Research,* vol. 30, no. 2, pp. 145–150.

Nygaard, R., M. Gutierrez, R.K. Bratli and K. Høeg. (2006). Brittle-ductile transition, shear failure and leakage in shales and mudrocks. *Mar. Petrol. Geol.*, vol. 23, no. 2, pp. 201–212.

Spencer, C.W. (1989). Review of characteristics of low-permeability gas-reservoirs in Western United-States. *AAPG Bull.*, vol. 73, no. 5, pp. 613–629.

Thomsen, L. (1986). Weak elastic-anisotropy. *Geophysics*, vol. 51, no. 10, pp. 1954–1966.

Geomechanics and Geodynamics of Rock Masses – Litvinenko (Ed.)
© 2018 Taylor & Francis Group, London, ISBN 978-1-138-61645-5

Geomechanical characterization of the upper carboniferous under thermal stress for the evaluation of a High Temperature-Mine Thermal Energy Storage (HT-MTES)

Florian Hahn
International Geothermal Centre—GZB, Bochum, Germany

Theresa Jabs
Ruhr Universität Bochum—RUB, Bochum, Germany

Rolf Bracke
International Geothermal Centre—GZB, Bochum, Germany

Michael Alber
Ruhr Universität Bochum—RUB, Bochum, Germany

ABSTRACT: The goal of this R&D project is to create a technically and economically feasible conceptual model for a High Temperature-Mine Thermal Energy Storage (HT-MTES) for the energetic reuse of abandoned collieries based on the example of the former Dannenbaum colliery in Bochum, Germany. During summer non-used surplus heat from solar thermal power plants, garbage incineration, Combined Heat and Power plants (CHP) or industrial production processes can be stored within the mine water of former drifts and mining faces. During the winter season, this surplus heat can be extracted from the mine water and then directly utilized for heating purposes of commercial and/or residential areas. For the evaluation of such a HT-MTES within a former colliery, the corresponding geomechanical parameters of the Upper Carboniferous under thermal cyclic loading need to be evaluated. Therefore, the main rock types of the Upper Carboniferous (claystone, mudstone and sandstone) are subject to a geomechanical characterization before and after thermal cyclic loading of temperatures up to 90°C. Almost 200 abandoned collieries, just within the Ruhr area of North Rhine-Westphalia, represent a vast potential for this new type of thermal energy storage.

Keywords: high temperature mine thermal energy storage

1 INTRODUCTION

At the moment a seasonal underground heat storage within an abandoned colliery has not been realized in Germany. Therefore the HT-MTES (High Temperature-Mine Thermal Energy Storage) project of the International Geothermal Centre (in cooperation with the Stadtwerke Bochum Holding and delta h Ingenieurgesellschaft mbH) would lead the way within the sector of renewable energy storage systems. The aim of this project is to design a technically and economically feasible pilot plant for a HT-MTES for the energetic reuse of the abandoned Dannenbaum colliery, which is located below the premises of the former Opel plant 1 in Bochum, Germany. The abandoned Dannenbaum colliery operated between 1859–1958.

The conceptual model (see Fig. 1) is based on the storage of seasonal unutilized surplus heat during the summer from solar thermal power plants, industrial production processes or

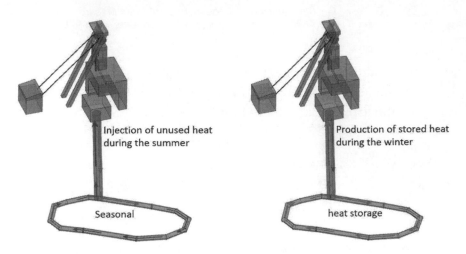

Figure 1. Conceptual model of seasonal underground heat storage in an abandoned coal mine (GZB).

CHP plants within the mine layout and to utilize the stored heat for industrial and/or commercial purposes, e.g. through the distribution of a district heating grid during the winter, when there is a high heat demand.

The two main extraction shafts of the Dannenbaum colliery are located in the northwest of the former Opel plant in Bochum. The mine layout was developed down to a depth of −696 m NHN (Huske 2006). After the colliery was shut down, the two shafts were backfilled. Presently, the mine is flooded above the 4th level, up to −190 m NHN. It is assumed that within the mine building, an undisturbed temperature level of approx. 36°C (Leonhardt 1983) can be anticipated on the 8th level at a depth of −693 m NHN.

In a later test operation, the mine layout of the former Dannenbaum colliery is to be developed by a production and an injection well, using directional drilling technology. For this purpose, the production well should be drilled into the 8th level at −693 m NHN and the injection well into the 4th level at −227.6 m NHN. Once the feasibility study has been successfully completed, a pilot plant is planned to be implemented promptly. This would aim at the establishment of a central heating system with solar collectors and a heat pump for the innovative heat supply (with a heat output of 1 MW$_{th}$) of the new settlements on the former Opel premises, which is now marketed under the acronym Mark 51°7.

2 CURRENT STATE OF TECHNOLOGY

The idea of obtaining thermal energy from an inoperative colliery has already been pursued for a long time, although to a comparatively limited extent. Up to this point a pilot plant has not been established, in which the possibility of thermal energy storage in a former colliery has been considered. Well-known executed projects concerning the utilization of mine water include:

- The Mijnwater project in Heerlen (Netherlands), whereby an already completely flooded and no longer accessible mine structure was accessed through directional drilling technology.
- The building of the School of Design at the Zeche Zollverein in Essen, which is heated by 28°C mine water from a depth of about 1.000 m (originating from the mine drainage of the RAG AG).
- The utilization of mine water of the former Robert Müser colliery in Bochum as an energy source for the heat supply of two schools and the main fire station in Bochum. Within this pilot plant the 20°C warm mine water, which originates from the mine drainage of the RAG AG from a depth of 570 m, is being used.

One way of increasing the efficiency of these systems is to enhance the temperature of the mine water through the storage of seasonal surplus heat in the mine layout, which has not been performed yet. Currently merely a few medium-depth hydrothermal aquifer storages are in the planning stage or in operation, which are similar in regard of the temperature and the layer depth. Worth mentioning here are a project of the BMW Group at the Dingolfing plant (planning stage), the Neubrandenburg deep storage (in operation) and the Spreebogen energy concept (in operation).

3 SAMPLE ACQUISITION AND PROCESSING

For the rock mechanical investigations, existing core samples from the Heinrich-Robert colliery in Hamm were utilized, as it was not possible to obtain any samples from the abandoned Dannenbaum colliery. The cores were originally drilled in 2006 near the longwall S109 at a depth of approx. −1090 m NHN. All cores were drilled with an inner diameter of 60 mm and stored dry at room temperature. These samples were considered favourable, as they originate from the same geological setting around the Sonnenschein coal seam, which is present in both collieries within the Bochum main basin. Altogether three core sections were obtained exhibiting the following lengths and origins:

- Core section I was drilled 90 m into the hanging wall at the beginning of the longwall;
- Core section II was drilled 70 m upwards at the end of the coal discharge section;
- Core section III was drilled 30 m downwards at the end of the discharge section.

Mainly all specimen were sampled from the second core section (see Fig. 2). Only a small amount of claystone samples were obtained from the first core section. The third core section was not used. During the first examination, all 23 core boxes of the second core section were screened for potential specimens with a diameter of 40 mm and a length of 80 mm for the medium- and coarse-grained sandstone. In order to enhance the possible outcome of claystone and mudstone specimens, it was decided to prepare the samples with a diameter of 30 mm and a length of 60 mm, due to their brittle behavior during preparation. All specimens were drilled parallel to the core axis and therefore perpendicular to the bedding. The

Figure 2.   Core section II during examination (GZB).

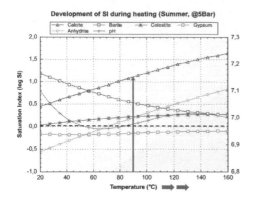

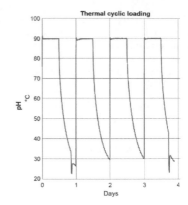

Figure 3.   PHREEQC mine water model (GZB).   Figure 4.   Thermal cyclic loading rates (RUB).

specimens prepared out of the Upper Carboniferous core sections were differentiated based on their grain sizes according to the nomenclature of Potter et al. 1980 to the following rock types: laminated siltstone, mudstone and claystone. Additionally, the sandstone was classified after the DIN 4022 (German Industry Standard) into medium- and coarse-grained sandstone.

4   METHODOLOGY

All specimens were saturated with mine water from the Robert Müser colliery, as it was the closest opportunity to obtain mine water samples (from the existing mine water utilization pilot plant), which were similar in respect of chemical properties to the anticipated mine water in the former Dannenbaum colliery. In Fig. 3, it can be seen that calcite will precipitate (SI > 1), if the mine water is heated above 90°C. Based on this PHREEQC model, it was determined to heat all saturated specimens up to a maximum temperature of 90°C for 12 hours per day (see Fig. 4), in order to simulate the effect of thermal cyclic loading. The geomechanical characterization of the Upper Carboniferous under thermal cyclic loading was performed after 10, 20 and 30 days with 3 specimens per rock type at each test interval. At each measuring cycle, the following parameters were measured: ultrasonic p-wave velocity, thermal conductivity, tensile, uniaxial and triaxial strength. Furthermore, the initial strength of the rock samples was tested before the heating cycles were started, which additionally included thin section analysis and three density measurements.

5   RESULTS

In Fig. 5 the ultrasonic p-wave velocities (Vp) have been correlated with the buoyant density of the mine water saturated specimen (in accordance to the Archimedes' principle), in order to have a better comparison between the properties of the different rock types. For the mud- and claystone, a clear linear trend can be seen between the density and the p-wave velocities, which is mainly based on the mineralogical composition and the low porosity of these rock types. The same effect cannot be observed for sandstone, as they display a higher porosity and therefore the p-wave velocities are more dampened due to the higher water content in the specimen. In Fig. 6 the correlation of the thermal conductivity of the Upper Carboniferous specimens to the increasing thermal cyclic loading rates can be observed. The thermal conductivity decreases slightly diametrically opposed to the increasing thermal loads for all displayed rock types, which is in accordance to Mücke (1962). The following thermal conductivity values were obtained for claystone in the range

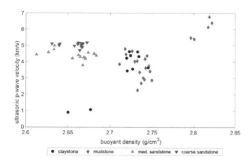
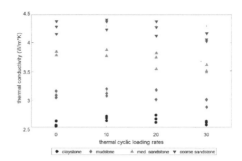

Figure 5. P-wave velocity subject to buoyant density (RUB).

Figure 6. Effect of thermal cyclic loading on thermal conductivity (RUB).

Table 1. Failure criteria (RUB).

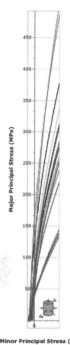

| Rock type | Thermal cyclic loading rates | Mohr-Coulomb | | Hoek-Brown | |
		Cohesion c(MPa)	Internal friction $\varphi(°)$	Uniaxial compressive strength σc(MPa)	Constant mi [–]
Claystone	0	7,47	31,44	43,11	4,20
	10	6,65	38,76	40,00	4,03
	20	5,65	39,47	38,37	5,60
	30	6,18	51,78	80,29	14,43
Mudstone	0	11,41	61,70	90,87	16,74
	10	10,86	50,73	81,28	9,54
	20	7,00	53,44	70,21	14,69
	30	8,76	49,75	44,11	23,00
Medium-grained sandstone	0	11,43	63,45	97,18	18,95
	10	8,87	55,91	78,77	16,43
	20	10,78	53,69	63,99	25,93
	30	10,85	53,52	69,72	22,69
Coarse-grained sandstone	0	8,65	51,49	65,42	8,07
	10	10,70	52,60	128,18	18,06
	20	10,50	60,48	162,35	25,25
	30	8,99	57,97	98,26	50,00

of 2,5–3,3 W/(m*K), for mudstone in the range of 2,8–3,5 W/(m*K), for medium-grained sandstone in the range of 3,4–4,1 W/(m*K) and for coarse-grained sandstone in the range of 4,0–4,4 W/(m*K).

In order to determine the Mohr-Coulomb and Hoek-Brown failure criteria all the results of the tensile, uniaxial and triaxial strength tests were summarized (see Table 1) using the software program RocData 5.0. All specimen were treated as intact rock material and therefore allocated with a GSI of 100.

6 DISCUSSION

Based on the failure criteria, it can be stated that, with the exception of claystone, all of the other tested rock types exhibit a decrease in their rock strength between 28 and 51%, regarding the initial and final measurements after 30 heating cycles. The results of the coarse-grained sandstones have not been taken into consideration for this evaluation, as the variances in their results do not display any clear trends. Interestingly, the claystone shows an increased rock strength of up to 46% after the conclusion of the test series. This can be explained by the fact that the thermal cyclic loading induces shear stress within the sample and hence a reduction of the attractive forces between the clay particles takes place, resulting in compaction. Overall, this leads to a structure densification, which enables the claystone to withstand higher loading rates (Cekerevac and Laloui 2004).

Thermal conductivities were measured at room temperature in the range of 2,5–4,4 W/(m*K) parallel to the sample stratification. After the tests were completed, a slight decrease of the thermal conductivity for the tested rock samples was observed within the range of 4–8%. This is significantly lower than the suggested magnitude of up to 15% by Mücke (1962), which would have a great impact on the efficiency of the mine thermal energy storage system.

7 CONCLUSION

This paper represents the first results of the laboratory studies on the geomechanical characterization of the Upper Carboniferous under thermal cyclic loading. In order to simulate the effect of thermal cyclic loading, the specimens, which were saturated with mine water, were heated up to 90°C for 12 h each day for a total duration of 30 days. Altogether four main test campaigns were conducted after 0, 10, 20 and 30 days. It could be observed that the different rock types show a diverse response to the cyclic heating procedures, as for instance the compressive strength of claystone increases, whereas it decreases for the medium-grained sandstone. Therefore, further tests under high thermal cyclic loading rates need to be conducted, in order to have a clear understanding of the effect on the surrounding rock of the stone drifts, which finally should evolve in long-term stability predictions for upcoming underground thermal energy systems.

The development of diversified storage capacities will have a great impact on the future promotion of renewable energies. In the Ruhr area unused mine structures in combination with available unutilized surplus heat from power plants and industrial processes, constitute a vast potential for large heat storage capacities. Out of this reason, fundamental research in the field of seasonal heat storage in abandoned mines has to be conducted, so that this technology can be further developed and established in the near future.

REFERENCES

Cekerevac, C.; Laloui, L. (2004): Experimental study of thermal effects on the mechanical behaviour of a clay. In: *Int. J. Numer. Anal. Meth. Geomech.* 28 (3), S. 209–228. DOI: 10.1002/nag.332.
Huske, J. (2006): Die Steinkohlenzechen im Ruhrrevier. Daten und Fakten von den Anfängen bis 2005. 3., überarb. und erw. Aufl. Bochum: Dt. Bergbau-Museum (Veröffentlichungen aus dem Deutschen Bergbau-Museum Bochum, Nr. 144).
Leonhardt, J. (1983): Die Gebirgstemperaturen im Ruhrrevier 90 (2), S. 218–230.
Mücke, Gerhard (1962): Untersuchungen über die Wärmeleitfähigkeit von Karbongesteinen und ihre Beeinflussung durch Feuchtigkeit im Zusammenhang mit der Wärmeübertragung des Steinkohlengebirges an die Grubenwetter. Aachen, T.H., F. f. Bergbau u. Hüttenw., Diss. v. 26. Juni 1962, Aachen.
Potter, Paul Edwin; Maynard, J. Barry; Pryor, Wayne Arthur (1980): Sedimentology of shale. Study guide and reference source. New York: Springer-Verlag.

Geomechanics and Geodynamics of Rock Masses – Litvinenko (Ed.)
© 2018 Taylor & Francis Group, London, ISBN 978-1-138-61645-5

Investigating the mechanism contributing to large scale structurally driven hangingwall instabilities on the UG2 reef horizon

A.G. Hartzenberg & M. du Plessis
Lonmin Marikana, Marikana, South Africa

ABSTRACT: The mine design and layout of a Shaft should be based on the geotechnical parameters of the specific mining block. This should include the impact of both regional and secondary geological structures exposed in the underground mining operations. Anomalous or unexpected conditions encountered could contribute to large-scale Falls of Grounds (FOG's), unstable beam behaviour, unpredictable pillar behaviour or support failure. All of these incidents could significantly impact on the health and safety of mine workers as well as cause detrimental financial losses as a result of the instability risk and loss in production.

On many operations, mining layouts are inherited from best practices on a mine or to accommodate a specific mining method. The consequence of mining in unfavorable conditions is often only realized once significant instabilities are experienced. A case study is presented in this paper based on instabilities experienced on Lonmin. The contributing geological structures, associated behaviour and resulting failure mechanisms are discussed. The intent is to share the design strategies and the early identification of areas with a potential for a fall of ground risk ahead of mining to prevent occurrences.

Keywords: regional geological structures, mine design and layout, span analysis

1 INTRODUCTION

Eastern Platinum Limited (EPL) 2 Shaft is an inclined shaft extracting the UG2 Chromitite Reef. Figure 1 show the location of this shaft where a geotechnical investigation was conducted. The resource block is structurally complex, where major dykes, faults and secondary geological structures dissect the ore body.

Anomalous conditions were reported on the Shaft. This included:

- Unstable beam behaviour with separation occurring along low-angle geological features,
- Buckling of the pre-stressed elongates,
- Pillar slabbing experienced along the rigid in-stope dip pillars.

An assessment of the general mining layout, support and geological environment was conducted in the panels. The prevailing conditions were compared to similar studies on adjacent operations to ensure that the most appropriate preventative layout and design strategies could be implemented.

2 GEOGRAPHIC LOCATION OF THE STUDY SITE

EPL2 Shaft (Figure 1) is located on the eastern side of Lonmin's Marikana property. The UG2 Reef strikes E-W and dips at approximately 10° towards the North. Geological features, including potholes, rolling reef, faults, dykes and joints disrupt the reef horizon. The UG2A Markers (thin Chromitite layers) are located at a maximum of 8.5 m in the hangingwall

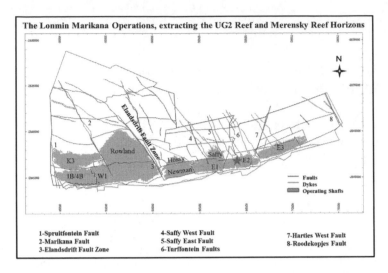

Figure 1. Major geological features exposed along the Lonmin Marikana lease area. The study site (EPL2 Shaft) is indicated by the red star.

of the UG2 Reef. These layers act as natural parting planes in the hangingwall. The layer above the UG2 Reef up to the UG2A Markers is referred to as the critical beam thickness. The Hangingwall 1A/Hangingwall 2 contact is located up to a maximum of 15.0 m in the hangingwall. A thin chromitite layer (natural parting plane) can also be present along this contact. Alteration at the UG2A Markers or HW1A/HW2 contact can contribute to large scale hangingwall instability (Hartzenberg and Du Plessis, 2014). The stope panels are mined at a maximum inter-pillar span of 30.0 m. A rigid, non-yield pillar system with a minimum Factor of Safety of 1.6 (Ryder and Jager, 2002) is currently applied in the layout design. A long panel dip mining method is used with raise lines situated alongside the in-stope dip pillars. In-stope support consist of 1.6 m long rock bolts to cater for the 95% fall-out thickness of 1.2 m (Ryder and Jager, 2002 and Du Plessis, 2015), pre-stressed elongates to support up to the UG2A Markers (8.5 m) and grout packs as supplementary back area support to prevent the occurrence of backbreaks.

3 EPL2 SHAFT SITE DESCRIPTION

The EPL2 Shaft mining personnel reported the instabilities at 10cW33 and 10cW34 Raise lines (Figure 2a) to the Rock Engineering Department. Figure 2b shows the geological features exposed in the panel, including NNW-SSE striking, near-vertical joints and faults (J1), WNW-ESE striking, low-angle joints and faults (J2) as well as interlinking joints and faults which alternate in strike direction from E-W to NE-SW (J3). Figure 2b shows the geological features that were exposed in the raise line and panel where structural mapping was done. The structural interpretations discussed in the remainder of the paper identify the potential causes that contribute to the hangingwall instabilities.

The slabbing experienced along the rigid non-yield pillars is as a result of both the joint orientation and altered material present along the reef top and/or bottom contacts. Site 1 in Figure 2b, show the orientation of the NNW-SSE striking joints in relation to the dip pillars. The joints intersect the pillars at oblique angles (Figure 3). These slabs defined by the joint planes are mobilised where altered material is present along the pillar top- or bottom contacts of the UG2 Reef. This results in the observed "pillar slabbing" experienced.

The zone/location of alteration has proven to be problematic at other investigations sites. If this zone is located along the top- or bottom contact of the UG2 Reef, the anomalous

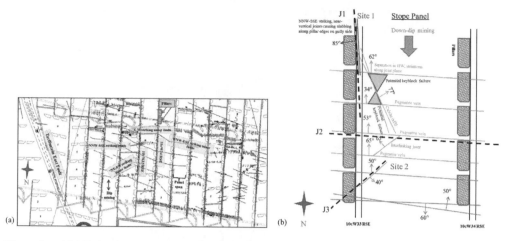

Figure 2.   (a) Mining layout at EPL2 Shaft, (b) Geotechnical mapping of the long down-dip panel.

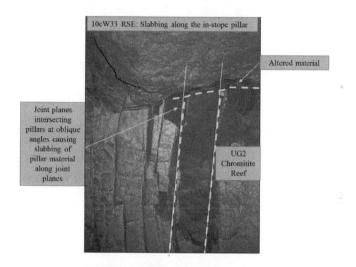

Figure 3.   Slabs defined by joint planes at 10cW33 RSE.

pillar behaviour or immediate unravelling of the hangingwall can be identified (Hartzenberg and Du Plessis, 2017). If it migrates to the UG2A Markers or HW1A/HW2 contact, large-scale instabilities can occur. The occurrences and risk associated needs to be taken into consideration when designing the layout and support.

The buckling elongates observed, indicated potential overloading. Underground HW instability was identified. This instability could be as a result of either beam separation along the UG2A Markers (maximum of 8.5 m in the hangingwall) or along the HW1A/HW2 contact (maximum of 15.0 m in the hangingwall). "Regional", large scale, low-angle geological structures were identified in this area. These structures dip towards the South (Figure 2b) and cut across the hangingwall layers. Consequently, the hangingwall beam is cut-off at the one end, creating a cantilever beam. The defined beam can detach from the potential parting planes (UG2A Markers or HW1A/HW2 contact). The effect (beam bending) is only realised when the in-stope pre-stressed elongates show signs of buckling and/or failure due to the loading capabilities being exceeded when a critical mining span is reached. In some instances, the beam bending causes a higher compressional stress along the edges of the underlying pillars. This can be observed as anomalous stress fracturing conditions.

Figure 4a and b. Cantilever beam with the buckling pre-stressed elongates situated on the down-dip side (defining the beam edge).

Site 2, Figure 2b, is represented by Figure 4a and b, where "regional", large scale, low-angle geological structures are exposed. Previous case studies identified this to usually occur when the panel has advanced for approximately twice the panel span (e.g. 60 m). This behaviour results in large scale (panel size) backbreaks.

4 FINDINGS AND REMEDIAL ACTIONS

To circumvent the risk of slabbing pillars inuring people, a split-dip layout (Figure 5b) as well as the pinning and strapping of pillars can be considered.

Where the buckling of pre-stressed elongates has been identified, changes in terms of the mining span, mining direction and in-stope support were considered. Cantilever analysis indicated a maximum inter-pillar span of 27.0 m should be implemented for the prevailing geotechnical conditions. Post implementation and follow-up visits confirmed stable conditions.

Van Zyl (2011) indicated through joint modelling (JBlock) that the instability risk can also be minimized by considering a change in the mining layout. For the prevailing structure, the following can be considered:

• Split-dip layout, least risk with the western face leading (Figure 5b),
• Breast mining (Figure 5a) mining towards the east (low risk),
• A breast panel mining towards the west (high risk),
• Single up-dip panel (high risk), the hazard extends well into the back area.

Figure 5a and b, show the most suitable (least risk) mining layouts and mining directions pertaining to the intersection of the "regional", large scale, low-angle geological structures. In Figure 5a, the structure cross-cuts the mining direction obliquely. Only a portion of the feature is exposed at a time (after each blast). The structure and associated behaviour can therefore be controlled by the appropriate support. The cantilever span will only reach its maximum once the structure is fully exposed. In Figure 5b only a half panel is mined at a time, reducing the overall mining span. Once the structure is fully exposed, the associated behaviour can be controlled by the appropriate support.

Grout packs should be considered to support the thickness of the potentially unstable layers. The grout pack should provide high early strength close to the face and prevent backbreaks.

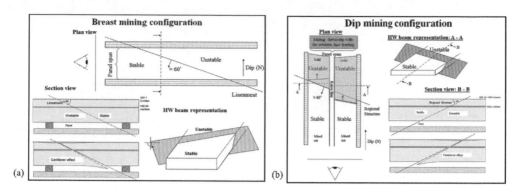

Figure 5. (a) Recommended breast mining configuration in an eastern direction, (b) Split down-dip mining configuration, with the western face leading.

5 CONCLUSIONS

Instabilities identified could be attributed to the intersection of "regional", large scale, low-angle geological structures and alteration layers. The resulting beam behaviour can be prevented through appropriate beam analysis and favourable mining layouts. The preventative strategies include decreasing the panel span to cater for the beam behaviour, a change in the mining layout and mining direction to ensure the most stable mining layouts and to implement a stiff, active grout pack support. The objective of this paper is to share the design strategies and learnings to promote early identification of areas with a potential for a fall of ground risk ahead of mining and to prevent occurrences.

ACKNOWLEDGEMENTS

The findings presented in this paper were conducted as part of the author's MSc. studies. The project is scoped as part of on-going research at Lonmin.

REFERENCES

Du Plessis, M. (2015). Marikana Mine Design Criteria Rock Engineering Guideline. *Appendix E: Tendon Design Methodology.*

Hartzenberg, A.G. and Du Plessis, M. (2017). Unravelling the structural mysteries of the 'Bermuda Triangle' at Lonmin's Saffy Shaft. *AfriRock 2017 Rock Mechanics for Africa.* The Southern African Institute of Mining and Metallurgy. Volume 2, pg. 933–945.

Hartzenberg, A.G. and Du Plessis, M. (2014). The influence of regional structures associated with the Bushveld Complex on the mechanism driving the behaviour of the UG2 hangingwall beam and in-stope pillars at Lonmin's Marikana Operations. *The 6th International Platinum Conference.* The Southern African Institute of Mining and Metallurgy, 2014.

Ryder, J.A. and Jager, A.J. (2002). *A textbook on Rock Mechanics for Tabular Hard Rock mines.* SIMRAC, Johannesburg.

Van Zyl, J.C. (2011). *Joint Modelling to determine the optimal mining layout, orientation and support strategy.* Internal Lonmin document.

Geomechanics and Geodynamics of Rock Masses – Litvinenko (Ed.)
© *2018 Taylor & Francis Group, London, ISBN 978-1-138-61645-5*

Strength estimation of fractured rock using compression—a specimen with spherical indenters

V.A. Korshunov, D.A. Solomoychenko & A.A. Bazhukov
Saint-Petersburg Mining University, Saint-Petersburg, Russia

ABSTRACT: New methods for limit and residual strength certificates construction were developed. The results of tests which including loading monolithic solid specimens of irregular shape by spherical indenters are needed.

The method of construction the strength certificate of intact rock is based on an estimate of the inhomogeneous stress state when it is loaded with spherical indenters. Tensile and compressive components of cohesive shear strength were taken for definition characteristics of specimen failure.

Residual rock strength certificate is constructed according to the data on the strength of solid rock, in accordance with the established dependencies on the rock brittleness index and taking into account its lithological composition.

It is proposed to approximate the envelope curve of Mohr's stress circles by linear segments corresponded to stable types of destruction, and transitional curvilinear segments for which the type of failure have the probabilistic nature.

A complex method for estimating the strength of fractured rocks is proposed. It is based on results of mechanical testing of specimens with spherical indenters. Strength of the rock is determined by comparing the indicators of the limit and residual strength of solid specimens, taking into account the parameters of fracture in the natural conditions.

Correlations which depend of the strength parameters of fractured rocks on the stress level and rock brittleness index are set.

Keywords: rocks, irregular shape lump sample, spherical indenters, compressive and tensile stresses, tearing, shearing, brittleness, strength certificate

1 COMPUTATIONAL METHOD FOR STRENGTH CERTIFICATE CONSTRUCTION

It's recognized that plotting an empirical envelope curve based on Mohr's stress circles (strength certificate) allows researcher to make a complete representation about rock's strength characteristics in different stress conditions. Thus the computational method for strength certificate construction was developed in Saint Petersburg Mining University. This method is based on estimation of rock's anisotropic stress condition and on estimation of specimen complex failure mechanism caused by loading it with counter moving spherical indenters (Korshunov, 2010). In the course of further research, the method is improved.

The refined destruction failure design of indenter loading process was approved (Fig. 1a). In this design three different types of destruction are introduced (Fig. 1b):

1. plastic behavior under uneven volume compression with high stresses in proximity to indenters;
2. failure by tearing in the plane of loads;
3. shearing along the intermediate areas.

For definition of specimen failure two components of cohesive shear strength C_0 were accepted: tensile component σ_t and compressive component p:

$$C_0 = \sqrt{\sigma_t p} \qquad (1)$$

The interconnection of σ_t and p with compressive strength σ_c was derived:

$$\sigma_c = p + C_0 \qquad (2)$$

a – specimen loading diagram; b – development of residual deformations under stress

There are two options for representing the rock strength criteria: in the Mohr's coordinate system ($\sigma - \tau$ axis) an in the Heigh coordinate system (principle stress axis $\sigma_3 - \sigma_1$).

In the Mohr's coordinate system, it is proposed to approximate the enveloping curve by straight lines. Each rectilinear segment corresponds to stable types of failure (tear, shear and ductile failure in conditions of non-uniform compression under high level of stresses).

Also, transitional curvilinear segments are used for destruction areas of a probabilistic nature (Fig. 2).

The transition destruction areas of a probabilistic nature correspond to the main components of cohesive shear strength C_0 (σ_t; p) and maximum shear strength τ_{max} ($\sigma_3{}^M$ and $\sigma_1{}^M$):

$$\sigma_3^M = \sqrt{K}\left(\frac{\sigma_c}{2} - 2\sigma_t \right) \qquad (3)$$

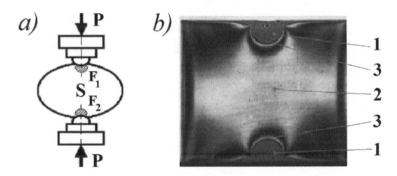

Figure 1.   Spherical indenter test design.

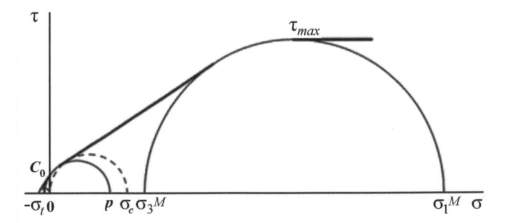

Figure 2.   Strength certificate construction in Mohr's coordinate system.

$$\sigma_1^M = \sigma_c + K(\frac{\sigma_c}{2} - 2\sigma_t)$$ (4)

where K – brittleness index, numerically equal to the ratio p/σ_t.

In the Heigh coordinate system strength modules assume a simple mathematical form:

for range of stresses $-2\sigma_t \leq \sigma_3 \leq -\sigma_t : \sigma_1 = K\sigma_3 + 2p$ (5)

for range of stresses $-\sigma_t \leq \sigma_3 \leq \sigma_3^M : \sigma_1 = \sqrt{K}\sigma_3 + \sigma_c$ (6)

for range of stresses $\sigma_3 \geq \sigma_3^M : \sigma_1 = \sigma_3 + \frac{\sigma_c}{2}(K - \sqrt{K} - 2) + 4C_0$ (7)

The deviation of calculated strength parameters from the values calculated by the standard method in accordance with state standard (GOST 21153. 8–88) does not exceed 10%.

2 COMPUTATIONAL METHOD FOR RESIDUAL STRENGTH CERTIFICATE CONSTRUCTION

It is known that at the level of τ_{max} (σ_3^M; σ_1^M) limit strength certificate and residual strength certificate are equal (Tarasov, 1991). A comprehensive study was conducted to assess the possibility of constructing the entire residual strength certificate using data of point load tests. This research consists of matching the results of different rock tests: indentation tests with spherical indenters and triaxial compression tests.

The graph (Fig. 3) shows resulting data of triaxial and indenter tests of sandstone and siltstone specimens. At the graph the triaxial compression test results are shown as dots and indenter test average results are shown as line graph. All data on graph is plotted in the Heigh coordinate system.

The method is proposed for constructing a residual strength certificate using data of spherical indenters tests of solid specimens.

It is proposed to approximate the certificate of residual strength in Heigh coordinate system the same way as ultimate strength certificate "1". The segments of approximated graph correspond to tear, shear and ductile failure (Fig. 4).

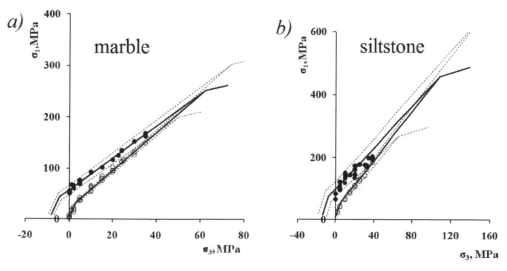

Figure 3. Limit and residual strength certificate of marble (a) and siltstone (b) in Heigh coordinate system.

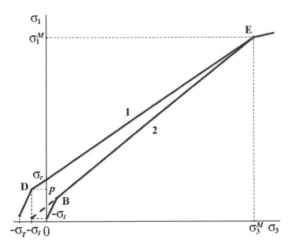

Figure 4. Residual strength certificate construction in Heigh coordinate system.

The segment plotted in shear failure range includes points corresponding to simple shear level $(-\sigma_t; \sigma_t)$ and maximum shear strength τ_{max} $(\sigma_3^M; \sigma_1^M)$. In tear failure range the graph can be approximated by a segment parallel to the analogous segment "1" of the envelope of ultimate strength.

As a limit of residual strength σ_R under compression, the average tensile stress σ_t, which is determined from spherical indenters test data, can be adopted in the first approximation:

$$\sigma_R \approx \sigma_t \tag{8}$$

The following values of the principal stresses correspond to transition region of residual strength with probabilistic tear or shear nature of failure:

$$\sigma_3 = \frac{\sigma_t(\sigma_1^M - \sigma_t)}{K(\sigma_3^M + \sigma_t) - (\sigma_1^M - \sigma_t)} \tag{9}$$

$$\sigma_1 = \frac{\sigma_1^M - \sigma_t}{\sigma_3^M + \sigma_t} \sigma_3 + \sigma_t \frac{\sigma_1^M + \sigma_3^M}{\sigma_3^M + \sigma_t} \tag{10}$$

The application field of calculation methods is hard rock formations with brittleness ratio (σ_c/σ_T) more than 5. To ensure that the spherical indenters test results are consistent with standard methods, restrictions on the shape and dimensions of the samples are placed. It is recommended to use specimens from 20 to 50 mm thickness with the expected surface area of the rupture in the range from 5 to 25 cm^2.

3 ESTIMATING OF FRACTURED ROCK STRENGTH

The experimental approach to estimating the strength of a fractured massif is based on the establishment of criteria linking the rock strengths in the massif and in the specimens, taking into account fracture parameters in the field conditions (Fissenko, 1972, Bieniawski, 1975). Usually formulas describe two main trends: decreasing in strength in the range from the maximum strength of solid specimens to the minimum values corresponding to the fractured rock; and reduction of the influence of cracks with an increase in the stress level under conditions of triaxial compression (Protodyakonov, 1964, Chirkov, 1974).

The comparative tests of rocks on the splitting of solid specimens by indenters and on compression in the regime of controlled deformation of structurally disturbed samples were carried out. In the specimens, a block structure was simulated with an intensity of cracks w equal to 1, 2, and 5 (Fig. 5).

The results of the study illustrate experimental points obtained from following data: results of triaxial tests of marble and sandstone and certificates of limit and residual strength constructed by using result data of indenter tests of specimens (Fig. 5).

It is established that envelope curves of limiting states, constructed using triaxial compression tests of homogeneous samples with different crack intensities, can be approximated similarly to the limit and residual strength certificates by rectilinear segments corresponding to tear and shear failure. It is justified to approximate it by rectilinear segments parallel to analogous segments of the envelope of residual strength. In this case, the conjugations of the segments characterizing the transition states lie on the line passing through the points corresponding to the limit strength for uniaxial compression of intact rock and to the analogous transition state of the destroyed rock.

A complex method for estimating the strength of fractured rocks is proposed. It is based on results of mechanical testing of solid specimens with spherical indenters. Strength of the rock is determined by comparing the indicators of the limit and residual strength of solid specimens, taking into account the parameters of fracture in the natural conditions. For this purpose, the main parameters of the fracture systems in the array are pre-established—the average distance between the cracks and their orientation.

Then, specimens are tested by spherical indenters. According to the test results data the main components of C_0 (σ_i; p) are calculated. Next, the construction of passports of limit 1 and residual 2 strengths is performed (Fig. 6). Enveloping curves 3 and 4 of the limiting stress state of the rock with different fracture character occupy an intermediate position between them.

It is proposed to approximate the envelope curve of the fractured rock certificate as similar to the limit and residual strength certificate by rectilinear segments corresponding to: tear failure $\sigma_w N$, shear failure (NA and AE) and ductile failure in conditions of non-uniform compression under high level of stresses.

Construction of the strength certificate of fractured rock is carried out in the following sequence.

First, from the values of the limiting σ_c and the residual σ_c strength under uniaxial compression, taking into account the chosen parameters of the crack systems, the limit strength is determined by the correlation dependence for uniaxial compression of the fractured rock σ_w:

$$\sigma_w = \sigma_R + (\sigma_c - \sigma_R) \cdot C_w \cdot C_\alpha \qquad (11)$$

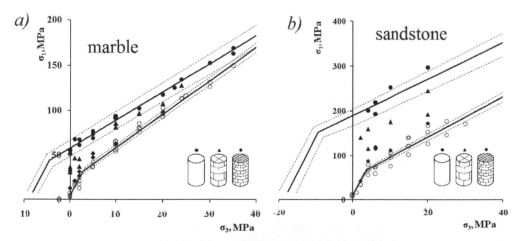

Figure 5.   Determination of the strength of fractured marble (a) and sandstone (b) specimens.

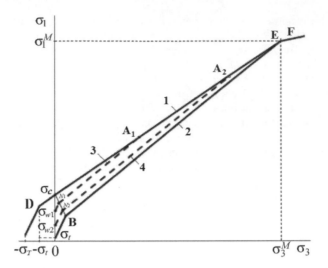

Figure 6.   Construction of fractured rock strength certificate.

where C_w and C_α – modules of fracture character and intermediate position.

The following empirical formulae is proposed to assess the strength of a rock weakened by normally secant fracture system:

$$\sigma_w = \sigma_t + (\sigma_c - \sigma_t) \cdot e^{-0,08 \, K_f \cdot (w-1)} \tag{12}$$

where K_f – brittleness coefficient, numerically equal to the ratio σ_c/σ_T.

The value of the indicator C_α for a rock weakened by a complex system of cracks tends to zero. In this case, when estimating the strength of fractured rock, as σ_w, the value σ_t can be taken:

$$\sigma_w \approx \sigma_t \tag{13}$$

Next the tear and shear failure segments of strength certificate are constructed. First, from the point σ_w, on X axis a straight line $\sigma_w N$ is drawn that should be parallel to the initial segment $\sigma_t B$ of residual strength certificate. This straight line ends at the point N located on a segment $\sigma_c B$ that connects the points corresponding to σ_c and the transition state of the residual strength B. Secondly, from the point N a straight line parallel to the BE is drawn to intersect with the envelope 1 of the ultimate strength at point A.

Fractured rock strength certificate after touching the envelope 1 of the ultimate strength at point A coincides with it (section AE).

The strength condition of a fractured rock in the indicated stress ranges assumed the following form:

$$\text{for range of stresses} \, 0 \le \sigma_3 \le \sigma_{3N} : \sigma_1 = \sigma_w + K\sigma_3 \tag{14}$$

$$\text{for range of stresses} \, \sigma_{3N} \le \sigma_3 \le \sigma_{3A} : \sigma_1 = \sigma_{3B} + (\sigma_3 - \sigma_{3B}) \cdot \frac{\sigma_1^M - \sigma_t}{\sigma_3^M + \sigma_t} \tag{15}$$

4   SUMMARY

New methods for limit and residual strength certificates construction were developed. The results of tests including loading solid specimens of irregular shape by spherical indenters are needed.

The method of construction the strength certificate of intact rock is based on an estimate of the inhomogeneous stress state when it is loaded with spherical indenters. Tensile and compressive components of cohesive shear strength were taken for definition characteristics of specimen failure.

Residual rock strength certificate is constructed according to the data on the strength of solid rock, in accordance with the established dependencies on the rock brittleness index and taking into account its lithological composition.

It is proposed to approximate the envelope curve of Mohr stress circles by linear segments corresponding to stable types of destruction, and transitional curvilinear segments for which the type of failure have the probabilistic nature.

A complex method for estimating the strength of fractured rocks is proposed. It is based on results of mechanical testing of specimens with spherical indenters. Strength of the rock is determined by comparing the indicators of the limit and residual strength of solid specimens, taking into account the parameters of fracture in the natural conditions.

Correlations which depend of the strength parameters of fractured rocks on the stress level and rock brittleness index are set.

The study was carried out at the expense of a grant from the Russian scientific Foundation (project № 17-77-10101).

REFERENCES

Bieniawski Z.T., van Heerden W.L. The significance of in situ tests on large rock specimens. Int. J. Rock Mech. Min. Sci., 1975, vol. 12, № 4, p. 101–113.

Bieniawski Z.T. Estimating the strength of rock materials. – J.S. Afr. Min. Metall., vol. 74, 1974, pp. 312–320.

Chirkov S.E. Influence of stress state and scale effect on rock strength and deformability. M., Institute of mining named after A.A. Skochinski, 1974. – 32 c.

Fissenko G.L. Methods of quantitative estimation of structural weakening in connection with the analysis of their stability. Modern problems of rock mechanics. L., Nedra. – 1972, c.21–29.

Korshunov V.A., Kartashov. Yu.M., Kozlov V.A. Determination of indices of strength certificate of rocks using the method of specimens failure with spherical indenters//Problems in geomechanics of technogenious rock mass/Proceedings of the mining Institute. V.185. Saint-Petersburg, 2010. pp. 41–45.

Protodiakonov M.M. Методы оценки трещиноватости и прочности горных пород в массиве Methods of assessment of fracture and rock strength in massif. M., Institute of mining named after A.A. Skochinski, 1964. – 32p.

Tarasov B.G. Laws of deformation and fracture of rocks at high pressures. Saint-Petersburg, Mining institute, 1991. – 46 c.

Geomechanics and Geodynamics of Rock Masses – Litvinenko (Ed.)
© *2018 Taylor & Francis Group, London, ISBN 978-1-138-61645-5*

Measurements of thermal properties of rock samples under high temperature conditions

Weiren Lin
Katsura Campus, Kyoto University, Nishikyo-ku, Kyoto, Japan

Osamu Tadai
Marin Works Japan Ltd., Technical Researcher, Monobe-otsu Nankoku, Kochi, Japan

Tatsuhiro Sugimoto
Katsura Campus, Kyoto University, Nishikyo-ku, Kyoto, Japan

Takehiro Hirose & Wataru Tanikawa
JAMSTEC, Senior Scientist, Monobe-otsu Nankoku, Kochi, Japan

Yohei Hamada
JAMSTEC, Scientist, Monobe-otsu Nankoku, Kochi, Japan

ABSTRACT: To investigate temperature effects on the thermal transport properties of several typical rocks, thermal conductivity, thermal diffusivity and specific heat capacity or volumetric heat capacity of dry granite, gabbro, marble and sandstone samples and a fused silica sample were measured under high ambient temperature conditions from room temperature up to 160°C and under atmospheric pressure condition. For this purpose, we developed a new measurement system for the three thermal transport properties of dry rock samples under high ambient temperatures up to 180°C. The results of our experiments conducted in this study clearly showed that the thermal properties of dry rocks change with ambient temperature changes, in addition, the thermal property change patterns were different for different rock types. For example, the thermal conductivities of granite, marble and sandstone decreased with temperature, however that of gabbro kept almost constant.

1 INTRODUCTION

Thermal transport properties are necessary for understanding heat flow and temperature structure in the crust. Usually, thermal properties of rock formations are measured by using rock core samples retrieved from a great depth where temperature is higher than on the ground. To exactly evaluate thermal transport properties of a rock sample retrieved from a deep borehole at its original in-situ conditions, the high temperature effects must be taken into consideration. To realize this purpose, we developed a simple new measurement system for the three thermal transport properties including thermal conductivity, thermal diffusivity and volumetric heat capacity and/or specific heat capacity of rock specimens under high ambient temperatures. This system consists of a commercial thermal constants analyzer TPS 1500 (Hot Disk, Gothenburg, Sweden), a high temperature chamber in which the rock samples and the sensor of thermal property measurements were installed. The highest temperature limit of the measurement system is 180°C. In this study, we used this new system and examined the effects of high temperature on thermal conductivity, thermal diffusivity and specific heat capacity for dry samples of four terrestrial rock types and an artificial material of fused silica.

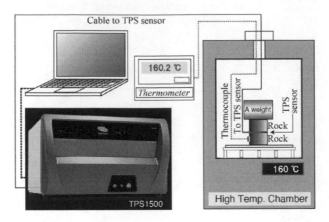

Figure 1. Schematic diagram of the apparatus used in this study for measurements of thermal conductivity under high temperature and under atmospheric pressure conditions (modified from Lin et al., 2018).

2 MEASUREMENT SYSTEM OF THERMAL PROPERTIES UNDER HIGH TEMPERATURE CONDITIONS

We have established a new, simple thermal property measurement system for rock samples under high ambient temperature and atmospheric pressure (Figure 1).

This simple system consists of a high temperature chamber (maximum temperature 270°C, real temperature control accuracy ~1°C), a thermal constants analyzer TPS 1500 to measure thermal conductivity and thermal diffusivity simultaneously, and a thermometer and its sensor (thermocouple), which measures the temperature of the rock sample rather than the air temperature in the chamber. The transient plane heat source sensor of the TPS 1500 is sandwiched between two pieces of the same rock, and a weight was put on the top of them for good contact (Lin et al., 2014; Tanikawa et al., 2016).

This thermal property measuring system TPS 1500 is based on the transient plane heat source method (ISO, 2008) to measure thermal conductivity, λ, and thermal diffusivity, α, simultaneously. From them, the volumetric heat capacity (heat capacity per unit volume), C_{vp}, or/and the specific heat capacity (heat capacity per unit mass), C_p, can be determined by the following equation:

$$\alpha = \lambda/C_{vp} = \lambda/\rho C_p \qquad (1)$$

where ρ is bulk density of the test rock sample. The thermal constants analyzer TPS 1500 has two measurement modes. In this study, we applied the bulk mode, which assumes the test sample is isotropic and calculates bulk thermal conductivity and bulk thermal diffusivity.

The maximum temperature of the high-temperature chamber is 270°C, but the limit of high-temperature resistance for the TPS sensor cable is 180°C. Therefore, the applicable highest temperature of this system is 180°C. We set the ambient temperature conditions of the thermal property measurements as room temperature (between 21 and 25°C), 40, 60, 80, 100, 120, 140, and 160°C. Thermal property measurements were conducted under atmospheric pressure.

3 ROCK SAMPLES

To test if our thermal property measurement system works correctly, a standard sample made of the artificial material of fused silica was prepared because its thermal conductivity is known (Horai and Susaki, 1989; Abdulagatov et al., 2000; Lin et al., 2011). Its bulk density is ~2.21 g/cm³, and porosity is zero. As typical terrestrial rocks, four rock samples of a granite, a gabbro, a marble and a sandstone were collected. They are the fine-grained Aji Granite (porosity 0.43%, dry bulk density 2.66 g/cm³, wet bulk density 2.66 g/cm³ determined by the buoyancy method) from Kagawa, Japan; Indian Gabbro (0.16%, 2.98 g/cm³, 2.99 g/cm³) from India; Carrara Marble (0.30%, 2.71 g/cm³, 2.72 g/cm³) from Italy, and Shirahama Sandstone (12.2%, 2.29 g/cm³, 2.41 g/cm³) from Wakayama, Japan.

Under atmospheric pressure conditions, it is difficult to keep the rock sample in the water-saturated state at a high temperature around or higher than the boiling point of water. Thus, we measured the thermal properties under high temperatures up to 160°C and at atmospheric pressure for dry rock samples.

As mentioned above, one pair of two cylindrical pieces of each rock are necessary for the measurements by hot disk method. The size of each piece is ~5 cm in diameter, and 4–5 cm in length. Basically, we prepared dry samples for the four terrestrial rocks dried in an oven at 110°C for more than 24 hours and then moved into a dried desiccator for cooling to room temperature and keeping dry state until thermal property tests.

4   RESULTS AND DISCUSSIONS

Figure 2 shows the thermal conductivity of the various dry rock samples and fused silica measured in this study. It is obvious that thermal conductivity changes with temperature. The granite, marble and sandstone samples showed a clear decreasing trend of thermal conductivity with increasing ambient temperature; but the gabbro kept almost constant; and in contrary the fused silica sample showed a increasing trend. The decreasing trend is common in many natural rocks including sedimentary, crystalline, magmatic, and metamorphic rocks, as shown by numerous previous research articles and review papers (e.g., Heuze, 1983; Schön, 1998, Abdulagatov et al., 2006). In general, an inverse correlation between temperature and thermal conductivity is usually understood in terms of phonon conduction. An increasing trend with temperature, however, seems to be much rarer than a decreasing trend, but has been found in a few studies, e.g., in fused silica (Abdulagatov et al., 2000), in pyroxene-granulite (Abdulagatov et al., 2006), and in basalt (Lin et al., in press).

By applying the hot disk thermal property measurement method (ISO, 2008), we can obtain thermal conductivity and thermal diffusivity simultaneously, and can estimate volumetric heat capacity and/or specific heat capacity using equation (1).

Thermal diffusivity of the dry samples of the granite, marble and sandstone decreased steeply with increasing ambient temperature, but the gabbro and fused silica decreased gently

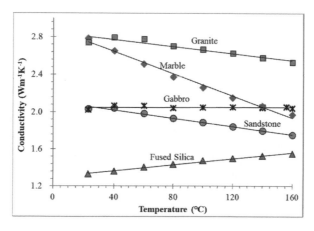

Figure 2.   Relationships between measured thermal conductivity and ambient temperature of the dry samples including Aji Granite, Indian Gabbro, Carrara Marble, Shirahama Sandstone, and fused silica.

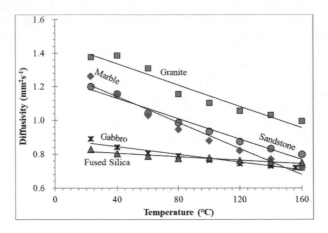

Figure 3. Relationships between measured thermal diffusivity and ambient temperature for the same dry samples shown in Figure 2.

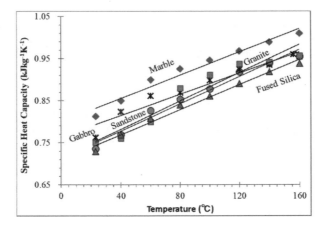

Figure 4. Relationships between measured volumetric heat capacity and ambient temperature for the same dry samples shown in Figures 2 and 3.

(Figure 3). Specific heat capacity of all the samples showed a similar increasing trend with ambient temperature, in addition the values of all the samples at the same temperature are within a narrow range (e.g., 10–20%) (Figure 4).

5 CONCLUSIONS

To investigate temperature effects on the thermal transport properties including thermal conductivity, thermal diffusivity and specific heat capacity of four dry rock samples and a fused silica sample were measured under high temperature conditions from room temperature to 160°C and under atmospheric pressure. As results of the measurements, temperature effects of thermal conductivity clearly depended on rock types. Particularly, the thermal conductivity decreased with temperature for granite, marble and sandstone; whereas slightly increased for fused silica; but kept almost the constant for the gabbro.

The thermal diffusivity decreased for all the rock types and the fused silica tested in this study, although the decreasing rate of the gabbro and the fused silica were much less than those of the other rocks. The specific heat capacity of all the rocks and fused silica increased

310

at a similar increasing rate with temperature, in addition the values of all the samples at the same temperature are within a narrow range. The results of our experiments clearly show that the thermal properties of rocks change with ambient temperature changes, in addition, the thermal property change patterns were different for different rock types.

ACKNOWLEDGMENTS

Part of this work was supported by Grants-in-Aid for Scientific Research 16H04065 of the Japan Society for the Promotion of Science (JSPS), Japan.

REFERENCES

Abdulagatov, I.M., Emirov, S.N., Tsomaeva, T.A., Gairbekov, Kh.A., Askerov, S.Y., and Magomedova, N.A., 2000, Thermal conductivity of fused quartz and quartz ceramic at high temperatures and high pressures, *J. Phys. Chem. Solid*, Vol. 61, pp. 779–787, doi:10.1016/S0022-3697(99)00268-1.

Abdulagatov, I.M., Emirov, S.N., Abdulagatova, Z.Z., and Askerov, S.Y., 2006, Effect of Pressure and Temperature on the Thermal Conductivity of Rocks, *J. Chem. Eng. Data*, Vol. 51, pp. 22–33, doi:10.1021/je050016a.

Heuze, F.E., 1983, High-temperature mechanical, physical and thermal properties of granitic rocks— A review, *Int. J. Rock Mech. Min. Sci. & Geomech. Abstr.*, Vol. 20, pp. 3–10.

Horai, K., and Susaki, J., 1989, The effect of pressure on the thermal conductivity of silicate rocks up to 12 kbar, *Phy. Earth Planetary Interiors*, Vol. 55, pp. 292–305.

ISO, 2008, Plastics—Determination of thermal conductivity and thermal diffusivity—Part 2: Transient plane heat source (hot disc) method, *International Standard ISO* 22007-2.

Lin, W., Tadai, O., Hirose, T., Tanikawa, W., Takahashi, M., Mukoyoshi, H., Kinoshita, M., 2011, Thermal conductivities under high pressure in core samples from IODP NanTroSEIZE drilling site C0001. *Geochem. Geophys. Geosyst.*, Vol. 12, No. Q0AD14, doi:10.1029/2010GC003449.

Lin, W., Fulton, P.M., Harris, R.N., Tadai, O., Matsubayashi, O., Tanikawa, W., and Kinoshita, M., 2014, Thermal conductivities, thermal diffusivities, and volumetric heat capacities of core samples obtained from the Japan Trench Fast Drilling Project (JFAST), *Earth, Planets and Space*, Vol. 66, No. 48, doi:10.1186/1880-5981-66-48.

Lin, W., Tadai, O., Kinoshita, M., Kameda, J., Tanikawa, W., Hirose, T., Hamada, Y., Matsubayashi, O., 2018, Thermal conductivity changes of subducting basalt, Nankai subduction zone, SW Japan: An estimation from laboratory measurements under high-pressure and high-temperature conditions, *in* T. Byrne, D. Fisher, L. McNeill, D. Saffer, K. Ujiie, M. Underwood, and A. Yamaguchi, eds., *GSA Book SPE534: Geology and Tectonics of Subduction Zones: A Tribute to Gaku Kimura*, Geological Society of America Special Paper 534, pp. 1–16, https://dx.doi.org/10.1130/2018.2534(XX).

Schön, J.H., 1998, Ch. 8: Thermal properties of rocks, in *Physical properties of rocks, Second Ed. - Handbook of geophysical exploration, Seismic exploration*, Vol. 18, Pergamon, p. 323–378.

Tanikawa, W., Tadai, O., Morita, S., Lin, W., Yamada, Y., Sanada, Y., Moe, K., Kubo, Y., Inagaki, F., 2016, Thermal properties and thermal structure in the deep-water coalbed basin off the Shimokita Peninsula, Japan, *Marine and Petroleum Geology*, Vol. 73, pp. 445–461.

Geomechanics and Geodynamics of Rock Masses – Litvinenko (Ed.)
© 2018 Taylor & Francis Group, London, ISBN 978-1-138-61645-5

Static and dynamic effective stress coefficient of St. Peter sandstone during depletion and injection

Xiaodong Ma
Department of Geophysics, Stanford University, Stanford, US
State Key Laboratory of Geomechanics and Geotechnical Engineering, Institute of Rock and Soil
Mechanics, Chinese Academy of Sciences, Wuhan, China
SCCER-SoE and Department of Earth Sciences, ETH, Zürich, Switzerland

Mark D. Zoback
Department of Geophysics, Stanford University, Stanford, US

ABSTRACT: A medium-to-high porosity St. Peter sandstone was subjected to hydrostatic confining pressure $P_c (= S)$ under fully-drained conditions, simulating the stressed rock *in situ* during depletion or injection. We recorded variations of static strain and dynamic velocities with confining pressure and pore pressure to estimate the corresponding effective stress coefficient. For the static deformation data, α is clearly less than unity, ranging between 0.3 and 0.7 for any tested stress conditions. The effective stress coefficient is dependent on P_p during depletion but not so during injection. Given the same stress condition, the effective stress coefficient during injection is consistently higher than during depletion. The dynamic effective stress coefficient for V_p is generally close to unity when σ is less than 20 MPa. During depletion, α for V_p first decreases with σ when the latter is up to 30 MPa. However α jumps to unity for $\sigma \geq 30$ MPa. The effective stress coefficient for V_s increases significantly with σ, regardless of the loading path. It is important to note that the effective stress coefficient with respect to velocities is different from that of static deformation, in both the magnitude and the dependency on P_c and P_p.

Keywords: sandstone; poroelasticity; effective stress coefficient; seasoning; ultrasonic velocity, deformation

1 INTRODUCTION

The production of hydrocarbons typically causes the reservoir to contract. Significant deformation can cause wellbore failure, surface subsidence and production loss. On the other hand, hydraulic fracturing stimulation and enhanced oil recovery inject large volumes of fluid into the reservoirs, which can potentially trigger faulting or earthquakes. What accompanies such deformation is the stress changes around and inside the reservoir, which may enhance or inhibit the rate of deformation. Thus, it is imperative to fundamentally understand the change in stress and strain as the reservoir depletes or undergoes injection. Commonly, the stress changes associated with depletion and injection are considered as poroelastic and treated as such. The effective stress coefficient α, as a key parameter in poroelasticity, is used to evaluate the relative contribution of total stress S and pore pressure P_p on rock properties (*Biot*, 1962). While there is a continuing interest in the effective stress coefficient, the experimental studies on α has been scarce. Motivated by the effective stress alteration due to depletion and subsequent fluid injection in conventional reservoirs, we characterized a St. Peter sandstone in order to obtain its dependencies of α on S and P_p with respect to volumetric deformation and velocities. We present our preliminary test results here.

2   MATERIAL AND METHODS

We tested a St. Peter sandstone with medium-to-high porosity (~19%), nearly monomineralic lithology (98% pure quartz) and rounded grains (0.1–0.7 mm). The tests are configured to subject the specimen (prepared into 25.4 mm length and 25.4 mm diameter cylinder) to hydrostatic confining pressure $P_c (= S)$ under fully-drained conditions, simulating the stressed rock *in situ* during depletion or injection. The jacketed specimen is sandwiched by two core holders (Figure 1). The assembly is then housed inside a pressure vessel to subject to P_c. Pore pressure (P_p) is applied as the compressed argon (Ar) gas is injected from both ends of the specimen.

Two pairs of electrical-resistance strain gages are epoxied directly on the specimen to measure its axial and radial deformation, which allow for calculating volumetric strain ε_v. Ultrasonic velocities ($P/S1/S2$) along the axial direction were measured by the coupled wave emitter and receiver embedded in the core holders. We recorded variations of static strain and dynamic velocities with confining pressure and pore pressure to estimate the corresponding effective stress coefficient. The variations of P_c and P_p followed a predetermined loading path. Under constant P_p, P_c is cycled between P_p plus 10 MPa and a maximum of 50 MPa. Then, P_p is elevated to a higher level and P_c is cycled again. The condition of loading P_c with constant P_p simulates the scenario of depletion, and unloading P_c for injection. We waited sufficiently long after each step of P_c and P_p for the pore pressure to fully equilibrate.

3   DEFORMATION AND VELOCITY DATA

The deformation data under the designated loading path were analyzed under the context of volumetric strain variation with P_c and P_p. The stress-strain response of the St. Peter sandstone specimen under different constant pore pressure is shown in Figure 2. Constant P_p data series are highlighted and fitted with contour lines of the corresponding color. These lines of data illustrate the effect of P_c. Constant simple effective stress $\sigma (= P_c - P_p)$ data series are fitted with dashed black contour lines. These lines of data reveal the counteraction between P_c and P_p. Clearly, P_p relieves the confinement of P_c on the rock frame, as higher P_p decreases

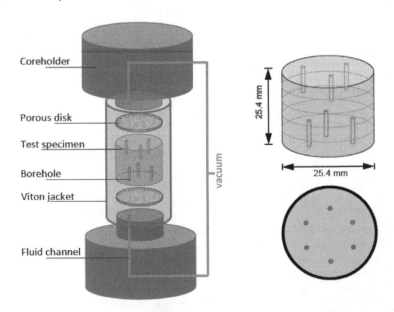

Figure 1.   Left: Illustration of the experimental specimen-coreholder assembly housed inside a pressure vessel. Right: Dimensions of the specimen and the configuration of boreholes drilled inside the specimen.

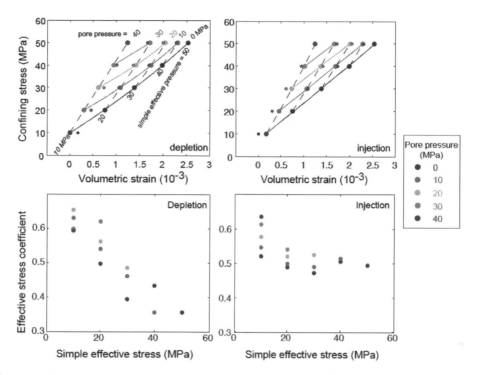

Figure 2.   Upper row: confining pressure—volumetric strain relationships for constant pore pressures. Lower row: Variations of Effective stress coefficient (for volumetric strain) with simple effective stress for constant pore pressures under both depletion and injection cases.

the volumetric strain. However, an inclined constant σ curve suggests P_p does not completely cancel the effect of P_c; in other words, the effective stress coefficient for static volumetric deformation is less than unity. Figures 1 detail the scenarios of depletion (loading P_c) and injection (unloading P_c), respectively. The discrepancy between the two scenarios is generally regarded as the hysteresis between loading and unloading material behavior in geomaterials.

The velocity data under the designated loading path were analyzed in a way similar to the deformation data. The response of the specimen under different constant pore pressure is shown in Figure 3, which depicts the scenarios of depletion and injection for both Vp and Vs. Regardless of loading path and wave type, the effect of P_p on velocities appears to strengthen when confining stress is held constant. This is evident as the horizontal distances between adjacent data points increases from right to left.

4   EXPERIMENTALLY-DERIVED EFFECTIVE STRESS COEFFICIENT

We follow *Todd and Simmons* (1972) to derive α via:

$$\alpha = 1 - \frac{\partial Q / \partial P_p \big|_\sigma}{\partial Q / \partial \sigma \big|_{P_p}} \tag{1}$$

where Q is any measured physical quantity, and $\sigma = S - P_p$ is the simple effective stress σ. Eq. (1) had been used by a few other researchers in measuring velocities (*Christensen and Wang*, 1985; *Hornby*, 1996; *Sarker and Batzle*, 2008).

Since Q represents either volumetric strain or P-/S-wave velocity, it is important to not mix the effective stress coefficient α for different physical quantities. According to Eq. (1), α

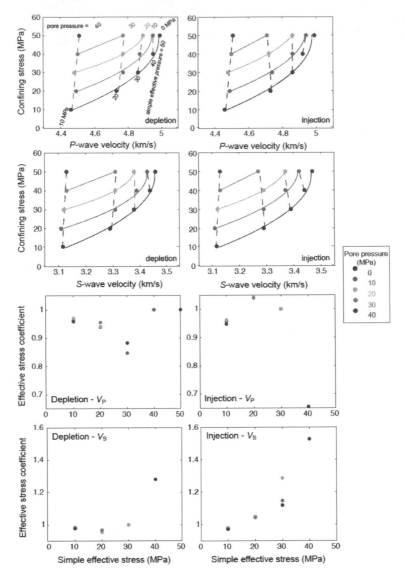

Figure 3.   Upper two row: confining pressure—ultrasonic velocity relationships for constant pore pressures. Lower two row: Variations of Effective stress coefficient (for ultrasonic velocity) with simple effective stress for constant pore pressures under both depletion and injection cases.

is computed using the local tangent to the curves of constant σ and P_p at any given (P_c, P_p) magnitudes.

For the static deformation data, α is clearly less than unity, ranging between 0.3 and 0.7 for any tested stress conditions (Figure 2). One can evaluate the P_p effect by examining the trend of α for constant σ. In general, α increases with P_p when σ is held constant, and the highest α is obtained when P_p and σ reaches its maximum and minimum value we applied. The effective stress coefficient is dependent on P_p during depletion but not so during injection. Given the same stress condition, the effective stress coefficient during injection is consistently higher than during depletion.

The dynamic effective stress coefficient for V_p is generally close to unity when σ is less than 20 MPa (Figure 3). During depletion, α for V_p first decreases with σ when the latter is up to 30 MPa. However α jumps to unity for $\sigma \geq 30$ MPa, the significance of which is questionable since the derivation of α is based on fewer data points. The effective stress coefficient for V_S increases significantly with σ, regardless of the loading path. For example, α for V_S reaches as high as 1.5. Regardless of the wave type, the effect of P_p α is insignificant.

5 DISCUSSION AND CONCLUDING REMARKS

We have performed hydrostatic experiments on a St. Peter under drained, jacketed conditions, and derived the effective stress coefficient for both static deformation and dynamic velocities. With respect to static deformation, the effective stress coefficient is less than unity, and typically varies between 0.3 and 0.7 under different P_c and P_p. The variation of α with P_c and P_p implies the modeling of reservoir stress change using a constant α could be erroneous.

It is important to note that the effective stress coefficient with respect to velocities is different from that of static deformation, in both the magnitude and the dependency on P_c and P_p. Although the transmitting waves through the rock also cause vibration, or strain, the amount of the induced strain is a few orders less than produced by the static measurements. In addition, the dynamic measurements only sample elastic strain, which does not allow for the inelastic deformation. As the deformation associated with depletion or injection generally involves both elastic and plastic strain, we consider the effective stress coefficient derived from static tests is more suitable in the analysis of reservoir stress changes.

It is particularly intriguing that the effective stress coefficients with respect to velocities increase significantly with effective stress and go beyond unity (Figure 3). This is rather surprising as normally we expect the opposite, which has been identified in several sedimentary rocks (*Todd and Simmons*, 1972; *Christensen and Wang*, 1985; *Ma and Zoback*, 2017). It is difficult to compare the effective stress coefficients with respect to different physical quantities and different rocks as the underlying mechanism differs. Nonetheless this intriguing phenomenon is perhaps associated with issues such as inadequate equilibrium of pore pressure when high-frequency wave passing, pore fluid stiffness variation with pressure, and the microstructure alteration and even damage under stress.

ACKNOWLEDGEMENT

This work was supported by the Stanford Rock Physics and Borehole Geophysisc Project (SRB) and the Open Research Fund of State Key Laboratory of Geomechanics and Geotechnical Engineering, Institute of Rock and Soil Mechanics, Chinese Academy of Sciences (Grant NO.Z015002).

REFERENCES

Biot, M.A. (1962). Mechanics of deformation and acoustic propagation in porous media, Journal of Acoustic Society of America, 28: 168–191.

Christensen, N.I. and Wang, H.F. (1985). The influence of pore pressure and confining pressure on dynamic elastic properties of Berea sandstone: Geophysics 50(2): 207–213.

Hornby, B.E. (1996). An experimental investigation of effective stress principles for sedimentary rocks: SEG 1996 annual meeting.

Ma, X. and Zoback, M.D. (2017). Laboratory experiments simulating poroelastic stress changes associated with depletion and injection in low-porosity sedimentary rocks, Journal of Geophysical Research-Solid Earth.

Sarker, R. and Batzle, M. (2008). Effect stress coefficient for North Sea shale—An experimental study: SEG 2008 Annual meeting.

Todd, T. and Simmons, G. (1972). Effect of pore pressure on the velocity of compressional waves in low-porosity rocks: Journal of Geophysical Research, 77, 3731–3743.

Geomechanics and Geodynamics of Rock Masses – Litvinenko (Ed.)
© *2018 Taylor & Francis Group, London, ISBN 978-1-138-61645-5*

Jointed rock mass characterization using field and point-cloud data

Miloš Marjanović
Faculty of Mining and Geology, University of Belgrade, Belgrade, Serbia

Marko Pejić
Faculty of Civil Engineering, University of Belgrade, Belgrade, Serbia

Jelka Krušić & Biljana Abolmasov
Faculty of Mining and Geology, University of Belgrade, Belgrade, Serbia

ABSTRACT: This research addresses a rockslope on the motorway No. 228, labeled as a 2A category road, according to the official road authority—Public Enterprise Roads of Serbia, which links Prolom Banja with a higher-category road network. Prolom Banja is an uprising touristic destination, famous for its spa and wellness resort, which shows an increasing trend of visits (over 10k tourists per month in the peak-season). As the road 228 is the only link to the resort, there is a raising concern about the road safety for about 650 vehicles per day (on average) along its many critical road cuts. One such rockslope (Lat 43.039850, Lon 21.380114) in weathered, jointed andesite is exampled in this research, using both terrestrial and airborne systems for reconstructing surface point-clouds, as well as the abovementioned index field and lab tests. These allowed quantifications of the rock joint sets and their kinematic stability, joint spacing, blocks volumes, and finally, profiling the critical blocks for potential rockfall development. It is planned to annually monitor the pilot site in the future, so that all changes, i.e. detached blocks and fragments will be identified.

Keywords: rockslope, kinematic analysis, index properties, LiDAR, point-cloud

1 INTRODUCTION

Point-cloud-based analyses are becoming widespread in rockslope engineering practice (Slob and Hack, 2004; Sturzenegger and Stead, 2009; Sturzenegger et al., 2011), especially in transportation (Lato et al., 2009) in the last couple of decades. Terrestrial laser (LiDAR) scanning (TLS) and digital photogrammetry using airborne UAV systems both provide such point-clouds, and nowadays, these techniques are reaching a sub-cm resolution and accuracy, which can be well exploited for reconstruction of the rock face geometry, mapping of its discontinuities and its further characterization. In combination with rock index properties, obtained by field testing (such as joint roughness estimation, Schmidt hammer test, lab tests of other index properties), these digital analyses give a thorough and reliable quantification of a rock mass (Gigli and Casagli, 2011; Abellán et al., 2013).

In this work, we demonstrated one such case study along the motorway No. 228, near Prolom banja spa in Serbia (Fig. 1). This 2A category road (a second-level road according to the official road labeling legislation in Serbia) is the only link from Prolom banja to a higher-level road network (1 A and 1B category). Inconveniently, the 228 road is cut through the rocky Prolom river gorge (in fact, the name Prolom refers to an abyss or a sinkhole in

Figure 1. From the left: site location in relation to the national road network of Serbia; geological setting of the wider area along the Prolomska river; rockslope site and scan station dispositions; site photograph with the Leica ScanStation P20 instrument (photo M. Marjanović).

Figure 2. Panoramic photograph of the slope (joint sets are color coded, but their daylighting is affected by panoramic projection), with a detail on the fallen boulder in the incept to the right (photo M. Marjanović).

Serbian, suggesting a hostile terrain). Scarce protective measures (only road mashes locally) witness of the poor road condition (especially after heavy rain or a snowmelt) as its culverts and gutters are often covered with rock shatter and even larger blocks (Fig. 2). On the other hand, the spa resort itself, as well as the associated water bottling factory have a rather high traffic demand. Statistics from 2016 (www.putevi-srbije.rs) shows that there are 637 vehicles per day on average, wherein 87% are passengers cars, around 7% cargo (mostly heavy-duty trucks to and from the factory), and remaining 6% (or about 35) are shuttles that commute to and from the spa resort. It is safe to say that all these circumstances make sufficient motive to investigate critical sites along the route 228.

2 CASE STUDY AREA

The chosen critical rockslope site is located along the 228 road at Lat 43.039850, Lon 21.380114 (Fig. 1). It was chosen as the most critical due to the current evidence of the rockslope activity that occasionally impedes the traffic, especially along its left lane (next to the rockslopes, opposite to the river), by introducing rock shatter and larger blocks that can reach well into the left lane of the carriageway.

Wider area (Fig. 1) is located along the right valley side of the Prolom river, a relatively short (16 km) river, with the watershed of around 65 km^2 squeezed between the Sokolovica (1050 m), Radan Arbanaška (1128 m) and Prolom Mountain (1370 m). It flows westward, slightly arching around the former caldera rim (Sokolovica), and in the largest part cuts through a rugged hilly-mountainous landscape (elevation range between 450 and 1400 m), locally with steep (>45°) slopes (gorge-like parts of the valley). Bedrock geology is relatively uniform and tightly related to the Lecki volcanic massive of Oligocene age, i.e. its northernmost caldera. It is mainly composed of andesite—fresh (αh), and hydrothermally altered (αΘ), with subordinate appearances of tuffs and breccia (ω). Volcanites penetrate the Cretaceous basal sandstones and conglomerates, that are subsequently buried beneath andesitic lava flows and associated pyroclastites. The tectonic setting is complex, but mainly dominated by the caldera's outer rim, which is indicated by a set of concentric and radial faults (estimated) that occasionally intersect. Our critical rockslope site is located precisely at one such juncture, which generally represents the weakest and most altered and weathered parts of the andesite rock mass (Fig. 1).

The slope is engineered for the road construction purposes, to fit the geometry of the local road curve. Evidently, larger rock volume that was plunging towards the river (some 50 m away) was removed. Drilling traces found on the slope, as well as the traces along the other nearby rockslopes indicate that systematic blasting technique was used to remove the rock mass. Removed material (probably conveniently fragmented due to blasting) was used for constructing a rest area pocket along the external outline of the curve. The resulting slope is about 14.5 m high (with reference to the road level), and about 45 m long, with relatively flat continual face, with steep plunge (Figs. 2–4).

Figure 3. The point-cloud with manually selected and color-coded joint sets with the global kinematic stability on a stereonet in the incept to the right.

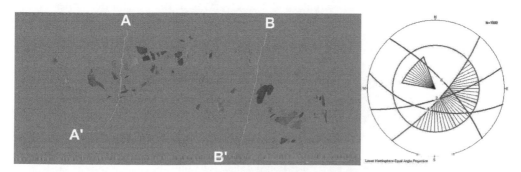

Figure 4. Plane sets extracted from the point cloud, and appropriate local profiles with the local kinematic stability conditions of A-A′ on a stereonet in the incept to the right.

3 METHODOLOGY

The proposed approach involved conventional field investigation of the rockslope face, and point-cloud analyses. These were undertaken simultaneously to maintain the consistency of all conditions.

Field investigation included the petrological recognition of the rock (mineralogical content, structure, texture), detection of discontinuity sets and related properties (orientation, spacing, block size, JCS freshness), percolating water, fill, and other general conditions. Most of these could be reliably mapped and estimated only for the lower, accessible part of the slope.

Point-cloud analyses included direct mapping of different joint sets (their orientation and spacing), and subsequent kinematic conditions for block, wedge and toppling failure. The first step was performed in COLTOP3D software (some preprocessing steps were done in CloudCompare software), by a supervised selection method, while the second step used the outcomes of the first one to establish kinematic conditions using the DipAnalyst software. Although the joint set analysis was performed in the field investigation part as well, the point-cloud approach allowed observation of the entire slope, not only the accessible, lower section, thereby making the kinematic analyses more consistent. The coupled field investigation joint set analyses were therefore used only for control.

4 DATA COLLECTION

Point-cloud data were collected using the Leica ScanStation P20 terrestrial LiDAR scanner. There was no positioning to a national reference system, so the subsequent analyses were performed in the local coordinates of the scanner. The scanning was performed from 5 stations with different angles and positions (from a standard tripod) to avoid all possible shadowing effects. All 5 scans were acquired by using 6.3 mm/10 m scanning mode. Given that the stations were located across the road, some 20 m away from the slope, the effective point cloud resolution, after applying uniform point standard in LeicaCyclone software, was 3 cm, sufficient to support all planned analyses. Previously, all 5 point-clouds were co-registered into one, using 3 fixed tie points (markers). Co-registration accuracy was 2 mm. Resulting point-cloud contains 5.8 M of points. Since the scans were taken in November 2017, well after the vegetative season, and the slope face was clear of any screens, no additional post-processing and filtering was necessary.

Other, conventional, field data were collected by standard field investigation procedures and tools (geological compass, measuring tape, N type Schmidt hammer, etc.).

5 RESULTS

The rockslope can be separated into several zones, varying in the general properties of the joint system and index properties of rock. One of the typical classification schemes, such as Rock Mass Rating-RMR, is applicable.

Automatic mapping of joint sets on the point-cloud by using $\pm\alpha/\nu$ tolerance in COLTOP3D did not prove useful, since too many noisy points intertwine, whereas, some planar surfaces that evidently belong to the set were not included, because they fall outside the reasonable tolerance range. The reason is a slight radial arching of joint sets, which changes their orientation and/or dip which is typical for volcanic rocks. In addition, the middle part of the slope is dominated by a fault zone, which symptomatically altered spatial orientation of the primary (pre-faulting) joint sets, and created additional tension cracks as the faulting propagated. For this reason, joint set mapping on the point-cloud was performed manually (Fig. 3). Mapping of these joint sets also allows for a conventional kinematic analysis, i.e. profiling (Fig. 4). For instance, a typical local profile A-A' is showing how Set 3 and Set 2 form the local face of the slope, and can cause potential wedging (Figs. 3–4).

The weakest are the parts belonging to the fault zone in the middle, and the weathering zone atop the slope face. Four major joint sets were determined by mapping their orientation on the point-cloud and controlling it by using field data (especially since the acquisition coordinate system was relative), but the last one, the Set 5, was mapped only on point-cloud, since it was not accessible on site:

- Set 1: Gentle dipping to sub-horizontal discontinuities with average elements 193/54°; split into left and right flanks by a fault zone (the Set 4); on the right flank spacing is evidently smaller, 15 cm on average, whereas the left flank has larger spacing, 90 cm on average; aperture is small, and the walls are flat; it is probably a generic (concentric) jointing crack system that is formed concentrically around the magma extrusion center, as the andesitic lava cooled; relatively fresh with JCS of about 42 MPa.
- Set 2: Also split by the Set 4, and significantly differ at right and left flank in orientation/dip, but average are 162/72° elements; penetrative across the entire slope and form most of the slope face; they match the orientation of the principal faulting along the river valley (Fig. 1) perpendicular to the faulting plane of the Set 4, so it is likely created by the erosional decompression along the valley, which is also suggested by spacing and apertures, but it might be also of radial polygonal jointing origin; spacing determined on a point-cloud 40 cm on average, while aperture is in cm order of magnitude; not as fresh as the walls of Set 1 – JCS of about 34 MPa.
- Set 3: Isolated, difficult to track along the entire slope, except for the left flank middle where they dominate, and evidently generate large unstable wedges; probably primary (pre-faulting) structure broken by Set 4; average orientation and dip 299/78° (confirmed also on the point-cloud at inaccessible parts); large spacing of about 150 cm (determined on the point-cloud) wavy to rough wall surface, with calcareous fill, large aperture (dm order of magnitude), inaccessible for further index analysis.
- Set 4: Fault zone set, vertically penetrative across the entire slope, slightly arched – 47/72°, with tight spacing (25 cm on average, but in many places <5 cm); heavily weathered (JCS < 20 MPa), with changes of color from fresh grey to light orange; warped and rough wall surface; apertures mm-cm order of magnitude; walls covered with white calcite screens.
- Set 5: Inaccessible to index field investigation, orientation determined on the point-cloud (350/77°), as well as its spacing (60 cm on average); tight, flat walls, fresh color.

Kinematic conditions were analyzed by approximating the entire slope face as a continual surface with elements 130/68°. The 68° angle was an average measured inclination for 10 regularly spaced profiles (e.g. A-A′ in Fig. 4) along the slope. Since the slope geometry is relatively regular and flat no further faceting was necessary. The following global kinematic conditions for the slope apply:

- Planar block failure is not globally (for the entire slope) relevant, as neither of the averaged planes do not meet the Panet/Markland's criteria. However, due to the local arching and divergence of planes close to the fault zone, singular local planar failures are possible.
- Wedge failure occurs between the Sets 1 and 4, as well as 2 and 4 (to a lesser degree), and globally influence the stability of the slope, as well as locally (Fig. 4).
- Toppling is possible with the Set 3, both globally and locally in some profiles, wherein the Set 1 plays an important role in undercutting the slabs for potential toppling.

6 CONCLUSION

The need of intervention along the present slope is evident. Relatively large blocks (as much as 1.5 m³) can be seen covering the gutters or even the road lanes. The analysis in this work showed that the preliminary stability concern is originating from a combination of topples of the Set 3, and wedge failures between the Set 1 and 4, and subordinately Set 2 and 4.

Further research will include annual monitoring of the slope, as this first scanning epoch will be its base for comparison of all further surface models (Bogdanović et al. 2015). This follows the principal motive and objective of the research, which is to propose some efficient protective measures on this particular slope along the relatively important touristic route.

ACKNOWLEDGEMENTS

This work was supported by the project of Ministry of education of the Republic of Serbia—TR36009 and the grant of the "Start up for Science" foundation of Phillip Morris International – the "MEDJA" project.

REFERENCES

Abellán A, Oppikofer T, Jaboyedoff M, Rosser N, Lim M, Lato M.J. (2013) Terrestrial laser scanning of rock slope instabilities. Earth Surf. Process. Landforms, 39/1:80–97.

Bogdanović S, Marjanović M, Abolmasov B, Đurić U, Basarić U. (2015) Rockfall Monitoring Based on Surface Models. In: Surface Models for Geosciences (eds. Růžičková K, Inspektor T), 37–44.

Gigli G, Casagli N. (2011) Semi-automatic extraction of rock mass structural data from high resolution LIDAR point clouds. International Journal of Rock Mechanics and Mining Sciences, 48/2: 187–198.

Lato M, Diederich MS, Ball D, Harrap R. (2009) Engineering monitoring of rockfall hazards along transportation corridors: using mobile terrestrial LIDAR. Nat Hazards Earth Syst Sci 9:935–946.

Slob S, Hack R. (2004) 3D Terrestrial laser scanning as a new field measurement and monitoring technique. Eng Geol Infrastruct Plann Europe Lecture Notes Earth Sci 104:179–189.

Sturzenegger M, Stead D, Elmo D. (2011) Terrestrial remote sensing estimation of mean trace length, trace intensity and block size/shape. Eng Geol 119:96–111.

Sturzenegger M, Stead D. (2009) Close range digital photogrammetry and terrestrial laser scanning for discontinuity characterization on rock cuts. Eng Geol 103:17–29.

Geomechanics and Geodynamics of Rock Masses – Litvinenko (Ed.)
© *2018 Taylor & Francis Group, London, ISBN 978-1-138-61645-5*

New failure criterion for rocks by using compression tests

Celestino González Nicieza, Martina Inmaculada Álvarez Fernández,
Carmen Covadonga García Fernández, Román Fernández Rodríguez &
Juan Ramón García Menéndez
Department of Exploitation and Prospecting Mines, University of Oviedo, Oviedo (Asturias), Spain

ABSTRACT: In the present research a new non-lineal failure criterion in rocks called GIT criterion is established. The new criterion has been developed by the Ground Engineering Group from the University of Oviedo (Spain), and it can be expressed in terms of both normal and shear stresses (σ_n, τ), and principal stresses (σ_3, σ_1). Three characteristic parameters define the criterion, which are estimated applying nonlinear least squares regression methods, using the Levenberg-Marquardt (LM) algorithm. The novelty of this new criterion is that its determination can be carried out only by using uniaxial compression tests, which are more economic and easier to implement than triaxial tests. The theoretical formulation is accompanied by set laboratory tests, and the results obtained are compared with the classical criteria employed in rock mechanics, such as those of Mohr-Coulomb and Hoek-Brown.

Keywords: failure criterion, non-lincal, rocks, compression tests, theoretical, laboratory, Levenberg-Marquardt

1   INTRODUCTION

When a system of loads is applied to an elastic material, a stress state is created and generates a state of deformation, both related by Hooke's Law. But when the material exhibits an elastic-plastic behavior, once the load exceeds a threshold, the material ceases its elastic behavior meaning that the material does not return to its original size if the load is removed.

During the loading process, the characteristic values of the stress state as well as the variation in the distance between atoms will increase until to break the bonds between the atoms that constitute the internal structure of the material. Then, two situations can take place. One consists in the sustaining of the cohesion due to the formation of new bonds that substitute the original ones, which produces a plastic permanent deformation. The other situation implies that the cohesion is not sustained, in which case the failure of the bond is definitive and rupture is produced.

The onset of permanent deformations will produce qualitative variations in the properties of the material and, even more so, the failure of the material accompanied by the possibility of its collapse. For this reason is essential to understand the process as well as the causes that initiate the inelastic deformations. Many researchers have studied in detail these questions in order to determine the stress-strain state for which the elastic regime of the material ceases and thus, to establish the law which defines the limit of the elastic behavior of the material. This is known as the strength criteria which allow describing the state of stresses in materials at failure.

An ideal strength criterion needs to closely fit test data with acceptable accuracy over the stress state expected in practice. Some of the most important studies were carried out by

Balmer (Balmer, 1952), Sheorey (Sheorey,1997); Mohr Coulomb (Mohr & Welche, 1900); or Hoek-Brown (Hoek & Brown, 1982), who formulated parabolic criteria in the principal stress plane. The work by Pincus (Pincus, 2000) is also worth highlighting. This author found exact linear expressions for failure criteria, both parabolic and hyperbolic, formulated in the plane of normal and shear stresses. In all these cases, the input data come from triaxial tests in rock samples or from empirical criteria.

This study attempts to verify a new criterion named 'GIT criterion' which parameters may be obtained from simple uniaxial compression tests. By carrying out laboratory tests, the proposed criteria will be compared to the most widely employed failure criteria in rock mechanics: the Mohr Coulomb criterion and the Hoek–Brown criterion.

2 THEORETICAL CONSIDERATIONS

2.1 *Parabolic intrinsic curve*

Mohr's theory admits the uniqueness of the enveloping curve of the Mohr's circles for the limit states and its independence of intermediate principal stress. The form of this intrinsic curve is a mechanical characteristic of the material that depends on its physical properties, and thus the envelope may be obtained from laboratory tests using optimization techniques.

The proposed criterion by the authors comes from the consideration of the Mohr's circles that define the stress state in a material. Eq. (1) is considered for representing the intrinsic curve in the plane (σ_n, τ):

$$\sigma_n = -A + B\tau^2 \tag{1}$$

where A and B are the characteristic parameters of the criterion. Moreover, the following hypothesis is formulated: "*the common point (σ_n, τ) of the envelope and any of the Mohr's circles is a first-order contact; meaning that the tangent to the envelope and to the circle are the same*".

Otherwise, the equation of Mohr's circle is given by:

$$(\sigma_n - S)^2 + \tau^2 = D^2 \tag{2}$$

where S and D are defined such as the abscissas of the centre of the circle and its radius, respectively, as it is shown in Figure 1.

From Figure 1, it is verified that:

$$\tan\alpha = \frac{S + \sigma_n}{\tau} = \frac{1}{\dfrac{d\sigma}{d\tau}} = \frac{1}{2B\tau} \tag{3}$$

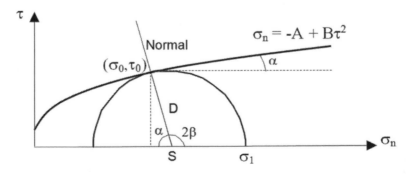

Figure 1.   Parabolic criterion.

From Eq. (1) it is obtained:

$$\tau^2 = \frac{2B(S+A)-1}{2B^2} \tag{4}$$

Substituting in Eq. (2), as well as $S = \frac{\sigma_1 + \sigma_3}{2}$ and $R = \frac{\sigma_1 - \sigma_3}{2}$:

$$\sigma_1 = \frac{2B(B\sigma_3 + 1) + \sqrt{4B^2(B\sigma_3 + 1)^2 - 4B^2(B^2\sigma_3^2 - 2B\sigma_3 - 4AB + 1)}}{2B^2} \tag{5}$$

dividing by $2B$ and developing the discriminate, finally Eq. (6), which represents the intrinsic curve formulated in the principal stress plane (σ_3, σ_1), is reached:

$$\sigma_1 = \sigma_3 + G + 2\sqrt{G(\sigma_3 + A)} \tag{6}$$

where $G = \frac{1}{B}$. Thus, the characteristic parameters of the criterion in the principal stresses plane are (A, G) and in the plane of normal and shear stresses (A, B). This envelope has been called the GIT criterion (from the Spanish name for the research group: *Grupo de Ingeniería del Terreno*, at the University of Oviedo), which is consistent with the two systems of coordinates (σ_3, σ_1) and (σ_n, τ).

2.2 Parameters of the intrinsic curve

By considering Figure 1, the equation for the normal straight line common to the Mohr's circle and the parabolic envelope is given by:

$$\tau = \tau_0 - \frac{1}{\tau_0'}(\sigma_n - \sigma_0) \tag{7}$$

where according to Eq. (3):

$$\tau_0' = -\frac{1}{2B\tau_0} \tag{8}$$

The shear point of the normal with respect to the σ_n axis is (S,0). Thus, it is verified that:

$$\tau_0 - \tau_0' = S - \sigma_0 \tag{9}$$

By taking into account the generic formulation of the Mohr Coulomb failure criterion $\tau = C + \sigma_n \tan\varphi$, being C the cohesion of the material and φ the internal friction angle, as well as the fact that $\cos 2\beta = -\cos\alpha$, the following expression is reached:

$$\alpha = \cos^{-1}\frac{G}{\sigma_1 - \sigma_3} \tag{10}$$

By substituting Eq. (6) in Eq. (10):

$$\alpha = \frac{180}{\pi}\cos^{-1}\frac{G}{G + 2\sqrt{G(\sigma_3 + A)}} \tag{11}$$

where α is expressed in degrees. Eq. (11) can be expressed in terms of the angle β, as a function of the confinement stress σ_3:

$$\beta = 90\left(1 - \frac{1}{\pi}\cos^{-1}\frac{G}{G + 2\sqrt{G(\sigma_3 + A)}}\right) \tag{12}$$

being β the angle between the normal to the shear (or failure) plane and the direction of the maximum principal stress. Thus, taking into account that in a compression test is verified that $\sigma_1 = \sigma_c$ and $\sigma_3 = 0$, Eq. (13) are reached, which allow to determine the characteristic parameters of the parabolic criterion from simple compression tests:

$$G = \sigma_c \cos 2\beta_c; \quad A = \frac{(\sigma_c - G)^2}{4G}; \quad B = \frac{1}{G} \tag{13}$$

It will be recommendable to carry out a number of tests to estimate the average values of σ_c and β_c, being β_c the angle between the failure plane and the direction of the applied load in a compression test. In this study, the fitting of the parameters A and G is made on the principal stress plane by means of nonlinear least squares regression methods that use the Levenberg-Marquardt (LM) algorithm (Levenberg 1944; Marquardt, 1963). The LM algorithm is an iterative technique that locates the minimum of a multivariable function expressed as the sum of squared nonlinear functions of the actual variable.

3 RESULTS

In order to validate the parabolic criterion developed, experimental tests—which include both triaxial and uniaxial compression tests—were carried out in two materials. Results obtained in the fitting of the characteristic parameters of the GIT criterion will be compared with those obtained for the two classical criteria in rocks: Mohr Coulomb criterion and Hoek Brown criterion, which were established by using the classical formulation (Mohr & Welche, 1900; Hoek & Brown, 1982), while for the parabolic criterion developed in this work, Eq. (1) and Eq. (6) will be used, fitting the parameters of the criterion A, G and B by using Eq. (13).

It should be stated that the fitting of the parameters of the Hoek-Brown criterion with values $s \neq 1$ was not possible with the algorithm used here. Accordingly, the value $s = 1$ has been used in the fittings of the parameters of this criterion.

3.1 Results in homogeneous material

A set of experimental tests were carried out in samples of a homogeneous and uniform material (such as mortar), since in rock materials the inherent heterogeneity is a variable that can lead to confusion when the results are analyzed.

Specimens of mortar with proportions cement:sand of 1:1 and 1:3 were tested by triaxial and uniaxial compression tests, and characteristic parameters of the failure criteria of Mohr Coulomb (cohesion C and friction angle φ) and Hoek Brown (m) were established. Moreover, the compression tests were analyzed in detail in order to determine the characteristic parameters of the GIT criterion. This detailed analysis included both the calculus of the mean value of compression strength, and especially, the determination of the failure angle β_c. Results obtained for each proportion tested are summarized in Table 1.

Table 1. Fitting of the characteristic parameters in mortar.

	GIT					HOEK-BROWN		MOHR-COULOMB		
	β_c (°)	σ_c (MPa)	G	A	B	σ_c (MPa)	m	σ_c (MPa)	C (MPa)	φ (°)
1:1	11.00	13.60	12.61	0.019	0.079	15.26	10.37	14.15	3.05	43.35
1:3	21.00	5.70	4.23	0.127	0.236	6.23	5.29	5.80	1.49	35.55

It is important to highlight that the failure angle was carefully calculated by scanning of the "wedge" failures using the device PIX-30 of ROLAND PICZA, which lets to obtain a digital image 3D with a maximum accuracy of 0.05 mm, as well as being exported to its treatment in common formats (such as *stl, *dxf, *igs, *txt,*vrml, etc.). In this work, the calculus of the angle β_c in mortar was achieved by using an algorithm created in MATLAB, which lets to generate profiles along the longitudinal section in the wedge scanned. So, for each wedge several profiles were obtained, in which the angle between the failure plane and the vertical direction was calculated. Finally, a mean value of β_c was considered for each proportion. Figure 2 shows the comparison between a photo of a "wedge" failure obtained after a test in a sample of mortar 1:3 and its respective scanned image.

Figure 3 represents the fitted envelopes of the three criteria (GIT, Mohr Coulomb and Hoek Brown) in principal stresses for both proportions. Also, in the plot are represented the experimental points cloud obtained, which were used as data for the fitting. A good agreement of the results was reached.

The GIT envelope is more conservative (further from safety requirements) than the classical criteria, but presents the advantage of not requiring triaxial test to determine the characteristic parameters, since the simple compression test is sufficient. Finally Figure 4 shows the

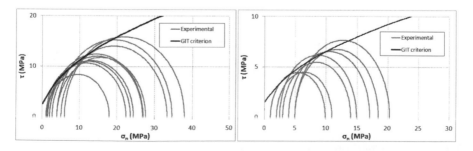

Figure 2. Comparison between a "wedge" failure (left) and its respective image scanned (right).

Figure 3. Fitted envelopes of the three criteria in the mortar 1:1 (left) and the mortar 1:3 (right).

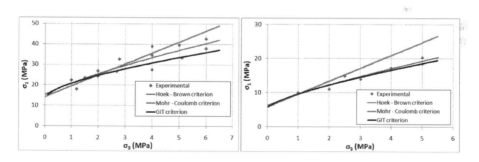

Figure 4. Fitted envelope in the stress plane $(\sigma n, \tau)$ in the mortar 1:1 (left) and the mortar 1:3 (right).

Table 2. Fitting of the characteristic parameters in slate.

	GIT					HOEK-BROWN		MOHR-COULOMB		
	$\beta_c\,(°)$	$\sigma_c\,(MPa)$	G	A	B	$\sigma_c\,(MPa)$	m	$\sigma_c\,(MPa)$	$C\,(MPa)$	$\varphi\,(°)$
Slate	17.80	178.50	145.13	1.917	0.007	179.78	7.50	188.60	36.20	48.00

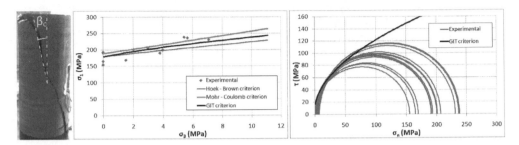

Figure 5. Pattern failure in slate (left); fitted envelopes of the three criteria in the slate in principal stress (middle); and in terms of normal and shear stress (right).

fitted GIT envelopes in the stress plane (σ_n, τ), as well as the Mohr's circles corresponding to the experimental tests for each material. A good agreement of the envelope was achieved again, and the condition of tangency with respect to the majority of circles is verified.

It is not intended of the authors to state that GIT criterion improves the classical criteria, but proving that the proposed criterion shows a good adjustment with the experimental tests, so it may be used in order to establish the stress limit of materials in a simple way.

3.2 Results in foliated material

In order to check the applicability of the GIT criterion in rock, laboratory tests were carried out in slate from NW of the Iberian Peninsula (Spain). Slate is a foliated metamorphic rock characterized by a set of parallel well-defined discontinuities, called cleavage or foliation, which constitute planes of relative weakness in the rock. Triaxial and compression tests were carried out by applying the maximum principal stress in the perpendicular direction to the discontinuity planes. In the same way that the methodology exposed above, the equivalent parameters for the GIT criterion were calculated; by analyzing in detail both the compression strength and the failure angle β_c. Results are shown in Table 2.

Failure in slate is characterized by only a single fracture, which was seen in all the samples tested, as it is shown in Figure 5, in which one of set samples tested is exposed with the angle β_c highlighted. Figure 5 also represents both the fitted envelopes of the three criteria and the experimental points cloud in terms of principal stress, as well as the fitted and approximated GIT envelopes in the stress plane (σ_n, τ), and the Mohr's circles corresponding with each test. Again, the good agreement of the results can be observed, so the accuracy of the GIT criterion is verified even in rock.

According to the experimental results, it can be stated that the parabolic GIT criterion guarantees an accuracy calculus of the stress limits, at least for the material tested, both in terms of principal stress and in terms of normal and shear stress.

4 CONCLUSIONS

A novel failure criterion formulated both in the normal and shear stress plane and in principal stresses plane has been presented. Estimation of the characteristic parameters of the envelope was established by applying the Levenberg-Marquardt algorithm. Laboratory tests were carried out in order to verify the criterion: firstly, by testing homogeneous material such as mortar, and secondly, in slate, which is a rock material characterized by the presence of discontinuity planes. Experimental results obtained were compared with those obtained for classical criteria such as Mohr-Coulomb and Hoek-Brown, proving the proximity between the proposed criterion and the classical criteria. The novelty of the GIT criterion lies in the fact that only is necessary to carry out unixial compressive tests to determine both the compression strength and the angle between the failure plane and the maximum principal stress (or direction of the applied load). This supposes a major economic advantage respect to the procedures traditionally used up (such as triaxial tests) in order to characterize rock at failure.

ACKNOWLEDGEMENTS

The authors of this paper would like to acknowledge the financial support of the PhD fellowship Severo Ochoa Program of the Government of the Principality of Asturias (PA-14-PF-BP14-067).

REFERENCES

Balmer, G.A., 1952. A general analytical solution for Mohr's envelope. Am Soc For Testing and Materials, Proceedings. 52, 1260–1271.

Hoek, E., Brown, E.T., 1982. Underground Excavations in Rock. London: IMM.

Levenberg, K., 1944. A Method for the solution of Certain Non-linear Problems in Least Squares. Quarterly of Applied Mathematics, 2(2), 164–168.

Marquardt, D.W., 1963. An Algorithm for the Least-Squares Estimation of Nonlinear Parameters. SIAM Journal of Applied Mathematics, 11(2), 431–441.

Mohr, O., Welche., 1900. Umstände bedingen die Elastizitä tsgrenze und den Bruch eines Materials. VDI-Zeitschrift. 45, 1524–1530.

Pincus, H.J., 2000. Closed-form/leas-squares failure envelopes for rock strength. International Journal of Rocks Mechanics and Mining Sciences, 37, 763–785.

Sheorey, P.R., 1997. Empirical Rock Failure Criteria. A.A. Balkema.

Geomechanics and Geodynamics of Rock Masses – Litvinenko (Ed.)
© *2018 Taylor & Francis Group, London, ISBN 978-1-138-61645-5*

Ore strength property evaluation in the design of ore preparation cycles

N.V. Nikolaeva, T.N. Aleksandrova & A.M. Elbendari
Mineral Processing Department, Saint-Petersburg Mining University, St. Petersburg, Russia

ABSTRACT: In order to analyze and reveal mathematical relations between the drop weight parameters and Bond work indexes, a data base on laboratory testing of various type ore grindability has been collected. However, it must be considered that impact dominates when performing drop weight test (test to study abrasive destruction is carried out separately), meanwhile impact and abrasion destructions occur in a ball mill of Bond. The destruction process depends on form and structure of sample during destruction of large pieces using installation of drop weight (the physical and mechanical properties). Destruction process begins to take place at micro level, and often ore becomes more resistant to breakage. It can be associated with the mineral composition of ore in the first place. Consequently, if the ore has dynamic of uniform destruction at impact and abrasion, it can be claimed that there is correlation dependence, which can be used to express strength parameters of ore through Work Index of Bond. This data was used for statistical analyzes, plotting and for deriving mathematical relations, which showed good convergence of characteristics of impact strength, wear capacity and ball mill work index. The research findings allow passing between the results of different tests without additional studies, which considerably shortens the period of analysis of ore characteristics. Furthermore, the research has shown feasibility of having the detailed and extensive knowledge on strength characteristics of ore as early as the stage of planning of ore pretreatment, which enables improvement of reliability and efficiency of design solutions within the entire system of ore processing.

Keywords: Ore preparation, the strength properties of ores, Bond test, drop weight test, parameters of impact fracture

1 INTRODUCTION

Crushing and grinding processes are one of the energy-intensive and costly processes of ore processing, as is known. They determine enrichment efficiency indicators largely. Research in the area of ore dressing processes optimization and modernization is focused on development of scientifically-based methods of designing and managing processes. The establishment of quantitative relationships between the main parameters of processes is necessary for this. They are determined by means of special calculations or simulations and are based on the revealed regularities of the processes of desintegration and separation of products by their size during laboratory tests [1, 3–5, 7, 9].

Research on the determination of various physical and mechanical characteristics of mineral raw materials, such as hardness, abrasiveness, mineralization, product size of material (which promotes the maximum unlocking) and so forth, is necessary when selection and design of ore preparation flowsheets. The first three characteristics are taken into account when choosing dimensions of equipment and required drive power to obtain of required product size. At the same time, it is necessary to take into account that the desintegration processes in ore-preparation devices have different action mechanisms on the processing mineral raw materials. Consequently, devices and procedures that have a similar process of destruction with an industrial device should be used in study of mineral raw materials grindability. This will provide more accurate predicted indicators.

2 THE RESEARCH

Establishing of relationship between the strength parameters (A · b) and Bond Ball Mill Work Index (BWI) for various ore types of has been main objective of our research. This will allow to pass from one test results to another results of without additional research. Strength parameters are obtained as a result of Drop Weight Test. Parameters A and b characterize strength of material under impact effect (statistical record has form: A · b), parameter t_{10} characterizes granulometric composition of product at specific crushing energy (E_{cs}). Parameter (ta) characterizes the ability to autogenous grinding of material. However, impact fracture prevails in carrying out test on JK drop weight tester (Fig. 1), and impact and abrasive fracture occurs in Bond mill (Fig. 2). Therefore, process of deterioration depends on the habit and structure of sample when large pieces are destroyed in drop weight tester (physical and mechanical properties) [2]. When the Bond Test, the feed is smaller particle size (−3.35 mm + 0) than in the drop weight tester (−63 + 13.2 mm). The process of destruction begins to occur at the micro level, and often the ore becomes more resistant to grinding. First of all, this can be linked with the mineral composition of ore. Hence, if the mineral raw material has the dynamics of uniform fracture during impact and abrasion, it can be argued that there is a correlation dependence that will allow the ore strength parameters (A · b) to be expressed through the Bond Work Index (BWI). Internal structure affects the strength of mineral raw materials in addition to physical and mechanical properties. For example, the porosity factor is often determinant factor in interpreting the strength characteristics of rock and its behavior in process of deterioration. Since the destruction occurs along the most weakened zones—the boundaries of the intergrowths of individual mineral phases, cracks, etc. [6].

A database of laboratory studies results on the matter of grindability of various ore types was collected for analysis and identification of mathematical relationships between the parameters of Drop Weight Tests and energy Work indexes of Bond (Table 1).

Figure 1. JK drop weight tester.

Figure 2. Bond mill LM-BM1000.

Studies conducted on various types of mineral raw materials in the period from 2014 to 2017 were the sources of replenishment of this database. The data were averaged over mineralogical types of ores. Statistical analyzes were performed, the graphs were constructed and the equations of dependencies were obtained using this data.

The dependencies $BWI = f(A \cdot b)$ and $BWI = f(t_a)$ for 13 pairs (13 different mineralogical mineral types) are shown in Figures 3 and 4.

Energy requirements of crushing can be estimated based on results of Drop Weight Test, and energy requirements of autogenous/semiautogenous grinding (AG/SAG) may be determined by results of simulation. Analysis of regression equations shows good correlation between parameters which characterize the impact strength (A·b) and Bond Ball Mill Work Index (Figs. 3 and 4).

But it should be noted that in some cases, rebound of the drop weight was observed as a result of the impact of maximum specific energy during the test. It causes a repeated destruction of the material and consequently leads to overestimation of the energy parameters. An example of destruction of single sulfide ore particles with sufficiently high density (more than 4 tons/m³) is considered. Rebound of the drop weight was observed as a result of the impact of maximum specific energy in four kinds of factions during the test (Table 2). Rebound of each impact was recorded on video to determine average values of rebound heights. It is necessary to more accurately determine of specific energy of destruction.

Such effect causes overgrinding of material, and increases of values (A·b) by 4–6 units. It was also noted that there was a dependence of mechanical parameters of layered rocks on the direction of the compressive force (parallel or perpendicularly (Fig. 5)).

Table 1.   Values of energy parameters of various ore types at destruction process.

Ore	Value A · b	Value t_a	Value BWI, kW·h/t
Copper ore	59.7	0.61	12.73
Gold ore	34.9	0.39	18.93
Hematite ore	50	0.62	13.4
Lead ore	63.5	0.57	11.9
Lead-zinc ore	67.8	0.58	10.93
Lead-zinc ore	55.6	0.59	13.65
Pyrite ore	76.8	0.85	8.93
Tin ore	68	0.63	10.9
Zinc ore	65	0.56	11.56
Titanium ore	61.5	0.54	12.33
Quartz	55.9	0.6	13.57
Pyroxenite	32.5	0.38	23.8
Serpentinite	27.2	0.28	27.8

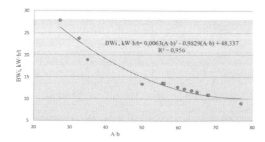

Figure 3.   Relationship between Bond Ball Mill Work Index (BWI) and ore destruction parameter (A · b).

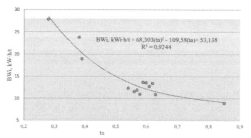

Figure 4.   Relationship between Bond Ball Mill Work Index (BWI) and the ore abrasion parameter (ta).

Table 2. The height of weight rebound and specific energy of secondary impact.

Size, mm	Specific energy of impact, kW·h/t	Height of rebound of weight, cm	Specific energy of secondary impact, kW·h/t
−63 + 53	not observed		
−45 + 37.5	0.71	3.1	0.02
−31.5 + 26.5	2.2	5.8	0.12
−22.4 + 19	2.5	6.3	0.17
−16 + 13.2	2.5	5.7	0.26

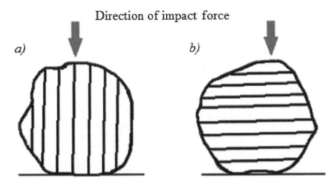

Figure 5. Direction of impact force is referenced to weakening planes: *a)* parallel; *b)* perpendicularly.

Table 3. Mineral composition effect on strength characteristics.

Size, mm	Type of ore	Ore minerals, %	Barren minerals, %	The average weight of particles, g	Specific energy, kW·h/t
−45+37.5	1	35	65	195.9	0.7
				193.7	0.25
				192.2	0.1
	2	18	82	193.7	0.7
				191.3	1
				193.2	0.25

Note: The ore minerals are represented by pyrrhotite, chalcopyrite, etc. Barren minerals are represented by feldspar, pyroxene, quartz, chlorite, biotite, etc.

Detached test for "narrow" sets of size −45 + 37.5 mm was carried out additionally. The sets were divided into two types with respect to mineral composition of ores. Initial data for test are shown in Table 3. The destruction degree (t_{10}) for two types, depending on specific crushing energy (E_{cs}), is determined by the results of sieve analysis. The test result is shown in Figure 6 from which it follows that parameter A·b (type 1) is greater than the parameter A·b (type 2).

Impact strength of sulphide-containing ore samples under impact flatwise, having a layered structure was higher in the vast majority of cases than with parallel impact of impact force about lamination. This is due to the fact that weak streak are kept from splitting by solid layer in case crosscut impact. It increases the resistance to impact destruction. When impact load is applied along bedding, impact strength is determined mainly by the strength of weak bind upon which rock is splitted [9].

Analyzing the existing methodology for conducting of Drop Weight Test is referenced to impact destruction, it is necessary to take into account circumstances that reduce the error of the test being performed. Rational selection of specific fracture energy allows a more detailed analysis of the nature of the particles fracture.

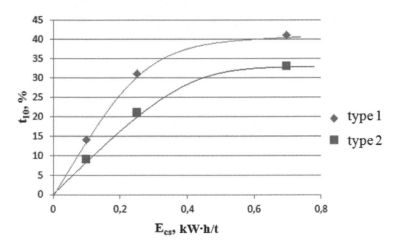

Figure 6.   Curves E_{cs}-t_{10} for two types of ore.

3   CONCLUSION

Desintegration processes in ore-preparation devices have different action mechanisms on the processing mineral raw materials. Consequently, devices and procedures that have a similar process of destruction with an industrial device should be used in study of mineral raw materials grindability. The direction of research relationship amongst mineral composition, structural and textural characteristics and form of particles with strength properties of ore can contribute more accurate selection and calculation cycles of AG/SAG using laboratory equipment. Executed studies have proved possibility of correct transition from results of one test to the results of another without additional research, expenses of time, as well as financial costs reduction. This allows obtaining more detailed and comprehensive information about strength properties of ore at stage of study and design of ore dressing cycles.

ACKNOWLEDGEMENT

The work is carried out under financial support of the Ministry of education and Science of the Russian Federation, the project RFMEFI57417X0168.

REFERENCES

[1] Andre McKen and Steve Williams. An overview of the small-scale tests available to characterize ore grindability. International Autogenous and Semiautogenous Grinding Technology/Proceedings of SAG conference Held in Vancouver, B.C., September 23–27, 2006, Vol. 4, p. 315–330.

[2] Andreev, E.E., Dokukin, V.P. and Nikolaeva, N.V. Estimation of feed size effect in designing and modeling of autogenous and semi-autogenous grinding mills. Obogashenie rud. 2009. №1. pp. 14–16.

[3] Gupta, A. and Van, D.S. Mineral Processing Design and Operations. An Introduction.//ELSEVIER. 2006, pp. 65–76.

[4] John Mosher and Tony Bigg. Bench—Scale and Pilot Plant Test for Comminution Circuit Design. Mineral Processing Plant Design, Practice and Control Proceedings/Published by the Society for Mining, Metallurgy and Exploration, Inc. Edited by Andrew L. Mular, Dough N. Halbe and Derek I. Barrat. 2002, Vol. 1, p. 123–135.

[5] John Starkey and Glenn Dobby. Application of the Minnovex SAG Power Index at Five Canadian SAG Plants. International Autogenous and Semiautogenous Grinding Technology/Proceedings of SAG conference Held in Vancouver, B.C., October 6–9, 1996, Vol. 1, p. 345–360.

[6] Melnikova, T.N., Yatlukova, N.G. and Litvinova, N.M. To the question of optimizing the process of ore crushing//Obogashchenie Rud (1), c. 44–48. 2006. № 4. P. 5–7.

[7] Morrell, S. Predicting the specific energy of autogenous and semi-autogenous mills from small diameter drill core samples//Minerals Engineering 17. 2004.

[8] Talovina, I.V., Aleksandrova, T.N., Popov, O. and Lieberwirth, H. Comparative analysis of rocks structural-textural characteristics studies by computer X-ray microtomography and quantitative microstructural analysis methods. Obogashchenie Rud. Issue 3, 2017, Pages 56–62. DOI:10.17580/or.2017.03.09.

[9] Taranov, V.A. Increase of efficiency of ore preparation of gold recovery factories on the basis of optimization of technological schemes. Thesis for the degree of candidate of technical sciences/ National Mineral Resources University (University of Mines). St. Petersburg, 2016. – 181 p.

[10] Verret, F.O., Chiasson, G. and Mcken, A. SAG mill testing—an overview of the test procedures available to characterize ore grindability. SGS MINERALS SERVICES. TECHNICAL PAPER 2011-08.

Geomechanics and Geodynamics of Rock Masses – Litvinenko (Ed.)
© *2018 Taylor & Francis Group, London, ISBN 978-1-138-61645-5*

Geomechanical behaviour of a rock barricade and cemented paste backfill: Laboratory experiments on a reduced-scale model

M. Nujaim & C. Auvray
CNRS, CREGU, Université de Lorraine, Vandœuvre-lès-Nancy, France

T. Belem
Institut de Recherche en Mines et en Environnement Rouyn Noranda, Université du Québec en Abitibi-Témiscamingue, Québec, Canada

ABSTRACT: The GeoRessources Laboratory (University of Lorraine, France) in collaboration with the Research Institute on Mines and the Environment (RIME-University of Québec in Abitibi-Témiscamingue, Canada), have developed an innovative experimental set-up to allow us reproduce a rock barricade within a drift leading into an exploitation chamber (called stope). For this study, we reproduce the mining-operations scheme at the LaRonde mine in Québec (Canada) by our 1:50 scale model. The objective of the experimental program is to measure the barricade-backfill and barricade-gallery interactions during emplacement of the backfill. The barricade is observed with a high-definition camera to determine the eventual displacements of the barricade using digital image correlation method (DIC). Also, mini pressure sensors are placed on the upstream side of the barricade to measure the pressure applied by the backfill.

1 INTRODUCTION

In recent decades, the use of mining backfill in underground stope for mine has become an increasingly employed approach in mining operations. Underground backfilling with waste-rock and tailings requires the construction of a retaining structure, known as a barricade, which is installed in a drift in order to hold the backfill in place. Barricades must be properly designed to prevent failure that can slow down mining production. Failure is often due to the high pressure induced by the backfill that has been conducted too quickly. Several documented failures indicate that barricade design is still a major challenge (Yumlu and Guresci 2007; Helinski et al. 2006; Sivakugan et al. 2006a, b, 2013; Grice 1998, 2001; Kuganathan 2001, 2002; Bloss and Chen 1998; Soderberg and Busch 1985). Barricades are generally constructed from timber, permeable bricks allowing easy drainage; concrete blocks and reinforced shotcrete with drainage pipes are also used to construct barricades (Hughes et al. 2010; Grabinsky 2010; Sivakugan 2008; Yumlu Guresci 2007 and Sivakugan et al. 2006b). Barricades with waste rocks are lower cost because these readily available and produced underground. Appropriate tools are therefore needed to evaluate the effective and total stresses in backfilled stopes in order to model the geomechanical responses and optimize the backfilling procedure. In order prevent failure, the barricades that maintain the backfill in place must be designed according to the state of stress, which depends on the properties of the backfill, the geometries of the openings, and the sequence of backfilling.

For the present study, we use an innovative experimental set-up to investigate the behaviour of barricades and backfill in the laboratory. Different parameters were studied, which can influence the behaviour of a barricade/backfill complex. These parameters included the backfilling rate, the water content of the backfill, and the position and grain-size of the barricade.

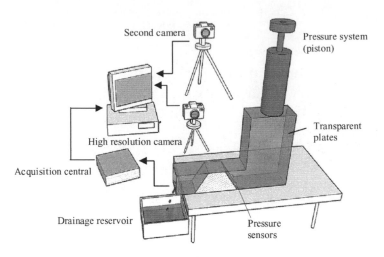

Figure 1.   Experimental set-up with associated metrology.

2   PRESENTATION OF THE EXPERIMENTAL SET-UP

The model was constructed at the 1:50 scale (Figure 1) to reproduce the exploitation operations scheme at the LaRonde mine, with a vertical section of 90 × 90 mm, length of 630 mm for the lower gallery, horizontal section of 90 × 200 mm and a height of 1000 mm for the stope. The model was designed using different PMMA-type polymer plates, which have sufficient mechanical properties (stiffness and resistance to traction and compression) for this application. The main advantage of this type of polymer is its transparency, which allows the barricade-backfill interaction to be observed during the tests. The pressures on the barricade were measured by mini-pressure sensors (0–50 kPa with 1% accuracy) positioned at different locations on the upstream side of the barricade. The set-up is equipped with a pressure system (piston) to increase the pressure of the backfill on the barricade until the failure. The emplacement of the backfill and displacements of the barricade were observed using a Prosilica GE2040 high-resolution camera (2048 × 2048 resolution), equipped with a 60 mm Nikon-AF sensor. A second camera was used to monitor the overall backfill emplacement procedure.

3   EXPERIMENTAL PROTOCOL

After having checked the seals and the mechanical resistance of our scale model, a protocol for installing the barricade and backfill was established in order to ensure the reproducibility and the repeatability in the experiments. The protocol used for positioning the barricade was: (i) fabrication of a mould in the chosen dimensions; (ii) positioning of the mould in the drift; (iii) filling the mould with dry gravels of 0 to 8 mm grain-size in accordance with the 1:50 scaling; (iv) removal of the barricade mould.

A protocol for preparing and emplacing the backfill was established. In order to properly respect the in situ conditions, in the first tests, we used calcareous sand (0/400 μm grain-size) mixed with 7% cement and 30% water, which allows rapid drainage during the simulation. For the other simulations, clay (kaolin) with a water content of 110% was used. This percent of water was necessary to obtain a homogeneous mix during the backfill emplacement and to avoid the difficulties resulting from sedimentation of sand grains. In order to understand the mechanism of failure barricades and to avoid the sliding, shims (pieces of plastic) have been fixed at the base of the gallery to increase the friction between the barricade and the gallery walls. During injection of the backfill into the stope (Figure 2a), the barricade was observed with the high-definition camera. These recordings were used to determine the eventual displacements of the barricade. In addition, a reservoir was positioned in the downstream part of the

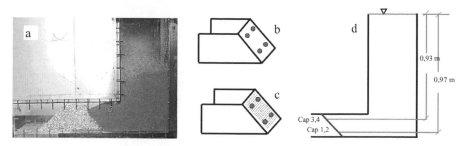

Figure 2: a) Emplacement the backfill, b) closed plate, c) perforated plate, d) position of the plate with its sensors of pressure in the set-up.

drift gallery to allow the fluids passing through the barricade to be recuperated. The volumes collected were used for subsequent drainage rate calculations. Tests were carried out on plates, representing the barricade surface, to study the backfill behavior of the two materials used in the experimental tests. To study the behavior of the barricade without drainage, a closed plate was used (Figure 2b). However, holes (1 mm of diameter) are perforated in the plate to study it with drainage. These holes allows us to drain the water without passing the solid grains (Figure 2c). In order to measure the pressures on the plate, four sensors were fixed (two in the upper part placed at 0.93 m from the top of the filling chamber) and two in the lower part (placed at 0.97 m) as shown in the Figure 2d. These pressure sensors should be calibrated using water before each test.

4 EXPERIMENTAL TESTING PROGRAMME

4.1 *Simplified preliminary tests (without pressure sensors)*

Before enrichment the model with mini-sensors of pressure, a series of 6 tests was carried out using the mixture (sand, cement and water) to study the influence of several parameters on the behavior of the barricade. The parameters during the first test were by default: backfilling rate (335 cm/h), water percentage (30%), grain seize (0/8 mm), distance between the barricade and the stope (100 mm), type of the barricade (trapezoidal). However in each new test, one of the parameters was changed to: backfilling rate (120 cm/h), water percentage (27%), grain seizes (1/8 mm), distance between the barricade and the stope (200 mm), barricade (triangular barricade with 210 mm wide at the base). Tests with the changed parameters were finally compared to the test with all parameters in default value.

4.2 *Instrumented tests (with mini-pressure sensors)*

A series of four tests was conducted to study the behavior of the two backfill used. Two of them were to study the behavior without drainage (by using a closed plate) and the other two to study the behavior with drainage (by using a perforated plate). A series of five tests was carried out using the clay on barricades, located at 150 mm from the access of the gallery, in order to understand the barricades failure mechanism. Two tests were performed without shims at the base of the gallery. The other tests were carried out with shims to prevent the sliding of the barricade and to connect the resistance of the barricade with the friction between its grains. Table 1 shows a summary of all the tests performed.

5 RESULTS

5.1 *Simplified preliminary tests (without pressure sensors)*

The main results of the tests 1 to 6 demonstrated that: a high backfilling rate would likely generate overpressures that could then induce large displacements of the barricade (the barricade was displaced by 600 μm for a backfilling rate of 335 cm/h whereas the displacement was zero

Table 1. Summary of tests.

		Test no.	Type retained structure	Type of backfill	Objective
Without shims at the base of the gallery	Without sensors	1 to 6	Barricade	Calcareous sand (0/400 µm) + 7% cement + 30% water	Study of the influence of several parameters on the behavior of the barricade: the filling rate, the percentage of water of the backfill, the granulometry, the position and the size of the barricade.
	Instrumented tests with sensors	7 & 8	Plate (closed/ perforated)	Calcareous sand (0/400 µm) + 7% cement + 30% water	Compare the behavior of two different materials used in experimental tests.
		9 & 10	Plate (closed/ perforated)	Clay (kaolin) + 110% water	
With shims		11 & 12	Barricade	Clay (kaolin) + 110% water	Predict the resistance of the barricades and their failure mechanism with different grain size classes.
		13 to 15	Barricade		

for a rate of 120 cm/h). A triangular barricade has a resistance lower than a trapezoidal but it allows a rapid drainage (the measured displacement was zero for the triangular barricade and 600 µm with the trapezoidal barricade, during increasing of the pressure on the barricades, the triangular one dissipated the pores water pressure faster but it collapsed first). Increasing the water content of the backfill increases the barricade drainage (600 µm of barricade displacement was measured with backfill have 30% of water content, when the water content was reduced to 27%, the measured displacement was zero). Increasing the distance between the barricade and the stope reduces the amount of barricade displacement (zero displacement was observed for a barricade located 200 mm whereas 600 µm was recorded for a barricade positioned at 100 mm). A dense barricade containing fine elements resists better than a barricade without fine elements (a barricade with particles size 1/8 mm was displaced by 1 mm. In contrast, the barricade containing fine material (0/8 mm) was displaced by 600 µm).

5.2 *Instrumented tests (with mini-pressure sensors)*

The second tests campaign were realized on plates. Pressure sensors were required on the plates to estimate the pressure applied by the backfill during a continuous filling. Calibration the pressure sensors used was validated (100%) by comparing the weight of a 1 meter water column (multiplying the density by the height) to the measured pressure by these sensors.

The tests 7 and 8 were performed on plates using clay. These tests showed that: the pressure exerted by the clay (density of 1.35 g/cm³) is the same with the closed plate or with the perforated plate. For both plates, the overburden (density of the backfill multiplied by its high) in the lower part was 127.5 µbar whereas 126 µbar of pressure was measured by the sensors. In the upper, 120 µbar of pressure was recorded whereas its overburden was 122 µbar. It has also been noted that the filling rate (continuous filling) and the position of the barricade with respect to the access of the filling chamber have no effect on the pressure.

The tests 9 and 10 were performed on plates using the other backfill (sand, water and cement). In these tests, a rapid sedimentation for the solid particles and separation of water from the mixture was noted. The overburden in the lower and upper parts was 178, 170 µbar respectively. We found pressures twice as low as the overburden for the closed plate. For the perforated plate, the pressure was decreased to 88% at the lower part (sensors 1 & 2) and to 97% at the upper part (sensors 3 & 4).

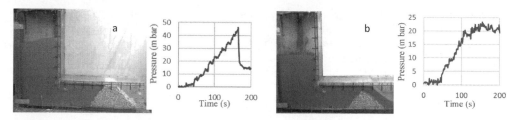

Figure 3.   a) A trapezoidal barricade without shims, b) A triangular barricade without shims.

Grains size 0 / 2 mm   Grains size 2 / 6.3 mm   Grains size 6.3 / 8 mm

Figure 4.   Trapezoidal barricades (of different particles size) with shims at the base of the drift.

Tests 11 to 15 were performed in order to understand the barricade's response during the filling. The pressure was measured by a sensor located at 0.0 m from the base and 140 mm from the entrance of the stope. The tests 11 and 12 were carried out on barricades (1/8 mm grains size) without shims at the base of the gallery. Both tests showed that: a pressure (4.63 kPa) applied on a trapezoidal barricade (1.8 kg of weight) produced a full sliding (Figure 3a). Other pressure (2.1 kPa) applied on a triangular barricade (1.2 kg of weight) led to a collapse for its upper part (Figure 3b). Tests 13 to 15 were carried out with shims at the base of the drift to increase the barricade-wall friction. The results of three tests with different gradations of the barricade show that: barricades with a grain size class 0/2 and 2/6.3 can support a backfill column equal to 2 times the main height of the filling chamber (1 m) without failure. A barricade (grain size 6.3/8 mm) was collapsed when the level of backfill was near to the top of the filling chamber (pressure 12.3 kPa). It has been observed that, the backfill was deep in the porous net of the barricade which does not contain fine elements (Figure 4).

6   INTERPRETATIONS

The laboratory tests on the barricade have allowed us to better understand the effect of the main parameters that influence the mining operations such as filling rate, the position and the dimensions of the barricade.

It has been found that a rapid filling can produce large interstitial pressures at the bottom of the stope, this is reflected directly on the displacement of the barricade. This remark agrees with previous studies (Nujaim et al 2016, El-Mkadmi et al 2014). Increasing the distance between the barricade and the gallery access has an influence on the constraints in the stope, the displacements seem to increase with the decrease of the distance from the extraction chamber. A triangular barricade with respect to a trapezoidal barricade allows a rapid drainage and its resistance is lower than the trapezoidal. The grain size of the barricade plays a very important role in the resistance. A dense barricade, which contains fine elements, resists better the displacement or the failure. Continuous filling can sweep the upper grains when there is no good connection between the barricade roof and the wall of the gallery, which confirms the need to support the upper part of the barricade (by a layer of shotcrete).

The experimental tests performed on the plate demonstrated the importance of understanding the backfill behavior during backfilling instantly. Clay retains water for a long time (drainage is very slow during filling) and applies pressure on the barricade corresponding to the overburden. With this behavior, the filling rate and the position of the barricade do not significantly influence the measured pressures. A rapid drainage with grains sedimentation led to significant pressure decreases for the backfill (sand with 7% cement and 30% water). Thus, it is important to define the suitable behavior of the backfill that takes into account the different phenomena before realization of numerical modelling.

7 PERSPECTIVES

Obtained results are consistent with many previous studies. Additional tests will be conducted with respecting the simulation rules on barricades with different particles size. We will increase the pressure on the barricades, using the piston, in order to measure their maximum resistance. Also, the failure mechanism will be studied using the digital image correlation method (DIC). The results will be compared to numerical simulations and future in situ data. Therefore, we can enhance the precision in calculation of dimensions of barricades and the increase of filling rate which in turn will improve mining production and ensure the safety of equipment and workers.

ACKNOWLEDGMENTS

The authors acknowledge the financial support from the ASGA (Scientific Association for Geology and its Applications, France).

REFERENCES

Bloss, M.L., and Chen, J. (1998). "Drainage research at Mount Isa Mines Limited 1992–1997." Proc., 6th Int. Symp. on Mining with Backfill, M. Bloss, ed., AusIMM, Carlton, Australia, 111–116.

El Mkadmi N, Aubertin M, Li L (2014). "Effect of drainage and sequential filling on the behavior of backfill in mine stopes". Can Geotech J; 51(1):1–15.

Grabinsky, M.W. (2010). "In situ monitoring for ground truthing paste backfill designs." Proc., Paste 2010, Australian Centre for Geomechanics, Crawley, Australia, 85–98.

Grice, T. (1998). "Stability of hydraulic backfill barricades." Proc., 6th Int. Symp. on Mining with Backfill, M. Bloss, ed., AusIMM, Carlton, Victoria, Australia, 117–120.

Helinski, M., Fourie, A., and Fahey, M. (2006). "Mechanics of early age CPB." Proc., Symp. at the 9th Int. Seminar on Paste and Thickened.

Hughes, P.B., Pakalnis, R., Hitch, M., and Corey, G. (2010). "Composite paste barricade performance at Goldcorp Inc. Red Lake Mine, Ontario, Canada." Int. J. Min, Reclam., Environ., 24(2), 138–150.

Kuganathan, K. (2001). "Mine backfilling, backfill drainage and bulkhead construction-a safety first approach." Aust. Min. Mon., 58–64.

Kuganathan, K. (2002). "A model to predict bulkhead pressures for safe design of bulkheads." Proc., Filling with Hydraulic Fills Seminar, section 6, Australian Centre for Geomechanics, Perth, Australia.

Nujaim M, Auvray C, Belem T (2016). "Comportement géomécanique d'une barricade rocheuse et du remblai en pâte cimenté: expérimentations sur un modèle réduit" RST 2016 conference, Caen.

Sivakugan, N. (2008). "Geotechnical issues of mining with hydraulic backfills." Elect. J. Geotech. Eng., Special Volume: Bouquet 08.

Sivakugan, N., Rankine, K.J., and Rankine, R.M. (2006b). "Permeability of hydraulic fills and barricade bricks." Geotech. Geol. Eng., 24,661–673.

Sivakugan, N., Rankine, R.M., Rankine, K.J., and Rankine, K.S. (2006a). "Geotechnical considerations in mine backfilling in Australia." J.Cleaner Prod., 14(12–13), 1168–1175.

Soderberg, R.L., and Busch, R.A. (1985). Bulkheads and drains for high sandfill stopes, U.S. Dept. of the Interior, Bureau of Mines, Washington, DC.

Tailings, R. Jewell, S. Lawson, and P. Newman, eds., Australian Centre for Geomechanics, Perth, Australia, 313–322.

Yumlu, M., and Guresci, M. (2007). "Paste backfill bulkhead monitoring-A case study from Inmet's Cayeli mine." Proc., 9th Int. Symp. in Mining with Backfill (CD-ROM), Canadian Institute of Mining, Metallurgy and Petroleum, Montreal.

Geomechanics and Geodynamics of Rock Masses – Litvinenko (Ed.)
© *2018 Taylor & Francis Group, London, ISBN 978-1-138-61645-5*

Multivariate Artificial Neural Network (ANN) models for predicting uniaxial compressive strength from index tests

Burkan Saeed Othman
Department of Applied Geology, College of Science, Kirkuk University, Kirkuk, Iraq

Nazlı Tunar Özcan, Aycan Kalender & Harun Sönmez
Department of Geological Engineering, Hacettepe University, Ankara, Turkey

ABSTRACT: Uniaxial Compressive Strength (UCS) of rock material is an important parameter used as input for rock engineering applications. However, preparation of standard test samples to determine UCS from some rock materials such as thinly bedded, weak and jointed rock masses are almost impossible. Therefore some index and indirect tests such as Block Punch Index (BPI), point load index (I_{s50}) and Brazilian tensile strength have been proposed in order to estimate the UCS of such rocks. In the literature, the relations between UCS and index tests were extensively investigated by using statistical models. In fact, different rock materials with similar index test results may have different UCS's depending on their failure envelopes. By considering this fact, multivariate equation models were developed in this study to estimate UCS values. A large database was prepared with data compiled. In addition to developed simple empirical equations, Artificial Neural Network (ANN) method was used to develop models for prediction of UCS considering BPI, $I_{s(50)}$, tensile strength (σ_t), γ and m_i parameters. Although the prediction models developed in this study may be considered for practical purposes, the results indicated that the models using BPI as an input parameter have higher prediction performance than others.

Keywords: Uniaxial compressive strength, index tests, multivariate artificial neural network

1 INTRODUCTION

The uniaxial compressive strength of rock material is determined by conventional laboratory tests employed on high quality core samples in accordance with ISRM (2007). However, for some rocks such as laminated and fragmented rock materials, preparation of high quality cores is almost impossible. Therefore some index and indirect tests such as block punch index (BPI), point load index (PLI) and Brazilian tensile strength have been used in order to predict the UCS of such rocks. However, simple relations between index parameter and UCS may not reflect sufficiently type of rock due to the use of unique input parameter. In this study, multivariate prediction models were investigated by Artificial Neural Network (ANN) after some simple regression analyses.

2 ESTABLISHMENT OF DATABASE

The database established by Othman (2012) was checked for this study. In addition to laboratory tests results for 14 different rock types and datasets compiled by Othman (2012), some new datasets was also searched from additional literature survey. The studies considered for the establishment of the database were given in the reference list. The database was composed of rock type, unit weight (γ), point load index (I_{s50}), block punch index (BPI) and Brazilian tensile

Table 1. Statistical summary for the database.

Parameter	Uniaxial compressive strength, UCS (MPa)	Brazilian tensile strength, σ_t (MPa)	Point load index, I_{s50} (MPa)	Block punch index, BPI (MPa)	Unit weight, γ (kN/m³)	m_i parameter (from Hoek, 2007)
Minimum	2.2	0.4	0.1	0.7	13.1	4
Maximum	355	28.03	16.21	42.9	29.2	32
Average	90.42	9.83	4.95	13.48	24.20	18.91
Standard Deviation	65.40	5.69	3.30	9.70	3.13	8.62
Number of data	528	225	334	190	305	485

strength (σ_t) values as inputs and uniaxial compressive strength (UCS) values as output. The m_i parameter of Hoek and Brown (HB) failure criterion was also taken into consideration as a new input parameter in multivariate models. As known, the m_i parameter is mainly control the general slope of the nonlinear envelope for rocks in the HB criterion. The general inclination of curve type failure envelope play important role on the value of UCS. By considering this possible effect, Sönmez and Tunusluoğlu (2008) developed an analytical equation including m_i parameter together with BPI for predicting of UCS. Although the precise value of m_i parameter can be obtained from HB equations by using triaxial compression test results, in Hoek's some studies as originator of the Hoek and Brown failure criterion the possible value of m_i parameter just considering type of rock were available for practical purposes. In this study the m_i values for different type of rocks listed by Hoek (2007) were used. The summary information of database is given in Table 1. While the number of UCS value is 387 complied by Othman (2012), the established database include total 528 dataset after additional literature survey performed by this study. The datasets were obtained from 41 different type of rocks. In this study, while the parameters may be directly linked to the UCS such as Brazilian tensile strength, point load index and block punch index were evaluated as primary input parameters, the unit weight and m_i parameter were used as calibration (secondary) inputs together with primary inputs. In other words multivariate models investigated using two inputs composed of one primary and one calibration input parameters. Unfortunately whole parameter for each dataset may not be available for the collected data from literature. Therefore the number of data used in the multivariate models are different from each other. The number of data used for multivariate models for Model_1a (σ_t;m_i), Model_1b (σ_t;γ), Model_2a (I_{s50};m_i), Model_2b (I_{s50};γ), Model_3a (BPI;m_i) and Model 3b (BPI;γ) are 216, 128, 296, 217, 185 and 51, respectively. The number of data set for the model (3b) which considers BPI and γ as inputs is 51. Therefore the Model_3b was not used in multivariate ANN.

3 SIMPLE REGRESSIONS

As a statistical method, regression analyses widely used for investigation of prediction equations. Because the predictions can be done by handy calculator easily by regression based empirical equations. However regression analyses may not be sufficient to solve nonlinear problems. To overcome this problem soft computing tools have been preferred in many rock mechanics problems. The simple relations between UCS and primary inputs such as σ_t, I_{s50} and BPI were investigated to determine type and degree of relations that can be seen from Figure 1. Similar multipliers for predicting of UCS from BPI were obtained from the studies performed by van der Schrier (1988), Gokceoglu (1997), Ulusay and Gokceoglu (1997), Sulukcü and Ulusay (2001) and Ulusay et al. (2001). The relations and the cross check graphs between measured and predicted UCS was given in Figure 1. As seen from Figure 1, while the best prediction obtained from BPI as input, the relation of I_{s50} is better than that of σ_t. The prediction capacity of the primary inputs together with calibration (secondary) inputs such as γ and m_i were investigated by multivariate ANN in the next section.

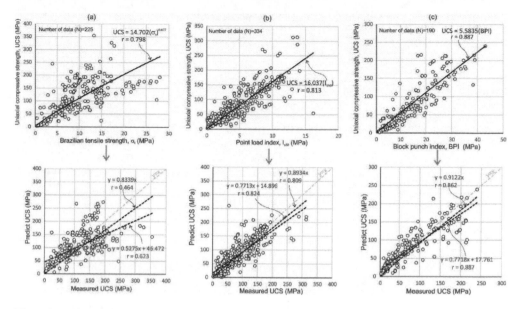

Figure 1. The relations between measured UCS and primary inputs and their prediction performance on y = x cross graphs for primary inputs as (a) σ_t, (b) I_{s50} and (c) BPI.

4 MULTIVARIATE ARTIFICIAL NEURAL NETWORK (ANN) ANALYSES

Soft computing techniques have been widely used as a popular tool in rock mechanic applications. Artificial Neural Network (ANN) is one of the soft computing methods to be used for solving nonlinear and multivariate problems. Sonmez et al. (2006) developed ANN based chart for selection of intact elastic modules considering unit weight and uniaxial compressive strength as inputs. In this study, ANN was used for the prediction of UCS considering primary and calibration parameters as inputs. The studied models were mentioned above.

ANN models were analyzed with the computer codes ANNES developed by Dr. H. Sönmez. The back-propagation ANN was used with possible simple structure including one hidden layer with two neurons. Although simple ANN structure was preferred due to the quite simple models defined in this study, of course complex ANN architecture may be preferred for complex problems. The sigmoid transfer function was used to transform the weighted sum of all signals impinging on a neuron. The learning rate and momentum coefficient are two parameters which control training rate of the ANN model. In general, small learning rate may cause slow training phase, the large learning rate may cause oscillations in training phase. While learning rate was set to 0.1, momentum coefficient was used as 0.95 as preferred in literature (Sonmez et al., 2006). Root mean square error (RMSE) less than 0.001 or maximum training cycle (epoch) equal to 10000 were used for stop criteria. While 70% of dataset was used as train dataset, rest of them used as test data. Logically the UCS must be equal to zero for the zero values of BPI, I_{s50} and σ_t. To assign this logic assumption in the regression analysis, the regression equation can be fixed to zero practically. However, it is not possible to do for ANN. Therefore a series of logical dataset were included to the train datasets such as pairs of BPI = 0 (similarly for I_{s50} and σ_t) and the possible nine values of m_i (similarly for γ) between maximum and minimum ranges for output is UCS = 0. These data were not considered on the performance evaluations by cross correlation between measured and predicted UCS's.

The criteria for selection of optimum training cycle (epoch) proposed by Basheer and Hajmeer (2000) was preferred to avoid overlearning of the ANN model. While the prediction performances of the ANN models with comparison simple regression are given in Table 2, the cross correlations between measured and predicted UCS values from the ANN models

Table 2. The prediction performances of the ANN models with comparison simple regression.

Simple regressions		ANN models		
Regression relation	Cross correlation between measured and predicted UCS	Inputs	Selected optimum training cycle	Cross correlation between measured and predicted UCS
$UCS = 14.702(\sigma_t)^{0.877}$ $r = 0.798$	$y = 0.5275x + 46.472$ $(r = 0.623)$	σ_t, m_i	~278	$y = 0.4584x + 63.995$ $(r = 0.475)$ $y = 0.8787x$ $(r: \text{not determined})$
	$y = 0.8339x$ $(r = 0.464)$	σ_t, γ	~200	$y = 0.5488x + 44.981$ $(r = 0.742)$ $y = 0.8684x$ $(r = 0.557)$
$UCS = 16.037(I_{s50})$ $r = 0.813$	$y = 0.7713x + 14.896$ $(r = 0.824)$	I_{s50}, m_i	~70	**$y = 0.7361x + 26.944$** **$(r = 0.843)$** **$y = 0.9494x$** **$(r = 0.789)$**
	$y = 0.8934x$ $(r = 0.809)$	I_{s50}, γ	~50	$y = 0.6794x + 32.168$ $(r = 0.823)$ $y = 0.9541x$ $(r = 0.722)$
$UCS = 5.5835(BPI)$ $r = 0.887$	$y = 0.7718x + 17.761$ $(r = 0.887)$	BPI, m_i	~50	**$y = 0.87x + 17.021$** **$(r = 0.904)$** **$y = 1.0058x$** **$(r = 0.885)$**
	$y = 0.9122x$ $(r = 0.862)$	BPI, γ	Not considered due to the limited number of data for ANN	

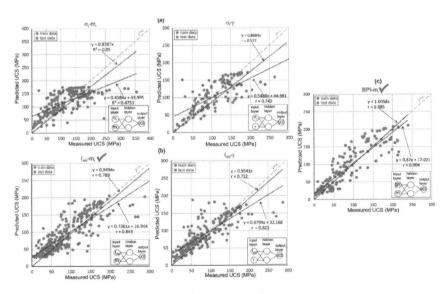

Figure 2. The prediction performance of the ANN models for primary inputs as (a) σ_t, (b) I_{s50} and (c) BPI.

are shown in Figure 2. By considering their high prediction performance of the ANN models on y = x cross graphs, prediction charts of Model 2a (for inputs of I_{s50} and m_i) and Model 3a (for inputs of BPI and m_i) were prepared for the optimum learning cycle of the ANNs. For this aim input matrices for both ANN models was prepared considering possible range of I_{s50}, BPI and m_i. Then UCS values as output were determined by using the weights and threshold values for the optimum learning cycles of two ANN models separately. Finally, input matrices (for I_{s50}-m_i and BPI-m_i) and UCS output files used together for preparation of the prediction charts given in Figure 3.

As seen in Figure 3, particularly for weak to moderate strength rock materials (predicted UCS is less than about 50 MPa), the use of ANN based prediction charts may overestimate because the zero intercept could not be satisfied for both BPI = 0 and I_{s50} = 0. To minimize overestimation on predicted UCS value for weak to moderate strength rock material, a correction procedure were proposed using possible error on measured UCS values for both ANN models as illustrated in Figure 4. The prediction performances were also slightly increased after corrections applied on ANN results (Figure 4).

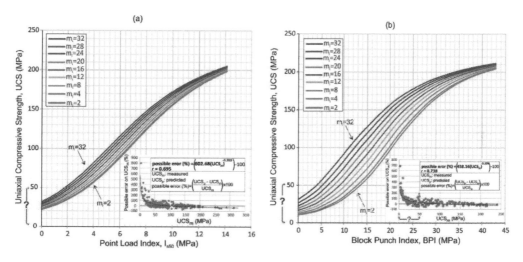

Figure 3. The multivariate ANN based prediction charts for (a) I_{s50} and (b) BPI as primary inputs and m_i as secondary input.

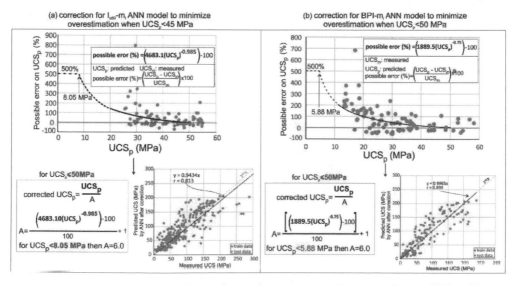

Figure 4. A correction procedure to minimize overestimation on predicted UCS value for weak to moderate strength rock material for both ANN models.

5 RESULTS AND CONCLUSIONS

It is very difficult to prepare standard UCS test samples from thinly bedded, weak and jointed rock masses. The prediction tools have been widely preferred in engineering applications for practical purposes to overcome difficulties encountered direct measurements. However, the generalization capacity of the prediction tool have crucial importance for reliable outputs. In this study by considering established database, Brazilian tensile strength (σ_t), point load index (I_{s50}) and block punch index (BPI) were selected as primary inputs and the prediction capacities were improved by secondary inputs such as γ and m_i by using ANN. The ANN models which include I_{s50} and BPI have better prediction performance than Brazilian tensile strength. In addition the use of m_i parameter in the ANN models have more contribution on performance for BPI when compared I_{s50}. The UCS prediction tool investigated in this study such as simple equations and the ANN models were developed by empirically based on the established database. It should be underlined that the established database mainly composed of comprehensive literature survey, therefore the possible uncertainties may be expected in the results.

The models for predicting of UCS were investigated for wide range of UCS values from minimum 2.2 to 355 MPa. Although the overall prediction capacity may be evaluated as sufficient on y = x cross graphs with r = ~0.9, the prediction error increases for lower predicted UCS values. However to minimize this overestimation a procedure was instructed, nevertheless new models for weak or moderate strength rock material may be re-investigated. Although soft computing methods are powerful tool to be considered in engineering application, it should be carefully evaluated to be sure about generalization capacity satisfied. Soft computing methods such as ANN based models should not be forced to yield highest prediction to avoid overlearning.

REFERENCES

Adriansyah, Y., Zakaria, Z., Muslim, D., Haryanto, I., and Hutabarat, J., 2017. Determining of Coefficient Correlation between UCS and PLI data for Various Rock Types at Batu Hijau Mine PT Amman Mineral Nusa Tenggara. J. of Geo. Sci. and App. Geo., (2017) V.2. No.1.

Basheer, I. and Hajmeer, M., 2000. Artificial neural networks: fundamentals, computing, design, and application. J. of Microbiological Methods, 43(1):3–31.

Bell, F.G., and Lindsay, P., 1999. The petrographics and geomechanical properties of some sandstones from the Newspaper Member of the Natal group near Durban, South africa. Engineering Geology, 53, 57–81.

Bell, F.G., Entwisle, D.C., and Culshaw, M.G., 1997. A geotechnical survey of some British coal measure mudstones, with particular emphasis durability. Engineering Geology, 46, 115–129.

Bell, F.G., 1981a. Geotechnical properties of some evaboritic rocks. Bulletin of the Int. Assoc. of Eng. Geology, 24, 137–144.

Bell, F.G., 1981b. A survey of the physical properties of some carbonate rocks. Bulletin of the Int. Assoc. of Eng. Geology, 24, 105–110.

Bell, F.G., 1978. The physical and mechanical properties of the Fell sandstones. Northurmberland, England. Engineering Geology, 12, 1–29.

Bieniawski, Z.T., 1975. The point load test in geomechanical practice. Engineering Geology, 9, s1–11.

Butenuth, C., De Freitas, M.H., Park, H.D., Schetelig, K., Van Lent, P., and Grill, P., 1994. Observation on the use of the hoop test for measuring the tensile strength of anistropic rock Int. J. of Rock Mech. and Min. Sci. and Geomech. Abstr., 31, 733–741.

Cargill, J.S. and Shakoor, A., 1990. Evaluation of empirical method for measuring the uniaxial compressive strength of rock. Int. J. of Rock Mech. and Min. Sci. and Geomech. Abstr., 27(6), 495–503.

Christaras, B., 1991. Weathering evaluation method and changes in mechanical behaviour of granites Northern Greece. Bulletin of Int. Assoc. of Eng. Geology, 43, 21–26.

Dalgıç, S., 2002. A comparsion of predicted and actual tunnel behaviour in the İstanbul Metro, Turkey. Engineering Geology, 63, 69–82.

Del Greco, O., Ferrero, A.M., and Oggeri, C., 1993. Eperimental and analytical interpretation of the behaviour of laboratory tests on composites specimens. Int. J. Rock Mech. Min. Sci. and Geomech. Abstr. 30(7), 1539–1543.

Diamantis, K., Gartzos, E., and Migiros, G., 2009. Studay on uniaxial compressive strength, point load strength index, dynamic and physical properties of serpentinites from central Greece: Test results and empirical relations. Engineering Geology, 108, 199–207.

Endait, M. and Juneja, A., 2015. New correlations between uniaxial compressive strength and point load strength of basalt, Int. J. of Geotechnical Eng., 9:4, 348–353.

Ghosh, D.K. and Srivstava, M., 1991. Point load strength. An indeks classification of rock material. Int. Assoc. Bull. Eng. Geol. 44, 233.

Gokceoglu, C. 1997. The approaches to overcome the difficulties encountered in the engineering classification of clay-bearing densely jointed and weak rock masses. PhD thesis, Hacettepe University, Ankara, [in Turkish].

Gokceoglu, C. and Aksoy, H., 2000. New approaches to the characterization of clay-bearing, densely jointed and weak rock masses. Engineering Geology, 58, 1–23.

Gunsallus, K.L. and Kulhawy, F.H., 1984. A comparative Evaluation of rock strength measures. Int. J. of Rock Mech. and Min. Science. Abstract, 21(5), 233–248.

Hoek E. Practical rock engineering. 2007. http://www.rocscience.com/roc/Hoek/Hoek.htm.

ISRM, 2007. The complete ISRM suggested method for rock characterization, testing and monitoring, 1974–2006. R Ulusay and JA Hudson (Eds.), Compilation Arranged by the ISRM Turkish National Group, Kozan ofset, Ankara, Turkey, 628 pp.

Karakul, H., Ulusay, R. and Isik N.S. 2010. Empirical models and numerical analysis for assessing strength anizotropy based on block punch index and uniaxal compression tests. International Journal of Rock Mech & Min. Sci. 47, pp. 657–665.

Karakus, M. and Fowell, R.J., 2005. Back analysis for tunnelling induced ground movments and stress redistribution. Tunnelling and Underground space Technology. 20, 514–524.

Mishra, D.A. and Basu, A., 2013. Estimation of uniaxial compressive strength of rock materials by index tests using regression analysis and fuzzy inference system. Engineering Geology 160 (2013) 54–68.

Othman, B., 2012. Investigation of the Estimation of Uniaxial Compressive Strength from Multi Parameter Models. Unpublished PhD Thesis, Hacettepe University, Ankara, 87 p (in Turkish).

Roghanchi, P. and Kallu, R.R., 2014. Block Punch Index (BPI) Test—A New Consideration on Validity and Correlations for Basalt and Rhyolite Rock Types. Journal of Mining Science, 2014, Vol. 50, No. 3, pp. 475–483.

Russell, A.R. and David Muir Wood, D.M. 2009. Point load tests and strength measurments for brittle spheres. Int. J. Rock Mech. Min. Sci. 46, 272–280.

Sabatakakis, N., Koukis, G.,Tsiambaos, G., Papanakli, S., 2008, İndex properties and variation Controlled by microstructure for sedimentry rocks, Engineering geology, 97, 80–90.

Sair Kahraman, S., Fener, F. and Kılıc C.O., 2016. Preliminary study on the conversion factor used in the prediction of the UCS from the BPI for pyroclastic rocks. Bull Eng Geol Environ (2016) 75:771–780.

Schrier van der, J.S., 1988. The block punch index test. Bulletin of the International Association of Engineering Geology, 38, 121–126.

Sönmez, H., Tunusluoglu, C., 2008. New consideration on the block punch index for prediction of uniaxial compressive strength of rock material, Int. J. of Rock Mech & Min Sci., 45 (2008) 1007–1014.

Sönmez, H., Gökçeoğlu, C., Nefeslioğlu, H.A., Kayabaşı, A., 2006. Estimation of rock modulus: For intact rocks with an artificial neural network and for rock masses with a new empirical equation. Int. J. of Rock Mech. & Min. Sci. 43(2), pp. 224–235.

Sulukcu, S., Ulusay, R., 2001. Evaluation of the block punch index test with particular reference to the size effect, failure mechanism and its effectiveness in predicting rock strength. Int J Rock Mech Min Sci 38(8), 1091–1111.

Topal, T., 2000. Problems faced in the applications of the point load index test, Jeoloji Mühendisliği Dergisi, 24, 73–86. (in Turkhish).

Topal, T., and Duyuran, V., 1997. Engineering geological properties and durability assessment of the Cappadocian tuff. Engineering Geology, 47: 175–187.

Tuğrul, A., and Zarif, I.H., 1999. Correlation of mineralogical and textural characteristics with engineering properties of selected granitic rocks from Turkey. Engineering Geology, 51, 303–317.

Ulusay, R. and Gokceoglu, C., 1997. The modified block punch index test. Canadian Geotech. Journal, 34(6), 991–1001.

Ulusay, R., Gokceoglu, C., and Sulukcu, S., 2001. Draft ISRM suggested method for determining block punch strength index (BPI). Int. J. of Rock Mech. & Min. Sci., 38, 1113–1119.

Ulusay, R., Tureli, K. and Ider, M.H., 1994. Prediction of Engineering properties of a selected Litharenite sandstone from its petrographic characteristics using correlation and multivariate statistical techniques. Engineering Geology. 38, 135–157.

Villeneuve, M.C., 2008. Examination of geological influence on machine excavation of highly stressed tunneles in massive haed rock. Ph.D. thesis, Queen's University Kingston, Ontario, Canada, p. 788.

Yagis, S., 2009. Assessment of brittleness using rock strength and density with punch penetration test. Tunnelling and Underground Space Technology. 24, 66–74.

Investigation of scale effects on uniaxial compressive strength for Kars-Kagızman Rock Salt, Turkey

Ihsan Ozkan & Zahir Kızıltaş
Mining Engineering Department, Selçuk University, Konya, Turkey

ABSTRACT: The Uniaxial Compressive Strength (UCS) test results carried out for intact rocks under laboratory condition are used widely in rock mechanics analyses. However, if the rock engineers performs UCS testing program to evaluate the impact of shape and size on rock materials, unfortunately, they can encounter with some surprising results, particularly for special materials like coal and rock salt. In addition, the grain size (d) is effective on UCS test results. In this study, the change in UCS attributable to scale was determined for Kars-Kagızman rock salt, and the results are presented to emphasize the importance of the scale effect on failure phenomena in rock salt under uniaxial compression. With statistical analysis, the critical Height/Diameter (H/D) and Height/Width (H/W) ratios were determined to be 2.0 for core specimens and 2.0 for prismatic specimens.

1 INTRODUCTION

In design studies of mining and the other rock engineering structures, the determination of the in situ strength parameters of a rock mass is difficult, expensive, and time consuming. Therefore, designers have to prefer empirical approaches, which has been developed based on laboratory studies. The uniaxial compressive strength (UCS) test is widely used to determine mechanical properties of the intact rock in rock engineering studies. It is assumed by designers that the UCS used in various design approaches is an important input parameter. Therefore, test procedures published by the ISRM (1978, 2007) and the ASTM (1986) specify specimen dimensions and loading rate. The specified loading rate based on area is 0.5 to 1 MPa/s. However, especially, the suggested specimen dimensions are generally valid for much encountered rock types such as granite, basalt, limestone, sandstone, marl ect. Its validity should be discussed for special rock types such as coal and rock salt. Therefore, rock mechanic studies are required to investigate the scale effects for coal and rock salt.

It is accepted by authoritarians of rock mechanics that the scale effect on UCS is an important consideration (Hoek and Brown, 1980; Bieniawski, 1984; Hawkins, 1998; and Thuro et al., 2001). The scale effect in the literature is defined as different geometric forms and their different sizes, such as core samples and prismatic samples. The UCS test results for both geometric forms can be different because of the shape effect. Most UCS testing on rock materials is carried out on cylindrical rock core samples, whereas prismatic samples are used less than them (Hawkins, 1998). The effect on UCS of variations in the height/diameter (H/D) ratios of cylindrical specimens and the height/width (H/W) ratios of prismatic samples are known as scale effects.

The UCS test is also commonly used to determine the strength of rock salt. The test results are to be used in design studies for open-pit or underground salt mines, also caverns used as natural gas storages. In fact, the UCS of rock salt depends on a number of factors including the specimen shape-size, porosity, moisture, mineral composition, grain size, temperature, and clamping effect. Variations in the strength properties of rock salt can occur between various salt deposits. It can be due to its different types and structures of the mineral components (Jeremic, 1994). Impurities in the samples (anhydrite, clay) may also have an important effect on the strength parameters of rock salt, especially in bedded salt deposits (Aksoy et al., 2006; Liang et al., 2007). In addition, grain size also has an important effect on the UCS of rock salt samples (Lux, 1984). Some empirical equations explaining scale effect on UCS are given in Table 1. As a result, it should be emphasized in particular that the permanent UCS values independent of the scale effect must be used by designers. Therefore, in this study, we investigated the scale effect on the UCS values of the Kars-Kagızman rock salt located in Turkey. The permanent UCS values, independent of the scale effect, were determined for underground mine design applications.

2 KARS-KAGIZMAN MINE SITE

Thick natural salt deposits are found in certain parts of Turkey. The main salt deposits are located in central Anatolia (Cankiri and Nevsehir regions), eastern Turkey (Erzurum region), southeast Anatolia (Cukurova region), Kars-Kagızman located in east black sea region of Turkey and the Salt Lake region. Salt rocks of the Tertiary period are relatively widespread in Turkey.

In mine, the underground room-and-pillar mining method was applied. The bedded rock salt deposit contains approximately 92–95% NaCl, 2–3% gypsum, and 3–5% other impurities. The density of the rock salt has been determined to be 2.07 g/cm^3, and the salt grain size (d) is approximately 7 mm. The other information about rock salt underground mine: depth is 150 m, salt thickness is 230 m, salt reserves is 200 million tons, pillar dimensions (H × W_1 × W_2) are 10 m × 10 m × 10 m, and H/W ratio is 1.

3 SPECIMEN PREPARATION AND TESTING

In this study, first of all, 30 rock salt blocks, which are the approximately 6 tons, have been transplanted to Selcuk University-Konya from Kars-Kagızman rock salt mine. In fact, this project study have been focused on determination of creep behavior of rock salt material. In order to goal, this project study were considered in three stages. They are: (i) to determine scale effect on UCS and the optimum dimensions (H/D and H/W) based on it, (ii) to determine the permanent UCS values independent of the scale effect, (iii) to determine the creep behavior of rock salt material based on the critical H/D and H/W ratios. In this paper, the research results carried out in the first and second stages were presented.

A total of 45 cores and 60 prismatic specimens were extracted from these blocks. Fifteen cube specimens were also prepared to assess the effect of the condition H/W = 1. The dimensions of the specimens used in this study are presented in Tables 2 and 3. In the laboratory experiments carried out to determine the UCS of rock salt, the method suggested by ISRM (1978, 2007) were used. The specified loading rate based on area has been selected as 0.75 MPa/s. Then, the tests were carried out at constant room temperature with standard steel plates. A hydraulic press, which has a servo control facility and a loading capacity of 3000 kN, were used for extensive laboratory tests.

Table 1. The empirical equations explaining scale effect on UCS.

The researcher	The suggested equation	The special condition
ASTM, 1986	$K = [UCS_{(D/H)\neq0.5}]/[UCS_{(D/H)=0.5}] = 0.88 + 0.24(D/H)$	for various rocks
Anonymous, 2004	$K = [UCS_{(D/H)\neq0.5}]/[UCS_{(D/H)=0.5}] = 0.65 + 0.70(D/H)$	0<D/H≤0.5 for rock salt
Anonymous, 2004	$K = [UCS_{(D/H)\neq0.5}]/[UCS_{(D/H)=0.5}] = (2D/H)^{0.5}$	D/H≥0.5 for rock salt
Zhigalkin et al., 2005	$K = [UCS_{(D/H)\neq0.5}]/[UCS_{(D/H)=0.5}] = 0.775 + 0.45(D/H)$	for red sylvinite
Zhigalkin et al., 2005	$K = [UCS_{(D/H)\neq0.5}]/[UCS_{(D/H)=0.5}] = 0.52 + 0.96(D/H)$	for carnallite
Hoek and Brown, 1980	$K = [UCS_{D\neq0.5}]/[UCS_{D=0.5}] = (50/D)^{0.18}$	for crystalline, intact rocks

Table 2. Dimensional specifications of the prepared core specimens of rock salt and sample number (#).

H/D	D = 42 mm H (mm) / #	D = 54 mm H (mm) / #	D = 93 mm H (mm) / #
0.5	21 / 3	27 / 3	46.5 / 3
1.0	42 / 3	54 / 3	93 / 3
1.5	63 / 3	81 / 3	139.5 / 3
2.0	84 / 3	108 / 3	186 / 3
2.7	113.4 / 3	145.8 / 3	251.1 / 3

Table 3. Dimensional specifications of the prepared prismatic rock salt specimens and sample number (#).

H/W	W = 25 mm H (mm) / #	W = 50 mm H (mm) / #	W = 75 mm H (mm) / #	W = 100 mm H (mm) / #	W = 125 mm H (mm) / #
0.5	12.5 / 3	25 / 3	37.5 / 3	50 / 3	62.5 / 3
1.0	25 / 3	50 / 3	75 / 3	100 / 3	125 / 3
1.5	37.5 / 3	75 / 3	112.5 / 3	150 / 3	187.5 / 3
2.0	50 / 3	100 / 3	150 / 3	200 / 3	250 / 3
2.5	62.5 / 3	125 / 3	187.5 / 3	250 / 3	312.5 / 3

Table 4. UCS test results for core specimens based on the height-to-diameter (H/D) ratio.

H/D	D = 42 mm, UCS (MPa)	D = 54 mm, UCS (MPa)	D = 93 mm, UCS (MPa)
0.5	29.92 ± 2.02	36.33 ± 1.43	36.44 ± 9.22
1.0	15.26 ± 2.93	20.88 ± 2.24	25.91 ± 4.75
1.5	8.67 ± 0.00	18.12 ± 4.51	23.59 ± 2.99
2.0	12.4 ± 3.17	20.85 ± 5.26	15.4 ± 2.79
2.7	20.03 ± 4.53	14.87 ± 4.63	19.91 ± 1.84

Table 5. UCS test results for prismatic specimens based on the height-to-width (H/W) ratio.

H/W	W = 25 mm UCS (MPa)	W = 50 mm UCS (MPa)	W = 75 mm UCS (MPa)	W = 100 mm UCS (MPa)	W = 125 mm UCS (MPa)
0.5	40.32 ± 5.43	42.97 ± 8.32	40.23 ± 5.62	39.54 ± 2.54	35.54 ± 6.48
1.0	25.44 ± 10.04	31.88 ± 3.78	22.89 ± 6.11	26.17 ± 4.51	21.58 ± 1.95
1.5	14.88 ± 6.56	24 ± 1.58	19.51 ± 1.04	19.15 ± 2.85	17.76 ± 3.48
2.0	21.6 ± 1.81	17.28 ± 6.79	16.96 ± 0.23	16.15 ± 5.47	19.64 ± 2.237
2.5	6.35 ± 1.60	16 ± 0.73	21.75 ± 9.93	19.71 ± 1.94	15.31 ± 3.15

4 INVESTIGATION OF THE SCALE EFFECT

The tests were performed on core and prismatic rock specimens. A database was prepared to using the UCS test results given in Tables 4 and 5 to determine scale effect. In the first stage of this study, Figs. 1 and 2 were prepared based on database. However, the assessments made from Figs. 1 and 2 seemed to be inadequate for determining a critical value for the turning point of the curves relative to the x-axis. In order to determine the permanent UCS values independent of the scale effect, the critical turning point in x-axis must be defined. Because

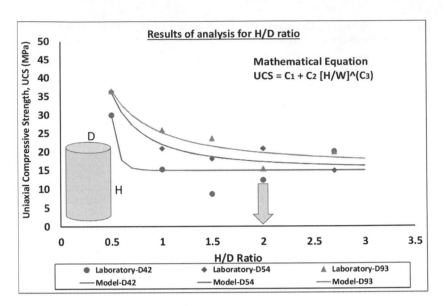

Figure 1.   Relationship between the height/diameter (H/D) and UCS values for core specimens.

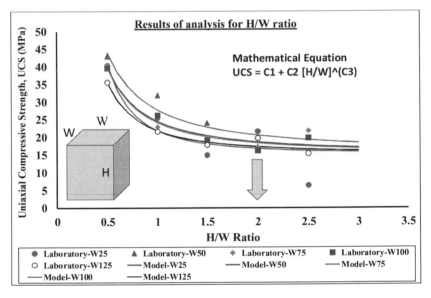

Figure 2.   Relationship between the height/width (H/W) ratios and UCS values for prismatic specimens.

this point that will be H/D or H/W ratio values required for the permanent UCS values independent of the scale effect. Therefore, the statistical software package SPSS (2004) was used to construct mathematical equations based on the test results. The nonlinear behavior for core and prismatic forms was defined by the same type of equation which is similar to model suggested by Ozkan et al. (2009).

$$\sigma_C = C_1 + C_2 \,[H/D]^{C3} \quad \text{for core specimens } (0.5 \leq H/D \leq 3.0) \tag{1}$$
$$\sigma_C = C_1 + C_2 \,[H/W]^{C3} \quad \text{for prismatic specimens } (0.5 \leq H/W \leq 3.0) \tag{2}$$

where C_1, C_2, and C_3 are statistical constants dictated by the H/D or H/W ratio and σ_C is the UCS.

Table 6. Constants C_1, C_2, and C_3 for UCS equations based on the height-to-diameter (H/D) ratio.

Sample type	Statistical constants			R²	Mathematical model
	C_1	C_2	C_3	R^2	
D42	15	0.024	−9.304	0.73	$\sigma_c = 15 + 0.024\,(H/D)^{-9.304}$
D54	15	6.993	−1.597	0.94	$\sigma_c = 15 + 6.993\,(H/D)^{-1.597}$
D93	15	10.158	−1.095	0.90	$\sigma_c = 15 + 10.158\,(H/D)^{-.095}$

Table 7. Constants C_1, C_2, and C_3 for UCS equations based on the height-to-width (H/W) ratio.

Sample type	Statistical constants			R²	Mathematical model
	C_1	C_2	C_3	R^2	
W25	15	6.764	−1.931	0.78	$\sigma_c = 15 + 6.764\,(H/W)^{-1.931}$
W50	15	12.679	−1.195	0.92	$\sigma_c = 15 + 12.679\,(H/W)^{-1.195}$
W75	15	9.095	−1.454	0.93	$\sigma_c = 15 + 9.095\,(H/W)^{-1.454}$
W100	15	9.478	−1.386	0.96	$\sigma_c = 15 + 9.478\,(H/W)^{-1.386}$
W125	15	6.802	−1.591	0.97	$\sigma_c = 15 + 6.802\,(H/W)^{-1.591}$

The constant values and regression coefficients obtained by SPSS are given in Tables 6 and 7. According to Figs. 1 and 2, the constant UCS value on the y-axis was about 15 MPa for the core and prismatic specimens. It was noted that these constant UCS values were equal to the C1 values given in Tables 6, 7 and Eqs. 1, 2. The constant C_1 can be considered to be the permanent UCS, independent of the scale effect, as shown in Eq. 3. The values for the permanent UCS do not change on the y-axis, because the curves become parallel to the x-axis at a specific point (Figs. 1 and 2).

In addition, if the H/D and H/W ratios in Eqs. 1 and 2 are infinite, sub-equations, which are second parts of equation, will be nil with the statistical constants C_2 and C_3. So, Eqs. 1 and 2 will be equal to C_1, which is the permanent UCS (σ_{Cp}). As a result, the permanent UCS for both geometric forms can be defined as follow.

$$C_1 = \sigma_{Cp} \tag{3}$$

It is seen that the changes of H/D and H/W ratios are only affected the second parts of Eqs. 1 and 2. Therefore, Eqs. 4 and 5 will represents the scale effects defined in Eqs. 1 and 2. UCS with the scale effect for core and prismatic specimens can be written as below.

$$C_2\,[H/D]^{C_3} = \sigma_{Cse} \tag{4}$$
$$C_2\,[H/W]^{C_3} = \sigma_{Cse} \tag{5}$$

Consequently, Eqs. 1 and 2 can also be generalized. In other words, total of the permanent UCS (σ_{Cp}) and the scale-affected UCS (σ_{Cse}) can be written, as shown in Eq. 6. As a results, the general equation for both geometric forms can be rewritten as follow.

$$\sigma_C = \sigma_{Cp} + \sigma_{Cse} \tag{6}$$

5 RESULTS AND DISCUSSION

In this study, a power-type equations (Eqs. 1 and 2) based on the scale effects for both core and prismatic rock salt samples were derived by the graphical evaluations and statistical analysis. After a critical H/D ratio of ≥ 2.0 for core samples and H/W ratio of ≥ 2.0 for prismatic samples (Figs. 1 and 2), the permanent UCS is achieved.

The design engineers may want to predict the strength of different pillar dimensions for underground rock salt mines utilized room-and-pillar mining method. In order to predict permanent pillar strength for the selected pillar dimensions, the engineers use widely the empirical equations, which are based on laboratory UCS test results. Therefore, the standardized calculation of a correction factor (K) for their test results UCS values between different sizes is required. As a result, a correction factor (K) is proposed for both geometric forms of Kars-Kagızman rock salt, as given below.

$$K = 1 + \sigma_{Cse} / \sigma_{Cp} \qquad \text{for general equation} \qquad (7)$$
$$K = 1 + (C_2 [H/D]^{C3} / C_1) \quad \text{for core specimens} \qquad (8)$$
$$K = 1 + (C_2 [H/W]^{C3} / C_1) \quad \text{for prismatic specimens} \qquad (9)$$

To use in Kars-Kagızman rock salt region, the approaches given in Eqs. 8 and 9 can be generalized by specimens with $D = 93$ mm and $W = 100$ mm. It is considered that this choice is fit for ISRM (1978, 2007) suggestion. Because, It is seen that the selected dimensions for both geometric forms are greater than 10 times of grain size ($d = 7$ mm). In conclusion, the correction factor (K) for core and prismatic geometric forms to use in pillar design studies are given in Eqs. 10 and 11. So the engineers can be calculated the permanent UCS value (σ_{Cp}) using Eq. 12.

$$K = 1 + 0.677 (H/D)^{-1.095} \quad \text{geometric shape: cylinder} \qquad (10)$$
$$K = 1 + 0.632 (H/W)^{-1.386} \quad \text{geometric shape: prismatic} \qquad (11)$$
$$\sigma_{Cp} = \sigma_C / K \qquad \qquad \text{for both geometric forms} \qquad (12)$$

In addition, the average conversion factor (CF_{av}) to convert the strength of a core to a cube are presented. As it is known, UCS tests are widely carried out on core and cube specimens prepared in laboratory. Therefore, Eq. 13 was suggested under condition presented below (Tables 4 and 5).

$$CF_{av} = [\sigma_{C\,core}] / [\sigma_{C\,cube}] \approx 0.71 \text{ for } D \geq 50 \text{ mm and } H/D = 2, W \geq 50 \text{ mm and } H/W = 1 \quad (13)$$

6 CONCLUSIONS

The results of this study confirm that the scale of the rock specimen significantly affects the UCS value. The following conclusions are obtained from this study: (i) a mathematical model was developed that the permanent and scale-affected values for UCS for core and prismatic specimens can be calculated separately, (ii) the permanent UCS value is determined as 15 MPa for both core and prismatic specimens, which is under condition be 2.0 of the critical H/D and H/W ratios, (iii) to estimate their permanent UCS values, the correction factors (K) for core and prismatic specimens were developed for the Kars-Kagızman rock salt deposit, (iv) in addition, to turn the UCS results determined for cube specimens (H/W = 1) to the strength of core specimens (H/D = 2), an average conversion factor (CF_{av}) of 0.71 was defined, i.e. shape factor.

ACKNOWLEDGMENTS

This study was supported by Crystal Salt Mine Enterprise in Kars-Kagızman. The authors are also indebted to the reviewers for their valuable comments.

REFERENCES

Aksoy, C.O., Onargan, T., Yenice, H., Kucuk, K., Kose, H., 2006, Determining the stress and convergence at Beypazari trona field by three dimensional elastic-plastic finite element studies:

A case study, *International Journal Rock Mechanics Mining Sciences*, Vol. 43, No. 2, pp. 166–178.

American Society of Testing and Materials (ASTM), 1986, Standard Test Method for Unconfined Compressive Strength of Intact Rock Core Specimens, Report D 2938, pp. 390–391.

Anonymous, 2004, Guides on Flooding Prevention in Mines and Undermined Objects Protection in the Upper Kama Potash Deposits [in Russian], Saint Petersburg, Russia.

Bieniawski, Z.T., 1984, Rock Mechanics Design in Mining and Tunneling, A.A. Balkema, Rotterdam, The Netherlands, 272 p.

Hawkins, A.B., 1998, Aspects of rock strength, *Bulletin Eng. Geology Environment*, Vol. 57, No.1, pp. 17–30.

Hoek, E. and Brown, E.T., 1980, Underground Excavations in Rock, *The Institution of Mining and Metallurgy*, Chapman & Hall, London, United Kingdom, 527 p.

International Society for Rock Mechanics (ISRM), 1978, Suggested methods for determining the uniaxial compressive strength and deformability of rock materials, *Commission on Standardization of Laboratory and Field Tests: International Journal Rock Mechanics Mining Sciences Geomechanics Abstracts*, Vol. 16, pp. 135–140.

International Society for Rock Mechanics (ISRM), 2007, The Complete ISRM Suggested Methods for Rock Characterization, Testing and Monitoring: 1974–2006, Ulusay, R. and Hudson, J.A. (Editors), ISRM Turkish National Group, Ankara, Turkey, 628 p.

Jeremic, M.L., 1994, Rock Mechanics in Salt Mining: A.A. Balkema, Rotterdam, The Netherlands, 532 p.

Liang, W., Yang, C., Zhao, Y., Dusseault, M.B., and Liub, J., 2007, Experimental investigation of mechanical properties of bedded salt rock, *International Journal Rock Mechanics Mining Sciences*, Vol. 44, No. 3, pp. 400–411.

Lux, K.H., 1984, Gebirgsmechanischer Entwurf und Felderfahrung im Salzkavernenbau, *Enke Verlag*, Stuttgart, Germany, 360 p.

Ozkan, I., Ozarslan, A., Genis, M., Ozsen, H., 2009, Assessment of Scale Effects on Uniaxial Compressive Strength in Rock Salt, *Environmental & Engineering Geoscience*, Vol.XV, No.2, pp. 91–100

SPSS, INC., 2004, Statistical Package for the Social Sciences: SPSS, Inc., Chicago, IL.

Thuro, K., Plinninger, R.J., Zah, S., and Schutz, S., 2001, Scale effect in rock strength properties. Part 1: Unconfined compressive test and Brazilian test, *Proceedings of the ISRM Regional Symposium EUROCK 2001*, Espoo, Finland, pp. 169–174.

Zhigalkin, V.M., Usoltseva, O.M., Semenov, V.N., Tsoi, P.A., Asanov, V.A., Baryakh, A.A., Pankov, I.L., and Toksarov, V.N., 2005, Deformation of quasi-plastic salt rocks under different conditions of loading. Report I: Deformation of salt rocks under uniaxial compression, *Journal Mining Science*, Vol. 41, No. 6, pp. 507–515.

Geomechanics and Geodynamics of Rock Masses – Litvinenko (Ed.)
© 2018 Taylor & Francis Group, London, ISBN 978-1-138-61645-5

Practical aspects of boundary condition selection on direct shear laboratory tests

Timothy R.M. Packulak
Department of Geological Sciences and Geological Engineering, Queen's University, Kingston, Ontario, Canada

Jennifer J. Day
Department of Earth Sciences, University of New Brunswick, Fredericton, Canada

Mark S. Diederichs
Geological Sciences and Geological Engineering, Queen's University, Kingston, Ontario, Canada

ABSTRACT: While laboratory direct shear testing of rock fractures in constant normal stress (CNL*) conditions are well established in practice, testing in Constant Normal Stiffness (CNS) boundary conditions is less common and few recommendations on CNS testing are available regarding appropriate laboratory testing conditions. This paper provides practical guidelines needed to design an effective suite of CNS test conditions on rock joints and other fractures that can be applied to a variety of rock types based on intact elastic properties. The results of 42 direct shear tests on NQ and NQ3 size granitic drill core under both CNS and CNL* boundary conditions are analyzed and discussed in this study. Boundary conditions were found to have minimal effect on pre-yield sample deformation properties including normal stiffness and shear stiffness, but have significant influence on peak shear strength and post-peak residual shear strength and dilation. Finally, sources of error that may arise in direct shear testing that are relevant to boundary conditions are discussed and potential solutions are presented.

1 INTRODUCTION

Rockmass discontinuities govern rockmass behaviour around rock engineering projects such as natural slopes, surface excavations, and underground openings such as tunnels or caverns. Geotechnical laboratory tests such as direct shear provide a means to determine stiffness, strength, and post-peak material properties. Direct shear tests are typically conducted under constant normal stress conditions, which are considered to represent fracture behaviour in slopes and other near surface gravity-driven environments, or constant normal stiffness conditions, which are considered to better reflect behaviours of rock fractures near underground excavations, such as tunnels, mines, and nuclear waste repositories, where in situ and induced stresses dominate (Indraratna et al., 1999). This paper presents practical guidelines needed to design an effective suite of constant normal stiffness test conditions that can be applied to a variety of rock types based on intact elastic properties.

1.1 *Boundary conditions*

The shear strength of a discontinuity has the largest influence on the potential failure of the discontinuity and can be determined through direct shear testing. Fracture shear behaviour is dependent on the level of confinement stress ranging from no external stiffness (constant normal stress) to infinite external stiffness (constant normal stiffness). Both of these real world conditions can be simulated in the lab, as shown in Figure 1. These real world conditions

361

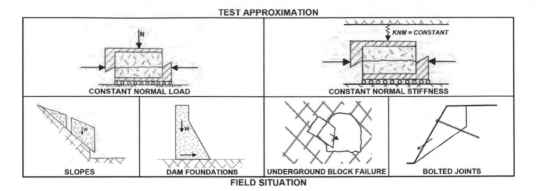

Figure 1.   Examples of joint boundary conditions in situ and their lab equivalent.

include sliding failure on slopes, sliding under dams, and above ground and underground excavations (tunnels, nuclear repositories, and mines).

It is important to note the constant normal stress boundary condition used in this study differs from the conventionally considered constant normal load, where constant normal stress is achieved with a servo-controlled direct shear machine that adjusts the applied normal load as the cross-sectional sample area under load reduces during shear displacement. In a conventional constant normal load test, the load is maintained while the effective sample area decreases. Both conditions, however, allow the fractures to dilate freely during the test so there is no feedback between dilation during shear and the normal stress (Indraratna et al., 1999), which simulates the shearing behaviour of sliding under dams and slips on rock slopes (Figure 1). For this study, the abbreviation CNL* refers to the constant normal stress boundary condition.

The constant normal stiffness (CNS) boundary condition applies a constant stiffness, instead of a constant stress, to the specimen during the shear stage of the test. Constant stiffness restricts the dilation of joints during shearing by increasing the normal stress during dilation to maintain the preset machine normal stiffness (KNM) input. Since joint shear strength is dependent on applied normal stress, the increase in normal stress increases the shear strength response. The CNS boundary condition is intended to simulate conditions observed in underground environments where rock blocks are confined and stable due to the confinement applied by the surrounding rockmass and in situ ground stresses (Figure 1).

2   LABORATORY TESTING

2.1   *Creating a direct shear testing program*

Prior to the development and use of servo-controlled direct shear machines, the constant normal stiffness boundary condition was applied by springs with a known stiffness. The use of springs limited the testing under constant normal stiffness to springs that could be produced and procured. Servo-controlled direct shear machines allow for greater freedom in choosing a stiffness condition that is more representative of the underground environment of interest that can be described by the global normal stiffness. The first developments of equations to numerically estimate the global normal stiffness of a rockmass are attributed to work by Johnston and Lam (1989) on the dilation and shear behaviour exhibited by a concrete pile. However, while these equations are applicable to concrete piles near ground surface, they do not fully capture the behaviour of planar rockmass fractures in underground environments. Additional work by Skinas et al. (1990) and Thirukumaran and Indraratna (2016) informed the development of the equation for machine normal stiffness presented in Figure 2. Based on the consideration of a rockmass system of intact rock (defined by E and v) cut by infinitely repeating parallel joints with a spacing of L that simultaneously dilate under identical

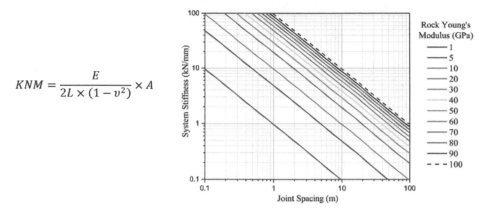

$$KNM = \frac{E}{2L \times (1 - v^2)} \times A$$

Figure 2. (Left) Direct shear machine normal stiffness (KNM) equation, A = sample area, L = rockmass joint spacing; (Right) KNM (for a 48 mm diameter sample, Poisson's ratio = 0.25) vs. in situ joint spacing.

boundary conditions, the relationship in Figure 2 estimates the machine normal stiffness input parameter (KNM), which is the simulated stiffness of the machine for CNS tests performed with springs; this would be the spring constant.

Once the machine stiffness has been determined based on sample geometry and material properties, an appropriate range of confining stresses for the testing program is needed. A test program should contain at least three different applied normal stress conditions. According to the ISRM guidelines (Muralha et al., 2013), between three and five specimens should be tested at each boundary condition on fractures from the same joint or test horizon. This is only feasible for oriented drill core; for unoriented core where it is difficult to identify joint sets based on their orientations, test specimens should have similar surface characteristics such as roughness, alteration and mineral coating.

2.2 Laboratory testing program

Forty-two direct shear tests were conducted on hard rock samples with pre-existing fractures in the Queen's University Advanced Geomechanics Testing Laboratory in Kingston, Canada, under the boundary conditions listed in Table 1. Three boundary conditions were tested, CNL* (equivalent to CNS with a stiffness of 0 kN/mm), CNS with a stiffness of 6 kN/mm, and CNS with a stiffness of 12 kN/mm. All tests were conducted using a GCTS RDS-200 Servo-Controlled Rock Direct Shear System. The machine features electro-hydraulic closed-loop digital servo control of the shear and normal loads for test automation, where the normal actuator has a load capacity of 50 kN and the shear actuator has a capacity of 100 kN. Tests were completed using the ISRM suggested method (Muralha et al., 2013) with a shear velocity of 0.2 mm/min for a shear displacement of 10 mm. Prior to shearing, the samples were subjected to a three cycled normal loading-unloading stages at a rate of 10 kPa/second (zero shear loading). Upon reaching the preset applied normal stress on the third loading cycle, the machine began the shearing stage, where in CNS conditions the normal stress increased as needed as the discontinuity dilated.

These tests were conducted on a combination of machine breaks and natural fractures through NQ (47.6 mm diameter) and NQ3 (45 mm diameter) size drill core of gneissic tonalite, pink granite, and leucogranite units of the Canadian Winnipeg River Complex within the Pointe Du Bois Batholith. Machine breaks and joints were distinguished by the characteristics of the fracture; natural joints typically had a trace mineral coating of calcite or iron oxide while machine breaks had rough, irregular profiles and fresh fracture faces. The tonalites and granites are equigranular, medium to coarse grained and are unfoliated to weakly foliated. The joints are smooth to semi-rough, planar to sub-planar, and are generally fresh with trace mineral coating. Laboratory unconfined compressive strength (UCS) and indirect Brazilian

Table 1. Boundary conditions tested and number of tests run per boundary condition.

Initial normal stress, σ_{n0} [MPa]	Machine normal stiffness, KNM [kN/mm] (# of Tests)
1	0 (4), 6 (3), 12 (3)
2	0 (4), 6 (3), 12 (3)
4	0 (3), 6 (3), 12 (3)
6	0 (2)
8	0 (5), 6 (3), 12 (3)

Table 2. Intact rock properties determined by UCS and Brazilian tensile laboratory testing.

Rock type	Tonalite gneiss	Pink granite	Leucogranite
Number of tests	4	5	3
Young's Modulus [GPa]	83.2–84.6	73.9–79.6	78.6–81.1
Poisson's Ratio	0.25–0.27	0.26–0.32	0.22–0.23
UCS [MPa]	154–275	241–349	262–319
Tensile Strength [MPa]	8.6–14.6	14.8–18.0	14.4–15.2

tensile strength tests of intact rock samples were used to measure the UCS, tensile strength, Young's Modulus, and Poisson's Ratio of the intact rock (Table 2).

3 TEST RESULTS AND ANALYSIS

3.1 Pre-peak deformation behaviour: Joint normal stiffness

To estimate joint normal stiffness (K_n), all test results from the normal loading stage (zero shear load) were analyzed using both linear (Hungr and Coates, 1978) and semi-logarithmic (Zangerl et al., 2008) closure relationships between normal displacement and normal stress. A compilation of normal stiffness results for all the Pointe Du Bois Batholith samples range from 4,000 to 20,000 MPa/m when analyzed using a linear relationship, and ranged from 3,000 to 15,500 MPa/m when analyzed using a semi-logarithmic relationship (Figure 3). The normal stiffness does not appear to be dependent on normal stress, however testing at lower normal stresses is recommended to confirm this behaviour.

3.2 Pre-peak deformation behaviour: Joint shear stiffness

To determine joint shear stiffness (K_s), the slope of the shear stress versus shear displacement curve is determined between the start of shearing and the peak shear strength. Values presented in Figure 3 were determined using guidelines proposed by Day et al. (2017) for the determination of secant (K_{SS}, zero to peak) and tangent (K_{ST}, at 50% of peak shear strength). Joint shear stiffness is dependent on normal stress, as shown in Figure 3, although different boundary conditions (CNL* versus CNS) do not appear to influence shear stiffness. In the Pointe Du Bois hard rock material, the tangent shear stiffness measurements are higher than the secant shear stiffness.

3.4 Peak and post-peak behaviour: Shear strength

Fractures tested under CNL* conditions show two distinct failure envelopes: yield strength and residual strength. Under CNS conditions, there is a third strength envelope between yield and residual, which is referred to as maximum strength. The shear strength results (Figure 4)

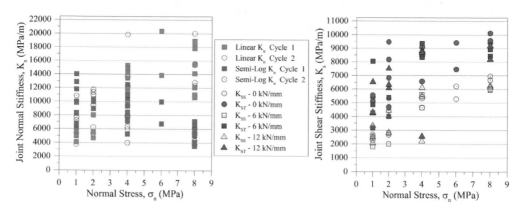

Figure 3. (Left) Joint normal stiffness (K_n) for loading cycle 1 and cycle 2; (Right) Secant (K_{SS}) and tangent (K_{ST}) shear stiffness results for CNL* and CNS boundary conditions.

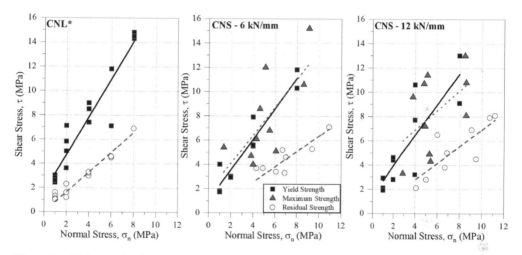

Figure 4. Mohr-Coulomb strength envelopes for all boundary conditions (properties are in Table 3).

Table 3. Summary of Mohr-Coulomb shear strength parameters for all boundary conditions.

Boundary condition	Number of tests	Yield Shear Strength		Maximum Shear Strength		Residual Shear Strength	
		φ_y [°]	c [MPa]	φ_m [°]	c [MPa]	φ_r [°]	c [MPa]
CNL*	16	57.3 ± 7.7	1.6 ± 0.6	N/A	N/A	36.4 ± 2.7	0.4 ± 0.2
CNS-6	12	51.6 ± 7.6	1.0 ± 0.6	48.3 ± 14.1	4.3 ± 1.2	31.0 ± 7.7	3.1 ± 0.7
CNS-12	12	51.5 ± 14.5	1.4 ± 1.1	38.2 ± 17.0	5.7 ± 1.3	17.9 ± 13.5	4.8 ± 1.2

are sorted by boundary condition (CNL*, CNS – 6 kN/mm, and CNS – 12 kN/mm) and plotted with linear best fit Mohr-Coulomb strength envelopes (Mohr-Coulomb properties are listed in Table 3). Residual shear strength data points for tests with post-yield interference are not included in these best fit calculations.

3.5 Post peak behaviour. Dilation

The dilation angle is measured using a secant dilation angle. An example measurement of the secant dilation angle is shown in Figure 5 (left). First, the shear strength yield point and

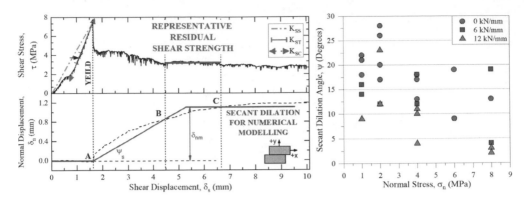

Figure 5. (left) Example shear stress (τ) and normal displacement (δ_n) results with a secant dilation angle (ψ) measurement; (right) compilation of secant dilation angle measurements.

a zone of shear displacement that is representative of residual shear strength are determined. The secant dilation angle (ψ) is the angle from horizontal that is formed from drawing a line from the yield point on the normal displacement curve to the start of residual shear strength zone on the normal displacement curve. Specimens that experienced grout interference during the post peak portion of the direct shear stage have been filtered out of the data set. The secant dilation angle decreases with increasing normal stress and is further dampened under CNS conditions (Figure 5 right).

3.6 *Potential sources of error*

Prior to any shear displacement, the applied normal and shear loads are assumed to be uniform across the specimen. However, once the specimen has become unmated after shear displacement, an overturning moment develops as the normal load moves away from the centre of the sample. Capturing the effect of the moment arm on the test through additional LVDTs may improve the understanding of dilation and sample behaviour. The characterization of sample rotation, if present, may also explain certain phenomenon during shearing such as tensile fracturing on the edge of the sample in the direction of shear. This failure decreases the effective fracture contact area in the sample, which is not accounted for in stress calculations by the machine. This may result in the machine outputting a lower stress measurement then what is physically occurring on the sample. By capturing sample rotation and tensile fracturing, post-peak data can be corrected, giving a greater understanding of post-peak shearing behaviour.

4 CONCLUSIONS

The equation for machine normal stiffness (KNM) proposed in this study provides a means to target appropriate CNS boundary conditions in a direct shear test program, which better reflects rockmass behaviour around an underground excavation. The proposed secant dilation angle provides an improved approach for numerical modelling applications. Pre-yield parameters such as joint normal stiffness and shear stiffness are independent of the test boundary condition, as the CNS component does not take full effect until yield shear strength. Post-peak shear strength and dilation are dependent on boundary condition. As confinement increases, the dilation angle is reduced in CNL* and even more in CNS tests.

ACKNOWLEDGEMENTS

The Natural Sciences and Engineering Research Council of Canada, the Nuclear Waste Management Organization of Canada and the Tunnelling Association of Canada have financially supported this research. Manitoba Hydro and KGS Group Consulting Engineers supplied the rock core samples.

REFERENCES

Day, J.J., Diederichs, M.S. and Hutchinson, D.J. 2017. New direct shear testing protocols and analyses for fractures and healed intrablock rockmass discontinuities. *Engineering Geology*, 229: 53–72.

Hungr, O. and Coates, D.F. 1978. Deformability of joints and its relation to rock foundation settlements. *Canadian Geotechnical Journal*, 15(2): 239–249.

Indraratna, B., Haque, A. and Aziz, N. 1999. Shear behaviour of idealized infilled joints under constant normal stiffness. *Geotechnique*, 49(3): 331–355.

Johnston, I.W. and Lam, T.S. 1989. Shear behaviour of regular triangular concrete/rock joints – analysis. *Journal of Geotechnical Engineering Division, ASCE*, 115(5), 711–727.

Muralha, J., Grasselli, G., Tatone, B., Blumel, M., Chryssanthakis, P., and Yujing, J. 2013. ISRM suggested method for laboratory determination of the shear strength of rock joints: Revised version. *Rock Mechanics and Rock Engineering*, 47: 291–302.

Skinas, C.A., Bandis, S.C. and Demiris, C.A. 1990. Experimental investigations and modelling of rock joint behaviour under constant stiffness. In *Rock Joints*, Barton & Stephansson (eds), Loen: 301–308.

Thirukumaran, S. and Indraratna, B. 2016. A review of shear strength models for rock joints subjected to constant normal stiffness. *J. Rock Mech. Geotech. Engrg.*, 8(3): 405–414.

Zangerl, C., Evans, K.F., Eberhardt, E. and Loew, S. 2008. Normal stiffness of fractures in granitic rock: a compilation of laboratory and in situ experiments. *Int. J. Rock Mech. Min. Sci.*, 45(8): 1500–1507.

Geomechanics and Geodynamics of Rock Masses – Litvinenko (Ed.)
© *2018 Taylor & Francis Group, London, ISBN 978-1-138-61645-5*

Experimental studies of saliferous rock direct tension

I. Pankov & V. Asanov
Mining Institute of the Ural Branch of the Russian Academy of Sciences, Moscow, Russia

V. Kuzminykh & I. Morozov
Perm National Research Polytechnic University, Russia

ABSTRACT: The main direct tensile problems of rock specimens have been analyzed at the laboratory conditions. Based on the laboratory and numerical experiments the geometry of saliferous specimen has been suggested considering the stress state mode arising under loading process, grain size, bracing influence. Rectangular samples sized $250 \times 50 \times 50$ mm were used for this study. Before testing the sample has been cemented in metal clips to be used for clamping in the reversing device or in the grippers of the testing machine. The cementation of the samples is carried out using the magnesia cement mortar based on saturated carnallite brine. The experiments were carried out using the electromechanical testing machine Zwick/050. A constant traversing speed of 0.1 mm/min was maintained during the test. The measurement of longitudinal deformations was carried out on the surface of the sample with the help of three remote LVDT sensors with a measurement accuracy of 0.0005 mm arranged in an equilateral triangle. The "sensor-sample" contact was carried out with the help of plates fixed on the sample surface. The experiment continued until the sample was completely failure, i.e. when it was divided into two pieces. The obtained results were presented in the form of stress-strain diagrams, which were used to determine strength and strain parameters. Variation intervals of strength, failure strain, and deformation modulus and energy capacity of fracture under extension are established for the samples with different layering orientations.

The obtained results are intended to determine the mechanical salt rock characteristics of Verchnekamskoe potassium salts deposit under direct tensile loading.

Keywords: tensile strength, failure strain, deformation modulus, specific deformation energy, saliferous rocks, direct tensile, numerical modeling

1 INTRODUCTION

Development of the Verkhnekamskoye potassium salts deposit depends on difficult mining and geological occurrence conditions of the mineral. Underground water-bearing horizons lying above the saliferous layers are the reason for the biggest accidents at the deposit. As the result, two mines were flooded in 1986 and 2006. Yet water started to penetrate into another mine in 2014. The water inflow continues until the present day. The reason for water inflow into the mine openings is vertical fissures in water-blocking layers (rock mass) situated between the openings and water-bearing horizons. The fissures are the result of tensile deformation of the saliferous rock mass. In that regard, the studies of tension deformation process [1] arouse special interest. Experimental identification of tensile strength and deformation criteria of saliferous rock tensile fracture is one of the geomechanical problems oriented to mining works safety assessment.

It is known that tensile strength of rocks is weaker than their compressive strength. Ratio of the values changes from 2.7 to 50 [2, 3, 4]. There are direct and indirect methods of tensile strength tests. The indirect methods are: fissuring by compressive loads; bending; breaking by inside pressure [5].

The most common method of the rock tensile strength test is the method where cylindrical samples are fractured along the generatrix using compressive loads. This method is internationally known as the Brazilian method [6]. The sample is fractured by tensile stress generated in its central part. The distribution pattern of the tensile stress is highly variable and is determined by the way of compressive load transmission to the sample [7, 8]. Therefore, the resulting value of indirect tensile strength for one and the same rock may vary over a wide range [7]. Researchers suggest that indirect tensile strength tested by the Brazilian method depends not only on the way of load transmission, but also on the transmission speed, structure features of the rock, temperature and shape of the sample [9–15].

The most common bending methods are the methods of three-point or four-point bending of the beam-shaped samples with rectangular or round cross sections. The beam bending methods are considered in details in the following papers [15–18]. Many researchers state that the values of breaking tensile stress in the three-point bend are higher than in the four-point bending. Experimental data show that it is not always true [15]. Bending fracture behavior is determined not only by the rock properties, but also by the size of the sample as well as by the way and speed of load transmission [15–18]. The methods of rock fracture by inside pressure are not covered by the paper because of the specifics of the experiment and limited application of these methods.

Despite a relatively simple implementation of the considered indirect methods, they have series of major deficiencies. The indirect methods allow to perform only a rough estimate of tensile strength. Results of different indirect methods cannot be compared both to each other and to the results of direct tests [4, 7, 13–16, 19–22]. The main objective of the considered indirect methods is the evaluation of tensile strength. Tensile stress does not allow to obtain other mechanical characteristics (energy capacity properties, deformation properties, stiffness properties). Often the application requires a comparison to direct tension data. Broadly speaking, this fact casts doubt on the practicability of indirect researches of the matter.

Direct tension tests are standard reference research methods designed for identification of strength and tensile deformation properties. Current recommended practices for this type of tests, described in [2, 5, 6], are quite general and give no distinct methods to solve practical issues arising during laboratory studies of a certain rock. In general, the available studies of direct tension, including [7, 23], are focused on investigating solid and brittle rocks. In addition, there are almost no data on direct tension of saliferous rock.

2 TESTING EQUIPMENT AND SAMPLES UNDER INVESTIGATION

The direct stress test was performed using the multipurpose electromechanical test machine Zwick Z050 with a maximum axial force of tension/compression of 50 kN (Fig. 1). Precision of axial force measurement is ± 5 N, of mobile traverse movement it is − ± 5 μm. The machine allows to adjust load transmission speed in the range from 0.001 mm/min to 100 mm/min. The machine is equipped with three console sensors with a precision of ± 0.5 μm located in its working (middle) part to measure longitudinal (axial) deformation of the sample. Steel casings with the samples were fixed by the wedge clamps (Fig. 2).

Usually a direct tension test of artificial materials is performed using samples with a narrow test section (dumbbell shape). It is very difficult to create a sample of such a complex shape from brittle saliferous rock, so the shape of the sample should be as simple as possible. One should take into account that the ratio of the cross-section area of the sample to the cross-section area of the grain should be no less than 15 [24] to decrease the influence of a grain size on the values of the mechanical properties. A sample's height should provide both a sufficient size of the test section and a sufficient area for a firm fixation of the sample in the casings. The height to width ratio of the sample should be as high as possible to minimize the influence of buttend fixation on stressed and deformed condition of the test section. At the same time, undue elongation of the saliferous rock samples increases the possibility of their breakdown at the stage of manufacturing. With that in mind, it was decided to manufacture samples of a prismatic shape with the dimensions of 250 × 50 × 50 mm.

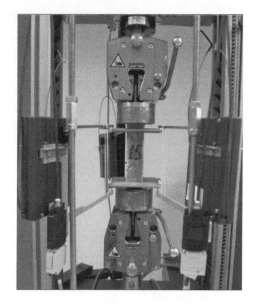

Figure 1. Electromechanical test machine Zwick Z050.

Figure 2. Direct tension test of saliferous rock samples.

The samples were manufactured from sylvinite monoliths of an even-grained and stratified structure taken from the productive formations of the Verkhnekamskoye potassium salts deposit. The average size of the sylvinite grains was less than 3–5 mm. The length of the sample test section was 110 mm. The length of the fixation area was 40 mm, and the transition area was 30 mm. Direct fixation of the sample by the clamps of the test machine is impossible because of a low contact strength of saliferous rock. Transmission of tensile stress to the sample was performed using special steel casings—heavy wall tubes with the inside diameter and height sufficient for a firm fixation of the sample. The lower part of the casing is designed to be equipped with the steel lead with a cylindrical shank to fix the casing in the wedge clamps of the stress machine and transfer tensile stress to the sample. The magnesian phosphate cement was used to fix the sample in the casings. The cementing agent is applied for cementation of the oil wells drilled through saliferous horizons of the Verkhnekamskoye potash-magnesium deposit and exhibit high adhesive and strength properties. The sample was cemented in the steel casings clamped in the test machine to prevent the generation of the bending moment resulting from the coaxial misalignment of the steel casings.

The dynamic modulus of elasticity and Poisson's ratio were defined before the test for each of the samples according to the method described in [25]. The samples with a markedly different elastic behavior were rejected. The samples with visually defined structure features different from the total sample, defects, and damages were rejected as well. Load transmission was performed with the constant deformation speed of $1.7{\cdot}10^{-5}$ s^{-1} on the basis of 110 mm up to the ultimate failure. During the test process the researchers recorded the longitudinal force, traverse location, and readings of the three console sensors for longitudinal deformation measurement installed according to the equilateral triangle design.

3 RESULTS OF LABORATORY AND NUMERICAL STUDIES

24 samples were tested according to the method described in Section 1. The test results were used to construct a deformation diagram with the coordinates of "longitudinal stress (σ_1) - relative longitudinal deformation (ε_1)" for each sample. The Fig. 3 shows an example of the deformation diagram for the saliferous rock samples stressed along and across the banding with brittle and ductile fracture.

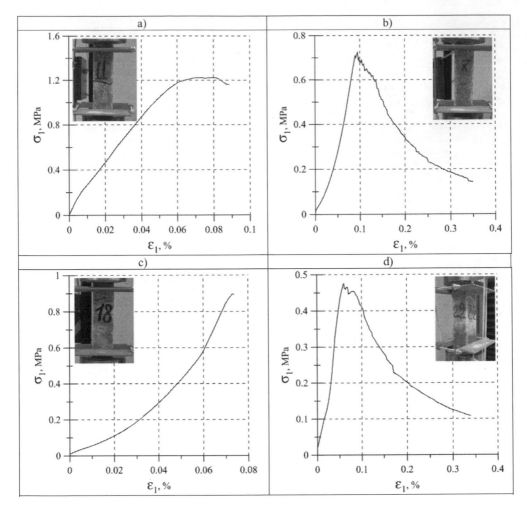

Figure 3. Tensile deformation diagrams and photos of the samples after the test: a is along the banding (brittle fracture); b is along the banding (ductile fracture); c is across the banding (brittle fracture); d is across the banding (ductile fracture).

The complex of mechanical metrics is defined (Fig. 4) on the basis of the results of the tests performed according to the method described in [26]: tensile strength (σ_{1str}); longitudinal breaking strain ($\varepsilon 1st_r$); tangent modulus of prelimiting deformation (M_1); tangent modulus of superlimiting deformation (M_2); energy intensity of deformation at the breaking point (W). The results of the experimental studies are presented in Table 1.

The analysis of the stressed and deformed condition of the sylvinite samples was performed using the finite-element method. The geometry of the steel casings and cement fixation complied with the geometry of the laboratory experiments. The problem was solved with the help of the elastic approach. The sample was modeled as a homogeneous isotropic medium. Elastic behavior of the materials used for calculations is presented in Table 2.

Numerical studies of direct tension were performed in the three dimensional symmetrical setup for the one fourth part of the full-scale model. Movements of the lower casing basis were prohibited. The upper casing basis was moved along its positive Z-axis. Fig. 5a shows the distribution of longitudinal stress for the outside edge of the sample. Fig. 5b shows the distribution of longitudinal stress for the cross-section driven though the axis of symmetry of the sample and center points of the opposite edges.

Table 1. Results of direct tension tests of sylvinite samples.

Sample No	σ_{1str}, MPa	ε_{1str}, %	M_1, hPa	M_2, hPa	W, kJ/m³
Along the banding					
1	1.06	0.09	1.36	–	0.61
2	0.63	0.11	1.36	–	0.34
3	1.15	0.17	1.27	–	0.64
4	1.27	0.07	2.33	–	0.39
5	1.58	0.15	1.35	–	1.00
6	2.09	0.15	2.00	–	1.34
7	0.73	0.10	1.20	0.28	0.29
8	0.59	0.07	1.27	0.15	0.24
9	1.52	0.15	1.46	–	1.06
11	1.22	0.08	2.01	–	0.63
12	1.05	0.07	1.77	–	0.33
13	1.44	0.11	1.93	–	0.73
16	1.37	0.16	1.72	–	0.78
Average	**1.21**	**0.11**	**1.62**	**–**	**0.64**
Coefficient of variation,%	**34.6**	**33.4**	**22.5**	**–**	**52.1**
Across the banding					
10	0.25	0.09	0.61	0.09	0.16
14	0.84	0.09	1.14	–	0.35
15	1.01	0.14	0.93	–	0.69
17	0.54	0.07	1.05	–	0.23
18	0.90	0.07	1.75	–	0.28
22	0.48	0.06	1.20	0.26	0.14
Average	**0.67**	**0.09**	**1.11**	**–**	**0.31**
Coefficient of variation,%	**34.6**	**35.8**	**28.4**	**–**	**68.5**

Table 2. Elastic behavior of materials.

Material	Modulus of elasticity, hPa	Poisson's ratio
Steel	200.0	0.3
Sylvinite	1.0	0.35
Magnesian phosphate cement	10.0	0.1

4 ANALYSIS OF RESULTS AND CONCLUSION

Laboratory studies of direct tension along and across the banding of sylvinite samples of prismatic shape showed no obvious differences in their fracture mechanism. On both accounts, the deformation of the most saliferous rock samples is close to the elastic deformation. Loss of sustaining capacity was determined by a fast propagation of the main crack in the central part of the sample perpendicularly to its axis (Fig. 3). The analysis of the experimental data (Table 1) showed that the sylvinite samples tested along the banding have better mechanical properties. In most cases, tension of saliferous samples showed brittle fracture (Fig. 3, a, c). At the same time, several cases showed ductile fracture with an apparent superlimiting area (Fig. 3, b, d) that allowed to define the tangent modulus of the superlimiting deformation (Table 1, samples 7, 8, 10, and 22). The analysis showed that the superlimiting area is present primarily due to low tensile strength properties of the saliferous rock and consequently low level of elastic energy saved in the "test machine-sample" system that defines ductile or brittle fracture of the rock. The obtained results are highly variable in values which is one of the property measurement features of granular and laminated saliferous rock under tension. When considering low mechanical properties showed by the tests across the banding, the

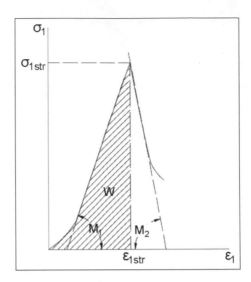

Figure 4.   Determining the mechanical metrics for tension using the deformation diagram.

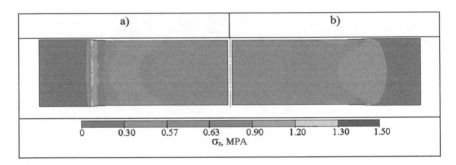

Figure 5.   Longitudinal stress array for prismatic samples.

evaluation of the mine roof stability should be given special attention, as it is the area of tension stress concentration.

The results of numerical modeling (Fig. 5) show that the stress in the central part of the prismatic samples is close to homogeneous. At the same time, the fixation area of the samples experiences a minor stress concentration. To reduce its level, it is practical to cement the area of 5–7 mm.

The research results serve to the parametric control of geomechanical assessment of mining works safety at the Verkhnekamskoye potassium salts deposit.

REFERENCES

[1] Panteleyev I.A., Plekhov O.A., Naimark O.B., Evseev A.V., Pankov I.L., Asanov V.A. Features of deformation localization under tension of sylvinite//Bulletin of Perm National Research Polytechnic University. Mechanics. 2015. – No. 2. – p. 127–138.

[2] Kartashov Yu.M., Matveev B.V., Mikheev G.V., Fadeyev A.B. Strength and deformability of rocks. Moscow: Nedra, 1979. p. 269.

[3] Pankov I.L., Garaeva Yu.I. Investigation of mechanical properties of salt rocks under tension//Mining Information and Analytical Bulletin (scientific and technical journal). 2011. – No. 9. pp. 154–157.

[4] Cai M. Practical Estimates of Tensile Strength and Hoek–Brown Strength Parameter m_i of Brittle Rocks // Rock Mechanics and Rock Engineering. 2010. – Vol. 43, № 2. – pp. 167–184. doi: 10.1007/s00603–009–0053–1.

[5] Ilnitskaya E.I., Teder R.I., Vatolin E.S., Kuntysh M.F. Properties of rocks and methods for their determination. Moscow: Nedra, 1969. p. 392.

[6] ISRM. International society for rock mechanics. Suggested methods for determining tensile strength of rock materials//Int J Rock Mech Min Sci Geomech Abstr. 1978. – Vol. 15. – pp. 99–103.

[7] Erarslan N., Williams D.J. Experimental, numerical and analytical studies on tensile strength of rocks//International Journal of Rock Mechanics & Mining Sciences. 2012. – Vol. 49. – pp. 21–30.

[8] Ye Jianhong, Wu F.Q., Sun J.Z. Estimation of the tensile elastic modulus using Brazilian disc by applying diametrically opposed concentrated loads//International Journal of Rock Mechanics & Mining Sciences. 2009. – Vol. 46. – pp. 568–576.

[9] Hui Zhou, Yanshuang Yang, Chuanqing Zhang, Dawei Hua. Experimental investigations on loading-rate dependency of compressive and tensile mechanical behaviour of hard rocks // European Journal of Environmental and Civil Engineering. 2015. – Vol. 19, No. 1. – pp. 70–82. doi: 10.1080/19648189.2015.1064621.

[10] Sopon Wisetsaen, Chaowarin Walsri, Kittitep Fuenkajorn. Effects of loading rate and temperature on tensile strength and deformation of rock salt//International Journal of Rock Mechanics & Mining Sciences. 2015. – Vol. 73. – pp. 10–14.

[11] Xueliang Xu, Shunchuan Wu, Yongtao Gao, Miaofei Xu. Effects of micro-structure and micro-parameters on brazilian tensile strength using flat-joint model//Rock Mechanics and Rock Engineering. 2016. – Vol. 49, № 9. – pp. 3575–3595. doi: 10.1007/s00603–016–1021–1.

[12] Abbass Tavallali, André Vervoort. Behaviour of layered sandstone under Brazilian test conditions: layer orientation and shape effects.//Journal of Rock Mechanics and Geotechnical Engineering. 2013. – Vol. 5, No. 5. – pp. 366–377.

[13] Rafiei Renani, H., Martin, C.D. The direct and Brazilian tensile strength of rock in the light of size effect and bimodularity//48th US Rock Mechanics/Geomechanics Symposium 2014: Rock Mechanics Across Length and Time Scales: conference paper. - USA, Minneapolis, Minnesota, 2014 – pp. 1052–1060.

[14] Tolooiyan, A., Mackay, R., Xue, J. Measurement of the Tensile Strength of Organic Soft Rock // Geotechnical Testing Journal. 2014. – Vol. 37, No. 6. – pp. 991–1001. doi: 10.1520/GTJ20140028.

[15] Efimov V.P. The rock strength in different tension conditions//Journal of Mining Science. 2009. – Vol. 45, № 6. – pp. 569–575. doi: 10.1007/s10913–009–0071–0.

[16] Efimov V.P. Rock tests in nonuniform fields of tensile stresses // Journal of Applied Mechanics and Technical Physics. 2013. – Vol. 54, № 5. – pp. 857–865. doi: 10.1134/S0021894413050192.

[17] Mehdi Galouei, Ali Fakhimi. Size effect, material ductility and shape of fracture process zone in quasi-brittle materials // Computers and Geotechnics. 2015. – Vol. 65. – pp. 126–135.

[18] Yuichi Nakamura, Sang Ho Cho, Katsuhiko Kaneko, Shoji Kajiki, Yoshio Kiritani. Dynamic fracture experiments of mortar using a high-speed loading apparatus driven by explosives // Science and Technology of Energetic Materials. 2012. – Vol. 41, № 5. – pp. 136–141.

[19] M. Cai. Practical Estimates of Tensile Strength and Hoek–Brown Strength Parameter mi of Brittle Rocks//Rock Mechanics and Rock Engineering. 2010. – Vol. 43, № 2. – pp. 167–184. doi: 10.1007/s00603–009–0053–1.

[20] Zhigalkin, V.M., Rychkov, B.A., Usol'tseva, O.M. et al. Estimation of strength properties of rock samples in terms of calculated Mohr's envelopes//Journal of Mining Science. 2011. – Vol. 47, № 6. – pp. 714–721. doi: 10.1134/S1062739147060025.

[21] Asanov V.A., Pankov I.L., Evseev V.S. Evaluation of strength and deformation properties of salt rocks under tension//Geology, geophysics and development of oil and gas fields. 2010. – No. 12. pp. 65–66.

[22] Pankov I.L., Gushchina K.S., Bogdanova A.S. Results of complex determination of strength and deformation properties of salt rocks under tension//Mining echo. 2012. – No. 3. pp. 15–18.

[23] Baoyun Zhao, Dongyan Liu, Qian Dong. Experimental research on creep behaviors of sandstone under uniaxial compressive and tensile stresses//Journal of Rock Mechanics and Geotechnical Engineering. 2011. – Vol. 3. Supp. No. 1, – pp. 438–444.

[24] Proskuryakov N.M., Permyakov R.S., Chernikov A.K. Physical and mechanical properties of salt rocks. – L.: Nedra, 1973. p. 271.

[25] Lomtadze V.D. Physical and mechanical properties of rocks. Methods of laboratory research: Textbook for high schools. – 2nd Edition, revised and supplemented – L.: Nedra, 1990. p. 328.

[26] Baryakh A.A., Asanov V.A., Pankov I.L. Physical and mechanical properties of salt rocks of the Verkhnekamskoye potassium deposit: Textbook. - Perm: PSTU Publishing House, 2008. p. 199.

Geomechanics and Geodynamics of Rock Masses – Litvinenko (Ed.)
© 2018 Taylor & Francis Group, London, ISBN 978-1-138-61645-5

Understanding tilt-test results on saw-cut planar rock surfaces from a statistical perspective

I. Pérez-Rey & L.R. Alejano
Department of Natural Resources and Environmental Engineering, University of Vigo, Spain

J. Martínez
Centro Universitario de la Defensa de Marín, CUD, Spain

M. Muñiz
Centro de Estudios y Experimentación de Obras Públicas, CEDEX, Ministerio de Fomento, Spain

J. Muralha
National Laboratory for Civil Engineering, LNEC, Portugal

ABSTRACT: Tilt tests have been used *in situ* and in laboratory as a practical technique to estimate the basic frictional component of rock discontinuities. A number of studies on tilt test results carried out by several authors revealed an unexpected variability of the results. To clear this issue, a statistical approach intended to allow a better understanding of basic friction angle results is presented. Three different parameters potentially affecting the behaviour of a sliding rock slab were considered: base slenderness, tilting rate and saw blade used to cut the slabs. Nearly three hundred tests with different combinations of these parameters were carried out on granite rock slabs. General trends observed for results were analysed by means of a descriptive statistical analysis, histograms and fitted probability distributions. One–way analyses of variance (ANOVA) were performed. Occurrence of abnormal values within datasets was also detected by a thorough assessment of outliers. From this study, it could be concluded that tilt test results may be considered as reproducible values to estimate basic friction angle of rock sliders, since identified variability was within the range of other common geotechnical parameters.

Keywords: tilt test; normality; reproducibility; friction; discontinuity

1 INTRODUCTION

The definition of the frictional behavioural component of a rock discontinuity and, particularly, its experimental determination has been been subject of discussion from the beginnings of rock mechanics science (Barton 1973; Hencher 1976) to nowadays (Alejano et al. 2017). Classic configurations proposed to obtain this frictional component correspond to direct shear tests, pull/push tests and tilt tests, not only performed on saw-cut rock surfaces but also on rock cores.

The idea of this study is to perform an experimental analysis and a statistical assessment of results based on a good number of tests, carried out for 27 different combinations of three relevant factors (length-to-thickness ratio, selected saw blade and tilting rate of the testing platform) reported to have potential influence on tilt-test results (Hencher 1976; Alejano et al. 2017).

2 MATERIALS AND METHODS

A medium-grained granite, known as *Blanco Mera*, was selected to obtain all rock specimens. This rock presents a mean UCS of about 120 MPa and an approximate elastic modulus of 36.5 GPa.

Rock specimens were obtained and tests were carried out by varying three controllable features: the length-to-thickness ratio (l/h), the tilting rate of the testing machine and the type of saw blade. The selected tilting rates were 5, 10 and 25°/min, covering an acceptable range found in bibliography (Alejano et al. 2017). The saw blades (named as A, B and C, respectively) used to prepare the specimens were a 350 mm-diameter, a bronze-alloy matrix jagged disk; a 600 mm-diameter, bronze/cobalt alloy matrix jagged disk and a 300 mm-diameter, continuous-rim disk. The selected l/h ratios were 5, 3 and 1.67.

In any experimental study, it is critical to define the of sample size number required to assure statistical robustness to the results. In this case, the minimum number of runs or tilt tests to be carried out (n) was calculated considering Equation 1.

$$n = \frac{Z_\alpha^2 \sigma}{d^2} \tag{1}$$

where Z_α is the coefficient of the assigned level of confidence α (for this case, $\alpha = 0.95$ and, accordingly, $Z_\alpha = 1.96$); σ is the standard deviation considered as 2.2° from previous studies and d is the desired/imposed precision of the experiment (in this case, $d = 0.2°$ was selected). By equating previous figures, a number of experiments n = 212 should be performed. The present study considered 27 possible combinations of the three selected variables and 11 tilt tests were carried out for each series, thus generating a total amount of 297 results and fulfilling the minimum number of tests condition.

Results of all tilt tests are presented in Table 1. First letter of series' titles corresponds to the type of saw blade (A, B or C); first number is the used l/h ratio of the slab – 5 (l/h = 5), 3 (l/h = 3) and 17 (l/h = 1.67) – and last number corresponds to the tilting rate in °/min.

Table 1. Raw values from all tilt tests (in degrees).

Series	ϕ_{b1}	ϕ_{b2}	ϕ_{b3}	ϕ_{b4}	ϕ_{b5}	ϕ_{b6}	ϕ_{b7}	ϕ_{b8}	ϕ_{b9}	ϕ_{b10}	ϕ_{b11}
a_5_5	33.2	30.4	32.4	31.3	31.6	32.4	32.8	31.6	31.0	30.7	30.9
a_5_10	35.7	34.4	34.3	32.9	32.3	32.0	32.0	32.5	32.6	31.9	31.8
a_5_25	29.9	27.9	28.2	28.4	28.0	28.2	27.5	29.2	26.9	29.1	28.6
a_3_5	28.5	27.0	27.0	27.3	25.4	27.1	26.2	26.9	26.2	26.8	27.8
a_3_10	29.3	29.9	28.4	29.0	27.6	28.1	29.8	27.6	28.7	28.0	28.2
a_3_25	29.7	28.4	28.5	28.5	27.5	29.0	28.3	29.9	28.9	29.7	29.1
a_17_5	32.2	32.1	31.4	31.8	31.0	30.5	31.1	30.4	30.4	30.5	31.8
a_17_10	35.2	27.1	33.1	29.9	27.6	25.9	28.4	29.3	26.7	25.9	26.3
a_17_25	32.5	30.7	31.6	31.0	31.4	31.6	30.4	30.3	29.6	30.3	30.6
b_5_5	30.4	28.2	27.5	28.6	30.8	27.9	30.0	27.9	27.9	27.5	28.6
b_5_10	29.4	25.6	27.0	27.4	26.3	25.0	22.5	24.5	25.5	23.5	25.5
b_5_25	36.0	30.8	31.3	30.7	29.4	30.0	30.4	30.0	30.7	32.0	29.8
b_3_5	30.1	27.2	28.6	29.1	27.9	27.9	28.5	28.0	28.0	32.4	28.2
b_3_10	30.6	28.2	28.3	28.4	28.9	26.7	28.2	28.4	27.8	27.9	29.9
b_3_25	32.5	32.2	29.8	28.8	29.7	28.7	28.8	29.1	27.7	28.9	28.5
b_17_5	30.9	27.7	25.7	26.8	28.3	27.7	27.0	26.6	27.2	28.3	27.4
b_17_10	24.0	24.2	26.3	23.7	25.2	27.3	24.2	23.5	23.8	24.8	25.3
b_17_25	31.1	31.5	30.6	28.7	29.0	28.5	29.6	28.3	29.5	27.9	29.2
c_5_5	32.0	31.2	29.4	29.9	30.0	28.9	29.9	29.8	29.1	28.3	29.1
c_5_10	32.5	30.4	30.3	30.0	29.5	31.1	30.1	29.2	29.8	33.1	28.9
c_5_25	32.1	32.4	31.1	31.1	30.5	32.1	30.7	31.8	30.5	30.6	31.1
c_3_5	28.9	27.9	27.2	27.8	27.8	27.1	27.2	27.1	27.0	27.2	26.9
c_3_10	30.6	26.4	27.3	28.2	28.0	28.7	27.9	28.3	27.7	28.9	29.3
c_3_25	32.2	31.2	31.4	30.8	31.3	31.2	31.0	29.9	30.1	29.8	29.3
c_17_5	32.4	29.2	29.7	30.1	29.9	28.9	28.3	27.7	29.1	27.3	27.5
c_17_10	31.7	30.6	30.6	30.7	28.9	29.2	29.7	28.3	28.3	29.5	29.8
c_17_25	33.5	31.2	32.1	32.3	29.5	28.5	28.3	27.9	29.1	28.0	28.6

3 STATISTICAL ASSESSMENT OF RESULTS

Results obtained in this experimental study were analysed by means of different statistical tests and techniques described in the following sections.

3.1 *Descriptive analysis*

Some statistical descriptors have been studied for the complete dataset (297 results). The kurtosis coefficient $\gamma_2 = 0.40$ indicates that data distribution can be considered quite close to a normally distributed one, supported by a skewness $\gamma_1 = -0.05$, which is almost null. Range value becomes logically greater than that observed for each individual series, since this dataset includes all tests. Mean and median are very close amongst each other (variation of 0.1°), as in other previously published results (Alejano et al. 2017).

3.2 *Histograms and boxplots*

Histograms are used to show representations of the distribution of numerical data and to infer the probability distribution of a continuous variable. To provide an overview of the distribution of results obtained in the present study, all 297 basic friction angle values were plotted by means of a histogram (Figure 1). A normal fit for the considered data is also provided, superposed to the histogram bars.

The histogram presented in Figure 1 can be considered reasonably symmetric and unimodal. The data arrangement appears quite similar to a normal distribution.

If data are clustered according to selected parameters (different length-to-thickness ratios, used saw blades or tilting rates), histograms like those shown in Figure 2 (a1, b1, and c1) can be obtained. With the aim of a better understanding of results, boxplots for the same groups of data are provided below the histograms (Figure 2 a2, b2 and c2). Each boxplot displays median (50th percentile) and 25th and 75th percentiles by the central mark and box edges, respectively. Those values located out of the whiskers, if applicable, are considered outliers and are identified with red crosses.

Histograms for different l/h ratios in Figure 2(a1) show that the most homogeneous results are those corresponding to l/h = 3 (all-3 series). This can be attributed to a relatively small size of the block, which allows a more flat and homogeneous cutting than that for larger blocks (presenting ratios l/h = 5, in this study).

Nevertheless, for the smaller l/h ratio l/h (l/h = 1.67) compressive strength is not completely distributed on the contact surfaces (Hencher 1976), generating more scattered results than

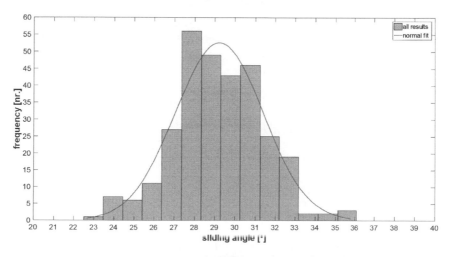

Figure 1. Histogram representing all 297 results and normal fit (mean = 29.21° and standard deviation = 2.19°).

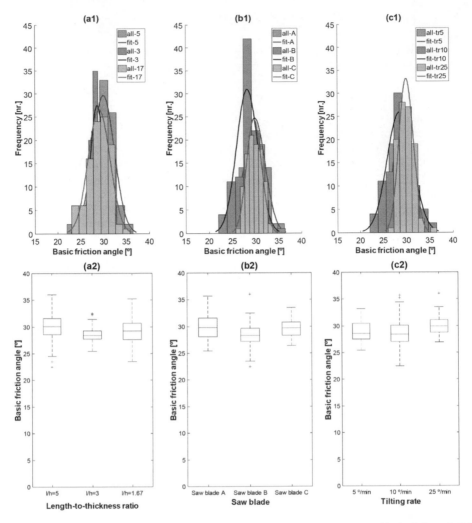

Figure 2. Histograms for data clustered by length-to-thickness ratio (a1), saw blade (b1) and tilting rate and boxplots for the same groups, respectively (a2, b2 and c2).

for the other cases, despite rock specimens are thought to present flatter surfaces. A somewhat low contact surface may also contribute to increase scattering.

For the second case (Figure 2(a2)) it has been observed that specimens cut with sawblade C (a continuous-rim sawblade) yielded tilt test results with less dispersion than the other ones. This fact can be associated to surfaces presenting less roughness than those cut with the other jagged sawblades.

For the last parameter, tilting rate (Figure 2(c2)), mean and median values for the three studied histograms are located very close and scattering becomes lower for test results carried out at 25 °/min. Although it produces similar mean and median results, this somewhat high velocity introduces an effect that makes the beginning of the sliding dependent on the mobilizing component besides of the friction of the material or other features.

A group of boxplots for each test series is provided in Figure 3. As can be appreciated, median of these raw values fall within a range of 25° to 32° approximately, and some variability is observed. Those values considered outliers (displayed out of the whiskers) were identified as the first sliding of each series for the majority of datasets. This may be due to the condition of the fresh saw-cut surface at the first stage that may display some small grains and unevenness that only contribute to the friction resistance in the first sliding since they have not yet been worn.

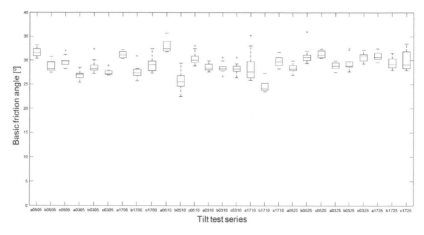

Figure 3. Boxplots for each carried out series (x-axis code: a, b, c are the corresponding saw blades; first two numbers are the l/h ratio and last two numbers are the tilting rate (°/min).

Table 2. ANOVA results for different datasets.

Dataset analysed	p-value	F	Acceptance of null hypothesis
All 27 series	<0.05	21.14	Rejected
Means of series, classified by type of sawblade A, B, C	0.12	2.29	Accepted
Medians of series, classified by type of sawblade A, B, C	0.08	2.74	Accepted
Means of series, classified by l/h ratio: 5, 3 and1.67	0.28	1.34	Accepted
Medians of series, classified by l/h ratio: 5, 3 and1.67	0.31	1.24	Accepted
Medians of series, classified by tilting rate: 5, 10 and 25 [°/min]	0.26	1.42	Accepted
Medians of series, classified by tilting rate: 5, 10 and 25 [°/min]	0.25	1.49	Accepted

3.3 One-way analysis of variance (ANOVA)

A one-way analysis of variance was applied to the different datasets. First, all data was jointly analyzed and then it was divided into different groups, according to the three factors being studied (sawblade, l/h ratio and tilting rate) and the ANOVA test was also run. These arrangements were studied for two cases: mean and median of each series. Results for these analyses are presented in Table 2.

Despite the null hypothesis is rejected for all data, it is possible to conclude that those datasets classified by type of saw blade, l/h ratios and tilting rates all belong to the same statistical population or, in other words, they do not depend on any factor beyond the inherent frictional component of the contact surfaces. Moreover, the analysis does not depend neither on mean or median values, being results very close in terms of p-value.

3.4 Evolution of the median

Median value of a tilt test series seems to be a good estimator of the basic friction angle of a planar sawcut rock surface. A criterion, as presented in Equation 2, has been proposed. This equation allows to decide whether a certain number of tests can be representative to obtain a reliable value of the basic friction angle of a couple of rock surfaces.

$$t_i = \mathrm{abs}\left[\mathrm{median}\left\{ \phi_{b_j} \right\}_{j=1}^i - \mathrm{median}\left\{ \phi_{b_j} \right\}_{j=1}^{i-1} \right] < 0.2 \text{ for } i = 6, 7..., 11 \tag{2}$$

When t_i value becomes less than a threshold value, set as 0.2 (in this case, 10% of the standard deviations has been deemed appropriate), median can be considered stabilized and

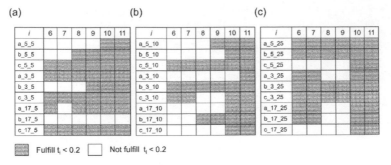

Figure 4. Graphical description of median evolution by series, regarding equation (2).

the number of carried out tests. This equation has been applied to all performed series in this study (11 tests per series) and results are described in Figure 4 graphically.

4 CONCLUSIONS

27 series of 11 tilt tests on granite rock slabs have been performed. Samples presented three possible values of three different parameters, namely length-to-thickness ratio of the specimen base, type of sawblade used and tilting rate of the testing platform. The obtained basic friction angle values for the studied rock slightly vary, independently of the type of block tested, type of sawblade or tilting rate.

Though reduced in number, outliers are almost always referred to the first test of each series, which leads to think that it is the condition of the intact rock (immediately after being cut) that determines these abnormally high values more than other aspects.

Median has been considered a sufficiently robust estimator to evaluate results of this test (at least better than mean, since it mathematically takes into account the outliers). Regarding this analysis, it is recommended to always carry out a minimum number of tests between 5 and 7 repetitions. Nevertheless, as can be appreciated for the present case, the median of the values completely stabilizes for a larger number of tests, commonly 11.

All in all, this study suggests the basic friction angle can be considered not only an inherent characteristic of this *Blanco* Mera granite, but also a reproducible parameter as long as certain factors are controlled when performing laboratory tests. This conclusion holds for the parameter variations considered in the study.

ACKNOWLEDGEMENTS

Co-authors I. Pérez-Rey and L.R. Alejano would like to acknowledge the Spanish Ministry of the Economy, Industry and Competitiveness for funding their research studies under Contract Reference No. BIA2014-53368P, partially financed by ERDF funds from the EU.

REFERENCES

Alejano, L.R., Muralha, J., Ulusay, R., Li, C.C., Pérez–Rey, I., Karakul, H., Chryssanthakis, P., Aydan, Ö., Martínez, J., and Zhang, N. "A benchmark experiment to assess factors affecting tilt test results for sawcut rock surfaces," (Technical Note) *Rock Mech. Rock Eng.* Vol. 50, No. 9, pp. 2547–2562, https://doi.org/10.1007/s00603-017-1271-6.

Barton, N., "Review of a new shear-strength criterion for rock joints," *Eng. Geo.*, Vol. 7, No. 4, 1973, pp. 287–332.

Hencher, S.R., Discussion of: "A simple sliding apparatus for the measurement of rock joint friction" by D.C. Cawsey and N.J. Farrar. The Article Was Published in Géotechnique, Vol. 26, No. 4, 1976, pp: 641–644.

Geomechanics and Geodynamics of Rock Masses – Litvinenko (Ed.)
© *2018 Taylor & Francis Group, London, ISBN 978-1-138-61645-5*

The use of InSAR as a tool to manage precursor ground displacement in rock masses

Josep Raventós, Ciscu Sánchez & Aritz Conde
TRE ALTAMIRA, Barcelona, Spain

ABSTRACT: The slope stability of altered rocks that can behave as soils (generally referred to as "soft rocks") can be a major concern when these slopes are surrounding key infrastructures. This paper describes the role of satellite technologies, and specifically InSAR (Interferometry of Synthetic Aperture Radar), as a tool for monitoring precursor ground displacement in the case of a massive slope in Northern Spain affecting a major European link highway. The slope failed at least two times in the period between March and November 2015. The first failure resulted in the highway blocked and the necessity of implementing an emergency project and remediation works. During the remediation works a second failure occurred.

The geological profile of the site is made of alluvial soils, altered rock mainly due to previous landslide in the same area, and rock mass, also affected by faulting, made by breccia, limolite and mylonite.

A stack of images from the European Space Agency satellite Sentinel was processed for a period starting before the first failure and up to several months after the highway reopening. This ground motion analysis aimed to help understand the mechanism of failure as well as the precursory movements before the failures.

Keywords: InSAR, landslide, displacement monitoring, accuracy, rock mass, soft rock, precursory movement

1 INTRODUCTION

The study here presented corresponds to the surroundings of a highway in Spain. In March 2015, after a long rainy period, there was a landslide affecting the north area of the highway. The affected slope was 110 m long and 60 m height.

In this context, with the objective of study the stability of the slopes surrounding the highway after the event before mentioned, an InSAR (Interferometric synthetic aperture radar) analysis using radar satellite images have been achieved. Images provided by the Sentinel-1 ESA's satellite acquired from July 2016 to October 2017 have been used.

This paper describes the results obtained on the area of study and shows the capability of InSAR techniques to monitor landslides using radar images of the past.

2 SITE DESCRIPTION: GEOTECHNICAL CONTEXT AND FAILURES

The design phase previous to the construction of the motorway highlights the existence of a mobilized mass of rock in this area. The rock mass consists in mica siltstone dark grey coloured with centimetric intercalations of grey sandstones. The rock does not have significant stratification or joints, and when this is the case, they are closed or filled. The rock mass is jointed with centimetric openings. This stratification has a dip/slip orientation E008/30–70, so it is sub-parallel to the track of the road. This geometry helps the flexural toppling problems to occur.

From the structural point of view, the rock mass is faulted in two main directions: NW-SW and SW-NE, with the first group oriented parallel to the existing thrusts.

At the beginning of March 2015, a major landslide occurred forcing a traffic block of the road. An image of the landslide can be seen in Figure 1, with a strong activity at the feet of the slope. A technical note was proposed with several construction activities related to the excavation of the mobilized rock mass and also a new support designed to stabilize the slope. Among others, three anchored micropiles walls were prosed. The construction of the support starts at the end of March 2015.

In June 2015, after the construction of the upper level of the support and during the construction of the intermediate level, significant deformation over the intermediate level and a new support, behind this intermediate level occurred. After the construction of these two intermediate levels and also a rainy period, a new instability occurred in the unsupported rock mass. After that a third anchored level in the intermediate wall was constructed.

At the beginning of July, the lower wall level started its construction in junction with the anchor construction in the intermediate level. During its construction, and again after a rainy period, the anchors of the lower level started to fail and a collapse of the slope.

3 METHODOLOGY

The presented results have been obtained by processing one datastack with advanced PSI techniques: images from the Sentinel-1 satellite acquired in descending geometry from July 2016 to October 2017. PSI techniques can retrieve a high number of measurements points from natural reflectors located on the Earth's surface when no big changes occur during the period of study. The combination of different SAR processing techniques guarantees the best measurement precision for all ranges of motion, from millimetric to centimetric magnitudes. All measurements provided are taken in the Line-of-Sight (LOS) direction meaning that measurements are a projection of the real motion into the detection vector or looking vector of the satellite.

3.1 Technical basis of DInSAR

Differential Interferometry Synthetic Aperture Radar (DInSAR) is a remote sensing technique able to measure surface deformation on the ground using complex satellite radar image datasets. The main information used by DInSAR analysis corresponds to the interferometric phase obtained by the computation of the phase difference between a pair of SAR images, the interferogram. The phase information coming from an interferogram is related to the topography of the area observed and to the deformation occurred in the time between the

Figure 1.   Collapse of the slope. March 2015.

two images used (Lanari et al., 2004). The first description of DInSAR technique using SEASAT SAR images corresponds to Gabriel (Gabriel et al., 1989).

Advanced DInSAR techniques applied to a set of interferograms (redundancy) are the ones known as advanced PSI techniques. PSI techniques are related to the detection of millimetric and centimetric displacements on the ground surface caused by underground activity or low-motion instability (Adam et al., 2009). The PSI family of algorithms can identify high quality reflectance points (high coherence) and obtain its temporal evolution of the displacement (time series). The basis of the PSI techniques consist of the separation of each component: motion, topographic error, atmospheric artefacts and signal noise from the input interferogram phases. In a step by step procedure, using a stack of interferograms coming from a minimum image dataset of more than 20 images, atmospheric artefacts are estimated and compensated by using low pass filters and deformation and topographic contribution values are extracted through a high resolution analysis. The PSInSAR application (Ferretti et al., 2000) was able to retrieve milimetric accuracies in motion estimation with very good results especially in urban areas by removing atmospheric noise.

PSI techniques such as PSInSAR identifies individual pixel with strong signal returns referred as Permanent Scattereres in order to measure surface movement occurring through the time (Ferretti et al., 2001). These PS are limited in distribution to rocky terrain or more urban settings. To improve the technique under other surface conditions, such as those found in and around mining operations a more complex scattering model was needed. Distributed scatterers (DS) are low-amplitude but coherent returns that are identified on a pixel-by-pixel basis. The better temporal and spatial resolution of the new satellites improved the signal-to-noise ratio to the point where these DS became a significant contributor to the deformation monitoring signal. Advanced processing that optimizes the PS and DS returns, such as SqueeSAR™, were developed to provide displacement information not only on manmade structures or exposed rocks, but on outcrops and thinly vegetated areas (in Colombo et al., 2015). Further details of advance DInSAR techniques can be found in Rosen et al. (2000), Hanssen (2001) and Ferretti (2014).

4 InSAR STUDY

Ground displacement information around the failure area has been obtained by processing a Sentinel-1 dataset. Table 1 shows the characteristic of the image dataset while Figure 2 shows the temporal distribution of the Sentinel-1 acquisitions, most of the period has acquisitions every 6 days although at the beginning there are gaps of 12 and 24 days.

4.1 Results

An area of 2.8 km² around a highway has been studied where 980 measurements points have been obtained providing ground deformation data over 2.6 km². This measurement points can be divided in two sets: the 313 persistent scatters (PS) which represent 25,070 m² of the area and the 4,667 distributed scatters (DS) representing of 2,591,541 m².

Table 1. Dataset used for an historical study.

Satellite	Geometry	Number of images	Time spam
Sentinel-1	Descending	67	8th July 2016–13rd October 2017

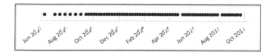

Figure 2. Temporal distribution of the SENTINEL dataset used for this study.

Figure 3 shows the results obtained over the area. The measurement points are color coded by using the average displacement. The 90% of the measurement points show deformation between −5 and +5 mm/years which can be considered stable measures due to the ±5 mm/year of precision. Nevertheless, some motion patterns are detected in the highway surrounding slopes. Figure 3 shows the location of those motion patters which are mostly located over excavated highway slopes. Two white squares highlight the main areas commented hereafter.

Figure 4 shows a close up on the western area of study. The thin black circles around the measurements represent the approximate area represented by each measure. All the slope located at the north of the highway show different kind of motion. In the central part of this area the measurement points show smooth deformation with values around −6 mm/year, while in the western area the detected motion is stronger. The thick black circle in Figure 4 shows an area with non-linear behavior: almost stability from July 2016 to the end of March 2017 and deformation from 2017 spring on. During this last 7 months −15 mm/year of deformation has been detected. All these aforementioned measurements are located in the same natural slope and they are part of the same geodynamic system.

Figure 5 shows the average displacement obtained around the highway. The thin black circles around the measurements represent the approximate area represented by each measure.

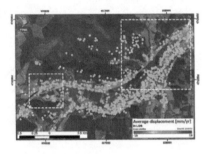

Figure 3.   Ground deformation results over the site from July 2016 to October 2017. White squares highlight the main areas.

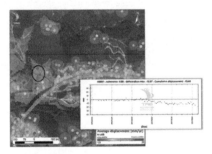

Figure 4.   Western area with an unstable slope.

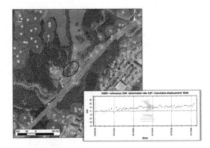

Figure 5.   Results over a representation of DS computing area.

386

The highlighted area shows measures located over an excavated slope with accumulated motion up to 23 mm during the period of study. The time series shows the mean behavior of the blue measures, reaching 9.8 mm/year of deformation with a linear behavior. The surrounding measures located over natural slopes show stability.

5 CONCLUSIONS

InSAR was used in this project to provide information of the stability of several slopes surrounding a main roadway in Spain. Some of these slopes were excavated in a rock mass made of weathered siltstone. One major slope failure occurred in March 2015, and during the construction of a new support for this slope, several anchors failed and the structure collapsed. The main conclusion, in terms of stability, is that the failed slope is stable, but the surrounding slopes show non-linear displacement patterns, with accelerations on the spring time and velocities up to 1.5 cm/year.

The main advantages of the use of this technology in this kind of projects are:

- The use of remote sensing techniques allows us to measure instability without installing any device on site.
- It provides a general overview of the stability and helps to understand the (un)linear pattern of the displacement and correlate this behavior with construction and/or weather changes.
- It helps to manage and to improve the installation and measurement ratios of in site geotechnical monitoring.

REFERENCES

Adam N., Parizzi A., & Crosetto, M. (2009). Practical Persistent Scatterer Processing Validation in the Course of the Terrafirma Project. Journal of Applied Geophysics, 69: 59–65.

Berardino, P., Fornaro, G., Lanari, R. & Sansoti, E. (2002), "A new algorithm for surface deformation monitoring based on small baseline differential SAR interferograms". IEEE Transactions on Geoscience and Remote sensing, 40(11), 2375–2383.

Colombo, D. & MacDonald, B. (2015). Using advanced InSAR techniques as a remote tool for mine site monitoring. In Proceedings, Slope Stability Congress, Cape Town, South Africa.

Ferretti, A., Prati, C. & Rocca, F. (2000). Nonlinear subsidence rate estimations using permanent scatteres in differential sar interferometry. IEEE Transactions on Geoscience and Remote Sensing, 38(5): 2202–2212.

Ferretti, A., Prati, C. & Rocca, F. (2001). Permanent scatterers in SAR interferometry. IEEE Trans. Geosci. Remote Sens. 39(1): 8–20.

Ferretti, A. (2014). "Satellite InSAR data. Reservoir Monitoring from space". EAGE publications.

Gabriel, A.K., Goldstein, R.M. & Zebker, H.A. (1989). "Mapping small elevation changes over large areas: differential radar interferometry". J. Geophys. Res., 94(B7), 9183–9191.

Hanssen, R. (2001). "Radar interferometry". Kluwer Academic Publishers, Dordrecht, The Netherlands.

Lanari, R., Mora, O., Manunta, M., Mallorquí, J.J., Berardino, P. & Sansosti, E. (2004). "A small-baseline approach for investigating deformations on full-resolution differential SAR interferograms". IEEE Transactions on Geosciences and Remote Sensing, 42(7), 1377–1386.

Mora, O., Mallorquí, J.J. & Broquetas, A. (2003). "Linear and nonlinear terrain deformation maps from a reduced set of interferometric SAR images". IEEE Transactions on Geosciences and Remote Sensing, 41(10), 2243–2253.

Rosen, P.A., Hensley, S., Joughin, I.R., Li, F.K., Madsen, S.N., Rodríguez, E. & Goldstein, R.M. (2000). "Synthetic Aperture Radar Interferometry". Proc. of the IEEE, 88(3), 333–382.

Werner, C., Wegmüller, U., Strozzi, T. & Wiesmann, A. (2003). "Interferometric point target analysis for deformation mapping". Proceedings of IGARSS, 2003, 4362–4364.

Geomechanics and Geodynamics of Rock Masses – Litvinenko (Ed.)
© 2018 Taylor & Francis Group, London, ISBN 978-1-138-61645-5

Correlations of geomechanical indices for Andean environments

Sergio Sánchez Rodríguez, José Diego López Valero & Carlos Laina Gómez
Geoconsult España Ingenieros Consultores, Madrid, Spain

ABSTRACT: The use of geomechanical classifications, especially RMR, GSI and Barton Q, applied to underground civil works is very common in Andean areas where important infrastructures are being built in recent times and many others projected related to tunnels and hydroelectric projects. The use of different classifications is needed to characterize the rock masses, but their individual use reveals inconsistencies between the results of these classifications implying a deficit in the characterization of rock masses affected. In this paper new criteria of correlation between these geomechanical classifications, RMR, GSI, and Barton Q, associated with Andean contexts are presented, in order to obtain an optimum characterization of the rock mass affected by the infrastructure.

Keywords: Andean enviroments, Geomechanical classification, RMR, GSI, Q

1 INTRODUCTION

Nowadays geomechanical classifications are very much used in practical Rocks Mechanics, especially in the Andean region where, at this moment, a wide development of infrastructures are taken place.

The engineer usually uses different geomechanical indices, in combination with data of 'in situ' and laboratory tests, to obtain engineering calculation parameters. However, there is a risk of making inaccuracies or obtaining results which can differ in some order of magnitude.

Therefore this study deepens on correlations GSI-RMR and Q-RMR indices, which are commonly used in almost every geological engineering study. This study is an update of previous results obtained by the authors [12].

2 GEOMECHANICAL CLASSIFICATIONS

Geomechanical classifications allow the characterization of rock masses from simple tests and field observations. There are many geomechanical indices in use today, among which are: RMR, Q, GSI SMR, RQD, RSR, RMi and so on.

The present study focuses its analysis on Andean rock masses which have been characterized by Rock Mass Rating (RMR), Geological Strength Index (GSI) and Tunneling Quality Index (Q).

2.1 *Rock Mass Rating (RMR)*

Rock Mass Rating (RMR) index was developed by Bieniawski (1976) [3]. It is one of the most used geomechanical classifications for the characterization and description of rock masses.

The RMR, value from 0 to 100, is obtained as the total sum of parameters related with strength of material, rock fracturing, discontinuities and groundwater conditions. In the specialized bibliography two different versions can be found, depending on the scoring ranges assigned to each of the parameters previously defined (RMR_{76}, RMR_{89}) as shown in Table 1.

Table 1. Differences between RMR$_{76}$, RMR$_{89}$ and GSI.

Parameter	RMR$_{76}$	RMR$_{89}$	GSI
Uniaxial compressive strength of rock material	15	15	0
RQD and Spacing of discontinuities	50	40	50
Condition of discontinuities	25	30	50
Groundwater conditions	10	15	0

2.2 Geological Strength Index (GSI)

Geological Strength Index (GSI) was developed by Hoek et al. in 1995 [6]. Unlike RMR and Q, it is an index that is obtained in a more simplified and visual way from two parameters: the structure and the superficial state of the rock mass. GSI considers a final score within the range 0–100, having the two parameters considered on it the same importance or weight. Table 1 summarizes the main ideas showing the differences between RMR$_{76}$, RMR$_{89}$ and GSI.

As it is extracted from the previous paragraphs, GSI introduces a greater empiricism in an empirical classification by itself. However, it is the nexus of union between the visual description made in the field works and the constitutive model of non-linear failure developed by its authors and implemented in a wide range of software used in the design of engineering works.

2.3 Tunneling Quality Index (Q)

Tunneling Quality Index (Q) was proposed by Barton et al. (1974) [1] with the aim of determinate rock mass characteristics and tunnel support requirements based on a large number of case histories of underground excavations. Index Q varies on a logarithmic scale from 0.001 to 1000 and its numerical value is the product of only three parameters:

- Block size $\left(\frac{RQD}{J_n}\right)$,
- Inter-block shear strength $\left(\frac{J_r}{J_a}\right)$ and
- Active stress $\left(\frac{J_w}{SRF}\right)$.

Equation (1) shown below represents Index Q proposed by Barton et al. in 1974 [1]:

$$Q = \left(\frac{RQD}{J_n}\right) \cdot \left(\frac{J_r}{J_a}\right) \cdot \left(\frac{J_w}{SRF}\right) \tag{1}$$

where RQD is the Rock Quality Designation, J_r is the joint roughness number, J_a is the joint alteration number, J_w is the joint water reduction factor and SRF is the stress reduction factor.

2.4 Existing correlations between RMR, GSI and Q

Authors consider further studies on RMR-Q-GSI correlations are of high interest. Some reasons are:

- RMR and Q classifications have been used in studies of geological engineering for a long time. In consequence, there are remarkable empirical knowledge, large bibliographical references related to correlations between other parameters and great of data collected during the development of engineering projects existing.
- GSI classification, despite being more recent, has a nonlinear failure criterion associated with it, that it is implemented in many rock engineering softwares.

These were the main reasons why in recent years several authors have invested great efforts in developing relationships between these geotechnical classifications. Among which are:

- Hoek et al. (1995) [6] proposed the following general mathematical expression is:

$$GSI = RMR' - 5 \quad if \ RMR' > 23$$
$$GSI = (*) \qquad\quad if \ RMR' < 23 \tag{2}$$

where RMR′ is the value of RMR in dry conditions (with 15 points for groundwater condition) and uncorrected by orientation of the joints
(*) The RMR′ should not be used as an indirect parameter in obtaining the GSI.

- Bieniawski (1976) [3] correlated RMR and Q with the well-known general proposal:

$$RMR = 9 \cdot \ln(Q) + 44 \tag{3}$$

RMR value is uncorrected for the orientation of the joints.

3 STUDY FOR ANDEAN ENVIRONMENTS

3.1 *Goals of the study*

Currently a very important part of the global investment on civil and construction works is located in Latin American countries.

Consequently, this study aims to obtain correlations between the Rock Mass Rating (RMR), Geological Strength Index (GSI) and Quality Index (Q), depending on the different types of rocks existing in the Andean environment. Secondly, correlations obtained shall be compared with results obtained previously by other authors, such as Hoek et al. (1995) [6], Bieniawski (1976) [3] and so on.

All of this with the ultimate goal of being able to use, with the best accuracy, all the tools, applications and correlations associated with the different geomechanical classifications regardless the indices obtained during field investigation.

3.2 *Description of the database*

The characterization and description of Andean rock masses is based on 298 outcrop mapping measurements and 61 tunnel face mapping results, which were executed in 10 projects located in 4 Latin American countries where the authors of this publication participated recently (Bolivia, Ecuador, Colombia and Peru).

All of the outcrop mapping and tunnel face mapping have been classified according to the type of rock. In addition, pairs of RMR-GSI and Q-RMR values have been obtained in all of them.

3.3 *Working methodology*

The first step of this study consisted on compilation of data and its classification based on the lithology, type of measurement (tunnel face or outcrop) and geomechanical index used.

Basic information regarding the compiled data is shown in Table 2. As can be seen, a total of 359 pairs of RMR-GSI for 6 types of rocks were considered, whilst the number of RMR-Q values were 136.

Secondly, a statistical study of the data series described above and based on the type of rock was carried out. The main target of this was obtaining the best fit using the least squares method.

3.4 *Results obtained*

3.4.1 *Correlations RMR'-GSI and RMR-Q*

The results obtained from this study are shown below. After analyzing different mathematical expressions, it was confirmed that the best statistical fit that correlates the RMR′ and GSI is the linear. RMR′ is the uncorrected value for dry conditions.

Table 2. Number of pair data RMR-GSI and RMR-Q depending on the type of rock.

Type of rock	RMR-GSI			RMR-Q		
	Outcrop Mapping	Face Mapping	N° of data	Outcrop Mapping	Face Mapping	N° of data
Coarse grained sedimentary rocks	105	31	136	0	31	31
Fine grained sedimentary rocks	51	11	62	0	11	11
Chemical sedimentary rocks	1	1	1	0	0	0
Metamorphic rocks	69	12	81	67	12	79
Plutonic rocks	21	7	28	7	7	14
Volcanic rocks	51	0	51	1	0	1
Entire of database rocks	298	61	359	75	61	136

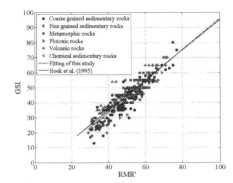

Figure 1. Statistical fit RMR'-GSI for the entire simple of data considered in the study.

Table 3. Correlations GSI = GSI (RMR') depending on the type of rock.

Type of rock	Equation	R^2
All data rocks	$GSI = RMR' - 6$	0.82
Coarse grained sedimentary rocks	$GSI = RMR' - 7$	0.81
Fine grained sedimentary rocks	$GSI = 1.1 \cdot RMR' - 12.5$	0.87
Metamorphic rocks	$GSI = RMR' - 4$	0.68
Plutonic rocks	$GSI = 1.15 \cdot RMR' - 15$	0.76
Volcanic rocks	$GSI = 0.95 \cdot RMR'$	0.87

Figure 1 shows the correlation obtained for the entire sample data considered regardless of lithology, correlating the GSI index compared to RMR', i.e., the RMR uncorrected and with the highest score for water condition.

In further analysis, RMR-GSI has been correlated to different lithologies of the rock masses studied, including: sedimentary rocks of coarse grain (sandstone and conglomerates), fine-grained sedimentary rocks (siltstones, shales), metamorphic rocks (schists, slates), plutonic rocks (granites, granodiorites) and volcanic rocks (andesites, basalts). Results obtained are shown in Table 3.

As can be seen, correlation is based on a range of RMR' data between 30–80 points, and provides a reasonable correlation coefficient of 0.82, for the expression of GSI = RMR'-6, which is close to the proposal of Hoek et al. (1995) [6].

However, depending on the type of material or lithology this study has shown discrepancies or differences above 5 points of GSI depending on the nature of the rocks. For example, one of these cases is related with fine grained sedimentary rocks, which can be clearly unconservative in case of weak rocks masses.

The same analysis was made with pairs of RMR-Q values. Figure 2 and Table 4 show the results obtained along with some of the traditional correlations proposed in specialized bibliography. A general conclusion of this study is that results were consistent with those proposed by several authors.

3.4.2 New proposal for GSI estimation based on RQD and Jcond
Similarly to Hoek et al. (2013) [7], it has been obtained an expression for estimating the GSI as a function of the RQD and the J_{cond}. Subsequently, GSI values obtained by this new

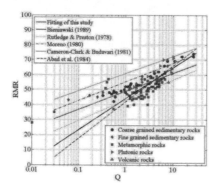

Figure 2. Statistical fit RMR-Q for the entire simple of data considered in the study.

Table 4. Correlations RMR = RMR (Q) depending on the type of rock.

Type of rock	Equation	R^2
All data rocks	$RMR = 5 \cdot LnQ + 49$	0.55
Coarse grained sedimentary rocks	$RMR = 7 \cdot LnQ + 50$	0.93
Fine grained sedimentary rocks	$RMR = 5 \cdot LnQ + 54$	0.88
Metamorphic rocks	$RMR = 4 \cdot LnQ + 47$	0.48
Plutonic rocks	$RMR = 9 \cdot LnQ + 48$	0.78

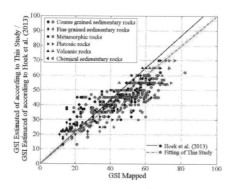

Figure 3. Statistical fit GSI Mapped-GSI Estimated for the entire simple of data considered in the study.

Table 5. Correlations depending on the type of rock of GSI Estimated = $a \cdot J_{cond}$ +b·RQD.

Type of rock	Equation
All data rocks	$GSI = 1.28 \cdot J_{cond} + 0.48 \cdot RQD$
Coarse grained sedimentary rocks	$GSI = 0.76 \cdot J_{cond} + 0.53 \cdot RQD$
Fine grained sedimentary rocks	$GSI = 0.81 \cdot J_{cond} + 0.59 \cdot RQD$
Metamorphic rocks	$GSI = 1.99 \cdot J_{cond} + 0.41 \cdot RQD$
Plutonic rocks	$GSI = 1.47 \cdot J_{cond} + 0.45 \cdot RQD$
Volcanic rocks	$GSI = 0.62 \cdot J_{cond} + 0.57 \cdot RQD$

expression and GSI mapped data available have been compared, showing that obtained results differ from the ones deduced by the use of Hoek's proposal. Additionally, dependency on lithology has been proved as can been seen in the Figure 3.

4 CONCLUSIONS

This study analyzed large number of pairs of data GSI-RMR corresponding to different projects and works located in the Andean region. These have been studied in order to shed light on the correlation between these two indices so frequently used in Rock Mechanics. The main conclusion of this study refers to the goodness of the generally used correlation proposed by Hoek.

However, depending on the type of material or lithology, it is strongly recommended to obtain site specific correlations for each case or project; especially in the case of detailed studies, since this study has shown discrepancies or differences above 5 points of GSI depending on the nature of the rocks.

Here, a new expression to obtain GSI from RQD and J_{cond} has also been proposed. This provides a best fit with available data visually measured on site by the authors.

Finally, a new correlation between RMR and Q was also obtained, although this should be improved with more data in the range of Q < 1.

ACKNOWLEDGEMENTS

Workers involved in works and projects that allowed Geoconsult compilation of the data used to carry out this study.

REFERENCES

[1] Barton, N.R., et al. (1974). Engineering classification of rock masses for the design of tunnel support. Rock Mech. 6(4), 189–239.

[2] Bhawani, S., Ranjnish, K.G. (2006). Tunneling in weak rocks. El-sever, London.

[3] Bieniawski, Z.T. (1976). Rock mass classification in rock engineering. In Exploration for rock engineering, proc.of the symp., ed. Z.T. Bieniawski, 1, 97–106. Cape Town: Balkema.

[4] Bieniawski, Z.T. (1989). Engineering rock mass classification. New York: Wiley Interscience.

[5] Ceballos, F., Olalla, C. and Jiménez, R. (2014). Relationship between RMRb and GSI based on in situ data. Rock Engineering and Rock Mechanics: Structures in and on Rock Masses, pp. 375–380

[6] Hoek, E., Kaiser, P.K. and Bawden. W.F. (1995). Support of underground excavations in hard rock. Rotterdam: Balkema.

[7] Hoek, E., et al. (2013). Quantification of the geological strength index chart. In 47th US rock mechanics/geomechanics symposium. American Rock Mechanics Association.

[8] Marinos, P., & Hoek, E. (2000). GSI: A geologically friendly tool for rock mass strength estimation. In proceeding of the GeoEng2000 at the international conference on geotechnical and geological engineering, Melbourne, Technomic publishers, Lancaster, pp. 1422–1446.

[9] Marinos, V., Marinos, P. and Hoek, E. (2005). The geological strength index: applications and limitations. Bull Eng Geol Environ, 64(1), pp. 55–65.

[10] Palmström, A. and Stille, H. (2003). Classification as a tool in Rock Engineering. Tunneling and Underground Space Tech., vol 18, no. 4, 331–345.

[11] Priest, S.D. and Hudson, J.A. (1976). Discontinuity spacings in rock. Int. J. Rock Mech. Min. Sci. & Geomech. Abstr. Vol. 13, pp. 135–148.

[12] Sánchez Rodriguez, S., López Valero, J.D. & Laina Gómez, C. (2016) Correlaciones entre clasificaciones geomecánicas en ambientes andinos. In ISRM 2nd International Specialized Conference on Soft Rocks. International Society for Rock Mechanics.

Geomechanics and Geodynamics of Rock Masses – Litvinenko (Ed.)
© 2018 Taylor & Francis Group, London, ISBN 978-1-138-61645-5

Thermal behavior of Indian shale rock after high temperature treatment

Sahil Sardana, A.K. Verma, Manish Kumar Jha & Pushpendra Sharma
*Department of Mining Engineering, Indian Institute of Technology (Indian School of Mines),
Dhanbad, India*

ABSTRACT: Thermal properties of rock, more specifically thermal conductivity is a critical property and is useful in different engineering and scientific applications. In the present study, an attempt has been made to analyze the effect of high temperature on thermal conductivity and induced damage of Jhiri shale rock. Shale samples have been kept at a designated temperature from room temperature to 900°C at an interval of 100°C. Thermal conductivity and P-wave velocity have been measured at room temperature after thermal treatment. Steady state method has been used for the measurement of thermal conductivity of Jhiri Shale.

Assessment of induced damage at different temperatures has been done and correlated with the thermal conductivity of shale rock. It has been observed that thermal conductivity decreases from 4.36 to 1.77 W/m.K as temperature changes from 25°C to 900°C. P-wave velocity also decreases from 4.33 to 1.52 Km/s with the increase in temperature up to 900°C. SEM analysis has also been carried out to visualize the change in microstructure caused by thermal treatment. It has been observed that up to the temperature of 300°C, there is a very small change in measured properties.

1 INTRODUCTION

Thermophysical properties of rock after high temperature treatment provides valuable knowledge for various science and engineering related application. The information of such properties of rocks after high temperature treatment are of extreme importance in the development of deep geological repository for high level nuclear waste, economical and sustainable extraction of natural gases, exploration of geothermal energy, storage and protection of underground energy and restoring of important buildings exposed to fire (Zhang et al. 2001; Hajpal, 2002; Emirov et al. 2013; Sun et al., 2016). Extensive studies have been carried out to analyse the thermal properties. Somerton and Boozer have conducted experimental study to determine the thermal properties of rocks, especially for sedimentary rocks. The study was focused to measure thermal conductivity and thermal diffusivity of sedimentary rocks using un-steady state method and observed that both properties found to be decreasing with increase in temperature up to 900°C. Vosteen and Schellschimdt and Hanley et al. have measured the thermal conductivity of rocks at high temperature and found that thermal conductivity of rocks decreases with increase in temperature. The exposure to elevated temperature results in the considerable change of microstructure and mineralogical composition in the rocks. Such changes cause variation in thermal and physical properties of rocks specifically in high temperature environment. As rocks are composed of different minerals having varying chemical composition along with the dissimilar degree of crystallization. Due to such variations, the thermal properties of rocks considerably depends on the microstructure, grain size, mineralogical composition, pore geometry (shape and size of pores), dimension and volume of existing cracks and cracks density of rock. The mechanism of heat transfer in a porous medium (rock) is extremely complex and hence, difficult or in some cases even impossible to predict the thermal behavior of rocks. Therefore, the experimental study for thermal behavior of rock at high temperature is essential. In the present study, an attempt has been made to investigate the thermal behavior of shale rocks from India after high temperature treatment.

2 EXPERIMENTAL PROGRAM

In the present study cylindrical sample of 50 mm diameter and 60–115 mm thickness, have been prepared from collected shale rock mass from the field (Fig. 1). Thermal conductivity along with porosity, density and P-wave velocity has been measured after treating the samples at designated temperatures. The samples have been kept for more than 24 hrs. at the temperatures ranging from room temperature to 900°C at an interval of 100°C.

The thermophysical properties of each sample have been measured in the laboratory after cooling samples into the furnace. The time for thermal treatment has been chosen to ensure the penetration of heat into the each grain of sample. Similarly, cooling has been done into a furnace to void the chance of thermal shock by putting the sample from high temperature environment to the ambient temperature condition. More than five samples have been tested for measurement of each property to ensure the reliability of results.

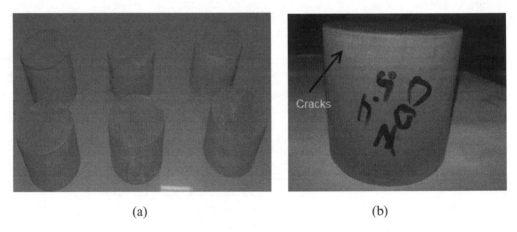

(a)   (b)

Figure 1.   (a) Cylindrical samples of Jhiri Shale (b) Sample after thermal treatment at 700°C.

Table 1.   Thermophysical properties of Jhiri Shale (Jha et al. 2016).

Temperature (°C)	Density (Kgm^{-3})	Porosity (%)	P-wave velocity (Kms^{-1})	Thermal conductivity (Wm^{-1}K^{-1})
Air Dried	2818.97	0.8	4.33	4.36
100	2658.44	1.1	4.16	3.49
200	2667.92	1.45	4.01	2.97
300	2653.70	1.8	3.95	2.74
400	2506.82	1.99	3.39	2.33
500	2491.60	3.38	2.36	2.22
600	2417.99	3.42	1.91	2.16
700	2271.98	3.87	1.87	2.12
800	2208.30	3.94	1.68	1.80
900	2190.65	4.10	1.52	1.77

3 PROPERTIES OF SHALE ROCK AFTER THERMAL TREATMENT AT DIFFERENT TEMPERATURE

The elevated temperature condition causes significant variation in the existing microstructure which causes the change in the porosity of the rock. Sufficient micro-cracks essentially exists in the rocks and due to the thermal treatment more micro-cracks originates and the density of cracks increases drastically with increase in temperature level. On the other hand, the widening of new and pre-existing micro-cracks and joining of these cracks take place. The extensive experimental work has been done to obtain the reliable results of each property (Table 1). In order to ensure the accuracy of result each sample has been tested 3–4 times for each property and then average data has been presented.

4 RESULTS AND DISCUSSION

4.1 *Thermal conductivity and induced damage after thermal treatment*

The thermal conductivity of rock largely depends upon the physical properties such as density, porosity and compressional wave velocity. Damage coefficient has been estimated from P-wave velocity measurement. Thermal damage caused by the temperature has been estimated from measured P-wave velocity as

$$D = 1 - \left(\frac{V_{P(T)}}{V_{P(25)}}\right)^2 \tag{1}$$

In Eq. 1, D is thermal damage coefficient, $V_{P(T)}$ is P-wave velocity at designated temperature and $V_{P(25)}$ is P-wave velocity at room temperature (25°C) (Km/s). P-wave velocity is a property that can be used to estimate the induced damage in the rock sample due to heating at high temperature. Thermal treatment of rocks results in development and accumulation of micro-cracks and hence, the increases the density of microcracks/cracks. High micro crack/crack density forces the damage of grain boundaries. Such events yield to lower resistance against grain sliding induced by frictional wear of debonded grains. An adequate amount of thermal stresses developed in rock samples during the exposure to elevated temperature conditions which causes the premature debonding of rock grains (Gautam et al. 2016).

P-wave velocity and thermal conductivity have been measured after thermal treatment at a designated temperature. The results obtained shows that both properties decrease with increase in temperature (Fig. 2). P-wave velocity does not decrease significantly up to 300°C. Up to this temperature the damage is dominant and only evaporation of water takes place. On the other hand thermal conductivity decrease as water has high thermal conductivity as compared to air. Beyond this temperature range, P-wave velocity decreases drastically up to 600°C. Thermal conductivity also decreases considerably in this temperature range. FESEM images also validate the damage in this temperature range. Further increase in temperature causes more damage and some of the samples got severe damage due to the occurrence of a circular and large number of irregular cracks in samples. Both properties decrease up to the temperature of 900°C.

Damage coefficient at room temperature is lowest and highest at maximum temperature (Fig. 3). The value of damage coefficient increases with increases in temperature up to 0.88. At the same time, thermal conductivity follows the reverse trend as of damage coefficient. The thermal conductivity of tested rock samples decreases with increase in temperature. Such behavior can be the result of the change in physical properties with temperature. Initially, up to 200°C, the different type of water present in the rock get evaporated and air replaces the water which causes the reduction in thermal conductivity. Above this temperature and up to 600°C, the damage also becomes severe and more air filled in newly formed cracks and contribute to lowering the thermal conductivity. Beyond this temperature, shale rock samples show severe physical damage and many cracks observed on the surface of the sample. The coefficient of damage and thermal conductivity shows inverse trend and it can be explained as more damage causes more air to present within the rock sample and hence, lowers the thermal conductivity. The behavior of Jhiri shale rock follows an approximate similar trend, except at some points where the variation of thermal conductivity and damage shows a considerable difference (Fig. 3).

FESEM analysis of Jhiri shale has been done to observe the change in its microstructure due to thermal treatment (Fig. 4). It can be observed that the cracks are not visible on the samples at room temperature and as temperature increases the cracks becoming visible after the 300°C. The cracks become wider and clearly observed after 600°C. Up to this

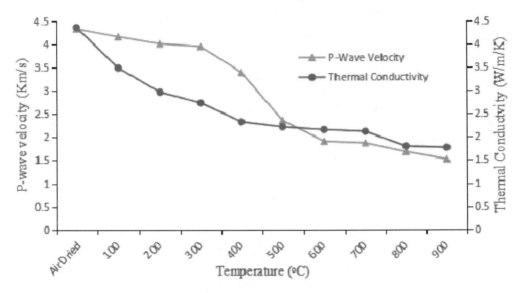

Figure 2.   Thermal conductivity and P-wave velocity of Jhiri shale.

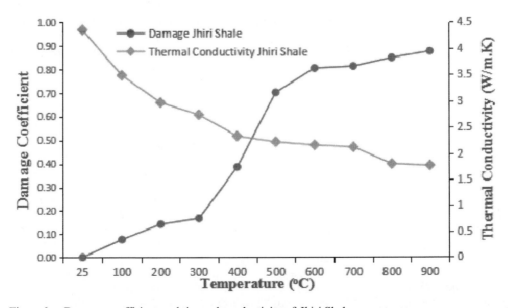

Figure 3.   Damage coefficient and thermal conductivity of Jhiri Shale.

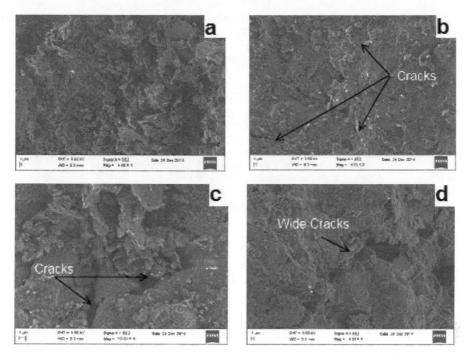

Figure 4.   FESEM images of Jhiri shale at (a) Room Temperature (b) 300°C (c) 600°C and (d) 900°C (Jha et al. 2016).

temperature, Jhiri shale shows severe damage due to thermal treatment. These changes are well related to the change in thermal conductivity and estimated damage. As thermal conductivity of Jhiri shale shows considerable changes after 300°C and 600°C temperatures and thermal conductivity decreases drastically. The damage coefficient also validates the change in microstructure with an increase in the temperature.

5   CONCLUSIONS

The study was focused on the determining thermal conductivity, density, P-wave velocity and porosity or Jhiri shale. The estimation of induced damage has been done and its effect on thermal conductivity has been explained. It has been observed that thermal conductivity decreases considerably (4.36 to 1.36 W/m.K) with an increase in temperature, whereas damage increases (0 to 0.88) significantly with temperature. Induced damage is an important parameter which affects the thermal conductivity of Jhiri shale rock. It has been found that the coefficient of damage shows opposite trend that of thermal conductivity with increase in temperature. Scanning electron microscopy (SEM) analysis has also been carried out to visualize the change in microstructure caused by thermal treatment. It has been observed that up to the temperature of 300°C, there is a very small change in measured properties. Such small change in the behavior of Jhiri shale shows that it can be one of the potential candidates as a suitable host rock for the nuclear waste repository.

REFERENCES

Emirov S.N., Ramazanova A.E., Zarichnyak Y.P. (2013) Contribution of thermal radiation in measurements of thermal conductivity of sand stone. Phys Solid-state. 55(12): 2463–2441.

Gautam P.K., Verma A.K., Jha M.K., Sarkar K., Singh T.N. (2016) Study of Strain Rate and Thermal Damage of Dholpur Sandstone at Elevated Temperature, Rock Mech Rock Eng, DOI 10.1007/s00603-016-0965-5.

Hajpál M.(2002) Changes in sandstones of historical monuments exposed to fire or high temperature. Fire Tech. 38: 373–382.

Hanley E.J., Dewitt D.P., Roy R.F. (1978) The thermal diffusivity of eight well-chracteried rocks for the temperature range 300–1000 K. Eng Geol. 12: 31–47.

Jha M.K., Verma A.K., Maheshwar S., Chauhan A. (2016) Study of temperature effect on thermal conductivity of Jhiri shale from Upper Vindhyan, India, Bull Eng Geol Env, 1–12. DOI 10.1007/s10064-015-0829-3.

Sun, Q., Lü, C., Cao, L., Li, W., Geng, J. and Zhang, W. (2016) Thermal properties of sandstone after treatment at high temperature. Int. J. Rock Mech. Min. Sci., Vol. 85, pp. 60–66.

Vosteen H.D., Schellschmidt R. (2003) Influence of temperature on thermalcon-ductivity, thermal capacity and thermal diffusivity for different types of rock. Phys ChemEarth. 55(12): 499–509.

Zhang Z.X.,Yu J., Kou S.Q., Lindqvist P.A. (2001) Effects of high temperatures on dynamic rock fracture. Int J Rock Mech Min 38(2): 211–225.

Swelling pressures of some rocks using different test procedures

Lena Selen & Krishna Kanta Panthi
*Department of Geosciences and Petroleum, Norwegian University of Sciences and Technology (NTNU),
Trondheim, Norway*

Maximilliano R. Vergara
Karlsruhe Institute of Technology (KIT), Karlsruhe, Germany

ABSTRACT: There are no clearly defined rules for the investigation procedures of swelling rocks. Difficulties are generally met for characterization and testing of swelling rocks and for prediction of the response to tunnel excavation. The waterway tunnels of hydropower projects and other water tunnels are even special since these tunnels are persistently exposed to the moisture changes caused by flowing water. Reported case histories have shown that severe stability problems have been experienced during the operation of hydropower plant caused by swelling of rocks. This important issue needs to be investigated thoroughly. Different laboratory oedometer testing methodologies have been developed and proposed in recent years by different institutions to carry out testing of swelling of the rocks. However, there is no standard methodology to assess the swelling behavior under moisture changes nor the application of the results obtained from powder tests.

The site-specific swelling rock potential including moisture changes should be investigated with effective diagnostic methods, which provide reliable data for the considerations and decisions to be made. The main aim of this manuscript is to provide insight and qualitative description of the laboratory methods in operation at two leading laboratories in rock mechanics; i.e. Norwegian University of Science and Technology (NTNU) of Norway and Karlsruhe Institute of Technology (KIT) of Germany, in order to determine the swelling potential of various rocks through cyclic testing. Oedometer swelling tests on pulverized material as well as intact rock discs have been performed at these two laboratories; i.e. at NTNU and KIT. The manuscript highlights the results of the swelling tests from both laboratories, and comparisons and discussions of the testing methodology at these two institutions are made.

Keywords: Water tunnels, hydropower, swelling of rocks, stability issues

1   INTRODUCTION

The determination of the swelling potential of rocks is necessary to make adequate choices on the dimensioning of the tunnel support. However, from swelling tests it is difficult to estimate the in-situ swelling pressure necessary in the design phase of a project (Galera et al. 2014). In the case of hydropower tunnels, the surrounding rock mass will be exposed to cyclic wetting and drying processes during the life-time of the hydropower-project. In several cases, swelling zones or rock mass containing swelling minerals have caused tunnel collapse, which has resulted in considerable additional construction costs and delays in completion of the projects (Selmer-Olsen & Palmstrøm 1989). The cyclic moisture changes and swelling of rocks may be replicated in the laboratory and the main patterns of the swelling behavior can be assessed.

The laboratory work presented in this manuscript was performed at two different universities; i.e. at the Norwegian University of Science and Technology (NTNU) and Karlsruhe

Institute of Technology (KIT); with an aim to compare the methodologies used in determining the swelling capacity of the intact rock and rock powder. The results of the tests performed on the same rocks are presented and compared. It is highlighted on the importance of the set-up at each laboratory and the knowledge needed to interpret the results obtained from the test.

2 THE CASE PROJECT AND MATERIAL

The Alimit hydropower plant is currently in its feasibility stage and is located at Ifugao, North Central Luzon in the Philippines. The rocks in the area are primarily volcanic rocks of basaltic and andesitic origin, which have undergone hydrothermal alteration or metamorphic transformation processes and can be found in different weathering stages (SN Aboitiz/Stache 2015). The tested material was obtained from borehole core samples. The rocks were categorized prior to laboratory investigation as *"strong"* and *"weak"* based on the visual inspection and assessment.

2.1 *Sample description*

The main characteristics for the rock group categorized as *"strong"* are intact core-lengths over 15 cm and low degree of visible disintegration. The majority of the intact cores show appearance similar to the assumed andesitic rock type as of AD-02 (box 12) shown in Table 1. The distribution of grain size and minerals appear as uniform throughout the samples, and the color is medium grey with shades of green (Table 1). The main characteristics of the *"weak"* group of the rock samples are heterogeneity regarding grain sizes and color, and high degree of disintegration. The samples break easily by hand force (Table 2).

Table 1. Samples of the category "strong" (Selen 2017).

AD-02, box 12 ~40.35–44.05 m	AD-06, box 25 ~86.30–87.30 m	AD-07, box 12 ~38.3–41.00 m	AQD-02, box 12 ~42.25–45.60 m

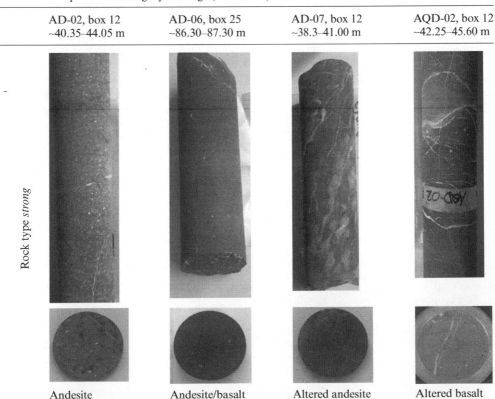

Andesite	Andesite/basalt	Altered andesite	Altered basalt

Rock type *strong*

Table 2. Samples of the category "weak" (Selen 2017).

AQD-02, box 5 ~16.70–20.60 m	AQD-02, box 6 ~20.60–24.60 m	APH-02, box 18 ~59.80–63.40 m

Rock type *weak*

No disc able to be prepared

Volcanic agglomerate/breccia	Volcanic agglomerate/breccia	Altered andesite/basalt

3 THE OEDOMETER SET-UP AND PROCEDURES AT NTNU AND KIT

Different variations of oedometer-tests are being used around the world, whereas many of them are based on the work performed by Huder and Amberg (1970) and Grob (1972) (Wittke-Gattermann & Wittke 2004). The oedometer swelling tests method described by ISRM (1977 and 1989) were updated several times (Madsen 1999). The suggestions include recommendations on preparation, apparatus configuration, procedures and reporting of the results. The maximum swelling pressure test suggested by ISRM (1977 and 1989) are the modified swelling tests of the method suggested by Huder and Amberg (1970). The oedometer swelling tests can be performed both on pulverized and intact rock samples. However, there is no methodology to assess the swelling behavior of the intact rock from the results obtained from powder tests. No consistent procedures on the preparation and testing of intact rock specimen are found. This may be due to the fact that different institutes construct the odeometer with some modifications of known standards, in this case the ISRM standards. No characterization system for swelling test results on intact rock specimen is developed yet. The results obtained are therefore difficult to compare directly, since the test results are influenced by both the apparatus used and the methodological constituency. This issue also applies to the results obtained from the tests performed at NTNU and KIT.

3.1 Oedometer test configurations

The swelling apparatus configuration should allow to obtain either swelling pressure or swelling strain (deformation) or combination of both. To obtain the characteristic swelling strain-pressure relationship, measurements on both strain (deformation in axial direction) and swelling force are needed. Some frequently used test-configurations are described in Table 3.

Table 3. Overview of conditions under which radially constrained swelling pressure tests may be performed in oedometers.

Method variations	Test configuration	Output
1) Zero volume change/zero deformation (ISRM, 1977 and 1989) (maximum swelling pressure tests)	Single test (one wetting phase)	Maximum swelling pressure in axial direction under constant volume.
2) Swelling under constant load followed by unloading stages in stress control (ISRM, 1999)	Single test (one wetting phase)	Swelling stress-strain relationship
3) Zero volume change followed by unloading stages in stress or strain control (Updated ISRM 1989 by Pimentel, 2007)	Single test (one wetting phase)	Maximum swelling pressure in axial direction Swelling stress-strain relationship First stage differs from method 2
4) Cyclic tests with controlled axial deformation (Updated ISRM 1989 by Vergara et al. 2014) (cyclic swellintests)	Multiple tests in cycles, often starting with one or more wetting and drying cycles allowing zero deformation (as for 1). Further unloading stages can be performed. The deformation allowed is fixed in each cycle (as for 2 or 3).	Swelling stress-strain-relationship is obtained Changes in swelling capacity between cycles may be evaluated.

3.2 The NTNU method

There exists a long tradition of testing the swelling potential of rock material at NTNU, due to the well documented problems with swelling gouge in weakness zones in different types of engineering projects (Nilsen 2016). The established form of oedometric swelling test at NTNU is maximum swelling pressure tests under conditions of zero volume change (Table 3, method 1a). The majority of tests have been carried out on the swelling potential of gouge material, and to some extent pulverized and compacted mixed soil/rock samples. In some cases, intact rock structure specimen has also been assessed, as in the study of Skippervik et al (2014).

The principle of the swelling pressure test developed at NTNU follows the method for determining maximum axial swelling stress for swelling rocks as suggested by ISRM (1977). The swelling pressure-swelling strain relationship cannot be obtained by the use of the current version of the oedometer at NTNU. Hence, single swelling pressure tests are performed on both powder and intact rock specimen. The procedure is similar for both powder- and intact rock structure specimen and is based on the ISRM (1977). The swelling test principle is illustrated in Figure 1.

3.3 The KIT method

At KIT, the ISRM suggested methods of 1989 is used for the laboratory set-up and the apparatus and methodology was modified several times. The latest changes on the oedometer apparatus configuration at KIT was made based on Pimentel (2004) and Vergara et al. (2014). This oedeometer allows to control the deformation or the load on the specimen in order to perform stress or strain swelling tests. In addition, cyclic swelling tests can be performed at KIT where the sample is subjected to dry and wetting cycles. These tests are performed under controlled axial deformation of the intact rock specimen (Table 3, method 4), and data on both swelling stress and swelling strain are obtained. The oedometer configuration is illustrated in Figure 3.

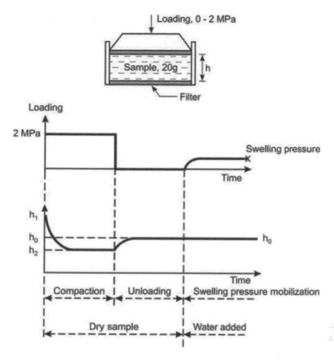

Figure 1.   Principle for testing swelling pressure at constant volume (Nilsen 2007).

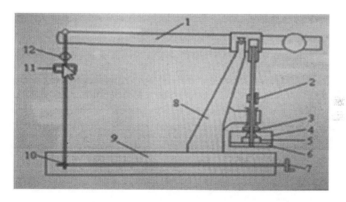

Figure 2.   Apparatus at NTNU. 1) Balance lever, with the ratio of 1:10. 2) Dial gauge with a sensitivity of 0.001 mm to measure the height (volume) of the specimen. 3) Adjustment screw. 4) Container. 5) Cylindrical test cell. 6) Steel base plate of the container. 7) Wheel. 8) Frame. 9) Base. 10) Worm gear. 11) Pressure ring. 12) Dial gauge.

The procedure of the cyclic swelling tests is summarized as following:

The axial swelling pressure developed by the specimen under conditions of zero volume change is recorded over the time, and after no noticeable change in the pressure is observed, the first cycle is assumed completed and the water is removed. As an effect of drying, the axial pressure produced by swelling decreases until a constant value is reached, marking the "finish-point" of the drying process. In case if swelling pressure does not increase after about 3 cycles, further cycles are performed with an increased deformation to see if it is an effect on the swelling behavior. Cycling is repeated until the pressure is stabilized. The adjustments to keep the desired volumes during the tests and the recording of swelling pressures are performed manually. The oedometer configuration is illustrated in Figure 3.

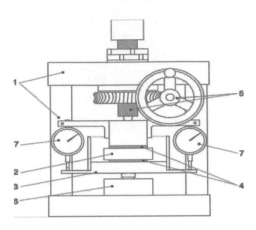

Figure 3. Modified oedometric apparatus at KIT (Vergara et al. 2014 and 2015). 1) rigid frame, 2) ring with rock specimen, 3) watering cell, 4) porous metal plates, 5) spindle, 6) load cell and 7) dial gauges.

Table 4. Main differences in methodology between NTNU and KIT (Selen 2017).

	NTNU	KIT
Preparation of rock specimen	– Core drilling (overcoring) – Use of trimming ring to fit specimen to oedometer ring	– Keep the core diameter if possible and remove the external surface by a lathe – Use of lathe to fit ring to specimen
Preferred test configuration (cf. Table 3)	Method 1)	Method 1) and 4).
Placement of dial gauges	– One dial gauge placed about 20 cm above the specimen. – Limited correction of the deformation of apparatus components between the dial gauge and specimen during the tests.	– Two dial gauges placed at opposite diameter ends of the loading plate. – Deformation of apparatus is avoided by abutting of dial gauges and sample, and by manual corrections during the tests.
Administration during tests	– Automatic volume control. – Automatic recording of swelling displacement and pressure.	– Manual volume control by reading the dial gauges and manually increasing/decreasing the load. – Manual recording of swelling displacement and pressure.
Sample size/mass (dry condition)	Powder – Mass: 20 g – Height: not measured – Diameter: 20 mm Rock – Mass: not measured – Height: ~5 mm – Diameter: 35,7 mm	Powder – Mass: 100 g – Height: ~18 mm – Diameter: ~60 mm Rock Mass: ~135 g Height: ~18.5 mm Diameter: ~60.5 mm
Pre-loading before tests	Yes, on both powder and intact rock structure samples with 2 MPa.	No, except 0.1 kN in order to ensure contact.
Number of wetting (and drying) cycles	Normally one.	Normally three or more.
Swelling stress and strain relationship	None	Yes, by allowing deformation (volume expansion in axial direction) in a stepwise manner in the conventional or in the cyclic tests.

3.4 *Main differences between NTNU and KIT*

The methodologies in operation at NTNU and KIT differ in several ways. The main differences include on the method, and on the internal modifications on apparatus and test procedures. Table 4 summarizes the main differences. Figure 2 and Figure 3 show the apparatus configurations.

The main difference in the methodologies in the powder tests, is the sample preparation. The sample dimensions are 2–4 times higher at KIT compared with the procedure used at NTNU. In addition, there is no pre-loading on the procedure used at KIT while a pre-loading of 2 MPa is carried out at NTNU. It is not known to which degree these differences appear in the results, but it is assumed that the influence is significant.

The main difference in the apparatus set-up is the placement of the dial gauge(s). At NTNU, the dial gauge is placed about 20 cm above the specimen, so it is unknown to which degree the deformation due to swelling pressure is absorbed in the apparatus between the specimen and the dial gauge. At KIT, the dial gauges are placed only a few millimeters from the specimen assuming that most of the swelling pressure induced by the specimen are detected by the gauges.

4 COMPARISON OF TEST RESULTS

The results comprise the comparison of oedometer swelling test results between NTNU and KIT. The swelling pressure test results on the powder material, whereby the tests were performed under conditions of zero volume change at both NTNU and KIT, are shown in Table 5. Since cyclic tests are not performed at NTNU, there is no comparable results for this type of test.

Table 5. Maximum swelling pressures in oedometer powder tests (Selen 2017).

Samples		Maximum swelling pressure (powder), MPa	
		NTNU	KIT
"strong"	AD-02, box 12 (1)	0.33	4.88
	AD-06, box 25	0.06	0.38
	AD-07, box 12	0.10	2.42
	AQD-02, box 12 (1)	0.10	0.41
«weak»	AQD-02, box 5	0.43	3.03
	AQD-02, box 6	0.35	2.87
	APH-02, box 18	0.12	0.82

Table 6. Overview of tests on discs. All values are given in MPa (Selen 2017).

	NTNU	KIT							
	–	1.cycle	2.cycle	3.cycle	4.cycle	5.cycle	6.cycle	7.cycle	8.cycle
Disc	$\varepsilon = 0$	$\varepsilon = 0$	$\varepsilon = 0$	$\varepsilon = 0$	$\varepsilon = 0.5\%$	$\varepsilon = +0.5\%$	$\varepsilon = +0.5\%$	$\varepsilon = +0.5\%$	$\varepsilon = +0.5\%$
AD-02, box 12	1.33	**2.08**	1.68	1.58	1.74	1.91	1.85	0.61*	–
AD-06, box 25	0.01	**0.05**	0.04	**0.05**	–	–	–	–	–
AD-07, box 12	0.22	0.13	0.14	0.15	0.17	**0.18**	–	–	–
AQD-02, box 12	0.09	0.04	0.04	0.04	–	–	–	–	–
AQD-02, box 5	–	**0.38**	0.26	0.25	0.13*	0.20*	–	–	–
AQD-02, box 6	0.08	0.17	0.17	0.18	0.09*	0.14*	**0.19***	0.18*	0.05*
APH-02, box 18	0.04	0.49	0.48	0.52	0.17*	0.52*	0.48*	0.57*	**0.74***

* = Controlled deformation allowed by reducing the load acting on the specimens.

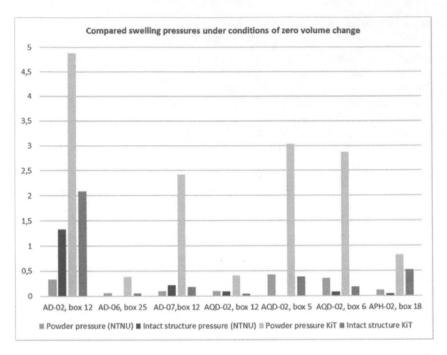

Figure 4. Compared swelling pressures in MPa of powder samples and intact rock samples (Selen 2017).

An overview of the intact rock structure (disc) swelling results is given in Table 6. The results from NTNU are obtained by single "zero volume change" tests. The results from KIT are obtained by cyclic tests, where some samples underwent cycles under conditions of "controlled deformation" as marked with "*". The maximum swelling pressure obtained in each cycle is presented. The highest value obtained by the cyclic tests is marked with bold text.

The obtained swelling pressure magnitudes from KIT are 2 to 4 times higher than for the corresponding tests carried out at NTNU. The differences in the magnitudes between powder tests and intact rock structure tests is also found higher at KIT. Figure 4 shows a comparison of the swelling pressures under conditions of zero volume change measured at both NTNU and KIT. The highest swelling pressures from the cyclic tests under conditions of *zero volume change* are representing the values obtained at KIT. Powder sample results are colored light blue (NTNU) and light orange (KIT), while intact rock structure results have corresponding strong blue and orange colors. The vertical axis shows the swelling pressure in MPa.

As can be seen from Figure 4, the NTNU method gave much lower swelling pressure potential of most of the intact rock samples compared KiT. Moreover, the swelling pressures obtained from the powder samples are also lower than that of the KIT results.

5 CONCLUSIONS

The comparison of the oedometer testing methodologies in operation at NTNU and KIT uncovered important differences. The deviations apply on both the apparatus used, and the procedures of swelling tests that is being practiced. The differences include the version of the ISRM suggested methods (1977 and 1989), and internal modifications on both apparatus and procedures made at each institution. The fact that different institutes operate with intern modifications of methodologies and apparatus configurations, leads to different results for similar rock types.

It is difficult to conclude on which of the methods is closest to the real swelling potential of the rock mass, since there exists no data on the in-situ swelling behavior. It is important to underline that the swelling pressure test results achieved from each laboratory cannot directly represent the swelling potential of the in-situ rock mass, and therefore must be used as indicative indexes. Each testing institution should classify indexes accordingly based on their data base recorded from actual projects and test results from the lab. This will help to understand on in what swelling pressure category the test results belong to. In the future, if laboratory data could be correlated with in-situ data, it would be possible to make an approach to a methodology appropriate to detect the swelling potential of the actual rock mass. However, it is reasonable to imply that the swelling measurements should be performed by dial gauges as close to the sample as possible. It is also clear that different institutions should agree on a standard method and a standard configuration of the apparatus, so comparisons of laboratory data can be made with lesser degree of uncertainty in the future. For projects where the rock mass is exposed to humid variations, as in hydropower projects, cyclic swelling tests should be included in the testing procedure. By allowing the specimen to deform during the test, data on both swelling stress and swelling strain can be obtained, and a simulation of the rocks interaction with support in a tunnel is possible to evaluate.

Further, it is important to highlight that there is a need for closer communication and cooperation among the institutions so that the swelling pressure test are standardized and differences in the test results are known, as have been made between NTNU and KIT. This cooperation will help to develop an index on the swelling pressure potential of both powder and intact rock discs, that could be translated in the consideration of potential swelling problems in the project. If comparable data on powder samples and intact structure specimen are systematically collected by synchronized test procedures, it may be possible to predict the intact rock behavior from powder test results in cases where the rock quality prevents preparation of intact rock specimens. Evaluations of laboratory testing will then be of higher value in the design of rock support on the tunnels passing through swelling rock mass.

REFERENCES

Galera, J., Paredes, M., Menchero, C. & Pozo, V. 2014. Risk Associated with Swelling Rocks in Volcanic Formations in the Design of Hydro-Tunnels. ISRM Regional Symposium-EUROCK 2014. International Society for Rock Mechanics.

ISRM. 1977. Suggested Methods for Determining Water Content, Porosity, Density, Absorption and Related Properties and Swelling and Slake-Durability Index Properties. Test Standards by International Society for Rock Mechanics, pp. 89–92.

ISRM. 1989. Suggested Methods for Laboratory Testing of Argillaceous Swelling Rocks. International Journal of Rock Mechanics, Mining Sciences and Geomechanics Abstracts, vol. 26 (5), pp. 415–426.

Madsen, F.T. 1999. International Society for Rock Mechanics Commission on Swelling Rocks and Commission on Testing Methods, *International Journal of Rock Mechanics and Mining Sciences*, p 291–306.

Nilsen, B. 2007. *Reliability of Swelling Pressure Testing for Tunnel Support Evaluation.*

Skippervik, C.P., Panthi, K.K. and Dahl F. 2014. Study on the Swelling Potential of some Selected Rocks. Proceedings: Norwegian Tunneling Conference, Oslo 2014.

Selen 2017. Study on material properties and testing of various rock types, development of investigation procedure and test methodology for future projects. NTNU, Institutt for geologi og bergteknikk.

Selmer, R. & Palmstrom, A. 1989. Tunnel collapses in swelling clay zones. *Tunnels & tunnelling.*

SN Aboitiz/Stache M. 2015. *Geological Report, Olilicon and Alimit HPP's and Alimit pump storage schemes, Feasibility phase* (Unpublished).

Vergara. 2016. Oral discussion.

Vergara, M.R., Balthasar, K. & Triantafyllidis, T. 2014. Comparison of experimental results in a testing device for swelling rocks. *International Journal of Rock Mechanics and Mining Sciences*, p 177–180.

Vergara, M.R. & Triantafyllidis, T. 2015. Swelling behavior of volcanic rocks under cyclic wetting and drying. *International Journal of Rock Mechanics and Mining Sciences,* 80, p 231–240.

Wittke-Gattermann, P. & Wittke, M. 2004. Computation of strains and pressures for tunnels in swelling rocks. *Tunnelling and Underground Space Technology,* 19, 422–423.

Xanthakos, P. *Ground Anchors and Anchored Structures*, John Wiley & Sons, 1991, Technology and Engineering, 686 p., p 432.

Geomechanics and Geodynamics of Rock Masses – Litvinenko (Ed.)
© 2018 Taylor & Francis Group, London, ISBN 978-1-138-61645-5

The implementation of de-stress gold mining technique along complex geological structures and heavily fractured ground conditions

F. Sengani
South Deep Gold Mine, South Africa
School of Mining Engineering, University of the Witwatersrand, Johannesburg, South Africa

T. Zvarivadza
School of Mining Engineering, University of the Witwatersrand, Johannesburg, South Africa

ABSTRACT: De-stress mining is one of the techniques used in mechanized, semi-mechanized and conventional deep to ultra-deep gold mines in South Africa. This mechanism involves the use of face-perpendicular preconditioning practice that transfers induced stress far ahead from the mining faces. This paper aims to highlight the effectiveness of face perpendicular preconditioning practice when mining heavily fractured ground. The trial was conducted in two sections of a deep level gold mine where the mining faces intersected the major faults. The ground conditions in the vicinity of the faults were heavily fractured, the previous geotechnical strategic designs made in order to mine the areas were not successful, leading to some of the ore being left unmined. Eventually, face perpendicular preconditioning practice was developed for such special areas. The developed face-perpendicular practice involved the use of five 4 m long drilled and blasted face-perpendicular preconditioning holes coupled with an increase in support density and with the use of 1.5 m long production holes. During the trial; no face burst/pillar bursting were reported, ground conditions on the hangingwall, sidewall and face were found to improve. A borehole camera was used to validate fracture frequency ahead of the mining faces while a Ground Penetrating Radar (GPR) was used to generate depth of fracturing ahead of the mining faces. Numerical modeling and microseismicity monitoring were also used to verify the results.

Keywords: Face-perpendicular preconditioning, destressing mining, fractured ground, hangingwall fracturing, borehole log, microseismicity monitoring, numerical modelling

1 INTRODUCTION

Preconditioning or de-stress blasting was initially started at East Rand Proprietary Mines (ERPM) with the guidance of CSIR in the early 1950's (Roux. et al. 1957). The argument for this was based on the concept that, if the holes drilled at right angles into the face were blasted, they would advance the depth of fracturing and in so doing transfer the high-stress zone further away from the face into the solid, should sudden failure occur in the high-stress zone, only limited damage would result, because of the cushion effect of the 'distressed' zone ahead of the face (Topper, et al., 2000). During the testing period, incidences of rockburst per area mined reduced by 36%, severe rockburst events by 73% and on-shift events dropped to almost zero. This was however not accepted by the mines as a viable and safe mining method and subsequently stopped (Roux et al., 1957). Safety in Mines Research Advisory Committee (SIMRAC) re-investigated pre-conditioning in the late 1980's. Long hole, face parallel pre-conditioning tests started at West Driefontein and at Blyvooruitzicht GM in 1990. This was not successful from a production viewpoint as it caused production delays. Tests on face perpendicular pre-conditioning in a Longwall started at Mponeng (then Western Deep Levels South) in 1994. Although successful it was not pursued after the testing was completed. Various attempts were made to do preconditioning mine-wide but it was only in the beginning of the 2000's that it was rolled out and enforced mine-wide (Toper et al., 2000; Mahne, 2004).

This study investigates the effectiveness of face-perpendicular preconditioning practice on mining ground conditions that consist of heavily fractured ground and complex geological structure. The study started with an extensive literature review on preconditioning practice, rockburst, instrumentation and monitoring systems related to the application of face-perpendicular preconditioning practice in heavily fractured and complex geological structures. In order to procure an increased understanding of the effectiveness face perpendicular preconditioning practice, the trial was carried out in two sections where most of the ore were left un-mined due to multiple occurrences of face-burst and pillar bursting. Underground observations (Visual examination), recordings and data collection were conducted within the sections. Microseismic monitoring, Ground Penetrating Radar (GPR) and Borehole cameras were used for monitoring the effectiveness of face perpendicular preconditioning practice. Abaqus Explicit numerical modeling software was also used to simulate the sigma 1 ahead of the mining face and in the vicinity of the in-stope pillars.

2 FACE PERPENDICULAR PRECONDITIONING METHODS

Face-perpendicular preconditioning practice was designed with the following descriptions: five drilled and blasted face-perpendicular pre-conditioning holes, drilled with a drill bit of 51 mm five, face-perpendicular preconditioning holes were charged up with emulsion and a 30 cm gassing gap was created. The rest of the hole was tamped by appropriate methods and equipment. Detonation of the production and five face-perpendicular pre-conditioning holes were sequenced with 1 millisecond delays chronologically as follows; detonate face-perpendicular preconditioning holes below the grade line, detonate face-perpendicular pre-conditioning holes above the grade line, detonate the cut and then the rest of the production holes, and lastly, alternate the positions of the face-perpendicular pre-conditioning holes after each blast. The preconditioning and production holes layout is as given in Figure 1.

3 STUDY RESULTS

This section provides results on rock mass fracturing, Ground Penetration Radar monitoring, seismicity monitoring and stress simulated ahead of the mining faces. The comparison on the rock mass fracturing was based on the way in which the rockmass responds and behaves after effective and ineffective face perpendicular preconditioning practices. Hangingwall, sidewall and blastholes fracturing were considered during the evaluation of the results. In order to validate the results of the study, GPR scans were conducted along different mining faces, some ineffectively preconditioned and some effectively preconditioned. Further studies

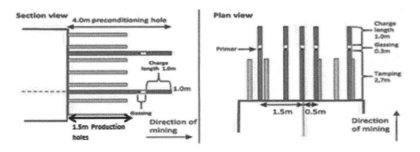

Figure 1.   Preconditioning and production holes layout.

on the fracture frequency were conducted using borehole camera. Lastly, Abaqus Explicit numerical modeling was used to simulate sigma $_1$ ahead of the mining faces before and after the implementation of face perpendicular preconditioning practices.

4   ROCK MASS FRACTURING

According to Toper (2003), one of the beneficial side effects of preconditioning is an improvement in the ground conditions of the hanging wall. During the implementation of the face-perpendicular preconditioning methods, hanging wall conditions were evaluated before the practice and during the practice. The data obtained on the intensity of fracturing on the hanging wall has shown that the hangingwalls were heavily fractured before the implementation of the practice. Furthermore, observation has shown that most of the hangingwalls were associated with bulging of the crushed and large rocks contained between support tendons and welded mesh. Owing to that, most of the roof bolts were noted to be poorly installed due to the heavily fractured ground, most of the roof bolts were not flushed against the hanging wall. Moreover, several mining faces were also found to be affected by face-bursting when small and large seismic events occur ahead or far from the mining faces.

In order to extract/mine out ore within the complex geological structures and heavily fractured ground, a strategic design which involves the use of five face-perpendicular preconditioning holes, increased support density and reduction of the production hole length to 1.5 m was implemented. The results of the study have indicated that the intensity of fracturing on the hangingwalls was rapidly improved as compared to the previous situation. However, minor fractures on the hangingwalls, with no difficulties on the installation of support system were found to be the results of the implementation of the practice. Owing to that, this practice has shown great improvement on the hanging wall fracturing, allowing for the development of minor fractures on the hanging wall and there was no face-burst reported during this practice. Although extensive scaling of the mining faces was achieved during this practice, the extensive scaling might be influenced by the number of preconditioning holes drilled and blasted as well as the ground conditions within the areas.

5   GROUND PENETRATING RADAR RESULTS

Five face-perpendicular pre-conditioning practice was found to produce a significant difference in the nature of fracturing ahead of pre-conditioned faces as compared to the previous situation where face-perpendicular preconditioning was found to be ineffective. This is evident in the GPR images (see Figure 2A and Figure 2B). Before the implementation of the practice, fractures ahead of the mining face were found to extend from 0 m to less than 3 m in depth while the fracturing frequency analysis after the practice was observed between 0 m to

 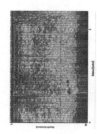

Figure 2A.   GPR scan during ineffective preconditioning practice, 2B GPR scan during effective preconditioning.
Note that the bright red color on the GPR image indicates stronger reflections within the material under test, the reflections are caused by discontinuities in the material and change in material properties.

3.5 m. The density of open fractures was also much higher during the implementation of the practice than before the implementation. This can be observed from Figure 2A and 2B where the high intensity of red color was observed during the implementation of the practice and scattered intensity of red color was observed before the implementation.

6 BOREHOLE CAMERA RESULTS

Borehole periscope surveys were conducted daily within the sections and the analysis of the study was based on the fracture frequency standard developed by Sengani and Zvarivadza, 2017 (see Table 1).

The results of the study have shown that preconditioning holes were found to be fractured for the first 1 m beyond the face and the ground was solid up to the end (4 m) of the holes before the implementation of the practice. Previous work has shown that a face with such short depth of fracturing is prone to face bursting (Sengani and Zvarivadza, 2017). After the implementation of the practice, most of the face-perpendicular preconditioning holes were noted to have extensive fracturing between 0–3 m and the fracturing decreased towards the end (4 m) of the preconditioning blast hole. Owing to that, there was no face burst or pillar burst reported during this practice.

7 SEISMICITY RESULTS

The seismicity, as expected, was clustered around the mining panels. It was noted that high seismic activities with many large events occurred before the implementation of the practice. Most of the large magnitude events occurred along geological structures, events with small magnitudes were reported during the implementation of the practice (Figures 3A and Figure 3B) and Different colors were used to differentiate the time at which the event occurred and the size of the circle represents the magnitude of the event.

Table 1. Standard for fracture frequency analysis (after Sengani and Zvarivadza, 2017).

Fracture frequency/meter	Risk profile	Color coding
5	High strainburst risk, rockmass not fractured/yielded	
5–10	Ff/m between 5–10: Medium strainburst risk, rockmass beginning to fracture/yield	
10–20	Low strainburst risk, rockmass has fractured/yielded	
>20	Very low strainburst risk, rockmass highly fractured/ yielded	

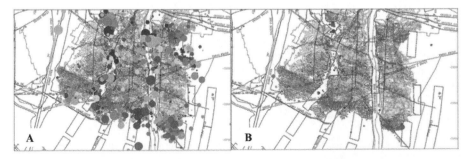

Figure 3A. Seismic plots before the implementation of the practice, 3B) Seismic plots after the implementation of the practice.

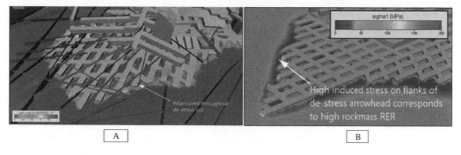

Figure 4A. and 4B.   Modelled induced stresses ahead of distress faces and along the yield pillar.

8   STRESS SIMULATION ALONG THE INSTOPE PILLAR

In order to understand the behavior of the distribution of stress ahead of the mining faces, numerical analysis was carried out to simulate the magnitudes of stress ahead of the mining faces and at the vicinity of the in-stope pillar using Abaqus Explicit software. Only σ_1 was generated and it was found that σ_1 ahead the face and at the vicinity of the in-stope pillar were about 150 MPa to 200 MPa (see Figure 4A and 4B). However, a gradual reduction of stress ahead of the mining faces was noted from Figure 4B to Figure 4A. This has shown that high stress was simulated before the implementation of the practice and intermediate to moderate stresses were simulated during the implementation of the practice.

9   CONCLUSIONS

It was concluded that the implementation of face-perpendicular preconditioning practice when mining along heavily fractured ground conditions comes with several benefits such as; provision of smooth hangingwalls with moderate fractures on the sidewalls of the excavation, provision of extensive fracturing on the mining faces which also lead to adequate face scaling and reduce the chance of face burst occurrence, ability to install support systems with no difficulties.

As a result, borehole log results before the implementation of the face-perpendicular preconditioning practice has shown that only first 1 m of the preconditioning holes consist of more than 20 fractures and a rapid decrease in fracture frequency was noted from 1.5 m towards the end (4 m) of the preconditioning hole. After the implementation of the practice, most of the preconditioning holes were noted to have extensive fracturing between 0–3 m and poorly developed dog-earing was noted from 3.2 m to 4 m. It was then concluded that the implementation of effective face-perpendicular preconditioning holes helps in the development of intensive fracturing ahead the mining faces to the depths of 3.5 m to 4 m depending on ground conditions.

Ground Penetrating Radar survey images before the implementation of the practice have shown a weak reflection within 1 m to 3 m. This was an indication of poor fracturing ahead the mining faces and therefore the rock ahead of the mining face was prone to face bursting. The Ground Penetrating Radar image taken during the implementation of the practice has indicated stronger reflections ahead of the mining faces, significant less possibility of the occurrence of face bursting. It was then concluded that the implementation of the practice has produced stronger reflection than before the practice. The density of open fractures was also much higher during the practice than before.

High seismic activities with many large events were noted before the implementation of the practice as compared to during the practice. However, most of the large events were found to occur along the geological structures. The stress simulated ahead of the mining faces has shown gradual decrease after the implementation of the practice compared to before the implementation of the practice. This could be attributed to deeper fracturing produced by face-perpendicular preconditioning practice, playing a major role in transferring stresses far ahead from the mining faces.

REFERENCES

Adams, D.J., Gay, N.C. and Cross, M. (1993). Preconditioning—a technique for controlling rock bursts. Proceedings of the 3rd International Symposium on Rockbursts and Seismicity in Mines, Young, R.P. (ed.), Balkema, Rotterdam.

Adams, D.J. and Geyser, D. (1999). Preconditioning of 43 Hangingwall Haulage at Kloof No. 4 Shaft. Proceedings of the 2nd Southern African Rock Engineering Symposium, SARES99. Hagan, T.O. (ed.), 13–15 September 1999, Johannesburg Facilities.

Cook, N.G.W., Hoek, E., Pretorius, J.P.G., Ortlepp, W.D. and Salamon, M.D.G. (1966). Rock mechanics applied to the study of rockbursts. Journal of the South African Ins. of Min. and Metall. May 1966. pp. 435–528.

Cook, N.G.W. (1983). Origin of rockbursts. Proc. of the Symposium on Rockbursts: prediction and control. IMM and IME, London, Oct. 1983. pp. 1–9.

De Kock, W.P. (1964). The geology and economic significance of west wits line. Geology of Some Ore Deposits in Southern Africa. (S.H. Haughton, Ed.). Geol. Soc. of S. Afr. Johannes-burg.

Giltner, S.G. (1992). Proposed Method of Preconditioning Stope Faces with Small Diameter Blastholes. COMRO Internal Note No: 06/92.

Hill, F.G. and Plewman, R.P. (1957). Destressing: A means of ameliorating rockburst conditions, Part 2 – Implementing destressing with a discussion on the results so far obtained. Journal of SAIMM, October 1957, pp. 120–127.

Legge, N.B. (1987). The incidence and location of rockbursts and rockfalls in gold mines as indicated by historical and contemporary accident data. Chamber of Mines Research Organisation, COMRO, research report, No. 33/87.

Roux, A.J.A., Leeman, E.R. and Denkhaus, H.G. (1957). Destressing: A means of ameliorating rockburst conditions, Part 1 – The conception of destressing and the results obtained from its application. Journal of SAIMM, October 1957. pp. 101–119.

Sengani, F. and Zvarivadza, T. (2017). Review of pre-conditioning practice in mechanized deep to ultra-deep level gold mining. 26th International Symposium on Mine Planning and Equipment Selection (MPES2017). 29–31 August 2017, Luleå, Sweden. pp. 113–127.

Toper, A.Z., Kabongo, K.K., Stewart, R.D. and Daehke, A. (2000). Mechanism, optimization and effects of pre-conditioning. Journal of South African Institute of Mining and Metallurgy. pp. 7–16.

Toper, A.Z. (2003). The effect of blasting on the rockmass for designing the most effective preconditioning blasts in deep level gold mines. PhD thesis, University of the Witwatersrand Johannesburg, school of mining engineering, South Africa.

Wilson, N.L., Oosthuizen, D.H., Brink, W.C.J. and Toens, P.D. (1964). The geology of Vaal Reef basin in the Klerksdorp area. Geology of Some Ore Deposits in Southern Africa. (S.H. Haughton, Ed.). Geol. Soc. of S. Afr. Johannesburg.

Geomechanics and Geodynamics of Rock Masses – Litvinenko (Ed.)
© *2018 Taylor & Francis Group, London, ISBN 978-1-138-61645-5*

The use of face perpendicular preconditioning technique to destress a dyke located 60 m ahead of mining faces

F. Sengani & T. Zvarivadza
School of Mining Engineering, University of the Witwatersrand, Johannesburg, South Africa

ABSTRACT: Face perpendicular preconditioning technique has become a useful tool for destressing geological structures such as dykes that are located 20 m and more away from the mining faces. In this study, a trial on destressing a dyke that was located 60 m ahead of mining faces was conducted through drilling face perpendicular preconditioning holes that intersected and passed the dyke with two meters. Drilled cores were taken for laboratory testing to confirm the stiffness and strength of the dyke. The results from laboratory tests indicated high stiffness and strength of the dyke material. From the results of the superior stiffness of the dyke, it was noted that stresses were concentrated on the dyke. When the strength of the dyke material is exceeded, the violent bursting of the dyke material may occur resulting in seismicity along the feature. A sticky emulsion (Ug101 s) with the density of 1.1 g/cc, charging a mass of 30.4 kg, powder factor of 1.3 kg/tonne, with gassing gap and 5 m long tamping material was used to destress the dyke. After detonating the holes, a small seismic magnitude was noted. However, as mining progressed towards the dyke and passing through the dyke, good ground conditions were observed. Less face, sidewall and hangingwall damage were noted due to the good ground conditions as compared to the previous situations. Numerical modelling was also conducted to simulate Rate of Energy Release (RER) and it was found that RER gradually reduce after preconditioning the dyke.

Keywords: Face perpendicular preconditioning, de-stressing, RER, hangingwall fracturing, microseismicity, numerical modelling, dyke

1   INTRODUCTION

Historically, in the mid-1980s, the Chamber of Mines Research Organisation (COMRO) initiated a research programme which was to investigate active rockburst control techniques in South African mines, specifically gold mines. The investigation was aimed at developing effective preconditioning methods that can transfer stress far ahead from the mining faces and also assess the potential of these technologies to be implemented in deep gold mines (Brummer, 1988). According to Rorke et al (1989), the COMRO's involvement in preconditioning began in 1987 with experimentation at West Driefontein Gold Mine, where the 32–12 W stope was being mined into a large remnant along the Western Deep Levels (WDL) boundary. The technique that was implemented made use of long, face-parallel holes drilled along the length of the 30 m panels. The 76 mm diameter holes were positioned between 2.5 m and 3.5 m ahead of the face and drilled within a shift. The panels were mined beyond the line of the previous preconditioning holes and, hence, the spacing between preconditioning holes averaged about 8 m (Toper, 2002).

Following a trial period of five months of test drilling, preconditioning was implemented in two panels. Once the technique had been optimised, all five panels within the stope were preconditioned. A total of 18 preconditioning blasts were carried out in the 11-month period of the project. Convergence of 5 mm to 40 mm associated with a preconditioning blast, face scaling of up to 300 mm, and a lack of damage at the face following large seismic events were

reported. An observed reduction in shallow-dipping fracturing compared to non-precondi-
tioned panels resulted in improved hangingwall stability, which may partly account for the
lack of seismic damage. However, the project was terminated when the technique could not
be integrated into the new layout that was required, as the stope was approaching a seismically hazardous structure.

A further preconditioning experiment was initiated on Blyvooruitzicht Gold Mine
(BGM) in the 18–13 W stope, an up-dip panel along a protection pillar adjacent to a seismically active fault (Rorke *et al.*, 1990). A series of 30 m long, 76 mm diameter holes fanned
out from the dip gullies was planned to be drilled into the entire pillar, with the intention of
"preconditioning" the pillar with one blast. Eventually, only the holes on the edge of the pillar could be drilled to nearly their full length. The other holes were drilled to only 10 m. Difficulties experienced with drilling into the core of the pillar provided some insight regarding
the condition of the pillar. Despite the seismicity from the adjacent fault, the pillar was
eventually mined out without incident by fanning the preconditioning holes drilled from the
up-dip gullies at five different face positions. Improved hangingwall stability was reported to
be due to steeper extension fractures following the introduction of preconditioning.

Extraction of a dip pillar with preconditioning at BGM began in mid-1990 (Lightfoot *et al.*,
1996). The 30–24 W stope is situated at the southern extent of BGM near the boundary pillar
to Western Deep Levels (WDL). This dip pillar was 40 m wide and 150 m long with the top of
the pillar terminating on a stabilizing pillar. Initially, the pillar was conventionally mined but,
after problems with consistently poor ground and several large seismic events, the mine decided
to implement their own preconditioning project in mid–1990. They requested COMRO to
monitor the project. The method of preconditioning was similar to that used at the 18–13 W
stope: 10 m long holes fanned out from the dip gullies. Difficulties with drilling and frequent
damage to support and the collar area of the holes all resulted in production delays.

Toper (2003) pointed out that in late 1994, it was recognized that, although face-parallel
preconditioning appeared to be well suited to the mining of long and narrow strikes pillars,
it would be difficult to implement in a normal deep-level longwall production environment
without imposing considerable delays in the mining cycle. Due to that reason, the new experimental site was opened on a deep-level longwall of WDL South Mine. The Experiments
involved drilling short face-perpendicular preconditioning holes as a standard addition to
every production blast. The result pointed out that the implementation could be practiced in
a deep-level longwall mining environment without significant disruption to the mining cycle
(Lightfoot *et al.*, 1996).

2 MOST RECENT EXPERIMENTS

In 1999, the studies conducted by Adams and Geyser outline a successful preconditioning
experiment which was carried out in a development end at Kloof gold mine No.4 shaft. In
their study, it was pointed out that a haulage used for trimming was experiencing instability
and strain burst. However, this problem was associated with the tunnel obliquely intersecting
a geological structure (dyke) which was weaker than the surrounding lava and assumed to be
seismically active (Toper, 2002). A preconditioning pattern designed with 14 preconditioning
blasts, the constant number of development blast in conventional mine, was implemented. The
tunnel was then driven through the dyke and was found to be safer, without experience injuries
and face burst problems. After this experience, the preconditioning technique on this site was
found to be very successful in reducing face burst as well as eliminating injuries (Toper, 2002).

3 RESEARCH APPROACH

In September 2016, the Grey Ghost Dyke was intersected by one of the section at the mine.
Immediately after intersecting the dyke, a rapid increase in seismicity was experienced at the
mining faces, and ahead of the mining faces. As described earlier in this paper, it was postulated

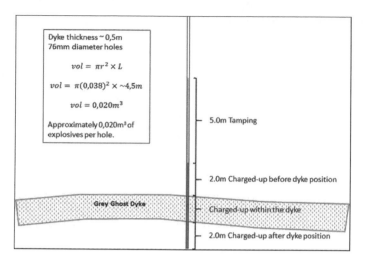

Figure 1. Plan view of how the dyke was de-stressed.

that the dyke contains a build-up of stress as the mining front approaches it. This built up stress eventually exceeds the strength of the dyke material and results in increased seismic activity. The dyke also acts as a barrier for stress to propagate further south and thus result in increased seismicity between the dyke and the mining front. For this reason, it was decided to implement long hole preconditioning in order to fracture the dyke and thus negate the ability of the dyke to build-up stress as well as to allow stress to naturally propagate south of the structure (see Figure 1).

As a result, there were a number of strategies which were employed together with long hole preconditioning, which includes the following:

➢ Precautionary Zones
The seismically active geological structure is demarcated on the plan and a precautionary area, 15.0 m ahead, and 15.0 m beyond the structure (including the actual structure) is highlighted in orange. This precautionary zone indicates the onset of a number of special precautions to be taken in order to safely negotiate the geological anomaly.
➢ Face Advance
Once in the demarcated precautionary zone, face advance is limited to a maximum of 1.5 m. This ensures unsupported spans are limited to a minimum following the blast and assists with swift cleaning and re-supporting of the blasted face.
➢ Additional Support
Although the support system in general caters for the anticipated seismicity expected at the deep level gold mine, additional support is required within the precautionary zone. The normal square pattern of support is supplemented with an additional bolt in the middle of every four bolts installed, effectively doubling up on the support requirements.

4   RESEARCH RESULTS

This section provides results on the seismic monitoring, rock mass fracturing and the Rate of Energy Release (RER) before and after the application of the method. Seismic monitoring results were based on the recorded events that occurred before and after the practice. Visual observations were also conducted to compare the rock mass behaviour in the vicinity of the de-stress cuts, these include hangingwall and sidewall fracturing. Lastly, the RER evaluation was based on the Peak Particle Velocities (PPVs) of all the seismic events that took place before and after the application.

5 SEISMICITY MONITORING RESULTS

The results of the study were based on the seismic event that took place in the vicinity of the dyke, this was done before preconditioning the dyke and after preconditioning the dykes. As it was expected the results of the study have shown a large quantity of seismic events taking place before the long hole preconditioning blast. However, this was based on the six months seismic plots along the dyke. Six months later after preconditioning the dyke, seismic plots were also conducted and it was noted that the seismicity along the dyke was rapidly reduced, no face bursting reported within the sections (see Figure 2). This method also brought some of the benefits, such as solid hangingwalls with minor fractures on the sidewalls. Besides that, the advance of the mining faces was improved rapidly.

6 ERR RESULTS

The ERR of the mine was simulated using a non elastic model called Abaqus Explicit. This was conducted three months before the implementation of the techniques for de-stressing dyke and three months after de-stressing the dyke to compare the ERR of the mine. However, in this analysis, the dark red color was used to denote high ERR and light grey was used to denote low ERR. Three months before the practice, higher ERR was simulated along the mining faces, regional pillars, and the dyke. Three months after the implementation of the practice, the ERR rate decreased gradually, with low ERR simulated along the regional pillars and scattered high ERR rate in some portion of the dyke as shown in Figure 3. It was then noted that the gradual decrease of ERR was influenced by the implementation of long hole preconditioning blast through de-stressing and transferring stress far ahead of the mining faces and the vicinity of the dyke.

7 ROCK MASS FRACTURING

Toper (2002) studies pointed out that one of the beneficial side effects of preconditioning is an improvement in the ground conditions of the hangingwall. The comparison of ground fracturing on both hangingwall and mining faces as well as sidewalls were compared. The data obtained from the hangingwall assessments have shown a smoother hangingwall on the mining faces after de-stressing the dyke. Intermediate to minor fracturing on the hangingwall was

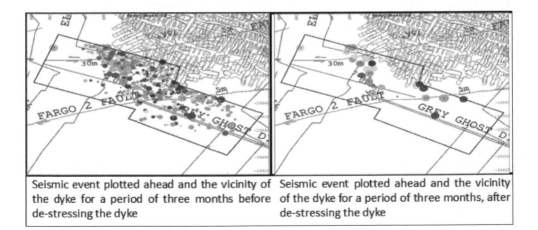

| Seismic event plotted ahead and the vicinity of the dyke for a period of three months before de-stressing the dyke | Seismic event plotted ahead and the vicinity of the dyke for a period of three months, after de-stressing the dyke |

Figure 2.   Seismic events plotted in the vicinity of the dyke and ahead of the mining faces before and after de-stressing the dyke.

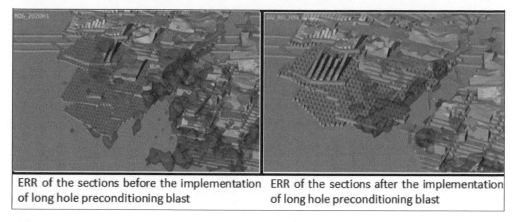

ERR of the sections before the implementation of long hole preconditioning blast	ERR of the sections after the implementation of long hole preconditioning blast

Figure 3. ERR simulated before and after the implementation of long hole preconditioning blast.

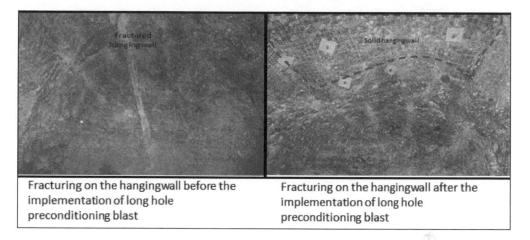

Fracturing on the hangingwall before the implementation of long hole preconditioning blast	Fracturing on the hangingwall after the implementation of long hole preconditioning blast

Figure 4. Hangingwall fracturing before and after implementation of long hole preconditioning hole blast.

observed before the implementation of the practice. On the other hand, the fracturing at the vicinity of the mining faces was observed to improve with extensive fracturing on the mining faces, especially around the normal preconditioning holes. This extensive fracturing was only observed after de-stressing the dyke, in addition to that, there was no face burst reported for a period of three months along all the mining faces intersected by the dyke. Before the implementation of the practice, the mining faces were observed to be fractured but several face bursts were reported along the mining faces that intersected the dyke (see Figure 4).

8 CONCLUSIONS

Rockbursts and Falls of Ground related fatalities have been the most common problematic issues faced by deep to ultra-deep gold mines in South Africa. Most of these fatalities were seismic related due to sudden release of energy from the rock. It has been reported that the most dangerous seismic events are the ones related to seismically active geological structures such as dykes. Historically, several rockburst committees were appointed to resolve the issue, but still all the solutions provided could not be able to stop the occurrence of rockburst events due to seismically active structures within the mine. This study attempted to provide clarity on how to control the behavior of geological structures within the vicinity of underground workings. To achieve the objective of the study, a detail research on destressing a dyke was investigated. After detailed research on the impact of de-stressing the dyke using long hole preconditioning blast, it was concluded that the practice of de-stressing the dyke was successful. This was validated through the improvement on the hangingwall fracturing, side-wall fracturing and extensive fracturing on the mining faces. As a result, seismic events plotted ahead of the mining faces and in the vicinity of the dykes have shown gradual decrease on recorded events along the dyke and ahead of the mining faces. Furthermore, analysis has shown that the ERR along the sections has shown a gradual decrease after de-stressing the dyke. It can be concluded that the use of long hole preconditioning blast for de-stressing seismically active geological structure was successful.

REFERENCES

Brummer, R.K., (1988). Active methods to combat the rockburst hazard in South African gold mines. Conference on Applied Rock Engineering, (CARE), IMM, Newcastle-upon-Tyne, London. pp. 35–43.

Kullmann, D.H., Stewart, R.D. and Grodner, M., (1996). A pillar preconditioning experiment on a deep-level South African gold mine. 2nd North American Rock Mechanics Symposium, Montreal, Canada.

Lightfoot, N., Kullmann, D.H., Toper, A.Z., Stewart, R.D., Grodner, M., Janse van Rensburg, A.L. and Longmore, P.J., (1996). Preconditioning to reduce the incidence of face bursts in highly stressed faces. CSIR Division of Mining Technology, SIMRAC Final Report, Project GAP 030.

Rorke A.J., Brenchley P.R. and Van Rensburg A.J., (1989), Preliminary Preconditioning Results Obtained at West Driefontein. COMRO Internal Report No: 548.

Rorke, A.J., Cross, M., Van Antwerpen, H.E.F. and Noble, K., (1990). The mining of a small up-dip remnant with the aid of preconditioning blasts. International Deep Mining Conference, Technical Challenges in Deep-level Mining. SAIMM. Johannesburg.

Toper, A.Z. (2002). Destressing/Preconditioning to control rock bursts in South African deep-level gold mines. Int. Seminar on Deep and High-Stress Mining. Australian Centre for Geomechanics. Perth, November 2002.

Geomechanics and Geodynamics of Rock Masses – Litvinenko (Ed.)
© 2018 Taylor & Francis Group, London, ISBN 978-1-138-61645-5

Prospects of the physical model-based study of geomechanical processes

A.N. Shabarov
Geomechanics and Mining Production Problems Scientific Centre, St. Petersburg Mining University, Sankt-Peterburg, Russia

B.Yu. Zuev & N.V. Krotov
St. Petersburg Mining University, Sankt-Peterburg, Russia

ABSTRACT: In the current conditions of the intensive Mineral Deposit (MD) mining and the growing complexity of the mining and geological setting, geomechanical processes, while continually developing in space and in time in structurally damaged rocks under the effect of various natural and man-caused factors, are taking expressly non-linear forms appeared as a change over time in different structural parameters of rock mass, and its Stress-Strain Behaviour (SSB) both in quasi-static and dynamic modes. The modelling method based on Equivalent Materials (EM) is one of the most effective study methods for deformations and damage in large rock masses at the MD mining which allows the most adequate simulation and study of their physical behaviour. The new simulation methodology developed on its basis and the study of different structures, physical fields and their evolution enabled to study the physical principles of the reviewed geomechanical processes more deeply, to enhance the reliability of a prediction on their basis and the prevention of dangerous dynamic phenomena. The developed innovative modelling technology allowed us to solve a number of perspective fundamental and application problems in the mining geomechanics through the obtainment of the new data on the physical processes happening in space and in time.

Keywords: geomechanical processes, modelling, equivalent materials, rock mass, physical fields, deformations

1 INTRODUCTION

By now the world practice of mineral deposit (MD) mining has gained the vast experience of geomechanical processes control, i.e. the purposeful modification of stress-strain behaviour (SSB) of rock masses to ensure efficiency and safety of mining.

Nevertheless, there are many events and unsuccessful efforts occurred in respect to control of geomechanical state of rock masses which cause the destruction of mines, disastrous rush of ground waters and soils, destruction of mining plants, formation of holes in the land, dynamic phenomena, induced earthquakes and other effects. This problem is extensively connected with the lacking study and incomplete scientific notations regarding the physics of a number of existing deformation processes in a block rock mass at the modern mining geomechanic development.

In the current context of the intensive MD mining and the growing complexity of mining and geological setting, the processes continually developing in space and in time in structurally damaged rock masses under the effect of various natural and man-caused factors are taking expressly non-linear forms. These processes in scale, duration and frequency may be of the most various forms: regional, local, dynamic, static, impulse, long, and appear in different combinations, occur independently of each other, or may be cause-and-effect linked.

The modelling method based on equivalent materials (EM) is one of the most effective study methods of non-linear geomechanical processes of deformations and destructions in large rock masses at the MD open pit and underground mining which allows the most adequate simulation and study of their physical behaviour.

At the same time, the conventional EM modelling method was not capable to achieve its potentials to the full. The new simulation methodology developed on its basis and the study of different structures, physical fields and their evolution at MD mining enabled to study the physical principles of the reviewed geomechanical processes more deeply, and to enhance the reliability of a prediction on their basis and the prevention of dangerous dynamic phenomena.

2 THE NEW METHODOLOGY INVOLVES A NUMBER OF NEW DEVELOPMENTS

- a unified comprehensive condition for similarity of stresses, deformations, main unit components of the energy balance in the event of concurrent "characteristic similarity criterion" processing for EM strength and deformation parameters [1];
- new EMs meeting the unified comprehensive condition for similarity [1,2];
- the automated system and special software for formation of a wide range of boundary conditions at the periphery of the modelled area of rock mass [1,2,3];
- simulation methods for various rock mass structures: single and systems of tectonic faults, block-hierarchy structures, free-form MD, stratas of varying thickness [1,2];
- a complex study technique for stress fields, deformations, temperatures, components of energy balance, acoustic and electromagnetic fields [4].

The developed innovative modelling technology allowed us to solve a number of perspective fundamental and application problems in the mining geomechanics. A series of studies have been carried out in the recent years, which enables to obtain new data on non-linear physical processes in the mining:

- redistribution of stresses in block structures in the light of natural and man-made factors depending on their hierarchy, geometry, and conditions of contacts between blocks;
- study of dynamic high-frequency geomechanical processes at the fall of a roof demonstrated by: the change of stress with time and in space, displacement of roof materials, their gradients, speeds and accelerations on different levels; emission of heat energy at non-reversible deformations in mass elements being destructed, change in parameters of acoustic signals;
- study of regularities in the change of the SSB and strike shifting of tectonically destructed rock mass at the extraction of ore deposit with goaf stowing;
- study of lamination parameters and development of cross fractures in the zone of fracture effect at the modelling of extraction of coal seams superimposed by several faces;
- study of regularities in the change of the SSB and the development of fractures in a rock mass at gas drainage from coal seams;
- study of formation processes of slides at reverse falling of seams in permafrost conditions.

The research results were stated in detail in a number of articles, and reports at international conferences [1–4].

3 LET US PROVIDE BELOW THE MOST MEANINGFUL RESEARCH RESULTS OF THE FUNDAMENTAL AND APPLICATION IMPORTANCE

1. Regularities in redistribution of stress in a block-hierarchy mass of three hierarchy grades are found at removing works (linear scale: 1:500, mass blocks from durable sandstone: R_{comp} = 40–50 MPa, depth: 400–1000 meters, wall resistance factor 0.3–1,0):

- The available tectonic fault system and the generated fragments are one of the determining factors which form extremely non-uniform SSB with normal compressing stresses in the centre of blocks with maximal concentrations reaching values: $K_{max} = 9$;
- With the increase in the hierarchy grade of blocks $c \rightarrow b \rightarrow a$ the growth of stress concentration factors in their centres is found;
- Average stress concentration factors (K_c and K_b) in the centre of different-hierarchy blocks (c and b) increase with the growth of weighted friction factor along their borders;
- The pattern of stress distribution in blocks is attributed with their location in the bigger blocks of the given hierarchy system, which is confirmed by existing trend to symmetrical distribution of stress relatively to their centres;
- During removing works in a block mass, stresses are redistributed essentially differently to the known in the layered-fractured mass—towards the larger stress concentration and the zone of mining effect (Fig. 1).
- Extreme values of stresses are determined more by block-hierarchy structure of a mass, than by the bearing pressure zone, and are of the asymmetrical and poorly predicting nature;
- A block mass is nonlinearly receptive to weak impacts, which appears as a change of stress concentration factors K in particular blocks at the distances 3–5 fold more the similar changes of K in masses not having block structures.

2. Study of dynamic processes at extraction of the suite of coal seams in typical conditions of OAO SUEK-Kuzbass mines.

The study was carried out at modelling of superimposed coal seams with the complete falling in the extracted space, in scale 1:87.5. Standard parameters for modelled rock material: thickness of each of two coal seams, m = 3.5 m; summary thickness of interseam and modelled roof: 25 and 58 m, correspondingly; thickness of the roof rock: 1.5–3.5 m; compression strength and modulus of elasticity for coal seams and roof rocks: 20–88 MPa and 7–22 MPa.

As a result of the study, the new data were obtained critical for the estimation of potential impacts from dynamic phenomena:

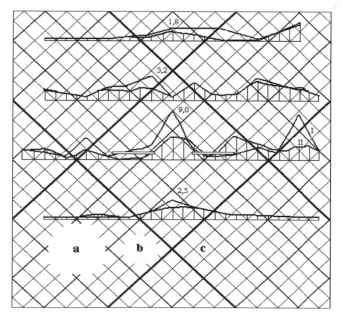

Figure 1.   Distribution of stress concentration factors along different levels prior to (I) and after (II) mining operations.

- At the primary roof subsidence during the excavation of the lower seam starting from a holing chute BE (Figs. 2 and 3), oscillation processes take place in coal seam roof with significant variations in acceleration values: 10 Hz to 4000 Hz, corresponding to life-size accelerations 3–63 Hz;
- Change in frequency of oscillation processes has the complex regularity, featured by a constant alternation of high and low frequencies;
- Variation of the acceleration amplitude, both in the model and in life, is 0 to 7.4 g (Figs. 4 and 5);
- Maximal accelerations are recorded at time interval of the first fall of the main roof (Fig. 5).

3. Study of the destruction of a rock mass affected by mining operations in regard to the development of lamination processes and evolution in the growth of cross fractures during the consequent extraction of 3 faces and development of these processes in typical conditions of OAO Vorkutaugol mines, in model scale 1:200.
   Upon the studies the following is found:

- The developed modelling methods of removing works at the extraction of superimposed seams across the strike by two or more faces (Fig. 6) allow to obtain the new data on parameters of the destruction zone of a rock mass affected by mining operations, in regard to the development of lamination processes and evolution in the growth of cross fractures during the consequent extraction of faces and the development of these processes in time;
- Upon the research, the new data is found regarding the potential appearance of the complex spatial system of lamination in the roof rock and through gas conductive cracks spreading for overlying rock up to 280 m (i.e. to 70 summary thicknesses of extracted seams);
- A fact is ascertained that the processes of displacement and lamination, layering and cracking in rock roof layers are developed in time in the area of falls at a distance over 400 meters behind the stall (Fig. 7), which allows to take a fresh look at the activation processes of strike shifting at the repeated mining of a rock mass appeared as secondary destructions and solidifying of the fallen rock in the filled stope mines.

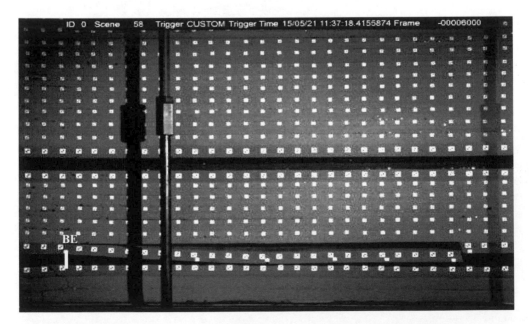

Figure 2.   Photo of the rock mass model after falling of the lower seam immediate roof.

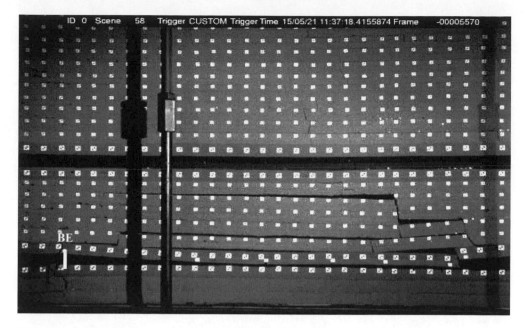

Figure 3.   Photo of the rock mass model after falling of the lower seam main roof.

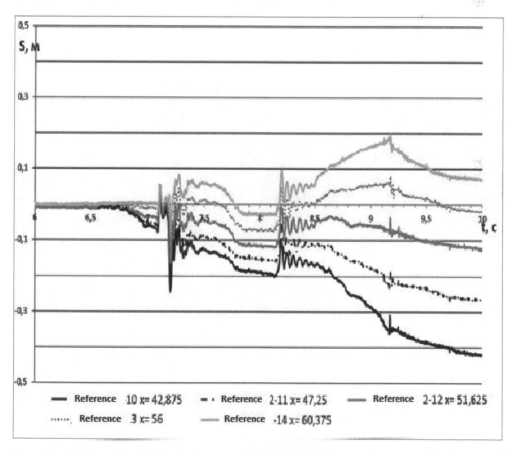

Figure 4.   Displacements (S) in the lower seam roof depending on the distance x(m) from a holing chute BE, modified to life-size conditions.

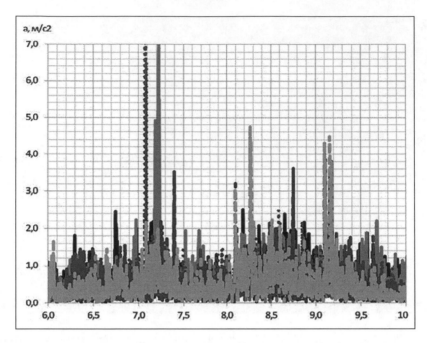

Figure 5. Change in acceleration (a) within the interval of 6–10 sec during the fall of the lower seam main roof.

Figure 6. Photo of the model after the consequent extraction of 3 faces (L1, L2, L3).

Further perspectives of the application of the EM modelling method are related to the solution of current complicated mining problems, as well as to fundamental studies of non-linear geomechanical processes at the evolution of structural parameters of rocks and SSB affected by various natural and man-caused factors in order to enhance the reliability of the prediction of dangerous dynamic phenomena and to elaborate preventive measures.

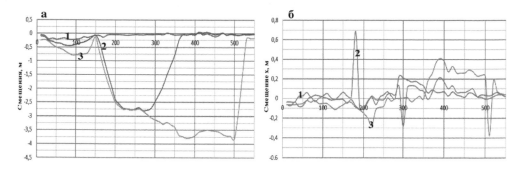

Figure 7.   Vertical (a) and horizontal (б) displacements of the immediate roof of 4 m thick seam, where x-coordinate, m; 1,2,3-dependences after extraction of 1,2,3 faces, correspondently. Displacement, m.

REFERENCES

[1]   Zuev B.Yu. The physical modelling of geomechanical processes in block-hierarchy masses based on the unified comprehensive condition for similarity. Mining Information-Analytical Bulletin. No.4/2014 pp. 356–360.
[Zuev B.Yu. Fizicheskoe modelirovanie geomekhanicheskikh protsessov v blochno-ierarkhicheskikh massivakh na osnove edinogo kompleksnogo usloviya podobiya. Gornyy informatsionno-analiticheskiy byulleten'. №4/2014s.356–360].

[2]   Zuev B.Yu. Development of methods and engineering means for the determining of static and dynamic stresses in physical models of laminated and block-hierarchy rock masses. Mining Information-Analytical Bulletin. No. 4/2014 pp. 356–360.
[Zuev B.Yu. Razrabotka metodov i tekhnicheskikh sredstv opredeleniya staticheskikh i dinamicheskikh napryazheniy v fizicheskikh modelyakh sloistykh i blochno-ierarkhicheskikh gornykh massivakh. Gornyy informatsionno-analiticheskiy byulleten'. №4/2014s.356–360].

[3]   Zuev B.Yu., Paltsev A.I. The research-and-methodological fundamentals of the physical modelling of geomechanical processes at underground mining. Mining Information-Analytical Bulletin No. 5, M., 2010, pp.29–37.
[Zuev B.Yu., Pal'tsev A.I. Nauchno-metodicheskie osnovy fizicheskogo modelirovaniya geomekhanicheskikh protsessov pri podzemnoy razrabotke poleznykh iskopaemykh. Gornyy informatsionno-analiticheskiy byulleten' № 5, M., 2010, s.29–37].

[4]   Zuev B.Yu., Krotov N.V., Shabarov A.N., Borisov A.V., Meshkov A.A. Results and perspectives of the study of geomechanical processes activation at the intensive mining of suites of coal seams on physical models. Mining in the XXI century: technology, science, education: Report abstracts from International scientific-practical conference dedicated to the 185th anniversary of the Department "Mining craft". St. Petersburg: Mining University, 2017. pp. 37–39.
[Zuev B.Yu., Krotov N.V., Shabarov A.N., Borisov A.V., Meshkov A.A. Rezul'taty i perspektivy issledovaniy aktivizatsii geomekhanicheskikh protsessov pri intensivnoy razrabotke svit ugol'nykh plastov na fizicheskikh modelyakh. Gornoe delo v XXI veke: tekhnologii, nauka, obrazovanie: Tezisy dokladov Mezhdunarodnoy nauchno-prakticheskoy konferentsii, posvyashchennoy 185-letiyu kafedry «Gornoe iskusstvo». Sp-b: Gornyy universitet. 2017. s.37–39].

Geomechanics and Geodynamics of Rock Masses – Litvinenko (Ed.)
© *2018 Taylor & Francis Group, London, ISBN 978-1-138-61645-5*

Long term creep pressuremeter tests in soft rock (St. Petersburg, Russia)

Anna Shidlovskaya
Department of Hydrogeology and Engineering Geology, Mining University, Saint-Petersburg, Russia

Jean-Louis Briaud, Mabel Chedid & Somayeh Tafti
Zachry Department of Civil Engineering, Texas A&M University, College Station, Texas, USA

ABSTRACT: Pressuremeter test is used worldwide in geotechnical engineering particularly useful for the design of foundations. It consists of inflating a cylinder in a borehole. The expansion is typically done in equal pressure steps that last one minute. Special pressuremeter tests with up to 20 minutes long pressure steps were conducted for the foundation of a tall building on soft clay rock in St. Petersburg, Russia and are presented in this paper. The behavior of the soft clay rock under creep loading is modeled with a power law model. Recommendations are made to optimize the length of the PMT test while retaining much of the needed information.

Keywords: pressuremeter, soft rock, hard clay, standard, creep settlement, modulus of deformation, limit pressure

1 INTRODUCTION

The pressuremeter (PMT) test is used worldwide in geotechnical engineering. It is particularly useful for the design of foundations on stiffer soils and soft rocks especially if samples are difficult to obtain. The PMT standards include ASTM in the USA, NIIOSP in Russia, and AFNOR in France. These standards do not match. For example the ASTM standard allows for equal volume increments lasting 15 seconds while the NIIOSP standard (GOST) requires equal pressure steps that last until the increase in radius has decreased to 0.1 mm in 30 minutes step which leads to very time-consuming tests. Such tests were conducted for the foundation design of one of the tall buildings in St Petersburg, Russia and are presented and analyzed.

2 THE PRESSUREMETER AND SITE GEOLOGY

The pressuremeter used was a TEXAM pressuremeter (Briaud, 1992). It is made of a monocellular probe which is 500 mm long and 75 mm in diameter inflated with water by a pump which can develop up to 14000 kPa. The borehole was drilled with the wet rotary method using a 76 mm drill bit and was advanced only far enough to run one test at a time. The site is located in St Petersburg on the northern bank of the Neva River delta (Shidlovskaya, 2016). From the geological point of view, the two main strata in St Petersburg are surficial non-lithified saturated sand-with-clay deposits of quaternary age which are about 400,000 years old underlain by an older lithified clay strata called the Upper Vendian clay which is about 600 million years old. The quaternary soils were deposited during a series of glacial, interglacial, and postglacial periods. At the site, the thickness of the quaternary deposits is

Table 1. Index properties of the Vendian clay.

| PMT | Borehole # | Depth (m) | Grain size distribution | | | Unit weight, kN/m³ | Water content, % | PI, % | Void ratio, e |
			% clay particles	% silt particles	% sand particles				
1	1	25.5	23	43	34	20.6	19	18	0.560
3		30.4	32	50	18	20.8	15	17	0.500
4		32.9	38	49	13	21.2	14.5	16	0.460
5		35	20	52	28	21.8	14	16	0.413
6		40	25	50	25	22.0	13	17	0.385
7	2	26	36	51	13	20.6	19	18	0.560
8		28	18	51	31	20.8	16	19	0.508
9		32.5	34	40	26	21.0	15	18	0.475
10		36	22	48	30	21.9	14	17	0.406
11		40	27	38	35	22.2	12	17	0.363
12	3	25.5	30	30	40	20.6	18	19	0.525
13		28	30	48	22	20.8	16.5	19	0.508

about 24 m with a peat layer from 7.0 to 8.5 m. The top of the Upper Vendian clay is about 24 m deep. The Upper Vendian clay was considered as the bearing layer for the pile foundation of the tall building. The index properties of the Vendian clay are shown in Table 1. The unit weight of solids was measured and averaged 27.0 kN/m³.

3 THE PRESSUREMETER TEST PROCEDURE

The tests were performed in three boreholes in the Upper Vendian clay and followed the Russian standard (GOST 20276-12) which is quite different from the ASTM standard (ASTM D4719-07). The Russian standard calls for a relatively slow test for tall buildings (height > 100 m) with pressure controlled steps lasting 30 min or until relative stabilization of the radial deformation in clays with a liquidity index (LI) larger than 0.25. The criteria for the relative stabilization of radial deformation is set as a rate of increase in the radius of the borehole not to exceed 0.1 mm in 30 min. Fast pressuremeter tests are allowed when at least two additional correlations between slow pressuremeter tests and fast pressuremeter tests are performed for a given type of soil within the area of the site investigation (GOST 20276-12). Generally, 100 kPa pressure steps are required for testing clays with LI ≤ 0.25 and a void ratio e ≤ 0.8. For tests in soft clay rocks, this may lead to 30 pressure steps or more and tests which typically last several hours as in the case reported here. This is quite different from the ASTM standard which calls for tests with a total test duration of the order of 10 minutes or 10 to 30 times shorter.

4 THE PRESSUREMETER TEST RESULTS

A total of 12 PMT tests were performed at depths varying from 25 m to 40 m below the ground surface. They are numbered from 1 to 13 in Table 1 because PMT test #2 could not be performed due to an error in drill bit size used. An example curve is shown in Fig. 1. It reveals the many pressure steps accomplished according to the Russian standard. In the softer rocks (Fig. 1) the whole curve could be defined making it possible to obtain a good estimate of the limit pressure. However, in the stronger rocks, only the modulus could be determined due to the pressure limits of the equipment. The parameters that can be obtained from a pressuremeter curve are summarized in Table 2 and are discussed in the following sections.

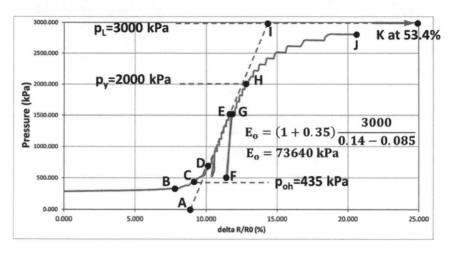

Figure 1. PMT test #1 in borehole 1 at a depth of 25.5 m.

Table 2. Pressuremeter results.

PMT test no.	Borehole	Depth (m)	Modulus E_o (MPa)	Total horizontal stress at rest p_{oh} (kPa)	Coefficient of earth pressure at rest K_o	Yield pressure p_y (kPa)	Limit pressure p_L (kPa)	Viscous exponent n
1	1	25.5	74	435	0.72	2000	3000	0.015
3	1	30.4	608	1000	2.18	5727	12000	0.018
4	1	32.9	519	677	1.03	5192*	10385*	0.016
5	1	35	628	667	0.89	6279*	12558*	0.016
6	1	40	533	1207	1.88	5329*	10658*	0.016
7	2	26	38	294	0.18	1529	2765	0.017
8	2	28	145	441	0.59	3794	6200	0.014
9	2	32.5	771	733	1.22	7714*	15429*	0.018
10	2	36	643	882	1.39	6429*	12857*	0.017
11	2	40	228	771	0.89	4000	9507*	0.019
12	3	25.5	55	500	0.96	1717	3200	0.017
13	3	28	234	771	1.70	3771	7000	0.019

Note. Numbers were obtained based on the correlations of other tests.

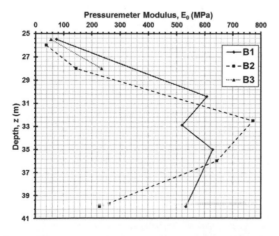

Figure 2. PMT modulus profiles.

433

5 THE PRESSUREMETER MODULUS E_o

The pressuremeter modulus E_o is obtained from the following equation:

$$E_o = (1 + \vartheta) \frac{\Delta p}{\Delta \left(\dfrac{dr}{r_o} \right)} \qquad (1)$$

where E_o is the first load pressuremeter modulus, v is Poisson's ratio estimated as 0.35 for the calculations, and $\Delta p / \Delta (dr/r_o)$ is the slope of the PMT curve (e.g.: slope DE = AI on Fig. 1). The first load PMT modulus values are listed in Table 1 and shown as profiles in Fig. 2. The PMT modulus E_o corresponds to a strain level of about 1% (Briaud, 2013b), a stress level associated with the middle of the linear portion of the PMT curve or one half of the yield pressure p_y, a rate of loading corresponding to reaching failure in 10 minutes, and zero cycles since it is associated with the first monotonic loading during the test.

6 THE LIMIT PRESSURE p_L

The limit pressure p_L is read on the pressuremeter curve at a relative increase in cavity volume equal to the initial cavity volume (doubling the initial cavity volume). This corresponds to a relative increase in cavity radius equal to 0.41 times the initial cavity radius. Such large expansion is very rarely obtained during PMT testing because of the volume expansion and pressure limits of typical PMT equipment. Practically, the limit pressure is the maximum pressure that the soil can resist and is obtained by manual extrapolation of the curve to an asymptotic value. In order to obtain an estimate of the limit pressure for the tests where only the modulus could be defined, the following procedure was adopted. For all the tests where the limit pressure could be determined with reasonable confidence, the ratio between the modulus and the limit pressure was calculated (see question marks on the Figure 3). This ratio was then used to obtain the limit pressure for the tests in soils with similar properties and where only the modulus could be determined from the test curve. These extrapolation ratios were in the range of 15 to 50. These limit pressure values are shown followed by a question mark in Table 1 and in the profile of Fig. 3.

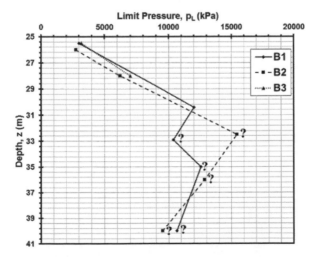

Figure 3.   Limit pressure profiles.

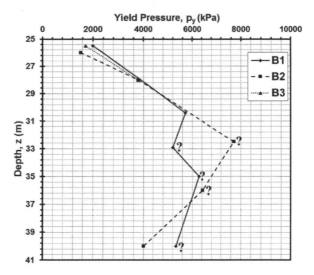

Figure 4. PMT yield pressure profiles.

7 THE YIELD PRESSURE p_Y

The yield pressure p_y is defined as the pressure on the pressuremeter curve at the end of the straight portion of the curve used to obtain the modulus. On Fig. 1 the yield pressure is found at point H. This pressure corresponds, as will be shown later, to the point at which the creep exponent n and associated creep rate starts to increase. It is also known as the creep limit. This pressure is particularly useful in design if the engineer wishes to minimize the accumulation of creep movement versus time. The yield pressure could not be read directly on the curve when the pressuremeter expansion was limited. The ratio of the modulus over the yield pressure was calculated for the PMT tests where the yield pressure could be precisely defined, and was then used to estimate the yield pressure. These extrapolation ratios were in the range of 30 to 100. These yield pressure values are shown followed by a question mark in Table 1 and Fig. 4.

8 THE CREEP EXPONENT n

During each pressure step, the radius of the probe increases slightly as a function of time while the pressure is kept constant. This displacement may be due to radial consolidation (movement due to water stress dissipation) or to creep (movement due to plastic sliding between particles at constant water stress). The pressuremeter is best suited to testing stronger soils such as unsaturated soils, sands, stiff clays and soft rocks because it is easier in those soils to prepare a high quality borehole. Thus, the discussion below is focusing on the case where creep settlement is predominant although it has been shown that the following model can also represent the consolidation phase (Bi, 2016). The creep exponent profile is shown on the Figure 5.

The creep model is written as follows

$$\frac{E_t}{E_{to}} = \left(\frac{t}{to}\right)^{-n} \tag{2}$$

where E_t is the PMT secant modulus corresponding to an elapsed time t after the start of the chosen pressure step (Fig. 6), E_{to} is the PMT secant modulus corresponding to a reference time $t_o < t$ after the start of the chosen pressure step, and n is the creep exponent. For the

435

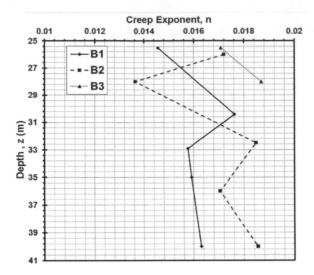

Figure 5. PMT creep exponent profiles.

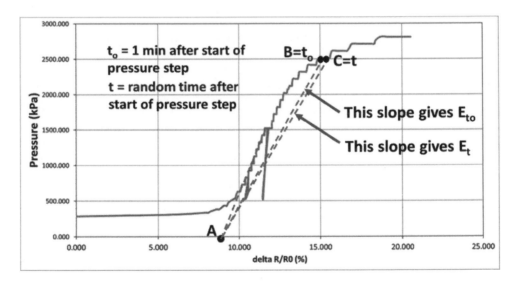

Figure 6. Definition of terms for the creep model.

tests reported here the reference time t_o was chosen as 1 minute. This model has been shown to describe well the movement of foundations as a function of time as will be presented later.

$$n = -\frac{\log\left(\dfrac{E_t}{E_{to}}\right)}{\log\left(\dfrac{t}{to}\right)}$$ (3)

The values of n can be obtained for each pressure step. Because of the unique aspect of the Russian tests reported here, many pressure steps were conducted and many associated n values can be obtained for corresponding values of p (Fig. 7). Fig. 8 shows that the n value is

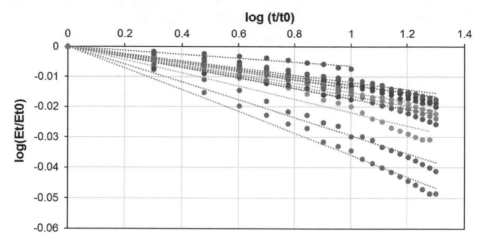

Figure 7. Regressions for all creep exponent determinations in PMT test #1 in borehole 1 at a depth of 25.5 m.

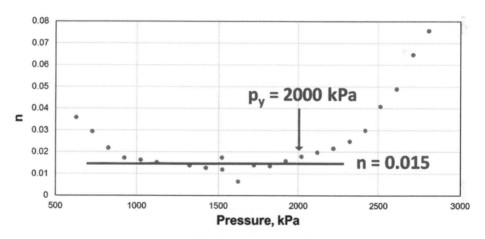

Figure 8. Creep exponent n vs. pressure for PMT test #1 in borehole 1 at a depth of 25.5 m.

high (high creep rate) at lower pressures towards the beginning of the test, lowest in the mid pressure range (low creep rate) and rises again as the pressure approaches the limit pressure (high creep rate).

At low pressures, there is a lot of creep because of the disturbance created by the drill bit during the preparation of the borehole. As the pressure rises and the probe expands that zone is recompressed, the effect of disturbance decreases and the true value of n is reached. The value of n increases again as the pressure increases past the yield pressure and up to the limit pressure. It is recommended that the creep test be performed towards the end of the linear part of the pressuremeter test curve. The mean values of n obtained from the linear part of the PMT curve for all the tests reported here are shown in Table 1 and on the profile of Fig. 5. The value of n has been found to range from 0.01 to 0.03 in sand and from 0.03 to 0.06 in clays (Briaud, 2013a). The tests reported here extend the known values of n to the case of a soft rock. This creep exponent can be used to calculate the long-term settlement of a foundation due to creep.

9 CONCLUSIONS AND RECOMMENDATIONS

The following conclusions and recommendations are made but are limited by the amount of data presented here.

1. The pressuremeter test gives a measure of the creep movement in soils. The characteristic parameter is the creep exponent n defined as the slope of the modulus vs time on a log-log scale. This plot is approximated by a straight line though some curvature is observed. Further work may be necessary to study the value of n over longer creep periods.
2. There is no need to perform many creep tests at all pressure levels. One single creep pressure test at the end of the linear part of the pressuremeter curve is sufficient and gives a reasonable value of the creep exponent n.
3. It is recommended that the creep pressure step last 10 minutes with readings every minute.
4. The n value obtained in large scale footing tests and large scale pile load tests under horizontal loads show a good comparison between the pressuremeter n values and the foundation elements n values.

ACKNOWLEDGEMENTS

The authors wish to acknowledge the contributions of Denis Michailov for his help in performing the long PMT tests in St Petersburg.

REFERENCES

ASTM D4719-07. USA Standard Test Methods for Prebored Pressuremeter Testing in Soil.
Bi G., 2016, "A power law model for time dependent behavior of soils", PhD dissertation, Zachry Dpt of Civil Engineering, Texas A&M University, College Station, Texas.
Briaud, J.-L. (2013a), "Geotechnical Engineering: Unsaturated and Saturated Soils". John Wiley and Sons Publishers, New York.
Briaud, J.-L. (2013b), "The Pressuremeter: Expanding its use", The Menard Lecture, Proceedings of the 2013 International Conference on Soil Mechanics and Geotechnical Engineering, Presse des Ponts et Chaussees, Paris, France.
Briaud, J.-L., 1992, "The Pressuremeter", Taylor and Francis.
GOST 20276-12. Soil. Field methods for determining the strenght and strain characteristics. NIIOSP. 2012.
Shidlovskaya, A.V. (2016). "Pressuremeter testing in the boreholes for a complex of the buildings in St. Petersburg". Report. Saint-Petersburg, Russia. May 2016.

Geomechanics and Geodynamics of Rock Masses – Litvinenko (Ed.)
© 2018 Taylor & Francis Group, London, ISBN 978-1-138-61645-5

Reviewing length, density and orientation data of fractures in a granitic rock mass

Gábor Somodi
Geographer, Kőmérő Ltd., Hungary

Ágnes Krupa
Geotechnical Engineer, Civil Engineer, Kőmérő Ltd., Hungary

László Kovács
Mining and Environmental Engineer, Kőmérő Ltd., Hungary

Gábor Szujó
GIS Geographer, Kőmérő Ltd., Hungary

ABSTRACT: In aspect of underground excavations geological and geotechnical investigation take into consideration the 3D geometry of fracture network according to the size of the excavation geometry. Fracture density and orientation distribution can differ due to different investigation scale and purpose. Research area is situated in the host rock of Hungarian National Radioactive Waste Repository (NRWR), which is a fractured granitic body. Previous field observations and model results suggested that the host rock formation is hydraulically strongly compartmented. Each block and the boundaries characterized by different fracture orientation also. The repository chambers of the NRWR were built in monzogranite block which have NE to SW strike and bordered by highly fractured zones and hybrid or monzonite rocks. This preferred zone has fracture sets with distinctly different orientation. Spatial distribution map of fracture density shows a pattern EN to NS strike, and it coincides with the strike of the main shear zones and with the main rock domains. Although field mapping of rock mass classification show that the most frequent fracture orientation in the repository area is perpendicular to this direction in scale of tunnel size. Our results can help for in-situ monitoring installation and numerical modelling works also.

Keywords: fractures, granite, RQD, radioactive waste repository, 3D visualization

1 INTRODUCTION

Construction of a radioactive waste disposal should have a wider-range investigation and design program than an ordinary underground construction. Besides serving the static design, the hydro-mechanical correspondences have to be explored circumstantially. Because of the transport of radionuclides through groundwater systems is the most important factor of environmental impact, the identification of possible contamination pathways is required (IAEA, 2003, Kovács et al., 2014). In a fractured rock researchers have to get acquainted with fracture network explicitly. In aspect of underground excavations geological and geotechnical investigation take into consideration the 3D geometry of fracture network according to the size of the excavation geometry. Investigation scale will define fracture density and orientation distribution. In most cases joint length can be crudely quantified by observing the discontinuity trace lengths on surface exposures (Palmström, 1995).

The research area is in a fractured granitic body which is the host rock of the Hungarian Radioactive Waste Repository (NRWR) near the village of Bátaapáti. Up to now, more than

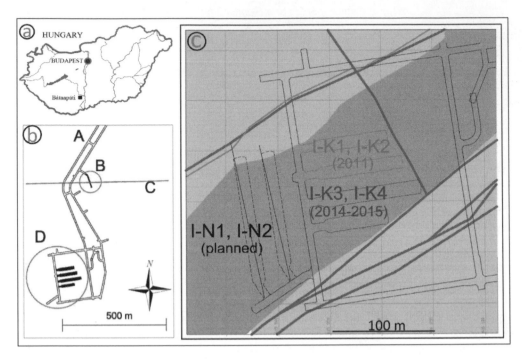

Figure 1. **a** Schematic view of Bátaapáti, Hungary, location of the NRWR, **b** tunnels reviewed in this paper (A: Access tunnels, B: 3rd research tunnel 3, C: Fault zone, D: Repository chambers), **c** Repository chambers with the year of construction. Green: Monzonite and hybrid rocks. Purple: Monzogranite rock. Red lines: Main shear zones with clay gauges. Green line: Basaltic dyke.

6 km long tunnel system was constructed and all the tunnel faces were documented (Deák et al. 2014, Kovács et al. 2015). The geographical position and the schematic view of the tunnel system are shown in Figure 1. From geological aspect, three main rock types can be distinguished in the Carbonian granite formation: monzogranite, monzonite, hybrid rocks. This granitic body is transected by Cretaceous trachyandesite dykes with NE-SW strike and randomly distributed aplitic veins also. The four repository chambers were excavated in a rock mass which is mainly composed of monzogranite with aplitic veins and scarce monzonite enclaves.

Results of previous field observations and models suggested that the granite formation is hydraulically strongly compartmented, dividing the underground flow system into several blocks of limited hydraulic connection. Each block and the boundaries are characterized by different fracture orientation also. Based on field observations, it was distinguished to the so-called more transmissive zones and the less transmissive blocks. The repository for low- and medium level nuclear waste disposal is placed in a less transmissive hydraulic compartment (Benedek et al., 2009, Benedek & Molnár, 2013). Geotechnical data from core loggings and tunnel face documentation show a clear connection between rock mass classification and rock type (Kovács et al., 2016, Vásárhelyi et al., 2016). In some part of the works geotechnical explanations of the rock mass quality distribution indicated a clear divergence from hydrogeological models. The goal of this paper is to create a usable model for geotechnical interpretation of the fracture system of the radioactive waste repository. The work will help the future numerical modelling works of the in situ stress and deformation around the tunnels.

2 GEOTECHNICAL SURVEY DATA AND APPLIED METHODS

During the excavation of the NRWR for all tunnel faces RMR and Q values were determined and from 2011 GSI determination have been brought into practice (Deák et al., 2014, Kovács et al., 2015, Vásárhelyi et al., 2016). During field work stereoscopic JointMetriX3D/Shap-eMetriX3D (JMX/SMX) system was used, which are able to generate real, precise, high reso-lution 3D images of rock surfaces (Gaich et al., 2007, Deák et al., 2014, Gaich et al., 2017). New complex system, the Advanced Survey has been developed for supporting other fields of surveying and creating directly usable database for GIS applications (Kovács et al., 2017).

For local or general modelling, engineering design and construction, it is highly important to recognise the standard features of fractures and fracture network. Geotechnical survey, field report, and 3D geotechnical advance models were summarized into an extensive data-base. In this database along with systematized orientation data, a number of other parameters describing fracture systems were saved also. These data are as follows: type of the fractures, distance between the fractures, planarity, the surface roughness, planarity and aperture, the type and thickness of the filling material, the surface weathering index. Due to these data it could be feasible to verify or reinterpret rock mass condition for obtaining new observations.

During data evaluations simple methods were used. Orientation distribution analysis was made by using Dips software. Homogeneity test was made in Statgraphics Centurion and parameter maps were made in Surfer software using Simple Kriging method. Fracture length were revised in ArcGIS where 3D fracture map was created from ShapeMetrix3D model data on the basis of an earth scientific database (Kovács et al, 2017).

3 ANALYSIS OF FRACTURE NETWORK PROPERTIES—LENGTH, ORIENTATION, DENSITY

The competent size of fractures define mostly the rock failures in tunnels excavated in fractured rock, but real sizes, real length data can reveal after construction the whole tunnel. In other hand determination of orientations and fracture densities can be determined more accurate from the first phase of investigation, however investigation scale will define fracture density and orientation distribution. In most cases joint length can be crudely quantified by observ-ing the discontinuity trace lengths on surface exposures. It is often an important rock mass parameter, but is one of the most difficult to quantify in anything but crude terms (Palmström, 1995). In aspect of underground excavations the geological and geotechnical investigation brings into focus the fracture geometry according to the size of the excavation geometry. In this investigation those fractures which are in the size of tunnel scale are important.

In our complementary studies re-evaluation of the geotechnical information of the reposi-tory was prepared, the complete database of NRWR research was analyzed, and it was focus-ing on the fracture orientations of the monzogranite rock zone which contains the repository chambers. Detailed analysis was performed based on the following principles: 1. Data were sorted out separately by rocks. The boundaries beetween different rock types were deter-mined on the basis of the field mapping. 2. Separate data were investigated from the point of view of dip/dip direction. The analysis showed that the main fracture orientations don't coincide with the strike direction of monzogranite domain and the main water-conducting structures. The strikes of the fractures are predominantly in NW-SE, their dip varies, but generally north. The steep (>75°) fractures are dominant. It reveals a difference beetween geotechnical and hydrogeological aspects.

Length of fractures is described proper by power function (Benedek and Bíró, 2010), but it suggested that fracture density analysis depends on the scale of the analysis, and according to detailed documentation the monzogranite compartment and the boundaries characterized by different fracture orientation (Benedek & Molnár, 2013). Using 3D database the distribution of fractures with NW to SE can be followed more precisely, and they have higher persistence than it was supposed in discrete fracture network modelling approaches (Benedek & Molnár, 2013), but confirm with wedge analysis presume (Kovács et al., 2016).

To characterize fracture density properties of the site RQD data was used. In the project RQD was initiated during rock core logging, but later was applied to the tunnel documenta-tion also in accordance with the recommendation of Palmström (2005). According to the homogeneity test of RQD values we can divide the rocks into four groups. Based on the mean

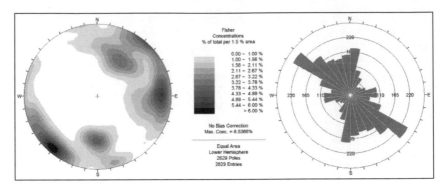

Figure 2. Distribution of fracture orientation in three repository chambers on stereonet plot with Fisher method, and relating rosette plot.

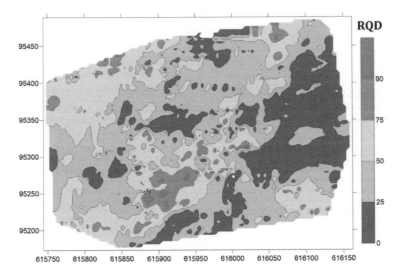

Figure 3. Distribution of RQD values in depth of the repository area. A visualisation with simple kriging method based on exploratory borehole data.

values of fracture density the monzonite, monzogranite, hybrid types can be separated and the forth group consist of aplitic and trachyandesite dykes. On Fig. 3 spatial distribution of RQD shows a very noisy pattern, characterised by NE to SW trend. It can be seen a correlation beetween good conditions and the monzogranite zone, degree of fracture density and lower RQD is typical for the shear zones.

4 CONCLUSIONS

Characterization of the assigned area of the next repository chambers on the basis of 3D fracture mapping results and new correlations was created.

Strikes of main shear zones have a pattern NE to SW strike, and it coincides with the strike of main rock domains. However field mapping of rock mass classification shows that the most frequent fracture orientation in the repository area is perpendicular to this direction in the scale of tunnel size. By means of geo-database length of these fractures was analysed and higher persistence was found than it was supposed in the earlier studies.

Good rock quality is pertained to the monzogranite zone while boundaries have much greater fracture density and lower RQD. It shows clearly that good rock conditions connected to the monzogranite zone while boundaries have much greater fracture density and lower RQD.

These investigated aspects are essential in underground analysis like numerical modelling of in situ stress and deformation or wedge analysis. Results are also useful for geotechnical monitoring installation. In further analysis block volume and fractal geometry will be examined.

ACKNOWLEDGEMENT

This paper has been published with the permission of Public Limited Company for Radioactive Waste Management (PURAM). We would like to thank all our colleagues in RockStudy Ltd for their help.

REFERENCES

Benedek K., Bőthi Z., Mező Gy., Molnár P. (2009): Compartmented flow at the Bátaapáti site in Hungary. – Hydrogeology Journal 17 (5), 1219–1232.

Benedek K., Molnár, P. (2013): Combining hydrogeological and structural data: A conceptualization of a fracture system. Engineering Geology, Vol. 163, 1–10, https://doi.org/10.1016/j.enggeo.2013.05.018.

Bíró L., Benedek K. (2010): A repedések hossz-eloszlásának vizsgálata. – Kézirat, RHK Kft., Paks, RHK-K-092/10.

Darcel, C., Davy, P. (2013): Development of the statistical fracture domain methodology—application to the Forsmark site. Swedish Nuclear Fuel and Waste Management Co. R-13-54.

Gaich A., Pötsch M., Schubert W. (2007): High resolution 3D imaging for site characterization of nuclear waste repository. Taylor and francis. 1st Canada – US Rock Mechanics Symposium, Vancouver, pp. 69–75. DOI 10.1201/NOE0415444019-c9.

Gaich, A., Pötsch, M. and Schubert, W. (2017): Digital rock mass characterization 2017 – Where are we now? What comes next? Geomechanics and Tunnelling, 10: 561–566. https://doi.org/10.1002/geot.201700036.

International Atomic Energy Agency, (2003): Scientific and Technical Basis for the Geological Disposal of Radioactive Wastes. IAEA Technical Report Series Number 413. Vienna.

Kovács L., Mészáros, E., Somodi, G., (2015): Rock Mechanical and Geotechnical Characterization of a Granitic Formation Hosting the Hungarian National Radioactive Waste Repository at Bátaapáti In: Procs. Eng. Geol. for Soc. Territory – 6. 915–918.

Kovács L., Kádár B., Krupa Á., Mészáros E., Pöszmet T., Rátkai O., Somodi G., Amigyáné dr. Reisz K.; Vásárhelyi B. (2016): The revision and upgrade of Geotechnical Interpretative Report. (In Hungarian) Manuscript. Puram (RHK Kft.), RHKK-028/16.

Kovács L., Krupa Á., Schön R., Gaburi I., (2017): "Data-Mine" software: Complex earth scientific documentation of excavations with a uniform, real 3D background. Proceeding book of 20th Hungarian and 9th Croatian-Hungarian Geomathematical Congress. Pécs, Hungary, 11–13. May 2017. 60.

Palmström, A., (1995): RMi – a rock mass characterization system for rock engineering purposes. PhD thesis, Univ. Oslo, Norway, 400, www.rockmass.net.

Palmström, A., (2005): Measurements of and Correlations between Block Size and Rock Quality Designation (RQD). Tunnels and underground Space Technology Vol. 20, Issue 4, (2005) 362–377. https://doi.org/10.1016/j.tust.2005.01.005.

Palmström, A., (2009): Technical note: Combining the RMR, Q, and RMi classification systems. Tunnels and underground Space Technology, Volume 24, Issue 4, 491–492. https://doi.org/10.1016/j.tust.2008.12.002.

Somodi G., Istovics K., Kovács L., M. Tóth T., (2017): Relationship between geotechnical parameters and discrete fracture network simulation results in Bátaapáti National Radioactive Waste Repository. Proceeding book of 20th Hungarian and 9th Croatian-Hungarian Geomathematical Congress, Pécs, Hungary, 11–13. May 2017. 242.

Vásárhelyi, B., Somodi, G., Krupa, Á., Kovács, L., (2016): Determining the Geological Strength Index (GSI) using different methods. Proc.: Rock Mechanics and Rock Engineering: From the Past to the Future: Eurock 2016. Cappadocia, Turkey, 1049–1054.

Geomechanics and Geodynamics of Rock Masses – Litvinenko (Ed.)
© 2018 Taylor & Francis Group, London, ISBN 978-1-138-61645-5

Change in elastic properties of hard rocks passing from thawed to frozen state

S.V. Suknev

Chersky Institute of Mining of the North, Siberian Branch, Russian Academy of Sciences, Yakutsk, Russia

ABSTRACT: Elastic properties of host rocks of Yakutia's diamond deposits were studied using STO 05282612-001-2013 internal standard. The standard was developed based on an original procedure that enables determination of static elastic properties (modulus of elasticity, Poisson's ratio) under variation of temperature and water content in a specimen, which is neglected by the effective Russian and international standards but is of critical importance in mine construction in permafrost zone. The testing includes multiple loading of a specimen within the range of low reversible strains, which enables highly accurate measurement of strain and physically correct estimation of temperature effect on change in properties of the material passing from thawed to frozen state. Air-dry and wet specimens of the rocks (limestone, siltstone) were tested under uniaxial compression. Specimen with extensometers attached to it was placed in a temperature control chamber and tested sequentially under room temperature and, then, under 0°C, −20°C, and −40°C. It was found that as the temperature is lowered, the elasticity modulus of the specimen grows and Poisson's ratio remains unaltered. To study the effect of moisture, a specimen was placed in water until complete saturation. The tests were carried out in the course of natural drying of the specimen at certain time intervals, and the values of the elastic modulus and Poisson's ratio were evaluated based on the resultant stress-strain diagrams. It was observed that the change in elastic properties as a function of water saturation has significantly nonlinear character.

Keywords: hard rock, compression, elastic modulus, Poisson's ratio, low temperature, water content

1   INTRODUCTION

Elasticity is a fundamental physical property of rock that specifies its mechanical behaviour. The elastic modulus of intact rock is used as an important input parameter in any analysis of rock mass deformations and for many rock engineering projects such as underground mines, tunnels, slopes, and foundations (Jaeger, 1979; Hoek and Diederichs, 2006; Sonmez et al., 2006). Poisson's ratio is equally important for the elastic deformation of rock mass and in geotechnical applications (Gercek, 2007). The standard test methods (such as ASTM D7012, 2010 or DIN EN 14580, 2005) or ISRM suggested methods (ISRM, 2007) are used to determine of elastic properties of rock under uniaxial compression. However, these methods are not suitable for testing and determination of elastic properties of the material immediately during the process of changing its state (temperature, water saturation). Another difficulty is that in order to determine the Poisson's ratio it is necessary to measure small (1–10 µm) transversal displacements of rock sample with high accuracy, which is a rather complicated technical problem (Suknev, 2012). With this in mind, a method for determination of the static modulus of elasticity and the Poisson's ratio has been developed. It enables to determine correctly the elastic properties of rock, including when its temperature and saturation are changing (Suknev, 2016). Based on this method the Chersky Institute of Mining of the

North SB RAS has developed and put into operation STO 05282612-001-2013 internal corporate standard (Suknev, 2015). The aim of the study is to establish regular patterns in changing of static elastic properties (elastic modulus, Poisson's ratio) of host rocks with varying water content within the range of testing temperature (from +20°C to –40°C) typical for permafrost zone, where the diamond deposits of Yakutia are located.

2 EXPERIMENTAL PROCEDURE

Rock specimens were tested in accordance with STO 05282612-001-2013 using a UTS 250 testing machine equipped with a temperature control chamber (Fig. 1).

Core samples of host rocks (limestone, siltstone) from the Botuobinskaya pipe were used to make square section prismatic specimens with side length of 50 mm and height of 150 mm. The specimens were mounted with axial and transverse Toni Technik 0712.001 and 0712.004 extensometers, or Epsilon 3542RA and 3975 extensometers. These extensometers provide correct and highly accurate measurements of deformations of specimens. A specimen with extensometers attached to it was placed in the temperature control chamber (Fig. 2) and tested as follows: at room temperature and at temperatures of 0°C, –20°C, and –40°C. The testing includes multiple loading of a specimen within the range of low reversible strains, which enables highly accurate measurement of strain and physically correct estimation of temperature effect on change in properties of the material passing from thawed to frozen state. At all times prior to a test a specimen would be necessarily pre-compacted (2 loading cycles); during subsequent loading cycles axial and transverse deformations would be measured. The upper and the lower limits of the loading range were set based on the conditions of reversibility and linearity of deformations. Values of elastic modulus and Poisson's ratio were evaluated by axial and transverse deformation diagrams.

To study the effect of moisture on elastic properties of rocks, a specimen would be saturated to a maximum level in a water bath, then it would be removed from the bath, installed with extensometers, and placed on a testing machine. Testing was carried out in the course of natural drying of a specimen at certain time intervals, and values of elastic modulus and Poisson's ratio were evaluated from stress-strain diagrams. Test cycle of one specimen made 28 days. During this period the specimen remained on the testing machine. Water saturation

Figure 1.   Testing machine equipped with a temperature control chamber.

Figure 2.   Rock specimen in temperature control chamber.

and water loss diagrams would be plotted for a specimen beforehand, through measuring its weight in process of water saturation and then in process of drying. A specimen would be dried naturally, at room temperature, as well as subsequently tested in the same conditions.

3   EXPERIMENTAL RESULTS AND DISCUSSION

Fig. 3 shows elastic modulus and Poisson's ratio behavior curves of three specimens of siltstone, numerated 1, 2, 3, in air-dry state, depending on the temperature. Each specimen was tested three times. To prevent moisture condensation on the surface of specimens, provisions were made to waterproof them. Before and after each test, water content in the specimen would be controlled by weighting of it. The follow-up test would be done a few days after the previous one, to make sure that the air-dry state of a specimen had stabilized. The following repeatability of elastic properties determination results was obtained: above 99% for the elastic modulus, and above 94% for the Poisson's ratio.

Fig. 3 shows that as temperature drops the elastic modulus of siltstone specimens grows linearly and Poisson's ratio remains unaltered (within the experimental error). The rate of elastic modulus increase against decreasing temperature ranged from 17 to 38 MPa/degree.

Fig. 4 elastic modulus and Poisson's ratio behavior curves of three samples of limestone in air-dry state depending on temperature.

Same as with siltstone, elastic modulus demonstrated a tendency to increase as the specimen's temperature was decreasing, while the value of Poisson's ratio maintained constant. The rate of elastic modulus increase against decreasing temperature ranged from 31 to 149 MPa/degree.

Specimens of siltstone were also tested in wet state at low and moderate levels of water saturation. Fig. 5 shows the elastic modulus and the Poisson's ratio curves of siltstone specimen as a function of temperature in air-dry (straight lines 1) and partially saturated (straight lines 2 and 3) states at two levels of water saturation. The degree of water saturation was 15% and 41% of the maximum water content that a specimen contained in fully saturated state. As water content increased, a decrease in elastic modulus and an increase in Poisson's ratio were observed throughout the entire temperature range. The behavior of elastic properties did not show any qualitative change: the rate of elastic modulus increase against decreasing

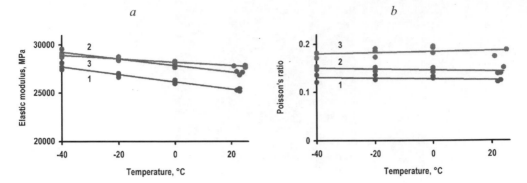

Figure 3. Variations of elastic modulus (*a*) and Poisson's ratio (*b*) with temperature for three specimens of siltstone.

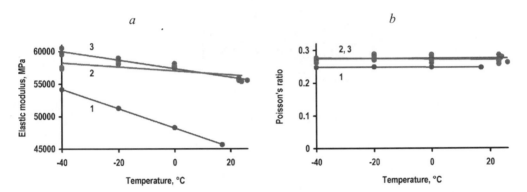

Figure 4. Variations of elastic modulus (*a*) and Poisson's ratio (*b*) with temperature for three specimens of limestone.

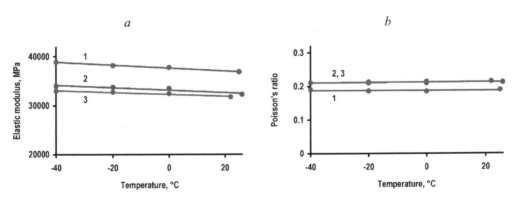

Figure 5. Variations of elastic modulus (*a*) and Poisson's ratio (*b*) with temperature for air-dry (line 1) and wet (lines 2 and 3) specimens of siltstone.

temperature as well as consistency of Poisson's ratio remained the same in the investigated temperature range.

It is noteworthy that values of elastic modulus of wet specimens were very close, and the Poisson's ratios were practically identical. That is, variation of water content in wet specimens in the given range has little effect on their elastic properties. This indicates a significantly nonlinear behavior of elastic properties of material, depending on the water content.

448

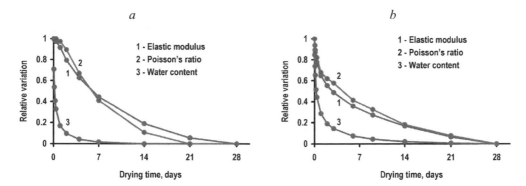

Figure 6. Relative variations of elastic modulus (curve 1), Poisson's ratio (curve 2), and water content (curve 3) with drying time for limestone (*a*) and siltstone (*b*).

Investigation of the moisture effect on elastic properties of rocks was carried out in the course of natural drying of saturated specimens. Fig. 6 shows curves of relative variation of elastic modulus $\overline{\Delta E}$, Poisson's ratio $\overline{\Delta \nu}$, and water content η as a function of the specimens' drying time. $\overline{\Delta E}, \overline{\Delta \nu}$, η are defined as follows:

$$\overline{\Delta E} = \frac{E_{wet} - E_{dry}}{E_{sat} - E_{dry}}, \overline{\Delta \nu} = \frac{\nu_{wet} - \nu_{dry}}{\nu_{sat} - \nu_{dry}}, \eta = \frac{P_{wet} - P_{dry}}{P_{sat} - P_{dry}},$$

where E_{wet}, E_{dry}, E_{sat}, ν_{wet}, ν_{dry}, ν_{sat}, P_{wet}, P_{dry}, P_{sat} are elastic modulus, Poisson's ratio, weight of the specimen in wet, air-dry, and saturated state respectively. The maximum water content of the specimens was: limestone – 1.6%, siltstone – 4.4%. Elastic moduli of limestone and siltstone in water saturated state made 34360 MPa and 13970 MPa respectively, and Poisson's ratios made 0.266 and 0.206 respectively.

The curves shown in Fig. 6 allow us to compare the change rates of water content in specimens and their elastic properties. As can be seen from Fig. 6, recovery of elastic properties occurs much more slowly than drying of specimens. As a result, elastic modulus and Poisson's ratio change slowly (relative to the drying rate of specimens) at high degrees of water saturation, and quite quickly at low degrees of water saturation, when the rate of water content changing in specimen is slow. This confirms the conclusion made above that variation of elastic properties of rocks, depending on their saturation, is significantly nonlinear.

The reason for such behavior of properties is a decreasing effectiveness of moisture impact on the rock skeleton as it becomes saturated with water. Since this impact occurs through the surface of pores, it is most effective at low water saturation levels. As a pore is filled with water, the effectiveness of moisture impact on rock skeleton is reduced, since some part of water located in the inner regions of the pore does not participate in this process.

4 CONCLUSIONS

In this paper static elastic properties (modulus of elasticity, Poisson's ratio) of host rocks of the Botuobinskaya diamond pipe were investigated within temperature range of the test (from +20°C to –40°C), typical for permafrost zone, which hosts diamond deposits of Yakutia. A series of static experiments were carried out on limestone and siltstone specimens with different water contents, and effects of testing temperature and degree of saturation on the elastic properties of rocks were experimentally studied. As a result of undertaken studies, the following have been established:

1. When temperature decreases while the material passes from thawed to frozen state, elastic modulus of the studied rock specimens grows linearly and Poisson's ratio remains

unaltered (within the experimental error). The rate of siltstone elastic modulus increase against decreasing temperature ranged from 17 to 38 MPa/degree, the value for limestone ranged from 31 to 149 MPa/degree.

2. As water content of specimens increases, elastic modulus demonstrates a decrease, and Poisson's ratio demonstrates an increase throughout the temperature range. The behavior of elastic properties does not show a qualitative change: the rate of elastic modulus increase against decreasing temperature and consistency of Poisson's ratio remain the same within investigated temperature range.

3. The change in elastic properties of rocks as a function of water saturation has significantly nonlinear character. Elastic modulus and Poisson's ratio change slowly at high degrees of water saturation, and quite quickly at low degrees when the rate of water content changing in specimen is slow.

ACKNOWLEDGMENTS

This work was supported by the Siberian Branch of the Russian Academy of Sciences, project no. 0382-2016-0002, and by the Russian Foundation for Basic Research, project no. 15-45-05014.

REFERENCES

ASTM D7012, 2010. Standard test method for compressive strength and elastic moduli of intact rock core specimens under varying states of stress and temperatures. West Conshohocken: ASTM International, 9 p.

DIN EN 14580, 2005. Prüfverfahren für Naturstein – Bestimmung des statischen Elastizitätsmoduls. Berlin: Deutsches Institut für Normung e.V., 15 p. (in German).

Gercek, H., 2007. Poisson's ratio values for rocks. Int. J. Rock Mech. Min. Sci., 44(1), 1–13.

Hoek, E., Diederichs, M.S., 2006. Empirical estimation of rock mass modulus. Int. J. Rock Mech. Min. Sci., 43(2), 203–215.

ISRM, 2007. The Complete ISRM Suggested Methods for Rock Characterization, Testing and Monitoring: 1974–2006. Suggested Methods Prepared by the Commission on Testing Methods, International Society for Rock Mechanics, Ulusay, R. & Hudson, J.A. (editors), Compilation Arranged by the ISRM Turkish National Group, Ankara, Turkey, 628 p.

Jaeger, C., 1979. Rock mechanics and engineering. Cambridge: Cambridge University Press, 523 p.

Sonmez, H., Gokceoglu, C., Nefeslioglu, H.A., Kayabasi, A., 2006. Estimation of rock modulus: For intact rocks with an artificial neural network and for rock masses with a new empirical equation. Int. J. Rock Mech. Min. Sci., 43(2), 224–235.

Suknev, S.V., 2012. Using of circular and diametrical extensometers for determination of Poisson's ratio in compression. Gornyi Informatsionno-analiticheskii Byulleten', no. 12, 22–27 (in Russian).

Suknev, S.V., 2015. Experience practice of development and application of internal standard for determination of elastic properties of rocks. Gornyi Zhurnal, no. 4, 20–25 (in Russian).

Suknev, S.V., 2016. Determination of elastic properties of rocks under varying temperature. J. Min. Sci., 52(2), 378–387.

Geomechanics and Geodynamics of Rock Masses – Litvinenko (Ed.)
© *2018 Taylor & Francis Group, London, ISBN 978-1-138-61645-5*

Tensile behavior of rock under intermediate dynamic loading for Hwangdeung granite and Linyi sandstone

Yudhidya Wicaksana & Seokwon Jeon
Department of Energy Systems Engineering, Seoul National University, Seoul, Republic of Korea

Gyeongjo Min & Sangho Cho
Department of Mineral Resources and Energy Engineering, Chonbuk National University, Jeonju, Republic of Korea

ABSTRACT: Intact rock is most prone to tensile failure. However, the behavior of tensile rock failure between static loading and dynamic loading is unlike. Research on the effect of load/strain rate on the behavior of rock and rock-like material has been attempted massively by many investigators. However, the data in the range between static and dynamic, i.e., intermediate strain rate (ISR), remains unexplored. To determine tensile strength and observe its behavior under intermediate strain-rate loading, specialized loading device using an explosive-driven piston was designed. Compared to Split-Hopkinson Pressure Bar (SHPB), which is commonly used in dynamic testing, the device adopted by this study produces a lower loading rate that is in the range of ISR loading. Two types of rock, namely Hwangdeung granite and Linyi sandstone, was collected for the test. The results indicate that the tensile strength increases with increasing strain rate. The increasing slope is more significant in the higher strain rate than in the lower strain rate. Failure behavior corresponds to a numerical simulation performed by ANSYS AUTODYN. This finding is essential for better understanding of real engineering problem such as rock and cutting tool interaction during rock excavation process.

1 INTRODUCTION

Tensile strength is one of the fundamental properties that is widely used in various engineering application including mining and civil projects. Tensile strength can be measured by the indirect tensile test, or Brazilian, from a disc-shaped core specimen (ISRM, 1978). Standard Brazilian tensile strength is commonly obtained from the static test that means the specimen is subjected to a relatively slow and constant loading. Recently, many researchers discuss the importance of tensile strength under dynamic loading (Xia et al., 2017; Zhang and Zhao, 2014) which means the specimen is forced by an instantaneous load. The reason is that many engineering practices include dynamic event such as impact action, blasting shock wave, earthquake, and so forth.

The standard method to measure dynamic tensile strength by Brazilian method has been established by ISRM (Zhou et al., 2012). However, the suggested method only covers high strain rate (HSR) testing that utilizes Split-Hopkinson pressure bar (SHPB). In the ISR loading state, dynamic Brazilian test using hydro-pneumatic loading machine has been conducted to obtain the tensile strength of Bukit Timah granite (Zhao and Li, 2000). Similar loading generation mechanism was also utilized to perform direct tension test (Asprone et al., 2009; Cadoni, 2010). Zhu et al. (2015) investigated failure pattern of Brazilian disc specimen of granite using numerical code. The simulation was then validated by a series of Brazilian test utilizing a pendulum hammer-driven SHPB.

This paper aims to study rock tensile strength under intermediate strain-rate loading. We utilized an explosive-driven loading device to generate the ISR loading. We compared the dynamic property with the property obtained from quasi-static loading for granite and sandstone rock types. We also plotted the results with others' findings and verified the dynamic test with numerical simulations. This contribution is essential for better understanding tensile behavior on real engineering problems including rock and cutting tool interaction during excavation process in the mechanized tunneling project.

2 LABORATORY EXPERIMENTATION

Brazilian test is a standard test to measure the tensile strength of rock indirectly. In this study, the experiment included two different rock types which are granitic rock (from Hwangdeung, Korea) and sandstone (from Linyi, China). The rock specimens were tested in two loading modes: dynamic and quasi-static loading.

2.1 Sample preparation

In this study, typical Korean granitic rock called Hwangdeung Granite and Chinese sandstone named Linyi sandstone were used. They were initially received as block samples and cored with the size of about 54 mm in diameter. The specimens were then cut with a diamond saw cutting machine to shape them into disc-shape specimens. The ratio of diameter to thickness of the disc specimen was about 2.00–2.25. Finally, the surfaces were polished with a grinding machine to ensure their parallelisms.

2.2 Dynamic test

A dedicated device was used to produce intermediate dynamic loading. The device consists of a frame, a loading piston, a reaction chamber, a load cell, a set of platens including a spherical seat and a stopper with rubber cushion. To produce the load, the NRC was used. The NRC stands for non-explosive rock cracker which is basically similar to the general explosive but provides less energy and lower velocity of detonation. It consists of metal powder and magnesium sulfate-hydrate mixture that can react when it is ignited. Based on the technical sheet, when thermite reaction occurs, it generates roughly 409 kcal/kg energy with chemical reaction velocity about 200–300 m/s. The amount of gas that can be produced is around 350 l/kg. It is usually used at the vibration and noise sensitive area such as construction site in the metropolitan city. In mining, this powder is commonly used as a secondary blaster to fracture boulder fragments from the primary blasting operation.

Twenty gram of NRC powder was placed in a sealed synthetic bag together with an electric detonator. The detonator was connected to an electronic discharge controller (EDC). The EDC sent the electric pulse to trigger the detonator as well as the NRC powder. The explosion created pressure inside the reaction chamber and subsequently pushed the loading piston down and impacted to the specimen. The testing data was measured and recorded by a system consisting of a load cell, an amplifier, and an oscilloscope. The voltage-time curves were recorded in the oscilloscope. The load cell was calibrated with the servo-controlled loading machine prior to the test which resulted the conversion factor that can transform voltage to force output (Min et al., 2017). A high-speed camera was adopted to capture images during the test. All information from the load cell and the high-speed camera were transferred to a personal computer. To protect investigators and devices from rock burst and flying objects, a thick acrylic glass was placed in front of the loading device. The schematic diagram of dynamic testing is illustrated in Figure 1.

Three specimens for each rock type were tested. Typical force-time curve and broken samples resulted from the dynamic tests can be seen in Figure 2.

2.3 Quasi-static test

To compare the result of dynamic test, a series of quasi-static tests were also carried out using an MTS 816 hydraulic servo-controlled loading system. Force, axial displacement and testing

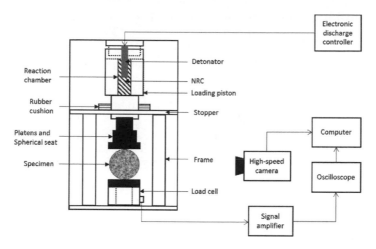

Figure 1. Schematic diagram of dynamic testing using NRC reaction-driven loading device and its connectivity with other devices.

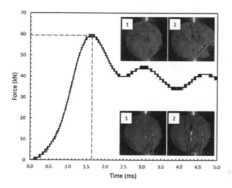

Figure 2. Typical force-time curve in dynamic test and specimen states before (1) and after (2) the test.

time were recorded by the system during the experiment. For the quasi-static test, two displacement rate scenarios were set in the program: 0.01 mm/s and 10 mm/s. Ten and 17 specimens of Hwangdeung granite and Linyi sandstone, respectively, were tested for both scenarios.

2.4 *Test Result*

Failure force was determined from the force history curve. The peak force value was used to calculate tensile strength as mentioned in the equation (1).

$$\sigma_t = \frac{2P}{\pi DT} \tag{1}$$

where σ_t is tensile strength, P is peak load, D is specimen diameter, and T is specimen thickness.

Corresponding strain rate was calculated by the dividing the stress rate with the specimen Young's modulus. The equation can be seen in the equation (2).

$$\sigma_t = \frac{\Lambda \sigma_t}{\Delta t}, \dot{\varepsilon} = \frac{\dot{\sigma}_t}{E} \tag{2}$$

where $\dot{\sigma}_t$ stress rate, Δt is time duration until failure, $\dot{\varepsilon}$ is strain rate, and E is Young's modulus.

453

Three groups of data represent three different loading rates: two groups from quasi-static testing (QS_{low} and QS_{high}) and one group from ISR testing (ISR). The average tensile strength and its strain rate for all testing scenarios and rock types can be seen in Table 1. The range of strain rate (on average) resulted from this experiment was ranging from 6.0×10^{-6} s^{-1} to 4.5×10^{-1} s^{-1}. The averaged tensile strength of QS_{low}, QS_{high}, and ISR for Hwangdeung granite were 9.15 MPa, 11.66 MPa, and 26.44 MPa while for Linyi sandstone were 4.00 MPa, 5.43 MPa, and 10.23 MPa. Overall, the result showed that tensile strength increased with increasing of strain rate for both rock types.

We found that the increasing gradient levels in quasi-static and dynamic loading were unlike for all rock types. The rising slope of quasi-static was represented by the regression between QS_{low} and QS_{high} while increasing slope of intermediate dynamic loading was deter-

Table 1. Averaged strain rate and tensile strength under quasi-static and dynamic loading conditions.

Loading Condition	Hwangdeung Granite		Linyi Sandstone	
	Average $\dot{\varepsilon}$ (s^{-1})	Average σ_t (MPa)	Average $\dot{\varepsilon}$ (s^{-1})	Average σ_t (MPa)
QS_{low} (1)	6.0×10^{-6}	9.15	1.0×10^{-5}	4.00
QS_{high} (2)	3.2×10^{-3}	11.66	3.7×10^{-3}	5.34
ISR (3)	4.5×10^{-1}	26.44	4.5×10^{-1}	10.23
Average Strength Increasing (%)				
(1) $\rightarrow$ (2)		27.44		33.50
(2) $\rightarrow$ (3)		126.67		91.78

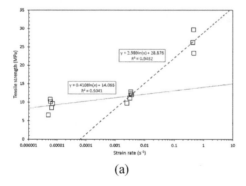

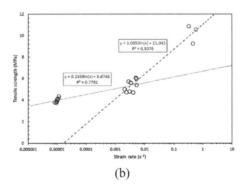

Figure 3. Tensile strength over strain rate; (a) Hwangdeung Granite, (b) Linyi Sandstone.

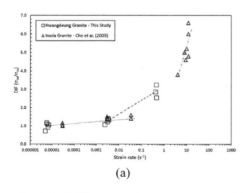

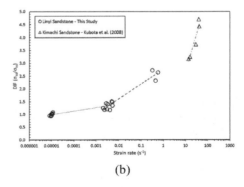

Figure 4. DIF over strain rate; (a) Granite, (b) Sandstone.

mined by the relationship between QS_{high} and ISR. The gradient was much higher in the ISR compared to the slope in the quasi-static as depicted in Figure 3. It supports the general statement that there are two distinct trends, i.e., modest increase in the low strain rate region and sharp rise in the high strain rate region (Asprone et al, 2009).

To indicate the increased strength under the higher strain rate loading, the results were processed in terms of dynamic increase factor (DIF) which is defined as the ratio of the dynamic strength to the referenced static strength. The data was plotted together with the data from others as can be seen in Figure 4. It showed that the slope is significantly higher at the higher strain rate. Cho et al. (2003) conducted a series of spalling test to determine the dynamic tensile strength of Inada granite. Their finding showed that tensile strength is significantly higher in dynamic loading. Kubota et al. (2008) estimated the dynamic tensile strength of Kimachi sandstone by using underwater explosive-driven method under 10–40 s^{-1} of strain rate. They claimed that the increase of dynamic strength and its strain rate is $\sigma_d = 4.78\dot{\varepsilon}^{0.333}$

3 NUMERICAL SIMULATION

Numerical simulation was done by ANSYS AUTODYN. A disc specimen with 54 mm diameter and 27 mm thickness was made. Two plates were positioned on the upper and lower relative to the specimen. Dynamic loading was applied to the specimen downward from the top plate parallel to the z-axis (see Figure 5a). RHT concrete strength criterion was adopted to represent the rock model. Contact force in z-direction was monitored on the upper plate then the maximum value of the contact force was taken for tensile strength calculation.

Figure 5b shows the contact force history and damage distribution of the specimen resulted from the simulation. The orientation of damage distribution was parallel to the

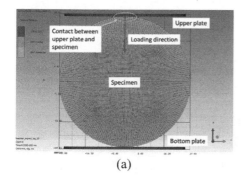
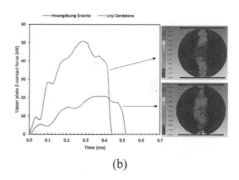

(a)　　　　　　　　　　　　　　　(b)

Figure 5.　Numerical simulation using ANSYS AUTODYN; (a) simulation model, (b) Z-contact force history of the upper plate and damage distribution of the specimen model.

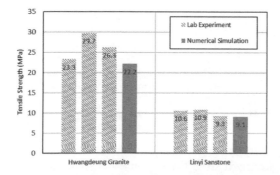

Figure 6.　Dynamic tensile strength comparison between laboratory experiment and numerical simulation.

direction of loading which was predominantly located in the center of the specimen along the z-axis. The maximum contact forces of Hwangdeung granite and Linyi Sandstone were 50.9 kN and 20.8 kN and referred to the equation (1), the dynamic tensile strength magnitudes were 22.2 MPa and 9.1 MPa, respectively. The tensile strength values resulted from numerical simulation are acceptable compared to the results from the experimental work as presented in Figure 6.

4 CONCLUSIONS

In this study, a specialized device based on thermite reaction of a non-explosive powder was successfully adopted. A series of Brazilian tests in quasi-static and dynamic loading were carried out for Hwangdeung granite and Linyi sandstone. The loading machine generated loading that falls in the strain rate between 10^{-1} s^{-1} and 10^{1} s^{-1} which is considered as an intermediate strain rate (ISR) loading condition. The result exhibits that tensile strength increases with increasing strain rate. The gradient becomes sharper as strain rate increases. It agrees with the previous findings that showed similar results. Additionally, we reproduced the dynamic loading by numerical code and showed that the tensile strength and fracture behavior between laboratory experiment and numerical model are comparable. It should be noted that the number of specimens tested in the ISR dynamic loading is limited due to some technical setbacks. Further experiments need to be addressed to give more comprehensive outcomes.

ACKNOWLEDGEMENT

This research is supported by Korea Energy and Mineral Resources Engineering Program (KEMREP), Republic of Korea. The first author is indebted to the LPDP Scholarship of Indonesian Government for providing financial support to study at Seoul National University.

REFERENCES

Asprone, D., Cadoni, E., Prota, A., and Manfredi, G. (2009). "Dynamic behavior of a Mediterranean natural stone under tensile loading." Int J Rock Mech Min Sci, Vol. 46, No. 3, pp. 514–520.

Cadoni, E. (2010). "Dynamic characterization of orthogneiss rock subjected to intermediate and high strain rates in tension." Rock Mech Rock Eng, Vol. 43, No. 6, pp. 667–676.

Cho, S.H., Ogata, Y., and Kaneko, K. (2003). "Strain-rate dependency of the dynamic tensile strength of rock." Int J Rock Mech Min Sci, Vol. 40, No. 5, pp. 763–777.

ISRM (1978). "Suggested methods for determining tensile strength of rock materials." Int J Rock Mech Min Sci Geomech Abstr, Vol 15, No. 3, pp. 99–103.

Kubota, S., Ogata, Y., Wada, Y., Simangunsong, G., Shimada, H., and Matsui, K. (2008). "Estimation of dynamic tensile strength of sandstone." Int J Rock Mech Min Sci, Vol. 45, No. 3, pp. 397–406.

Min, G.J., Oh, S.W., Wicaksana, Y., Jeon S., and Cho, S.H. (2017). "Development of the strain measurement-based impact force sensor and its application to the dynamic Brazilian tension test of the rock." Explosives and Blasting, Vol. 35, No. 3, pp. 15–20.

Wicaksana, Y., and Jeon, S. (2017). "Dynamic rock strength under intermediate strain rate loading." Proceeding of Korean Society of Explosives and Blasting Engineering Fall Conference, pp. 47–50.

Xia, K., Yao, W., and Wu, B. (2017). "Dynamic rock tensile strength of Laurentian granite: Experimental observation and micromechanical model." J Rock Mech Geotech Eng, Vol. 9, No. 1, pp. 116–124.

Zhang, Q.B. and Zhao, J. (2014). "A review of dynamic experimental techniques and mechanical behaviour of rock material." Rock Mech Rock Eng, Vol. 47, No. 4, pp. 1411–1478.

Zhao, J., and Li, H.B. (2000). "Experimental determination of dynamic tensile properties of a granite." Int J Rock Mech Min Sci, Vol. 37, No. 5, pp. 861–866.

Zhou, Y.X., Xia, K., Li, X.B., Li, H.B., Ma, G.W., Zhao, J., Zhou, Z.L., and Dai, F. (2012). "Suggested methods for determining the dynamic strength parameters and mode-I fracture toughness of rock materials." Int J Rock Mech Min Sci, Vol. 49, pp. 105–112.

Zhu, W.C., Niu, L.L., and Li, S.H. (2015). "Dynamic Brazilian test of rock under intermediate strain rate: pendulum hammer-driven SHPB test and numerical simulation." Rock Mech Rock Eng, Vol. 48, No. 5, pp. 1867–1881.

Development of a rock blasting management system

Masahito Yamagami

*TAISEI Corporation, Geotechnical Research Section, Infrastructure Technology Research Department,
Technology Center, Nase-cho, Totuka-ku, Yokohama, Japan*

Saburou Katayama

*TAISEI Corporation, Construction Technology Development Section, Advanced Technology
Development Department, Technology Center, Nase-cho, Totuka-ku, Yokohama, Japan*

ABSTRACT: Concrete aggregate production in dam construction is planned according
to the distribution of good quality rocks estimated by preliminary surveys such as borehole exploration and geophysical survey. The accuracy in estimating the distribution of good
quality rocks depends on the quantity of survey and is generally low. This may lead to a
limited enhancement in the aggregate production efficiency. In addition, rock quality classification is based on the results of visual observation on blasted rocks distributed near the
ground surface. Since it is impossible to inspect inside the ground, the accuracy in estimating
the rock quality for the solid portion of the rock mass is therefore also limited. Furthermore, works related to drilling for blasting such as determination of the drilling positions
and lengths are normally carried out manually. Under these conditions, a labor-saving means
by mechanization has been desired. To resolve the problem, we developed a rock blasting
management system, which consists of rational construction techniques with an Information
and Communication Technology (ICT). In this report we will describe the summary of the
system and the results of a verification experiment on the drilling survey and of a demonstration experiment at an aggregate production site.

1   INTRODUCTION

Concrete aggregate production in dam construction is planned according to the distribution
of good quality rocks estimated by preliminary surveys such as borehole exploration and geophysical survey. The accuracy in estimating the distribution of good quality rocks depends
on the quantity of survey and is generally low. This may lead to a limited enhancement in
the aggregate production efficiency. In addition, rock quality classification is based on the
results of visual observation on blasted rocks distributed near the ground surface. Since it is
impossible to inspect inside the ground, the accuracy in estimating the lated to drilling for
blasting such as determination of the drilling positions and lengths are normally carried out
manually. Under these conditions, a labor-saving means by mechanization has been desired.
To resolve the problem, we developed a rock blasting management system named "T-iBlast
DAM", which consists of rational construction techniques with an information and communication technology (ICT).

The "T-iBlast DAM" consists of two subsystems, which are named "the intelligent crawler
drill system (ICDS)" (Fig. 1) and "the ground evaluation system (GES)". ICDS has a
logging-while-drilling and the drilling guidance functions. The former function can estimate
rock quality using specific drilling energy during drilling works of a crawler drill used for
blasting, while the latter function can guide drilling position, orientation and depth using the
Global Navigation Satellite System (GNSS). GES can manage rock quality information in
three dimensions. The method of applying the drilling logging results to the informational

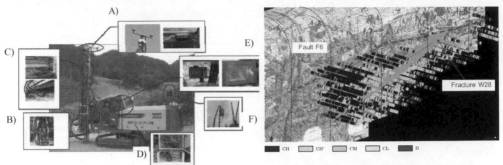

Figure 1.   Intelligent crawler drill.

Figure 2.   A case applying the drill logging to the underground cavern excavation.

construction has already been verified in a large-scale underground cavern (Takeda et al., 2012.) The method was then referred to in our research (Fig. 2).

In this report, we describe the outline of the T-iBlast DAM and the results of a verification test on the logging-while-drilling function and of a demonstration experiment conducted at the Gokayama Dam aggregate production work site (located in Fukuoka prefecture of Japan). If the rock quality inside the ground can be evaluated, it will be possible to collect good rock aggregates with higher precision. Also if the management of accumulated survey data is integrated in three dimensions all at once, it is possible to reduce many waste rock materials and as much processing loss as possible in the entire construction work. At the end of the report, we present a rationalization technique using an ICT in future dam aggregate manufacturing work combined with an "on-site analysis method" (Ichiki et al., 2015) which is separately under development.

2   T-ɪ BLAST DAM

The intelligent crawler drill system (ICDS) and the ground evaluation system (GES) which constitute the T-iBlast DAM are described below.

2.1   Intelligent Crawler Drill System (ICDS)

2.1.1   System overview
This system is a fusion of positioning guidance technology using a GNSS surveying system (Trimble DPS 900 Nikon Trimble Co., Ltd.) and rock quality evaluation technology with the use of drilling energy. Drilling work is usually performed at an interval of about 3 m, so high accuracy rock quality determination can be expected from the acquisition of high-density rock quality information. The main components mounted on the crawler drill and their roles are listed below and are shown in Fig. 1:

A. GNSS antenna and Receiver: Acquisition of coordinate and orientation
B. Inclinometer: Measurement of tilt angle of the mast
C. Leach sensor: Measurement of drilling length
D. Hydraulic force meter: Calculation of drilling energy
E. Tablet PC: Monitoring for guidance and rock quality determination
F. Radio equipment: Real Time Kinematic Positioning and Receiving GNSS correction information

2.1.2   GNSS machine guidance function
Bench drilling creates a drilling surface in a staircase shape, and natural grounds are blasted so that the elevation of the bench surface becomes uniform in the blasting operation. It is necessary to drill to the same elevation to keep a constant slope grade in the same direction.

Table 1. Comparison of ICDS with the conventional work flow.

	Worker	Position measurement	Guidance	Drilling	Checking
ICDS	Operater: 1	**No need**	Display on monitor in the operator room	Display on monitor in the operator room	**No need**
Conventional work	Operater: 1 Assistant: 1	Using measurement instrument and marking	Visual confirmation	Drilling by sense of operator	Using measurement instrument

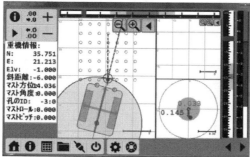

Figure 3. Machine guidance monitor in the opera-tor room.

Figure 4. Drilling survey monitor in the operator room.

We compared the conventional construction method with that used in the workflow of this function (Table 1). The greatest feature of the such as "measurement" and "examination". Moreover, setting the "elevation of the end point" instead of the "drilling length" to a cer-tain value, even when the undulation of the current bench surface is large, the elevation of the hole bottom can be kept constant, and the bench surface can be easily aligned horizon-tally. Another feature is the direction and angle guidance function. If it is possible to always arrange the crawler drill facing the inclination direction of the hole, it could be adjusted by the inclinometer guide attached to the crawler drill and by the assistant worker. In reality, it is difficult to adjust the crawler drill. On the other hand, regardless of the direction of the crawler drill, this function enables to guide the drilling rod to the hole bottom at desired posi-tion and inclination that match the designed drilling plan (Fig. 3).

2.1.3 *Logging-while-drilling function*
Drilling by a crawler drill is performed with a hydraulically driven rock drilling machine. The drill bit penetrates the rock with rotational, hammering and feeding pressures (last being the force to press the drill bit against the rock) applied to it through the rod. The logging-while-drilling function can calculate the drilling energy which is defined as the rotational and ham-mering pressures and the drilling rate to drill a unit of volume. Investigating the correlation between the drilling energy and rock class in advance, it becomes possible to classify the rocks according to their drilling energy. In addition, an operator can evaluate the rock condition in real time. Fig. 4 shows an example of the logging-while-drilling function in the operator's room.

2.2 *Natural Ground Evaluation System (GES)*

Using Geo-Graphia (GEOSCIENCE Research Laboratory) which is a three-dimensional integrated visualization software, it is possible to display the result of the logging-while-drill-ing as 3D contour display using geo-statistics method, and volume ratio for each rock class.

3 VERIFICATION EXPERIMENT ON LOGGING-WHILE-DRILLING FUNCTION

Prior to evaluation of relationship between specific energy while drilling by crawler drill and rock classification, a fundamental verification experiment was conducted by drilling layered concrete with known strengths.

3.1 Outline of experiment

To evaluate the logging-while-drilling function, a concrete block cast in four layers as an artificial rock was drilled. Every layer comprised 50 cm thick, 200 cm long and 200 cm wide. The concrete strengths were respectively 50, 100, 30 and 50 N/mm² from the bottom layer. The reason for setting the bottom and the top layers to 50 N/mm² is to confirm the reproducibility of the data. The layered concrete block was drilled after backfilling with soil and sand to simulate the conditions at the real drilling site (Fig. 5, Photo 1).

3.2 Experimental results

The experimental results are shown in Fig. 6 and Fig. 7. The relationship between specific drilling energy and compressive strength was recognized as given by (Eq. 1).

$$\text{Compressive strength} = 0.0862 \times (\text{specific drilling energy})^{1.5557} \tag{1}$$

Since the compressive strength of medium hard rock (equivalent to rock class CM) is approximately 30 N/mm², this function is conclusively possible to determine which portion of rock can be used for concrete aggregate.

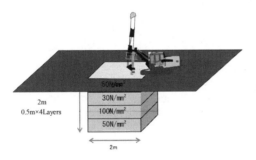

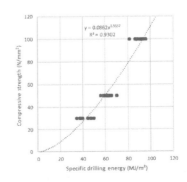

Photo 1. Experimental condition of a layered concrete block. (left: test block, right: drilling state).

Figure 5. Schematic chart of the drilling survey for a layered concrete block.

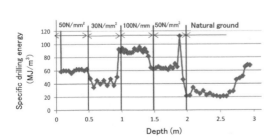

Figure 6. Experimental result of specific.

Figure 7. Schematic chart of the drilling survey for a layered concrete block.

4 DEMONSTRATION EXPERIMENT ON T-I BLAST DAM

We conducted a demonstration experiment to confirm the effectiveness of the system at the Gokayama Dam aggregate production work site ordered by the Fukuoka prefecture. The outline of the experiment will be reported below.

4.1 Experiment site overview

The Gokayama Dam is a concrete gravity dam in Fukuoka prefecture, constructed for flood control and secure water supply. The Gokayama Dam aggregate production work was a construction project to prepare concrete aggregates for the dam body. Distributed rock are biotite granite. As for the rock classification for aggregate, the target rock materials are I-1 (class B-CH) and I-2 (class CH-CM) while the waste materials are II (class CL) and III (class D). Typical construction cycle consisted of blasting at noon, transportation of blasted rock in the afternoon and on, and drilling work for charging and blasting preparation in the next morning.

4.2 Outline of experiment

The T-iBlast DAM was applied to one cycle of the blasting work, the ground condition was evaluated and the test results were verified.

4.2.1 Threshold setting

To set the threshold value for separating good rock from waste rock material using the specific drilling energy obtained by logging-while-drilling, we performed the experiment at a position 50 cm away from where a core drilling had already been drilled during the preliminary investigation. From the relationship between drilling energy and rock class at the corresponding depth, distributions of frequency for each rock class were created (Fig. 8), and the specific drilling energy at which each rock class was dominant was set as the threshold values. As a result, class D or CL was less than 35 MJ/m³, class CM was between 35 and 70 MJ/m³, and class CH more than 70 MJ/m³ (Fig. 9).

4.2.2 Data acquisition by logging-while-drilling function

At a bench section scheduled to be blasted on the following day at noon, the logging-while-drilling was carried out in 15 places (drilling length of 11 m with an interval of 4 m) by the intelligent crawler drill from the afternoon on that day to the time for blasting in the following day.

4.2.3 Analysis by Ground Evaluation System (GES) (Fig. 10)

Using a three dimensional visualizing software, we showed the color-coded specific drilling energy data by the threshold values first. Using the geostatistical method, we then showed the

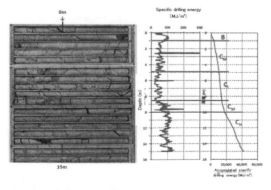

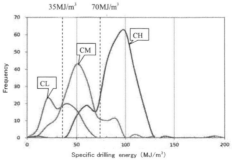

Figure 8. Correspondence between specific drilling energy and rock classes.

Figure 9. Frequency distribution of specific drilling energy and determination of threshold for each rock class.

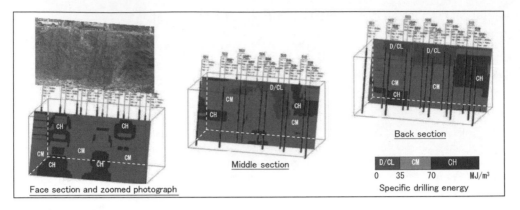

Face section and zoomed photograph

Middle section

Back section

D/CL	CM	CH
0	35	70	MJ/m³

Specific drilling energy

Figure 10. Analysis results of the drilling survey (from left to right, face section and zoomed photograph, middle section and back section).

· Waste rocks mainly composed of D and CL class are distributed.

· Rocks are so soft that can be broken by a light blow of a hammer because of hydrothermal alteration

Figure 11. Verification result of the waste rock area estimated by the drilling survey.

· Good rocks mainly composed of CM class are distributed.

· Rocks are hard, so can be broken by a strong blow because of non or week alteration.

Figure 12. Verification result of the good rock area estimated by the drilling survey.

zones in colored contours and created the main sectional views (front, middle and back sections). This way, we could confirm the distribution profile of the waste rock materials. Observation of the cross sections before blasting confirmed good rock materials (classes CM to CH) at the face section, whereas waste rock materials (class CL or poorer) could be assumed in the middle and back cross sections inside the blasting target block.

4.2.4 *Verification of results*
After blasting, we confirmed the distribution of the waste rock materials inside the blasted block. It was possible to confirm the waste rock materials which deteriorated due to hydrothermal alteration at the expected position (Fig. 11). Likewise, high quality rocks of class CM could be confirmed at the expected position next to the waste rock materials (Fig. 12).

5 CONCLUSIONS

We developed a rock blasting management system named "T-iBlast DAM", which consists of rational construction techniques with an ICT. Using machine guidance function, we could reduce monitoring time and human cost necessary for adjusting drill positions. Regarding the logging-while-drilling function, we could obtain the correlation between concrete strength and specific drilling energy. In addition, as a result of applying the T-iBlast DAM to the aggregate production work site, it was found possible to evaluate the inside of the rock block to be

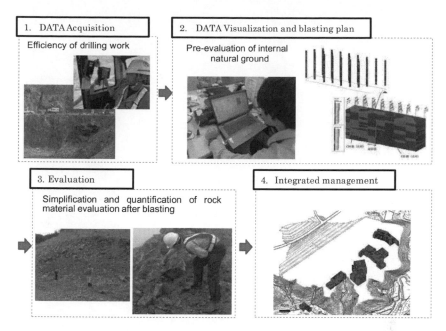

Figure 13. Flow of intelligent construction procedure with ICT for dam concrete aggregate production in the future.

blasted, which could not be evaluated by the conventional visual observation method. If this evaluation is given to work related persons before blasting, it is possible not only to grasp the amount of the waste rock in advance but also to make a reasonable aggregate production plan. Moreover, accumulating such results every time will be very useful for management of aggregate production. From these experiments, we confirmed the validity of the T-iBlast DAM.

In addition to the T-iBlast DAM reported here, we are developing a method of evaluating rock materials easily and quickly at dam construction site. In this method, the rock material is quantitatively evaluated from correlation of dry density with moisture content using a portable susceptibility meter, colorimeter, fluorescent X-ray analyzer (on-site analyzer) (Ichiki et al., 2015). Fig. 13 shows a rationalized construction plan using an ICT in the dam aggregate production work that integrates T-iBlast DAM together with an evaluation method using onsite analysis devices.

REFERENCES

Ichiki T., Yamagami M., Takemoto T. 2015: Simple and quick field evaluation of rock material quality for dam construction-examples of granite and andesite, report of taisei technology center No. 48, 26.
Takeda N., Nishimura T., Yamagami M. 2012: Applicability of drilling survey for informative construction on the underground excavation, electric power civil engineering, No. 359.

Geomechanics and Geodynamics of Rock Masses – Litvinenko (Ed.)
© 2018 Taylor & Francis Group, London, ISBN 978-1-138-61645-5

Experimental study on damaged zone around an opening due to thermo-mechanical loading

Nan Zhang & Seokwon Jeon
Department of Energy Systems Engineering, Seoul National University, Seoul, Republic of Korea

Shuhong Wang
School of Resources and Civil Engineering, Northeastern University, Shenyang, P.R. China

ABSTRACT: The influence of the elevated temperature on the evolution of the Excavation Damaged Zone (EDZ) in the rock mass is an important issue for the performance assessment of radioactive waste repositories. Thermal stress produce additional damage and may increase the extent of the EDZ resulting in the degradation of mechanical properties of rock mass around the deposition openings which provides a preferential pathway for radionuclides to migrate. In this research, cubic specimens with central circular opening were casted using cement mortar. Biaxial compression tests were firstly performed to generate damaged zone around the opening. Acoustic Emission (AE) technique was applied to monitor and characterize the brittle failure process around the circular opening. The experimental observations of the biaxial compression tests indicated that brittle failure around a circular opening was mainly a process of progressive spalling and finally resulted in a v-shaped failed zone. Laboratory scale heater tests were conducted under confined condition using intact and damaged cement mortar specimens to investigate the effect of elevated temperature on the evolution of the damaged zone. The results of the heater test using intact specimen showed that no clear damage was found on the surface of the opening. In contrast, results of heater test using damaged specimen revealed that fractures tended to initiate on the surface of the opening and no obvious further spalling was identified. These experimental findings are consistent with the field observations from the in situ single hole heating damage test.

Keywords: Thermo-mechanical, Damaged zone, Heater test, Spalling, Acoustic emission

1 INTRODUCTION

The thermo-mechanical response of rock is a critical research issue in many fields such as geological sequestration of CO_2 (Shiu et al., 2011), enhanced geothermal systems (Rutqvist et al., 2015) and nuclear waste repository (Tsang et al., 2012) which are usually subject to complex thermo-mechanical conditions. One of the key issues when assessing the performance of the nuclear waste repository is related to the effects of the temperature elevation due to the heat emitted from the placement canister (Pusch, 2008). The elevated temperature can cause expansion of rock mass and the resulting thermal stress will be concentrated around the boundary of the deposition holes and cause an increase in the tangential stresses. Hudson et al. (2009) emphasized that thermal stresses can have high magnitudes and extensive damage in the EDZ will be resulted. Jansen et al. (1993) and Ishida et al. (2004) carried out laboratory heater tests to study thermally-induced cracking processes using cubic granite specimen. However, it should be noted that these tests were conducted under unconfined condition which is not the realistic underground conditions. In this research, biaxial compression tests were firstly performed to generate damaged zone around the opening. The brittle failure process around the opening was investigated. Furthermore, laboratory heater tests were conducted under confined condition to study the effect of elevated temperature on the evolution of damaged zone.

2 LABORATORY EXPERIMENTAL STUDY

2.1 *Specimen and its properties*

Cubic specimens with side length of 170 mm and central circular opening (50 mm in diameter) were casted using ultra-rapid hardening cement mortar. The mixing ratio of cement to water as 2:1 by weight. Specimens were kept in the molds and covered with plastic sheets for 24 hours, then demolded and cured in water tank at room temperature. On the curing age of day-7, specimens were removed from water tank and placed in drying oven, dried under 105°C for four days until a constant weight was obtained. The physical and mechanical properties of the cement mortar specimen were summarized in Table 1.

2.2 *AE measurement system*

AE technique was applied to monitor and characterize the brittle failure process around the circular opening under thermo-mechanical loading condition. The AE measurement system was manufactured by Physical Acoustic Corporation (PAC). AE signals were detected through six PICO sensors and amplified by six PAC-1220A preamplifiers with 60 dB gain due to the high attenuation of signal transmission in cement mortar material. High vacuum silicone grease exhibits excellent thermal stability at temperatures up to 200°C and was used as acoustic couplant between the sensor and the specimen. Three AE sensors were attached to the front surface and back surface of the specimen, respectively. All sensors were attached to the left portion of the specimen as shown in Figure 1, based on two main concerns. Firstly, thermocouples were installed at the right portion of the specimen. Secondly, to prevent the

Table 1. Physical and mechanical properties of cement mortar.

Properties	Value	Unit
Uniaxial compression strength, σ_c	46.2	MPa
Brazilian tensile strength, σ_t	4.1	MPa
Young's modulus, E	8.2	GPa
Poisson's ratio, ν	0.28	–
Friction angle, Φ	31	°
Cohesion, c	10.3	MPa
Density, ρ	1950	kg/m³
Porosity, Φ	11	%
P-wave velocity, V_p	4010	m/s
S-wave velocity, V_s	2105	m/s

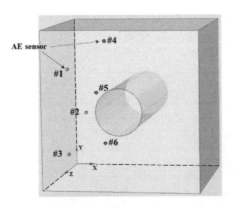

Figure 1. Schematic layout of AE sensors.

interference of acoustic wave due to the presence of the circular opening. In heater tests, specimen will be heated for several hours, it is critical to keep good coupling between specimen and AE sensor. Two sensor holders were applied to insure the adequate and stable coupling between the sensor and the specimen. In addition, pencil lead break tests under different sensor layouts were carried out at known positions on the specimen. The source locations were compared with the real locations and the sensor layout which has the lowest error was decided.

2.3 Experimental set-up and procedure

Two kinds of laboratory heater tests (i.e. #CM-M-TM and #CM-TM) were conducted under confined condition using intact and damaged specimen, respectively. In heater test #CM-M-TM, specimen was firstly damaged through biaxial compression test, then it was used in heater test under confined condition. In heater test #CM-TM, specimen was used directly in heater test under confined condition. The experimental apparatus for heater test is primarily composed of loading system (including a loading machine with a capacity up to 200 ton and a hydraulic pump unit which can apply confining pressure through a pair of flat jacks), AE monitoring system, DAQ system and the heating unit as shown in Figure 2. Temperature evolution was monitored by six thermocouples (see position in Figure 3). Cartridge heater was placed into the central opening and the interfacial gap between the cartridge heater and the surface of the opening was filled with standard sand. In the heat test, the confining pressure

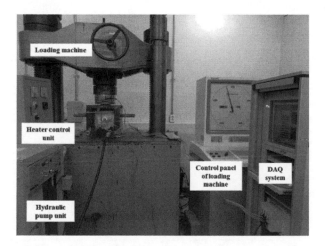

Figure 2.   Experimental setup of heater test.

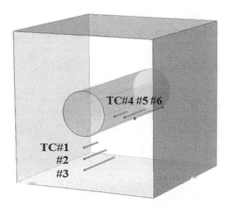

Figure 3.   Schematic view of thermocouple (TC) position.

was simultaneously applied to the specimen by the loading machine in vertical direction and hydraulic pump in horizontal direction up to 1 MPa. Then, the specimen was heated from 22°C (room temperature) to 350°C and keep at 350°C for two hours. The low heating rate of 3°C/min was applied to minimize the transient thermal effects (i.e. thermal shock) as much as possible.

2.4 *Results of heater tests*

AE sources locations in heater test #CM-M-TM are shown in Figure 4. The sources occurred in different time periods are depicted in different colors. AE sources started to cluster at the sidewall of the opening after heating about 50 minutes. These results were proved by the experimental findings from the test. Fractures initiated on the surface of the opening and some fractures propagated and coalesced with the pre-existing failed zone. The relatively scattered distribution of the AE sources can be explained by the presence of the opening which causes the reflection of the AE wave. Similar phenomenon was also found in previous studies by Bae (2005), Golshani et al. (2005), Wang et al. (2009) and Zhao et al. (2014). The temperature monitored in heater test #CM-TM was presented in Figure 5. It was found that

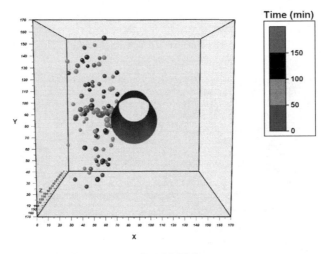

Figure 4.   AE source locations in heat test (#CM-M-TM).

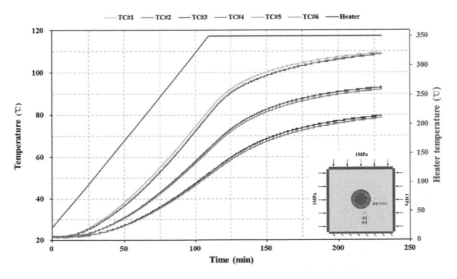

Figure 5.   Temperature evolution at six temperature monitoring points in heater test (#CM-TM).

468

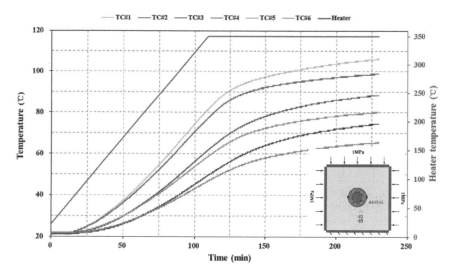

Figure 6. Temperature evolution at six temperature monitoring points in heater test (#CM-M-TM).

the temperature evolution of TC#1 was almost the same with TC#4 during the whole heating period. Similar tendency was also found between TC#2 and TC#5, TC#3 and TC#6. However, obvious temperature difference was found between each pairs of thermocouples in heater test #CM-M-TM (Figure 6). This can be explained by the presence of damaged zone which located at the side wall of the opening. Abundant cracks with low thermal conductivity were formed in this area during the biaxial compression test.

3 CONCLUSIONS

The experimental observations of biaxial compression tests indicated that brittle failure around a circular opening was mainly a process of progressive spalling and finally resulted in v-shaped failed zone. The results of the heater test using intact specimen showed that no clear damage was found on the surface of the circular opening after heating. In contrast, the results of heater test using damaged specimen revealed that fractures tended to initiate on the surface of the circular opening and no obvious further spalling was identified. These experimental findings are consistent with the field observations from the in situ heating damage test (i.e. the third phase of the POSE experiment at the ONKALO URL). It is believed that the results of this research can contribute to the understanding of thermo-mechanical behavior in the damaged zone around underground opening under the context of deep geological disposal of high-level radioactive waste.

ACKNOWLEDGEMENTS

This work was supported by National Natural Science Foundation of China (Grant Nos. 51474050 and U1602232), the Program of High-end Foreign Experts which initiated by State Administration of Foreign Experts Affairs (SAFEA) of China.

REFERENCES

Bae, S.H., 2005, Characteristics of initial rock stress state in Korean tectonic provinces by hydraulic fracturing stress measurement, Ph.D. dissertation, Seoul National University, South Korea. (In Korean).

Golshani, A., Okui, Y., Oda, M., Suzuki, K., 2005, Simulation of damage around a circular opening in rock, 11th International Conference of IACMAG, Turin, Italy.

Hudson, J.A., Bäckström, A., Rutqvist, J., Jing, L., Backers, T., Chijimatsu, M., Christiansson, R., Feng, X.T., Kobayashi, A., Koyama, T., Lee, H.S., Neretnieks, I., Pan, P.Z., Rinne, M., Shen, B.T., 2009, Characterising and modelling the excavation damaged zone in crystalline rock in the context of radioactive waste disposal, Environmental Geology, Vol. 57, No. 6, pp. 1275–1297.

Ishida, T., Kinoshita, N., Wakabayashi, N., 2004, Acoustic emission monitoring during thermal cracking of a granite block, heated in a center hole, Proceedings of the ISRM International Symposium, 3rd ARMS, pp. 133–138.

Jansen, D.P., Carlson, S.R., Young, R.P., Hutchins, D.A., 1993, Ultrasonic imaging and acoustic emission monitoring of thermally induced microcracks in Lac du Bonnet Granite, Journal of Geophysical Research: Solid Earth, Vol. 98, No. B12, pp. 22231–22243.

Rutqvist, J., Dobson, P.F., Garcia, J., Hartline, C., Jeanne, P., Curtis M., Oldenburg, C.M., Vasco, D.W., Walters, M., 2015, The northwest Geysers EGS demonstration project, California: Pre-stimulation modeling and interpretation of the stimulation, Mathematical Geosciences, Vol. 47, No. 1, pp. 3–29.

Shiu, W.J., Dedecker, F., Rachez, X., Peter-Borie, M., 2011, Discrete modeling of near-well thermo-mechanical behavior during CO2 injection, 2nd International FLAC/DEM Symposium, Melbourne, Australia.

Tsang, C.-F., Barnichon, J.D., Birkholzer, J., Li, X.L., Liu, H.H., Sillen, X., 2012, Coupled thermo-hydro-mechanical processes in the near field of a high-level radioactive waste repository in clay formations, International Journal of Rock Mechanics and Mining Sciences, Vol. 49, pp. 31–44.

Valli, J., Hakala, M., Wanne, T., Kantia, P., Siren, T., 2013, ONKALO POSE Experiment – Phase 3: Execution and Monitoring, POSIVA Working Report 2013–41, Eurajoki: Posiva Oy.

Wang, S.H., Lee, C.I., Ranjith, P.G., Tang, C.A., 2009, Modeling the Effects of Heterogeneity and Anisotropy on the Excavation Damaged/Disturbed Zone (EDZ), Rock Mechanics and Rock Engineering, Vol. 42, No. 2, pp. 229–258.

Zhao, X.D., Zhang, H.X., Zhu, W.C., 2014, Fracture evolution around pre-existing cylindrical cavities in brittle rocks under uniaxial compression, Transactions of Nonferrous Metals Society of China, Vol. 24, No. 3, pp. 806–815.

Geomechanics and Geodynamics of Rock Masses – Litvinenko (Ed.)
© 2018 Taylor & Francis Group, London, ISBN 978-1-138-61645-5

Ground closure monitoring systems on trial in deep to ultra-deep mechanized gold mining

T. Zvarivadza & F. Sengani
School of Mining Engineering, University of the Witwatersrand, Johannesburg, South Africa

ABSTRACT: As mining progresses towards deep to ultra-deep levels, rockbursts, support failure, stope closure, and stress fracturing become more pronounced. Electronic monitoring cable anchor devices, extensometers, closure meters, strain cells and FOG lights are currently employed in deep to ultra—deep level mechanized gold mining in South Africa for stability monitoring. The main objective of the study was to monitor the impact of blasting longhole stopes within development cuts on ground deformation or ground closure. This is meant to improve safety of employees that connect from one point to another performing different duties in the mine. This was achieved through installation of ground closure monitoring systems that measure the strain or ground deformation as mining progresses. However, the monitoring systems were performing differently based on their effectiveness. It was found that both electronic monitoring cable anchors and extensometers were more sensitive as compared to FOG lights through detecting micro strain or deformation. Numerical modelling was also conducted to determine the rate of ground closure influenced by blasting longhole stopes within development cuts. The model indicated that the ground closure rate ranged from 400 mm to 600 mm. The electronic Monitoring cable anchors were only installed along the ramp and haulages, strain measurements were found to range from -1×10^{-7} to 1.5×10^{-7}. The extensometers were installed within development ends where longhole stopes were blasted, the closure was found to range from 200 mm to 500 mm. It was noted that rapid blasting of longhole stopes within development cuts led to rapid ground closure and rockmass instability as well as massive Falls of Ground.

Keywords: Electronic monitoring cable device, extensometer, FOG light, strain measurements, longhole stope, distress cut, ground closure

1 INTRODUCTION

Rock related fatalities are still considered to be the most problematic issues faced by South African deep to ultra-deep gold mines (Toper, 2003; Sengani and Zvarivadza, 2017; Zvarivadza et al. 2017). Although many expensive and varied research projects have been completed in recent years, a significant decrease in the number of rock related fatalities in deep to ultra-deep gold mines has not been achieved to date (Kaiser et al. 1996; Karampinos et al. 2015; Potvin and Hadjigeorgiou, 2016). One of the possible factors contributing to this apparent failure has been discussed by Karampinos et al. 2015; they have indicated that lack of instrumentation for continuous monitoring on the behavior of the excavation has played a major role on the increase in number of rock related fatalities. Their discussion further outlines that although seismic systems play a very important role in ground motion, the system does not provide any information about the hangingwall stability, the risk of falls of ground or stope closure rates needed for effective support design. The current studies conducted by Kaiser et al. 1996; Mercier-Langevin and Hadjigeorgiou, 2011; Sweby et al, 2011; Potvin and

Wasseloo, 2013; Potvin and Hadjigeorgiou, 2015; Karampinos et al. 2015; Potvin and Haji-georgiou, 2016, have shown many hard rock mines around the world are still facing problems associated with ground closure, however only few cases have been documented. Looking at documented studies conducted in South Africa, it has been indicated that continuous stope closure measurements might be a useful base on the following aspects that were pointed out by different studies (Sengani and Zvarivadza, 2017; Zvarivadza et al, 2017):

- Identification of different geotechnical areas
- Identification of areas with high face stresses and therefore prone to face bursting
- Identification of areas with a large rock mass mobility leading to unstable hangingwall conditions
- Estimation of closure rates for different mining rates, for effective support design
- Assessment of the effectiveness of preconditioning
- Assessment of the effect of seismicity on stope closure

This study focuses more on the ground closure monitoring during the extraction of long hole stopes. In order to achieve the objectives of the study, several ground monitoring systems were installed along the main access drive, strike access drive and strike drive. This monitoring system includes extensometers, ground closure meters, Falls of Ground lights (FOG) and Electronic monitoring cable devices.

2 DE-STRESS MINING TECHNIQUE

The de-stress cuts are mined through the strata with a maximum span of about 180 m and a 20 m overlap between successive cuts (see Figure 1). The mining configuration of de-stress cuts has the shape of an arrow-head to account for proper leads and lags in the high stress conditions. These de-stress cuts are mined at a stoping height of 5.5 m, the stoping widths are about 5 m wide and 15 m long. Subsequently 10 m of the 15 m is backfilled to create a 5 m wide Strike Access Drive (SAD), and then the adjacent excavation is mined in the same direction (see Figure 1). In essence, the de-stressing reduces the vertical stress so that the major principal stress is in the horizontal direction. The de-stress cuts are supported with 2.5 m long Garford hybrid bolt with 3 m × 1.5 m rectangular weld-mesh on both sidewalls and hangingwalls.

3 LONG HOLE STOPING

Immediately after the completion of de-stress cuts, long hole stopes are then developed to extract the ground left between the two de-stress cuts (middling). This extraction is achieved through the development of a slot at the end of long hole stoping, after the extraction of a slot, a sequence of hangingwall ripping are then designed one after another. Eventually the

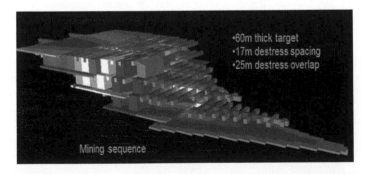

Figure 1. An oblique view of de-stress cut.

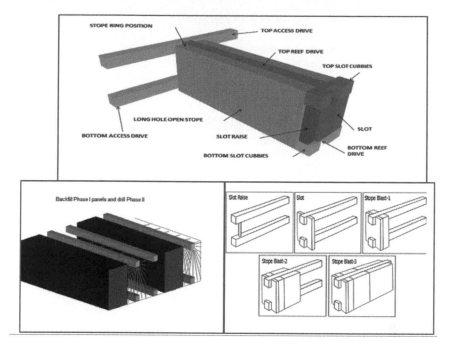

Figure 2.   Longhole stoping process.

extracted long hole stope is then bulk-filled with backfill. Figure 2 shows the cycle of long hole stoping.

4   GROUND CLOSURE

The Electronic Monitoring Cable Anchor Device (EMCAD), extensometers, closure meters and Falls of Ground (FOG) lights were used to monitor the ground closure at the back area and 30 m away from the face. Extensometers were installed on the hanging wall at 90°, 15 m and 20 m long extensometers were installed along the main access drive. EMCAD were also installed along the access drive, intersections and at the bullnoses of long-hole stopes. Closure meters of about 5.5 m height were also installed at the back area and 30 m away from the faces. Lastly, FOG light was installed along the main access drive and intersections.

The ground closure monitoring equipment was installed with the aim of identifying the ground closure generated by the extraction of long-hole stoping and its impact on in-stope pillars. The final results were based on the ground closure measurements before extraction of long-hole stopes and after, as well as ground closure during the extraction of mining stopes.

The closure monitoring systems were found to detect the ground closure differently depending on their sensitivity. The extensometer could identify closure ranging from 10 mm to 500 mm, during the investigation. The closure was found to increase rapidly after extraction of multiple long-hole stopes within the sections. The closure ranged from 200 mm to 500 mm just after extraction of long-hole stopes. These measurements were only measured using extensometers. At the same period, closure meter measurements were also reviewed and it was found that the closures of between 50 mm to 160 mm were recorded from different closure meters installed. On the other hand most of the FOG lights showed some indication of ground closure. Lastly EMCAD measurements on strain were also collected; they were found to range from -1×10^{-7} to 1.5×10^{-7}. It was then concluded that rapid blasting and failure of backfilling long-hole stopes had direct impact on the closure of the ground and also influences in-stope pillar scaling due to redistribution of stress along the in-stope pillars after blasting long-hole stopes.

5 NUMERICAL MODELING ON GROUND CLOSURE

The Abaqus Explicit software was used to simulate ground closure during the investigation. A strain-hardening material model was used through the application of the Hoek-Brown yield criterion and faults were represented using a discontinuum formulation using cohesive finite elements. To determine an accurate virgin (or primitive) stress tensor, stress measurements were carried out using strain gauge cells on 95 Level in the vicinity of one of the Shafts. The results of this exercise are presented in Table 1. Further analyses on stress simulation were conducted using Abaqus Explicit software, the simulation was focusing on sigma 1 only within the de-stress section. Sigma 1 values ranging from 150 MPa to 200 MPa were simulated within the de-stress cuts. As a result, the sigma 1 value simulated from the model within the back areas was higher than the unconfined compression strength of the hosting and country rock. Due to that, higher stress in back areas were also considered to be the problematic issue concerning ground closure (see Figure 3).

Figure 4 illustrates the ground deformation during de-stress development. The deformation was found to be approximately 300 mm. Several factors were also noted to influence the ground closure during de-stress development. The depth at which mining was taking place (+/−3500 m), the middling between two de-stress cuts (+/−15 m) and also high vertical stresses were all influencing the closure during de-stress development.

Table 1. Stress measurements results.

Principal stresses	Magnitude	Bearing	Dip
Major	80.8 MPa	144°	48°
Intermediate	58.9 MPa	359°	20°
Minor	36.3 MPa	265°	20°

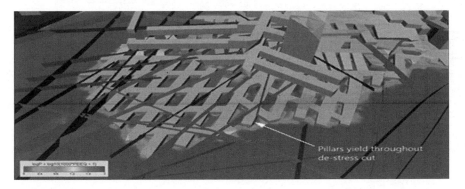

Figure 3.   Stress simulated within the de-stress cut.

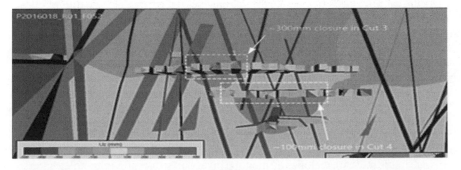

Figure 4.   Ground closure simulated in de-stress cuts before longhole stoping commences.

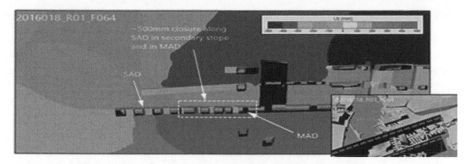

Figure 5.   Ground closure simulated in de-stress cuts when longhole stoping commences.

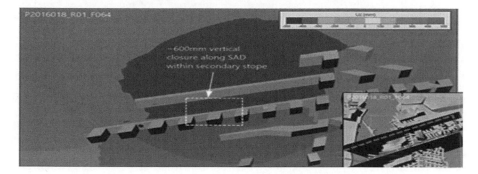

Figure 6.   Ground closure simulated in de-stress cuts when multiple extraction of longhole stoping commence.

Figure 5 illustrates simulated data for ground deformation using numerical modeling. The deformation condition was simulated when long-hole stoping started at the mine. It was found that the ground closure could be approximately 500 mm within de-stress cuts. However, the deformation was estimated to occur during extraction of secondary stoping (long-hole stope) along the strike access drive and main access drive.

Figure 6 illustrates the deformation conditions of the ground during multiple extraction of secondary access within de-stress cuts. The model estimated the vertical closure of approximately 600 mm. However, several factors were also stated which might influence the closure. These include, high release of strain energy, failure to install backfill, multiple large voids in de-stress cuts which were not backfilled.

The results simulated by the model indicated that the ground closure was ranging from 300 mm to 600 mm. It was also found that if the rate of mining remains the same, the mine might expect high ground closure due to insufficient backfill, and rapid blasting of long-hole stopes. However, these could be controlled by backfilling extracted long-hole stopes.

6   CONCLUSIONS

When mining at great depth, the occurrence of ground closure is expected to occur at high rate. It is absolutely crucial to monitor the ground closure within such conditions. The objective of the study was to understand some of the factors that influence ground closure, through the use of different ground closure monitoring systems and observations underground. The results of the study has shown that the ground closure in de stress cuts were ranging from 100 mm to 600 mm depending of the activities that were taking place during that period. Further analysis has indicated that the extraction of longhole stoping at faster rate has influence on the rapid increase in ground closure. As a result, insufficient backfill and rapid blasting of long-hole stopes were found to influence the rapid increase in ground closure of the study area.

REFERENCES

Kaiser, P.K., McCreath, D.R. & Tannant, D.D. 1996, Canadian Rockburst Support Handbook, Canadian Mining Industry Research Organization, Sudbury.

Karampinos, E., Hadjigeorgiou, J. & Turcotte, P. 2015b, 'Management of squeezing ground conditions at Laronde Mine', Proceedings of the 13th ISRM International Congress on Rock Mechanics, International Society for Rock Mechanics, Lisbon.

Mercier-Langevin, F. & Hadjigeorgiou, J. 2011, 'Towards a better understanding of squeezing potential in hard rock mines', Mining Technology, Transactions of the Institutions of Mining and Metallurgy: Section A, vol. 120, no. 1, pp. 36–44.

Potvin Y. & Hadjigeorgiou J. 2016, 'Selection of ground support for mining drives based on the Q-system', in E. Nordlund, T.H. Jones & A. Eitzenberger (eds), Proceedings of the 8th International Symposium on Ground Support in Mining and Underground Construction, Luleå University of Technology, Sweden.

Potvin, Y. & Hadjigeorgiou, J. 2015, 'Empirical ground support design of mine drives', in Y. Potvin (ed.), Proceedings of the International Seminar on Design Methods in Underground Mining, Australian Centre for Geomechanics. Perth, pp. 419–430.

Potvin, Y. & Wesseloo, J. 2013, 'Towards an understanding of dynamic demand on ground support', Journal of the Southern African Institute of Mining and Metallurgy, vol. 113, no. 12. pp. 913–922.

Sengani, F. and Zvarivadza, T. (2017). Review of pre-conditioning practice in mechanized deep to ultra-deep level gold mining. 26th International Symposium on Mine Planning and Equipment Selection (MPES2017). 29–31 August 2017, Luleå, Sweden. pp 113–127.

Sweby, G.J, Dight, P.M., Potvin, Y. & Gamble, N. 2011, 'An instrumentation project to investigate the response of a ground support system to stoping induced deformation', in E Nordlund, TH Jones & A Eitzenberger (eds), Proceedings of the 8th International Symposium on Ground Support in Mining and Underground Construction, Luleå University of Technology, Sweden.

Toper A.Z. 2003. The effect of blasting on the rockmass for designing the most effective preconditioning blasts in deep level gold mines. PhD thesis, University of the Witwatersrand Johannesburg, school of mining engineering, South Africa.

Zvarivadza, T., Sengani F. and Adoko A.C (2017). In-stope pillar scaling and fracturing in Southern African deep level gold mines. 26th International Symposium on Mine Planning and Equipment Selection (MPES2017). 29–31 August 2017, Luleå, Sweden. pp. 379–388.

Geophysics in rock mechanics

Geomechanics and Geodynamics of Rock Masses – Litvinenko (Ed.)
© 2018 Taylor & Francis Group, London, ISBN 978-1-138-61645-5

Study of factors behind rock slope displacement at Higashi Shikagoe limestone quarry, Japan

Chimwemwe Nelson Bandazi
Department of Mines, Malawi

Rith Uy, Jun-ichi Kodama, Yoshiaki Fujii & Daisuke Fukuda
Hokkaido University, Hokkaido, Japan

Hiroshi Iwasaki & Sanshirou Ikegami
Nittetsu Mining, Japan

ABSTRACT: Ensuring safe and productive operations in open pit mining entails undertaking displacement measurement of rock slopes to determine their stability. At Higashi Shikagoe limestone quarry in Hokkaido, Japan, Automated Polar System (APS) was installed to monitor rock slope displacement. The site is geologically complex and is subject to a wide variety of weather conditions due to its location in a cold region of Japan where winter temperatures can drop to −18°C and snow depths of about 80 to 90 cm are often measured. In this study, APS data was used to characterize rock slope deformation and identify its driving forces. Analysis of displacement data revealed changes in tendencies of displacement during warm and cold periods. Numerical analysis using linearly elastic and isotropic models of the quarry showed that excavation and backfilling were not the main causes of measured displacement. It was found that snowfall and snowmelt had an influence on rock slope displacement.

Keywords: rock slope, automated polar system, displacement measurement, excavation and backfilling, cold region, snowfall, snowmelt

1 INTRODUCTION

Displacement measurement in open pit mine operations is a vital part of ensuring safe and productive operations (Tarosen et al., 2016; Dick et al., 2015; and Brown et al., 2007). At Higashi Shikagoe limestone quarry located in Minami Furano, Hokkaido, Japan, an automated polar system (APS) was set up to monitor rock slope deformation. The system consists of a laser beam generator (at the base point) and 16 mirror points (Figure 1). 12 of the mirrors are installed on the rock slope of the mountain type quarry. In this paper data collected between April 2014 and April 2017 is used to clear the characteristics of rock slope deformation. Then, some of the factors affecting the rock slope deformation such as mining activities, snowfall and snowmelt are considered.

2 CHARACTERISTICS OF CHANGE IN DISTANCE

Change in distance data is presented in Figure 2 which shows steady decrease in distance between rock slope and base point over the entire time under consideration. The rate of change in distance is almost constant with a few exceptions where there are some micro-accelerations. The most noticeable irreversible acceleration occurred in the summer period of 2016 when a typhoon reached this part of Hokkaido. These tendencies are seen at all points.

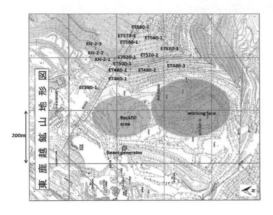

Figure 1. Mine map and APS layout.

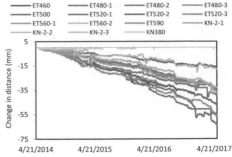

Figure 2. Change in distance.

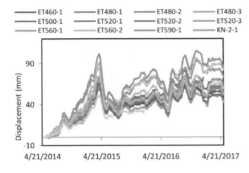

Figure 3. X displacement.

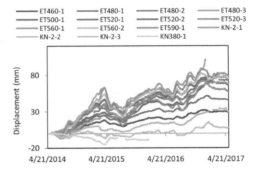

Figure 4. Y displacement.

3 CHARACTERISTICS OF X, Y AND Z COMPONENTS OF DISPLACEMENT

Figures 3 to 5 show the resolved components (x, y, z respectively) of the APS data. Regarding the x direction (sideways movement, where positive displacement represents movement to the southward direction of the site), Figure 3 shows that overall displacement is towards the southern part of the quarry with rates being steady after 2015.

In the y direction (forward and backward movement), forward displacement is observed over the three-year period under consideration with greater magnitudes in 2014 (Figure 4). Accelerated displacement rates are observed during warm periods.

In Figure 5, it is observed that z direction displacement (vertical movement) shows a cyclic pattern of upward and downward displacement depending on season. Although most of this displacement is reversible, a trend of overall downward displacement can be observed following each cycle of movement.

4 EFFECTS OF EXCAVATION AND BACKFILLING

Numerical analysis was used to determine the effects of excavation and backfilling (Kodama et al., 2008; Najib et al., 2015; and Obara et al., 2000). Three-dimensional finite element models of the quarry were built based on site topography and differences in self-weight of rock due to gravitational force were calculated to obtain displacement field.

Assumptions of homogeneity, isotropicity and linear elasticity were made, and unit weight of 26.2 kN/m³ with Young's modulus of 1.0 GPa were assigned to the modeled rock material. Boundary conditions were set by zero displacement vectors on model side and basal surfaces. Element size was based on dimensions of a cube of 10 m × 10 m × 10 m to correspond with

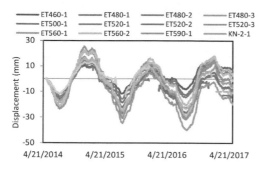

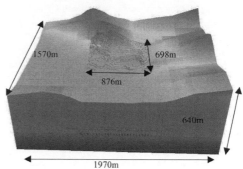

Figure 5.   Z displacement.

Figure 6.   Numerical model for elastic analysis.

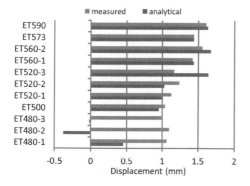

Figure 7.   Comparison between measured and analytical data in x direction.

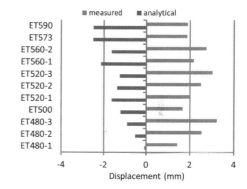

Figure 8.   Comparison between measured and analytical data in y direction.

bench height on site. The initial model had 1,540,689 nodes and 1,776,068 elements. One of the full numerical models is shown in Figure 6 below.

Results of displacement in three directions relative to the base point and analytical data from numerical analysis conducted at Poisson's ratio of 0.1 are shown from Figure 7 to Figure 9 for May 2014 to May 2015. The results show good agreement between measured and analytical data in the x direction (Figure 7) for points at the middle and upper part of the slope. The lower part of the slope shows little agreement. The results show no agreement between measured displacement and analytical data in y and z directions. From this it is concluded that the measured displacement cannot be explained by excavation and backfilling only.

5   EFFECTS OF SNOWFALL AND SNOWMELT

Located in a cold region of Japan where winter temperatures can drop to −18°C and snow depths of up to 90 cm are recorded, it is expected that displacement on-site is influenced by snowfall and snowmelt. It has already been described that displacement data shows different tendencies in summer and winter seasons. Figures 10 to 12 show the relationship between change in distance and snow depth. There is a reduction in rates of change in distance during the peak of winter in 2014/15 and 2016/17 (Figures 10 and 12 respectively). For 2015/16 season (Figure 11), change in distance data shows decreased rates and slightly reversing tendencies momentarily at the onset of the continuous snowfall period. Apart from that, there is no clarity in the relationship between snow depth and change in distance. In all cases, rates of change in distance increase after the disappearance of snow cover. From this, it can be considered that snowfall creates a loading effect on the rock slope that suppresses the rates of

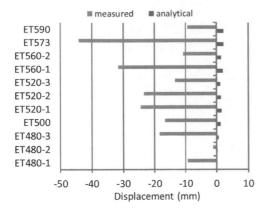

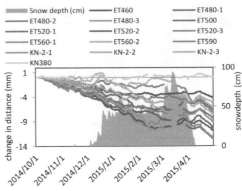

Figure 9.   Comparison between measured and analytical data in z direction.

Figure 10.   Change in distance and snow depth 2014/15.

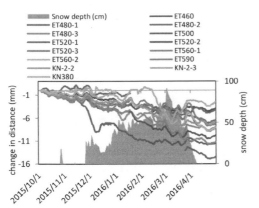

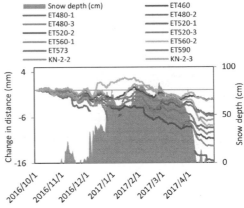

Figure 11.   Change in distance and snow depth 2015/16.

Figure 12.   Change in distance and snow depth 2016/17.

change in distance. Snowfall can induce an increase in normal stress to a sliding plane when dip angle of the sliding plane is low. This is in addition to its loading effect that can suppress displacement due to the added weight on top of the slope (Matsuura et al., 2017).

The increase in displacement rates after snow cover could be explained by snow load release and infiltration of water from the melting of snow. Snowmelt was calculated using the improved degree-hour method presented by Yoshino et al. (2017). The method determines snowmelt Q (cm), according to:

$$Q = k_q T (T > 0°C) \tag{1}$$

where T is temperature (°C) and k_q (0.0248 cm/°C/hour) is a snowmelt coefficient determined from the relationship between cumulative snowmelt and temperature. The method does not calculate snowmelt when temperature is below 0°C because the water is in a frozen state. Figures 13 to 15 show change in distance and snowmelt during the snowmelt periods of 2015, 2016 and 2017 respectively. In 2017, there is a response in displacement to snowmelt. For the 2015 and 2016 data, there are relationships during certain periods. However, the strongest response is recorded in the 2017 season. Effects of snowmelt are compounded with unloading effects resulting in accelerated displacements at the end of the snowmelt period.

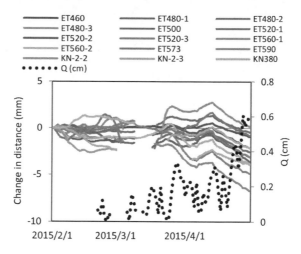

Figure 13.　Change in distance and snowmelt (2015).

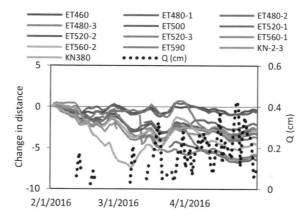

Figure 14.　Change in distance and snowmelt (2016).

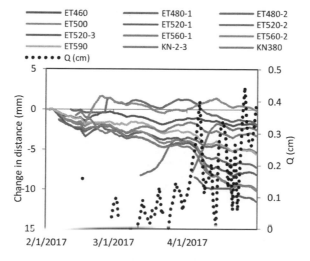

Figure 15.　Change in distance and snowmelt (2017).

6 CONCLUSION

APS data shows that rock slope deformation at Higashi Shikagoe quarry is characterized by forward and downward displacement towards the southern direction of the site. The measured displacement cannot be explained by the processes of excavation and backfilling only. The data shows signs of effects of snowfall and snowmelt on the rock slope.

REFERENCES

Brown, N., Kaloustian, S., & Roeckle, M., (2007), Monitoring of open pit mines using combined GNSS satellite recievers and robotic total stations. *2007 International Symposium on Rock Slope Stability in OPen Pit Mining and Civil Engineering, Perth*, 417–429.

Dick, G.J., Eberhardt, E., Cabrejo-Lievano, A.G. Stead, D. & Rose, D.N., (2015), Development of an early-warning time-of-failure analysis methodology for open-pit mine slopes utilizing ground-based stability radar monitoring data. *Canadian Geotechnical Journal*, 52, 515–529.

Kodama, J., Nishiyama, E., & Kaneko, K., (2008), Measurement and interpretation of long term deformation of a rock slope at the Ikura limestone quarry, Japan. *International Journal of Rock Mechanics and Mining Sciences*, 46, 488–500.

Matsuura, S., Okamoto, T., Asano, S., Osawa, H. & Shibasaki, T., (2017), Influences of the snow cover on landslide displacement in winter period: a case study in a heavy snowfall area of Japan. *Environmental Earth Sciences*, 76: 362. https://doi.org/10.1007/s12665-017-6693-7.

Najib, Fukuda, D., Kodama, J. & Fujii, Y., (2015), The deformation modes of rock slopes due to excavation in mountain-type mines. *Materials Transactions*, 56, 1159–1168.

Obara, Y., Nakamura, N., Kang, S.S. & Kaneko, K., (2000), Measurement of local stress and estimation of regional stress associated with stability assessment of an open-pit rock slope. *International Journal of Rock Mechanics and Mining Sciences*, 37, 1211–1221.

Tarosen, K., Gardhouse, G., & Ran, J., (2016), Lessons in slope stability management from Kinross' Tasiast mine, Mauritania. *3rdInternational Conference on Mine Safety, Science and Rock Engineering, ISMS, Montreal*, 114–120.

Yoshino, K., Agui, K., & Kurahashi, T., (2017), Review of estimation of amount of snowmelt - towards evaluation index for road slope. *2017 International Joint Symposium between CERI (Japan) and IEGS (Korea), Sapporo*, 161–164.

Geomechanics and Geodynamics of Rock Masses – Litvinenko (Ed.)
© *2018 Taylor & Francis Group, London, ISBN 978-1-138-61645-5*

Seismic measurements to recognize rock mass damaging induced by recurrent vibrations

Danilo D'Angiò & Roberto Iannucci
Department of Earth Sciences and Research Center for Geological Risks (CERI),
"Sapienza" University of Rome, Rome, Italy

Luca Lenti
French Institute of Science and Technology for Transport, Development and Networks (IFSTTAR) –
Paris East University, Paris, France

Salvatore Martino
Department of Earth Sciences and Research Center for Geological Risks (CERI),
"Sapienza" University of Rome, Rome, Italy

Antonella Paciello
Italian National Agency for New Technologies, Energy and Sustainable Economic Development,
(ENEA)—Casaccia Research Center, Rome, Italy

ABSTRACT: Recurrent train-induced vibrations can affect the long-term stability of rock slopes located in proximity of railways, where rock falls can cause high-risk conditions for transit and passengers. Seismic monitoring of the unstable rock walls was experienced as a risk-mitigation strategy, in order to detect variations in the vibrational behaviour over time that can be related to microcracking able to modify the pre-existing crack-net. This effect, known as "rock-mass damaging", can justify changes of rock mass mechanical parameters and consequently of rock mass rheology and they can lead toward slope failures (i.e. rock falls or slides). By deriving mean ambient noise levels from signals recorded over a representative time window and by comparing time histories and related Fourier spectra, potential changes in the rock mass vibrational trend can be observed. Moreover, by spectral analyses of seismic ambient noise records it is possible to distinguish the contributions to vibration due to natural sources (generally generating low-frequency signals), including wind, thermal effects or earthquakes, respect to the ones induced by artificial sources (generally generating high-frequency signals), including the train transit.

1 INTRODUCTION

Rock-slides and rock-falls represent one of the most hazardous natural events because of the short time available for taking actions in case of exposed infrastructures due to their rapid evolution as well as their hardly detectable precursors. A recent approach devoted to risk prevention consists in performing ambient vibration studies on potentially unstable jointed rock masses (Got et al., 2010; Levy et al., 2010; Bottelin et al., 2013), in order to detect irreversible changes in vibrational behavior that can be related to microcracking, i.e. the generation of new joints at the rock mass scale (Eberhardt et al., 2004; Stead et al., 2006). Seismic noise is influenced by several factors, both natural (earthquakes, rainfalls, wind speed and direction, air and rock mass temperature) and anthropic (car traffic, quarry activities). In case of rock masses located close to railways or highways, an important contribution to ambient vibration is provided by recurrent and often frequent solicitations generated by trains transit.

In order to investigate variations in the seismic noise, after setting a seismic monitoring in continuous acquisition mode, it is possible to disaggregate and analyse the records with the aim of distinguishing contributions in different frequency ranges, that can be related to natural (low frequency) or anthropic (high frequency) vibration sources. Moreover, the observation of permanent changes in the vibrational behaviour of the monitored rock mass can be related to a rock mass damaging and, hence, interpreted as a precursor of slope failures.

This methodology was applied on seismic record datasets collected in two natural-scale test sites, located in central Italy, the first one in the abandoned quarry of Acuto and the second one along the Terni-Giuncano railway. At Acuto, a protruding and potentially unstable rock block is monitored to detect possible effects due to natural vibrations induced by wind or man-induced vibrations, while at Terni-Giuncano, a rock wall is monitored to record solicitations generated by hourly transit of different trains.

2 SITES DESCRIPTION AND SEISMIC MONITORING

2.1 *Acuto test site*

The abandoned quarry of Acuto (Frosinone, central Italy) is located in the carbonatic Monti Ernici ridge. This quarry was chosen on Autumn 2015 as test site for the installation of a multi-sensor monitoring system on a rock block prone to failure, to investigate long-term rock mass deformations due to temperature, wind and rainfalls (Fantini et al., 2017). The multi-sensor monitoring system consists in: one thermometer for the rock mass temperature; six strain-gauges installed on micro-fractures of the rock mass; four extensimeter installed on open fractures; one optical device for the detection of rock fall events on a railway track, posed to reproduce hazard scenarios, and two weather stations, installed at the foot and the top of the slope wall, equipped with an air-thermometer, an hygrometer, a pluviometer and an anemometer for wind speed and direction. The sub-vertical quarry wall has a height ranging from 15 m up to 50 m and is composed of Mesozoic wackestone with rudists (Accordi et al., 1986). A geomechanical characterisation of the rock mass led to the identification of

Figure 1.   Acuto test site. Dashed limits show the positioning of the accelerometers. a) Zoom and ID number of the accelerometers. b) Main fracture that separates the rock block from the rock wall. c) Weather station at slope top.

four joint sets (Fantini et al., 2016), here indicated according to dip direction/dip convention: S0 (130/13) corresponding to the limestone strata, S1 (270/74), S2 (355/62) and S3 (190/64). The monitored sector is located in the NW portion of the 500-m-long quarry wall and is characterised by the presence of a 64 m³ intensely jointed protruding block, separated from the quarry wall by a main opened joint (oriented 115/90).

On March 2017, a three-days seismic monitoring was carried out to evaluate the vibrational response of the rock block and of the rock mass walls both to natural and to man-induced vibrations. At this aim, three 1-component FBA11 accelerometers were fixed on the monitored rock block and three on the back rock wall; the accelerometers were connected to a Kinemetrics K2 datalogger set for acquisition in continuous mode with a sampling frequency of 250 Hz.

2.2 *Terni-Giuncano test site*

The second experimental test-site is located along the Terni-Giuncano railway which runs along the Serra River valley in correspondence to a canyon cut by the river in the southern sector of the Martani Mounts, NE of the Terni geological basin. This area is characterised by folded and jointed rock masses involved in thrusting and faulting (e.g. Monte Torricella thrust and Battiferro fault) originated by the Appennine chain genesis (Calamita and Pierantoni, 1994; Bruni et al., 1996); in particular, the Scaglia Rossa Formation outcrops in the test site.

Several geomechanical surveys were carried out along the railway line between January and March 2017, to select the rock mass wall more suitable for installing seismic devices. The site was selected close to a tunnel, where no retaining nets were deployed to equip the rock walls and a visible block-size slope debris testifies recurrent rock falls and slides from the wall. The geomechanical characterisation, performed following ISRM (1978), provided J_v: 7–10 and I_b: 0.3–0.5.

A seismic monitoring of the rock mass was performed from 20th April to 15th June 2017. As shown in Fig. 2, the monitored sub-vertical rock wall is 4.4 m distant from the railway and its slope face (oriented 260/80 in dip direction/dip convention) is parallel to the railway track that runs along an almost N-S direction. Four 1-component FBA11 accelerometers were fixed on the rock wall, in particular two at its base (0.5 m from ground) and two at 1.5 m from ground, arranged for measuring along NS and UP directions. In addition, two 3-component

Figure 2.   Terni-Giuncano railway test site. Location and ID number of accelerometers (green circles), velocimeters (yellow circles) and K2 datalogger (red circle) are indicated. The REF TEK dataloggers were within an impermeable box visible at the right bottom of the picture.

velocimeters Lennartz LE-3Dlite MkII were installed at the bottom of the rock mass and as near as possible to the railway track respectively, in order to evaluate the soil attenuation and the amount of vibrations reaching the rock mass wall. The accelerometers were connected to a Kinemetrics K2 datalogger set in trigger mode, while the velocimeters were connected to REF TEK 130S-01 dataloggers set in continuous mode; the dataloggers recorded with a sampling frequency of 250 Hz.

3 DATA ANALYSIS

Although in the Terni-Giuncano test site the accelerometric records do not include all the trains transited during the experimentation period, because of a limited storage capacity, the collected data are reliable to perform preliminary analyses and to test the methodology of seismic record disaggregation at different frequency ranges. Moreover, in the Terni-Giuncano test site no weather station was installed and data taken from the closest meteorological stations are not suitable to be transposed to the experimental site. On the contrary, at Acuto test site, despite the shortness of the ambient vibration records obtained during the experiments performed so far, it is possible to relate the seismic records with temperature, wind speed and wind direction values that were recorded every minute by the in-site weather station.

The accelerometric records were filtered from 4 to 60 Hz and integrated to obtain velocity records comparable with those recorded by Lennartz velocimeters. Fast Fourier Transform (FFT) and spectrograms were then computed, in order to evaluate the frequency response of the investigated rock mass. In the Terni-Giuncano test site we observed that the transit of different train vehicles (i.e. freight or passenger ones and with different number of wagons) generates seismic vibrations with peaks around 50 Hz and a FFT amplitude that put in evidence an heterogeneous response of the vibrational motion observed for the different sensors installed; recorded motion can variate up to one order of magnitude (i.e. comparing sensors 5, 6, 12 with sensor 8 in Fig. 2).

A further analysis consisted in ambient noise disaggregation: the seismic records were filtered in three different frequency ranges (low, medium, high) chosen after consulting FFT and spectrograms results in order to isolate the effects induced by environmental conditions at Acuto and by train transits at Terni-Giuncano. In particular, for the Acuto test site the selected frequency ranges were 0.1–15 Hz, 15–30 Hz, 30–60 Hz while for the Terni-Giuncano test site they were 0.1–30 Hz, 30–50 Hz, 50–60 Hz. Amplitude values of the records were averaged in time windows of one minute. This analysis generates mean values for a frequency band as functions over the time. It was possible to compare the cumulated mean amplitude values with the external inputs received by the rock mass (i.e. train transits, rainfalls, ventilation), in order to measure the acting solicitations, and the detectable vibrational response of the monitored rock mass.

4 PRELIMINARY RESULTS

Fig. 3 shows the preliminary results obtained following the above-described methodology and regarding 72 hours of seismic monitoring at Terni-Giuncano test site. Along this time interval 92 trains transited on the railway, especially in early morning and in the afternoon, while between midnight and 5 a.m. the railway traffic was always suspended. Data related to the accelerometer 5 (cf. Fig. 2), show a main response of the rock mass to induced vibrations at the medium (30–50 Hz) and high (50–60 Hz) frequency ranges.

By observing the cumulated curves in each frequency range reported in Fig. 3, it is possible to note marked transient rate variations during hours of daylight (i.e. in correspondence with trains transit), while a flatten response is shown during late evening and night, when the traffic is reduced. In fact, a mean of 19 trains transit between 5:00 and 13:00 and between 13:00 and 21:00, while only 5 trains transit between 21:00 and 5:00. Moreover, by considering the rate of the curves in the same time interval (e.g. between 12:00 and 18:00), on 29th April an

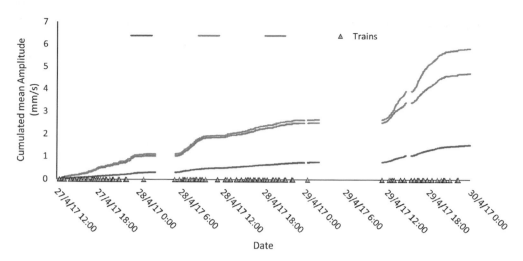

Figure 3. Ambient noise disaggregation of three-days seismic monitoring on Terni-Giuncano railway line. Coloured lines indicate different frequencies bands; interruptions indicate no-acquisition periods.

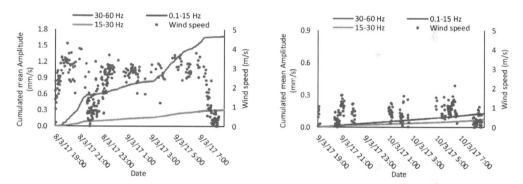

Figure 4. Ambient noise disaggregation for the accelerometer 12 during two nights (8–9th and 9–10th March 2017, left and right graph respectively) characterised by different wind intensity (purple circles). Coloured lines indicate different frequencies bands.

incremented rate is observed in respect to that showed on 27th and 28th April: at this regard, considering the similar amount of trains transited, this marked variation in the vibration rate could be referred to changes in environmental conditions. Beyond these transient changes in vibrational rates, no permanent changes in the long-term vibrational trend of the rock mass has been noted, because of the limited duration of the accelerometric records analysed.

At the Acuto test site, the ambient noise disaggregation was performed on the records obtained from an accelerometer fixed on the rock block during two nights (8–9th and 9–10th March 2017 respectively) characterised by comparable site conditions (air and rock temperature, vehicle traffic) except for wind intensity (Fig. 4). During the first night (8–9th March), characterised by intense and persistent gusts of wind, a strong vibrational contribution at the high frequency range can be detected, while during the second night (9–10th March), with negligible wind, the contribution to vibrations mainly results at the low frequency range.

5 CONCLUSIONS

The data collected so far demonstrate that the here presented methodology is able to detect different sources of vibration over short to middle time measurements. In fact, at the Terni test site it was possible to detect a main contribution to vibrations always at frequencies between 30 and 60 Hz, which testifies the major effect related to the train transits. On the other side, at the Acuto test site it was possible to observe a vibrational response at different frequencies of a rock block related to meteorological conditions since in case of stronger wind, a marked vibrational response is evident at frequencies ranging from 30 up to 60 Hz. These preliminary results encourage to perform this seismic monitoring over longer time to highlight irreversible changes in the vibrational behaviour that can be related to rock mass damaging and can be regarded as precursors for rock mass failures. As a future perspective, this methodology can be devoted to landslide risk management strategy, increasing the advice-time before rock falls or slides and avoiding accident during the train traffic. In order to improve the methodology performance, further experiments will be planned consisting in monthly-to-yearly long seismic measurements and/or considering more detailed frequency ranges.

ACKNOWLEDGMENTS

The Authors wish to thank the Municipality of Acuto for the authorization provided to the experimental activities carried out in the abandoned quarry; RFI (Rete Ferroviaria Italiana) for allowing the in situ test along the Terni-Giuncano railway as well as for providing data on train transits; the DTP (Direzione Territoriale Produzione) of Foligno for the logistic support to device installation and maintenance. This research is carried on in the framework of the Ph.D. project of Danilo D'Angiò (Department of Earth Sciences of the University of Rome "Sapienza").

REFERENCES

Accordi, G., Carbone, F., Civitelli, G., Corda, L., De Rita, D., Esu, D., Funiciello, R., Kotsakis, T., Mariotti, G. & Sposato, A. 1986. *Lithofacies map of Latium-Abruzzi and neighbouring areas.* Quaderno C.N.R. "La Ricerca Scientifica", Roma, 114 (5), 223.

Bottelin, P., Jongmans, D., Baillet, L., Lebourg, T., Hantz, D., Lévy, C., Le Roux, O., Cadet, H., Lorier, L., Rouiller, J.D., Turpin, J. & Darras, L. 2013. *Spectral analysis of prone-to-fall rock compartments using ambient vibrations.* Journal of Environmental and Engineering Geophysics, 18(4), 205–217.

Bruni, F., Calamita, F., Maranci, M., Pierantoni, P.P. 1996. *Il controllo della tettonica giurassica sulla strutturazione neogenica dei Monti Martani meridionali (Preappennino umbro).* Studi Geologici Camerti, Volume Speciale 1995/1, 121–135.

Calamita, F., Pierantoni, P.P. 1994. *Structural setting of the southern Martani mountains (Umbrian Apennines: central Italy).* Memorie della Società Geologica Italiana, 48, 549–557.

Eberhardt, E., Stead, D. & Coggan, J.S. 2004. *Numerical analysis of initiation and progressive failure in natural rock slopes—the 1991 Randa rock slide.* International Journal of Rock Mechanics & Mining Sciences 41, pp. 69–87.

Fantini, A., Fiorucci, M. & Martino, S., 2017. *Rock Falls Impacting Railway Tracks: Detection Analysis through an Artificial Intelligence Camera Prototype.* Wireless Communications and Mobile Computing, vol. 2017, Article ID 9386928, 11 pages, 2017. doi:10.1155/2017/9386928.

Fantini, A., Fiorucci, M., Martino, S., Marino, L., Napoli, G., Prestininzi, A., Salvetti, O., Sarandrea, P. & Stedile, L., 2016. *Multi-sensor system designed for monitoring rock falls: the experimental test-site of Acuto (Italy).* Rendiconti Online della Società Geologica Italiana, 41, 147–150.

Got, J.-L., Mourot, P. & Grangeon, J., 2010. *Pre-failure behaviour of an unstable limestone cliff from displacement and seismic data.* Nat. Hazards Earth Syst. Sci., 10, 819–829.

ISRM, 1978. *Suggested methods for the quantitative description of discontinuities in rock masses.* International Journal of Rock Mechanics and Mining Sciences and Geomechanics Abstracts, 15, 319–368.

Lévy, C., Baillet, L., Jongmans, D., Mourot, P. & Hantz, D. 2010. *Dynamic response of the Chamousset rock column (Western Alps, France).* Journal of Geophysical Research, 115, F4.

Stead, D., Eberhardt, E. & Coggan, J.S. 2006. *Developments in the characterization of complex rock slope deformation and failure using numerical modelling techniques.* Engineering Geology 83, pp. 217–235.

Geomechanics and Geodynamics of Rock Masses – Litvinenko (Ed.)
© 2018 Taylor & Francis Group, London, ISBN 978-1-138-61645-5

Dynamic investigations of EDZs from Bátaapáti radwaste repository based on passive seismoacoustic measurements

Ferenc Deák
Department of Engineering Geology and Geotechnics, Engineering Geologist—Mecsekérc Ltd., Pécs and Budapest University of Technology and Economics, Hungary

István Szűcs
Department of Environmental Engineering, Faculty of Engineering and Information Technology, LADINI Engineering Co. Ltd., Geophysicist—University of Pécs, Hungary

ABSTRACT: The program for the final disposal of low and intermediate level radioactive waste was established by Paks Nuclear Power Plant, Hungary. Preparation of final disposal has been done as part of a national program since 1993. The Central Nuclear Financial Fund and the Public Limited Company for Radioactive Waste Management (PURAM) have been established to coordinate organizations and activities for all tasks in connection, with nuclear waste treatment. The selected potential host rock is a granite complex in the Mórágy Granite Formation in the south-western part of Hungary, close to the village of Bátaapáti.

During the construction of the Bátaapáti radioactive waste repository the drill and blast excavation method was used. During the periodic verification of the blasting process, passive seismoacoustic monitoring systems were applied. Based on these measurements it was possible to set and supervise accurately the blasting quality in order to minimize the damage to the rock in vicinity of the tunnel contour. Cautious blasting is needed to minimize the Highly Damaged Zone.

The aim of this paper is to present the measuring analysis methodologies of the acquired data. Details will be explained on the dynamic numerical modeling results, on the site-specific attenuations and the blasting effects on the extension and behavior of Excavation Damage Zones.

Keywords: blasting, shock wave, attenuation, numerical modeling, EDZs

1 INTRODUCTION

Since the first excavation blast rounds in the Bátaapáti radwaste repository tunnels minimizing the blasting damage in vicinity of the excavation surface has been a priority. During the blasting design one of the most important points is the prediction of blast induced vibration (Lu et al., 2011). When explosive is detonated, shock waves are generated and propagated from the source. The dissipated energy content of the elastic waves is exposed to absorption of intensity as a function of the distance from the blast holes.

The characteristic shock wave propagating in rock mass can be estimated from attenuation relationships or during numerical modeling. There are numerous attenuation relationships, (relating the peak particle velocity or PPV with scaled distance defined as the ratio of distance from the charge point, to the square root of charge mass, expressed in TNT net equivalent charge weight), which are most often used to predict the amplitude of vibration (Dowding, 1984 and Kumar et al., 2016). Typical vibration measurements are carried out far from the blast source, hence the rock mass inside the zone of measurement are not well defined in terms of attenuation, for instance near field vibration is even more difficult to be characterized. In addition, the attenuation curves used in practice cannot be used to predict the vibrations close to the detonation.

Dynamic properties of rocks include Young's modulus, Poisson's ratio and a damping ratio. These properties of the rock mass are not well documented in the scientific literature and are typically selected based on experience. However, the lack of data is even more sparse for a jointed rock mass, with the damping ratio having the highest uncertainty (Ahn et al. 2016).

In this study we simulate the effect of a tunnel face blast, with an advance of 2.0 m in length using a 21 m² tunnel cross section. 2D finite element modeling was performed with the RS2 Rocscience software by applying both the static loads and the dynamic loads. During the 2D plane modeling it is difficult to determine the damping ratio, because it is rally a 3D problem. The damping ratio has to be additionally adjusted for the different attenuation characteristics of a 2D analysis since the spherically induced blast load propagates through the 3D environment.

2 PASSIVE SEISMOACOUSTIC MEASUREMENTS

During the excavation of the inclined twin access tunnels of the repository numerous seismoacoustic measurements were carried out to determine the blasting effects on the neighboring tunnel.

From the measurements we chose one, in which case the actual tunnel face coincided with one of the measurement points. The blasting was executed in the Western access tunnel, while the measurement points were situated in the Eastern tunnel. At the measurement location we were using 6 mounted accelerometers on the top of the 1 m long rock bolts inside the rock mass which were installed at a height of 1–2 m from the tunnel floor into the side wall. The direction of sensitivity of the accelerometers was perpendicular to the tunnel wall. In the Eastern access tunnel, point 4. mp was the only sensor which measured the acceleration in one direction (horizontal) in line with the Western access blasted tunnel face (Figure 1). Taking this advantageous position into the consideration, we present these results.

During the measurement we carried out one record, while the conventional work was interrupted in both tunnels, the ventilation system was turned off and the overall conditions could be considered "seismically silent".

Main technical parameters: sampling density – 3000 samples/sec/channel; frequency range – 0–1200 Hz; minimal measurable value (basic sensitivity) – 0.001 m/s²; maximum measurable

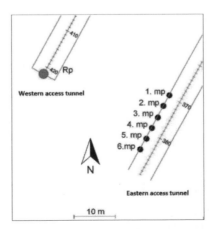

Figure 1.   Location of the executed seismoacoustic measurements (the actual blasted tunnel face is indicated by the red circle and the measurement points by the black points).

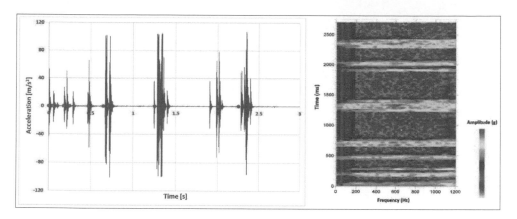

Figure 2. Recorded accelerogram at the measurement point 4. mp—delays are assigned by red points, at right figure is shown the time-varying frequency components (spectogram) of the 4. mp measurement (after Gacsályi et al. 2005).

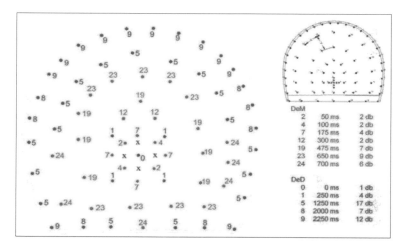

Figure 3. Schematic view on the blast holes of the experimental blasting, by used 122 kg explosives which produced 2.0 m tunnel advance—at the right corner is shown the standardized tunnel section with the blast holes from Bátaapáti radwaste repository access tunnels, in case of rock support class II (Deák et al. 2014).

value – 100 m/s². By using high sampling density, it was also possible to examine the whole blasting process, and to record the high frequency components. Different delays can be easily identified in the data series. The 280 ms incoming signal cannot be coupled to any blasting delay time preset. Figure 2 shows the recorded accelerogram at the 4. mp point with the assigned delay times. The Figure 3 shows positions of the blast holes used in this experimental tunnel face blasting.

During the wave spectral analysis, on the records, a time window of 26 ms width was run through 13 ms steps. The amplitude spectrum characteristic of the frequency component of the recording part of the window was determined (Figure 2). Each such spectrum is depicted in one row, the color scale shows the magnitude of the markers. For the better representation of the spectrum, the base 10 logarithm of the amplitude values was depicted. Each scale is relative, the presented maximum is the maximum of the actual channel. It was characteristic, that high amplitude signals were detected on multiple channels (as in the case of the presented channel 4.) in the frequency range above 1000 Hz.

3 NUMERICAL ANALYSIS

The numerical investigation was carried out by using RS2 (Rocscience) software taking advantage of the dynamic solutions.

The amplitude of vibration induced by blasting decreases with increasing distance from source. Due to attenuation, decay of vibration is produced by two phenomena, which are geometrical spreading and material damping (Dowding, 1996). The geometrical damping is caused by the expansion of the surface over which the vibration energy is transmitted. The material damping is caused by the nonlinear hysteretic behavior of the geologic media (Park et al. 2015).

During the preliminary work simplifications were made, such as the equations only described the amplitude and not the time series of the propagated motion. The registered amplitudes were amplified and then the dynamic load normal to the tunnel wall was defined. In the preliminary study no distinction between space and time was made for the different delays. As a further investigation, it's not difficult to specify the accurate time delays, because we have the registered times and the used delay times during the blasting process, accordingly in this case the question can be solved with a simple time correction.

For the attenuation, the following equation was used (Kim and Lee, 2000):

$$A_2 = A_1 \left(\frac{r_1}{r_2} \right)^m$$

where A_1 is the amplitude of vibration at distance r_1 from the source, A_2 is the amplitude of vibration at distance r_2 from the source and m is a geometric coefficient (for underground blasting, m = 1.0).

In Figure 4 the accelerogram is shown, which is the amplified result of the originally recorded values.

From the previous investigations and from the scientific literature, it can be assumed that blasting creates a fractured zone around the blast holes and the elastic waves are propagated beyond this crushed zone. This zone was assigned to be a circle with a radius 1 m (starting from that, we assumed a 1 m thick crushed zone started from the tunnel wall, around the tunnel section). During the amplification we used $r_2 = 1$ m, and the blast load was applied normal to the excavation surface.

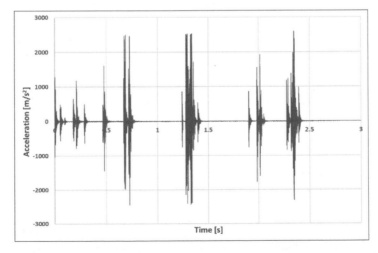

Figure 4.   The amplified accelerogram which was used as normal dynamic load on the tunnel wall.

Damping was solved by using Rayleigh damping. With this type of damping, the damping matrix that relates the damping force and velocity of the system is expressed in terms of the stiffness and mass matrix of the system. The damping becomes proportional to the mass and stiffness of the system.

$$[C] = \alpha[M] + \beta[K]$$

where $[C]$ is the damping matrix, $[M]$ is the mass matrix, $[K]$ is the stiffness matrix, ξ is the damping ratio, α and β are Rayleigh coefficients that determine the frequency dependence of the damping formulation. The damping of the soils and rocks is independent of the loading frequency. By contrast Rayleigh damping formulation is frequency dependent and adequate to solve numerical damping (Ahn et al. 2016).

Use of Rayleigh damping in the numerical model allows the user to define the damping ratio for two frequencies. Generally, the frequency range between the two defined threshold frequencies, have a damping ratio lower than the specified damping ratios and frequencies outside this range are damped more heavily. Numerous variations the two frequencies and damping ratios were modelled and a more appropriate result to the measured one. Based on the mentioned settings the software automatically calculates the α and β values.

The size of the elements was selected as suggested by (Lysmer and Kuhlemeyer, 1969)

$$\Delta l \leq \frac{\lambda}{10}$$

where Δl is the element size, λ is the wave length associated with the highest frequency component that contains appreciable energy. The typical frequency range of underground induced vibrations are between 50 and 100 Hz.

With this range in mind, the following calculation for the maximum frequency that still maintains the accuracy of the model:

$$C_s / \lambda$$

where C_s is shear wave speed and λ is the wave length. Based on the mesh density (maximum element length 0.1 m) the maximum frequency was found to be, $\frac{61\,m/s}{1\,m} = 61\,Hz$ (Figure 5).

From the numerical calibration work, specific model time series results, which were very similar to the originally measured one, were determined (Figure 6).

The best match was achieved by the following damping settings:

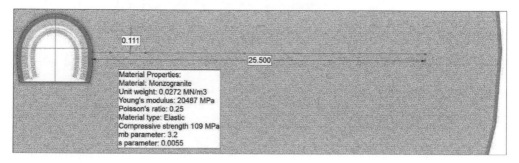

Figure 5. The used geometry and boundary conditions due to modeling (we used several variations, where both tunnels were defined, in these cases we found amplitude uncertainties which will be handled in the future).

Table 1. Dynamic settings for the adequate used damping.

$\alpha[M]$	$\beta[K]$
0.0314076	8.3763310
Frequency 1	**Frequency 2**
0.05 Hz	570 Hz
Damping ratio 1	**Damping ratio 2**
0.05	0.15

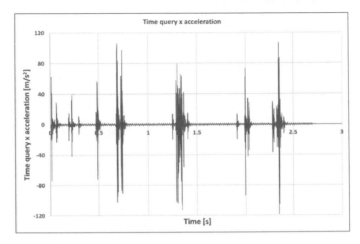

Figure 6. Modeling results at the point which distance from the Western access tunnel matched to the measured 4. mp accelerometer sensor.

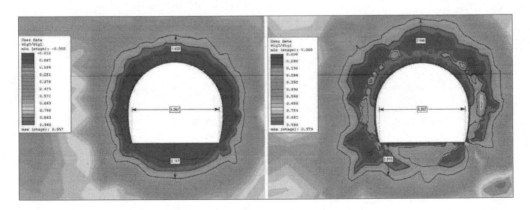

Figure 7. Extension of EDZs after the excavation: left figure without dynamic loads, at right figure with the blasting dynamic extra load—stage time: 3 sec (iso lines: HDZ-red, EDZ$_i$-violet, EDZ$_o$-black).

4 DYNAMIC EXAMINATION OF THE EDZs

The tunnel excavation process affects the rock mass producing zones of damage. These are the Excavation Damaged Zones (EDZs). Moving away from the excavation, the EDZs can be distinguished as the Highly Damaged Zone (HDZ), the inner EDZ (EDZ$_i$) and the outer EDZ (EDZ$_o$). Beyond the EDZ is a stress and/or strain influence zone that involves only

496

elastic change, the Excavation Influence Zone (EIZ) (Perras et al. 2012, Deák et al. 2013, Deák et al. 2016).

As a further experiment the amplified acceleration load (shown in Figure 4) normal to the excavated tunnel walls was modelled. The preliminary modelling was carried out by using three stages: in the first stage the rock mass was defined with no excavation, in the second stage the tunnel was excavated and in the third stage the dynamic load was applied. The resulting zones became much larger in space and their shape was also compared to the static load stage (Figure 7).

5 CONCLUSIONS

The work presented here highlighted the importance of complex numerical investigations for drilling and blasting excavation effect on the rock mass. A correct method of examination of the EDZs is required to use static and dynamic approaches in parallel and to incorporate these experiences into the design.

Due to calibrated modeling of existing acceleration measurements it was possible to define the site-specific attenuations. It was shown that the near field vibration can be accurately captured by the models, by using a very dense mesh, while the accuracy of the far field attenuation is influenced by the Rayleigh damping.

A further investigation will continue to use this work by separation of all blasted regions based on the different delays and the tunnel sections will be excavated in this way by an appropriate number of stages in the new numerical models.

During the characterization of the rock mass damage zones at the vicinity of the underground caverns, the presented approach has a great significance in the design of the engineered barriers for radioactive waste repositories.

ACKNOWLEDGEMENT

The authors acknowledge the permission of PURAM to publish this paper. The first author is grateful to Dr. Matthew A. Perras for the suggestions and comments.

REFERENCES

Ahn J.-K., Park D., Yoo J.-K. (2016): Estimation of damping ratio of rock mass for numerical simulation of blast induced vibration propagation, Japanese Geotechnical Society Special Publication, vol. 2, pp. 1589–1592.

Deák F., Kovács L., Vásárhelyi B. (2013): Modeling the Excavation Damaged Zones in the Bátaapáti radioactive waste repository, Rock Mechanics for Resources, Energy and Environment—Kwasniewski & Łydzba (eds) 2013 Taylor & Francis Group, London, pp. 603–608.

Deák F., Kovács L., Vásárhelyi B. (2014): Geotechnical rock mass documentation in the Bátaapáti radioactive waste repository, Central European Geology, Vol. 57/2, pp. 193–207.

Deák F., Szűcs I. (2016): Examination of a granitic host rock behavior around underground radwaste repository chambers based on acoustic emission datasets, Rock Mechanics and Rock Engineering: From the Past to the Future—Ulusay et al. (Eds) 2016 Taylor & Francis Group, London, pp. 1243–1248.

Dowding C. (1996): Construction vibrations, Prentice Hall.

Gacsályi M., Jenei A. (2005): Passive seismoacoustic measurements in rock mass classified as rock class 2. from Bátaapáti low- and intermediate level radioactive waste repository, Geopard Ltd. Report (in Hungarian language), 26 p.

Kim D.S., Lee J.S. (2000): Propagation and attenuation characteristics of various ground vibrations, Soil Dynamics and Earthquake Engineering, 19(2), pp. 115–126.

Kumar R., Choudhury D., Bhargava K. (2016): Determination of blast-induced ground vibration equations for rock using mechanical and geological properties, Journal of Rock Mechanics and Geotechnical Engineering, 8 (2016), pp. 341–349.

Lu W., Yang J., Chen M., Zhou C. (2011): An equivalent method for blasting vibration simulation, Simulation Modeling Practice and Theory, 19(9), pp. 2051–2062.

Lysmer J., Kuhlemeyer R. (1969): Finite element model for infinite media, Journal of Engineering Mechanics Division. ASCE, 95, pp. 859–877.

Park D., Ahn J.-K. (2015): Estimation of Attenuation Relationship Compatible with Damping Ratio of Rock Mass from Numerical Simulation, Journal of the Korean Geotechnical Society, vol. 31, pp. 23–26.

Perras. M., Langford. C., Ghazvinian. E. & Diederichs, M. (2012): Numerical deliniation of the excavation damage zones: From rock properties to statistical distribution of the dimensions. ISRM, EUROCK2012, Stockholm, 14 p.

Geomechanics and Geodynamics of Rock Masses – Litvinenko (Ed.)
© *2018 Taylor & Francis Group, London, ISBN 978-1-138-61645-5*

Reactivation of the old landslide caused by the land development—the case study

Lucyna Florkowska & Izabela Bryt-Nitarska
The Strata Mechanics Research Institute of the Polish Academy of Sciences, Cracow, Poland

Rafał Gawałkiewicz
Faculty of Mining Surveying and Environmental Engineering, University of Science and Technology, Cracow, Poland

Ryszard Murzyn
Geological and Engineering Company GEO-INŻ-BUD, Wiśniowa, Wiśniowa, Poland

ABSTRACT: Although landslides are a group of natural hazards, the factor causing their activation is often human activity. The construction of buildings and roads on slopes has an especially unfavourable impact on its stability. Additional load and disturbance of natural hydrological conditions are factors that may cause the reactivation of old landslides.

This paper contains a case study of the activation of landslide processes, which was caused by human activity. In the area of the old and inactive landslide an educational complex was built. The complex consisted of four buildings located in one line across the slope. An additional, very unfavourable factor was the high level of groundwater, which was rising up to the ground surface temporarily. At the time of long-term rainfall activation of landslide processes in the substrate of the building occurred. These processes are characterized by slow changes, causing progressive destruction of buildings.

The geological structure of the slope was determined by six core holes with a depth of about 20 m. On this basis, the geological layers, hydrological conditions and slide surface position were determined.

Deformations of the slope and the building were monitored by using integrated survey techniques: precision leveling, electronic techeometry, traversing (angular-linear), static GNSS and laser scanning.

The work includes the analysis of initial geological and hydrological conditions, the analysis of the causes and the course of deformation processes in the substructure, the results of the measurements and the analysis of the deformation of buildings caused by a landslide.

Keywords: landslides, natural hazards, flysch, source sink, fault, glide plane breccias, human activity, building, damage, core drilling, survey monitoring

1 INTRODUCTION

Erecting buildings in areas where geodynamic processes can occur carries a very high safety risk. If the geological situation is not assessed appropriately and if necessary security measures (geotechnical and structural) are not introduced at the design making stage, a state of emergency or a collapse of a building may occur. The basic conditions for landslide prevention are: not erecting any buildings on slopes and protecting the slopes against excessive water infiltration (Burrel et al., 2017, Kyoji et al., 2013). This paper is a case study in which neither of the conditions was met. Additional unfavorable factors which increased the risk were in this case: exceptionally unfavourable hydrological conditions, non-homogenous substrate, unfavourable geometry of the shape and the design solutions used in building construction.

2 THE GEOLOGICAL SITUATION

The area this paper deals with is part of a large geological unit called flysch Western Carpathians. The geological structure of its part close to the border is so exceptionally complex and so hard to decipher that it would be difficult to find a similar area anywhere else in the world. Flysch deposits were moved here from the south (over 50 km) to its foreland made mainly of mesozoic structural platforms (mainly Jurassic and Cretaceous period deposits) and Miocene period deposits lying on them in disconformity (Burtan J., 1984).

2.1 Geological structure

The building we are analysing in this paper was erected on the colluvia of an old landslide, directly below two source sinks, where ponds were situated in earlier times (Fig. 1). The concave slope shape indicated landslide origins, however the geologist did not participate in the designing process and the geotechnical engineer failed to notice the obliterated old form. Now the width of the colluvia of the old landslide was assessed based on core drilling (with a double tube core) – it is approximately 21.9 m near the school (Figures 2 and 3). The colluvia are composed of tectonic breccia, assize shales and sandstones, weathering waste and fragments of the Aeolian layer (loess-like deposits) carried by the wind to the concave slope shape (Fig. 2). The presence of boggy ground of approximately 3 m in thickness under the corner of the school points to the existence of constant anaerobic conditions and local bog sedimentation processes. Research shows that building the school caused landsliding processes to be activated in the old landslide. Pressure exerted by the colluvia on the building, uneven subsidence of the building/the ground exerting upward pressure/, creep movements of the entire building together with the ground were observed. The deposits which occur at the depth of the building's foundations have stiff, firm and soft consistencies.

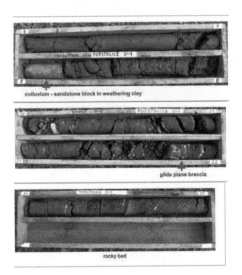

Figure 2.   Cores from borehole O-1.

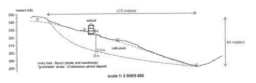

Figure 1.   The location of the building.

Figure 3.   Cross-section through the slope.

2.2 *Hydrological conditions*

The landslide headwall starting scar consists of two source sinks with constant water outflow. In normal conditions, the school's existing drainage system carries away water at the rate of approximately 3 l/s (10.8 m³/h). It was also found that groundwater flows from several other places on the slope. The existence of the sources is strictly connected with a fault (the fault area was identified on the basis of drilling). Erecting the school and building roads traversing the slope, which carry surface run-off and groundwater outflows, led to the shifting of the slope's natural water balance. The transverse orientation of the building on the slope led to a compression of natural migration pathways of groundwater within the colluvia and at the same time leads to the surface run-off accumulating at the building. The system of roads on the slope causes run-off to accumulate in the area of the colluvium—directly below the school building. The existing hydrogeological and hydrological conditions are the main factors leading to increased landsliding processes. Increased landsliding processes took place in 2010 during the millennium flood. Increased landsliding processes were also observed in 2013 due to a failure of the school's sewage treatment facility and in 2015 following a failure of the school's storm sewer system (damaged as a result of previous landslide movements).

3 DESCRIPTION OF THE BUILDING

The building was designed in 1991 as a complex of three functionally-linked segments (Fig. 1). The construction was planned in two stages. Stage 1 – erecting the building of the school (S1) and the link (S2), Stage 2 – erecting the sports and entertainment hall (S3) with amenities (S4). Stage 1 was completed at the beginning of the 1990s. The sports hall with the amenities was built in 2011, according to a design which was based on different assumptions than those planned for the original design (Florkowska et al., 2017).

The school building (S1) with the dimensions of the layout of approximately 33.5 × 15.5 m and the link building (S2) with dimensions of approx. 11.0 × 15.0 m were built on a system of reinforced concrete strip footing without a basement. The school segment has two full aboveground storeys with a habitable attic. The sports and entertainment hall segment (S3) with general layout dimensions of approx. 33.0 × 18.0 m was erected together with the amenities (S4) with dimensions of approx. 18.0 × 18.0 m. The gymnastics hall and the amenities were supported on a high raft foundation. The above-grade walls are made of ceramic blocks and are stiffened by means of a wet-poured post and beam structure.

Already at the stage of erecting the building the actual foundation conditions proved more difficult than planned. It can be proved by the fact that the drainage system is considerably more extensive than originally planned and by the decision not to build the gymnastics hall. From the moment it started to be used (in 1997) until the flood in 2010 the building interacted with the ground properly. No cracks were detected on the structure. In 2009 a decision was taken to build the gymnastics hall and amenities segment that had previously been abandoned.

During the flood in May 2010 landsliding processes intensified, which resulted in damage to the building's main segment (S1). In 2011, when works on the foundations for the gymnastics hall with amenities were in progress (according to a new design), a decision was taken to remove two reinforced concrete pillars strengthening the link building's foundation due to a geometric clash with the newly designed foundations. As a result of this decision the link building separated from the school building S1 and processes of deformation and damage to the entire structure (all the segments) started and have been continuing until today. Destruction processes were particularly severe during the thaws in 2016 in the area of the south-eastern corner of the building (segment S1), which was connected with the blocking of the pipes carrying away run-off.

4 GEODETIC RESEARCH FIELD

The following were used in integrated measurements taken to describe spatial changes to the building's shape and the ground surface of the surrounding area: GNSS satellite measurements (reference points), traversing (angular-linear measurements of the grid) and precision leveling. The planned control grid assumed the division of points into two categories (Fig. 1) (Baryła et al., 2012; Gawałkiewicz et al., 2011):

- single purpose points i.e.:
  - o wall benchmarks to determine vertical dislocations W_H (57 5F-75STK-type points);
  - o 13RM21 wall targets, to measure horizontal displacements W_{XY} (83 points);

- double-purpose, in the form of ground points (in the ground – 18 points) or floor points (in floors – 18 points), Plastmark 50 PP-PLAS50 modular type or 5F-75STK-type or 10TK-45-type survey steel nails with a marked centre, used to determine the values of both horizontal W_{XY} and vertical W_H displacements.

Geodesic monitoring of the slope and the educational complex based on discrete points (fixed in the ground, building floors and walls) was carried out from February 2017 at three-month and one-month intervals. As micro-movements of the substructure and the structure of the building were expected, achieving the highest possible precision of positioning the control grid points required the use of a complex set of measuring instruments with the best possible accuracies. For this reason the following were used as part of integrated technologies:

- the GNSS satellite system consisting of an R8s aerial and a Trimble TSC3 controller, used for horizontal and vertical positioning of reference grid points (static measurement on a point for the duration of 1 h 20 min) in connection with reference stations of the ASG-EUPOS system;
- TC 1800 Leica and 5503 DR Trimble electronic total stations to determine horizontal displacements of the points filling the triangulation grid;
- a Leica Na 3003 precision digital level to determine vertical displacements of points.

5 RESULTS OF OBSERVATIONS

Surveying measurements have a clear tendency to translate all the measured points with an increasing tendency to rotate the building in the same direction as the resultant of the slope's inclination (Fig. 4). The map of vectors of horizontal displacement of the points of the

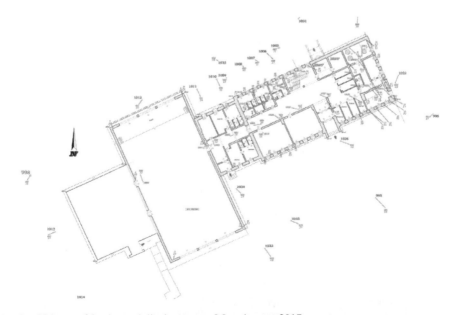

Figure 4.   Vectors of horizontal displacements May-August 2017.

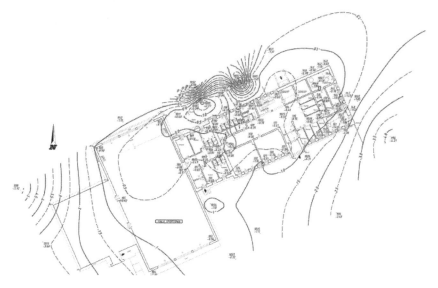

Figure 5.   Map of vectors of vertical displacements May-August 2017.

monitored grid displays high variability of the magnitudes and directions of the vectors in particular segments of the building and in its nearest surroundings. The rate of the changes measured during one year of observations varied and strongly correlated with increased intensity of precipitation and with spring thaws. In the most severely damaged fragment of the building comprising the eastern part of the S1 segment, especially the south-eastern corner, the rate of horizontal displacements averaged 10 mm/quarter.

The changes in subsidence in the period of May-August are shown in Fig. 5. The rate of subsidence in the south-eastern corner of S1 was on average 1 mm/3 months, with the highest value for the period of May-August, when it reached 1.4 mm/3 months. On the surface of the area in the northern part local zones of elevation connected with the colluvium exerting pressure on the building were recorded.

6   SUMMARY

Landsliding processes led to the structure of the building being damaged and caused a pre-emergency state in the structure. The geodesic observations showed that the deformation processes in the building are slow but progressive. These interactions are further exacerbated by the non-homogenous substrate. An additional significant factor are the hydrological conditions connected with the close proximity of source sinks. The areas where the main defects occur and their patterns indicate a tendency of the school segment to slide down and rotate. This tendency of the ground floor to dislocate, augmented by the effect of the colluvia exerting pressure on the subsided fragments of the segment, causes horizontal shear planes to arise in the most weakened wall cross-sections. These planes manifest themselves as cracks running through external walls—mainly on pier and window header links on the southern wall. The edges of these cracks are characterised by a dislocation "from the wall's plane". The tendency of the school building to slide down in the south-western direction is also indicated by the fact that the floor separated from the northern external wall of S1. The floor separated from the wall and was moved away by 20 mm, pulling the partition walls, the cladding of which was deformed at the area of contact with the external wall (Florkowska et al., 2017).

The newly build segment S3 is exposed to the pressure of the link building from the east and to the pressure from the colluvia from the north, caused by a significant subsidence and landsliding processes. The vertical cracks which correspond to the position of reinforced concrete columns indicate a loss of contact between the load-bearing structure and the infill elements. The patterns of horizontal and oblique cracks on external walls of segment S3 perpendicular to the northern wall indicate that the building moves in the southern direction.

The destruction process of the building is complex and multi-faceted. The impact of activated colluvium masses on individual segments of the building, the shifting of the natural water balance on the slope, the impact of the non-homogenous substrate as well as loading the slope with the weight of the building and its non-uniform subsidence in catastrophic conditions (the loss of the load-bearing capacity of the substrate) all overlap here. Errors in design and construction are also significant factors contributing to the existing state.

The case study presented in this paper confirms the principle which states that landslide-prone areas must not be developed without appropriate geotechnical solutions. Slopes which are temporarily stable can easily become destabilized and geodynamic processes can occur as a result of the load from the weight of a building and as a result of disturbing the hydrological conditions by introducing an impermeable obstacle in the form of foundations.

REFERENCES

Baryła R., Paziewski J., (2012): *Principles of the ground deformation monitoring technology based on GPS satellite measurements in control network.* Bulletin of the Military University of Technology, vol. LXI, Issue 2. Warszawa (in polish).

Burrell E. Montz, Graham A. Tobin, Ronald R. Hagelman III: Natural Hazards: Explanation and Integration. The Guilford Press. 2017.

Burtan J., 1984. *The tectonics of the flysh Carpathians south of Wieliczka.* (Tektonika Karpat fliszowych na południe od Wieliczki). Biul. Inst. Geol. 273 (in polish).

Florkowska L., Bryt-Nitarska I., Murzyn R.: *Diagnostics of the state of deformation and damage to the multi-segment building, which is subjected to influence of mass movements.* Awarie budowlane 2017, p. 521–532.

Gawałkiewicz R., Szafarczyk A.: *The application of GPS technology in the inventory of repositories and waste heaps.* Reports on Geodesy, 2011, Issue 1/90, p. 131–139.

Kyoji Sassa, Badaoui Rouhban, Sálvano Briceño, Mauri McSaveney, Bin He (eds.): Landslides: Global Risk Preparedness; Springer, 2013.

Geomechanics and Geodynamics of Rock Masses – Litvinenko (Ed.)
© *2018 Taylor & Francis Group, London, ISBN 978-1-138-61645-5*

Rock stress assessment based on the fracture induced electromagnetic radiation

V. Frid
Sami Shamoon College of Engineering, Ashdod, Israel

S.N. Mulev
VNIMI, Saint-Petersburg, Russia

ABSTRACT: An appearance of underground openings alters a stress field around them that sometimes provokes rock failure. Rock stress state is usually assessed at tunnel face or within boreholes using different types of electromechanical sensors that is laborious, expensive, and sometimes dangerous. Electromagnetic Radiation (EMR) caused by micro-fracturing is a non-destructive method of rock stress assessment. It is caused by the organized oscillation of charges on newly created surfaces in a frequency range kHz–MHz. Since the EMR in the MHz diapason appears much earlier than in the kHz range, it enables a short-term observation of failure transformation. The EMR parameters are related to the fracture length and width while its intensity is associated with rock elastic properties and stress state level. Despite of this comprehensive knowledge, quantitative unbiased criteria of stress assessment in the underground tunnels based on the EMR parameters are lacking yet. Here we consider several aspects related to the development of EMR quantitative criterion as follows: a. pre-calculation of the targeted EMR signal based on rock physical properties, b. regularity of the EMR attenuation based on the frequency of EMR signals, c. the symptoms of stress transformation from the steady to dangerous level.

Keywords: rock stress assessment, non-destructive methods, electromagnetic radiation, rock fracturing

1 INTRODUCTION

An increase in the stress intensity around a mine opening leads to crack enlargement, rock deformation and failure. Rock fracturing occurs in two main phases: the first one consists of an accumulation of micro-cracks and increase their density, while the second phase consists of micro-crack clustering (fracture nucleation) and final failure (Hallbauer et al. 1973; Kuksenko et al. 1985; Lockner and Reches 1994). Such two-phase mechanism was found to be valid for all scale levels: from rock samples up to the crust fracture (Kuksenko et al. 1985). It was shown by a wide variety of laboratory studies that the EMR is associated with micro-cracking and hence it appears prior to rock collapse (Khatiashvili 1984; Bahat et al. 2005 and refs. therein; Frid et al. 2003; Rabinovitch et al. 2002, 2007; Mavromatou et al. 2004; Fukui et al. 2005; Wang et al. 2012; Song et al. 2016; Carpinteri and Borla 2017; Potirakis and Mastrogiannis, 2017). This feature of EMR anomaly is indeed authentic both for the laboratory and in situ levels including rockburst and earthquake (Khatiashvili 1984). Careful investigations of EMR features during fracturing of rocks, glass and glass-ceramic materials resulted in the understanding that the oscillating electrical dipoles created on the crack surfaces comprise the EMR sources without regard to material chemical/mineralogical constitution and/or the mode of applied load (Frid et al. 2003; Bahat et al. 2005; Rabinovitch et al. 2007, 2017a). Utilization of the EMR method was shown to be highly useful for the identification of active faults (Akawwi 2008) and evaluation of stress direction around the tunnels (Greiling and Obermeyer 2010). The widest acceptance of the EMR method has been received to assess the intensity of stress in underground openings and tunnels (Frid 2001; Liu and He 2001; Frid and Vozoff 2005; Wang et al. 2011, 2012; Lu and Dou 2014; Song et al. 2016; Qui et al. 2017). Our overview indicates that despite of the EMR is actually a multi-scale phenomenon intensively employed for the stress assessment in underground openings, the studies of EMR features associated with its utilization in the underground conditions are mostly experimental. The last attempt to develop a theoretical EMR criterion for the rock stress assessment was carried out more that 15 years ago (Frid 2001). This paper consists of examining several aspects that are of great importance for the EMR application in the underground condition. It is mostly based on the knowledge accumulated in the field over the past fifteen years and aimed at making the method to be more accessible for the researchers.

2 THE METHOD OF PRE-CALCULATION OF THE TARGETED EMR SIGNAL BASED ON ROCK PHYSICAL PROPERTIES

The main problem in the EMR application is the lack of measuring equipment, the parameters of which will be theoretically justified. Hereafter several such parameters will be considered.

a. Frequency range: Based on the EMR mechanism (Frid et al. 2003; Bahat et al. 2005; Rabinovitch et al. 2007, 2017a) the frequency of EMR signals and the crack width are interconnected via the following relationships:

$$\omega = \frac{\pi v_R}{b} \tag{1a}$$

or

$$f = \frac{v_R}{2b} \tag{1b}$$

where ω is the angular frequency, f is the reciprocal frequency, b is the crack width, and v_R is the Rayleigh wave speed. Song et al. (2016) transformed Eq. 1b relating the reciprocal frequency f of the measured EMR signal with the crack length L, density ρ, elastic modulus E, Poisson's ratio μ and the uniaxial rock strength σ_c as follows:

$$f = \frac{1}{L} \frac{1}{32\sqrt[4]{2}} \frac{(0.87 + 1.12\mu)}{(1-\mu)\sqrt{(1+\mu)^5}} \frac{\sqrt{E^3}}{\sigma_c \sqrt{\rho}} \tag{2a}$$

hence:

$$L = \frac{1}{f} \frac{1}{32\sqrt[4]{2}} \frac{(0.87 + 1.12\mu)}{(1-\mu)\sqrt{(1+\mu)^5}} \frac{\sqrt{E^3}}{\sigma_c \sqrt{\rho}} \tag{2b}$$

Figure 1 portrays the calculated relationships between crack length and the measured frequency for several rock types. Our experience shows that the range of crack lengths relevant for the short-term stress assessment in the underground opening changes from the units of millimeters to the dozens of centimeters implying that the frequency range of the equipment suitable for the EMR registration has to be of the order 10 kHz–1 MHz or slightly wider.

b. Sensitivity: The sensitivity of the equipment suitable for the EMR registration in underground conditions can be estimated using the relationship between EMR field intensity

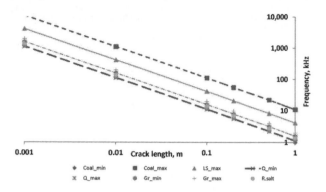

Figure 1. The relationships between crack length and the measured frequency: for Coal, Limestone (Ls), Quartzite (Q), Granite (Gr) and Rock salt (R.salt) (Dortman, 1984).

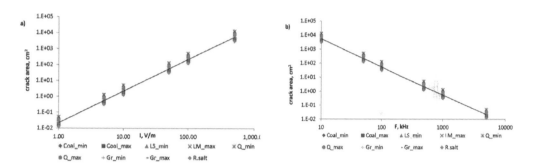

Figure 2. The relationships between EMR field intensity (V/m) and crack area (cm²) (a), EMR frequency (kHz) and the crack area (cm²) (b). The calculation is performed for Coal, Limestone (Ls), Quartzite (Q), Granite (Gr) and Rock salt (R.salt) (Dortman, 1984).

and measured frequency as follows (Frid et al. 2003; Bahat et al. 2005; Rabinovitch et al. 2007): $I = \frac{5*10^6}{f}$, where I is the electromagnetic field intensity (V/m). The sensitivity diapason of the measuring equipment has to be at least one order less than the lower calculated limit (the highest frequency limit) and at least one order higher than the upper calculated limit (the lowest frequency limit). Note that this calculation is for the fracture appearance at the working face.

It could be useful to estimate fracture areas based on the calculated EMR field intensity and frequency. It can be done by the superposition Eqs. 1b-2b and well known relationships between elastic wave velocities and elastic constants (e.g. Song et al. 2016). The procedure yields:

$$S = \frac{I^2}{2.5*10^{13}} \frac{1}{64\sqrt[4]{8}} \frac{(0.87+1.12\mu)^2}{(1-\mu)(1+\mu)^4} \frac{E^2}{\sigma_c\rho}$$ (3a)

or

$$S = \frac{1}{f^2} \frac{1}{64\sqrt[4]{8}} \frac{(0.87+1.12\mu)^2}{(1-\mu)(1+\mu)^4} \frac{E^2}{\sigma_c\rho}$$ (3b)

Figure 2 shows the results of the calculation. Note, an estimation of the intensity of the EMR field and the crack region was made assuming that a crack is created on the face/wall of the tunnel. In order to take into account the creation of cracks in the inner zone of the rock massif, one can consider the attenuation of the electromagnetic wave in rocks.

507

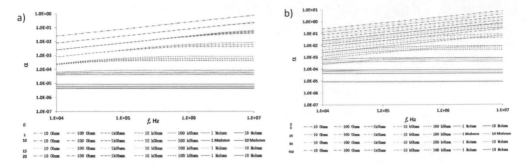

Figure 3. The results of the calculation of attenuation coefficient α. a) attenuation coefficient α vs. EMR frequency for ε_r range 5–20 while $\chi = 0$, b) attenuation coefficient α vs. EMR frequency for the χ range 0–100 while $\varepsilon_r = 5$ (modified from Rabinovitch et al. 2017b).

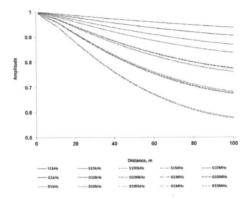

Figure 4. The results of calculation of EMR intensity attenuation vs. distance (in m) for the frequency range 1 kHz–10 MHz. Blue - Sandstone; Red - Granite; Green - Basalt. Full line - 1 kHz; Dashed line - 10 kHz; Dot-Dashed - 100 kHz; Dashed-Dot-Dot-Dashed - 1 MHz; Extended Dashed - 10 MHz. For the rock electromagnetic parameters see Dortman 1984, Telford et al. 1990) (Modified from Rabinovitch et al. 2017b).

3 REGULARITY OF THE EMR ATTENUATION IN ROCKS BASED ON SIGNAL FREQUENCY

The comprehensive investigation of the attenuation of electromagnetic waves in different rocks was performed by Rabinovitch et al. (2017b). Here we provide only the partial results, which are relevant to the restricted frequency diapason obtained above. The attenuation coefficient of EM waves 'α' (1/m) is known to be (Rgevsky and Novik 1978; Rabinovitch et al. 2017b):

$$\alpha = \omega \sqrt{\frac{\mu \varepsilon}{2} \left(\sqrt{1 + \left(\frac{\sigma}{\omega \varepsilon} \right)^2} - 1 \right)} \qquad (4)$$

where μ is the rock magnetic permeability (H/m), ε is the rock permittivity (F/m), σ is the rock conductivity (S/m). Note that $\mu = \mu_0 (1 + \chi)$, where χ is the rock magnetic susceptibility, $\mu_0 = 4\pi \times 10^{-7}$ H/m, while $\varepsilon = \varepsilon_0 \varepsilon_r$, where ε_r is the rock dielectric constant, $\varepsilon_0 = 8.84 \times 10^{-12}$ F/m. Figure 3 shows the results of α calculation for the wide range ε_r and χ (for typical values of rock electromagnetic parameters see e.g. Dortman 1984, Telford et al. 1990). The decrease in the intensity of the electromagnetic energy with increasing distance from the radiation point takes place exponentially: $I = I_0 e^{-\alpha R}$, where I is the EMR intensity (V/m) at the distance R (m) from the radiation point, where the EMR intensity is I_0(V/m). The results of calculating the decrease in the EMR intensity for several types of rocks are shown in Fig. 4. The size of the zone of tunnel influence (where rock fracturing can be the source of EMR) is restricted by the several meters from the tunnel face/wall. Hence, the attenuation of EMR signals will not be higher than 5–10%. To obtain a more accurate estimation, the attenuation must be calculated for each specific case.

4 THE SYMPTOMS OF STRESS TRANSFORMATION FROM A STEADY TO DANGEROUS LEVEL

a. EMR activity: the simplest attribute of increasing the intensity of stress field is an increase of the EMR activity (the number of measured EMR signals per unit of time). The application of this parameter naturally follows the two-phase mechanism of rock failure i.e. crack density increase prior to the rock coalesce. This parameter being easy for the measurement has been intensively utilized (Frid 2001; Liu and He 2001; Frid and Vozoff 2005; Wang et al. 2011, 2012; Lu and Dou 2014; Song et al. 2016; Qui et al. 2017), but our experience indicates that it is not enough for the accurate stress state assessment.
b. The EMR field intensity and frequency: An increase in the EMR intensity along with a decrease in the EMR frequency (Figs. 1–2) indicates an increase in the size of the cracks and, consequently, an increase in the intensity of stress level around the underground object. This conclusion is consistent with the mechanism of EMR, as well as with the two-phase model of rock failure and is confirmed by the numerous experimental results (e.g. Frid et al. 2003; Rabinovitch et al., 2007, 2017b; Song et al. 2016).
c. Failure statistics: Gutenberg-Richter relationship was found to be a fundamental statistical law valid both for seismic events and acoustic emission caused by rock sample fracturing (Rabinovitch et al 2002b and refs therein). The b-value behavior for EMR is similar to its features for seismic/acoustic emission, e.g. Rabinovitch et al (2002b) showed that b-value prior to rock collapse drops to about 2/3. Carpinteri et al (2012) found that at the period of microcracks coalesce the b-value decreases below 1.5. Yakovlev and Mulev (2014) revealed significant decrease in the b-value prior to rock collapse. The problem of uncertainty in the behavior of the b-value is well known in seismology: the b-value falls significantly prior to failure, but sometimes falls without it (Contoyiannis et al. 2010). Since, the EMR vs. b-value studies are rare in scientific literature, in order to obtain more repeatable and definite results, much more parametric studies are required. An alternative approach to the statistical consideration of EMR behavior is the use of Benioff strain release parameters. An appearance of irreversible changes in the slope of EMR-Benioff strain release diagrams was found to occur much earlier than rock collapse take place and to be independent from the failure scale (Rabinovitch et al. 2002; Frid and Vozoff 2005; Frid et al. 2011). Figure 5 shows the example of Benioff strain release behavior of EMR signals measured prior to roof fall in coal mine.

5 COMPARISON OF THE EXISTING EQUIPMENT FOR THE EMR MEASUREMENT

Figure shows four examples of the equipment used for the EMR measurement. Description of the equipment available in the scientific literature is very poor. However, even from the available data, it can be seen that although the sensitivity of the equipments is similar to the requirements calculated above, its frequency range should be extended for the successful stress assessment based on the EMR properties. Moreover, since utilization of a single EMR parameter causes uncertainty in the stress state assessment, the functionality of all types of equipment has to be modernized on the basis of the modern principles. As was shown above,

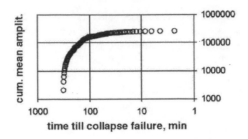

Figure 5.   The Benioff strain release diagrams of EMR prior to roof fall in coal mine.

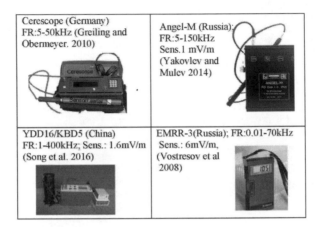

Figure 6.   Four examples of the EMR equipment.

only the integrated use of the entire complex of EMR characteristics (EMR activity, EMR field intensity, reciprocal frequency of EMR signals, and statistical characteristics ("b-value" and "Benioff-parameters") will allow an accurate assessment of the stress state in underground tunnels.

REFERENCES

Akawwi E. 2008. Investigate the active faults, sinkholes at the eastern shores of the Dead Sea by using electromagnetic radiation (EMR). WSEAS Int. Conf. on Engineering mechanics, structures, engineering geology (EMESEG '08), Heraklion, Crete Island, Greece, July 22–24.

Bahat D., Rabinovitch A., Frid V. 2005. Tensile fracture in rocks. Springer, Heidelberg.

Carpinteri A., G. Lacidogna, A. Manuello, G. Niccolini, A. Schiavi, A. Agosto 2012. Mechanical and Electromagnetic Emissions Related to Stress-Induced Cracks. Experimental Techniques 36, 53–64.

Carpinteri A., O. Borla. 2017. Fractoemissions as seismic precursors. Engin. Fracture Mech. 177, 239–250.

Contoyiannis Y.F., C. Nomicos, J. Kopanas, G. Antonopoulos, L. Contoyianni. K. Eftaxias. 2010. Critical features in electromagnetic anomalies detected prior to the L'Aquila earthquake. Physica A389, 499–508.

Dortman. N.B. 1984. Physical properties of rock and economical materials. Nedra. Moscow.

Frid V. 2001. Calculation of electromagnetic radiation criterion for rockburst hazard forecast in coal mines, Pure and Applied Geophysics 158, 931–944 (and refs therein).

Frid V., Rabinovitch A., Bahat D. 2003. Fracture induced electromagnetic radiation. J. Phys. D. 36, 1620–1628 (and refs therein).

Frid V. and Vozoff K. 2005. Electromagnetic radiation induced by mining rock failure. Int. J. Coal Geol. 64(1–2), 57–65.

Frid V., Goldbaum J., Rabinovitch A. and Bahat. 2011. Time dependent Benioff strain release diagrams. Phil. Mag. 90(12), 1693–1704.

Fukui, K., S. Okubo, T. Terashima. 2005. Electromagnetic radiation from rock during uniaxial compression testing: The effects of rock characteristics and test conditions. Rock mech. rock engng. 38(5), 411–423.

Greiling R.O., H. Obermeyer. 2010. Natural electromagnetic radiation (EMR) and its application in structural geology and neotectonics. J. Geolog. society of India. 75, 278–288 (and refs therein).

Hallbauer D.K., H. Wagner, N.G.W. Cook. 1973. Some observations concerning the microscopic and mechanical behaviour of quartzite specimens in stiff, triaxial compression tests. Int. J. Rock mech. min. sciences & geomechanics abstracts. 10(6), 713–726.

Khatiashvili, N. 1984. The electromagnetic effect accompanying the fracturing of alkaline halide crystals and rocks. Phys. Solid Earth 20, 656–661.

Kuksenko V.S., Manzikov V., Mansurov V.A. 1985. Regularities in the development of microfocal rupture. Izv., Earth Phys. 21(7), 553–556.

Lockner D., Z. Reches. 1994. Nucleation and growth of faults in brittle rocks. J. Geophys. Res.-Solid earth. 99 (b9) 18159–18173 (and refs therein).

Lu Cai-Ping, Dou Lin-Ming. 2014. The relationship between vertical stress gradient, seismic, and electromagnetic emission signals at Sanhejian coal mine, China. Int. J. rock mech. min sciences 70, 90–100.

Mavromatou C., V. Hadjicontis, D. Ninos, D. Mastrogiannis, E. Hadjicontis, K. Eftaxias. 2004. Understanding the fracture phenomena in inhomogeneous rock samples and ionic crystals, by monitoring the electromagnetic emission during their deformation. Physics and Chemistry of the Earth 29, 353–357.

Potirakis S.M., D. Mastrogiannis. 2017. Critical features revealed in acoustic and electromagnetic emissions during fracture experiments on LiF. Physica A 485, 11–22.

Qiu Liming, Enyuan Wang, Dazhao Song, Zhentang Liu, Rongxi Shen, Ganggang Lv, Zhaoyong Xu. Measurement of the stress field of a tunnel through its rock EMR. J. Geophys. Eng. 14 (2017) 949–959.

Rabinovitch A., Frid V., Bahat D. 2002. Gutenberg-Richter type relation for laboratory fracture induced electromagnetic radiation. Physical Review E65, 011401–011404.

Rabinovitch A., Frid V., Bahat D. 2007. Surface oscillations-A possible source of fracture induced electromagnetic radiation. Tectonophysics 431, 15–21 (and refs therein).

Rabinovitch A., Frid V., Bahat D. 2017a. Directionality of electromagnetic radiation from fractures. Intern. J. Fracture. (in press).

Rabinovitch A., Frid V., Bahat D. 2017b. Use of electromagnetic radiation to predict earthquakes. Geol. Magaz. (in press).

Rgevsky V.V., Novik G.Y. 1978. Foundations of rock physics. Nedra. Moscow.

Song D., E. Wang, X. Song, P. Jin, L. Qiu. 2016. Changes in frequency of electromagnetic radiation from loaded coal rock. Rock Mech Rock Eng. 49, 291–302.

Telford L.P., Geldart L.P., Sheriff R.E. 1990. Applied Geophysics. Cambridge University Press.

Vostresov A.G., A.V. Krivetskii, A.A. Bizyaev, G.E. Yakovitskaya. 2008. New methods and instruments in mining. J. Mining Science 44(2), 215–222.

Wang Enyuan, Xueqiu He, Xiaofei Liu, Wenquan Xu. 2012. Comprehensive monitoring technique based on electromagnetic radiation and its applications to mine pressure. Safety Science 50, 885–893.

Yakovlev D.V., S.N. Mulev. 2014. Experience in the use of multifunctional geophysical equipment Angel-M in the coal and ore industry. Coal. October. 14–19.

Geomechanics and Geodynamics of Rock Masses – Litvinenko (Ed.)
© 2018 Taylor & Francis Group, London, ISBN 978-1-138-61645-5

Modern principles of nondestructive stress monitoring in mine workings—overview

V. Frid
Sami Shamoon College of Engineering, Ashdod, Israel

A.N. Shabarov
Saint Petersburg, Mining University, Saint-Petersburg, Russia

ABSTRACT: The problem of assessing the state of stress in mine workings is vital for safe underground mining. The main source of information about a local stress field is usually obtained from measurements on the walls of underground galleries or in boreholes. An alternative to these methods are non-destructive short-term geophysical methods calibrated with a small amount of information from drilled wells and used to monitor stress conditions. Of this category, the greatest attention was paid to the method of acoustic emission and the method of electromagnetic radiation. Both of these phenomena are caused by the rock fracturing. This paper considers the physical basis for the application of acoustic emission in underground conditions based on a modern understanding of the phenomenon and its features.

1 INTRODUCTION

The release of high concentration of stress in the form of collapsed rocks towards the underground galleries is the main cause of accidents during mining (Zhang et al. 2017a), since the natural distribution of underground stress is significantly altered by the additional stress superimposed by the mining operations, and substantial decrease in strength of rocks (He et al. 20017). Hence, the problem of assessing the state of stress in mine workings is vital for safe mining. The manifold methods have been utilized to monitor the potential for the stress concentration (Zhang et al. 2017b). Hereafter we merely concern the local short-term methods. The main source of information about a local stressed field is usually obtained from measurements the walls of underground galleries or in drilled wells. An alternative to these methods are non-destructive short-term geophysical methods, i.e. acoustic emission (AE) or/ and electromagnetic radiation (EMR) caused by rock fracturing. Acoustic emission (micro-seismic) has been used since the 1920s to assess the underground stress conditions in Poland, South Africa, Canada, the U.S., Australia, China, etc. (He et al. 2017 and references therein). Numerous studies of the AE phenomenon make it possible to understand that the signals AE are a small-scale phenomenon whose properties are analogous to the large-scale radiation of an elastic wave caused by rock-bursts and earthquakes (Kuksenko et al. 1985; Lockner and Rehez 1994; Guha 2001 and references therein; Lei et al. 2003; Thompson et al. 2009; Johnson et al. 2013; Goebel et al. 2014; McLaskey and Lockner 2016 and references therein). The formation of cracks and/or macroscopic fractures in rocks is the main cause of the AE excitation, whose energy and spectrum depend on the mechanisms of their generation and rock properties (Michlmayr et al. 2012). Monitoring of AE excitation is applicable for the detection of crack sources, hazard assessment in mines, tunnels, etc (Kim et al. 2015; Kuksenko and Makhmudov 2017 and references therein). It was shown that the AE time-sequence parameters correspond well with the evolution of the rupture events of the rock material at different stages and hence can provide the precursor for the rock damage (Kuksenko et al. 1985; Lockner and Rehez 1994; Rehes 1999; Liang et al. 2017). For example, Cai et al. (2001) characterized rock damage near excavation using AE monitoring. It was shown that the combined AE and EMR methods it possible to accurately assess the risk of high stress accumulation (Frid and Vozoff 2005; Lu et al. 2015; Jiang et al. 2016). The experimental results obtained at the San Pietro gypsum mine, Prato Nuovo, underscore the close correlation between AE and EMR activity, while it was noted that both types of emissions preceded a failure event for approximately one day and 3–4 days, respectively (Carpinteri and Borla 2017). Henceforth, we consider several aspects of the phenomenon of AE, which are important for its application for local short-term stress assessment while features of EMR phenomenon were examined in detail by Frid and Mulev (2018 and references therein).

2 AE FREQUENCY VS. CRACK SIZE

The relation between the crack length and the corner frequency AE is known to as follows (Gibowicz and Kijko 1994; Cai et al. 2001; Shearer 2009):

$$l = \frac{kv_s}{f}(\text{m}) \tag{1}$$

where v_s is the shear wave velocity, f is the corner frequency and k = 0.21 for the shear wave (Shearer 2009). Fig. 1 shows the range of AE frequency for the v_s equal to 500–4500 m/s (Rgevski and Novik 1978).

The application of AE for the short-term stress assessment is based on the two-phase mechanism of rock failure while the first phase consists of an accumulation of micro-cracks and increase their density, while the second phase consists of micro-crack coalesence (fracture nucleation) and final failure (Hallbauer et al. 1973; Kuksenko et al. 1985; Lockner and Reches 1994). Therefore, for the successful use of the AE method for the short-term stress assessment in underground conditions (which implies the possibility of estimating the stress field much earlier than the rock collapse occurs), the range of crack lengths to be identified by AE measurements should be from a few millimeters to tens of centimeters. Hence, the frequency range of equipment suitable for measuring AE should be of the order of 1 kHz to 10 MHz, depending on the properties of rocks. This conclusion is consistent with those of Guha (2000) and Carpinteri and Borla (2017) who noted that the frequency range of acoustic emission lies between 0.1 to 10^3 kHz.

3 AE AMPLITUDE

For a circular fracture the relation between stress drop, seismic moment and crack length is as follows (Shearer 2009):

$$\Delta\sigma = \frac{7}{16}\frac{M_0}{l^3}(\text{Pa}) \tag{2}$$

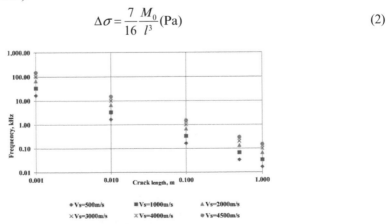

Figure 1.   The range of AE frequency vs. crack length.

514

where $\Delta\sigma$ is the stress drop value, M_0 is the seismic moment, the typical stress drop values for the rock sample cracking/failure and pillar destruction/rockburst in mines was found to be in the range 0.2–20 MPa (Rabinovitch et al. 1995; Guha 2000; Shearer 2009; Goodfellow and Young 2014; McLaskey and Lockner 2016; Li et al. 2017). The radiated seismic/AE energy from newly created surface area of crack can be calculated as follows:

$$E_s = 0.5\Delta\sigma l (J/m^2) \tag{3}$$

Figure 2 shows the results of calculation of the seismic moment M_0 and radiated AE energy for the wide range of crack lengths. As it is seen, the seismic moment and radiated energy values change broadly 10^{-5}–10^4 N*m ($-9 \leq M_w \leq -3.5$) and 10^1–10^6 J/m^2, respectively. These results are mainly consistent with those of McLaskey and Lockner 2016 ($-8 < M_w < -3.5$) and those presented by Rehes (1999) ($10^4 < E_s < 10^6$ J/m^2). who noted that the value of radiated AE energy from the newly created surfaces is about 10^4–10^6 J/m^2. The magnitudes calculated above are much lower than ones corresponding to the big scale failure phenomena in mines known in scientific literature $0 < M_w < 4.5$ (se e.g. Guha 2000; Frid and Vozoff 2005; Li et al. 2007). However, as we noted above, in order to assess the stress state in mine openings much earlier, than crack clustering occurs, it is necessary to investigate the failure events of small sizes. Ponomarev et al. (1997) shows that the AE amplitude being related to $\sim\sqrt{E_s}$ in the close vicinity of the crack creation can be calculated as follows:

$$U(mV) = \left(\frac{l}{n}\right)^{\frac{3}{2}} \tag{4}$$

where l is the crack length (mm), n = 0.17. Substituting Eq. 1 into Eq. 4 we can get the AE amplitude to be:

$$U(mV) = \left(\frac{kv_s}{nf}\right)^{\frac{3}{2}} \tag{5}$$

Figure 3 shows the relationships of the estimated AE amplitude vs. the corner frequency.

Note that the coefficient 'n' proposed by Ponomarev et al. (1997) does not takes into account amplification of the acquisition system (not described in Ponomarev's et al. paper). However, assuming the gain of AE system usually used in the laboratory experiments to be 40–60dB (McLaskey and Lockner 2016), it can be estimated that the amplitude of the AE in the immediate vicinity of the newly created crack should be of the order of tens micro-volts (for millimeter-length cracks) to several volts (for cracks of tens of centimeters in length).

This estimate can be checked in a different way. The mean energy density of AE signal can be calculated as follows (Shearer 2009; Grishin 2012):

$$E = \frac{\rho U^2 (2\pi f)^2}{2} (J/m^3) \tag{6}$$

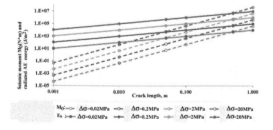

Figure 2. The seismic moment M_0 and radiated energy E_s for the wide range of crack lengths and stress drops.

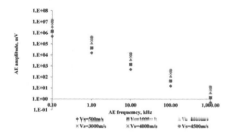

Figure 3. The AE amplitude (mV) vs. AE corner frequency (kHz).

515

where ρ is the rock density. Hence, the radiated seismic/AE energy from newly created surface area of crack can be found as follows:

$$E_s = \frac{\rho U^2 (2\pi f)^2 l}{2} (\text{J/ m}^2)$$
(6'),

and after transformation we get the AE amplitude:

$$U \approx \sqrt{\frac{2E_s}{l\rho(2\pi f)^2}} \underset{Eq.1}{\Rightarrow} \sqrt{\frac{E_s l}{2\rho(\pi k v_s)^2}} (V)$$
(7)

Substituting $E_s = 10^1 - 10^5$ J/m² (Fig. 2), rock density ρ = 1700–3500 kg/m³ and shear wave velocity v_s = 500–4000 m/s, it can be seen that the AE amplitude caused by the crack, length of which changes between 1 mm and 10 cm should be from a few microvolts to tens of millivolts. Such an assessment essentially agrees with the one presented above.

4 ATTENUATION

The comprehensive investigation of the attenuation of acoustic waves in granite and sandstone was performed by Rabinovitch et al. (2017). It was shown that the AE amplitude (A) at a distance r from the crack radiating the AE amplitude A_0 is determined as follows:

$$A(f,r) = A_0 \exp(-\alpha f r) \text{ or } A(f,r) = A_0 \exp(-\Theta r)$$
(8)

where $\Theta = \alpha * f$ is the attenuation factor, $\alpha = \pi/Q v_p$ (m⁻¹), and v_p (m/s) is the compression wave velocity, while Q is the quality factor. Typically, the attenuation factor ranges between 0.15 and 10. Figure 4 shows the changes of AE amplitude calculated using Eq. 8 for different values of the attenuation factor.

The analysis of Figs. 1, 3, and 4 shows that cracks with a length of about 1 mm can be detected only under very favorable conditions and only from a distance of several meters from the cracking point (Fig. 4a). Cracks of about 1 cm in length can be identified if the attenuation coefficient Θ does not exceed 0.5 and from a distance less than 50 m (Θ > 0.15, Fig. 4a, b). A longer crack can be recognized in more or less unfavorable conditions, but from a distance of several meters (Fig. 4c, d). Note that this conclusion implies a gain of the receiving

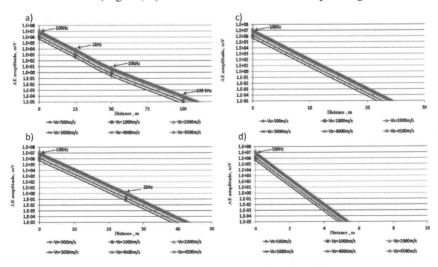

Figure 4. Attenuation of the amplitude AE and the distance (m) for different values of the attenuation factor: a) Θ = 0.15; b) Θ = 0.5; c) Θ = 1; d) Θ = 5.

system of at least 60 dB. To overcome this limitation, it is highly desirable to integrate the measurements of AE and EMR (Frid and Mulev 2018).

5 STATISTICS

The problems associated with AE statistics are very diverse, including: the lifetime, the decay laws, the b-value, and so on. They are comprehensively considered in the scientific literature, while the conclusions are sometimes contradictory. For example, it is commonly assumed that b-value reflects spatial variations in stress, e.g. Medincki (2016) suggested using this parameter to evaluate the stiff/nonstiff loading mode. Numerous studies have shown regular changes in the b-value before the failure of rocks (e.g. Lockner et al. 1991 and references therein; Thompson et al. 2006). On the other hand, the uncertainty problem in the behavior of the b-value is well known in seismology: the value of b significantly drops prior to a failure, but sometimes falls without it (Contoyiannis et al. 2010). The problem of the lifetime or, in other words, the precursory-period is also comprehensively studied, e.g. Guha (2000 and references therein) demonstrated that the precursory period τ (s) and focal area A ($A = \frac{\pi}{4}l^2$, cm) are related via the following relationship:

$$\tau = 2.43 * 10^{-4} A \tag{9}$$

A higher intensity of failure of rocks corresponds to a longer precursory period, and therefore knowledge of this value can be practical for stress assessing. In addition, a very useful estimate is the type of distribution AE, which, as shown, obeys the law of exponential distribution (in the first phase of destruction), whereas in the second stage (during cracks coalescence/prior to failure) it satisfies the power law (Damaskinskaya ct al. 2017).

6 CONCLUSION

Several aspects of the AE phenomenon that are important for its application for local short-term stress assessment are considered in the article, including the frequency of AE in comparison with the size of the crack, amplitude and energy, attenuation and the features of AE statistics (very briefly). Our overview indicates that to ensure the underground application of AE to be a successful story, a thorough preliminary analysis of the rock properties, the features of fracturing and elastic wave attenuation has to be performed as well as the possibility of the necessary arrangement of recording equipment. Moreover, to overcome the intrinsic limitations of AE, it is highly desirable to integrate its utilization with EMR.

ACKNOWLEDGMENT

We are very grateful to Mr. Mulev for his very useful comments and assistance during the preparation of the paper.

REFERENCES

Cai M., Kaiser P.K., Martin C.D. 2001. Quantification of rock mass damage in underground excavations from microseismic event monitoring. Int. J. Rock Mechanics & Mining Sciences 38, 1135–1145.
Carpinteri A., Borla O. 2017. Fracto-emissions as seismic precursors. Engineering Fracture Mechanics 177, 239–250.
Contoyiannis Y.F., C. Nomicos, J. Kopanas, G. Antonopoulos, L. Contoyianni. K. Eftaxias. 2010. Critical features in electromagnetic anomalies detected prior to the L'Aquila earthquake. Physica A389, 499–508.
Frid V., Vozoff K. 2005. Electromagnetic radiation induced by mining rock failure. Int. J. Coal Geol. 64(1–2), 57–65.
Damaskinskaya E., Frolov D., Gafurova D., Korost D., Panteleev I. 2017. Criterion for fracture transition to critical stage. Interpretaion. 11, sp1–sp8.

Frid V., Mulev S. 2018. Rock stress assessment based on the fracture induced electromagnetic radiation. Geomechanics and geodynamics of rock masses. ISRM. Eurorock 2018.

Gibowicz S.J., Kijko A. An introduction to mining seismology. New York: Academic press, 1994.

Goodfellow S.D., Young R.P. 2014. A laboratory acoustic emission experiment under in situ conditions. Geophysical Research Letters. 41, 3422–3430.

Goebel T.H.W., Schorlemmer D., Becker T.W., Dresen G., Sammis C.G. 2013. Acoustic emissions document stress changes over many seismic cycles in stick-slip experiments. Geophysical research letters. 40, 2049–2054.

Grishin E.N. 2012. Physics in tables and schemes. Pheniks. 2012.

Guha S.K. 2000. Induced earthquakes. Kluwer Academic Publisher. 313pp.

Hallbauer D.K., H. Wagner, N.G.W. Cook. 1973. Some observations concerning the microscopic and mechanical behaviour of quartzite specimens in stiff, triaxial compression tests. Int. J. Rock mech. min. sciences & geomechanics abstracts. 10(6), 713–726.

He J., Doub L.,Gong S., Lia J., Ma Z. 2017. Rock burst assessment and prediction by dynamic and static stress analysis based on micro-seismic monitoring. Int. J. Rock Mechanics & Mining Sciences 93, 46–53.

Jiang B., Wang L., Lu Y., Wang C., Ma D. 2016. Combined early warning method for rockburst in fully mechanized caving face in a Deep Island. Arab J Geosci. 9, 743–749.

Kim J.-S., Lee K.-S., Cho W.-J., Choi H.-J. Cho. 2015. G.-C. A Comparative Evaluation of Stress-Strain and Acoustic Emission Methods for Quantitative Damage Assessments of Brittle Rock. Rock Mech. Rock Eng. 48, 495–508.

Kuksenko V.S., Manzikov V., Mansurov V.A. 1985. Regularities in the development of microfocal rupture. Izv., Earth Phys. 21(7), 553–556.

Kuksenko V.S., Makhmudov K.F. 2017. Fracture in heterogeneous materials: experimental and theoretical studies. Russian Geology and Geophysics 5, 738–743.

Lockner D., Byerlee J.D., Kuksenko V.S., A.V. Ponomarev, A. Sidorin 1991. Quasi-static fault growth and shear frac-ture energy in granite. Nature 350, 39–42.

Lockner D., Z. Reches. 1994. Nucleation and growth of faults in brittle rocks. J. Geophys. Res.-Solid earth. 99 (b9) 18159–18173 (and refs therein).

Li T., Cai M.F., Cai M. 2007. A review of mining-induced seismicity in China. Int. J. Rock Mechanics & Mining Sciences 44, 1149–1171.

Liang Y., Li Q., Gu Y., Zou Q. 2017. Mechanical and acoustic emission characteristics of rock: Effect of loading and unloading confining pressure at the postpeak stage. Journal of Natural Gas Science and Engineering 44, 54–64.

Lu C-P., Liu G-J, Liu Y, Zhang N., Xue H-J., Zhang L. 2015. Microseismic multi-parameter characteristics of rockburst hazard induced by hard roof fall and high stress concentration. Int. J. Rock Mech. &Min. Sciences 76, 18–32.

McLaskey G.C. Lockner D.A. 2016. Calibrated Acoustic Emission System Records M 23.5 to M 28 Events Generated on a Saw-Cut Granite Sample. Rock Mech Rock Eng 49, 4527–4536.

Mendecki A.J. 2016. Mine Seismology Reference Book: Seismic Hazard, The Institute of Mine Seismology. 88pp.

Michlmayr G., Cohen D., Or D. 2012. Sources and characteristics of acoustic emissions from mechanically stressed geologic granular media — A review. Earth-Science Reviews 112, 97–114.

Ponomarev A.V., Zavyalova A.D., Smirnov V.B., Lockner D.A. 1997. Physical modeling of the formation and evolution of seismically active fault zones. Tectonophysics 277, 57–81.

Rabinovitch A., Bahat D., Frid V. 1995. Comparison of electromagnetic radiation and acoustic emission in granite fracturing. Intn. J. Fract. 71(2), r33–r41.

Rehes Z. 1999. Mechanisms of slip nucleation during earthquakes. Earth and Planetary Science Letters. 170, 475–486.

Rgevsky V.V. Novik G.Y. 1978. Foundations of rock physics. Nedra. Moscow.

Thompson B.D., Young R.P., Lockner D.A. 2006. Fracture in Westerly Granite under AE Feedback and Constant Strain Rate Loading: Nucleation, Quasi-static Propagation, and the Transition to Unstable Fracture Propagation. Pure appl. geophys. 163, 995–1019.

Shearer P.M. 2009. Introduction to Seismology. Cambridge University press. 2009. 396pp.

Zhang C., Canbulat I, Faham T., Hebblewhite B. 2017a. Assessment of energy release mechanisms contributing to coal burst. Int. J. Mining Science and Technology. 27, 43–47.

Zhang C., Canbulat I, Hebblewhite B., Ward C.R. 2017b. Assessing coal burst phenomena in mining and insights into directions for future research. Int.J. Coal Geology 179, 28–44.

518

Empirical-Statical-Dynamic (ESD) methodology for extrapolation in rock mechanics

Milorad Jovanovski, Igor Peshevski & Jovan Papic
Faculty of Civil Engineering, University Ss. Cyril and Methodius, Skopje, Macedonia

ABSTRACT: This article presents the basics of Empirical-Static-Dynamic (ESD) methodology for extrapolation of parameters in Rock Mechanics. The prerequisite for its successful application is to have enough data for reliable rock mass classification; to have enough testing data for deformability with static tests in a large scale and all zone of interaction between the structure and the rock mass to be covered with geophysical seismic tests. Applied tests must be performed in a manner that will ensure reliable data to establish correlations between necessary parameters and that will enable their extrapolation for each quasi-homogenous zone of the rock massif of interest. The analyses are given based on the results from investigations on location for arch dam "Sveta Petka" and number of tunnels and bridges for large infrastructure projects in Republic of Macedonia. Several regressive models between rock mass quality, statical (D) and dynamical modulus of deformability (Edyn) with velocity of longitudinal seismic waves (V_l) are presented in the paper, in order to illustrate the necessary steps to be undertaken in the methodology.

Keywords: ESD methodology, deformability tests, seismic tests, regressive models, extrapolation

1 INTRODUCTION

Succesful geotechnical and numerical modelling in the rock masses is possible only if the estimation of the input parameters is reliable, correct, and as much as posssible, close to the real natural state and propreties of the rock masses. The key problem is how to extrapolate the parameters from the zone of testing to the whole area (volume) that is of interest for interaction analyses of the system: rock mass-artificial structure. The extrapolation procedure is related to succesfull definition of the scale effect, deformability, shear strength of the rock masses and rock mass quality in an apppropiate way, and, to present data in Engineering Geological Section—EGS and Engineering Geological Model—EGM. The extrapolation methods are firstly developed for design problems at large dams by (Kujundžić, 1973), and (Kujundžić and Petrović, 1980). Shortly, their methodology is named as combined Static-Dynamic methodology for extrapolation, because the mesaured data from in situ static loading deformability tests with refraction seismic methods are combined using direct or indirect correlations between static (D) and dynamic modulus of deformability (Edyn) with velocity of longitudinal seismic waves (V_l). After that, there are many atetmps to use similar approach for other civil or mining structures (Jovanovski, 2001), (Jovanovski, Gapkovski, 2002). Classification systems developed in the field of rock mechanics gave another impuls in this area. The Geomechanical Classification—Rock Mass Rating system (Bieniawski, 1993); Q system—Rock Mass Quality (Barton, Lien and Lunde, 1974), multi-parameter classification system which can be used for main type of excavation problems—Excavation Rock Mass Rating (Jovanovski, 2001) can be mentioned in this context. Important improvement in this field is achieved by Barton (2002; 2008), with an empasis to the ways of exptrapolation of seismic data with rock mass quality, porosity, fracturing etc., mainly for tunneling problems. Beside numerous attempts, it is still evident that systematization of knowledge in the mentioned field is necessary. Thus, the present article describes a methodology that shows how it is possible to integrate several approaches in solving complex geotechnical problems in a "simple" way.

2   BASIC THEORETICAL SETUP ON THE PROBLEM OF EXTRAPOLATION

The proposed methodology can be shortly defined as **Empirical-Static-Dynamic** (ESD) methodology of extrapolation. In fact, this is an approach where static-dynamic approach given by Kujundžić (1980) is used as a starting point, but later integrates all known methods for defining of deformability and shear strength of rock masses. General description of the methodology is illustrated on a Figure 1. Following this procedure, extrapolation of the parameters can be done using defined correlative relations from the area of testing to the whole rock mass volume.

The central point in the methodology is to separate the investigated geological environment so-called quasi-homogeneous zones as basic constitutive elements of geological model. Inside such zones, rock mass characteristics are considered as similar in every point, and very different outside the zone. To this end, very usefull are considered the geophysical seismic methods, because of large possibilities (refraction seismic, cross-hole, parallel mesaurements to static deformability tests, etc). Some used combinations of such measurements and typical results are given in Figure 2 and Figure 3.

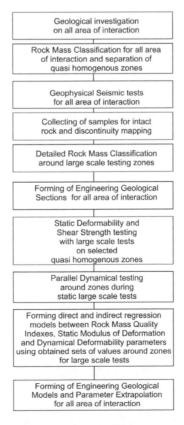

Figure 1.   General presentation of necessary steps for parameter extrapolation.

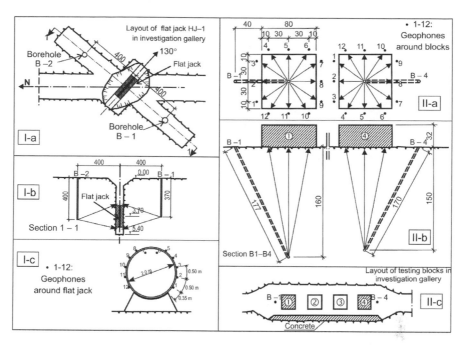

Figure 2. Disposition of parallel static and dynamic testing around zones for large scale deformability and shear strength testings on a profile for arch dam "Sveta Petka" in R. Macedonia.

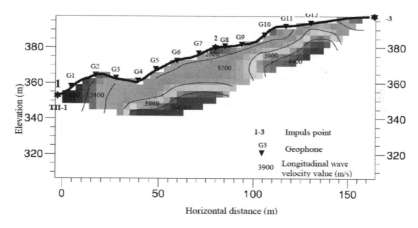

Figure 3. Example of geophysical section defined with refraction seismic test along diversion tunnel on a profile for arch dam "Sveta Petka" in R. Macedonia.

Figure 2 presents the procedure of micro-seismic cross-hole testing around flat jack and block scheme deformability tests in R. Macedonia. During each phase of loading and unloading, the seismic waves are produced from boreholes, in order to define simultaneously values of seismic longitudinal (V_p) and transversal (V_t) seismic waves. Figure 3 shows one profile along diversion tunnel for dam "Sveta Petka", using refraction seismic tests. Collected data were used in a first phase for preparation of geological models. To illustrate the methodology, cases for simplified engineering geological sections and models are presented in Figure 4 and Figure 5.

As a result, Figure 5 presents an approach how the rock mass deformability parameters are extrapolated using correlation with longitudinal wave velocities. Of course, there are large

521

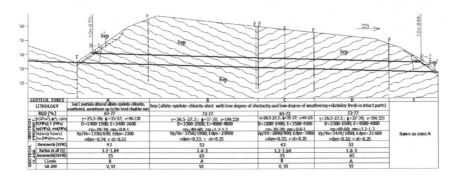

Figure 4. Engineering geological section of railway tunnel No. 4 on Corridor VIII, section Kriva Palanka—Deve Bair.

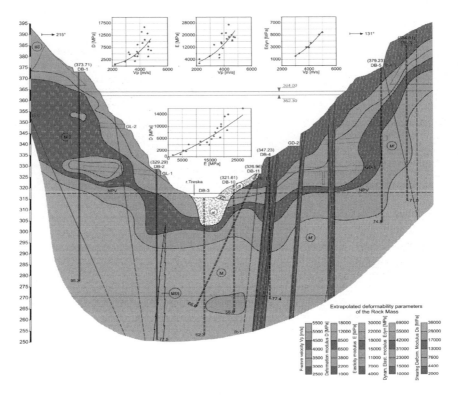

Figure 5. Engineering Geological Model (EGM) per deformability on a profile for arch dam "Sveta Petka".

possibilities for different relations between geological elements for each structure, so in every practical problem, there are always some solutions that can be adopted during investigation and design phases.

3 SOME DIRECT AND INDIRECT CORRELATIONS BETWEEN ROCK MASS PARAMETERS

With detailed analysis of collected tests' data, a numerous regression models can be obtained in order to fulfill necessary criteria for extrapolation. Based on major review by the authors,

Table 1. Obtained correlations between rock mass quality and seismic waves velocity and deformability fo cases in the R. Macedonia.

Case study	Established correlation	Determination coefficients (R^2)
European corridor X, Highway section Demir Kapija—Smokvica	$Vp \approx 057 \ln Q + 2.79$	0.83
R. Macedonia	$Vp \approx 055 \ln Q + 2.62$	0.85
R. Macedonia	$D = 1.69^{-6} RMR^{3.9}$	0.82
Dam "Sveta Petka"	$D = 0,1104 \ e^{0,0703 \ RMR}$	0.84
European corridor VIII, tunnels in shists	$D = 516.42 e^{0.0532 \ RMR}$	0.90
European corridor X, Highway section Demir Kapija—Smokvica	$RMR = 0.0161 Vp + 1.4973$	0.79
R. Macedonia	$RMR = 0.0229 \cdot Vp^{0.9573}$	0.70

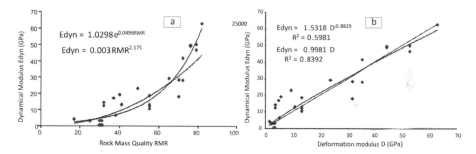

Figure 6. Correlation between quality of rock mass by the RMR classification and dynamical elasticity modulus for tunnels of the highway Demir Kapija-Gevgelija (Edyn) (a) and between deformation modulus (D) with Edyn (b).

several examples of regression links between velocity and rock quality, deformability and velocity are developed and demonstrated in Table 1. Thus, correlations between the quality of rock massif (RMR, Q indexes), dynamic (Vp, E_{dyn}) and static (D, E) characteristics of rock masses are prepared using results of the detailed classification of the rock massif around the measuring point with dilatometer tests. Two specific examples are shown in Figure 6.

Analysing all regression models, it is obvious that determination coefficients (R^2) indicate strong connection between examined parameters, and that for different structures and geo-tectonic regions, the correlation always had some differences.

4 CONCLUSION

The presented Empirical-Static-Dynamic methodology of data extrapolation is useful tool in a preparation of engineering-geological models for design analyses in Rock Mechanics. Because of its verification, suggested methodology with time must have critical re-examination in terms of possibilities to apply it on other locations and facilities in different geological media.

In addition, it can be concluded that the process of modelling must be harmonized with research and design phases. It is normal that simple approaches are used in initial phases, which meet current quality and quantity of available data. Results of such kind of initial models, especially for complex facilities, can indicate to need for now tests that will enable re-interpretation of existing data. This, in the other hand, influences the improvement of models or leads to new ideas for new model types. The used data in the article originate from sites in R. Macedonia, which does not mean that similar methodology can't be applied to other localities.

REFERENCES

Barton, N. (2002). Some new Q-value correlations to assist site characterisation and tunnel design. Int. J. Rock Mech. & Min. Sci. Vol. 392: 185–216.

Barton, N. (2006). Rock Quality, Seismic Velocity, Attenuation and Anisotropy. Taylor & Francis, UK & Netherlands, 729 p.

Barton, N., Lien, R. & Lunde, J. (1974). Engineering classification of rock masses for the design of tunnel support. Rock Mechanics, 1974; 6(4): 189–236.

Bieniawski, Z.T. (1993). Classification of Rock Masses for Engineering: The RMR System and future trends. Comprehensive Rock Engineering, 1993.

Jovanovski, M. (2001). Prilog kon metodologija na istražuvanje na karpestite masi kako rabotna sredina, Doktorska disertacija, Univerzitet Sv. Kiril i Metodij, Gradezen fakultet, Skopje.

Jovanovski, M., Gapkovski, N. & Ilijovski, Z. (2002). Correlation between Rock Mass Rating and deformability on a profile for arch dam Sveta Petka. 10-th International Conference of the DGKM, Ohrid.

Kujundžić, B. & Petrović, Lj. (1980). Korelacija statičkih i dinamičkih karakteristika deformabilnosti krečnjačkih stenskih masa. V Simpozij JDMSPR, 1, 5–12, Split.

Kujundžić, B. (1973). Sadržina i metodika izrade inženjersko-geoloških preseka i inženjersko-geoloških i geotehničkih modela. Saopštenja IX kongresa Jugoslovenskog komiteta za visoke brane, Zlatibor.

Geomechanics and Geodynamics of Rock Masses – Litvinenko (Ed.)
© *2018 Taylor & Francis Group, London, ISBN 978-1-138-61645-5*

State-of-the-art damage assessment methods for brittle rock using digital image correlation and infrared thermography

Murat Karakus, Selahattin Akdag, Lovepreet S. Randhawa,
Yiming Zhao & Zhenlong Cao
School of Civil, Environmental and Mining Engineering, University of Adelaide, Adelaide, SA, Australia

ABSTRACT: The microstructures of rocks are affected by mechanical loading causing damage in the rock leading to failure. Therefore, understanding the damage evolution in rock is of great significance to stability of rock engineering structures. For the damage evolution process the state-of-the-art Digital Image Correlation (DIC) and Infrared Thermography (IRT) systems were used. Complimentary measurements integrating the use of strain gauge and Acoustic Emission (AE) were also performed. Series of indirect tensile strength (Brazilian) tests were conducted for the damage evolution. DIC and IRT provide high resolution strain mapping across the surface of Brazilian discs throughout the test. Based on acoustic emission, full field stress-strain characteristics, and temperature change, damage evolution for brittle rock were analysed. Experimental results showed that the non-contact technique DIC and IRT have major advantages as an auxiliary to the conventional external strain measurements. DIC can be used as a better technique for not only tracing the full-field deformation behaviour but also strain localisation. IRT was also found to be effective method for detecting failure process. However, since quasi-static loading cause dissipation of heat quickly, IRT can be more effective for high strain rate loading cases.

Keywords: Digital Image Correlation (DIC), Thermal camera, Acoustic Emission (AE), Brittle rock, Damage evolution

1 INTRODUCTION

It is well known that gradual damage evolution, crack initiation and propagation cause deformation and failure of rock. Microcracking process leads to the alteration in the mechanical properties of rock which results in a decrease in the elastic stiffness and cohesive strength of rock (Eberhardt et al. 1999). The progressive propagation of microfractures contributes to damage evolution in rock.

Since brittle rock material is much weaker in tension, tensile failure of rock occurs more frequently in underground rock engineering projects. Various methods have been suggested to determine the tensile strength of rocks. Since performing a direct uniaxial tensile test on a rock sample is associated with many difficulties such as experimental setup, stress concentration and high testing cost, several indirect tensile testing methods have been proposed as alternatives to assess tensile strength (Mellor and Hawkes 1971; Hudson et al. 1972; Bieniawski and Hawkes 1978; Coviello et al. 2005). The conventional Brazilian test which was proposed by the International Society for Rock Mechanics (ISRM) is widely used in engineering practices to indirectly determine the tensile strength of rock materials (ISRM, 2007). This test is performed by applying a compression load to a circular disc sample until failure. Tensile stresses normal to the vertical diameter are induced by the compression assuming fracturing initiates from the central part of the disc where the failure is induced by maximum tangential stress.

Numerous experimental studies have been performed to investigate the rock damage evolution and failure process using different techniques. Eberhardt et al. (1999) investigated and quantified stress-induced fracture damage on Lac du Bonnet granite with the combined use of strain gauge measurements and acoustic emission (AE) monitoring during uniaxial compression tests. They found that damage and the deformation can be quantified by normalising the stresses and strains needed to pass from one stage of crack development to another. Karakus et al. (2016) conducted quantitative damage assessment on Hawkesbury sandstone under uniaxial cyclic loading conditions using AE. Akdag et al. (2018) proposed a method to quantify the thermal damage caused by temperature using cumulative amount of AE energy. Scanning election microscope (SEM) analysis was conducted to investigate the development of microcraking and spatial evolution of anisotropic damage was quantified using the crack density (Wu et al. 2000). Digital image correlation (DIC) is a robust technique which has been used to determine the displacement and strain fields on the surface of wide variety of materials (Sutton et al. 2009; Lin and Labuz 2013). Full-field displacement distribution and tensile strain on the surface of granite specimen were determined by DIC and damage was characterised by gradient of the tensile strain under tensile loading (Yang et al. 2015). Pre-peak and post-peak stress-strain characteristics of sandstone under uniaxial compression was studied via non-contact DIC method (Munoz et al. 2016) proving that DIC is a best tool to capture post-peak behaviour. Therefore, this study aims to investigate the whole failure process, damage evolution and localisation by adopting different non-destructive techniques such as AE, DIC and infrared thermography (IRT) as an auxiliary to the conventional external strain measurements rock testing.

2 EXPERIMENTAL STUDY

Iranian granite was used in the experiments. Mineral composition of the granite is composed of coarse grained (>5 mm) with a subhedral granular texture and porphyritic fabric including quartz (25%), orthoclase (25%), plagioclase (25%), biotite (20%), and minor minerals (5%) (Momeni et al. 2015). The diameter of the specimens is 54 mm and the thickness is 27 mm (see Fig. 1a). The end faces and sides of the samples were prepared smooth and straight within 0.25 mm according to the ISRM standard (ISRM 2007). The granite specimens were instrumented locally by strain gauges (Fig. 1b). Axial displacement was measured externally by a pair of LVDTs (see Fig. 1c).

DIC is a non-contact optical particle tracking technique dealing with real-time displacement and full-field strain fields on the surface of materials. Surface deformation is calculated by tracking the same pixel points and comparing the positional change of an undeformed

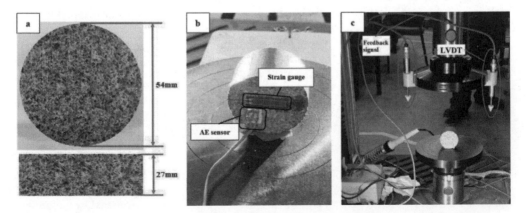

Figure 1. (a) Dimensions of Brazilian disc (b) Strain gauge and AE sensor instrumentation arrangement (c) Testing system and extensometer arrangement.

image with a deformed image. It relies on maximising the correlation coefficient which is determined by examining subsets at the considered points in the undeformed and their corresponding locations in the deformed images (Eqs. 1–2).

$$C(X) = \frac{\sum_{i=1}^{m} \sum_{j=1}^{m} \left[f(x_i, y_i) - \overline{f} \right] \cdot \left[g(x_i^*, y_i^*) - \overline{g} \right]}{\sqrt{\sum_{i=1}^{m} \sum_{j=1}^{m} \left[f(x_i, y_i) - \overline{f} \right]^2 \cdot \sum_{i=1}^{m} \sum_{j=1}^{m} \left[g(x_i^*, y_i^*) - \overline{g} \right]^2}} \tag{1}$$

$$x^* = x + u + \frac{\partial u}{\partial x} \Delta x + \frac{\partial u}{\partial y} \Delta y; \quad y^* = y + v + \frac{\partial v}{\partial x} \Delta x + \frac{\partial v}{\partial y} \Delta y \tag{2}$$

where $X = (u, v, \partial u / \partial x, \partial u / \partial y, \partial v / \partial x, \partial v / \partial y)$ containing six deformation parameters, $f(x, y)$ is the grey level value at coordinate (x, y) for the reference image, $g(x^*, y^*)$ is the grey level value at coordinate (x^*, y^*) for the target image, $\overline{f}$ and $\overline{g}$ are the average values of the image $f(x, y)$ and $g(x^*, y^*)$, respectively.

As shown in Fig. 2a, two digital high-resolution monochrome stereo cameras (i.e. Fujinon HF75SA-1, 1:1.8/75 mm, 5 megapixels resolution) which were set to point at the Brazilian disc specimen from two different directions were used. The angle between the cameras was kept below 60 degrees and they were programmed to capture 10 images per second until the end of the tests. Two goose-neck halogen light were also used to provide steady and uniform illumination across the whole Brazilian disc specimen. Since speckle pattern located on the sample has great importance in the DIC method, a black paint was sprayed on the white basecoat of the rock sample and then black speckles were created for high resolution (Yang et al. 2015; Munoz et al. 2016). Prior to Brazilian testing, each camera was calibrated by taking 50 image pairs at the calibration target. The calibration target was positioned and oriented differently such as tilted and rotated as well as in and out of the specimen plane (see Fig. 2b-c).

Captured images were computed and analysed via Vic-Snap photogrammetry software. A standard deviation of residuals of the stereo calibration was determined as 0.020 (in pixels) which suggests that the calibration process was conducted adequately for measurements (Sutton et al. 2009). Additionally, the loading rate was conducted under displacement control with a rate of 1 mm/min using a servo-controlled hydraulic compressive system of MTS Criterion which has a maximum load capacity of 300 kN.

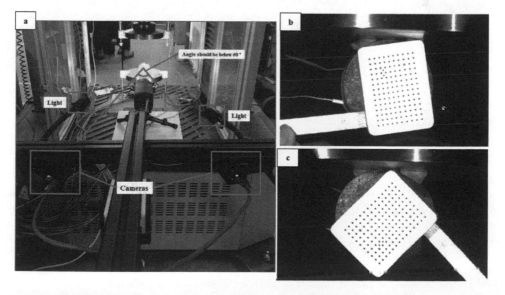

Figure 2. (a) DIC system instrumentation (b) Calibration target left camera (c) Calibration target right camera.

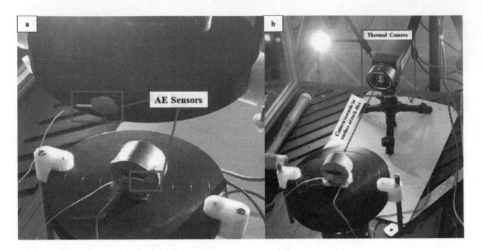

Figure 3.   (a) AE sensors instrumentation (b) thermal camera setup.

In the present study, two AE sensors were used to monitor micro crack initiation the damage mechanism of Brazilian discs. The AE transducers were attached on the rock specimen (see Fig. 3a). The output voltage of the AE was amplified by a gain of 60 dB. Additionally, thermal images were acquired using a thermal camera placed on opposite side of the specimen (see Fig. 3b). The temperature changes in the region of interest (RIO) of the specimen were monitored and analysed to investigate the strain localisation.

3   EXPERIMENTAL RESULTS AND DISCUSSION

The detailed results of indirect tensile tests on four Iranian granite disc specimens (B1, B2, B3, and B4) are given in Table 1. The tensile elastic modulus, E_t of the rock specimens was calculated using the recently proposed measuring method (Jianhong et al. 2009). Splitting elastic modulus which is the slope of line section of the stress-strain curve in the elastic stage recorded during the indirect tensile test was obtained first and then E_t was determined using Eq. 3 as below.

$$E_t = E_s \left\{ \left(1 - \frac{D}{L} arctan \frac{2L}{D}\right)(1-v) + \frac{2D^2(1+v)}{4L^2 + D^2} \right\}$$

(3)

where E_t is the tensile elastic modulus, E_s is the splitting elastic modulus, D is the diameter of Brazilian disc, L is the half-length of strain gauge and v is the Poisson's ratio of rock which were determined by uniaxial compression test. As can be seen from Table 1, experiment results indicate that E_t of the granite specimens are all less than their compressive elastic modulus, E_c, the average value of which was calculated as 46.5 GPa from conventional uniaxial and triaxial compression tests. The results show good reliability as their ratio is generally between 0.6 and 0.9.

Stress-strain relations of each granite specimen using DIC and strain gauge are compared in Fig. 4a. It can be seen that DIC strain measurements stops earlier because propagation of fracturing could not be captured due to the high velocity of fracturing (see Fig. 4). This is consistent with work conducted by Stirling et al. (2013) using DIC for investigating the influence of loading geometry on strain localisation in sandstones. Strain measured by DIC method are independent from bedding errors which also does not include the strains associated with localised crushing at the loaded surfaces or machine compliance.

The image at time, $t = 0$, was selected as a reference image, and then all the deformation fields at various load levels (where $t > 0$) on the Brazilian specimen was obtained. The lateral strain, ε_L fields reveal the evolution process of damage and deformation from the elastic stage until rock failure (see Fig. 5). The development of the full-field deformation can be qualitatively divided into three stages: elastic deformation stage, damage localisation stage and failure stage. In the initial stage, any obvious strain localisation region on the surface of the specimen was not observed, thus this stage can be considered as elastic stage.

Table 1. Summary of indirect tensile test results.

Sample no	Tensile strength, σ_t (MPa)	Splitting elastic modulus, E_s (GPa)	Tensile elastic modulus, E_t (GPa)
B4	5.61	22.53	36.23
B5	5.13	21.27	30.34
B6	5.72	23.73	40.56
B7	5.54	20.23	33.65
Average	5.49	21.94	35.20

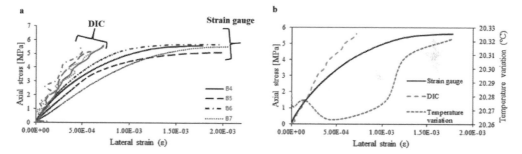

Figure 4. (a) Stress-strain curves for granite specimen from DIC and strain gauge measurements during Brazilian disc test (a) Temperature variation compared with stress-strain relation.

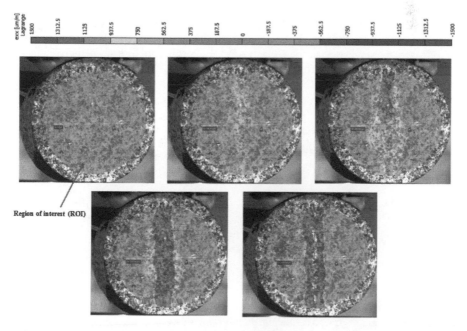

Figure 5. Evolution of tensile strain and damage accumulation.

In the second stage, crack initiation was observed in two strain localisation regions near the loading points. It is believed that the fracturing emanated from the loading ends rather than at the specimen centre when the total strain in the rock exceeds the critical extensional strain (Li and Wong 2013). Therefore, our experimental observations (see Fig. 5) show consistent results with this research.

Once the cracking began, a progressive damage accumulation was observed towards the centre of the specimen along the central diameter of the disc surface which is an indicator of the degradation of the rock. Upon the increasing load, the strain localisation and damage became more obvious when the stress concentration expanded towards the centre of the disc specimen. This localisation which can be associated with the propagation of fracture indicates that a damage zone occurred due to the displacement discontinuity and thus load carrying capacity of the rock started to decrease. At the failure stage, a damage band towards the centre of the specimen formed and with further loading, accumulated damage transformed to rapid tensile crack propagation which eventually resulted in a macro-crack due to the coalescence of linked strain localisation areas. Thus, this cracking lead to the failure of specimen as disc splitting where tensile stresses exceeded the load carrying capacity of the rock.

Thermal images were captured along the whole test on the Brazilian disc specimens and trend of temperature changes due the stress concentration were assessed to better detect strain localisation and damage accumulation, as shown in Fig. 6. All Brazilian discs exhibited temperature localisation when they were about to fracture which also enables to get the information about the crack propagation path. However, since the tests were conducted under quasi-static loading, the temperature distribution had a tendency to decrease as it allowed more time for heat dissipation. Temperature variation during the deformation stages in a Brazilian test is presented in Fig. 4b. Prior to the final failure stage, a sharp rising trend was observed due to the rapid damage which resulted in a macro fracture by the crack through the disc. The dominant fracture is also accompanied by secondary fracture at the points of loading (see Fig. 6).

AE signals generated by the microcraking provide useful information about deformation and failure characteristics of rocks. While AE hits is related to the number of cracks during loading, the size or magnitude of the cracks is related to the AE energy of each AE event. Fig. 7a-b shows the AE energy and hits characteristics during Brazilian test in which deformation stages can be observed. Trends of cumulative AE energy and hits and the damage evolution which exhibit a flat-to-sharply rising trend are shown in Fig. 7c-d. At the initial stage, damage in the rock is low in which the rock remained in a relatively stable condition. When the stress level reached criticality, the accumulated elastic energy within the rock was rapidly released leading to a sudden jump of AE energy and hits with increasing amplitudes. As tensile stresses exceed the load carrying capacity, a sharp rise in AE parameters was observed since damage started to get intensified leading to a major microcrack that caused the failure of the rock.

Figure 6. Comparison of the experimental failure and thermal image showing the deformation localisation.

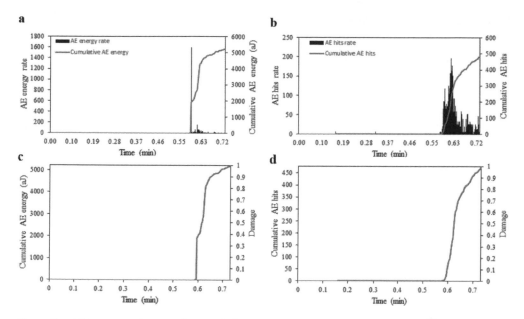

Figure 7.   AE energy and count characteristics and damage evolution.

4   CONCLUSION

Failure process of brittle rocks is closely related to crack initiation, crack propagation and damage localisation. In this paper, evolution of damage, deformation field and strain localisation of the Iranian granite using Brazilian test were investigated by the state-of-the-art DIC, IRT and AE monitoring systems. The experimental results demonstrated that DIC is an effective particle tracking technique to observe the evolution of the displacement and strain fields in rock which can describe the whole damage accumulation process and crack initiation and propagation of rock under Brazilian testing. The deformation process was divided into three stages based on the full-field strain. DIC proved that onset of crack initiation occurs near the loading ends and then localised progressively with the macrocrack nucleation along the central diameter of the specimen which led the failure of the specimen. Correlation of IRT with DIC also improved the reliability of exhibited strain and damage localisation of the surface of the Brazilian disc specimens where temperature localisation was observed prior to fracturing and failure of rock in the centre of the disc. However, due to lower loading rate, temperature distribution was tend to decrease due to rapid heat dissipation. Therefore, this monitoring method could be used to better identify the temperature trend of rock and predict failure under high strain rate loading cases. It was also concluded that complete evolution of damage in the rock during rock failure can be successfully characterised using the AE parameters i.e. AE energy and AE hits which identify damage state of the rock.

REFERENCES

Akdag, S., Karakus, M., Taheri, A., Nguyen, G., & Manchao, H. (2018). Effects of thermal damage on strain burst mechanism for brittle rocks under true-triaxial loading conditions. Rock Mech Rock Eng, (Accepted Manuscript).

Bieniawski, Z., & Hawkes, I. (1978). ISRM commission on standardization of laboratory and field tests suggested methods for determining tensile strength of rock materials. Int J Rock Mech Min Sci & Geomech Abstr, 15(3), 99–103.

Coviello, A., Lagioia, R., & Nova, R. (2005). On the measurement of the tensile strength of soft rocks. Rock Mech Rock Eng, 38(4), 251–273.

Eberhardt, E., Stead, D., & Stimpson, B. (1999). Quantifying progressive pre-peak brittle fracture damage in rock during uniaxial compression. Int J Rock Mech Min Sci, 36, 361–380.

Hudson, J., Rummel, F., & Brown, E. (1972). Controlled failure of rock discs and rings loaded in diametral compression. Int J Rock Mech Min Sci, 9(2), 241–248.

ISRM. (2007). The complete ISRM suggested methods for rock characterization, testing and monitoring, 177–184.

Jianhong, Y., Wu, F., & Sun, J. (2009). Estimation of the tensile elastic modulus using Brazilian disc by applying diametrically opposed concentrated loads. Int J Rock Mech Min Sci, 46, 568–576.

Karakus, M., Akdag, S., & Bruning, T. (2016). Rock fatigue damage assessment by acoustic emission. In P. Ranjith, & J. Zhao (Ed.), International Conference on Geo-mechanics, Geo-energy and Geo-resources, IC3G, (pp. 9–82–88). Melbourne, Australia.

Lin, Q., & Labuz, J. (2013). Fracture of sandstone characterized by digital image correlation. Int J Rock Mech Min Sci, 60, 235–245.

Mellor, M., & Hawkes, I. (1971). Measurement of tensile strength by diametral compression of discs and annuli. Eng Geol, 5(3), 173–225.

Momeni A., Karakus M., Khanlari G.R., Heidari M., 2015. Effects of cyclic loading on the mechanical properties of a granite, International Journal of Rock Mechanics and Mining Sciences 77, 89–96.

Munoz, H., Taheri, A., & Chanda, E. (2016). Pre-peak and post-peak rock strain characteristics during uniaxial compression by 3D digital image correlation. Rock Mech Rock Eng, 49, 2541–2554.

Stirling, R., Simpson, D., & Davie, C. (2013). The application of digital image correlation to Brazilian testing of sandstone. Int J Rock Mech Min Sci, 60, 1–11.

Sutton, M., Orteu, J., & Schreier, H. (2009). Image correlation for shape, motion and deformation measurements. Springer.

Wu, X., Baud, P., & Wong, T. (2000). Micromechanics of compressive failure and spatial evolution of anisotropic damage in Darley Dale sandstone. Int J Rock Mech Min Sci, 37, 143–160.

Yang, G., Cai, Z., Zhang, X., & Fu, D. (2015). An experimental investigation on the damage of granite under uniaxial tension by using a digital image correlation method. Opt Las Eng, 73, 46–52.

The increasing of exploitation safety of potassium salt deposit based on geological-geomechanical simulation

Yu. Kashnikov, A. Ermashov, D. Shustov & D. Khvostantcev
Perm National Research Polytechnic University, Perm, Russia

ABSTRACT: Major accidents at underground salt mines of PJSC «Uralkali», in most cases, lead to their total flooding. As a result, the need of detailed study of Verkhnekamskoe salt deposit geological features, in respect to oversalt rocks, appeared. First of all, abnormal complex zones in oversalt rocks and, primarily, in waterproof covering must be determined. A detection and investigation of local decompression zones, excessive fissuring and permeability of the rocks is significantly important while studying waterproof covering.

The paper introduces the method of weakened zones determination in waterproof covering. The method based on the joint application of geophysical and geomechanical approaches. The main idea is to obtain correlation dependencies between static and dynamic geomechanical properties, adjustment of dependencies using well logging data, to get distribution of compression wave velocity and acoustic impedance on the ground of 2D and 3D seismic data processing and, in final part, definition of zones with various consolidation degree in waterproof covering and salt rocks. Following this approach, it is possible to determine the values of uniaxial compressive strength, coefficient of elasticity, Poisson ratio and other characteristics in any point of waterproof covering.

The further geomechanical modeling of waterproof layers stress-strain state uses available values of physical and mechanical properties of rocks. The progression of ground movements and breaking zones in most weakened part of the mine at Verkhnekamskoe salt deposit is considered. The specific rock deformation rheological model implemented with finite element method was used. Physical and mechanical properties of salts and waterproof covering rocks defined from combination of geomechanical and geophysical methods were applied in geological-geomechanical simulation.

It is still not clear how dangerous weakened zones in waterproof covering and in salt rocks. From the usual point of view, weakened zones in waterproof covering are the reason of mines flooding. However weakened zones in salt rocks cause abrupt increasing of ground subsidence and possible destruction of hard rocks in waterproof covering as a result of higher loading of safety blocks. From the other side, even with low loading degree of safety blocks and solid rocks in salt layers, water-conducting cracks may appear due to weak strength properties of waterproof covering rocks.

1 GEOLOGICAL-GEOMECHANICAL SIMULATION

Presence of overlying waterlogged rocks is a geological feature of most potassium salt deposits in the world. Rocks between potassium layers and waterlogged rocks are called waterproof covering (WPC). At Verkhnekamskoe potassium salt deposit in Western Ural region water-tight rock section has thickness from 50 to 140 meters and located between production potassium layers and weakly-mineralized aquifer which has active water exchange with overlying freshwater layers with total seams height from 150 to 350 meters (Kudrjashov, 2001)

Through weakened zones in WPC, water invasion in mines is possible during deposit exploitation. The method based on the joint application of geophysical and geomechanical approaches was used to determine weakened zones in WPC and in productive potassium layers. The research main point was to obtain correlation dependencies between static and dynamic geomechanical properties, to adjust received dependencies using well sonic logging data, to get distribution of compression wave velocity and acoustic impedance based on 2D and 3D seismic data processing and, in final part, to define zones with various consolidation degree in waterproof covering and salt rocks. Actually, the values of uniaxial compressive strength, elastic modulus, Poisson ratio and other geomechanical characteristics can be determined in any point of waterproof covering. Available values of physical and mechanical properties of rocks from productive layers and waterproof covering were used in further geomechanical modeling of WPC stress state.

Geological structure of the deposit, 3D (2D) seismic data and well logging data processing results, geomechanical dependencies received from core sample test results are the foundation of the represented method. The method allows allocating zones with different physical and mechanical properties in waterproof covering and production layers. All that information is the basis of geological-geomechanical model of waterproof covering rocks and potassium salt layers. Reliable data on development of deformation processes in waterproof covering and salt rocks can be obtained through geological-geomechanical simulation considering different parameters of deposit exploitation.

Thereby, aggregate of geological, geophysical and geomechanical information is the base of geological-geomechanical model. The model is oriented towards solving the problem of waterproof covering safety evaluation with specific parameters of deposit exploitation (Figure 1).

Three-dimensional distribution of the static geomechanical parameters is the ground of geological-geomechanical simulation. 3D distributions are widely used to detect consolidated and unconsolidated zones in hydrocarbon reservoirs (Kashnikov et al., 2012, Shustov et al., 2015, M.D. Zoback, 2007). Correlation between static and dynamic geomechanical properties is one of the most important parts in geological-geomechanical modeling. Static and dynamic properties were obtained from core sample tests (M.D. Zoback, 2007, H. Sone, 2013, V. Brotons et al., 2016). Afterwards, static characteristics can be calculated from dynamic characteristics of the rock mass using 2D or 3D seismic data.

Monoliths from well of mine shaft were selected and 45 core samples were made from them. Dimensions of core samples were 30 × 60 mm. Geomechanical properties such as static and

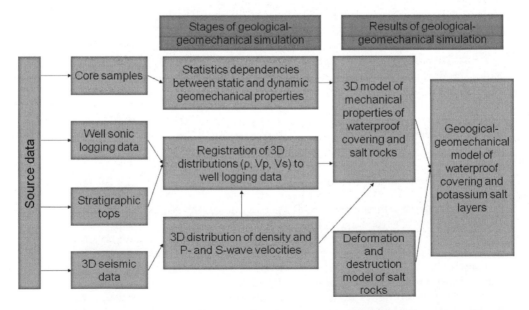

Figure 1. Block-diagram of geological-geomechanical simulation of waterproof covering and potassium salt layers.

dynamic values of elastic modulus, Poisson ratio, uniaxial compressive strength and strength envelopes in triaxial compression were determined using PIK-UIDK/PL equipment.

Figure 2 illustrates the dependence of uniaxial compressive strength from acoustic impedance. Correlation ratio is very high and the dependence can be used in creation of geological-geomechanical model. In total, following dependences were obtained:

– Static elastic modulus in atmospheric and reservoir conditions on dynamic elastic modulus, P-wave velocity, acoustic impedance;
– Uniaxial compressive strength on P-wave velocity and acoustic impedance;
– Tangent and secant deformation moduli and decline modulus on P-wave velocity.

After test results were tied with well-log data, similar dependencies for sonic logging data were obtained.

Next step was devoted to compute dynamic elastic characteristics of salt and overlying rocks based on velocities of longitudinal and transverse waves (Vp, Vs) and volume density ρ. 3D seismic survey data made for oil field situated under potassium salt deposit and well logging data were used to achieve needed results. Sonic logging data of four oil wells and well of mine shaft was used for stratigraphic interpretation and for creation of velocity inversion model. Interpolation of well logging data based on 3D distribution obtained from elastic inversion was made to get 3D distribution of density and 3D distribution of P- and S-wave velocities.

Analysis of wavefield's velocity properties is a base of qualitative detection of weakened zones in waterproof covering and potassium layers. Dependencies between static and dynamic geomechanical properties were obtained from core sample test results which were adjusted with sonic logging data. The obtained dependencies were used for quantitative evaluation of mechanical and deformation properties distribution. In fact, the dependencies of uniaxial compressive strength and elastic modulus on P-wave velocity and acoustic impedance were used in the research.

Figure 3 illustrates combined cross section of waterproof covering and potassium layers and distribution of uniaxial compressive strength along 61–60–107г-3 well line. The cross section shows presence of the solid layer of salt rocks (ПКС) with uniaxial compressive strength of about 22–25 MPa and many weak layers (Д-Г, Г-В, В-Б) with values of uniaxial compressive strength of 5–10 MPa. Similar cross sections with elastic and deformation properties

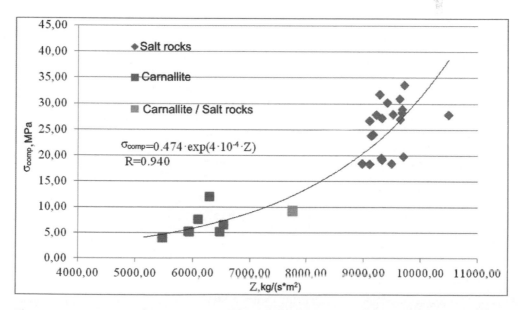

Figure 2. Correlation between uniaxial compressive strength and acoustic impedance.

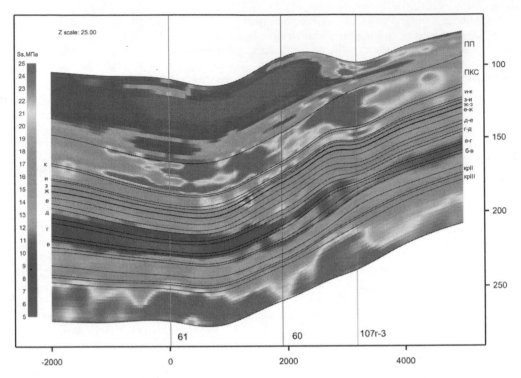

Figure 3. Distribution of uniaxial compressive strength along 61–60–107г-3 well line.

distribution were obtained as well. Weakened zones in potassium layers KpII and KpIII in work site's northwestern part were determined. The value of uniaxial compressive strength in a zone of 600×800 meters in potassium layer KpII is much lower than in mine exploitation project. As a result, safety block loading degree is 0.6 and even more with regulation value of 0.4. If so, it can lead to accelerated subsidence of earth surface and to discontinuity of water-tight rocks. The layer KpIII has the same situation.

2 GEOMECHANICAL SIMULATION OF WEAKENED ZONES

The next step is geomechanical simulation of part of mine field that takes into account parameters of deposit exploitation and obtained physical and mechanical properties of rocks. Deformation and failure rheological model is a ground of stress state calculation of rock mass. The model considers primary, secondary and tertiary creep deformations separately. It also makes provision for weakening and failure of rocks with appearance of dilatancy (W. Wittke 1999, Doering, T. & Kiehl, J.R., 1996). Viscoplastic theory is the base of the model. Viscoplastic deformation increment corresponds to the plasticity theory:

$$\left\{ d\varepsilon^{vp} / dt \right\} = \left\{ \dot{\varepsilon}^{vp} \right\} = \begin{cases} 0, & F \leq 0 \\ \dfrac{1}{\eta} F \{ \partial Q / \partial \sigma \}, & F > 0 \end{cases} \tag{1}$$

where F – flow function;
$\dot{\varepsilon}^{vp}$ – viscoplastic deformation velocity.

The rheological model was implemented into ANSYS Software using the initial strain method. Figure 4 shows earth surface subsidence with time due to deposit exploitation.

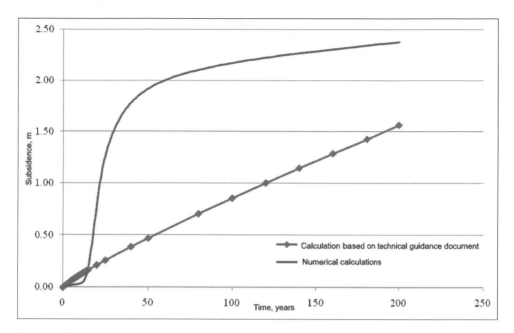

Figure 4.   Increase of earth surface subsidence because of part of deposit exploitation.

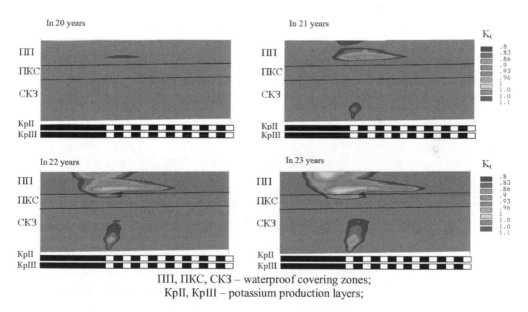

ПП, ПКС, СКЗ – waterproof covering zones;
KpII, KpIII – potassium production layers;

Figure 5.   Development of failure zones with time (K_t – Drucker-Prager criterion).

Physical and mechanical properties of potassium layers and waterproof covering for upper line were received from numerical simulation and for bottom line based on demands of technical guidance document. It can be seen that the difference in nature of subsidence increase is significant.

Figure 5 illustrates development of weakened zone in waterproof covering. Deformation processes begin only in 20 years from mine site development. Failure zone has an intensive development during 3 years and then the process practically stops. Thereby, the most dangerous situation probably should be in 23–24 years after beginning of mine site exploitation and this correspond to plot shown in Figure 4.

It is still not clear how dangerous weakened zones in waterproof covering and in salt rocks are. From the usual point of view, the weakened zones in waterproof covering are the reason of mines flooding. However, weakened zones in potassium layers cause abrupt increase of ground subsidence and possible destruction of hard rocks in WPC because of higher loading of safety blocks. From the other side, even with low loading degree of safety blocks and with the presence of solid rocks in salt layers, water-conducting cracks may appear due to weak strength properties of waterproof covering rocks.

3 CONCLUSION

Geological-geomechanical simulation of waterproof covering and potassium production layers of part of Verkhnekamskoe potassium salt deposit was based on sonic logging data processing, 3D seismic data processing and core sample test results.

Conducted research has shown that the increase of exploitation safety of potassium salt deposit should be based on an integrated application of geomechanical and geophysical methods. Different mechanical properties of rock mass should be taken into account by reliable geomechanical modelling of deformation and failure of waterproof covering and potassium layers. Geophysical survey in combination with geomechanical methods allows allocating zones with different mechanical properties in rock masses.

REFERENCES

Brotons, V., R. Tomás, S. Ivorra, A. Grediaga, J. Martínez-Martínez, D. Benavente, M. Gómez-Heras. Improved correlation between the static and dynamic elastic modulus of different types of rocks. Materials and Structures, August 2016, Volume 49, Issue 8, pp. 3021–3037.

Doering, T; Kiehl, J.R. Das primaeren, sekundaeren und tertiaelen Kreichen von Steinsalz -ein dreidimensionales rheologes Stoffgesetz. Geotechnik. Nr. 3. 194–199, 1996.

Kashnikov Yu. A., Gladyshev S.V., Shustov D.V., Yakimov S. Yu., Komarov A. Yu, Tinakin O.V. Geological and geomechanical model for Astrakhan gas/condensate field. Gas industry of Russia, March 2012, pp. 29–33.

Kudjashov A.I. Verkhnekamskoe potassium salt deposit.—Perm, MI UB RAS, 2001, pp. 429.

Shustov D.V., Kashnikov Yu. A. Development of geological-geomechanical model of some part of Urubcheno-Tokhomskoe oil field. Geology, Geophysics and development of oil and gas fields, April 2015, pp. 26–31.

Sone H., Zoback M.D. Mechanical properties of shale-gas reservoir rocks—Part 1: Static and dynamic elastic properties and anisotropy. Geophysics, vol. 78, no. 5 (September–October 2013); p. D381–D392.

Wittke W. Tunnelstatik. Verlag Glueckauf GmbH. Essen. 1999. S.408.

Zoback, M.D., Reservoir Geomechanics. Cambridge. University Press.—2007. - pp. 449.

Geomechanics and Geodynamics of Rock Masses – Litvinenko (Ed.)
© 2018 Taylor & Francis Group, London, ISBN 978-1-138-61645-5

Effect of viscosity of fault filling on stick-slip dynamics of seismogenic fault motion: A numerical approach

Srđan Kostić
Institute for Development of Water Resources "Jaroslav Černi", Belgrade, Serbia
Faculty of Mining, University of Banja Luka, Prijedor, Bosnia and Herzegovina

Nebojša Vasović
Faculty of Mining and Geology, University of Belgrade, Belgrade, Serbia

Igor Franović
Institute of Physics, University of Belgrade, Belgrade, Serbia

ABSTRACT: In present paper authors examine the conditions for which the fault motion exhibits stick-slip like dynamics, under the main assumption that a fault filling at least along a certain part of the fault possesses viscous properties. Analysis is conducted for the mechanical spring-block model of fault motion, suggesting that two neighboring parts of a fault, whose friction could be described by Dieterich-Ruina law, are separated by a viscous part of a fault filling, due to increased temperature, fluidization and melting. This initial assumption follows the original suggestion of Burridge and Knopoff, but with the introduced time delay in transition of the motion between the two blocks. Analysis is conducted using local bifurcation analysis, and by numerically solving the observed system of delay differential equations using Runge-Kutta 4th order method. Starting system, without the introduced time delay, exhibits only the first direct and inverse supercritical Andronov-Hopf bifurcation. Results obtained indicate that the effect of nonstationary time delay force the system under study to exhibit stick-slip like dynamics, which corresponds qualitatively well to full seismic cycle, with alternation of inter-seismic and co-seismic fault motion. Results obtained emphasize the role of viscous regions along the fault on seismogenic dynamics.

Keywords: fault motion, stick-slip dynamics, rock friction law, time delay, seismic cycle

1 INTRODUCTION

It has been long recognized that laboratory mechanical models of stick-slip motion provide results which are relevant from the viewpoint of real tectonic movement and earthquake nucleation (Burridge and Knopoff, 1967). The main contributing factor in such models is the specific nature of rock friction, which is usually observed either as only rate-dependent (Vieira, 1999) or both rate- and state-dependent (Erickson, 2008). These friction laws are explicitly or implicitly introduced in mathematical equations which describe the observed motion, and which are very convenient for numerical simulation of the real tectonic movement. Benefit of using such models lies in one's ability to introduce different factors which could significantly affect the fault motion, such as background seismic noise (Vasović et al., 2017), delayed interaction among the adjacent blocks (Kostić et al., 2017) and variable frictional strength (Kostić et al., 2016). Goal of such analyzes is simple: one needs to determine the conditions which enable transition of spring-block dynamics from the stable equilibrium state to the unstable dynamical regime. Instability, as an occurrence of regular periodic or irregular aperiodic behavior, is commonly treated as the onset of seismic motion in real conditions. Such way of interpreting different dynamical behavior of spring-block models is fully justified from the viewpoint of pure dynamics. Nevertheless, when it comes to the real fault movement, especially the seismogenic one, onset of regular periodic oscillations could hardly be assigned to the earthquake nucleation process. From the viewpoint of seismology, only stick-slip dynamics, as alternation of moving phase (slip) and standing or stationary moving phase (stick), could be relevant for the analysis of the mechanism behind the seismogenic fault motion. For very nice review of fault geomechanics and related dynamical behavior, one should refer to (Kocharyan, 2016). In order to obtain such dynamics, a researcher needs to cope with three challenging tasks. First of all, one should find an appropriate model which could simulate stick-slip motion for a certain parameter range. Secondly, one should determine values of the control parameters, for which the observed system exhibit stick-slip. And, finally, one needs to find physically justified motivation for an assumption that will lead to the onset of stick-slip behavior. In present paper, authors introduce the following assumption: *there is a time-dependent delayed interaction among the adjacent blocks in spring-block model*. Existence of delayed interaction is not new, and it goes back to the pioneer work of Burridge and Knopoff (1967), who shrewdly spotted a delayed interaction among two end group of blocks along a single chain, representing two different parts of a single fault. These groups are separated by blocks in a middle, providing a delaying interaction which is an effect that could be assigned to the increased viscosity along a certain part of the fault. Although many years have passed since then, this interpretation of the laboratory analysis of stick-slip motion is still intuitively clear and appears as possible explanation of the real fault movement. However, one could ask himself: is it justified to assume the constant value of time-delay in interaction, or this factor represents a variable that is evolving in time? If one compares the relevant time scales, i.e. the returning period of the large earthquakes and the duration of seismic event, it could be hard to estimate the constant value of time delay in interaction among the adjacent blocks. More probable assumption is that the viscous effect of fault filling is slowly changing over time, probably due to temporal variations in temperature, pore-pressure, different consolidation and lithification conditions along the fault, caused by variable normal stress and nearby tectonic activity.

In presented research, authors analyzed the dynamics of spring-block model composed of 2 interconnected blocks, with time-dependent delayed interaction, which are also attached to an upper plate driving the whole system in one direction with a constant velocity. Assumed friction law is rate- and state-dependent. Aim of the performed research is to determine the conditions for which spring-block model starts to behave in a stick-slip fashion, which is claimed to be relevant from the seismological viewpoint. Such approach is new and, as authors are aware, this is the first time such analysis is done for the spring-block model of earthquake nucleation.

2 APPLIED METHODOLOGY

In present paper authors start from the following initial dynamical system:

$$\dot{\theta} = -\nu(\theta + (1+\varepsilon)\log(\nu))$$
$$\dot{u} = \nu - 1 \qquad\qquad (1)$$
$$\dot{\nu} = -\gamma^2 \left[u + (1/\xi)(\theta + \log(\nu)) \right]$$

where $\varepsilon = B/A - 1$ measures the sensitivity of the velocity relaxation, $\xi = kL/A$ is the nondimensional spring constant, and $\gamma = (k/M)^{1/2}(L/v_0)$ is the nondimensional frequency (Erickson et al., 2008). One should note that system (1) is dimensionless, meaning that no real values of measurements could be directly incorporated into model (1), but this model enable qualitative analysis in order to capture the main dynamical mechanisms. Parameter M is the mass of the block and the spring stiffness k corresponds to the linear elastic properties of the rock mass surrounding the fault (Scholz, 2002). According to Dieterich and Kilgore (1994) the parameter L corresponds to the critical sliding distance necessary to replace the population of asperity contacts. The parameters A and B are empirical constants, which depend on material properties. Variables u and v represent displacement and velocity, while θ denotes the state variable describing the state of the rough surface along which blocks are moving (Clancy and Corcoran, 2009). Parameter v_0 represents the constant background velocity of the upper plate. As it was previously shown (Kostić et al., 2014), a supercritical direct Andronov-Hopf bifurcation curve occurs for the following parameter values $\varepsilon = 0.27$, $\xi = 0.5$ and $\gamma = 0.8$, indicating a transition from equilibrium state to regular periodic oscillations. In present paper, authors examined the interaction of two blocks in a spring-slider model, whose dynamics is governed by the following system of equations:

$$\dot{\theta}_1 = -V_1 \cdot \left(\theta_1 + (1+\varepsilon)\ln V_1\right)$$

$$\dot{U}_1 = V_1 - 1$$

$$\dot{V}_1 = \gamma_1^2\left(-U_1 + c\left(U_2(t-\tau) - U_1(t)\right) - \left(\frac{1}{\xi}\right)\left(\theta_1 + \ln(V_1)\right)\right)$$

$$\dot{\theta}_2 = -V_2 \cdot \left(\theta_2 + (1+\varepsilon)\ln V_2\right)$$

$$\dot{U}_2 = V_2 - 1$$

$$\dot{V}_2 = \gamma^2\left(-U_2 + c\left(U_1(t-\tau) - U_2(t)\right) - \left(\frac{1}{\xi}\right)\left(\theta_2 + \ln(V_2)\right)\right)$$

(2)

where c is the measure of frictional strength (k_1/k_2) and τ is interaction delay, Equation (2) was solved numerically using Runge-Kutta 4th order differentiation method, while the relevant bifurcation curves were obtained using the package DDE-BIFTOOL in Matlab (Matlab, 2017).

3 RESULTS

3.1 Impact of constant time delay in interaction of adjacent blocks

In the first phase of the analysis, authors assume that time delay in interaction between the adjacent blocks is not changing over time. Numerical bifurcation analysis of the system (2) without the introduced time-dependent character of time delay indicate the appearance of the inverse and direct supercritical Andronov-Hopf bifurcation, i.e. a transition from regular periodic oscillations to stable equilibrium state, and, again, to regular periodic oscillations (Figure 1a). Obtained time series before and after the bifurcation curve (point 1, 2, 3 and 4 in Figure 1a) are given in Figure 2. Results obtained indicate interesting dynamical behavior. In particular, it appears that the state with frictional strength $c > 1$ and time delay $0.2 < \tau < 0.3$ (area constrained by bifurcation curves) should be treated as a regime of stable aseismic creep along the fault (Figure 2c). Further change of time delay in interaction induces the following scenario. Decrease of τ induces the occurrence of high-amplitude long-period regular periodic oscillations (Figure 2a and 2b), while increase of τ leads to the occurrence of small-amplitude and short-period oscillations (Figure 2d). From the seismological viewpoint, these three distinctive regimes could be ascribed to the following modes of fault movement (Figure 1b): regime of potential aseismic creep along the fault, regime of potential strong earthquakes and a regime of potential medium to small earthquakes. Nevertheless, as one can see from Figure 2a and 2b, although regular periodic oscillations are treated as unstable dynamical regime from the viewpoint of pure dynamics, such motion could not be directly prescribed to the real fault movement during a seismic event. For this reasons, authors assume the slow temporal changing of delayed interaction among the interacting blocks, which is justified considering the existence of two relevant time scales: long, denoting the return period of large earthquakes, and short, which corresponds to the short duration of a seismic event.

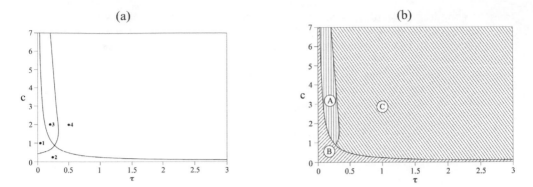

Figure 1.　(a) Bifurcation diagram τ(c), for the fixed values of parameters $\varepsilon = 0.3$, $\xi = 0.5$ and $\gamma = 0.5$ (regime of regular periodic oscillations of the starting system). Andronov-Hopf bifurcation curve denotes the transition from regime of regular periodic oscillations to stable equilibrium and again to regime of periodic oscillations. (b) Different dynamical regimes of potential relevance from the seismological viewpoint: A—regime of stable aseismic creep, B—regime of potential co-seismic fault movement—strong earthquakes, C—regime of possible co-seismic fault movement—medium to weak earthquakes.

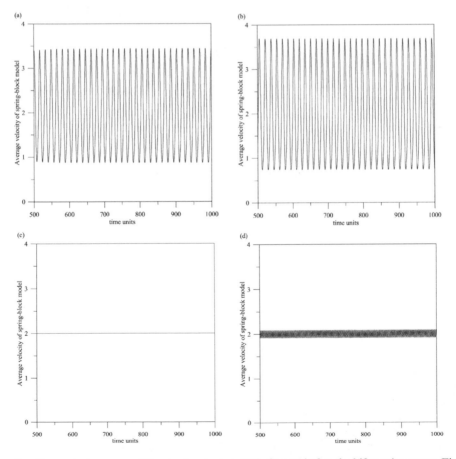

Figure 2.　Dynamical regimes of the starting system (2) before and after the bifurcation curve. Figures in (a), (b), (c) and (d) correspond to the points 1, 2, 3 and 4, in Figure 1, respectively.

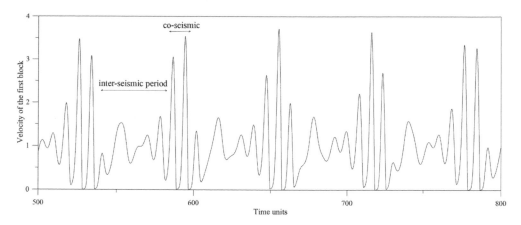

Figure 3. Characteristic time series illustrating the effect of nonstationary time delay: $\omega = 0.1$, $b_0 = 1.2$, $b_1 = 1$, $\varepsilon = 0.3$, $\xi = 0.5$, $\gamma = 0.5$, $c = 2$. Qualitatively similar dynamics is obtained for the second block of the observed model (2).

3.2 *Impact of nonstationary time delay*

In the second phase of the analysis, authors assume that time delay in interaction between the adjacent blocks is evolving in time in a slow sinusoidal manner, $\tau = b_0 + b_1 * sin(\omega t)$. In present case, frequency ω and amplitude of the time delay are chosen to be much smaller in comparison to the oscillation frequency and amplitude of model (2) in regime C ($\omega = 0.3$ and $A = 2.7$). Reason for this lies in the fact that expected changes of viscosity of fault filling are slow and with small amplitude, due to the effect of temperature oscillations, influence of pore pressure, variable normal stress and, hence, different consolidation and lithification processes in different parts of the fault. One should note that analysis is conducted for model (2) in regimes A and C due to the fact that regime B is rather narrow for higher values of frictional strength, while no effect of nonstationary time delay is observed for frictional strength below $c = 0.5$. Reason for this lies in the following. Firstly, when $c < 0.5$ coupling between the two blocks is too loose for time delay to have any effect on the dynamics of the observed model. Secondly, oscillation frequency in this regime is so small ($\omega = 0.07$), that one needs to assume extremely low-frequent oscillations of time delay which cannot be "felt" in the dynamics of the observed model.

If one assumes frequency of non-stationary time-delay $\omega = 0.1$ and $b_0 = 0.25$, while $b_1 = 0.05$, meaning that dynamics of the observed model (2) stays within regime A, no instability occurs. However, if one increases that observed range for time delay, $b_0 = 1.2$, while $b_1 = 1$, including also the regime C, than stick-slip-like behavior is obtained, as it is shown in Figure 3. Here one can distinguish rather long periods of quasi-stationary movement (inter-seismic periods) and short periods of bursting (co-seismic events). Oscillations in Figure 3 are irregular and nearly-periodic.

It should be noted that stick-slip dynamics shown in Figure 3 is not obtained values of ω much smaller or much higher than the assumed one, $\omega = 0.1$. In both cases of very small or very high value of ω, irregular oscillations also occur, but without the pronounced alternation between the period quasi-stationary motion and bursting intervals.

4 CONCLUSIONS

In present paper authors analyze the dynamics of fault motion by assuming non-stationary time delay among the adjacent parts of the fault zone. Analysis is conducted by modeling the dynamics of spring-block model, which is considered as the phenomenological model of fault motion. Authors conducted a numerical research, by solving the system of delay differential equations using the 4th order Runge-Kutta method, while bifurcation curves are obtained using DDE-BIFTOOL in Matlab.

Analysis of the effect of stationary (constant) delay in interaction of two adjacent blocks indicates rather interesting behavior. It appears that one needs to assume the existence of time-delay in the stable equilibrium regime of aseismic creep (regime A). Reduction of time delay leads to the onset of high-amplitude long-period oscillations, which could correspond to potential co-seismic dynamical regime of large earthquakes (regime B). On the other hand, increase of time delay indicates a transition to small-amplitude and short-period oscillations, denoting the probable co-seismic regime of medium to small earthquakes (regime C).

If the non-stationary delayed interaction among the blocks is assumed, than one could capture the stick-slip like behavior, when oscillation amplitude and frequency of time delay are significantly smaller in comparison to the oscillation of the observed system in regime B. In that case, on could easily distinguish betwen the long interseismic period and rather short co-seismic bursting. Effect of non-stationary time delay was not examined for the regime B, since oscillations in this regime are with so small frequency that any assumed frequency of time delay in interaction could not be "felt" by the dynamics of the observed model.

Although stick-slip-like behavior is captured when nonstationary time delay is introduced, one needs to note that occurrence of bursting events is still nearly periodic (please refer to Figure 3), which is only the case for large earthquakes with the strongest magnitudes. As it is well-known, temporal distribution of earthquakes by magnitudes is random (Kostić et al., 2014), even for earthquakes recorded along a single fault. Hence, authors need to establish a convenient controlling factor which would enable random bursting in the suggested fault model. One way of doing this is by assuming that nonstationary time delay is described by a more complex irregular function. Another way is by introducing the stronger effect of background seismic white noise. Such model, with random distribution of bursting events, would provide almost perfect qualitative description of seismogenic fault motion.

ACKNOWLEDGMENTS

This research was partly supported by the Ministry of Education, Science and Technological Development of the Republic of Serbia (No. 176016 and 171017).

REFERENCES

Burridge, R., Knopoff, L.1967. Model and theoretical seismicity. Bulletin of the Seismological Society of America 57, 3, 341–371.

Erickson, B., Birnir, B., Lavallee, D. 2008. A model for aperiodicity in earthquakes. Nonlinear processes in geophysics 15, 1–12.

Kocharyan, G.G. 2016. Geomechanics of Faults. Institute of Geosphere Dynamics, Russian Scientific Foundation, Moscow, 422 p.

Kostić, S., Vasović, N., Franović, I., Todorović, K., Klinshov, V., Nekorkin, V. 2017. Dynamics of fault motion in a stochastic spring-slider model with varying neighboring interactions and time-delayed coupling. Nonlinear Dynamics 87, 2563–2575.

Kostić, S., Vasović, N., Franović, I., Todorović, K. 2013. Dynamics of simple earthquake model with time delay and variation of friction strength, Nonlinear Processes in Geophysics 20, 857–865.

Kostić S., Vasović N., Perc M. 2014. Temporal distribution of recorded magnitudes in Serbia earthquake catalog. Applied Mathematics and Computation 244, 917–924.

Matlab R2017b. Mathworks.

Vasović N., Kostić S., Franović I., Todorović K. 2016. Earthquake nucleation in a stochastic fault model of globally coupled units with interaction delays, Communications in Nonlinear Science and Numerical Simulation, 38, 117–129.

Geomechanics and Geodynamics of Rock Masses – Litvinenko (Ed.)
© *2018 Taylor & Francis Group, London, ISBN 978-1-138-61645-5*

Integrated seismic measurement techniques to determine the velocity distribution near underground drifts at the Sanford Underground Research Facility (SURF)

W.M. Roggenthen

Research Science, Department of Geology and Geological Engineering, South Dakota School of Mines and Techonology, Rapid City, SD, USA

ABSTRACT: The Sanford Underground Research Facility (SURF) is located in the central part of the USA in a former deep gold mine. The primary purpose of the facility is to provide a host for physics experiments that require shielding from cosmic radiation, but it also offers an excellent deep testbed for geological and geoengineering experimentation. The facility is expected to be in operation for tens of years and it already has experienced an extended history as a mining operation. Determination of seismic velocities of the rock can provide information regarding the current geomechanical condition of the host rock and also will provide information for the design of geophysical monitoring in support of future hydraulic fracturing experiments. Measurement of seismic velocities on recently acquired core produced reproducible seismic profiles with lower seismic velocities close to a drift that is located 1.5 km below surface and that has been open for over 40 years. These core measurements, however, were acquired on unconfined cores, therefore, a series of seismic surveys to determine *in situ* velocities were undertaken in order to evaluate the effect of stress. These seismic surveys produced inferred seismic velocity profiles that were similar to those derived from core measurements, although with greater seismic velocities due to high rock stress.

Keywords: mining, seismic, rock quality, creep, physics, Sanford

1   UNDERGROUND LABORATORIES

Scientific and engineering basic research using underground laboratories has increased substantially in recent years to support studies in particle physics and waste disposal (Wang, 2010). The Sanford Underground Research Facility (SURF), developed in a repurposed gold mine, is a laboratory (Fig. 1) with a primary purpose of supporting physics experimentation (Heise, 2015). The maximum depth currently available for experimentation is 1.5 km below the surface on what is known as the 4850 Level. This part of the facility was developed nearly 40 years ago as part of the mining operation. Although the primary purpose of SURF is to support the physics community, the facility also is an excellent test-bed for geological, geobiological, and geoengineering studies. Detailed geotechnical engineering work was performed prior to installation of the present suite of experiments (Hladysz and others, 2011) and geological, geoengineering, and geobiological studies are ongoing.

The rock in which the laboratory is developed consists of a series of Proterozoic metasedimentary rocks (upper Poorman Fm.) overlying a thick amphibolite unit (Yates Member). The metasedimentary rocks are complexly folded and the overall fold system is steeply plunging to the south (Caddy and others, 1991). The current study was performed in the rocks of the Poorman Fm. which consists primarily of a graphitic carbonate mica phyllite and is highly anisotropic in that thin, more carbonate-rich layers alternate with more graphitic units.

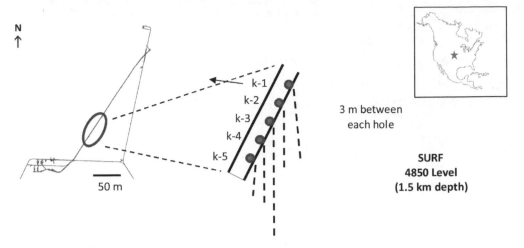

Figure 1. The location of SURF is shown in the insert in the upper right of the diagram. The experiment support area on the 4850 Level (1.5 km below the surface) is the area where the five holes of the kISMET site were drilled. The holes are located in a line along the drift with a separation of three meters. All holes are cased in their top ~3 m, and they are 50 m in depth with the exception of k-3 which is 100 m deep. Holes k-2, k-3, and k-4 are vertical and holes k-1 and k-5 are slightly slanted so as to form a five-spot pattern at a depth of 50 m.

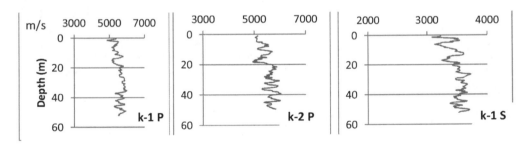

Figure 2. P-wave velocities on core from boreholes k-1 and k-2 and S-wave velocities are from k-1.

2 SEISMIC WORK AT SURF

Although direct measurement of ground stability is a necessity for safety of personnel and equipment, additional information can be gained by the use of seismic techniques. Previous seismic studies during the gold mining era were primarily concerned with location of microseismicity to identify areas of potential ground collapse (e.g. Filigenzi and others, 1995; Girard and others 1995; Friedel and others, 1996). Other, more recent studies initiated after conversion of the mine for scientific purposes include installation of broadband seismometers in the underground (Acernese and others, 2009) and a tomographic study of the condition of a crucial rock pillar on the 4850 Level (Roggenthen and Berry, 2016).

The kISMET project (Permeability [k] and Induced Seismicity Management for Energy Technologies) is a program to evaluate the ability to develop designed fractures, establish the stress conditions, and evaluate methods of geophysical monitoring during after the fracturing process (Oldenburg and others, 2017). Figure 1 shows the distribution of a series of five holes drilled on the 4850 Level (four holes are 50 m deep and k-3 is 100 m deep).

Seismic velocities of cores from k-1 and k-2 from the kISMET site in Figure 1 were determined using a 150 kHz ultrasonic source. Four measurements were made at each sampling point along the cores to determine the maximum velocities to account for core velocity anisotropy. Figure 2 shows the results of the unconfined measurements of maximum P and S velocities from k-1 measured previously (Roggenthen, 2017). The additional P-wave velocities from k-2 show a similar pattern as that determined previously in k-1 in that they both have a slower upper section and an increase in velocities toward the bottom of the hole. The finer scale structure of the velocity profile shows other important features, such as the velocity decrease at the top of the hole, which may be due to long-term, near-drift damage. The S-wave pattern is also shown for each of these holes and, in general, does not have as much variation down the hole.

3 IN SITU SEISMIC STUDY

A series of vertical seismic surveys were performed in borehole k-2 (Fig. 1) to provide constraints on the acoustic velocity profile downwards into the drift under confined stress conditions in the vicinity of the kISMET site on the SURF 4850 Level. The purpose of the study was to provide information regarding the integrity of the rock near the drift, which has been open for more than 40 years, and to provide estimates of the bulk acoustic velocities that will assist in the design of seismic monitoring during future hydraulic fracturing experiments currently being planned. The k-2 hole was chosen because it is a straight borehole as opposed to kISMET-001 and 005 which have a slight deviation from vertical. A three-component geophone with spring clips to force the geophone against the uncased portion of the borehole wall was used in k-2 beginning at a depth of 40 m. A large hammer was used to produce the seismic signal and surface offsets between the source and the collar location of k-2 were 30.5 m (100 ft), 22.9 m (75 ft), 15.2 m (50 ft), 7.6 m (25 ft), and 3.5 m (10 ft). The geophone was moved up the borehole at 1 m intervals for each of the source offsets. A Bison 24-channel seismograph was used to the record the signal with a 125 msec sampling interval. Because the rock velocity is high and the offsets small, it was necessary to use a reference geophone at the source measure the difference in time between the hammer strike and arrival times at the geophone station.

The geophone response is shown in Fig. 3 for the vertical geophone (Fig. 3A) and one of the horizontal geophones (Fig. 3B). Time is plotted on the vertical axis and the geophone stations in the boreholes are shown on the horizontal axis. The P and S arrivals are clearly identifiable in the panels of Figure 3. The prominent tube wave (T) in the vertical panel but not the horizontal geophone response panel arises due to a reflection from the bottom of the hole.

For each of the geophone stations a slant velocity was calculated, which represents the amount of time for a wave to travel from the source to the downhole geophone. This wave traverses varying distance through materials with varying velocities. By making the

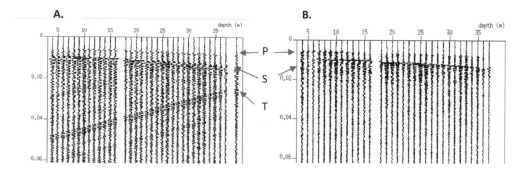

Figure 3. A and B showing the vertical and h2 (horizontal) geophone responses. The P and S arrivals are clearly indicated along with a prominent tube wave (T) that occurs due to a reflection from the bottom of the hole. Geophone depths in the borehole are shown on the horizontal axis and time (s) is on the vertical axis.

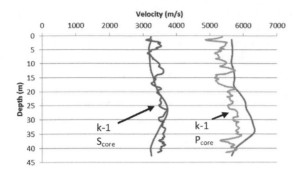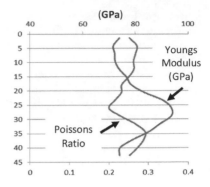

Figure 4. P and S velocity profiles from k-2 produced by a seismic survey with an offset of 30.5 m (smooth curve). The core velocity measurements from k-1 are plotted for comparison.

assumption of lateral homogeneity, it was possible to invert the slant times using a tomographic software package which yielded estimates of velocities with 2.9 m cell sizes. The assumption of lateral homogeneity may or may not be justifiable although it was possible to test it by establishing additional shot points at varying distances from the borehole. The resulting inferred vertical profile for both the P and S velocities along with the measurements of velocities on the core from k-1 is shown in Figure 4 for a source offset of 30.5 m.

Although in general the P and S velocities from the seismic surveys increase toward the bottom of k-2 in Figure 4, the location of the highest velocities for the P and S-waves, as well as the velocity gradients, are somewhat different between the two seismic phases. Based upon the inferred velocities in Figure 4 (smooth curves), it is possible to calculate Poissons Ratio and Young's Modulus, and those values are shown in Figure 4.

4 DISCUSSION

Comparison between the core velocities and the *in situ* velocities shows that a marked difference in P-wave velocity exists between the core measurements and the inferred velocities using a downhole geophone. S-wave arrivals were difficult to distinguish for near offsets. This means that usable S-wave velocities could only be found for longer offsets which resulted in averaging larger rock volumes. The *in situ* seismic velocities are higher, as would be expected, because they were derived from rocks that are under elevated confining stresses, the velocities are well within the values expected for these types of rocks. The near-drift seismic survey velocities are lower as are the core measurements, although the core measurements show much more small-scale variations. The lower velocities may be due in part to the damaged zone associated with the drift due to the long time that the drift has been open. Previous studies (Roggenthen, 2017) documented a decrease in the core velocities extending to a distance of ~2.5 m from the drift although the lower velocities in Figure 4 extend much further. This type of decrease in velocity also has been shown to occur in seismic tomography studies of pillar areas on the 4850 Level, as well (Roggenthen and Berry, 2016), although these studies were conducted in the amphibolite in the lower part of the section and not the metasedimentary rocks currently being considered. The S-wave velocities inferred from the *in situ* seismic survey do not differ significantly from velocities derived from core measurements although the velocities from core show more small-scale variation.

Measurements were also conducted at offset distances of 3.5 m, 7.6 m, 15.2 m and 22.9 m. Although a general pattern of increasing velocities with distance away from the bottom of the drift was noted, the resulting velocities were locally variable with smoother profiles at longer offsets, presumably due to averaging of the local velocity variations.

5 CONCLUSIONS

The use of the borehole seismic survey appears to be a method to determine seismic properties under stressed conditions. Useful geomechanical properties can then be calculated using the velocities determined in this manner. This method also allows an estimate of velocity to be gained even though the top part of hole is cased. Both core measurements and direct methods of velocity determinations, such as downhole logging, are constrained in that they only investigate close to the borehole, but the seismic surveys used here can provide additional information regarding the distribution of velocities away from the immediate vicinity of the borehole. Whereas velocity measurements on core velocity measurements are labor intensive and downhole logging may involve considerable expense, this downhole seismic survey technique is easily accomplished and cost effective.

ACKNOWLEDGEMENTS

The research supporting this work took place in part at the Sanford Underground Research Facility in Lead, South Dakota. Funding for this work is supported by Subcontract 7257287 from Lawrence Berkeley National Laboratory under the Dept. of Energy Prime Contract No. DE-AC02-05 CH1 1231. The help of Sterling Richard during the data acquisition was appreciated. The assistance of the Sanford Underground Research Facility and its personnel in providing physical access and general logistical and technical support is acknowledged.

REFERENCES

Acernese, Fausto, Rosario De Rosa, Riccardo De Salvo, Gerardo Giordano, Jan Harms, Vuk Mandic, Angelo Sajeva, Thomas Trancynger, and Fabrizio Barone, 2009, Long term seismic noise acquisition and analysis in the Homestake mine with tunable monolithic sensors, Proc. of SPIE Vol. 7478 74782 K-1, 8 p.

Caddey, S.W., Bachman, R.L., Campbell, T.J., Reid, R.R., and Otto, R.P., 1991. The Homestake Gold Mine, an Early Proterozoic Iron-Formation-Hosted Gold Deposit, Lawrence County, South Dakota. U.S. Geological Survey Bulletin 1857-J, Geology and Resources of Gold in the United States, 67 p.

Filigenzi, M.T. and J.M. Girard, 1995. Seismic Studies and Numerical Modeling at the Homestake Mine, Lead, SD, in Proceedings: Mechanics and Mitigation of Violent Failure in Coal and Hard-rock Mines, Special Pub. 01–95, U.S. Bureau of Mines, 347–355.

Friedel, M.J., Scott, D.F., Jackson, M.J., Williams, T.J., and Killen, S.M., 1996. 3-D tomographic imaging of anomalous stress conditions in a deep US gold mine: J Appl Geophys, 36: 1–17.

Girard, J.M., T.J. McMahon, W. Blake, and T.J. Williams, 1995. Installation of PC-based Seismic Monitoring Systems with Examples from the Homestake, Sunshine, and Lucky Friday Mines, in Proceedings: Mechanics and Mitigation of Violent Failure in Coal and Hard-rock Mines, Special Pub. 01–95, U.S. Bureau of Mines, 303–312.

Heise, J., 2015, The Sanford Underground Research Facility at Homestake, arXiv:1503.01112 [physics. ins-det].

Hladysz, Z.J., Callahan, G.D., Popielak, R., Weinig, W., Randolph-Loar, C., Pariseau, and W.G., Roggenthen, W., 2011, Site investigations and geotechnical assessment for the construction of the deep underground science and engineering laboratory, T I Min. Metall. A., 330: 526–542.

Oldenburg, C.M., Dobson, P.F., Wu, Y., Cook, P.J., Kneafsey, T.J., Nakagawa, S., Ulrich, C., Siler, D.L., Guglielmi, Y., Ajo-Franklin, J., Rutqvist, J., Daley, T.M., Birkholzer, J.T., Wang, H.F., Lord, N.E., Haimson, B.C., Sone, H., Vigilante, P., Roggenthen, W.M., Doe, T.W., Lee, M.Y., Ingraham, M., Huang, H., Mattson, E.D., Zhou, J., Johnson, T.J., Zoback, M.D., Morris, J.P., White, J.A., Johnson, P.A., Coblentz, D.D., Heise, J., 2017, Overview of the kISMET project on intermediate-scale hydraulic fracturing in a deep mine (ARMA 17–780), presented at the 51st US Rock Mechanics/ Geomechanics Symposium held in San Francisco, California, USA, 25–28 June 2017.

Roggenthen, W.M., 2017, Geophysical and Geological Characterization of Core Materials in Support of the kISMET Experiment at the Sanford Underground Research Facility (SURF) (ARMA 17-0896), presented at the 51st US Rock Mechanics/Geomechanics Symposium held in San Francisco, California, USA, 25–28 June 2017.

Roggenthen, W.M. and K.M. Berry, 2016, Acoustic Velocities and Pillar Monitoring On the 4850 Level of the Sanford Underground Research Facility, 50th US Rock Mechanics/Geomechanics Symposium held in Houston, Texas, USA, 26–29 June 2016, ARMA 16-0411.

Wang, J.S.Y., P.H. Smeallie, X.-T. Feng, J.A. Hudson, 2010, Evaluation of underground research laboratories for formulation of interdisciplinary global networks, International Society of Rock Mechanics and Imperial College, London, U.K. ISRM International Symposium 2010 and 6th Asian Rock Mechanics Symposium—Advances in Rock Engineering, 23–27 October, 2010, New Delhi, India.

Geomechanics and Geodynamics of Rock Masses – Litvinenko (Ed.)
© 2018 Taylor & Francis Group, London, ISBN 978-1-138-61645-5

Geophysical monitoring as an inherent part of the technological process in deep open pits

Vadim V. Rybin, Viktor I. Panin, Mikhail M. Kagan & Konstantin N. Konstantinov
Mining Institute of the Kola Science Centre of the Russian Academy of Sciences, Mining Institute KSC RAS, Apatity, Russia

ABSTRACT: One of the most important tasks in deep open mining is to provide for the slope stability, which requires an operative diagnostics of its conditions. Such diagnostics can be performed by geophysical methods only which allow obtaining objective information about the properties and parameters of the rock mass state through discrete and continuous in time observations.

The evolution of geomechanical processes in open pit walls with open mining progress can be traced through a microseismic method. The operational control of actual wall sectors (groups of benches) is performed by ultrasonic logging, well logging and seismic tomography methods.

All of the above methods have been successfully applied for many years by the Mining Institute KSC RAS in the open pits of the Kola Peninsula; in particular at the Zhelezny open pit mine, JSC Kovdorsky mining-processing enterprise (Kovdorsky GOK), and Khibiny apatite mines.

1 INTRODUCTION

The construction of deep open pit walls with the use of high benches on the ultimate contour requires improving a technology and organizing works, developing advanced geomechanical approaches for theoretical verification of mining-engineering structures stability, as well as managing special monitoring systems for rock mass, including open pits. In the context of hard rock conditions, new geomechanical approaches are based on the representation of a rock mass as a hierarchically-blocked medium under the influence of gravitational-tectonic stress field (Kozyrev, 2002, Kasparian, 2006).

Countries with a developed mining industry pay considerable attention to monitoring of slope stability, job security and elimination of the consequences of instability of individual benches (Becerra Abregu, 2013, Escobar, 2013). At that, slope stability monitoring is carried out mainly with the use of various modern radar scanning technologies that allow controlling the displacement of sectors of the pit contour in real time. In some cases the control of the geomechanical processes occurring in a rock mass outside an open pit is carried out using microseismic methods (Lynch, 2004). In general, such a combination of radar scanning technologies and microseismic methods allows obtaining a fairly complete picture of the geomechanical processes occurring in the deep rock mass and on the pit boundaries. To the date, Russian specialists have had extensive best practices in the experimental definitions of a geomechanical situation in ore open pits of the Murmansk region (Kozyrev, 2013a), the most studied of which is the open pit of the Zhelezny mine.

The application of steep dip angles for the high benches, up to vertical ones and, respectively, increased general angles of slopes, requires compulsory continuous monitoring of stability to provide for secure mining operations. To control stability, a set of experimental methods is used such as a relief method (Turchaninov, 1970), seismic and ultrasonic methods, a method of television monitoring for wells (Kozyrev, 2012), etc.

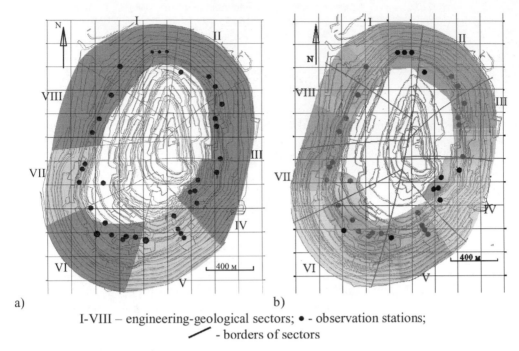

I-VIII – engineering-geological sectors; ● - observation stations;
╱ - borders of sectors

Figure 1.   Zoning the Zhelezny mine in terms of thickness of disturbed zone (a) and level of operating stresses (b).

2   ZONING OF AN OPEN PIT FIELD BASED ON THE RESULTS OF IN-SITU STUDIES

Taking into account great significance of adequate determination of slope angles in the ultimate position, Mining Institute KSC RAS had started in the early 1990s the research in this field, considering the actual stress-strain state of the host rock mass and excavation damage zone parameters in the rock mass surrounding the open pit (Kozyrev et al., 2001, Melnikov et al., 2004). To the date, a great number of experimental detections of a geomechanical setting in an open pit have been carried in the Zhelezny mine. This made it possible to carry out zoning of the open pit field in accordance with various factors affecting stability, in particular, in terms of the level of operating stresses and the thickness of a disturbed zone (Fig. 1).

On Figure 1 the green color shows the most favorable zones from the point of view of the factors under consideration, the yellow—less favorable, and the pale-red—unfavorable zones. The northern section of the pit wall appears to be quite favorable both from the point of view of the stress state and thickness of the disturbed zone, which is confirmed by the investigation results of 2017. Also, the rock mass that forms the deep mining levels at the pit wall's eastern part, in accordance with the data of several observation stations within engineering-geological sector IV, seems more favorable from the point of view of operating stresses than in the rock mass above the zero mark.

3   GEOPHYSICAL MONITORING OF STABILITY OF GROUP OF BENCHES

The increased probability of damage to stability of the benches and individual sections of the pit wall, as can be seen from Figure 1b, is forecasted on the eastern section of the pit wall at the levels above the zero mark from the day surface (Rybin, 2015). This is confirmed by the microseismic monitoring data organized on the eastern section of the Zhelezny mine open pit (Kagan, 2010), where continuous seismic activity has been registered since 2007 (Kozyrev, 2013b).

Seismic monitoring revealed a seismically active zone in the depth of the rock mass near the eastern section of the open pit. The zone reflects the formation of parallel adjacent cracks forming a weakening plane in 2010–2012, which, in terms of the parameters of occurrence relative to the open pit's final boundaries, may be of a certain danger for the stability of a group of benches. On Figure 2, the green intermittent line shows seismic events in 2010, the brown line—seismic events in 2012, and the blue line indicates a weakening zone according to the geological mine survey. A series of red lines detects the contours of faults after exploratory wells drilling.

This set of factors became the basis for the construction of a stationary geophysical testing site, where periodic seismic tomography measurements were carried out (Fig. 3) for three years (Rybin, 2017). The geophysical testing site was constructed in 2014 and consists of profiles of excitation and receiving of elastic fluctuations on level +94 m and +10 m, correspondently. In order to quickly obtain the dynamic elastic characteristics of the near-boundary rock mass in the vicinity of an ore crushing-conveyor plant on level +120 m, six measurement cycles were performed.

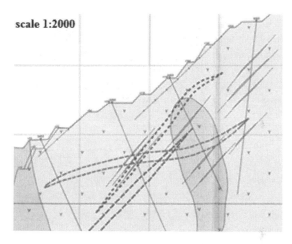

scale 1:2000

Figure 2. Seismic events localization zones. A fragment of a geological section of the Zhelezny open pit mine.

+94м

+10 м

Figure 3. Geophysical testing site. The eastern sector of the Zhelezny open pit mine (blue rectangle).

4 ANALYSIS OF THE STUDY RESULTS

An analysis of the seismic observation results has allowed the authors to trace dynamics of changes in the propagation velocities of longitudinal and transverse waves V_p and V_s in the rock mass and make an conclusion about general tendency of their decrease. These contrast changes are associated with both drainage-watering of the rock mass area under control and with increase of rocks fracturing due to mining-induced impact.

The Poisson's ratio μ distribution was calculated on the basis of the elastic waves' velocities measured, and the value is of interest within the controlled zone. Higher values of the ratio characterize the degree of weakening of structural bonds and increase of fracturing in the rock mass. Figure 4 shows the distributions of the Poisson's ratio values. Figure 4 demonstrates dynamics of changes of μ between the 5th and 6th cycles.

Comparison of data in Figures 3a and 3b shows more contrasting changes in the Poisson's ratio values between the 5th and 6th measurement cycles, and this fact indicates a high sensitivity of this parameter to dynamics of the controlled area.

It can be seen that in some areas there was a significant increase in this parameter, an average of 0.2 (Fig. 5). However, these planes are insignificant in area. The distributions have an asymmetry toward higher values, which characterizes the entire rock mass as fractured. In addition, the location of the geophysical testing site in the plan and comparison with the results of the Zhelezny mine open pit zoning in terms of the stress level indicates the unfavorable geomechanical state of this section of the pit wall (Figs. 1b & 6). Nowadays, the operations are being performed to make this rock mass sector safe.

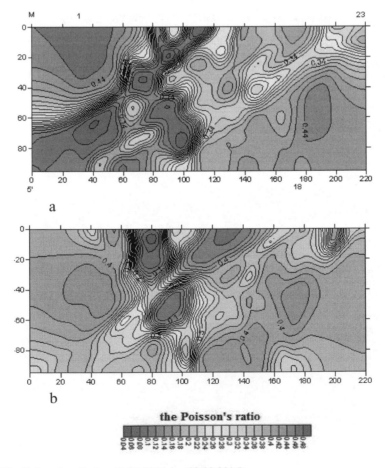

Figure 4.   The Poisson's ratio (a – 11.07.2016, b – 29.09.2016).

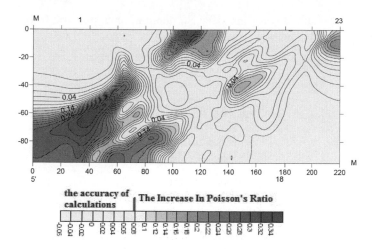

Figure 5. The Poisson's ratio dynamics, 29.09.2016–11.07.2016.

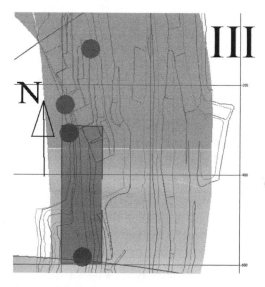

Figure 6. A fragment of mining plan and location of the geophysical testing site (blue rectangle).

5 CONCLUSION

1. As a result of zoning of the Zhelezny open pit mine, according to in-situ studies, the northern open pit sector seems to be quite favorable both from the point of view of the stress state and thickness of a disturbed zone. In addition, the geomechanical state of the rock mass in the deep levels within engineering-geological sector IV appears to be more favorable than in the rock mass of a level above the zero mark.
2. Seismic observation analysis within the geophysical testing site indicates unfavorable geomechanical setting of this pit wall sector.
3. The developed sequence of geophysical monitoring for critical sectors of the open pit walls is implemented on the basis of the principle "from general to particular" and reflects the hierarchical-blocked structure of the rock mass. The geophysical monitoring organization approach consists in continuous support of ongoing mining operations and can be applied to other similar mining industrial facilities, which in turn will allow more justified

approach to the issues of ensuring safety in mining operations and to open and underground mining technologies.

REFERENCES

[1] Kozyrev A. A., Reshetnyak S. P., Maltsev V. A., Rybin V. V. 2001. //Analysis of Stability Loss in Open-Pit Slopes and Assessment Principles for Hard, Tectonically Stressed Rock Masses, Slope stability in Surface Mining/edited by William A. Hustrulid, Michael K. McCarter, Dirk J.A. Van Zyl. – Littleton, Colorado, USA. –, 442 p. Chapter 27 in Section 2, pp. 251–256.

[2] Becerra Abregu G.A., Valencia Jeri E., Garcia E., Yuan P., Byers T. Case History: Deep-Seated Slope Failure in Weak Rocks, El Tapado Pit North Wall, Yanacocha Operation // Slope Stability 2013. Proceedings of the 2013 International Symposium on Slope Stability in Open Pit Mining and Civil Engineering, 25–27 September 2013, Brisbane, Australia, Publ. by Australian Center of Geomechanics – 2013.– P. 409–424.

[3] Escobar A.E.E., Farina P., Leoni L., Iasio C., Coli N. Innovative Use of Slope Monitoring Radar As A Support To Geotechnical Modelling of Slopes in Open Pit Mines // Slope Stability 2013. Proceedings of the 2013 International Symposium on Slope Stability in Open Pit Mining and Civil Engineering, 25–27 September 2013, Brisbane, Australia, Publ. by Australian Center of Geomechanics – 2013.– P. 793–802.

[4] Kagan M.M., Kozyrev A.A. Features of the seismic regime of the rock massif in a near-boundary zone of the open pit, Proceedings of VIII International School-Seminar "Physical Basis for Forecasting the Destruction of Rocks", May 24–29, 2010. – p. 42.

[5] Kasparian E.V., Kozyrev A.A., Iofis M.A., Makarov A.B. Geomechanics. – M.: Higher school. 2006. – 503 p.

[6] Kozyrev A.A., Kagan M.M., Chernobrov D.S. Results related Pit wall Microseismic Monitoring ("Zhelezny" mine, Kovdorsky GOK, JSC). Proceedings of 8th International Symposium on Rockbursts and Seismicity in mines. Perm. Mining Institute of RAS. 2013 – pp. 501–505.

[7] Kozyrev A.A., Panin V.I., Savchenko S.N. et al. // Seismicity in mining operations – Apatity, 2002. – 325 p

[8] Kozyrev, A.A., Rybin, V.V., Konstantinov, K.N. Field-Scale Investigations of the Stress Field and the Excavation Damaged Zone Extent, the Kola Peninsula, Russia // Book of Abstracts of 5th Jubilee Balkan Mining Congress, 18–21 September 2013, Ohrid, Macedonia, 2013, pp. 359–365.

[9] Kozyrev A.A., Rybin V.V., Konstantinov K.N. Evaluation of the geomechanical state of the rock massif in the open pit by a complex of instrumental methods // Mining information-analytical bulletin. – 2012.-No. 10. – pp. 113–119.

[10] Lynch R. Microseismic monitoring of open pit slope stability // Newsletter of Australian Centre for geomechanics, December 2004. – p. 13.

[11] Melnikov N. N., Kozyrev A. A., Reshetnyak S. P., Kasparian E. V., Rybin V. V. // Methodical approach to value of an optimal open pit slope angle in hard rocks, Proceeding of the ISRM International Symposium 3rd ARMS, Ohnishi & Aoki (eds); Millpress, Rotterdam, 2004, pp. 509–513.

[12] Guidance to measuring stresses in the rock massif by unloading (the end measurement method) // KF AS USSR, Mining Institute. Authors: Turchaninov I.A., Markov G.A., Ivanov V.I.—Apatity, 1970. – 48 p.

[13] Rybin V.V., Panin V.I., Konstantinov K.N., Startsev Yu.A. Results of geomechanical monitoring for the rock massif based on the use of geophysical research methods // Mining information-analytical bulletin. – 2015. – No. 11 (special issue 56). – pp. 193–200.

[14] Rybin V.V., Panin V.I., Konstantinov K.N., Startsev Yu.A., Kalyuzhny A.S. Monitoring of migration of water-saturated areas in the open pit wall for assessing its stability // Mining information-analytical bulletin. – 2017. – No. 4. – pp. 184–195.

Geomechanics and Geodynamics of Rock Masses – Litvinenko (Ed.)
© 2018 Taylor & Francis Group, London, ISBN 978-1-138-61645-5

Changes in the time of the properties of rock foundations of large hydraulic structures according to geophysical monitoring data

A.I. Savich, E.A. Gorokhova & M.M. Ilin
Geodynamic Research Centre—Branch of the Hydroproject Institute, JSC, Moscow, Russia

ABSTRACT: Operation of large hydraulic engineering structures such as the Sayano-Shushenskaya HPP (Eastern Siberia) and Inguri HPP (Western Georgia) and other high-pressure hydroelectric power plants causes significant changes in the initial rock properties, that is, the rock properties laid down in the project, as well as changes in the structure of the weakened zone and the discharge zone in comparison with the project. Long-term geophysical monitoring allow observe during significant time intervals, measured in tens of years, the changes in the properties and structure of the enclosing rock massif. Specified impacts are conditioned not only by rock excavation but also by the processes of their operation when due to periodical fluctuations of water level in the reservoir the significant changes of hydrogeological conditions and violent changes of force impacts transferred to the enclosing rock mass occur. At that, different natural and technogenic geodynamic processes development of which may cause hazard phenomena for the structures, such as damage of local parts of the foundation due to large cracks propagation, appearance of the zones of anomalously high seepage, collapse of separate blocks of the rock mass and others, are activated in the surrounding geological environment. Thus, geodynamic processes occurring during construction and operation of high dams result in significant changes of initial engineering and geological structure of enclosing rock mass.

The operation of large hydraulic structures such as Inguri HPP (Georgia), Sayano-Shushenskaya HPP (Eastern Siberia) and other high-head HPPs causes significant changes of initial, i.e. design properties of rock foundations (elastic, strain and strength indices), and changes in the structure of weak zone and zone of de-stressing in comparison with the design. Long-term geophysical investigations at monitoring allow observing the changes in properties of enclosing rock mass during the significant time intervals for the decades under the influence of seasonal fluctuations of the reservoir level and variations of stress-strain state of enclosing mass, water saturation, uplift pressure and factors that activate the modern natural and technogenic geodynamic processes in the near-surface parts of Earth's crust at dam section, and thereby increase the geodynamic risk in the area of the project.

1 INGURI HPP

Arch dam of Inguri HPP with a height of 271.5 m is located in the western part of Georgia near the town of Jvari on the Enguri River.

In the construction of arch dam of Inguri HPP during the investigations, as a result of detailed studies, the engineering-geological model of rock foundation was constructed on the basis of which a model of the stressed state of foundation as well as the models of elastic and strain indices were developed.

Geodynamic monitoring was regularly carried out at dam section of Inguri HPP during 1974–1992. Its purpose was to assess the change in physical and mechanical properties and rock mass condition at the foundation and abutments of a dam during the excavation of various underground workings and construction pit, when placing the concrete in the dam body, in the process of consolidation grouting, etc. as well as the monitoring for dynamics of strain processes of different scales in the mass during the impoundment and operation of the deep-water reservoir of Inguri HPP. For solving the assigned tasks a set of various methods were used at geodynamic monitoring, including engineering-geological, geodetic, geomechanical and geophysical studies (Mastitsky A.K., 1974 г.). Among the geophysical methods, seismic tomography is selected as the main one in the course of which the requirements for maximum identity of repeated measurements are mostly fulfilled, and the obtained data is related well to other types of studies.

Results of monitoring seismic tomography, obtained at the foundation of arch dam of Inguri HPP during the period from 1974 to 1992, proved the high efficiency of these methods in solving the above tasks of geomonitoring and providing the reliability and safety operation of the dam (Savich A.I., 1993 г., 1979 г., 1999 г.). The main results of completed studies are as follows:

- The established time variations of elastic waves velocities have the clear stages confined to certain stages of construction and operation of arch dam and the reservoir. Under the influence of the weight of concrete, placed in the dam body and consolidation grouting, the velocities V_p and V_s in the near-surface parts of the rock mass are increased again, and their values at this stage in many parts of the foundation exceed the initial (natural) level.
- Under the influence of the weight of concrete, placed in the dam body and consolidation grouting, the velocities V_p and V_s in the near-surface parts of the rock mass are increased again, and their values at this stage in many parts of the foundation exceed the initial (natural) level. The reservoir impoundment has led to the significant technogenic changes in the rock mass. Because of the high permeability of the rock foundation due to its fracturing, at the initial stage of the reservoir impoundment, the deconsolidation of the rock mass is occurred under the influence of hydrostatic forces of weighing and fracture seepage forces, which was accompanied by a significant decrease in the velocities of the elastic waves.
- Subsequently, after the full water saturation of the rock mass and its operation in the quasi-elastic regime during the reservoir operation the periodic seasonal fluctuations of velocity were observed in the dam abutments due mainly to the change of arch stresses transferred by the dam on the rock mass during rising and lowering of the reservoir level (Fig. 1). At rapid drawdown of the reservoir in the rock mass in zones of excess fracture pressure a sharp drop of velocity was recorded, caused by processes of deconsolidation and destruction of local parts of the rock mass (a phenomenon similar to hydraulic fracturing);
- At the initial stages of operation the information on the relative changes in the velocity of p waves was received depending on the reservoir level variation.
- Taking into consideration the data of geophysical observations the limit velocity of upstream water level variation (UWL) of the Inguri reservoir, equal to 1 m/day, was set.

After 1992 there was a break in the studies of riverside abutments of Inguri HPP. Some studies were conducted in 1996 and 2004. The monitoring network has been restored in 2016. Investigations are carried out in two stages: at the minimum and maximum UWL. The results of geophysical investigations are presented in the form of tomographic sections in isolines of the velocities of elastic waves (Fig. 2), and then recalculated in the total strain modulus by correlation dependences (Savich A.I., 1985 г.). These sections reflect the change in the velocity structure both during the operation period and during the variation of UWL, i.e. at the reservoir impoundment and drawdown. It should also be noted that the result of processing of the obtained data is not only seismic tomograms of the values of P-wave velocities, but also S-wave velocities. Thus, the complete analysis of the occurring changes is possible, i.e. it is possible to divide the compression and water saturation processes of the rock mass, because the S-wave velocities do not react to changes in the water saturation, however at increasing the rock mass compression the velocities are increased.

Subsequently, the cumulative curves for the values of p and s wave velocities are plotted and tables of average values are made. Hereinafter nine possible processes occurring in the rock mass were formulated. In this case the changes in s wave velocity are related to con-

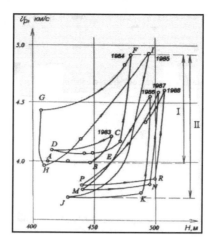

Figure 1. Changes in the velocity of elastic waves at the foundation of arch dam of Inguri HPP depending on the upstream water level (UWL). I - variations at normal discharge, II - variations at rapid and drawdown of large volume of water.

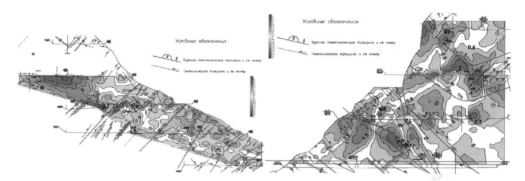

Figure 2. Tomographic sections along the axis of grouting galleries of left-bank and right-bank abutments of the dam, reflecting the change in the velocities of longitudinal elastic waves at UWL rise from el. 426 m up to el. 510 m.

solidation and deconsolidation of the rock mass (change in its fracturing), while the degree of water saturation of the rock mass also has a significant effect on the change in p wave velocity. The results show that the riverside abutments react considerably in different ways to change in the stress-strain state at UWL rise. So in the left-bank abutment the deconsolidation processes are dominated. At the same time, the consolidation process is observed in the whole right-bank abutment.

Analysis of the data on the change in the velocity structure in time from the construction commencement and for the operation period until 2016 shows that the largest changes of average values of p wave velocities both at low and high values of UWL are occurred for the period from 1974 to 2004–06 (Fig. 3). Moreover, in both abutments there was an increase in p wave velocity, which was 20% for the left-bank abutment and more than 40% for the right-bank abutment. The given data indicates to the compression (deconsolidation) of enclosing rock mass. However, for the period from 2004–06 to 2016 at high elevations of UWL a decrease in p wave velocity was observed in the right-bank abutment, it was 4%. At low elevations of UWL the decrease in p wave velocity was recorded in both abutments, which was about 1% for the left-bank abutment and more than 9% for right-bank abutment, reflecting a relative deconsolidation of enclosing rock mass.

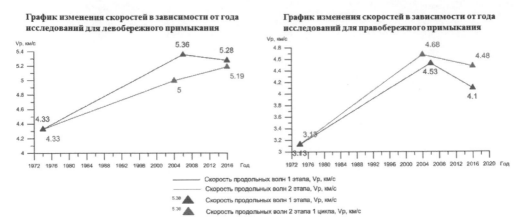

Figure 3. Curves of velocity changes depending on the year of observations for the left-bank and the right-bank abutments.

2 SAYANO-SHUSHENSKAYA HPP

Sayano-Shushenskaya HPP is the arch-gravity dam with a height of 242 m located on the Yenisei River. The physical prerequisite for the system development is the experimentally adopted changes in the indices of the physical and mechanical properties of the rocks including the velocities of elastic waves at the foundation and the riverside abutments of the dam as well as the experimental correlation dependences between p wave velocity and strength indices of enclosing rock mass: compression strength, cohesion and friction angle, obtained especially for the rock mass of Sayano-Shushenskaya HPP (Savich A.I., 1985 г.).

At Sayano-Shushenskaya HPP for the purposes of providing the safety operation of the project along with the monitoring of control parameters of the dam condition the system of geophysical observations was established for the indices of elastic, strain and strength properties of the rock foundation and the riverside abutments of the dam.

The rock mass, which is the dam foundation of Sayano-Shushenskaya HPP, is composed of metamorphic rocks of the Upper Proterozoic age, represented by ortho- and paraschist (Badukhin V.N., 1986 г.). The main factors that determine the variability of the properties of the rock mass are fracturing and tectonic disturbances as well as the processes of weathering and near-surface de-stressing. Under the influence of these factors the rock mass at the dam foundation and its abutments is divided into the number of geostructural blocks and zones that differ significantly in their physical and mechanical properties.

The character of the spatial variability of the properties of the test mass is visually represented by its 3D geomechanical model described in the document (Savich A.I., 2013 г.) (Fig. 4). The main feature of this model is distinguishing the rock layers of different rock quality, the thickness of which is assumed to be the same at the section of dam foundation in the downstream.

During the structure operation the rock mass at its foundation was experienced to various natural and technogenic factors, including additional variable technogenic loads and changes in the seepage regime, caused by seasonal fluctuations of UWL. As the result of these factors the rocks composing the rock mass underwent certain changes. In general, detailed investigations of the dam foundation and abutments confirmed the features of the above mentioned general geomechanical schematization are revealed a number of additional objective laws in the structure of the enclosing mass and variability of its physical and mechanical properties.

Data analysis obtained in 2009–2016 at the dam foundation shows that with some conventionality in terms of the absolute level of velocities Vp^c and the intensity of changes in this parameter, the modern rock foundation of the dam and power house can be divided into three subzones (Fig. 5):

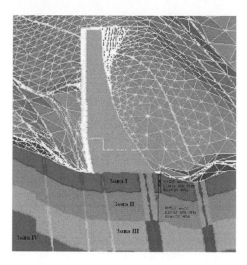

Figure 4.   Fragment of 3D geomechanical model, reflecting the initial zoning structure of the dam foundation.

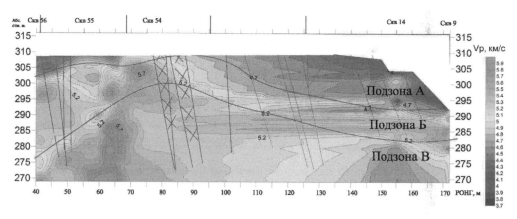

Figure 5.   Section in the isolines of p wave velocity reflecting the zoning structure of the dam foundation at the Section 18.

- «А» - is a near-surface weak subzone, characterized by low values Vp^c <5.2 km/s and too largely subjected to temporary changes;
- «Б» - is a transition (relatively intact) subzone with the average level values Vp^c = 5.0–5.6 km/s, where the occurring changes are less expressed;
- «В» - subzone with the values Vp^c > 5.5 km/s, where the changes in the properties of the rock mass are practically not shown.

It is significant that all these subzones are located within the upper and lower horizons of the zone of de-stressing singled out at the design stage (Savich A.I., 2015).

Thicknesses of «А» and «Б» subzones are not constant, they change as they are becoming further from the pressure face towards downstream. At the 1-st and 2-nd boundaries the base of the specified subzones is located at el. 297 m and 284 m respectively. At the 3-rd and 4-th boundaries of the base А and Б the boundaries of subzones rise to el. 309–310 m and 299–300 m. At the foundation of Unit No. 2 (ГА2) the boundary of subzone А is lowered to el. 282–285 m, the boundary between subzones Б and В is not singled out because of the insufficient degree of sophistication. Thus, it was established that the minimum thickness of indi-

561

cated subzones is reached at 3-rd and 4-th boundaries, and at the sections of 1-st boundary and, especially, at the section of power house the thickness of weak zones increases.

The similar nature of the velocity changes and the structure of indicated subzones show a change in time of the stress-strain state of the enclosing rock mass, which causes the consolidation of near-surface parts at the dam foundation during its operation, and the relative deconsolidation of the rock mass at the sections of upstream face of the dam as well as at the foundation of power house. According to the large number of subhorizontal cracks isolated by a complex of geophysical methods, this zone was formed as a result of partial de-stressing of the upper part of the rock mass at the foundation of power house under the influence of alternating seasonal dam impact in combination with the seepage processes determined by the downstream conditions.

3 CONCLUSION

High dams together with deep-water reservoirs impact significantly on enclosing rock masses in the zone of impact of these structures. Specified impacts are conditioned not only by rock excavation but also by the processes of their operation when due to periodical fluctuations of water level in the reservoir the significant changes of hydrogeological conditions and violent changes of force impacts transferred to the enclosing rock mass occur. At that, different natural and technogenic geodynamic processes development of which may cause hazard phenomena for the structures, such as damage of local parts of the foundation due to large cracks propagation, appearance of the zones of anomalously high seepage, collapse of separate blocks of the rock mass and others, are activated in the surrounding geological environment. Thus, geodynamic processes occurring during construction and operation of high dams result in significant changes of initial engineering and geological structure of enclosing rock mass.

REFERENCES

Badukhin V.N., Lutsevich O.A., Pirogov I.A. Sayano-Shushenskaya HPP dam on the Yenisei River// Geology and dams, Volume X, Energoatomizdat, Moscow, 1986, p. 6–42.

Mastitsky A.K., Dghigauri G.M., Kereselidze S.B. Engineering and geological peculiarities of the foundation of the Inguri arch dam//Hydraulic engineering. 1974. No. 6, p. 3–6.

Savich A.I., Basova A.B., Gorokhova E.A., Ilyin M.M. Existing conditions of the rock mass in the foundation of the Sayno-Shushenskaya HPP dam//Hydraulic engineering. 2015. No. 12, p. 2–12.

Savich A.I., Bronshtein V.I., Groshev M.E., Gaziev E.G., Ilyin M.M., Rechitsky V.I., Rechitsky V.V. Static and dynamic behavior of the Sayano-Shushenskaya HPP arch dam//Hydraulic engineering. 2013. No. 3, p. 2–13.

Savich A.I., Bronshtein V.I., Rechitsky V.I., Ilyin M.M. Problems of geodynamic safety of rock foundation and host rock mass of major power projects of Russia//International congress on rock mechanics, Paris, 1999, p. 465–470.

Savich A.I., Ilyin M.M., Elkin V.P., Rechitsky V.I., Basova A.B. Engineering-geological and geomechanical models of the rock mass in the foundation of the Sayno-Shushenskaya HPP dam// Hydraulic engineering. 2013. No. 1, p. 16–29.

Savich A.I., Ilyin M.M., Kozlov O.V. Different scale elastic wave velocity variation under technogenic impact//Scale effects in rock masses 93, ISRM, 1993.

Savich A.I., Koptev V.I., Ilyin M.M. Distinguishing features of deformation development in the foundation of the Inguri arch dam//ISRM, Switzerland, 1979, p. 585–588.

Savich A.I., Kuyundzhich B.D. Recommendations on application of engineering geophysics for studying strain characteristics of rock. Moscow-Belgrade 1985.

Geomechanics and Geodynamics of Rock Masses – Litvinenko (Ed.)
© 2018 Taylor & Francis Group, London, ISBN 978-1-138-61645-5

Recovery of in-situ orientation of drilled rock core samples for crustal stress measurements

Tatsuhiro Sugimoto
Kyoto University, Kyoto, Japan

Yuhji Yamamoto
Kochi University, Nankoku, Kochi, Japan

Weiren Lin
Kyoto University, Kyoto, Japan

Yuzuru Yamamoto
JAMSTEC, Yokohama-city, Kanagawa, Japan

Takehiro Hirose
JAMSTEC, Nankoku, Kochi, Japan

Nana Kamiya
Kyoto University, Kyoto, Japan

ABSTRACT: In many cases, the information about the in-situ orientation of rock core samples drilled from underground is not available although the core orientation is valuable for a wide range of geological researches (e.g. the determination of in-situ principal stress directions or the analyses of geological structures). In this paper, a method for recovering the in-situ orientation of drilled whole-round core samples is presented that is based on natural remanent magnetization of rocks. This method has been applied successfully to previous scientific drilling projects. As a case study, we applied this method to the 15 oceanic sedimentary rock samples collected at the toe of the Nankai trough, SW Japan, during International Ocean Discovery Program (IODP) Expedition 370. In this study, nine whole-round core samples were successfully reoriented based on two criteria adopted in this paper. On the other hands, six whole-round core samples were not able to be reoriented due to magnetic overprints associated with the drilling operations.

1 INTRODUCTION

The accurate information about the in-situ orientation of drilled whole-round core samples is essential for determining the in-situ principal stress directions from rock cores and for analyses of geological structures. There have been several methods to recover the orientation of the drilled whole-round core samples. They include the use of correlations between measurements of planar structures in a drilled whole-round core sample and those observed in spatially referenced images of borehole walls (Shigematsu et al., 2014) and the use of a mark scribed in a drilled whole-round core sample along its length within an oriented core barrel (Nelson et al., 1987). However, the first method can be only applicable when the planar structures are found in drilled whole-round core samples and clear borehole images are obtained. The second method is not always accurate owing to a number of potential mechanical errors during drilling operations.

In this study, we present another reorientation method which uses magnetization of drilled whole-round core samples. As a case study, we applied this method to whole-round core samples retrieved from the Nankai trough off Muroto, SW Japan. This method doesn't need any characteristic structures or scribed marks in drilled whole-round core samples. This method has been applied successfully to previous scientific drilling projects (e.g. Byrne et al., 2009; Yamamoto, et al., 2013).

2 CORE REORIENTATION BY REMANENT MAGNETIZATION

Rock core samples acquire natural remanent magnetizations (NRMs) which are parallel to the direction of the past geomagnetic field when the rocks were formed. In many cases of the sediments, their NRM directions are considered to correspond to geographic north and can be used to reconstruct the in-situ orientation of the samples. These NRMs, however, have been usually affected by later variations of the ambient magnetic field since the rock formations and have often resulted in partial acquisition of overprint magnetization. This magnetization, namely the secondary component of NRM, obscure the primary component of NRM. In order to reorient rock cores, the primary component which has been masked by the secondary component have to be determined. Generally, the secondary component is magnetically weaker than the primary component and can be eliminated selectively by demagnetization process. In this study, progressive alternating field demagnetization (AFD) was conducted on each sample by 80 mT to extract a primary component. We were able to determine the primary component of NRM direction of each specimen by applying the principal component analysis (Kirschvink, 1980) or the great circle analysis (McFadden et al., 1988) to the demagnetization results of NRM.

15 whole-round core samples were collected from Site C0023 (located at 32° 22. 00′ N, 134° 57. 98′ E, at the toe of the Nankai accretionary prism off Muroto, Japan) during IODP (International Ocean Discovery Program) Expedition 370. They are the Miocene—Quaternary sedimentary soft rocks with 40–50% porosity (collected from 429 mbsf to 1122 mbsf). For each whole-round core sample we fisrt marked a reference line and a disc-shaped sub-sample with a thickness of ~2 cm was cut from the whole-round core sample. The sub-sample was then further cut into nine specimens (Figure 1). To avoid drying of the specimens, they were subsequently wrapped by a parafilm.

3 RESULTS AND DISCUSSIONS

Nine whole-round core samples (ASR-1, 2, 3, 5, 7, 8, 11, 12, and 15) out of 15 samples were reoriented successfully. The declination and inclinations determined in this study are listed in

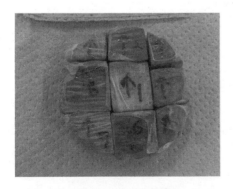

Figure 1. Nine specimens prepared from a parent sub-sample for demagnetization process. Red allows show the direction of the reference line of the whole-round core sample.

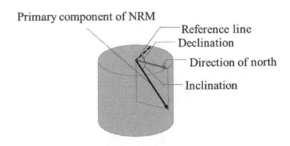

Primary component of NRM
Reference line
Declination
Direction of north
Inclination
Whole round core sample

Figure 2. The definitions of inclination, declination and reference line.

Table 1. The definitions of declination and inclination are shown in Figure 2. Reference lines were drawn arbitrary position in each whole-round core sample.

We adopted two criteria for the determination of the primary NRM directions. First, we checked if we can find a linear segment in a demagnetization result. Figure 3a is a Zijderveld diagram for the demagnetization result of a specimen from ASR–1. Solid black circles are projections of the end points of the NRM vector onto the horizontal plane and open circles are those onto a vertical plane oriented north-south. Gray circles (open and solid) are the secondary components of the NRM vector. Primary component of the NRM is determined successfully if the straight line through those points crosses the origin of the coordinate axes. The second criterion is the value of α_{95}, which is a confidence limit for the mean primary NRM direction

Table 1. The determined values of inclination and declination. Number of samples means the number of subsamples used for calculation of NRM mean direction of each sample.

	Depth (mbsf)	Number of samples	Inclination (°)	Declination (°)	A95 (°)
ASR-1	429.01	7	54.4	286.7	3.8
ASR-2	475.98	8	55.6	137.8	3.3
ASR-3	491.16	7	45.5	317.1	2.2
ASR-5	654.86	7	63.5	215.2	1.8
ASR-7	718.62	4	63.4	180.5	4.1
ASR-8	748.33	5	56.9	102.2	1.9
ASR-11	846.47	7	56.4	290.7	2.4
ASR-12	892.99	6	−54.4	117.5	3.8
ASR-15	1122.73	7	80.2	71.5	2.5

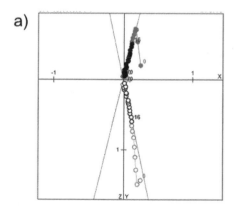

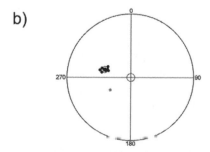

Figure 3. The demagnetization result of ASR-1. a) Zijderveld diagram b) Equal-area projections of the NRM directions for 8 specimens from ASR-1. For this result, $\alpha_{95} = 3.8°$.

calculated from all specimens for each sub-sample. For a directional data set obtained from each specimen, α_{95} means the angle within which the unknown true mean lies at confidence level 95% (Butler, 1992). In this study, $\alpha_{95} < 5°$ was adopted as a criterion. Figure 2b shows the directional data set of 8 specimens from ASR-1 (one outlier). In this figure, $\alpha_{95} = 3.8°$.

As to the other six whole-round core samples (ASR-4, 6, 9, 10, 13, 14), we were not able to determine the NRM directions due to magnetic overprints associated with the drilling operations.

4 SUMMARY

A core reorientation method using NRM is presented. As a case study, this method was successfully applied to nine whole-round core samples (ASR-1, 2, 3, 5, 7, 8, 11, 12, and 15) out of 15 samples based on the two criteria. Specimens from the other 6 cores (ASR-4, 6, 9, 10, 13 and 14) had heavy magnetic overprints associated with the drilling were failed to be reoriented.

ACKNOWLEDGMENTS

15 whole round rock core samples were provided by IODP Expedition 370. We gratefully acknowledge Verena Heuer, Fumio Inagaki, Yuki Morono and the other Expedition 370 scientists. This study was performed under the cooperative research program of Center for Advanced Marine Core Research (CMCR), Kochi University <16C002>.

REFERENCES

Butler, R.F. (1992). *PALEOMAGNETISM: Magnetic Domains to Geologic Terranes*. Boston: Blackwell Scientific Publications.

Byrne, T.B., Lin, W., Tsutsumi, A., Yamamoto, Y., Lewis, J.C., Kanagawa, K., ... Kimura, G. (2009). Anelastic strain recovery reveals extension across SW Japan subduction zone. *Geophysical Research Letters*, *36*(23), 1–6. http://doi.org/10.1029/2009GL040749.

Kirschvink, J.L. (1980). The least-square line and plane and the analysis of paleomagnetic data. *Geophysical Journal of the Royal Astronomical Society*, *62*, 699–718.

McFadden, P.L., & McElhinny, M. (1988). The combined analysis of remagnetisation circles and direct observation in palaeomagnetism. *Earth and Planetary Science Letters*, *87*, 161–172.

Nelson, R.A., Lenox, L.C., & Ward Jr, B.J. (1987). Oriented core: its use, error, and uncertainty. *American Association of Petroleum Geologists Bulletin*, *71*(4).

Shigematsu, N., Otsubo, M., Fujimoto, K., & Tanaka, N. (2014). Orienting drill core using borehole-wall image correlation analysis. *Journal of Structural Geology*, *67*(PB), 293–299. http://doi.org/10.1016/j.jsg.2014.01.016.

Yamamoto, Y., Lin, W., Oda, H., Byrne, T., & Yamamoto, Y. (2013). Stress states at the subduction input site, Nankai Subduction Zone, using an elastic strain recovery (ASR) data in the basement basalt and overlying sediments. *Tectonophysics*, *600*, 91–98. http://doi.org/10.1016/j.tecto.2013.01.028.

Geomechanics and Geodynamics of Rock Masses – Litvinenko (Ed.)
© 2018 Taylor & Francis Group, London, ISBN 978-1-138-61645-5

Monitoring and analysis of a large mass movement area in clay endangering a motorway in Bavaria, Germany

Lisa Wilfing, Claas Meier & Conrad Boley
Boley Geotechnik, Munich, Germany

Thomas Pfeifer
Autobahndirektion Nordbayern, Bayreuth, Germany

ABSTRACT: The federal motorway A70 in Bavaria crosses a mass movement area of at least 1.5 km length. Several costly reconstruction measures were conducted to remedy the damage, but they could only affect a deceleration of the deformation velocity. Until the relocation of the route is finished, a detailed monitoring program including risk levels acc. to the guidelines of Eurocode 7 (EC7) was installed. Topic of this paper is the interpretation of the complex ground conditions (disintegrated sandstone layer overlying massive clay layer) and corresponding movement process (block spreading). Therefore, measuring results (inclinometer, pore-water pressure and geodetic measurements) were combined with extensive laboratory tests on the clayey material that is decisive for the formation of a sliding surface. Monitoring results shown, that the clay has a high swelling potential and heavy rainfall trigger the movement. Laboratory tests reveal that the clay has distinct inhomogeneity with highly varying soil parameters. Therefore, statitical evaluation has been performed to determine representative input parameters for a slope stability analysis. The stability analysis combines the results of the field and laboratory tests to achieve a deep understanding of the landslide at the motorway A70.

1 INTRODUCTION

The movement at the federal motorway A70 has been noticed in 1991 for the first time. Several costly reconstruction measures have been conducted to remedy the caused damage. However, these measures could only affect a deceleration of the deformation velocity and up to now, the slope is still in movement causing severe damage to the motorway. Due to that, a relocation of the route is inevitable but the completion of road work will take several years. Until then, the safety of the motorway has to be guaranteed on basis of a detailed measuring and monitoring concept that follows the guidelines of the Eurocode 7 for real time surveillance. This includes the determination of thresholds for several risk levels with corresponding instructions.

1.1 *Project overview*

The motorway A70 is situated in the north of Bavaria between Bamberg and Bayreuth, Germany. It runs in east-west direction with a length of about 120 km and construction took place in 1981 with two lanes per direction. The active landslide area, which is near the municipality of Thurnau, is mainly forested and has an extent of about 1.5 km length. The terrain slope in the project area decreases from N/NW to S/SE for about 80 m.

1.2 *Geological and hydrogeological overview*

The project area is situated in the geological area of the *northbavarian platform*. The strata include Triassic to Cretaceous rock types covered by Quaternary sediments dipping with an angle of about 10°.

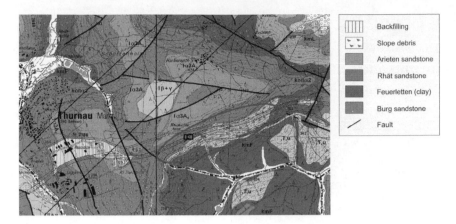

		Backfilling
		Slope debris
		Arieten sandstone
		Rhät sandstone
		Feuerletten (clay)
		Burg sandstone
	/	Fault

Figure 1. Geological overview of the project area (outlined in blue) with the crucial strata of Feuerletten (pink) and Rhätolias (rose). Map extract of the preliminary Geological Map of Bavaria (GK 25, Nr. 5934).

At the project site Thurnau, the Middle Keuper consists of sandstone (*Burgsandstein*) and clay (*Feuerletten*). The Upper Keuper includes sandstone with clay bands (*Rhätolias*). Crucial for the mass movement is the clayey *Feuerletten* and the overlying sandstone layer of *Rhätolias* (Figure 1).

The *Feuerletten* is a very fine-grained material of reddish color. Besides clay, the soft rock includes silty or sandy lenses. Furthermore, some layers are cemented and classified as claystone. This mixture leads to a very heterogeneous rock with varying rock mechanical characteristics. Due to the geological history, the material is overconsolidated with medium strength in an undisturbed condition. When exposed to water, the rock loses its strength quickly which leads to a high degree of disintegration. The surface of the fractures is often slickensided which is an indicator for dynamic processes.

The *Rhätolias* consists of medium-coarse grained, quarzitic sandstone with intersecting layers or lenses of fine-grained material (clay/silt). The clay bands are rich in Montmorillonite with a high swelling capacity and lead to a predestined sliding surface. Due to tectonic movement, the formerly compact sandstone bed is highly disintegrated in some parts into sandstone blocks or pillars. This is observed especially in the surrounding of the slope cuts where the contact to the intact rock mass is not given anymore.

According to the topography, the drain of the slope is in SE direction. In the project area, groundwater-bearing strata are the sandstone layers of the *Burgsandstein* and the *Rhätsandstein*. The fine-grained beds (clay and silt) within the *Rhätolias* lead to a locally decreased permeability. The clayey material of the *Feuerletten* serves as aquiclude so that the lower groundwater table within the *Burgsandstein* is artesian. In the case of tectonic stress, discontinuities may develop within the clay that lead to an increase of hydraulic permeability.

2 MONITORING PROGRAM ON-SITE

To guarantee the safety of the motorway, a detailed measuring and monitoring concept that follows the guidelines of the Eurocode 7 for real time surveillance was installed at the project site. This includes measuring devices like inclinometers, pore-water pressure sensors, groundwater wells and geodetic measurements (Figure 2). In total, about 25 inclinometers (1 automatic, 24 manual), 1 pore-water pressure sensor and 7 groundwater-measuring holes are monitored. The geodetic system controls all inclinometer heads and additionally 28 survey pillars along the motorway. At the time of the report, the geodetic system is manually measured every month. However, it is planned to enhance the accuracy by an automatically measuring total station. By means of this, all devices can be combined to an early-warning system for the slope.

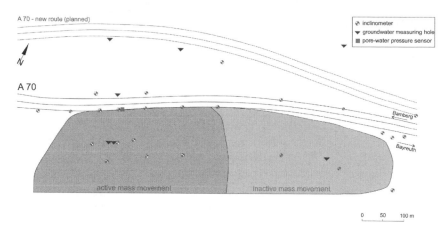

Figure 2.   Monitoring system on-site including inclinometer, pore-water pressure sensor and ground water measuring holes. The active mass movement area is shown in red while the inactive, but possibly reactivated, area is shown in green.

To categorize the risk potential for the motorway, a warning and emergency plan, that is based on the results of the monitoring system, was developed. The plan includes four risk levels with corresponding instructions to control the hazard optimally and quickest possible.

3   RESULTS

3.1   Field measurements

Results of the inclinometer demonstrate two different failure behaviors. Test sites that are located along the motorway (Figure 3, left) reveal movement from the bottom and no distinct shear plane. These data suggest multiple shear planes for test sites located in sandstone layer of the *Rhätolias* with its intersecting clay beds where each may act as plane of weakness within the strata. The deformation of the decisive axis (A-axis) is in the range of less than 1 cm. However, this magnitude is slowly increasing, causing damage to the motorway that is directly next to the test site.

In contrast, the slope-lower test sites show different results (Figure 3, right). Inclinometric measurements observe clearly one single shear plane and no movement from the bottom point. These sites are mainly located in the clay of the *Feuerletten* and the sandstone bed on top has already been eroded. The deformation of the decisive axis has been recorded with maximum 8 cm within one year. The deformation velocity for the lower slope is approximately 0.5 cm/month.

These findings contribute to an interpretation of the failure process that is described in chapter 4.

3.2   Laboratory testing

An extensive laboratory program has been conducted at the University of the German Armed Forces on about 200 samples taken out of the boreholes including the *Rhätolias* as well as the *Feuerletten* (UNIVERSITÄT DER BUNDESWEHR MÜNCHEN). Special focus has been placed on the determination of the shear parameters by triaxial and direct shear testing. These parameters are crucial to investigate the failure process and in a second step to conduct slope stability analysis.

In this paper, only the stratum of the *Feuerletten* is analyzed in detail since it is the influential layer for the movement. Laboratory tests reveal that the material is classified mostly as silty clay (siCl) acc. to DIN EN ISO 14688-1. However, the geomechanical properties vary extremely due to the inhomogeneity of the rock. Plastic, very soft clay lenses as well as cemented particles may occur within the generally semi-solid material.

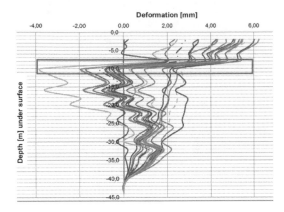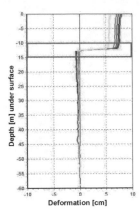

Figure 3. Inclinometric measurements (depth is plotted against measured deformation) at the active movement area for test sites along the motorway (left) and slope-lower sites (right). Different lines show different times of measurement.

One parameter that influences the soil properties is the water content. As the material consists of clay, increase of water content lead to distinct changes in the consistency accompanied with a drop of the cohesion. The dimension is significant. The stiff to semi-solid material reveals medium cohesion of 85 kN/m² whereas the plastic material only results in 30 kN/m².

Samples that contain cemented particles should result in higher shear parameters, as the rock is solid in some parts. However, this is not the case. The measured cohesion for cemented clay samples was only 25 kN/m². This is due to the fact that the solid particles lead to errors during the testing procedure since they affect the shearing process and cause highly fluctuating shearing resistance.

Focus of this chapter is the comparison of shear parameters for the same kind of soil, analyzed with different testing methods. These are mainly the direct shear test (DIN 18137-1) and the triaxal shear test (DIN 18137-2). The circular shear (DIN 18137-1) test is a special version of the direct shear test, where the shearing takes place circular, which allows the determination of the residual angle of friction.

Figure 4 shows the variation of the angle of friction and the cohesion for all three test methods. It is obvious that the **angle of friction** is more or less consistent and the mean values for each test vary in the range of maximum 2° (Figure 4, left). Highest variation within one test method is noticed for the triaxial test. Here, the angle of friction varies between 15° and 30°.

The **cohesion** results in an extreme fluctuation from minimum 10 kN/m² up to 115 kN/m². The results of the direct and circular shear tests are very similar (mean values of 74 kN/m² resp. 75 kN/m²). However, the triaxial test yields a mean value of only 40 kN/m². This is almost half of the direct shear test (Figure 4, right).

It seems that the triaxial test is not the right tool to analyze complex, heterogeneous material like the clay of the *Feuerletten*. The test determines shear parameter with high accuracy. However, the inhomogeneous material with cemented particles leads to distinct deviations of the stress-strain curve for the different loading levels. As small variations in the stress-strain curve have a significant impact on the parameters of φ and c, the test is very sensitive concerning heterogeneous material. Additional, the sample preparation is elaborately and needs quite an amount of sample material, which increases the risk of testing material with different soil characteristics. To sum up, the test basically reveals results with high accuracy but its sensitivity concerning the sample quality and preparation makes it inapplicable for this project.

The direct shear test is less sensitive to sample preparation and therefore more suitable for characterizing the complex material at the project site.

As the angle of friction and the cohesion are the parameters most influencing the shearing behavior of rocks, these parameters also control the failure behavior at the landslide in Thurnau. Major challenge is, to choose the 'correct' values for this heterogeneous material.

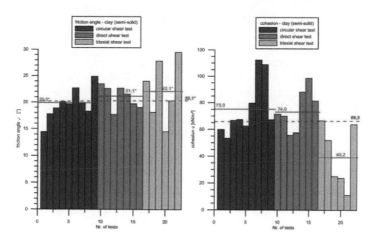

Figure 4. Angle of friction φ' and cohesion c for semi-solid clay samples of the *Feuerletten*. Compared are the results of three different testing procedures (circular shear test, direct shear test, triaxial shear test).

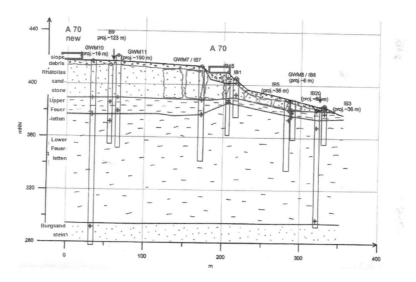

Figure 5. Schematic cross section through the active mass movement area with the geological strata of slope debris, Rhätolias sandstone, Feuerletten and Burgsandstein.

Results reveal that every stratum (*Rhätolias, upper/lower Feuerletten*) can be divided into several soil classes acc. to grain size and consistency category—each with its own shear parameter set. By means of this, the basic layer and planes of weakness within one stratum can be defined and further implemented into the soil model for a detailed slope stability analysis.

4 MOVEMENT PROCESS

The combination of the results from the field and laboratory testing lead to a profound understanding of the sub-surface conditions at the project site. By means of this, a detailed soil model and an interpretation of the failure mechanism is feasible. A schematic figure that includes the geological situation as well as the movement process is shown in Figure 5.

The area is a typical example for a mass movement where a 'hard slap' (the sandstone bed of *Rhätolias*) spreads on a 'soft ground' (the clay of *Feuerletten*). Due to anthropogenic changing of the slope geometry and loading situation (construction of the motorway), natural factors like erosion or weathering were amplified. Therefore, the once compact sandstone layer was disintegrated into sandstone blocks or pillars which shown no link to the compact rock mass any more. The loose blocks or pillars slowly sink in the soft clay layer and creep down the slope with a combined tilting-/spreading movement. The movement process is classified as block spread acc. to CRUDEN & VARNES (1996). It is a slow but continuously proceeding process with acceleration rate of about an mm/year up to an mm/day. The process is accompanied with a plastic deformation without shearing of the rock and a concrete shearing plane.

The process is affirmed by the results of the inclinometric measurements where the slope-lower inclinometer (IB5, IB20, IB3) reveal a concrete shear plane. These test sites are installed mainly in the layer of the Feuerletten and the sandstone bed is already eroded. In contrast, the inclinometers along the motorway show multiple shear planes that reflect the tilting of the sandstone pillars.

5 CONCLUSION

The detailed monitoring program in combination with the laboratory testing allows a profound knowledge of the geological and geomechanical characteristics at the landslide of Thurnau.

The laboratory and field measurements show that the geology of the project site is built up of several heterogeneous strata. Additional, each stratum itself is again divided into layers with different geomechanical properties. Layers with low shear strength parameters (angle of friction and cohesion) act as planes of weakness and facilitate the movement. This is the case for the *Rhätolias* where the fine-grained interlayers of clay and silt serve as lubricant. The inclinometric measurements along the motorway verify this hypothesis since they show multiple shear planes with varying movement directions that reflect the tilting of the sandstone pillars. However, this hypothesis could not be confirmed for the *Feuerletten*. The sliding surface at the slope-lower inclinometer sites (Figure 3) was determined in semi-solid clay which is the layer with higher shear parameters within the *Feuerletten*. It seems that the entire stratum of the *Feuerletten* can be activated as shear plane since the material is very sensitive to changes in the environment like hydrological balance. Thus, further investigation needs to be done on the swelling pressure of the material compared to the results of the pore pressure sensor to possibly find a critical threshold at which the clay changes its properties decisively causing a newly formed sliding surface and an increase in the movement velocity of the landslide.

REFERENCES

BAYERISCHES GEOLOGISCHES LANDESAMT (not yet published): Preliminary of Geological Map of Bavaria 1:25.000, Nr. 5934 Thurnau.

CRUDEN, D. & VARNES, D. (1996): Landslide types and processes—In Special report 247: Landslides: Investigation and Mitigation.

DIN 18137-1 (2010): Soil, investigation and testing—Determination of shear strength—Part 1: Concepts and general testing conditions.

DIN 18137-2 (2011): Soil, investigation and testing—Determination of shear strength—Soil, investigation and testing—Determination of shear strength—Part 2: Triaxial test.

DIN EN 1997-1 (2014): Eurocode 7: Geotechnical design—Part 1: General rules; German version.

UNIVERSITÄT DER BUNDESWEHR MÜNCHEN (2016–2017): Laboratory results of samples taken at bore cores of Thurnau, Chair of soil mechanics and geotechnical engineering.

Rock mass strength and failure

Geomechanics and Geodynamics of Rock Masses – Litvinenko (Ed.)
© 2018 Taylor & Francis Group, London, ISBN 978-1-138-61645-5

On a model of geomechanical effect of underground explosion in the massif of block-like structure and the mechanism of rock destruction

Vitalii V. Adushkin
Institute of Geosphere Dynamics RAS, Scientific Leader IDG RAS, Academician RAS, Moscow, Russia

Ivan V. Brigadin & Sergei A. Krasnov
Promstroyvzryv Ltd., Candidate of Technical Sciences, Saint Petersburg, Russia

ABSTRACT: The article is dedicated to challenging issues of drilling and blasting operations in mines at great depths from the point of the modern concept of hard rock fracture mechanisms and models. The diagram of two underground blasting development options in the block structure is presented. Basing on the analysis of the experimental data, the hard rock fracture mechanism at depth in the block structure is substantiated and the respective principles are formulated. The assumption is made with regards to the hypothetical existence of quasi-solid, which is an earlier unknown phase of the solid substance.

Keywords: fracture model and diagram, hypothesis, structural and tectonic block, fracture strength, phase transition pressure, drilling and blasting operations, fracture mechanism principles, great depth, hard rock

1   INTRODUCTION

To extract solid minerals and some non-metallic materials from underground, the blasting method of rock fracture is usually used with further processing of the geomaterial to obtain the required parameters.

The main topical issues and development directions of blasting operations in Russia are presented in the Article [18]. In case of underground mining, hard rock fracture is performed by drilling and blasting operations.

The blasting method of the underground mining is based on the classical concept of hard rock fracture patterns. The most comprehensive development results of the models for underground blasting mechanical action are presented in the monographies [14, 15].

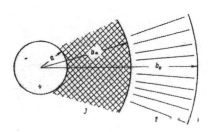

Picture 1.   Rock fracture diagram: 1 – elastic deformation zone; 2 – radial fracture zone (b_0); 3 – crushed zone (b_H); 4 – cavity (a).

The earlier monography [14] presents in general terms the extent, to which all main effects and fracture zones are interrelated in case of an underground blasting, on the basis of the rock racture diagram shown on Picture 1. Such diagrammatic model is based on the continual approach to description of the rock fabrics.

With accumulation of the experimental material and stricter requirements to accuracy of the forecasted underground blasting action parameters, the new model was proposed, which was described in the monography [15]. This model of a solid body with structure allows predicting development of deformation processes in the hard rock through time.

2 MODEL OF GEOMECHANICAL ACTION OF THE UNDERGROUND BLASTING IN THE BLOCK STRUCTURE

Theanalysis results of the experimental studies demonstrated that in a number of cases the respective hard rock fracture theory at great depths fails to explain the numerous experimental facts, while the applied technologies do not result in rock fracture at big distances from the tunnel, which are possible with the use of the modern drilling methods.

The continual approach has been used for a long time to describe the hard rock deformation and fracture processes. Description of the hard rock as a block and hierarchical environment has been widely used for the last three decades in different mining and geomechanical practices.

Analysis of the experimental data on results of examination of the central zones allows distinguishing two underground blasting development options [2, 5, 10, 11].

When the blast develops in a rather homogeneous structural and tectonic block, the rock does not break at strains less than the phase transition pressure (Picture 2). A cavity is formed in the middle. However, beyond the cavity no significant crushed zones are detected in the framework of the considered task.

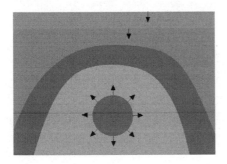 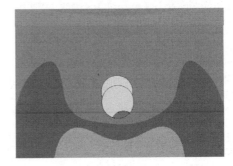

Picture 2. Blast diagram in single structural and tectonic block. (a – dynamic phase; b – quasi-static phase).

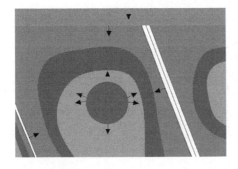 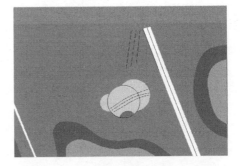

Picture 3. Diagram of typical blast in the rock consisting of several structural and tectonic blocks. (a – dynamic phase; b – quasi-static phase).

With cooling of the cavity, the break-down point of the rock decreases at the border with the main rock massif and slight fracture can take place due to the gravity pressure. Such effect is called "sloughing zone" in the mining practice [4, 9, 13].

The diagram of the blast option in block is shown on Picture 3.

In accordance with the typical scenario, one or several blocks break earlier, at the dynamic stage of compression wave emission. The main reason of fracture is the whole train of tension waves, including those reflected from the borders of the structural and tectonic block and land surface.

Phenomenologically, the breaking picture shown above can be explained based on development of the concept of the strong hard rocks fracture mechanism by underground blasting. The first systematic research performed on this issue is provided in the study [3].

3 SUBSTANTIATION OF THE ROCK FRACTURE MECHANISM

Ultimate compressive strength B_{com}, resistance to direct pull B_{pull}, ultimate shear stress B_{shear}, as well as phase transition pressure P_{ph} are used as hard rock strength parameters [1–3, 6–8, 13–15, 19, 20].

Description of the hard rock as a block and hierarchical environment has been widely used for the last three decades in different mining and geomechanical practices [10, 11, 16, 17].

At the initial development stage of drilling and blasting operations, the length of blast holes (up to 2÷5 m) was comparable with the unloading (sloughing) zone in the bottom hole. In this situation the blast hole was filled with the explosive material throughout its length and the bottom hole moved in proportion to the blast hole efficiency ratio during one cycle. When the holes with the length of up to 10 m started to be used, the blast hole efficiency ratio reduced considerably and the efficiency of drilling and blasting operations decreased as well.

Reduced efficiency of drilling and blasting operations and increased specific charge of the blasting agent may be caused by increased rock strength in case of triaxial compression—the so-called "compressed rock" [6, 9, 13].

It is also obvious that the compressed environment is stronger than the sample. Besides it was found that with increased depth of the hard rock and growth of overburden pressure, the general rock strength increases as well (against all types of stress). This fact was confirmed by special experiments and underground development of ore bodies, when both increase and reduction of the blasting agent specific charge could be observed [1].

At the same time, there are facts, which cannot be explained with the modern theory. These include rock strength against blast load in case of long holes, i.e. when this length exceeds considerably the typical size of underground mining. That is why a free hole is made in the middle of the cutting zone in case of tunnel driving by long holes [4, 9, 13].

The anomalous fracture effect seems to have been noted for the first time in the study by V.M. Kuznetsov [12].

Cavity strength in case of relatively small power of the underground nuclear explosion (UNE) shall be recognized as even bigger paradox. Specifically, melted rock can be observed both in the lens located at the bottom and on the walls [5]. The stress level in this zone is so high that the rock should have been broken up to fines [8]. A comparison was made in a number of cases between the properties of adjacent strata on samples [11] and it was found that the strength of the rock and massif in general at relatively small distances from the explosion cavity did not change significantly, although the registered stress values reached 50 hPa.

Unlike the fracture mechanism described in the study [14], the real process appears to be somehow different.

An expansion wave reflected from some non uniformity results in asymmetry of stress parameters at the border of cavity and ground. At some limit values the symmetry of cavity expansion is broken and break-away blocks of different sizes are formed. These blocks can move rather independently, including into the cavity. The cavity with moderate asymmetry

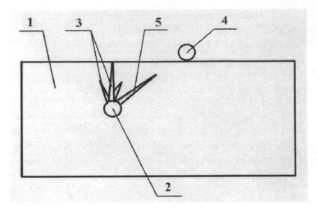

Picture 4.   Diagram of comparative experiments on different materials. 1 – block, 2 – cavity, 3 and 5 – cracks, 4 – charge.

parameters remains stable for a long period of time, with this fact having been observed in a number of experiments.

Special comparative experiments were carried out on different materials (brick, concrete, hard rock) to clarify the nature and mechanism of solid body fracture. The diagram of these experiments is shown on Picture 4.

A charge of several detonating cord threads was exploded in the cylindrical cavity 2 in block 1. As a result of the detonating cord explosion, the block material towards the open air was broken into separate fragments with several radial cracks 3, extending towards the open air. At explosion of the detonating cord in the cavity 2 and the surfaced charge 4 from one detonating cord thread, one main crack 5 was observed strictly towards the charge 4. At that, no cracks were detected towards the open air.

Thus, rock fracture in these tests is detected either by sufficient intensity of the explosive force, or by the reflected expansion wave, or by the counter-running compressive wave.

4   FRACTURE MECHANISM PRINCIPLES

Basing on the summary of experimental facts it is clear that the solid body of mainly crystal structure has the property of gaining the ordered structure without significant density changes, resulting in transfer into the new phase, for which not the compression strength B_{com}, but the phase transition pressure P_{ph} is typical.

In this regard, change (increase) of the hard rock strength from the initial value on land surface B_{com} to the phase transition pressure P_{ph} can be detected with moving away from the border of the structural and tectonic block (increased depth). Such fracture mechanism can be hypothetically explained by existence of an earlier unknown phase condition of the substance, which can be called "quasi-solid".

This even can be reported to have a hidden nature—when the rock sample is taken from the massif depth onto the land surface, its physical and mathematical properties gain slightly different values, received in laboratories.

The following main provisions of the blasting destruction model at big depth under development, can be formulated on the basis of received experimental results [3]:

1. Rock fracture strength in compression, cleavage and tension is a value changing within the limits of the block massif of any level.
2. Rock strength in compression within the block limits increases from edges to the middle up to the phase transition pressure P_{ph}.

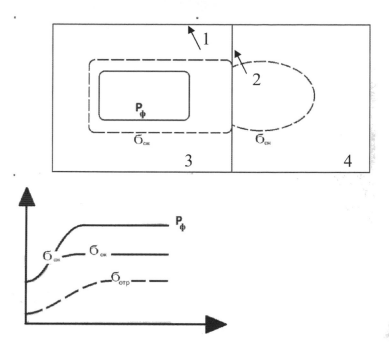

Picture 5. Diagram and nature of massif resistance changes. 1 – surface, 2 – fault, 3 and 4 – structural and tectonic blocks.

3. Strength at the block edge inside the massif may fail to correspond to the strength at the border with open air.
4. After the massif is broken into smaller blocks, they may move in an asymmetric, random manner.

When the mine tunnel is driven to such place in the block massif, where the strength reaches the phase transition pressure values P_{ph}, blasting destruction of the rock beyond the limits of the sloughing zone becomes impossible. The nature of strength changes is outlined on Picture 5.

Compressive strength B_{com} and resistance to direct pull B_{pull} close to the open air 1 or fault 2 of the structural and tectonic blocks 3 and 4 are close to traditional (laboratory) values. With increased distance from the borders of the blocks both resistance values increase with B_{com} being able to reach or not to reach P_{ph} value.

5 CONCLUSIONS

1. Modern concept of the hard rock fracture mechanism cannot explain some experimental facts obtained in the course of drilling and blasting operations at big depths.
2. One of the reasons may be the fact that most data, which form the basis for the fracture criteria and camouflet explosion development diagrams, are based on experimental data received on samples, during laboratory tests or explosions performed at shallow depth.
3. The rock fracture model with underground explosion in the block structure is proposed.
4. Main provisions of the rock fracture mechanism at big depths in the block structure are formulated in this study.
5. As a general result the assumption is made about hypothetical existence of "quasi-solid" as the new phase condition of the substance, along with the known conditions: gas, liquid, solid body and plasma.

REFERENCES

[1] Adushkin V.V., Spivak A.A. 2001, Destructive Effect of Explosion in Pre-stressed Environment// Proceedings of the Mining Academy, St. Petersburg, 2001, v. 148(1), pp. 21–32.

[2] Adushkin V.V., Spivak A.A. 2007, Underground Explosions, M., "Nauka", 2007.

[3] Adushkin V.V., Kocharyan G.G., Brigadin I.V., Krasnov S.A. 2015. Concerning the Hard Rock Fracture Mechanism by Underground Blasting; MIAB №7, pp 344–349, 2015.

[4] Baron V.L. 1989. Methods and Technology of Explosive Works in the USA. Nedra.

[5] Greschilov A.A, 2008, Nuclear Shield – M.: Logos; pp. 239–241.

[6] Dzhigrin A.V. High-Speed Tunnel Driving Technology by Hard Rocks using Drilling and Blasting Method. St. Petersburg. Blasting Work Challenges.; M, MGGU, 2002.

[7] Dugartsyrenov A.V. Physical Nature and Fracture Mechanism of Hard Rock in case of Camouflet Explosion. Collected Book "Blasting Work", issue No. 106/63, M.: "Interdepartmental Commission for Blasting Work under the Mining Academy".

[8] Samychljev B.V. 1990. Dynamic Hard Rock Deformation and Fracture Models; M., "Nauka", pp. 32–75.

[9] Isakov A.A. 1986. Strata Pressure Impact on Hard Rock Fracture Direction by Blasting; Collected Book "Blasting Work", issue No. 106/63, M.: Nedra.

[10] Kocharyan G.G., 1990, Rock Massif Retrograde Deformation Model; Collected Book "Blasting Work", issue No. 90/47, M.: Nedra.

[11] Kocharyan G.G., Spivak A.A. 2003, Deformation of Hard Rock Block Massifs, M. Publishing House "Akademkniga".

[12] Kuznetsov V.M. On Hard Rock Fracture Mechanism by Blasthole and Borehole Explosive Charges; Collected Book "Blasting Work", issue No. 83/40, M.: Nedra, 1982.

[13] Mosinez V.N. Hard Rock Fracture; M.: Nedra, 1975.

[14] Rodionov V.N. Mechanical Effect of Underground Blasting; M.: Nedra, 1971.

[15] Rodionov V.N. Fundamentals of Geomechanics; M.: Nedra, 1986.

[16] Sadovskiy M.A. Natural Granulometric Composition of Hard Rock. Reports of the Academy of Sciences of the USSR; 1979, t.247, No. 4, pp. 829–831.

[17] Sadovskiy M.A, Kocharyan G.G., Rodionov V.N. On Mechanics of Hard Rock Block; Reports of the Academy of Sciences of the USSR; 1988, t.301, No.2, pp. 306–307.

[18] Trubetskoy R.Y., Viktorov S.D. Blasting Work Development Challenges on Land Surface; Collected Book "Blasting Work", issue No. 101/58, M.: "Interdepartmental Commission for Blasting Work under the Mining Academy"; 2009, pp. 3–24.

[19] Shemyakin E.I. Seismic Effect of Massive Underground Explosion. Collected Book "Blasting Work", issue No. 94/51, M.: "Interdepartmental Commission for Blasting Work under the Mining Academy"; 2004, pp. 10–21.

[20] Shemyakin E.I. Strain Waves at Underground Blasting; Collected Book "Blasting Work", issue No. 93/50, M.: "Interdepartmental Commission for Blasting Work under the Mining Academy"; 2000, pp. 4–12.

Geomechanics and Geodynamics of Rock Masses – Litvinenko (Ed.)
© 2018 Taylor & Francis Group, London, ISBN 978-1-138-61645-5

Effect of thermal damage on strain burst mechanism of brittle rock using acoustic emission

Selahattin Akdag, Murat Karakus, Abbas Taheri & Giang Nguyen
School of Civil, Environmental and Mining Engineering, University of Adelaide, Adelaide, SA, Australia

He Manchao
State Key Laboratory for Geomechanics and Deep Underground Engineering, Beijing, China

ABSTRACT: Influence of thermal damage on strain burst characteristics and damage stress of granitic rocks were investigated in a series of strain burst tests. Granite samples were exposed to various temperatures ranging from 25°C to 150°C. A true-triaxial loading-unloading system was used to replicate in-situ stress condition taking into account creation of excavation process. During true-triaxial loading-unloading process, damage evolution of the rocks was examined using Acoustic Emission (AE). While the damage at the onset of Critical Strain Burst Stress (CSBS) was found to be around 35% in relation to cumulative AE hits, the damage starts at about ~15% according to the analysis based on cumulative AE energy at temperature 25°C. This suggests that damage evaluation with cumulative AE hits is inaccurate as it does not account for the size of the micro cracks which is related to the AE energy. Based on AE energy approach, thermal damage caused an approximately 55% decrease in the onset of strain burst stress when the temperature increased from 25°C to 100°C. A gradual increase was observed at temperatures 100–150°C but it was still less than that of at room temperature which showed more intense strain burst behaviour.

Keywords: Strain burst, Thermal damage, Acoustic Emission (AE), Brittle rock, Damage evolution, Temperature, True-triaxial loading

1 INTRODUCTION

Strain burst that has been the most frequently encountered rock burst type in deep underground engineering projects is a serious problem in the rock mass. It can be characterised by the violent ejection of rock blocks or rock fragments when the excavation-induced stresses exceed the rock mass strength in which accumulated strain energy within the rock is suddenly released. Strain burst poses a serious threat to the safety of deep underground excavations which also drastically increases with the increasing excavation depth due to high-geo-stress conditions.

Rock is a naturally occurring geological material composed of minerals, micro cracks and weak interfaces that is generally subjected to various environmental impacts such as high pressure and temperature. Due to the increasing depth, underground structures are becoming vulnerable to the effects of high ground temperature triggering strain burst. At elevated temperatures, physical and mechanical properties of rock were altered due to the change in the density of micro cracks (Heuze 1983). For instance, compressive strength, elastic modulus, stress thresholds, failure mode and colour of Australian granite were significantly influenced by the effect of temperature under unconfined stress conditions (Shao et al. 2015). To date, a number of comprehensive experimental research has been carried out to investigate the influence of temperature on the variation of mechanical and physical characteristics of various rocks subjected to different temperature treatments (Heuze 1983; Dwivedi et al. 2008; Tian et al. 2014; Liu and Xu 2015; Shao et al. 2015; Zhang et al. 2016, Xiaoli & Karakus 2018).

Under high temperatures, not only different rock types but also a same rock type, exhibit different mechanical characteristics due to the variations in mineral composition, microstructure and grain size. Some recent research works (Ranjith et al. 2012; Shao et al. 2015; Peng et al. 2016) have shown that there are temperature thresholds at which the mechanical and physical properties significantly change. Therefore, it is essential to better understand the temperature dependent mechanical and physical behaviour of rocks for the long-term stability design of rock structures and safety of underground excavation projects. All of these existing studies focused on the mechanical characteristics of thermally-damaged rocks under uniaxial compression or triaxial compression. However, studies considering the mechanism and deformation characteristics of strain burst in thermally-damaged rocks under true-triaxial loading-unloading conditions are limited.

In order to provide insight into the strain burst mechanism, extensive experimental works have been performed under uniaxial compression, triaxial compression and true-triaxial compression conditions (Wang and Park 2001; Hua and You 2001; Nasseri et al. 2014). However, these tests neither realistically replicate the polyaxial stress conditions nor unloading process near excavation boundary which is closely related to strain burst occurrence. Some recent research has been focused on studying strain burst mechanism experimentally under true-triaxial loading/unloading conditions to simulate and capture the complex stress distribution at a tunnel boundary (He et al. 2010; Gong et al. 2012; Zhao et al. 2014). Akdag et al. (2018) also investigated the influence of thermal damage on strain burst mechanism of brittle rock under true triaxial loading/unloading conditions. They found that while strain burst stress, the velocity of the initial fragments and kinetic energy decreased from 25°C to 100°C, a gradual increase in strain burst stress associated with more intense strain burst behaviour was observed in the range of 100°C to 150°C.

In the present study, a number of true-triaxial loading-unloading strain burst test was conducted on granite samples subjected to various temperatures ranging from 25°C to 150°C and micro-seismic activities were monitored. Acoustic response from each sample was analysed in terms of AE event and AE energy to assess the damage accumulation so that the critical strain burst damage level can be identified.

2 EXPERIMENTAL STUDY

The granite samples used in this study were sourced from a borehole at a depth around 1300 m in South Australia. The selected granite is the coarse-grained type of granite with weak to moderate alteration and occasional weak gneissic foliation. The tested rectangular prismatic granite specimens were prepared from 63 mm diameter drill cores. As can be seen from Fig. 1, the dimension of a rectangular prism of granite samples is 125 mm × 50 mm × 25 mm for the strain burst tests (Akdag et al. 2016). The geometry of the rectangular specimens is shown

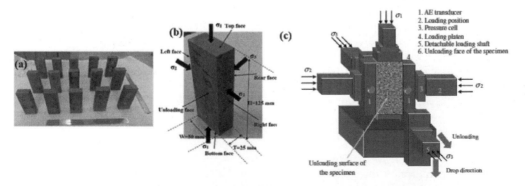

Figure 1. **(a)** Overview of tested granite specimens **(b)** Schematics of a specimen **(c)** Schematic illustration of the testing machine (He et al. 2010).

in detail in Fig. 1. The specimens were heated up to their target temperature of 50°C, 75°C, 100°C, 125°C and 150°C using a high-temperature furnace at a heating rate of 5°C/min, with the exception of 25°C at which the specimens were tested at room temperature. This low rate was employed to avoid microstructural damage due to thermal shock associated with the rapid heat development and to keep the thermal gradient as small as possible. They were then kept at their target temperature for 12 h, to ensure temperature uniformity. Finally, the specimens were placed in a room temperature environment for natural cooling to the room temperature.

Strain burst tests were conducted using the Modified True-Triaxial Apparatus (MTTA) developed by He et al. (2010). This true-triaxial strain burst test system consists of a loading/unloading device, a hydraulic-pressure controlling unit, and data acquisition system, linear variable differential transformer (LVDT) deformation sensors, a high-speed camera system and an AE monitoring system (see in Fig. 1c). The loading is applied to the six faces of prismatic specimen independently in three principal stress directions (σ_1, σ_2, σ_3) through the rigid solid plates. One surface of the sample is unloaded abruptly by dropping the loading bar and the plate suddenly which replicates the creation of a tunnel/excavation in the underground. Meanwhile, a continuous acoustic emission monitoring was conducted to understand the initiation and accumulation of micro cracks and damage mechanism of granite treated to temperatures from 25°C to 150°C. The AE sensors used in this study (provided by the American Physical Acoustics Corp.) are 18 mm in diameter with a resonance frequency of 125 kHz. Operating frequency of the sensors is in the range of 100 kHz to 1 MHz and a sampling rate used is 10 Msps (million samples per second). The output voltage of the AE was amplified by 40 dB gain and the threshold amplitude was set to 35 dB. The AE sensors were in direct contact with the rock specimen in the direction of σ_2 and a petroleum jelly was applied to the sensors and the steel plates to ensure a good acoustic coupling.

In the true-triaxial strain burst tests, the rock samples were loaded with 43 MPa, 23 MPa, and 11 MPa in the direction of in-situ stresses σ_1, σ_2, σ_3 respectively and these stress levels were retained for 5 minutes to ensure the stress uniformity in the rock. Then the loading plate in the direction of σ_3 was immediately dropped, creating an unloading free surface while keeping σ_2 constant. Following this, σ_1 was increased until strain burst occurred. Meanwhile, the high-speed camera was used to capture the strain burst process for further kinetic energy analysis.

3   EXPERIMENTAL RESULTS AND DISCUSSION

The AE monitoring method is a powerful non-destructive technique to investigate the micro cracking process and damage evolution in brittle rocks. The wave generated by each micro crack during loading was recorded as AE hit. Therefore, the number of AE hits can be related to the number of micro cracks generation and micro crack openings. Fig. 2a shows a representative cumulative AE hits response, damage evolution and assessment of the onset of CSBS of the rock at a temperature of 25°C.

At the early stage of the loading, cumulative AE hits exhibited a rapid increase due to the closure of micro cracks and voids. Then, the cumulative AE hits rate decreased at higher stress level due to the end of crack closure (end of the elastic stage). In the subsequent stress increment, the slight change in the cumulative AE hits may represent that there was not any significant micro-fracturing. After the unloading of σ_3 and increasing σ_1, AE hit rate drastically increased indicating the propagation and coalescence of micro-cracks and finally forming macro cracks. In the present study, this transition point in cumulative AE hits was used to identify the onset of CSBS. It was found that the damage at the onset of CSBS was around 35% based on the cumulative AE hits evolution at temperature 25°C (see Fig. 2a). Moreover, the onset of CSBS decreased approximately 63% as the temperature increased from 25°C to 100°C, as shown in Fig. 3a. The different rock minerals may be influenced by the thermal contrast, thereby decreasing the load bearing capacity and lower strain burst stress compared to the rocks at room temperature (25°C). When the temperature increased from 100°C to 150°C, strain burst started to occur at relatively higher stress levels (see Fig. 3a). This can be attributed to the reduction in micro crack generation as the internal

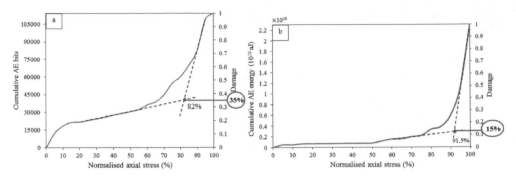

Figure 2. Damage evolution and assessment at the onset of CSBS by **a** cumulative AE hits and **b** cumulative AE energy at temperature of 25°C.

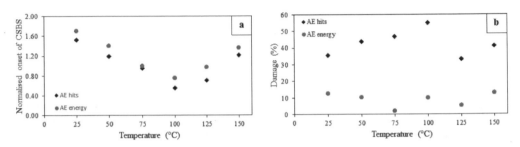

Figure 3. **a** Influence of thermal damage on the onset of CSBS **b** Difference of damage assessment by cumulative AE hits and cumulative AE energy.

mineral grains expand and improve the densification and compactness of the granite specimens and increased the onset of CSBS.

The size or magnitude of the cracks can be related to the AE energy of each AE event and as a result of our analysis, the use of AE energy as an indication of damage in rock can be more accurate than considering AE events/hits. In the present study, cumulative AE energy evolution was thus used to assess the thermally-induced damage evolution and damage at the onset of CSBS of the granite specimens subjected to various pre-heated temperatures. Fig. 2b shows that damage at CSBS according to cumulative AE energy characteristics at temperature of 25°C initiated around 15% of total damage which suggests that cumulative AE energy behaviour is more representative for the evaluation of damage as the size of the micro cracks is more related to the AE energy. According to the AE energy approach, thermal damage caused a 55% drop in the onset of CSBS as the temperature increased from 25°C to 100°C (see Fig. 3a). When the temperature was increased from room temperature (25°C), rock matrix was accompanied by the increased micro crack activities which resulted in a decrease in AE events during loading prior to bursting. Due to the temperature induced pre-damage, the cumulative AE energy was low and the strength of the rocks was deteriorated. Between the temperature of 100°C and 150°C, a slight rise in the onset of CSBS was observed due to the enhanced compactness of the specimens leading to intense strain burst, as depicted in Fig. 3a. It can be seen in Fig. 3b, there is a discrepancy in the damage assessment in relation to cumulative AE hits and cumulative AE energy which indicates that AE energy can be better to use as a damage indication.

4 CONCLUSION

In this study, thermal damage influence on the onset of CSBS and strain burst behaviour of granite samples were experimentally investigated using a unique true-triaxial strain burst testing system. Based on the cumulative AE hits and cumulative AE energy characteristics, damage evolution and the stresses at the strain burst initiation of the rocks subjected to different temperatures ranging from 25°C to 150°C were studied. It was found that damage at the onset of CSBS at the temperature 25°C was around 15% in relation to cumulative AE energy and this provided more accurate results on the strain burst mechanism compared to the use of cumulative AE hits. The onset of CSBS showed a decreasing trend up to 100°C due to rendering the rocks relatively reduced resistance to deformation. At temperatures from 100°C to150°C, the onset of CSBS gradually increases but it is still less than that of at room temperature. The results of this experimental research clearly demonstrate that strain burst stress for granite is dependent on temperature-induced pre-damage as supported by acoustic emission energy analysis.

ACKNOWLEDGEMENT

The authors gratefully acknowledge the financial support from the Australian Research Council (ARC-LP150100539), OZ Minerals, and the principal geotechnical manager, David Goodchild.

REFERENCES

Akdag, S., Karakus, M., Nguyen, G., & Taheri, A. (2017). Influence of specimen dimensions on bursting behaviour of rocks under true-triaxial loading conditions. In J. Wesseloo (Ed.), *8th International Conference on Deep and High Stress Mining* (pp. 447–457). Perth: Australian Centre for Geomechanics.

Akdag, S., Karakus, M., Taheri, A., Nguyen, G., & Manchao, H. (2018). Effects of thermal damage on strain burst mechanism for brittle rocks under true-triaxial loading conditions. *Rock Mech Rock Eng*, (Accepted Manuscript).

Dwivedi, R., Goel, R., Prasad, V., & Sinha, A. (2008). Thermo-mechanical properties of Indian and other granites. *Int J Rock Mech Min Sci*, 303–315.

Gong, Q., Yin, L., Wu, S., Zhao, J., & Ting, Y. (2012). Rock burst and slabbing failure and its influence on TBM excavation at headrace tunnels in Jinping II hydropower station. *Eng Geol, 124*, 98–108.

He, M., Miao, J., & Feng, J. (2010). Rock burst process of limestone and its acoustic emission characteristics under true-triaxial unloading conditions. *Int J Rock Mech Min Sci, 47*, 286–298.

Heuze, F. (1983). High-temperature mechanical, physical and thermal properties of granitic rocks— A review. *Int J Rock Mech Min Sci, 20*(1), 3–10.

Karakus, M., Akdag, S., & Bruning, T. (2016). Rock fatigue damage assessment by acoustic emission. In P. Ranjith, & J. Zhao (Ed.), *International Conference on Geo-mechanics, Geo-energy and Geo-resources, IC3G*, (pp. 9-82-88). Melbourne, Australia.

Liu, S., & Xu, J. (2015). An experimental study on the physico-mechanical properties of two post-high temperature rocks. *Eng Geol, 185*, 63–70.

Peng, J., Rong, G., Cai, M., Yao, M., & Zhou, C. (2016). Physical and mechanical behaviors of a thermal-damaged coarse marble under uniaxial compression. *Eng Geol, 200*, 88–93.

Ranjith, P., Viete, D., Chen, B., & Perera, M. (2012). Transformation plasticity and the effect of temperature on the mechanical behaviour of Hawkesbury sandstone at atmospheric pressure. *Eng Geol, 151*, 120–27.

Shao, S., Ranjith, P., Wasantha, P., & Chen, B. (2015). Experimental and numerical studies on the mechanical behaviour of Australian Strathbogie granite at high temperatures: an application to geothermal energy. *Geothermics, 54*, 96–108.

Tian, H., Ziegler, M., & Kempka, T. (2014). Physical and mechanical behavior of claystone exposed to temperatures up to 1000°C. *Int J Rock Mech Min Sci, 70*, 144–153.

Wang, J., & Park, H. (2001). Comprehensive prediction of rockburst based on analysis of strain energy in rocks. *Tunn Undergr Sp Technol, 16*(1), 49–57.

Xiaoli X., Karakus M., (2018). A coupled thermo-mechanical damage model for granite. *Int J Rock Mech Min Sci. (Accepted)*.

Zhang, W., Zhang, W., Sun, Q., Hao, S., Geng, J., & Lv, C. (2016). Experimental study on the variation of physical and mechanical properties of rock after high temperature treatment. *Appl Therm Eng, 98*, 1297–1304.

Zhao, X., Wang, J., Cai, M., Cheng, C., Ma, L., Su, R., & Li, D. (2014). Influence of unloading rate on the strainburst characteristics of Beishan granite under true-triaixal unloading conditions. *Rock Mech Rock Eng, 47*, 467–483.

Geomechanics and Geodynamics of Rock Masses – Litvinenko (Ed.)
© *2018 Taylor & Francis Group, London, ISBN 978-1-138-61645-5*

Gel explosives—a tool to improve the efficiency of drilling and blasting operations

Vladimir Eduardovich Annikov & Nikolay Ivanovich Akinin
Department of Technosphere Safety, D. Mendeleev University of Chemical Technology of Russia (MUCTR), Moscow, Russia

Vladimir Arnoldovich Belin
National Research Technological University "MISIS", Mining Institute, "Physical Processes of Mining and Geocontrol", Moscow, Russia

Denis Igolevich Mikheev
Department of Technosphere Safety, D. Mendeleev University of Chemical Technology of Russia (MUCTR), Moscow, Russia

Stanislav Ivanivich Doroshenko
Baltic State Technical University, «VOENMEH» of D.F. Ustinov, Candidate of Technical Sciences, Russia

Ivan Vladimirovich Brigadin
Promstroyvzryv Ltd., Scientific Consultant (Senior Researcher), Candidate of Technical Sciences, Saint Petersburg, Russia

Vitaliy Melissovich Gubaidullin
Promstroyvzryv Ltd. Director General, The Russian Federation, Saint Petersburg, Russia

Aleksandr Valerevich Shirokov
Candidate of Technical Sciences, The Russian Federation, Saint Petersburg, Russia

Alexei Nikolaevich Hasov
Promstroyvzryv Ltd. Researcher, The Russian Federation, Saint Petersburg, Russia

Vladimir Mikhailovich Mytarev
Gefest-M Ltd., The Russian Federation, Rezh, Russia

ABSTRACT: Results of the comparative analysis of measurements of impact parameters of explosions of several types of explosives in various environments are presented. The dependence of change of relative operability of water and gel structures was established. Results of experimental studies of influence on detonation characteristics of a compounding of water gels and their contents in explosives based on utilized single base propellant are presented. Influence of composition of water gel on detonation characteristics of explosive structures was established. On the base of results of studies by the electromagnetic method the features of detonation of water gel explosives based on expired single base propellants has revealed and interpretation of its mechanism was offered. It is shown that application of gunpowder water gel explosives for cutting non conditional rock formations and for using as a part of the combined charges at mass explosions allows to increase efficiency of drilling-and-blasting works considerably.

Certainly, the leading role in the modern field of destruction of rocks belongs to the emulsion explosives. In the same time increasing of efficiency and profitability of drilling-and-blasting works can be reached by complex application of emulsion explosives in a combination with traditional (regular) and/or new types of explosives.

Essentially new type of industrial explosive is Gelpor, based on single base propellant (pyroxyline gunpowder) taken from utilized ammunition.

Water gel explosives based on single base propellants was developed [1] and improved [2–4] in Russia. Single base propellants extracted from expired ammunition can be applied as components of the industrial explosives. Application of similar industrial explosives has shown increase in efficiency of conducting mining and also in destruction of concrete designs and other engineering works with application of explosion [8].

Keywords: pyroxyline powder, smokeless gunpowder, single base propellant, explosives, detonation, large-scale explosion, borehole charge, gel explosive, slurry, water gel, emulsion explosive, bottom amplifier, economic effect, borehole grid

1 EFFICIENCY OF APPLICATION OF WATER GEL EXPLOSIVES

On the mining pits of the Karelian Isthmus comparative tests on destruction of rocks by water gel, emulsion and regular industrial explosives were carried out.

For ensuring correctness of comparison of influence of a Gelpor with emulsion explosive in specific conditions of a pit at mass explosion the following methodology were accepted.

– boreholes with a Gelpor were placed between boreholes with emulsion explosive for an exception of edge effects;
– the boreholes which are completely equipped with Gelpor settled down nearby and compactly for concentration of the researched effect;
– boreholes with bottom amplifiers based on Gelpor were also placed compactly and adjacently with the boreholes equipped with Gelpor.

In case of that, methodical approach effects of application of Gelpor shall be the most expressed.

After excavation of rock mass visual survey in a foot wall of the unit obviously tracks the under mark of a face toe from 0,5 to 0,7 m. – in the location of the boreholes equipped with Gelpor or with the ground intensifier based on Gelpor. This fact demonstrates the reinforced destroying action of the Gelpor by itself and increasing of the destroying action of the main charge.

Taking into account the increased working capacity and bulk density of Gelpor, its integral efficiency in 2,0–2,5 times is higher, than emulsion explosives.

The main part of economic effect of application of Gelpor in case of mass explosions on mining pits consists in extension of a grid of boreholes with same diameter and depth of boreholes. For the conditional unit from 100 boreholes equipped with emulsion explosive using ground intensifier based on Gelpor charges, there is enough 80–85 boreholes.

At such approach, the efficiency of drilling-and-blasting works increases by 10% minimum.

For conditions of main types of granites of the Karelian Isthmus at the planned trench driving under a roadbed in Table 1 an assessment of relative efficiency of application of standard industrial and water gel explosives is given. The disposable driving of the site 500 m long and of 2800 m³ is assumed as a basis.

Emulsion explosive "Yarit" and powder ammonite are accepted as standard as the most widespread more preferably in the case under consideration. Generally the efficiency is understood as drilling-and-blasting works productivity.

In calculations the condition is accepted that standard industrial explosives are applied in a charges, and water gel explosive through a filling method. In the general assessment of efficiency also costs of equipment of boreholes are considered.

Table 1. Calculated values of general parameters of drilling-and-blasting works.

No	Explosive	Specific consumption, kg/m³	Mass of Explosive, kg	Diameter of borehole, mm	Number of boreholes	Relative Efficiency
1	2	3	4	5	6	7
1	Yarit	3,5	9750	70	1300	1
2	Ammonite	3,5	9750	70	1300	1
3	Gelpor	2,0	5600	70	330	4,1
4	Gelpor	1,9	5320	60	430	3,2
5	Gelpor	1,8	5040	50	590	2,5

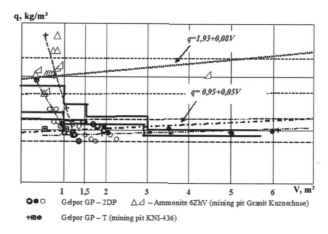

Figure 1. Dependences of a specific expense of explosive on non-conditional rock volume.

Important indicator of efficiency of Gelpor is the specific expense when cutting non-conditional rocks.

It is clearly possible to allocate two sites on Figure. 1. On the first site (up to 1,5 m³) the specific consumption of q is sharply cut, on the second (volume up to 6 m³) – insignificant, but noticeable growth is observed.

Reduction of q with growth of volume of non-conditional rocks V on the first site makes the "practical" sense connected with discretization of mass of charges of explosives.

Step approximation of dependence of q = f (v) for Gelpor is more convenient for practical estimates.

The dependence of q = f (v) and for ammonite has similar character: at the volume up to 1 m³ a specific expense more than 2,5 kg/m³, at non conditional rocks volumes about 3–5 m³–2,1 kg/m³.

The "physical" sense of a specific expense is revealed by dependence of q = f (v) for conditions of the minimum mass of a charge.

On the overall average effectiveness, application of Gelpor is better, than ammonite for 50–75%.

2 FEATURES OF A DETONATION AND OPERABILITY OF WATER GEL EXPLOSIVES

Water gel explosives based on expired single base propellants (SPB) structurally represent by mix of gunpowder elements (granulated gunpowder and/or their particles) with the water gel filling space between those elements. In that study seven-channel single base propellant was used. As the filler different gelled water solutions (water gels) on the base of polyacrylamide (PAA) were used. Composition of water gels are presented in Table 2.

Assessment of influence of composition of water gel on detonation ability is carried out. Results of experimental studies are given in Table 3.

The structure with inert water gel (Comp No. 3) steadily detonates in charges of 20–22 mm. Adding of oxidizer as a part of water gel (Comp No. 2) reduces the critical diameter of a detonation of water gel explosives based on single base propellant twice. The greatest detonation ability is received while using the water gel containing both oxidizer, and fuel (Comp No. 1).

Joint brining in of fuel and oxidizer to composition of filler provides maintaining detonation ability at considerable reduction of content of single base propellant in explosive. The received results of experimental studies have shown participation of water gel filler in detonation process.

For assessment of a mechanism of influence the research of detonation process by means of an electromagnetic method was conducted [11].

The standard profile of a detonation wave for regular explosive structure of Ammonite No. 6ZhV is presented in the Figure 2a, and characteristic dependence of change of pressure in a detonation wave for PVGS on the base of the used water gels of compositions No. 1, No. 2 and No. 3 respectively in Fig. 2b.

The parameters of a detonation wave for each structure average at least by three experiments are presented in Table 4.

The analysis of results of the experiments allows to offer the sequence of the processes happening in a detonation wave of PVGS charge.

The structure of a charge represents composition of two components—grains of the piroxyline gunpowder capable in certain conditions to a detonation, and the power-intensive water gel containing oxidizer and, in certain cases, fuel. Under the influence of a shock wave in grains of gunpowder there is a detonation extending on powder elements with a speed which isn't depending on composition of gel water solution. After achievement of a certain temperature and pressure in a detonation wave a part of energy of the wave is absorbed by water from gel that is expressed in delay or the termination of increase in pressure in the front of a detonation wave. At the same time the possibility of reaction of oxidizer with fuel within a detonation wave is provided that leads to increase in time of action of high pressure at an objects of destruction. Features of detonation processes are transformed to PVGS and in particular explosive characteristics.

Table 2.   Compositions of water gels.

| Component | Content, % mass. | | |
	Comp. №1	Comp. №2	Comp. №3
NH_4NO_3	45	50	–
$NaNO_3$	15	20	–
N_2H_4CO	10	–	–
H_2O	30	30	61,34
$CaCl_2$	–	–	38,66
PAA*	1,2	1,2	1,2

*Over 100%.

Table 3.   Critical diameter of water gel explosive in dependence from filler content.

Filler	Density of filler, g/cm³	Critical diameter, mm
Comp. №1	$1,38 \pm 0,02$	8–10
Comp. №2		10–12
Comp. №3		20–22

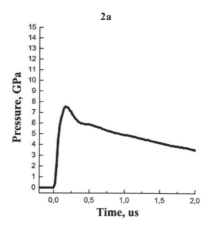

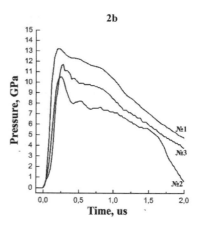

Figure 2. Standard profiles of change of pressure in a detonation wave: 2a—Ammonite No. 6ZhV; 2b – №1 (SBP/Comp. №1), №2 (SBP/Comp. №2), №3 (SBP/Comp. №3).

Table 4. Parameters of a detonation of PVGS depending on the used filler.

Filler	ρ, g/cm³	d₃, mm	D, km/s	P_{max}, GPa
Comp №1	1,42	20,4	6,56	15,17
Comp №2			5,86	10,28
Comp №3			5,67	10,29

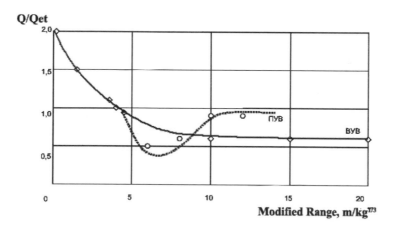

Figure 3. Changes in relative performance with distance.

Results of researches of specific explosive characteristics of charges of water gel explosives are stated in a number of articles [4–10]. In the Figure 3 changes of relative operability of Gelpor with distance is presented (Q – operability of Gelpor, Qet – operability of reference explosive).

The comparative nature of impact of explosions of charges of standard and water gel explosives in zones of destruction and negative action is presented in works [4–10].

3 CONCLUSIONS

1. Results of a research of detonation ability and parameters of a detonation of PVGS taking into account a compounding of water gel and its contents in PVGS are presented.

Influence of brining in of oxidizing and fuel components as part of water gel on detonation ability and parameters of detonation wave is shown.

2. The phenomenological mechanism of detonation process in PVGS that explaining the change of pressure in a detonation wave of similar explosive is offered.

3. Features of explosive characteristics of PVGS are revealed and the consistent pattern of change of relative operability of water gel explosives is determined.

4. Application of water gel explosives at mass explosions, special works and cutting of non-conditional rocks allows to increase efficiency of drilling-and-blasting works from 10% to 400%. It is necessary to recognize such growth of efficiency revolutionary.

REFERENCES

[1] Annikov V.E., Belin V.A., Smagin N.P. and others Vodosoderzhashhij porohovoj sostav (Water-containing gunpowder composition) *Patent of the Russian Federatiom*, № 2183209 since 26.12.2000.

[2] Annikov V.E., Oleynikov V.A., Sposob utilizacii (Method of utilization) *Patent of the Russian Federation*, № 2232739 since 27.06.2003.

[3] Annikov V.E., Kondrikov B.N., Oleynikov V.A. and others Sposob izgotovlenija porohovogo vodosoderzhashhego sostava (Method for manufacturing of gunpowdered water-containing composition) *Patent of the Russian Federation*, № 2253642 since 05.12.2003Г.

[4] Annikov V.E., Akinin N.I., Mikheev D.I., Soboleva L.I., Derzhavets A.S., Brigadin I.V., Doroshenko S.I. Ob osobennostjah detonacii i vzryvnogo vozdejstvija na gornye porody porohovyh vzryvchatyh veshhestv na gelevoj osnove (About features of detonation and impact on rock by powder water gel explosives) *Gorny informatsionno-analiticheskiy byulleten (Mining informational and analytical bulletin)*, Moscow, 2015, vol. 12, pp. 318–324.

[5] Belin V.A., Smagin N.P., Doroshenko S.I. Jeksperimental'nye issledovanija harakteristik PVM na gelevoj osnove (Experimental studies of the characteristics of industrial explosive materials on a gel basis) V*zryvnoe delo: Sbornik nauchnyh trudov/ Otdel'nyj vypusk Gornogo informacionno-analiticheskogo bjulletenja (Explosive case: Collection of scientific papers Separate issue of the Mountain Information and Analytical Bulletin)*, is.8, M.: Publishing house "World of Mining Books", 2007, с.143–148.

[6] Belin V.A., Doroshenko S.I. and others., Fizicheskie osnovy, tehnologicheskie shemy i jekonomicheskie pokazateli primenenija gelevyh PVV (Physical bases, technological schemes and economic indicators of application of gel industrial explosives) *Kompleksnaja utilizacija obychnyh vidov boepripasov: Sbornik dokladov. (Complex utilization of conventional types of ammunition: Collection of reports)*, – M.: Publishing house "Arms and Technologies" 2007, pp. 216–220.

[7] Brigadin I.V. and others. GEL'POR – mechta gornjaka?! Nekotorye rezul'taty ispytanij (gelpor - the dream of the mining?! Some test results) *Sbornik trudov chetvertoj mezhdunarodnoj nauchnoj konferencii «Fizicheskie problemy razrushenija gornyh porod»(Collection of Proceedings of the 4th International Scientific Conference "Physical Problems of Rock Destruction", IPCON RAS)*, M., 2005, pp. 391–394.

[8] Doroshenko S.I., Mikhailov N.P. and others Jeffektivnost' primenenija PVM na gelevoj osnove v inzhenernom dele (Efficiency of using industrial explosive materials on a gel basis in engineering) *Pjataja mezhdunarodnaja nauchnaja konferencija «Fizicheskie problemy razrushenija gornyh porod». Zapiski Gornogo institute (The Fifth International Scientific Conference "Physical Problems of Rock Destruction". Notes of the Mining Institute)*, is.171, 2007. – SPb.: SPSMU (TU), pp. 150–152.

[9] Doroshenko S.I., and others Nekotorye osobennosti parametrov podvodnyh vzryvov PVM na gelevoj osnove (Some features of the parameters of underwater explosions of industrial explosive materials on a gel basis) *Sbornik trudov chetvertoj mezhdunarodnoj nauchnoj konferencii «Fizicheskie problemy razrushenija gornyh porod», IPKON RAN (Collection of Proceedings of the Fourth International Scientific Conference "Physical Problems of Rock Destruction", IPCON RAS)*, M., 2005, с.394–397.

[10] Doroshenko S.I. Model' jenergovydelenija pri vzryve PVM na gelevoj osnove (Model of energy release in the explosion of industrial explosive materials on a gel basis) *Pjataja mezhdunarodnaja nauchnaja konferencija «Fizicheskie problemy razrushenija gornyh porod». Zapiski Gornogo instituta (The 5th International Scientific Conference "Physical Problems of Rock Destruction". Notes of the Mining Institute)*. is.180. SPb.: SPSMU (TU) 2009, pp.125–129.

[11] Zajcev V.M., Pohil P.F., Shvedov K.K. Jelektromagnitnyj metod izmerenija skorosti produktov vzryva (The electromagnetic method of measuring the velocity of explosion products), *Doklady AN SSSR (Reports of the USSR Academy of Sciences)*, 1960, vol. 132(6), pp. 1339–1340.

Geomechanics and Geodynamics of Rock Masses – Litvinenko (Ed.)
© 2018 Taylor & Francis Group, London, ISBN 978-1-138-61645-5

Dynamical destruction of rock mass due to excavation of a coal seam

A.S. Batugin
NUST MISIS, Moscow, Russia

V.N. Odintsev
IPKON RAS, Moscow, Russia

K.S. Kolikov
NUST MISIS, Moscow, Russia

Eu.I. Hotchenkov
SGM RAS , Moscow, Russia

ABSTRACT: The paper presents results of field and theoretical studies of geomechanical state of rock mass in a coal mine prone to rock bursts. The filed study was based on the method of geodynamical zoning and involved analysis of block structure of the mine field, natural and technogenic cracks in overburden rocks. In addition the study involved analysis of the open crack generation from the day surface into the rock mass with respect to the sequence of mining works and used a geomechanical model based on the FEM and the theory of cracks. It is shown that tensile crack development should proceed in a quasi-static mode as coal excavation progresses and the span of excavated space increases. It is also shown that the situation is inevitable when one of the tensile cracks provokes further rock fracture from this crack towards the excavation space by a shear mechanism. The shear fracture is unstable and must manifest itself as a rock burst leading to hard "landing" of a large block of rock mass on the ruined rock and soil in the excavated seam under its own weight.

Keywords: coal seam, overburden rocks, rock bursts, geodynamical zoning, field observations, theoretical assessments, tensile cracks, shear fracture, dynamical destruction

1 INTRODUCTION

The problem of rock bursts remains urgent for many countries worldwide in spite of the advance in this field (Petukhov I.M., 2004; Lasocki S. et al., 2017; Seedsman R.W., 2017; van Aswegen G., 2017; Lan Tianwei et al., 2015). It is recognized that rock mass structure and its stress state play the leading role in the mechanism of strong rock bursts (Petukhov I.M., 2004; Batugina I.M., Petukhov I.M. 1990; Lizurek G., Rudziński L., 2015; Brzovic. A, 2017; Tarasov, B.G. 2014; Lan Tainwei et al., 2012). In Russia a method of geodynamical zoning was developed more than 30 years ago to assess the geomechanical rock mass state and to identify dangerous sites in mines (Batugina I.M., Petukhov I.M., 1990; Petukhov I.M., Batugina I.M., 1999). According to the concept of geomechanical zoning hazardous zones arise in a mineral deposit area due to interaction between geodynamically active earth crust blocks of different hierarchical ranks generated as a result of global geodynamical processes.

During 2010 to 2012 a study based on geodynamical zoning was performed in the Huafeng mine in China (Qiao Jianyong et al., 2016). The mining of an inclined coal seam of 6 m thickness and 30° angle of incidence at a depth of about 1000 m (Fig. 1a) was associated with strong rock bursts of up to 10^6 J and generation of wide open, long cracks. The mine field is

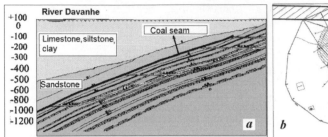

Figure 1.   Deposit geological section (a) and the plan of excavation works (b) with location of open cracks on the earth surface: 1 – block border; 2 – surface cracks; 3 – coal mine borderline; 4 – surface water drains; 5 – the number of the observation point.

located in a south limb of the Singmen syncline characterized by dislocations with breaks in continuity of an up to 28 m amplitude. The seam was excavated by system of longwalls along the strike. The roof rock is composed of aleurolite, sand stone and stone conglomerates. Power class and number of the rock bursts was increasing as mining works progressed deeper (Qiao Jianyong et al., 2016). Tens of rock bursts of higher than 1 magnitude occurred every month. The mining was also accompanied by abnormal rock movement and sudden inflow of underground water. Field and analytical studies of deformations and destruction of rock mass involving the method of geodynamical zonation were performed to understand these hazardous events and to reduce their negative effects.

2   FIELD OBSERVATIONS

Our field observations addressed block structure of the mining field, natural and techno-genical cracks in the undermined strata of rocks and interaction of earth surface subsidence with rock bursts (Qiao Jianyong et al., 2016). It was found that the earth surface underwent severe deformation under the effect of excavation of the coal seam with a 4 m maximum subsidence in the displacement syncline. When the earth surface deformation is greater than 2,8 mm/m, long open cracks looking like faults start to develop on the surface. The largest cracks demonstrated opening greater than 1 m with some fragments reaching several tens of meters in length.

The long cracks in the rock mass occur in front of the excavation and are oriented almost in parallel to the front (Fig. 1b). They have an apparent depth above 10 m, maximal depth of penetration into the undermined strata of rocks may reach several hundred meters. In par-ticular, well drilling in the area of such a crack has demonstrated that the technogenic zone of broken continuity of the rock covers the distance from the earth surface to the excavated space (Jing Jidong et al., 2006).

Microseismicity control was made using a Polish ARAMIS M/E recording station. More than 20 thousand seismic events of a 0.5 or higher magnitude were recorded with this station during 1995 through 2001. For example, Fig. 2a shows rock burst epicenters at the North-Eastern flank of the mining field.

It is also shown that the open cracks are associated mainly with natural large-block faults in the rock, which are characterized by a steep fall of the natural tectonic cracks.

Surveyors' measuring has also demonstrated that increased earth surface subsidence rate leads to strong rock bursts resulting in abnormal earth surface elevation at some areas of the mining field (Fig. 2b). In this figure the points in the column on the right side are points of observation.

Analysis of the field observations resulted in certain conclusions concerning characteristic features of geomechanical processes during coal excavation, though appropriate theoretical study is needed to understand these processes in a greater detail. A special model has been

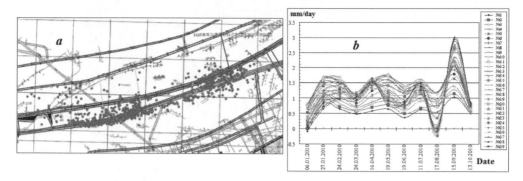

Figure 2.   Rock burst epicenters at the North-Eastern flank of mining field (a) and velocity of earth surface subsidence (b) along line No. 1 for 2010.

developed for this purpose that reflects principal conclusions made on the basis of the field observations. In particular, deep open cracks occur in the area of tensile strain in front of the mining and are oriented along the strike. Horizontal tensile strains in the upper part of rock mass are located at weakened areas, i.e. along block borders, which leads to relative block displacement and development of long surface cracks. Angle of incidence of the tensile cracks penetrating into the rock mass corresponds to the angle of incidence of natural cracks and is near 90°.

3   THEORETICAL ASSESSMENTS

The theoretical assessments used the elastic model of overburden rocks. Since the length of the excavated space along the strike resulting from the technology of coal extraction is much greater than the span of the longwall, we used the plain strain condition for active rock displacement area (Fig. 3a). This area includes the excavated space, has a great length underneath and on the left, is limited by the day surface from above and by the border of the "landing" block, on the right. One series of calculations involves the boundary condition of no horizontal displacements at the borderline of this block while the other one uses the condition of horizontal stresses. The Figure 3 shows small but the most relevant fragments of the large real computation area.

The purpose of the calculations was to assess changes in the strain-stress state of the rock occurring as the span was increasing and to assess conditions of tensile crack development. The finite element method was used in the calculations. The condition for crack growth was determined according the Griffith-Irwin criterion, and the calculation procedure was similar to that used in (Odintsev V.N., Miletenko N.A., 2015). The natural rock mass was under the load of its own weight, natural horizontal stresses were determined by the coefficient of the horizontal stress of 0.5, which corresponded to the geodynamical situation in the region in question. Averaged elastic modulus of overburden rocks was 5×10^3 MPa, Poisson's coefficient was 0.2. Since the tensile crack development occurred along natural contacts, the fracture toughness K_{1C} was determined by contact properties and was assumed 0.25 MPa·m$^{1/2}$.

As demonstrated by the calculations, a tensile stress area is generated in overburden rocks due to coal excavation. We considered a variety of numbers of cracks growing from the day surface into the rock mass and analyzed model conditions for their optimal growth. As demonstrated by the modeling a less favorable situation of crack development was observed in case of assuming the boundary condition of no horizontal displacements along the borderline of "landing" block.

Fig. 3a is one of many situations. It shows isolines of main tensile stresses involved in the tensile crack development criterion. The figure demonstrates a model situation occurring after landing of a large block of overburden rocks onto ruined rock and seam soil. The set of

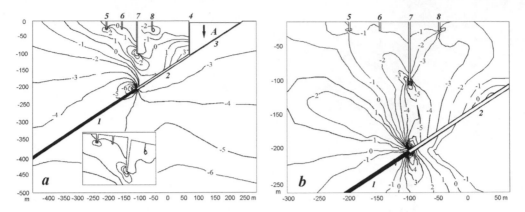

Figure 3.   Isolines of the stresses (MPa, tensile stresses are assumed to be negative): the least main stresses (a) and tangent stresses (b) in the rock mass with the span of excavated space of 200 m and mining depth of 200 m; 1 – coal seam, 2 – excavated space, 3 – "landing" block, 4 – technogenic fault, 5–7 – open tensile cracks, 8 – closed tensile crack.

such diagrams may help to determine crack condition in rock mass before the landing of the next large block due to coal excavation.

The figure inset shows schematically comparison of crack openings. Maximal opening is observed in the crack located on the line of stope projection on the day surface. In the case in question, the crack opening is about 0.5 m and depth of penetration into the rock mass is 100 m. Interestingly that the tensile crack number 8 started to develop when the roof span was 50 m, closed when the span increased to about 100 m. Similar calculations for other situations like this demonstrated that open cracks always occurred ahead of the excavation front. Tensile cracks located on the projection line of the stope have maximum length. Cracks in hanging rock mass near day surface become closed.

Fig. 4a shows a situation for a mining depth of 600 m and the span of 200 m. In this case, the crack in hanging rock mass was not taken into account. Similarly, to the previous example, the greatest crack opening was seen for the crack coaxial with the projection line of the excavation front. For this case, the opening was about 1 m, which corresponded to field data by the order of magnitude. Depth of crack penetration into the rock mass was about 330 m.

For comparison Fig. 5a shows a situation with five tensile cracks, none of them coaxial to the stope projection on the day surface. The calculations demonstrated that the deepest crack might reach a 310 m depth for a 600 m mining depth. The left neighbor crack (number 7) had a zero crack resistance in the calculations, which was a model situation with extremely weak natural possible joint contact. For this reason this crack grew to almost the same depth as the right one and had about the same surface opening of 1.2 m.

Basing on the calculation series concerning tensile cracks the following conclusions may be made. Open tensile cracks may grow for several hundred meters into the rock mass. The growth depth depends upon the tensile strength more accurately fracture toughness of possible joint contact. The open tensile cracks located on the day surface start to grow far before the excavated front approaches. Crack opening width depends on crack length: for mining depth of 600 m maximal opening width of cracks seen on the surface is about 1 m which corresponds to field observations. The open tensile crack in the hanging rock mass becomes a closed crack.

It follows from the calculations that the tensile crack should develop in a quasi-static mode, i.e. the crack grows slightly if the span of free excavated space increases but slightly. The crack stops to grow when crack growth criterion becomes invalid. In case the span of roof increases, a situation inevitably occurs when one of the existing tensile cracks provokes further fracture from this crack to rock mass by a shear mode. Figs. 3b and 4b show that when the tensile crack is approximately coaxial to the stope projection on the day surface, shear stresses are generated in the area between the stope and the tensile crack that are compatible with the shear strength

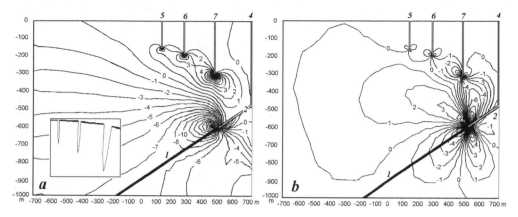

Figure 4. Isolines of the least main stresses (a) and tangent stresses (b) in the rock mass with the span of excavated space of 200 m and mining depth of 600 m: 1 – coal seam, 2 – excavated space, 4 – technogenic fault, 5–7 – open tensile cracks.

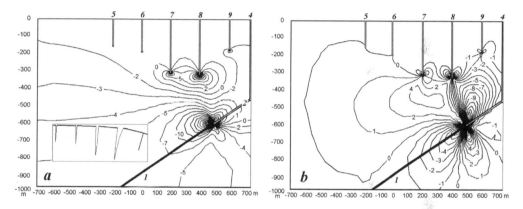

Figure 5. Isolines of main stresses (a) and shear stresses (b) in the rock mass at the span 200 m and mining depth of 200 m.

of rock mass. So, in this case a situation with further fracture is observed along the possible joint contact by the shear mechanism.

This situation was studied basing on assessment of validity of the Coulomb-Mohr criterion on the plane of the possible natural joint. The area of shear destruction was simulated as a cut with closed sides undergoing shear stress equal to rock slippage along the contact. The assessment showed that in case of shear fracture, the situation with development of this process becomes more favorable. Therefore, the process of shear fracture of rock mass is unstable and must occur in a dynamical mode. In practice, such shear fracture should lead to dynamical displacement of the large block of rock mass under the effect of its own weight, i.e. a strong rock burst.

It should be emphasized that the situation with the tensile crack stopping and further development by shear mechanism occurred with different task definitions. They include the number of initial cracks on the surface, consideration of local tensile cracks above the excavated space, setting of boundary conditions for stresses at the borderline of the landing block. Therefore, the conclusion about changing the mode of rock fracture obtains a theoretical basis. It should be noted that in case when the tensile crack is not coaxial to the stope projection on the day surface (Fig. 5b), the block landing is impossible due to kinematic condition, i.e. due to the block wedging when it sinks, though conditions for stresses of shear destruction are valid.

597

4 DISCUSSION OF RESULTS

The main conclusion of our study is the field and theoretical substantiation of the fracture mechanism of overburden rocks, i.e. transition of the fracture from one type to another (from tensile to shear fracture) accompanied by a change in the destruction mode from quasi-static to dynamical. The generated open technogenic tensile cracks close when rock mass is undermined. They develop further as shear cracks, which are much less dangerous in terms of serving as channels for penetration of surface water into underground space. A system of such closed cracks is clearly seen in analysis of periodicity of induced deformations in the undermined strata of rocks (Iophis M.A. et al., 2007).

As follows from our study a strong rock burst is a vertical dynamical displacement under the effect of the own weight of a large undermined block of rock mass limited from sides by techno-genic cracks and possible natural joints. This dynamical displacement ends with a hard landing of the block on the ruined rock in the excavated space. The relative elevation of earth surface after the block landing may be explained by unloading and vertical elastic response of a part of the rock mass adjacent to the plane of the technogenic displacement. The process of sequential generation and dynamical displacement of large blocks may explain systematic occurrence of rock bursts as the coal seam is excavated. Our in-field and theoretical studies helped understand better the observed geodynamical characteristics of Huafeng mine explotation and make proper prognostic assessment based on the method of geodynamical zoning up to a 1500 m depth.

REFERENCES

[1] Batugina I. M., Petukhov I. M., 1990. "Geodynamic Zoning of Mineral Deposits for Planning and Explotation of Mines, Oxford & IBH Publishing Co. Pvt. Ltd., New Delhi, 159 p.

[2] Brzovic A., Skarmeta J., Blanco B., Dunlop R., Sepulveda M. P., 2017. "Sub-horizontal Faulting Mechanism for Large Rock Bursts at the El Teniente Mine", Proceedings of the Ninth International Symposium on Rock Bursts and Seismicity in Mines, Nov. 15017, Santiago, Chile, pp. 124–132.

[3] Iophis M. A., Odintsev V. N., Blokhin D. I., Sheinin V. I., 2007. "Experimental Investigation of Spatial Periodicity of Induced Deformations in a Rock Mass", Journal of Mining Science, vol. 43, N2, pp. 125–131.

[4] Jing Jidong, Shi Longqing. Li Zilin et al., 2006. "Mechanism of Water Inrush from Roof in Huafeng Mine", Journal of China University of Mining & Technology, N5, pp. 642–647.

[5] Lan Tianwei, Zhang Hongwei, Han Jun, et al., 2012. "Study on Rock Burst Mechanism Based on Geo-stress and Energy Principle", Journal of Mining & Safety Engineering, vol. 29, N6, pp. 840–845.

[6] Lan Tianwei, Zhang Hongwei, Li Sheng, Han Jun, Song Weihua, Batugin A.S., Tang Guoshui, 2015. "Numerical Study on 4-1 Coal Seam of Xiaoming Mine in Ascending Mining". The Scientific World Journal, Volume (2015), Article ID 516095, 4 pages. DOI: 10.1155/2015/516095.

[7] Lasocki S., Orlecka-Sicora B., Mutke, G et al., 2017. "A Catastrophic Event in Rudna Copper-ore Mine in Poland on 29 November, 2016: what, how and why", Proceedings of the Ninth International Symposium on Rock Bursts and Seismicity in Mines, Nov. 15017, Santiago, Chile, pp. 316–324.

[8] Lizurek G., Rudziński L., 2015. "Mining Induced Seismic Event on an Inactive Fault", J. Acta Geophysica, vol. 3, no. 1, pp. 176–200. DOI: 10.2478/s11600-014-0249-y.

[9] Odintsev V. N., Miletenko N. A., 2015. "Water Inrush in Mines as a Consequence of Spontaneous Hydrofracture", Journal of Mining Science, vol. 51, N3, pp. 423–434. DOI: 10.1134/S1062739115030011.

[10] Petukhov I. M., 2004. "Rock Burst in Coal Mines", MNC VNIMI, Saint Petersburg, 237 p.

[11] Petukhov I. M., Batugina I.vM., 1999. "Geodynamics of Earth Interior", Nedra Communication, Moscow, 288 p. (in Russian)

[12] Qiao Jianyong, Batugin A. S., Batugina I. M., Yu Lijiang, Zhao Jingli, 2016. "The Conditions of Geodynamic Phenomena at Huafeng Mine in China", Sputnik+, Moscow, 144 p. (in Russian).

[13] Seedsman R. W., 2017. "Application of Rock Burst Concepts to Understanding the Sudden Collapse of Ribs in Coal Mines", Proceedings of the Ninth International Symposium on Rock Bursts and Seismicity in Mines, Nov. 15017, Santiago, Chile, pp. 297–303.

[14] Tarasov, B. G. 2014. "Hitherto Unknown Shear Rupture Mechanism as a Source of Instability in Intact Hard Rocks at Highly Confined Compression", J. Tectonophysics, vol. 621, pp. 69–84.

[15] van Aswegen G., 2017. "Seismic Sources and Rock Burst Damage in South Africa and Chile", Proceedings of the Ninth International Symposium on Rock Bursts and Seismicity in Mines, Nov. 15017, Santiago, Chile, pp. 72–86.

Geomechanics and Geodynamics of Rock Masses – Litvinenko (Ed.)
© *2018 Taylor & Francis Group, London, ISBN 978-1-138-61645-5*

Quantitative assessment of variability in values of Geological Strength Index (GSI)

A. Bedi & M. Invernici
Bedi Consulting Ltd., London, UK

J.P. Harrison
University of Toronto, Toronto, Canada

ABSTRACT: This paper presents a new approach for quantitatively incorporating variability of rock mass conditions in assessments of Geological Strength Index (GSI) within the generalised Hoek-Brown strength criterion (Hoek *et al.*, 2002). This paper presents a fully quantitative relationship to calculate GSI using the Rock Quality Designation (RQD) rating to characterise block volume and the Barton-Bandis joint shear strength parameters to define discontinuity surface condition. All of the parameters proposed in the new relation may be objectively measured using simple index tests, and so uncertainty in their characterisation defined probabilistically.

Keywords: Geological Strength Index (GSI), probabilistic, Hoek-Brown criterion, Limit State Design

1   INTRODUCTION

The generalised Hoek-Brown strength criterion (Hoek *et al.*, 2002) is widely used to predict the peak strength of jointed rock masses, with the parameters of the criterion often obtained in terms of GSI. A semi-quantitative approach to obtaining GSI values is available (Hoek *et al.*, 2013), and later work (Morelli, 2015; 2017) proposed techniques for including uncertainty through a combination of empirical relations, objectively measured values of rock mass structure and subjective assessments of discontinuity surface condition. The inclusion of subjective assessments in both these approaches indicates that uncertainty is being considered in an epistemic sense (Bedi & Orr, 2014; Hudson & Feng, 2015).

Probabilistic methods, which form the basis of Limit State Design (LSD) codes, are finding increasing application in rock engineering design as a means of quantifying risk and uncertainty. These methods require that uncertainty in design variables follows the aleatory, i.e. stochastic model, and as such need variability to be quantified by objective measurement. Given the widespread use of GSI as a design variable, it is necessary that variability in its assessment be in aleatory and not epistemic terms.

In this paper we present a fully quantitative relationship to calculate GSI that uses the RQD rating to characterise rock mass structure and joint shear strength parameters to define discontinuity surface condition. As these parameters may be objectively measured using simple index tests, uncertainty in their characterisation becomes aleatory and thus accords with LSD principles. The paper concludes with a case study showing the utility of our approach, and in particular its ability to quantify variability in GSI.

2   UNCERTAINTY IN ROCK MASS STRENGTH

In the equations above, GSI is the governing parameter defining the rock mass strength and is often estimated by comparing a linguistic description of blockiness and the surface condition

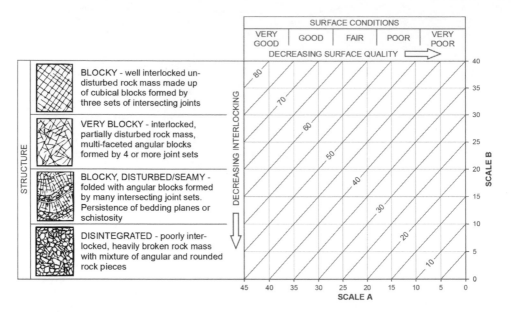

Figure 1. The basic structure of the *GSI* chart and possibilities for quantification (after Hoek *et al.*, 2013; Carter & Marinos, 2014).

of discontinuities to a tabulated range (see Fig. 1) (Carter and Marinos, 2014). This qualitative and subjective nature of assessment introduces epistemic uncertainty. In an attempt to reduce this epistemic uncertainty, several authors suggested methods aimed at making the assessment of GSI more quantitative. Indeed, Hoek *et al.* (2013) presented an updated GSI chart, amenable to quantification (see Fig. 1, below).

2.1 *Existing semi-quantitative methods*

Various semi-quantitative relationships to evaluate GSI are available in the current literature (Morelli, 2015, 2017; Hoek *et al.*, 2013; Russo, 2009; Cai *et al.*, 2004). All of these methods evaluate GSI through two sets of parameters on two scales. The parameters on the horizontal axis (scale A) characterise joint structure and conditions of discontinuity surfaces in the rock mass, while those on the vertical axis (scale B) define the lithology and blockiness (Carter and Marinos, 2014).

Cai *et al.* (2004) suggested calculating GSI based on block volume (V_b) and a joint condition factor (J_c), which itself is evaluated by combining three ratings from the Q (Barton *et al.*, 1974) and RMi (Palmström, 1995) systems. A similar approach was followed by Russo (2009), who suggested using V_b and a joint condition factor (jC) from the *RMi* system. Hoek *et al.* (2013) suggested assigning a rating based on RQD (Deere, 1963) and the joint condition index J_{Cond89} (after Bieniawski, 1989), or a correlation based on RQD and two indices from the Q system (Barton *et al.*, 1974). All these correlations rely on subjectively assessed parameters—which are expressed through qualitative and descriptive scales—to rate joint structure and conditions that form scale A, thereby introducing epistemic uncertainty.

3 PROPOSED FULLY QUANTITATIVE ASSESSMENT OF GSI

Hoek *et al.* (2013) proposed a semi-quantitative correlation,

$$\text{GSI} = \frac{52\left(J_r/J_a\right)}{1+\left(J_r/J_a\right)} + \frac{\text{RQD}}{2} \tag{1}$$

that uses the quotient J_r/J_a to characterise the shear strength of the discontinuities (Barton et al., 1974) on scale A, with a function of RQD to characterise block volume on scale B. These authors suggest that the parameters J_r and J_a be assessed subjectively using the Q system, which would introduce epistemic uncertainty.

Through a back-analysis of shear strength test data Barton (1973) confirmed that the function $\tan^{-1}(J_r/J_a)$ agreed well with the total friction angles (i.e. combined cohesion and friction) objectively measured using tilt tests. Later, Barton et al. (1974) noted that the angles given by the function $\tan^{-1}(J_r/J_a)$ closely resemble the actual frictional shear strength of a rock joint that results from joint wall roughness and alteration. Thus, we obtain

$$\tau = \sigma_n \tan(\phi_a) = \sigma_n (J_r/J_a) \tag{2}$$

where ϕ_a is the apparent friction angle.

The shear strength of discontinuities may also be predicted using the Barton-Bandis criterion (Barton & Choubey, 1977),

$$\tau = \sigma_n \tan\left[JRC \log_{10}(JCS/\sigma_n) + \phi_r \right], \tag{3}$$

and so a comparison of Eqs. (2) and (3) shows that

$$\tan(\phi_a) = (J_r/J_a) = \tan\left[JRC \log_{10}(JCS/\sigma_n) + \phi_r \right]. \tag{4}$$

Eq. (4) thus shows that Eq. (1) may be written as

$$GSI = \frac{52 + \tan\phi_a}{1 + \tan\phi_a} + \frac{RQD}{2}, \tag{5}$$

which is a new quantitative expression for GSI. We see that Eq. (5) expresses scale A using $52\tan\phi_a/(1+\tan\phi_a)$ and scale B using $RQD/2$, with ϕ_a, the apparent friction angle, being given as

$$\phi_a = JRC \log_{10}(JCS/\sigma_n) + \phi_r. \tag{6}$$

As shown in Fig. 2, the apparent friction angle given by Eq. (6) is entirely defined in terms of quantitative parameters that can be objectively measured using simple index tests.

When a sufficient number of tests are performed to quantify the parameters on Scale A and Scale B, an aleatory model may be fitted to quantify the uncertainty in each of these scales. Thus, epistemic uncertainty is removed and using Eq. (5), GSI can be calculated and an aleatory model fitted to characterise the uncertainty resulting from inherent variation.

The new quantitative relation proposed here is subject to the limitations of both the Barton-Bandis criterion and of the use of GSI to characterise peak strength of fractured rock masses in general. A summary is as follows:

- For intact massive or very sparsely jointed rock, the GSI chart should not be used for input into the Hoek-Brown criterion because there are too few pre-existing discontinuities to satisfy the conditions of homogeneity and isotropy which is a key assumption of the criterion (Hoek et al., 2013).
- For unfilled discontinuities the roughness and compressive strength of the walls are important, whereas in the case of filled discontinuities the physical properties of the material separating the discontinuity walls are of primary concern. Barton's criterion is thus only valid for those cases where the opposing discontinuity surfaces are in contact (Wines & Lilly, 2003).
- Within the JRC range of between 0 and 8, JRC values can be reliably obtained using tilt tests. Rougher surfaces, up to JRC values of about 12, can be tested with push/pull tests (Barton & Choubey, 1977).

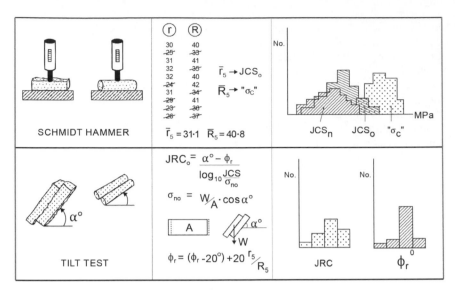

Figure 2. Objective test methods for index testing of discontinuity surfaces, and typical resulting frequency distributions for use with aleatory models (from Barton, 2013).

- To extend the application of the proposed quantitative GSI relation, additional work is required to investigate the applicability of various test methods over the whole range of JRC.

4 EXAMPLE APPLICATION OF FULLY QUANTITATIVE GSI RELATION

Applicability of the proposed quantitative GSI relation is demonstrated using data from the nuclear waste repository site at Oskarshamn, in Sweden. Borehole logging and tilt tests were carried out at the Swedish National Testing and Research Institute (SP) for the Swedish Nuclear and Fuel Waste Management Company (SKB). All test results were obtained from publically available data reports posted online at http://www.skb.se.

As part of the ground investigation regime to assess feasibility of an appropriate site, a 1000 m deep borehole was drilled through the granite/granodiorite rock mass. RQD was measured directly from the core, with J_r and J_a being measured through a total of 47 tilt tests. As all the parameters required by Eqs. (5) and (6) have been objectively measured, it is appropriate to produce frequency histograms to quantitatively characterise variability. Both the empirical and fitted cumulative distribution functions for Scale A and Scale B are presented in Fig. 3.

Using the objectively measured CDFs shown in Fig. 3, a Monte-Carlo simulation was used with Eq. (5) to obtain values of GSI. The resulting distribution of GSI was found to follow a Beta distribution, as illustrated in Fig. 4. As a result of obtaining this CDF for GSI, we are now able to make probabilistic statements regarding the occurrence of particular GSI values. Thus, we see that GSI has a mean value of 77 and a standard deviation of 8, and that there is 95% probability that GSI will be greater than 65.

Finally, in order to verify the appropriateness of the proposed quantitative method to calculate GSI, Fig. 5a presents a comparison of the GSI distribution determined from a Monte-Carlo simulation of Eq. (5) using aleatory PDFs fitted to the objective test data and the distribution determined from Monte-Carlo simulations using subjective estimates of J_r and J_a determined by a qualified geologist. Fig. 5b presents an analysis of the difference in GSI values obtained between GSI logged by the geologist and that obtained using the quantitative method presented in this paper. It is evident that the two distributions correspond closely, with the maximum difference in GSI value being approximately 4.

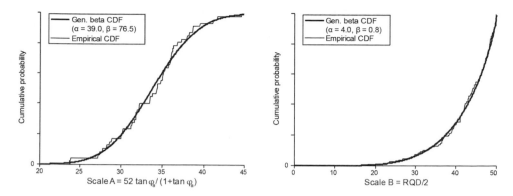

Figure 3.   Cumulative distribution functions fitted to data obtained by SKB.

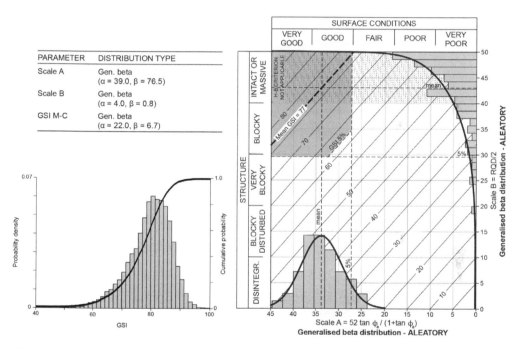

PARAMETER	DISTRIBUTION TYPE
Scale A	Gen. beta (α = 39.0, β = 76.5)
Scale B	Gen. beta (α = 4.0, β = 0.8)
GSI M-C	Gen. beta (α = 22.0, β = 6.7)

Figure 4.   PDF and CDF of GSI, resulting from aleatory models fitted to GSI Scale A and Scale B.

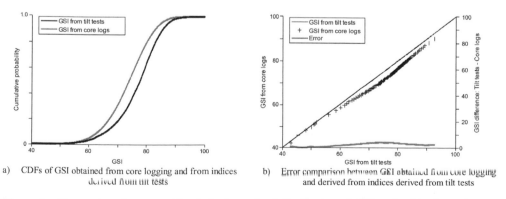

a)   CDFs of GSI obtained from core logging and from indices derived from tilt tests

b)   Error comparison between GSI obtained from core logging and derived from indices derived from tilt tests

Figure 5.   Comparison between objectively determined distribution of GSI and subjective estimates obtained by a field geologist at this site.

5 CONCLUSIONS

A new relation for GSI has been presented that uses objectively measured rock mass properties. This use of objective measurements allows a probabilistic interpretation of GSI and thus renders it suitable for use in probabilistic design. Using a case example from the SKB Oskarshamn site in Sweden, we show that at this site the distribution of GSI, and the surface condition and structure scales can all be characterised by a beta distribution.

We suggest that this quantitative means of assessing GSI may be applied by logging RQD and undertaking tilt tests on joints from cores obtained during early site investigation. This would allow a quantitative means of determining GSI during the early stages of design. Of particular importance, the applicability of probabilistic distributions to characterise GSI means that this critical rock mass parameter may be applied using Limit State Design principles.

REFERENCES

Barton N. (1973). Review of a new shear strength criterion for rock joints. *Eng. Geol.*, 7:28–332.
Barton N, Choubey V. (1977). The shear strength of rock joints in theory and practice. *Rock Mech.*, 10(1–2):1–54.
Barton N, Lien R, Lunde J. (1974). Engineering classification of rock masses for the design of tunnel support. *Rock Mech.*, 6:189–236.
Barton N. (2013). Shear strength criteria for rock, rock joints, rockfill and rock masses: Problems and some solutions. *J. Rock Mech. Geotech. Eng.*, 5:249–261.
Bedi & Orr (2014). On the applicability of the Eurocode 7 partial factor method for rock mechanics. L.R. Alejano, A. Perucho, C. Olalla, R. Jiménez (Eds.), Rock engineering and rock mechanics: structures in and on rock, Proceedings of Eurock 2014, ISRM European Regional Symposium, Vigo, CRC Press/Balkema, Leiden. pp. 1517–1523.
Bieniawski ZT. (1989). *Engineering rock mass classifications*. John Wiley & Sons.
Cai M, Kaiser PK, Uno H, Tasaka Y, Minami M. (2004). Estimation of rock mass deformation modulus and strength of jointed hard rock masses using the GSI system. *Int. J. Rock Mech. Min. Sci.* 41:3–19.
Carter TG, Marinos V. (2014). Use of GSI for Rock Engineering Design. 1st Int. Conf. Applied Empirical Design Methods in Mining, Lima 2014.
Deere DU. (1963). Technical description of rock cores for engineering purposes. *Rock Mech. Eng. Geol.*, 1(1):16–22.
Hoek E, Carranza Torres C, Corkum B. (2002). Hoek–Brown failure criterion—2002 edition. Proc. 5th North American Rock Mech. Symp., Toronto, Canada, vol. 1, 2002. p. 267–73.
Hoek E, Carter TG, Diederichs MS. (2013). Quantification of the Geological Strength Index chart. 47th US Rock Mechanics/Geomechanics Symposium (ARMA 13-672), San Francisco, CA, USA.
Hudson, JA & Feng, XT. (2015). *Rock Engineering Risk*. CRC Press/Balkema, Taylor & Francis Group.
Morelli GL. (2015). Variability of the GSI Index Estimated from Different Quantitative Methods. *Geotech. Geol. Eng.*, 33:983–995.
Morelli GL. (2017). Alternative quantification of the Geological Strength Index chart for jointed rocks. *Geotech. Geol. Eng.*, 35:2803–2816.
Palmström A. (1995). RMi – A Rock Mass Characterisation System for Rock Engineering Purposes. PhD thesis, University of Oslo, Norway.
Russo G. (2009). A new rational method for calculating the GSI. *Tunn. Undergr. Space Tech.*, 24:103–111.
Wines DR, Lilly PA. (2003). Estimates of rock joint shear strength in part of the Fimiston open pit operation in Western Australia. *Int. J. Rock Mech. Min. Sci.*, 40:929–937.

Geomechanics and Geodynamics of Rock Masses – Litvinenko (Ed.)
© *2018 Taylor & Francis Group, London, ISBN 978-1-138-61645-5*

The effect of stress level on the compressive strength of the rock samples subjected to cyclic loading

Melek Hanım Beşer & Kerim Aydiner
Karadeniz Technical University Mining Engineering Department, Trabzon, Turkey

ABSTRACT: The effect of stress level on the compressive strength of rock samples is researched using Tephrite samples subjected to cyclic loading. Four sets of stress level (% 20, %30, %40 and %50 of uniaxial compressive strength) were applied for cyclic loading tests on dry and saturated samples by 200 tons servo-hydraulic testing rig. The compressive strength of the rock by different stresses were determined under constant frequency and number of cycles at failure. The results of the cyclic loading tests indicate that stress level has significant influence on the uniaxial compressive strength. An increase in stress resulted in a decrease in compressive strength proportionality. In the case of the saturated samples, it was found that compressive strength reduced by approximately 25 per cent.

Keywords: Stress Level, Cyclic Loading, Rock Fatigue, Uniaxial Compressive Strength

1 INTRODUCTION

Rock units and tunnels, dams, roads, bridges and underground structures created in rock masses are exposed to static and dynamic loads (Bagde and Petros, 2009; Momeni et al., 2015). Static loads are the weight of the rock mass on any rock unit or structure. Dynamic loads are defined as loads that change with time and are caused by earthquakes, drilling and blasting, loading, mechanical excavation and heavy traffic. The understanding of mechanical properties of rock under dynamic loads is more crucial than under the static load.

Different rock materials exhibit different behavior under dynamic loading conditions. Some becomes stronger and ductile, but some becomes weaker and more brittle (Stavrogin and Tarasov, 2001; Liu and He, 2012). Limited numbers of past research on the topic have revealed that the cyclic loads cause weakening of rock mechanical properties. This has been called "rock fatigue" (Singh, 1989; Eraslan, 2011). Firstly, Burdine (1963) showed the weakening of the material under compressive cyclic loading. The detailed literature research on fatigue behavior in rocks has already been presented by Bagde and Petros (2005). Thereafter, many studies are reported about the effects of the cyclic loading. Bagde and Petros (2009) studied the effects of uniaxial cyclic compression tests on sandstone and conglomerate rock samples in different amplitudes and frequencies. These researchers reported that fatigue failure is influenced by the petrographic, physical and mechanical properties of these rocks. Liang et al. (2012) studied the effect of cyclic load on samples of tenardite, glauberite and gypsies under uniaxial compressive strength and founded that the uniaxial compressive strengths of of tenardite, glauberite and gypsies decreased by 34%, 19% and 35%, respectively. Khanlari and Momeni (2014) investigated the fatigue behavior of Monzogranite subjected to various loading levels (85%, 90% and 95% of the strength of the rock) at amplitude %70 with 1 Hz cyclic loading frequency. They observed that different loading levels have effect on the fatigue behavior of the rock. For the first few cycles, it concluded that the rock displays elastic behavior and elasto-plastic behavior with increasing cycle numbers. As the result of studies, the yield stress level reduces, and the plastic behavior of the rock is dominant in each cycle with the increasing cycle numbers. Taheri et al. (2016) showed peak strength variations of Hawkesbury sandstone under cyclic loading and detected that fatigue influenced the peak strength of the sandstone. It is reported that fatigue failure occurs during cyclic loading if the stress level in cyclic loading is equal or higher than 94% of rock peak strength. All of these studies revealed that different materials showed different behaviors in cyclic conditions. There is some missing about the fatigue behavior of the dry and saturated rocks for low cyclic loads. The study was aimed to fill this gap. In this paper, different stresses that are one of the important variables of fatigue mechanism were performed in the experiments and the effect of the different stress levels on the cyclic loading of the rock samples for dry and saturated condition was investigated. Also, the compressive strength of the rock under cyclic compressive loading conditions is evaluated for Tephrite samples. This paper presents the preliminary results of an ongoing research. The number of the samples and the amount of data presented are relatively limited. However, the rock response under the cyclic loading may be seen explicitly, when especially combined with the literature work.

2 EXPERIMENTAL SET-UP AND ROCK PROPERTIES

2.1 Rock samples

Thin sections were prepared to analyze the mineralogical and petrographic properties of the rock. According to the results of analyses, this rock generally shows porphyritic, microlitic porphyritic and hyalo-microlitic porphyritic texture. It is observed that main components of most of these rocks are leucite, calcitic plagioclase, pyroxene (%20), amphibole (3–5%), biotite (3–5%) and opaque (2%) minerals. The secondary minerals are chlorite, calcite, zeolite and clay minerals. As a result of microscopic examination, this rock is located tephrite/basanite groups. Also, this rock is named as tephrite due to the olivine absence.

Rock samples were cut to a length to diameter ratio of 2.5–3.0 with a diameter of NX core size, approximately 54 mm. The ends and sides of the specimens were prepared to ISRM (2007) testing procedures. The physical and mechanical properties of the rock samples were determined (see Table 1).

2.2 Testing methodology

A series of uniaxial cyclic loading tests were performed on the core specimens using a servo-hydraulic Besmak Machine. Four different stress levels (%20, %30, %40 and %50 of uniaxial

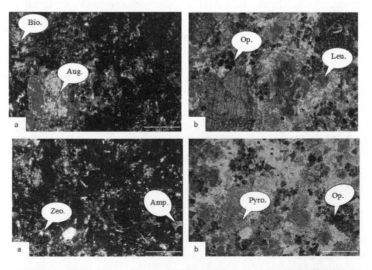

Figure 1.   Microscopic view of the rock (Pyro: Pyroxene, Amp: Amphibole, Bio: Biotite, Ze: Zeolite, Leu: leucite, O: Opaque mineral, Aug: Augite).

Table 1. Physical and mechanical properties of the rock samples.

Material properties	Trabzon rocks
Density (g/cm³)	2.59
Uniaxial Compressive Strength (MPa)	98.81
Tensile Strength (MPa)	6.42
Apparent Porosity (%)	3.74

Figure 2. Cracks features in specimen after failure.

compressive strength (UCS) are applied under constant frequency (0.5) and number of cycles (100). Each experiment was repeated three times, and their average was used as the result of the test. These experimental procedures were implemented for dry and saturated rocks. The saturated samples were heated, dried and cooled to room temperature and then saturated through free immersion. These samples were tested after 2 days of saturation in water. After cyclic loading until 100 cycles, UCS of these specimens is immediately determined again.

3 EXPERIMENTAL RESULTS AND DISCUSSION

Experiments were performed to determine the effect of the stress level on the strength of the rock. The strengths of the rock obtained after cyclic loads are evaluated by comparing with the static uniaxial compressive strength (98.8 MPa). After cyclic loading tests, the fractures occur, propagate and these fractures coalesce. These three steps have also been reported by many researchers (Xiao et al., 2010; Liu and He, 2012; Momeni et al., 2015). The Tephrite specimen that failed under uniaxial compression cyclic loading is shown in Figure 2. The applied stress level has effect on the percentage decrease in uniaxial compressive strength (Ray, 1999; Bagde and Petros, 2009). In this study, the graph was plotted to investigate the effect of the applied stress level on the strength of the rock at fixed frequency (0.5 Hz) and number of cycles (100). It shows that the strength of the rock reduces with increasing the stress level and also, relation between UCS of the rock and stress level is inversely proportional as shown in Figure 3. Momeni et al. (2015) stated that the deformation development in the rocks

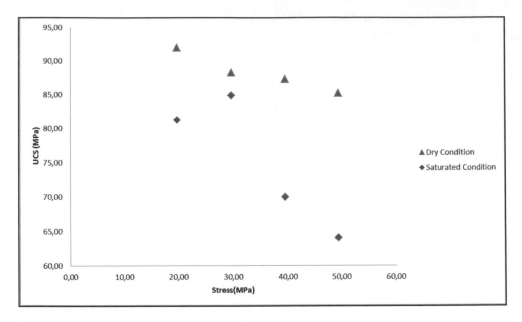

Figure 3.   The graph versus UCS and load under cyclic loading.

in the fatigue process started with the formation of capillary cracks in the rock mass. While the stress level continues, the cracks propagate and this causes rock failure. Hence, the UCS of the rock decreases with increasing of the applied stress level. It was determined that the uniaxial compressive strength of the saturated rock samples decreased by 13%. The decrease in fatigue strength (nearly %25 of UCS of the rock) in the case of saturation was also obvious and related to pore pressure effect and internal structural changes of the rock specimens as seen in the Figure 3. Similar observations were made by Burdine (1963) and Tien et al. (1990). While stress level is at 20 MPa for saturated condition, there is an increase in uniaxial compressive strength due to machine behavior and the testing conditions.

4   CONCLUSION

The work presented here was performed to investigate the effects of stress level on strength of dry and saturated rock in uniaxial compression subjected to cyclic loading. The following conclusions were drawn from this research.

1. The fatigue strength of the rock generally reduced by 10–15% on average under cyclic loading compared to static strength of the rock. Also, it was determined that uniaxial compressive strength for saturated rock samples reduced by %13 of uniaxial compressive strength in terms of the water pressure in fissure. Also, fatigue strength decreased by % 25 of uniaxial compressive strength.
2. The linear relation between stress level and strength of the rock are observed. The compressive strength decreases under constant cycle number and constant frequency, by an increase of the stress level.
3. The fatigue mechanism is affected by mineralogical and petrographic properties and loading variables such as frequency, amplitude and loading cycles. Therefore, these parameters need to be considered to understand the fatigue mechanism of the rock. As a result, more studies should be conducted to understand the rock behavior under cyclic loading.

ACKNOWLEDGEMENTS

The authors would like to thank to the Scientific Research Fund of Karadeniz Technical University for financial support for this research (FYL-2016-5658).

REFERENCES

Bagde, M.N. and Petros, V., 2005. Fatigue properties of intact sandstone samples subjected to dynamic cyclical loading, Int. J. Rock Mech. Min Sci. 42, 237–250.

Bagde, M.N. and Petros, V., 2009. Fatigue and dynamic energy behavior of rock subjected to cyclical loading, Int. J. Rock Mech. Min. Sci. 46, 200–209.

Burdine, N.T., 1963. Rock failure under dynamic failure conditions, Soc. Petr Eng J. 3, 1–8.

Eraslan, N., 2011. Static and Cyclic Laboratory Testing of Brisbane Rocks, PhD Thesis, School of Civil Engineering, The University of Queensland, Australia.

ISRM. Ulusay, R. and Hudson, J.A. (eds)., 2007. The complete ISRM suggested methods for rock characterization, testing and monitoring, 1974–2006. Suggested methods prepared by the commission on testing methods. Compilation arranged by the ISRM Turkish National Group. ISRM, Ankara, 87–88.

Khanlari, Gh. and Momeni, A.A., 2014. Assessment of fatigue behavior of Alvand Monzogranite Rocks, Journal Geology Engineering, Vol. 8, No. 1.

Liang, W., Zhang, C., Gao, H., Yang, X., Xu, S. and Zhao, Y., 2012. Experiments on mechanical properties of salt rocks under cyclic loading, J Rock Mec. and Geo. Engng, 4 (1), 54–61.

Liu, E. and He, S., 2012. Effects of cyclic dynamic loading on the mechanical properties of intact rock samples under confining pressure conditions. Engineering Geology 125, 81–91.

Momeni, A., Karakus, M., Khanlari, G.R. and Heidari, M., 2015. Effects of cyclic loading on the mechanical properties of a granite, Int. J. Rock Mech. Min. Sci. 77, 89–96.

Ray, S.K., Sarkar, M. and Singh, T.N., 1999. Effect of loading and strain rate on the mechanical behaviour of sandstone, Int J Rock Mech. Min. Sci. 36, 543–549.

Singh, S.K., 1989. Fatigue and strain hardening behaviour of greywacke from the flagstaff formation, NSW, Engineering Geology, 26, 171–179.

Stavrogin, A.N. and Tarasov, B.G., 2001. Experimental physics and rock mechanics, Balkem (Roterdam), 356.

Taheri, A., Royle, A., Yang, Z. and Zhao, Y., 2016. Study on variations of peak strength of a sandstone during cyclic loading, Geomech. Geophy. Geo-energy Geo-resour. 2, 1–10.

Tien, Y.M., Lee, D.H. and Juang, C.H., 1990. Strain, pore pressure and fatigue characteristics of sandstone under various load conditions. Int J Rock Mech Min Sci Geomech Abstr, 27(4), 283–9.

Xiao, J.Q., Ding, D.X., Jiang, F.L. and Xu, G., 2010. Fatigue damage variable and evolution of rock subjected to cyclic loading, Int. J. Rock. Mech. Min. Sci. 47, 461–468.

Geomechanics and Geodynamics of Rock Masses – Litvinenko (Ed.)
© *2018 Taylor & Francis Group, London, ISBN 978-1-138-61645-5*

Normandy cliff stability: Analysis and repair

Jean-Louis Briaud

Distinguished Professor, Texas A&M University, USA

ABSTRACT: The cliffs and beaches of Normandy in France were the scene of the D-Day invasion on 6 June 1944 during World War II. One of the landing site was Pointe du Hoc where the German defense was set up on top of 30 m high limestone cliffs. The infrastructure of this defense included canon bunkers back from the front of the cliff and an observation post near the edge of the cliff. In 1944, the Observation Post was 20 m away from the edge but in 2004 that distance was down to 10 m. The reason was the continued erosion process by the sea and associated failures of very large blocks of cliff. The situation became precarious enough that the French administration closed the Observation Post to visitors.

In 2006, the American Battle Monument Commission cooperated with the French administration to slow down the erosion process in order to save the historic monuments from falling into the sea. The study conducted by Texas A&M University included field tests with geophysical equipment, coring and sampling of the rock, laboratory testing, and numerical simulations. The study showed that the failure mechanism was the development of deep caverns at the base of the cliffs followed by collapse of the overhang when the depth of the caverns reached a point where the tensile strength of the rock mass was insufficient to carry the rock mass in cantilever. The repair scheme consisted of two parts: backfilling of the caverns and tying the periphery of the foundation of the observation post to deep micropiles.

In 2011, the repairs were effected using esthetic concrete to backfill the caverns and deep micropiles around the Observation Post. The Observation Post was reopen to visitors after a ceremony in the presence of D-Day invasion soldiers as well as US and French dignitaries.

Keywords: cliff, erosion, limestone, stability, failure

1 INTRODUCTION

The Pointe Du Hoc site, in Normandy, France was host to one of the most important battles of D-Day on 6 June 1944. The Pointe du Hoc cliffs are being eroded by the waves of the Channel between France and England, especially during winter storms. The Observation Post (O.P.), for one, located at the northern-most position of the site (Cliff Head), appears most vulnerable due to its proximity to the edge of the cliff and was closed to tourists in 2004 (Fig. 1).

2 LITERATURE REVIEW

The Channel is the part of the Atlantic Ocean that separates the island of Great Britain from northern France and joins the North Sea to the Atlantic. Duperret *et al.* (2004) reported nine different chalk units along the coastline of NW France (120 km from upper Normandy to Picardy). The chalk sea cliffs retreat along the Normandy coast of France over decadal time scales, is ranging from effectively stable to landward retreat between 0.1 to 0.5 m/yr with a mean value of 0.23 m/yr (Duperret *et al.* 2004). The retreat usually takes place by successive local collapse. Many authors (e.g. Emery & Kuhn 1982; Sunamura 1992) propose that marine parameters acting at the toe of rock cliffs are responsible for under cutting the cliff, which leads to rock failures. The coastal monitoring program at the Sandown Bay, UK, which is

Figure 1.   Pointe du Hoc, Normandy, France.

Figure 2.   Massive collapse about 300 m west
of the O.P.

Figure 3.   Caverns near the Observation Post (West
side).

about 140 km north of Pointe du Hoc, gives a range of maximum wave height, H_{max}, between
5 and 6 m during the storm season. Benumof and Griggs (1999) also state that waves are one
of the leading forcing mechanisms of sea-cliff erosion, secondary only to the material properties of the rock itself. The typical geology of the Pointe du Hoc area consists of the following
sequence. At the top is a layer of sediments, which have been deposited in the Bajocien—
Bathonien period. Below is a limestone layer of "Calcaire de St Pierre du Mont" also called
"Calcaires du Bessin". This limestone layer is based on a marl layer of "Marnes de Port en
Bessin". This marl is apparent in the nearby small harbor of Port en Bessin under the form of
slopes created by the erosion on both sides of a small river that reaches the sea at this location.

3   SITE DATA COLLECTION

The site reconnaissance revealed the following observations: The observation post (O.P.)
which is nearest to the cliff edge appears to be intact; however, some cracks in the O.P. walkway and entrance stairs were identified. Massive rock failures (Fig. 2) were observed along
the beach, all of them had a vertical plane of failure. The rock bedding is nearly horizontal,
dipping down towards the north with an approximate angle of 5 degrees δ. Fresh water seepage from the cliff face was noticed in several places at a level of approximately 2–3 m above
the base of the cliff and sometimes higher. The observed joints' spacing ranged between 1
and 2 m. On the East side of the Pointe, just under the O.P., two extended joints appeared at
the cliff face with the possibility that they coincide with the O.P. walkway cracks. Some large
size undercutting and caverns were found to have heights up to 3.5 m and depth (i.e. perpendicular to the cliff face) from 3 to 4 m (Fig. 3). The fact that no caverns were found with depth

deeper than 4 m indicates that this depth may be a limit for the cliff rock mass and lead to the collapse of the cliff overhang. The high tide level may reach up to 3 m from the cliff base at some locations and the waves may add another 6 m of water attack above that. Calculation of the erosion rate over the years from aerial photos indicate that at the most aggressive locations, the Pointe Du Hoc site experienced about 10 m of erosion due to collapse over the last 56 years (from 1944 to 2000), or an average erosion rate of 0.18 m/year. This erosion rate is consistent with the numbers found in the literature.

4 GEOTECHNICAL BORINGS

The site investigation included drilling six bore holes numbered B1 to B6 to a depth of 30 meters. B1 and B3 were equipped with piezometers and B4 was equipped with a settlement benchmark. B1 and B5 were core drilling boreholes, and the rest of the boreholes were destructive drilling. In the destructive boreholes, the following drilling parameters were recorded during the drilling: instantaneous penetration speed, pressure on tool, rotation torque pressure, and drilling fluid injection pressure. One part (12 samples) of the intact samples taken during soundings B1, B1', and B5 was sent to the FUGRO Geotechnical Laboratory in France. The second part (10 samples) was sent to Texas A&M University, College Station, Texas, USA. The boring logs led to the stratigraphy shown in Fig. 4.

5 TESTING AND TEST RESULTS

Rock tensile strength, soil and rock erosion tests, suction measurements in the soil cover were carried out at Texas A&M University. The average rock tensile strength was found to be 3.36 MPa in the limestone, 4.55 MPa in the sandstone, and 4.52 MPa in the marly limestone. The suction values in the soil cover ranged from –5 to –160 kPa corresponding to a water content ranging from 49.4% to 31.8%. Soil erosion tests showed that the top soil has a high to medium erodibility, yet the rock erosion tests showed that the erodibility of the intact rock mass is almost negligible.

Index properties (unit weight, water content, Atterberg limits, grain size distribution), direct shear tests, and unconfined compression tests were carried out by Fugro France.

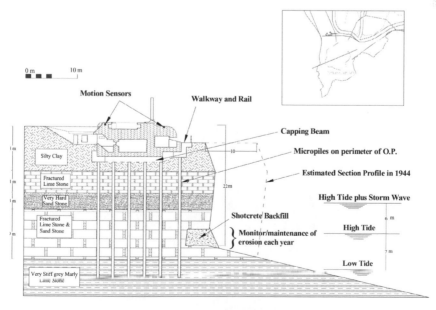

Figure 4.   Stratigraphy, observation post, and repair scheme.

The average shear strength parameters from drained direct shear tests gave cohesion intercept which varied from 7 to 42 kPa and friction angles which varied from 15 to 35 degrees. The average unconfined compression strength of the rock varied from 67.5 MPa in the limestone/sandstone down to 315 kPa in the shale below.

6 ANALYSIS OF THE FAILURES

Several analyses were carried out to understand the failure mechanism that led to the observed collapse and failures at the site. A wedge analysis of the soil cover showed that for the 9 chosen study cases, and all ϕ' values (from 5~45), the factor of safety for dry suction conditions was greater than 1.0, yet the factor of safety for wet suction conditions was less than 1.0 for many cases. This analysis indicates that the likely increase in water content of the soil after a heavy rain may trigger wedge failures in the upper zone of the cliff. A Finite Element Analysis (FEA) was carried out to study the stress distribution in the rock mass created by the caverns and the associated overhang. The FEA showed that the cliff material is too strong to fail in tension if it is assumed that the cliff material is acting as a continuum (i.e. ignoring the role of rock discontinuities). The maximum tension calculated by the FEA simulation was about 40 kPa for a cavern depth of 4 m which seems to be the maximum tolerable by the cliff rock mass before failure of the overhang occurs. This would indicate that the rock mass tensile strength is only 1/100th of the intact rock tensile strength which varies between 3 and 5 MPa. The influence of the water pressures on the rock block stability due to high tides and storm waves was studied next assuming that the rock blocks were 1 to 2 m in size as observed. The analysis showed that a wave head of 2 to 3 m can suck out the rock blocks out of the rock mass upon retreat of each wave. Fig. 5 shows an overhang with a nearly complete separation of the rock block from the entire rock mass.

7 FAILURE MECHANISM

The most likely failure mechanism is as follows: The first part of the mechanism is that the fresh water seeps through the soil cracks in the top soil layer and through the joints network in the rock layers. This seepage of water causes erosion of the soil filling the rock joints and some dissolution of the limestone rock which is heavily jointed. The geophysical test performed but not reported here corroborated by the borings B1 (void from 19.7 to 21.1 m) and B2 (void from 19.7 to 20.6 m) indicate that caverns do exist at depth within the mass of the limestone. The flow process added to the leaching by the waves leads to a rock mass towards the bottom of the cliff which becomes an assembly of preexisting caverns and dissociated rock blocks. These blocks are of the order of 1 to 2 m in size. The second part of the mechanism is the

Figure 5.   Overhang with loose rock blocks.

formation of extended vertical cracks at locations of weak planes due to frost heave activities, horizontal decompression, or bending of the rock mass. Water flows towards the cliff edge through the joints network to find an exit out of the cliff; it gets subjected to very low temperatures close to the edge especially in the winter. Freezing and consequently expansion may take place. This phenomenon results in the pre-definition of vertical planes of potential failure. The third part of the mechanism starts with the sea water providing buoyancy for the lower blocks and continues with the wave action removing the rock blocks thereby creating caverns or deepening pre-existing internal caverns at the bottom of the cliffs. Indeed the Channel (La Manche) is a very active sea with very high tide fluctuations and very large storm waves during the winter. These waves, with the buoyancy provided by high tides, facilitate the loosening of the rock blocks, which are then sucked out from the cliff base by the increase of the sea water head. The depth of the caverns progresses and increases the overhang, which creates tensile stresses in the rock mass until these stresses reach the level of the rock mass tensile strength. Failure of the rock supporting the soil cover follows as shown in Fig. 2.

8 GENERAL REMEDIATION

Two phases of remediation are recommended (Fig. 4). The first phase consists of completely filling the caverns and associated joints with shot-crete (Phase I-A), or concrete (Phase I-B) to support the ceiling of the overhang and protect the cliff base. Filling the caverns with rocks would not be a good remediation as the ceiling of the overhang needs to be well supported. An overall concern for the whole cliff height is that any remediation must be esthetically pleasing. The second phase consists of stabilizing the Observation Post by strengthening and deepening its foundation. This is done by placing micropiles below the O.P. as shown in Fig. 4. This foundation system will significantly increase the probability that the O.P. would survive a cliff failure ahead of the O.P. This type of underpinning will make use of drilled and grouted steel bar or pipe micropiles. The micropiles will be drilled directly through the existing concrete Observation Post, or new concrete pile caps could be attached to the existing structure. Both vertical and battered (inclined) micropiles can be installed.

9 REOPENING OF THE OBSERVATION POST

As recommended by Texas A&M University research investigators, the caverns were back-filled with aesthetic concrete and architectural rock blocks and micropiles were installed around the Observation Post. On 6 June 2011, 67 years after D-Day, the Observation Post was reopen to the public in a ceremony which took place on top of the O.P. The ceremony was attended by D-Day surviving soldiers and many American and French dignitaries (Fig. 7).

Figure 6. Backfilled caverns.

Figure 7. Reopening ceremony.

REFERENCES

Benumof, B.T. & Griggs, G.B. (1999) "The Dependance of Sea-cliff Erosion Rates on Cliff Material Properties and Physical Processes: San Diego County, California". Journal of Shore and Beach, V. 67, No. 4, October 1999, pp. 29–41.

Briaud, J.-L., Nouri, H.R., Darby, C., 2008, "Pointe du Hoc Stabilization Study: geotechnical Report", Zachry Dpt. of Civil Engineering, Texas A&M university, College Station, Texas, USA, pp. 149.

Duperret, A., Genter, A., Martinez, A. & Mortimore, R.N. (2004). "Coastal chalk cliff instability in NW France: the role of chalk lithology, fracture pattern and rainfall". In: Mortimore, R.N. & Duperret, A. (eds) Coastal Chalk Cliff Instability. Geological Society, London, Engineering Geology Special Publication, 20, pp. 33–55.

Emery, K.O. & Kuhn, G.G. (1982). "Sea cliffs: their processes, profiles, and classification". Geological Society of American Bulletin, 93, pp. 644–654.

Sunamura, T. (1992). "Geomorphology of Roack coasts". Wiely, Chichester.

Geomechanics and Geodynamics of Rock Masses – Litvinenko (Ed.)
© 2018 Taylor & Francis Group, London, ISBN 978-1-138-61645-5

Analytical formulation of stand-up time based on 1989 Beniawski's chart

E. Estébanez & A. Lage
TUNELSOFT S.L, Madrid, Spain

ABSTRACT: The stand-up time is a main topic in the field of rock mechanics. It represents the time that an unsupported underground excavation with a given dimension remain stable. In tunneling, it leads to assess the available time to install rock support, and thus, the length of the advance step, which implies a high economic impact.

Complexity of rock masses makes the derivation of predictive equations difficult. Some authors have analyzed the main factors involved, and have proposed analytical approaches to assess the stand-up time. Nevertheless, until now, the empirical approximatios are the methods currently used to solve this issue. Probably, the most widespread nowadays is the 1989 Bieniawski's chart. This approach allows one to relate the stand-up-time to the dimensions of the excavation (active span) and to the rock mass quality, estimated through the RMR (Rock Mass Rating).

An analytical solution based on this approach is proposed here. The aim of this work is to extract the underlying mathematical law that govern the phenomenon and, furthermore, to permit to assess the stand-up time in an easier manner, keeping the reliability of the original chart.

Keywords: Stand-up time, active span, rock mass quality, self-supported excavation, analytical solution, tunnel stability

1 INTRODUCTION

The stand-up time is the period of time that a tunnel will stand unsupported after excavation. This concept was introduced by Lauffer [6] in his geomechanical classification, published in 1958. It should be noted that a number of factors may affect the stand-up time, as illustrated in Fig. 1. This figure shows Lauffer's relationship between active span and stand-up time for different classes of rock mass: A – very good rock, G – very poor rock.

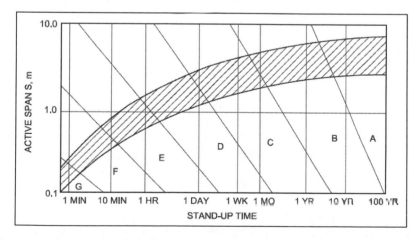

Figure 1. Lauffer's classification (Lauffer, 1958).

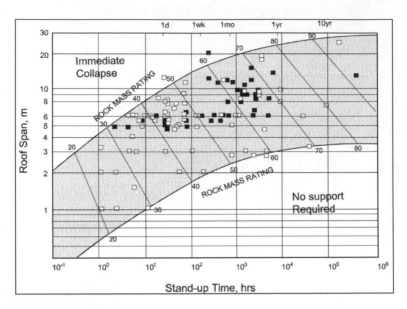

Figure 2.   Beniawski's chart (1989).

The active unsupported span is the width of the tunnel or the distance from the face to the support if this is less than the tunnel width.

Since then, several authors have proposed different approximations to assess the stand-up time from both empirical and analytical points of view. The most relevant are depicted below.

2   EVOLUTION OF MAIN EMPIRICAL APPROACHES

Lauffer's classification was based in the ideas of Stini (1950) [11], who analysed the influence of span and rock structure in the stability of underground openings. In this way he elaborated a rock classification based on its consistency, establishing eight categories and the main issues associated with its excavation.

Since then, other authors have proposed different approaches as the modification of Pacher et al. (1974) [9], and that led to the development of the New Austrian Tunneling Method. Also *Barton* (1976) [2], proposed one during the development of his geomechanical classification [1], by studying 29 historic cases. Within he related the quality index Q of his classification, for different span values between less than 2 metres to 100 metres (range of span lengths that correspond to the registered historic cases). It was however, in his own words, "a preliminary attempt". Even Lauffer (1988) [4], adapted his original classification linking it with RMR parameter and considering the excavation with TBM.

Bieniawski (1979) [3] published a chart based on 49 historic cases that modified later with more data. This last approach, *Beniawski (1989)* [4], is probably the most used to determine the stand-up time of an underground excavation. Of the 351 historic cases, originating from civil construction and mining, on which the author based his geomechanical classification, 123 registered the stand-up time. Correlating this variable with the active span and the RMR, he obtained the chart showed in Fig. 2.

3   ANALYTICAL FORMULATIONS

The afore-mentioned approaches only consider two variables in the estimation of stand-up time: rock mass quality and length of the span.

Nevertheless the stand-up time is affected, in a greater or lesser degree, by other factors as the shape of the opening, the excavation method (drill & blast, mechanical excavation, TBM) or the advancing rate (Myer et al. 1981 [7]). Also it seems that not all rock mass parameters have the same influence and specific weight.

Ramamurthy (2007) [10], proposed a relation that includes the main factors that he consider are decisive in the value of the stand-up time. In the expression below the stand-up time is related with the modulus ratio of rock mass ($k_s M_{rj}$), the effective span (S_u), the in-situ stress (p_0) and the seepage pressure (u).

$$t_f = \frac{k_s M_{rj}}{S_u(p_0 + u)} \tag{1}$$

After checking this expression in several actual cases the author concluded that the results are acceptable and are within the limits established by Beniawski (1989) [4].

From other perspective, Nguyen V.-M and Nguyen Q.-P. (2015) [8] proposed a "rheological deformational creep approach". The authors consider that the terrain surrounding the excavation suffer a deformation delayed in the time. In order to model that behaviour and predict the stand-up time they utilise a rheological deformational model known as "Abel's Creep Kernel". Below the proposed formulation is showed:

$$L^* t^{*(1-\alpha)} = \frac{2E}{3\gamma Ha} \frac{(1-\alpha)}{\delta} u_l^* \tag{2}$$

Within the expression the stand-up time is related with the active span (L), the admissible displacement of the wall (u_l), the depth (H), the Young modulus (E) and density (γ) of the rock, time dependent constants of the rock mass (α, δ) and tunnel face distance (l) and radius (R).

After verifying the expression for different cases, the authors concluded that the results show a similar trend that Beniawski's and Lauffer's empirical approaches.

4 ANALYTICAL APPROACH TO BIENIAWSKI'S CHART

Within Beniawski's Chart a relationship is established between three variables: active span, stand-up time and RMR. The domain where this relation is applicable is limited by two curves which define the borders where the excavation is stable or collapse immediately. Below, an analytical approach to the law expressed graphically on the chart is developed [5].

4.1 Starting point

For constant RMR values, the Bieniawski's chart shows a lineal relation between the logarithms of stand-up time (T) and active span (S)

$$\log(T) = A + B \cdot \log(S) \tag{3}$$

That means that T and S are related by a power law

$$T = 10^A \cdot S^B \tag{4}$$

Taking into account that for each value of RMR we can get a pair of values for A and B, a general relationship between S and T can be expressed as:

$$T = \alpha(\text{RMR}) \cdot S^{\beta(\text{RMR})} \tag{5}$$

where

$$\alpha = 10^A \qquad (6)$$
$$\beta = B \qquad (7)$$

4.2 Estimation of α(RMR) and β(RMR)

Values of A and B for different values of RMR have been obtained from the Bieniawski's chart. The corresponding values of α and β expressed in function of RMR, have been adjusted to achieve a minimum coefficient of determination of 0,99. The result is shown in Fig. 3.
    We get the following expressions:

$$\alpha(RMR) = 0,04 \cdot e^{0,27RMR} \qquad (8)$$
$$\beta(RMR) = -0,81Ln(RMR) + 1,35 \qquad (9)$$

4.3 Estimation of the relation between span and RMR for the boundaries

From the chart, the values of span corresponding to each RMR value in the boundaries have been obtained. By adjustment of these data, we get the correlations shown in Fig. 4.
    For self-supported boundary we get:

$$S = 0,0104 \cdot RMR^{1,377} \qquad (10)$$

For immediate collapse boundary we get:

$$S = 0,0249 \cdot RMR^{1,574} \qquad (11)$$

4.4 Proposed formulation

By replacing (8) and (9) in (5) we get:

$$T = 0.004 \cdot e^{0.27 \cdot RMR} \cdot S^{-\left(0.81 \cdot \ln(RMR) - 1.35\right)} \qquad (12)$$

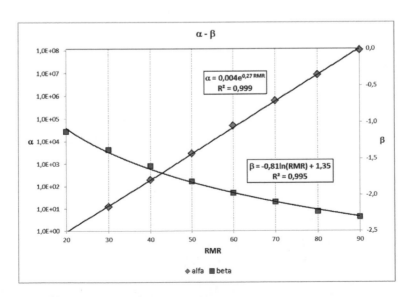

Figure 3.   α and β in function of RMR.

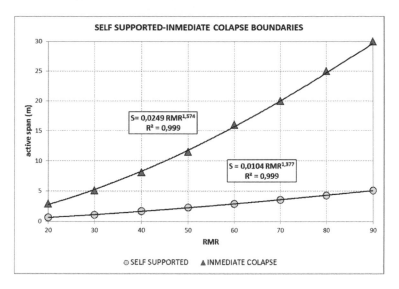

Figure 4. Correlation between span and RMR in the boundaries.

By elimination of RMR between (10) and (12) we get for the self-supported boundary:

$$T_E = 0.004 \cdot e^{7.44 \cdot S^{0.726}} \cdot S^{-(1.34 + 0.59 \cdot \ln S)} \tag{13}$$

By elimination of RMR between (11) and (12) we get for the immediate collapse boundary:

$$T_C = 0.004 \cdot e^{2.82 \cdot S^{0.635}} \cdot S^{-(0.55 + 0.51 \cdot \ln S)} \tag{14}$$

For a given span and RMR:
if $T > T_E$: self-supported excavation
if $T < T_C$: immediate collapse

4.5 *Derived charts*

Equation (12) defines a surface in a three dimensional space. This surface is showed in Fig. 5, limited by (13), (14) and planes corresponding to RMR equals to 20 and 100.

When projecting lines for a given RMR on the plane T-S we get the chart shown in Fig. 6. The RMR curves are obtained by particularizing the expression (12) for the corresponding value. The boundary curves are expressed by equations (13) and (14).

Projecting lines for a given T on the plane RMR-S we get the chart shown in Fig. 7.

The curves corresponding to a specific time are obtained expressing S in equation (12) as a function of T and RMR and particularizing for T.

$$S = \left(\frac{T}{0.004 \cdot e^{0.27 \cdot RMR}} \right)^{\frac{-1}{0.81 \cdot \ln RMR - 1.35}} \tag{15}$$

The boundary curves are expressed by:

$$\text{Self-supported } S_E = 0.01 \cdot RMR^{1.34} \tag{16}$$

$$\text{Immediate collapse } S_C = 0.025 \cdot RMR^{1.57} \tag{17}$$

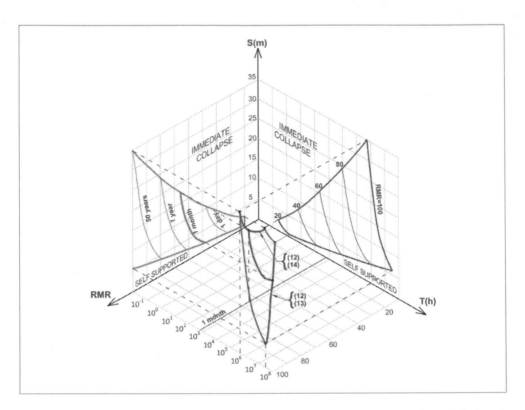

Figure 5.   Graphical representation of equation (12) and its projection over T-S and RMR-S planes.

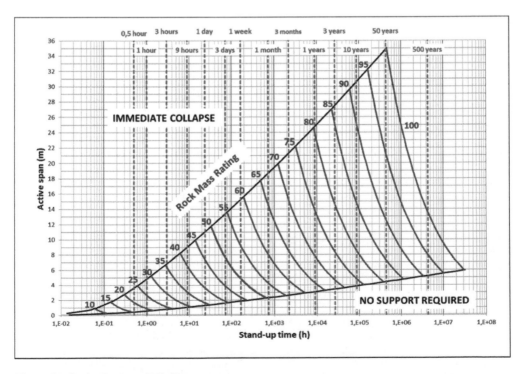

Figure 6.   Projection over T-S plane.

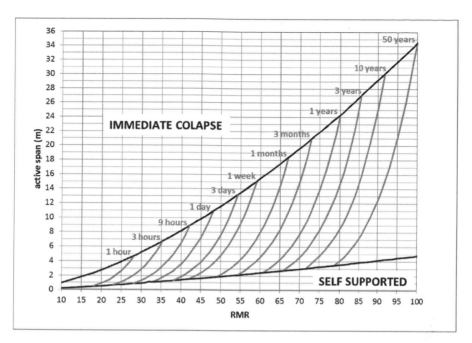

Figure 7. Projection over RMR-S plane.

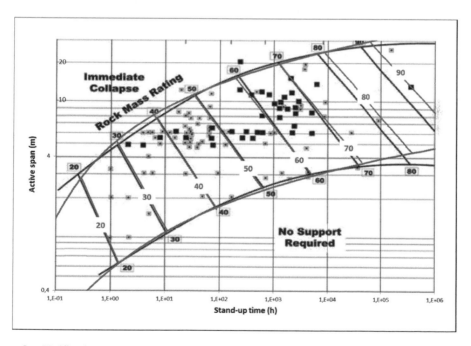

Figure 8. Verification.

5 VERIFICATION

The curves constructed from equations (12), (13) and (14) are depicted in Fig. 8. (red lines) over the Bieniawski's chart. The line corresponding to the first equation has been particularized for values of RMR from 20 to 90 every 10 points. It can be seen that the proposed solution provides an accurate fit to the original diagram.

6 CONCLUSIONS

Since Lauffer published his geomechanical classification in 1958, several empirical and analytical approximations to assess the stand-up time of an unsupported underground excavation have been made. Nowadays, the empirical Bieniawski's 1989 approach seems to be the most widespread.

In an attempt to assess the stand-up time in an easy way, but keeping the reliability that provide an empirical approach, a new analytical formulation is proposed in this research. This formulation, based on the 1989 Beniawski's Chart and consistent with it, is a mathematical approximation to the law described graphically there, and can be a contribution in order to provide a theoretical basis.

REFERENCES

[1] Barton, N. 1976. Unsupported underground openings. Rock Mechanics Discussion Meeting, Befo, Swedish Rock Mechanics Research Foundation, Stockholm, pp. 61–94.
[2] Barton, N; Lien, R; Lunde, J; 1974. Engineering classification of rock masses for the design of tunnel support. Rock Mechanics 6, pp. 183–236.
[3] Bieniawski, Z.T. 1979. The Geomechanics Classification in rock engineering applications. Proceedings, 4th International Congress on Rock Mechanics, International Society for Rock Mechanics, Montreux, A.A. Balkema, Rotterdam, Vol. 2, pp. 51–58.
[4] Bieniawski, Z.T. 1989. Engineering Rock Mass Classificatons. New York. John Wiley & Sons. pp. 251.
[5] Estébanez, E., Lage, A. 2017. Formulación analítica del tiempo de autoestabilidad de una excavación subterránea basada en el ábaco de Bieniawski de 1989. Ed. C Lopez Jimeno. Ingeotúneles 25 (pp. 39–50). Madrid.
[6] Lauffer, H., 1958. Gebirgsklassifizierung für den stollenbau. Geologic und Bauwesan. 24, 46–51.
[7] Myer, L.R., Brekke, T.L., Dare, C.T., Dill, R.B., Korbin, G.E., 1981. An investigation of stand-up time of tunnels in squeezing ground. In: Rapid excavation and tunneling conference proceedings, San Francisco. California, pp. 1415–1433.
[8] Nguyen, V.M.; Nguyen, Q.P., 2015. Analytical solution for estimating the stand-up time of the rock mass surrounding tunnel. Tunnelling and Underground Space Technology 47. Hanoi, Viet Nam. Elsevier. pp. 10–15.
[9] Pacher, F., Rabcewicz, L., and Gosler, J., 1974. Zum Derseitigen Stand der Gebirgsklassifizierung in Stollen-und Tunnelbau. Proceedings, XXII Ceomechanics Colloquium, Salzburg, pp. 51–58.
[10] Ramamurthy, T., 2007. A realistic approach to estimate stand-up time. In: 11th Congress of the International Society for Rock Mechanics, Lisbon, Taylor & Francis Group, London, pp. 757–760.
[11] Stini, J. 1950. Tunnelbaugeologie: Die geologischen Grundlagen des Stollen—und Tunnelbaues. Vienna. Springer-Verlag. pp. 366.

Assessing the influence of discontinuities and clayey filling materials on the rock slope instability

Davood Fereidooni
School of Earth Sciences, Damghan University, Damghan, Semnan, Iran

ABSTRACT: For performing the research, seven rock slopes along the margin of Ganjnameh-Shahrestaneh Road, Hamedan Province, western Iran, were selected, and the physical and mechanical properties of their rocks and discontinuities were determined. Rock slope stability analysis has been performed using kinematic and limit equilibrium methods so that safety factors for the rock slopes can be calculated. Also, sampling of filling materials and XRD tests have been done to identify the clay minerals in the filling materials. The lithologies of the studied rock slopes are granite, diorite, and hornfels. The presence of discontinuities and weakness planes with different orientations and clay minerals in filling materials of discontinuities are effective factors that cause plane, wedge, and toppling failures in the rock slopes. Clay minerals as filling materials of discontinuities in the studied rock slope facilitate their instability by two different methods. First, absorption of water by infilling clay minerals causes the friction angle of discontinuity surfaces that leads to plane and wedge failures to be reduced. Second, water absorption causes the swelling of clay infilling minerals that leads to toppling failure.

Keywords: Rock slope, clay mineral, discontinuity, plane failure, wedge failure

1 INTRODUCTION

The instability of a slope is depended to many factors including the gradient and height of the slope, the geotechnical properties of the material involved, cohesion, degree of weathering, and the presence of induced discontinuities and inherent weakness planes. The effects of mentioned factors on instability of rock slopes have been studied by many researchers (e.g. Eberhardt et al. 2005; Shi 2014; Indraratna 2014; Nkpadobi et al. 2015a; Sainsbury et al. 2016). The stability of rock slopes depends kinematically on the orientation of the discontinuities (Priest 1985). The stability of rock slopes, therefore, depends to the intersection between the orientation of discontinuities and the direction of slopes, and whether the dip direction of discontinuities is parallel with or perpendicular to the direction of slopes. Zhang et al. (2015) found that the potential instability of a rock slope is more sensitive to major geological discontinuities such as fault, fractured or weak zone, and strata interface at a large scale.

Discontinuities are often filled with different materials which may be detrital material or gouge. The filling material may be in the form of partially—to completely-loose cohesive or noncohesive weathered materials deposited in open joints or faults (Sinha and Singh 2000). On the other hand, the increase in the degree of weathering causes an increase in the aperture that will then be filled with materials, such as clay, sand, and plant roots. In discontinuities water can be trapped and cause the surrounding minerals to change as chemical decomposition occurs. During the decomposition, large amounts of clay minerals are produced (Agustawijaya 2003). Clay minerals play an important role in the instability of rock slopes (Eberhardt et al. 2005; Yokota and Iwamatsu 1999) and may accelerate time-dependent deformation of slopes, and their contributions to slope evolution and failure of slopes are notable. Distribution of clay minerals is controlled primarily by the rock type and climate in their source regions and by the transport agents and directions (Cagatay et al. 2002). Another parameter that influences the stability of jointed rock masses is the shear strength of discontinuities which will in turn depend on the roughness of the discontinuities (Priest, 1993).

2 METHODS AND MATERIALS

Seven rock slopes along the margin of Ganjnameh-Shahrestaneh Road, Hamedan Province, western Iran, were selected, and field surveys and laboratory investigations were conducted; including scanline surveys, mineralogical and lithological studies, determination of physical properties, discontinuities and filling materials, and direct shear tests. Scanline surveys were conducted on the rock faces of the slopes. Measurements of discontinuity conditions such as dip, dip direction, aperture, spacing, roughness, and infilling material were then conducted along this line. Mineralogical and lithological properties of the rock samples and filling materials were performed by thin section and XRD analyses. Direct shear tests of discontinuities were conducted on discontinuity samples taken from the seven rock slopes, and parameters of the shear strength of discontinuities were evaluated. Orientation data of discontinuities were analyzed using Dips software and kinematical analysis could consequently be conducted for evaluating the stability of the slopes. Also, rock slope stability analyses were performed by the limit equilibrium method.

3 RESULTS

3.1 Geometric and engineering properties of the rock slopes

Different characteristics of rock slopes and their rock compositions include lithology, dry unit weight, slope height, average angle of slope, friction angle, and cohesion of joint surfaces were evaluated. Rock lithologies were determined by thin section studies in accordance with the ISRM (1978) suggested method. Evaluation of slope height was carried out using GPS. Dry unit weight of the rocks was evaluated in accordance to the ISRM (1972) suggested method. Spacing of discontinuities, based on the scanline method, was determined as suggested by ISRM (2007). The results are presented in Table 1.

3.2 Clay minerals filling rock discontinuities

In order to identify clay minerals in filling materials of discontinuities in the studied rock slopes, X-ray diffraction (XRD) analysis was applied according to ISRM (2007). The results (Fig. 1) show that the clay minerals have mineralogical correlations with the rocks composing the rock slopes. The most important identified clay minerals are in the chlorite and illite-mica group.

Table 1. Geometric and engineering properties of the studied rock slopes.

Rock slope	Lithology	Slope height (m)	Slope angle (Deg.)	Unite weight (Kg/m³)	Average spacing of discontinuities (m)
1	Diorite	15	71	2925	0.11
2	Granite	11	61	2657	0.20
3	Hornfels	10	75	2802	0.10
4	Granite	18	67	2657	0.18
5	Hornfels	17	65	2762	0.15
6	Hornfels	10	49	2695	0.18
7	Hornfels	11	72	2680	0.16

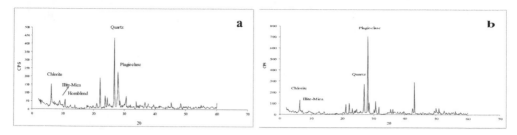

Figure 1. The diffractogram of analyzed filling materials obtained from the rock slopes No. 1 and 2 as representative rock slopes.

3.3 Shear strength of discontinuities

To evaluate shear strength parameters of discontinuities (cohesion and friction angle), some rock cores were obtained from rocks composing the rock slopes. Direct shear tests were carried out according to the ISRM (2007) suggested method in both normal and reconstructed conditions. In the first condition, after obtaining rock cores from the rocks containing natural discontinuity, a direct shear test was carried out on the rock cores at three different normal forces, 10, 15 and 20 KN. In the second condition, the amount of saturated mixture of filling materials passing through a sieve No. 40, was applied between two walls of sawed discontinuities of each rock core, and also a direct shear test was measured at the mentioned normal stresses. This second condition indicates discontinuities that are filled with the materials containing clay minerals in a saturation state (reconstructed or filled condition). The consequences of the direct shear test are shear stress versus shear displacement and shear stress versus normal stress shown in Fig. 2 for the rock slope No. 1 as a representative sample of the studied rock slopes.

Fig. 2a shows that the tested specimens after a semi linear shear displacement reach a peak in the curves (peak shear strength). Then, they are ruptured and continue to be displaced at constant shear stress (residual shear strength). If peak and residual shear strengths are plotted in the shear stress versus normal stress coordinate system in both normal and reconstructed test conditions, Fig. 2b is achieved. Intercept and slope of each graph in this figure are the cohesion and friction angle, respectively. The values of shear strength parameters are given in Table 2.

3.4 Kinematic slope stability analysis

Kinematic slope stability analysis was done using Dips software. Fig. 3 shows contour diagram and major discontinuity planes of the rock slopes No. 1 as a representative rock slope. Failure types for each rock slope are recognizable from these figures. The required geometrical conditions for occurring different types of failures are presented by Wyllie and Mah (2004). The obtained results show that some of the studied rock slopes are stable whereas, some of them have the potential of failure as presented in Table 3.

3.5 Limit equilibrium slope stability analysis

In limit equilibrium method, the factor of safety (F) for plane failure is determined by resolving all forces acting on the slope into components parallel and normal to the sliding plane:

$$F = \frac{cA + \left(W\,Cos\psi_p - U - V\,Sin\psi_p\right)\tan\phi}{W\,Sin\psi_p + V\,Cos\psi_p}$$ (1)

where c is cohesion of the plane of failure, A is the area of the sliding plane, W is the weight of the sliding block, U and V are the water forces acting on the sliding plane and in the

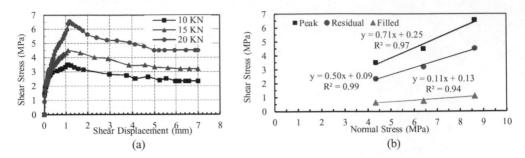

(a)　　　　　(b)

Figure 2.　a) Shear stress versus shear displacement curve and b) Shear stress versus normal stress curve for discontinuities of the rock slope No. 1.

Table 2.　The values of shear strength parameters for the studied rock slopes.

Rock slope	Normal condition				Reconstructed condition	
	c (MPa)	c_r (MPa)	ϕ (Deg.)	ϕ_r (Deg.)	ϕ_f (Deg.)	c_f (MPa)
1	0.254	0.089	35.451	26.776	6.271	0.132
2	0.382	0.056	36.815	27.420	5.162	0.114
3	0.229	0.017	32.729	23.446	4.649	0.059
4	0.382	0.055	36.825	27.320	6.509	0.101
5	0.993	0.063	39.242	29.610	6.318	0.109
6	0.879	0.001	34.431	26.730	5.310	0.043
7	0.327	0.002	30.715	23.258	4.721	0.054

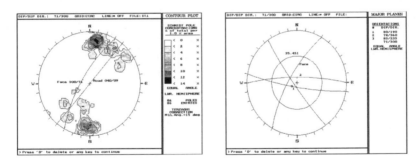

Figure 3.　Contour diagrams and major discontinuity planes of the rock slopes No. 1.

Table 3.　Stability or instability conditions of the studied rock slopes.

Rock slope	Stability or instability conditions and type of failure
1	Wedge failure along the line of intersection between discontinuities 2 and 3
2	Wedge failure along the lines of intersection between discontinuities 1 and 5 and discontinuities 2 and 3
3	Toppling failure due to the presence of discontinuities 2
4	Stable
5	Stable
6	Stable
7	Wedge failure along the line of intersection between discontinuities 1 and 2

Table 4. Rock slope stability analysis for plane failure.

Condition	Rock slope No.	Critical discontinuity	A (m²)	W (Kg)	V (Kg)	U (Kg)	ψ_p (Deg.)	ϕ (Deg.)	c (MPa)	F
Normal	6	3	0.032	15.49	0	0	51	34.43	0.879	238.82
Reconstructed	6	3	0.032	15.49	56.32	103.00	51	5.31	0.043	−0.27

Table 5. Rock slope stability analysis for wedge failure.

Condition	Rock slope No.	Critical discontinuities	β (Deg.)	ξ (Deg.)	ϕ (Deg.)	ψ_i (Deg.)	F
Normal	1	2 and 3	84	68.5	35.45	58	0.39
	2	1 and 5	58	64	36.82	44	1.24
	2	2 and 3	77	68	36.82	56	0.88
	7	1 and 2	77	68	30.72	51	0.84
Reconstructed	1	1 and 2	89	22	6.27	31	0.95
	2	2 and 5	88	54	5.16	32	0.32
	7	2 and 3	84	64	4.72	13	0.68
	7	3 and 4	72.5	41	4.72	11	0.78

tension crack, respectively. ψ_p is angle of the sliding plane and ϕ is the friction angle of the sliding plane.

Observations of the studied rock slopes show that plane failure can occur in the rock slope No. 6 along discontinuity 3 in the reconstructed (filled and saturated) condition. But, in normal (dry) condition, the rock slope is stable. The results of rock slope stability analysis for plane failure are presented in Table 4.

The factor of safety for wedge failure is calculated from the following equation:

$$F = \frac{Sin\beta}{Sin\frac{1}{2}\xi} \times \frac{\tan\phi}{\tan\psi_i} \qquad (2)$$

where β is the angle between the bisector of wedge angle and the horizontal plane, ξ is the angle between two planes of failure, ϕ is average friction angle of two planes of failure and ψ_i is dip of the line of intersection of two planes of failure. Observations of the studied rock slopes show that wedge failure can occur in rock slopes No. 1, 2 and 7. The results of stability analysis for wedge failure are given in Table 5.

On the basis of the analyses, toppling failure can occur in the rock slope No. 3 due to the presence of discontinuities 2, but circular failure has not occurred in the studied rock slopes.

4 DISCUSSIONS

The kinematic analyses have demonstrated three joint or discontinuity sets in the rock slope No. 1. Wedge failure has occurred because of discontinuities 2 and 3 in dry conditions. Also, filling materials of discontinuities are clay minerals, such as those that occur in the chlorite and illite-mica group where the clay minerals have high abundance. If the clay minerals become saturated, the friction angle of the failure plane can be highly reduced. In this case, the crescent of instability in the stereogram becomes larger, and the intersection point of discontinuities 1 and 2 lies within this area (Fig. 3). This intersection means that wedge failure can occur along the mentioned discontinuities. This possibility was confirmed by reaching a factor of safety less than one (0.95) in the limit equilibrium analysis.

In rock slope No. 2, under normal conditions wedge failure can occur along discontinuities 1 and 5 and discontinuities 2 and 3. Also, in filling materials of the discontinuities, the clay minerals of the chlorite and illite-mica group are identified. If these minerals become saturated, wedge failure can occur along the intersection line of discontinuities 2 and 5. Because, with reducing friction angle of the failure plane, the crescent of instability in the stereogram becomes larger, and the intersection point of the discontinuities lies within this area. The occurrence of wedge failure was confirmed by limit equilibrium analysis in filled and saturated condition of discontinuities.

Due to the different lithology of rock slope No. 3, in comparison to the previous rock slopes, the clay minerals of the chlorite and illite-mica group are more abundant than in the other studied rock slopes. Also, the occurrence of metamorphic minerals in the filling materials is another characteristic of rock slope No. 3. On the basis of kinematic analysis, toppling failure can occur due to the presence of discontinuity 2. On the other hand, the presence of expansion clay minerals in rock discontinuities causes a compressive force to be induced in the wall surface of toppling blocks that promotes toppling failure.

The orientation of discontinuities in rock slope No. 4 makes this rock slope stable. The mineralographical studies of filling materials of the rock slope indicate a small percentage of clay minerals in the filling materials. Therefore, clay minerals have not produced any influence on the instability of the rock slope.

According to the presence of a discontinuity set with minimal orientation in the rock slope No. 5, this rock slope is stable. The XRD analyses show small percentage of clay minerals in the filling materials that have not any influences on the instability of the rock slope.

According to the orientation of the discontinuities in the stereogram, the rock slope No. 6 is stable. This was confirmed by limit equilibrium analysis in dry condition. But, in filled and saturated condition of discontinuities, the plane failure can occur along discontinuity 3 in the rock slope. In the field investigations, the presence of water has been proven in the discontinuities of the rock slope. Therefore, clay minerals in the rock slope discontinuities are the most important parameters that cause plane failure.

The orientation of four sets of discontinuities in the rock slope No. 7 provides a condition for the occurrence of wedge failure in the dry state in this rock slope. This possibility is confirmed by kinematic and limit equilibrium analyses. In addition, the presence of clay minerals of the chlorite and illite-mica group in the rock discontinuities causes friction angle of the failure plane to be reduced. This helps wedge failure to occur along intersection lines of discontinuities 2 and 3 and discontinuities 3 and 4, because the factor of safety becomes less than one.

5 CONCLUSIONS

Based on Kinematic slope stability analysis, the orientation of discontinuities is so that some of the studied rock slopes are stable whereas, some of them have the potential of failure. In spite of roughness joint surfaces, the presence of filling materials will greatly contribute to instability in the rock slopes No.1, 2, 3, 6 and 7. The orientation of discontinuities in the rock slopes No. 1, 2 and 7 has provided the conditions for the occurrence of wedge failure when these rock slopes are dry. Instable conditions for the rock slopes are demonstrated by factors of safety less than one. Furthermore, if saturated filling materials exist in the rock discontinuities, wedge failure will occur along intersection lines of the other discontinuities. This failure indicates the important role of filling materials of discontinuities on rock slope instability. The occurrence of toppling failure is certain due to the presence of discontinuity 2 in the rock slope No. 3. The clay minerals with absorption of water induce a compressive force to the walls of critical discontinuities and increase the potential of toppling failure in this rock slope. The presence of a discontinuity set parallel to the slope face in the rock slope No. 6 is one of the most important factors for plane failure occurrence when the discontinuities are filed by clay minerals. The presence of filling materials in discontinuities of the studied rock slopes helps to create their instabilities by two different ways. First, absorption of water by filling materials reduces the friction angle of the failure planes that leads to plane and wedge failures. Second, the expansion of clay minerals causes a compressive force to be induced to the walls of critical discontinuities that leads to toppling failure.

REFERENCES

Agustawijaya, D.S., 2005. The influence of rock engineering properties on slope stability at Senggigi Resort area. Lombok Island, Journal of Teknik Sipit, 6(1): 92–97.

Cagatay, M.N., Keigwin, L.D., Okay, N., Sari, E., and Algan, O. 2002. Variability of clay-mineral composition on Carolina Slope (NW Atlantic) during marine isotope stages 1–3 and its paleoceanographic significance. Marine Geology, 189: 163–174.

Eberhardt, E., Thuro, K., Luginbuehl, M., 2005. Slope instability mechanisms in dipping interbedded conglomerates and weathered marls–the 1999 Rufi landslide, Switzerland. Eng Geol, 77: 35–56.

Indraratna, B., Premdasa, W., Brwon, E.T., Gens, A., Heitor, A., 2014. Shear strength of rock joints influenced by compacted fill. Int. J Rock Mech. Min. Sci., 70: 296–307.

ISRM, 1972. Suggested methods for determining water content, porosity, density, absorption and related properties and swelling and slake-durability index properties, 36pp.

ISRM, 1978. Suggested methods for determining petrographic description of rocks. International Journal of Rock Mechanics and Mining Sciences, 15: 43–45.

ISRM, 2007. The Blue Book: The complete ISRM suggested methods for rock characterization, testing and monitoring. 1974–2006. Ulusay R., Hudson J.A., (editors), Compilation arranged by the ISRM Turkish National Group, Ankara, Turkey. Kazan Offset Press, Ankara.

Kuhlemann, J., Lange, H., Paetsch, H., 1993. Implications of a connection between clay mineral variations and coarse grained debris and lithology in the central Norwegian-Greenland Sea. Mar. Geol, 114: 1–11.

Markland, J.T., 1972. A useful technique for estimating the stability of rock slopes when the rigid wedge sliding type of failure is expected. Imp. Coll. Rock Mech. Res., 19: 10pp.

Nkpadobi, J.I., Raj, J.K., Ng, T.F., 2015a. Failure mechanisms in weathered meta-sedimentary rocks. Environmental Earth Sciences, 73: 4405–4418.

Nkpadobi, J.I., Raj, J.K., Ng, T.F., 2015b. Influence of discontinuities on the stability of cut slopes in weathered meta-sedimentary rocks. Geomechanics and Geoengineering, 10(4): 290–302.

Priest, S.D., 1985. Hemispherical Projection in Rock Mechanics. George Allen & Unwin, London, 124 pp.

Priest, S.D., 1993. Discontinuity Analysis for Rock Engineering. Chapman & Hall, London, 473 pp.

Sainsbury, D.P., Sainsbury B. L., Sweeney E., 2016. Three-dimensional analysis of complex anisotropic slope instability at MMG's Century Mine. Mining Technology, DOI:10.1080/14749009.2016.1163918.

Shi, G.H., 2014. Application of discontinuous deformation analysis on stability analysis of slopes and underground power houses. Geomechanics and Geoengineering, 9(2): 80–96.

Sinha, U.N., Singh, B., 2000. Testing of rock joints filled with gouge using a triaxial apparatus. International Journal of Rock Mechanics and Mining Sciences, 37: 963–981.

Wyllie, D.C., Mah, C.W., 2004. Rock Slope Engineering. Spoon Press, Taylor and Francis Group, London, UK, 429 pp.

Yokota, S., Iwamatsu A., 1999. Weathering distribution in a steep slope of soft pyroclastic rocks as an indicator of slope instability. Engineering Geology, 55: 57–68.

Zhang, K., Cao P., Ma, G., Ren, F., Li, K., 2016. Stability Analysis of Rock Slope Controlled by Major Geological Discontinuities Based on the Extended Kinematical Element Method. Rock Mech Rock Eng, 49(7): 2967–2975.

Geomechanics and Geodynamics of Rock Masses – Litvinenko (Ed.)
© *2018 Taylor & Francis Group, London, ISBN 978-1-138-61645-5*

Laboratory investigation of crack initiation on hourglass-shaped granite specimens

L. Jacobsson & J.E. Lindqvist
RISE Research Institutes of Sweden, Borås, Sweden

ABSTRACT: Laboratory experiments on axially compressed hourglass-shaped specimens of medium to coarse grained granite specimens were conducted. A tangential stress is generated in the circular notches which is intended to initiate surface spalling similar to what can be seen at circular openings in a rock mass. Specimens of three different sizes were tested with notch radii 98.4, 225 and 375 mm, which are equivalent to a hole diameter of 197, 450 and 750 mm. The spalling initiation and progress in the notches were monitored by acoustic emission and digital correlation measurements. From the acoustic emission measurements it could be seen that the tangential stress at spalling initiation decreased with increasing notch radius. Results from digital image correlation show how the surfaces in local zones in the notch were pushed outwards due to subsurface cracking parallel to the notch surface. Analyses of thin sections and polished slabs taken in the notch area showed that the cracking depth increased with increasing notch radius.

1 INTRODUCTION

The understanding of spalling around circular openings in a rock mass is of interest in rock mechanics applications. Laboratory and in-situ experiments have shown that the tangential stress level at which spalling initiates depends on the diameter of the circular opening from experiments in the in the same material (e.g. Martin 1994, 1997). Previous laboratory experiments were mainly on axially compressed blocks which had a central drilled hole through the block in order to generate a tangential stress in the circumferential direction on the hole wall. The block sizes and required forces increase rapidly if blocks with large diameter holes are going to be investigated. A new type tests on hourglass-shaped specimens, where the hole wall is resembled by circular notches on the specimen, were previously reported by Jacobsson et al. (2015). This specimen design allows inducing a tangential stress on a simulated hole wall at a lower force as compared with tests on blocks containing a hole for a given hole diameter.

In this paper we present results from a new series of tests which were conducted on hourglass-shaped samples made from cylindrical cores of quarried medium to coarse-grained granite. Samples of three sizes were tested. The specimen geometries were uniformly scaled in all dimensions. The effect of the uniform scaling can be seen as the grain size is varied in a specimen for a given stress field. The aim with the tests was to investigate at which tangential stress spalling initiates for a range of radii (hole diameters) which previously have not been explored by laboratory experiments. The fractures were investigated visually on polished slabs and by microscopy investigations on thin sections using fluorescence technique to visualize the cracks. The crack analysis yields an insight of the crack formation process for this particular specimen geometry and material. Besides acoustic emission and strain measurements, the deformation at the notches was monitored using digital image correlation (DIC).

2 EXPERIMENTS

2.1 *Specimens*

Cylindrical specimens were manufactured from a granite block obtained from the Flivik quarry in southeast of Sweden. Flivik Granite is a porphyritic granite with 5–20 mm large euhedral K-feldspar phenocrysts, set in a medium-grained (2–4 mm) matrix. Main minerals are K-feldspar, quartz and plagioclase, with minor biotite and titanite. K-feldspar is red, plagioclase greenish yellow from low-grade alteration (sausseritization) and quartz light purple. The rock is structurally isotropic and shows no signs of metamorphic processes. The uniaxial compressive strength, determined from six 47 mm cores, was 205.1 ± 1.6 MPa. The crack initiation and crack damage stress were 107.8 ± 2.4 MPa and 160.2 ± 9.4 MPa.

Twelve large cores of three different diameters were drilled from the quarried block perpendicular to the rift plane. Four identical cylindrical specimens of each diameter with machined parallel end surfaces were manufactured. Planar circular notches were machined on opposite sides of the specimens yielding an hourglass shape, see Figure 1. The notch intended to resemble the wall on a circular opening. The notches were carefully grinded to yield minimum surface damage and roughness. The notch radii 98.4, 225 and 375 mm on the three specimen sizes were chosen (Figure 1). The specimens were uniformly scaled in all dimensions to obtain the same shape of the stress field in the different specimens. The same type of specimen design was used by Jacobsson et al. (2015) in experiments on Äspö diorite. Table 1 shows the specimen dimensions and the calculated axial load that yields a tangential stress of 200 MPa in the notch centre.

2.2 *Experimental set-up*

The specimens were instrumented with four strain gauges and eight acoustic emission (AE) sensors on each specimen to make it possible to monitor deformations, crack noise emission levels and localisation of crack events (Figures 1 and 2). Micro30 sensors were used on the R98 specimens and R15 sensors on the R225 and R375 specimens for the AE measurements. The AE sensors, acquisition system and software were from Physical Acoustics. The deformation field on the notch surfaces was measured using digital image correlation (DIC) on the two smallest specimen sizes. The DIC system Aramis 12 M was used to measure the displacement field on one

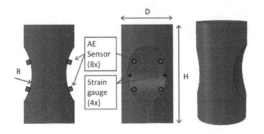

	R98	R225	R375
R (mm)	98.4	225	375
D (mm)	105	240	400
H (mm)	201	460	767
F_{200MPa} (kN)	1028	5376	14929

Figure 1. Left: Specimen geometry and locations of strain gauges and AE-sensors; Right: Dimensions for the different specimen sizes and axial force yielding 200 MPa tangential stress in the notch centre.

Table 1. Maximum axial load F_{max} and equivalent tangential stress σ_t in the notch centre. Observations on the notches are fracture (fr), spalling (sp) and possible spalling (ps).

	R98-1	R98-2	R98-3	R98-4	R225-1	R225-2	R225-3	R225-4	R375-1	R375-2	R375-3	R375-4
F_{max} [kN]	1230	1217	1302	1407	6408	6369	6538	7152	17570	18150	17410	18600
σ_t [MPa]	239	237	253	274	238	237	243	266	235	243	233	249
Notch 1			fr					fr	ps	sp	fr	fr
Notch 2			fr						ps	ps	ps	ps

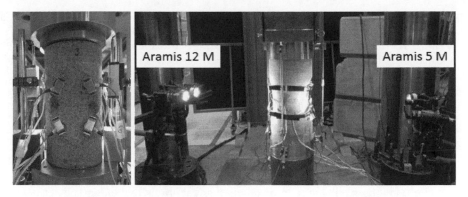

Figure 2. Set-up for the tests of the specimens R98 (left and middle) and R225 (right).

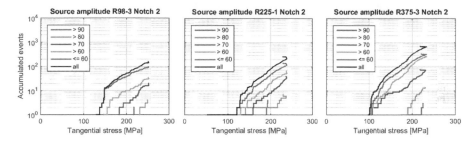

Figure 3. Accumulation of localized events of different source amplitude (dB).

of the notches on the R98 specimens and two DIC systems, Aramis 5M and Aramis 12M, were used to measure the displacement field on one notch each during the tests of the R225 specimens. The calibration volumes were approximately 85 mm × 65 mm on the R98 specimens and 260 mm × 200 mm on the R225 specimens. The total deformation between the loading platens using linear resistive gauges was also measured, but the results are not shown here.

The specimens were uniaxially compressed using two different servo-hydraulic equipment, one high-stiffness load frame used for rock mechanics testing with a loading range up to 1,600 kN and a general purpose four-column load frame with a loading capacity up to 20,000 kN. Pictures of the test set-ups are shown in Figure 2.

3 RESULTS

3.1 *General*

The specimens were loaded as high possible without causing a complete specimen failure. This upper load limit was difficult to know and some specimens did fail by shear-tension and slab types of pieces were formed. No specimen failed completely. The maximum applied load, equivalent maximum tangential stress in the notch centre and observations from the notch surfaces are shown in Table 1. The notch surfaces were visually examined directly after a completed test. Spalling was seen in one case, where a surface crack could be seen along with an outward deformation of the surface. Possible spalling in the notches was identified by feeling a slight outward surface deformation using finger tips.

3.2 *AE*

The amplitude and accumulated number of localised events for three different specimen sizes are shown in Figure 3. It can be seen that the trend is that crack events occur at a lower stress when the radius increases. Looking for example at events with a source amplitude >70 dB, we see that activities starts around 182 MPa for specimen R98-3, 158 MPa for R225-1 and 108 MPa for R375-3. There is a scatter between results of different specimens and notches (not shown here), but a general trend can be clearly be seen. Figure 4 shows the maximum tangential stress at AE initiation for different amplitudes. The ratio of the maximum tangential stress over the compressive strength obtained from the uniaxial compression tests versus twice the notch radius (equivalent hole diameter) are shown in Figure 4.

3.3 DIC

The out-of-plane deformations on specimen R98-4 and R225-4, measured using the Aramis 12 M system, are shown in Figure 5. Local peak values "hot spots" (red) with size of c 5–8 mm could be observed. This is probably individual grains that either split or detach at grain boundaries. By looking at the average displacement on a small region (c 4 mm) within a hot spot relative to neighboring regions (c 4–5 mm) versus stress it becomes evident when the displacements locally starts to deviate from a linear response (Figure 6). For two selected hot

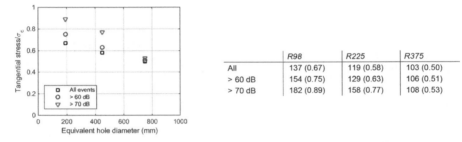

	R98	R225	R375
All	137 (0.67)	119 (0.58)	103 (0.50)
> 60 dB	154 (0.75)	129 (0.63)	106 (0.51)
> 70 dB	182 (0.89)	158 (0.77)	108 (0.53)

Figure 4. Left: Ratio of the maximum tangential stress at AE initiation for different amplitudes over compressive strength σ_c versus equivalent hole diameter; Right: Observed initial tangential stress in MPa at initiation of acoustic emission (Accumulated events = 2). The values within the parentheses denote tangential stress/σ_c.

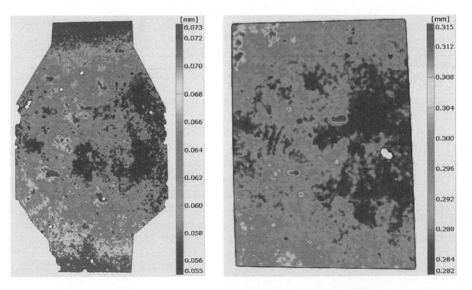

Figure 5. Out-of-plane displacement at notches measured by DIC. Left: R98-4 (c 55 mm picture width); Right: R225-4 (c 88 mm picture width).

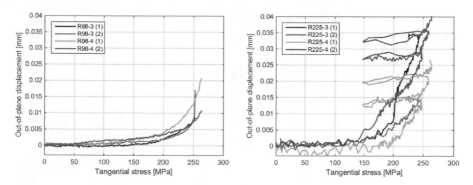

Figure 6. Out-of-plane measurements of local regions with possible spalling. The specimen R225-4 was partially unloaded twice until loading up to fracturing.

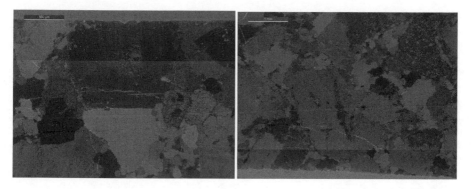

Figure 7. Thin sections impregnated with fluorescent epoxy. Left: R98-1 (3); Right: R98-3 (3).

spots each in the notches monitored by DIC on specimens R98-3, R98-4, R225-3 and R225-4 it can be noted that the out-of-plane displacement successively starts to deviate from a linear response at 160-190 MPa on the R98 specimens and around140-190 MPa on the R225 specimens. Sudden increase of the out-of-plane displacement can be seen on the R225 specimens. Moreover, it can be seen that the noise and out-of-plane displacement is larger on the R225 specimens. The larger noise in the R225 results is due to less measurement points in the averaging in the selected areas.

3.4 Crack observations

Thin sections (R98) and polished slabs (R225 and R375) with the rock material impregnated with fluorescent epoxy showing a cross section over the notch were prepared for some specimens (Figures 7 and 8). The cracks and fractures were studied and photographed in fluorescence light. Shallow cracks, subparallel to the notch surface in the direction of the largest compressive stress, were formed at several locations. Some grains, somewhat with a deeper location, were also split at some locations in the same direction. These types of results were also seen on experiments on diorite (Jacobsson et al. 2015). It can be seen that the subparallel cracks to the notch surface are formed more shallowly in the R98 specimens and the depth increases with increasing radius. The mineral micro structure plays a crucial role for the initiation, shape and propagation of these near surface cracks. Further crack investigations using thin sections will be carried out to study the crack paths in relation to the minerals and crystallographic directions.

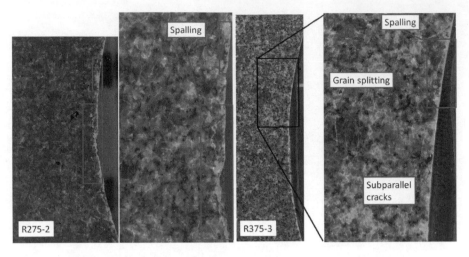

Figure 8.   Polished slabs impregnated with fluorescent epoxy.

4   CONCLUSIONS

The mechanical experiments on specimens with notches resembling the hole geometry around circular openings show a trend that cracks form at a lower stress when the notch radius increases, which is in correspondence with previous investigations (Martin 1993, 1997). A combination of AE and deformation field measurements was used to investigate the crack initiation in the notches. The out of-plane deformations reveal local dilations on up to 0.01–0.02 mm. The measured localized AE-events and out-of-plane displacements correlate to each other for one of the spots.

ACKNOWLEDGEMENTS

The work presented in this paper has been funded by Swedish Nuclear Fuel and Waste Management Co (SKB). Assistance during measurements by Torsten Sjögren, Mathias Flansbjer and staff in the laboratory at RISE is gratefully acknowledged. The experimental work was carried out during 2014–2015 at SP Technical research Institute of Sweden in Borås, which since then changed name to RISE in January 2016.

REFERENCES

Jacobsson, L., Appelquist, K. & Lindqvist, J.E., 2015. Spalling experiments on large hard rock specimens. *Rock Mechanics and Rock Engineering* 48(4):1485–1503, doi: 10.1007/s00603-014-0655-0.

Martin, C.D., 1993. The strength of massive Lac du Bonnet granite around circular openings. PhD-thesis, University of Manitoba, Canada.

Martin, C.D., 1997. Seventeenth Canadian Geotechnical Colloquium: The effect of cohesion loss and stress path on brittle rock strength. *Canadian Geotechnical Journal*, 34(5), 698–725. doi:10.1139/t97-030.

Geomechanics and Geodynamics of Rock Masses – Litvinenko (Ed.)
© 2018 Taylor & Francis Group, London, ISBN 978-1-138-61645-5

Comparison of different approaches to predict the shear strength of large rock discontinuities

M. Jeffery, L. Lapastoure Gritchou & A. Giacomini
Priority Research Centre for Geotechnical Science and Engineering, University of Newcastle, Callaghan, NSW, Australia

V. Griffiths
Department of Civil and Environmental Engineering, Colorado School of Mines, Golden, CO, USA

O. Buzzi
Priority Research Centre for Geotechnical Science and Engineering, University of Newcastle, Callaghan, NSW, Australia

ABSTRACT: Reliable prediction of shear strength for engineering scale in-situ discontinuities is still problematic. There is currently no consensus or a satisfactory method to estimate shear strength, account for surface variability and manage the effects of the recognised 'scale effect' phenomenon. Shear behaviour of a discontinuity rock mass greatly depends upon the rock joint roughness, which is generally concealed within the rock mass. As such, the limited amount of accessible surface information complicates even further the exercise of shear strength prediction. Often, small size specimens (e.g. rock core) are recovered to conduct experimental tests and predictions can be made from analysing traces. In this paper, three different shear strength prediction approaches were followed and their relative performances were compared. The results reveal that the prediction of shear strength from recovered sub samples can be significantly variable. The application of Barton's empirical model to four selected traces, produced a large scattering of results, with the prediction highly dependent on the trace used. Conversely, the application of Casagrande and co-workers' stochastic approach on the same traces, produced the least scatter and provides statistical data to quantify variability and uncertainty.

1 INTRODUCTION

The prediction of shear strength is an important component in large scale engineering rock mass stability analysis and design applications. It is well known that roughness is a key factor influencing the mechanical response of a discontinuity and its peak shear strength. There is currently no consensus or a satisfactory method to estimate shear strength of a large discontinuity, whilst accounting for surface roughness variability and managing the effects of the recognised 'scale effect'. As such, reliable prediction of the shear strength of large in situ discontinuity is still problematic.

Extensive work has been undertaken over the last four decades to characterise joint surface information (Ge et al, 2015). However, the first difficulty with estimating shear strength of a large in-situ discontinuity, stems from the limited amount of available surface information. Due to convenience and a lack of reliable alternatives, researchers and engineers typically employ empirical approaches, or use small scale specimens and either numerical or experimental methods to predict a shear strength (Buzzi et al, 2017).

Approaches for estimating shear strength of large discontinuities generally rely on the assumption that obtained roughness data, is representative and shares some characteristics of the whole surface (Fardin et al, 2001, Buzzi et al, 2017). If this assumption is incorrect or disregarded, having an understanding of the potential variability is critical.

Since the 1960's numerous studies have investigated the scale effects on roughness, shear strength development and deformability. With respect to the shear strength studies, opposite conclusions have been drawn, regarding the need for positive or negative upscaling requirements.

Given a laboratory size sample, it is extremely difficult to determine the degree of scaling required (Giani, 1992). One of the most used empirical model for shear strength, has been modified to try and capture the scale dependence (Barton and Bandis, 1980). More recently a new stochastic approach for predicting peak and residual shear strength was developed by Casagrande et al. (2017). This approach is applicable at field scale and has the potential to minimise or entirely circumvent the scale effect issue.

This paper compares the results obtained from three shear strength prediction approaches, to gain an understanding of prediction performance with respect to prediction behaviour, magnitude and variability as well as managing scale effects.

2 OBJECTIVES

Assessing the shear strength of a large in situ discontinuity is a non-trivial task. Amongst the different options, it is possible to core the rock to obtain small size specimens (here referred to as "sub-sample") that will be tested in the laboratory or to estimate the strength from the roughness information that can be collected from visible traces. Consequently, this paper compares the shear strength predictions obtained by following three different approaches:

1. Using sub-samples of 100 mm per 100 mm.
2. Using Barton's model (Barton and Choubey, 1977), applied to a single trace.
3. Using a new stochastic approach proposed and validated by the authors, also applied to a single trace.

3 STUDY SURFACE

This study is based on a natural discontinuity located in a former sandstone quarry, in the Pilkington Street Reserve in Newcastle (NSW, Australia). The surface was accurately surveyed using 16 coded ground control points (GCP) that were evenly placed on the wall with their coordinates measured twice using a reflectorless total station (Leica TPS1205). More than 250 photographs were taken (Canon E0S 7D camera) at a distance of 1 m and the commercial software called "Structure for Motion Agisoft Photoscan" was used for a 3D reconstruction. The point cloud was structured into a grid with a 1 mm spatial increment. More information on the reconstruction process, is presented in Casagrande et al. (2017). The reconstructed surface is presented in Figure 1.

4 APPROACHES TO PREDICT SHEAR STRENGTH

4.1 *Using sub-samples*

For this first approach, the surface was sub-divided into 400 sub-samples, 100 mm × 100 mm domains as shown in Figure 1a. All 400 samples were analysed using the semi analytical shear strength model developed by Casagrande et al. (2017), which can predict peak and residual shear strength of rough, perfectly matched discontinuities for given shear direction, normal stress (0.02, 0.1, 0.25 and 0.5 MPa) and material properties (unaxial compressive strength (UCS)/joint compressive strength (JCS) value of 39.67 MPa, basic/residual friction angle (ϕ_b/ϕ_r) of 35°, cohesion (c) value of 4.74 MPa). Note that the values of normal stress and material properties given above were used for all three methods.

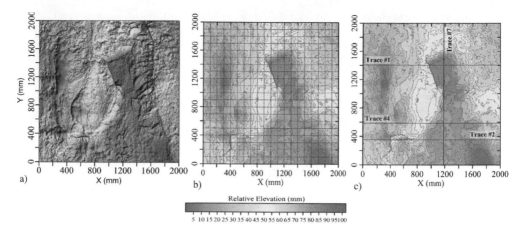

Figure 1.   Reconstructed Pilkington study surface (a), contour map with (b) domains of the 400 sub-surface and (c) selected traces for shear strength prediction.

4.2 *Barton's empirical model*

Barton's empirical failure criterion (Barton and Choubey, 1977) is commonly employed to predict the shear strength of discontinuities. The model accounts for the effect of roughness via the Joint Roughness Coefficient (JRC):

$$\tau_{peak} = \sigma_n \tan\left(\phi_r + JRClog_{10}\left(\frac{JCS}{\sigma_n}\right)\right) \tag{1}$$

where σ_n is the normal stress acting on the discontinuity, JCS is the Joint Compressive Strength and ϕ_r is the residual friction angle. The residual strength is obtained by equating the JRC to 0 in Equation 3.2, i.e.:

$$\tau_{residual} = \sigma_n \tan(\phi_r) \tag{2}$$

The JRC value is here back calculated from the surface descriptor Z_2 and Yang et al (2001) equation;

$$JRC = 32.69 + 32.98 \, log_{10}Z_2 \tag{3}$$

where Z_2 is defined as:

$$Z_2 = \left(\frac{1}{L.\Delta x^2}\sum_{i=1}^{L}\left(z_{\left(y_{(i+1)}\right)} - z_{\left(y_{(i)}\right)}\right)^2\right)^{\frac{1}{2}} \tag{4}$$

Equations (1) to (4) were applied to the four traces highlighted in Figure 1c.

4.3 *Stochastic approach*

Casagrande et al. (2017) developed a new stochastic approach to predict the peak and residual shear strength of rock discontinuities. This approach is applicable at field scale and has the potential to minimise or entirely circumvent the scale effect issue. The foundation of the method is to use the statistics of a visible trace, to form a large number of statistically similar synthetic surfaces via the rigorous application of a random field model. Shear strength distributions of shear strength (peak and residual) are obtained through analysing all of the synthetic surfaces using a semi analytical shear strength model. Note that this is the same semi-analytical model that is used for the approach described in section 4.1. The prediction is expressed in terms of mean shear strength with an error bar equal to the standard deviation. The reader is invited to **refer to** Casagrande et al. (2017) for more information on the details of the stochastic method.

The statistics of the four traces highlighted in Figure 1c were used as an input of the random field model and the stochastic approach. 100 synthetic surfaces were created for each trace.

4.4 *Reference shear strength*

In absence of experimental data on the surface used here (it is located in a protected reserve and cannot be tested), the predictions of the three approaches described above will be compared to the peak and residual shear strength pertaining to the whole surface and obtained via the semi-analytical model described in Casagrande et al. (2017). The same material properties than those given in section 4.1 prevail.

5 RESULTS

5.1 *Prediction of shear strength using sub-samples*

The cumulative distribution of strength (obtained on the 400 sub-samples) shows a considerable variability under all five normal stresses (see Figure 2a). For example, under 0.02 MPa, there is a factor 10 between the lowest and the highest predicted value of peak shear strength, and 50 for the residual strength.

Comparing the mean shear strengths (peak and residual) obtained from the 400 sub-samples to the peak and residual shear strength of the whole surface (obtained from the semi-analytical model, and referred to as "deterministic") reveals a clear negative scale effect (Figure 2b). Indeed, the whole surface consistently displays a higher strength than the

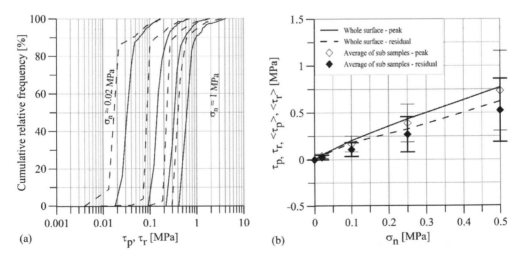

Figure 2. Cumulative relative frequency distribution of peak (continuous line) and residual (dashed line) shear strength for the 400 sub surfaces (a) and a comparison of mean peak and residual shear strength for the 400 sub surfaces and deterministic peak and residual shear predictions on the whole surface (b).

sub-samples tested. At this stage, this is not fully explained (it has not been fully analysed) but is believed to be a matter of the roughness features captured in the each sub-sample.

5.2 Prediction using Barton's model

The JRC values for traces 1, 2, 4 and 7 were back-calculated as 19.12, 14.63, 7.01 and 5.52, respectively. Figure 3a compares the deterministic criterion of the whole surface (peak and residual) to the predictions obtained from Barton's model. Based on the trace selected, it is possible to under-estimate (by as much as 40%) or over-estimate (by as much as 60%) the peak shear strength of the surface. The failure criterion obtained for trace #1 is quite questionable: the peak shear strength is negative at low normal stress before increasing significantly and over-predicting the peak shear strength (by about 50% at 0.5 MPa of normal stress). This behaviour is due to equating the tangent of a combined friction angle value that is close to $\pi/2$ (98.41° at 0.02 MPa and 85.05° at 0.1 MPa), before multiplying by the normal stress to determine the peak shear strength. In the instance of trace 1, the large JRC value contributed to the combined angle terms to be greater than $\pi/3$, for the considered normal stresses. The JRC being nil for the residual strength prediction, all traces have the same residual failure criterion (Figure 3b), which falls under the deterministic value (under estimation ranging from 40% to 70%).

5.3 Prediction using the new stochastic approach

The peak and residual failure criteria obtained by the stochastic method, presented in Figure 4, show a good agreement with the deterministic failure criteria. Note that the approach relies on creating synthetic surfaces created from the statistics of the seed trace. In that sense, the new approach does not use the shear strength of the whole surface. At low normal stresses, the difference between the predicted and deterministic criterion is only in the order to −25% to 15%. Under 0.5 MPa, the difference between the highest and lowest shear strength increases to 0.21 MPa and 0.173 MPa for peak and residual shear strength, respectively. Note that for the sake of clarity, the error bars are not shown.

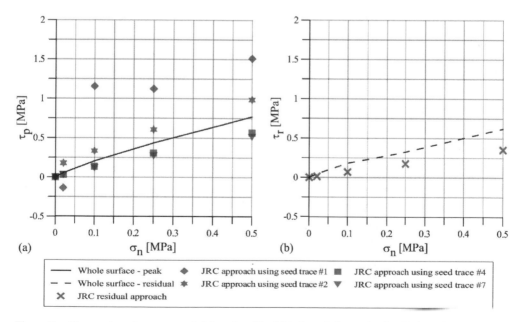

Figure 3. Comparison between peak (a) and residual (b) shear failure criteria obtained from Barton's model (using traces 1, 2, 4 and 7) and deterministic criteria pertaining to the whole surface.

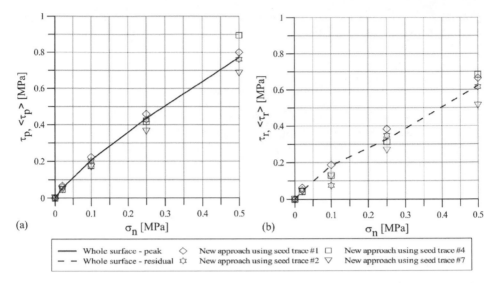

Figure 4. Comparison between peak (a) and residual (b) shear failure criteria obtained from the new stochastic approach (using traces 1, 2, 4 and 7) and deterministic criteria pertaining to the whole surface.

6 DISCUSSION

The distributions of shear strength obtained for the 400 subsamples presented in Figure 2a, highlight one of the main disadvantages of testing sub-samples, namely the possibility of large variability. The extent of variability can be correlated to the degree of morphology variability of the parent surface, which cannot be easily assessed if the surface is contained within a rock mass. The surface studied here is quite variable with some marked roughness features (such as steps), resulting in a wide distribution of predictions.

Figure 2 also confirms another long-known problem: the existence of a scale effect. Published results in the literature demonstrate that there is no such thing as a systematic scale effect. Positive (REF) and negative effects (REF) have been observed as well as no scale dependence (REF). To date, there are no reliable methods that can help characterise, qualitatively or quantitatively, the scale effect.

The application of the JRC approach for the four traces tested produced mixed results. The degree of variability between the four failure criteria and the unusual failure criterion associated to trace #1 is concerning. The failure criteria obtained under and overestimate the strength by about −40 to +60%. Note that the reference strength (deterministic value) is obtained from the semi-analytical model used for the other two approaches, so there is a slight bias in the results. However, the focus is here more on the scattering of results between the different approaches rather than reaching a slightly biased target value.

The variability of these results can be attributed to the effect of surface morphology on the evolution of the Z2 and JRC value. Figure 5a and b, show traces 1 and 2 encounter a morphological feature characterised by a significant and abrupt increases in elevation. The evolution of Z2 and JRC along the traces, presented by Figure 5c and d respectively, illustrates that the magnitude of these values, is increased by an order of 2–3 times after the feature is encountered. These figures also show, the values of Z2 and JRC to gradually reduce in magnitude away from the crest, but appear to be still affected by the initial encounter at the end of the trace length. In contrast, despite different elevation profiles, traces 4 and 7, produced Z2 and JRC evolution profiles that show only minor localised variations values along the traces length. From this it can be concluded that morphological features such as steps can affect the evolution of Z2 and JRC values, which will in turn effect the prediction of shear strength and contribute to an increase in prediction variability.

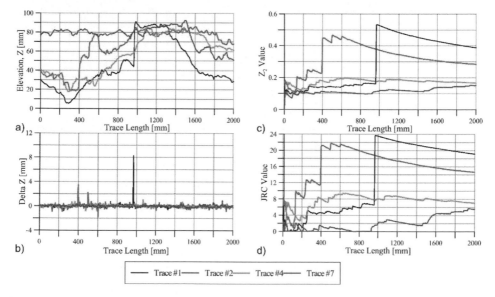

Figure 5. Comparison of Z2 and JRC value realisation for the 4 traces: trace elevation profiles (a), trace profile incremental elevation difference (b), evolution of Z2 values (c) and evolution of JRC value (d).

The prediction of both peak and residual shear strength using the new stochastic approach produced results with much more reduced scattering. It has to be reminded here that the rationale of this approach is that there is enough information on a visible trace to create a range of realistic discontinuities that underpin the stochastic prediction of shear strength. As such, the full surface does not constitute an input of the new stochastic approach. The new approach does not rely on a scale dependent descriptor (such as JRC) and hence circumvents the scale effect, the approach produces mean values of shear strength with reduced scattering (despite using four different traces, and unlike the JRC approach) and a standard deviation, which can be used to compute confidence intervals.

7 CONCLUSIONS

This paper compares three approaches to predict shear strength of a large discontinuity (2m per 2m) coming from a natural reserve in the vicinity of Newcastle, NSW, Australia. The first approach consists of testing sub-samples (from cored rock). The study revealed that the results variability can be very large, with as much as a factor 50 between one prediction and another. Also, the occurrence of a scale effect was confirmed, meaning that the results obtained in the laboratory cannot easily be extrapolated to the whole surface.

The second approach consists of applying Barton's model, based on back calculated JRC, to a selection of traces. A large scattering of results was observed, with the prediction being highly dependent on the trace used. One trace produced an inconsistent failure criterion with a negative value of shear strength. It was concluded that the JRC approach generally produces conservative predictions of shear strength. The study noted that abrupt morphological features can detrimentally effect the evolution of Z2 calculated JRC values.

The third approach tested is that developed by Casagrande et al. (2017) which relies on random field model and stochastic analysis via a semi-analytical model for shear strength. It was found that this approach produced the most accurate predictions of shear strength and the least scatter. The obtained peak and residual shear strength distributions can be used to attain statistically confident mean estimates and an indication of confidence interval via the standard deviation. One of the advantages of this approach is that scale free surface characteristics are obtained at the intended scale and therefore the scale effect can potentially be avoided.

ACKNOWLEDGEMENTS

The authors would like to acknowledge the financial contribution received from Pells Sullivan Meynink, Engineering Consultants, Sydney.

REFERENCES

Barton N. (1976). The Shear Strength of Rock and Rock Joints. International Journal of Rock Mechanics and Mining Science & Geomechanics Abstracts, 13, 255–279.

Barton N., Choubey V. (1977). The shear strength of rock joints in theory and practice. Rock Mechanics and Rock Engineering, 10, 1–54.

Barton, N., Bandis, S. (1980). Some effects of scale on the shear strength of joints. International Journal of Rock Mechanics and Mining Sciences & Geomechanics Abstracts, 17, 69–73.

Buzzi O., Casagrande, D., Giacomini, A., Lambert, C. Fenton, G. (2017). A new approach to avoid the scale effect when predicting the shear strength of large in situ discontinuity. In: Proceeding of 70th Canadian Geotechnical Conference, Ottawa, Canada.

Casagrande, D., Buzzi, O., Giacomini, A., Lambert, C., Fenton, G. (2017). A New Stochastic Approach to Predict Peak and Residual Shear Strength of Natural Rock Discontinuities. Rock Mechanics and Rock Engineering, 1–31.

Fardin, N., Stephansson, O., Jing, L. (2001). The scale dependence of rock joint surface roughness. International Journal of Rock Mechanics and Mining Sciences, 38 (5), 659–669.

Ge, Y., Tang, H., Eldin, M.A.M.E., Chen, P., Wang, L., Wang, J. (2015). A Description for Rock Joint Roughness Based on Terrestrial Laser Scanner and Image Analysis. Scientific Reports, 5, 16999. http://doi.org/10.1038/srep16999.

Giani, G.P., Ferrero, A.M., Passarello, G., Reinaudo, L. (1992). Scale effect evaluation on natural discontinuity shear strength. In Proceedings of the international congress fractured jointed rock mass. Lake Tahoe, California, 456–462.

Maerz, N.H., Franklin, J.A. and Bennett, C.P. (1990). Joint roughness measurement using shadow profilometry. International Journal of Rock Mechanics and Mining Sciences & Geomechanics Abstracts 27(5): 329–343.

Yang, Z.Y., Lo, S.C., Di, C.C. (2001). Reassessing the joint roughness coefficient (JRC) estimation using Z2. Rock Mechanics and Rock Engineering, 34(3), 243–251.

Geomechanics and Geodynamics of Rock Masses – Litvinenko (Ed.)
© 2018 Taylor & Francis Group, London, ISBN 978-1-138-61645-5

Effects of excavation damage on the electrical properties of rock mass

Pekka Kantia
Geofcon, Rovaniemi, Finland

Risto Kiuru & Mikael Rinne
Department of Civil Engineering, School of Engineering, Aalto University, Espoo, Finland

ABSTRACT: Electrical and Electromagnetic (EM) methods have been used for characterisation of Excavation Damage Zone (EDZ) in conjunction with long-term safety evaluation of geological disposal of spent nuclear fuel. Physical properties of the rock have been tested in laboratory to characterise the property changes related to EDZ. EM research has focused around Ground Penetrating Radar (GPR) utilizing GPR EDZ method, developed for excavation quality control in means of EDZ extend. For 20 specimens, resistivity, relative dielectric permittivity (ε_r) as well as high frequency scattering parameters (electrical conductivity and ε_r) were measured. Induced Polarization (IP) values were calculated from resistivity data. Resistivity, IP and ε_r, was produced to support the use of GPR image analysis and the GPR EDZ method. Results were analysed to reveal links between ε_r and conductivity, and to analyse possible depth dependencies. A positive association between electrical conductivity and ε_r was observed. Changes in the electrical properties linked to shallow depths and visual EDZ features were observed, which seems to validate the theoretical basis of the GPR EDZ method. Electrical property data allows further modelling, development and assessment of the GPR EDZ method.

Keywords: Excavation damage, scattering parameters, electrical resistivity, electrical conductivity, induced polarisation, relative dielectric permittivity, network analyser

1 INTRODUCTION

Excavation damages the rock in the tunnel vicinity and this weakens the rock mass. Formed damage zone in the tunnel vicinity, microscopic as well as open fractures, are referred to as Excavation Damage Zone (EDZ). Understanding the formation and physical characteristics of excavation damage is critical for the long-term safety evaluation of deep geological disposal of spent nuclear fuel. More intense and deeper the EDZ penetrates, the higher is the risk for formation of pathways for the radio nuclides to the organic nature. As the EDZ is seen as a long-term disposal safety issue, it has to be controlled. Methods for controlling EDZ are Drill and Blast (D&B) method related (drilling, charging and workmanship, e.g.) but a measurement method for EDZ is needed as well. The Finnish company Posiva has been developing high frequency Ground Penetrating Radar (GPR) method for the task since 2008, resulting in a GPR signal frequency content analysing method called GPR EDZ method (Kantia et al., 2012; Kantia et al., 2013; Kantia et al., 2016a). It appears that the GPR EDZ method can be used in EDZ characterization (Heikkinen et al., 2010) as well as in EDZ extent measurements (Kantia et al., 2016b). In this work, electrical properties of intact and damaged rock specimens were determined to verify the feasibility of the GPR EDZ method and allow theoretical modelling of the GPR signal in the rock mass.

2 EXCAVATION DAMAGE ZONE

Terminology used in EDZ investigations is based on a paper by Dinis da Gama and Torres (2002) which defines several levels of distinctly different damaged zones: (1) zone of crushing, (2) zone of radial cracking, (3) zone of extension and expansion of fractures and (4) elastic zone where no cracks are formed (Figure 1). Siren et al. (2014) separate the most important formation mechanisms of EDZ as (1) EDZ_{CI} – Construction induced excavation damage zone, which is instantly formed by the construction method and (2) EDZ_{SI} – Stress-induced excavation damage zone, which is consequence of the redistribution of the stress field in the rock (Figure 2). EDZ influences the physical and mechanical properties of the rock mass and EDZ effects are different in different rock types. For example, granitic pegmatoid (PGR) with

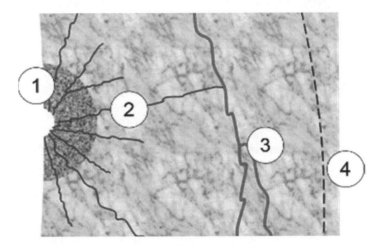

Figure 1.   Schematic presentation of different type of damages near the blasting hole and tunnel surface as classified by Dinis da Gama and Torres, 2002: (1) zone of crushing, (2) zone of radial cracking, (3) zone of extension and expansion of fractures and (4) elastic zone, where no cracks are formed. Adapted from Kantia et al. (2016).

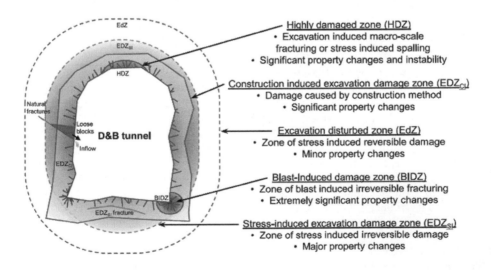

Figure 2.   Definitions of the different damage zones used in this work. Modified from Siren et al. (2014).

intense micro fracturing differs from veined gneiss (VGN) characterised by clear individual fractures that penetrate deeper.

3 MEASURED ELECTRICAL PROPERTIES AND OBSERVED CHANGES

For 20 specimens, electrical resistivity, induced polarisation (IP effect, describes chargeability of the material) and relative dielectric permittivity (describes attenuation of an electromagnetic wave in the material) values as well as high frequency scattering parameters (electrical conductivity and relative dielectric permittivity) were determined. Resistivity was first measured at 0.1 Hz, 10 Hz and 500 Hz using a proprietary measurement system developed by the Geological Survey of Finland. From the measured resistivity values, IP effects (0.1 Hz/10 Hz and 0.1 Hz/500 Hz) were calculated. Relative dielectric permittivity was measured using an Adek v.7 percometer operating in the 40–50 MHz range (for 19 of the 20 specimens). Description of the resistivity, IP effect and relative dielectric permittivity measurements can be found in Kiuru (2017). High frequency scattering parameters (electrical conductivity and relative dielectric permittivity) were measured for all specimens both saturated and dry at 2 GHz and 3 GHz using an Agilent network analyser. Results are shown in Table 1.

Results were analysed to reveal associations between electrical parameters and their possible depth dependencies. A general increase in relative dielectric permittivity was observed with increasing resistivity (Figure 3), while scatter of the observed values also increases. At the low frequencies (0.1 Hz, 10 Hz and 500 Hz) resistivity is generally higher in the first 20 cm of the excavated surface, and shows higher variation than deeper (Figure 4, left). A similar effect was also observed at higher (2 GHz and 3 GHz frequencies), if not as pronounced (Figure 4, right).

Table 1. Results from the measurements.

Sample	Rock type	Depth m	Porosity %	R0.1	R10	R500	2G	3G	ε_r	$\varepsilon_r'^1$	$\varepsilon_r'^2$
						Ohm.m					
ED123	VGN	0.054	0.81	11900	11200	10200	67.0	47.4	6.30	6.65	6.58
ED124	VGN	0.135	0.60	18200	16700	14500	114.9	87.3	6.44	6.15	6.10
ED131	VGN	0.115	0.24	34600	31300	27300	188.0	151.8	6.17	5.53	5.51
ED132	VGN	0.170	0.25	21500	20100	18400	264.8	247.5	6.35	5.19	5.17
ED141	VGN	0.590	0.45	8540	7810	7080	118.7	103.1	6.40	6.28	6.19
ED142	VGN	0.645	0.55	7190	6990	6640	87.8	71.6	6.26	5.99	5.91
ED144	PGR	0.200	0.41	5050	5190	5140	200.5	189.6	5.32	5.39	5.34
ED145	PGR	0.255	0.36	6010	6450	6340	155.9	142.1	5.28	5.28	5.24
ED146	PGR	0.310	0.27	14600	14300	13700	107.9	93.9	5.81	5.51	5.43
ED147	PGR	0.365	0.32	9850	9470	9060	105.9	94.5	5.71	5.47	5.41
ED152	PGR	0.530	0.32	8100	7990	7740	189.7	179.8	5.59	5.31	5.26
ED153	PGR	0.585	0.40	5920	5740	5480	180.4	166.3	5.61	5.35	5.31
ED154	PGR	0.640	0.26	7370	7240	7000	184.7	170.4	5.22	5.41	5.38
ED165	PGR	0.115	0.31	4780	4640	4500	199.2	190.9	5.56	5.23	5.20
ED166	PGR	0.170	0.28	6260	6150	5960	180.8	169.8	5.48	5.30	5.26
ED170	PGR	0.390	0.41	9900	9500	8910	188.0	181.3		5.28	5.24
ED172	VGN	0.055	0.22	16400	15300	13800	214.6	173.9	5.99	5.59	5.54
ED173	VGN	0.110	0.23	24000	21800	18500	87.7	65.8	6.31	7.02	6.95
ED174	VGN	0.165	0.26	36300	31500	25800	117.7	91.5	6.38	6.52	6.47
ED175	VGN	0.220	0.20	38700	34300	28900	210.7	179.1	5.85	5.54	5.50

R0.1 is measurement at 0.1 Hz, R10 at 10 Hz and R500 at 500 Hz.
2G is the measurement at 2 GHz and at 3 GHz.
ε_r is the measurement at 40–50 MHz, $\varepsilon_r'^1$ at 2 GHz and $\varepsilon_r'^2$ at 3 GHz.

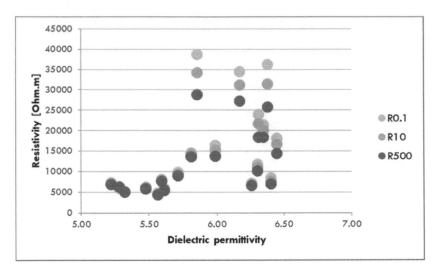

Figure 3.   Relative dielectric permittivity appears to increase with increasing resistivity. R0.1, R10 and R500 correspond to measurements at 0.1 Hz, 10 Hz and 500 Hz, respectively.

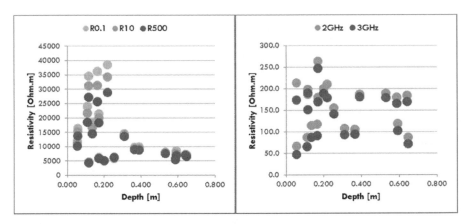

Figure 4.   Resistivity vs depth from excavated surface. At low frequencies (0.1 Hz, 10 Hz and 500 Hz, left figure), resistivity variation within the first 20 cm is much higher than deeper from the excavated surface. R0.1, R10 and R500 correspond to measurements at 0.1 Hz, Hz and 500 Hz, respectively. Similar effect is observed with high (2 GHz and 3 GHz) frequencies as well (right figure).

Relative dielectric permittivity shows higher values and variation near the excavated surface as well (Figure 5). Changes in the electrical properties linked to shallower depths could in some cases also be linked to visually observed EDZ fractures in core samples (Kiuru et al., this publication), which seems to validate the theoretical basis of the GPR EDZ method.

4   CONCLUSIONS

Changes in the electrical properties linked to shallow depths and visually observed EDZ features were observed, which seems to validate the theoretical basis of the GPR EDZ method. Knowing the electrical properties of intact as well as damaged rock allows assessing the feasibility of the method and enables theoretical modelling of the GPR signal behaviour in the site-specific rock mass. Improved understanding of the correlation between the geophysical and mechanical parameters of the rock mass provides better capability to detect and model the development of the excavation damage zone with help of the GPR EDZ method.

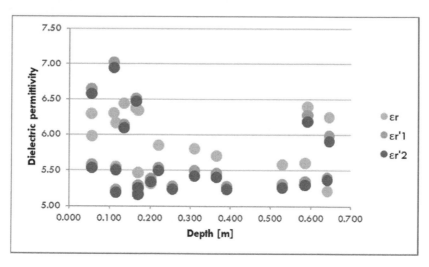

Figure 5.   Higher relative dielectric permittivity is observed in the surface layer, regardless of the measurement frequency. ε_r is the measurement at 40–50 MHz, $\varepsilon_r'1$ at 2 GHz and $\varepsilon_r'2$ at 3 GHz.

DISCLAIMER AND ACKNOWLEDGEMENTS

The views expressed are those of the authors and do not necessarily reflect those of Posiva. We would like to thank Academy of Finland for funding (grant 297770), Posiva Oy and Geofcon for providing the possibility to publish the data, VTT Technical Research Centre of Finland and the Geological Survey of Finland for help with the measurements and Sanna Mustonen, Eetu Pussinen, Eero Heikkinen, Noora Riihiluoma, Johannes Suikkanen, Lasse Koskinen and Eeva Käpyaho for cooperation on the EDZ research.

REFERENCES

Dinis da Gama, C. and V.F. Navarro Torres, 2002. *Prediction of EDZ (excavation damaged zone) from explosive detonation in underground openings.* In Proceedings of the ISRM International Symposium on Rock Engineering for Mountainous Regions—Eurock 2002 Funchal, Portugal, 26–28 November 2002. Eds. C. Dinis da Gama and L. Riberia E Sousa.

Heikkinen E., Kantia P., Lehtimäki T., Silvast M. and Wiljanen B., 2010 *EDZ Assessments in Various Geological Environments Using GPR Method.* Posiva Working Report 2010–04.

Kantia, P., Mustonen, S. and Mellanen, S., 2016b. *EDZ Control in Nuclear Waste Disposal Facility in Finland.* ISEE 42nd Annual Conference on Explosives and Blasting Technique, Las Vegas, Nevada, Jan. 31st—Feb. 3rd 2016.

Kantia, P., Mustonen, S., Kouvonen, T., Lehtimäki, T. and Olsson, M., 2016a. *Excavation damaged zone research in Tampere test mine Finland.* ISRM International Symposium 2016. Cappadocia, Turkey.

Kantia, P., Heikkinen E., Mustonen S., Lehtimäki T. and Silvast M., 2012. *Excavation Damage Zone Mapping Using EDZ GPR Method.* EAGE Near Surface Geoscience 2012–18th European Meeting of Environmental and Engineering Geophysics.

Kantia P., Heikkinen E., Mustonen S., Mellanen S., Lehtimäki T. and Silvast M., 2013. *Quality control of drill and blast excavated tunnels using GPR EDZ method.* In: World Tunnel Congress 2013 Geneva. Eds.: G. Anagnostou & H. Ehrbar.

Kiuru, R., 2017. *EDZ Study Area in ONK-TKU-3620: Association Analysis of Petrophysical and Rock Mechanics Data.* Posiva Oy, Working Report 2016-42. In press.

Siren, T., Kantia, P. and Rinne, M., 2014. *Considerations and observations of stress-induced and construction induced excavation damage zone in crystalline rock.* International Journal of Rock Mechanics & Mining Sciences. Elsevier Ltd 2014.

Geomechanics and Geodynamics of Rock Masses – Litvinenko (Ed.)
© *2018 Taylor & Francis Group, London, ISBN 978-1-138-61645-5*

Study of physical-mechanical properties of hard rocks under water-saturated conditions

Nikolai Kuznetcov, Iuliia Fedotova & Alexander Pak
MI KSC RAS, Apatity, Russia

ABSTRACT: The development of mineral deposits both by open and underground mining methods is always accompanied by the mining-induced stress field formation in the rock mass. The specificity of such stresses occurs in the form of their concentration or release in the geological medium structures. The rock mass in such conditions is subjected to permanent loadings and unloadings. This influence also results in the change of rock properties.

The hard rock mass is characterized by high strength and deformation parameters. However, the watered areas (aquifers), the structural discontinuities of different rank and the water-soluble minerals along with permanent loadings and unloadings in the rock mass can cause dynamic fracturing of rocks. In this regard it is important to study the variation of rock properties under the change of loading character under the water-saturated conditions.

The paper presents the results of laboratory studies on hard rock samples (carbonatite, fenite and gneiss) from the Kovdor deposit under uniaxial stress state with the change of their water-saturation time (from 1 to 13 days). A series of deformation tests in the form of loading and unloading up to a value of 30% of the compressive strength with subsequent fracturing of samples were performed simultaneously with the water-saturation process.

The dependence between the change of physical-mechanical properties of studied rocks and the quasistatic alternating loadings was established by comparing the test results. The authors have determined a degree of influence of water-saturation time increase on the acceleration of rock samples' fracturing and the possibility of dynamic rock pressure occurrences in the hard rock mass.

1 INTRODUCTION

The rock mass is a complex hierarchically-blocked geological environment subjected to various man-made and nature factors (Turchaninov, 1977). One of the possible sequences from such influences is the occurrence of alternating loads in the rock mass due to the stress concentration and unloading (Oparin, 2011). The long-term action of such processes can promote the rock softening which, in turn, can lead to the formation of joints and rock mass fracturing. The alternating loads can trigger the dynamic failure of a brittle hard rocks.

In general, it is considered that loading and unloading of hard rocks up to 30% of their compressive strength occur in the elastic area and don't cause the rock failure (Stavrogin, 2001; Suknev, 2012). However, additional factors can decrease the elastic strength of such rocks. One of such factors is water saturation at a part of the rock mass (Grebenkin, 2010). The water saturation mainly effects on the water-soluble minerals and fractured rocks. Nevertheless, it should also take into account its contribution to the decrease of strength properties of the hard rocks having high-order structural discontinuities. This is related to the fact that the hierarchically-blocked rock mass characterized by different-scale fractures is exposed to the over-saturation of near-surface rocks during the intensive snow melting and heavy raining. In this case the increase of rock mass fracturing and water saturation decreases the cohesion of interblock links, which can disturbed the open-pit slope stability (Fedotova, 2013; Fedotova, 2015).

One of the main parameters characterizing the physical-mechanical properties of hard rocks is their deformability depending on the applied loads. In the case when the loads are alternating and act together with the water saturation, the change of deformation values of various rocks can be different. Consequently, the proneness of such rocks to dynamic failure under given conditions will also vary. In this regard, it is important to study the deformability parameters of hard rocks under water-saturated and alternating loading conditions.

2 OBJECT AND METHODS OF THE STUDY

The study object is the Kovdor rock massif, a complex multiphase intrusion of the central type made of different rocks and ores. Its annular structure is explained by gradual injection of various intrusive rocks attracted to a single center and their intensive metasomatic modification. The most ancient rocks are the olivinites composing the rock massif's core. The marginal zone is made of alkaline rocks of ijolites and turiyaites. Their injection at the contact of olivinites with embedded gneisses was accompanied by active change in both types of rocks. As a result, gneisses have turned into alkaline rocks—fenites, and olivinites—into pyroxenites with more or less quantity of phlogopite, melilithic and monticellite metasomatites. One of the final formation stages was formation of numerous carbonatite stockworks. These rocks are very different and represent the greatest industrial interest because of the association with baddeleyite-apatite-magnetite, rare metal, carbonate and apatite-carbonate ore deposits. The outcrop of the main ore body before the exploitation was observed as a brow at the Pilkoma-Selgi slope. At the present time on its place there is a dip open-pit, a rounded pit bowl of about 2 × 1.5 km across.

The results of a hydrogeological research carried out in the open-pit of the Zhelezny mine, JSC Kovdorsky GOK, have revealed groundwater inflows filtered by the slopes of the upper benches to the lower levels, up to the bottom open-pit. At the same time the values of water inflows in different parts of the open-pit are different throughout the year. This is connected with the physical-mechanical properties of the different rocks composing the rock massif (Reshetnyak, 2008). To determine the properties, the core samples of the hard rocks were taken from various parts of the deposit mined. Twenty-seven samples were made with the height-to-diameter ratio as 1:1 (Fig. 1) and divided into 5 groups by dedicated lithotypes. The amount of samples in each group was as follow: 6 for both types of fenite and 5 for the rest of the rocks.

The research was carried out in four stages. At the first stage several samples had been selected from each lithologic group to determine an average compressive strength in each group. At the second stage the samples were subjected to 5 cycles of alternating loads up to 30% of an average compressive strength value and the values of longitudinal deformations were recorded. A testing method is presented in (Kuznetcov, 2015). At the third stage the samples were kept under water-saturated conditions for 6 days and after that were loaded and unloaded according to the scheme presented in the second stage. At the fourth stage the samples were kept under water-saturated conditions for 13 days. After that they were loaded and unloaded according to the scheme of the second stage and then were destroyed.

Figure 1.   Samples: a – carbonatite, b – fenite amphibole-biotite fine-grained, c – fenite amphibole-feldspar medium-coarse-grained, d – gneiss biotite, e – gneiss biotite with streaks of fine-grained micaceous carbonatite.

3 RESULTS AND DISCUSSION

At the first research stage the rock samples were exposed to the uniaxial compression tests and their average compressive strength values were determined (Table 1). The maximum values of compressive strength found corresponding to two types of gneiss, while the minimum ones—to carbonatite and fenite amphibole-feldspar. The character of the rock samples failure was also different. The gneiss samples were failed dynamically, with a wide dispersion of rock particles. The carbonatite and fenite amphibole-feldspar samples were subsided under the load with a slight dispersion of rock particles.

Based on the further study results, the graphs of variations of average values of relative longitudinal deformations in the rock samples under dry and water-saturated conditions were plotted. According to obtained data, three deformation types of samples were identified.

The first deformation type under uniaxial compression (Fig. 2) was determined for two types of gneiss and fenite amphibole-biotite, the strongest rock samples among the studied ones. The deformation is characteristic by the increase of the longitudinal deformation values both on each loading cycle and with increase of the water saturation time. In the case of absolutely elastic deformation without influence of the water saturation, the sample deformation graph wouldn't have changed. According to the results, under the permanent cyclic alternating loads and water-saturated conditions the relative deformation of samples increases and doesn't return to zero. This indicates the occurrence of micro-fracturing processes. If we compare the compressive strength values of gneisses and fenite amphibole-biotite under dry and water-saturated conditions on the 13th day (Table 1), we can observe the

Table 1. The strength values of the rock samples under compression under dry and water-saturated conditions on the 13th day.

Rock	Average compressive strength value under dry conditions, $<\sigma_{cd}>$ (MPa)	Average compressive strength value under water-saturated conditions, $<\sigma_{cw}>$ (MPa)
Carbonatite	43	39
Fenite amphibole-biotite	106	71
Fenite amphibole-feldspar	50	24
Gneiss biotite	143	132
Gneiss biotite with carbonatite	136	95

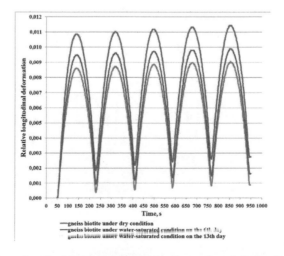

Figure 2. Deformation of gneiss biotite samples during five cycles of alternating loads under dry and water-saturated conditions on the 6th and 13th day.

decrease of the values under the latter conditions. For the gneiss biotite, the average strength value decreases by 8%, for the remaining two rocks—by 30%.

Therefore, we can suppose that the considered factors can lead to the failure of the studied rocks. In this case it should be noted that the long-time alternating loads and water saturation will contribute to the decrease of the rock strength. Also these factors can reduce the possibility of rocks' dynamic fracturing but increase the probability of stability loss. In the case when the alternating loads are more intensive, the possibility of dynamic fracturing of the studied rocks may increase.

The second deformation type (Fig. 3) was identified for the carbonatite samples. It is characterized by the fact that the increase of the alternating loading cycles leads to the insignificant rise of the longitudinal deformation values of the samples. Also, the deformation curve under water-saturated conditions on the 13th day almost repeats the deformation curve under dry conditions. It is also interesting that the average compressive strength values of the carbonatite samples under dry and water-saturated conditions on the 13th day don't differ (Table 1).

According to the test results, we can make the conclusion that the applied alternating loading under water-saturated conditions was insufficient to initiate significant failure. Therefore, it can be supposed that if an alternating load is small (less than 30% of the compressive strength) and the rocks have the highest-order structural discontinuities only, the rock failure under specified conditions may not occur or occur after a long period of time.

The third deformation type (Fig. 4) was determined for fenite amphibole-feldspar samples. Its characteristic feature is a steep increase of longitudinal deformations in the samples under water-saturated conditions. At the same time, the change of the deformation values on the 13th day of the water saturation differs slightly from the values on the 6th day. Also, it should be noted that an average compressive strength value under water-saturated conditions on the 13th day has decreased by half in comparison to a value under dry conditions (Table 1).

The change of the relative longitudinal deformation values under water-saturated conditions for fenite amphibole-feldspar samples is mainly connected to streaks and shallow fractures. At the same time, according to the graph (Fig. 4), further water saturation didn't influence the deformation mode of the samples studied. Nevertheless, it is important that the active water inflows in autumn and spring periods can significantly decrease the strength parameters of rocks having similar structural discontinuities. The influence of alternating loads in these periods can lead to loss of stability of such rocks.

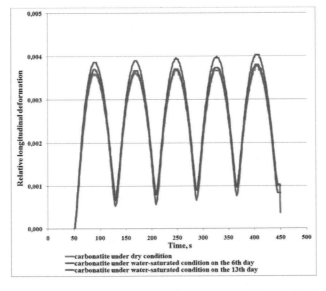

Figure 3.   Deformation of carbonatite samples during five cycles of alternating loads under dry and water-saturated conditions on the 6th and 13th day.

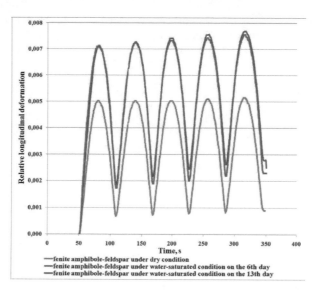

Figure 4. Deformation of fenite amphibole-feldspar samples during five cycles of alternating loads under dry and water-saturated conditions on the 6th and 13th day.

4 CONCLUSION

Based on the experimental study results, three deformation types of rock samples from the Kovdor rock massif have been identified.

The first deformation type is characterized by the gradual increase in the longitudinal deformation values of the rock samples with increasing the alternating loading cycles both under dry and water-saturated conditions. In this case, the long-time alternating loads under water-saturated conditions will decrease the rock strength properties which can lead to the rock instability. The increase of intensity of quasi-static alternating loads, in turn, can lead to dynamic failure of the rocks.

The second deformation type is typical for rocks having the highest order of structural discontinuities (micro-cracks and grain contacts). In this case water saturated conditions don't result in the water saturation and alternating loads don't lead to the fissure propagation. However, as for the previous deformation type, the increase in the load intensity can result either in the rise of the probable dynamic failures in such rocks or in the loss of their stability.

The third deformation type is associated with a steep increase of longitudinal deformations in the rocks having streaks and shallow fractures under water-saturated conditions. The influence of alternating loads during active water inflows can result in the significant decrease of the strength properties of the rocks and to the stability loss.

From the physical point of view, the considered deformations and failure can be explained as follows.

Firstly, in the hard rock mass without water-soluble inclusions the stresses are concentrated in individual blocks because of the pore pressure increase in the rocks under water-saturated conditions and alternating loads. This results in dynamic fracturing. The deformation velocity rises abruptly up to dynamic failure under the pore pressure and increasing alternating load velocity, as the rock mass doesn't have time to unload from the excess stresses. Secondly, the mechanical properties of the hard rock mass with the water-soluble inclusions change when the inclusions dissolve under the abundant water inflows and simultaneous effect of alternating loads. This, in turn, leads to the formation of shear planes due to more intensive rock mass fracturing within the boundaries of the blocks and, eventually, to the instability of individual structural blocks. This process is characterized by permanent static loading velocity.

Thus, the study of the deformation mode under alternating loadings and water-saturated conditions allows establishing the proneness of hard rocks to different dynamic failures or loss of stability in the form of rock falls under the given conditions.

REFERENCES

Fedotova Iu.V., Zhukova S.A. Natural and Man-made Factors Influence on Seismicity Changing of Hard Rock//Proceeding of the 23rd International Mining Congress of Turkey. April 16–19, 2013. Antalya, Turkey. Ed.Ilkay Celik & Mehtap Kilic. Publ. TMMOD Maden Muhendisleri Odasi Selanic Cad. 19/14 Kizilay-Ankara. – Pp. 2111–2120.

Fedotova Iu.V., Zorin A.V. Analysis of meteorological data from the monitoring system for the inner open-pit atmosphere and estimation of microseismic activity in the open-pit slope of the Zhelezny mine, JSC Kovdorskii GOK//Deep mining: Mining informational and analytical bulletin, № 11, 2015 (special issue 56). – Moscow: Mining Book, 2015. – Pp. 294–309.

Grebenkin S.S., Pavlysh V.N., Samoilov V.L., Petrenko Iu.A. Rock mass state management. – Doneck: VIK, 2010. – 193 p.

Kuznetcov N.N., Pak A.K., Fedotova Iu.V. Study of deformation behavior and energy intensity of hard rock samples from the Kovdor deposit//Deep mining: Mining informational and analytical bulletin, № 11, 2015 (special issue 56). – Moscow: Mining Book, 2015. – Pp. 286–292.

Oparin V.N., Tanaino A.S. Canonical scale of hierarchical representations in rock science. – Novosibirsk: Nauka, 2011. – 259 p.

Reshetnyak S.P., Melikhova G.S., Fedotova Yu.V., Melikhov M.V. Problems of Deep Open Pits Closure in the Kola Peninsula//Proceedings of Mine Water and Environment. – Ostrava, Czech Republic, 2008. – Pp. 171–174.

Stavrogin A.N., Tarasov B.G. Experimental physics and mechanics of rocks. – Saint-Petersburg: Nauka, 2001. – 343 p.

Suknev S.V., Fedorov S.P. Standard methods for determining the elastic properties of rocks//Mining informational and analytical bulletin, № 12, 2012. – Moscow: Mining Book, 2012. – Pp. 17–21.

Turchaninov I.A., Iofis M.A., Kasparyan E.V. Fundamentals of rock mechanics. – Saint-Petersburg: Nedra, 1977. – 503 p.

Geomechanics and Geodynamics of Rock Masses – Litvinenko (Ed.)
© *2018 Taylor & Francis Group, London, ISBN 978-1-138-61645-5*

On the effectiveness of rocks and materials destruction based on shock-wave cutting technology

Nikolay P. Mikhailov, Evgeniy A. Znamenskiy, Stanislav I. Doroshenko & Yurii A. Telegin
Baltic State Technical University «VOENMEH» named after D.F. Ustinov, Saint-Petersburg, Russia

Ivan V. Brigadin
Promstroyvzryv Ltd., Saint Petersburg, Russia

Vitaliy M. Gubaidullin
Promstroyvzryv Ltd. The Russian Federation, Saint Petersburg, Russia

Gennadii P. Paramovov
National mineral resources University «MINING UNIVERSITY», St. Petersburg, Russia

ABSTRACT: The paper assesses the effectiveness of shock-wave technology cutting for rocks, based on irregular modes of shockwaves interference. That modes occurs when angle in the waves collision greater then critical angle. Shock waves forms in the material upon synchronous detonation of bilinear charges, that designed in the BSTU "VOENMEH". A theoretical and experimental comparison of materials cutting technologies using a shock-wave and a shaped charge is done. The usage of cumulative cutting in emergency rescue operations is limited by the presence of a hazardous effect on objects behind a barrier. Unlike the shaped charge, the shock-wave cutting destruct materials without mass transfer, and does not damage objects behind the barrier. A reduced in 2...5 times consumption of explosives and higher efficiency of the shock-wave cutting technology is experimentally shown. As the result of significant reduction in consumption of explosives reduces lead times, transportation costs and environmental damage. It follows that the shock-wave cutting technology can be successfully used in mining and emergency rescue operations.

Keywords: shock-wave charge, bilinear charge, mathematical model, Mach wave, experiment, rock, concrete block

1 INTRODUCTION

Emergency situations irrespective of their nature and causes, require complex engineering tasks associated with large amount of rescue and other emergency operations.

Complexity, danger and strict deadlines of such works dictate the need for widespread usage of explosion energy. Structural divisions of various departments involved in liquidation of emergency situations should be equipped with modern innovative explosive technologies enabling them to effectively tackle these tasks.

Currently, the primary means of materials (metals) cutting by explosion are shaped charges, which are based on the high-speed removal of material from the cut cavity, which volume is roughly proportional to the second degree of the barrier thickness. The same dependence is preserved between the weight of the shaped charge explosive and the thickness of the barrier. With through cutting of barriers by high-speed jet its particles affect the objects behind barrier. This fact limits the applicability of shaped charges cutting technology.

Unlike the shaped charge, the shock-wave cutting is based on extreme (Mach) modes of shock waves interference, which are formed in the material upon synchronous detonation of parallel to the barrier's surface charges.

The destruction happens almost without mass transfer due to the interaction of three rarefaction waves behind the front of Mach wave. As the result, the barrier is cut following the least resistance line by flat cracks. A linear dependence between shock-wave charge mass and the barrier thickness was found.

2 THEORETICAL STUDY

In order to assess the applicability of shock-wave cutting to the rocks, an analytical calculation of kinematic parameters of the shock waves collision in granite coming from parallel charges was done.

As it's known, the velocity of shock waves (D_y) in granite remains equal to the speed of sound (C) up to the pressure $p \leq 37$ GPa [1]. In the general case, the formation of conically and cylindrically shaped shock waves is provided in term $D_H > D_v = C$, where D_H – detonation speed of the charges.

The main parameter that determines the mode of waves' interference is their collision angle (α). Irregular (Mach) mode occurs when $\alpha \geq \alpha_{cr} \approx 68...74°$ (Orlenko, 2002).

The dependence of the angle α in the conical waves' collision ($D_H \geq C$) is defined analytically as (Mikhailov, 2012):

$$\sin(\alpha/2) = \sqrt{1 - \frac{\cos^2 \beta}{1 + y/L}},$$ (1)

where $\beta = \arcsin(D_v/D_H) = \arcsin(C/D_H)$, L – distance between the charges; y – distance from the surface of the barrier to the point of the waves' collision. For cylindrical waves, $\beta = 0$, and from (1) we obtain

$$\sin(\alpha/2) = \sqrt{1 - L/(y + L)}.$$ (2)

As follows from (2) for the irregular mode of cylindrical waves collision is realized in the wide range of bilinear charges' parameters. That ensures its applicability for breaking rocks, concrete and other crystalline materials.

As the result of theoretical and experimental research performed in the BSTU "VOEN-MEH", a number of shock-wave charges models has been developed.

In order to improve the shock-wave cutting technology the theoretical studies of barriers destruction by implosive (symmetrically converging) shock waves were carried out. Physical and mathematical models described by systems of equations (Brigadin, 2014) were developed.

3 EXPERIMENTAL STUDY

Tests of rocks and concrete blocks cutting by implosive shockwaves were performed on the "Promstroyvzryv" premises. Bilinear charges made of emulsion explosives Nitronit-P and Gelpor (Brigadin, 2014) were used for cutting.

Experiments on the bilinear charges were conducted in order to remove the piece of rock hanging above "St. Petersburg-Sortavala" road. The dimensions of that piece were $3.1 \times 1.3 \times 1.5$ m. The cut area was up to 3.5 m². Charges were located on the block surface in two parallel rows along the cut line. Distance between the rows was 100 mm.

As the result of the explosion, separation of the block was observed in the bilinear charge plane of symmetry. Specific consumption of explosives was 1.26 g/cm^2, which is 2...5 times less comparing to a separation by concentrated charge.

4 CONCLUSIONS

Shock-wave cutting of rocks is an innovative technology, for significantly reducing the explosive consumption. The positive effect is achieved through the usage of bilinear charges, implementing cutting by Mach waves. As the result of reduction in explosives consumption reduces lead times, transportation costs and environmental damage. The shock-wave cutting technology can be successfully used in mining and emergency rescue operations.

REFERENCES

[1] Brigadin I.V. [and others]. On the issue of effective elimination of natural disasters on the basis of innovative technologies of shock wave cutting: Problems of ensuring security in the aftermath of emergencies: a Collection of articles of III all-Russian NPK with international participation. Voronezh, December 19, 2014 Voronezh: Voronezh Institute of state fire service of EMERCOM of Russia; 2014. pp. 246–249.
[2] Mikhailov N.P. Fundamentals of mathematical simulation of explosion and impact: Textbook/N.P. Mikhailov. Balt. state. tech. un-t. SPb., 2012. 202 p.
[3] Orlenko L.P. [and others] Physics of explosion./Under edited by L.P. Orlenko. Ed. 3rd, revised, in 2 Vol. M.: FIZMATLIT, 2002.

Geomechanics and Geodynamics of Rock Masses – Litvinenko (Ed.)
© *2018 Taylor & Francis Group, London, ISBN 978-1-138-61645-5*

Modelling of fault reactivation in applications of mining and petroleum industry

Roberto Quevedo & Cristian Mejia
Institute Tecgraf/PUC-Rio, Rio de Janeiro, Rio de Janeiro, Brazil

Deane Roehl
Institute Tecgraf/PUC-Rio, Rio de Janeiro, Rio de Janeiro, Brazil
Civil and Environmental Engineering Pontifical Catholic University of Rio de Janeiro,
Rio de Janeiro, Brazil

ABSTRACT: The seal of critically stressed faults can be compromised by deformation/ stress changes induced by external solicitations. The stress relief due to excavations or reservoir pressure changes, for example, can create favorable conditions for reactivation of faults by shear or tensile modes. Among several problems, fault reactivation may result in slope detachment, the generation of a fault scarp along the ground surface, oil exudation and seismicity. This study presents a methodology for the evaluation of fault reactivation using numerical modelling. With the finite element method, it is possible simulate the effect of both, excavations and reservoir pressure changes over geological faults. Those faults are introduced in the models using zero thickness interface elements. Some 2D and 3D models are analyzed with focus on mining and petroleum engineering applications. In relation to the mining industry, fault reactivation caused by open-pit and underground mines are taken into account. Regarding the oil industry, a case of fault slip due to hydrocarbon production is also presented. The results call attention to the different causes and effects of fault reactivation and provide some clarity to the understanding of this phenomenon.

Keywords: Fault reactivation, numerical modelling, mining, underground excavation, petroleum

1 INTRODUCTION

In the mining industry, the stress relief owing to superficial or underground excavations usually induces relative displacements of the rock masses along pre-existent discontinuities or faults. This phenomenon is known as fault reactivation and can be responsible for several geotechnical problems such as, tunnel collapse, slope instability, mining shaft deformation, topographic scarp and surface subsidence (Jeon et al. 2004, Donnelly 2006, Donnelly 2009, Zhao et al. 2013). In a similar way, in hydrogeology and the petroleum industry, the strain owing to the pressure changes in reservoirs compartmentalized by geological faults can reactivate them, triggering potential geomechanical problems, such as seismicity, well collapse, loss of reservoir sealing and exudation (Baranova et al. 1999, Wiprut and Zoback, 2002).

Focusing on those geotechnical and geomechanical applications, three different activities can induce fault reactivation: slope excavation, mineral extraction and water/hydrocarbon production. The use of numerical modelling to identify potential fault reactivation is widespread in the mining (Bruneau et al. 2003, Zhao et al. 2013) and the petroleum industry (Rutqvist et al 2013, Quevedo et al. 2017). However, numerical simulations of fault reactivation in the three activities previously mentioned have not been performed in a single study for discussion and comparison purposes. Observe that some issues, such as geometric shapes, loadings, and stress and displacements fields can create different reactivation mechanisms over the fault planes. Probably, difficulties associated with different methodologies, complex model building and large computational effort have not allowed the comparison of those mechanisms of fault reactivation. In this paper, we introduce a numerical methodology based on the Finite element Method for representation of different situations in which geological faults are reactivated. This methodology was implemented in an in-house code for modelling of 2D and 3D problems. Numerical simulations of open-pit mining process and mineral and hydrocarbon extraction were carried out. The results call attention to the different causes, mechanisms and effects of fault reactivation and provide some clarity to the understanding of this phenomenon.

2 METHODOLOGY

Using the traditional Galerkin Finite Element method, the governing equation of mechanical process can be written in an incremental way as follows:

$$\int_{\Omega} \boldsymbol{B}^T \boldsymbol{D} \boldsymbol{B} d\Omega \cdot \Delta u = \Delta F_{ext}^E + \Delta F_{ext}^R \tag{1}$$

where Δu represents the variation of nodal displacements, $\boldsymbol{B}$ is the matrix that relates the strains and the displacements, $\boldsymbol{D}$ is the stiffness matrix and Ω is the spatial domain of the model.

ΔF_{ext}^E represents the nodal external forces due to the material removal in open-pit slope and underground mining, this vector is given according to the next relationship:

$$\Delta F_{ext}^E = \int_{\Omega E} \boldsymbol{B}^T \boldsymbol{\sigma} d\Omega_E + \gamma \int_{\Omega E} \boldsymbol{N}^T d\Omega_E \tag{2}$$

in which Ω_E is the excavated domain, σ is the stress tensor of the elements in the excavated zone, N is the element shape functions and γ is the material unit weight.

In turn, ΔF_{ext}^R represents the nodal external forces due to pressure changes in water or oil reservoirs, this vector is assessed according to the following relationship:

$$\Delta F_{ext}^R = \int_{\Omega R} \boldsymbol{B}^T \alpha m d\Omega_R \cdot \Delta p \tag{3}$$

where Ω_R is the reservoir domain, α is the Biot coefficient, Δp is the nodal pressure change within the reservoir and m is a vector that introduces the effects of that pressure change.

The rock layers are assumed to have elastic behavior. Thus, only the Young elastic modulus (E) and the Poisson ratio (ν) define the stiffness matrix $\boldsymbol{D}$. In turn, the geological faults are represented trough zero-thickness interface elements. In this case, the tangential (k_s) and the normal (k_n) elastic stiffness are considered in the assessment of $\boldsymbol{D}$.

3 NUMERICAL SIMULATIONS

3.1 Open-pit mining model

This section deals with the reactivation of faults adjacent to open-pit mines. According to several geological investigations conducted in open-pit mines (Zhao et al. 2013), when steep faults are located in the upper part of excavated slopes, fault reactivation usually results in a downward movement of the hanging wall relative to the footwall, regardless of the trend and the original properties of the fault. Figure 1 presents two idealized scenarios to analyze fault reactivation, two in open-pit mining.

The initial stresses before excavation consider homogeneous rock unit weight of 27 kN/m³. At this step, the horizontal stresses were taken equal to the vertical stresses. The elastic modulus and the Poisson ratio have the values 0.6GPa and 0.4, respectively. For the geological faults, the tangential and normal stiffness are 0.1 MPa/m and 0.1GPa/m, respectively.

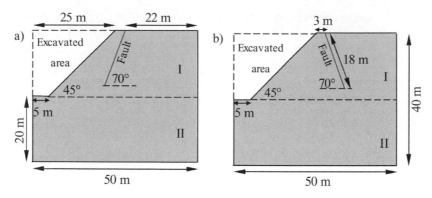

Figure 1. Models used for simulation of fault reactivation induced by an open-pit excavation over a) the hanging wall and over b) the footwall.

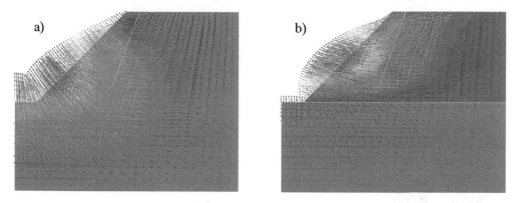

Figure 2. Displacement vectors for an open-pit excavation in the hanging wall considering a) a single set of elastic parameters and b) two sets of elastic parameters.

In a first simulation, those elastic parameters were adopted in the whole model of Figure 1a. Figure 2a shows the corresponding vector of displacements found after the excavation. Different from field observations, the slope shows upward movements over the entire region of the hanging wall. Zhao et al. (2013) found similar results. Before the excavation, all elements suffer volume compression and store elastic strain energy that is released when some elements are removed. For this reason, in a second simulation, the slope configuration in Figure 1a has a stiffer layer in the bottom of the slope (region II) with 10 times the original elastic modulus. Two main arguments support this assumption. First, deeper layers are prone to be stiffer due to the higher stresses in which they are subjected. Second, deeper layers are less exposed to environmental factors that degrade and affect the properties of superficial rock layers. Figure 2b shows the results corresponding to the use of a stiffer layer in the bottom of the model. Now, the top of the slope in the hanging wall follows a downward movement, while the middle and inferior regions follow movements towards the pit.

Figure 3 shows the corresponding contour map of the vertical displacements in the models of Figure 1 considering stiffer rock layers in the bottom. A normal fault-style movement is present in both cases, characterized by downward movements of the hanging wall relative to the footwall. Those results indicate the reactivation of faults induced by the tensile stresses over the fault planes due to the open-pit slope excavation.

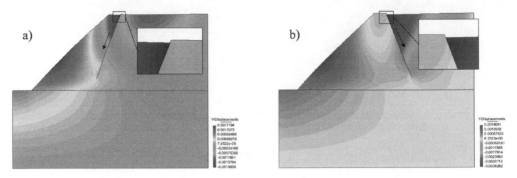

Figure 3.   Contour map of vertical displacements and fault reactivation due to open-pit excavation in a) the hanging wall and b) the footwall.

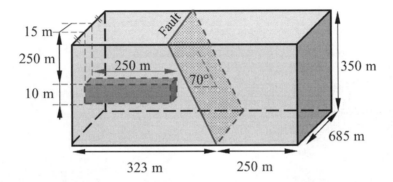

Figure 4.   Simulation model for fault reactivation induced by underground mining excavation.

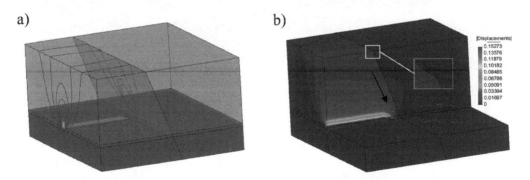

Figure 5.   Results for the underground mining excavation at the hanging wall a) Vertical displacements and subsidence bowl. (b) Contour map of vertical displacements and fault reactivation.

3.2   *Underground mining*

Fault reactivation induced by underground mining excavation is the focus of this section. In this case, the excavated zone is located at an average depth of 255 m. Figure 4 shows the corresponding model and its dimensions. Before the excavation, initial stresses are due to the rock unit weight of 22 kN/m³, and the horizontal stresses are equal to the vertical ones. From the surface to the depth of 250 m, the elastic modulus is 0.6GPa; from 250 m to 260 m it is 1.0GPa; and, from 260 m to 310 m it is 5.0GPa. The Poisson ratio is equal to 0.3 in all layers. For the geological faults, the tangential and normal stiffness are 3.0 MPa/m and 0.5GPa/m, respectively. The excavated region is located over the footwall and defines a parallelepiped with the dimensions 250 m×10 m×15 m.

Figure 5 shows the corresponding results after the excavation. Figure 5a shows the contour lines of the resulting displacements. The maximum values are concentrated around the excavation region. The vertical convergence in the gallery, i.e. the difference between the largest displacement of the floor and that of the roof of the excavated zone, was around 0.13 m. Over the surface of the model, there is a maximum subsidence of 1 cm. Figure 5b shows the contour plots of the displacements on a cutting surface. It shows a reverse fault-style movement characterized by downward movements of the footwall relative to the hanging wall. According to the results, fault planes close to the gallery reactivate due to the tensile stresses. However, in regions near the surface, reactivation is due to shear stresses.

3.3 Hydrocarbon production

This last section presents a simulation of fault reactivation induced by hydrocarbon production. Differently from the previous examples, field deformations occur because of pressure reductions inside the reservoir. The model considers plane strain conditions, a common assumption when the reservoir is laterally extended in the horizontal directions. Before reservoir depletion, the initial stresses reflect a vertical effective stress gradient of 12 kN/m^3. At this step, the initialization of the horizontal stresses takes into account a lateral stress coefficient of 0.75. In the Overburden, Sideburden, Reservoir and Underburden layers, the elastic modulus are 1.0GPa, 5.0GPa, 2.0GPa and 5.0GPa, respectively. The Poisson ratio is equal to 0.25 in all layers. The tangential and normal stiffness are 0.2GPa/m and 5.0GPa/m, respectively. The reservoir was depleted in 10 MPa with a Biot coefficient equal to 1.0.

Figure 6 shows the contour map of the vertical displacements, showing downward movements concentrated above the depleted reservoir. The maximum reservoir compaction is about 0.63 m while the maximum surface subsidence is 0.48 m. In the hanging wall, where the reservoir is located, relative displacements to the footwall are present, indicating the reactivation of the fault in a normal fault-style movement. The results indicate that fault planes

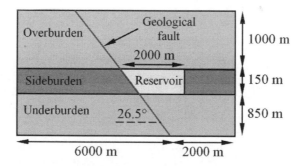

Figure 6.   Model for the simulation of fault reactivation induced by hydrocarbon production.

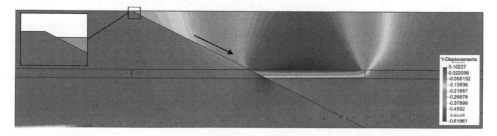

Figure 7.   Contour map of vertical displacements and fault reactivation in a simulation of hydrocarbon production.

around the reservoir reactivate by tensile stresses. On the other hand, fault planes located in the overburden reactivate by shear stresses. In such situation, the high depletion of the reservoir and the induced fault reactivation increase risks for loss of reservoir sealing integrity, seismicity and exudation.

4 CONCLUSIONS

A methodology for the assessment of fault reactivation induced by mining operations and hydrocarbon production was introduced in this study. Synthetic scenarios with different conditions of geometry, loadings and contrast of properties provide the test cases for the methodology. The analysis of the results of the test cases provided better understanding of the mechanisms of fault reactivation. In addition, the results highlight the applicability of numerical modelling in the assessment of geological fault reactivation and its usability for the forecasting of potential geotechnical and geomechanical problems.

ACKNOWLEDGEMENTS

The authors gratefully acknowledge support from Shell Brasil through the "Coupled Geomechanics" project at TecGraf Institute (PUC-Rio) and the strategic importance of the support given by ANP through the R&D levy regulation.

REFERENCES

Baranova V., Mustaqeem A., Bell S. 1999. "A Model for Induced Seismicity Caused by Hydrocarbon Production in the Western Canada Sedimentary Basin." Canadian Journal of Earth Sciences 36 (1): 47–64. doi:10.1139/e98-080.

Bruneau GDB, Hadjigeorgiou T., Potvin Y. 2003. "Influence of Faulting on a Mine Shaft a Case Study Part II Numerical Modelling." International Journal of Rock Mechanics and Mining Sciences 40 (1): 95–111. doi:10.1016/S1365-1609(02)00115-6.

Donnelly L.J. 2009. "A Review of International Cases of Fault Reactivation during Mining Subsidence and Fluid Abstraction." Quarterly Journal of Engineering Geology and Hydrogeology 42 (1): 73–94. doi:10.1144/1470-9236/07-017.

Donnelly L.J. 2006. "A Review of Coal Mining Induced Fault Reactivation in Great Britain." Quarterly Journal of Engineering Geology and Hydrogeology 39 (1): 5–50. doi:10.1144/1470-9236/05-015.

Jeon S., Kim Y., Seo Y., Hong Ch. 2004. "Effect of a Fault and Weak Plane on the Stability of a Tunnel in Rock - A Scaled Model Test and Numerical Analysis." International Journal of Rock Mechanics and Mining Sciences 41 (SUPPL. 1): 1–6. doi:10.1016/j.ijrmms.2004.03.115.

Quevedo R.J., Ramirez M.A., Roehl D. 2017. "2d and 3d Numerical Modeling of Fault Reactivation." 51st U.S. Rock Mechanics/Geomechanics Symposium, 25–28 June, San Francisco, California, USA.

Rutqvist J., Rinaldi A., Cappa F., Moridis G.J. 2013. "Modeling of Fault Reactivation and Induced Seismicity During Hydraulic Fracturing of Shale-Gas Reservoirs" 44: 31–44.

Wiprut D., Zoback M.D. 2002. "Fault Reactivation, Leakage Potential, and Hydrocarbon Column Heights in the Northern North Sea." Norwegian Petroleum Society Special Publications 11 (C): 203–19. doi:10.1016/S0928-8937(02)80016-9.

Zhao H., Ma F., Xu J., Guo J. 2013. "Preliminary Quantitative Study of Fault Reactivation Induced by Open-Pit Mining." International Journal of Rock Mechanics and Mining Sciences 59. Elsevier: 120–27. doi:10.1016/j.ijrmms.2012.12.012.

Geomechanics and Geodynamics of Rock Masses – Litvinenko (Ed.)
© 2018 Taylor & Francis Group, London, ISBN 978-1-138-61645-5

Acoustic emission precursor criteria of rock damage

Alexander O. Rozanov, Dmitri N. Petrov & Aleksei M. Rozenbaum
Saint-Petersburg Mining University, Saint-Petersburg, Russia

Andrei A. Tereshkin
Khabarovsk Mining Institute, The Russian Academy of Sciences, Khabarovsk, Russia

Michael D. Ilinov
Saint-Petersburg Mining University, Saint-Petersburg, Russia

ABSTRACT: Such processes as production of ore deposits and mining excavation are associated with the phenomenon of Acoustic Emission (AE). Changes of in situ stress conditions can result in sudden movements along pre-existing faults or can generate new fractures which in turn induce the emission of acoustic energy. Monitoring and analysis of AE events can detect dangerous areas in minings and predict rockbursts during actual mining procedure. A huge AE data have been acquired under different mining conditions in the PhosAgro apatite mines. The measurements have been carried out with the help of one-channel multifunctional device «Prognoz-L» developed in the Mining Institute of Khabarovsk, Russia. The spectral correlation analysis combined with conventional event statistics were used to gain reliable AE criteria of rockbursts. Verification of these criteria has been carried out in laboratory on samples of three main types of rock contained in massif.

1 INTRODUCTION

Now different geophysical safety systems are widely introduced in mining. One of them is geoacoustics. Geoacoustic systems are developed in order to provide a remote permanent control of rock massif state under mining. These systems are based on acoustic emission (AE) phenomenon and relate to non-destructive kind of material tests. A conclusion on rock state is provided by intellectual software solution developed on wide range AE data analysis. For instance, nowadays new technology of geoacoustic systems is developed in the Mining Institute of Khabarovsk, Russia. These are a system of mine section control (Prognoz-ADS) and a system of mine local control (Prognoz-L). Both systems have been successfully deployed in PhosAgro ore mines for AE data acquisition procedure. Now an intensive AE data processing is carried out both by scientific center of Saint-Petersburg Mining University and the Mining Institute of Khabarovsk in order to advance rock burst precursor algorithms.

2 RESULTS

In this work a problem of precursor criteria is attacked from a new standpoint. *A.A. Griffith* showed in 1921 that inherent material inhomogeneities such as microcracks substantially reduce the strength of material (Griffith A.A., 1921). Moreover it is derived from lab triaxial tests of hard rocks that the strength strongly depends on confining pressure. That is we may encounter various manifestations of rock pressure activity depending on strain energy level in the vicinity of mine excavation. We classify AE signatures of fracture process and identify each signature to a certain stage of fracturing. A developed package of spectral-correlation algorithm is used to analyze AE signatures (Rozanov A.O., 2003, Rozanov A.O., 2012).

To study fracture process in rocks ten triaxial compression tests were performed on samples with different content of P_2O_5 – samples with P_2O_5 ~ 12–14%, with P_2O_5 ~ 4–6%, and with P_2O_5 < 4%. The experimental setup we used consists of the loading frame (servo-controlled MTS 815 frame, 4600 kN), and the AE acquisition system ErgoTech. One series of samples was tested under 20 MPa of confining pressure, and another one under 40 MPa. All experiments were fixed displacement rate tests (0.01 mm/min). To map the hypocenters of AE events a commercial code package ASC has been used. For spectral correlation analysis a self-developed code written in Pascal is used.

We calculated two main AE spectral parameters in order to qualify fracturing process in rock. The first one is so called median frequency which depends on the ratio of high and low frequencies in AE signal spectrum. And another one is standard deviation which is proportional to signal duration and amplitude square. Thus AE median frequency characterizes a value proportional to crack growth rate and inverse proportional to crack growth length. Meanwhile AE standard deviation characterizes crack radiated energy.

A typical stress-strain diagram from hard-rock lab triaxial test on PhosAgro ore mine samples is shown in Figure 1. Four stages of fracturing and a rupture formation are depicted there to point out different AE manifestations usually observed during fracturing. Stage A characterizes a process of linearization at the very beginning of loading. Stage B indicates the yield response of rock. Then comes stage C of quasi-plastic behavior. When the strain exceeds the strength of rock a *rupture* occurs and a stress drop is observed. This is a tensile *Griffith's* fracture mode. During the post-strength stage D the rock deforms as two parts of it shear along the rupture surfaces. In seismology this stage is usually called a *stick-slip* frictional mode. The earthquake they consider to happen is the "slip" along an existing fault, and the "stick" is the interseismic period of elastic strain accumulation.

Figure 2 shows the results of AE signal analysis obtained during mining excavation. AE median frequency trend (black line) and AE energy trend (grey line) evidently indicate the transition point from the stable process of strain energy accumulation to the unstable process of rupture formation as it is observed during lab experiments (Figure 1). Such kind of AE parameter behavior specifies a state of rock massif as *dangerous*.

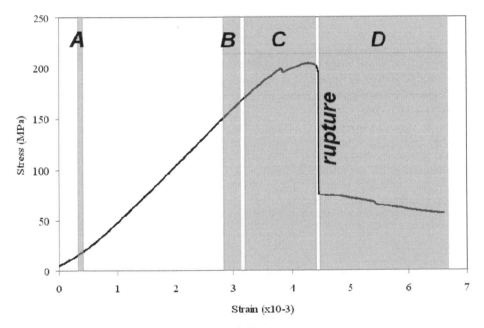

Figure 1.   The stress-strain diagram of lab triaxial compression test of urtite sample. The picture performs four stages of AE radiation and a *rupture* occurrence.

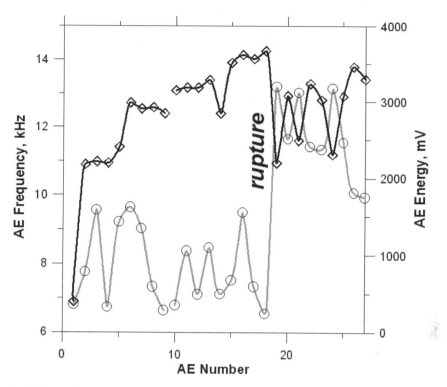

Figure 2. The *AE* frequency (*black line*) and energy (*grey line*) spectral parameters perform strain energy accumulation and a *rupture* occurrence during mining excavation.

Figure 3. The typical hypocenter display of 677 *AE* events in pegmatoid urtite ($P_2O_5 \sim 4\text{–}6\%$).

Figure 3 shows a sample of pegmatoid urtite with $P_2O_5 \sim 4\text{–}6\%$ after a test under 40 MPa of confining pressure, and a typical *AE* hypocenter distribution. One can see that nucleation process develops along a plane making an angle of about 30° with respect to the maximum compression axis.

3 CONCLUSIONS

Thus we can make the following conclusions:

1. The *AE* median frequency and standard deviation are explicitly indicate an in situ stress change in mines.
2. *AE* hypocenter distribution delivers information about nucleation process of damage in rocks.
3. *AE* signatures provides a perfect approach to real stress state in mines and can deliver relevant criteria for rock state characterization.

REFERENCES

Griffith A.A. The Phenomena of Rupture and Flow in Solids, Philosophical Transactions of the Royal Society of London. Series A, Containing Papers of a Mathematical or Physical Character. 1921, Vol. 221. – P. 163–198.

Rozanov A.O. Spectral and Correlation Analysis of Acoustic Emission Waveforms and Failure Micromechanics in Rocks, Int. Geophys. Conf. & Exhibition "Geophysics of 21 Century—The Leap into Future", Extended Abstracts, Session OS22, 2003.

Rozanov A.O. Microseismic Event Spectrum Control and Strain Energy Release in Stressed Rocks, GEO 10th Middle East Geosciences Conference & Exhibition, 2012.

Geomechanics and Geodynamics of Rock Masses – Litvinenko (Ed.)
© *2018 Taylor & Francis Group, London, ISBN 978-1-138-61645-5*

Numerical simulation of stress distribution within a rock discontinuity asperity

Atsushi Sainoki
International Research Organization for Advanced Science and Technology, Kumamoto University, Japan

Yuzo Obara
Department of Civil and Environmental Engineering, Kumamoto University, Japan

Hani S. Mitri
Mining and Materials Engineering, McGill University, Canada

ABSTRACT: The present study investigates the stress distribution and the extent of failure within rock discontinuity asperities. The fractal geometry of the discontinuity surface is generated with successive random addition and mid point method, based on which numerical analyses are quasi-statically performed with the three-dimensional discrete element method (3DEC). The analysis results show that extremely high compressive stress is locally generated by the collision of asperities, while tensile stress fields are extensively produced in the surrounding region because the asperity collision draws the surrounding rock. The generation of the large compressive stress implies the occurrence of tertiary creep, resulting in asperity comminution and the increase in the contact area. On the other hand, the tensile stress field is assumed to form extension fractures with a length being the order of mm or less.

1   INTRODUCTION AND NUMERICAL MODEL DESCRIPTIONS

The deformational behaviour of a rock discontinuity and its shear strength is of paramount importance in various engineering projects. Over the past decades, significant efforts have been made to elucidate the mechanical behaviour and strength of rock discontinuities (Barton, 1977). In the developed frictional resistance models, the influence of the surface asperities is considered implicitly by proposing representative parameters, while to explicitly model the surface asperity abrasion, numerical simulation techniques have been employed (Elmo, 2006; Bahaaddini et al., 2013). These studies showed fracture formation, propagation and its coalescence within the surface asperities during the shear behaviour of a rock joint. However, the detailed analysis of the stress distribution within the asperities has not yet been extensively conducted as the previous studies mainly focused on "fracture development". When the stress state of a rock discontinuity is below its critical stress state, asperity creep behaviour is assumed to play a critical role in its deformation behaviour, e.g. the contact area would increase with asperity deterioration due to the occurrence of tertiary creep. Such time-dependent behaviour is crucial when estimating the long-term stability of faults and jointed rock masses. As the first step to achieve the goal, this study aims to better understand the stress distribution within the surface asperities of a rock joint as well as their failure modes when subjected to high confining stress.

2 NUMERICAL MODEL DESCRIPTION

This study employs 3DEC software, which is based on the three-dimensional distinct element method, to analyze the stress distribution and resultant failure within joint surface asperities. Figure 1 shows the numerical model generated. The discontinuity geometry of the upper block is the same as that of the lower block, but the upper block is slightly displaced to replicate a non-interlocking discontinuity. It is well-known that the surface geometry of a fault or natural rock joint is fractal. To consider such fractal characteristics, the surface geometry of the numerical model was generated with the successive random addition and midpoint method proposed by Voss (1985). In this method, a key parameter is Hurst exponent, which is related to fractal dimension with the following equation in the case of two dimensions.

$$D = 2 - H$$

where D and H denote fractal dimension and Hurst exponent. The detailed mathematical concept and implementation of the method are found in the study (Ozdemirtas et al., 2009). According to Eq. (1), a rougher discontinuity surface is formed with a lower H. We generated four numerical models while varying H from 0.6 to 0.9 as shown in Figure 2. It is found that the surface with H = 0.9 is the smoothest, while the model with H = 0.6 has the irregular surface with many asperities. The range corresponds to joint roughness coefficients from 10 to 19 (Jia, 2011).

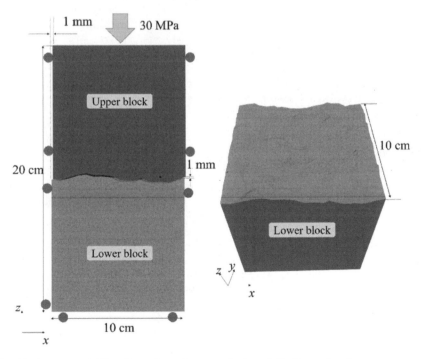

Figure 1.   Basic geometrical configuration of a numerical model to be analyzed.

(a)   (b)   (c)   (d)

Figure 2.   Asperity geometry on the cross-section at y = 5 cm generated with different fractal dimensions.

Table 1. Mechanical properties of the rock for the numerical model.

Rock type	E (GPa)	σ_c (MPa)	$\Phi(°)$	v	γ (kN/m³)	σ_T (MPa)	$\psi(°)$
Granite	31	240	63	0.26	25.5	11	8.75

Table 2. Mechanical properties of a joint between the upper and lower blocks.

k_s (GPa/m)	k_n (GPa/m)	$\phi(°)$	C (MPa)	σ_T (MPa)
12800	31000	30	1	1

3 MECHANICAL PARAMETERS AND SIMULATION PROCEDURE

This study focuses on a rock joint formed in a hard rock. Accordingly, the mechanical properties of granite are extracted from previous study (Melek et al., 2015) and are applied to the numerical model. The parameters are summarized in Table 1. As the numerical simulation is performed with the purpose of investigating the stress distribution and resultant failure of the asperities when subjected to only normal stress, fracture generation due to shearing is not considered in the present study. Thus, the upper and lower blocks are considered to be a continuum material, while contacts are produced on the boundary between the blocks. The mechanical properties of the contacts are listed in Table 2. The contact stiffness was calculated according to the equation ($k_n = E/L$ and $k_s = G/L$). L is a thickness of an interface, and G is the shear modulus. In the calculation, we assumed L = 0.001 m since the upper block has a contact with the lower block. The other parameters do not have large influences on the analysis result.

Numerical simulation is carried out while applying a normal stress of 30 MPa on the top boundary of the upper block under static condition. The lateral boundaries of the upper and lower blocks as well as the bottom of the lower block are constrained in the direction perpendicular to the boundaries. The analysis is continued until unbalanced force in the numerical model reaches a sufficiently small value.

4 ANALYSIS RESULTS

4.1 Stress distribution of the simulated rock joint

Figures 3 shows the maximum and minimum principle stress fields of the lower block obtained from the static analysis. Note that in the model the negative values represent compression. As the upper block is displaced by 1 mm as shown in Figure 1, the asperities of the lower block do not perfectly interlock with those of the upper block, thus generating the high compressive stress in the contact regions. It is further found from the result that as the surface becomes rougher with a high fractal dimension (low Hurst exponent), the stress distribution becomes more complex. And, interestingly, not only compressive stress but also tensile stress is produced in the extensive area of the discontinuity. The high compressive stress is obviously attributed to the small contact area between the lower and upper blocks generated by displacing the upper block by 1 mm in the x-direction. However, the generation of the tensile stress field cannot be readily explained from the contact of asperities subjected to normal stress. To elucidate the mechanism of the tensile stress field, the stress tensor orientation obtained from the model with $H = 0.9$ is plotted in Figure 4. In the figure, the lines represent the orientation of stresses; for each zone, three lines are drawn, and the longest line corresponds to the orientation of the maximum principle stress. It is found from the figure that the tensile stress is oriented in the direction parallel to the discontinuity surface, whereas compressive stress is oriented perpendicular to the surface. From this result, it can be conjectured that the tensile stress field is generated by the asperity penetration that draws the surrounding rock into the contact region. The mechanism is further explored in the following section.

In order to verify the postulation, a simple numerical simulation is carried out to replicate an asperity penetrating to the other block. Figure 5 shows the model geometry as well as the

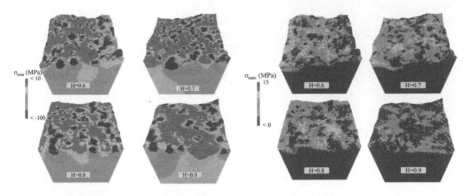

Figure 3. Minimum and maximum compressive stress distribution (compression as negative).

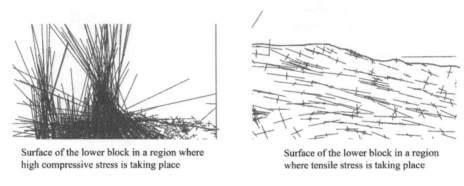

Surface of the lower block in a region where high compressive stress is taking place

Surface of the lower block in a region where tensile stress is taking place

Figure 4. Stress tensor orientation on the cross section at y = 5 cm in the model with $H = 0.9$.

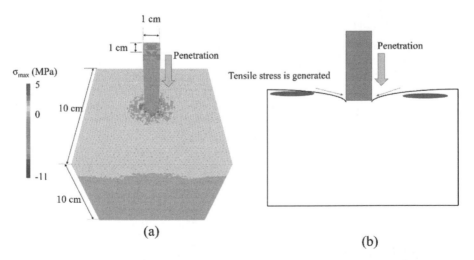

Figure 5. Numerical simulation of a block penetrating into the lower block: (a) Model geometry and analysis result, (b) Schematic illustration showing tensile stress generation caused by the block penetration dragging the surrounding region into the inside of the lower block.

analysis result. As can be seen, the model is composed of two blocks. The lower block has the same geometry and the same boundary conditions as that in Figure 1, while the upper block is a cuboid, which is quasi-statically moved down towards the lower block by applying a normal stress of 30 MPa on the top boundary of the cuboid. As expected, the upper block (cuboid) is compressed because of the generation of the contact with the lower block. More importantly, it can be seen that in the extensive area on the upper surface of the lower block, the tensile stress field is formed. A plausible explanation for this is that the surface region is pulled into the contact area because of the penetrating upper block; as the lateral boundaries are fixed, this leads to the generation of tensile stress. Hence, this would be the plausible mechanism of the tensile field generation observed in Figure 3. It is to be noted that even if the lateral boundaries are not fixed, the tensile stress field can be produced when there are multiple asperity contacts, i.e. each contact causes the surrounding region to be pulled into the contact areas. Thus, the region located at the midpoint of the contact would undergo tensile stress.

4.2 *Failure of asperities subjected to high normal stress*

The extent of failure is shown in Figures 6 and 7. In these figures, green, sky blue, red colors represent shear failure, shear and tensile failure, and tensile failure that took place during the static analyses, respectively. Figure 6 indicates the complex distribution of shear and tensile

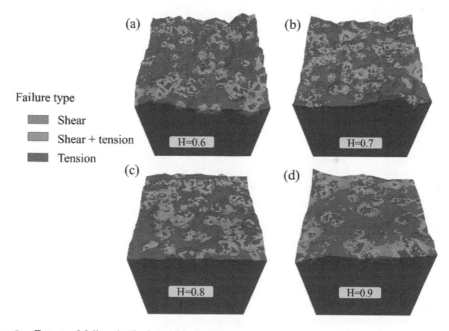

Figure 6.   Extent of failure in the lower block.

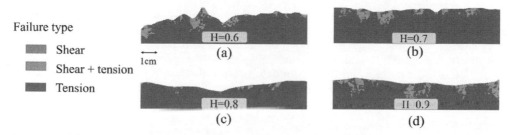

Figure 7.   Extent of failure on the cross-section at y = 5 cm.

677

failure on the rock discontinuity subjected to high normal stress. It seems that the smooth surfaces (high H) generate a more extensive tensile failure area on their surfaces than the rough surfaces. Specifically, in the model with $H = 0.6$, the red-colored surface area approximately accounts for less than one-third of the entire surface, while more than half of the surface undergoes tensile failure for the model with H = 0.9. It is speculated that the uniform stress condition formed on the smooth surface results in the extensive tensile failure area. This can be justified by the result shown in Figure 5, where the entire surface area except the contact region uniformly undergoes tensile stress because of the block penetrating into the completely planar surface. The model with a high Hurst exponent is presumed to yield a similar stress condition because of the non-irregular surface geometry. On the other hand, the irregular, rough discontinuity surface would create a more complex stress field due to the irregular geometry, i.e. the presence of a number of asperities and apertures.

In terms of the depth of failure from the discontinuity surface, it is found from Figure 7 that the shear failure caused by the collision of asperities reaches approximately 1 cm from the surface, although the depth varies place to place, while the damage depth related to the tensile failure ranges from 0.5 mm to about 8 mm. Compared to the shear failure-induced damage, the damage caused by tensile failure is limited to the region in the vicinity of the discontinuity surface.

5 DISCUSSION

To author's knowledge, previous studies predominantly investigated asperity degradation while simulating the shear movement of a rock discontinuity, of which initial surface asperities are fully interlocked (e.g. Bahaaddini et al., 2013), and do not consider the complex stress distribution of the rock discontinuity surface. The present study has characterized the stress state on the rock discontinuity while varying fractal dimension from 1.1 to 1.4. Consequently, it was found out that extremely high compressive stress conditions are generated at locations where the collision of asperities takes place, while the tensile stress field is extensively formed around the contact area. The high compressive stress state is deemed to have a strong relationship with the time-dependent increase in frictional resistance. It has been reported in a number of studies that the frictional resistance of a fault increases with time. It was then assumed that the creep behaviour of fault surface asperities is responsible for the time-dependent resistance increase, which was implemented into the well-known rate- and state-dependent law. It is to be noted, however, that the constitutive equation was empirically formulated based on experimental results. The result obtained from the present study contributes to explicitly modelling the creep behaviour of surface asperities. As shown in Figure 6, the complex stress field is formed on the discontinuity surface, which in tern results in complex time-dependent behaviour causing the degradation of the asperities with the increase in contact area. Most likely, the regions with high compressive stresses in Figure 6 would undergo creep deformation, starting from primary creep. When tertiary creep is initiated, the asperities would experience sever damage associated with comminution, eventually alleviating the stress concentration. The high compressive stress distribution simulated in the present study thus agrees well with the empirical knowledge accumulated with the experiments in previous studies.

What is unclear is the implication of the tensile stress field caused by the penetration of the asperities explored in Figures 3 and 5. Presumably, the tensile stress generates a number of extension fractures in the region. Based on the extent of tensile failure in the depth direction shown in Figure 7, the extension fracture would be the order of mm or less, if fracture generation is taken into account in the simulation. It would be future study to investigate the influence of the micro- and mesoscopic extension fractures on the frictional behaviour of a rock joint surface, including its verification.

6 CONCLUSIONS

The present study investigates the stress distribution and the extent of failure within rock joint surface asperities subjected to high confining stress with the aim of gaining fundamental knowledge of the stress state of rock discontinuities before being sheared. Fractal surfaces are generated with successive random addition and midpoint method, and numerical simulation was quasi-statically performed with 3DEC employing the three-dimensional discrete element method. The analysis revealed that a complex stress field is formed on the surface of the rock joint, producing extremely high compressive stress in the asperity contact regions as well as moderate tensile stress in the surrounding region. It was demonstrated that the tensile stress is caused by asperity collision that draws the surrounding rock into the contact area. The high compressive stress regions are assumed to be associated with the creep behaviour of asperities that results in the time-dependent increase in frictional resistance found in the empirical friction models, such as rate- and state-dependent friction laws. The tensile stress field is, on the other hand, assumed to produce extension fractures with a length being the order of mm or less on the surface, estimating from the extent of tensile failure. The influence of the extension fractures produced in the tensile stress region on the frictional resistance and/or the mechanical behaviour of a rock discontinuity needs to be studied in the future.

REFERENCES

Bahaaddini, M., Hagan, P., Mitra, R. & Hebblewhite, B.H. Numerical investigation of asperity degradation in the direct shear test of rock joints. *In:* Kwasniewski & Lydzba, eds. Eurock, 2013 Poland. Taylor & Francis Group, 391–397.

Barton, N. & Choubey, V. 1977. The shear strength of rock joints in theory and practice. *Rock Mechanics and Rock Engineering,* 10, 1–54.

Elmo, D. 2006. *Evaluation of a hybrid FEM/DEM approach for determination of rock mass strength using a combination of discontinuity mapping and fracture mechanics modelling, with paricular emphasis on modelling of jointed pillars.* PhD, University of Exeter.

Jia, H. Q. 2011. *Experimental research on joint surface state and the characteristics of shear failure.* Master of Science, Central South University.

Malek, F., Suorineni, F. T. & Vasak, P. Geomechanics Strategies for Rockburst Management at Vale Inco Creighton Mine. *In:* Diederichs, M. & Grasselli, G., eds. Rockeng 09, 2009 Toronto.

Ozdemitras, A., Babadagli, T. & Kuru, E. 2009. Effects of fractal fracture surface roughness on borehole ballooning. *Vadose Zone Journal* 8, 250–257.

Voss, R.F. 1985. Random fractal forgeries. *In:* Earnshaw, R.A. (ed.) *Fundamenatl algorithms for computer graphics.* Berlin: Springer.

Geomechanics and Geodynamics of Rock Masses – Litvinenko (Ed.)
© *2018 Taylor & Francis Group, London, ISBN 978-1-138-61645-5*

Thermo-temporal behaviour of uniaxial compressive strength of a fine-grained Indian sandstone

Nikhil Sirdesai & Vinoth Srinivasan
Department of Earth Sciences, Indian Institute of Technology Bombay, Mumbai, Maharashtra, India

Rajesh Singh
Department of Geology, University of Lucknow, Uttar Pradesh, India

T.N. Singh
Department of Earth Sciences, Indian Institute of Technology Bombay, Mumbai, Maharashtra, India

ABSTRACT: Several energy recovery processes such as Underground Coal Gasification (UCG), Enhanced Geothermal Systems (EGS) and Enhanced Oil Recovery (EOR), and processes involving the disposal of nuclear waste in Deep Geological Repositories (DGR) involve the interaction of host rocks with temperature. However, a large variance can be observed in the pattern and nature of thermal interaction within these processes. Since the success and efficiency of the process rely largely on the state of the nearby strata, it is imperative to understand the behaviour of the geotechnical properties under high temperature conditions. Although, lot of research discuss the effect of temperature on the geotechnical properties, studies on effect of duration of thermal treatment are relatively fewer in number. Therefore, in this study, the effect of duration of thermal treatment on the Uniaxial Compressive Strength (UCS) of the rock has been analysed. Samples of fine-grained Dholpur sandstone were treated and tested up to 500°C. The effect of duration of thermal treatment was studied by varying the exposure time from 5 to 30 days. Additionally, the effect of thermal condition of the sample at the time of mechanical tests, namely, hot or cool, was also analysed. The results suggest that a distinct relation exists between the UCS and duration of thermal treatment. The study can be help add upon the existing pool of knowledge on the effect of heat on geomaterials.

Keywords: Sandstone Thermal treatment Strength Time Microcracks

1 INTRODUCTION

Detailed information of rocks and their geotechnical properties is imperative for the success of any civil, mining or energy recovery process. However, studies conducted by several researchers suggest that the geotechnical properties are susceptible to large variations when exposed to high temperatures (Sharma et al. 2018; Somerton 1992; Tian et al. 2012). Rocks in processes such as underground coal gasification (UCG), enhanced geothermal systems (EGS), enhanced oil recovery (EOR) and disposal of nuclear wastes in deep geological repositories (DGR) are exposed to high temperatures during the operational stages (Siratovich et al. 2015; Sirdesai et al. 2015). Additionally, in the case of fires in buildings, tunnels and mines, the rocks and other geo-materials are subjected to varying degree of thermal treatment (Das et al. 2017; Mahanta et al. 2017). However, the nature and magnitude of thermal profile differs substantially from one process to the other. Such large variations in the thermal treatment profile (maximum temperature and the duration of exposure) makes it imperative to study the geomechanics of these processes, individually, to ensure operation success. Several morphological and mineralogical changes occur in rocks when exposed to high temperatures. As observed in all physical material, rocks and rock-forming minerals expand upon heating. However, as studied by Clark (1966), the thermal expansion behaviour of various minerals is anisotropic in nature. Additionally, the thermal expansion along the various crystallographic axes of a mineral is inconsistent. This induces large amounts of thermal stress within the mineral assemblage, which subsequently alters the microcrack network. Besides morphological changes, several chemical reactions occur in a rock at elevated temperatures (Hajpál and Török 2004). The α-β transition of quartz that occurs at temperatures around 573°C, is associated with a volumetric (2%) and linear (0.7%) expansion of the quartz grain (Schacht 2004). The expansion accelerates the process of thermal microcracking. Further, studies suggest that the effect of transition is completely reversible at slow cooling rates (Kerr et al. 2004). Higher cooling rates induces thermal shock within the microstructure, thereby accelerating the process of microcracking. Therefore, the rate of cooling also plays an important role in the morphological transformation of the rock. The closure, formation and coalescence of the microcracks subsequently alters the physico-mechanical response of the rocks (Sirdesai et al. 2017c). The creation of microcracks in any of the above-mentioned processes can lead to catastrophes such as subsidence and groundwater contamination, as seen at the Linc Energy's UCG project at Chinchilla in the state of Queensland, Australia (The Courier Mail 2017). Since the nature of thermal treatment varies across these process, it becomes imperative to perform detailed study of the thermo-mechanical performance of the rocks to ensure safe and successful operation. Therefore, in this study, the mechanical response of an Indian sandstone has been analysed under varied durations of thermal treatment.

2 MATERIAL AND METHODS

2.1 Sample characterisation and preparation

Specimen of fine-grained sandstone belonging to the Upper Bhander Group of the Vindhyan Supergroup, were collected from the Dholpur district of Rajasthan, India (Figure 1). The sandstone is rich in quartz and feldspar, and the mineral grains are held together by a siliceous cement. Dholpur sandstone is a major construction material in India, and in the past, it has been widely used to build some of the major political and cultural monuments (DMG-Rajasthan 2006). Additionally, the sandstone shares similar geological features to those present in the coal/lignite seams that have been chosen for the Indian UCG trials (Ministry of Coal 2015). Diamond core-bits were used to recover cylindrical specimen from the

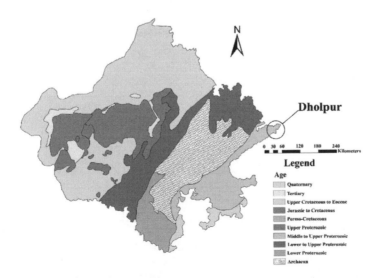

Figure 1. Geological map of Rajasthan, India.

Figure 2. Experimental setup.

blocks of Dholpur sandstone. The cores were recovered perpendicular to the bedding plane, and thereafter, they were cut to attain a length-to-diameter ratio of 2:1.

2.2 *Thermal treatment*

Parameters such as rate of heating, duration of exposure, type and rate of cooling have a large impact on the physico-mechanical response of a rock. Additionally, high rates of heating or cooling induce thermal shock subsequently leading to accelerated microcracking (Den'gina et al. 1993; Tian et al. 2015). Therefore, in this study, the samples were heated to the target temperatures (50–500°C, every 50°C) at a rate of 5°C/min. The samples were exposed to the target temperatures for a duration ranging from 5 to 30 days. Further, in order to analyse the effect of thermal condition at the time of testing, two sets of samples, each containing three specimens, were treated simultaneously at each temperature level. While the first set was tested at heated condition, the other set of samples was allowed to cool, at room temperature conditions, for the exact duration as that of the heating. The samples that were tested at heated condition were named '*NC*' to represent '*No Cooling*', while the samples that were tested after cooling were named '*WC*' to represent '*With Cooling*'. The scheme of thermal treatment has been enlisted in Table 1.

2.3 *Methodology for strength tests*

The thermally-treated samples were tested for their UCS in accordance to the standards mandated by ASTM (2014). The strength tests were performed in a universal testing machine (UTM) under a constant loading rate of 0.1 mm/min. In the case of NC samples, an environmental chamber was used to provide heating during the strength test (Figure 2). The samples were allowed to soak at the target temperature in the chamber in order to compensate for the loss of heat. The surface temperature of the sample was continuously monitored with the help of thermocouples.

3   RESULTS AND DISCUSSION

The strength of non-treated (NT) and thermally treated samples have been enlisted in Tables 2a and 2b, respectively. The cumulative effect of time and temperature on the compressive strength of hot (NC) and heat-treated (WC) samples has been illustrated in Figure 3, wherein, the trends of strength of NC and WC for all every treatment duration have been plotted for better comparison.

As seen in Figure 3, the strengths of all the specimens decrease at the onset of thermal treatment (50–100°C). This can be attributed to the exposure of the inherent pores and microcracks due to

Table 2a. Strength of non-treated samples.

Sample	UCS (MPa)	Average UCS (MPa)
NT1	54.98	56.00
NT2	56.24	
NT3	56.78	

Table 2b. Average compressive strength of treated samples.

Temp	5NC	5WC	10NC	10WC	15NC	15WC	20NC	20WC	25NC	25WC	30NC	30WC
50	52.91	46.95	50.04	55.11	54.54	45.63	49.67	45.33	49.26	42.83	55.26	52.24
100	54.87	48.64	53.69	57.80	57.31	65.16	62.20	61.94	48.89	45.41	61.92	60.21
150	63.86	68.01	72.13	72.91	70.59	70.10	68.27	69.81	75.10	73.97	55.22	60.15
200	70.60	75.53	75.53	72.80	59.81	62.16	61.72	60.81	64.02	65.49	67.01	65.13
250	52.03	48.56	48.74	47.65	54.08	47.93	55.83	53.57	53.02	50.63	63.64	62.06
300	52.50	44.27	52.77	48.82	53.62	55.64	49.62	56.10	50.53	55.13	47.87	45.91
350	48.93	42.39	51.19	45.02	43.71	53.29	51.84	45.40	50.52	44.94	53.61	51.20
400	45.06	50.59	37.82	51.74	57.63	50.79	47.61	55.84	44.24	51.04	44.15	45.05
450	49.08	57.19	48.33	55.02	57.08	56.20	55.09	51.69	51.36	52.37	46.53	48.50
500	58.72	53.27	60.39	59.72	49.96	52.98	55.64	54.16	59.82	55.49	55.98	52.35

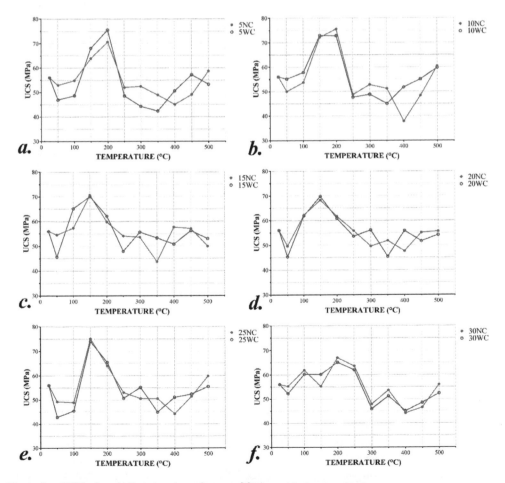

Figure 3. UCS of variedly treated specimens with respect to temperature.

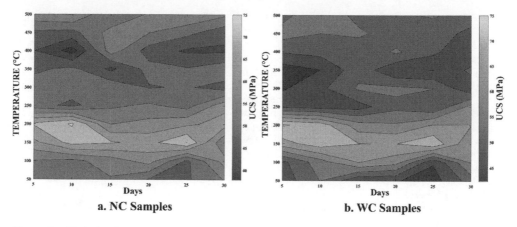

a. NC Samples　　　　　　　　　**b. WC Samples**

Figure 4.　Variation in UCS as a function of duration and treatment temperature.

the evaporation of free-water. On further heating, the mineral grains begin to expand which subsequently leads to the closure of the pores and microcracks. This increases the compaction of the rock, thereby leading to an increase in strength. However, on further heating, the expansion of the mineral grains continues, thereby leading to the formation of new microcracks along the grain boundaries. The formation of microcracks reduces the compaction of the rock, thereby causing the sample to fail. The phenomenon is consistent for all the samples, and can be observed at temperatures over 200°C. The temperature-range between 150–200°C serves as the inflection point for strength, and is known as the *critical temperature zone* (CTZ). Additionally, Figure 3 suggests that the strength of WC samples is less than that of the NC sample. This can be attributed to the state of microcracks within a specimen. Since the WC sample are cooled at room temperature conditions, the resultant thermal shock results in the creation of microcracks, thereby causing an irreversible reduction in strength. Similar results have been reported by various researchers (Brotóns et al. 2013; Sirdesai et al. 2017a; Sirdesai et al. 2016; Sirdesai et al. 2017b; Sirdesai et al. 2017d). In order to analyse the strength as a function of both, duration and temperature, the results of NC and WC samples were studied using MATLAB software to obtain contour plots (Figure 4). The plots confirm the presence of CTZ between 150–200°C. Additionally, it can be observed that the NC specimens, which have been treated for a duration of 10 days, exhibit the highest strength. On the other hand, the strength of the WC samples is highest after 5 days of thermal treatment. This variation can be attributed to the phenomenon of accelerated microcracking in the WC samples due to thermal shock. The contour plot helps in visualising the thermo-temporal behaviour of strength of the fine-grained Dholpur sandstone.

4　CONCLUSIONS

In this study, cylindrical specimens of thermally-treated, fine-grained Dholpur sandstone were examined for their uniaxial compressive strength (UCS). The target temperature and the duration of exposure at the target temperature were varied in order to observe the thermo-temporal behaviour of UCS. The samples were treated at temperature between 50–500°C for 6 durations ranging between 5–30 days. Additionally, the effect of thermal condition at the time of testing was also considered. The results suggest that the mechanical response of the samples varies significantly with the change in duration, temperature and/or thermal condition. The primary cause of the variation can be attributed to the change in the nature and volume of microcracks. Additionally, the results suggest that the highest strength of all the samples can be observed between 150–200°C, thereby, suggesting the presence of a *critical temperature zone* (CTZ). The results of this study will help in understanding the physico-mechanical response of strata in in-situ coal gasification, nuclear waste disposal and enhanced geothermal systems.

REFERENCES

ASTM. 2014. D7012-14E: Standard Test Methods for Compressive Strength and Elastic Moduli of Intact Rock Core Specimens under Varying States of Stress and Temperatures. ASTM International.

Brotóns, V., Tomás, R., Ivorra, S. & Alarcón, J.C. 2013. Temperature influence on the physical and mechanical properties of a porous rock: San Julian's calcarenite. Engineering Geology, **167**, 117–127.

Clark, S.P. 1966. Handbook of physical constants. Geological Society of America.

Das, R., Sirdesai, N. & Singh, T. 2017. Analysis of deformational behavior of circular underground opening in soft ground using three-dimensional physical model. *51st US Rock Mechanics/Geomechanics Symposium*. American Rock Mechanics Association.

Den'gina, N.I., Kazak, V.N. & Pristash, V.V. 1993. Changes in Rocks at High-Temperatures. Journal of Mining Science, **29**, 472–477.

DMG-Rajasthan. 2006. Sandstone - Rajasthan. World Wide Web Address: www.dmg-raj.org/sandstone.html.

Hajpál, M. & Török, Á. 2004. Mineralogical and colour changes of quartz sandstones by heat. Environmental Geology, **46**.

Kerr, R., Needham, J. & Wood, N. 2004. Science and Civilisation in China: Volume 5, Chemistry and Chemical Technology, Part 12, Ceramic Technology. Cambridge University Press.

Mahanta, B., Sirdesai, N., Singh, T. & Ranjith, P. 2017. Experimental study of strain rate sensitivity to fracture toughness of rock using flattened Brazilian disc. Procedia Engineering, **191**, 256–262.

Ministry of Coal, G.o.I. 2015. Steps for Development of Underground Coal Gasification Technology. World Wide Web Address: http://pib.nic.in/newsite/PrintRelease.aspx?relid=132935.

Schacht, C. 2004. Refractories handbook. CRC Press.

Sharma, L.K., Sirdesai, N.N., Sharma, K.M. & Singh, T.N. 2018. Experimental study to examine the independent roles of lime and cement on the stabilization of a mountain soil: A comparative study. Applied clay science, **152**, 183–195.

Siratovich, P.A., Villeneuve, M.C., Cole, J.W., Kennedy, B.M. & Bégué, F. 2015. Saturated heating and quenching of three crustal rocks and implications for thermal stimulation of permeability in geothermal reservoirs. International Journal of Rock Mechanics and Mining Sciences, **80**, 265–280.

Sirdesai, N., Mahanta, B., Ranjith, P. & Singh, T. 2017a. Effects of thermal treatment on physico-morphological properties of Indian fine-grained sandstone. Bulletin of Engineering Geology and the Environment, 1–15.

Sirdesai, N., Mahanta, B., Singh, T. & Ranjith, P. 2016. Elastic modulus of thermally treated fine grained sandstone using non-contact laser extensometer. *Recent Advances in Rock Engineering (RARE 2016)*, Bangalore.

Sirdesai, N., Singh, R., Singh, T. & Ranjith, P. 2015. Numerical and experimental study of strata behavior and land subsidence in an underground coal gasification project. Proceedings of the International Association of Hydrological Sciences, **372**, 455.

Sirdesai, N., Singh, T., Ranjith, P. & Singh, R. 2017b. Effect of varied durations of thermal treatment on the tensile strength of Red Sandstone. Rock Mechanics and Rock Engineering, **50**, 205–213.

Sirdesai, N.N., Singh, A., Sharma, L.K., Singh, R. & Singh, T. 2017c. Development of novel methods to predict the strength properties of thermally treated sandstone using statistical and soft-computing approach. Neural Computing and Applications, 1–27.

Sirdesai, N.N., Singh, T.N. & Gamage, R.P. 2017d. Thermal alterations in the poro-mechanical characteristic of an Indian sandstone–A comparative study. Engineering Geology, **226**, 208–220.

Somerton, W.H. 1992. Thermal properties and temperature-related behavior of rock/fluid systems. Elsevier.

The Courier Mail. 2017. Contamination 'much worse than thought'. World Wide Web Address: http://www.couriermail.com.au/news/queensland/chinchilla-contamination-worse-than-thought-warning-on-mount-isa-lead-levels/news-story/a19a56e1289c98023438fce45a28c681.

Tian, H., Kempka, T., Xu, N.-X. & Ziegler, M. 2012. Physical Properties of Sandstones After High Temperature Treatment. Rock Mechanics and Rock Engineering, **45**, 1113–1117.

Tian, H., Kempka, T., Yu, S. & Ziegler, M. 2015. Mechanical Properties of Sandstones Exposed to High Temperature. Rock Mechanics and Rock Engineering, **49**, 321–327.

Geomechanics and Geodynamics of Rock Masses – Litvinenko (Ed.)
© 2018 Taylor & Francis Group, London, ISBN 978-1-138-61645-5

Acoustic and failure behaviour of Gondwana shale under uniaxial compressive and indirect Brazilian tensile loading—an experimental study

Ashutosh Tripathy, Vinoth Srinivasan, Krishna Kumar Maurya,
Nikhil Sirdesai & T.N. Singh
Department of Earth Sciences, Indian Institute of Technology Bombay, Mumbai, Maharashtra, India

ABSTRACT: Shale has become a principle source of future energy in India. Therefore, research on mechanical and fracture behavior of shale rock has proven imperative in rock engineering. The present study investigated significance of acoustic behavior in shale rock under constant uniaxial compression and tensile Brazilian loading. The XRD and SEM analysis revealed that the investigated shale samples exhibited natural flaws at micro levels with higher percent of flaky micaceous minerals. The compaction of laminations and microstructural flaws is evident during the compression than in tensile condition. This was marked by gradual acoustic rate increase during the preliminary stages in compression. However, the rate of acoustic emission in medial time sequence is feeble under both loading conditions. This indicated the damage accumulation with associated deformation. The stress increase seems to have initiated propagation of fractures leading to failure of rock witnessed by steep increase in acoustic data at final stages. Likewise, tensile failure under Brazilian loading was observed with burst of acoustic signals. The acoustic patterns revealed that layering in shale failure during compression and tensile loading. The present study tried to understand the failure behavior of shale through acoustic emission parameters to infer fracture initiation and propagation stages. The research can be extended to study the fracture stimulation in shale gas exploration and CO_2 injection as an aid to understand shale behavior under compression and tensile loading.

Keywords: Acoustic Emission, Shale, AE Rate, AE Energy, Fracture, Damage accumulation

1 INTRODUCTION

Shale has become a key rock in many rock engineering investigations in India. The relative importance of shale is magnified in the recent years due to its abundance in both natural gas extraction and as an important coal measure rock in Indian subcontinent conditions. Therefore, much focus has been diverted on inferring the mechanical behavior of shale rock. However, the lower permeability and complex fracture properties of the shale deposits enriched with natural gas has been the area of greater concern. This added uncertainty and difficulty in understanding its deformation and their failure behavior great depths. Hence, a complete knowledge of rock stability and its damage conditions of shale rock proved important. This may add sufficient confidence for planning deeper exploration and to arrest any unexpected geo hazards. The exploration of shale gas principally involves injection of hydraulic fluid into the stratum and inducing artificial fractures, thereby stimulating the flow of gas through the rock. Also, shale can be utilized as ideal cap rock for CO_2 sequestration into the deep unmined coal measure rocks. Controlled fracture stimulation at greater depth requires sound knowledge on rock mass behavior while fracturing. Also, these activities will be carried out at very deeper depth, where in-situ estimation of rock behavior is highly difficult. Under the circumstances, laboratory investigations on shale behavior during fracturing and failure proves vital.

The concept of AE monitoring has opened window for wider perspective of rock engineering studies in the recent years. The method is based on the principle that, rocks, when loaded externally will develop micro-cracks due to the stress accumulation and the subsequent re-distribution leading to failure. This will lead to the generation of acoustic signals, which can be captured through a serious of sensors attached to the rock specimen during loading. On processing and inferring those acoustic emission signals effectively, will be help to understand the deformation and the damage in rocks and to locate pre-failure zones with stress redistribution. The application of AE technique for inferring rock damage and deformation through stress release and subsequent generation of microfractures and their propagation has been plentiful. Various researchers have discussed and reported the successful application of AE technique for rock damage assessment and deformation forecasting. Notable work in the recent years include those of (Al-Bazali et al. 2008; Fortin et al. 2009; Stanchits et al. 2009, 2014; Ishida et al. 2012; Inserra et al. 2013; Khazaei et al. 2015; Stoeckhert et al. 2015; Xiao et al. 2016; Rodríguez et al. 2016; Kong et al. 2017; Zhang et al. 2018) and much more.

In the present study, the mechanical and acoustic behavior of shale rock from deep underground coal mine with potential natural gas resource is studied. The specimens were loaded both under compression and Brazilian conditions, so as to study the effect of layering on the acoustic behavior of the shale samples. The change in AE features were directly correlated with the damage conditions and failure characteristics of the loaded specimens. Another significant aspect of the study is to infer the effect of layering on the acoustic behavior of rocks using parametric analysis.

2 MATERIALS AND METHODS

2.1 Sample description

The shale samples tested in the present study were collected from boreholes of Jamadoba coal mine in Jharia coal fields of Damodar river valley belonging to the Gondwana formation in the North East Peninsular Indian State of Jharkhand (Fig. 1). The Jharia coal field is an oval shaped coal field with areal distribution of about 458 sq·km and is located in the eastern end of the damodar valley basin. The damodar basin forms a part of east-west trending the Saptupra—Damodar with a thick sedimentary sequence of about 2900 m of shallow water, fluviatile, lacustrine and glacial environment ranging from Carboniferous to Permian (Verma et al. 1979; Saikia and Sarkar 2013). Sand stones, shales and sandy shales are the principal

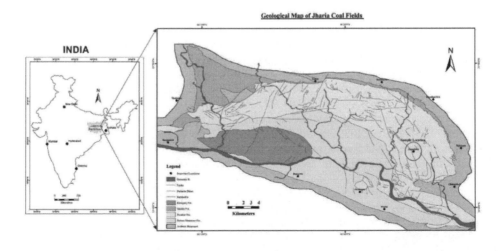

Figure 1.   Location and geological map of Jharia Coal Field.

rocks encountered in the region with intrusion of dolerite and mica-peridotite dykes. Structurally, the region consists of some normal-tensional fault with several strike-slip faults and is surrounded on all sides by crystalline geneiss of Archean age. Also, the region is highly fire prone, the hot fumes extending up to several kilometers with some normal.

The samples were dark black in color with visible layered sedimentation. The core logs collected from the field were cut in the smaller specimens of variable length with the constant diameter of 44 mm. The samples were prepared following the ISRM specifications by ensuring the length to diameter ratio of 2.0 to 2.5. Also, the specimen sides were polished to ensure the tolerance limit is within the prescribed in the standards. Before performing experiments, the samples were analyzed for inferring their compositional and microstructural characteristics. The microstructural characteristics were inferred using scanning electron microscopic analysis and the results are furnished in Fig. 2a. The mineralogical composition of the sample was determined using XRD analysis from the powered samples sieved down to 75-micron mesh (Fig. 2b). The XRD analysis suggested that the samples consisted predominantly of non-siliceous minerals (Fig. 2a). This was strongly supported by SEM imaging results gave insights about the flaky mineral layers of micaceous in nature (Fig. 2b). These bedding layers constituted natural micro-structural weak planes of the rock. The compressional wave velocities suggested that the shale is densely compacted possessing good strength with increase in depth (Fig. 2c). Experiments were carried out in cylindrical samples with height varying from 90 to 104 mm and tensile disc specimens with dimension of 1:1.25 as per the ISRM standards (ISRM 1978; ASTM 1988).

2.1 Experimental setup

The schematic representation of AE setup used in the present study is shown in Fig. 2. The instrumentation for recording acoustic signals consisted of two piezoelectric transducers (R6D Type) with acoustic data logger from Physical Acoustics Corporation (MISTRAS) connected to a PC. The transducers were connected with a pre-amplifier of 40 dB to the 60 dB front amplifier. The signals were filtered to the frequency range of 20 kHz to 3 MHz. The signal threshold of the AE system was set to 45 dB with the sampling frequency of 1 million samples per second (MSPS). The AE experiments were performed as suggested by standards in ISRM (Ishida et al. 2017).

The samples were then loaded using semi servo-controlled loading system having fixed upper platen with force applied by movement of the lower platen. In order to minimize the defects in recorded acoustic signals, a thick glue was applied between the contact of sensor and the samples. The cylindrical samples were loaded perpendicular to the sedimentation layer and the Brazilian discs were loaded parallel to the bedding planes of the shale.

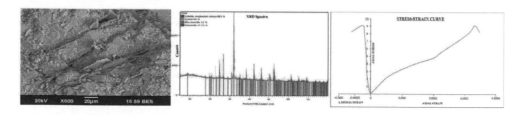

Figure 2.   (a) SEM image, (b) XRD Spectra and (c) Stress Strain Data Curve of the Jharia shale sample.

Table 1.   Important properties of Jharia shale used in the study.

Property	V_P (m/sec)	V_S (m/sec)	BTS (MPa)	UCS (MPa)	E (GPa)	ν
Shale	3344	1319	4.54	9.11	21.4	0.19

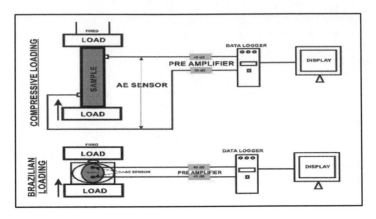

Figure 3.   Schematic of AE Monitoring and Loading System used in the experiment.

3   EXPERIMENTAL RESULTS ANALYSIS OF ACOUSTIC CHARACTERISTICS

Acoustic Emission data will give insight information on the damage initiation and failure precursor of the samples under load (Ohnaka and Mogi 1982). As Acoustic data is rapid release of stored stress of a loaded material through fracture channels, they will help to understand the rock deformation and natural flaws present in rocks. This may be useful in infer the pre-fracture damage and failure characteristics of rock while planning deep underground structures in shales such as in natural gas extraction and in coal mines. In the present study, the acoustic characteristics such as AE Count, Cumulative AE Count (∑AE Count), AE Energy, Cumulative AE Energy (∑AE Energy) were correlated with load.

3.1   *Uniaxial compressive loading*

Although many research on effect of uniaxial compression with acoustic monitoring studies are reported, rarely found research studying the shale behavior with acoustic monitoring. This is due to fact that, shales are highly anisotropic and inhomogeneous in nature with natural bedding layers. Therefore, the present study aims to infer the acoustic behavior of Gondwana shale rock under compression. The results of rate and energy characteristics of acoustic monitoring for cylindrical specimen under compressive loading is given in Fig. 4 and Fig. 5 in correlation with applied load. The results suggested that Acoustic emission attained its peak, just prior to the rock failure, depicted from abnormal increase in both acoustic counts and associated energy. The initial excitation in acoustic signals suggested the closing of the micro-cracks with release of feeble energy. The acoustic signals attained peak after every fracture initiation leading to failure. Since, shale tested in this study is of relatively low strength, the fracture peaks were least prominent from the acoustic data recorded as depicted in Fig. 4.

The correlative plot of studied acoustic parameters with the load is presented in Fig. 5. Both acoustic count and energy levels displayed similar trend throughout the experiment cycle. Also, the cumulative plot of acoustic event count and energy suggested that the samples exhibited three phase of acoustic behavior under compressive loading (Fig. 5).

The first phase of rock deformation is represented with a gentle increment in number of acoustic counts having feeble energy. Furthermore, the slope of the cumulative acoustic data flattens signifying further consolidation of weak layers within shale with damage accumulation. The third phase perceived a drastic increase in intensity of acoustic signals in terms of both number and energy preceding the failure of the sample. This phase is associated with both accumulated stress release linked with fracture propagation.

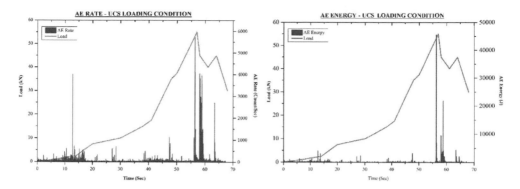

Figure 4. Results of AE parameter behavior under Uniaxial Compression. Correlating Load with (a) AE Count (b) AE Energy.

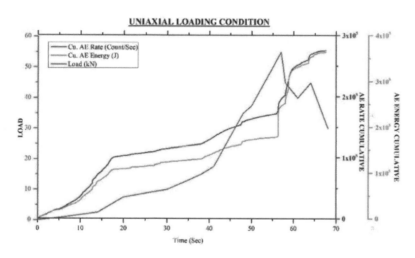

Figure 5. Correlation of Cumulative AE count and Cumulative AE Energy with load.

3.2 *Tensile loading under indirect brazilian condition*

The tensile strength of rock is very important in any excavation processes and failure under tensile loading is one of the predominant failure mode in deep underground excavations (Zhang et al. 2018). One of the principle aim of the present study is to understand the failure behavior of shale under tensile loading and to infer their acoustic behavior. The tensile disc samples were loaded parallel to the bedding layer orientation with simultaneous monitoring of the acoustic signals. The time sequence plot of acoustic data revealed the recording of weak energy associated with rock deformation. The acoustic data projected more count peak with lower energy peaks. The high acoustic peak revealed generation of microcracks during rock deformation. However, the recorded energy in contrast exhibited very low peaks signifying the distribution of stress along the microfractures prior to failure.

The correlation between the cumulative acoustic count and cumulative energy with load is presented in Fig. 7. Unlike uniaxial compression, tensile loading in Brazilian disc exhibited relative inconsistent pattern between acoustic rate and subsequent energy. A gradual increase in acoustic rate was witnessed against low acoustic energy release. However, relative near the peak strength there has been a drastic release of acoustic signals prior to the failure. This drastic increase in acoustic behavior may be due to the sudden release of accumulated stress during the middle period of deformation. The results suggested that the stress

691

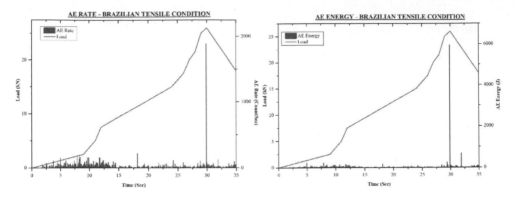

Figure 6.   Results of AE parameter behavior under Brazilian Tensile Loading. Correlating Load with (a) AE Count (b) AE Energy.

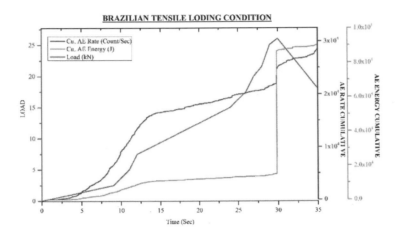

Figure 7.   Correlation of Cumulative AE count and Cumulative AE Energy with load.

accumulation was feeble during loading which dissipated along micro-cracks generated and sudden increase in energy was due to the major crack resulting in the failure of the sample.

4   CONCLUSION

In the present study, the acoustic emission datasets generated during compressive and tensile loading in shale were investigated to understand their fracturing behavior during deformation and associated failure. The shale samples from deep underground coal mine belonging to Gondwana formation were studied. A parametric analysis is carried out with acoustic rate and acoustic energy correlated with load. During compressive, shale sample underwent three stages of deformation evidenced from cumulative curves of acoustic parameters. Both the acoustic rate and energy release followed a significantly similar trend under compression. The first state is marked with compaction of existing natural microcracks, followed by reduced acoustic data representing the damage accumulation leading to fracturing. At the last stage, there is gradual increase in frequency and rate of acoustic signals which signified the propagation of fractures with increased acoustic rate leading to failure. However, the tensile loading resulted in contrast acoustic emission behavior. The tensile loading is marked with gradual increase in acoustic rate with minimal energy at initial stages, and a sudden burst of acoustic energy prior to the failure. This represented the generation of microcracks as evidenced from acoustic rate peak leading to the dispersion of accumulated energy and stress. However, in both the loading conditions, the acoustic emission attained its peak prior to the failure. These results suggested the potential application of acoustic monitoring for predicting rock deformation behavior at loading and failure. The results from the present study could to extended for controlled stimulation of fracture in shale reservoirs during hydraulic fracturing, inferring the potential of shale as a cap rock in CO_2 sequestration site in unmineable coal resources and to determine their injection rate.

REFERENCES

[1] Al-Bazali T, Zhang J, Chenevert ME, Sharma MM (2008) Factors controlling the compressive strength and acoustic properties of shales when interacting with water-based fluids. Int J Rock Mech Min Sci 45:729–738. doi: 10.1016/j.ijrmms.2007.08.012.

[2] Fortin J, Stanchits S, Dresen G, Gueguen Y (2009) Acoustic emissions monitoring during inelastic deformation of porous sandstone: Comparison of three modes of deformation. Pure Appl Geophys 166:823–841. doi: 10.1007/s00024-009-0479-0

[3] Inserra C, Biwa S, Chen Y (2013) Influence of thermal damage on linear and nonlinear acoustic properties of granite. Int J Rock Mech Min Sci 62:96–104. doi: 10.1016/j.ijrmms.2013.05.001

[4] Ishida T, Aoyagi K, Niwa T, et al (2012) Acoustic emission monitoring of hydraulic fracturing laboratory experiment with supercritical and liquid CO_2. Geophys Res Lett 39. doi:10.1029/2012GL052788

[5] Ishida T, Labuz JF, Manthei G, et al. (2017) ISRM Suggested Method for Laboratory Acoustic Emission Monitoring. Rock Mech Rock Eng 50:665–674. doi: 10.1007/s00603-016-1165-z

[6] Khazaei C, Hazzard J, Chalaturnyk R (2015) Damage quantification of intact rocks using acoustic emission energies recorded during uniaxial compression test and discrete element modeling. Comput Geotech 67:94–102. doi: 10.1016/j.compgeo.2015.02.012

[7] Kong B, Wang E, Li Z, et al. (2017) Acoustic emission signals frequency-amplitude characteristics of sandstone after thermal treated under uniaxial compression. J Appl Geophys 136:190–197.

[8] Ohnaka M, Mogi K (1982) Frequency characteristics of acoustic emission in rocks under uniaxial compression and its relation to the fracturing process to failure. J Geophys Res 87:3873–3884.

[9] Rodríguez P, Arab PB, Celestino TB (2016) Characterization of rock cracking patterns in diametral compression tests by acoustic emission and petrographic analysis. Int J Rock Mech Min Sci 83:73–85.

[10] Saikia K, Sarkar BC (2013) Coal exploration modelling using geostatistics in Jharia coal field, India. Int J Coal Geol 112:36–52.

[11] Stanchits S, Fortin J, Gueguen Y, Dresen G (2009) Initiation and propagation of compaction bands in dry and wet bentheim sandstone. Pure Appl Geophys 166:843–868.

[12] Stanchits S, Surdi A, Gathogo P, et al. (2014) Onset of hydraulic fracture initiation monitored by acoustic emission and volumetric deformation measurements. Rock Mech Rock Eng 47:1521–1532.

[13] Stoeckhert F, Molenda M, Brenne S, Alber M (2015) Engineering Fracture propagation in sandstone and slate e Laboratory experiments, acoustic emissions and fracture mechanics. J Rock Mech Geotech Eng 7:237–249. doi: 10.1016/j.jrmge.2015.03.011

[14] Verma RK, Bhuin NC, Mukhopadhyay M (1979) Geology, Structure and Tectonics of the Jharia Field, India-A Three-Dimensional Model. Geoexploration 17:305–324.

[15] Xiao F, Liu G, Zhang Z, et al. (2016) Acoustic emission characteristics and stress release rate of coal samples in different dynamic destruction time. Int J Min Sci Technol 26:981–988.

[16] Zhang SW, Shou KJ, Xian XF, et al. (2018) Fractal characteristics and acoustic emission of anisotropic shale in Brazilian tests. Tunn Undergr Sp Technol 71:298–308.

Geomechanics and Geodynamics of Rock Masses – Litvinenko (Ed.)
© 2018 Taylor & Francis Group, London, ISBN 978-1-138-61645-5

Rocks drillability classification based on comparison of physico-mechanical properties with drilling rate timing

Andrey Trofimov, Alexandr Rumyantsev, Vladislav Vilchinsky & Konstantin Breus
LLC "Institute Gipronikel", Saint-Petersburg, Russia

Andrey Skokov
Department Industrial Assets, PJSC "MMC "Norilsk Nickel", Moscow, Russia

ABSTRACT: The article considers various approaches to the assessment of rock drilling. Domestic classifications and methods for estimating the drillability according to physicomechanical properties are given. Based on the analysis of the results of international practice and various studies, the use of strength properties as a characteristic of drillability is justified. To determine the local productivity of drilling rigs, a video-timing of the pure drilling rate and a complex of mechanical tests are performed. Timing was carried out on various models of the Atlas Copco and Sandvik drilling rigs and various drilling diameters. The best correlation between the pure drilling rate is established with the strength of the rocks for uniaxial tension in accordance with the procedure of GOST 21153.3 paragraph 2. To account for the diameter of drilling and the power of the drummer, a transition to the conditional specific volume rate of drilling is performed. To account for the diameter of drilling and the power of the hummer, a transition to the conditional specific volume rate of drilling is performed. As a result of the generalization of the results, dependencies have been obtained that make it possible to determine the pure drilling speed of drilling rigs, depending on the power of the hummer, the drilling diameter and the tensile strength. Dependencies are presented separately for face and long-hole drilling rigs. For the long-hole, the coefficient of reducing the drilling speed from depth is given.

Keywords: drillability, drilling rate, drilling rigs, mechanical tests of rock, tensile strength

1 INTRODUCTION

The most widely used in the domestic mining industry has got the classification of drillability in which rocks are classified according to the duration of the regular time of drilling one meter hole of different machines and tools such as SNiP IV-2-82 classification (11 classes) and Uniform classification of rocks drillability (20 classes). Scales of classifications are not comparable, since they are constructed from the time of drilling by different machines. However, the Standard developed by Soyuzdorstroy provides a reference table of different drillability classification. A different approach to the classification of drilling rocks is proposed to be evaluated as a function of the UCS, the shear strength and the density of the rocks. A comparison of domestic scales, criteria and methods for estimating the drillability of rocks is discussed in detail by A. Tanayno. In modern conditions, up to 10 models of different generation drilling rigs and jumbos and hammer power can be applied to one type of drilling operations and also the drilling tool used from different suppliers and diameters. In this regard, the assignment of the category of rocks in the classical sense of a certain reference drilling speed is very conditional. For the most accurate description of drillability, it is necessary to characterize the rocks by the mechanical property responsible for the resistance of the material to the formation and propagation of cracks when a concentrated dynamic load is applied.

In international practice, rocks drillability characterize the mechanical properties defined in accordance with the standards ASTM or ISRM, or by specific tests, such as index DRI drillability—drilling rate index (Figure 1).

From the mechanical properties, we can attach the USC and Brazilian UTS method, shore hardness test and Point load index (Figure 2). The obvious disadvantage of special tests to determine the drillability is the large complexity in the preparation of samples and tests, also required a unique laboratory stands with a large number of given initial parameters, which affects the accuracy of test and repeatability of results. In scientific works compared, the index of DRI and the mechanical properties of rocks and with the above properties there is a good correlation.

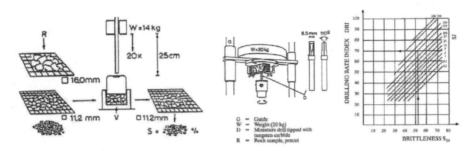

Figure 1.   Determination of DRI index.

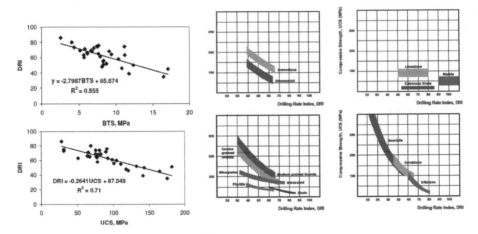

Figure 2.   Correlation between DRI index, UCS and BTS test (Kelessidis, 2011; Yarali and Soyer, 2011).

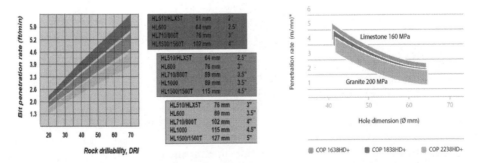

Figure 3.   Sandvik and Atlas Copco recommendations on the choice of the hammer type.

Figure 4. Drilling rate timing.

Figure 5. Mechanical test of rocks.

In the works [Heinio 1999; Thuro 1997; Bilgin and Kahraman 2003], as well as the estimation of rock drilling by physico-mechanical characteristics. Similarly, the recommendations of the leading manufacturers of the drilling rigs and jumbos (Sandvik and Atlas Copco) on the choice of the hammer type are also based on the rock properties (Figure 3). Thus, the use of mechanical properties of rocks to assess the drillability is justified by international practice.

2 RESEARCH METHOD AND MECHANICAL TESTS OF ROCK

To determine local performance of drilling rigs from the physico-mechanical properties of rocks was made pure drilling rate timing (Figure 4). During the timing, the following s models were tested: face drilling jumbos—Boomer M2C, L2D, DD 420–60, 421–60, Axera D5–140, Minimatik D07 260 C and long-hole rigs—Simba M7C, M6C, L6C, SOLO 7–7F, DL 421–15, 321–7, 420–7, 411–15, 420–10, DS 421. The following drilling diameters were tested: 43, 48, 64, 76, 89, 102, 115 mm.

Based on drill timing had gotten drilling speed for different diameters and lithological types of rocks. With the aim of obtaining correlations between pure drilling rate and strength and deformation properties of a complex mechanical test of rocks (Figure 5) that have been taken on the venue of drill timing.

3 RESULTS PROCESSING

Based on the correlation coefficients highlighted the dependence of the conditional volume of the drilling speed of the tensile strength, since this properties most accurately describes drilling of rocks. The investigated differences are represented by sulphide rich, cuprous and

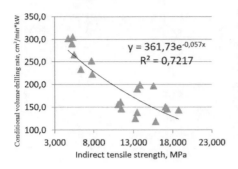 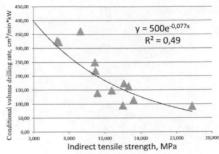

Figure 6. The dependence of the conditional volume drilling rate of the indirect tensile strength, for face drilling rigs (left), long-hole drilling rigs (right).

disseminated ores, hornfelses and metasomatites, diabases, anhydrite, limestone, peridotite, serpentinite. In order to operate with a large amount of statistical data and establish a more accurate relationship, a change is made from the drilling speed in m/min for each drilling diameter to the volume drilling speed m³/min. In order to exclude the influence of the power of the hammer drill, in the presented dependences the volume drilling speed is presented in the dimension m³/min*kW. This approach allows us to estimate the specific energy consumption for rock destruction. Such an approach makes it possible to estimate specific energy costs for the destruction of rock and to determine a single dependence on all sampling points. According to the best correlation is shown in Figure 6.

To determine the pure drilling rate of face drilling rigs, the empirical formula is used:

$$v_{face\_drill} = \frac{361{,}73 \cdot \exp(-0{,}057 \cdot \sigma_{utt}) \cdot W}{\pi / 4 \cdot d^2 \cdot \kappa_r}, \, \text{m/min} \tag{1}$$

were: σ_{utt} – indirect uniaxial tensile strength determined in accordance with GOST 21153.3 paragraph 2, MPa.

W – drilling hummer power, kW;

d – borehole diameter, mm;

K_r – coefficient for drilling in the roof and sides, $K_r = 1{,}135$.

For long-hole drilling rigs:

$$v_{long\_hole} = \frac{500 \cdot \exp(-0{,}077 \cdot \sigma_{utt}) \cdot W}{\pi / 4 \cdot d^2 \cdot}, \, \text{m/min} \tag{2}$$

The effect of the depth of the long-hole $l_{long\text{-}hole}$ on the drilling rate is determined by the formula:

$$v_l = v_{long\_hole} \cdot k_{long\text{-}hole} \tag{3}$$

$$k_{long\text{-}hole} = 1{,}082 - 0{,}0161 \cdot l_{long\text{-}hole} \tag{4}$$

4 CONCLUSION

Based on the results of the study, a dependence has been obtained that makes it possible to estimate the pure drilling rate in rocks of different strengths (tensile strength) for drilling rigs with a varied impact power and drilling diameters. The results obtained are valid only for drilling with top-hammers percussive drilling rigs in the range of sizes of drill bits 43–115 mm.

REFERENCES

Bilgin. N., S. Kahraman. Drillability Prediction in Rotary Blast Hole Drilling. 1a" International Mining Congress and Exhibition ot Turkey-IMCET 2003.

Kelessidis, V.C. Rock drillability prediction from in situ determined unconfined compressive strength of rock. The Journal of the Southern African Institute of Mining and Metallurgy. June 2011. Rock Excavation Course Notes—Spring 2003 University of Arizona Mining and Geological Engineering.

Matti Heinio. Rock Excavation handbook. Sandvik Tamrock Corp. 1999.

Olgay Yarali and Eren Soyer. The effect of mechanical rock properties and brittleness on drillability. Scientific Research and Essays Vol. 6(5), pp. 1077–1088, 4 March, 2011.

SNiP IV-2-82, Rules for the development and application of elemental estimates for building structures and applications. Collections of elemental estimates for building structures and work. Volume 1. Collection 3. Blasting and exploratory work (not cited).

Thuro K. Drillability prediction—geological influences in hard rock drill and blast tunnelling. Geol Rundsch (1997) 86: pp. 426–438.

Geomechanics and Geodynamics of Rock Masses – Litvinenko (Ed.)
© 2018 Taylor & Francis Group, London, ISBN 978-1-138-61645-5

About specific energy intensity behavior under multistage triaxial compression of sandstone specimens

Pavel Aleksandrovich Tsoi
Chinakal Institute of Mining of the Siberian Branch of the RAS, Novosibirsk, Russia
Novosibirsk State Technical University, Novosibirsk, Russia

Olga Mikhailovna Usol'tseva
Chinakal Institute of Mining of the Siberian Branch of the RAS, Head of the Shared Use Center of Geomechanical, Geophysical, and Geodynamic Measurements, Novosibirsk, Russia

Vladimir Nikolaevich Semenov
Chinakal Institute of Mining of the Siberian Branch of the RAS, Novosibirsk, Russia

ABSTRACT: There are cases when it is very difficult to extract enough unbroken cores for the aims of geo-engineering works and studies. Therefore (in the case of core deficiency) the required (by standards) number of laboratory conventional mechanical tests can't be carried out. To overcome this obstacle, a compromise settlement based on the use of a multi-stage compression test was used. The data of such the experiments were used in this work. On the basis of the Instron-8802 testing machine, multistage tests were performed under triaxial compression of cylindrical sandstone specimens. The "axial stress-axial strain" diagrams were obtained for the six tested specimens. Based on the diagram data, the specific energy intensities for each of the five loading stages were determined. It is proposed to assess the predisposition to the brittle or plastic behavior of the studied rock from the change in specific energy intensity during the transition from one loading stage to another.

Keywords: multistage, triaxial compression, specific energy intensity, sandstone, specimen, lab testing

1 INTRODUCTION

The construction of mining objects deals with the natural or technogeneous activities (involving regular loads-unloads) and leads to the mechanical degradation of rocks. Rock mass microfissures induced by development, weakening and corresponding changes in the initial stress state can cause changes of strength and elastic properties [Song et al., 2016]. Therefore calculating techniques based on the use of conventional mechanical properties may be the cause of inaccurate estimation of stress-strain state. There is also complication in extracting the undisturbed core from petroleum deposits. Thus, to characterize the strength parameters, a scant amount of specimens must be tested. Carrying out the multistage triaxial compression tests make it possible to get Mohr-Coulomb envelope involving single specimen data [Myers et al., 2015]. The lab tests data that composed the presented paper are based on multistage triaxial compression of the sandstone, the lateral pressure was varied depending on the stage of the loading program. A number of the international research studies are devoted to such or related topic [Gatelier et al., 2002, Youshinaka et al., 1998, Youn et al., 2010, Bro, 1997]. The aim of the present paper was to consider the assessing of specific energy intensity and its possible interdependence with plastic or brittle behavior during the multistage triaxial compression.

2 EXPERIMENT TECHNIQUE AND ACQUIRED DATA ANALYSIS

The sandstone cores were chosen from the rock collection of the Shared Use Center of Geo-mechanical, Geophysical and Geodynamic Measurements, Siberian Branch, Russian Academy of Sciences. Six cylindrical specimens of the standard diameter (30 mm) were made using the CPM 400 Coretest System. To be sure the parallelism of the end faces and required length (60 mm) were reached, drilled specimens were sawed on the DTS-430 high precise apparatus. The sandstone specimens (Figure 1) were tested under the multistage triaxial compression.

During the multistage test the axial stress value reached to the compressive elastic limit at each loading stage at defined confining pressures. Their values were 0.5, 1.5, 7, 9.5, 11 MPa, respectively, for 1st, 2nd, 3rd, 4th, 5th testing stages.

Each specimen was put into the stabilometer. The specimens were wrapped by special rubber shell to prevent confining liquid (machine oil) from penetration into the sample pores during tests. The oil is used to for even distribution of confining pressure. The axial load was applied to the specimen ends.

The technique for testing rock samples under the multistage triaxial compression consists of several steps [Myers et al., 2015]. The first confining pressure value is applied, and then the axial load (by given loading path) increases until the plastic strain begin. After this, given unloading of the specimen up to the equality value of confining and axial pressures is carrying out. Then the confining pressure increases up to the second value of lateral pressure, the axial load reaches the second value of plastic strain. This procedure is repeated several times (it depends on given quantity of stages). The specimen is brought to disintegration at the experiment completion. The typical instance of obtained «Stress–strain» curve is shown in Figure 2.

The value of specific energy intensity (kJ/m²) was estimated for each loading stage. Specific energy intensity is the energy value accumulated in the specimen at each loading stage of loading up to compressive elastic limit value. This characteristic is described by the definite integral below:

$$W = \int_{u_{z1}^{i}}^{u_{z2}^{i}} \left(\sigma_z^i \left(u_z^i \right) - \sigma_\phi^j \right) \cdot du_z,$$ (1)

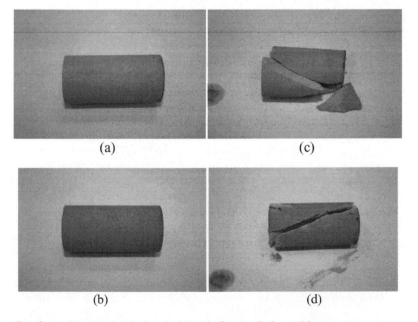

(a)   (c)

(b)   (d)

Figure 1.   Specimens No 1–3, 2–2 before (a, b) and after (c, d) the multistage tests.

where $\sigma_z^i\left(u_z^i\right)$ is the polynomial dependence of the axial stress σ_z^i on the axial displacement u_z^i up to elastic limit value before unloading; σ_φ^i is the value of the confining pressure; u_{z1}^i and u_{z2}^i are related to the range of integration; index i means the i-th (i = 1, ..., 5) loading stage.

The interesting task was to trace the changes of the value of from one stage to another in order to describe the sample energy behavior within the framework of its multistage degradation. As can be seen, there is no clear regularity in the specific energy intensity variation

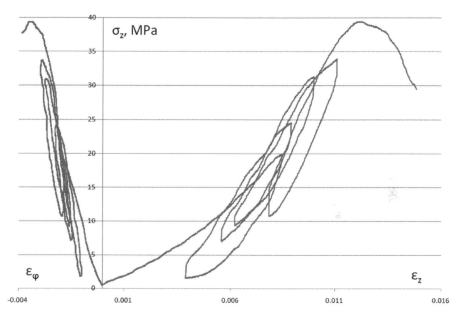

Figure 2.   «Stress-strain» curve for sandstone specimen No 2-1.

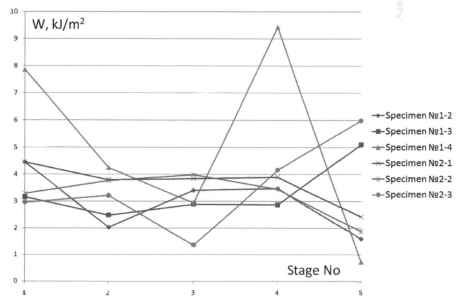

Figure 3.   The plots of dependence of the specific energy intensity on the stage number for tested sandstone samples.

703

(Figure 3). It can be noted (from Figure 3) that for samples 1–3 and 2–3 during the last (5th) stage, the specific energy intensity increased dramatically, while for the remaining four samples, 2-1, 2-2, 1-2, 1-4 decreased. The first observation, apparently, is due to the fact that the material has not yet fully accumulated enough energy for ultimate destruction (it looks like the behavior of a brittle material), and the second observation can be explained by the fact that as plastic deformation preceded the destruction of the sample, a certain amount of the accumulated energy has already been released (for example a sample 1–4 had the highest value on the 4th stage; on the 5th stage the material behaved as plastic).

3 CONCLUSION

In this study, multistage triaxial compression tests on sandstone samples were carried out with the making emphasis on the result related to a specific energy intensity behavior.

Based on the processed experimental data, the following conclusions can be drawn:

- In the case of each tested sample the specific energy intensities for each of the five loading stages were determined.
- It was revealed that there is no common clear pattern of variation in the specific energy intensity (weak dependency on stage number).
- But in the same time a tracing the changes of the value of (from one stage to another) can give some apparent explanation of material behavior. That is, depending on the sample and loading stage, geomaterial demonstrates more brittle (material has not yet fully accumulated enough energy for ultimate destruction) or more plastic (a certain amount of the accumulated energy has already been released) state.

ACKNOWLEDGMENTS

We appreciate the contribution of the Shared Use Center of Geomechanical, Geophysical and Geodynamic Measurements, Siberian Branch, Russian Academy of Sciences to the present research work (State registration No AAAA-A17-117121140065-7).

REFERENCES

Bro A. Analysis of multistage triaxial test results for a strain-hardening rock // International Journal of Rock Mechanics & Mining Sciences. Vol. 34, iss. 1, pp. 143–145, 1997.

Gatelier N., Pellet F., Loret B. Mechanical damage of an anisotropic porous rock in cyclic triaxial tests // International Journal of Rock Mechanics & Mining Sciences. Vol. 39, iss. 3, pp. 335–354, 2002.

Myers M.T. and Sharf-Aldin M.H. Comparison of Multistage to Single Stage Triaxial Tests // 2015 Proc. ARMA-2015-767, 49th U.S. Rock Mechanics, 11 p.

Song H., Zhang H., Fu D., Zhang Q. Experimental analysis and characterization of damage evolution in rock under cyclic loading // International Journal of Rock Mechanics & Mining Sciences. Vol. 88, pp. 157–164, 2016.

Yoshinaka R., Tran T.V., Osada M. Non-linear, stress- and strain-dependent behavior of soft rocks under cyclic triaxial conditions // International Journal of Rock Mechanics & Mining Sciences. Vol. 35, iss. 7, pp. 941–955, 1998.

Youn H., Tonon F. Multi-stage triaxial test on brittle rock // International Journal of Rock Mechanics & Mining Sciences. Vol. 47, iss. 4, pp. 678–684, 2010.

Geomechanics and Geodynamics of Rock Masses – Litvinenko (Ed.)
© 2018 Taylor & Francis Group, London, ISBN 978-1-138-61645-5

The relationship between strain, microstrain, temperature fields and microseismic emission parameters in geomodels with hole under uniaxial and biaxial loading

Vladimir Ivanovich Vostrikov
Chinakal Institute of Mining of the Siberian Branch of the RAS, Head of Mining Geophysics Laboratory, Novosibirsk, Russia

Olga Mikhailovna Usol'tseva
Chinakal Institute of Mining of the Siberian Branch of the RAS, Head of the Shared Use Center of Geomechanical, Geophysical, and Geodynamic Measurements, Novosibirsk, Russia

Pavel Aleksandrovich Tsoi
Chinakal Institute of Mining of the Siberian Branch of the RAS, Novosibirsk, Russia
Novosibirsk State Technical University, Novosibirsk, Russia

Vladimir Nikolaevich Semenov
Chinakal Institute of Mining of the Siberian Branch of the RAS, Chief Specialist, Novosibirsk, Russia

Olga Alekseevna Persidskaya
Chinakal Institute of Mining of the Siberian Branch of the RAS, Lead Engineer, Novosibirsk, Russia

ABSTRACT: In this research the authors have carried out uniaxial and biaxial compression tests of artificial cube specimens with hole and argillite to study the process of deformation prior to failure using the multiparametric equipment designed for synchronous recording of physical fields of stresses, macrostrains, Microseismic Emission (MSE) and microstrains by speckle method. The complex data of evolution of microseismic emission signals, temperature field and microdeformation field under uniaxial and biaxial loading prior to the destruction of prismatic samples from artificial geomaterial allowed to establish the time-space relationship between the features of signal changes depending on loading level.

The evolution of deformation process, development of microdamages and the formation of main rupture fracture lead to significant transformation of spectral composition of microseismic emission signals, microdeformation field, and the temperature field. In the region of future main discontinuity the temperature increases, localization of maximum microdeformations occurs and velocities of microdeformations components increase. Generation of powerful low-frequency harmonics at loads approaching the peak, can serve as a precursor of a rupture on the surface and, consequently, destruction of the geomaterial.

Keywords: laboratory test, stress, strain, microseismic emission, speckle method, temperature

1 BRIEF INTRODUCTION

The processes of crack formation, discontinuity propagation and generating fields of various physical natures under the influence of critical loads in materials are related to: deformation, microseismic and electromagnetic emission, and temperature. The task of adequate description of mechanical behavior of various materials, rocks and massifs under the influence of different

Figure 1. General view of the experiment: 1 – optical/tv measurement complex ALMEC-tv; 2 – argillite specimen; 3 – compression grips of the Instron-8802; 4 – artificial geomaterial sample.

types loading in underground construction and mining requires the investigation these field regularities, the identification of intercorrelation between them, and, ultimately, the development of new assessment parameters of geo-environment disruption, which can be used to predict the destruction of rock massifs (dumps, rock blows, technogenic earthquakes). Over the past year the review of the literature has shown that there are a number of papers devoted to the study of the features of physical fields of a different nature—the field of microdeformations by the speckle method (Wang, 2015, Kim, 2015, Shi, 2010], acoustic emission signals (Nejati, 2014, Kim, 2015, Shkuratnik, 2014, Zuev, 2014], the temperature field (Shi, 2010] at deformation of rocks. In (Nejati, 2014], the characteristics of acoustic emission signals are correlated with the deformation of various rocks with their fragility and durability. In [Hedayat, 2014] seismic wave transmission and digital image correlation were employed to study slip processes along frictional discontinuities in biaxial compression experiments were performed on gypsum specimens composed of two blocks with non-homogeneous contact surfaces. In (Kim, 2015, Shkuratnik, 2014, Zuev, 2014], the temperature field (Shi, 2010] in the deformation of rocks. In (Nejati, 2014] experimental studies and theoretical justification of the evolution of acoustic emission signals appearing in different genotypes rock samples under their mechanical loading are presented. Studies using the speckle method make it possible to study in detail at the micro level the process of formation and development of microdamages in a rock under the action of stresses (Wang, 2015, Shi, 2010]. The purpose of this study was to establish the features of the change and the relationship between stresses, deformations, microdeformations, temperature field and parameters of microseismic emission signals (MSE) under uniaxial and biaxial loading of geomaterial and rock samples. Figure 1 illustrates general view of the experiment.

2 EXPERIMENTAL PROCEDURE AND MEASUREMENT EQUIPMENT

Two series of experiments were carried out on the following samples: 1) prismatic samples of argillite by dimensions of $50 \times 50 \times 20$ mm³ with a hole diameter of 15 mm in the center; 2) samples from an artificial geomaterial ("Neolit" glue – 1 part, calibrated sand with particle size $0.25 \div 0.315$ mm – 3 parts, cement – 1 part, water), which were cubes by 200 mm edge with the cylindrical hole diameter 20 mm in the cube center. The average strength limit was 39.1 MPa for argillite prismatic samples under uniaxial compression, 20.1 MPa for artificial geomaterial samples under uniaxial compression and 20.3 MPa – under biaxial compression.

Tests of argillite samples were carried out under uniaxial compression and tests of artificial geomaterial samples– under uniaxial and biaxial compression prior to failure on Instron 8802 servo-hydraulic press. Movement and force in the axial (vertical) direction was recorded by the measuring system of the press Instron 8802 and was saved to the computer file.

The moving speed of the movable gripper was varied in the range 0.1 ÷ 10 mm/min for various experiments. To implement the biaxial loading, the special device was used, which made it possible to create an additional side load, independent of the press, on the prismatic sample, which was also continuously recorded the computer file. Four microseismic KD 91 sensors were installed on 4 lateral faces of the cube to record the MSE signals. The measurement of the temperature field was carried out using by computer thermal imager TKVr-SVIT 101, the accuracy of the measurement is 0.03°. Figure 1 illustrates general view of the experiment. Microstrains were recorded using automated digital speckle photography analyzer ALMEC-tv at frequency of 27 frames per second and spatial resolution not less than 1 μm. The processing output is the coordinates and displacements of the specimen surface points and timing, which allows calculating strain tensor components (Shi et al., 2010).

Figure 2a illustrates the diagram «stress σ/σ^{lim} – strain ε» obtained for artificial geomaterial sample under uniaxial and biaxial compression (σ^{lim} is the average strength limit under uniaxial compression), Figure 2b illustrates the diagram «stress σ/σ^{lim} – time t» under uniaxial compression.

A large number of microseismic signals were recorded under uniaxial and biaxial compression prior to destruction of samples. As it turned out, the process of deformation can be conditionally divided into three stages, for which the same patterns of change are characteristic. Figure 3 shows the characteristic values of parameters of signals of microseismic emission

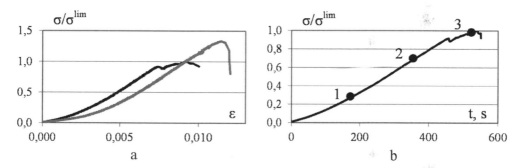

Figure 2.   Diagram «stress σ/σ^{lim} – strain ε» for artificial geomaterial sample under uniaxial (black) and biaxial (red) compression (σ^{lim} is the average strength limit under uniaxial compression) – a; diagram «σ/σ^{lim} – time t».

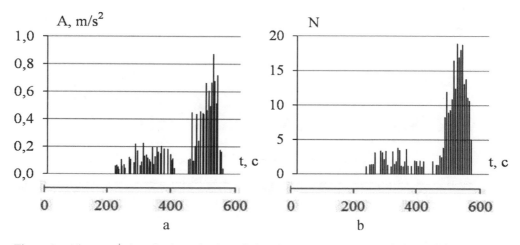

Figure 3.   Characteristics of microseismic emission signals averaged over 10 secondsinterval, under uniaxial loading of artificial geomaterialsample: a – acceleration amplitudevalues A (m/s²); b – number of signals N.

obtained for the three isolated loading stages of the diagram «stress $\sigma/\sigma^{\lim}$ – time t». The data obtained under uniaxial compression loading is shown in Fig. 2b. Figure 3 illustrates the dependence of characteristics of microseismic emission signals on time under uniaxial loading of cubic samples from an artificial geomaterial, averaged over an interval of 10 seconds: 3a—values of the acceleration amplitude A (m/s^2); 3b – number of MSE signals N; in accordance with the diagram «$\sigma/\sigma^{\lim}$ – t» – 3d. The insignificant number of small amplitude MSE signals fixed before levels $\sigma/\sigma^{\lim} = 0{,}3$ is not shown in the diagrams, since these signals are most likely the cause of the alignment of the roughness on the side faces of the sample and are not related to the formation of the main discontinuity. At stress levels $\sigma/\sigma^{\lim} = 0{,}3 \div 0{,}4$ (point 1), the first MSE signals are recorded, the acceleration magnitude is $0{,}1 \div 0{,}2$ m/s^2, the broadband frequency signal is $8 \div 24$ kHz. Then, as the load increases, the number of signals increases too, distributing chaotically evenly in the volume of the sample. At stress levels $\sigma/\sigma^{\lim} = 0{,}5 \div 0{,}8$ (point 2), the amplitude of the MSE signal increases, its magnitude reaches a value of $0.3 \div 0{,}35$ m/s^2. At the same time, the frequency spectrum is somewhat narrowed and shifted to low frequencies of $f = 12 \div 18$ kHz, the distance between the hypocenters decreases, which indicates the localization of microdefects. At values of stresses close to the ultimate strength, the certain period of MSE absence before the formation of the main discontinuity. At the last deformation stage at a stress close to $\sigma/\sigma^{\lim} = 0.8 \div 1$ (point 3), the number of microdamages and MSE signals increases significantly, their energy increases, the frequency spectrum is further narrowed and shifted to low frequencies, up to $f = 8 \div 10$ kHz. A powerful low-frequency signal of $8 \div 12$ kHz is generated at the time of the main discontinuity initiation.

The analysis of microdeformation fields for specimens from argillite and artificial geomaterials under uniaxial and biaxial compression has shown that plastic deformation is inhomogeneous from the beginning of the loading, which is related to the mineralogical and structural inhomogeneities of test samples. Figure 4 (a, b, c) illustrates deformation mapping shots of scanned surface of the part of working surface of cubic sample for deformation component in the x-direction (perpendicular to the load axes) at three strain levels. Green color of deformation component corresponds to positive values (increasing in size), red color – decreasing in size, black color – zero deformations, white color – deformation exceeding the value of 0.007. Despite the fact that the given kind of loading of rock sample is with constant speed compression, nevertheless, both the shortening and elongation regions are present in the space-time field of microdeformations scanned surfaces. At the first stage of loading (point 1), microdeformations field is chaotically inhomogeneous, elongation-shortening zones are randomly distributed over sample surface, the vibrations of microdeformation components are practically absent. At the 2nd stage of deformation (point 2), microdeformations field becomes more inhomogeneous, maximum microdeformation zones occur, the values of which exceed the average values over sample surface, the amplitudes of the oscillations of microdeformations and their velocities in the region of future destruction increase. At the

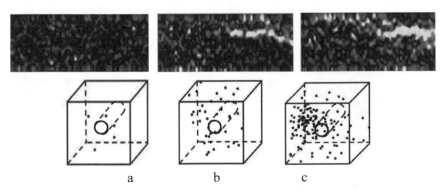

a   b   c

Figure 4.   Shots of deformation mapping of scanned surface of working surface part of cubic sample for deformation component in the x-direction at three strain levels: point 1 (a), point 2 (b), point 3 (c).

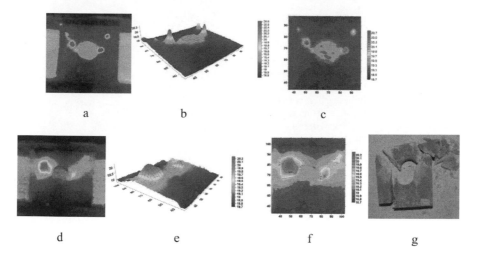

Figure 5. Temperature field on the sample surface at loading moments t = 0,8 t_m (a, b, c) and t = 0,9 t_m (d, e, f), photograph of the corresponding destructed rock sample (g).

third stage of deformation (point 3), maximum microdeformations zones are localized in the certain volume of the sample, which indicates the beginning main discontinuity formation. Maps of the space-time distribution of the MSE signals are schematically shown in Figure 4 (d, e, f), in accordance with three stages of loading.

Analysis of temperature data shows that at the beginning stage of loading, temperature field and have no obvious change (Figure 5). With the loading development, strains focus around the hole gradually, over and under of the hole appear tensile strain along the direction of loading and grow gradually.

At the values t = 0,8 t_m the temperature increases in the region with an X-shaped pattern near the hole. When the loading time is t = 0,9 t_m, the red area to the left of the hole increases in size. On the diagram of y-component of the microdeformation there is a correspondence—this is the region of higher values of tensile microdeformations. With further increase in the load, a main crack appears at the top left of the hole, then from the top right, at the last moment—a vertical crack under the slope and the sample is completely destructed. The temperature rise in the zones of localization of the maximum microdeformation is 1.5°. With an increase of loading rate from 0.1 mm/min to 6 mm/min, these patterns of temperature field variation manifest more contrastively.

3 CONCLUSION

The complex analysis of evolution of microseismic emission signals, strain, stress, microdeformation field and temperature field made it possible to determine the regularities of the change of these parameters and their relationship with the loading stage of geomaterials and rocks samples, which makes it possible to predict the region of the main rupture appearance on the sample surface at loads less than the peak value, when the sample still maintains its visible intact state.

ACKNOWLEDGMENTS

This study has been carried with partial financial support of the Russian Foundation for Basic Research, projects nos. 16-05-00992, using equipment of the Shared Use Center for Geomechanical, Geophysical and Geodynamic Measurements, Siberian Branch, Russian Academy of Sciences.

REFERENCES

Hedayat A., Pyrak – Nolte L.J., Bobet A. Multi – Modal Monitoring of Slip Along Frictional Disconti-nuities. 2014, Rock Mechanics and Rock Engineering, vol. 47, issue 5, pp. 1575–1587.

Kim J.S., Lee K.S., Cho W.J., Choi H.J., Cho G.C. A Comparative Evaluation of Stress–Strain and Acoustic Emission Methods for Quantitative Damage Assessments of Brittle Rock. 2015, Rock Mechanics and Rock Engineering, vol. 48, issue 2, pp. 495–508.

Reza Nejati H., Ghazvinian A. Brittleness Effect on Rock Fatigue Damage Evolution//Rock Mechanics and Rock Engineering, 2014, Volume 47, Issue 5, pp. 1839–1848.

Shi Y., He Q., Liu S., Wu L. The time – space relationship between strain, temperature and acous-tic emission of loaded rock. 2010 Progress In Electromagnetics Research Symposium Proceedings, Xi'an, China, pp. 114–118.

Shkuratnik V.L., Novikov E.A., Oshkin R.O. Experimental analysis of thermally stimulated acoustic emission in various—genotype rock specimens under uniaxial compression. 2014 Journal of Mining Science, vol. 50, issue 2, pp. 249–255.

Wang L., Bornert M., Heripre E., Chanchole S., Pouya A., Halphen B. The Mechanisms of Deforma-tion and Damage of Mudstones: A Micro—scale Study Combining ESEM and DIC. 2015, Rock Mechanics and Rock Engineering, vol. 48, issue 5, pp. 1913–1926.

Zuev L.B., Barannikova S.A., Nadezhkin M.V., Gorbatenko V.V. Localization of deformation and prog-nostibility of rock failure. 2014 Journal of Mining Science, vol. 50, issue 1, pp. 43–49.

Geomechanics and Geodynamics of Rock Masses – Litvinenko (Ed.)
© *2018 Taylor & Francis Group, London, ISBN 978-1-138-61645-5*

Monitoring of coal pillars yielding during room and pillar extraction at the great depth

Petr Waclawik, Radovan Kukutsch, Petr Konicek & Vlastimil Kajzar
The Czech Academy of Sciences, Institute of Geonics, Ostrava-Poruba, Studentska, Czech Republic

ABSTRACT: A considerable amount of coal reserves are located in protection pillars that lie under built-up region in active mining areas at the Czech part of the Upper Silesian Coal Basin. The commonly used controlled caving longwall mining method is not applicable in these areas because significant deformation of the surface is not permitted. For this reason the room and pillar method with stable coal pillars has been tested in order to minimise subsidence of surface.

Stress-deformation monitoring was essential as this was the first application of the conventional room and pillar mining method within the Upper Silesian Coal Basin mines. More than six kilometres of roadways were driven within two panels during last three years. To determine pillar stability, vertical stress and horizontal displacement of coal pillars were measured in coal pillars which are located within a row of pillars forming the panels. Two monitored pillars diamond in shape and slightly irregular sides have been observed into the first mined panel "V" and three monitored pillars have been observed into the second panel "II". To measure the increase in vertical stress due to mining, hydraulic stress cells were installed in each coal pillar. The 5-level multipoint rib extensometers measured displacements of all sides within each monitored pillar. The results of stress-deformation monitoring allowed pillar loading and yielding characteristics to be described.

Keywords: stress-deformation, monitoring, room and pillar, coal pillar, yielding, displacement

1 INTRODUCTION

The pilot project of the mining modified room and pillar method with stable pillars has been running since 2014. The method was tested within the shaft protective pillar located in CSM-North Mine (Karvina coal sub-basin in the Czech part of the Upper Silesian Coal Basin—USCB) coal seam No. 30, where the risk of rockbursts was low and roof conditions were acceptable for bolting reinforcing. However, the variable geology and several faults of regional importance complicated the mining conditions.

The project includes the detailed geomechanical monitoring of stress and deformation in the driven roadways and the surrounding rock mass. Monitored pillars with 3.5 m in high and different sizes were selected to determine stress-deformation characteristics under different geotechnical conditions. Two monitored pillars diamond in shape and slightly irregular sides were approximately 860 m² and 1200 m² in size into the first mined panel "V" (locality A) and three monitored pillars were approximately 590 m², 590 m² and 730 m² in size into the second panel "II" (locality B). Mining depth of room and pillar trial ranged from 700 to 900 m, being perhaps the deepest room and pillar mining in the world coal mines.

Monitored data and other analyses are essential to establishing procedures for a safe room and pillar method of mining within USCB. The results are also important for worldwide mining, for the largest coal producers will reach higher mining depth in near future.

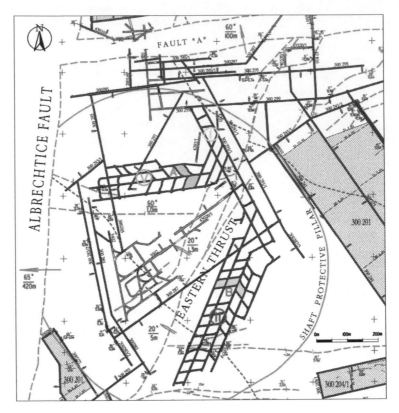

Figure 1. Tectonic situation and position of monitored pillars in panel V and II.

2 GEOLOGICAL AND MINING CONDITIONS

The targeted coal seam (No. 30) is at a depth of approximately 700 m to 900 m below the surface. The thickness of coal seam is extremely variable (from 180 to 520 cm) within the proposed mining area. The thickness of coal seam ranges 300 cm to 350 cm in monitored pillars. The strata dip oriented in the north-east direction ranges from 8° to 17°. Occasionally the dip of the coal seam can reach up to 20°. There are several faults of regional importance in the area of the CSM-North shaft protective pillar (see Fig. 1). The significant regional tectonic fault zone "Eastern Thrust" (Waclawik et al. 2013, Grygar & Waclawik 2011) divides the area of the protective pillar into two separate blocks with different geotechnical conditions. The immediate roof above concerned coal seam No. 30 consists of a thin 0.1 m thick sandy claystone layer. This layer is relatively weak and disturbed with slickensides present on the surrounding bedding planes. Above this is 5 m thick siltstone overlain with 6 m thick medium-grained sandstone and 0.3 m thick coal seam. The immediate floor below mined seam No. 30 consists of 0.5 m thick siltstone underlain by 0.6 m thick coal seam No. 31. The interbedded siltstone and sandstone layers follow down to coal seam No. 32 located around 10 m below the seam No. 30. More details about natural and mining conditions can be found in previous published papers (e.g. Waclawik et al. 2016, Waclawik et al. 2017).

3 DESIGN OF GEOMECHANICAL MONITORING

Monitoring of stress and the deformation state of rock mass is an essential requirement for the design of safe and successful room and pillar method that can be applied in the Czech part of the USCB. In the context of stress and deformation, the monitoring are covering deformability of rock overlaying the room and pillar roadways, measuring pre-mining stress and stress change monitoring in rock and coal during mining, deformability of coal pillars, load on the installed cable bolts, roadway convergence monitoring. On top of all that the seismology and seismo-acoustic monitoring were carried out to characterize fracturing of rock mass during mining.

The instrument locations are shown in Figures 2 and 3. To monitor roof deformation, 5-level multipoint extensometers monitored roof displacements (VE1 to VE14 in locality A; IIE1 to IIE19 in locality B) and strain gauged rockbolts (VS1 to VS11 in locality A) were installed at various locations. The 5-level multipoint rib extensometers (VEH1 to VEH8 in locality A; IIEH1 to IIEH12 in locality B) measured displacements of all sides within each monitored pillar were installed. Vertical and horizontal displacements together with the convergence measurements (VP1 to VP9 in locality A; IIP1 to IIP12 in locality B), changes in vertical pillar loads and the periodic 3D laser scanning of the overall roadway displacements (roof, rib and floor heave) provided data to evaluate coal pillars deformability. To describe pre-mining stress-state condition of coal pillars area 3-dimensional CCBO stress overcoring cells (Obara & Sugawara, 2003; Stas, Knejzlik & Rambousky, 2004) were used (VCCBO1, VCCBO2 in locality A; IICCBO1, IICCBO2 in locality B) and 3-dimensional CCBM stress change monitoring cells (Stas, Knejzlik, Palla, Soucek & Waclawik, 2011; Stas, Soucek, & Knejzlik, 2007) were installed to measure stress changes during mining (VCCBM1 to VCCBM8 in locality A; IICCBM1 to IICCBM3 in locality B). The 1-dimensional hydraulic stress monitoring cells were installed at various depths in each pillar to measure vertical stress (VSC1 to VSC8 in locality A; IISC1 to IISC5 in locality B), seven hydraulic dynamometer load cells measured the cable bolt

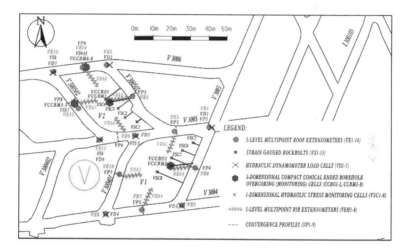

Figure 2. Positions of the monitoring equipment in locality A.

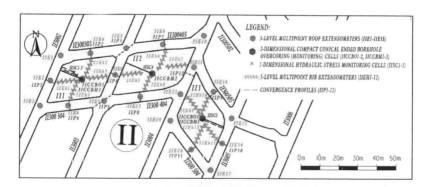

Figure 3. Positions of the monitoring equipment in locality B.

713

loads installed at the roadway intersections (VD1 to VD7 in locality A). The monitoring equipment were reduced in locality B due to monitored results from locality A. Due to minimal roof displacement during the whole time of monitoring in locality A, the strain gauged rockbolts, hydraulic dynamometers have not been installed in locality B. The vertical 5-level multipoint extensometers were substitute by the cheaper 3-level multipoint extensometers.

4 RESULT AND DISCUSSION

The displacement and deformation development are determinative factor for assessment of coal pillar stability. The results of pillar displacement and loading monitoring allowed the monitored pillars deformation characteristics to be described. The data showed that the monitored coal pillar ribs displaced into the roadway mainly due to a large vertical stress and the presence of weak slickensides layers above and below the seam. This mechanism has caused large floor heave, rib convergence, therefore weakening the coal and causing the pillar to yield (Waclawik et al. 2016).

In locality A, due to the incorrect stress cell installation within the larger coal pillar (V1), the loading results were limited to the smaller pillar (V2) only. The stress cells installed in coal pillar V2 gave the information about stress changes in coal pillar, depending on coal pillars forming. The maximum vertical load of 49 MPa on the stress cell VSC4 was registered. Considering the position of the load cell, and the registered values of the other stress cells, it was the extreme short-time load of the central part of the coal pillar V2. On basis of data from stress cells (vertical load) and horizontal extensometers (displacement) we can define development of yielding zones in coal pillar (Waclawik et al. 2016).

In locality B, the maximum vertical load of 39 MPa was registered on the stress cell IISC2 (monitored pillar II1). The maximum vertical load of 16 MPa in monitored pillar II2 and 14 MPa in monitored pillar II3 were recorded only. These relatively small values of maximum vertical load were influenced by more rapid yielding of coal pillar to the depth in locality B (compare results from horizontal extensometers – see Figs. 5, 6). The smaller size of coal pillars, added number of pillars in row and presence of overthrust above monitored pillars significantly affected rate of coal pillars yielding.

The displacements of the coal rib recorded by horizontal extensometers are comparatively different within the monitored pillars. In locality A, the larger displacements were recorded by horizontal extensometers installed in monitored pillar V2. The values of displacement ranged between 212 mm to 300 mm in monitored pillar V2 (see Fig. 5). The displacement coal ribs of monitored pillar V1 ranged between 59 mm to 223 mm. These values indicate that the displacement of the coal pillar V2 as large as monitored coal pillar V1, caused by higher area loading. From the results recorded by the horizontal extensometers in location B, it is evident that the maximum horizontal displacement is 478 mm (IIEh5) in the monitored pillar II2. Also, in the monitored pillar II1, the relatively higher values are recorded (468 mm – IIEh1, 379 mm – IIEh3). Even in the monitored pillar II3, which was last formed, the values of displacements of around 300 mm (IIEh9 – 344 mm, IIEh11 – 353 mm) are reached.

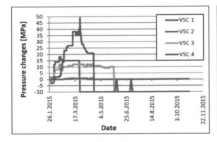 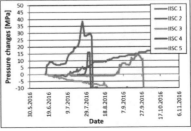

Figure 4. Pillar load results in monitored pillar V2 – locality A (on the left) and load results in monitored pillars II1, II2 and II3 – locality B (on the right).

In locality A, the major strata displacement zone occurred in the area 1.5–5 m from the pillar side (see Fig. 5). The displacement at the depth of 0–1.5 m was much smaller due to the efficiency of rock bolts. In two cases (extensometers VEH7, VEH8) had no influence of the rock bolts and the maximum strata displacements occurred at the depth of 0–3 m into the pillar. The reason for this was considered to be the primary pillar damage by fractures in highly stressed ground. In locality B, the major strata displacement zone occurred mainly at the depth of 5–8.5 m from the side (see Fig. 6). The significant strata separation was recorded at the deepest monitored zone 8.5–12 m, which indicated that the monitored pillars were totally fractured. In most cases there was no measured rockbolt influence on coal behaviour and the significant strata separation occurred at the depth of 0–1.5 m into the pillar (see Fig. 6). In addition to absolute values of pillars displacement, data from horizontal extensometers provided important information about the dynamics of displacement of coal pillars.

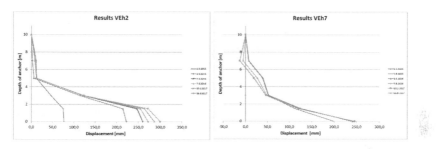

Figure 5.   Example of horizontal displacement indicated by rib extensometers in monitored locality A.

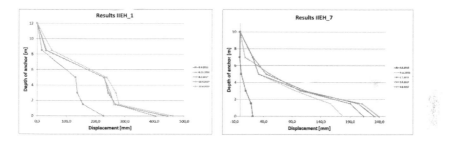

Figure 6.   Example of horizontal displacement indicated by rib extensometers in monitored locality B.

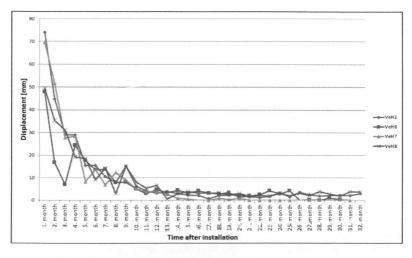

Figure 7.   Rate of the V2 coal pillar displacements versus time.

Concerning the dynamics, we could see a decrease of displacements during the whole evaluation period (see Fig. 7). The monthly gain of displacements reduced up to 25 times during monitoring period of 32 months in locality A. The monthly gain of displacements stabilised at 3 mm/month during the last eighteen month of monitoring period. Continuous deformation processes indicate, that yielding of coal pillars is still in progress therefore the long-term stability of coal pillars has not been established yet. Similar dynamic of displacement changes was recorded in locality B during monitoring period of 15 months only.

5 CONCLUSION

Evaluation from presented part of monitoring contribute to better knowledge of yielding assessment of coal pillar during room and pillar mining in great depth and difficult natural and mining conditions. Based on the long-term stress-deformation monitoring (up to 32 months), it was possible to determine the operating stability and long-term stability of coal pillars. Safe operating stability has been proven for all coal pillars formed within the trial operation of room and pillar method. The long-term stability of coal pillars have not been confirmed yet because the deformation processes of coal pillars are still in progress

The room and pillar method was trialled in the shaft protective pillar at the CSM Mine located in the Upper Silesian Coal Basin. Coal pillar monitoring was essential as this was the first application of the conventional room and pillar mining method in USCB mines. Based on the measurements, numerical modelling and other analyses were possible to assess stability of the coal pillars at the great depth. The results are also important for global mining, for the largest coal producers will reach higher mining depth in near future.

ACKNOWLEDGMENTS

This article was written in connection with the Project Institute of Clean Technologies for Mining and Utilization of Raw Materials for Energy Use—Sustainability Program (reg. no. CZ.1.05/2.1.00/03.0082 and MSMT LO1406), which is supported by the Research and Development for Innovations Operational Programme financed by the Structural Funds of the European Union and the Czech Republic project for the long-term conceptual development of research organisations (RVO: 68145535).

REFERENCES

Grygar, R and Waclawik, P, 2011. Structural-tectonic conditions of Karvina Subbasin with regard to its position in the apical zone of Variscan accretion wedge, Acta Montanistica Slovaca, Vol. 16, No. 2, pp. 159–175.
Obara, Y and Sugawara, K, 2003. Updating the use of the CCBO cell in Japan: overcoring case studies, International Journal of Rock Mechanics and Mining Sciences, Vol. 40, pp. 1189–1203.
Stas, L, Knejzlik, J, Palla, L, Soucek, K and Waclawik, P, 2011. Measurement of stress changes using a Compact Conical-ended Borehole Monitoring. Geotechnical Testing Journal, Vol. 34, No. 6, p. 685–693.
Stas, L, Soucek, K and Knejzlik, J, 2007. Conical borehole strain gauge probe applied to induced rock stress changes measurement. In Proceedings of 12th International Congress on Energy and Mineral Resources, p. 507–516.
Stas, L, Knejzlik, J, and Rambouský, Z, 2004. Development of conical probe for stress measurement by borehole overcoring method. Acta Geodyn. Geomater, Vol. 1, No. 4.
Waclawik, P, Snuparek, R and Kukutsch, R, 2017. Rock bolting at the room and pillar method at great depths. Symposium of the International Society for Rock mechanics, 20–22 June 2017, Procedia Engineering. Volume 191, pp. 575–582.
Waclawik, P, Ptacek, J, Konicek, P, Kukutsch, R, and Nemcik, J, 2016. Stress state monitoring of coal pillars during room and pillar extraction, Journal of Sustainable Mining, Vol. 15, Issue 2, pp. 49–56.
Waclawik, P, Ptacek, J and Grygar, R, 2013. Structural and stress analysis of mining practice in the Upper Silesian Coal Basin, Acta Geodyn. Geomater, Vol. 10, Issue 2, pp. 255–265.

Nonlinear problems in rock mechanics

Geomechanics and Geodynamics of Rock Masses – Litvinenko (Ed.)
© *2018 Taylor & Francis Group, London, ISBN 978-1-138-61645-5*

Ultimate bearing capacity analysis of foundation on rock masses using the Hoek-Brown failure criterion

A.A. Chepurnova

Gersevanov's Research Institute of Bases and Underground Structures (NIIOSP)—JSC Research Center of Construction, Moscow, Russia

ABSTRACT: Among many issues to be solved in the engineering structures design, one of the most controversial is estimation the strength and stability of rock masses. Rock mass is a discrete heterogeneous anisotropic medium, which properties are rather difficult to reflect correctly within the used models. In this paper, the possibility of determining the ultimate bearing capacity of foundation resting on naturally discontinuing rock mass with the Hoek-Brown failure criterion (Hoek, 2007) implemented in the geotechnical software Optum CE (Krabbenhoft, 2016) is presented. Numerical solutions of the bearing capacity are performed by the finite element method with determination of the upper and lower bounds of the ultimate load. The results show good agreement with analytical solution results. Alternatively, an approach to estimate the bearing capacity of rock masses by fitting the linear Mohr-Coulomb relationship to the curved Hoek-Brown solution is observed.

1 INTRODUCTION

Rock masses, in general case, are characterized by substantial heterogeneity at the microscopic and macroscopic level, anisotropy of properties, multiphase. Herein, estimation of a fractured rock mass strength is considered, with no dominant systems of cracks and inhomogeneities of a known direction, but multiple fracturing allows consider it as a homogeneous isotropic mass which properties are different from intact structure.

In this case, analyzing the bearing capacity of closely fractured or very weak rock mass can be performed in accordance with the fracture mechanism, similar to soil mechanics (Wyllie, 1999, Zertsalov, 2014). The simplified analysis assumes straight lines for the failure surfaces and ignores the weight of the rock in the foundation as well as the shear stresses that develop along the vertical interface between two wedges. Shear strength parameters along vertical shear planes can be assumed the same as for a rock mass. The analysis is based on the assumption that active and passive pressure wedges (zones) are formed in the rock under the footing, defined by straight lines, and the shear strength parameters of these surfaces corresponds to those of the rock mass.

Figure 1 shows a strip footing bearing on a horizontal rock surface under conditions of plane deformation; zone A experiences a triaxial contraction. The major principal stress in zone A is determined by the footing pressure, if the weight of rock beneath the footing is neglected with the following relation:

$$\sigma_{1A} = q_u \tag{1}$$

Zone B also undergoes a triaxial compression with the major principal stress acting horizontally, and the minor principal stress acting vertical; with the foundation position at the ground surface $\sigma_{3B} = 0$. At the moment of foundation failure both zones shear simultaneously

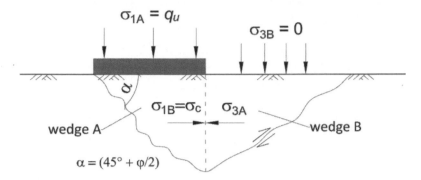

Figure 1. Analysis of bearing capacity of rock mass (after Wyllie, 1999).

and the minor principal stress in zone A, σ_{3A} equals the major principal stress in zone B, σ_{1B} where passive pressures are realized. The minor principal stresses in zone A is produced by the resistance of zone B to be compressed and equal to the compressive strength of the rock mass σ_c.

There are several approaches to determining the strength of a rock mass, in this paper the empirical failure criterion of Hoek-Brown, which takes into account fractures of rock and incorporated into the Optum CE software, is considered.

2   HOEK-BROWN FAILURE CRITERION

The Hoek-Brown Failure Criterion, HBFC, (Hoek, 2007) is an example of a nonlinear criterion for shear strength, developed specifically for fractured rock mass. The starting point for the criterion was the Griffiths theory for brittle fracture, however, the process of deriving the criterion was based on trial and error method. The empirical criterion is obtained as a result of processing triaxial tests of rock samples. As a result of the almost twenty-year history of the criterion development, a nonlinear relationship between the major and minor principal stresses was recorded, written in the following form:

$$\sigma_1 = \sigma_3 + \sigma_{ci}\left(m_b \cdot \frac{\sigma_3}{\sigma_{ci}} + s\right)^{\alpha} \tag{2}$$

where σ_{ci} is the uniaxial compressive strength of the intact rock, in MPa, m_b, s, α are constants for the rock mass which depends on the rock type, its quality and in the process of developing the criterion undergone alteration. Table 1 shows the dependencies for their determination, developed for the criterion version of 1997 and 2002 years.

In the above mentioned equations of Table 1: m_i is the material constant (i = intact), GSI is the Geological Strength Index; D is a factor that depends on the degree of disturbance to which the rock mass has been subjected by blast damage and stress relaxation and vary from D = 0 for undisturbed in situ rock mass to D = 1 for disturbed rock mass properties. The GSI was introduced by Hoek and in many respects is similar to the Rock Mass Raiting (RMR) developed by Z. Bieniawski (Bieniawski, 1976). Parameter D is introduced into the criterion for a smoother transition from very good quality to extremely poor ($GSI < 25$). An intact rock mass is counted with parameters: $m_b = m_i$, $s = 1$, $\alpha = 0,5$.

In general, the criterion parameters m_i, GSI, D are empirical constants assigned in accordance with experimental field and laboratory data, which purpose is to relate the physical and mechanical properties of rock mass to its structural discontinuities, degree of fracturing, etc. Its determination in accordance with authors (Sas, 2015) in accordance with the requirements of the Russian' Code of Regulation is possible to achieve, and is not the subject of discussion in the article.

Table 1. Estimation of Hoek-Brown parameters.

Parameter		Generalized HBFC, Hoek et al., 1997	Generalized HBFC, Hoek et al., 2002
m_b		$m_b = m_i \, exp\left(\dfrac{GSI - 100}{28}\right)$	$m_b = m_i exp\left(\dfrac{GSI - 100}{24 - 14D}\right)$
s	$GSI > 25$	$s = exp\left(\dfrac{GSI - 100}{9}\right)$	$s = exp\left(\dfrac{GSI - 100}{9 - 3D}\right)$
	$GSI < 25$	$s = 0$	
α	$GSI > 25$	$\alpha = 0{,}5$	$\alpha = \dfrac{1}{2} + \dfrac{1}{6}\left(e^{-GSI/15} - e^{-20/3}\right)$
	$GSI < 25$	$\alpha = 0{,}65 - \dfrac{GSI}{200}$	

The unconfined compressive strength is obtained by setting $\sigma_3 = 0$ in Eq. (2), giving

$$\sigma_c = \sigma_{ci} s^{\alpha} \qquad (3)$$

The tensile strength at $\sigma_1 = 0$ from (2) is expressed as follows:

$$\sigma_t = -\frac{s\sigma_{ci}}{m_b} \qquad (4)$$

It is important to note again that the HBFC assumes isotropic rock and rock mass behavior and extends to those rock masses in which there are a sufficient number of closely spaced discontinuities with similar surface characteristics. Thus, it is possible to assume isotropic behavior and failure through discontinuities. Where the block size is of the same order as the analyzed system (i.e. "footing-rock mass"), or when one of the discontinuities sets is significantly weaker than the others, the HBFC is not applicable (Hoek, 2007). In these cases, the stability of the structure should be analyzed by considering failure mechanisms associated with sliding or rotation of blocks and wedges defined by intersecting structural features.

Despite the popularity of the HBFC and its obvious advantages, practical analysis of rock masses (using numerical methods) in most cases is carried out using a linear Mohr-Coulomb failure criterion. Equivalent Mohr-Coulomb strength parameters c', φ' of rock with specified characteristics (σ_{ci}, m_i, GSI, D) can be obtained by fitting a linear relationship to the curved generated by Eq. 2 for a range of minor principal stress values defined by $\sigma_t < \sigma_3 < \sigma'_{3max}$ (Hoek, 2007). This results in the following equations:

$$\varphi' = \sin^{-1}\left[\frac{6\alpha m_b(s + m_b\sigma'_{3n})^{\alpha-1}}{2(1+\alpha)(2+\alpha) + 6\alpha m_b(s + m_b\sigma'_{3n})^{\alpha-1}}\right] \qquad (5)$$

$$c' = \frac{\sigma_{ci}\left[(1+2\alpha)s + (1-\alpha)m_b\sigma'_{3n}\right](s + m_b\sigma'_{3n})^{\alpha-1}}{(1+\alpha)(2+\alpha)\sqrt{1 + \left(6\alpha m_b(s + m_b\sigma'_{3n})^{\alpha-1}\right)/\left((1+\alpha)(2+\alpha)\right)}} \qquad (6)$$

$$\text{where } \sigma'_{3n} = \sigma'_{3n}/\sigma_{ci} \qquad (7)$$

Note, that precise recommendations for upper limit of confining stress (σ'_{3max}) for bearing capacity analysis are not given. In any case, stress state over which the relationship between the Hoek-Brown and the Mohr-Coulomb criteria is considered has to be determined for each individual case and particular problem. Thus, for example, an upper bound estimate of the stress state can be found from an elastic stress analysis for opening and slope (Sjöberg, 1997).

This approach gives slightly lower friction angle and slightly higher cohesion, depending on the curvature of the actual Hoek-Brown failure envelope.

For the generalized HBFC criterion written in the Eq. 2, from experience and trial and error, Hoek and Brown suggest a value of $\sigma'_{3max} = \sigma_{ci}/4$ that will provide consistent result (Hoek, 2007).

3 STATEMENT OF A PROBLEM

The plane strain bearing capacity problem of a strip footing of width B resting on jointed rock mass is illustrated in Fig. 2. The ultimate capacity can be expressed by analogy with formula (1) by the following relationship:

$$q_u = N_{\sigma_0} \cdot \sigma_{ci} \qquad (8)$$

where $N_{\sigma 0}$ is a dimensionless bearing capacity factor, depending on the values of the weight-less ($\gamma = 0$) Hoek-Brown material m_i, GSI, D. Accordingly, the HBFC implies an increase in the bearing capacity with increasing values of the parameters m_i, GSI (for $\sigma_{ci} = const$). Problem formulation in Eq. (8) is a convenient way of expressing the ultimate bearing capacity as a function of uniaxial compression strength of a sample and agrees with the work of other authors. The strength of rock mass is determined by the Hoek-Brown criterion (2), which establishes the major principal stress, indicating the intensity of the foundation pressure on the base. For the case where footing is located on the rock surface, the minor principal stress at the moment of failure is determined by Eq. (3).

The failure mechanism of rock masses is determined in most cases by discontinuities and their location, which allows to be conditionally assigned into three structural groups schematically shown in Fig. 2 (Merifield, 2006). Rock masses suitable for the description of group I (intact tock) and group III (heavily jointed with "small spacing" between discontinuities so that, on the scale of the problem, it can be regarded as an isotropic assembly of interlocking particles) are applicable to the Hoek-Brown failure criterion and considered as a homogeneous and isotropic mass.

Further, the ultimate bearing capacity of rock herein is estimated for a practical range of values m_i, GSI at $\sigma_{ci} = 10$ MPa, $\gamma = 0$, $D = 0$. It should be noted that in connection with some uncertainties in the Hoek-Brown model parameters designation including lack of certain recommendations for their determination and taking into account national peculiarities,

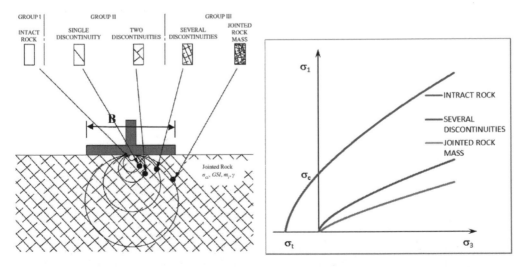

Figure 2.   Hoek-Brown failure criterion applicability for shallow foundations (after Merifield, 2006).

analysis of rock masses stability and bearing capacity in the Russian Federation more often is produced by means of the Mohr-Coulomb model. In this connection, ultimate bearing capacity of rock is also determined through the equivalent strength parameters c', φ', defined for two ranges of the minor principal stress. The upper estimate of the ultimate load capacity is given for $0 < \sigma_3 < \sigma^e_{3max}$ (hereinafter referred to as MC (1)), and the lower limit for $0 < \sigma_3 < 0.25\sigma_{ci}$ (denoted as MC (2)).

4 HOEK-BROWN SOIL MODEL IN OPTUM CE

The Optum CE is a relatively new geotechnical software complex appeared on the international market. The plain strain version G2 (Krabbenhoft, 2016) is available for free download from the web-site http://optumce.com/, 3D beta version is under active testing now. Among the variety of geotechnical problems OptumG2 solves the system of equations of limiting equilibrium method. The key features of the computational core of the program are:

- Strength condition is considered in the inequality mode;
- Mathematics optimization problem is solved, namely, applied load is maximized if equality and inequalities are observed;
- Implementation of an arbitrary set of boundary conditions;
- Automatic adaptive mesh refinement to maximize accuracy while keeping the computational cost at a minimum (see Fig. 3).

Concerning the limit analysis of the strip footing in general, OptumG2 implements features that are fundamentally different from other finite element programs, namely:

1. A cohesion identically equal to zero can be used (actual for cohesionless soils);
2. Definition of the upper and lower bounds, which gives a direct measure of the error in the numerical solution. In addition, it is often observed that the mean between the upper and lower bounds gives a good estimate of the exact solution—even if the gap between the boundaries is significant;
3. The singularity at the footing edge may be handled using a special tool (so-called the Mesh Fan). This feature constructs a fan of elements around the singularity which often leads to improved solutions, especially for lower bound elements, which leads to more precise solutions, especially for the elements of the lower boundary (Krabbenhoft, 2016).

During verification OptumG2, the program developers simulated various situations, including the problem of determining the ultimate bearing capacity of rock using the HBFC (2). The maximum strength of rock foundation is determined by calculating the upper and lower bounds of the footing pressure for the range of rock parameters of the Hoek-Brown

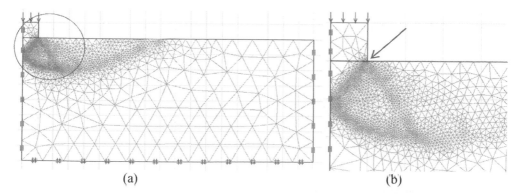

(a)   (b)

Figure 3.   Limited analysis solution for a strip footing resting on rock mass in OptumG2: mesh adaptation to calculate the lower bound (a) and an enlarged fragment of a Mesh Fan at footing edge (b).

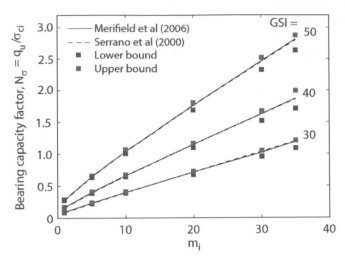

Figure 4. Bearing capacity factors for strip footing on weightless Hoek-Brown material (Krabbenhoft, 2016).

model m_i, *GSI* at $\gamma = 0$, $\sigma_{ci} = 10$ MPa (Krabbenhoft, 2016). Limited analysis solution is shown in Fig. 3. The results are presented as a relation of bearing capacity factors $N_{\sigma 0}$ on parameter of undistributed material m_i for geological strength index range *GSI* = 30,40,50 and are shown in Fig. 4. As can be seen from the graph, numerical results determined with OptumG2 are within the close proximity to the exact solutions (obtained by Merifield and Serrano).

Numerical simulation of limit analysis in OptumG2 using the Hoek-Brown model resulted, as expected, in an increase in the ultimate bearing capacity for a given *GSI* with increasing m_i. The Fig. 4 indicates that the bearing capacity factor $N_{\sigma 0}$ increases non-linearly with m_i and *GSI*. The upper bounds tend to be somewhat more accurate than the lower bounds and the accuracy decreases with increasing material strength. Calculation results of rock with lower value of m_i tend to be more accurate.

5 RESULTS OF NUMERICAL SIMULATION

In OptumG2, as well as in some others, for example, FLAC (Itasca, 2015), is implemented algorithm that allow to "convert" a curved Hoek-Brown failure envelop to a linear Mohr-Coulomb envelop by determining the tangent to the curved envelop at the calculation stress in each element and for each calculation step. The tangent at each section has its own pair of equivalent strength parameters (5) and (6), which vary throughout the model. However, for many practical applications it is necessary to approximate the curved envelop to with one set of equivalent strength parameters c', φ' or the straight line Mohr-Coulomb envelope. Therefore, a methodology for determining one set of such parameters is needed.

Further, Table 2 shows the results of rock strength (in terms of bearing capacity factor $N_{\sigma 0}$ from (8)) obtained in OptumG2 for various parameter group of the Hoek-Brown model (m_i, *GSI*) and for the Mohr-Coulomb model with equivalent strength parameters defined for $0 < \sigma_3 < \sigma_{3max}^e$ where σ_{3max}^e is a maximum value of the minor principal stress determined by an elastic analysis (MC (1)); and for $0 < \sigma_3 < 0.25\sigma_{ci}$ (MC (2)). Bearing capacity factor for three serious of calculation is carried out by determining the upper and lower boundaries and Table 2 shows the mean values of $N_{\sigma 0}$ (note, that the error in determining the mean is within 4%).

Fig. 5a shows the ultimate bearing capacity solutions of fractured rock *GSI* = 25 for different values of m_i, obtained for both the Hoek-Brown (solid line) and Mohr-Coulomb (dashed lines) criteria. Strength envelops for the same parameters are shown in Fig. 5b.

Table 2. Results for the bearing capacity factor $N_{\sigma 0}$ of the fractured weightless rock.

Hoek-Brown			Mohr-Coulomb					
		$N_{\sigma 0}$ OptumG2 (Krabbenhoft, 2016)	MC (1) $0 < \sigma_3 < \sigma_{3max}^e$		$N_{\sigma 0}$ OptumG2 (Krabbenhoft, 2016)	MC (2) $0 < \sigma_3 < 0.25\sigma_{ci}$		$N_{\sigma 0}$ OptumG2 (Krabbenhoft, 2016)
GSI	m_i		c', kPa	φ', degree		c', kPa	φ', degree	
10	1	0,022	21	11,45	0,019	−15%	63	5,96	0,043	91%
	7	0,283	254	12,46	0,243	−14%	163	15,42	0,184	−35%
	10	0,392	373	13,21	0,371	−5%	192	17,98	0,251	−36%
	15	0,556	570	14,2	0,599	8%	229	21,19	0,367	−34%
	17	0,623	649	14,51	0,694	11%	242	22,24	0,415	−33%
	25	0,857	966	15,55	1,097	28%	284	25,65	0,616	−28%
25	1	0,117	111	9,96	0,093	−21%	125	9,16	0,100	−14%
	7	0,616	678	12,35	0,644	5%	275	20,35	0,418	−32%
	10	0,794	927	13,05	0,915	15%	315	23,07	0,571	−28%
	15	1,060	1312	13,94	1,361	28%	336	26,38	0,768	−28%
	17	1,174	1475	14,14	1,545	32%	383	27,44	0,945	−20%
	25	1,504	2040	15,08	2,253	50%	439	30,8	1,409	−6%
50	1	0,380	407	9,51	0,332	−13%	273	13,67	0,279	−27%
	7	1,265	1664	13,29	1,664	32%	447	27,53	1,112	−12%
	10	1,625	2222	13,94	2,304	42%	498	30,52	1,562	−4%
	15	2,144	3172	14,45	3,383	58%	563	34	2,374	11%
	17	2,358	3444	14,87	3,759	59%	585	35,08	2,716	15%
	25	2,990	4644	15,76	5,331	78%	656	38,45	4,179	40%
65	1	0,669	778	9,39	0,630	−6%	505	15,88	0,584	−13%
	7	1,948	2712	13,8	2,790	43%	596	31,76	2,067	6%
	10	2,469	3618	14,4	3,846	56%	643	34,89	2,924	18%
	15	3,320	5056	15,07	5,581	68%	707	38,47	4,510	36%
	17	3,666	5647	15,21	6,281	71%	729	39,57	5,205	42%
	25	5,007	7932	15,67	9,096	81%	804	42,93	8,306	66%

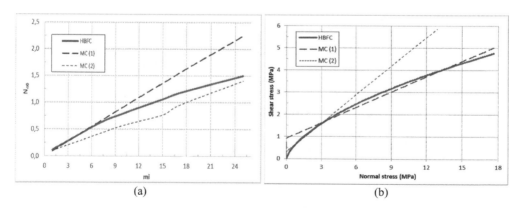

(a)   (b)

Figure 5. Bearing capacity factor $N_{\sigma 0}$ of fractured rock determined in OptumG2 (a) and a plot of normal-shear stresses (based on the values from RocLab) (b) simulated using the HBFC of $m_i = 10$, $GSI = 25$ and Mohr-Coulomb equivalent parameters of MC (1) and MC (2).

As can be seen from Table 2, determination of the ultimate bearing capacity of rock masses through equivalent Mohr-Coulomb parameters, given by one set of compression stress, provides overestimated values with increasing GSI, m_i. Wherein definition of equivalent parameters by limiting the minor principal stresses $0 < \sigma_3 < \sigma_{3max}^e$ (MC (1)) almost always gives too high results. Nevertheless, the second approach (MC (2)) can be considered as the lower limit for determining the ultimate bearing capacity for $GSI \leq 25$. In other cases, the error in determining the bearing capacity is up to + 66% (except for $m_i = 1$, $GSI = 10$ where the error is 91% and need to be specified). It should be noted, that the error of determining bearing capacity for the Mohr-Coulomb model in OptumG2 was also verified using the exact solution implemented by Martin C. in his program (ABC) and shows correlation with good proximity (not presented in this article).

6 CONCLUSIONS

The Hoek-Brown failure criterion was used to determine the ultimate pressure on a rock foundation using the finite element method realized by the original solution in OptumG2. Upper and lower bounds of the bearing capacity factor $N_{\sigma0}$ for weightless media are in good agreement with the results of the exact solutions obtained by Merifield, 2006 and Serrano, 2000. Ignoring the self-weight of rock gives a slightly conservative estimate of the bearing capacity.

It is established, that estimation of the ultimate bearing capacity of rock mass using one set of equivalent strength parameters of the Mohr-Coulomb failure criterion can significantly distort the actual bearing capacity. One of the reasons for this may be called ambiguity in limiting the minor principal voltage. The limitation of the minor principal stress as a function of the uniaxial compression strength $\sigma'_{3max} = \sigma_{ci}/4$ gives a lower estimate of rock strength for $GSI \leq 25$.

Numerical results of bearing capacity estimation by means of the Hoek-Brown failure criterion for fractured rock masses are well corresponded with analytical solutions in all range of material parameters.

REFERENCES

Bieniawski Z.T. Rock mass classification in rock engineering, 1976. Symposium: Exploration for rock engineering, vol. 1. Cape Town, p. 97–106.
Hoek E., Marinos P. A brief history of the development of the Hoek-Brown failure criterion, 2007.
Itasca. FLAC version 8. Manual, 2015. Minneapolis.
Krabbenhoft K., Lyamin A., Krabbenhoft J. OptumG2, 2016. http://optumce.com.
Merifield R., Lyamin A., Sloan S. Limit analysis solutions for the bearing capacity of rock masses using the generalised Hoek–Brown criterion, 2006. Int. J. Rock Mech. & Min. Sci. Vol. 43, pp. 920–937.
Sas I.E., Bershov A.V. On features of the Hoek-Brown rock behaviour model and determining its initial parameters, 2015. Inzhenernie iziskaniya, №13, pp. 42–47 (in Russian).
Sjöberg J., Estimating rick mass strength using the Hoek-Brown failure criterion and rock mass classification—a review and application to the Aznalcollar pit, 1997. Internal report of division of rock mechanics, Lulea University of Technology, p. 49.
Wyllie D. Foundations on rock, 1999. E&FN SPON, London and New York, p. 401.
Zertsalov M.G. Geomechanics. Introduction into the rock mechanics, 2014. ACB, Moscow, p. 351. (in Russian).

Geomechanics and Geodynamics of Rock Masses – Litvinenko (Ed.)
© 2018 Taylor & Francis Group, London, ISBN 978-1-138-61645-5

Conception of highly stressed rock and rock mass—as the step to theory of hierarchical cracking mesostructures

M.A. Guzev & V.V. Makarov
Far Eastern Federal University, Primorskii krai, Vladivostok, Russia

V.N. Odintsev
Research Institute of Comprehensive Exploitation of Mineral Resources RAS, Moscow, Russia

ABSTRACT: Conventional geomechanics bases on principles of the classical theory of continuum mechanics. However, new experimental results such as zonal disintegration around deep openings and reversible deformations of highly stressed rock samples cannot be described accurately in its terms. A new approach to geomechanical mathematical models involves non-Euclidian modelling to describe abnormal experimental findings. As a result, a new concept of the theory of geomechanics, i.e. geomechanics of highly compressed rock and rock mass is developed that improves considerably prediction of geomechanical phenomena. This paper describes principles of highly compressed rock and rock mass geomechanics. The use of non-Euclidian modelling is demonstrated at two hierarchical levels of block geomedium, i.e. a rock sample and rock mass around the openings.

Keywords: high compression, reversible deformation, zone failure, non-Euclidian model

1 INTRODUCTION

Experimental study of deformation relationships in rock and rock mass under high (greater than half of strength) compressing stresses have demonstrated two abnormal effects, such as reversible deformation of rock samples [1–3] and zonal deformation and failure of rock mass around underground openings [4–6].

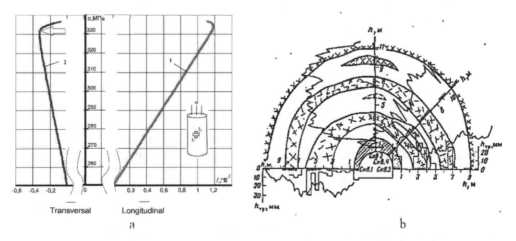

Figure 1. Abnormal phenomena of highly stressed rock and rock mass: a, reversible deformation of rock samples [3]; b, zonal failure around deep underground openings [4].

These two new phenomena cannot be explained in terms of classical geomechanics. The attempts of theoretical description of individual effects were not a success [4, 5]. There is still no general concept to explain all experimental data from a common position.

Rock mass failure under high stress can be described as thermodynamically non-equilibrium processes [7]. On the other hand, methods of the gauge theory for representation of defective continuums were introduced into the mechanics of continuous media, which allowed transition to a new class of models where the defective media were replaced by equivalent continuous ones in order to introduce the statements about non-Euclidian model space [8].

Investigation of defect formation processes, in turn, led to emerging of mesomechanics that focused on formation and development of various mesostructures, which were areas of preparation of macrodefects [9–11]. This approach described rock mass at a great depth as an open thermodynamically non-equilibrium defective medium [12–14]. Therefore, development of new models of rock mass at a great depth requires a theoretical basis.

2 NON-EUCLIDIAN MODEL TO SIMULATE THE GEOMEDIUM ABNORMAL PHENOMENA

The approach has been realized in non-Euclidian model of the zonal failure phenomena at the hierarchical scale level of the mining opening in block geomedium [15]. The zonal structure of the stress field around a cylindrical opening is considered under conditions of plane strain and stationarity. For the non-Euclidean model, the classical equations of equilibrium and the boundary conditions remain true:

$$\frac{\partial \sigma_{ij}}{\partial x_j} = 0. \tag{1}$$

In this model, the stress σ_∞ is specified at infinity under the condition $\sigma_r = 0$ at the boundary $r = r_0$. According to linearity of the equations (1) it is possible to present the required solution as the sum of classical Σ_{ij} field and additional field T_{ij}:

$$\sigma_{ij} = \Sigma_{ij} + T_{ij}. \tag{2}$$

The T_{ij} field is reflected by the parameter of incompatibility R. In a plane strain approximation, for the deep opening with a circular cross section we have R-function:

$$\frac{R}{2} = 2\frac{\partial^2 \varepsilon_{12}}{\partial x^1 \partial x^2} - \frac{\partial^2 \varepsilon_{11}}{\partial x^2 \partial x^2} - \frac{\partial^2 \varepsilon_{22}}{\partial x^1 \partial x^1}, \tag{3}$$

that characterizes the incompatibility of the components ε_{ij}. In polar coordinates we have the following expression for the stress components [15]:

$$\sigma_r = \sigma_\infty\left(1 - \frac{r_0^2}{r^2}\right) + (\gamma + \beta - \alpha)\frac{\partial^2 R}{\partial r^2} + \frac{\gamma}{r}\frac{\partial R}{\partial r}, \quad \sigma_\theta = \sigma_\infty\left(1 + \frac{r_0^2}{r^2}\right) + (\gamma + \beta)\frac{\partial^2 R}{\partial r^2} + \frac{\gamma}{r}\frac{\partial R}{\partial r}, \tag{4}$$

where α, β, γ are constant coefficients.

This model (1)–(4) can describe the periodicity of the stress field around a cylindrical opening. The results of boundary problem solution give formulas for the stress components:

728

$$\sigma_{rr} = \sigma_\infty \left(1 - \frac{r_0^2}{r^2}\right) - \frac{E}{2\left(1-\nu^2\right)\gamma^{3/2}} \cdot \frac{1}{r} \times \left[aJ_1\left(\sqrt{\gamma}r\right) + bN_1\left(\sqrt{\gamma}r\right) + cK_1\left(\sqrt{\gamma}r\right)\right];$$

$$\sigma_{\varphi\varphi} = \sigma_\infty \left(1 + \frac{r_0^2}{r^2}\right) - \frac{E}{2\left(1-\nu^2\right)\gamma} \times \left[aJ_0\left(\sqrt{\gamma}\cdot r\right) + bN_0\left(\sqrt{\gamma}\cdot r\right) - cK_0\left(\sqrt{\gamma}\cdot r\right)\right] + \tag{5}$$

$$+ \frac{E}{2\left(1-\nu^2\right)\gamma^{3/2}} \cdot \frac{1}{r}\left[aJ_1\left(\sqrt{\gamma}\cdot r\right) + bN_1\left(\sqrt{\gamma}\cdot r\right) + cK_1\left(\sqrt{\gamma}\cdot r\right)\right],$$

where r is the radius-vector, which characterizes the zones position.

Calculation with the failure criteria [20] demonstrates that the zone characteristics depend but slightly on the deformational parameters in the model E, ν (elastic modulus and Poisson ratio). The model was shown effective on a broad range of properties of solid rock mass ($\sigma_c = 150$ MPa) and for weak rock ($\sigma_c = 15$ MPa). We developed algorithms and programs to calculate expressions of incompatibility $R(r)$, stresses, and the criterion function $K(r)$ (see Makarov et al. in the EUROCK 2018 "Proceedings...").

Experimental results of zonal failure were obtained at the Nikolaevsky ore mine (Dalnegorsk, Russia). The correlation between theory and experiment findings was assessed by comparing *in situ* measurements of the radial displacements near the openings at great depth (Nikolaevsky ore mine) with the model calculations (Fig. 2). As demonstrated by the comparison, the difference between the predicted and measured values was not more than 47%.

Comparison of analytical and experimental findings for weak rock also showed good compliance. The analysis for both weak and solid rock demonstrates that the number and radial width of the failure zones increase with depth.

The abnormal strains in rock samples (Fig. 1a) can also be described by the non-Euclidian model [16]. The extensive field and laboratory experimental research at some world laboratories allow us to make the conclusion about inhomogeneous space-time distribution of longitudinal and lateral deformations of local fields of the sample (Fig. 3).

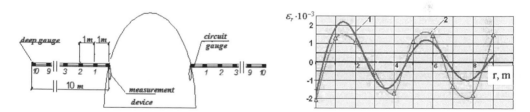

Figure 2. Comparison of theoretical (1) and experimental (2) values of radial deformations near the opening (right) and research station (left).

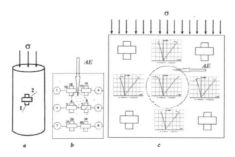

Figure 3. Mesocrack structure of the source type: a, scheme of loading (1, 2, linear deformation gauges); b, source area location (1–9, AE sensors); c, reversible deformation location in the near source area (AE).

The boundary problem of determining the stress distributions σ_{ij} in a cylindrical sample is formulated in stationary formulation (1). Let external force $\boldsymbol{P} = (0,0,P)$ act on sample ends, then in the balance, it is compensated by the force of internal stresses, and boundary conditions are satisfied at the ends $(S_\pm)$ and on the lateral surface (Γ) of the sample:

$$\int_{S_\pm} \sigma_{ij} n_j dS = 0, \quad \int_\Gamma \sigma_{zj} n_j dS = \int_\Gamma P dS. \tag{6}$$

The requirements (6) are satisfied under the following conditions:

$$\Sigma_{zz}\big|_{z=\pm h} = P, \quad \Sigma_{zr}\big|_{z=\pm h} = 0, \quad \Sigma_{z\varphi}\big|_{z=\pm h} = 0,$$
$$\Sigma_{rr}\big|_{r=R} = 0, \quad \Sigma_{r\varphi}\big|_{r=R} = 0, \quad \Sigma_{rz}\big|_{r=R} = 0, \tag{7}$$

$$\int_0^R r dr \int_0^{2\pi} d\varphi T_{iz}\bigg|_{z=\pm h} = 0, \quad \int_{-h}^h dz \int_0^{2\pi} d\varphi T_{ir}\bigg|_{r=R} = 0, \tag{8}$$

where R is the radius, $2h$ is the height of the sample. The conditions (7) comply with classical ones locally in each point. It follows from (8) that components T_{ij} are self-balanced fields.

To construct T_{ij} one has to introduce stress function g:

$$T_{ij} = \delta_{ij}\Delta g - \frac{\partial^2 g}{\partial x^i \partial x^j} \tag{9}$$

As well known, the classical theory requires $\Delta g = 0$ (here Δ is the Laplace operator), then T_{ij} satisfies the compatibility conditions. The non-Euclidian model results in equation

$$\Delta g = -\gamma g \tag{10}$$

with a phenomenological parameter γ. In the cylindrical coordinates we have the following expression for the stress components (9):

$$T_{rr} = \frac{1}{r}\frac{\partial g}{\partial r} + \frac{1}{r^2}\frac{\partial^2 g}{\partial \varphi^2} + \frac{\partial^2 g}{\partial z^2}, \quad T_{r\varphi} = -\frac{\partial}{\partial r}\left(\frac{1}{r}\frac{\partial g}{\partial \varphi}\right), \quad T_{rz} = -\frac{\partial}{\partial r}\frac{\partial g}{\partial z},$$
$$T_{\varphi\varphi} = \frac{\partial^2 g}{\partial r^2}, T_{z\varphi} = -\frac{\partial}{\partial z}\left(\frac{1}{r}\frac{\partial g}{\partial \varphi}\right), \quad T_{zz} = \frac{\partial^2 g}{\partial r^2} + \frac{1}{r}\frac{\partial g}{\partial r} + \frac{1}{r^2}\frac{\partial^2 g}{\partial \varphi^2}. \tag{11}$$

The function g is expressed through cylindrical function $Z_n(kr)$, which is the solution of the equation (12). Then the components of field T_{ij} are given by formulae:

$$T_{rr} = \left(\frac{1}{r}\frac{dZ_n(kr)}{dr} - \frac{n^2}{r^2}Z_n(kr) - q^2 Z_n(kr)\right)\cos(n\varphi + \varphi_0)\cos qz,$$

$$T_{r\varphi} = n\frac{d}{dr}\left(\frac{Z_n(kr)}{r}\right)\sin(n\varphi + \varphi_0)\cos qz,$$

$$T_{rz} = q\frac{d}{dr}(Z_n(kr))\cos(n\varphi + \varphi_0)\sin qz, \tag{12}$$

$$T_{\varphi\varphi} = -\left(\frac{1}{r}\frac{dZ_n(kr)}{dr} - \frac{n^2}{r^2}Z_n(kr) + k^2 Z_n(kr)\right)\cos(n\varphi + \varphi_0)\cos qz,$$

$$T_{z\varphi} = -qnZ_n(kr)\sin(n\varphi + \varphi_0)\sin qz,$$

$$T_{zz} = -k^2 Z_n(kr)\cos(n\varphi + \varphi_0)\cos qz.$$

The parameters q, k cannot be arbitrary. The boundary condition at the ends for $T_{z\varphi}$, T_{zr} demands sin $qh = 0$. This results in a discrete set for the parameter q: $q_m = \pi m/h$, $m \geq 1$. The integrated condition (8) for T_{zz} contains cylindrical function $Z_0(kr)$. Then the condition may be written as

$$\int_0^R rdr \int_0^{2\pi} d\varphi T_{zz}\bigg|_{z=\pm h} = -2\pi(-1)^m k^2 \cos\varphi_0 \int_0^R rdr J_0(kr) = 2\pi k R J_1(kR)(-1)^{m+1}\cos\varphi_0 = 0. \quad (13)$$

It follows that the parameter k is defined by roots of the Bessel function $J_1(z)$. Comparison of experimental and analytical findings shows a full qualitative and maximum 23% quantitative difference [16]. The details see at (Golosov et al.) in these Proceedings.

The new approach to the modelling of rock mass at great depth requires new conceptual principles including principle of rock mass characterization, principle of modelling and description of the subject and method of a new conception.

3 THE BASIC CONSEQUENCES

The basic consequences of the experimental data analyses and the non-Euclidian approach to modelling may be formulated as the following scientific positions. The rock mass represents hierarchical block medium with crystal-like formations (minerals) located at the lowest level, and the highest level is limited to the scale of the objects considered within the crust. Dissipative mesocrack structure (DMS) is considered as a set of areas of active (increase) and passive (decrease) defect formation. At each hierarchical level, it is reasonable to consider three scale levels of defects, i.e. micro- meso- and macrodefects, which makes difficult the analysis of such media and construction of adequate mathematical models. Macrodefects of the lowest hierarchical level can be considered mesodefects at the next higher hierarchical level (macro-meso-transition).

Conception of "high compression of rocks and rock mass" characterizes formation and development of the dissipative mesocrack structures (DMS) of the corresponding hierarchical level. So, subject of the geomechanics of highly compressed rocks and rock mass is the rock phenomena generated by hierarchical rock mass at each level of geomedium, Method of geomechanics of 'highly compressed rocks and rock mass' is presentation of hierarchical block geomedium by a system of non-Euclidian models. Thus, we replace mesodefective structures at each hierarchical level of the defective medium by the models of the continuous medium. Therefore, main principle of geomechanics of highly compressed rock and rock mass is the principle of non-Euclidian hierarchy of geomedium, i.e. replacement of block hierarchical medium by continuous hierarchically structured medium with the non-Euclidian metrics, where transition between structural levels (models) is defined by condition criteria corresponding to the meso-macro-meso process. The conditions of compatibility of elastic deformations are not satisfied in each point in the defective medium, while conditions of compatibility of general deformations for the corresponding hierarchical level of blocks remain satisfied (the monolithic block principle).

Some conditions of the geomechanics of highly compressed rocks and rock mass applicability must be added to the principles described above. These are: the condition of the geomedium being far from thermodynamic balance, the condition of transference from one structural level of the hierarchical block medium to another, and the condition of criteria satisfaction [21–22].

4 CONCLUSION

The conception of the geomechanics of highly compressed rocks and rock mass has been developed on the basis of principles of non-Euclidian hierarchy of geomedium and the monolithic block. It was shown that abnormal phenomena of 'zonal disintegration' in rock mass and 'reversible deformation' of rock samples could be simulated successfully by the same model. And the theory of geomechanics of highly compressed rocks and rock mass can be constructed on this approach as an answer on the new experimental results, which are appeared to be as anomalous phenomena from position of temporary geomechanics.

REFERENCES

[1] Seldenrath, Th. R., J. Gramberg. (1958) Stress-strain relations and breakage of rocks. In: Mechanical Properties of Non-Metallic Materials. ed. Walton W.H.L., Butterworths, pp. 79–102.

[2] Guzev M. A, Makarov V.V. (2007) Deformability and failure of highly stressed rocks around openings, - Vladivostok: Dalnauka, 2007. – 231 P. (in Russian).

[3] Guzev M. A, Makarov V. V, Ushakov A.A. (2005) Modelling of elastic behavior of samples of the compressed rocks in predestroying area//Journal of Mining Science. № 6. pp. 3–13.

[4] Effect of zone disintegration of rocks round underground openings (1986)/Shemjakin E.I., Fisenko G. L, Kurlenja M. V, Oparin V. N, etc.//Proc AS. V. 289. № 5. pp. 1088–1094. (in Russian).

[5] Chanyshev А.И. (1988) To research of the phenomenon of zone disintegration of rocks//Intense-deformed condition of a rock massive. Novosibirsk: Mining Institute, pp. 3–8.

[6] Vladimir V. Makarov, Mikhail A. Guzev, Vladimir N. Odintsev, Lyudmila S. Ksendzenko (2016) Periodical zonal character of damage near the openings in highly-stressed rock mass conditions. J. Rock Mech. and Geotech. Eng. V. 8, N 2, pp. 164–169. doi:10.1016/j.jrmge.2015.09.010.

[7] Prigogine I., Kondepudi D. (1998) Modern thermodynamics. Published in 1998 by John Wiley & Sons Ltd, 1998, 506 P.

[8] A. Kadiʹc, D.G.B. Edelen, (1983) A Gauge Theory of Dislocations and Disclinations, In: Lectures Notes in Physics, Vol. 174, Springer, Berlin.

[9] Panin V.E. Structural levels of plastic deformation and destruction (1990) /V.E. Panin, J.V.Grinjaev, etc. Novosibirsk: Science, 1990. 255 p. (in Russian).

[10] The Physical mesomechanics and computer designing of materials: in 2 v. (1995) / under the editorship of V.E. Panin. – Novosibirsk: Science, V. 1. 297 p.; V. 2. 320 p. (in Russian).

[11] D.A. Lockner, J.D. Byerlee V. Kuksenko, A. Ponomarev, A. Sidorin (1991) Quasi-static fault growth and shear fracture energy in granite. Nature, V. 350, N 7, pp. 39–42.

[12] Guzev M.A, Miasnikov V.P. (1998) Thermomechanical model of elastic-plastic material with defects//News RAS. Mechanics of a solid body. 1998. № 4. pp. 156–172. (in Russian).

[13] C.R. De Groot, P. Mazur (1964) Non-Equilibrium Thermodynamics. – M: World, 524 p.

[14] Sadovsky M.A. (1979) Natural blocky of rock//Proc. AS USSR, V. 247, № 4. (in Russian).

[15] M.A. Guzev, (2014) Non-classical solutions of a continuum model for rock descriptions, Journal of Rock Mechanics and Geotechnical Engineering, v. 6, N. 3, pp. 180–185.

[16] Guzev, M., Makarov, V., & Ksendzenko, L. (2015, January). Non-Euclidean Model of High Stressed Rocks and Rock Masses. In *ISRM Regional Symposium-EUROCK 2015*. International Society for Rock Mechanics. – Salzburg, Austria, pp. 979–984.

[17] Goldin S.V. (2005) Macro - and mesostructures of source earthquake areas// Physical mesomechanics – V. 8, № 1. – pp. 5–14 (in Russian).

[18] Guzev M. A, Paroshin A.A. (2000) Non-Euclidian model of zone disintegration of rocks around underground openings//PMTF. № 3. pp. 181–195. (in Russian).

[19] Guzev M. A, Miasnikov V.P., Ushakov A.A. (2004) Field of the self-counterbalanced pressure in continuous//Acta media. V. 45, № 4. pp. 121–130.

[20] Odintsev V.N. (1996) Tension destruction of a massive of brittle rocks. M: IPKON, RAS.

[21] Makarov, VV, Ksendzenko, LS, Golosov, AM & Opanasiuk (2017) Mesocracking structures of the 'source type' in highly stressed rocks. In J Wesseloo (ed.), *Proceedings of the Eighth International Conference on Deep and High Stress Mining*, Australian Centre for Geomechanics, Perth, pp. 403–411. https://papers.acg.uwa.edu.au/p/1704_28_Makarov/.

[22] Makarov, V.V, Ksendzenko, LS, Opanasiuk, NA & Golosov, AM (2017). Zonal failure structure near the deep openings. In J Wesseloo (ed.), *Proceedings of the Eighth International Conference on Deep and High Stress Mining*, Australian Centre for Geomechanics, Perth, pp. 423–432. https://papers.acg.uwa.edu.au/p/1704_30_Makarov/.

Geomechanics and Geodynamics of Rock Masses – Litvinenko (Ed.)
© 2018 Taylor & Francis Group, London, ISBN 978-1-138-61645-5

Zonal type mesostructures around single openings in deep rock mass

Lyudmila S. Ksendzenko & Vladimir V. Makarov
Laboratory of Highly Compressed Rock and Rock Mass, Far Eastern Federal University, Vladivostok, Russia

ABSTRACT: The phenomenon of rock mass zonal failure (disintegration) has been known lately. For the safety mining operations it is necessary to know the regularities of zonal mesostructures behavior around deep mine openings. In this paper dependences from influencing factors of the first failure zone radial extent, of the depth of the contour zone and the first failure zone merging, and the depth of the second failure zone appearance are investigated. In that, the non-Euclidean model of a rock mass zonal failure around deep openings and the Odintsev's criterion of the failure under compression were used. The influencing factors are the uniaxial compression strength of the rock σ_c, MPa; the opening depth H, m; the relative rock tenseness $\gamma_r \times H/\sigma_c$; the ratio of empirical coefficients γ_3/γ_1; the values of the rocks elastic modulus, and the Poisson's ratio, v. The study was carried out for hard rocks. The regularities established for hard rocks can be shown to be also valid for low-strength rocks.

Keywords: rock mass, zonal failure, first zone of cracking, depth of zones merging

1 INTRODUCTION

The non-Euclidean model of a continuous medium turned out to be the most effective for studying this phenomenon (Guzev, and Paroshin, 2001; Guzev, and Makarov, 2007; Guzev, et al., 2015). Refusing the deformations compatibility condition, the authors were able to take into account the defects in the rock material internal structure. The ideas of the works were developed in the papers of Qian, and Zhou (2011), Zhou, and Shou (2013), Qian, Zho, & Xie (2012) and others. Zonal type mesostructure in the rock mass is a collection of alternating disturbed and relatively undisturbed zones, surrounding deep opening. Investigation of zonal mesostructures in a deep rock mass was carried out by Makarov, Guzev, et al. (2015, 2016, 2017) in close connection with the study of the sample source type mesostructures, which are formed, when the load on the sample reaches a certain critical value.

A rock mass under conditions of high compression is modeled by a continuous medium with defects. The gauge field is constructed (Guzev, et al., 2015) from the conditions of thermodynamic irreversibility of plastic deformations caused by the failure.

The macro-crack of the tearing off is formed as a result of mesodefects confluence in the areas in which conditions of the Odintsev's fracture criterion (1996)—tearing off under compression—are carry out.

Under these assumptions, the mechanism of rock mass zonal failure phenomenon around underground openings was established by Makarov, & Guzev, (1999), and also the boundary-value problem for the mechanics of defective media was solved. A good correspondence between the results of theoretical studies and data of laboratory experiments VNIMI Institute was obtained.

Despite the advances in the modeling of zonal failure (disintegration) phenomenon (Qian, Zhou, Shou, Xie, and et al., 2011, 2012, 2013), the proposed models do not specify methodology for the model parameters determining, which would allow to study zonal mesostructures behavior patterns and predict rock mass behavior around the deep underground openings.

2 MATHEMATICAL MODEL OF ZONAL TYPE MESOSTRUCTURES

Mathematical model of the rock mass zonal failure at a great depth and the method for determining its parameters are presented in (Makarov, Guzev, et al., 2014, 2016). The task is considered to be as flat and stationary under the assumption of hydrostatic loading of an extended underground single cylindrical opening of a round cross section. The stress components in the rock mass around the underground mine opening are determined by the equalities (Makarov, et al., 2016, 2017):

$$
\sigma_r = \sigma_\infty\left(1 - \frac{r_0^2}{r^2}\right) - \frac{E}{2(1-v^2)\gamma^{1/2}} \cdot \frac{1}{r}\left[aJ_1\left(\sqrt{\gamma}\cdot r\right) + bN_1\left(\sqrt{\gamma}\cdot r\right) - cK_1\left(\sqrt{\gamma}\cdot r\right)\right],
$$

$$
\sigma_\theta = \sigma_\infty\left(1 - \frac{r_0^2}{r^2}\right) - \frac{E}{2(1-v^2)\gamma}\left[aJ_0\left(\sqrt{\gamma}\cdot r\right) + bN_0\left(\sqrt{\gamma}\cdot r\right) - cK_0\left(\sqrt{\gamma}\cdot r\right)\right] \tag{1}
$$

$$
+ \frac{E}{2(1-v^2)\gamma^{1/2}} \cdot \frac{1}{r}\left[aJ_1\left(\sqrt{\gamma}\cdot r\right) + bN_1\left(\sqrt{\gamma}\cdot r\right) - cK_1\left(\sqrt{\gamma}\cdot r\right)\right],
$$

where $\sigma_\infty = \gamma_r \times H$, γ_r is the unit weight of rock (kN/m³), and H is the opening depth(m); r_0 is the radius of opening (m); r is the distance from the center of opening to the current point of the rock mass (m); E is the modulus of elasticity (MPa); v is the Poisson's ratio; γ, c are the model parameters (m⁻²); J_0, N_0, K_0, J_1, N_1, K_1 are the Bessel, Neumann, Macdonald functions of zero and first orders, respectively. To determine the location of the failure zones around the underground opening, a force criterion of cracking under compression is used (Odintsev, 1996):

$$
K_1 = (\pi l)^{1/2}\left(\gamma_1\sigma_1^0 - \gamma_3\sigma_3^0\right) \le K_{1c}, \tag{2}
$$

where l is the half-length of the rock mass fracture faults and is assumed to be equal to the minimum half-length of a tensile macrocrack (m), which his unstable in stress conditions; σ_1^0 and σ_3^0 are the maximum and minimum principal stresses, respectively (MPa); γ_1 and γ_3 are the empirical factors (according to Odintsev, 1966, $\gamma_3/\gamma_1 = 0.8 - 0.9$); K_1 is the coefficient of stress intensity (МПа·м$^{1/2}$); K_{1c} is the fracture toughness of rock material (МПа·м$^{1/2}$). The stress intensity factor values in the rock mass at the time preceding the first failure zone formation are determined from the relation

$$
K_1(r) = \sqrt{\pi l_{\text{mezo}}^{\text{mass}}}\left(\gamma_1\sigma_\theta - \gamma_3\sigma_r\right) \le K_{1c}^{\text{mass}}, \tag{3}
$$

here the stresses σ_θ and σ_r are determined by the relations (1). We introduce the criterial function $K_r(r) = K_1(r)/K_{1c}$. If $Kr(r) <1$, then a fracture process at the meso-level develops around the opening; if $Kr(r) \ge 1$, formation of macro-cracks begins around the opening. The moment of the first failure zone appearance corresponds to the relative tenseness $\gamma_r \cdot H/\sigma_c^{\text{res}} = 1$–1.04 (Glushikhin, et al., 1991) (Figure 1). At the point P_1, the criterion function Kr (r) has reached the value 1 (Figure 1b). This indicates the appearance of the failure first zone at the existing level of the stressed state of the rock mass. The point P_1 also denotes the moment of the first intermediate (contour) zone formation, and also shows the order of the second failure zone appearance.

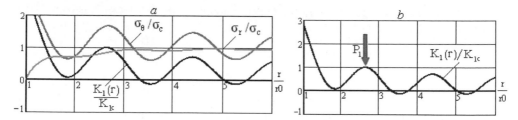

Figure 1. The state of the rock mass stresses around the unlined openings at the time of the first failure zone appearance (*a*); the criterion function at the time of the first fracture zone appearance (arrow P_1) (*b*).

Table 1. Physical and mechanical characteristics of the studied rocks.

Type of rocks	UCS (MPa)	Young's modulus (MPa)	Poisson ratio	Fracture toughness (Mpa × m^0.5)	Volume weight (KN · m^-3)
Dacite	100–150	17000–35000	0.15–0.18	1.25–2.5	26
Rhyolite	130–180	20000–30000	0.26–0.29	1.25–2.5	27
Diorite	150–280	26000–37000	0.24–0.30	1.25–2.5	28

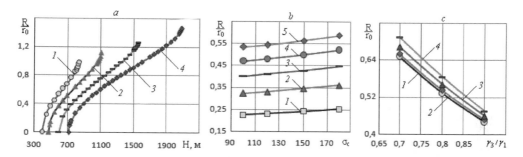

Figure 2. Dependence of the first failure zone radial extent on: *a*) the opening depth at different values of the uniaxial compression strength of the rock (MPa): *1* – 100; *2* – 120; *3* – 150; *4* – 180; *b*) the uniaxial compression strength of the rock at the rock mass values relative tenseness: *1* – 1.1; *2* – 1.2; *3* – 1.3; *4* – 1.4; *5* – 1.5; *c*) the ratio of the empirical coefficients for the values of the rock uniaxial compression strength (MPa): *1* – 100; *2* – 120; *3* – 150; *4* – 180.

3 THE RESULTS OF THE FIRST FAILURE ZONE RADIAL EXTENT DEPENDENCE ON THE INFLUENCING FACTORS FOR HARD ROCKS STUDIES

The first failure zone radial extent dependence on the influencing factors of rocks at Nikolaevskij mine (Dalnegorsk, Primorsky Krai, Russia) was investigated (Table 1).

With an increase of mining depth, the radial extent of the first failure zone increases. At a fixed depth, with an increase in the strength of the rock on uniaxial compression, the radial extent of the first failure zone decreases (Figure 2*a*). As the relative tenseness of a rock mass increases, the radial extent of the first failure zone increases (Figure 2*b*) and with an increase in the ratio of empirical coefficients γ_3/γ_1 – decreases (Figure 2*c*).

It is found out that the deformation modulus E and the Poisson ratio have practically no effect on the radial extent of the first failure zone.

4 THE INVESTIGATION RESULTS OF THE CONTOUR MERGING DEPTH DEPENDENCE AND THE FIRST FRACTURE ZONE, AND THE SECOND FRACTURE ZONE DEPTH OF APPEARANCE FROM THE INFLUENCING FACTORS FOR HARD ROCKS

Along with the increase in the radial extent of the first failure zone with the growth of the openingdepth and the corresponding increase in the gravitational stresses $\gamma_r \times H$, the radial extent of the contour zone also increases, and the radial extent of the first intermediate zone decreases to zero, which leads to the merging of the contour zone and first zone of failure (Figures 3–4).

As the mining depth increases, the appearance of the second and subsequent failure zones is observed, as well as the merging of the contour and first failure zones (Figure 4).

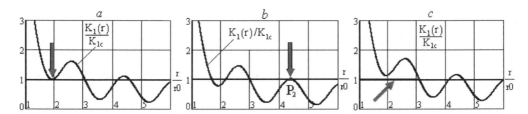

Figure 3. Dependence of the contour radial extension and first intermediate zone on the gravitational stress: a – an increase in the radial extent of the contour zone with increasing gravitational stress; b – decrease in the radial extent of the first intermediate zone with increasing gravitational stress.

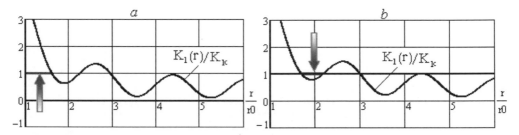

Figure 4. Reduction to zero of the first intermediate zone radial extent with increasing gravitational stress (a); the second failure zone appearance (arrow, point P_2) (b); merged contour and the first failure zones (c).

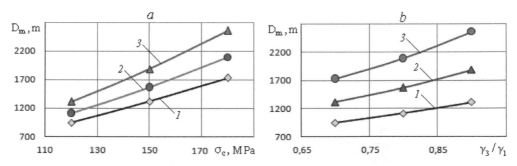

Figure 5. Graphs of contour merging depth and first fracture zones dependence on: a) the uniaxial compression strength of the rock σ_c at the values γ_3/γ_1, respectively: $1 - 0.7$; $2 - 0.8$; $3 - 0.9$; b) ratio of empirical coefficients γ_3/γ_1 at the values uniaxial compression strength of the rock σ_c (MPa): $1 - 120$; $2 - 150$; $3 - 180$.

The patterns of realizing the conditions for the merging of the first fracture zone with the contour zone are established for different values of the uniaxial compression strength, and the ratio of the empirical coefficients γ_3/γ_1 (Figure 5). It has been established that with the increase in the rock uniaxial compression strength the merging of the contour zone and the first fracture zone occurs at a greater depth (Figure 5a). With an increase in the empirical coefficients ratio γ_3/γ_1, the depth of the zones merging also increases (Figure 5b). A similar dependence was obtained for the second zone of failure depth appearance. When calculating the deep-lying mining opening support parameters, it is important to know not only the depth at which the contour and first fracture zones merge, but also the depth at which the second zone of failure appears.

5 CONCLUSION

1. The hard rock massfailure around underground openings in conditions of great depths has a zonal oscillating character, where fractured zones and non-fractured zonesalternate in the direction from the outline to the depth of the mass.
2. With increasing depth, the radial extent of the first failurezone increases, up to the merging of neighboring zones; the radial extent of the first failure zone decreases with the increase in the uniaxial compression strength of the rock; with an increase in the ratio of empirical coefficients γ_3/γ_1, the radial extent of the first failure zone also decreases.
3. With the increase in the uniaxial compression strength of the rock, the merging of the contour zone and the first failure zone occurs at a greater depth; with an increase in the ratio of empirical coefficients γ_3/γ_1, the zones merging depth also increases.
4. With the increase in the uniaxial compression strength of the rock,the second failure zone appearance depthalso increases. With an increase in the ratio of empirical coefficients γ_3/γ_1, the second failure zone appears at a greater depth.

REFERENCES

[1] Adams, G.R., & Jager, A.J. (1980). Petroscopic observation of rock fracturing ahead of stop faces in deep-level gold minees. *J. South African Inst. Mining and Metallurgy*, *80(6)*, 204–209.
[2] Glushikhin, F.P., Kuznetsov, G.N., Shkliarsky, M.F., Pavlov, V.N., & Zolotnikov, M.S. (1991). Modeling in geomechanics, M. Nedra, 240 p.(in Russian).
[3] Guzev, M.A., & Makarov, V.V. (2007). Deforming and failure of the high stressed rocks around openings. Vladivostok: RAS Edit., 232 p. (in Russian).
[4] Guzev, M.A., & Paroshin, A.A. (2001). Non-Euclidean model of the zonal disintegration of rocks around an underground working. *Journal of Applied Mechanics and Technical Physics*, *42(1)*, 131–139. DOI: 10.1023/A:1018877015940.
[5] Guzev, M.A., Makarov, V.V., & Ksendzenko, L.S. (2015). Non-Euclidean model of high stressed rocks and rock masses//ISRM Regional Symposium «EUROCK 2015: *Future Development of Rock Mechanics».- EUROCK 2015 & 64th Geomechanics Colloquium*. Schubert & Kluckner (ed). ©ÖGG.October 7–10. 2015. Salsburg, Austria, pp. 979–984. ISBN 978-3-9503898-1-4.
[6] Ksendzenko, L.S., Makarov, V.V., Opanasyuk, N.A, & Golosov, A.M. (2014). The patterns of deformation and failure of highly compressed rocks and Masss: monograph. [Electronic resource]. Far Eastern Federal University, School of Engineering. Vladivostok, Far Eastern Federal University, [192 p.]; 1 CD. (Series "Geology and exploration of mineral resources", ISSN 2305-4158). ISBN 978-5-7444-3415-1 (in Russian).
[7] Makarov, V.V., & Guzev, M.A. (1999). The Mechanism of zonal failure and deformation of rocks around underground openings/Geodynamics and the Earth Stress Condition. Novosibirsk; Siberian Department of the Russian Academy of Sciences, Mining Institute Press: 120–125.
[8] Makarov, V.V., Guzev, M.A., Odintsev, V.N., and Ksendzenko, L.S. (2016). Periodical zonal character of damage near the openings in highly-stressed rock mass conditions. *Journal of Rock Mechanics and Geotechnical Engineering*, *8(2)*, 164–169. doi.org/10.1016/j.jrmge.2015.09.010.

[9] Makarov, VV, Ksendzenko, LS, Opanasiuk, NA & Golosov, AM. (2017). Zonal failure structure near the deep openings, in J. Wesseloo (ed.),*Proceedings of the Eighth International Conference on Deep and High Stress Mining,* Australian Centre for Geomechanics, Perth, pp. 423–432.

[10] Odintsev, V.N. (1996). Rupture destruction of brittle rocks mass. Moscow: IPKON, the Russian Academy of Sciences, 166 p.

[11] Oparin, V.N., Tapsiev, A.P., Rozenbaum, M.A., Reva, V.N., Badtiev, B.P., Tropp, E.A., & Chanyshev, A.I. (2008). Zone disintegration of rocks and stability of underground workings, the Russian Academy of Sciences. Siberian Branch. Institute of Mining, 278 p.

[12] Qian, Q., Zhou, X., & Xie, E. (2012). Effects of the Axial In Situ Stresses on the Zonal Disintegration Phenomenon in the Surrounding Rock Masses around a Deep Circular Tunnel, *Journal of Mining Science, 48(2),* 276–285.

[13] Qian, Q.H., & Zhou, X.P. (2011). Non-Euclidean continuum model of the zonal disintegration of surrounding rocks around a deep circular tunnel in a non-hydrostatic pressure state, *Journal of Mining Science, 47(1),* 37–46. DOI:10.1134/S1062739147010059.

[14] Shemyakin, E.I., Fisenko, G.L., Kurlenya, M.V., Oparin, V.N., Reva, V.N., Glushihin, F.P., Rozenbaum, M.A., Tropp, E.A., & Kuznetsov Yu.S. (1986). Zonal disintegration of rocks around underground workings. Part. I. The data of based on field observations. *Journal of Mining Science,* 22(3):157–68.

[15] Shemyakin, E.I, Kurlenua, L.V., Oparin, V.N., Reva, V.N., & Glushihin, F.P. (1992). USSR Opening no. 400. The phenomenon of zonal disintegration of rocks around underground workings, *Byul. Izobr.,* 1992, no. 1, p.3.

[16] Zhou X.P., & Shou,Y.D. (2013). Excavation induced zonal disintegration of the surrounding rock around a deep circular tunnel considering unloading effect, *International Journal of Rock Mechanics & Mining Sciences, 64,* 246–257. DOI:10.1016/j.ijrmms.2013.08.010.

Geomechanics and Geodynamics of Rock Masses – Litvinenko (Ed.)
© 2018 Taylor & Francis Group, London, ISBN 978-1-138-61645-5

Models of strength and fracture of rocks

Anatoliy Grigorievich Protosenya & Maxim Anatolievich Karasev
Construction of Mines and Underground Structures Department, Saint-Petersburg Mining University, Saint Petersburg, Russia

ABSTRACT: The paper outlines a view on the mechanisms of deformation and fracture of rocks, the studies of which were carried out in the laboratories of the St. Petersburg Mining University. The concept of rock fracturing is presented, which is based on mixed fracturing of the medium by shearing and splitting and a physical model of rocks deformation and fracture is proposed. The results of laboratory studies were summarized in the form of a rock strength criteria, which later was called the Stavrogin strength criterion. This criterion included a number of analytical equations for limiting elastic, peak and residual states in the form of equations of exponential type. Implementation of the Stavrogin physical model in the framework of the method of finite-discrete elements is presented. Analytical and numerical methods for solving elastoplastic problems using exponential and other conditions of the limiting state of rocks are presented.

Keywords: rock, deformation, fracturing, limit state zones, excavations, stability, strength criteria, formation and propagation of micro cracks

1 INTRODUCTION

The heterogeneity of the rock structures is the reason for the specific behavior of rocks when they are deformed and fracturing in conditions of complex stress states. The most significant features in the behavior of rocks are the effect of increasing the volume (dilatancy) and the presence of a maximum in the process of irreversible deformation under conditions of a triaxial compression and a softening in the stress-strain diagram.

2 EXPERIMENTAL RESEARCH OF ROCK DEFORMATION AND FRACTURING

2.1 *General mechanical behavior of rock*

A large number of works have been devoted to the fundamental problem of the strength and fracturing of rocks [1–6]. The stress-strain diagram is the main experimental material, which is subjected to further analysis in order to obtain more complete information on the mechanical properties of the materials under study. The nature of rock behavior under loading condition can be divided into several stages. The first stage of deformation is characterized by a linear relationship between stresses and strains. At this stage, mainly elastic deformations are realized, and the formation of new microcracks and the development of existing ones are of limited nature. In the second stage, hardening of the rock is observed, the relationship between stresses and deformations is nonlinear, the process is accompanied by a significant growth of existing cracks and the formation of new microcracks both on the surface and in the volume. The second stage of deformation is completed by reaching the ultimate strength of the rock. Further deformation is accompanied by weakening of the rock and is characterized by the avalanche-like growth of microcracks, the destruction of residual bonds between the mineral particles of the rock. At the end of the third stage, one or more main macro cracks are formed, and the strength of the rock is reduced to a residual one. At small values of lateral stress, a pronounced maximum is observed, followed by a descending branch. In this area, the material completely loses its adhesion, and further deformation is accomplished by sliding of two or more samples of rock. As the lateral compression increases, the outermost part of the curve becomes more shallow, irreversible deformations increase the strength and the residual strength increases, the value of which at high pressures becomes equal to the ultimate strength and completely determined by internal friction.

The results tests of various types of rocks in triaxial stress condition show that the envelope of Mohr circles of maximum rock stresses has a nonlinear form [2–6] and A.N. Stavrogin and his co-workers carried out fundamental experimental studies of the strength, deformation and fracturing of the main types of rocks in a wide range of regimes and types of loading.

On the basis of stress-strain diagrams, in addition to elastic constants and softening characteristics, it is possible to obtain the conditions of three limiting states: the conditions of the limits of elasticity, the conditions of the peak strength, and the conditions of the residual strength. In these papers the analytical representations of all three kinds of limiting states in the form of equations of exponential type is proposed:

$$\tau_y = \tau_y^0 e^{BC}; \tag{1}$$

$$\tau_\Pi = \tau_\Pi^0 e^{AC}; \tag{2}$$

$$\tau_o = \tau_o^0 e^{OC}, \tag{3}$$

where equation (1) – elasticity limit state. Equation (2) – peak strength condition. Equation (3) – residual strength condition. In this equations $\tau_y = (\sigma_1^y - \sigma_3)/2$, $\tau_\Pi = (\sigma_1^\Pi - \sigma_3)/2$, $\tau_o = (\sigma_1^o - \sigma_3)/2$ – respectively elastic limit, peak and residuals strengths; τ_y^0, τ_Π^0, τ_o^0 – constants, determined elastic limit, peak and residual strengths in uniaxial strength conditions; B, A, O – constants, relate strength of rocks to hydrostatic pressure; $C = \sigma_3/\sigma_1$ – parameter, considering the mode of stress state.

The values of the parameters for various types of rocks are given in [2–6]. The development of other areas of research is given in [7–12].

A system of differential equations for the plane and axisymmetric problems of the theory of the limiting state under the condition of strength (2) is investigated. A.G. Protosseny obtained analytical solutions on the formation of the zone of the limiting state in the vicinity of the excavations located in various mining and geological conditions, the strength of which is described by the condition (2). It is established that the zone of the limiting state of the rocks around the outline of the circular outline has the shape of an ellipse elongated in the horizontal direction. The presence of dilatancy and its influence on the deformation of rock mass in limiting zone around excavation is investigated.

It should be noted that the use of conditions (1)–(3) causes complexity in getting nonlinear solution Numerical solution of practical problems using the Stavrogin criterion is presented in [11–12], where solutions of inhomogeneous elastic-plastic problems are considered concerning the size of the region of the limiting state around the excavations of non-circular shape and various conditions of the limiting state of rocks. The model is implemented as a user model via the UMAT interface into Abaqus/Standard.

2.2 Physical model of deformation and fracture of rocks

The mechanism of deformation and fracturing of rock [2,3,6] can be represented as the process of formation and development of microcracks in the body of the rock. Graphically, the process of deformation and fracturing of rock under conditions of uniaxial compression can be represented as the formation of microcracks of shearing and detachment, the growth of which propagates through the rock from the center to the edge parts.

The initiation of cracks is related to the heterogeneity of the rock or the existing of micro-defects. The formation of cracks also affects the nature of deformation of the rock. Thus, with the appearance and growth of new microcracks, the body is weakened in the considered rock region and the relationship between stresses and deformations is no longer linear. When the number of microcracks reaches a critical value, the stress in the rock ceases and the ulti-mate strength is reached. Further deformation of rock leads to the destruction of residual bonds along the forming sliding or tearing surface and the yield to the residual strength, which is determined by friction along the surfaces of weakening.

That is, the relationship between individual particles of the rock is conducted through con-tact surfaces, which before the time of complete destruction are connected surfaces. Moving of such surfaces relative to each other is impossible until the moment of complete destruc-tion of these bonds. The physical model considered above was used as a basis for the devel-opment of numerical models of rock deformation and destruction in the framework of the method of finite-discrete elements.

As in the physical model, the process of deformation and fracturing of rock will be associ-ated with the formation of microcracks inside the body of the rock. Formation of a cracks inside the body of the rock is determined by the tensile and shear strength of bonds.

3 NUMERICAL MODELING OF ROCK DEFORMATION AND FRACTURE PROCESSES

3.1 *Model of deformation and fracture of rocks in the framework of finite-discrete elements method*

The numerical model of deformation and fracture of rocks within the framework of the method of finite-discrete elements [13] includes two types of elements—solid and cohesive elements. Elements of the theory of elasticity, the theory of plastic flow, the theory of viscous and viscoplastic flow, and also their combinations can be used to describe the mechanical behavior of solid elements. The cohesive elements provide links between solid elements and allow modeling crack initiation and propagation due to the reduction of their deformation characteristics and their subsequent removal from the numerical model. In this paper, a medium modeled by solid elements was considered as a linear deformable isotropic or transversally isotropic medium. Some non-linearity of deformation at the pre-limit stage, achievement of the ultimate strength and softening of the medium are realized through a change in the stiffness of the cohesive elements. Mechanical behavior of cohesive bonds must be divided into two components: behavior in the pre-limiting and out-of-bound stages of deformation.

Cohesion bonds between individual elements of the medium can be damaged by exceeding their tensile or shear strengths. The criterion for achieving the ultimate strength of a cohesive material can be generalized as

$$\max \left\{ \frac{\langle t_n \rangle}{t_n^0}, \frac{t_s}{t_s^0}, \frac{t_t}{t_t^0} \right\} = 1,$$

где t_n^0, t_s^0, t_t^0 – respectively, the tensile strength of the cohesive bonds in the n direction, the shear strength of the cohesive bonds in the direction s and t.

In order to assess the degree of damage, function D that shows at what rate the medium's stiffness degrades after reaching the ultimate stress state damage is introduced. To describe the development of the damage function D, various dependencies are used, expressed in linear form, power or logarithmic form, and others. In general, the damage function can be represented as

$$D = f\left(u_p, u_r, G\right),$$

где u_p – magnitude of the displacements corresponding to the moment of reaching the limiting stress state; u_r – displacement value corresponding to full crack opening; G – fracture energy.

The Mohr-Coulomb strength criterion for describing the strength of a bond under the action of shear stresses and the Ranke strength criterion for describing the strength of a bond under the action of normal tensile stresses is adopted in the paper. The process of damage to the medium is specified through the accumulation of fracture energy in the cohesive elements [14]. A quantitative estimation of the fracture energy can be expressed in terms of the area under the softening function of the medium. Fracture energy G^c, which depends on stress direction of loading, convenient to write through the law M.L. Benzeggagh—M. Kenane [15].

3.2 Implementation of finite-discrete elements method into Abaqus/Explicit

The construction of a numerical model within the framework of finite-discrete elements method involves several basic steps. First, the constitutive models for both solid and cohesive elements should be work out. Next stage involves the generation of an element grid where the solid elements are separated from each other by cohesive elements, i.e., the nodes of the solid elements are not connected to each other, and the transfer of forces from one solid element to the adjacent is effected through the cohesive elements. Such an approach ensures the continuity of displacements until the moment of crack formation. Next stage involves the development of a model of contact interaction between the faces of solid elements, which allows contact between the surfaces of solid elements after the formation of the weakening surface (exclusion of the cohesive element from model). This step is the most difficult, since the search algorithm for the contact between the newly formed surfaces must track their position in the process of solving the problem and in automatic mode enable or disable contact interaction.

The existing algorithms for generating a finite-element grid perform its construction in such a way that the interconnection between adjacent elements is carried out through common nodes, which ensures the continuity of displacements. In finite-discrete elements method it is necessary to separate the solid elements so that each of them does not have common nodes with each other. Solid elements usually have the form of a triangle (plane strain condition), a wedge (generalized plane strain condition) or a tetrahedron (3d condition), which makes it possible to form a maximally unstructured grid. When performing the simulation, their geometric thickness of cohesive elements is assumed to be zero, and the physical thickness is determined through an internal parameter that does not change in the solution process.

Due to the fact that existing programs for the generation of an element grid allow the formation of only joint meshes where solid elements are interconnected through adjacent node points, algorithms for generating elemental grids for the conditions of planar deformation and solving spatial problems suitable for their subsequent application in the framework of the method of finite-discrete elements. Based on the proposed algorithm, a software solution was developed that made it possible to generate an elemental grid for elements of the first and second order under both planar deformation conditions (triangular or rectangular elements) and for solving spatial problems (tetrahedral, prismatic or wedge). The output file included node points and elements in the format adopted in the Abaqus/Explicit software package. The mechanism of deformation and damage of cohesive elements is realized through the user procedure VUSDFLD, which allows you to redefine the parameters of the environment behavior model in the process of solving the problem. During the solution, the change in average stresses at each point of integration of the cohesive elements was tracked and the strength indices of the medium were corrected.

742

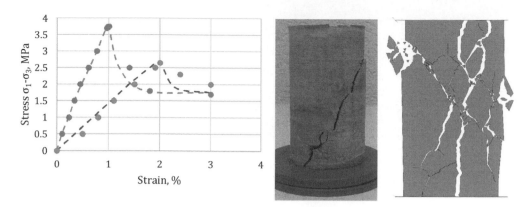

Figure 1. Comparison of laboratory tests and numerical simulation of britle clay: dots—data from laboratory tests; dashed lines—numerical simulation.

3.3 *Examples of modeling deformation and fracture at the level of rock samples*

As the material under investigation, solid argillite-like clay rocks that undergo a brittle fracture character and can be satisfactorily described within the framework of the physical model proposed by Stavrogin in the framework of the method of finite-discrete elements is taken.

Strength and deformation characteristics of rocks obtained on the basis of numerical simulation, correspond to the results of laboratory tests (Figure 1). It has been established that the nature of rock fracture obtained on the models allows to obtain both the brittle (formation of the main cracks of separation) and the plastic (formation of the main fracture of the shear) character of the rock failure, which varies depending on the magnitude of the all-round reduction.

The numerical simulation of the rock test under conditions of uniaxial compression, volume compression, shear and uniaxial stretching made it possible to build a passport for the strength of the rock. The envelope of the strength of the rock as a result of modeling takes a nonlinear form, which is well correlated with the Stavrogin strength condition and also this type of envelope was obtained on the basis of the results of laboratory tests [16].

The presented concept of numerical modeling of the behavior of rocks with the introduction of some changes in the formulation of the model and the algorithm for the formation of an element grid [17] can be successfully used to study the processes of deformation and destruction of layered rocks. The obtained model allows an explicit description of the anisotropic structure of layered media, and the simulation results correspond to complete deformation diagrams of such a rock taking into account the direction of application of the load.

4 CONCLUSION

The paper presents some results of rock tests carried out in the laboratories of the Mining University. The generalization of laboratory studies made it possible to form an idea of the strength of rock as a process that occurs at the micro level and is expressed in the form of fracture formation and shear. The strength of the rock at various stages of loading can be described by three strength conditions (strength at the limit of elasticity, ultimate strength, residual strength), which later were called the Stavrogin strength condition. On the basis of the Stavrogin strength condition, the authors of the paper obtained analytical solutions of boundary-value problems of geomechanics. The presented physical model of deformation and fracturing of the rock formed the basis for the development of numerical models in the framework of the finite-discrete elements method, the implementation of which was carried out in the form of user procedures in the Abaqus/Explicit code. The numerical calculations based on the models formed in the framework of the finite-discrete element method have shown the promise of such an approach for the modeling of deformation and fracture of rocks.

ACKNOWLEDGEMENTS

The research was carried out at the expense of a grant from the Russian Science Foundation (project №16-17-00117).

Исследование выполнено за счет гранта Российского научного фонда (проект № 16-17-00117).

REFERENCES

[1] Kwasniewski, M. True triaxial testing of rocks/M. Kwasniewski, Xiaochum Li, Manabu Takahashi// CRC Press/Balkema. 2013. P.367.

[2] Stavrogin, A.N. Experimental physics and rocks mechanics/A.N. Stavrogin, B.G. Tarasov// A.A. Balkema. 2001. P. 356.

[3] Ставрогин А.Н. Экспериментальная физика и механика горных пород/А.Н. Ставрогин, Б.Г. Тарасов/- СПб.:"Наука", 2001 г., 343 с.

[4] Ставрогин А.Н. Пластичность горных пород. Монография/А.Н. Ставрогин, А.Г. Протосеня. - М.: Недра, 1979г., 302 с.

[5] Ставрогин А.Н. Прочность горных пород и устойчивость выработок на больших глубинах/А.Н. Ставрогин, А.Г. Протосеня/ М.: Недра, 1985г., 270 с.

[6] Ставрогин А.Н. Механика деформирования и разрушения горных пород. Монография/А.Н. Ставрогин, А.Г. Протосеня/М.: Недра, 1992г., 222 с.

[7] Протосеня А.Г. О построении модели смешанного разрушения горных пород и твердых тел/А.Г. Протосеня, В.А. Александров//ФТПРПИ № 3, 1986. С. 39-46.

[8] Протосеня А.Г. Запредельное деформирование вокруг выработки в негидростатическом поле напряжений/А.Г. Протосеня, Г.Н. Журов, В.А. Александров//ФТПРПИ № 2, 1983. С. 10-17.

[9] Протосеня А.Г. К определяющим уравнениям состояния при деформировании горных пород в запредельной области/А.Г. Протосеня, А.Н. Ставрогин, А.К. Черников, Б.Г. Тарасов//ФТПРПИ №3, 1981. С. 33-42.

[10] Ставрогин А.Н. Пластичность горных пород в условиях переменных скоростей деформирования/ А.Н. Ставрогин, А.Г. Протосеня//ФТПРПИ № 4, 1983. С. 3-13.

[11] Протосеня А.Г. Разработка численной модели прогноза предельного состояния массива с использованием критерия прочности Ставрогина/А.Г. Протосеня, М.А. Карасев, Н.А. Беляков// Физико-технические проблемы разработки полезных ископаемых. 2015, №1, с. 40-48.

[12] Протосеня А.Г. Упругопластическая задача для выработок различных форм поперечных сечений при условии предельного равновесия Кулона/А.Г. Протосеня, М.А. Карасев, Н.А. Беляков// Физико-технические проблемы разработки полезных ископаемых. 2016, №1, с. 71-81.

[13] Munjiza A. The combined finite-discrete element method. Chichester, UK: John Wiley & Sons Ltd., 2004. P. 352.

[14] Mahabadi O.K. A novel approach for micro-scale characterization and modelling of geomaterials incorporating actual material heterogeneity/O.K. Mahabadi, N.X. Randall, Z. Zong, G. Grasselli// Geophysical Research Letters. 2012. Vol. 39(1). P. 532–541.

[15] Benzeggagh M.L. Measurement of Mixed-Mode Delamination Fracture Toughness of Unidirectional Glass/Epoxy Composites with Mixed-Mode Bending Apparatus/M. Kenane//Composites Science and Technology. 1996. Vol. 56. P. 439–449.

[16] Protosenya A/G. Investigating Mechanical Properties of Argillaceous Grounds in Order to Improve Safety of Development of Megapolis Underground Space./A.G. Protosenya, M.A. Karasev, D.N. Petrov//International Journal of Applied Engineering Research. 2016, vol. 11, pp. 8849–8956.

[17] Lisjak A. Continuum-discontinuum analysis of failure mechanisms around unsupported circular excavations in anisotropic clay shales/A. Lisjak, G. Grasselli, T. Vietor//International Journal of Rock Mechanics and Mining Sciences. 2014. Vol. 65. P. 96–115.

Geophysics in rock mechanics

Geomechanics and Geodynamics of Rock Masses – Litvinenko (Ed.)
© 2018 Taylor & Francis Group, London, ISBN 978-1-138-61645-5

Tunnel restoration in unstable rock masses: Numerical analysis and validation of monitoring data from innovative instrumentation

Andrea Segalini, Andrea Carri, Roberto Savi & Edoardo Cavalca
DIA, Università di Parma, Parma, Italy

Carlo Alessio & Georgios Kalamaras
AK Ingegneria Geotecnica S.r.l., Torino, Italy

ABSTRACT: The increase of global temperatures caused shortages of water supply in several Italian cities. One of the challenges of responsible institution is how to maintain and restore the old aqueducts facilities. In particular, the most critical problem is related to the water losses along the aqueduct pipes and tunnels due to the aging of the structures and the damages caused by rock mass instabilities.

This paper deals with one of these cases, where an aqueduct tunnel, damaged by the progressive movement of a rockslide, underwent to a complete reconstruction. The restoration design required the installation of CIR-Array, an innovative automated monitoring system, which would enable to monitor the surrounding rock mass, as well as the provisional reinforcement and the final lining, during both construction and operational phases. This case has been the first in situ application of this instrumentation and it represents an interesting test, complementary to the previous laboratory one.

Construction works are still in progress, as well as the installation of monitoring instrumentations. In this paper, the preliminary results are presented, described and then compared with the expected behavior of the rock mass predicted by the design numerical model.

Keywords: Tunneling, Monitoring, Back analysis, Early warning, Numerical model, Convergence

1 INTRODUCTION

The increase of global temperatures around the globe caused unique shortages of water supply in several Italian cities, especially during the last few summers. The institutions that are responsible for the drinking water supply in some Italian metropolis are facing two different challenges: the first is how to increase the available water supply, and the second is how to maintain and restore or adapt the old aqueducts facilities in order to comply with the increased demand. One of the most critical problem, which affects both aspects of water shortage, is related to the water losses along the aqueduct pipes and tunnels due to both, the aging of the structures and the damages caused by rock mass instabilities.

This paper deals with one of these cases, where an aqueduct tunnel in the center of Italy, damaged by the progressive movement of a rockslide in which it was originally built, underwent to a complete reconstruction. In particular, this aqueduct has a leading importance for the water supply for millions of people of one of the major city of the area.

The restoration design required the installation of an automated monitoring system, which would enable to monitor the rock mass surrounding the tunnel as well as the provisional reinforcement and the final lining, during both the restoration works and afterwards, when the tunnel will return operational. The CIR-Array monitoring system (Carri *et al.*, 2017) is able to detect automatically the convergence of a tunnel section. It has been designed to give an appropriate tool for the observational method philosophy (Peck, 1969), the "design as you monitor" approach of NATM (Rabcewicz, 1964) and the "manual discrete monitoring approach" of ADECO-RS (Lunardi, 1988), based on distometers and topography. Carri *et al.* (2017) have suggested a new approach, defined as "automated semi-continuous", and it is based on the use of automated monitoring devices, using different technologies, each having its own limitations in terms of sensitivity, reliability, full scale and application.

In this paper, the initial data retrieved by the in situ application of the monitoring system are presented, described and then compared with the expected behavior of the rock mass predicted by the design numerical model.

2 CIR-ARRAY

CIR-Array is a chain of nodes (links), called "Tunnel Link", located at known distances and equipped with 3D MEMS sensors (accelerometer and thermometer). A fiberglass rod links the nodes preserving the alignment while a quadrupole cable provides power and retrieve data. An aramidic fiber cable replaces the fiberglass rod in the invert. Each sensor provides its relative position in the space, referring to the previous one, starting from a fixed point. By cumulating all the results, it is possible to reconstruct the section shape. The control unit automatically queries each link at defined time intervals that can be varied according to the monitoring needs. The instrument is installed on the bare rock just after the excavation and provides results both during construction and usage without becoming a hindrance during the works.

A previous paper (Carri *et al.*, 2017) studied the instrumentation through a series of laboratory tests where a real scale tunnel section was reproduced and known deformation applied. An algorithm (called CSC) was developed in order to elaborate the raw data of CIR-Array. During the mentioned tests, the device provided good agreement with topography and photogrammetry results. The case study analyzed in this paper represents the first on site application of this type of instrumentation and highlights some unexpected behaviors. These unpredicted results required an improvement on the software.

The tunnel monitoring plan required seven CIR-Array sections, five of which have been installed starting from May 2017. Every convergence array is equipped with 20 links located at an interspace of one meter along the tunnel boundary, except for the four links in the invert, which are positioned at variable distances. In order to reproduce the convergence star typically used for the topographic surveys, the authors identified seven segment of convergence, as in Figure 1a.

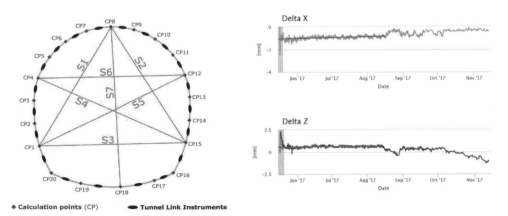

Figure 1.   a) position of nodes in the tunnel section and definition of segments of convergence, b) Noise due to vibrations and not solved by the de-spike filter (area in light blue) and trend of data (line in red).

2.1 Calculation algorithm

The CSC algorithm was designed in order to provide optimal performances in terms of calculation time, and results reliability. The first issue outlined just after the installation was related to the influence of works and vibrations, that could jeopardize the accuracy, precision and repeatability of measures. Moreover, after the first month of monitoring, the large amount of data related to the sample period of 15 minutes, caused a progressive increase of the calculation time, which soon became unsustainable for a near-real time monitoring.

In order to solve the first problem, a new de-spike filter has been developed and applied successfully. In particular, this subroutine excludes spike noise using phase-space method, based on modified Goring and Nikora (2002) algorithm developed by Mori et al. (2007) and then modified by Ulanowski (2014). This algorithm identifies and deletes the noise due to vibrations, using an adequate sample period and data set. The filter is sensitive to border effects and therefore should not be applied to the first N data of the data set. The noise due to vibrations will affect them: this only determines punctual unreliable results but does not influence the trend of displacements (Figure 1b). After a series of parametric analysis, a value of N equal to 40 was found appropriate for the described monitoring sample period.

The second problem has been solved improving the algorithm and the interconnection with the database where the data are stored. In the first version of the software, all the data set starting from the reading of zero were considered in order to apply the de-spike algorithm, while only the new results were written in the database. This procedure and in particular the DB read and write operation caused an unsustainable calculation time. The newly developed approach recognizes automatically which is the last elaborated data that can be considered as a new reference reading, taking into account the border effect of de-spike filter as well. Then CSC algorithm downloads the raw data starting from this reference reading and elaborates the relative displacements, then adds them to the displacements previously recorded. This solution has greatly improved the elaboration time, which now is less than a minute if we consider a data set of one week (current interval of manual transmission of data to the server) with a sample period of 15 minutes and could be much faster with a daily, hourly or instantaneous transmission, such in the case of real-time monitoring.

The new approach unexpectedly improved also data stability, which is now higher. This is probably related to an unknown behavior of the de-spike filter, which did not performed at its best with the large data set previously required. Using the appropriate size of the input data set, the filter is able to better recognize the spikes and delete them. Figure 2 shows an example of this behavior, applied at the convergence segment S6: after the convergence effect, which lasted between three weeks and one month with a total displacement of 53 mm, is possible to identify stable conditions in two main periods. The first one has been elaborated with the original version of the algorithm and highlights a stability of ± 0.85 mm of displacement around the mean value, while the second one, elaborated with the improved version of CSC,

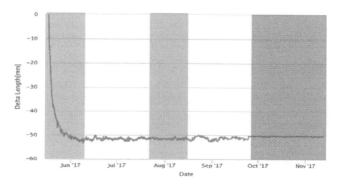

Figure 2. Convergence after the excavation (first identified area, light blue), data stability with the original version of the software (second identified area, green) and data stability with the introduction of the new version of the software (last identified area, red).

evaluates a stability of ± 0.30 mm. This effect can be found in each calculation point or segment of convergence.

3 NUMERICAL MODEL

In order to validate the results provided by the instrumentation on site, it was decided to compare them with the displacements forecasted by the original numerical model developed during the design. AK Ingegneria Geotecnica developed such a preliminary numerical model using RS2® (Rock and Soil bi-dimensional) software produced by Rocscience® software house. This 2D software is based on finite elements formulation to evaluate displacements, stresses and potential failures of an equivalent continuum material surrounding the tunnel. The geometry has been set up considering a planar extension of 100 m and a height of 50 m in order to avoid boundary effects (Figure 3a). The tunnel section has a diameter of 6.87 m and the plasticization area surrounding the tunnel extends for 15.00 m at the sides and around 28 m in height (Figure 3b). The grid has 15235 elements, 30686 calculation nodes and constrains used are rollers for all the boundaries.

Due to the presence of highly fragmented rock, the design of the restoration works forecasted to enlarge the section using umbrella supports. This methodology was abandoned during the excavation because an injected concrete layer of an adequate thickness has been identified around the old tunnel section. This layer is able to entirely contain the enlargement excavation, and therefore a full section excavation technique was adopted without the advancing support. The preliminary numerical model did not take into account the injected concrete layer and this original assumption has been maintained due to the lack of knowledge regarding its thickness and geometrical distribution. Numerical simulation has been divided into seven stages to simulate the progressive perturbation of the rock mass state of stress induced by the original tunnel excavation (stage 1, 2, 3 and 4) carried out some 80 years ago and the currently undertaken enlargement (stage 5 and 6). Table 1 reports the failure

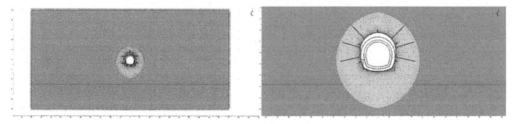

Figure 3. a) Planar extension of the model and tunnel section, b) Detail of the tunnel section and the plasticization area.

Table 1. Failure criterion, stiffness and mechanical parameters.

Stage	1	4	Stage	1	4
Failure criterion	Mohr-Coulomb	Mohr-Coulomb	Cohesion (residual)	1.90 MPa	0.05 MPa
Tensile Strength (peak)	1.33 MPa	0.01 MPa	Dilation Angle	0°	0°
Friction Angle (peak)	54.5°	50.0°	Young's Modulus	4658 MPa	50 MPa
Cohesion (peak)	1.90 MPa	0.05 MPa	Poisson's Ratio	0.25	0.30
Tensile Strength (residual)	0 MPa	0 MPa	Type	Isotropic	Isotropic
Friction Angle (residual)	54.5°	50.0°			

Table 2. Maximum variations of the values recorded by pressure cells.

Pressure cell	DT0040 Δ_{max} [KPa]	DT0054 Δ_{max} [KPa]	DT0059 Δ_{max} [KPa]
CP-01	40.91	46.75	22.09
CP-02	14.84	101.28	28.88
CP-03	52.80	1.41	6.95

criterion, the stiffness and the mechanical parameters for the undisturbed rock mass (stage 1) and the plasticization area (stage 4).

During the modeling phase of the tunnel enlargement, both the presence of spritz-beton and metal ribs were taken into account. Three pressure cells are positioned in the extrados of the metal rib in correspondence of the CIR-Array. During the monitoring period, pressure data did not highlighted load increments over the supports. Table 2 shows the maximum variations of the values recorded by pressure cells of the three sections considered in this paper.

4 RESULTS

The original monitoring plan foresaw three main section types related to differences in the lithology of the rock mass surrounding the excavation. Three of the five CIR-Array (DT0040, DT0054 and DT0059) that have been already installed are located in the same section type, while the other two are positioned in different ones and have been installed more recently.

For this reason in this paper are presented the displacements provided by the model for the first section type and compared with the convergence segments defined in Figure 1a and Table 3. In order to do so the same segments were defined and calculated in the numerical model, knowing the design position of CIR-Array calculation points.

It is necessary to underline that the displacements recorded could be related to two different causes: the convergence after the excavation and the potential slow moving rockslide affecting the tunnel. Since the design model is isotropic and does not take into account the rockslide effect, the first month of the monitoring data set for each CIR-Array has been considered in order to consider exclusively the effects of convergence. The zero reading has been set up just after the installation for DT0054 and DT0059, while DT0040 considers the reference position five days after the installation. This is because each sensor has self-control parameters able to identify malfunctions or unreliable data. During the first days, accelerometers of DT0040 installed in the invert highlighted a behavior highly influenced by temperature variation related to the concrete aging. This aging most probably induced movements of the steel net where sensors were inappropriately fixed during installation (in the other parts of the section, sensors were correctly fixed directly to the rock mass) causing the anomalies. After five days, data became stable and reliable. This behavior was not found in the other two CIR-Array previously mentioned, since the installation occurred properly. The comparison identifies the same qualitative trend between the CIR-Arrays recordings and the numerical design model, except for the DT0059, that shows a different behavior (Figure 4). In particular, segment S3 records the maximum displacement, while secondary maximums are on segments S2 and S4. All these vectors have in common the calculation point 15 and probably the displacements recorded are located there. Segment S7 is the more stable, while the model overestimates the S1 vector. A quantitative analysis highlights the same order of magnitude in the displacements recorded by the instrumentations and forecasted by the model (Table 4).

The mean differences between CIR-Array and the model are respectively of 5.9 mm for DT0040, 3.0 mm for DT0054 and 9.6 mm for DT0059, while the mean differences between the DT0040 and DT0054 is of 5.7 mm. This is probably related to the observed changes of geology and lithology around the considered tunnel sections, which are located at different progressives (DT0059 is 26 m far from DT0054, which in turn is 52 m far from DT0040).

751

Table 3. Design length and calculation points used for the detection of the segments of convergence.

	S1 [m]	S2 [m]	S3 [m]	S4 [m]	S5 [m]	S6 [m]	S7 [m]
Design length	5.981	5.953	6.126	6.789	6.830	6.139	7.011
Calculation points	1–8	8–15	1–15	4–15	1–12	4–12	8–18

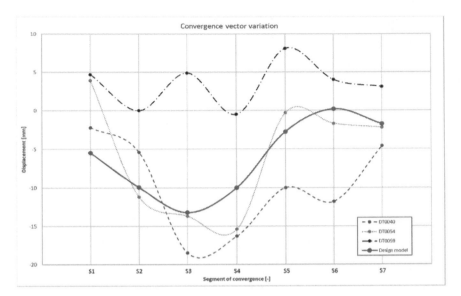

Figure 4. Comparison between the displacements recorded by the three CIR-Array studied and forecasted by the design model.

Table 4. Comparison between displacements recorded by CIR-Array installed in the first section type and design model.

	Date of installation [dd-mm-yy]	Zero reading [dd-mm-yy]	Δ S1 [mm]	Δ S2 [mm]	Δ S3 [mm]	Δ S4 [mm]	Δ S5 [mm]	Δ S6 [mm]	Δ S7 [mm]
DT0040 Section B3	15-05-2017 (cast of shotcrete during the 16-05-2017)	20-05-2017	−2.2	−5.4	−18.5	−16.3	−10	−11.8	−4.6
DT0054 Section B4	07-08-2017	08-08-2017	3.9	−11.2	−13.7	−15.4	−0.3	−1.7	−2.2
DT0059 Section B5	18-09-2017 (cast of shotcrete during the 19-09-2017)	20-09-2017	4.7	0.0	4.9	−0.5	8.1	4.0	3.1
Design model	–	–	−5.5	−9.9	−13.2	−10.0	−2.8	0.2	−1.8

5 CONCLUSIONS

The paper deals with the monitoring of restoration works of an aqueduct tunnel. This tunnel is of primary importance for the water supply of one of the main Italian city and was damaged by a slow moving rockslide.

The restoration design required the installation of CIR-Array, an innovative automated monitoring system, which would enable to monitor the surrounding rock mass, as well as the provisional reinforcement and the final lining, during both construction and operational phases. This case represents the first in situ application of this kind of instrumentation.

After the first period of monitoring, it was necessary to improve the CSC algorithm, in order to achieve sustainable elaboration time and correctly identify the noise due to vibration induced by construction works.

The displacements recorded by three different CIR-Array, installed in the same section type, have been compared with the movements forecasted by the design numerical model, using the traditional segments of convergence. The comparison identifies the same qualitative trend between two of the three investigated CIR-Arrays and the design model, highlighting the expected rock mass behavior. A quantitative analysis shows a good agreement between the displacements recorded by the instrumentations and forecasted by the model. The CIR-Array tool was able to follow the evolution of the rock mass deformation induced by the tunnel excavation and, coupled with a provisional numerical model, could provide a valuable solution for the application of the observational method, control the tunnel excavation progress and evaluate its safety conditions.

REFERENCES

Carri A., Chiapponi L., Savi R. and Segalini A. (2017). Innovative technologies for monitoring underground excavations during construction and usage. *ISRM International Symposium "Rock Mechanics for Africa"*. Cape Town. Volume: Symposium Series S93.

Goring D.G. and Nikora V.I. (2002). Despiking Acoustic Doppler Velocimeter Data. *Journal of Hydraulic Engineering*. Vol. 128 Issue 1 – January 2002. https://doi.org/10.1061/(ASCE)0733-9429(2002)128:1(117).

Lunardi P. (1988). ADECO-RS Analisi delle deformazioni controllate nelle rocce e nei suoli. *Seminar on "Design and Construction of Tunnels"*. ISMES. Bergamo.

Mori N., Suzuki T. and Kakuno S. (2007) Noise of Acoustic Doppler Velocimeter Data in Bubbly Flows. *Journal of Engineering Mechanics*. Vol. 133 Issue 1 – January 2007. https://doi.org/10.1061/(ASCE)0733-9399(2007)133:1(122).

Peck R.B. (1969). Advantage and limitations of the observational method in applied soil mechanics. *Géotechnique*, 19 (2), pp. 171–187.

Rabcewicz L. (1964). The New Austrian Tunnelling Method, Part one. *Water power*. Nov. pp. 453–457.

Rocscience. RS2 Manual. https://www.rocscience.com/rocscience/products/rs2 [last accessed 14 Nov. 2017].

Ulanowski J. (2014). Goring and Nikora method. Modified to remove offset in output. Mathworks Community Forum. http://it.mathworks.com/matlabcentral/fileexchange/15361-despiking?focused=3838229&tab=function [last accessed 16 Nov. 2017].

Geomechanics and Geodynamics of Rock Masses – Litvinenko (Ed.)
© 2018 Taylor & Francis Group, London, ISBN 978-1-138-61645-5

Forecast of rock mass stability under industrial open pit mine facilities during the open pit deepening. A case study of the Zhelezny open pit, JSC Kovdorsky GOK

Ivan M. Avetisian & Inna E. Semenova
Mining Institute of the Kola Science Center of the Russian Academy of Sciences, Russia

ABSTRACT: The paper presents the results on numerical modeling of stress-strain state in the rock mass adjacent to an eastern wall of the Zhelezny open pit, JSC Kovdorsky GOK. The authors estimated influence of gravitational-tectonic stress field and fault structures on wall stability. The work reveals specifics of the stress field transformation of a deepened pit wall. The authors have determined parts of the open pit wall with the compression level to be sufficient for the rockbursts (according to forecasting) and the parts with a probable germination of tensile cracks due to high tensile stresses.

Based on the numerical modeling results, the rock mass stability under the industrial facilities located near the eastern open pit wall has been estimated. The calculations were performed using Sigma GT software, developed at the Mining Institute of the Kola Science Centre of the Russian Academy of Sciences.

1 INTRODUCTION

The past few decades demonstrate a steady world trend to deepen mining operations. One of mining methods to excavate deep reserves is the deepening of existing open pits through increasing a slope angle. The design of open pit walls or change of its project configuration requires geomechanical verification of its parameters and structural elements. To do this, it is necessary to determine the behavior of stress redistribution in the rock mass adjacent to a pit wall and identify potentially hazardous areas. Numerical modeling methods are now widely used and successfully applied to forecast the slope stability [Hamman, 2007; Stead, 2007]. Geomechanical 3D numerical modeling allows the forecast of stress-strain state in the rock mass taking into account the main geological and mining factors.

2 SUBJECT AND METHODS OF THE RESEARCH

The Zhelezny open pit, JSC Kovdorsky GOK, is located in the south-west of the Kola Peninsula, Russia. To the moment, the depth of the open pit is about 450 m.

In 2011–2012, a small-scale numerical model of the rock mass in the vicinity of baddeleite-apatite-magnetite, apatite-shtafelyte and apatite-carbonate ores was designed within the studies on the slope stability forecast for the Zhelezny open pit. The model includes a surface relief, geological environment objects (ore bodies, tectonic faults, and weakened zones), a tectonic stress field, actual and designed open excavations and designed underground stoping chambers [Kozyrev, 2015]. To substantiate a physical model of the environment, specialists from the Mining Institute KSC RAS analyzed available data on strength and elastic characteristics of ores and rocks in the vicinity of the Zhelezny open pit. Based on the analysis of both the absolute values of parameters and their changes under loading, a conclusion was made that most rocks were strained elastically until failure. This fact has allowed adopting an elastic model as a model of the environment.

The boundary conditions in the model were set taking into account the field stress measurements by a relief method and analysis of the curvature of long exploration wells [Rybin, 2009]. The study's results indicate the correspondence of the rock mass stress state to a gravitational-tectonic type if tectonic forces act sub-parallel to a long axis of the open pit.

Recently, due to revision of a designed outline of the open pit, it has become necessary to study in more detail the stress-strain state in the rock mass adjacent to the eastern open pit wall. To solve this problem, a large-scale model has been designed, which represents a part of the rock mass adjacent to the eastern wall. The model was obtained by cutting out a local area from the small-scale model with a four-fold compaction of the volumetric finite element mesh. As a result, the model size was $2,600 \times 2,280 \times 1,900$ m^3. The total number of elements is about 3.2 million. The minimum element size is $10 \times 12 \times 5$ m^3. A general view of the 3D model is shown in Figure 1.

When designing the model, the authors took into account various rocks (host rocks, ore body and fault structures), the surface relief and three pit outlines (an actual outline and two versions of the projected outline made by the Giproruda Institute). The boundary conditions were set in the form of nodal displacements taken from the small-scale model. The physical-mechanical properties of the rocks considered in the modeling are presented in Table 1.

The stress-strain state was calculated using SigmaGT software, developed at the Mining Institute KSC RAS [Kozyrev, 2012].

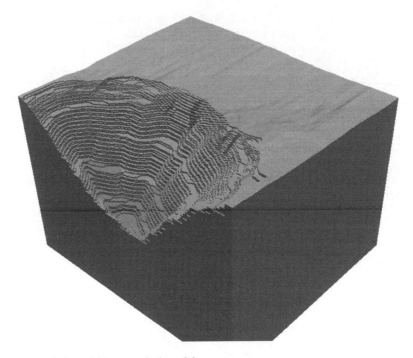

Figure 1. General view of 3D numerical model.

Table 1. Physical-mechanical properties of rocks.

Rock type	Elastic modulus E, MPa	Poisson's ratio ν	Volume weight γ, g/cm^3
Host rocks	$1 \cdot 10^5$	0.2	3.0
Ore body	$0.5 \cdot 10^5$	0.35	3.5
Fault structures	$0.1 \cdot 10^5$	0.4	3.0

3 RESULTS

Let's consider the stress-strain state of the rock mass in the vicinity of a projected open pit (variant 1). As it can be seen from Figure 2, a pronounced concave area has been designed in the eastern open pit wall, in the vicinity of which a stress concentration zone σ_{max} is formed. In this zone the values of σ_{max} reach 50 MPa at the zero mark, and 65–70 MPa at –220 m mark. The studies on physical-mechanical properties of rocks were conducted by the specialists of the Mining Institute KSC RAS and results have revealed the ultimate compressive strength (UCS) of rocks in the vicinity of the eastern wall (fenites, ijolites) to be 144–165 MPa. Thus, the stresses σ_{max} in the concentration zone approach a value of 0.5 UCS, which can cause dynamic failures. In other parts of the eastern pit wall the values of σ_{max} vary from 25 to 45 MPa.

In addition, on the flanks of the concave area in the eastern wall, the tensile zones with σ_{min} values reaching 2 MPa are formed. According to numerical modeling results, they cover

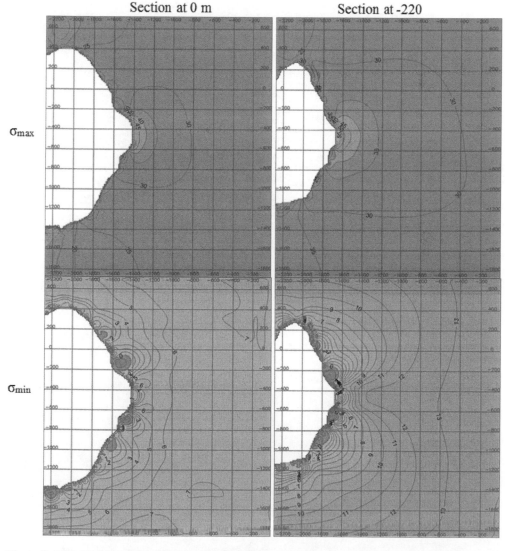

Figure 2. Distribution of main stresses in the vicinity of the projected eastern pit wall (variant 1) in horizontal sections.

extended parts of the wall. In this case, the stresses σ_{min} act in such a way that the planes, on which tensile cracks can be formed, cut the open pit wall. This fact is unfavorable for its stability.

Let's consider the stress-strain state of the rock mass adjacent to the projected open pit (variant 2). As it can be seen from Figure 3, the pit wall has more curving shape in plan than in variant 1. So, concave and convex areas alternate at level −220 m. In the concave areas the zones of σ_{max} concentration occur, reaching 60–70 MPa and approaching a value of 0.5 UCS, which can cause dynamic failures. In the convex areas the tensile stress concentration zones σ_{min} arise, reaching 3 MPa. The direction of the planes, where tensile cracks can be formed in the convex areas, is similar to variant 1 and is unfavorable, as it cuts the open pit wall.

As experience shows, slope stability is greatly influenced by undercutting fault structures. During the study two typical fault structures cutting the actual pit wall were modeled: structure R62 in the eastern part of the wall and structure R2 in the south-eastern part.

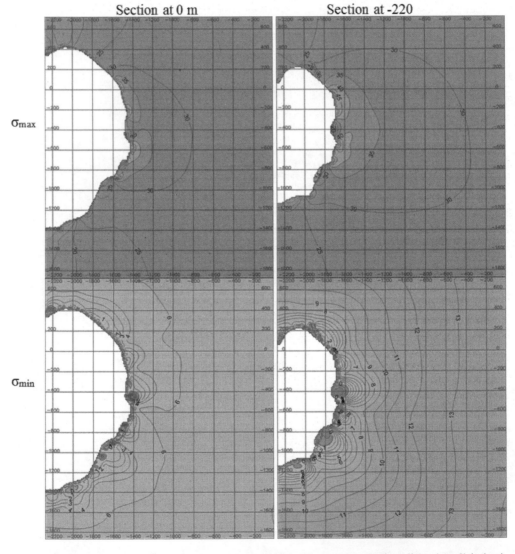

Figure 3.   Distribution of main stresses in vicinity of the projected eastern pit wall (variant 2) in horizontal sections.

As it can be seen from Figure 4, the account of fault structures in the stress-strain state calculations allows observing additional tensile zones which occur in the boundaries between the structure and host rocks. The absolute values of tensile stresses σ_{min} in these zones are higher than in a homogeneous rock mass and reach 6–8 MPa in local areas. This is comparable to the tensile strength of rocks in a sample. In addition, the stress planes σ_{min} cut the open pit wall and almost coincide with dip of fault structures in direction. Thus, it has been found that in the parts of the actual open pit undercut by fault structures R62 and R2, probable failures can occur due to germinations and further opening of tensile cracks.

The calculation results of stress-strain state were confirmed in practice. So, on 24.08.2015 a rock slide occurred on the eastern wall part, undercut by fault structure R62. Although the mine personnel was timely evacuated, the operating regime at this part was violated, and safety hazard for mining operations was created [Melnikov, 2015].

As for the projected outline of the open pit, modern geological data have not revealed any fault structures undercutting the pit wall in the ultimate position.

The authors have also evaluated the influence of the deepened pit wall on industrial site facilities, in particular, the potential formation of mining-induced cracks in the eastern pit wall with subsequent germination beyond the open pit outline to the industrial site. Figure 5 shows the distribution of stresses σ_{min} in a near-surface zone under the surface facilities for the actual and projected pit walls. The analysis has revealed formation of slight tensile zones with values of σ_{min} not exceeding 0.1–0.2 MPa in the vicinity of the actual pit wall, at the roughness of the surface relief. They are two orders of magnitude lower than the tensile strength. With the deepening of the open pit, tensile stresses don't increase in the near-surface zone under the surface facilities. Therefore, it can be concluded that most of the surface industrial facilities will remain stable during the deepening of the open pit to the project depth.

Rock mass without accounting fault structures Rock mass taking into account fault structures

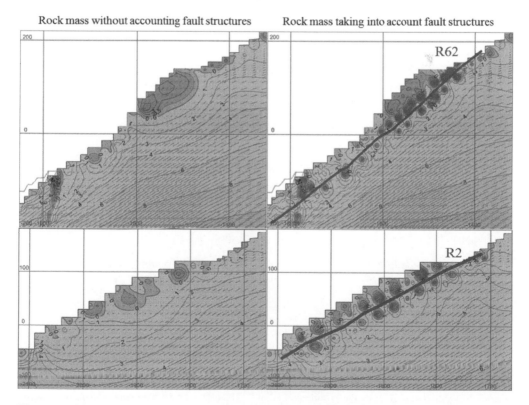

Figure 4.   Distribution of σ_{min} in the vicinity of the actual eastern pit wall.

Actual pit wall   Projected pit wall

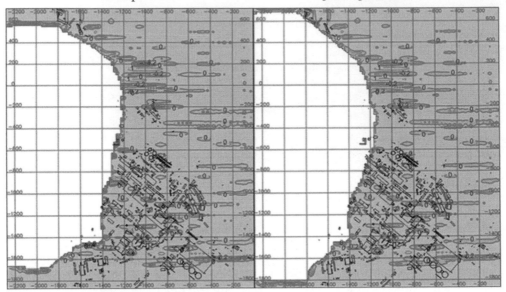

Figure 5.   Distribution of stresses σ_{min} in the near-surface zone under surface facilities.

4 CONCLUSIONS

The analysis of the stress-strain state in the rock mass adjacent to the projected open pit walls, taking into account tectonic stresses, has revealed general stability of the eastern part. There is no significant increase in tensile stresses in the near-surface zone under the surface industrial facilities. The loss of stability is probable in concave areas of the open pit wall, where, according to forecast, the compression level is sufficient for dynamic rock pressure, and in convex areas with a probable germination of tensile cracks. In this regard, it is recommended to correct a projected wall of the deep open pit in order to obtain a smoother outline without typical convex and concave parts concentrating stresses.

REFERENCES

Hamman E.C.F., Coulthard M.A. Developing a Numerical Model for a Deep Open Pit/Slope Stability 2007. Proceedings of the 2007 International Symposium on Rock Slope Stability in Open Pit Mining and Civil Engineering, 12–14 September 2007, Perth, Australia, Publ. by Australian Center of Geomechanics – 2007. pp. 225–237.

Kozyrev A.A., Panin V.I., Semenova I.E. Experience of using the expert systems to estimate the stress strain state in the rock mass to choose safe mining methods. Mining Institute Papers. 2012. V.198. pp. 16–23.

Kozyrev A.A., Semenova I.E., Rybin V.V., Avetisian I.M. Stress redistribution in deep open pit mine Zhelezny at the Kovdor iron ore deposit/Journal of mining science. № 4. 2015. pp. 659–665.

Melnikov N.N., Kozyrev A.A. Changes in geodynamic behavior of geological environment at large-scale deep open-pit mining/Mining Informational and Analytical Bulletin, 2015, № 11 (special issue 56). – pp. 7–23.

Rybin V.V., Kozyrev A.A., Danilov I.V. Determination of the parameters of the stressed state of the contour rock mass in the Kola Peninsula open pits // Mining Informational and Analytical Bulletin, 2009. № 10. pp. 402–405.

Stead D., Coggan J.S., Elmo D., Yan M. Modelling Brittle Fracture in Rock Slopes—Experience Gained and Lessons Learned/Slope Stability 2007. Proceedings of the 2007 International Symposium on Rock Slope Stability in Open Pit Mining and Civil Engineering, 12–14 September 2007, Perth, Australia, Publ. by Australian Center of Geomechanics – 2007. pp. 239–252.

Geomechanics and Geodynamics of Rock Masses – Litvinenko (Ed.)
© 2018 Taylor & Francis Group, London, ISBN 978-1-138-61645-5

Benefits and limitations of applying directional shear strengths in 2D and 3D limit equilibrium models to predict slope stability in highly anisotropic rock masses

Neil Bar
Gecko Geotechnics, Cairns, Australia

Geoffrey Weekes
Red Rock Geotechnical, Perth, Australia

Senaka Welideniya
Gecko Geotechnics, Willetton, Australia

ABSTRACT: The bedded iron ore deposits in Western Australia are hosted by highly anisotropic rock masses that typically comprise strong banded iron formation discretely interbedded with very weak shales. Slope instability mechanisms within these bedded units generally involve sliding along bedding planes combined with joints or faults acting as release surfaces. Slope stability modelling techniques have significantly developed over the years from basic kinematic analysis in the 1990s to complex two-dimensional limit equilibrium analysis and numerical modelling in the 2000s. Limit equilibrium analysis software now offers the possibility of modelling the behaviour of anisotropic rock masses in either 2D or 3D. The results obtained by these different modelling methods can vary significantly. Whilst the choice of either a 2D or 3D modelling code will generally be dictated by the geometry of the situation, it has been found that selecting either inappropriate anisotropic shear strength models for a given rock mass or using poorly calibrated models can result in overly conservative slope designs, regardless of the modelling code used. This paper presents case studies which illustrate the importance of geological interpretations and correct constitutive model selection.

1 ANISOTROPIC ROCK MASSES

Iron ore deposits in the Pilbara region of Western Australia occur within the banded iron formations of the Hamersley Group which is made up of Archaean to Proterozoic marine sedimentary and volcanic rocks. Geological structures are the primary control for the location, size, geometry and preservation of high grade iron ore bodies. The structural evolution of the Hamersley province is complex but well understood. It comprises normal faulting and thick-skinned tectonics in the west near Tom Price and intense folding and minor thrust faulting with possible thin-skinned tectonics in the east near Newman (Dalstra, 2014). The iron ore province is geographically extensive, spanning over 200 kilometers.

The stratigraphic units of economic interest to the iron ore mining industry consist of banded iron formation (BIF) with interbedded shales and carbonates. BIF thickness can vary due to differing amounts of carbonate dissolution and silica replacement during iron ore enrichment phases. The BIF sequences usually contain thick interbedded shale bands, some of which are useful stratigraphic marker horizons in the mining areas as they are very persistent across hundreds of kilometers (Harmsworth et al. 1990).

Rock mass and bedding shear strengths in the Pilbara region are typically well understood due to a combination of: a) mining in over 250 individual open pits in similar stratigraphic units since the 1970s; b) remarkably limited variation between individual deposits or open pits within the same stratigraphic unit (Bar et al. 2016). Intact rock strength varies significantly with the degree of weathering, and to a lesser extent, alteration. Rock mass shear strengths usually correlate quite well with intact rock strength and weathering, particularly in BIF-dominated stratigraphic units. Maldonado & Haile (2015) stated that no significant difference exists between the shear strength of bedding planes of shale and BIF units across the Pilbara. Maldonado & Mercer (2015) further indicated that the shear strength of shale bedding planes is independent of weathering grade. Site specific drilling, mapping, lab testing and failure back-analyses (when possible) provide a means of assessing local variation in material density, rock mass and discontinuity characteristics and shear strengths.

In iron ore deposits, two scales of anisotropy exist as shown in Figure 1:

1. Bedding scale—between individual bedding planes (e.g. micro to meso scale BIF-BIF or shale-shale bedding planes).
2. Banding scale—between known specific bands within stratigraphic layers (e.g. macro scale banding comprising 16 recognized shale bands and BIF bands in the Dales Gorge Member).

Planar sliding along unfavorably oriented shale bedding is the most common failure mechanism in the Pilbara region and occurs on all scales from bench failures to overall slopes in open pit mines and in nature (Bar, 2012; Seery, 2015). Figure 2 presents examples of

A: Bedding scale anisotropy (individual bedding planes)   B: Banding scale anisotropy (discrete shale bands)

Figure 1.   A: Intricately folded bedding scale anisotropy—individual BIF-BIF bedding planes; Gently folded banding scale anisotropy - white-light yellow layers are discrete 0.5 m to 1.0 m thick shale bands within anisotropic banded iron formation in the Dales Gorge Member of the Brockman Iron Formation.

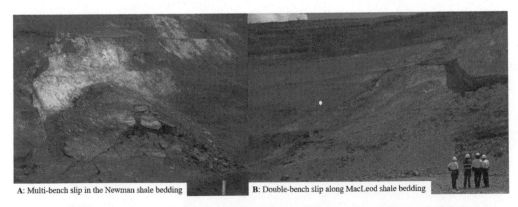

A: Multi-bench slip in the Newman shale bedding   B: Double-bench slip along MacLeod shale bedding

Figure 2.   Planar sliding along shale bedding—most common failure mechanism in the Pilbara region; A: 45 m high slip along Newman shale bedding; B: 30 m high slip along MacLeod shale bedding.

interesting but troublesome near-perfectly planar sliding slope failures induced by locally unfavorable bedding orientation and the presence of shale bands. Joass et al. (2013) provide a detailed account of the management and remediation of a 90 m high planar sliding failure along a shale band within the Newman Member of the Marra Mamba Iron Formation at West Angelas mine in Western Australia.

Slope failures in the Pilbara region are generally quite brittle in that only tens of millimeters of deformation are required prior to failure. Risk to personnel and equipment in the mines is typically managed with the use of alarmed and automated, near real-time monitoring instrumentation including automatic total stations and slope stability radar (Saunders & Nicoll, 2016; Baczynski & Bar, 2017).

2 2D AND 3D MODELS WITH DIRECTIONAL SHEAR STRENGTHS

Directional shear strength models require four key components:

1. Rock mass shear strength (usually described using the Hoek-Brown failure criterion).
2. Anisotropy shear strength (bedding, foliation, pervasive joints, etc).
3. Transition of shear strength between the rock mass and anisotropy.
4. Orientation of the anisotropy plane or planes.

Limit equilibrium analysis software such as *SLIDE* and *SLIDE³* of Rocscience (Canada) is well suited for the analysis of open pit slopes in bedded rocks, and allows users to select a variety of directional shear strength models using linear and non-linear shear strengths. Where applicable, the use of non-linear shear strengths is important in slope stability analyses as simple Mohr-Coulomb approximations in most cases, over-estimate shear strength at low stresses in any slip-circle analysis and have the potential to provide higher than normal factors of safety.

Limit equilibrium analysis using directional shear strength models is generally undertaken with the selection of a 2D cross-section of a slope chosen by a geotechnical engineer as being 'critical' or 'representative' for assessing slope stability. Anisotropy is inherently almost always represented by 'apparent dips' rather than 'true dips'. This must be accounted in the selection of an appropriate strength model (typically using an anisotropic strength model for situations in which bedding strikes less than 30° to the strike of the slope). The dip of anisotropy is typically modelled in 5–10° increments depending on the complexity of the available geological model. The selection of several cross-sections for analysis is currently considered standard practice.

With the recent developments in 3D limit equilibrium software, a 2D cross-section can easily be extruded laterally into 3D-extruded models to provide an indication of the possible extent of a potential failure.

Full 3D models of open pits or an area of interest in an open pit are also possible and facilitate the analysis of complex three-dimensional slope geometry and geological conditions as shown in Figure 3. Anisotropy is fundamentally represented with 'true dips' and the resolution of the model is dependent on the anisotropy wireframes rather than user-input (i.e. it is inherently more reliable than in 2D cross-sections which use 5–10° incremental zones).

A comparative assessment between 2D and 3D limit equilibrium methods was performed using a preliminary pit design for an unmined iron ore deposit in the Marra Mamba Iron Formation.

Anisotropic shear strengths were applied to both 2D and 3D models for bedded (anisotropic) rock masses. In the upper quaternary sediments, isotropic shear strengths were applied. Critical slip circles in both the 2D and 3D analyses were located within the Newman Member and had similar mechanisms to the planar sliding example illustrated in Figure 2A. The Newman Member was represented using non-linear shear strength envelopes using the Barton-Bandis failure criterion (Barton & Bandis, 1982) for shale bedding anisotropic direction and the Generalized Hoek-Brown failure criterion for the remainder of the rock mass as illustrated in Figure 4. Also illustrated is the transition relationship between bedding

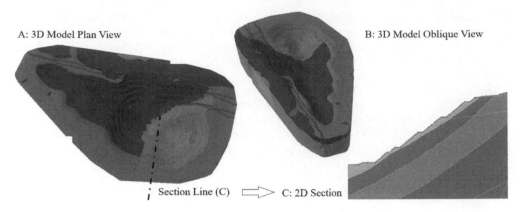

A: 3D Model Plan View

B: 3D Model Oblique View

Section Line (C) ⟹ C: 2D Section

Figure 3. A: Plan view of 3D pit model with location of cross-section (C) for 2D model; B: Oblique view of 3D pit model; C: 2D slope cross-section.

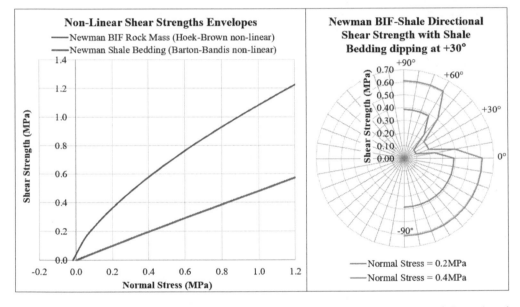

Figure 4. Left: Non-linear shear strength envelopes for Newman BIF rock mass (Hoek-Brown) and Newman Shale Bedding (Barton-Bandis); Right: Direction shear strength interpretation of non-linear shear strengths at normal stresses of 0.2 and 0.4 MPa assuming bedding (anisotropy) is dipping at +30° ± 5° with a linear transition requiring 30° between the bedding and rock mass.

(anisotropy) and rock mass shear strength using the anisotropic linear directional shear strength model for shale, with parameters, A = 5° and B = 30°.

The modelling results obtained from 2D and 3D analyses were quite different. The 2D limit equilibrium analysis suggested the entire slope design was likely to be unstable with a factor of safety (FS) of approximately 0.9 (i.e. less than equilibrium) as illustrated in Figure 5. Anisotropy was modelled using 10° increments and assuming that no lateral variability occurs out-of-the-plane (which of course is not realistic). Essentially, several aspects of the model are conservative. However, the benefit is that over 20,000 individual non-circular slip surfaces were evaluated for the cross-section.

The 3D limit equilibrium analysis for the region in which the two-dimensional cross-section was adopted from identified triple bench instability with a FS of approximately 0.99 (Figure 6). The failure mechanism was similar to the 2D analysis; however, local variability

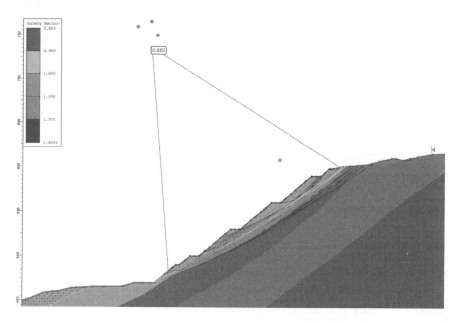

Figure 5. 2D limit equilibrium analysis results—lowest FS = 0.88 for overall slope.

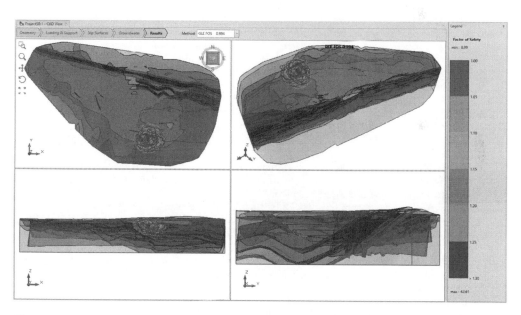

Figure 6. 3D limit equilibrium analysis results—lowest FS = 0.99 for triple bench (similar to failure mechanism in 2D analysis).

in slope geometry (pit curvature) and anisotropy in both the direction of the slip surface and laterally (out-of-plane with respect to the 2D analysis) facilitated higher resisting forces, and therefore, a higher factor of safety. Contrary to the 2D analysis, the 3D analysis only analyzed 800 individual surfaces for the region but each with several iterations.

Although neither 2D or 3D analyses in this case study support the use of this particular slope design geometry since factors of safety are below equilibrium, the 3D analysis will support the use of a steeper slope angle by accounting for local variations in both 3D slope geometry and anisotropy. These aspects are assumed to be constant in a 2D cross-section.

3 DISCUSSION

The benefits of applying directional shear strengths in 2D and 3D limit equilibrium analyses to model anisotropy are quite easily identifiable in that they provide a more realistic and sophisticated representation of plausible failure mechanisms, compared to using isotropic strength models with downgraded shear strengths.

Three-dimensional limit equilibrium models generally require a little more time to construct, particularly when wireframes contain minor geometric errors, and more time to compute the same number of slip surfaces as their 2D counterparts. However, 3D models can still be constructed in the space of a few hours, and not days. Whilst the simplicity of 2D models makes them perfectly suited to reasonably linear slope alignments, 3D modelling would certainly be encouraged where the slope is concave (or convex) along strike and where a more complex interaction with bedding is to be experienced, in addition to the effects of lateral confinement.

The authors recommend the use of both 2D and 3D models, as appropriate, to gain a better understanding of anisotropic rock mass behavior, and ultimately to optimize slope designs to increase ore recovery and reduce waste removal requirements in open pit mines. The economic benefits of slope optimization in the mining industry usually range from millions to hundreds of millions of dollars (Bar et al. 2016).

REFERENCES

Baczynski NRP, Bar, N (2017) *Landslide monitoring and management challenge in remote Papua New Guinea*, Advancing Culture of Living with Landslides, Volume 4, Springer, Switzerland: 343–354.

Bar N (2012) *Performance driven slope management, design and optimization at Brockman Operation, Rio Tinto Iron Ore*. In Proc. 9th Young Geotechnical Professionals Conference (9YPGC), St Kilda, 11–14 July 2012, Australian Geomechanics Society: 19–24.

Bar N, Johnson TM, Weekes G (2016) *Using directional shear stress models to predict slope stability in highly anisotropic rock masses*. In Ulusay et al. (eds.), Rock Mechanics and Rock Engineering: From the Past to the Future; Proc. ISRM int. symp. Eurock 2016, Cappadocia, 29–31 August 2016: 595–600.

Barton N, Bandis S (1982) *Effects of block size on the shear behavior of jointed rock*, J. Rock Mechanics vol. 23: 739–760.

Dalstra HJ (2014) *Structural evolution of the Mount Wall region in the Hamersley Province, Western Australia and its control on hydrothermal alteration and formation of high-grade iron deposits*. Journal of Structural Geology Volume 67, Elsevier: 268–292.

Harmsworth RA, Kneeshaw M, Morris RC, Robinson CJ, Shrivastava PK (1990) *BIF-derived iron ores of the Hamersley Province*. Geology of the Mineral Deposits of Australia and Papua New Guinea Volume 1, Australasian Institute of Mining and Metallurgy Monograph 14: 617–642.

Joass, GG, Dixon R, Sikma T, Wessels SDN, Lapwood J, de Graaf PJH (2013) *Risk management and remediation of the north wall slip, West Angelas Mine, Western Australia*, In Proc. Slope Stability 2013, 25–27 September 2013, Brisbane, Australian Centre for Geomechanics: 995–1010.

Maldonado A, Haile A (2015) *Application of ANOVA and Tuckey-Cramer statistical analysis to determine similarity of rock mass strength properties across Banded Iron Formations of the Pilbara region in Western Australia*. In Proc. SAIMM Slope Stability 2015, 12–14 October 2015, Cape Town, South African Institute of Mining and Metallurgy: 15–33.

Maldonado A, Mercer KG (2015) *An investigation into the shear strength of bedding planes in shale materials from the Hamersley Group rocks in the Pilbara region of Western Australia*. In Proc. SAIMM Slope Stability 2015, 12–14 October 2015, Cape Town, SAIMM: 47–62.

Saunders P & Nicoll S (2016) *Slope stability radar alarm threshold validation and back analysis at Telfer Gold Mine*, In Proc. APSSIM 2016, 6–8 September, Brisbane, Australian Centre for Geomechanics.

Seery JM (2015) *Limit equilibrium analysis of a planar sliding example in the Pilbara Region of Western Australia—comparison of modelling discrete structure to three anisotropic shear strength models*. In Proc. SAIMM Slope Stability 2015, 12–14 October 2015, Cape Town, SAIMM: 681–696.

Geomechanics and Geodynamics of Rock Masses – Litvinenko (Ed.)
© *2018 Taylor & Francis Group, London, ISBN 978-1-138-61645-5*

Mathematical modelling of limit states for load bearing elements in room-and-pillar mining of saliferous rocks*

Alexander Baryakh
Perm Federal Research Center of Ural Branch of Russian Academy of Science, Perm, Russia

Sergey Lobanov, Ivan Lomakin & Andrey Tsayukov
Mining Institute of the Ural Branch of Russian Academy of Sciences, Perm, Russia

ABSTRACT: As a rule, saliferous rock and potash deposits are developed using the room-and-pillar method. Field studies demonstrate that the thin-layered structure of a saliferous rock deposit and presence of clay bands reduce the durability of stope roofs. Therefore, it increases the deformation rate of rib pillars and leads to a gradual loss of bearing capacity.

This work presents the results of experimental and numerical studies of the stress-strain behavior of the bearing elements of the room-and-pillar method. The theoretical description of deformation and fracture of saliferous rocks was based on the elastoplastic models of the medium. The deformation of contacts was determined using their full failure diagram. Instrumental monitoring results were used for the mathematical model calibration. The paper formulates the criterion of the chamber roof collapse based on the field observations and numerical study results. The critical rates of the transverse deformation of pillars were estimated. The verification of the modelling results for the deformation and fracture processes of bearing elements of the room-and-pillar method proved an appropriate conformity to the field data.

1   INTRODUCTION

Sylvinite AB and KrII strata of Verkhnekamsk saliferous rock deposit (VKSD) are developed using the room-and-pillar method which supports the overlying rock layers with stabilizing pillars of various sizes. Their stability depends on the physical and mechanical properties of rocks, as well as on technical and geological conditions of mining. The bearing capacity of the rib pillars greatly depends on the condition of the stope roofs and especially on those of AB-KrII interbed. It is especially relevant under the conditions of thin-layered structures of the industrial saliferous rock formation with a high content of clay. The presence of clay bands decreases the durability of the stope roofs and causes an intensive segregation [Toksarov, 2009; Asanov, 2012] which may even lead to the fracture of AB-KrII interbed. It also decreases the bearing capacity of rib pillars and sometimes interrupts stoping works. The experimental studies carried out by [Yudin, 1985; Nesterov, 1968; Serata, 1972; Marakov, 1978] revealed a number of patterns in the processes of deformation and fracture of bearing elements during the room-and-pillar mining and formed the basis for the engineering calculation method. Particularly, the load degree of the rib pillars is calculated using the Tournaire-Shevyakov method [Tournaire, 1884; Shevyakov, 1941] which has been modified with regard to the results of the experimental studies carried out at VKSD:

$$C = \xi \frac{\gamma(a+b)H_0}{bk_f \sigma_m} \tag{1}$$

*This study was conducted with the support from the grant of the Russian Foundation for Basic Research Nr. 17-45-590681.

where ξ is the coefficient of pillar load change due to various mining technical factors (salt tailing pile surcharge, bearing pressure, presence of inter-passage pillars, etc.); γ is the bulk density of rock; H_0 is the maximal distance from the stope roofs to the surface; a is the stope width; b is the width of the rib pillars; k_f is the pillar shape coefficient; σ_m is the in-situ rock durability.

Subject to restrictions specific for the Tournair-Shevyakov method, the engineering formula (1) provides the estimation precision acceptable only for simple geological and technical conditions of mining. Modern simulation-based approaches open up new horizons in analyzing limit states of bearing elements in the room-and-pillar method thus providing new opportunities for a more profound description of layered structures of saliferous deposits and considering the interrelated fracture processes occurring in stope edges and introducing the time factor that corresponds to the temporal scale of real geomechanical processes.

This paper estimates the critical values of pillar transverse deformation rates obtained by analyzing the change of their load degree within time.

2 MATHEMATICAL MODELLING METHOD

The analysis of the load value change of rib pillars within time was based on the mathematical modelling of the stress-strain behavior of a dual-completion chamber block [Baryakh, Shumikhina, 2011] with regard to the interrelated fracture processes of the development interbed and pillar edges. The principal computational model for the problem is presented in Fig. 1.

It is based on the example of a chamber block under the influence of bulk forces of γ_i (γ_i is the density of rocks). The distributed load γH was set on the upper horizontal boundary (with H being the distance to the surface). There were no horizontal shears on the side boundaries while there were no vertical shears on the lower ones. There were clay layers and clay bands in the interbed interval between the commercial beds.

The mathematical modelling of the interbed fracture process was made for plane deformation conditions. The stress condition of the chamber block was described as a perfect elastic-plastic medium where the relation between deformations and stresses at the pre-limit stage was determined by Hooke's law. Limit stresses in the compression area were calculated using the linear envelope of the Mohr's circles:

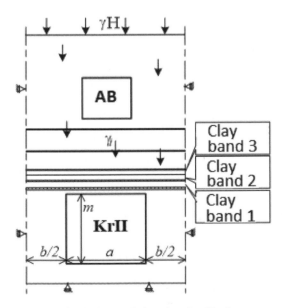

Figure 1. Computational model of a dual-completion chamber block.

$$\tau_{\max} = \tau_{pr} = C_m + \sigma_n \operatorname{tg}\varphi \tag{2}$$

where adhesion coefficient C_m and internal friction angle φ are determined with the strength of saliferous rocks under uniaxial compression σ_c and the ultimate tensile stress σ_t of rocks:

$$C_m = \frac{\sigma_t \sigma_c}{\sigma_t + \sigma_c}; \quad \varphi = \operatorname{arctg}\left(\frac{\sigma_c - \sigma_t}{\sigma_t + \sigma_c}\right) \tag{3}$$

In equation (2), $\tau_{\max}$ and σ_n are calculated via the principal stress values:

$$\tau_{\max} = (\sigma_1 - \sigma_3)/2; \quad \sigma_n = (\sigma_1 + \sigma_3)/2, \tag{4}$$

where the principal stresses for plane deformation are calculated using the formulas

$$\sigma_1, \sigma_3 = \frac{1}{2}\left[(\sigma_x - \sigma_y) \pm \sqrt{(\sigma_x - \sigma_y)^2 + 4\tau_{xy}^2}\right] \tag{5}$$

In the strain area the limit stress was within the ultimate tensile stress: $\sigma_1 = \sigma_t$.

The numerical implementation was carried out using the finite element method in displacements [Zenkevich, 1971] with the discretization of the considered area into first-order triangular elements. The finite element solution of the elastic-plastic problem was based on the initial stress method [Malinin, 1975].

During the numerical modelling of the clay bands' deformation between the layers, Goodman elements were used [Goodman, 1974]. The relation between normal stress (σ_n) and the corresponding deformation (δ_n) is presented with the linear equation

$$\sigma_n = k_n \delta_n \tag{6}$$

where k_n is the normal rigidity of the contact. When the value of $\delta_n > 0$, the contact was considered open, and $k_n = 0$ was admitted to (6).

For tangential stresses (τ_s) along the clay band line, the relation with the shear deformation (δ_s) was determined with the three-link piecewise linear approximation [Baryakh, Dudyrev, 1992]:

$$\tau_s = \begin{cases} k_s \delta_s & at\ 0 < \delta_s \le \delta_p, \\ \tau_p - k_m(\delta_s - \delta_p) & at\ \delta_p < \delta_s \le \delta^*, \\ \tau^* & at\ \delta_s > \delta^*, \end{cases} \tag{7}$$

where k_s is the shear rigidity of the contact, k_m is the shear rigidity of the contact in the softening area; τ_p is the peak strength of the contact; τ^* is the residual strength of the contact.

The ultimate shearing strength of the contact (peak strength) was calculated according to Coulomb equation:

$$\tau_p = C_k + \sigma_n \operatorname{tg}\varphi_k, \tag{8}$$

where C_k is the adhesion coefficient of the contact, and φ_k is its internal friction angle.

When the tensile strength influences the band, it is considered that its shear strength decreases to zero ($k_n = k_s = 0$).

The following conditions were admitted as the rock collapse criterion [Baryakh, Shumikhina, 2011; Baryakh, Fedoseev, 2011]: the baring of the tensile strength influence area ("masslf-chamber" boundary) and the shearing crack zone's reaching the zone of segregation through clay bands.

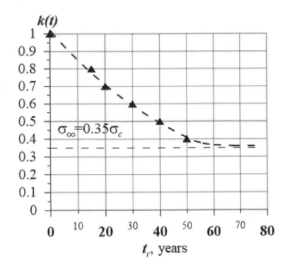

Figure 2.   Stress-rupture strength coefficient of the pillars based on the diagrams predicting ground settlement.

The numerical correspondence between the load degree of the rib pillars and invariants of their stressed condition is substantiated in the work by [Baryakh, Samodelkina, 2007] where the energy criterion is admitted as the stress condition of the saliferous rocks:

$$K = \sigma_i / \sigma_m \leq 1 \tag{9}$$

In expression (9), σ_m is the aggregate strength of rocks in the massif, $\sigma_i = \sqrt{I_2(D_\sigma)}$ is the intensity of the tangential stresses determined by the value of the second invariant of the stress deviator. The carried out multivariant calculations demonstrated a sustainable interconnection between the load degree of the rib pillar with the parameter K* operating along the whole width or height of the pillar [Baryakh, Samodelkina, 2007]:

$$C = K^* \tag{10}$$

Saliferous rocks are the geomaterials that reveal distinctive rheological properties under loading. Field observations show that the deformation of rib pillars within time occurs mostly in the creep process with a loss of bearing capacity [Baryakh, Asanov, 1998]. Meanwhile a decrease of structural and deformation strengths of saliferous rocks within time plays a dominant role in the display of rheological processes. For this reason, the developed procedure of variable deformation modules [Baryakh, Samodelkina, 2005] can be used to analyze the influence of time on the pillar loading degree; and the change of strength characteristics can be mathematically expressed in accordance with the stress-rupture strength curve based on the diagrams predicting ground settlement progress [Baryakh, Lobanov, Lomakin, 2016] (Fig. 2). According to the curve, the average value of the stress-rupture strength is $\sigma_\infty = 0.35\sigma_c$ which generally corresponds to the laboratory experiments [Baryakh, Asanov, Pankov, 2008].

3   RESULTS

The geomechanical computations are presented by four (equal to the number of variants of the selected parameters, Table 1) series of numerical experiments carried out as a step-by-step procedure. Each step determined a time segment within which the initial loading degree (C_0) of the rib pillars in KrII stratum increased by 0.1 due to the fracture of the technological interbed and edge areas of the pillars, as well as due to the decrease of their strength properties.

The calculation series are carried out until the value $C_{krII} = 1.0$ is reached, which means a complete loss of a pillar's bearing capacity in KrII stratum. At the overlying developed AB stratum, the loading degree remained unchanged and was approximately 0.2. During the mathematical modelling the developed interbed thickness h was considered to be equal to 8 m.

The basic parametric support is based on the representative physical and mechanical tests of saliferous rocks [Baryakh, 2008] and determinations of properties of clay layers and bands [Baryakh, Shumikhina, 2011].

Fig. 3 shows the computational results of how the load degree of rib pillars in KrII stratum changes within time during the fracture of its edge parts and technological interbed for the variant when $C = 0.4$. As we can see from (Fig. 3a), the load increases from 0.4 to 0.52 during $t_1 = 12$ years. Within this period the pillar width decreases by 1 meter and its height increases by 0.5 m.

Table 1.   Parameters of mining.

Variant of computation	Stratum	Axle spacing (l), m	Width of chamber (a), m	Width of pillar (b), m	Height of pillar (m), m	Load degree, C_0
1	AB	12.0	3.2	8.8	3.2	0.20
	KrII		6.0	6.0	7.0	0.40
2	AB	12.0	3.2	8.8	3.2	0.20
	KrII		7.0	5.0	7.5	0.50
3	AB	12.0	3.2	8.8	3.2	0.20
	KrII		7.5	4.5	8.0	0.60
4	AB	12.0	3.2	8.8	3.2	0.20
	KrII		7.8	4.2	8.5	0.70

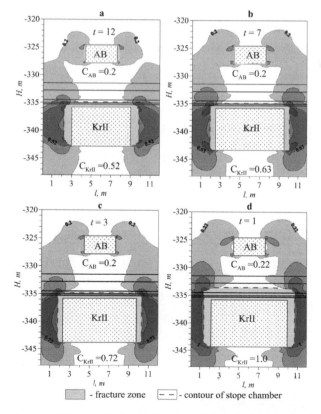

Figure 3.   Change of the load degree of rib pillars within time at the initial load degree $C_{krII} = 0.4$.

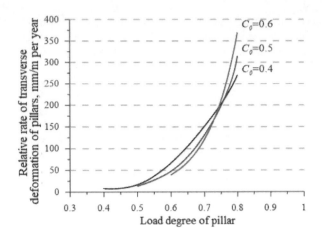

Figure 4.   Change of the relative rate of transverse deformation of pillars at various initial load degree.

The changed parameters of the bearing elements of KrII stratum room-and-pillar mining give the basis for the next modelling stage (Fig. 3b). The time period of the pillar load degree increase from 0.52 to 0.63 is $t_2 = 7$ years (Fig. 3b), while it took $t_3 = 3$ years to change it from 0.63 to 0.72 (Fig. 3c). From this moment its complete loss of bearing capacity ($C_{krII} = 1.0$) takes 1 year (Fig. 3d). So the total fracture time of the rib pillars in KrII stratum under the initial load degree $C_0 = 0.4$ and the geological conditions admitted for computations will amount to 23 years.

A similar mathematical modelling procedure was applied to other initial load degrees of pillars in KrII stratum. According to the obtained results, when $C_0 = 0.5$ the time of the pillar rigidity loss is 15 years, when $C_0 = 0.6$ it is 8.5 years and when $C_0 = 0.7$ it is 1.5 years.

According to the multivariant mathematical modelling, the average estimations of the relative rate of transverse deformation of pillars at different initial load degrees were made (Fig. 4). The data analysis shows that after the pillars reach the relative rate of transverse deformation of 50–100 mm/m per year, the relative rate of transverse deformation of pillars starts to increase rapidly. The excess of the said values may mean the loss of their bearing capacity and beginning of the pillars' progressing creeping stage. So the transverse deformation rate interval of 50–100 mm/m per year may be used as an indicator that stands for the ultimate rigidity of bearing elements in the room-and-pillar method. The provided estimations are approximate and may be subject to further specification.

4   CONCLUSION

The present study suggests an approach based on the mathematical modelling method that can be used to estimate the load degree of rib pillars within time. It considers stress-rupture strength of rocks and interconnected processes of fracture of bearing elements in the room-and-pillar mining. It estimates the time during which pillars lose their bearing capacity at different initial load values. It determines the critical rates of transverse deformation of pillars that indicate the limit state of bearing elements during the room-and-pillar method.

REFERENCES

[1] Asanov V.A., Toksarov V.N., Evseev A.V., Bel'tyukov N.L. Specific roof behavior in the southern wing of the upper Kama potash salt deposit/Journal of Mining Science. 2012. V. 48. № 1. P. 71–75.

[2] Baryakh A.A., Asanov V.A., Pankov I.L. Physical and mechanical properties of saliferous rocks of Verkhnekamsk potash deposit: course book. – Perm: Publishing house of Perm State Technical University, 2008.

[3] Baryakh A.A., Asanov V.A., Toksarov V.N., Gilev M.V. Evaluating the residual life of salt pillars/ Journal of Mining Science. 1998. V. 34. № 1. P. 14–20.

[4] Baryakh A.A., Dudyrev I.N., Asanov V.A., Pankov I.L. Interaction of layers in salt massif. Part 1. Mechanical properties of contacts/Journal of Mining Science. 1992. V. 28. № 2. P. 48–52.

[5] Baryakh A.A., Fedoseev A.K. Sinkhole formation mechanism/Journal of Mining Science. 2011. V. 47. № 4. P. 404–412.

[6] Baryakh A.A., Lobanov S.Yu., Lomakin I.S. Estimation of Salt Rocks' Long-Term Strength in Natural Conditions/Solid State Phenomena. – 2016. – Vol. 243. – Pp. 11–16.

[7] Baryakh A.A., Samodelkina N.A. On one approach to the rheological analysis of geomechanical processes/Journal of Mining Science. 2005. V. 41. № 6. P. 522–530.

[8] Baryakh A.A., Samodelkina N.A. On the calculation of pillar stability during room-and-pillar mining/ Journal of Mining Science. 2007. V 43. № 1. P. 8–16.

[9] Baryakh A.A., Shumikhina A.Yu., Toksarov V.N., Lobanov S.Yu., Evseev A.V. Fracture criteria and features of layerwise roof chambers during the development of Verkhnekamsk deposit of potash salts/Mining Journal – 2011. – Nr. 11. pp. 15–19.

[10] Goodman R. The mechanical properties of joins, Adv. Rock Mech., 1974, Vol.1, Pt A.

[11] Malinin N.N. Applied theory of plasticity and creep. – M.: Mechanical Engineering, 1975.

[12] Marakov V.E., Nesterov M.P., Neprimerov A. F. Stress changes in sylvinite pillars depending on their age and location in the worked-out area./Stressed state of massifs: Collection of scientific papers of Mining Institute of Siberian branch of Academy of Sciences of the USSR. – Novosibirsk, 1978.

[13] Nesterov, M.P. On engineering methods for calculating chain pillars/Mining Journal – 1968. – Nr. 9.

[14] Serata S. and Schults W. G. Application of stress control in deep potash mines, Mining Congress Journal, 1972, Nr. 58(11).

[15] Shevyakov, L.D., On the calculation of consistent sizes and deformation of pillars, Izvestiya of Academy of Sciences of the USSR, Engineering Supervision Department. – 1941. – Nr. 7–9.

[16] Toksarov V.N. Field studies of stope roof deformations under conditions of increased clay sealing/ Materials of the scientific conference: Strategy and processes of development of geo-resources. – Perm, 2009. P. 72–75.

[17] Tournaire. Des dimensions a donner aux pilliers des carriers et des pressions aux quelles les terrains sont soumis dans les profondeurs, Annales des mine, 8 series, 1884, T. V.

[18] Yudin R.E, Marakov V.E., Sivkov E.S., et al. Control of the stressed-strain state of the pillars and roof of stopes in the mines of Verkhnekamsk potassium deposit/Control, prediction and management of rock status in potash mines. – Leningrad: All-Russian Vedeneev Hydraulic Engineering Research Institute, 1985.

[19] Zienkiewicz O.C. The finite element method in engineering science. – London, 1971.

Geomechanics and Geodynamics of Rock Masses – Litvinenko (Ed.)
© *2018 Taylor & Francis Group, London, ISBN 978-1-138-61645-5*

Quantitative risk analysis of fragmental rockfalls: A case study

Jordi Corominas, Gerard Matas & Roger Ruiz-Carulla
Department of Civil and Environmental Engineering, Division of Geotechnical Engineering and Geosciences, Universitat Politècnica de Catalunya-BarcelonaTech, Barcelona, Spain

ABSTRACT: Rockfalls are frequent natural processes in mountain regions with the potential to produce damage. The Quantitative Risk Analysis (QRA) is an approach increasingly used to assess risk and evaluate the performance of mitigation measures. In case of fragmentation of the falling rock mass, the results of the QRA differ significantly because the number of new fragments generated increases the probability of impact while the kinetic energy of blocks and the runout is overestimated.

We have developed a procedure to account for the fragmentation, integrated in the RockGIS code (Matas et al. 2017), which is a rockfall propagation model. The procedure is applied at the Monasterio de Piedra, Spain as part of a QRA. The results show that for small-size rockfalls (<1 m³), fragmentation reduces risk to the visitors. For rockfall events >100 m³, fragmentation increases the overall risk due to the generation of multiple divergent trajectories and higher exposure of the elements at risk. For 10 m³-size rockfall events, the shorter runout and smaller kinetic energy compensates the effect of exposure. Furthermore, fragmentation makes feasible the implementation of protective measures.

Keywords: rockfall, fragmentation, Quantitative Risk Analysis, modelling, case study

1 INTRODUCTION

The quantitative analysis of risk of landslides (hereinafter QRA) has undergone remarkable development in recent years (Corominas et al. 2014). The objective of the QRA is to evaluate the consequences (damages, casualties) in case of a landslide and their probability. The QRA provides an objective evaluation of risk because the assumptions and uncertainties are declared. It yields reproducible results which allow the analysis of different scenarios and the comparison of their results. It also allows the interpretation of the results in terms of risk acceptability criteria.

For rockfalls, risk (R) is expressed as follows (modified from Agliardi et al. 2009):

$$R = \sum_{j=1}^{J} \sum_{i=1}^{I} N_i \cdot P(S/D)_i \cdot P(T/S)_j \cdot V_{ij} \tag{1}$$

where:

R the risk due to the occurrence of a rock fall of magnitude (volume) "i" on an exposed element "j" located at a reference distance S from the source,

Ni is the annual frequency of rockfalls of volume class "i",

P (S | D)$_i$ is the probability that the detached rock mass of the size class "i" reaches a point located at a distance S from the source,

P (T | S) is the exposure or the probability that an element "j" is located in the trajectory of the rock fall at the distance S, at the time of its occurrence,

V$_{ij}$ is the vulnerability of the element "j" in the case of being impacted by a block of magnitude "i".

The summation indicates that the expression of the risk must be calculated for the whole range of magnitudes (volumes) because each one is characterized by a probability of occurrence, runout, probability of impact and, therefore, different consequences.

Most of rockfalls become fragmented along the path. Fragmentation produces the separation of a rock mass into several smaller pieces upon the first impact(s) on the ground surface, which follow independent trajectories. Rockfall fragmentation has a direct effect on both hazard and risk as it causes the redistribution of the initial rock mass among the new fragments. The overall effect on the rockfall behavior is that the smaller mass of the new fragments makes them travel shorter distances and mobilize lesser kinetic energy, thus reducing $P(S \mid D)_i$. On the contrary, the probability of impact increases significantly with fragmentation

2 QRA OF FRAGMENTAL ROCKFALL EVENTS: A CASE STUDY IN SPAIN

A rock fall event of about 800 m^3 took place on February 17th, 2017 in the cliff above the Lago del Espejo (Mirror lake) in Monaterio de Piedra, Spain (Fig. 1). The cliff is composed of limestone layers, locally cavernous, that show dissolution features. The detached mass fell from a 60 m height and fragmented upon the impact on the ground. The debris extended downslope up to the lake, burying a stretch of the visitors trail. Several modules of the rockfall barrier of 1500 kJ were destroyed. The reach angle (H/L ratio) ranges between 49 and 53°, which is an abnormally high value for this volume. This is probably due to the high energy expenditure by fragmentation and breakage of the barrier. The event occurred early in the morning and no further damages were accounted.

After the event, a QRA was carried out to assess the degree of risk for different scenarios. We present here some of the results with emphasis on the influence of the fragmentation of the falling mass. The QRA analysis is performed at the reference variable distance S from the cliff, that corresponds to the visitor trail around the Lago del Espejo (Fig. 1).

The components of equation 1 are calculated for a range of rockfall volumes. The rockfall sources are assumed regularly distributed along the cliff (320 sources, one every meter). The frequency-magnitude relation for this section of the cliff was prepared using the inventory of

Figure 1. Partial view of the cliffs above the trail of the Lago del Espejo (Mirrow lake) at Monasterio de Piedra. At the foot of the cliff, the rockfall debris of February 2017.

rock blocks retained in the existing barriers over a period of 15 years (2003–2017) and completed with three large rockfall events (>500 m³) For the sake of brevity, here we present the frequency for 1, 10, and 100 m³ only, which are respectively 0.03, 0.004 and 0.0006 events/year.

The probability of reaching the trail is calculated using the simulation program RockGIS developed by our research group, whose details and characteristics are found in Matas et al. (2017). The program includes a fragmentation module based on a rockfall fractal fragmentation model (Ruiz-Carulla et al. 2017). The parameters of the model for both fragmentation and propagation were calibrated using the rockfall event of February 2017 and the location of a few fallen blocks that were removed from the cliff during prevention works carried out in March 2015.

Each source released 100 rock masses than remained intact along the path and 10 rock masses that fragmented, totaling 32,000 and 3,200 simulations respectively. The effect of fragmentation on the rockfall runout and impact probability is illustrated in Figure 2. For the sake of visualization, the figure only shows one trajectory of intact rock fall masses from a few selected detachment sources (top). For the same reason, only one fragmental rockfall event is shown as well (bottom). In the latter, the trajectories of the rock fragments are displayed. The simulations illustrate the effect of the topography in the generation of preferential trajectories, the effect of fragmentation on both runout and the kinetic energies of the blocks, and the efficiency of the barriers.

The simulation of the fragmented 10 m³ rockfalls generates a completely different scenario. First, fragmentation produces a high number of divergent trajectories. The width of the cone of block fragments increases with the distance from the impact points and with the number

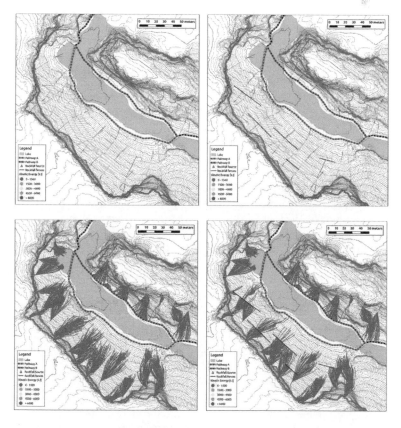

Figure 2. Top: trajectories of rockfall masses of 10 m³ without fragmentation on pathway A. Considering the presence (right) or absence of rockfall barriers; Bottom: trajectories of 10 m³ rockfall masses with fragmentation. Considering the presence (right) or absence of rockfall barriers; The existing rockfall barrier are represented by blue lines. The kinetic energies are displayed following a color code (from high to low: red orange, yellow and green).

of impacts. At the trail, the affected length may increase by a factor of between 9 and 18. This has a direct effect on risk, which increases in the trail sections that are close to the cliffs while it is significantly reduced in those sections located far away (see Fig. 2 bottom, left). However, one of the most important effects of fragmentation is the substantial reduction of the rockfall runout. This is clearly illustrated in Figure 2 bottom. A high number of block fragments either stop or are retained by the rockfall barriers. The simulation also shows that some trajectories may avoid the barriers by passing between them and/or by high bounces.

The risk for visitors to the lake, in this example, is evaluated exclusively for individuals (annual probability of loss of life) considering the annual frequency (Ni) of 1, 10 and 100 m³ rockfalls, and the runout probability P (S | D)$_i$ of reaching the trail.

The probability of impact P (T | S) takes into account two factors: the probability that the person is in the rockfall trajectory and the width of the section of the trail affected by the cone of rock fragments (W$_c$). For intact rock fall masses, it is assumed a cubic shape. For fragmental rockfalls W$_c$ is variable and depends on the distance to the source and on the characteristics of the path. It is determined by numerical modeling using the RockGIS program, considering different initial rockfall volumes and sources (some examples are shown in Figure 2 bottom) and summarized in Table 1.

The exposure, P (T | S) is based on the number of visitors. The last 15 years, the site has received an average of about 700 persons per day. In this example, the continuous flow of visitors is assumed. For people in movement, the probability of intersection with any rockfall is as follows (adapted from Nicolet et al., 2016):

$$P(T/S) = \frac{f_p \cdot (w_c + l_p)}{24 \cdot 1000 \cdot v_p} \tag{2}$$

where:
f$_p$: flow of visitors (persons/day)
W$_c$: width of the rockfall debris front (m)
l$_p$: width of the person (m)
v$_p$: is the mean velocity of persons (km/h)

A worked example is presented in Tables 2 and 3. The effect of fragmentation is nicely illustrated by the rockfall runout. For 1 m³ and 10 m³ rockfalls, the percentage of rockfall events that reach the trail has reduced from 62 to 3% and from 76 to 20%, respectively. This effect vanishes progressively as the rockfall volume increases. The reduction of the kinetic energy allows considering the feasibility of protection measures, such as the rockfall barriers. For 1 m³ intact blocks, the rockfall barriers are highly efficient allowing the reduction of P(S:D) from 62% to 13%. On the other hand, high energies developed in large unfragmented rockfalls make rockfall barrier inefficient, as shown by the P(S:D) values of Table 2. In case of fragmental rockfalls, 1500 kJ barriers can intercept a number of rockfall events of 10 and 100 m³. The comparison for 10 m³ rockfall is particularly relevant because the occurrence of fragmentation shows that the percentage of 10 m³ rockfall events that reach the trail is reduced from 75% in case of intact rock masses to 13% when fragmentation is accounted for. These results are a first estimation only because the multiple impact of blocks is not considered. It is also worth noticing that a percentage of events reach the trail because their trajectories avoid the barriers (Figure 2 bottom right) or by the high bounces of the blocks.

Table 1. Width W$_d$ (m) of the intact blocs and cone the rockfall fragments at the reference trail.

Volume (m³)	1	10	100
Intact block	1	2.2	4.6
Cone of fragments	17.5	20	40

Table 2. Annual probability of loss of life for intact (unfragmented) rockfall masses.

Pathway A (length 195 m) – contributing Cliff (length 320 m)

Without protection barriers

Class M_i (m³)	Ni	P(S:D)	P(T:S)	V	Risk R(A)
1	0.0331	0.6176	0.0219	1,0	4.50×10^{-4}
10	0.0043	0,7619	0.0394	1,0	1.30×10^{-4}
100	0.00057	0.8254	0,0744	1,0	3.52×10^{-5}

With protection barriers (1500 KJ)

Class M_i (m³)	Ni	P(S:D)	P(T:S)	V	Risk R(A)
1	0.0331	0.1273	0.0219	1,0	9.19×10^{-5}
10	0,0043	0,7472	0.0394	1,0	1.28×10^{-4}
100	0,0006	0.8254	0.0744	1,0	3.53×10^{-5}

Table 3. Annual probability of loss of life considering fragmentation of rockfalls.

Pathway A (length 195 m) – contributing Cliff (length 320 m)

Without protection barriers

Class M_i (m³)	Ni	P(S:D)	P(T:S)	V	Risk R(A)
1	0.0331	0.0322	0.0779	0,5	4.35×10^{-5}
10	0.0043	0.2035	0.1343	1,0	1.19×10^{-4}
100	0.00057	0.6240	0.2981	1,0	1.07×10^{-4}

With protection barriers (1500 KJ)

Class M_i (m³)	Ni	P(S:D)	P(T:S)	V	Risk R(A)
1	0.0331	0.0205	0.0973	0,5	3.29×10^{-5}
10	0.0043	0.1318	0.1461	1,0	8.36×10^{-5}
100	0.00057	0.5150	0.2973	1,0	8.80×10^{-5}

Fragmentation increases the impact probability due to the presence of a cone of fragments as shown by the comparison of values of P(T:S) in Tables 2 and 3. The risk values, in the worked example are of the same order of magnitude. However, in the case of Lago del Espejo, when considering fragmentation the feasibility of protection measures for mid-size events improves because risk may be reduced up to one order or magnitude when compared to unfragmented rockfalls.

REFERENCES

Agliardi F, Crosta GB, Frattini P. 2009. Integrating rockfall risk assessment and countermeasure design by 3D modelling techniques. Nat Hazards Earth Syst Sci 9:1059–1073.

Corominas, J; van Westen, C.; Frattini, P.; Cascini, L.; Malet, J.P.; Fotopoulou, S.; Catani, F.; Van Den Eeckhaut, M.; Mavrouli, O; Agliardi, F.; Pitilakis, K.; Winter, M.G.; Pastor, M.; Ferlisi, S.; Tofani, V.; Hervás, J. & Smith, J.T. 2014. Recommendations for the quantitative analysis of landslide risk. Bulletin of Engineering Geology and the Environment, 73: 209–263.

Matas, G., Lantada, N., Corominas, J., Gili, J.A., Ruiz-Carulla, R., Prades, A. 2017. RockGIS: a GIS-based model for the analysis of fragmentation in rockfalls. Landslides, 14: 1565–1578.

Nicolet, P. Jaboyedoff, M., Cloutier, C., Crosta, G., Lévy, S. 2016. Brief Communication: On direct impact probability of landslides on vehicles. Natural Hazards and Earth System Sciences 16, 995–1004.

Ruiz-Carulla, R., Corominas, J., Mavrouli, O. 2017. A fractal fragmentation model for rockfalls. Landslides, 14: 875–889.

Geomechanics and Geodynamics of Rock Masses – Litvinenko (Ed.)
© *2018 Taylor & Francis Group, London, ISBN 978-1-138-61645-5*

Experimental and numerical investigation the divergence of horizontal and vertical displacement in longwall mining

E.T. Denkevich
"TOMS-Project", Head of Mine Survey Team, Vasilevsky Island, St. Petersburg, Russia

O.L. Konovalov & M.A. Zhuravkov
Belarusian State University, Nezavisimosty av., Minsk, Belarus

Keywords: Viscoplasticity; InSAR; Displacement; Subsidence; Numerical simulation; Mixed meshing; Longwall-mining; Creep flow

Mining companies are looking for higher productivity solutions for low and medium-thickness potash seams. The suitable approach here is high-speed cutting longwall-mining. The extreme excavation process require the more frequency control over stress–strain state of rock's massif to reach safety of above process. Many specialists propose to use InSAR-technology to control the deformations of rock's massif surface (Figure 1).

Unfortunately, it is not possible to retrieve the full displacement vector from a single InSAR measurement. To overcome this there are proposed some heuristic methods based on the hypothesis of a physical relation between horizontal and vertical displacements. In above methods use the hypothesis that the horizontal displacements are proportional to the tilts (i.e. first spatial derivative of vertical deformation) [1].

During experimental study of subsidence process caused by longwall potash mining in Starobin Deposit of potassium salts, it was detected that Kratzsch's hypothesis [1] do not take place. The high accuracy GPS monitoring of subsidence process shows that point with maximum of absolute vertical displacement (MVD) are always falls behind from the point with zero horizontal displacement (ZHD). For 400-meter excavation depth, the divergence between above extremums had place in interval from 10 to 30 meters. However, according

Figure 1.   View of subsidence process on InSAR interferometry.

Kratzsch's hypothesis positions of MVD and ZHD should be equal. The experimental study shows also the dependence of divergence from speed of excavation.

To investigate the reason of above phenomena, the special numerical model was developed that provide to simulate high-speed longwall-mining. The goal of such developments was find the way for estimation MVD/ZHD divergence to improve InSAR deformation measurements. Take into account that we had not known the reason of the above phenomena, we had try to elaborate maximum universal and flexible numerical model. To describe hardening and softening as well as dilatancy and creep behavior of salt rocks, the numerical model with modified continuum damage model based on Mohr-Coulomb criteria was utilized.

We propose the following equation for the damage calculation:

$$D = \begin{cases} 1 - B/A & \text{if} \quad A > B \\ 0 & \text{else} \end{cases}$$

where $A = 1/2(\sigma_1 - \sigma_3) + 1/2(\sigma_1 + \sigma_3)\sin(\varphi)$ and $B = C \cdot \cos(\varphi)$. Here C is cohesion and φ friction angle.

Unfortunately numerical elastic model with described above continuum damage model do not able to reproduce the real form and dynamic of subsidence surface for Starobin Deposit. The reason of this is complexity of subsidence process. The form of subsidence surface are extremely dependent from size and dynamic of failure zone "1" (Figure 3).

For more accurate modeling of failure zone (zone "1") we will take into account the laminar structure of potash deposit. We propose to add in FEM grid special contact elements

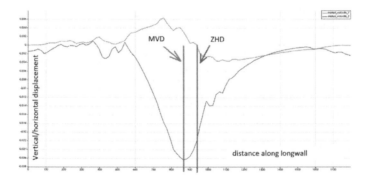

Figure 2.   MVD/ZHD divergence.

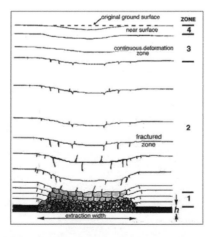

Figure 3.   Deformation and failure zones (1 – near zone; 2 – continous deformation zone; 3 – fracture zone; 4 – failure zone).

(truss element) with asymmetric behavior (Figure 4) to describe the layers behavior. During compressive loading, the Young coefficient of truss element is equal to appropriate coefficient of material. In case of tension, the Young coefficient are degraded to zero.

The damage model based on Mohr-Coulomb criteria and special contact elements for lamination give as possibility to reproduce form and dynamic of subsidence surface on active stage (Figure 5). However, the long-term modeling need to include in numerical model some rheology "submodel". We follow to approach, that for thin-layering salt massif, viscoelastic deformation can be modeling as planar viscous-flow [2].

The visco-creep flow of salt rocks is controlled by the Maxwell viscosity η_M

$$\dot{\varepsilon}(\sigma_{eff}) = \frac{\sigma_{eff}}{3\eta_M},$$

where $\dot{\varepsilon}$ is creep rate of deformation. We also consider visco-creep flow as planar process and restrict by criterion $\sigma_{eff} > \sigma^*_{eff}$.

To avoid numerical instability caused by extraction of some finite element from mesh, we use special technique of mixed meshing. Extracted potash seams are presented as grid of contact elements (truss element). Mixed meshing give us possibility to simulate excavation process as decreasing of length for some vertical truss elements. The example of mixed meshing is presented on Figure 6.

To investigate the reason of MVD/ZHD divergence, the series of numerical experiments based on described above model was implement. Three basic sets of rock's beds was included in FE network: sedimentary bed, argillo-marlaceous strata (AMS) and salt bed. The physics-mechanical properties of the selected sets of rock's beds are presented in Table 1.

Figure 4.   Layers of contact elements.

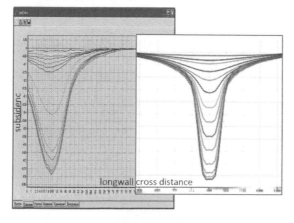

Figure 5.   Observed (left) and modeling (right) subsidence.

783

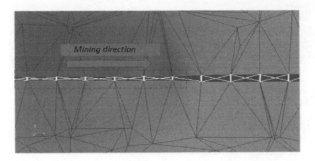

Figure 6. The potash seams are presented as grid of contact elements.

Table 1. Values of the physics-mechanical parameters.

Bed	Physics-mechanical parameters					
	$E(GPa)$	v	$C(MPa)$	$\varphi(rad)$	$\mu(MPa \cdot day)$	$\rho(kg/m^3)$
Sedimentary	0.7	0.3	1	0.6	10^{14}	2200
AMS	15	0.3	2	0.6	10^{14}	2400
Salt	20	0.3	2	0.77	10^{12}	2400

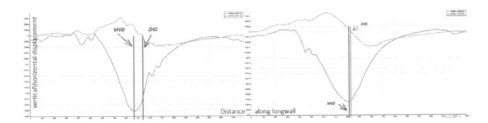

Figure 7. MVD/ZHD divergence for Maxwell viscosity value 10^{11} (left) and 10^{12} (right).

Extracted potash seams are presented as sub-grid of contact elements (truss element) inside FE network. Mentioned experiments showed that visco-creep flow is key process of VHE divergence. There are demonstration of MVD/ZHD divergence for different values of Maxwell viscosity of salt bed in Figure 7. The increase of viscosity lead to degradation of divergence.

In case of 10^8 viscosity, numeric model present 20 meter MVD/ZHD divergence. The results of modeling correlates with observed values.

The future improvement of presented numerical model can lead to accurate estimation of MVD/ZHD divergence and as result to increase accuracy of InSAR measurement for potash mining.

REFERENCES

[1] Kratzsch H., 1983, "Mining Subsidence Engineering", 543 pp.
[2] Kononova, N.S., "Geomechanical substantiation of stability of mine workings and bore holes in viscoplastic massifs", International Journal of Mining Institute, 152, 2002, pp. 129–132.

Geomechanics and Geodynamics of Rock Masses – Litvinenko (Ed.)
© *2018 Taylor & Francis Group, London, ISBN 978-1-138-61645-5*

Numerical analysis of the rheological behaviour of the Socompa debris avalanche, Chile

Federico Vagnon
Department of Earth Science, University of Turin, Turin, Italy

Marina Pirulli
Department of Structural, Geotechnical and Building Engineering, Politecnico of Turin, Turin, Italy

Irene Manzella
School of Geography, Earth and Environmental Sciences, Plymouth University, Plymouth, UK

Karim Kelfoun
Laboratoire Magmas et Volcans, OPGC, UMR Clermontn Université-CNRS-IRD, Clermont-Ferrand, France

Anna Maria Ferrero
Department of Earth Science, University of Turin, Turin, Italy

ABSTRACT: Socompa Volcano provides one of the world's best-exposed example of a sector collapse that generated debris avalanche deposit. The debris avalanche, occurred about 7000 years ago, involved 25 km^3 of fragmented rock that formed a thin but widespread (500 km^2) deposit.

Numerical model of this event was already performed using a shock-capturing method based on double upwind Eulerian scheme in order to provide information for investigating, within realistic geological context, its dynamic and run-out (Kelfoun and Druitt 2005).

This paper analyses an important aspect of the continuum numerical modeling of rapid landslides as debris avalanche: the interchangeability of rheological parameter values. The main question is: by using the same rheological parameter values, are the results, obtained with codes that implement the same constitutive equations but different numerical solvers, equal? Answering this question has required to compare the previous back analysis results with new numerical analyses performed using RASH3D code.

Different rheological laws were selected and calibrated in order to identify the law that better fits the characteristics of the final debris deposit of the Socompa landslide.

Keywords: Volcanic debris avalanche, Numerical modelling, Runout simulation, Depth-averaged equations, rheological laws

1 INTRODUCTION

The collapse of a giant sector of the Socompa Volcano caused a long runout debris avalanche, which represents one of the most critical and hazardous types of geological instability phenomena (Melosh, 1990). The potential for destruction of this type of flow-like landslides, due to the extremely rapid propagation velocity, requires reliable forecasting methods to predict their motion characteristics.

Continuum mechanics based numerical models (e.g. Savage and Hutter, 1989, O'Brien et al., 1993, Hungr, 1995, Iverson and Delinger, 2001, Mc-Dougall and Hungr, 2004, Pirulli, 2005, Pastor et al., 2009, Manzella et al., 2016) are useful tools for investigating, within realistic geological contexts, the dynamics of these phenomena.

The back analysis of real events is indispensable for the correct selection of the rheological laws and the calibration of the rheological parameters. Moreover, in order to perform robust numerical analyses, two aspects shall be considered: firstly, the use of more than one code and the comparison of results are recommended (Pirulli and Sorbino, 2010). Secondly, the interchangeability of rheological parameters should be evaluated. This aspect is particularly important because it helps users in the decisional process for assessing potential risks and evaluating/designing possible countermeasures (Vagnon, 2017).

The aim of this paper is to evaluate the interchangeability between calibrated values of rheological parameters comparing the simulation results of two different continuum-based numerical codes: VolcFlow (Kelfoun and Druitt, 2005) and RASH3D (Pirulli, 2005). In the next Sections, the codes are briefly described and used to back-analyse the Socompa debris avalanche. The obtained results are compared and discussed. Moreover, new simulations are carried out using Bingham rheology.

2 DESCRIPTION OF THE SOCOMPA AVALANCHE

Socompa Volcano is a stratovolcano located at the border between Chile and Argentina, in the Andes Mountains (Figure 1a). About 7000 years ago, the Chilean sector collapsed, generating a 40 km long debris avalanche that flowed into the flat and arid plan below before being deflected to northeast by a range of hills, forming a frontal lobe (Francis et al., 1985, Wadge et al., 1995, Van Wyk de Vries et al., 2001, Kelfoun and Druitt, 2005). The debris deposit covered an area of 500 km^2, forming a sheet of 50 m average thickness. The deposit has an estimated volume of about 36 km^3 and it results from the sum of two subsequent events. The first of 25 km^3 is analysed in the paper, while the second of 11 km^3 that gave origin to the Toreva blocks deposit (Figure 1b) is not analysed due to its negligible runout distance.

The avalanche deposit is characterized by a mixture of brecciated lavas and volcanoclastic deposit (Socompa Breccia Facies; SB) directly originated by the Socompa edifice itself and ignimbrites, gravels, sands and minor lacustrine evaporates from the Saline Formation (Reconstruited Ignimbrite Facies; RIF) of the volcano basement). Mostly of the deposit volume is constituted of RIF and only the 20% of SB.

The first avalanche was generated by a series of retrogressive failures that merged to form a single flowing mass (Wadge et al., 1995) that spread on a basal layer of RIF, characterized by very weak mechanical resistance (Van Wike de Vries et al., 2001). The deposit can be morphologically divided into two main zones by a median escarpment (ME), oriented NE-SW (Figure 1b), generated by secondary flow off the western and north-western basin margins.

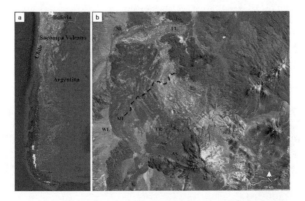

Figure 1. Location of Socompa Volcano (a) and aerial image of the avalanche deposit showing the avalanche scar (AS), the Toreva blocks (TB), the median escarpment (ME), the frontal lobe (FL) and northern and western levees (NL and WL) (b). In detail: the red dotted line surrounds the deposit limits; the blue continuous line draws the margins of Toreva deposit; the black dotted line highlights the medial escarpment.

3 NUMERICAL MODELING

Kelfoun and Druitt (2005), starting from geological investigations and morphological observations, reconstructed the original topography of the area before the collapse. Then, they performed several numerical simulations using VolcFlow code, testing different rheological laws in order to find the best model for obtaining the actual avalanche deposit configuration.

In this work, the Authors want to compare VolcFlow numerical results with those obtained with RASH3D code (Pirulli, 2005) for evaluating the interchangeability between codes of calibrated rheological values and providing new run-out simulations with a Bingham rheology.

3.1 *Basic equations*

The numerical simulation of rapid landslides is a common practice since in 1989, when Savage and Hutter firstly introduced the depth-averaged equations for the dynamic analysis of flowing mass. The hypotheses for applying depth-averaged equations to rapid landslides are:

- both thickness and length of flowing mass are assumed to exceed the size of single moving particles of several times;
- the flow thickness is considerably smaller than its length;
- the real moving mixture is replaced by an "equivalent fluid" whose properties approximate the bulk behaviour of the real mixture;
- the flowing mass is described as a single-phase, incompressible and homogeneous material;
- a kinematic boundary condition is imposed on free and bed surfaces;
- the rheological characteristics are all included in a single term acting at the interface between flow and terrain surface.

Under the above listed conditions, the motion is described by the equations of mass and momentum conservation:

$$
\begin{cases}
\dfrac{\partial h}{\partial t} + \dfrac{\partial (\overline{v_x}h)}{\partial x} + \dfrac{\partial (\overline{v_y}h)}{\partial y} = 0 \\[3mm]
\rho \left(\dfrac{\partial (\overline{v_x}h)}{\partial t} + \dfrac{\partial (\overline{v_x^2}h)}{\partial x} + \dfrac{\partial (\overline{v_x v_y}h)}{\partial y} \right) = -\dfrac{\partial (\overline{\sigma_{xx}}h)}{\partial x} + \tau_{Zx_{(z=b)}} + \rho g_x h \\[3mm]
\rho \left(\dfrac{\partial (\overline{v_y}h)}{\partial t} + \dfrac{\partial (\overline{v_y v_x}h)}{\partial x} + \dfrac{\partial (\overline{v_y^2}h)}{\partial y} \right) = -\dfrac{\partial (\overline{\sigma_{yy}}h)}{\partial y} + \tau_{Zy_{(z=b)}} + \rho g_y h
\end{cases}
\tag{1}
$$

where $\overline{v} = \left(\overline{v_x}, \overline{v_y} \right)$ denotes the depth-averaged flow velocity in a reference frame (x, y, z) linked to the topography, ρ is the bulk material density, h is the flow depth, τ is the shear stress in the x and y direction, $\overline{\sigma} = \left(\overline{\sigma_{xx}}, \overline{\sigma_{yy}} \right)$ is the depth-averaged stress and g_x, g_y are the projections of the gravity vector along the x and y direction.

The here applied VolcFlow and RASH3D codes differ in the numerical scheme adopted for solving the above equations. VolcFlow code uses a Eulerian explicit upwind scheme for solving the system of equations (1) where scalar quantities (thickness and terrain elevation) are evaluated at the centres of cells and vectors (velocity and fluxes) at the edges (Figure 2a). For a complete description of this method, see Kelfoun and Druitt 2005.

The RASH3D Eulerian code, developed by Pirulli (2005) uses a finite volume scheme for modelling rapid landslide run out problems. The system of equations (1) is discretized on an unstructured triangular mesh with a finite element data structure using a particular control volume, which is the median dual cell (Pirulli, 2005). Dual cells C_i are obtained by joining the centres of mass of the triangles surrounding each vertex P_i of the mesh (Figure 2b).

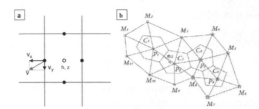

Figure 2. Definition of scalars and vector (a) in the numerical scheme of VolcFlow code (modified after Kelfoun and Druitt 2005) and triangular finite-element mesh and dual cells (C1, C2, C3, C4) in RASH3D code (b) (modified after Pirulli 2005).

3.2 Rheological laws

As stated above, the complex rheology of the flowing mass is incorporated in a single term (τ) that describes the frictional stress generated between terrain surface and flowing body.

In this paper, three rheologies were selected for the numerical back-analysis of Socompa avalanche:

1. Frictional rheology in which the resisting shear stress depends only on normal stress and it is independent of velocity.

$$\tau_{zi} = -\left(\rho \cdot g_z \cdot h \cdot tan\varphi_{bed}\right)\frac{v_i}{\|\overline{v}\|} \quad i = (x, y) \tag{2}$$

where φ_{bed} is the bulk friction angle.

2. Constant retarding stress in which the basal shear stress is constant and consequently independent by velocity, normal stress and frictional parameters.

$$\tau_{zi} = -const\frac{v_i}{\|\overline{v}\|} \quad i = (x, y) \tag{3}$$

These two rheological laws are implemented in both the presented codes and they were used to compare the RASH3D analyses with the already published VolcFlow simulations (Kelfoun and Druitt, 2005).

3. Bingham rheology combines plastic ad viscous behaviour, so that the flowing mass moves as a rigid body below a given threshold yield strength and then have a viscous behaviour above this threshold. The basal stress is determined solving the following equation:

$$\tau_{zi}^3 + \left(\frac{\tau_y}{2} + \frac{\mu_B \overline{v}_i}{h}\right)\tau_{zi}^2 - \frac{\tau_y^3}{2} = 0 \tag{4}$$

where τ_y is the Bingham yield stress and μ_B is the Bingham viscosity.

In RASH3D equation (4) is solved using polynomial economization technique proposed by Pastor et al. (2004). Bingham rheological law was selected to back-analyse Socompa avalanche since the type of material that characterized the deposit had a ductile behaviour (RIF) and behaved as a lubricant for the SB facies.

4 RESULTS

Numerical analyses were carried out following two different steps. Firstly, the VolcFlow numerical simulations (Kelfoun and Druitt, 2005) were replicated using RASH3D code for evaluating the interchangeability of rheological values. Then, once that RASH3D results were commented, a back-analysis using Bingham rheology was performed.

The goodness of numerical simulations is evaluated if the following conditions are satisfied:

1. best fit to the north-western margin
2. best fit to overall outline of the deposit
3. reproduction of the main structures, especially the median escarpment (cfr. Figure 1).

4.1   *Evaluation of the two codes interchangeability of rheological values*

Figure 3 compares VolcFlow (a and c) and RASH3D (b and d) simulations of the final ava-lanche deposit considering a frictional behaviour (model 1, Figures 3a and 3b) and a constant retarding stress rheological law (model 2, Figures 3c and 3d). The rheological values used for model 1 are $\varphi_{bed} = 2.5°$, in an isotropy condition of stresses, and a constant retarding stress equal to 52 kPa for model 2.

For each time step of the simulations, the thickness and the areal distribution of the deposit simulated by the two codes are satisfyingly comparable. In general, RASH3D simula-tions show a marked lateral spreading: however, the calculated thickness values at the margin of the simulated deposit are less than 10 cm. For what it concerns model 2, the conditions previously imposed for evaluating the goodness of the model (point 1 to 3, Section 4) were satisfied: the overall outline of the deposit was respected and the median escarpment, char-acteristic of this deposit, was well reproduced.

4.2   *Bingham rheology*

The Bingham rheology was never used before for simulating Socompa avalanche but, on the basis of previously discussed geological and geomorphological evidences, this rheology was adopted to evaluate thickness and velocity of the Socompa emplacement with the RASH3D code.

A large number of analyses was performed to obtain the combination of rheological val-ues that best simulate the deposit in terms of extension, thickness and escarpments. These conditions were satisfied considering the Bingham yield stress and the viscous coefficient respectively equal to 52 kPa and 10 kPa*s. Figure 4 shows the depositional height (a) and

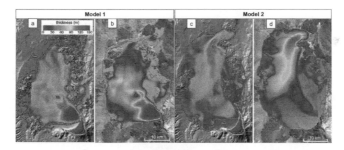

Figure 3.   Final deposit thickness of the Socompa avalanche considering frictional rheological law with $\varphi_{bed} = 2.5°$ and in an isotropy condition of stresses in VolcFlow code (a) and RASH3D code (b). Figures c and d show the obtained final deposit considering a constant retarding stress rheological law with $\tau = 52$ kPa in VolcFlow code (c) and RASH3D code (d).

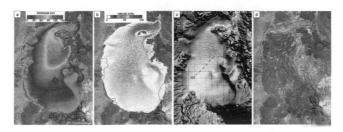

Figure 4.   Deposit thickness (a), flow velocity (b) and shaded relief map of the Socompa avalanche, simulated considering Bingham rheological law with $\tau = 52$ kPa and $\mu = 10$ kPa*s using RASH3D code and satellite image of the actual Socompa emplacement (d).

the flow velocity (b) of the simulated emplacement. The simulated final deposit (Figure 4a) remarkably well reproduces the real event. In particular, analysing Figures 4c and 4d that represent the shaded relief map of the simulated deposit and the satellite image, a topographic discontinuity is evident (red dotted line in Figure 4c) and it represents the median escarpment. Moreover, the presence of a frontal lobe can be clearly identified.

5 CONCLUSIONS

In this paper, the two codes VolcFlow and RASH3D, based on a continuum mechanics approach, were compared.

The results obtained from the carried out analyses have highlighted the good interchangeability of the rheological values between the presented codes. Moreover, the Bingham rheological law was applied to further simulate the avalanche emplacement: the results were satisfying both in terms of areal extension, depositional heights and topographical evidences (frontal lobe, median escarpment and well-defined lateral margins) compared to the actual morphological situation.

Further developments of this research will include the use of others numerical codes with different numerical scheme for solving mass and momentum conservation equations (e.g. Lagrangian code) for again evaluating and comparing these approach.

REFERENCES

Francis, P.W., Gardeweg, M., Ramirez, C.F., and Rothery, D.A. 1985. Catastrophic debris avalanche deposit of Socompa volcano, northern Chile, Geology, 13, 600–603.
Hungr, O. 1995. A model for the runout analysis of rapid flow slides, debris flows, and avalanches. Canadian Geotechnical Journal, 32(4), 610–623.
Iverson, R.M., and Denlinger, R.P. 2001. "Flow of variably fluidized granular masses across three-dimensional terrain: 1. Coulomb mixture theory." Journal of Geophysical Research: Solid Earth, 106(B1), 537–552.
Kelfoun, K., and Druitt, T.H. 2005. Numerical modeling of the emplacement of Socompa rock avalanche, Chile, J. of Geophysi. Res., 110, B12202, doi:10.1029/2005JB003758.
Manzella, I., Penna, I., Kelfoun, K., and Jaboyedoff, M. 2016. High-mobility of unconstrained rock avalanches: Numerical simulations of a laboratory experiment and an Argentinian event, in Landslides and Engineered Slopes. Experience, Theory and Practice, 1345–1352.
Mcdougall, S., and Hungr, O. 2005. Dynamic modelling of entrainment in rapid landslides. Canadian Geotechnical Journal, 42(5), 1437–1448.
Melosh, H.J. 1990. Giant rock avalanches, Nature, 348, 483–484.
O'brien, J.S., Julien, P.Y., and Fullerton, W.T. 1993. Two Dimensional Water Flood and Mudflow Simulation. Journal of Hydraulic Engineering, 119(2), 244–261.
Pastor, M., M. Quecedo, E. Gonzalez, M.I. Herreros, J.A. Fernandez Merodo, and P. Mira. 2004. Simple approximation to bottom friction for Bingham fluid depth integrated models. Journal of Hydraulic Engineering 130(2): 149–155.
Pastor, M., Haddad, B., Sorbino, G., Cuomo, S., and Drempetic, V. 2009. A depth-integrated, coupled SPH model for flow-like landslides and related phenomena. International Journal for Numerical and Analytical Methods in Geomechanics, 33(2), 143–172.
Pirulli, M. 2005. Numerical modelling of landslide runout, a continuum mechanics approach. Ph.D dissertation, Politecnico of Turin, Turin, Italy.
Pirulli, M., and Sorbino, G. 2008. Assessing potential debris flow runout: a comparison of two simulation models. Natural Hazards and Earth System Science, 8(4), 961–971.
Savage, S.B., and Hutter, K. 1989. The motion of a finite mass of granular material down a rough incline. Journal of Fluid Mechanics, 199(1), 177.
Vagnon, F. 2017. Theoretical and experimental study on the barrier optimization against debris flow risk. Ph.D dissertation, University of Turin, Turin, Italy.
Van Wyk de Vries, B., Self S., Francis, P.W., and Keszthelyi, L. 2001. A gravitational spreading origin for the Socompa debris avalanche, J. Volcanol. Geotherm. Res., 105, 225–247.
Wadge, G., P.W. Francis, and C.F. Ramirez (1995), The Socompa collapse and avalanche event, J. Volcanol. Geotherm. Res., 66, 309–336.

Numerical study on the strategies to reduce the risk of induced seismicity in an enhanced geothermal system

Wentao Feng
Energy Research Center of Lower Saxony (EFZN), Goslar, Germany

Zhengmeng Hou
Energy Research Center of Lower Saxony (EFZN), Goslar, Germany
Institute of Petroleum Engineering, TU Clausthal, Clausthal-Zellerfeld, Germany

Jianxing Liao & Patrick Were
Energy Research Center of Lower Saxony (EFZN), Goslar, Germany

ABSTRACT: Hydraulic fracturing technology is essential for the development of Enhanced Geothermal Systems (EGS) to increase the permeability of tight rock formations and hence the energy recovery from a petro geothermal reservoir. However, fracturing and the subsequent increase in energy production may pose the risk of induced seismicity. Previous studies established the mechanisms of microseismic events during hydraulic fracturing in the Deep Heat Mining project Basel using the numerical simulator FLAC3D^{plus}. In this paper, an innovative injection strategy using a linear increasing injection rate to reduce the maximum magnitude of microseismic events M_{max} has been proposed. In addition, the new EGS-strategy allows a combination of both the linear increased injection rate and the multiple fracture system to be studied numerically. Results show that the risk of induced seismicity can be considerably minimized by the proposed strategy. The simulation shows that increasing the number of fractures in the tight reservoir decreases M_{max} significantly. It is recommendable that multiple hydraulic fracturing technology be applied in the development of enhanced geothermal systems not only to minimize the risk of induced seismicity but also to increase the surface area for heat exchange and recovery efficiency.

1   INTRODUCTION

The exploitation of renewable energy has become a crucial method for mitigating global energy and climate crises (Gou et al., 2015). Compared with other types of renewable energy, e.g. PV or wind energy, deep geothermal energy offers more advantages including being both sustainable and economical (IEA, 2017). In an enhanced geothermal system, hydraulic fracturing is a key technology to increase formation permeability (Lu et al., 2015) and improve the energy recovery from reservoir. However, this technology has its own shortcomings. Hydraulic fracturing can trigger seismicity (Hou et al., 2012), which is one of the main obstacles hindering its industrial application in production from enhanced geothermal system.

In the previous study by Hou et al., 2013, the coupled Thermo-Hydro-Mechanica (THM) numerical simulator FLAC3D^{plus} was developed to investigate hydraulic fracturing in the Deep Heat Mining (DHM) project Basel. Meanwhile, the fundamental mechanisms of microseismic events during hydraulic fracturing have been studied in detail. In this paper, innovative injection strategies in combination with multiple hydraulic fracturing technologies have been proposed to reduce the risks of induced seismicity and to increase the heat exchange area in the reservoir and hence the efficiency of heat recovery. These parameters have been studied and tested based on the history-matched Basel model.

2 ASSESSMENT OF INDUCED SEISMICITY BASED ON NUMERICAL SIMULATION

In the developed numerical simulator FLAC3D^{plus}, the coupled HM responses were obtained by solving the basic equations of static mechanical equilibrium and geometry, (poroelasto-plastic) constitutive equations as well as the fluid motion equations. These equations have been solved using the explicit finite difference method (Itasca 2009). This paper also considers the enhancement permeability by hydraulic fracturing of a petrogeothermal reservoir. The seismic events can be evaluated based on the simulated results, including node displacements and velocities.

In seismology, the magnitude of a seismic event is commonly expressed as moment magnitude M_w, which can be calculated from the seismic moment M_0 using Eq. 1 (Hanks & Kanamori, 1979). Seismic moment M_0 is a measure of the total deformation energy released during an event. It is calculated using Eq. 2 (Kanamori & Anderson, 1975).

$$M_w = \frac{2}{3} log M_0 - 6.07 \tag{1}$$

$$M_0 = \sum_{i=1}^{n} G_i \cdot \Delta D_i \cdot A_i \tag{2}$$

where G_i is shear modulus [Pa], D_i is dislocation vector for the fractured element [m] and A_i the fracture area in the element [m^2].

Another important assessment parameter is the local magnitude M_L, which can be calculated using the empirical equation of Ahorner & Sobisch (1988) (Eq. 3). In FLAC3D^{plus} the released energy E would be computed as the kinetic energy from each grid point (Eq. 4) at a certain point in time.

$$log E = 3.81 + 1.64 M_L \tag{3}$$

$$E = \frac{1}{2} \sum_{i=1}^{n} m_i \left(v_{ix}^2 + v_{iy}^2 + v_{iz}^2 \right) \tag{4}$$

where E is the released kinetic energy [J], M_L is the local magnitude of the seismic event [–], v_{ij} is the velocity of grid point [m/s] and m_i the mass [kg].

3 NUMERICAL INVESTIGATION OF THE PROPOSED STRATEGIES TO REDUCE THE RISKS OF INDUCED SEISMICITY IN EGS

In this paper, numerical simulations were carried out using the history-matched model of the DHM project Basel from previous studies (Hou et al., 2013). The model was a ¼ symmetric model (Fig. 1a) and has been calibrated and verified relying on history matching (Fig. 2).

This model has a height of 1,179 m (in the z-direction, from –4,030 m to –5,209 m) and a breadth of 500 m (in the y-direction). Its length is 700 m (in the x-direction). The grid elements are divided into three zones according to their properties: "granite" (intact granite), "granite_frac" (naturally fractured granite) and "water" (naturally fractured granite), as well as the zone of injection. The "granite_frac" zone extends over a range from z = –4,030 m to –5,109 m with a width of 100 m (in the y-direction). Its strike direction has an orientation angle of ± 15° from the direction of the maximum horizontal stress σ_{Hmax} (in the x-direction). Compared to the zone "granite_frac" the zone "granite" possesses no pre-existing joints and hence characterized by a higher strength as well as a lower permeability. The zone "water" corresponds to the injection section from –4,630 m to –5,009 m (in the z-direction), i.e. a total height of ca. 380 m). Fig. 1b shows the initial stress state and Table 1 lists the mechanical and hydraulic parameters that are used in the simulation. The fluid applied in the stimulation possesses a bulk modulus of 2 GPa and a viscosity of 1$_{cp}$. The measured and simulated bottomhole pressure (BHP) with the corresponding injection rates are illustrated in Fig. 2. The

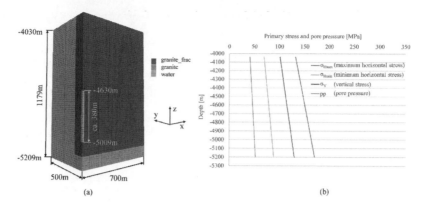

(a)   (b)

Figure 1. (a) Quarter of geometrical model and (b) initial primary stress and pore pressure distribution vs. depth.

Table 1. Parameters for simulation of hydraulic fracturing in the DHM project, Basel.

Parameters	Units	Joint surface	Granite_Frac	Granite
Density (ρ)	kg/m³	–	2500	2500
Young's modulus (E)	GPa	–	60	60
Poisson's ratio (υ)	–	–	0.2	0.2
Cohesion (c)	MPa	4	10	20
Friction angle (φ)	°	30	30	45
Tension strength (σ_t)	MPa	1	2.5	2.5
Porosity (n)	–	–	1%	1%
Permeability (k)	m²	–	4×10^{-17}	4×10^{-18}

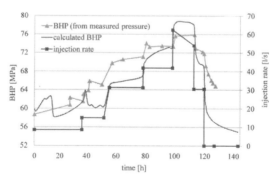

Figure 2. Treatment schedule and comparison of the bottomhole pressure (BHP) calculated from the measured pressure and simulated pressure vs. time.

simulated pressure is comparable to the measured pressure. This simulation was considered as a basis for the study of the proposed injection strategies and multiple hydraulic fracturing in this paper.

3.1 *Injection strategies with linear increasing injection rate*

A linear increasing injection method was proposed and simulated in the new strategy. To ensure the injected volume was the same as that in the basis situation, the maximum injection rate was raised to 48.58 l/s (see Fig. 3b). The results of the basis simulation are shown in Fig. 3a for comparison.

The new injection strategy yielded a final stimulated reservoir volume (SRV) of 7.011×10^7 m^3 (900 m $\times$ 100 m $\times$ 779 m). The maximum profile of the fractured zone is about 0.65 km^2. The bottomhole pressure attained a maximum level of 74.5 MPa (see Fig. 3b) comparable to that in the basis simulation (78.3 MPa, Fig. 2). However, a few fluctuations in the pressure curve are noticeable at the beginning of the injection (Fig. 3c). These oscillations can be attributed to the effect of "break down pressures". This means that a pressure drop occurs during the initiation of larger cracks, since the injected fluid has an increased accessible volume during this process. This effect is compensated for by continuous injection at an increasing rate. The seismic magnitude forms a corridor during the stimulation, i.e. the average level is relatively constant. The M_{max}, which also occurs after shut-in (see Figs. 3a & 3b), is clearly reduced from $M_L = 2.80$ and $M_w = 2.30$ to $M_L = 2.47$ and $M_w = 2.09$ (Figs. 3a & 3b), respectively.

Since the linear increasing injection method showed an immediate improvement in terms of reducing the seismic risk, further variants were also tested. In all the variants, the maximum injection rate was maintained at 46.58 l/s. However, the turning point at which the injection rate changed varied with variant. Variant 1 provided the same duration of increase and decrease (Fig. 3d), while Variant 2 consisted of a rapid rise and a slow decay (Fig. 3e). The resulting induced seismicity showed slight reductions in M_{max} ($M_{L,max}$) to 2.41 and 2.39, respectively.

Table 2 summarizes the results for different simulations. Variant 2 of the linear increasing injection method (with a rapid increase and slow decrease) appears to be the most advantageous. The generated fracture zone profile in this variant takes first place and simultaneously releases the largest total seismic energy during the simulation period. In spite of this, the maximum seismic magnitude is the lowest in comparison with the other cases. The seismic events appear to be well distributed during the simulation period, i.e. more small events appear.

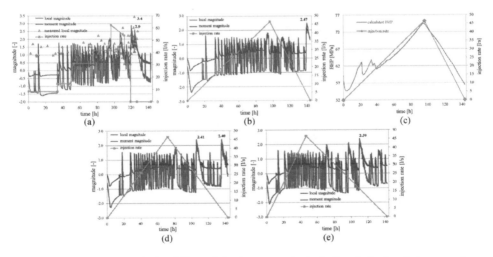

Figure 3. Treatment schedule for different injection strategies including (a, b, d and e) comparison of their calculated local and moment magnitude vs. time, (c) compares the calculated BHP and the linear injection rate vs. time.

Table 2. Results of the simulation cases with varying injection strategy.

Cases	Max. profile of the fracture zone [km^2]	Total seismic energy [J]	Calculated M_{Wmax} [−]
History matched DHM project	0.66	10.8×10^3	2.3
Basis linear injection method	0.65	9.49×10^3	2.09
Variant 1	0.69	9.83×10^3	2.03
Variant 2	0.71	10.2×10^3	2.03

3.2 *Injection strategies for linear increasing injection rates combined with a multiple fracture system*

In a further step, a combination of the modified injection strategy (Variant 2 of the linear increasing injection method, see Fig. 3e) and different variants with multiple fractures systems were numerically tested. For these tests, a fictitious horizontal bore at the depth of –4,825 m has been assumed, from which the transverse multiple fractures were generated (Fig. 4). It was presumed that the stimulations were carried out at the same time with the same injection strategy.

A model generation was performed in a similar pattern to that used in the DHM project, i.e. the models still possessed three zones ("granite", "granite_frac" and "water"). Only its construction was adapted. For reasons of symmetry, a ½ model was used, with dimensions of 350 m (y) × 679 m (z) (from –4,530 m to –5,209 m) × 1,200 m (y) for the 2-Frac model (Fig. 4, blue line), 1,600 m (y) for the 4-Frac model (Fig. 4, red line) and 2,000 m (y) for the 6-Frac model (Fig. 4, geometric model). The zone "granite_frac" was centrally positioned in a y-direction extending over the entire model width (in an x-direction). This zone lies between –4,530 m and –5,000 m. Two, four and six injection points at a horizontal distance of 200 m were defined at a depth of –4.825 m. This distance was chosen based on the results of the history matching, which ensures that no short circuit is generated between the multiple fractures, since the simulated seismic events are distributed at a width of about 80 m (y-direction in Fig. 1) in the basis simulation (DHM project).

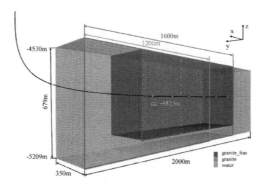

Figure 4.   Half of geometric model for testing a new EGS-strategy with 6 fractures (the red line demonstrates the dimension of a 4-Frac model and the blue line represents that of a 2-Frac model).

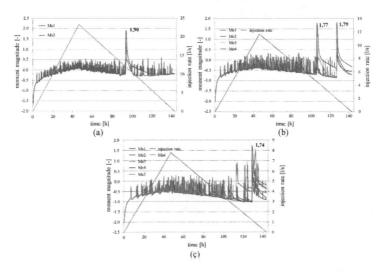

Figure 5.   Treatment schedule in different models and comparison of the calculated moment magnitude of each single fracture vs. time.

795

In the simulated results (Fig. 5), the strategy of the multiple fractures system combined with linear increasing injection method shows a more obvious improvement. The more fractures generated, the more seismic events happened. However, the calculated M_{Wmax} was reduced (from 1.90 to 1.79 and 1.74).

4 CONCLUSION AND OUTLOOK

Based on previous work innovative linear increasing injection strategies in combination with a multiple fracture system have been proposed and studied in order to optimize the hydraulic fracturing and reduce the risks of induced seismicity.

The simulated results show that the risks of induced seismicity can be reduced with the help of the new injection strategy (linear increasing injection). Variant 2 (a rapid rise and a slow decay) appeared to be more advantageous. Through this variant the seismic events gained a better distribution during the operation period, i.e. more small events in a long time interval, instead of large values at a certain point in time. In a further step, their combination with the multiple fractures system achieved a more significant improvement. The more fractures that were considered, the more seismic events happened, but the maximum magnitudes decreased.

Hence, from analysis of the simulation results it can be stated that

1. By releasing the energy at a more leisurely pace for a longer period of time, the M_{max} can clearly be reduced;
2. The same operation could be carried out from the perspective of space, i.e. with the help of a multiple fractures system the stimulation works will run as multiple fractures. Thus, the stimulation of each fracture can be more leisurely. At the same time, the fractured zone will become larger. The remained question is the influences between individual fractures.

Finally, this work confirms that injection strategy using a linear increasing injection rate could reduce the risk of induced seismicity in hydraulic stimulation. In addition, heat recovery efficiency would be further increased, when this injection strategy is applied in combination with the multiple fracturing technology.

REFERENCES

Ahorner L, Sobisch H-G (1988) Ein untertägiges Überwachungssystem im Kalibergwerk Hattorf zur Langzeiterfassung von seismischen Ereignissen im Werra-Kaligebiet. Kali und Steinsalz 10 (2): 38–49.
EIA (2017) Levelized Cost and Levelized Avoided Cost of New Generation Resources in the Annual Energy Outlook 2017. U.S. Energy Information Administration.
Gou Y, Zhou L, Zhao X, Hou MZ, Were P (2015) Numerical study on hydraulic fracturing in different types of georeservoirs with consideration of H²M-coupled leak-off effects. Environ. Earth Sci., 73(10):6019–6034. doi: 10.1007/s12665-015-4112-5.
Hanks TC, Kanamori H (1979) A moment magnitude scale. J. Geophys. Res. 84(B5): 2348–2350. doi: 10.1029/JB084iB05p02348.
Hou MZ, Kracke T, Zhou L, Wang X (2012) Rock Mechanical Influences of Hydraulic Fracturing Deep Underground the North German Basin: Geological Integrity of the Cap Rock Salt and Maximum Magnitude of Induced Microseismicity Based on the GeneSys Stimulation in May 2011. Erdöl Ergas Kohle 128(11): 454–460.
Hou MZ, Zhou L, Kracke T (2013) Modelling of seismic events induced by reservoir stimulation in an enhanced geothermal system and a suggestion to reduce the deformation energy release. Rock Dynamics and Applications: 161–175. doi: 10.1201/b14916-15.
Itasca (2009) FLAC3D Manual, Version 4.0. ITASCA Consulting Group, Inc.
Kanamori H, Anderson DL (1975) Theoretical basis of some empirical relations in seismology. B. Seismol. Soc. of Am. 65(5): 1073–1095.
Lu C, Guo J, Liu YX, Yin J, Deng Y, Lu QL, Zhao X (2015) Perforation spacing optimization for multi-stage hydraulic fracturing in Xujiahe formation: a tight sandstone formation in Sichuan Basin of China. Environ. Earth Sci., 73(10):5843–5854. doi: 10.1007/s12665-015-4366-y.

Geomechanics and Geodynamics of Rock Masses – Litvinenko (Ed.)
© *2018 Taylor & Francis Group, London, ISBN 978-1-138-61645-5*

The reasons of landslides activization at Sakhalin Island (on the example of landslide exploration at the river Lazovaya)

I.K. Fomenko, D.N. Gorobtsov, V.V. Pendin & M.E. Nikulina
Department of Geology, Russian State Geological Prospecting University n. a. Sergo Ordzhonikidze (MGRI-RSGPU), Moscow, Russia

Keywords: factors of the landslide process activation, stability calculations, probabilistic analysis, sensitivity analysis, inverse analysis, assessment of landslide hazard

1 ANNOTATION

Sakhalin area is characterized as one of the most problem in Russia. It connects with the amount of sliding phenomena expression. Understanding the reasons and mechanisms of landslides formation gives advantages in their increased activity forecast. It also helps in possible prevention of tragically accidents. Last year's landslides activation was registered in 36 localities, i.e. 10 were urban and 26 were rural. Landslides in bykovsky mudstones formation are of particular interest as properties of such rock are not fully explored.

At present sliding phenomena studying at Sakhalin area connects with the need of safety exploration the oil and gas pipelines. The section of the pipeline system, located in the Dolinsky and Makarov districts, passes through landslide areas. There take places mud and liquefaction slides with a capacity of 7,5 m and block slides characterized by 10,0 m capacity. That is likely to be hazardous to pipelines.

The considered object is situated on the west slope of dividing crest which is undermined by right river Lazovaya inflow.

According to the map of seismic zonation GSZ-97, the area seismic hazard is 8 points for the periodicities 500 years – map A. For the times of occurrences 1000 and 5000 years – maps B and C the area seismic hazard is about 9 points. Seismic impulses which are caused by earthquakes are the reasons of sliding processes.

The assessment of sliding riskiness was made by impact analysis the sliding processes factors on the stability coefficient's value of investigated slide. Among factors were considered seismic activity, the value of groundwater level as consequence the value of interstitial pressure and strength soil's properties.

Calculations of the stability were made using the Rocscience software. The faculty of engineering geology named in honour of F.P. Savarenski in MGRI-RSGPU is a part of Rocscience Education Program.

2 THE STUDY AREA DESCRIPTION

Landslide at the river Lazovaya is situated near 384 km of "Sakhalin-2" piping system at Makarovsky district, Sakhalin region (Figure 1).

The contemporary relief of studied territory is relatively young. Its formation has started at the end of Lower Cretaceous period after the mountain building process completion and the main watershed forming. Intensive erosion and denudation processes during Paleogene and Neogene periods contributed to purchasing by the relief the low and middle mountain identities. Gently undulating swelling surfaces of denudation levelling which bordered the lowlands were formed. Accumulating plain relief was formed at Quaternary period.

Geomorphologically the study area is situated near the east flank of West-Sakhalin anticlinorium. The area is represented by erosion and denudative landforms. These include low hills, V-shaped river valleys which are intensively divided. The slope steepness is 20–40°.

The considered object is situated on the west slope of dividing crest which is undermined by right river Lazovaya inflow. The highest degree of slope is 30°. This value is observed on sliding slope's cliff. Cliffs alternate with aligned grade levels. Slope consists of eluvial and deluvial deposits with capacity from 1,5 to 3 m. These deposits overlap by transported claystones and clays which belong to upper subsuite of bykovsky suite (K_2bk) (Sergeev, 1984). Claystones are weathered in the form of a grass and rock debris material with clayey aggregate. Such claystones are deposited at rock outcrops and sliding brow's disruptions. Claystones transform into bluish-grey clays of soft plastic consistency. It happens at the interaction zone

Figure 1.   Dividing crown and drawing to the river Lazovaya, 384 km of "Sakhalin-2" piping system.

Figure 2.   Brow failure between sliding steppes. Lack of wood vegetation is one of the landslides' activity indications.

Figure 3. Brook's bed. Abrupt banks are the evidence of active lateral erosion. Tree remnants in brook indicate on flow capacity at flood period.

with stratum and pore water which matches the weakness zones. These clays form the main deformed horizon of landslide's main bodies which are represented at the territory.

Landslides phenomenon on the studied area is represented by blocked landslides. There thickness is up to 15 m. Such landslides cover the whole slope from crown to thalweg's water cource.

Morphologically sliding body is in blocks. It looks like circus or frontal landslides which located along the slope. Such landslides have clearly expressed brow failures and sliding steppes (Figure 2).

Lateral erosion of watercourses bed significantly increase the affect degree and sliding processes intensity. Landslides growth has a regressive character which caused by undercutting the lower slope's parts. Starting near slope's foots such landslides eventually may reach the divide's crown (Figure 3).

The main reasons of nature landslides development at the studied area are: clay rocks at slope's structure which may transform into soft plastic consistency due to ground water, high slope's steepness, stream's erosion activity, high seismicity.

The assessment of sliding riskiness was made by impact analysis the sliding processes factors on the stability factor's value of investigated slide. Among factors were considered seismic activity, the value of groundwater level (as consequence the value of interstitial pressure) and strength soil's properties. (Zerkal, 2016; Pendin, 2015; Recommendations, 1984; Fomenko, 2012).

3   THE RESULTS OF STABILITY DESIGN

The considered landslide belongs to sliding landslip according to mechanism of sliding process. However such landslide has features of squeezing landslides. As this landslide is multicycle this takes the form of some sliding steppes so it should be allocated to complex landslides type (Bondarik, 2007).

The stability estimation of such landslides should be made per blocks (Pendin, 2015; Fomenko, 2011). Meanwhile it should be understood that sliding blocks are energetically connected. For this reason sliding slope based on morphological authorities was divided on two blocks—upper and lower. The stability estimation was made for each block.

Stability estimation was made by limit equilibrium methods (Morgenstern-Price, Bishop and Janbu simple). The feature of theme and maybe the main deficiency is the miss of correlation between stress and deformations (Krahn, 2004). For this reason soil's strength properties wasn't considered in the modelling.

Soil's strength properties (friction angle and cohesion) which are a part of Mohr-Coulomb strength criteria were estimated through triaxial test. It was accompanied by estimation the peak and residual strength in incorporated society "MOSTDORGEOTREST" laboratory.

The colour legend is shown on the Table 1. It was used in preparing geomechanical schemes.

Final geomechanical schemes with the results of slope's stability estimation according to Morgenstern-Price method are shown on Figure 4 and Figure 5.

The comparison between the results obtained due to various estimation versions and methods is shown at the Table 2.

The results of probabilistic analysis (Krahn, 2004; Zerkal, 2015; Zerkal, 2016; Pendin, 2015; Sisoev, 2011; Fomenko, 2012) according to Morgenstern-Price method are shown at the Table 3.

Integral function's distributions of the estimated modeling slopes' blocks stability factor are shown on Figure 6.

Predication slope's stability estimation on special combination of loads considering the seismic effect and ground water level lifting was made due to analysis of slope's stability factor sensitivity (S_f) and factors of sliding process activation (Krahn, 2004; Pendin, 2015; Fomenko, 2012). The results of analysis are shown on Figure 7 and Figure 8.

The seismic impact accounting was made on the base of pseudo static analysis (Krahn, 2004). Measurement ground water piezometric level limits were estimated according to monitoring.

Table 1. The legend to geomechanical schemes.

Material name	Color	Unit weight (kN/m³)	Angle internal friction (deg)	Cohesion (kPa)
2a		18.7	8.9	34.2
2b		18.7	23.0	40.5
3		19.3	38.8	23.9
4		18.3	7.4	18.0
5		19.6	7.6	37.4
6		19.6	40.4	17.7
7		22.1	63.0	30.5

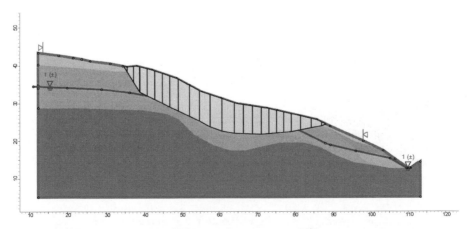

Figure 4. Geomechanical scheme with results of slope's stability estimation (M-P method), upper block.

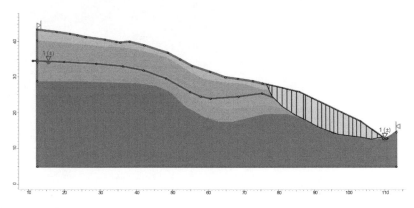

Figure 5. Geomechanical scheme with results of slope's stability estimation (M-P method), lower block.

Table 2. Slope's stability estimation.

Way of estimation				
Method	Morgenstern-Price (M-P)		Bishop	Janbu (simple)
1. Natural condition, upper block	1,45		1,41	1,36
2. Natural condition, lower block	1,02		1,02	0,99

The colour legend to Table 2		
Slope is stable	Slope is in limit equilibrium state	Slope is unstable

Table 3. The results of slope's stability probabilistic estimations (Morgenstern-Price method).

Way of estimation	Middle coefficient of slope's stability (K)	Standard deviation	Probabilistic parameters for the analysis of landslide formation risk			
			K_{min}	K_{max}	β	$f_k = 1$, %
Upper block	1,51	0,19	1,04	2,09	2,7	0
Lower block	1,06	0,13	0,68	1,46	0,47	32,5

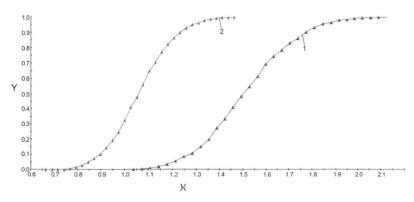

Figure 6. Integral function's distributions of the estimated modeling slopes' blocks stability factor. 1 – Upper block; 2 – Lower block; axis X – S_c (M-P method); axis Y – the sliding process development variety.

801

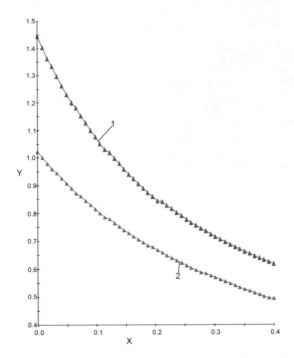

Figure 7.   Sensitivity of S_f to the horizontal seismic acceleration's value. 1 – Upper block; 2 – Lower block; axis X – horizontal seismic coefficient; axis Y – S_c (M-P method).

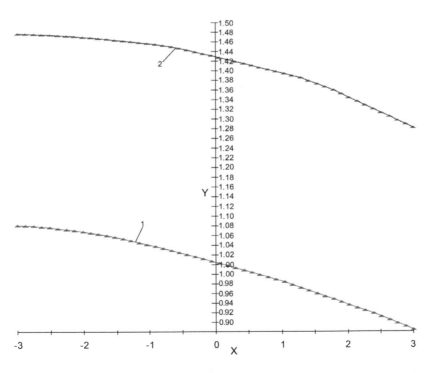

Figure 8.   Sensitivity of S_f to the piezometric level change. 1 – Upper block; 2 – Lower block; axis X – predictive change in ground water piezometric level (m) (0 – level on the moment of engineering and geological researches carrying out); axis Y – S_c (M-P method).

4 THE ANALYSIS OF QUANTITY SLOPE'S STABILITY ESTIMATION RESULTS

Estimations made by Morgenstern-Price (Morgenstern, 1965), Bishop (Bishop, 1955) and Janbu (Janbu, 1954) methods allow making a conclusion. In natural the upper block of the considered sliding area with given estimated indicators is stable. The lower block is in limit equilibrium state according to Morgenstern-Price and Bishop methods. The lower block is unstable according to Janbu method. It happens because simple Janbu method is more conservative. Thus estimated stability coefficients as a rule have minimum values in comparison with other limit equilibrium methods.

Estimated results allow making a conclusion the sliding phenomena progress on the slope has regressive character. So landslide develops from the bottom up. The lower block stability decline is determined by lateral erosion of the river Lazovaya right inflow.

The activation of sliding processes on the slope may happen by adverse engineering and geological situation. For example by increase the ground water piezometric level. The critical raising of ground water level is 0,5 m for lower block. The forecasting increase of the level for upper block higher than 3 m doesn't lead to the loss of stability.

Seismic effect of any intensity fewer than 6 strength may activate sliding processes at the lower block. Earthquake higher than 7 strength using Richter scale 64 is critical for upper block.

Probabilistic analysis shows that the variety of sliding process activation is 0% for upper block and 32,5% for lower block.

However it should be taken into account that sliding process development has a regressive character. Also the activation of landslide at the lower block will influence on the upper block. So localization activities including engineering survey should be organized for this slope.

Calculations of the stability were made using the Rocscience software. The faculty of engineering geology named in honour of F.P. Savarenski in MGRI-RSGPU is a part of Rocscience Education Program.

REFERENCES

Bishop A.W. The use of the slip circle in the stability analysis of slopes. Géotechnique, 1955. Y. 5. P. 7–17.

Bondarik G.K., Pendin V.V., Yarg L.A. Engineering geodynamic: textbook. M, 2007. pp. 439.

Fomenko I.K. "Modern trends in slope's stability assessment"//Engineering geology. 2012. № 6. pp. 44–53.

Fomenko I.K., Sirotkina O.N. Complex estimating methodology of slope's stability//Digest of scientific works on the international research conference materials "Modern trends of theoretical and applied researches". Odessa: Black Sea Region, 2011. pp. 88–96.

Krahn J. Stability modeling with SLOPE/W. An Engineering Methodology: First Edition, Revision 1. Calgary, Alberta: GEO-SLOPE International Ltd., 2004. 396 p.

Morgenstern N.R. and Price V.E. The analysis of the stability. of general slip surface//Géotechnique. 1965. V. 15. P. 70–93.

Pendin V.V., Fomenko I.K. The assessment and prediction methodology of sliding hazard. M.: Publishing house RF Lenand, 2015. p. 320.

Recommendations about quantity stability assessment of sliding slopes. PNIIS. M.: Stroiizdat, 1984. p. 80.

Sergeev K.F., "New data about the relationship between late Mesozoic and Cainozoe deposits of Western Sakhalin east slopes (Makarovsky district, i. Sakhalin) // Pacific geology, 1984, №1., pp. 99–103.

Sisoev J.A., Fomenko I.K. The sliding hazard variety analysis //Digest of scientific works on the international research conference materials "Scientific researches and its practical use. Current state and development trends". Odessa: Black Sea Region, 2011. pp. 125–129.

Zerkal O.V., Fomenko I.K. The variety slope's stability assessment and it's appliance in sliding hazard analysis // Analysis, prediction and management of natural risks in modern world: The materials of 9-th international research conferences "Georisk-2015" (October 13–14, 2015, Moscow) – Moscow, 2015.-V.1-pp. 225–231.

Zerkal O.V., Fomenko I.K. The influence of various factors on the results of sliding processes variety analysis//Engineering geology.–2016.-№1.-pp. 16–22.

Geomechanics and Geodynamics of Rock Masses – Litvinenko (Ed.)
© *2018 Taylor & Francis Group, London, ISBN 978-1-138-61645-5*

Prediction of rock movements using a finite-discrete element method

Bulat Ilyasov
TERETAU LLC, Bashkortostan Republic, Russia

Alexander Makarov
SRK Consulting (Russia), Principal Geotechnical Consultant, ISRM, Moscow, Russia

Ivan Biryuchiov
SRK Consulting (Russia), Geotechnical Consultant, Moscow, Russia

ABSTRACT: The aim of the study described in this article is to evaluate the suitability of a finite-discrete element method for predicting rock movement during underground mining of mineral deposits by caving systems. A comparison is made of the measured and modelled characteristics of deformation processes, which took place during mining at various deposits. The developed algorithms for simulating sub-level and block caving are described.

Keywords: fractured rock failure, disintegration and displacement, finite-discrete element method

1   INTRODUCTION

Predicting rock and ground surface movements during mining by caving methods is usually based on empirical methods, which cannot take into account all geological conditions. For this, it is necessary to simulate the processes of breaking fractured rock mass and the subsequent behaviour of disintegrated material: movement of rock blocks in the mined-out space and the caving zone.

Discrete elements methods have considerable potential for solving problems with disintegration. For their practical application, it is necessary to understand their specific features and the limits of applicability. The finite-discrete element method is one of the most advanced numerical methods in mechanics of a discrete medium (Munjiza, 2004). To perform geomechanical calculations by this method, Teretau develops Prorock software package. Software processor implements the algorithm of forced stabilization with Coulomb strength criterion. In the calculations, an extreme deformation by dilatancy is simulated, along with variability of strength characteristics, tectonic discontinuities and anisotropy of strength properties, plastic deformation of finite elements. To ensure the high speed, computations are performed on general-purpose processors (Ilyasov, 2016).

Prorock software was used to simulate movement of rock mass, ground surface and host rocks at mines using caving systems: Degtyarsky (Russia), Ridder-Sokolny (Kazakhstan) and Palabora (South Africa).

2 SUB-LEVEL AND BLOCK CAVING SIMULATION SCHEME

Caving simulation begins only after the model achieves a state of quiescence with the help of the forced stabilization algorithm (Ilyasov, 2016). This ensures a significant reduction of the impact of inertial oscillations in the system that occur after the simulation is started. After simulation is initiated, elastic modulus, Poisson's ratio and cohesion parameters of rocks in the mining area are gradually reduced to the values corresponding to the disintegrated rock mass (rubble). This is necessary to reduce the number of elements caused by the exclusion from calculations of unrealistic dynamic effects, consisting of high-amplitude oscillations of the system, which cause extensive disintegration around the area with excluded elements. Another consequence of excluding elements from the calculations is significant acceleration of nearby non-excluded finite elements and, consequently, an unnatural increase in the level of disintegration in the mining area. To reduce such effects, the algorithm of plastic deformation of finite elements has been developed and introduced into the program code (Ilyasov, 2016), and a dissipative impact model has been added (Mahabadi, 2012).

The values of the elastic modulus, Poisson's ratio and cohesion are reduced until either one of the two conditions is met:

$$S_c/S_0 < 0.18 n_{ts} > 7000$$

where S_c is the current area of the element, S_0 is the initial area of the element, n_{ts} is the number of time steps since the beginning of simulation of the current area. After the condition is met by at least one element, all elements are excluded from calculations.

For block caving, the undercut level is modelled first in the same way as for sub-level caving method. Next, to simulate the ore draw as it caves, elastic properties of the elements are changed, and the rate of change depends on the reduced distance l_i/L_0, where li is the distance from the centre of the element to the base of the undercut level. L_0 value is calculated depending on the average size of the element in the model. When the elastic modulus reaches 18% of the initial value, the element is excluded from calculations. This limit has been found empirically. Figure 1 presents a diagram explaining the ore draw modelling.

The above modelling scheme provides realistic simulation of ore mining by block caving method and stable operation of the processor. It should be added, however, that even with such analytical model, it is necessary to overstate strength properties of production level

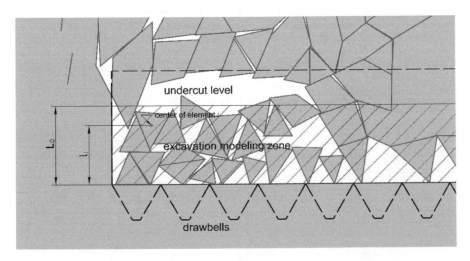

Figure 1.   Ore draw modelling diagram.

elements by several times in order to minimize their unnatural disintegration resulting from exclusion of elements from calculations.

3   CASE STUDY: DEGTYARSKY MINE

Degtyarsky copper deposit (Urals, Russia) is a steeply-dipping tabular lode with a thickness of 6 ÷ 25 meters, which was mined by sub-level caving method. During the development of the deposit, mine survey department monitored deformations of the ground surface (Kuznetsov, 1971). According to monitoring results, first caving of the ground surface occurred during extraction of the upper level (highlighted with violet shading in Figure 2a). This fact is used to back-calculate strength properties of the rock mass. Used rock mass properties are presented in Table 1.

Figures 2a and 2b show the results of modelling by mining stages.

Ground surface movement based on modelling results was close to the measured one. Estimated values of vertical displacements in the hanging wall are sometimes up to 100% higher than the measured ones. Based on survey data, horizontal displacements of the surface were 1.5–1.8 higher than vertical displacements, whereas in the model, this proportion ranges between 1.3 and 1.9. Significant differences in estimated and actual surface displacements in the footwall of the deposit are explained by presence of fragmented rock mass close to surface (Kuznetsov, 1971), which could not have been accounted for in simulation, as their parameters were unknown. Based on modelling data, hazardous impact angles β were plotted, that characterize the boundary of the ground surface zone with deformations exceeding 2 mm/m; the error of the estimate versus actual values amounted to 1.7° and 2.8°.

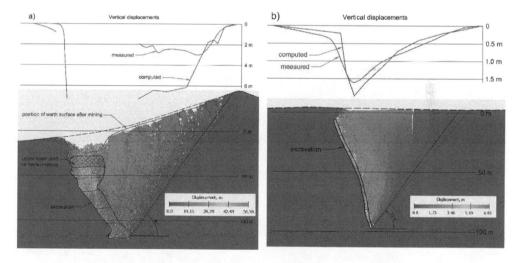

Figure 2.   Rock movement at Degtyarsky mine: a) cross-section 7, b) cross-section 9.

Table 1.   Degtyarsky deposit rock mass properties.

Rock type	Density (t/m³)	Elastic modulus (GPa)	Poisson's ratio	Cohesion (MPa)	Internal friction angle (deg)	Tensile strength (КРа)
Shales	2.8	12	0.29	0.2	38	5
Ore	3.6	12	0.29	0.2	38	5

According to I.A. Kuznetsov (Kuznetsov, 1971), displacements in the hanging wall in areas with significant thickness of mineralization happened in the form of slabbing and sagging of shale formation with subsequent caving of blocks sized 40 to 60 meters. Similar movement parameters were obtained through modelling by finite-discrete elements method in Prorock.

4 CASE STUDY: RIDDER-SOKOLNY MINE

Ridder-Sokolny polymetallic deposit (Kazakhstan) comprises about 10,000 discrete lenticular ore bodies. The deposit is mined using caving method with backfill (under protected areas).

Table 2 presents rock mass properties by geotechnical domains identified through statistical analysis. Cohesion values were obtained by means of back-calculation of the subsidence event shown in cross-section 5 (Figure 3a).

SRK Consulting reviewed cases of ground subsidence caused by extraction of individual ore bodies using caving methods (Makarov, 2017). In all cases, ore bodies had isometric shape in plan view.

Figures 3 a–c show the results of ground subsidence simulation. Simulation accounted for jointing and tectonic discontinuities and used the algorithm to incorporate horizontal tectonic stresses.

Simulation of a caving zone above one of the mined-out ore bodies was also performed. The actual contours of the caving zone were established through drilling observation holes. Figure 4 shows the mining outlines and the zone of disintegration based on measurements in observation holes and modelling. In addition to jointing, the influence of excavations located above was also taken into account.

Based on results of simulation it was concluded that 2D modelling of isometric ore bodies, dimensions of subsidence zones are determined with significant error. However, the very fact

Table 2. Ridder-Sokolny deposit rock mass properties.

Rock type	Density (t/m³)	Elastic modulus (GPa)	Poisson's ratio	Cohesion (MPa)	Internal friction angle (deg)	Tensile strength (KPa)
Aleuropelite, tuff	2.71	28.2	0.19	0.69	34	5
Microquartzite	2.72	39.6	0.2	1.14	34	5
Shale	2.72	19.0	0.19	0.75	36	5
Quartz formations	2.68	27.6	0.15	0.60	31	5
Quaternary soils	2.00	0.05	0.3	0.20	20	1

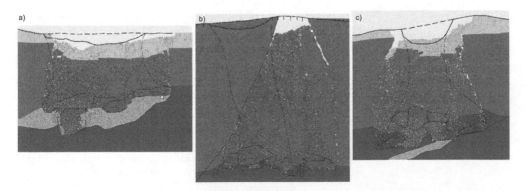

Figure 3, Modelled and actual outlines of ground subsidence at Ridder-Sokolny mine: a) cross-section 5, b) cross-section 3, c) cross-section 5a.

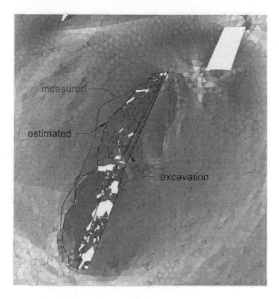

Figure 4. Outline of the subsidence above a stope (Ridder-Sokolny mine).

Table 3. Palabora deposit rock properties.

Rock type	Density (t/m³)	Elastic modulus (GPa)	Poisson's ratio	Cohesion (MPa)	Internal friction angle (deg)	Tensile strength (KPa)
Carbonatite	2.88	39.7	0.21	2.18	45	58
Foskorite	3.53	38.4	0.21	2.00	43	43
Micaceous pyroxenite	3.04	39.5	0.20	1.81	39	41
Feldspathic pyroxenite	3.10	11.0	0.24	1.30	37	40
Fenite	3.10	44.7	0.23	2.69	48	50

of subsidence was predicted correctly for all 6 models. Dimensions of the disintegration zone above a large stope were determined with an error of less than 40%.

5 CASE STUDY: PALABORA MINE

Palabora copper deposit (South Africa) was mined as an open pit to a depth of 800 meters, with subsequent underground block cave mining. The undercut level is located 400 meters below the bottom of the pit. A pit wall failure happened in late 2004 in the course of underground mining.

The dynamics of development of the disintegration zone above the draw level was studied (Severin, 2017). The development of the pit wall failure was also restored by years from public domain information. Indirect data on mine productivity was used to estimate approximate ore extraction volumes by years, which made it possible to compare production volumes and disintegration process under the pit floor and in the wall.

Deposit rock properties (Table 3) were adopted from Sainsbury (2016). Cohesion values were estimated by back-calculation, so that the modelled outline of the disintegration zone as of April 2003 coincided with the actual one. Calculations incorporate rock mass jointing and tectonic faulting as presented by Sainsbury (2016).

Figure 5 compares the results of disintegration simulation with the actual data. As can be seen, the simulated disintegration in the course of mining generally agrees with observation data.

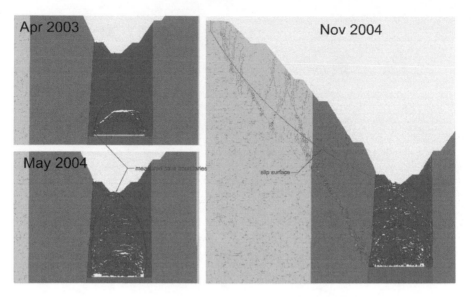

Figure 5.   Disintegration zones: modelled and actual.

6   CONCLUSIONS

It is possible to solve geomechanical practical problems related to rock movement in the course of mining by caving methods with the use of the finite-discrete element method implemented in Prorock software: predict zones of disintegration in the rock mass and on surface as mining operations progress, including in the walls of the pit in case of development of an underground mine underneath it. When extracting extended lodes, prediction of displacements in 2D can achieve an accuracy of a few meters. For isometric ore bodies, calculation errors can be much higher.

Prediction is possible if there is data for back-calculation for a particular deposit on a similar scale, e.g. development of a disintegration zone at early stages of mining. Back calculations are used to determine cohesion with the account for the scale effect.

REFERENCES

[1] B.T. Ilyasov (2016). Study of the kinetics of deformation of rock mass using finite-discrete element method. PhD Thesis. Ekaterinburg. P. 138.
[2] S.N. Ivanov, M.I. Merkulov (1937). Degtyarskoe pyrite deposit. Moscow. P. 124.
[3] M.A. Kuznetsov, A.G. Akimov, V.I. Kuzmin et al. (1971). Movement of rocks on ore deposits. Nedra, Moscow. P. 224.
[4] O.K. Mahabadi, A. Lisjak, A. Munjiza, and G. Grasselli (2012). Y-Geo: a new combined finite-discrete element numerical code for geomechanical applications. International Journal of Geomechanics, 12, pp. 676–688.
[5] A.B. Makarov, A.I. Ananin, D.V. Mosyakin (2017). Weakening of failed rocks and sinking conditions. Mining Journal, 3, pp. 32–36.
[6] A. Munjiza (2004). The combined finite-discrete element method. John Wiley & Sons Ltd. Chichester, UK. P. 350.
[7] D.P. Sainsbury, B.L. Sainsbury, H-D. Paetzold, P. Lourens, A. Vakili (2016). Caving-induced subsidence behaviour of lift 1 at the Palabora block cave mine. Proceedings Seventh International Conference & Exhibition on Mass Mining. Sydney, 9–11 may 2016. pp. 415–426.
[8] J.M. Severin (2017). Impact of faults and fault damage zones on large open pit slopes. PhD Thesis. Vancouver. P. 168.

Geomechanics and Geodynamics of Rock Masses – Litvinenko (Ed.)
© 2018 Taylor & Francis Group, London, ISBN 978-1-138-61645-5

Simulation of fracture propagation depth and failure in long hole open stoping

P.J. le Roux & K.R. Brentley
Brentley, Lucas and Associates, University of the Witwatersrand, Johannesburg, South Africa

ABSTRACT: There are numerous factors which affect open stope stability and often result in falls of ground. These falls of ground can be attributed to a number of factors such as beam failure due to a larger than normal roof area (hydraulic radius too large), adverse ground conditions, seismicity, the stress-strain environment, absence of support and poor drill and blast practices. The effect of fracture depth in open stope failure is sometimes under-estimated and relatively unknown. Actual data collected from open stopes and the analysis thereof is used for back analysis on the failure depth. The benefits of this analysis will result in improved understanding of fracture propagation in Long Hole Open Stoping at depths.

1 INTRODUCTION

The aim and objectives of this research was to develop a method to determine the expected failure depth into the hangingwall and sidewalls of large excavations with a good degree of certainty. With the current failure criteria currently available, this cannot be done with certainty. Making use of back analyses is one of the most important aspects in any engineering field. Compared with other engineering fields such as Aeronautical, Civil and Mechanical engineering, back analysis in Rock engineering is not always being utilized efficiently. Back analyses of open stope hangingwall and sidewall failure can yield an insight into the true behaviour of these excavations in the mining environment. Knowing that these stopes failed, the magnitude and mode of failure can prove extremely useful. Ultimately, the failure of these stopes should be "designed", and not be seen as "unexpected failure".

2 FAILURE CRITERIA USED IN EXCAVATION DESIGN

A failure criterion can be defined as the instance where the stress condition at which the ultimate strength of the rock is reached. Failure criteria can be expressed in terms of the major principal stress σ_1 that rock can tolerate for a given value of intermediate principal stresses σ_2 and minor principal stresses σ_3 (Ulusay and Hudson, 2007).

To understand the behaviour of the rockmass around open stopes, failure criteria are used. If expected failure can be calculated the amount of expected dilution or overbreak can be determined using numerical analyses. Some of the failure criteria being used in rock engineering will include the Mohr-Coulomb criterion, Hoek-Brown criterion, Zhang-Zhu Criterion, Pan-Hudson Criterion, Priest Criterion, Simplified Priest Criterion and Drucker-Prager Criterion. The Mohr-Coulomb criterion and Hoek-Brown criterion are two-dimensional criteria in which the intermediate principal stress value is ignored. Three-dimensional criteria such as 3D Hoek-Brown criterion, Zhang-Zhu Criterion, Pan-Hudson Criterion, Priest Criterion, Simplified Priest Criterion and Drucker-Prager Criterion, include the intermediate stress value. Using these three-dimensional criteria, the influence of the intermediate stress value can be taken into account.

2.1 Mohr-Coulomb failure criterion

The Mohr-Coulomb failure criterion is a set of linear equations in principal stress space describing the conditions for which an isotropic material will fail, irrespective of any effect from the intermediate principal stress σ_2 being neglected (Ulusay and Hudson, 2007). Mohr-Coulomb failure can be written as a function of major σ_1 and minor σ_3 principal stresses, or normal stress σ_n and shear stress τ on the failure plane (Jaeger and Cook, 1979).

In the investigations of retaining walls by Coulomb (Heyman, 1972), the following relationship was proposed:

$$|\tau| = S_O + \sigma \tan\varnothing \tag{1}$$

where S_O is the inherent shear strength, also known as cohesion, $\varnothing$ is the angle of internal friction, and the coefficient of internal friction $\mu = \tan\varnothing$. The criterion contains two material constants, $\varnothing$ and S_O. The representation of Equation (1) in the Mohr diagram is a straight line inclined to the σ-axis by the angle $\varnothing$ as shown in Figure 1.

Designing underground excavations utilizing numerical models can be difficult as they do not necessarily reflect the actual behaviour of the rock mass. In the case of brittle failure this is particularly true, the fundamental assumption of the Mohr-Coulomb criterion $|\tau| = S_O + \mu\sigma$, relating the cohesion S_O to a shear strength τ and a simultaneously acting frictional resistance $\mu\sigma$ not being valid according to Kaiser and Kim (2008). As intact rock is being strained, cohesive bonds start to fail, and only after this does frictional resistance develop. Damage initiation and propagation occur at different stress thresholds according to Diederichs (2003) and the propagation of tensile fractures depends on the level of confinement as established by Hoek (1968) and used to explain brittle failure. Wiles (2006) explains that the Mohr-Coulomb failure criterion can also be mathematically expressed as shown in Equation (2):

$$|\tau| = S_O + \sigma \tan\varnothing \tag{2}$$

where σ_1 and σ_3 represent, respectively, the major and minor principal stresses, C_o and q represent, respectively, the rock mass unconfined compressive strength and slope of the best fit-line as shown in Figure 2, where $q = \tan^2\left(45 + \frac{\varnothing}{2}\right)$; $\varnothing$ is the friction angle

2.2 Hoek-Brown failure criterion

The Hoek-Brown failure criterion follows a non-linear, parabolic form that separates it from the linear Mohr-Coulomb failure criterion. This criterion is an empirically derived relationship used to describe a non-linear increase in peak strength for isotropic rock with increasing confining stress. The criterion includes procedures developed to provide a practical means to estimate the rock mass strength from actual laboratory test values and underground observations (Ulusay and Hudson, 2007).

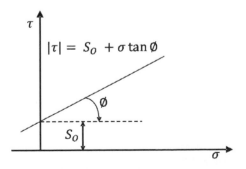

Figure 1.   The Mohr-Coulomb failure criterion for shear failure (Brady and Brown, 1985).

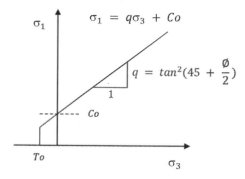

Figure 2. Alternative representation of the Mohr-Coulomb failure criterion (Wiles, 2006).

This criterion was developed as a means of estimating the rock mass strength by scaling the geological conditions present underground. Based on Hoek's (1968) experiences with brittle rock failure and his use of a parabolic Mohr envelope derived from Griffith's crack theory (Griffith, 1920, 1924) to define the relationship between shear and normal stress at fracture initiation, the criterion was conceived. Hoek and Brown (1980) proceeded through trial and error to fit a variety of parabolic curves to triaxial test data and associating rock failure and fracture initiation with fracture propagation, to derive their criterion (Ulusay and Hudson, 2007).

The non-linear Hoek-Brown failure criterion for intact rock (Hoek and Brown, 1980) was introduced as shown in Equation (3):

$$\sigma_1 = \sigma_3 + \sqrt{m\,C_{ucs}\,\sigma3 + s\,C_{ucs}^{\ 2}} \tag{3}$$

where m and s are dimensionless empirical constants and C_{ucs} is the uniaxial compressive strength (UCS) of rock in MPa. The parameter m is comparable to the frictional strength of the rock and s indicates how fractured the rock is, and is related to the rock mass cohesion (Ulusay and Hudson, 2007). The Hoek–Brown criterion has been updated several times to address certain practical limitations, and with experience gained with its use to improve the estimate of rock mass strength (Hoek and Brown, 1988; Hoek et al, 1992, 1995, 2002). It was assumed that the criterion was valid for effective stress conditions thus the principal stress terms in the original equation had been replaced earlier with effective principal stress, σ_1' and σ_3' terms (Hoek, 1983). One of the major updates was the reporting of the 'generalised' form of the criterion (Hoek et al, 1995):

$$\sigma_1' = \sigma_3' + C_{ucs}\left(m_b \frac{\sigma_3'}{C_{ucs}} + s\right)^a \tag{4}$$

For broken rock the term m_b was introduced. Hoek et al, (1992) reassessed the original m_i value and found it to be depending upon the grain size of the intact rock, mineralogy and composition. To address the system's bias towards hard rock and to better account for poorer quality rock masses by enabling the curvature of the failure envelope to be adjusted, particularly under very low normal stresses, the exponential term a was added (Hoek et al, 1992). As shown in Figure 3 the Geological Strength Index (GSI) was subsequently introduced together with several relationships relating m_b, a and s, with the overall structure of the rock mass and surface conditions of the discontinuities (Hoek et al, 1995).

A new factor D, also known as the blast damage factor, was introduced by Hoek et al. (2002), to account for near surface blast damage and stress relaxation in the rock mass. The factor D can range between 0 and 1 where D = 0 for undisturbed rock and D = 1 for highly disturbed rock mass. The m_b, a and s were reported as:

813

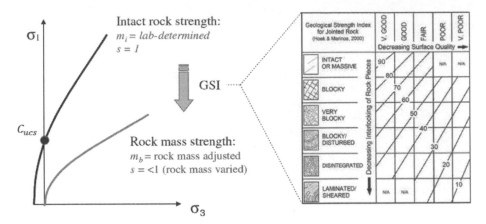

Figure 3. Scaling of Hoek-Brown failure envelope for intact rock to that for rock mass strength (Ulusay and Hudson, 2007).

$$m_b = m_i \exp\left(\frac{GSI - 100}{28 - 14D} \right) \tag{5}$$

$$s = \exp\left(\frac{GSI - 100}{9 - 3D} \right) \tag{6}$$

$$a = \frac{1}{2} + \frac{1}{6}\left(e^{-\frac{GSI}{15}} + e^{-\frac{20}{3}} \right). \tag{7}$$

where m_i is a curve fitting parameter derived from triaxial testing of intact rock. The parameter m_b is a reduced value of m_i, which accounts for the strength reducing effects of the rock mass conditions defined by GSI as shown in Figure 3 (Ulusay and Hudson, 2007).

2.3 Strain-based failure criteria

There are numerous strain based criteria such as the extension strain criterion after Stacey (1981), the direct strain evaluation technique after Sakurai (1981), Fujii et al. (1998) proposed the critical tensile strain criterion for brittle failure of rock and Kwaśniewski and Takahashi, (2010) considered the relationship between the octahedral shear strain and it was found that the mean normal strain yielded much better results than the mean normal strain at strength failure.

3 NUMERICAL MODELLING

Map3D is based on Banerjee and Butterfield (1981), a very efficient Indirect Boundary Element Method, and incorporates simultaneous use of both fictitious force and displacement discontinuity elements. Special boundary elements are incorporated for the thermal and non-linear analysis versions. This Boundary Element formulation offers many advantages over other stress analysis techniques. Direct Boundary Element formulations require approximately twice the computing effort to assemble and solve the boundary element matrix, compared to the indirect method used in Map3D (Wiles, 2006).

3.1 Input parameters for MAP3D

The rock mass in the numerical model is assumed to be homogeneous and isotropic to simplify numerical modelling (Wiles, 2006). MAP3D-SV was used to model the mining of the open stopes and to determine the strain and stress values. These stress values for σ1, σ2 and σ3 are used as inputs into the Mohr-Coulomb, Hoek-Brown, Zhang-Zhu, Pan-Hudson, Priest, Simplified Priest and Drucker-Prager Criteria to determine whether any of these criteria can be used for assessing failure around open stopes.

The following input parameters were used for MAP3D-SV:

Young's modulus	: 70000 MPa
Poisson's ratio	: 0.2
Density	: 2700 kg/m³
k-ratio	: 0.5

These input parameters for Young's modulus, Poisson's ratio and density were obtained from laboratory testing that was conducted at the University of the Witwatersrand by Le Roux (2004) for the Eldorado Reefs. The k-ratio is an estimate based on actual underground observations and back analyses.

4 OUTCOME OF THE APPLIED FAILURE CRITERIA

4.1 *Failure criteria applied to Map3D results*

Using the results obtained from Map3D on the hangingwall and sidewalls for the twenty-two case studies simulated, the failure criteria as discussed in section 2 will be applied. Figure 4 and Figure 5 show the results of application of the Mohr-Coulomb, Hoek-Brown, Zhang-Zhu, Pan-Hudson, Priest, Simplified Priest and Drucker-Prager Criteria to the median σ1, σ2 and σ3 results obtained from the Map3D analyses of the open stopes. Each of the criteria mentioned above either over or under estimate the failure around these case studies. The Drucker-Prager criterion does not fit the Map3D results and substantially overestimates the failure around open stopes. Thus it is not suitable for application to open stopes.

Using the stresses and strains determined with Map3D, the various stress-based failure criteria and strain-based failure criteria mentioned above were applied to predict failure depths into the hangingwall and sidewalls of the case study open stope as shown in Figures 6 to 8. These results show that the stress-based failure criteria and strain-based failure criteria either completely overestimate or under estimate the failure for most of the case studies. It can be concluded that these methods are not appropriate for accurate design of open stopes in the hard rock gold mining environment. In Table 1 the Predicted rock mass unconfined compressive strength C_{UCS} using different failure criteria is shown. Details of the method can be obtained from the original reference Le Roux (2015).

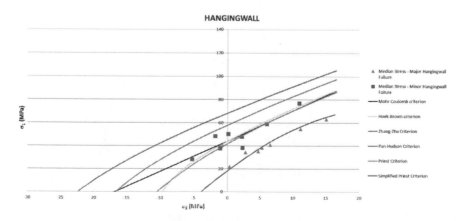

Figure 4. Graph showing the relation between various criteria used and obtained results for open stopes with hangingwall failure after Le Roux (2015).

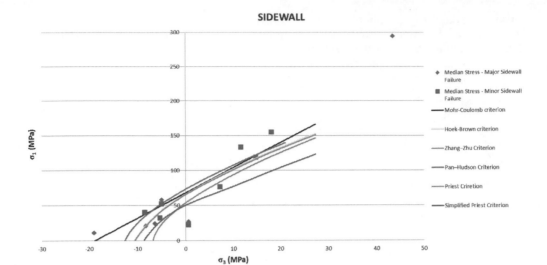

Figure 5. Graph showing the relation between various criteria used and obtained results for open stopes with sidewall failure after Le Roux (2015).

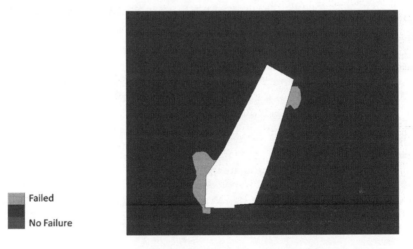

Figure 6. Application of the Mohr-Coulomb criterion after Le Roux (2015).

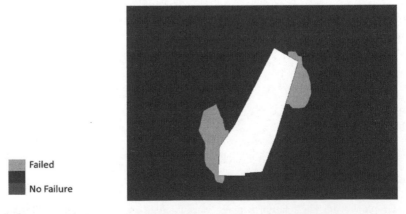

Figure 7. Application of the Hoek-Brown criterion after Le Roux (2015).

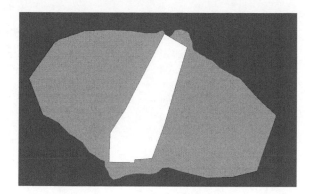

Failed

No Failure

Figure 8. Application of the extension strain criterion after Stacey, (1981) with a modulus of elasticity E = 70000 MPa after Le Roux (2015).

Table 1. Predicted rock mass unconfined compressive strength C_{UCS} using different failure criteria.

	Predicted Rock Mass UCS	
	Sidewall	Hangingwall
Mohr-Coulomb criterion	68 MPa	44 MPa
Hoek-Brown criterion	66 MPa	44 MPa
Zhang-Zhu Criterion	54 MPa	58 MPa
Pan-Hudson Criterion	73 MPa	68 MPa
Priest Criterion	65 MPa	44 MPa
Simplified Priest Criterion	50 MPa	18 MPa
Drucker-Prager Criterion	–	–

5 MEAN STRESS-VOLUMETRIC STRAIN CRITERION

Open stopes have a three-dimensional geometry and are created in a three-dimensional stress field. It is therefore to be expected that the stability of these stopes, and of course the potential dilution, will be dependent on the three-dimensional stress and strain conditions around these stopes. To take these three-dimensional conditions into account, the mean stress, σ_m, also known as the octahedral normal stress, was plotted against volumetric strain, ε_{vol}, in Figures 9 and 10 for open stopes with major failure, and minor failure, in the hangingwall and sidewalls respectively. These results indicate, as expected, a linear relation between the mean stress and volumetric strain-stress and strain are linked in the linear numerical model by constitutive behaviour known as Hooke's Law (Brady and Brown, 1985). This explains the linear relation between mean stress and volumetric strain.

Evaluating the stress-strain environment around these open stopes the following were observed from the numerical analyses. It would appear that there is a good relation between mean stress in MPa and volumetric strain in millistrains. A design criterion was proposed for open stopes allowing the prediction of the failure extent in the hangingwall and sidewalls of open stopes with accuracy. From the back analyses, it was found that for hard quartzite rock the tolerance for stress-strain changes in the immediate vicinity of the open stopes were very small. Mean stress σ_m and volumetric strain ε_{vol} can be mathematically expressed as follows:

$$\sigma_m = \frac{\upsilon_1 + \upsilon_2 + \upsilon_3}{3} \tag{8}$$

$$\varepsilon_{vol} = \varepsilon_1 + \varepsilon_2 + \varepsilon_3 \tag{9}$$

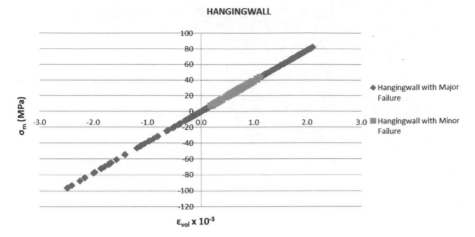

Figure 9. Graph showing the relation between mean stress and volumetric strain for open stopes with major and minor hangingwall failure after Le Roux (2015).

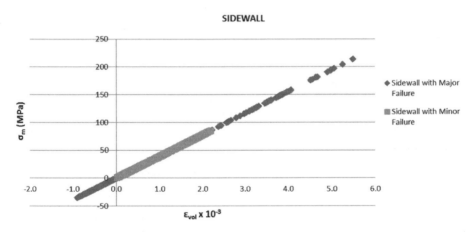

Figure 10. Graph showing the relation between mean stress and volumetric strain for open stopes with major and minor sidewall failure after Le Roux (2015).

As failure of these simulated open stopes is bounded by Hooke's Law, the Stress-Strain criterion (S_{sc}) also known as the Dilution Stress Strain Index (DSSI) as per Le Roux (2015), is the relation between mean stress and volumetric strain and can be mathematically expressed as follows:

$$S_{sc} = \frac{\sigma_m}{q\varepsilon_{vol}} \qquad (10)$$

where q in GPa, which is the slope of the linear trend line of Figures 9 and 10. The q-value can be different for each operation depending on the Young's Modulus (E) and Poisson's Ratio (v). With this criterion the expected failure depth in the hangingwall or sidewalls of excavations can be determined. In this method the assumption is made that if the volumetric strain exceeds the critical value for mean stress, failure will occur as shown in Figure 11 in light grey. This method considers all three Principal stresses and strains components, which agrees with the actual environment these open stopes are being excavated in. The contour range for plotting the Ssc design criterion was set to minimum 0 (zero) and the maximum to 1, with intervals of 1 in Map3D. This means that if the Ssc obtained value is > 1, it will be

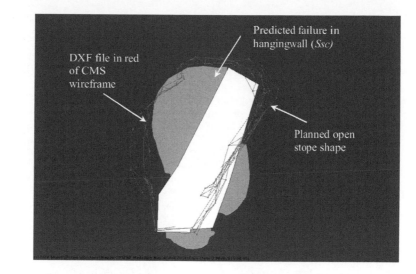

Figure 11. Application of the stress-strain criterion after Le Roux (2015).

indicated as light grey on the grid plane. The predicted failure corresponded very well with the actual observed failure in the hangingwall as shown by the CMS (Cavity Monitoring System) of the open stope plotted in red on Figure 11.

6 DISCUSSION OF RESULTS

It was illustrated that, even with very limited information available, relatively accurate results could be obtained for the open stope design. This is significant, since when a new mine is designed there is very limited information available, and the expected fracture depth is normally assumed to be within a certain value, which could completely underestimate or overestimate your support design. The design approach that has resulted from the research allows failure depth into the hangingwall and sidewalls of open stopes to be predicted accurately, and the fracture depth can be calculated for use in large excavation design with a high degree of certainty.

7 CONCLUSION

The objective of this research was to develop a method to determine the expected failure depth into the hangingwall and sidewalls of large excavations with a good degree of certainty. With the existing methods available, this could not be done with certainty, and a very large database is required (Capes, 2009). Rockmass properties, rockmass classifications, blast design, blast techniques, the stress strain environment and hydraulic radius all have some effect on, or play a part in the evaluation of dilution. It was found however, that the stress strain environment actually plays a significant role in the behaviour of open stopes at depth. Twenty-two case studies were selected with sufficient information for the research. The results of predictions of the extents of failure into the open stope hangingwall or sidewalls, based on application of the *Ssc* criterion, allow open stopes to be redesigned to "fail" up to the required stope shape and thus to reduce dilution.

ACKNOWLEDGEMENTS

The authors would like to thank Brentley, Lucas & Associates, Mining Consultants, for there assistance and it is greatly appreciated.

REFERENCES

Banerjee, P.K. and Butterfield, R. (1981) Boundary Element Methods in Engineering Science. McGraw-Hill Book Company (UK) Limited, London.

Brady, B.H.G., and Brown, E.T. (1985) Rock mechanics for underground mining. London: Allen and Unwin.

Capes, G.W. (2009) Open stope hangingwall design based on general and detailed data collection in rock masses with unfavorable hangingwall conditions, PhD. Thesis, University of Saskatchewan.

Diederichs, M.S. (2003) Rock fracture and collapse under low confinement conditions. Rock Mechanics and Rock Engineering 36 (5), pp. 339–381.

Fujii, Y., Kiyama, T., Ishijima, Y. and Kodama, J. (1998) Examination of a rock failure criterion based on circumferential tensile strain, Pure and applied Geophysica, Vol. 152, pp. 551–577.

Griffith, A.A. (1920) The phenomena of rupture and flow in solids. Philos Trans R Soc Lond Ser A Math Phys Sci 221(587), pp. 163–198.

Griffith, A.A. (1924) The theory of rupture. In: Biezeno CB, Burgers JM (eds) Proceedings of the First International Congress for Applied Mechanics. Delft. J. Waltman Jr, Delft, pp. 55–63.

Heyman, J. (1972) Coulomb's Memoir on Statics. Cambridge University Press, London.

Hoek, E. (1968) Brittle failure of rock. In: Stagg K.G., Zienkiewicz O.C. (eds) Rock mechanics in engineering practice. Wiley, New York, pp. 99–124.

Hoek, E. and Brown, E.T. (1980) Underground excavations in rock. The Institution of Mining and Metallurgy, London.

Hoek, E. (1983) Strength of jointed rock masses, 23rd Rankine Lecture. Geotechnique 33(3), pp. 187–223.

Hoek, E. and Brown, E.T. (1988) The Hoek–Brown failure criterion—a 1988 update. In: Curran J (ed) Proceedings of the 15th Canadian Rock Mechanics Symposium. University of Toronto, Toronto, pp.31–38.

Hoek, E., Wood, D., Shah, S. (1992) A modified Hoek–Brown criterion for jointed rock masses. In: Hudson JA (ed) Rock characterization: ISRM Symposium, Eurock '92, Chester, UK. Thomas Telford, London, pp. 209–213.

Hoek, E., Kaiser, P.K. and Bawden, W.F. (1995) Support of Underground Excavations in Hard Rock, A.A.Balkema, Rotterdam, Brookfield.

Hoek, E. and Brown, E.T. (1997) Practical estimates of rock mass strength. Int J Rock Mech Min Sci Geomech Abstr 34, pp. 1165–1186.

Hoek, E., Carranza-Torres, C.T. and Corkum, B. (2002) Hoek–Brown failure criterion—2002 edition. In: Hammah R., Bawden W., Curran J., Telesnicki M. (eds). Proceedings of the Fifth North American Rock Mechanics Symposium (NARMS-TAC), University of Toronto Press, Toronto, pp. 267–273.

Hoek, E. and Marinos, P. (2007) A brief history of the development of the Hoek-Brown failure criterion. Soils and Rocks, No. 2., November 2007.

Jaeger, J.C. and Cook, N.G.W. (1979) Fundamentals of rock mechanics. Chapman and Hall Ltd., London

Kaiser, P.K. and Kim, B-H. (2008) Rock Mechanics Advances for Underground Construction in Civil Engineering and Mining, Keynote lecture, Korea Rock Mechanics Symposium, Seoul, pp. 1–16.

Kwaśniewski, M. and Takahashi, M. (2010) Strain-based failure for rocks: State of the art and recent advances, Rock Mechanics in Civil and Environmental Engineering – Zhao, Labiouse, Dudt & Mathier (eds), pp. 45–56.

Le Roux, P.J. (2004) Project on Rock Mass Properties for the Free State, Mechanical Properties of Rocks and Rock Masses, University of the Witwatersrand, South Africa.

Le Roux, P.J. (2015). Measurement and prediction of dilution in a gold mine operating with open stoping mining methods. PhD thesis, University of the Witwatersrand, Johannesburg.

Le Roux, P.J. and Brentley, K.R. (2017). Time-dependent failure of open stopes at Target Mine. Afri-Rock Rock Mechanics for Africa, Cape Town, 2–7 October 2017, The Southern African Institude of Mining and Metallurgy, pp. 535–548.

Ulusay, R. and Hudson, J.A. (2007) The complete ISRM suggested methods for rock characterization, testing and monitoring: 1974–2006, pp. 971–1010.

Sakurai, S. (1981) Direct strain evaluation technique in construction of underground openings. Proceedings of the 22nd U.S. Symposium on Rock Mechanics, June 29 – July 2, 1981, Cambridge, MA, pp. 278–282.

Stacey, T.R. (1981) A simple extension strain criterion for fracture of brittle rock. Int. J. Rock Mech. Min. Sci. & Geomech. Abstr. Vol. 18, pp. 469–474.

Wiles, T.D. (2006) Course Notes, Mine Modelling Report.

Geomechanics and Geodynamics of Rock Masses – Litvinenko (Ed.)
© 2018 Taylor & Francis Group, London, ISBN 978-1-138-61645-5

Hydraulically fractured hard rock aquifer for seasonal storage of solar thermal energy

Mateusz Janiszewski
Department of Civil Engineering, School of Engineering, Aalto University, Finland

Baotang Shen
CSIRO Energy, Commonwealth Scientific and Industrial Research Organisation (CSIRO), Kenmore, QLD, Australia

Mikael Rinne
Department of Civil Engineering, School of Engineering, Aalto University, Finland

ABSTRACT: The intermittent nature of solar thermal energy derives from its oversupply during the low season and undersupply during the peak season. The solution is to accumulate and store the surplus energy that can be used in times of high demand and low supply. The HYDROCK concept is a method developed for seasonal heat storage in artificially fractured bedrock. This study aims to investigate the rock fracturing process in the construction of hydraulically fractured hard rock aquifer for seasonal storage of thermal energy. The primary objective of this study is to perform a sensitivity analysis of numerical simulations of rock fracturing processes that are taking place during the development of artificially fractured heat storage in hard rocks. Coupled hydro-mechanical numerical models are generated using rock fracture mechanics code FRACOD2D. The sensitivity of critical parameters is presented, and all relevant influencing factors are investigated. Suggestions for practical applications of HYDROCK are given.

Keywords: hydraulic fracturing, underground thermal energy storage, HYDROCK, numerical modelling, fracture mechanics

1 INTRODUCTION

The key of seasonal storage of solar thermal energy is to accumulate the surplus energy available in the low season to be used when the demand is high, and supply is low. The HYDROCK concept is a method for seasonal storage of thermal energy in artificially fractured hard rock aquifer, where the heat transfer takes place between the fluid and sub-horizontal fracture planes as depicted in Figure 1 (Larson, 1984; Hellström and Larson, 2001). The fracture planes are created by the use of hydraulic fracturing technique in boreholes, which is used commonly to increase the production rates in unconventional oil and gas reservoirs in tight shales and coal seams (Liu *et al.*, 2015). It has also been used successfully to increase the yield of water from boreholes in hard rock aquifer (Joshi, 1996) Additionally, the method is used commonly to measure the *in situ* stresses in rocks (Amadei and Stephansson, 1997).

In hydraulic fracturing, a liquid (most often water) is pumped into a sealed section of a borehole until the pressure reaches a level needed to initiate a hydraulic fracture. The orientation of hydraulic fracture is perpendicular to the least principal stress and parallel to maximum and medium principal stress. Hence, in reverse faulting stress regime which is typical for the Fennoscandian shield area, the hydraulic fracturing in vertical boreholes will result in sub-horizontal fracture planes. The first field experiment using HYDROCK method was conducted during 1982 in Bohus granite quarry in Rixö, Sweden (Larson et al. 1983,

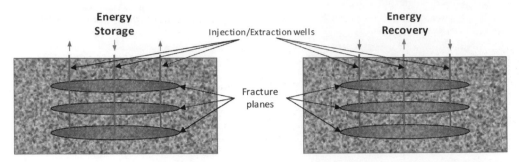

Figure 1. Schematic diagram of the HYDROCK method. During the energy storage phase (left), the heat carrier liquid is pumped into the central hole, flows through sub-horizontal fracture planes towards the peripheral wells and heats up the surrounding rock. During the energy recovery phase (right), the cycle is reversed and the cold fluid is pumped into the peripheral wells, flows through fractures and removes the heat from rock, and is extracted from the central well.

Eriksson et al. 1983, Larson 1984, Sundquist and Wallroth 1990). The test confirmed that hydraulic fracturing could produce sub-horizontal fractures in shield areas.

The HYDROCK is advantageous compared to other heat storage methods, e.g. borehole storage, as it requires fewer boreholes to be drilled and the investment cost is reduced. Latest HYDROCK field experiments have shown that reduction of 50% in the construction cost could be achieved if the energy extraction is larger than 105 MWh (Ramstad, 2004; Ramstad et al., 2007).

The HYDROCK method requires a sufficient fracture hydraulic conductivity and works best in homogenous rocks, but anisotropic and layered rock can also be utilised. Grouting may be needed if natural, vertical or steeply inclined fractures are present in the rock mass. Sufficient hydraulic conductivity in the fracture plane is required for fluid flow to connect the boreholes hydraulically. Hence, one of the most critical parameters for a successful operation of a fractured hard rock aquifer for seasonal thermal energy storage is the resulting aperture of the created fracture planes. Nordell *et al.* (1986) proposed that a fracture aperture of 1 mm is required for a proper water circulation. Results of HYDROCK *in situ* experiments indicated that proppants, such as quartz sand might be required to increase the hydraulic conductivity of fractures and to get a more controlled flow (Larson, 1984; Nordell *et al.*, 1986; Ramstad, 2004). Reaming of the borehole wall was also confirmed to reduce the impedance of the fracture (Larson, 1984).

The combined effects of explicit rock fracturing (mechanical), fluid flow (hydraulic), and temperature change (thermal) are essential to understand and forecast the rock behaviour in underground thermal energy storage in hard rocks. When constructing HYDROCK thermal storage, the hydro-mechanical coupling is most important. This study focuses on fluid flow in rock fractures because in low permeability rocks the fluid flows through fractures predominantly. The pressure of the fluid in fractures may cause movement and an increase of aperture and propagation of the fracture. The changes in fracture geometry will change the hydraulic conductivity of the fracture and create new flow paths, which will enhance the flow of fluid (Shen et al. 2014).

This study aims to investigate the rock fracturing process that is used in the construction of hydraulically fractured hard rock aquifer for seasonal storage of thermal energy. Coupled hydro-mechanical (HM) numerical model is prepared using the commercial FRACOD2D rock fracture mechanics code. The primary objective of this study is to perform a sensitivity analysis of the output values on varying input parameters and to give practical suggestions regarding the implementation of HYDROCK method.

2 METHODOLOGY

In this study, coupled hydro-mechanical numerical models were generated using rock fracture mechanics code FRACOD2D. FRACOD is a Boundary Element Method (BEM) and uses an indirect boundary element technique—Displacement Discontinuity Method (DDM) with fracture mechanics theory integrated into it. The model consisted of one injection borehole, where the fluid was injected under high pressure to produce a single horizontal hydraulic fracture propagating from the borehole (see Figure 2). The rock was assumed to be perfectly isotropic and homogeneous with no internal fractures present. The reverse faulting stress regime was assumed, with horizontal rock stresses being five times higher than vertical.

In the numerical model, a flow rate boundary condition was used to supply pressure into the injection borehole. The total flow rate in borehole was specified, and it remained constant throughout the whole process. The operational parameters of the hydraulic equipment applied in field experiments by Ramstad (2004) were employed.

First, a base case scenario of hydraulic stimulation was simulated using the base value input properties presented in Table 1. The properties were selected to represent a generic case of a hard crystalline rock, which is used in the HYDROCK method. Next, a sensitivity analysis of the output values on varying input parameters in hydraulic fracturing model in FRACOD2D was performed. The input parameters were varied by ±50% (see the list in Table 1). The maximum fracture aperture was measured as the primary output result. Additionally, the maximum flow rate and maximum vertical and horizontal induced stress were measured at a monitoring point positioned 0.1 m to the right of the injection borehole.

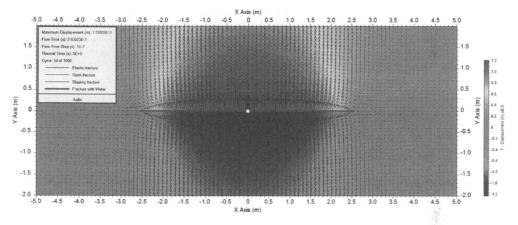

Figure 2. Vertical displacement of the rock mass during hydraulic fracturing in a borehole after 50 cycles of fracture propagation.

Table 1. Sensitivity analysis of the input parameters for hydraulic fracturing models in FRACOD2D.

Parameter	Symbol	Low value	Base value	High value	Unit
Poisson's ratio	ν	0.125	0.25	0.375	–
Young's modulus	E	18.75	37.5	56.25	GPa
Internal friction angle	ϕ	16.5	33	49.5	°
Cohesion	c	16.5	33	49.5	MPa
Tensile strength	σ_t	6.25	12.5	18.	MPa
Fracture normal stiffness	K_n	500	1000	1500	GPa/m
Mode II/Mode I toughness ratio	K_{IIc}/K_{Ic}	1	2	3	–
Flowrate at the boundary	$value\_h$	0.004	0.008	0.012	m³/s
Initial borehole (pump) pressure	$value\_pi$	12.5	25	37.5	MPa
Borehole volume	vol_hole	0.015	0.03	0.045	m³
Initial fracture aperture	e_{ini}	0.05	0.1	0.15	mm
Residual fracture aperture	e_{res}	0.05	0.1	0.15	mm
Rock porosity	ϕ	0.05	0.1	0.15	%
Horizontal rock stress	S_{xx}	2.5	5	7.5	MPa
Vertical rock stress	S_{yy}	0.5	1	1.5	MPa

3 RESULTS AND DISCUSSION

The result of the base case simulation is presented in Figure 2. The maximum fracture aperture of 2.5 mm was reached after 50 cycles and then decreased to 1.2 mm with decreasing fluid pressure as the fracture propagated further. The resulting fracture aperture is higher than the minimum width of 1 mm required for proper circulation of water that was suggested by Nordell *et al.* 1986. Nevertheless, using propping agents such as quartz sand may be required to increase the flow capacity of created fractures as suggested by Larson (1984), Nordel et al. (1986) and Ramstad (2004).

The fracturing process stopped after the fracture plane reached 15.6 m radius. As expected, the orientation of hydraulic fracture was perpendicular to the least principal stress and parallel to maximum and medium principal rock stress and is perfectly horizontal due to the homogenous rock without any internal discontinuities. This outcome corresponds well to the results of the first HYDROCK field experiment, where the resulting hydraulic fractures had at least a 10 m radius and were parallel to each other (Larson, 1983). However, in some *in situ* conditions, the hydraulic fracture may alter its orientation away from the drill hole if the local stress orientation is disturbed by discontinuities or flaws.

The results of the sensitivity analysis are plotted on tornado plots in Figure 3. The maximum aperture of the resulting hydraulic fracture (Figure 3a) was most sensitive to changes in the Young's modulus (+33.9% and −11.8% change in the aperture for −50% and +50% change in Young's Modulus, respectively). The bigger the Elastic modulus of the rock, the lower the resulting fracture aperture. Second input parameter with high influence on the fracture aperture was the borehole volume (11% and 29.9% change in the aperture for -50% and +50% change in borehole volume, respectively). This is dictated by the length of the borehole

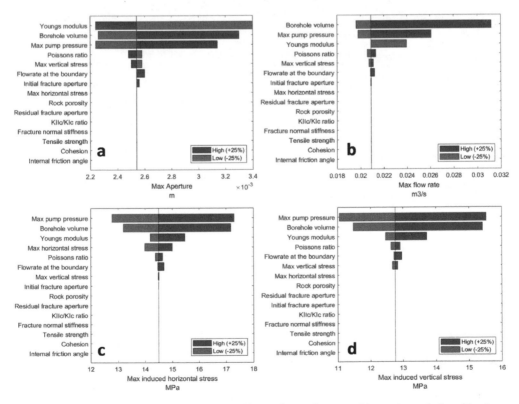

Figure 3.   The sensitivity of fracture aperture (a), maximum flow rate (b), maximum induced horizontal stress (c), and vertical stress (d) to ±50% change of the input parameters in the hydraulic fracturing numerical model in FRACOD2D.

sections at which it is pressurized. The longer the segment, the higher the borehole volume and the larger the resulting aperture of the hydraulic fracture. The maximum aperture of the fracture was also influenced by the maximum pressure of the pump used for hydraulic fracturing (−11.8% and 23.6% change in the aperture for −50% and +50% change in pump pressure, respectively). This is undoubtedly logical, as more pressure delivered by the pump results in higher pressures at the fracture tip available for fracture propagation. Other input parameters had a very low influence on the aperture (+1.6% change for −50% change in the vertical rock stress, and +1.6% change for −50% change in Poisson's ratio) or no influence at all. Surprisingly, the increase of flow rate had a relatively low influence on resulting fracture aperture (2.4% increase). Sundquist and Wallroth (1990) suggested that high flow rate (>10 l/s) should be used to increase the hydraulic conductivity of hydraulic fractures based on the field test results in granite. However, it should be noted that the hydrological behaviour of rock fracture is not only influenced by the mechanical aperture, but also other properties such as contact area, roughness, matedness, or presence of channels (Hakami, 1995).

The maximum flow rate at the monitoring point (Figure 3b) was most sensitive to the two operational parameters of the hydraulic fracturing used as a boundary condition in the model. By increasing the borehole volume and initial pump pressure by 50%, the flow rate was increased by 49.3% and by 24.8%, respectively. Interestingly, the decrease of Young's modulus resulted in higher flow rate in the fracture. However, the change was small (+12% change). Other parameters had low or no influence on the maximum flow rate at the monitoring point.

The max initial pump pressure and borehole volume were also influencing heavily the resulting induced rock stress at the monitoring point (Figure 3c,d) so that the 50% increase in their value resulted in 20% increase in both horizontal and vertical stress. The resulting rock stresses were also influenced by the elastic properties of the rock. However the increase was low (around 7% and 1% increase in stress with increased Young's modulus and Poisson's ratio, respectively).

The main limitation of the numerical modelling results presented in this study is that no natural fractures were included in the rock mass, which could influence the outcome and alter the path of the propagating fracture as observed by Ramstad (2004). In some cases, it can even lead to a situation where the hydraulic fracture is arrested on a natural fracture and stops propagating. However, the focus of this study was to investigate how the generic rock properties and the operational parameters of the hydraulic fracturing equipment affect the resulting fracture, without the influence of existing discontinuities that are very site dependent. This phenomenon will be investigated more closely in future studies, where a back-calculation of an HYDROCK field test will be performed taking into account the distribution of natural discontinuities in the rock mass.

4   CONCLUSIONS

The hydraulic fracturing procedure for constructing an artificially fractured hard rock aquifer was successfully simulated numerically using FRACOD2D. It was found out that the most influencing input parameters in hydraulic fracturing numerical model are the operational parameters of the hydraulic fracturing equipment used (i.e. the maximum pump pressure and the borehole volume). Therefore, the proper setting of the operational parameter of the fracturing equipment is crucial. Such finding is essential for the selection of appropriate hydraulic fracturing equipment for a construction of HYDROCK storage. Larger pump capacity can increase the resulting rock fracture aperture. Hence, the hydraulic conductivity of fracture planes will be higher, and the thermal performance of the system will improve.

The second group of input parameters that have a significant influence on the output are the elastic parameters (i.e. Young's modulus and Poisson's ratio). The higher the elastic properties, the more difficult is to create an open fracture with sufficient aperture and more pressure is required. This implies that selection of the site for HYDROCK construction and accurate site investigation will directly influence its performance. It is also crucial for numerical simulations of the hydraulic fracturing process as good quality laboratory data for the model input is needed.

In the future, a variety of scenarios will be tested to investigate the influence of different geological and geomechanical conditions, such as rock types, anisotropy, *in situ* rock stress, and presence of discontinuities. Additionally, a back-analysis of HYDROCK field test from the literature will be performed with an upgraded numerical model.

REFERENCES

Amadei, B. and Stephansson, O. (1997), *Rock Stress and Its Measurement*, Springer Netherlands, Dordrecht, the Netherlands.

Eriksson, K.G., Larson, S.Å. and Haag, O. (1983), HYDROCK—en ny metod att lagra värme i berg (HYDROCK—a new method of storing heat in the bedrock), Technical Report R105:1983, Statens råd för byggnadsforskning, Stockholm, 117–123.

Hakami E (1995) Aperture distribution of rock fractures. PhD thesis, Royal Institute of Technology, Stockholm, Sweden

Hellström, G. and Larson, S.Å. (2001), Seasonal thermal energy storage—the HYDROCK concept. *Bulletin of Engineering Geology and the Environment* 60(2):145–156. doi:10.1007/s100640100101

Joshi, V. (1996), Borehole rejuvenation for sustainability, *Proceedings of the 22nd WEDC Conference*, 193–194.

Larson, S.Å. (1984), Hydraulic fracturing in the Bohus granite, SW-Sweden. Test for heat storage and heat extraction. Geothermal Resources Council TRANSACTIONS 8:447–449. http://pubs.geothermal-library.org/lib/grc/1001215.pdf Accessed 7 October 2016.

Larson, S.Å., Fridh, B. and Haag, Ö. (1983), Hydrock—värmelager i berg. Anläggning av värmeväxlarytor med hjälp av hydraulisk uppspräckning; HYDROCK—metoden (Hydrock method—Heat Storage in Rock: The construction of heat exchanger surfaces by hydraulic fracturing), Technical Report Publ. B 222, Chalmers University of Technology/University of Götenburg.

Liu, H., Yang, T., Xu, T. and Yu, Q. (2015), A comparative study of hydraulic fracturing with various boreholes in coal seam, *Geosciences Journal*, 19(3), 489–502.

Nordell, B., Bjarnholt, G., Stephansson, O. and Torikka, A. (1986), Fracturing of a pilot plant for borehole heat storage in rock, *Tunnelling and Underground Space Technology*, 1(2), 195–208.

Ramstad, R.K. (2004), *Ground source energy in crystalline bedrock—increased energy extraction by using hydraulic fracturing in boreholes*. PhD thesis, Norwegian University of Science and Technology, Norway, http://hdl.handle.net/11250/235848. Accessed 7 October 2016.

Ramstad, R.K., Hilmo, B.O., Brattli, B. and Skarphagen, H. (2007), Ground source energy in crystalline bedrock-increased energy extraction using hydraulic fracturing in boreholes. *Bull Eng Geol Environ* 66: 493–503. doi:10.1007/s10064-007-0100-7.

Shen, B., Stephansson, O. and Rinne, M. (2014), *Modelling Rock Fracture Processes: a fracture mechanics approach using FRACOD*. Dordrecht, The Netherlands: Springer. doi:10.1007/978-94-007-6904-5.

Sundquist, U. and Wallroth, T. (1990), Hydrock—energilager i berg—slutrapport för etapp 1 & 2 (Hydrock—energy storage in rock: final report for phase 1 & 2), Technical Report Publ. B 349, Chalmers University of Technology/University of Götenburg.

Geomechanics and Geodynamics of Rock Masses – Litvinenko (Ed.)
© *2018 Taylor & Francis Group, London, ISBN 978-1-138-61645-5*

Peculiarities of numerical modeling of the conditions for the formation of water inflows into open-pit workings when constructing the protective watertight structures at the Koashvinsky quarry

Sergey Kotlov, Denis Saveliev & Artemiy Shamshev
Saint Petersburg Mining University, Saint-Petersburg, Vasilievsky Island, Russia

ABSTRACT: The article contains a detailed description of the peculiarities of the formation of the geofiltration regime on the northeastern side of the Koashvinsky quarry. In this area, the possibility of constructing a grout curtain is considered as one of the methods of controlling high inflows of water into the quarry. At the expectable construction site, the Quaternary age rocks, which are associated with the valley of the Vuonnemyok River, are developed and lying on the bedding crystalline rocks of ijolite-urtites; hydrogeological conditions are characterized as complex. Quaternary rocks, the influx from which is up to 2000 m^3/h in normal time and up to 3000 m^3/h in high-water periods, play the main role in the quarry watering. The water inflow into the quarry from crystalline rocks is approximately 500 m^3/h.

Formed at the time being in the open field, the filtration flow has a complex three-dimensional structure due to the presence of three aquifers in the section and a fairly close location of several boundaries feeding groundwater. The filtration calculations of the projected grout curtain with the use of analytical methods will not make it possible to take into account fully the complex hydrogeological structure of the northeastern side of the Koashvinsky quarry. The efficiency of the planned grout curtain can only be substantiated using geofiltration modeling.

To create a numerical geofiltration model of the Koashvinsky quarry area, the Visual MODFLOW software complex was used. The model takes into account the filtration flows formed both in Quaternary and in crystalline rocks, the feeding of aquifers by means of rivers and lakes and through the infiltration of atmospheric precipitation, as well as by virtue of discharge into the quarry and existing drainage wells.

In the course of numerical experiments, various alternate layouts for the grout curtain have been considered both in plain view and in section. It has been established that when the grout curtain is erected in the moraine aquifer that is the first from the surface, a decrease in the water inflow into the quarry will be insignificant in comparison with the total quarry drainage. In the case of the construction of the grout curtain in the artesian aquifer that is second from the surface, there will be an increase in the filtration rates in the upper part of strongly fractured and weathered crystalline rocks, the filtration properties of which have not been studied properly.

As a result of the study of geofiltration processes, the optimum position of the grout curtain has been determined with the use of numerical modeling, the structure efficiency has been assessed, and recommendations have been developed for the further study of the hydrogeological structure of the site in question.

Keywords: numerical geofiltration modeling, water inflows into the quarry, anti-filtration structures, hydrogeology

1 INTRODUCTION

The Koashvinskoye deposit of apatite-nepheline ores is confined to the southern part of the Khibin Massif and is located 14 km from the city Kirovsk. Absolute marks of the surface are in the range from +200 to +900 m. The deposit is confined to the intermountain valley of Elarge catchment area and favorable conditions for feeding aquifers due to a significant amount of precipitation.

According to hydrogeological stratification, two interconnected aquifers are distinguished: a complex of Quaternary deposits represented by sandy-argillaceous sediments, and a complex of crystalline intrusive rocks of Paleozoic age represented mainly by yolite-urtites. In the complex of Quaternary rocks, two aquifers are distinguished, the first from the surface is Ostashkovsky and the hydraulic head of Podporozhsky (Figure 3). Quaternary rocks, the influx from which is up to 2000 m³/h in normal time and up to 3000 m³/h in high-water periods, play the main role in the quarry watering. The water inflow into the quarry from crystalline rocks is approximately 500 m³/h.

In the immediate vicinity of the Koashvinsky Quarry, there is a large surface watercourse—Lake Porokyavr. Thus, the main inflow is formed from the northeastern side of the quarry edge, where Quaternary sediments are opened, as well as the Porokyavr Lake and Vuonne-myok River are located. On this area of the edge, filtration loss of a fine fraction of sandy rocks is registered, as a result of which sloughing tongues are formed. In order to reduce the negative impact due to high groundwater inflows, the design institutes proposed various solutions to reduce water inflows into the quarry, among which the construction of the grout curtain (GC) in the Ostashkov Aquifer was chosen as one of the methods for combating high water inflow into the quarry.

2 PROBLEM STATEMENT AND METHODS

In order to assess the efficiency of GC construction in this area, the experts of the Research Center GiPGP of the Saint Petersburg State Mining University performed work using numerical geofiltration modeling in the software application Visual MODFLOW. With the use of this software package, a model of the area of the northeastern side was created, which reveals water quarries of the Quaternary age that play the main role in the quarry watering (Figure 1). The use of the created model makes it possible to identify the main features of the formation of the geofiltration regime on the northeastern edge of the Koashvinsky Quarry, perform the forecast with regard to the change of water inflows into the quarry, and to assess the efficiency of various types of watertight structures and drainage measures.

Formed at the time being in the open field, the filtration flow has a complex three-dimensional structure due to the presence of three aquifers in the section and a fairly close location of several boundaries feeding groundwater. The filtration calculations of the projected grout curtain with the use of analytical methods will not make it possible to take into account fully the complex hydrogeological structure of the northeastern side of the Koashvinsky quarry. The efficiency of the planned grout curtain can only be substantiated using geofiltration modeling. The general structure of the numerical model is shown in Figure 2.

The created numeric geofiltration model takes into account the filtration flows formed both in Quaternary and in crystalline rocks, the feeding of aquifers by means of rivers and lakes and through the infiltration of atmospheric precipitation, as well as discharge into the quarry and existing drainage wells. An extensive regime network was created at the Koashvinskoye deposit, which consists of observation wells equipped for various aquifers. When creating the model, the results of plane surface surveys were used, while data from experimental filtration observations were used for model calibration. The calibration process involved comparing the real values of groundwater heads with the calculated levels obtained on the model; at the same time, the check was performed with respect to the consistency between the model costs and the actual productivity of the drainage system. In total, data from more than fifty wells was used; the values of level convergence for most of the wells are in the range of ± 2 m, for

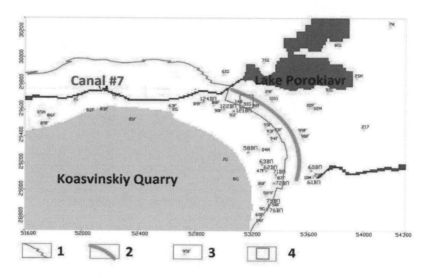

Figure 1. Layout of the modeled area. 1 – Contour of the Quarry at 2027 year; 2 – GC; 3 – drainage and observations wells; 4 – observed sectors.

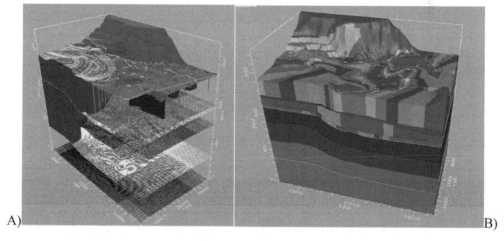

Figure 2. General structure of the numerical geofiltration model. A – profile breakdown into layers; B – distribution of the filtration coefficients over the model layers.

almost all the others—in the range of ± 5 m, a number of wells, approximately 8 pieces, stand out of the general range; however, despite this, we can talk about a fairly accurate mapping of the real geological and hydrogeological situation on the created numerical model, since this area is confined to the intermountain valley region. Due to a rather sharp change in the hypsometric marks of the surface, the actual values of the hydraulic heads for the given area reach 50 m–70 m, and sometimes even more. For such conditions, the results obtained with respect to level convergence are more than satisfactory.

In the course of numerical experiments, various alternate layouts for the grout curtain have been considered both in plain view and in section. Thus, the main options for comparison were the following situations: Option I – GC location in the Ostashkov Aquifer, which is the first aquifer from the surface (this was considered the main option among the possible design solutions aimed at reducing water inflows into the quarry); Option II – GC construction in the Pressure Podporozhsky Aquifer (Figure 3).

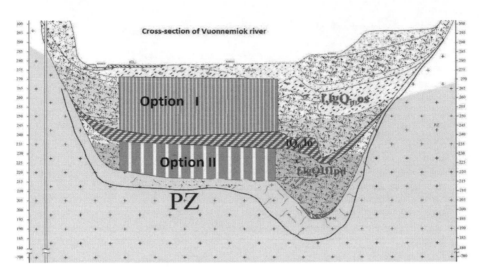

Figure 3.   Schematic geological section of the valley of the river. Vuonnemyok with options for modeling the location of the anti-filtration structures.

3   RESULTS

When considering Option I, one of the design institutes proposed the total GC length of 900 m; its erection will require to drill approximately 300 wells, and the base will rest upon a layer of relatively water-resistant banded clays. On a numerical model, a similar situation was shown, on which, according to the design decision, the GC was placed in the first calculated model layer (Figure 4).

Table 1 shows the distribution of water inflows into the quarry, on the northeastern edge of the quarry, before GC construction. When modeling Option I of the GC location, the results of water inflows into the quarry were obtained taking into account the change in the filtration regime of the area (Table 1). When analyzing these results, it was determined that the total reduction of water inflow into the quarry would be approximately 200 m³/h, which is less than 10% of the total value in the side-land period. Water inflow reduction occurs due to the redistribution of underground flows and changes in the hydrodynamic regime in the Ostashkov Aquifer.

The manifestation of negative phenomena and processes on quarry edges, such as sinking of the sides and filtration removal of the fine fraction, is mainly characteristic of the Ostashkov Aquifer, so GC erection in this element will undoubtedly reduce the activity of these processes; however, this design solution will not be able to affect in full a significant reduction of water inflows into the quarry and stop the edge slipping, as the value of the specific water inflow to the north-eastern edge will remain fairly high.

Since the rocks of the Podporozhsky Pressure Aquifer still play the main role in the quarry watering, a decision was made to consider the possibility of placing GC on the model in the rocks of this aquifer. Edge slipping is not typical for these sediments, despite the higher value of the specific water inflow, which is associated with the spread of a larger sandy, sometimes even gravel fraction.

When modeling this situation, it was found that the overall reduction in the inflow into the quarry will not exceed 180 m³/h, which is even lower than during GC erection in the Ostashkov Aquifer. This is due to the fact that during GC erection in the Podporozhsky Aquifer, more active filtration of groundwater will begin through the upper, more cracked and weatherworn part of crystalline rocks. As a result, the inflow from the Podporozhsky Aquifer will be almost halved, but at the same time, an inflow from the upper part of the crystalline rocks will increase by 30% of the initial value thereby reducing the overall efficiency of GC construction.

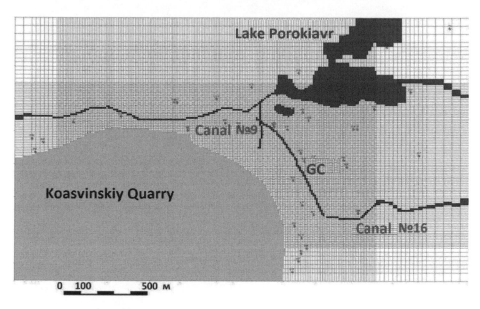

Figure 4. The first design layer of the model, indicating the zone construction of GC.

Table 1. The values of water inflows in the quarry, depending on the version of the simulation.

The values of water inflows from simulation results	Without GC	With GC in $Q_{III}os$ (Option I)	With GC in $Q_{III}pd$ (Option II)
Total inflows in the Koashvinsky quarry, m³/hour	**2465**	2270	2290
Total productions of drainage system, m³/hour	720	720	72
Inflow from Ostashkov aquifer, m³/hour	725	540	720
Summary inflow from Podporozskiy aquifer and upper, more permeable part of crystalline rocks, m³/hour	1100 (700 + 400)	1100 (700 + 400)	920 (390 + 530)
Inflow from crystalline rocks, m³/hour	510	510	530

4 DISCUSSION AND CONCLUSIONS

As a result of the modeling work, it was found that GC erection in the first aquifer from the surface or in the pressure aquifer will have little effect on the reduction of water inflows and only to a small extent will help reduce the occurrence of such negative processes as the slipping of the quarry edges in the north-eastern section of the edge. Using numerical geo-filtration modeling, the following solution was proposed to reduce water inflows into the quarry: extension and increase in the capacity of the encroachment line of dewatering wells (DWW). New DWWs were additionally set on the model (the specific yield of each of which is approximately 50 m³/h), and when analyzing the modeling results, it was concluded that the efficiency of the construction of five DWWs would outweigh the effect of GC construction (Figure 5). In case ten DWWs are constructed, the overall reduction of water inflow into the quarry from the north-eastern side will reach a value of 460 m³/h, which is much more efficient than the construction of a grout curtain, and, most importantly, requires much less capital construction costs.

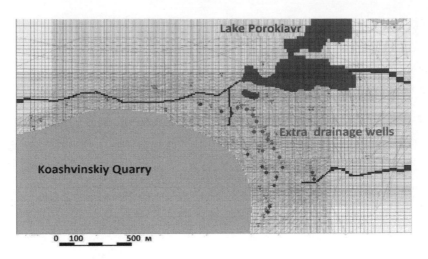

Figure 5. Project of extra drainage wells.

Thus, as a result of studying the geofiltration processes using numerical modeling, the efficiency of the erection of the grout curtain was determined, which turned out to be quite low; the overall decrease in water inflow into the quarry will only be equal to 10% of the initial value in the side-land period. Moreover, an option was considered on the model that involved the extension and capacity expansion of the DWW encroachment line, the efficiency of which exceeds the effect of GC construction by a huge ratio. To make DWW operation more efficient, it is also possible to consider on the model options with the different location of DWWs within the area under investigation with a view to determine their rational location in order not to reduce the water-absorbing capacity of each individual well.

Such numerical experiments make it possible to visually assess the efficiency of various methods of combating high values of water inflows into the quarry, consider the change in the existing hydrodynamic regime during the construction of protective structures, and help calculate in advance the efficiency and appropriateness of a certain method. Proceeding from these facts, contractors should remember that any design solution should be accepted and approved, based, among other things, on the results of numerical geofiltration modeling. Under the conditions of the development of rocks that are anisotropic and heterogeneous in terms of filtration parameters, any forecast estimates, as well as any recommendations regarding various methods aimed at combating high water inflow values should only be performed on the basis of numerical modeling results. Any numerical geofiltration model requires a large number of initial data to correctly display the real situation, but the opportunities that open up to the contractors, taking into account numerical modeling, make it possible to recoup unnecessary or inefficient design solutions performed without reliance on the modern methods of geofiltration modeling.

REFERENCES

[1] Gavich I. Theory and practice of using numerical modelling in hydrogeology. 1980, Moscow.
[2] Konosavskiy P., Soloveichik K. Matematical modelling of geofiltrational processes. Saint-Petersburg Еусртсфд university, 1988.
[3] Lomakin E., Mironenko V., Shestakov V. Numerical modelling of geofiltration, Moscow, 1988.
[4] Norvatov U. Studying and predictions of technogenic regime of groundwaters. Leningrad, 1988.
[5] Hill M.C. MODFLOW/P—A computer program for estimating parameters of a transient, three-dimensional, groundwater flow model using nonlinear regression, U.S. Geological Survey, Open-file report, 1992.

Geomechanics and Geodynamics of Rock Masses – Litvinenko (Ed.)
© 2018 Taylor & Francis Group, London, ISBN 978-1-138-61645-5

Numerical study of the hydraulic fracturing and energy production of a geothermal well in Northern Germany

Mengting Li
Energy Research Center of TU Clausthal (EFZ), Goslar, Germany

Michael Z. Hou
Energy Research Center of Lower Saxony (EFZN), Goslar, Germany
Institute of Petroleum Engineering, TU Clausthal, Clausthal-Zellerfeld, Germany

Lei Zhou
State Key Laboratory of Coal Mine Disaster Dynamics and Control, Chongqing University, China

Yang Gou
Energy Research Center of Lower Saxony (EFZN), Goslar, Germany

ABSTRACT: The hydraulic fracturing is an essential tool to increase the permeability of tight formations and to increase the petrogeothermal energy recovery. In this paper, a hydraulic fracturing model was developed based on the previous work and implemented in the coupled numerical simulator TOUGH2MP-FLAC3D^{plus}. It considers the stress redistribution due to fracture opening and hydromechanical effects in full three dimensions. The cubic law was implemented, so that the fluid flow in both porous media and fractures can be simulated at the same time. With the advantages of TOUGH2MP, it is also capable to track the migration and distribution of injected fracturing fluid in the reservoir formation. The model has been used to study the hydraulic fracturing in the geothermal well Gross Buchholz Gt1 in Hanover, Germany. The measured data during the hydraulic fracturing treatment has been matched. The simulated fracture geometry was comparable with that analyzed from the well test. The verified model has been used to study the geothermal utilization in the Detfurth sandstone formation.

1 INTRODUCTION

The development of renewable energy is of high priority due to the world's increasing energy consumption and climate change. As one of the most important members, the geothermal energy attracts a lot of attention because it is of huge amount, regenerable, and not dependent on weather. Especially in recent years, the development of deep geothermal energy is a hot topic (Kolditz et al. 2015). Currently, the development of deep geothermal energy is restricted by the current drilling technology, energy transition efficiency, environmental impacts etc. The deep geothermal energy is normally stored in tight sandstone or granite formations which has normally low or ultra-low permeability. This characteristic makes it difficult to produce the original fluid in place (for hydrothermal system) or injected fluid (for petrothermal system) in an economic rate. In order to increase the deep geothermal energy recovery, the hydraulic fracturing should be carried out to increase the permeability of tight formations. The hydraulic fracturing must be well designed and optimized, on one side, to maximize the productivity, and on the other side, to remove or minimize the related risks such as environment contaminations and induced micro earthquakes.

In this paper, a hydraulic fracturing model was developed based on the previous work (Zhou 2014, Gou et al. 2015) and implemented in the coupled numerical simulator TOUGH2MP-FLAC3D. It considers the stress redistribution due to fracture opening and hydromechanical effect in full three dimensions. The cubic law has been implemented, so that the fluid flow in both porous media and fractures can be simulated at the same time. With the advantage of TOUGH2MP, it is also capable to track the migration and distribution of injected fracturing fluid in the reservoir formation.

2   NUMERICAL MODELS

The hydraulic fracturing is a coupled hydomechanical geo-process combined with multiphase-multicomponent flow. Such processes have been considered in the developed model by Zhou (2014). However, it treated the flow in the fracture as incompressible and does not distinguish between the reservoir fluids and injected fluids. In order to overcome these shortcomings as well as consider more complicated fluid properties (e.g. the temperature—and pressure-dependent density, viscosity as well as enthalpy), TOUGH2MP was adopted to simulate the fluid flow in both fractures and reservoir formations.

In this study, the EOS7 (Zhang et al. 2008) was adopted to describe the fluid properties. Following the approach in Gou et al. (2015), the second component of EOS7 (originally brine) was switched to fracturing fluid with its own properties. It was assumed that the fracturing fluid is water-based fluid (miscible with water), so that it will not form the third phase and can be described by the mass fraction in the aqueous phase (as the second primary variable). At current stage, it is simplified that the fracturing fluid has constant density and viscosity.

The fluid flow in both fractures and reservoirs can be described with the following flow equation and mass conservation equation

$$\vec{F}_\beta = -k\frac{k_{r\beta}\rho_\beta}{\mu_\beta}\left(\vec{\nabla}p_\beta - \rho_\beta\vec{g}\right)$$ (1)

$$\frac{\partial\left(\phi\sum_\beta S_\beta\rho_\beta x_\beta^\kappa\right)}{\partial t} = -\vec{\nabla}\cdot\left(\sum_\beta \vec{F}_\beta x_\beta^\kappa\right) + q^\kappa$$ (2)

where F_β is the mass flow rate of phase β ($\beta = l$ for liquid, g for gas) in (kg/m²/s), k is rock intrinsic permeability in (m²), $k_{r\beta}$ is phase relative permeability in (–), ρ_β is phase density in (kg/m³), μ_β is phase viscosity in (Pa·s), p_β is phase pressure in (Pa), S_β is phase saturation in (–), x_β^κ is mass fraction of component j in phase β in (–), and q^κ is the sink/source in (kg/s). The fluid flow in the fracture is treated as flow between two parallel planes, which is normally described with the cubic law. So the fracture permeability is calculated as

$$k = \frac{(fw)^2}{12}$$ (3)

where the w is the fracture width in [m] and f is a parameter that reflects the influence of the roughness on the transmissivity ([–]).

The setup of the coupled non-linear equation systems is different for flow in the fracture and reservoir formations. For the reservoir formation, the element volume is constant. The original way in TOUGH2MP was adopted and the resulted set of coupled non-linear equations is

$$R_n^{\kappa,k+1} = M_n^{\kappa,k+1} - M_n^{\kappa,k} - \frac{\Delta t}{V_n}\sum_m A_{nm}F_{nm}^{\kappa,k+1} + \Delta t q_n^{\kappa,k+1} = 0$$ (4)

where $R^{\kappa,k+1}{}_n$ and $M^{\kappa,k+1}{}_n$ are the residuum and mass of component κ per unit volume in the n-th element at the time step $k+1$ in [kg/m³], respectively. Δt is the time step in [s], V_n is the volume of the n-th element, A_{nm} is the cross section area between the n-th and m-th element in [m²], $F^{\kappa,k+1}_{nm}$ is the flow term between the n-th and m-th element at the time step $k + 1$ in [kg/m²/s], and $q^{\kappa,kZ+1}_n$ is the sink/source term in the n-th element at the time step $k + 1$ in [kg/m³/s]. Eq. 4 is not applicable for the modelling of fracturing, because the fracture width (or volume) is dependent on the fracture pressure, which is the first primary variable, and will change with the time. In such a case, combining the mass per unit volume with the fracture width, the set of coupled non-linear equations for fracture element is modified as

$$R^{\kappa,k+1}_n = \left(wM\right)^{\kappa,k+1}_n - \left(wM\right)^{\kappa,k}_n - \frac{\Delta t}{A_n}\sum_m A_{nm}F^{\kappa,k+1}_{nm} + \Delta t\left(wq\right)^{\kappa,k+1}_n = 0 \qquad (5)$$

where $R^{\kappa,k+1}_n$ is the residuum of the width-mass per unit volume product in [kg/m²]. If two adjacent elements are both fracture element, A_{nm} is also dependent on the fracture pressure and is evaluated at $k+1$ time step for numerical stability. The Eq. 4 and Eq. 5 are solved by the AZTEC parallel linear solver using the Newton method, leading to

$$-\sum_i \left(\frac{\partial R^{\kappa,k+1}_n}{\partial x_i}\right)_p \left(x_{i,p+1} - x_{i,p}\right) = R^{\kappa,k+1}_n\left(x_{i,p}\right) \qquad (6)$$

where $x_{i,p}$ is the value of the i-th primary variable at the p-th Newton-Raphson iteration step.

3   CASE STUDIES

The developed model was applied in the study of the hydraulic fracturing in the GeneSys (Generated Geothermal Energy Systems) project. In this project, single well concept was developed to directly use the geothermal energy (Tischner et al. 2010). Two wells were involved in the GeneSys project, including the well Horstberg Z1 for the testing and concept development, and another well Gross Buchholz Gt1 in Hanover for the demonstration. More details can be found in Tischner et al. (2010). In this study, the hydraulic fracturing of the well Gross Buchholz Gt1 was investigated.

The well Gross Buchholz Gt1 was drilled in 2009. The well reached the depth of ca. 3900 m TVD and the target formation is Middle Buntsandstein. Hydraulic fracturing treatment was applied to create artificial fracture for heat exchange. According to Tischner et al. (2013), the hydraulic fracturing was carried out in 2011. A typical injection rate for the fracturing of tight sandstone formation, namely 90 l/s (5.4 m³/min), was adopted and no proppant was used. The whole operation lasted for 106 hours (23–28 May 2011) with 5 injection-pause-cycles, so that a final volume of 20,000 m³ fresh water was injected. After the fracturing treatment two low-rate injection tests were carried out in July and October 2011 (Pechan et al. 2014). The results of the pressure transient analysis show that the final fracture area was more than 0.5 km².

3.1   *Numerical simulation of the hydraulic fracturing in the well Groß Buchholz Gt1*

According to the geological and stratigraphic conditions, a 3D ¼ model was generated with FLAC3D^plus and used for the simulation of the hydraulic fracturing (Fig. 1a). The model has a dimension of 1,700 m × 350 m × 563 m and lies at the depth between −3,287 m and −3,850 m. It was discretized into 38,016 elements. The model considers the main sandstone layers, including Solling, Detfurth and Volpriehausen formations, several interlayers as well as rock salt as caprock. The rock mechanical parameters and in-situ stresses were adjusted based on the previous study in Hou et al. 2012 as well as Zhou 2014. The primary stress distribution was shown in Fig. 1b. The maximum principal stress was in vertical direction and calculated from the weight of the overburden. The minimum principal stress was estimated from the mini-frac tests. The ratio between the stress components shows an extensional stress regime. According to the in-situ measurement, the pressure decline at the post-injection phase showed that the reservoir is highly pressurized (Tischner et al. 2013). An initial pore pressure of 65 MPa was used in this study. For the initial temperature, a natural geothermal gradient of 0.03°C/m was used. The temperature at the surface was adjusted, so that the temperature at the depth of 3700 m was 165°C (Schäfer et al. 2012).

The simulation results are shown in the following figures. Fig. 2 shows the temporal evolution of the simulated bottom hole pressure (BHP). Fig. 2a shows the results during the stimulation phase. It can be seen that the simulated bottom hole pressure was slightly below 80 MPa. The simulated bottom hole pressure was comparable with the fracture pressure from Tischner et al. (2013), which was calculated from the measured well head pressure with consideration of well height and the flow friction. The pressure decline after shut-in was shown in Fig. 2b. The whole post-frac process from May until November 2011 was considered. The pressure decreased slowly after shut-in and reached 71 MPa at the end. The pressure decline after shut-in was also comparable with the measured value.

Fig. 3 shows the simulated fracture geometry at shut-in (t = 185 h). It can be seen that the final fracture had a half-length of 1,390 m and height of 358 m (Fig. 3a). This corresponds to a total area of 1.1 km² for the bi-wing fracture, which is enough for the geothermal utilization of this project. The upward fracture growth was restricted by the rock salt layers. The maximum fracture width was 2.5 cm at shut-in, while the leak-off ratio is 21% due to the low permeability

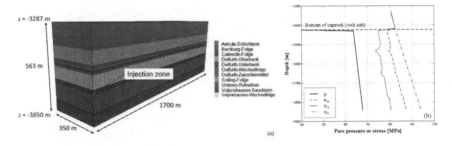

Figure 1. (a) 3D ¼ model geometry with stratigraphy; (b) initial conditions.

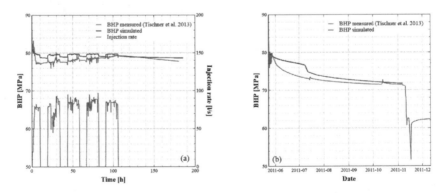

Figure 2. (a) Matching of the bottom hole pressure during the hydraulic fracturing operation; (b) Matching of the bottom hole pressure during the post-frac phase.

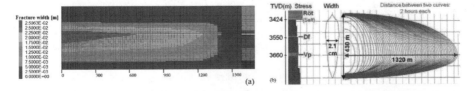

Figure 3. (a) The fracture geometry at shut-in (t = 106 h) from the full 3D simulation with TOUGH2MP-FLAC3D; (b) The simulation results from FIELDPRO (Tischner et al. 2013).

of the rock formations. For comparison, the simulation results with FIELDPRO based on semi-analytical solution (Tischner et al. 2013) were shown in Fig. 3b. The fracture geometry was more ideal with lower length and larger height, in comparison with the simulation with TOUGH2MP-FLAC3D[plus], which considers the full 3D stress redistribution. But the fracture opening was comparable.

3.2 Numerical simulation of the geothermal utilization

Based on the simulated fracture geometry, a 3D ½ model was generated for the simulation of the geothermal utilization. The model has a dimension of 3,500 m × 7,000 m × 563 m and lies at the depth between −3000 m and −4200 m (Fig. 4a). It contains all the relevant rock formations and considers the simulated hydraulic fracture in the previous section. TOUGH2MP is used for the coupled hydro-thermal simulation. Because of the low permeability of the formation, the hydro-mechanical effects should be taken into account, especially the fracture opening, the induced storage effect and permeability enhancement. In this study, a simplified approach was adopted, on one side to consider these effects and on the other side to avoid the time-consuming hydro-mechanical coupling. In this approach, the minimum horizontal stress was taken as input parameters and the fracture width was correlated to the effective stress.

Two different annual cyclic schemes were considered. For both cases, the annual injection/production volume is 50,000 m³ and the temperature of the injected water is 50°C. In case 1 the water injection took place from April to June (3 months), after which there is a 3-month pause. The water is then produced from October to March (6 months). In case 2 the cold water was injected from April to September (6 months) and hot water produced from October to March (6 months). The results are shown as follows.

Fig. 4b shows the pressure distribution after 5 years' operation. It can be seen that the pore pressure in the fractured zone reduced from 65 MPa to 50 MPa, although the annual injection volume was equal to the production volume. This is because some of the injected

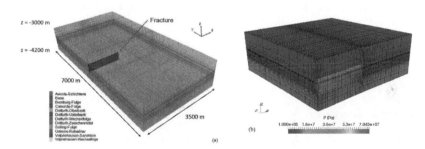

Figure 4. (a) 3D ½ model geometry for the simulation of geothermal utilization; (b) Simulated pore pressure distribution after 5 years' operation (case 1).

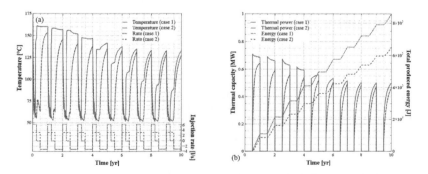

Figure 5. (a) Simulated temperature change in the injection/production zone during the operation; (b) Simulated geothermal capacity and total extracted geothermal energy in 10 years.

water leaked off into the tight formation and cannot be produced easily. Fig. 5a shows the temperature evolution of the injection/production zone. The simulation results indicate that the produced water has higher temperature in the cyclic scheme 1 due to the 3 months' pause period. This effect is obvious in the first 5–6 years. After that, the difference between the two cyclic schemes decreased. The geothermal capacity and total produced energy are shown in Fig. 5b. With the cyclic scheme, 32% more energy can be produced in 10 years' operation.

4 CONCLUSION AND OUTLOOK

In this paper, a hydraulic fracturing model was developed and implemented in the coupled numerical simulator TOUGH2MP-FLAC3D[plus]. The cubic law has been implemented, so that the fluid flow in both porous media and fractures can be simulated at the same time, with consideration of full 3D stress redistribution due to fracture opening and hydromechanical effect. The migration and distribution of injected fracturing fluid in the reservoir formation can be tracked.

As the case study, the hydraulic fracturing and geothermal production of the well Gross Buchholz Gt1 in Hanover in Northern Germany was investigated. The measured well pressure during stimulation as well as post-frac phase was matched. The fracture geometry was comparable with those simulated from commercial simulator. The simulated fracture geometry was then used in the subsequent simulation of geothermal utilization. Two different annual cyclic schemes were studied with coupled hydro-thermal simulation.

Based on the model, innovative cyclic schemes will be proposed and studied in the future. In addition, since the geothermal capacity of single well and single fracture was limited, the feasibility of multiple fractures with horizontal well will be studied.

REFERENCES

Gou Y, Zhou L, Zhao X, Hou ZM, Were P (2015) Numerical study on hydraulic fracturing in different types of georeservoirs with consideration of H^{2}M coupled leak-off effects. Environ Earth Sci. 73(10): 6019–6034. doi: 10.1007/s12665-015-4112-5.

Hou MZ, Kracke T, Zhou L, Wang X (2012) Rock Mechanical Influences of Hydraulic Fracturing Deep Underground the North German Basin: Geological Integrity of the Cap Rock Salt and Maximum Magnitude of Induced Microseismicity Based on the GeneSys Stimulation in May 2011. Erdöl Erdga Kohle, 128(11): 454–460.

Kolditz O, Xie H, Hou Z, Were P, Zhou H (Eds.) (2015) The Thematic Issue: Subsurface Energy Systems in China: production, storage and conversion. June 2015, Springer Publisher, Berlin Heidelberg, Germany. ISSN: 1866–6299.

Pechan E, Tischner T, Renner J (2014) Fracture properties after hydraulic stimulation in low-permeability sediments (GeneSys-project). ISRM Regional Symposium—EUROCK 2014, Vigo, Spain, 27–29 May.

Rioseco EM, Löhken J, Schellschmidt R, Tischner T (2013) 3-D Geomechanical modeling of the stress field in the North German Basin: case study GeneSys-borehole GT1 in Hannover Groß-Buchholz. Proceedings of the 38th Workshop on Geothermal Reservoir Engineering, Stanford University, Stanford, California, February 11–13.

Schäfer F, Hesshaus A, Hunze S, Jatho R, Luppold FW, Orilski J, et al. (2012) Kurzprofil der Geothermiebohrung Groß Buchholz Gt1. Erdöl Erdgas Kohle, 128(1): 20–26.

Tischner T, Evers H, Hauswirth H, Jatho R, Kosinowski M, Sulzbacher H (2010) New Concepts for Extracting Geothermal Energy from One Well: The GeneSys-Project. Proceedings World Geothermal Congress 2010, Bali, Indonesia, 25–29 April.

Tischner T, Krug S, Pechan E, Hesshaus A, Jatho R, Bischoff M, Wonik T (2013) Massive hydraulic fracturing in low permeable sedimentary rock in the GeneSys project. Proceedings of the 38th Workshop on Geothermal Reservoir Engineering, Stanford University, Stanford, California, February 11–13.

Zhang K, Wu YS, Pruess K (2008) User's guide for TOUGH2-MP—amassively parallel version of the TOUGH2 code. Earth Sciences Division, Lawrence Berkeley National Laboratory, LBNL-315E.

Zhou L (2014) New numerical approaches to model hydraulic fracturing in tight reservoirs with consideration of hydro-mechanical coupling effects. Cuvillier Verlag Göttingen. ISBN: 9783954046560.

Geomechanics and Geodynamics of Rock Masses – Litvinenko (Ed.)
© *2018 Taylor & Francis Group, London, ISBN 978-1-138-61645-5*

A mathematical approach for prediction of inclinometer measurements in open-pit coal mine slopes

Mehmet Mesutoglu & Ihsan Ozkan
Mining Engineering Department, Selçuk University, Konya, Turkey

ABSTRACT: The inclinometer measurement method is used widely to monitor unstable slopes encountered in open-pit mines. Although an indispensable method for determining the horizontal displacement, implementation of the method is quite expensive and difficult. In order to determine the deformation behavior, in situ measurements should be followed with frequent intervals. However, because of adverse weather conditions, inclinometer measurements sometimes cannot be performed. To predict the inclinometer measurement values that were not performed, a mathematical model was developed based on the time and rainfall. To test performance of the mathematical model, inclinometer measurement results obtained from 9 boreholes in Orhaneli-Turkey coal mine region were used. According to the data obtained from the in situ measurements monitored for approximately one year, in the mouth of the boreholes and also in the depth of the shear plane, horizontal displacement values in the north-west direction were determined as 19 mm and 20 mm, respectively. It was determined that the mathematical model results predicting the horizontal displacement values were strongly related to the in situ inclinometer measurement results. Thus, when in situ measurements cannot be performed, the engineers can interpret the graphical outputs that are prepared using the predicted displacement values.

1 INTRODUCTION

The slope design of an open pit is extremely important in terms of the safety of the enterprise. However, tension cracks in mine sites are an indicator of unstable slopes and must be monitored (Ozgenoglu, 1986). The stability of the slopes in open pit mines depend on geotechnical properties including rainfall, groundwater conditions, drilling-blasting, excavation, mine geometry and seismic activity. Depending on the open pit mine operations and effect of these parameters, slope movements increase with time and eventually reach a dangerous level (Ulusay et al., 2014). This unfavorable situation can endanger life and property.

There are two different methods to track time-dependent displacement behaviors of unstable slope. One of them uses the topographic surface, and the latter use a borehole. In the measurements taken from the topographic surface, the slope movements can be determined but the depth of the shear zone formed in the rock mass cannot be determined. The inclinometer measurements carried out in the borehole determines the depth of the shear zone in important projects (Dunnicliff, 1993). The horizontal displacement at every 0.5 m depth in the borehole is measured by an inclinometer probe that has 0.01 mm sensitivity (ISRM, 2007). The inclinometer measurement is expensive and laborious because it requires: perforation of boreholes; establishment of measurement systems; protection of boreholes; and time-consuming in situ measurements. Therefore, the in situ measurements must be carried out with small time intervals. However, this situation is not always possible in the winter climatic conditions (Ozkan et al., 2010).

In this study, a series of statistical analyses were performed to predict displacements when in situ measurements were not possible. As a result, a mathematical model was prepared based on time and rainfall. Thus the displacements not measurable by an inclinometer system can predict based on the developed equation. The equation was tested on the Orhaneli project results obtained from 9 boreholes (Mesutoglu, 2013 and Gokay et al., 2013). The graphical outputs based on performance results predicted displacement results for randomly selected time values and was compatible with real in situ measurements.

2   IN-SITU MEASUREMENTS

2.1   *Inclinometer measurement method*

The inclinometer system included an inclinometer casing, an inclinometer probe and cable, and an inclinometer readout unit. Inclinometer casing is typically installed in a vertical borehole drilled up to stable rock unit. The bottom of the casing is anchored on the stable ground. The initial borehole profile was recorded as a reference by the inclinometer probe. Slope movements cause movement away from the initial position of the casing. The rate, depth, direction, and magnitude of the slope movements are calculated by comparing the initial reference and the subsequent measurements.

2.2   *Orhaneli-Gümüşpınar open pit coal mine and measurement studies*

Orhaneli–Gümüşpınar coal basin is 55 km south of Bursa. It has been operated by Turkish Coal Enterprises (TKI) since 1979. From 1998 and 2012, there were many unstable slope problems. Serious slope failure problems occurred due to mining operations in 2012. In situ measurements for the unstable slope formed in the mine region were begun.

To measure the horizontal displacements, 9 inclinometer measurement boreholes were drilled at locations determined by preliminary studies like site investigation, geophysical, and geology. Boreholes were drilled between 17.5 m and 63 m to reach stable ground in region. Four boreholes located in bottom of open pit were cut coal seam which has approximate thickness of 7 m. The in situ measurement studies were performed one by one in the 9 boreholes. The measurements were carried out in sub-stations which were installed with 0.5 m intervals in each borehole. There were 733 sub-stations established in all boreholes. The inclinometer measurements were performed for 105 days and averaged one every 8–10 days in the boreholes. Measurements were performed 101 times for all boreholes.

The data received from 9 inclinometer measurement boreholes were transferred from the data logger to the computer via Smart software. The data were then processed by a second software program named INCLI-2 to prepare graphical outputs. A typical horizontal displacement and incremental displacements were given in Fig. 1 for OINK-1.

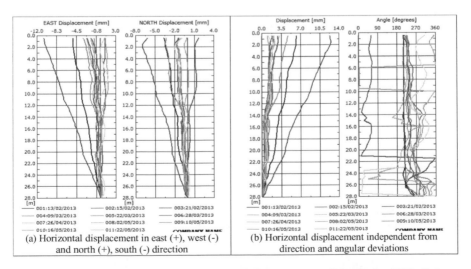

Figure 1.   Horizontal displacement values from in situ measurement (Mesutoglu, 2013, Gokay et al., 2013).

According to in situ measurements, in the mouth of the boreholes, the maximum horizontal displacement value was 19 mm in the north-west direction. In addition, the depth of the shear zone in the nine boreholes was determined as horizontal displacement graphs. At the critical depths, the maximum horizontal displacement value was approximately 20 mm in the north-west direction.

The width of the slope excavation in the original mine project is 800 meters. However, according to in situ measurement results, mine management changed the excavation method. In revised excavation, unstable slopes were separated from the narrow corridors with 60–80 meter intervals. After the coal production in the corridor, the next corridor excavation was carried out. Excavated mining pit was stopped by stripping material of next corridor. Slope movements were under control highly by new excavation method. However, it was observed that the displacements based on the new mine geometry still continued with low displacement rates.

3 MATHEMATICAL MODELLING

Inclinometer measurement systems were one of the in situ measurement methods. It is time consuming and expensive. However, shear zone depth and its thickness can only be used with borehole extensometers. Therefore, inclinometer measurement systems are used widely in important projects (Dunnicliff, 1993). Slope movements are based on rock and rock mass properties, mining activities, rainfalls, mine geometry, etc. (Ulusay et al., 2014). Thus, time intervals in inclinometer measurements are very important. Site engineers in project studies generally take measurements once or twice a month. Sometimes, measurement time intervals can be expanded due to the severe weather conditions. To predict measurement values for unmeasurable days, a series statistical analysis were carried out. In analysis, Orhaneli project measurement results were used. The statistical analysis and its interpretations can be considered in four stages. They are:

3.1 Stage-1 (calculation of ΣU from U)

The direct modelling of the behavior seen between the borehole depth and inclinometer measurements (Fig.1) are very difficult. That's why, an indirect way has been considered in this study. In analysis, first of all, time-dependent displacement (U) behavior for each sub-station in the boreholes were determined by inclinometer measurements (Fig. 2a). However, the displacement graph does not contain a systematic structure. Because deformations presented here are vector magnitudes. In addition, in each measurement time, the determined measuring value is equal to distance from reference point to the last position. In next measuring time, the last determined measuring value is this time equal to distance from reference point to the latest position. Therefore, to overcome this problem, the total displacement values (ΣU) were calculated (Fig. 2b). That is, the vectorial magnitudes were considered as the scalar values and then the vectorial magnitudes were collected one after the other. Finally, graphs as Fig. 3 for all inclinometer boreholes were prepared.

3.2 Stage-2 (determination of ΣU_p by statistical analysis)

The statistical analysis were carried out on data of these type of graphs (Fig. 3). In addition, in statistical analysis, total of the seasonal rainfall for a year in mine region was considered. The average rainfall values observed for the last two decade are recorded. The average rainfall per months is approximately 56.1 mm in the region. The following mathematical equation was developed by the statistical analyses conducted by SPSS V16.0.

$$\sum Up = C_1[1 - e^{(-t/C_2)}][1 + C_3(RF/(673 - RF))^{C_4}] \qquad (1)$$

Here, C_1, C_2, C_3 and C_4 are the statistical constants depending on sliding rock material, ΣU_p is the total predicted horizontal displacement (mm), t is time (day) and RF is total rain

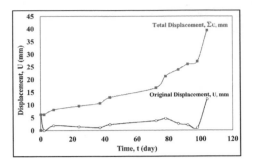

Figure 2.   The original and total displacement behavior in OINK-1/0.5 (Mesutoglu, 2013).

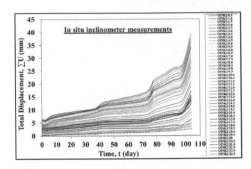

Figure 3.   The original total displacement behaviors for all of sub-stations in OINK-1 (Mesutoğlu, 2013).

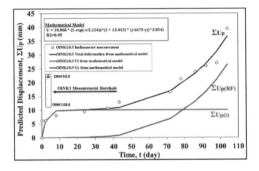

Figure 4.   A typical time-dependent displacement behavior predicted and in situ measurements for ONK-1/0.5 sub-station.

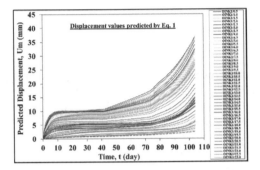

Figure 5.   The predicted displacement behavior in OINK-1 (Mesutoğlu, 2013).

amount (mm) which is at the moment of measurement. Also, 673 is the total rainfall in per year for Orhaneli region. It can change for different regions. The mathematical approach recommended in this study consists of basically two parts (Eq. 1). They are:

$$\sum Up_{(t)} = C_1[1 - e^{(-t/C_2)}] \qquad (2)$$

$$\sum Up_{(RF)} = C_1[1 - e^{(-t/C_2)}][C_3(RF/(673 - RF))^{C_4}] \qquad (3)$$

Eq. 2 depends on the time and Eq. 3 depends on the total amount of rainfall in mine region. As a result, the final equation can be rewritten as follows. The sum of the values found from the two equations gives the value of the total displacement.

$$\sum Up = \sum Up_{(t)} + \sum Up_{(RF)} \qquad (4)$$

A typical example related to the statistical results is presented in Fig. 4. The time-dependent (Eq. 2), rainfall-dependent (Eq. 3) and sum of the two data (Eq. 4 that is Eq. 1) were plotted on this graph. The graph shows that time-dependent behavior is low level and is an asymptote to the X-axis. But it is seen that the rain dependent curve is nonlinear. The sum of the time and rain dependent curves gives the total predicted displacement (ΣU_p, mm). Fig. 5 shows results predicted by the model. To determine the model performance, the mathematical model results (Fig. 5) can compare with the original measurement results (Fig. 3). This success of the model is valid for the other station points (from OINK-1 to OINK-9). The statistical regression coefficients (R^2) are generally 0.95 (Table 1).

Table 1. The statistical constants and R^2 values determined for OINK-1 (reduced data).

Depth (m)	Regression Coefficient				R^2
	C_1	C_2	C_3	C_4	
0.5	10.068	2.224	13.013	3.814	0.95
1.0	9.900	2.201	14.600	4.037	0.95
1.5	9.770	2.211	16.568	4.265	0.95
2.0	10.175	2.617	18.001	4.638	0.95
2.5	9.923	2.647	18.371	4.720	0.95
3.0	9.866	2.812	18.352	4.824	0.95
3.5	9.798	3.023	18.017	4.900	0.94
4.0	9.675	3.173	18.347	4.955	0.94
4.5	9.612	3.395	18.749	5.066	0.94
5.0	9.517	3.677	18.960	5.149	0.94

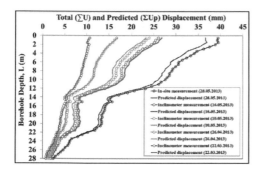

Figure 6. Success of Eq. 1 on the OINK-1 borehole with reduced data (Mesutoğlu, 2013).

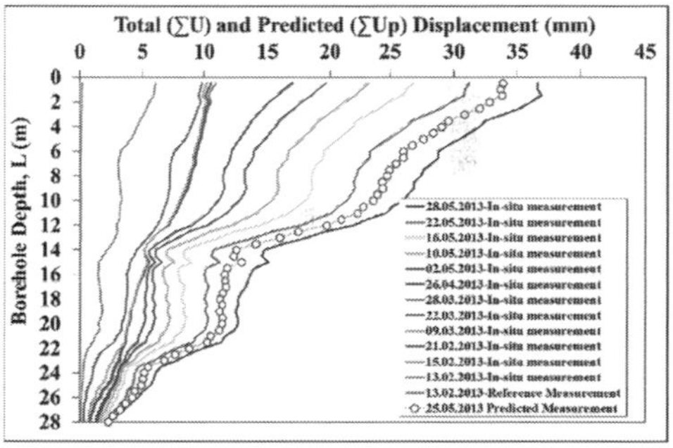

Figure 7. Performance of Eq. 1 on the OINK-1 borehole data (Mesutoğlu, 2013).

3.3 Stage-3 (a new graphical presentation for evaluations about inclinometer results)

We used Eq. 1 to predicted total displacement (ΣU_p) values of boreholes for all of measurement date. The database used in statistical analysis consist of time and total displacement (ΣU) values recorded for each sub-station. In other words, the time vs displacement (ΣU_p) behavior for each sub-station of borehole were determined by Eq. 1. As a typical graph, Fig. 5 was given. The total displacement (ΣU, ΣU_p) values defined in Fig. 3 and Fig. 5 were graphed for each sub-station which are located in each 0.5 m along borehole axis. The displacements (ΣU, $\Sigma U p$) measured and predicted for each sub-station in each measurement time can be prepared as a separate database. The new database would be based on borehole depth and displacement. In conclusion, borehole depth vs displacement behaviors can be graphed using this database. For example, in 28 sub-stations of OINK-1, the original total values (ΣU) and the predicted values (ΣU_p) by Eq. 1 were presented in Fig. 6. Of note, in Eq. 1, the total displacement values were considered (Fig. 2, 3 and 5).

In this stage, it would be appropriate that Fig. 1 and Fig. 6 are compared. In evaluations of inclinometer out graphs, the deviations formed in borehole axis and those depths are very important. Because shear plane will be formed at this depth. In Fig 1, it is seen that the critical depths were occurred in 21th, 14th and 9th meters. When the new graphical output (Fig. 6) declared in this paper are evaluated, it is seen that the same the critical depths (21th, 14th and 9th m) will be determined even easier.

3.4 Stage-4 (evaluations by the suggested approach for date that not carried out measurement)

The final result will actually bring the main a benefits. For dates that inclinometer measurements cannot be performed, the predicted displacement values can be determined. A typical example for this condition is prepared. It can be seen easily from Fig. 1 that in situ measurements for 25.05.2013 in timetable valid for OINK-1 could not be carried out. Note that this date is between two dates in which in situ measurement are carried out (22.05.2013 and 28.05.2013, in Fig. 1). By using the above mathematical approach and Table 1, the predicted displacements (ΣU_p) for the randomly selected date were calculated by Eq. 1. Then, Fig. 7 was prepared. Fig. 7 consists of inclinometer results for in situ measurements (ΣU) carried out in different dates and the displacement values (ΣUp) predicted for 25.05.2013. As a result, it was seen that the predicted values were located between the original measurement values.

4 CONCLUSIONS

The approach developed here can estimate the horizontal displacements. The constants of the mathematical models can be determined for different fields. Thus, the borehole profiles along vertical axis of borehole will be determined. If used by engineers, then the unmeasurable dates in the works conducted on unstable slopes will no longer be a problem. As a result, the interpretations and evaluations about unstable slopes will be carried out in a database.

ACKNOWLEDGMENTS

This study was supported by Selçuk University and TKI. The author is also indebted to the reviewers for their valuable comments.

REFERENCES

Dunnicliff, J. 1993. Geotechnical instrumentation for monitoring field performance. *New York NY: John Wiley & Sons*, 608p.
Gokay, M.K., Ozkan, I., Ozsen, H., Dogan, K., Mesutoglu, M. 2013. In-situ measurements and evaluations intended for slope stability analysis in TKI-GLI-BLI Gumuspinar open pit mining, *TKI Final Report,* Selcuk University, Department of Mining Engineering, Konya, 326p. (in Turkish).
ISRM (International Society for Rock Mechanics), 2007. The complete ISRM suggested methods for rock characterization, testing and monitoring: 1974–2006. In: Ulusay, R., Hudson, J.A. (eds). *Suggested Methods Prepared by the Commission on Testing Methods. International Society for Rock Mechanics, Compilation Arranged by the ISRM Turkish National Group*. Ankara, Turkey.
Mesutoglu, M. 2103. Mathematical analysis for in-situ deformations obtained from inclinometer measurement boreholes established inside rock and soil structures, *MSc Thesis, Selçuk University, Department of Mining Engineering*, Konya, 213p. (in Turkish).
Ozgenoglu, A. 1986. Slope stability analysis approaches in mining, Scientific Mining Journal, 25, 1.17–27.
Ozkan, I., Ozsen, H., Oltulu, F., Boztas, S. 2010. Mathematical model based on long term inclinometer measurements at an open pit limestone mine. *2nd Conference on Slope Tectonics*, Vienna, Austria, 1–5.
Ulusay, R., Ekmekci, M., Tuncay, E., Hasancebi, N. 2014. Improvement of slope stability based on integrated geotechnical evaluations and hydrogeological conceptualization at a lignite open pit. *Eng. Geol.* 181: 261–280.

Geomechanics and Geodynamics of Rock Masses – Litvinenko (Ed.)
© 2018 Taylor & Francis Group, London, ISBN 978-1-138-61645-5

Comparison of limit equilibrium and finite element methods to slope stability estimation

A.B. Makarov, I.S. Livinsky & V.I. Spirin
SRK Consulting (Russia), ISRM, Moscow, Russia

A.A. Pavlovich
*Center of Geomechanics and Issues of Mining Industry, Saint-Petersburg Mining University,
St. Petersburg, Russia*

ABSTRACT: The key aim of this study was to compare the slope stability estimation results by the limit equilibrium and numerical modelling methods on the examples of large open pit mines. This study compares factors of safety resulting from standard estimation methods for dry slopes and with account of water table, including impacts of large scale blasting and earthquakes, based on Mohr-Coulomb and Hoek-Brown criteria.

Keywords: slope stability, limit equilibrium method, finite element method, factor of safety, strength reduction factor

1 INTRODUCTION

With the increase of the open pit mining depths, the slope stability issues in complex mining and geological conditions are becoming more important.

In the clear majority of cases, open pit slope stability is estimated by limit equilibrium methods and, mainly, in two dimensions. These estimation methods have proven to be sufficiently reliable, however they have certain assumptions that can become very significant in deep pits, where even small changes in the slope angles can have a significant economic impact.

Numerical modelling, and primarily the finite element modelling method, is an increasingly common approach to slope stability estimation. Its main advantage is not only in estimating the slope stability, but also in estimating stress and strain distributions at each point of the rock mass.

The methods of limit equilibrium have been widely tested throughout the world, and specialists have learned to use them for various geomechanical conditions. The finite element method, despite the active development and implementation of newer software, requires sufficiently high qualification to obtain reliable results. However, as our experience shows, the methods of limit equilibrium and numerical modelling can complement each other.

To enable practical application of the finite elements method, it is necessary to test it and its algorithm of safety factor search, named SRF (Strength Reduction Factor), in comparison with the traditional Factor of Safety (FoS). For this purpose, the comparison analysis has been carried out. The results of two-dimensional analysis of different open pits by both methods have been considered, which allowed to estimate the differences and similarities of the finite element method compared to the classical limit equilibrium method (Figure 1). The analysis was carried out in Rocscience software: the method of limit equilibrium was tested in Slide v 6.0, and the finite element method was tested in Phase v 8.0.

A comparative analysis of the slope stability estimation by limit equilibrium and finite element methods in two dimensions will subsequently allow to summarize the experience and develop a methodology for slope stability estimation in three dimensions.

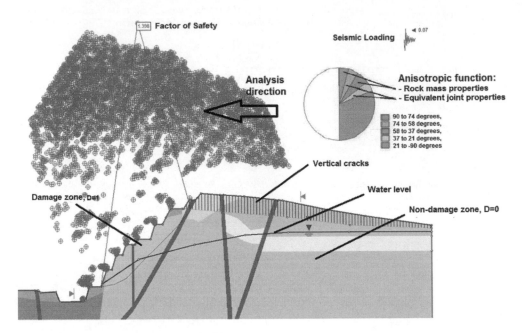

Figure 1.   View of Slide estimation model.

2   THE KEY PRINCIPLES OF GEOMECHANICAL SLOPE STABILITY MODELLING

For reliable slope stability estimation, a geomechanical model is required that will adequately reflect the actual conditions of slope failure and allow for possible unfavourable external impacts (John Read, 2009).

Depending on the complexity of the geomechanical conditions, a geomechanical model can include not only pit slope boundary, geological composition and geomechanical properties of the rock mass, but also, if necessary, groundwater level (or pore pressures in the joints) and the possible earthquake impact. However, in a number of cases, it is necessary to additionally simulate the zone disturbed by blasting operations, rupture cracks, structural features (jointing, cleavage, stratification, etc.) and other factors. Figure 1 presents a geomechanical model developed in Slide for limit equilibrium slope stability estimation.

Rock mass structure (jointing) must be taken into account in estimating the slope stability in hard rock material, as it predefines the failure mechanism.

To account for the jointing, Slide software has the Anisotropic Function, which reduces strength properties of the rock mass in the dip directions of the joint sets to reach the adhesion and friction angle along the joints (Figure 1). The anisotropic function is used to take into account the weakening impact of the joint sets with the same strike direction as the pit slopes (± 20°), including the joint sets dipping both into the pit and into the slope. Between the joint set directions, the estimations use the jointed rock mass properties. If data on the lengths of weakening surfaces is available, equivalent properties should be defined to consider the persistence of joints and the rock mass properties between the joints.

The finite element method (Phase2) allows to define the location of faults and discontinuities in the model itself. For a long time, modelling the rock mass structure was quite a difficult task using the finite element method, until the Joint Network tool for defining the rock mass composition was implemented in Phase2. This allows modelling of both continuous and intermittent joints by defining the persistence. Another jointing parameter to be defined in the software is joint spacing. Figure 2 shows an example of joint set modelling using the Joint Network module, which takes into account their statistical variability.

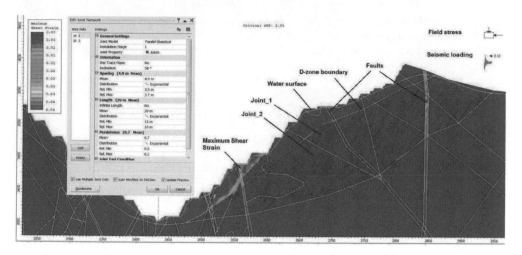

Figure 2.   Definition jointing network in Phase2.

Figure 3.   Comparing the safety factor by limit equilibrium method (LEM) and finite element method (FEM).

3   RESULTS OF SLOPE STABILITY ESTIMATION FOR ACTIVE OPEN PITS

Figure 3 is a comparative diagram of Factor of Safety (FoS) estimation by the limit equilibrium method and the finite elements method using the SRF procedure. The estimations were conducted for various geomechanical conditions and for a wide range of slope parameters (slope height was from few hundreds of meters to more than 500 m). Anisotropy of the jointed rock mass properties was defined. Some models also included the groundwater level, pore pressure and the seismic impact from earthquakes.

The comparison of the two fundamentally different slope stability estimation methods for 5 actual open pits showed their fairly good linear correlation with correlation factor of 0.97. The deviation of the trend line from the 45° line is 3° towards higher FoS estimated by the finite element method in comparison with the limit equilibrium method. In the range of FoS values between 1.0 and 2.0, which are important from practical standpoint, the estimated SRF values in most cases differ from the FoS by ±0.1–0.2.

4 COMPARISON OF ESTIMATION RESULTS FOR STANDARD ESTIMATION SCENARIOS

A comparative analysis was conducted to assess the similarity and difference of the limit equilibrium and finite element methods applied to various estimation scenarios (Table 1, Figure 4). The following main scenarios were assessed:

- homogeneous slope using the Mohr-Coulomb (MC) and Hoek-Brown (HB) strength criteria, considering water table (w) and seismic impacts from earthquakes (s);
- inhomogeneous slope of three lithological rock types (L), using the Mohr-Coulomb and Hoek-Brown strength criteria, and with zone disturbed by blasting operations;
- fault/weak contact (F) – a steeply dipping and a gently dipping contact / layer were considered, with broken sliding surface (the sliding surface partially or completely coincided with the weakening surface);
- anisotropic rock mass (A) – with different dip angles of joints, also several joint sets were defined;
- toppling (T) of a set of blocks formed by the planes of layers, cleavage or fracturing, with steep dipping of layers different from the slope angle, and with an underlying contact;
- the diagram in Figure 4 also includes data from the "Verification Manual..." (Rocscience, 2011), marked with an asterisk, to add to the statistics.

The comparative analysis allowed to estimate the variance between the two methods, which was determined as the percent difference in the factors of safety.

The following conclusions can be made following the comparison of the two methods for different estimation scenarios:

- In most cases, there is a slight difference, which confirms the applicability of both methods for assessing the slope stability.
- The differences in the two methods are observed primarily in inhomogeneous and anisotropic slopes.
- The greatest differences are observed with a gently dipping contact / layer or a flat joint set.
- The similarity of the results depends on the correct selection of the structural disturbance rank in the finite element method. Setting too small joint spacing can lead to local failures, and great spacing can lead to an overestimated Factor of Safety. In our opinion, setting the joint spacing should correspond to the hierarchical level of the scale of the estimated area.
- Traditional methods cannot estimate the slope stability with steeply dipping joints with bedding in the inverse direction to the slope. For such scenario, the Goodman-Bray method (Goodman, RE, 1976) produced a good correlation with numerical estimation methods.
- Numerical analysis should also be carried out to assess the stress-strain conditions of the rock mass, because these aspects are not considered in the limit equilibrium method.
- Slope stability analysis by limit equilibrium method should be carried out at all stages of pit design and mining operations. It is recommended to use numerical modelling to verify the limit equilibrium estimation results.

Table 1. The differences between the limit equilibrium and finite elements methods under different estimation scenarios.

Name of analysis	Description	Slope parameters		Factor of Safety (FOS)		Variance	Note
		Height, m	Angle, °	Slide (FS)	Phase 2 (SRF)		
CM	Simple slope, Mohr-Coulomb criterion	200	48	1.45	1.44	0.7%	
HB	Simple slope, Hoek-Brown criterion	200	48	1.345	1.36	1.1%	
CM_v	Simple slope, Mohr-Coulomb criterion, watered slope	200	48	1.23	1.20	2.5%	
CM_s	Simple slope, Mohr-Coulomb criterion, earthquake loading	200	48	1.319	1.29	2.2%	
L_CM	Lithologies slope (three lithologies), Mohr-Coulomb criterion	140	48	1.18	1.20	1.7%	
L_HB	Lithologies slope (three lithologies), Hoek-Brown criterion	140	48	2.714	2.80	3.1%	
L_HB-D	Lithologies slope (three lithologies), Hoek-Brown criterion, damage zone (D)	140	48	2.206	2.31	4.5%	
F_(70°)	Steep_Fault (dip 70°)	200	48	1.23	1.27	3.1%	
F_(14°)	Sloping_Fault (dip 14°)	200	48	1.162	1.03	12.8%	LEM (Slide) exaggerates FOS
F_(14°)_NC	Sloping_Fault (dip 14°). non-circular failure surface	200	48	1.117	1.03	8.4%	LEM (Slide) exaggerates FOS
A_(14°)	Steep_Anizotropic (dip 14°)	200	48	1.32	1.24	6.5%	LEM (Slide) exaggerates FOS
A_(14°)_NC	Steep_Anizotropic (dip 14°), non-circular failure surface	200	48	1.153	1.24	7.0%	
A_(30°)	Inclined_Anizotropic (dip 30°)	200	48	0.98	1.01	3.0%	
A_(70°)	Sloping_Anizotropic (dip 70°)	200	48	1.384	1.33	4.1%	
A_(70°)_NC	Sloping_Anizotropic (dip 70°), non-circular failure surface	200	48	1.296	1.33	2.6%	
A_(-40°)_NC	Steep back_Anizotropic (dip -40°), non-circular failure surface	200	48	1.405	1.39	1.1%	
A_2 se.s	Multijoint_Anizotropic with two sets (dips 20° & 70°)	200	48	1.236	1.08	14.4%	LEM (Slide) exaggerates FOS
A_2 se.s_NC	Multijoint_Anizotropic with two sets (dips 20° & 70°), non-circular failure surface	200	48	1.013	1.08	6.2%	
T_(-70°)	Toppling (dip -70°)	200	48	1.135	1.15	1.3%	Calculated by Goodman & Bray (RocTopple) due to LEM (Slide) does not suit for Back Anisotropy
T_2 ses	Toppling with basic joint (dips 20° & -70°)	200	48	0.983	0.98	0.3%	

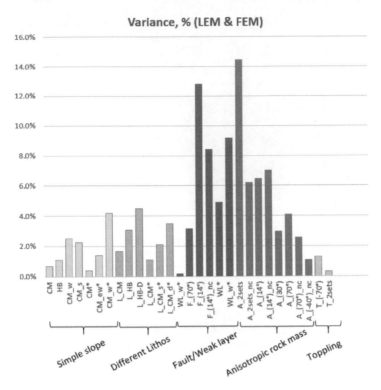

Figure 4. Variance of estimation results by the limit equilibrium and finite elements methods under various estimation scenarios.
* – according to Rocscience, 2011.

- A comparative analysis of the slope stability estimation by limit equilibrium and finite element methods in two dimensions will subsequently allow to summarize the experience and develop a methodology for slope stability estimation in three dimensions.

5 CONCLUSIONS

The comparison shows that the finite element method is acceptable for estimating the pit slope stability. The finite element method covers a wider range of slope stability estimation problems, and also provides the solution to the stress-strain modelling.

Despite the universality of numerical methods in assessing the pit slope stability, these methods are very time-consuming and require high qualification of specialists. Therefore, numerical methods are recommended for verification of the classical estimation methods, as well as for specific tasks associated with analysing the stress-strain state generated by open pit mining.

REFERENCES

[1] Goodman R.E. Toppling of rock slopes./R.E. Goodman, J.W. Bray // ASCE Specialty conference on rock engineering for foundations and slopes. Vol. 2. – 1976. pp. 201–234.
[2] Guidelines for open pit slope design/Editors John Read, Peter Stacey. – CRC Press/Balkema, 2009. – 509 p.
[3] Phase 2. Slope Stability Verification Manual. Part I. 1989–2011, Rocscience Inc.

Analyzing slope stability in bimrocks by means of a stochastic approach

Maria Lia Napoli, Monica Barbero & Claudio Scavia
Department of Structural, Geotechnical and Building Engineering, Politecnico di Torino, Torino, Italy

ABSTRACT: Bimrocks (block-in-matrix rocks) are chaotic geological formations defined as heterogeneous mixtures of hard rock blocks encased in a fine-graded matrix. The inherent geometric, lithological and mechanical variability of bimrocks imply a great challenge in their characterization and modeling.

A common practice when planning engineering works in these complex formations is to neglect the contribution of blocks and assign the strength and deformation properties of the weaker matrix to the whole rock mass. However, this assumption can lead to erroneous results and, consequently, to technical problems during construction works.

The aim of this study was to investigate stability of theoretical slopes in bimrocks using a stochastic approach, in order to consider the spatial and dimensional variability of rock inclusions. Many 2D stability analyses were performed on slope models with simple geometries, elliptical block shapes and variable block contents. The results were compared to those obtained in a previous work, where slope stability analyses were carried out on bimrocks with blocks of circular shape.

The findings of this research confirm that rock inclusions play an important role and strongly influence the slope stability of bimrocks. Furthermore, the advantages of using a stochastic approach when working with these heterogeneous materials are highlighted.

1 INTRODUCTION

The term *bimrock* (block-in-matrix rock) was defined by Medley (1997) to be a mixture of rocks "composed of geotechnically significant blocks within a bonded matrix of finer texture". Several geological formations, including melanges, breccias, weathered rocks, conglomerates and agglomerates, can be considered to be bimrocks (Medley, 1994; Haneberg, 2004; Wakabayashi and Medley, 2004). A reliable characterization and modeling of bimrocks is extremely complex due to the inherent spatial, lithologic and dimensional variability of rock inclusions. Hence, geotechnical engineers often plan engineering works in these challenging materials taking into account only the strength and deformation properties of the weaker matrix. However, based on many case histories reported in the literature, such a simplified approach can lead to wrong forecasts, instability problems, unexpected difficulties and delays during construction works (Medley and Zekkos, 2011; Afifipour and Moarefvand, 2014). Recently, a lot of research (laboratory tests, numerical analyses and in situ tests) has been conducted to investigate the mechanical properties of bimrocks, in order to correctly design civil engineering works in these complex formations. The main results of these studies are summarized below:

- bimrocks have scale independent (or fractal) block size distributions (Medley, 1994; Medley and Lindquist, 1995; Medley and Sanz, 2004; Sonmez et al., 2016);
- the block/matrix threshold, i.e. the smallest geotechnically significant block and the largest block size within a volume of bimrock, should be defined according to the scale of engineering interest, termed "characteristic engineering dimension", Lc. It could be the slope

height, the specimen diameter, the tunnel diameter, etc. (Medley, 1994; Wakabayashi and Medley, 2004);
- at the selected scale of interest, blocks can be considered all the inclusions with dimensions between 0,5·Lc (below which rock fragments are considered to belong to the matrix) and 0,75·Lc (Medley, 1994; Barbero et al., 2006; Medley and Zekkos, 2011);
- the overall strength of bimrocks is affected by many factors. The most important is the Volumetric Block Proportion (VBP), but orientation and spatial location of blocks, matrix strength, block size distributions, block count, block shapes, etc., play an important role, as well (Lindquist, 1994; Irfan and Tang, 1993);
- to be classified as bimrock, a sufficient mechanical contrast between blocks and surrounding matrix must be afforded by the material, so as to force failure surfaces to negotiate tortuously around the blocks. In particular, a minimum friction angle ratio (tanφblock/ tanφmatrix) of between 1.5 and 2 and a minimum stiffness contrast (Eblock/Ematrix) of about 2 have been suggested in the literature (Medley, 1994; Lindquist and Goodman, 1994; Barbero et al., 2007; Medley and Zekkos, 2011);
- an increase in the strength of bimrocks was registered for VBP between about 25% and 75% (Lindquist, 1994; Medley and Lindquist, 1995). In this range, researchers have observed an increase of both Young's modulus and friction angle, related to the increase in tortuosity of the failure surfaces, and a decrease in the cohesion, due to the poor mechanical properties of the matrix, where deformations develop;
- the presence of blocks (their position, shape and number) within slopes yields to irregular and tortuous sliding surfaces, far different from those obtained in homogeneous materials (Irfan and Tang, 1993; Medley and Sanz, 2004; Barbero et al., 2006). Greater safety factors have been found for higher VBP.

Some authors developed preliminary (simplified) strength criteria, which assume bimrocks to be homogeneous and isotropic masses (Lindquist, 1994; Sonmez et al., 2009; Kalender et al., 2014). Block proportions and matrix strength parameters are necessary in order to define the equivalent mechanical properties of the rock mass.

Lindquist (1994) proposed the empirical strength criterion reported in Eq. (1):

$$\tau_p = C_{matrix} \cdot (1 - VBP) + \sigma \cdot \tan\left(\varphi_{matrix} + \Delta_{matrix}(VBP)\right) \tag{1}$$

where t_p is the equivalent mass shear strength, c_{matrix} is the matrix cohesion (assumed to decrease with increasing VBP), φ_{matrix} is the internal friction angle of the matrix and $\Delta_{\varphi matrix}(VBP)$ is its increase, assumed by Lindquist to be, above 25% VBP, equal to 3° for every VBP increase of 10%.

The approach proposed by Kalender et al. (2014), which takes also into account contact strength between blocks and matrix, is reported in Eqs. (2)–(4).

$$\varphi_{bimrock} = \varphi_{matrix}\left[1 + \frac{1000\left[\dfrac{\tan(\alpha)}{\tan(\varphi_{matrix})} - 1\right]}{1000 + 5\left(\dfrac{100 - VBP}{15}\right)}\left(\dfrac{VBP}{VBP + 1}\right)\right] \tag{2}$$

$$UCS_{bimrock} = \left[\left(A - A^{\frac{VBP}{100}}\right) / (A - 1)\right] UCS_{matrix} \qquad 0,1 \le A \le 500 \tag{3}$$

$$c_{bimrock} = UCS_{bimrock}\left[1 - \sin(\varphi_{bimrock})\right] / \left[2\cos(\varphi_{bimrock})\right] \tag{4}$$

where α is the angle of repose of blocks, UCS is the uniaxial compressive strength and A is a parameter that can be defined according to both the compressive strength of the matrix and α.

Both the empirical approaches, as stated by the authors themselves, have some limitations and should be applied carefully and only in predesign stages of engineering applications.

2 2D STABILITY ANALYSIS OF SLOPES IN BIMROCKS

The aim of this study was to evaluate the effects of rock blocks on the stability of theoretical slopes in bimrocks, whose characteristic dimension, Lc, was their height. The slopes had an inclination of 30°, elliptical block shapes (with major axes inclined 90° to the vertical axis) and different block contents. In particular, 25%, 40%, 55% and 70% volumetric block proportions were examined.

To take the spatial and dimensional variability of the inclusions into account, the stochastic approach proposed by Napoli et al. (2017) was applied. In particular, a specific Matlab routine, performing numerical Monte Carlo simulations, was implemented to randomly generate elliptical blocks within the slope models according to specific statistical rules (Barbero et al., 2012). 15 extractions and, hence, 15 stability analyses were performed for each VBP considered, so as to achieve a statistical validity of the results. 0% VBP configurations (matrix only models) were also analyzed in order to evaluate potential inaccuracies that can be made designing without taking the presence of blocks into account.

Altogether, more than 120 slope stability analyses were carried out using both FEM and LEM methods, with Phase² and Slide computer codes (from Rocscience), respectively. Safety factors were evaluated and compared.

Table 1 shows the input parameters that were used in the stability analyses. Both matrix and blocks were assumed to have an elastic-perfectly plastic behavior and to follow the Mohr-Coulomb failure criterion. The empirical approach proposed by Lindquist was also applied, by way of comparison. Table 2 shows the input parameters that were used for analyzing these equivalent homogeneous slope models.

2.1 FEM analyses

Finite element (FE) slope stability analyses were conducted using the software Phase2 (vers. 8.0).

Six-node triangular elements were used and, in order to avoid stress modelling disturbance, an excavation process was simulated to reproduce the face geometry of the slopes.

Table 1. Input parameters for matrix and blocks of heterogeneous slope models.

	E [GPa]	ν [–]	γ [kN/m³]	c [kPa]	φ [°]
Matrix	0.04	0.25	22	30	24
Blocks	5.1	0.22	27	600	40

Table 2. Input parameters for equivalent homogeneous materials, according to the Lindquist criterion.

	LINDQUIST'S APPROACH			
VBP [%]	(1–VBH)	$c_{bimrock}$ [kPa]	$\Delta\varphi_{matrix}$ [°]	$\varphi_{bimrock}$ [°]
0	1	30	0	24
25	0.75	22.5	0	24
40	0.6	18	4.5	28.5
55	0.45	13.5	9	33
70	0.3	9	13.5	37.5

Table 3. Average safety factors and standard deviations obtained performing FEM analyses.

VBP [%]	Average SF	Standard deviation
0 (matrix-only)	0.80	–
25	0.79	0.036
40	0.91	0.094
55	1.10	0.144
70	1.41	0.189

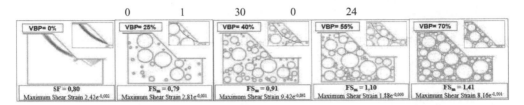

Figure 1. FEM analyses results for one of the bimrock configurations generated for each VBP considered: failure surfaces (in the magnified views), critical safety factors (SRF) and maximum shear strains.

The results, shown in Table 3, indicate that factors of safety increase significantly for higher VBP. This result can be ascribed to the increase of failure surface tortuosity with increasing VBP (Figure 1). Furthermore, the standard deviations reported in Table 3 indicate that a high variability in the results exists, and that it increases with increasing VBP. These results are in good agreement with previous findings reported in Medley and Sanz (2004), Irfan and Tang (1993), Barbero et al. (2006) and Napoli et al. (2017).

2.2 LEM analyses

Limit equilibrium analyses were carried out on the same extended slope models of the FEM analyses using the code Slide (vers. 5.0). The Simplified Bishop method was applied.

As shown in Table 4, a significant increase of the safety factors and standard deviations is achieved for higher VBP values, particularly for both 70% VBP configurations, according to FEM analyses results.

As shown in Figure 2, the tortuosity of failure surfaces is not taken into account, since they have circular shapes. The positions of critical surfaces are affected by the presence of blocks, having greater strength than the matrix. They tend to be, among those analyzed, the ones that encounter the lowest number of inclusions and are all quite superficial. Furthermore, the results are not representative of the real problem and significantly overestimate safety factors (SFs), with respect to FEM results.

2.3 Application of the Lindquist empirical strength criterion

The empirical strength criterion proposed by Lindquist was applied, by way of comparison, on the same slope models previously analyzed. The one proposed by Kalender et al. (2014) was not applicable, since the UCSmatrix was less than 0,1 MPa. For 25%VBP, Eq. (1) provided equivalent bimrock cohesion and internal friction angle basically coincident with those of the matrix (as reported in Table 2). Hence, only 40%, 55% and 70% VBP configurations were analyzed.

Table 5 compares the SFs obtained performing LEM and FEM analyses. It shows that SFs grow as the VBP increases. This trend is consistent with the one obtained assuming bimrocks

Table 4. Average safety factors and standard deviations obtained performing LEM analyses.

VBP [%]	Average SF	Standard deviation
0 (matrix-only)	0.83	–
25	0.90	0.065
40	1.16	0.205
55	1.57	0.279
70	2.24	0.391

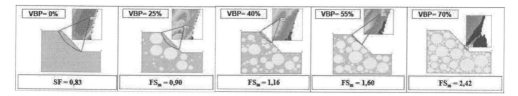

Figure 2. LEM analyses results for one of the fifteen bimrock configurations generated for each VBP considered: critical surfaces and minimum safety factors (SFs) provided by simplified Bishop's method.

Table 5. Safety factors obtained by FEM and LEM analyses applying the Lindquist criterion.

VBP [%]	Safety factors—Lindquist criterion	
	LEM Analyses	FEM Analyses
25	–	–
40	0.84	0.94
55	0.92	0.95
70	1.0	1.0

to be heterogeneous materials and with previous findings from Napoli et al. (2017), who analyzed slope stability in bimrocks using the same stochastic approach but rock inclusions of circular shape.

However, it is worth pointing out that assuming bimrocks to be homogeneous and isotropic materials does not allow the tortuosity of the slip surfaces to be taken into account and the critical slip surfaces to be correctly identified. This produces an underestimation of the rock volume involved in the instability.

3 COMPARISON OF RESULTS AND CONCLUSIONS

The average SFs, provided by the different approaches applied, are compared and reported in Figure 3.

The results show that:

– there is a clear trend toward increasing SF with increasing VBP, whatever the analysis performed. This trend, which is more evident for VBP greater than 25%, is in line with the findings of previous studies on slope stability (Irfan and Tang, 1993; Medley and Sanz, 2004; Barbero et al., 2006; Napoli et al., 2017);

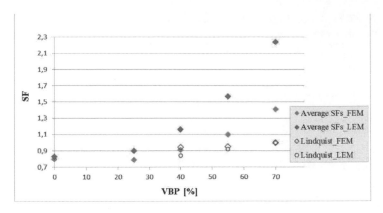

Figure 3.   Average SFs obtained for heterogeneous and homogeneous (Lindquist criterion) bimrock models.

– FEM analyses appear to be representative of the real behavior of these materials. As shown in Figure 1, this method allows potential slip surfaces to be correctly identified (Lindquist and Goodman, 1994; Medley and Sanz, 2004; Barbero et al., 2006; Napoli et al., 2017);
– when analyzing heterogeneous geomaterials such as bimrocks, LEM analyses should not be applied using the classic grid search method with circular failure surfaces. Such an approach does not allow the tortuosity of slip surfaces to develop within the matrix. Critical failure surfaces, indeed, are those encountering the lowest number of (stronger) inclusions and are usually located near the slope surfaces. This leads to overestimations of the SFs and to underestimations of unstable volumes involved, that increase with increasing VBP;
– the use of a matrix-only model (0% VBP), which does not take the presence of blocks into account, leads to a significant underestimation of the SFs, especially for high VBP. Furthermore, shapes and positions of failure surfaces are not correctly identified, since their tortuosity is not taken into account. These results are in good agreement with previous studies conducted on bimrocks (Lindquist 1994; Barbero et al. 2006);
– when applying the strength criterion proposed by Lindquist (1994), LEM and FEM analyses provide SFs close to each other (as shown in Table 5 and Figure 3) and quite similar to those yielded by FEM analyses in heterogeneous materials. Anyway, given the limited geometric configurations analyzed, further studies are still required to verify it. Furthermore, since the application of this approach does not allow tortuosity of sliding surfaces to be taken into account (because it assumes bimrocks as homogeneous continuous and isotropic rock masses), it seems to be acceptable if used in predesign stages only (Kalender et al., 2014);
– although further analyses will be performed to validate and generalize the findings of this paper (analyzing different block orientations and shapes, i.e. eccentricity, strength parameters and slope geometries), it appears that both elliptical and circular block shapes, that were analyzed in a previous work by Napoli et al. (2017), influence slope stability of bimrocks in a comparable way. However, failure surfaces obtained in this study show less tortuous paths with respect to those found by Napoli et al. (2017) for bimrocks with circular block shapes.

As illustrated in Table 3 and Table 4, both FEM and LEM results for heterogeneous materials are extremely variable, especially for higher VBP. The difference between the maximum and the minimum SF of slope models with a given VBP ranges from 0,13 (ΔSF25%VBP) up to around 0,60 (ΔSF70%VBP). This high variability can be ascribed to the different dimensions and locations of the rock inclusions within the slope models, which strongly affect the positions and shapes of sliding surfaces and the stability of the slopes. These findings, in accordance with previous results found by Napoli et al. (2017), demonstrate that blocks play an important role in slope stability and that their presence should not be neglected. Furthermore, when dealing with such heterogeneous materials, the use of a stochastic approach is highly recommended in order to achieve reliable results.

REFERENCES

Afifipour M, Moarefvand P (2014). Mechanical behavior of bimrocks having high rock block proportion. Int J Rock Mech Min Sci, 65, pp. 40–48.

Barbero M, Bonini M, Borri Brunetto M (2006). Analisi numeriche della stabilità di un versante in bimrock. In: Incontro Annuale Dei Ricercatori Di Geotecnica 2006—IARG 2006, Pisa.

Barbero M, Bonini M, Borri-Brunetto M (2007). Numerical Modelling of the Mechanical Behaviour of Bimrock. In: 11th Congress of the International Society for Rock Mechanics (ISRM 2007). Lisbon, Portugal: International Society for Rock Mechanics.

Barbero M, Bonini M, Borri-Brunetto M (2012). Numerical Simulations of Compressive Tests on Bimrock. EJGE, Vol. 15, pp. 3397–3399.

Haneberg WC (2004). Simulation of 3D block populations to charaterize outcrop sampling bias in bimrocks. Felsbau Rock Soil Eng J Eng Geol Geomech Tunneling, Vol. 22(5), pp. 19–26.

Irfan TY, Tang KY (1993). Effect of the coarse fractions on the shear strength of colluvium. Hong Kong Geotech. Eng. Off. Civ. Eng. TN 4/92.

Kalender A, Sonmez H, Medley E, Tunusluoglu C, Kasapoglu KE (2014). An approach to predicting the overall strengths of unwelded bimrocks and bimsoils. Eng Geol., 183, pp. 65–79.

Lindquist ES (1994). Strength and Deformation Properties of Melange. PhD dissertation. University of California at Berkeley.

Lindquist ES, Goodman RE (1994). Strength and deformation properties of a physical model melange. In: Nelson PP, Laubach SE, eds. Proc. 1st North America Rock Mech. Symposium. Austin, Texas, pp. 843–850.

Medley EW (1994). The engineering characterization of melanges and similar block-in-matrix rocks (bimrocks). PhD dissertation. University of California at Berkeley.

Medley EW, Lindquist ES (1995). The engineering significance of the scale-independence of some Franciscan melanges in California, USA. In: Daemen JJK, Schultz RA, eds. Rock Mechanics Proceedings of the 35th U.S. Symposium. Rotterdam: A.A. Balkema, pp. 907–914.

Medley EW, Sanz Rehermann PF (2004). Characterization of Bimrocks (Rock/Soil Mixtures) with Application to Slope Stability Problems. In: Eurock 2004 & 53rd Geomech. Colloquium, pp. 425–430.

Medley EW, Zekkos D (2011). Geopractitioner Approaches to Working with Antisocial Mélanges. Invited paper. In: Wakabayashi J, Dilek Y, eds. Mélanges: Processes of Formation and Societal Significance, 480. Geological Society of America Special, pp. 261–277.

Napoli, M.L., Barbero, M., Ravera, E., Scavia, C., 2017. A stochastic approach to slope stability analysis in bimrocks. Int. Journal of Roch Mech. And Min. Science. In printing.

Sonmez H, Ercanoglu M, Kalender A, Dagdelenler G, Tunusluoglu C (2016). Predicting uniaxial compressive strength and deformation modulus of volcanic bimrock considering engineering dimension. Int. J. Rock Mech Min Sci, 86, pp. 91–103.

Sonmez H, Kasapoglu KE, Coskun A, Tunusluoglu C, Medley E, Zimmerman RA (2009). Conceptual empirical approach for the overall strength of unwelded bimrocks. In: ISRM Regional Symposium, Rock Engineering in Difficult Ground Condition, Soft Rock and Karst. Dubrovnik, Croatia.

Wakabayashi J, Medley EW (2004). Geological Characterization of Melanges for Practitioners. Felsbau Rock Soil Eng J Eng Geol Geomech Tunneling, 22(5), pp. 10–18.

3D finite element modelling of fracturing in heterogeneous rock: From pure solid to coupled fluid/solid analysis

Ramin Pakzad, Shanyong Wang & Scott Williams Sloan
The University of Newcastle, Callaghan, NSW, Australia

ABSTRACT: The elastic-brittle-damage constitutive model for three-dimensional stress state is incorporated into the commercial finite element code ABAQUS. To take into account the heterogeneity effect, mesoscale elements in finite element analysis are assigned by different strength and stiffness properties according to the Weibull distribution function. A FORTRAN code is developed and employed within the static equilibrium equation to analyse three-dimensional progressive failure of rock masses. The numerical model is extended to the coupled hydro-mechanical problem by adding the fluid mass conservation equation to the governing equation system and relating hydraulic conductivity of the elements to their damage and hydrostatic stress states. The results for two typical problems with the pure dry and coupled solid/fluid conditions, respectively, are presented. Good agreement between the numerical results and previously findings in the literature verify such implementation and demonstrates its significance.

1 INTRODUCTION

Three-dimensional (3D) fracture patterns in quasi-brittle geo-materials such as rocks have exhibited to be tortuous even under simple loading. This tortuosity is associated to the heterogeneity of material properties. The numerical simulation of fracture propagation in hetero-geneous brittle materials has been paid much attention in recent years. With this regard, the smeared-cracking-like model has emerged as a powerful numerical procedure which is capable of capturing the macro-scale quasi-brittle fracturing response of rock formations from their micro-scale structure and meso-scale behaviour (Li and Tang, 2015). While most of the works done in this area are mainly related to two-dimensional (2D) analysis under dry conditions, just a few works are published concerning 3D analysis particularly for coupled fluid/solid analysis (Wang et al., 2014, Wang et al., 2013). In this paper, first, the 2D smeared-cracking-like model will be extended to 3D; and then the 3D smeared-like-cracking model along with a damage/stress-induced field variable will be incorporated into the coupled fluid/solid analysis of ABAQUS to simulate hydraulic fracturing. Such implementation is important because ABAQUS provides many built-in features making it possible for the users to apply their own code to more complex situation. For example, the probable partially-saturated condition in the hydraulic fracturing of low-permeability media can be simulated just by adding the partially-saturated option of ABAQUS to the model presented in this paper (Pakzad et al., 2017). The focus of this paper is mainly on the implementation aspects of the 3D modelling of fracturing in heterogeneous rock under either dry or fully-saturated conditions.

2 MATHEMATICAL FORMULATION

2.1 *Solid phase contribution*

In the framework of the finite element method, the equilibrium equation shown in Eq. (1) is implicitly solved by the ABAQUS/Standard solver to obtain the displacement degrees of freedom for every node in the model under static conditions. The strain components are then calculated from the displacement field (Eq. (2)) and sent to the UMAT subroutine, accompanied by other state variables, at the beginning of every increment.

$$\frac{\partial \sigma_{ij}}{\partial x_j} + F_j = 0 \tag{1}$$

$$\varepsilon_{ij} = \frac{1}{2}\left(\frac{\partial u_i}{\partial x_j} + \frac{\partial u_j}{\partial x_i}\right) \tag{2}$$

The user must use a constitutive model relating stresses to strains at the integration point(s) of each element. In this study, the elastic-brittle-damage constitutive model is employed as follows:

2.1.1 Elastic regime

Following Hookes' law, the principal stresses and strains are correlated through Eq. (3), in which $\lambda = \frac{vE}{(1-2v)(1+v)}$, $G = \frac{E}{(1+v)}$, v is Poisson's ratio and E is the current Young's modulus.

$$\sigma_{ij} = 2G\varepsilon_{ij} + \lambda\varepsilon_{jj} \quad (i, j = 1, 2, 3) \tag{3}$$

The stiffness begins to degrade gradually as soon as the stress state meets the damage surface. This occurs incrementally by the evolution of the damage parameter (D) in Eq. (4), where E and E_0 represent the updated and initial stiffness of the mesoscopic element, respectively.

$$E = (1 - D)E_0 \tag{4}$$

2.1.2 Damage initiation and evolution

The damage surface is defined by two individual criteria: the tensile failure criterion and the shear failure criterion, with priority given to the former. First, the maximum principal stress (σ_1) is compared with the uniaxial tensile strength of element (f_{t0}) according to Eq. (5), under the condition that the shear criterion is less critical than the tensile criterion.

$$\sigma_1 \geq f_{t0} \tag{5}$$

If the tensile failure criterion is met, the damage parameter is calculated by Eq. (6), depending on the value of equivalent strain ($\tilde{\varepsilon}$) at the end of the current increment.

$$D = \begin{cases} 0 & \tilde{\varepsilon} < \varepsilon_{t0} \\ 1 - \dfrac{f_{tr}}{\sigma_1^0} & \varepsilon_{t0} \leq \tilde{\varepsilon} < \varepsilon_{tu} \\ 1 & \tilde{\varepsilon} \geq \varepsilon_{tu} \end{cases} \tag{6}$$

In this equation, the equivalent strain at which the tensile damage surface is met for the first time is symbolized by ε_{t0}, and σ_1^0 denotes the maximum principal stress calculated by the initial elastic modulus. The residual tensile strength and ultimate tensile strain are defined as $f_{tr} = \gamma f_{t0}$ and $\varepsilon_{tu} = \eta\varepsilon_{t0}$, respectively. The equivalent strain corresponding to the tensile damage evolution is assumed to be a combination of the principal strains as follows:

$$\tilde{\varepsilon} = \sqrt{\langle\varepsilon_1\rangle^2 + \langle\varepsilon_2\rangle^2 + \langle\varepsilon_3\rangle^2} \tag{7}$$

The Macaulay brackets $\langle\ \rangle$ are defined by Eq. (8):

$$\langle x \rangle = \begin{cases} x & x \geq 0 \\ 0 & x < 0 \end{cases} \tag{8}$$

To be able to take into account the effect of intermediate principal stress as well, instead of Mohr-Coulomb strength criterion, the shear damage criterion is defined following the linear unified strength theory (Yu, 2006):

$$\psi \sigma_1 - \frac{b\sigma_2 + \sigma_3}{1+b} \geq f_{c0} \quad when \quad \sigma_2 \leq 0.5[(\sigma_1 + \sigma_3) - \sin\phi(\sigma_1 - \sigma_3)] \tag{9a}$$

$$\psi \frac{\sigma_1 + b\sigma_2}{1+b} - \sigma_3 \geq f_{c0} \quad when \quad \sigma_2 \geq 0.5[(\sigma_1 + \sigma_3) - \sin\phi(\sigma_1 - \sigma_3)] \tag{9b}$$

where $\psi = \frac{1+\sin\phi}{1-\sin\phi}$ and b is the coefficient of intermediate principal stress and takes a value between zero and one. ϕ, f_{c0}, σ_1, σ_2 and σ_3 are the internal frictional angle, uniaxial compressive strength, maximum; intermediate and minimum principal stresses, respectively.
When the shear criterion is satisfied, the damage parameter is calculated via Eq. (10):

$$D = \begin{cases} 0 & \tilde{\varepsilon} > \varepsilon_{c0} \\ 1 - \dfrac{f_{cr}}{\psi\sigma_1^0 - \dfrac{b\sigma_2^0 + \sigma_3^0}{1+b}} & \tilde{\varepsilon} \leq \varepsilon_{c0} \quad and \quad \sigma_2 \leq 0.5[(\sigma_1 + \sigma_3) - \sin\phi(\sigma_1 - \sigma_3)] \\ 1 - \dfrac{f_{cr}}{\psi\dfrac{\sigma_1^0 + b\sigma_2^0}{1+b} - \sigma_3^0} & \tilde{\varepsilon} \leq \varepsilon_{c0} \quad and \quad \sigma_2 \geq 0.5[(\sigma_1 + \sigma_3) - \sin\phi(\sigma_1 - \sigma_3)] \end{cases} \tag{10}$$

where $f_{cr} = \gamma f_{c0}$, $\tilde{\varepsilon} = \varepsilon_3$, ε_{c0} is the equivalent strain at the moment that the shear failure criterion meets the current strength of the element for the first time, and σ_1^0, σ_2^0 and σ_3^0 are the maximum, intermediate and minimum principal stresses of the intact element calculated by the initial Young's modulus. The convention of ABAQUS is followed in the abovementioned relationships such that the stress and strain components are negative in compression and positive in tension (Hibbitt et al., 2014).

2.2 Fluid phase contribution

To capture the interaction between the fluid and solid phases in a coupled problem, the transient mass conservation equation (Eq. (11)) needs to be simultaneously solved with the equilibrium equation (Eq. (1)).

$$\frac{\partial}{\partial t}(\rho n) + \frac{\partial}{\partial x_i}(\rho q_i) = 0 \tag{11}$$

In Eq. (11), q_i is the vector value of the pore fluid flux and ρ and n are the fluid density and porosity, respectively. According to this equation, the mass change of the stored fluid inside a RVE during a period (t) is equal to the net fluid flux crossing the RVE.
The fluid constitutive model has already been defined by ABAQUS/Standard and it should be appropriately selected by assigning values to the related material parameters. For instance, the flux parameter (q_i) in Eq. (11) is replaced by Eq. (12), which is known as Darcy's law:

$$q_i = \hat{k}_{ij} \frac{d}{\partial x_j}\left(z + \frac{p}{\rho g}\right) \tag{12}$$

where z and g represent an elevation above an arbitrary vector and gravitational acceleration, respectively. The second-order hydraulic conductivity tensor has units of velocity (length/time) and is the product of the hydraulic conductivity of a fully saturated porous medium (k) and a saturation-dependent factor (k_s), which has a value of one for fully saturated conditions.

In case of coupled analysis, the constitutive model presented above defines the effective stress (σ'_{ij}) instead of the total stress (σ_{ij}) which requires to be in equilibrium with the external load vector F_j. Following the Biot's theory of consolidation (Biot, 1941) the total stress is in relationship with the effective stress and pore pressure (p) through Eq. (13):

$$\sigma_{ij} = \sigma'_{ij} - sp\delta_{ij} \tag{13}$$

where s represents the degree of saturation and δ_{ij} is the Kronecker delta, which has a value of one for similar indexes and zero for dissimilar ones.

2.2.1 Evolution of permeability

The influence of damage and stress state on the hydraulic conductivity and thereby permeability of elements is included through Eq. (14):

$$k = \begin{cases} k_0 \exp\left[\beta\left(\dfrac{\sigma_{ii} + sp\delta_{ij}}{3} \right) \right] & D = 0 \\[3mm] \zeta k_0 \exp\left[\beta\left(\dfrac{\sigma_{ii} + sp\delta_{ij}}{3} \right) \right] & D > 0 \end{cases} \tag{14}$$

where k_0 and k are the initial and modified hydraulic conductivity under fully saturated conditions, respectively. Permeability is assumed to be isotropic and homogeneous for every element. The effect of damage is shown by the mutation coefficient of permeability ζ ($\zeta > 1$). The coupling coefficient β determines the intensity of the hydrostatic effective stress $\left(\sigma'_{ii} / 3 \right)$.

2.3 Material heterogeneity

The heterogeneity of rock is modelled statistically via the Weibull distribution function. To produce the random variable from the random number generated by the Monte Carlo method, the inverse of the Weibull cumulative probability function was used as follows:

$$f(u) = 1 - \exp\left[-\left(\frac{u}{u_0} \right)^m \right] \tag{15}$$

where u corresponds to the random material property of the element and m is the shape factor of the Weibull distribution function determining the dispersal of u around u_0. The parameter u approaches the value of u_0 as the value of m increases. In this study, the elastic modulus and strength are considered to vary among elements with the same homogeneity indexes (m) but different initial seeds.

3 INCREMENTAL PROCEDURE AND PARALLEL EXECUTION

Fig. 1 demonstrates the solution procedure of ABAQUS/Standard for numerical simulations. ABAQUS/Standard incrementally solves a system of equations to find the changes in the degrees of freedom (ΔU) which satisfy the governing equation(s) for the current increment of loading (ΔF). The residual vector (R) is calculated using the stresses (σ) calculated in UMAT which themselves as well as other solution dependent state variables ($SDV's$) are calculated based on the strain components (ε) and other input data from the former increment. The material tangent matrix ($\partial\Delta\sigma/\partial\Delta\varepsilon$) and stresses from UMAT are used for new approximation

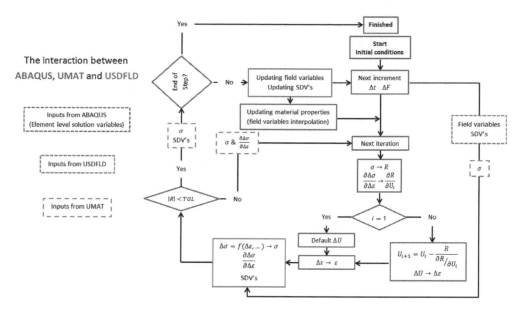

Figure 1. Incremental solution procedure of ABAQUS/Standard.

in the next iterations unless the residual is within a small tolerance (*TOL*). Once the solution is accepted, field variables are updated in USDFLD using the updated values of stress components and *SDV's*. This procedure is repeated until the end of analysis.

3D problems usually entail a huge number of elements, rendering the run time impractical particularly when nonlinear effects may result in so many iterations. Parallel execution is a technique based on which the model is decomposed into different domains an each domain is handled by a specific number of processors. In this work, MPI-based parallelization (Hibbitt et al., 2014) with 40 domains each with one processor is used to increase efficiency.

4 NUMERICAL EXAMPLES

Two numerical examples are presented in this section, one of which is related to the simulation of fracture propagation in a heterogeneous dry block with a pre-existing fissure under uni-axially displacement-control compression (Fig. 2). The results obtained from this simulation are consistent with the experimental observation of other scholars (Yang and Jing, 2011), verifying the numerical modelling. Two 3D wing cracks initiate at the two end corners of the pre-existing fissure, where the maximum tensile stress is critical. Secondary tensile cracks are generated near one of the corners of the fissure, followed by the appearance of a shear-damaged region linking the corner to the secondary tensile cracks. A new shear crack emerges at the other corner of the fissure concurrent with slow extension of the previous tensile cracks until the specimen splits at its top half section.

The second example represents hydraulically-induced fractures around a cavity inside a block pressurized by water and confined at its external boundaries (Fig. 3). According to Fig. 3, the fracture surfaces are generally perpendicular to the minimum far-field principal stress. Such influence of far-field principal stress on the direction of propagating hydraulic fractures has been observed and reported for physical modelling of hydraulic fracturing (Zhou et al., 2008). The asymmetric distribution of pore pressure illustrated in the right hand side of Fig. 3 is due to the asymmetric generation of damaged elements forming fractured area around the pressurized region of the cavity, indicating the proper function of our developed code for incorporating Eq. 14 in the solution procedure.

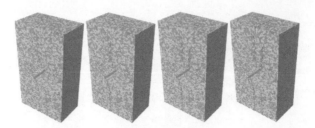

Figure 2.   Fracture propagation in a heterogeneous dry block with a pre-existing fissure under uni-axially displacement-control compression.

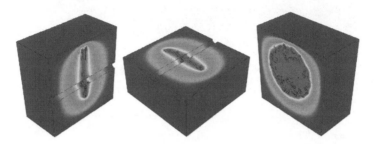

Figure 3.   Distribution of pore pressure throughout a hydraulically-fractured block pressurized by water at the central region of its cylindrical cavity (cross-sectioned views).

5   CONCLUSIONS

In this paper, the smeared-like-cracking model for 3D modelling of brittle failure of hetero-geneous material is implemented into the commercial finite element code ABAQUS through its UMAT user-subroutine interface. The incremental change of permeability due to damage and hydrostatic effective stress is incorporated into the USDFLD user-subroutine interface of ABAQUS to be used with the developed UMAT in coupled fluid/solid analysis. The MPI-based parallel execution technique is used for both dry and wet conditions to reduce run time which plays an important role in 3D problems. For the verification purpose, 3D fracture propagation in a dry block with an inclined fissure under uni-axial compression and hydrau-lic fracturing in a fully-saturated block pressurized by water at its internal boundary with a cylindrical cavity are simulated with promising results.

REFERENCES

Biot, M.A. 1941. General theory of three-dimensional consolidation. *J. appl. Mech.,* 12, 155–164.
Hibbitt, D., Karlsson, B. & Sorensen, P. 2014. ABAQUS, ABAQUS/Standard User's Manual, Inc.
Li, G. & Tang, C.A. 2015. A statistical meso-damage mechanical method for modeling trans-scale pro-gressive failure process of rock. *Int. J. Rock Mech. Min. Sci.,* 74, 133–150.
Pakzad, R., Wang, S.Y. & Scott, W.S. 2017. Numerical simulation of hydraulic fracturing in low/high permeability, quasi brittle, heterogeneous rocks. *Rock Mech. Rock Eng.,* submitted.
Wang, S.Y., Sloan, S.W., Fityus, S.G., Griffiths, D.V. & Tang, C.A. 2013. Numerical modeling of pore pres-sure influence on fracture evolution in brittle heterogeneous rocks. *Rock Mech. Rock Eng.,* 46, 1165–1182.
Wang, S.Y., Sloan, S.W., Sheng, D.C., Yang, S.Q. & Tang, C.A. 2014. Numerical study of failure behaviour of pre-cracked rock specimens under conventional triaxial compression. *Int. J. Solids Struct.,* 51, 1132–1148.
Yang, S.Q. & Jing, H.W. 2011. Strength failure and crack coalescence behavior of brittle sandstone samples containing a single fissure under uniaxial compression. *International Journal of Fracture,* 168, 227–250.
Yu, M.H. 2006. *Generalized plasticity,* Springer Science & Business Media.
Zhou, J., Chen, M., Jin, Y. & Zhang, G.-Q. 2008. Analysis of fracture propagation behavior and fracture geometry using a tri-axial fracturing system in naturally fractured reservoirs. *Int. J. Rock Mech. Min. Sci.,* 45, 1143–1152.

Geomechanics and Geodynamics of Rock Masses – Litvinenko (Ed.)
© *2018 Taylor & Francis Group, London, ISBN 978-1-138-61645-5*

Global sensitivity analysis on thermo-hydro-mechanical coupled processes in a low strength sedimentary rock

Samuel Parsons, Graham Stuart, Bill Murphy & David Price
School of Earth and Environment, University of Leeds, UK

ABSTRACT: Radioactive waste may be stored in deep geological disposal facilities. Heat-producing radioactive waste would drive coupled thermo-hydro-mechanical (THM) processes in the host rock mass. Rock mass uncertainty could affect predictions of observable properties that are perturbed by these THM processes. However, the contribution of the individual rock properties to the uncertainty is unknown and is critical for informing modelling and monitoring for reducing uncertainty. Here we rank the rock properties (model input factors) according to their contribution to uncertainty in the observable properties: temperature, pore pressure and displacement (model outputs). We identify non-influential rock properties which have a negligible effect on the uncertainty and should no longer be considered uncertain. We identify the thermal conductivity and permeability as the most influential rock properties which was expected because these are anisotropic. Interestingly we found that closer to the heat source these become more influential relative to the other rock properties. The ranking may be used to inform prioritizing further rock mass characterization efforts for uncertainty reduction. The spatial effect on thermal conductivity and permeability may be used to improve model calibration and back-calculations by counterintuitively using data proximal to the heat source. We demonstrate that global sensitivity analysis methods can be applied to geomechanical models to improve our understanding of the rock mass uncertainty. We present validated values for the number of models required for results to converge which should be used to estimate appropriate values for future sensitivity analyses on geomechanical models.

1 INTRODUCTION

Modelling and monitoring for reducing uncertainty in aspects of geological disposal of radioactive waste is a paramount aim for the community (IAEA, 2011). Often the problem for geologists is that rock masses are heterogeneous and characterized with a large degree of uncertainty. Underground research laboratories provide access to the in-situ rock mass and resultantly tend to be situated in the most characterized rock masses in the world (Jalali et al., 2017).

The complex thermo-hydro-mechanical processes occurring in the geosphere around heat-producing radioactive waste are simulated by in-situ heating experiments (e.g. Conil et al., 2012; Zhang et al., 2007). The observable physical properties are temperature, pore pressure and displacement. These data may be used in benchmark tests to validate coupled THM models (e.g. Wang and Kolditz, 2013; Gens et al., 2007; Zhang et al., 2007).

Despite being well characterized at test sites there is still significant uncertainty reported in rock mass properties. Gens et al. (2007) addresses this by running additional model evaluations with one-at-a-time changes to end-member values within the uncertainty for selected 'important' parameters. However, there has been a lack of investigation into the rock mass uncertainty beyond this.

This is diagnostic of a broader lack of investigation into rock mass uncertainty in geomechanical models. Notable cutting-edge work on rock mass uncertainty has been presented in

Jobmann et al. (2016). The authors use the software tool optiSLang for CAE-based sensitivity analysis and optimization that automatically fit parameters and analyse the importance of individual parameters for the general system development. However, no results are presented towards the robustness or validity of the sensitivity analysis and the sample size of 80 for 29 input factors is an order of magnitude less than the recommended values for similar aims (Sarrazin et al., 2016).

The high number of input factors for a geomechanical model is due to the THM coupling. The low number of model evaluations is because THM coupled finite element models are computationally expensive. Petropoulos and Srivastava (2016) present a credible, computationally efficient multi-method global sensitivity analysis approach for input factor ranking. Three approaches measure the sensitivity indices differently and presented together provide a more credible input factor ranking result. The approach is computationally efficient because the three approaches use the same generic input-output dataset and so no additional model evaluations are required.

We apply the multi-method global sensitivity analysis approach (Petropoulos and Srivastava, 2016) to a THM model to rank input factors according to estimated sensitivity indices. We identify the most important input factors contributing to uncertainty and the input factors which do not contribute to uncertainty in temperature, pore pressure and displacement. We benchmark our THM model against an in-situ heating test to ensure that it is a good representation of reality.

2 MULTI-METHOD GLOBAL SENSITIVITY ANALYSIS

Input factor ranking is achieved using global sensitivity analysis techniques in which the whole input factor uncertainty space is investigated using Monte Carlo style simulation. Sensitivity indices are estimated for each input factor and these indices are used to rank them.

The input factor uncertainty space (Table 1) is sampled using a Maximin Latin Hypercube. This sampling strategy provides enhanced coverage of the input factor space. The samples

Table 1. A table of the input factors investigated in the sensitivity analysis. The maximum and minimum values describe the uncertainty range in those input factors. The Opalinus clay rock mass parametrs are established by laboratory tests performed on normally-sized samples and confirmed by back-calculations of mock-up heating tests on large-scale samples taken from the test sites (Zhang et al., 2007).

Property		Minimim	Maximum	Recommended	Unit
Poisson's ratio	v	0.24	0.33	0.27	
Reference bulk modulus	B_{ref}	871.18	4564.20	3300	MPa
Cam Clay constant	κ	0.0030	0.0040	0.0035	
Porosity	ϕ_{init}	0.135	0.179	0.16	
Kozeny-Carmen constant	K_0	1.02E-19	1.02E-21	2.00E-20	m^2
Fluid density	ρ_f	971	1030	1000	kg/m^3
Grain density	ρ_g	2680	2740	2710	kg/m^3
Fluid stiffness	K_f	2000	2500	2222	MPa
Grain stiffness	K_g	20000	50000	40000	MPa
Biot constant	Biot	0.48	1.0	0.6	
Viscosity	μ	3.53×10^{-10}	1.12×10^{-9}	1.00×10^{-9}	MPa.s
Thermal conductivity	λ	1.0	2.1	1.7	W/m.K
Fluid heat capacity	c_f	3992	4182	4182	J/kg.K
Solid heat capacity	c_s	720	880	800	J/kg.K
Rock mass linear coefficient of expansion	α_{ref}	1.5×10^{-5}	1.9×10^{-5}	1.7×10^{-5}	K^{-1}
Fluid volumetric thermal expansion coefficient	α_f	2.07×10^{-4}	4.5×10^{-4}	3.4×10^{-4}	K^{-1}
Solid volumetric thermal expansion coefficient	α_s	4.5×10^{-6}	4.8×10^{-5}	4.5×10^{-6}	K^{-1}

form a k-by-n matrix (X) where k is the number of uncertain input factors and n is the number of samples. 500 models are built using the input factor samples. The models are evaluated and an output function is selected for each sample, forming a vector of length n (Y). The output functions investigated in this study are the temperature, pore pressure and displacement.

The global sensitivity analysis method used to evaluate X and Y for input factor ranking is the multi-method approach designed in Petropoulos and Srivastava (2016). The approach enhances the credibility of the study by using three sensitivity analyses to estimate three indices for each input factor instead of one. This is achieved without increasing the computational expense of the study because the methods use the same generic input-output dataset, therefore, no additional model evaluations are required.

The three sensitivity analyses are the Regional Sensitivity Analysis, PAWN (Pianosi et al., 2015), and an estimate of the main effects indices from the Variance-based Sensitivity Analysis (Petropoulos and Srivastava, 2016). SAFE Toolbox (Pianosi et al., 2015) contains Matlab functions for calculating the sensitivity indices of these approaches using X and Y.

3 THE MODEL

We use the finite element software ELFEN (Rockfield Software) for the THM modelling. We use fully-coupled thermal and porous flow fields within an implicit solution method, which are semi-coupled at regular intervals to the geomechanical field within an explicit solution method. The thermal field calculates bulk properties based on the grain and fluid properties to simulate conduction and advection (Equation 1). The porous flow field simulates Darcy fluid flow, consolidation and aqua-thermal pressure (Equation 2).

$$(\rho c)b\frac{\partial T}{\partial t} = div(K_b \nabla T) + \rho_f c_f q_f \cdot \nabla T \tag{1}$$

$$div\left(\frac{k(\phi)}{\mu_f(T)}(\nabla pf - \rho fg)\right) = \left(\frac{\phi}{K_f} + \frac{(\alpha - \phi)}{K_s}\right)\frac{\partial pf}{\partial t} - \frac{\alpha}{1-\phi}\frac{\partial \phi}{\partial t} + \beta_{s.f}\frac{\partial T}{\partial t} \tag{2}$$

where the subscripts b, f and s denote bulk, fluid and solid grain values, ρ is density, c is specific heat capacity, T is temperature, t is time, κ is thermal conductivity and q is the Darcy fluid flux, k is intrinsic permeability, μ is viscosity, p is pore pressure, g is acceleration due to gravity, K is stiffness, α is Biot's constant and β is thermal expansion coefficient.

We use a poro-elastic material model for the Opalinus clay with the Soft Rock (SR3) state boundary surface (Crook et al., 2003) for weakly cemented rock. A smooth hardening law approximates to the Cam Clay hardening model. We define non-linear permeability using the Kozeny-Carmen model and temperature-dependent viscosity for the fluid.

The THM coupled calculations include the following major assumptions. Heat transport is by conduction (Fourier's law) through porous medium and advection of liquid water. Fluid transport is controlled by liquid water advection (Darcy's law). A thermo-mechanical model is used for the description of the mechanical behaviour of the clay rock with the main features of thermal expansion. The clay rock is assumed to be isotropic and homogeneous. The modelled clay rock is isotropic because of the axisymmetric set-up. Since the aim of our work is not to fit parameters we are able to simplify the model to an axisymmetric representation, vastly reducing the computation time of one model evaluation to increase the total number of model evaluations.

The model (Figure 1) is allowed to reach equilibrium and then a heat flux is applied to simulate the heating. As in the original test the thermal load is 650 W for 92 days and then increased to 1950 W for 252 days. The temperature increases up to around 100°C. The model was initially run using the recommended parameters for the test field. The temperature and pore pressure results were accurately predicted up to the time at which the output was taken for the sensitivity analysis.

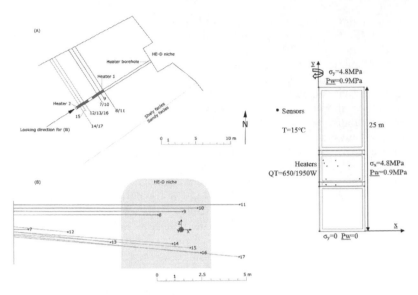

Figure 1. Left: Illustration of the experiment layout and instrumentation (adapted from Zhang et al., 2007). Sensor numbers are consistent with the original experiment. (A) Plan view. (B) Profile view. Right: An illustrative description of the axisymmetric model we used to simulate the in-situ heater experiment. The heater is centred on the symmetry axis. The gallery is represented by a zero-stress boundary condition whilst the mean in situ stress is applied to the other boundaries. The in-situ pore pressure is reduced from 2.2 MPa to 0.9 MPa because of the proximity to open tunnels (Gens et al., 2007).

4 RESULTS

The multi-method approach is applied to the generic input-output dataset for temperature, pore pressure and displacement. A sensitivity indices is estimated for each input factor and for each of the three sensitivity analyses. The analyses are repeated using the different locations of sensors from the original laboratory experiment to investigate the spatial effects on input factor sensitivity. Furthermore, the analyses are repeated throughout model time to investigate the temporal effects.

The spatial location affects the sensitivity of the thermal conductivity for temperature and the permeability and viscosity for pore pressure. These three input factors are the transport properties. The spatial analysis shows that the transport properties of the rock mass become more sensitive towards the heater. This is because changing the property next to the heater affects the amount of heat or fluid that is transported away from the heater, whereas away from the heater, changing the property affects the amount of heat or fluid that is transported away and towards the location which has a negative interference for the sensitivity index. This spatial effect is strong enough that within 1 m of the heater the transport properties dominate the resultant temperature and pore pressure. This enhanced understanding can improve back analyses by focusing only on the transport properties when back analysing for temperature or pore pressure proximal to a heat source.

The temporal analysis identified viscosity becoming more sensitive as the model advanced because the fluid becomes less viscous as its temperature increases. We also found the sensitivity analysis required a greater number of model evaluations to achieve convergence at the beginning of the model because the output perturbations were small. We observed no other temporal effects.

Figure 2 shows the results for a radial distance of 1.2 m from the heater at a time equal to the mean occurrence of peak pore pressure. The correlation between the indices predicted by the different sensitivity analysis methods demonstrates the importance in robust sensitivity analyses. We observe that the methods generally agree but with local conflicting ranking. For example, the relative ranking of Biot's constant and viscosity depends on the sensitivity analysis method. Therefore, we rank the input factors in groups.

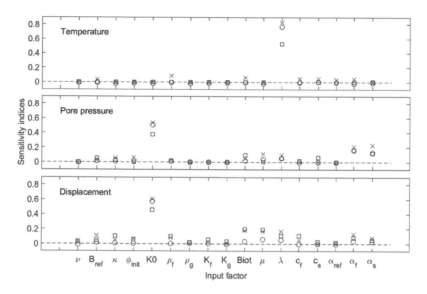

Figure 2. Multi-method sensitivity analysis results. Each subplot title indicates the model output for which the results were calculated. The y-axis values are the sensitivity indices for each input factor identified by the x-axis nomenclature. The nomenclature is defined in Table 1. Squares represent the indices calculated by the Regional sensitivity analysis, crosses represent the indices calculated by the PAWN sensitivity analysis and circles represent the indices calculated by the Variance-based sensitivity analysis.

For temperature, thermal conductivity is ranked first; grain heat capacity second; porosity third; and all other input factors are determined to have negligible uncertainty. For pore pressure, permeability is ranked first; grain and fluid thermal expansion coefficients second; viscosity and conductivity second; bulk modulus, Cam Clay constant, porosity, Biot's constant and solid heat capacity third; and all other input factors are determined to have negligible uncertainty. For displacement, permeability is ranked first; Biot's constant and fluid thermal expansion coefficient second; bulk modulus, viscosity, thermal conductivity and grain thermal expansion coefficient third; Cam Clay constant and porosity fourth; and all other input factors are determined to have negligible uncertainty.

Input factor ranking for temperature achieved convergence in as few as 40 model evaluations. Contrastingly, pore pressure and displacement required up to 420 model evaluations to achieve convergence of the input factor ranking. We interpret this to be because pore pressure and displacement are dependent on more input factors than temperature. Finally, we found that convergence required fewer model evaluations closer to the heat source for all outputs. Therefore, spatial analysis is critical for assessing the robustness of geomechanical sensitivity analyses.

When implementing the results it should be considered that we used a specific case study and experimental set-up. The case study and experimental set-up were selected to increase the applicability of the work. The case study was an in-situ heating experiment which provides a good representation of heat-producing waste packages in rock. The host rock was Opalinus clay which is a low permeability, lower strength sedimentary rock. The parameter ranges are based on uncertainty ranges in the characterisation of the experiment site, which are relatively well constrained for rock masses, as would be expected at a site for radioactive waste disposal.

5 CONCLUSIONS

We demonstrate that increasing computing power can be used for global sensitivity analyses on coupled geomechanical models to enhance our understanding of the modelled systems and their uncertainties.

Rock mass uncertainty at an underground laboratory in low strength sedimentary rock is investigated in the context of its effect on the range of temperature, pore pressure and displacement in forward modelling. Independent input factors that characterise the rock mass in the model are ranked according to their contribution to the range in the outputs using sensitivity indices. Uncertainty is case specific so the conceptualisation of the model is important for the applicability of the results. The results from this study can be applied to THM modelling of heating in low strength sedimentary rocks characterized at underground laboratories.

The result can be used to prioritize efforts for uncertainty reduction. Efforts should focus on the thermal conductivity to reduce temperature uncertainty. And on permeability, Biot's constant and thermal expansion coefficients of the grain and fluid to reduce pore pressure and displacement uncertainty.

Furthermore, the input factor ranking enhances our understanding of the dominant controls in the THM coupled processes. It indicates that uncertainty of how the pore pressure affects the pore space contributes more significantly to the displacement uncertainty than uncertainty in the thermal expansion coefficients. This should be considered if attempting to back analyse thermal expansion coefficients during heater tests. We find that a temperature dependent viscosity is required to capture realistic pore pressure evolution. Finally, we find that the spatial analysis of input factor ranking shows the best place to back analyse the thermal conductivity from temperature data is proximal to the heat source.

REFERENCES

Crook, T., Willson, S., Yu, J. and Owen, R. (2003). Computational modelling of the localized deformation associated with borehole breakout in quasi-brittle materials. *Journal of Petroleum Science and Engineering* **38**(3), 177–186.

Conil, N., Armand, G., Garitte, B., Jobmann, M., Jellouli, M., Filippi, M., De La Vaissière, R. and Morel, J. (2012). In situ heating test in Callovo-Oxfordian claystone: measurement and interpretation. In Proceeding of the 5th International meeting of Clays in Natural and Engineered Barriers for Radioactive Waste Confinement, Montpellier, October 22–25.

Gens, A., Vaunat, J., Garitte, B. and Wileveau, Y. (2007). In situ behaviour of a stiff layered clay subject to thermal loading: observations and interpretation. *Geotechnique* **57**(2), 207–228.

International Atomic Energy Agency (2011). Geological disposal facilities for radioactive waste: specific safety guide. IAEA Safety Standards Series No. SSG-14, IAEA.

Jalali MR, Gischig V, Doetsch J, Krietsch H, Amann F and Klepikova M. (2017). Mechanical, hydraulic and seismological behavior of crystalline rock as a response to hydraulic fracturing at the Grimsel Test Site. *American Rock Mechanics Association*; 51st U.S. Rock Mechanics/Geomechanics Symposium, 25–28 June, San Francisco, California, USA.

Jobmann, M., Li, S., Polster, M., Breustedt, M., Schlegel, R., Vymlatil, P. and Will, J. (2016). Using Statistical Methods for Rock Parameter Identification to Analyse the THM Behaviour of Callovo-oxfordian Claystone. *Journal of Geological Resource and Engineering* **3**, 125–136.

Petropoulos, G. and Srivastava, P.K. (2016). Sensitivity Analysis in Earth Observation Modelling. Elsevier Science and Technology Books.

Pianosi, F., Sarrazin, F. and Wagener, T. (2015). A Matlab toolbox for Global Sensitivity Analysis. *Environmental Modelling and Software* **70**, 80–85.

Sarrazin, F., Pianosi, F. and Wagener, T. (2016). Global Sensitivity Analysis of environmental models: Convergence and validation. *Environmental Modelling & Software* **79**, 135–152.

Wang, W. and Kolditz, O. (2013). High performance computing in simulation of coupled thermal, hydraulic and mechanical processes in transverse isotropic rock. Rock Characterisation, Modelling and Engineering Design Methods, 485–490.

Zhang, C., Rothfuchs, T., Jockwer, N., Wieczorek, K., Dittrich, J., Müller, J., Hartwig, L. and Komischke, M. (2007). Thermal effects on the Opalinus clay. A joint heating experiment of ANDRA and GRS at the Mont Terri URL (HE-D Project). Final report. Gesellschaft fuer Anlagen-und Reaktorsicherheit mbH (GRS).

The effect of rock mass stiffness on crush pillar behaviour

M. du Plessis
Lonmin Marikana, Marikana, North West, South Africa

D.F. Malan
Department of Mining Engineering, University of Pretoria, South Africa

ABSTRACT: Various parameters affect the behaviour of crush pillars in intermediate depth platinum mines. The final crush pillar dimension and overall mining layout is influenced by the original design methodology, mining discipline and the effect of unexpected geological losses. The combination of these factors can prevent the pillars from crushing and achieving the desired post-peak residual state. This may result in unpredictable pillar behaviour and damaging seismicity.

Effective crush pillar design will require the pillars being crushed while being formed at the mining face. A stability analysis has been recommended in the past as a method to design yield pillars at moderate mining depths. The methodology assumes that stable pillar crushing will occur if the local rock mass stiffness is greater than the post-failure stiffness of the crush pillar at the specific pillar location. This paper explores this design methodology and some preliminary numerical results are discussed.

Keywords: crush pillar behaviour, stiffness, numerical modeling

1   INTRODUCTION

Research conducted on the behaviour of crush pillars on the Merensky Reef highlighted the detrimental effect of oversized pillars (Du Plessis and Malan 2014). These intact pillars can result in violent failure with the associated seismic energy release in the back area of stopes. Du Plessis and Malan (2016) determined that the amount of convergence experienced in a crush pillar stope can be directly related to the pillar deformation processes. In a subsequent case study, Du Plessis and Malan (2017) described the effect of an intact oversized crush pillar has on the amount of convergence experienced in the stope. The pillar triggered a magnitude 1.9 seismic event which was followed by a substantial increase in convergence.

The objective of this paper is to explore a "stiffness model" to gain a better understanding of the interaction between the crush pillars and the surrounding rock mass. In future, this may provide an improved design methodology to reduce the number of cases of unstable pillar failure.

2   STIFFNESS MODELS

Cook (1965) proposed that rock bursts is a problem related to regional stability in mines and subsequently he discussed the significance of the post-peak behaviour of rock in compression (Cook, 1967). Salamon (1970) noted that the stability of a laboratory rock specimen under compression depends on the stiffness of the testing machine and the slope of the post failure behaviour of the specimen. If the stiffness of the testing machine (slope indicating soft loading system in Figure 1) is less than the slope of the post-peak load-deformation

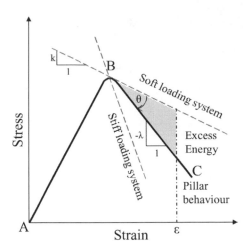

Figure 1. The effect of stable (stiff) and unstable (soft) system loading on pillar behaviour. Regions AB and BC represents the pre- and post-peak pillar strength respectively. If the angle θ increases, the magnitude of the eventual unstable pillar failure also increases.

relation of the specimen (pillar), the test will result in violent failure. Stable deformation in the post-failure region will, however, occur if the stiffness of the machine (slope indicating stiff loading system in Figure 1) is greater than the slope of the post-peak load-deformation relationship of the specimen.

Stiffness is defined as the ability of an object to resist deformation in response to an applied force (N/m). In an underground mining scenario, stoping operations will induce elastic deformation of the hangingwall strata. Pillars should ideally be designed to resist this deformation (loading) in a controlled and predictable manner. In the South African intermediate depth platinum mines significant convergence in the order of centimeters has been recorded in some areas (Malan *et al.*, 2007). The stiffness of the loading system (the surrounding rock mass) and the slope of the failed crush pillars are therefore of particular interest especially as the crush pillars, in some instances, contribute to unstable lading.

By applying the relationship between uniaxial rock testing and loading of an isolated pillar, Salamon (1970) showed that the equilibrium between a pillar being loaded and the post-peak behaviour is stable irrespective of the convergence experienced by the pillar if:

$$(k + \lambda) > 0 \tag{1}$$

The parameter k represents the stiffness of the loading strata and is understood as the force required to cause a unit increment in closure between the hangingwall and footwall at the pillar position. λ defines the post-peak pillar stiffness. The parameter k is positive by definition and λ is positive in the portion of the load-deformation curve where the pillar is intact. A negative λ in the post-failure portion may result in unstable failure, but only if the relationship in equation (1) is violated.

Ozbay and Roberts (1988) stated that if pillars were designed to be fractured during cutting by the face abutment stress, the pillars would already have yielded and reached a residual strength. Further deformation of the pillars would be associated with an increase in load (λ becomes positive), and according to equation (1), stability will be assured. This is, however, dependent on the assumption that the failed pillar material can be compacted so that the pillar would regenerate load. In contrast, if the pillars are intact when cut at the mining face and only fail later, λ will become negative once the pillar reaches peak strength and instability may occur (k + λ < 0).

The slope of the post-peak load deformation relationship levels off with increased pillar w:h ratios (λ tends to zero and may become positive). Ryder and Ozbay (1990) suggested that the post failure modulus of pillars with a w:h ratio >5 would be zero or positive and therefore such pillars would become "unconditionally" stable.

Pillar stiffness (λ) can be expressed by (Ryder and Ozbay, 1990):

$$\lambda = \frac{Force}{Displacement} = \frac{lwE_p}{S_m} \qquad (2)$$

where l is the pillar length, w is the pillar width, E_p is the post-failure modulus and S_m is the stoping width or pillar height.

The stiffness of the strata is proportional to Young's modulus. It is also strongly influenced by the mining geometry and the number of neighbouring pillars which may also be in a state of post-failure. The stiffness of the strata at an individual pillar location is termed the local stiffness (Ryder and Ozbay, 1990). The local strata stiffness can be determined through the application of a numerical model. In the model, the pillar is replaced with a variable probing force F. The change in closure ΔS resulting from a change in force ΔF gives the required local strata stiffness (k_L). When all the pillars in a layout are assumed to be in a failed state, the local stiffness will also define the critical stiffness λ_c (minimum value of k_L)

$$k_L = \Delta F / \Delta S \qquad (3)$$

The post-peak stiffness of pillars has predominantly been investigated through pillar-model tests conducted in laboratories or *in-situ* tests carried out on coal pillars. Based on estimates from this data, Ozbay (1989) assumed the relationship between normalized post-peak stiffness (λ/E) and width to height ratios of pillars which are in the range of $0.7 <$ w:h <5 to be:

$$\lambda/E = 0.16 \cdot (w/h) - 0.8, \text{with} \qquad (4)$$

where E is the initial elastic modulus, w is the width of the pillar and h is the height of the pillar. Ryder and Ozbay (1990) indicated the ratio of E_p/E to be highly associated with the pillar w:h ratio.

3 CRUSH PILLAR BEHAVIOUR

Crush pillars are typically cut at a w:h ratio of approximately 2:1 and a length to width (l:w) ratio of up to 4:1. Ideally, the pillar dimensions are selected such that the pillars are fractured while being formed at the mining face. Once crushed, the residual strength of the pillars fulfill a local support function in supporting to the height of the upper-most parting to prevent the occurrence of back-breaks (large scale instabilities).

Du Plessis and Malan (2014) demonstrated the effect of oversized crush pillars in the back area of a stope. The findings indicated that if an oversized pillar did not crush at the face, such a pillar will become highly stressed as it moves into the back area of the stope. The pillar may therefore either not crush or may fail violently. Of particular interest is to investigate the local stiffness of the loading strata close to the face and in the back area as a reduced stiffness k_L in the back area may contribute to violent pillar failure.

Equation (4) presents a possible contradiction. If it is accepted that this empirical relationship also holds for the crush pillar behaviour in platinum stopes, it implies that an oversized pillar with a larger w:h ratio will have smaller negative values of pillar stiffness λ. This may therefore contribute to increased stability according to equation (1). From observations it is known, however, that some oversized pillars that do not crush in the face may be prone to bursting. The reduction in the local strata stiffness k_L in the back area of stopes and the actual post failure pillar stiffness λ therefore needs to be investigated in more detail. As a first step, some modelling to investigate the effect of local strata stiffness as a function of distance to face was conducted. The results are illustrated below.

4 NUMERICAL MODELLING

For the numerical modelling study, the stope geometry described in Du Plessis *et al.* (2011) was used. For simplicity, the same elastic constants, mining depth and stress gradient were also used. The reader is referred to Figure 15 on page 882 of this reference for the particular geometry. The layout consists of two adjacent 30 m wide panels with a row of 4 m × 6 m crush pillars situated between the two panels. The mining height selected was 2 m resulting in the pillars having a w:h ratio of 2. The method of "probing" as described above was used to determine the local strata stiffness. Figure 2 illustrates the geometry simulated. The four probing positions, A, B, C, and D are shown with the sizes of the probes being similar to the sizes of the pillars at the corresponding positions.

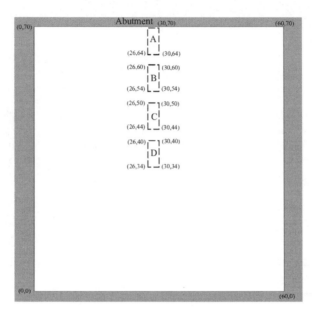

Figure 2.   Geometry simulated with the four probe positions indicated as A, B, C and D.

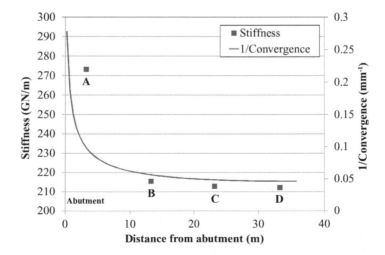

Figure 3.   Values of stiffness calculated at the four probe points. Note the high value of strata stiffness close to the abutment. The inverse of the convergence is plotted as the blue line.

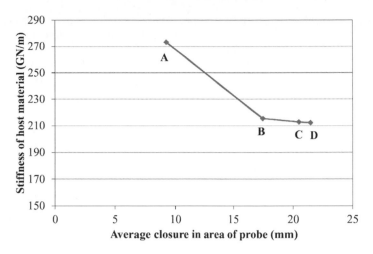

Figure 4. Values of stiffness as a function of closure in the area of the probe. There is not a linear trend between these parameters. There does, however, appear to be a change in the rate of convergence experienced with increased distance from the abutment.

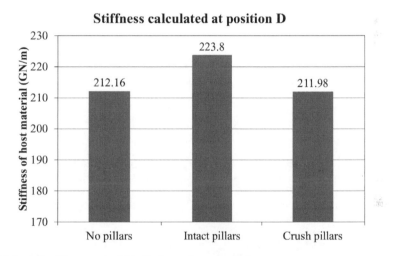

Figure 5. Values of stiffness at position D for various simulations.

Figures 3 and 4 considers the position of each probe (A, B, C and D) relative to its position from the abutment at y = 70. Figure 3 illustrates the decrease in stiffness as the distance to the abutment increases. This confirms the hypothesis that the stiffness is large close to the face and decreases into the back area. The inverse of the convergence simulated from the abutment into the back area of the stope (towards probe D position) is also plotted in Figure 3. In a crush pillar environment the majority of the convergence occurs close to the mining face and then levels off in the back area (Du Plessis and Malan, 2016). Although it seems tempting to use low convergence experienced close to the face compared to the cumulative effect experienced in the back area of a stope as indicators of stiffness, there is not a linear relationship between the two parameters as shown in Figure 4. It is therefore important to use the probe method in irregular geometries with different pillar sizes.

Two further simulations were conducted by including the entire row of pillars (A – G) from the original model and using the probe method to determine the stiffness at position D. If the pillars are simulated as intact, the stiffness is larger at a value of 223.8 GN/m compared 212.16 GN/m if no pillars are used. If crush pillars are used with a residual strength of approximately 50 MPa, the stiffness is approximately similar to the case where no pillars are used. This demonstrates that the function of crush pillars is not to control the regional rock mass, but rather act as a local support to the overlying potential unstable hangingwall strata. The result also signifies the impact of uncrushed pillars (intact) on the local strata stiffness.

5 CONCLUSIONS

The use of crush pillars in platinum mines can be problematic as pillar seismicity may be encountered for certain layouts and geometries. This paper assesses a stiffness criterion as a possible method to improve the design of these pillar layouts. The preliminary modelling confirmed the hypothesis that the local strata stiffness is high at a position close to the face. According to the stability criterion, it is preferable that the pillars crush close to the face. The methodology of using so-called yield pillars which only fail in the back area of the stopes is therefore questioned. From an instrumented crush pillar site not discussed in this paper, it appears that the transition from a stiff to a soft loading environment occurs at approximately 10 m behind the mining face. Further research is required to gain a better understanding of the local strata stiffness and the post-failure stiffness of these crush pillars.

ACKNOWLEDGEMENTS

The contribution and assistance of Prof. John Napier with regards to the development of the limit equilibrium model as well as the TEXAN code is greatly appreciated.

REFERENCES

Cook, N.G.W. (1965). A note on rockburst consideration as a problem of stability. *J. South. Afr. Inst. Min. Metall.*, vol. 65, pp. 437–446.
Cook, N.G.W. (1967). Contribution to discussion on pillar stability. *J. South. Afr. Inst. Min. Metall.*
Du Plessis, M., Malan, D.F. and Napier, J.A.L., (2011). Evaluation of a limit equilibrium model to simulate crush pillar behaviour. *J. South. Afr. Inst. Min. Metall.*, vol. 111, pp. 875–885.
Du Plessis, M. and Malan, D.F. (2014). Evaluation of pillar width on crush pillar behaviour using a limit equilibrium solution. *Proc. EuRock 2014*, Vigo, Spain.
Du Plessis, M. and Malan, D.F. (2016). The behaviour of Merensky crush pillars as measured at a trial mining site. *Proc. EuRock 2016*, Turkey.
Du Plessis, M. and Malan, D.F. (2017). Mining with crush pillars. *Proc. AfriRock 2017*, Cape Town, South Africa.
Malan, D.F. Napier, J.A.L. and Janse van Rensburg, A.L. (2007). Stope deformation measurements as a diagnostic measure of rock behaviour: A decade of research, *J. South. Afr. Inst. Min. Metall.*, vol. 107, pp. 743–765, 2007.
Ozbay, M.U. (1989). The stability and design of yield pillars located at shallow and moderate depths. *J. South. Afr. Inst. Min. Metall.*, 89 (3), pp. 73–79.
Ozbay, M.U. and Roberts, M.K.C. (1988). Yield pillars in stope support. *Proceedings of the SANGORM Symposium in Africa*, Swaziland. pp. 317–326.
Ryder, J.A. and Ozbay, M.U. (1990). A methodology for designing pillar layouts for shallow mining. *Proceedings of the International Symposium on Static and Dynamic Considerations in Rock Engineering*, Swaziland. ISRM. pp. 273–286.
Salamon, M.D.G (1970). Stability, instability and design of pillar workings. *International Journal of Rock Mechanics and Mining Sciences*, 7, pp. 613–631.

Geomechanics and Geodynamics of Rock Masses – Litvinenko (Ed.)
© *2018 Taylor & Francis Group, London, ISBN 978-1-138-61645-5*

Discrete element modelling of a soil-mesh interaction problem

A. Pol & F. Gabrieli
ICEA Department, University of Padova, Padova (PD), Italy

K. Thoeni
Centre for Geotechnical Science and Engineering, University of Newcastle, Callaghan, NSW, Australia

N. Mazzon
Maccaferri Innovation Center, M.I.C., Bolzano, Bozen (BZ), Italy

ABSTRACT: The design strategies of metallic cortical meshes used for slope protection are mostly linked to the application of empirical and semi-empirical methods. This limitation is due to the difficulties in representing such complex large-deformation rock-soil-structure interaction problems: the intrinsic complexities of these structures is combined with the difficulties in describing the geometry both the mechanical and physical properties of the rock and soil behind. The Discrete Element Method proved their effectiveness in describing the non-linear and large strain behaviour of the soil, and recently also the mechanical response of steel wire meshes which are represented as regular patterns of remote interactions between nodes. In this work, a discrete element model of a double-twisted hexagonal wire mesh is calibrated on the base of experimental tensile tests on two wire types. The validation of the model is performed using the results of a standard punch test. The same mesh is then loaded by a granular layer constituted by discrete element particles mimicking the soil earth pressure. A comparison between the two load types is discussed with reference to the force-displacement curve and the distribution of the tensile stresses on the mesh panel. The effect of the boundary conditions is also analysed.

Keywords: Discrete Element Method, soil-structure interaction, wire meshes, slope protection, hazard mitigation, punch tests, double-twisted hexagonal wire mesh

1   INTRODUCTION

Cortical meshes are common rockfall mitigation structures used for the protection of buildings, infrastructures and people in the proximity of hazardous mountains and mining areas. Their correct design is of great importance not only to maximize the effectiveness of the interventions but also for material and cost optimizations. However, up to now, the geometrical and mechanical complexities of these structures mostly confine their design to empirical and semi-empirical methods. Among all the possible numerical models the Discrete Element Method (DEM) (Cundall and Strack, 1979), which is extensively used to model granular materials, recently proved its effectiveness in representing the mechanical behaviour of different wire mesh types for problems involving complex failure mechanisms, large deformations, quasi-static and dynamic laboratory conditions (see, e.g., Nicot et al., 2001; Thoeni et al., 2014; Breugnot et al., 2016; Gabrieli et al., 2017).

2   DISCRETE ELEMENT MODELLING OF THE MESH

In this work, the mechanical behaviour of the double-twisted hexagonal steel wire mesh with wire diameter equal to 2.7 mm has been considered in the modelling process. The sample is

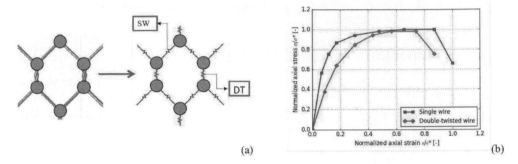

(a)   (b)

Figure 1. (a) Sketch of the discrete element model used for the double-twisted hexagonal wire mesh; (b) tensile stress-strain curves applied to the wire remote interactions, normalised on maximum stress and strain values of the single wire.

numerically schematized by means of spherical particles located at the nodes of the mesh (see Figure 1a). Each node is connected to the surrounding nodes by fictitious remote interactions which have no physical mass and volume, they only represent the specific "constitutive tensile model" of the corresponding wire. The size and the density of the nodal particles are selected to concentrate the mass of the wires and then to preserve the inertial properties of the mesh panel. Two types of stress-strain curves are considered to describe the mechanical behaviour of both single and double-twisted wires respectively. The two laws are depicted in Figure 1b. These tensile curves were directly derived from experimental tensile tests on single and double-twisted wires (Thoeni et al. 2013).

It should be noted that, due to the real production process, the single wire interaction law results moderately stiffer and with a higher maximum elongation value than that of the double-twisted wire.

3 PUNCH TESTS

The punch test is a common laboratory test used to compare the out-of-plane stress-strain performance of cortical meshes and it is standardized in UNI 11437-2012.

A square 3×3 m^2 mesh panel is unrolled horizontally and fixed at the edges of a larger steel frame. The shape of the punching element is standardized: it is a concrete dome with a diameter of 1 m, a curvature radius of 1.2 m and smoothed edges which have a curvature radius of 0.05 m (see Figure 2a). The punching element is vertically lifted with a constant velocity that should not be greater than 10 mm/s, while displacement and force are continuously monitored during the test.

The mechanical response of the mesh, obtained performing an experimental punch test, is represented by a single force-displacement curve (Figure 2b). The same test has been simulated, with a punching element body enveloped with a triangular mesh using the conditions used in the real test, i.e., constant velocity of the punching element and fixed nodes at the edges of the mesh panel (model 1). The good agreement, in terms of force-displacement curves, between numerical and experimental results (see Figure 2b) confirms the effectiveness of the adopted numerical approach to simulate this kind of structures in quasi-static loading conditions as reported with more details in Pol et al. (2017).

The tensile strain distribution before failure is depicted in Figure 3a in terms of normalized strain ε^* (ratio of the current strain value εi to the maximum strain value εmax supported by the wire).

In the first part of the test, roughly up to a value of $F/F^* = 0.2$, the mesh panel experiences a geometrical distortion of the wire hexagons without a relevant force level. Then, after this initial deformation, the contribution of the wire mechanical behaviour becomes higher and higher up to the yielding of the single wires above the punching element (see Figure 3a). Such a

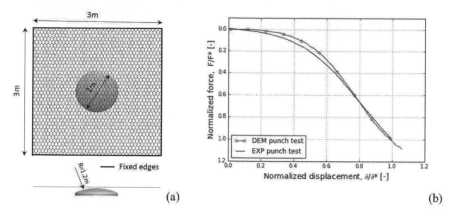

(a)   (b)

Figure 2.   (a) Layout of the punch test; (b) Mechanical response of the hexagonal double-twisted wire mesh panel for the laboratory and the numerical punch tests.

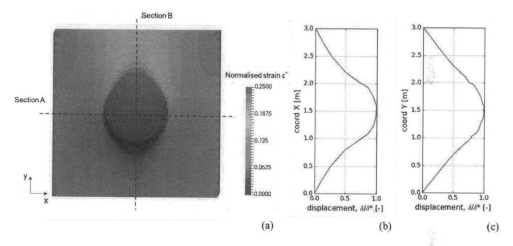

(a)   (b)   (c)

Figure 3.   (a) Interpolated normalized tensile strain distribution ($\varepsilon^* = \varepsilon i/\varepsilon max$) on the mesh panel just before failure and displacement profiles in the centre of the mesh panel, normalized on maximum experimental value of punch test, for the punch test problem: (b) section A and (c) section B.

behaviour, i.e. the initial deformation before the reaction of the mesh panel, is typical of passive structures when subjected to a puncture load (as rock blocks or fragments), whereas it tends to disappear for uniform loading conditions (as in the following soil-mesh problem). The final profiles of the panel along the two centred and orthogonal xz and yz section planes are shown in Figure 3b and Figure 3c respectively. The differences in tensile strain distribution and profiles along the two sections highlights the complex anisotropic behaviour of the mesh panel.

4   THE SOIL-MESH RETAINING MODEL

After the validation of the numerical approach with the punch test configuration, the modelling of the soil-mesh interaction problem can be tackled (model 2). The soil is represented by spheres with a mean diameter $d50 = 12$ cm. Considering a value larger than the mesh opening size of the panel is useful to prevent the possible loss of particles through the mesh during the simulation. Approximatively 84000spheres were randomly generated in a prismatic volume and settled under gravity, filling a final column box of size $3 \times 2 \times 20$ m^3 (see Figure 4a). The parameters of the spheres representing the soil are: solid density $\rho = 2900$ kg/m^3, interparticle friction coefficient $\mu = 0.61$, Young modulus $E = 1e8$ Pa, Poisson's ratio $\nu = 0.30$.

At the bottom of the box, on the xy-plane, the same 3×3 m^2 mesh panel used in the punch test simulation is attached and fixed at its edges as shown in Figure 4a. Then, in a second phase, the soil is let free to push on the testing mesh panel until failure.

During this second phase, the force and displacement of the mesh panel is monitored as well as the mechanical conditions of each wire. In such a way, it is possible to derive the force-displacement curve of the mesh panel (see Figure 4b) and the strain-stress path of each single element of the panel.

A comparison between the mechanical response obtained with the punch-test (laboratory conditions) and the self-weight soil-mesh retaining problem (approximation of the in-situ conditions) is also presented in Figure 4b. The force exerted by soil, in this case, is computed as the integral of tensile z-forces at the boundaries while the displacement is the maximum z-displacement of the mesh. The response of the mesh in the soil-mesh problem appears stiffer and presents a higher maximum force than in experimental and numerical punch tests. Such a behaviour can be attributed to the different loading distributions of the two analysed cases.

In the soil-mesh problem, the tensile loads are almost uniformly distributed on the mesh panel. However, important stress concentration and strain localization can be observed at the bottom corners (see Figure 5a), whereas in the punch test the tensile loads are mainly distributed above the punching element. Distributing the load on a larger area, i.e. on a larger number of wires, leads to an increment of the maximum force sustainable by the mesh panel. This is demonstrated in Figure 4b by the force-displacement curve obtained from a numerical punch test performed with a punching element 2.5 times larger than the standard one (model 3). Also in this case the strain is mostly distributed on the surface of the punching element but the number of collaborating wires increases of 2.5 times similarly to the maximum force experienced by the punching element. The higher and earlier stiffness obtained with the soil-mesh test (see Figure 4b) is ascribable to the proximity of the loading area to the fixed edges of the mesh panel. This second effect can be simulated also with a 2.5 m side square punching plate (model 4) as highlighted in Figure 4b. The maximum force reached by model 4 is approximately the same of model 3. Nevertheless, in model 4 only wires along the perimeter of the punching plate, and especially at its the corners, experience large strains and stresses (Figure 7a).

Finally, the most elongated wires are located at the bottom of the panel on model 2. This is due to the gravity field stress distribution in the soil-mesh problem. This also reflects in the lack of symmetry along the y-axis of symmetry of the panel (compare Figure 3b and Figure 5b).

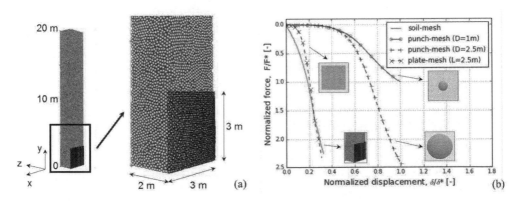

Figure 4. (a) Layout of the soil-mesh problem; (b) comparison of the mechanical response of the mesh for different sources of load.

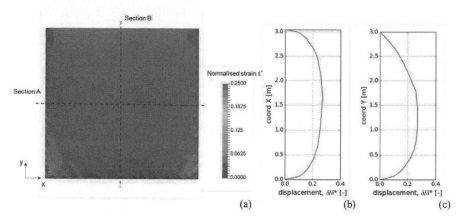

(a)   (b)   (c)

Figure 5.   (a) Interpolated normalized tensile strain distribution ($\varepsilon^* = \varepsilon i/\varepsilon max$) on the mesh panel just before failure and displacement profiles in the centre of the mesh panel, normalized on maximum experimental value of punch test, for the soil-mesh problem: (b) section A and (c) section B.

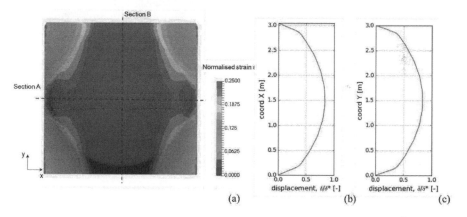

(a)   (b)   (c)

Figure 6.   (a) Interpolated normalized tensile strain distribution ($\varepsilon^* = \varepsilon i/\varepsilon max$) on the mesh panel just before failure and displacement profiles in the centre of the mesh panel, normalized on maximum experimental value of punch test, for the punch test with the 2.5 bigger punching element: (b) section A and (c) section B.

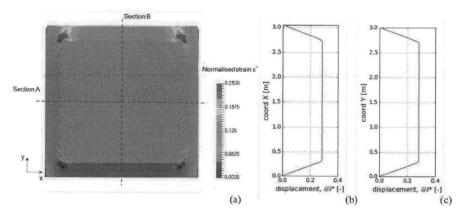

(a)   (b)   (c)

Figure 7.   (a) Interpolated normalized tensile strain distribution ($\varepsilon^* = \varepsilon i/\varepsilon max$) on the mesh panel just before failure and displacement profiles in the centre of the mesh panel, normalized on maximum experimental value of punch test, for the punch test with the large square plate: (b) section A and (c) section B.

5   CONCLUSIONS

The Discrete Element Method was proved to be a good candidate for the comprehension of the mechanical behaviour of in-situ cortical meshes dealing with large deformations and complex interaction problems. The capability of DEM to simulate the overall behaviour of a sample up to the failure has been summarized. Additionally, in this work the laboratory conditions of punch tests have been successfully extended to the problem of a simplified soil-mesh interaction. A difference in terms of overall load-displacement behaviour can be observed between the results related to the standard punch test (model 1) and those of a real loading condition (model 2). This could be attributed to the different mechanisms that rule the transmission of stresses from the external bodies to the mesh panel. When analysing the tensile stress or strain distributions it is evident that in the punch test conditions the load is concentrated in a central zone, while in the second case the soil earth pressure distributes the tensile stresses in a more homogeneous manner on a larger area of the mesh panel. This overall difference is also clearly highlighted by the different deformed shape of models 1 to 3. The application of the load in a more spread configuration (model 2) allows the overall sustained force to be increased more than two times. Furthermore, the analysis of the soil-mesh interaction highlights a stiffer overall behaviour, resulting in a deformation at failure almost three times lower than that resulting from punch simulations 2 and 3.

Finally, as a future development of this investigation, an in-depth study of the influence of the boundary conditions, soil types and parameters is required to improve and optimize the numerical predictions. Extending these simulations may also help the development and the design of new meshes as well as to find the solutions for an increase of their performances.

REFERENCES

Bertrand, D., F. Nicot, P. Gotteland and S. Lambert, "Modelling a geo-composite cell using discrete analysis", *Computers and Geotechnics* 32(8), 564–577 (2005).

Breugnot, A., S. Lambert, P. Villard, P. Gotteland, "A Discrete/continuous Coupled Approach for Modeling Impacts on Cellular Geostructures", *Rock Mechanics and Rock Engineering* 49(5), 1831–1848 (2016).

Cundall, P. and O.D.L. Strack, "A Discrete numerical model for granular assemblies", *Geotechnique*, 29(1), 47–65, (1979).

Gabrieli, F., A. Pol and K. Thoeni, "Comparison of two DEM strategies for modelling cortical meshes", *V International Conference on Particle-based Methods—Fundamentals and Applications, Particles 2017,* 489–496, (2017).

Nicot, F., B. Cambou, and G. Mazzoleni, "Design of Rockfall Restraining Nets from a Discrete Element Modelling" *Rock Mechanics and Rock Engineering* 34 (2), 99–118, (2001).

Pol, A., F. Gabrieli, K. Thoeni and N. Mazzon, "Discrete element modelling of punch tests with a double-twist hexagonal wire mesh", *J. Corominas, J. Moya, M. Janeras, editors, RocExs 2017, 6th Interdisciplinary Workshop on Rockfall Protection*, 145–148, (2017).

Thoeni, K., A. Giacomini, C. Lambert, S.W. Sloan and J.P. Carter, "A 3D discrete element modelling approach for rockfall analysis with drapery systems", *International Journal of Rock Mechanics and Mining Sciences* 68, 107–119, (2014).

Thoeni, K., C. Lambert, A. Giacomini and S.W. Sloan, "Discrete modelling of hexagonal wire meshes with a stochastically distorted contact model", *Computers and Geotechnics* 49, 158–169, (2013).

Geomechanics and Geodynamics of Rock Masses – Litvinenko (Ed.)
© 2018 Taylor & Francis Group, London, ISBN 978-1-138-61645-5

Numerical modelling of fracture processes in thermal shock weakened rock

Martina Pressacco & Timo Saksala

Laboratory of Civil Engineering, Tampere University of Technology, Tampere, Finland

ABSTRACT: This paper presents some preliminary results of a research project aiming at the simulation of thermal shock assisted percussive drilling. In the present study, a numerical model for transient thermal shock induced damage in rock is presented. This model includes a rock mesostructure description accounting for different mineral properties and a thermo-mechanical constitutive model based on embedded discontinuity finite elements. In the numerical simulations, the thermal shock induced damage process is first simulated. Then the uniaxial compression test on thermally affected numerical rock samples is carried out. The effect of thermal shock is demonstrated by comparison to uniaxial compression test simulation on intact rock. The results show that the thermal-shock assisted rock breakage is a feasible idea to be extended to percussive drilling as well.

1   INTRODUCTION

Weakening the rock by thermal shock is a promising method to facilitate mechanical breakage in traditional drilling or rock crushing. This method consists of damaging the rock first by application of high amplitude heat shock and then applying the mechanical loading. During the last 15 years, some encouraging results have been obtained in numerical modelling of microwave-assisted breakage (Whittles et al., 2003), thermal spallation (Walsh et al., 2013), and plasma torch shocked granite (Mardoukhi et al., 2017).

In this paper, some preliminary results on numerical modelling of heat shocked granite under mechanical loading are presented. For this end, we present a combined thermo-mechanical fracture model to simulate heat-shock induced damage and consequent mechanical breakage. More specifically, the constitutive model for rock is based on the embedded discontinuity finite elements (Saksala et al., 2015), where thermal coupling is taken into account through thermal strains. An explicit time-integration algorithm is presented for solving the global thermo-mechanical problem. The performance of this method is demonstrated in the numerical examples by simulating uniaxial compression tests on thermally affected numerical rock samples.

2   FINITE ELEMENT FORMULATION FOR THE UNCOUPLED THERMO-MECHANICAL PROBLEM

The aim of this analysis is to model the behavior of a rock sample exposed first to a thermal shock and then subjected to mechanical loading. If the external heat source (which in the present case is a flux **q** applied at the boundaries of the rock specimen) is significantly larger than the heat generation due to the mechanical response, the coupling term in the solution for the heat equation can be neglected. Therefore, for our purposes, the finite element formulation for the heat balance equation becomes

$$\mathbf{C}\dot{\theta} + \mathbf{K}_\theta \theta - \mathbf{f}_\theta = \mathbf{0} \qquad (1)$$

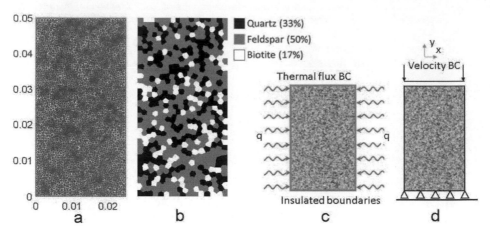

Figure 1. FE mesh and centroidal Voronoi tessellation for the numerical rock sample a-b), boundary conditions for the heat shock test c) and the compression test d).

where $\mathbf{C}$, $\mathbf{K}_\theta$ and $\mathbf{f}_\theta$ are the capacity matrix, the conductivity matrix and the external force, respectively, defined as follows:

$$\mathbf{C} = \int_V \rho c \mathbf{N}_\theta^T \mathbf{N}_\theta dV, \qquad \mathbf{K}_\theta = \int_V k \mathbf{B}_\theta^T \mathbf{B}_\theta dV, \qquad \mathbf{f}_\theta = -\int_S q n \mathbf{N}_\theta^T ds \qquad (2)$$

where ρ is the density, c is the specific heat capacity, θ is the temperature, $\mathbf{N}_\theta$ is the temperature interpolation matrix, k is the conductivity, q_n is the normal component of the external heat flux and finally $\mathbf{B}_\theta$ is the gradient of the temperature interpolation matrix.

In general, both the heat capacity matrix and the conductivity matrix depend on temperature. However, in this preliminary analysis the dependance of the material properties on temperature is neglected.

The mechanical problem is governed by the finite element discretized equation of motion

$$\mathbf{M\ddot{u}} + \mathbf{f}_{int} = \mathbf{f}_{ext}, \qquad \mathbf{f}_{int} = \int_V \mathbf{B}_u^T \boldsymbol{\sigma} dV \qquad (3)$$

where $\mathbf{M}$ is the lumped mass matrix, $\mathbf{f}_{ext}$ is the external force vector, $\mathbf{B}_u$ is the kinematic matrix and $\boldsymbol{\sigma}$ is the stress vector.

3 ROCK MESOSTRUCTURE DESCRIPTION

The rock mesostructure is represented as a random mineral grain distribution based on centroidal Voronoi tessellation. The tool adopted for this purpose is the `PolyMesher` Matlab code by (Talischi et al., 2012), which generates centroidal Voronoi tessellations, whereas the FE mesh consists of constant strain three-node triangles. The rock sample is composed of three main minerals, (i.e. quartz, feldspar and biotite), each of them having different mechanical and thermal properties. The FE mesh and the random mineral texture are shown in Figure 1a) and b). It should be reminded that the Voronoi cells (mineral grains) are meshed with the three-node elements.

4 ROCK CONSTITUTIVE MODEL

The model used here adopts the main idea developed in (Saksala et al., 2015), namely the representation of cracks by embedded discontinuity finite elements. According to the adaptation here, the material response is linear elastic until the tensile strength is reached. The fracture criterion adopted here is the Rankine criterion. A discontinuity (crack) is introduced when the first principal stress in the element exceeds the tensile strength. The normal vector $\mathbf{n}$ of the discontinuity is parallel to the first principal stress direction.

The model components, i.e. the loading functions, the evolution laws for discontinuity jump and the internal variable, and the Kuhn-Tucker conditions, can be summarized as:

$$\phi_t\left(t_{\Gamma_d},\kappa,\dot{\kappa}\right)= \mathbf{n}\cdot t_{\Gamma_d} -\left(\sigma_t + q\left(\kappa,\dot{\kappa}\right)\right), \qquad \phi_s\left(t_{\Gamma_d},\kappa,\dot{\kappa}\right)= \left|\mathbf{m}\cdot t_{\Gamma_d}\right| -\left(\sigma_s + \frac{\sigma_s}{\sigma_t}q\left(\kappa,\dot{\kappa}\right)\right) \quad (4)$$

$$\dot{\boldsymbol{\alpha}}_d = \dot{\boldsymbol{\alpha}}_I + \dot{\alpha}_{II} = \dot{\lambda}_t \frac{\partial \phi_t}{\partial t_{\Gamma_d}} + \dot{\lambda}_s \frac{\partial \phi_s}{\partial t_{\Gamma_d}} \tag{5}$$

$$\dot{t}_{\Gamma_d} =-E:\left(\nabla\phi \otimes \dot{\boldsymbol{\alpha}}_d\right)^{sym}, \quad \dot{\kappa} =-\dot{\lambda}_t \frac{\partial \phi_t}{\partial q} - \dot{\lambda}_s \frac{\partial \phi_s}{\partial q} \tag{6}$$

$$q = h\kappa + s\dot{\kappa}, \qquad h =-g\sigma_t \exp\left(-g\kappa\right) \tag{7}$$

$$\dot{\lambda}_i \geq 0, \quad \phi_i \leq 0, \quad \dot{\lambda}_i\phi_i = 0, \; i = t,s \tag{8}$$

where $\mathbf{n}$ and $\mathbf{m}$ are respectively the unit normal and tangent vectors for the crack, σ_t and σ_s are the elastic limits for the stress in tension and shear, and $\kappa,\dot{\kappa}$ are the internal variable and its rate which relate to the softening law for the discontinuity. The softening slope parameter g is defined with the mode I fracture energy G_{Ic} as $g = \sigma_t/G_{Ic}$, and s is the viscosity modulus. The total displacement jump $\boldsymbol{\alpha}_d$ is decomposed in two parts, $\boldsymbol{\alpha}_I$ and $\boldsymbol{\alpha}_{II}$, correspondent to mode I and II increments, while $\dot{\lambda}_t$ and $\dot{\lambda}_s$ represent the crack opening and sliding increments. The stress tensor assumes the following expression

$$\boldsymbol{\sigma}= E:\left(\boldsymbol{\varepsilon}_{tot} -\left(\nabla\phi \otimes \boldsymbol{\alpha}_d\right)^{sym} - \boldsymbol{\varepsilon}_\theta\right) \tag{9}$$

with E being the elasticity tensor and

$$\nabla\phi = \arg\left(\max_{k=1,2} \frac{\left|\sum_{i=1}^{k}\nabla N_i \cdot \mathbf{n}\right|}{\left\|\sum_{i=1}^{k}\nabla N_i\right\|}\right), \qquad \boldsymbol{\varepsilon}_\theta = \alpha\Delta\theta I \tag{10}$$

where ϕ is a function that restricts the effect of the displacement jump within the corresponding finite element so the essential boundary conditions remain unaffected. More details can be found in (Saksala et al., 2015). Moreover, $\boldsymbol{\varepsilon}_\theta$ is the thermal strain, with α being the thermal expansion coefficient.

5 TIME INTEGRATION SCHEME

By applying the forward Euler scheme $\dot{\boldsymbol{\theta}}_n = \left(\boldsymbol{\theta}_{n+1} - \boldsymbol{\theta}_n\right)/\Delta t$ to the FE discretized heat equation (1), the following explicit expression is obtained for solving the temperature

$$C_n\boldsymbol{\theta}_{n+1} =\left(C_n - \Delta t K_{\theta,n}\right)\boldsymbol{\theta}_n + \Delta t f_{\theta,n} \tag{11}$$

which is valid also for nonlinear cases where $\mathbf{C}$ and $\mathbf{K}_\theta$ depend on temperature.

Table 1. Solution procedure for the thermo-mechanical problem (computations during each time step n).

1. Solve for the temperature θ_{n+1} from equation (11)
2. By looping over all the elements, solve the model defined by equations (4)-(8) for displacement jump α_d and the internal variables, calculate the new stress σ by equation (9), and assemble the internal force vector (3)
3. Solve for the acceleration $\ddot{u}_n$ from equation (3)
4. Predict the mechanical response by equations (12)

Table 2. Material properties and model parameters used in simulations.

Parameter			Quartz	Feldspar	Biotite
Percentage in the sample		[%]	33	50	17
ρ	(Density)	[kg/m³]	2.65	2.62	3.05
E	(Elastic modulus)	[GPa]	80	60	20
v	(Poisson's ratio)		0.17	0.29	0.20
σ_t	(Tensile strength)	[MPa]	10	8	7
σ_s	(Shear strength)	[MPa]	50	50	50
φ	(Internal friction angle)	[°]	50	50	50
G_{Ic}	(Mode I fracture energy)	[J/m²]	40	40	28
α	(Thermal expansion coefficient)	[1/K]	1.60×10^{-5}	0.75×10^{-5}	1.21×10^{-5}
k	(Thermal conductivity)	[W/mK]	4.94	2.34	3.14
c	(Specific heat capacity)	[J/kgK]	731	730	770

The explicit modified Euler time integration scheme is chosen for the integration of the mechanical part. It is based on solving the acceleration from the equation of motion (3) and then predicting the response by

$$\dot{u}_{n+1} = \dot{u}_n + \Delta t \ddot{u}_n, \quad u_{n+1} = u_n + \Delta t \dot{u}_{n+1} \qquad (12)$$

where $\mathbf{u}$ is the displacement. The solution procedure for the uncoupled thermo-mechanical problem is sketched in Table 2.

6 SIMULATION RESULTS AND DISCUSSION

Here we present some representative simulations of compressive tests performed on heat shocked numerical rock samples (Figure 1a-b) generated with the method illustrated in Section 3. The boundary conditions for the thermal problem, i.e. the heat shock, are defined as a constant flux q at the left and right edges of the numerical sample. The initial temperature in the sample is 20°C. For the consequent uniaxial compression test, the mechanical boundary conditions characterize the sample as simply supported at the bottom, with a constant velocity of 0.1 m/s at the top edge (Figure 1a-b). The material properties are given in Table 2.

The mineral properties are mainly from (Park et al., 2015). It should be noted that the properties do not necessarily represent any single real rock, but still are representative for granitic rocks. Moreover, the temperature dependance is neglected. Finally, the viscosity value is set to 0.005 MPa.s/m.

In Figure 2 results for the heat-shock simulation are shown. While there is a substantial increase in temperature at the boundary nodes, most part of numerical sample remains at the reference temperature due to the very short duration of the heat shock. A significant amount of heat induced cracks can be observed at the boundaries of the specimen. However, the crack openings are quite small (Figure 2b).

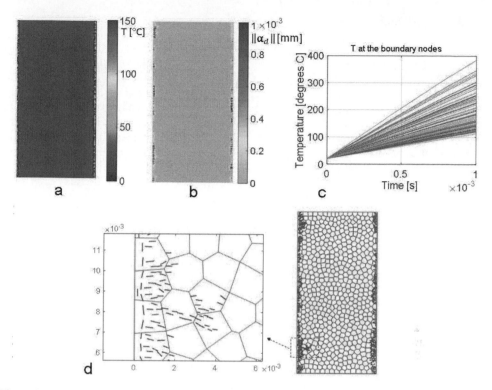

Figure 2. Simulation results for heat shock: temperature distribution a), displacement jump distribution b), temperatures at the boundary nodes c) and orientation of the thermal shock induced cracks in the sample d).

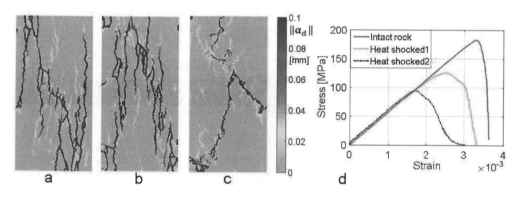

Figure 3. Simulation results for compression tests: final failures mode for intact rock a), heat shocked, cooled down case b), heat shocked and immediate compression c), average stress-strain curves for the three cases d).

Figure 3a-c) presents the failure modes in the uniaxial compression tests for three different cases: intact rock sample a); sample cooled down to room temperature b); sample mechanically loaded immediately after the heat shock c).

The failure modes in cases a) and b) are quite similar, but c) deviates substantially due to the highly disturbed initial stress state. Figure 3d) shows the average stress-strain curves for the corresponding three cases. The resulting compressive strength for case b) and c) show a reduction in strength of, respectively, 35% and 50%, compared to the strength of the intact rock (~180 MPa).

7 CONCLUSIONS

A 2D numerical approach for modelling the heat shock-induced damage in the granitic rock was introduced in this paper. As the model is based on the embedded-discontinuity finite elements, it can account for the microcrack orientations induced by the heat shock modelled as a lateral heat flux applied at the vertical boundaries of the numerical specimen. Despite the fact that the cracks were limited in a very narrow zones close to the boundaries, significant reduction (35% to 50%) in the compressive strength of the intact rock was observed in the samples. Therefore, the short-duration high-intensity heat flux provides an efficient means to enhance mechanical breakage of rock. This method will be applied in numerical modelling of heat shock enhanced percussive drilling. Moreover, the effect of temperature dependance of the mechanical and thermal properties of the rock constituting minerals needs to be addressed in the future studies of the method.

ACKNOWLEDGMENTS

This research was funded by Academy of Finland under grant numbers 298345 and 307105.

REFERENCES

Mardoukhi A., Saksala T., Hokka M. and Kuokkala V.-T., A numerical and experimental study in the tensile behavior of plasma shocked granite under dynamic loading. *Rakenteiden Mekaniikka (Journal of Structural Mechanics)*, vol. 50, pp. 41–62, 2017.

Park J.-W., Park C., Ryu D. and Park E.S., Numerical simulation of thermo-mechanical behavior of rock using a grain-based distinct element model. *In: Schubert & Kluckner (Ed.). Proceedings of EUROCK 2015 & 64th Geomechanics Colloquium.*, Salzburg, Austria, 2015.

Saksala T., Brancherie D., Harari I. and Ibrahimbegovic A., Combined continuum damage-embedded discontinuity model for explicit dynamic fracture analyses of quasi-brittle materials. *International Journal for Numerical Methods in Engineering*, vol. 101, pp. 230–250, 2015.

Talischi C., Paulino G.H. and Menezes I.F.M., PolyMesher: a general-purpose mesh generator for polygonal elements written in Matlab. *Structural and Multidisciplinary Optimization*, vol. 45, pp. 309–328, 2012.

Walsh S.D.C. and Lomov I.N., Micromechanical modeling of thermal spallation in granitic rock. *International Journal of Heat and Mass Transfer*, vol. 65, pp. 366–373, 2013.

Whittles D.N., Kingman S.W. and Reddish D.J., Application of numerical modelling for prediction of the influence of power density on microwave-assisted breakage. *International Journal of Mineral Processing*, vol. 68, pp. 71–91, 2003.

Geomechanics and Geodynamics of Rock Masses – Litvinenko (Ed.)
© 2018 Taylor & Francis Group, London, ISBN 978-1-138-61645-5

Static and dynamic analysis of a rock slope in Sikkim: A case study

Naveen Reddy Kallam & Murali Krishna Adapa
Department of Civil Engineering, Indian Institute of Technology Guwahati, Guwahati, India

ABSTRACT: Discontinuities in the form of joints, bedding planes and faults create anisotropy in the rock mass. Discontinuities are the weakest zones in the rock mass. The slope stability analysis of rock masses has been a challenging task due to the presence of discontinuities in various forms which result in different types of slope failures. Failure mechanism of a rock slope mainly depends on the characteristics of discontinuities. This paper deals with the stability assessment of a rock slope in Sikkim by static analysis and pseudo-static analysis using shear strength reduction technique coupled with finite element method and dynamic analysis using time response analysis in PHASE^2 software. The results obtained from the parametric analysis are presented in terms of the variations of horizontal and vertical displacements along the face of the slope corresponding to the different acceleration—time histories of different earthquakes. The results obtained from both static and pseudo-static analyses confirmed the global stability of the slope as the FOS in both the cases is more than permissible limit and the displacements observed in case of time response analysis are within the permissible limits.

Keywords: Rock Slope; Static Analysis; Pseudo-Static Analysis; Dynamic Analysis; PHASE^2; Sikkim

1 INTRODUCTION

The stability evaluations of natural rock slopes are very essential for any construction activity on the rock slope, especially, when the slopes were located in residential areas. Rock slope fails in one or more combination of failure mechanisms like circular failure, plane failure, buckling failure, toppling failure and wedge failure (Cai and Hori 1992). Slope instability is dictated by the presence of discontinuities in the form of joints, bedding planes and faults create anisotropy in the rock mass. Discontinuities are the weakest zones in the rock mass. Failure mechanism of a rock slope mainly depends on the characteristics of discontinuities. The stability analysis of a rock slope becomes very critical if the slope lies in earthquake prone areas. Landslides and slope failures are the most common natural hazards and mainly caused due to the earthquake induced ground shaking. Even earthquakes of very small magnitude can trigger failures of rock slopes in jointed rock masses which are perfectly stable. Hence the study of the behavior of rock slope in actual dynamic scenarios is to be checked.

Several researchers conducted the dynamic analysis of a jointed rock slope using different techniques. Gupta et al., (2016) conducted the stability analysis of Surabhi landslide in the Uttaranchal located in Mussoorie using the FEM-SSR technique in PHASE^2. Pal Shilpa et al., (2012) conducted the dynamic analysis of Surabhi landslide in the Uttaranchal located in Mussoorie using the distinct element method (DEM) in UDEC. Kainthola et al., (2015) conducted the stability analysis of cut slopes along State Highway-72, using the finite element method in PHASE^2. Liu Yaqun et al., (2014) on the seismic stability analysis of a layered rock slope using pseudo static analysis in UDEC. Tiwari et al., (2014) performed the stability analysis of Himalayan rock slope using continuum interface approach by FEM-SSR technique in PHASE^2. Hatzor et al. (2004) carried out dynamic 2D stability analysis of upper terrace of King Herod's Palace in Masada, which is a highly discontinuous rock slope. Latha and Garaga (2010) performed the seismic slope stability analysis of a 350-m-high slope using the equivalent continuum approach in FLAC. Kanugo et al., (2013) performed the stability analysis of rock slopes using finite element method in PHASE².

This paper presents the results with an emphasis of static and dynamic slope stability of a jointed rock mass using PHASE² (Rocscience 2016) through a case study in the Sikkim, India.

2 DESCRIPTION OF THE SITE

The case study considered for the study is the Theng rock slope which is located along North Sikkim Highway between Chungthang (27.60391° N, 88.64644° E) and Tung (27.54368° N, 88.64821° E). Teesta River flows nearly parallel to the highway in north to south direction. The corridor is major link that connects Chungthang to the rest of Sikkim. The area is found to have been subjected frequent rock falls and rock slides due to which several commuters are stranded especially during rainy seasons. The geology of the site is mainly composed of quartzose and quartzose felspathic gneiss. The rocks present at the site are heavily jointed. Previously, stability analysis on the rock slope based on Kinematic analysis and Slope Mass Rating (SMR) method had been conducted by Ghosh et al. (2014). Generalized Hoek–Brown failure criteria were used to define material characteristics of rock mass and Mohr-Coulomb failure criteria is used to define joints. Joints are introduced with a persistence of 0.5 and uniform spacing of 3 m. The summary of the joints present in the area and properties of the rock mass and joints are adapted from Ghosh et al., (2014) and are given in Table 1.

3 STATIC AND PSEUDO-STATIC SLOPE STABILITY ANALYSIS

Numerical model of a rock slope, located in Sikkim, India, was developed in an elasto-plastic finite element analysis program, PHASE² (Rocscience, 2016) as shown in Fig. 1(a). Joints are introduced in between intact rock as negligible thickness interface element formulated by Goodman *et al.* (1968), which connects two intact rock element. Shear and normal stiffness of the interface element governs the displacement of the jointed system. Generalized Hoek–Brown failure criteria were used to define material characteristics of rock mass and Mohr-Coulomb failure criteria is used to define joints. Finite element based Shear strength reduction (SSR) method has been used for the static and pseudo-static analysis of rock slopes to obtain factor of safety of the rock slope. The ratio of horizontal to vertical stress in the

Table 1. Properties of Joints and Rock mass (Source: Ghosh et al., 2014).

Joints	Dip (°)	Dip direction (°)	Normal Stiffness (GPa/m)	Shear Stiffness (GPa/m)	Joint Cohesion (KPa)	Joint Friction Angle (°)
J1	65	110	38	3.8	148	35
J2	50	55	23	2.3	148	35
J3	30	210	23	2.3	148	35
J4	75	342	23	2.3	148	35

	Elastic Modulus (GPa)	UCS (MPa)	Rock Mass Parameter			Poisson's Ratio	Unit Weight (kN/m³)
			m_b	s	a		
Rock Mass	26.25	70	8.314	0.023	0.502	0.3	27

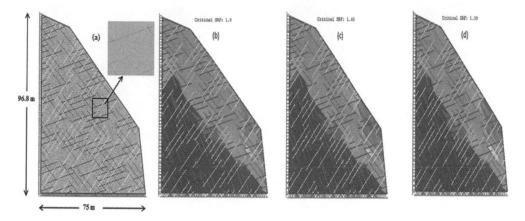

Figure 1. a) Numerical Model of rock slope b) Displacement response in static condition c) Displacement response in Pseudo-static condition (k_H) for Chamba Earthquake d) Displacement response in Pseudo-static condition (k_H & k_V) for Chamba Earthquake.

rock mass is considered to be 1.0 (Pal et al., 2012). In this study the intact rock is discretized using six nodded triangular meshing element with uniform gradient of 1.0 m. Fixed boundary condition is assumed at the base of the model which will restrict displacement in x and y direction at the base and roller boundary condition is assigned at the left side and right side of the slope which will allow movement in the vertical y direction.

In the pseudo-static analysis, the time-history sequence of the earthquake is considered in terms of a constant horizontal and vertical time-independent force generated due to the earthquake. The horizontal acceleration coefficients considered for the present study are the 'a_{max}' values of the three earthquakes Uttarkashi, Chamba and Chamoli i.e., 0.309, 0.146 and 0.359 respectively and the vertical acceleration coefficient is taken as two-thirds of the horizontal acceleration coefficient. The stability analysis of slope is carried out for three different cases, i.e., considering the static condition in the first case and k_H alone in the second case and both k_H and k_V in the third case.

Static slope stability analysis is carried out and FOS without earthquake loading was found to be 1.8 and the displacement (2.9 mm) response of the rock slope is shown in Fig. 1(b). By considering the horizontal acceleration coefficient alone, resulted FOS of 1.2, 1.05 and 1.48 and the displacements observed are 5.11, 7.5 and 4.31 mm for Uttarkashi, Chamoli and Chamba Earthquakes respectively. Similarly, by considering both horizontal and vertical acceleration coefficient, resulted FOS of 1.14, 1.01 and 1.39 and the displacements observed are 6.85, 7.25 and 5.25 mm and displacement responses of the rock slope for Chamba Earthquake are shown in Fig. 1(c) and Fig. 1(d). The results implies when the horizontal acceleration coefficient and vertical acceleration coefficient is more, then the factor of safety of a rock slope reduces. The slope is in marginally stable for Chamoli earthquake as FOS value of 1.0 is acceptable as per the guidelines of National Earthquake Hazards Reduction Program (NEHRP), US, for land sliding hazards.

4 DYNAMIC SLOPE STABILITY ANALYSIS

The pseudo-static approach for stability analysis is simple and straight forward but it cannot simulate the transient dynamic effects of earthquake shaking, because it assumes a constant unidirectional pseudo-static acceleration. Dynamic analysis is carried out on the rock slope by subjecting to base shaking corresponding to the Uttarkashi earthquake, Chamba earthquake, Chamoli earthquake recorded on 20 October 1991, 24 March 1995, 29 March 1999. The dynamic input can be applied by any one of the following ways: an acceleration history, a velocity history, a stress (or pressure) history or a force history.

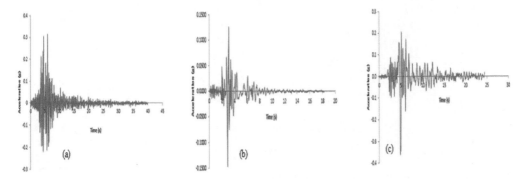

Figure 2. Corrected transverse component of acceleration–time history a) Uttarkashi Earthquake b) Chamba Earthquake c) Chamoli Earthquake.

The dynamic analysis option in PHASE[2] permits two-dimensional plane-strain dynamic analysis. The dynamic input applied is the transverse component of the acceleration—time history of the different earthquakes shown in Fig. 2 and is applied at the base of the slope. The displacements in both x and y directions are fixed at the base and displacements in x direction are fixed on the left side of the slope along y axis and slope is allowed free to move in both directions. Boundary conditions such as absorber and transmitter is applied at the base and left side of the slope. Rayleigh damping of 5% is chosen in the present study.

5 RESULTS AND DISCUSSIONS

The parametric analysis of rock slope is performed in the case of time response analysis to understand the effect of acceleration—time history on the overall stability of the rock slope. The effects of the acceleration—time response of the earthquake on the displacements of slope are observed by conducting the parametric analysis. Three earthquake scenarios namely Uttarkashi earthquake, Chamba earthquake and Chamoli earthquake occurred in India. The values of 'a_{max}' for the three earthquakes Uttarkashi, Chamba and Chamoli are 0.309, 0.146 and 0.359 respectively. The corresponding durations of these three earthquakes are 39.9, 18.3 and 24 s respectively. The slope under consideration is analyzed for all the three earthquake scenarios and the results are presented individually in terms of displacements in the slope. The properties of the rock mass were kept constant for all the cases. Figure 2 shows the acceleration–time history recorded for the earthquakes.

Numerical model of rock slope used for dynamic event and the variation of displacement of rock slope with time along the face of the slope during the earthquake is observed at points A (75, 0), B (70, 35) and C (20, 90) as shown in Fig. 3(a). Figure 3(b) shows the displacement response after the complete dynamic event subjected to Chamoli earthquake and Fig. 3(c) shows the displaced shape after the dynamic event subjected to Chamoli earthquake.

The variation of horizontal and vertical displacements at various points on the slope face from toe to crest, at the end of the dynamic event for the three earthquakes considered is shown in Fig. 4. A maximum horizontal displacement of 32.56 mm, 5.38 mm and 82.34 mm was observed at a distance of 50–60 m along the slope from the toe and similarly the maximum vertical displacement of 6.23 mm, 1.65 mm and 52.34 mm was observed at a distance of 50–60 m along the slope from the toe for Uttarkashi earthquake, Chamba earthquake and Chamoli earthquake and then it is slowly reduced towards the crest. It can be observed from the Fig. 4 that both horizontal and vertical displacements reached maximum at a distance of about 50–60 m along the slope from the toe for all the three earthquakes and stabilized afterwards and Fig. 3(b) depicts that the displacements are more at a distance from the toe of the rock slope. It is determined that a_{max} is not only the parameter that governs the displacements.

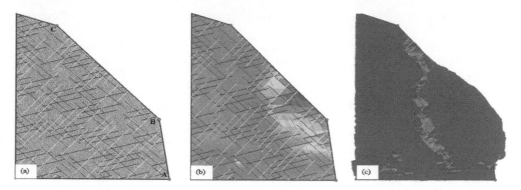

Figure 3. a) Numerical model in dynamic condition b) Displacement response after dynamic event for Chamoli Earthquake c) Displaced shape after dynamic event for Chamoli Earthquake.

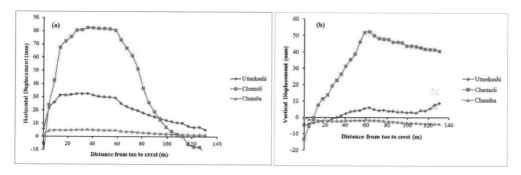

Figure 4. Comparison of Horizontal (a) and Vertical (b) displacements along the face of the slope from toe to crest for all the three earthquakes after dynamic event.

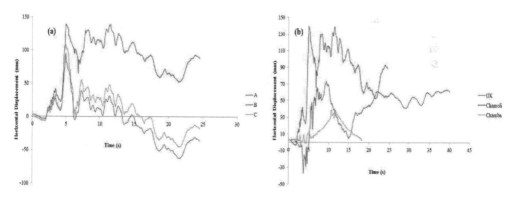

Figure 5. a) Horizontal displacement plots along slope face during the dynamic event b) Variation of horizontal displacements at point B with time for three earthquakes.

Along with a_{max} the frequency content and duration of earthquake also play a vital role in influencing the displacements during the event.

Figure 5(a) shows the variation of horizontal displacements with time at points A, B and C along the face of the slope during the Chamoli earthquake. It was observed that the maximum displacements are occurring at 4.94 s after the initiation of the event (at the point of application of peak amplitude). The horizontal displacements are plotted for all the three earthquake scenarios under consideration. It can be seen from the Fig. 5(b) that the Chamoli

893

earthquake has highest magnitude of all and hence maximum displacement is observed for this case.

The horizontal displacements observed are 139.42, 72.964 and 38.90 mm for Chamoli, Uttarkashi and Chamba earthquakes, respectively. These results indicate that the peak amplitude of an event has no significance on the deformation of slopes during earthquakes. It is the total acceleration—time response which includes amplitude, frequency and duration of the event that needs to be considered in the analysis.

6 CONCLUSIONS

Stability of a rock slope in Sikkim is studied using static, pseudo-static and time response analysis in PHASE^2 software. The rock slope is found to be stable in static and marginally stable in pseudo-static conditions, which are within permissible limits. The results from pseudo-static analysis implies when the horizontal acceleration coefficient and vertical acceleration coefficient is more, then the factor of safety of a rock slope reduces. Pseudo-static analyses, where only peak amplitude alone is considered, will lead to erroneous results while predicting the deformations. The response of the slope to different earthquake events is also studied. The variation of horizontal and vertical displacements observed in the slope for three earthquake scenarios are presented and are well within the permissible limits. It was observed that peak amplitude of the earthquake event has less influence on the dynamic response of the slope. It is the total acceleration—time response which includes amplitude, frequency and duration of the event that needs to be considered in the analysis.

REFERENCES

Cai, M., & Horii, H. (1992). A constitutive model of highly jointed rock masses. Mechanics of Materials 13: 217–246.

Ghosh, S., Kumar, A., & Bora, A. (2014). Analyzing the stability of a failing rock slope for suggesting suitable mitigation measure: a case study from the Theng rockslide, Sikkim Himalayas, India. *Bulletin of Engineering Geology and the Environment*, 73(4), 931–945.

Gupta, V., Bhasin, R.K., Kaynia, A.M., Kumar, V., Saini, A.S., Tandon, R.S., & Pabst, T. (2016). Finite element analysis of failed slope by shear strength reduction technique: a case study for Surabhi Resort Landslide, Mussoorie Township, Garhwal Himalaya. *Geomatics, Natural Hazards and Risk,* 7(5), 1677–1690.

Hatzor, Y.H., Arzi, A.A., Zaslavsky, Y., & Shapira, A. (2004). Dynamic stability analysis of jointed rock slopes using the DDA method: King Herod's Palace, Masada, Israel. *International Journal of Rock Mechanics and Mining Sciences, 41*(5), 813–832.

Kanungo, D.P., Pain, A., Sharma, S. (2013) Finite element modeling approach to assess the stability of debris and rock slopes: a case study from the Indian Himalayas. Journal of the International Society for the Prevention and Mitigation of Natural Hazards 69:1–24

Latha, G.M., & Garaga, A. (2010). Seismic stability analysis of a Himalayan rock slope. *Rock Mechanics and Rock Engineering*, 43(6), 831–843.

Liu, Y., Li, H., Xiao, K., Li, J., Xia, X., & Liu, B. (2014). Seismic stability analysis of a layered rock slope. *Computers and Geotechnics, 55*, 474–481.

Pal, S., Kaynia, A.M., Bhasin, R.K., & Paul, D.K. (2012). Earthquake stability analysis of rock slopes: a case study. *Rock Mechanics and Rock Engineering, 45*(2), 205–215.

Rocscience Inc. (2016). PHASE^2 Version 9.0 – Finite Element Analysis for Excavations and Slopes. www.rocscience.com, Toronto, Ontario, Canada.

Tiwari, G., Gali, M.L., & Rao, V.R. (2014). Finite Element Study of a Rock Slope Using Continuum-Interface Approach.

Geomechanics and Geodynamics of Rock Masses – Litvinenko (Ed.)
© 2018 Taylor & Francis Group, London, ISBN 978-1-138-61645-5

Finite-element analysis as a means of solving geomechanics problems in deep mines

Alexandr Evgenevich Rumyantsev, Andrey Viktorovich Trofimov &
Vladislav Borisovich Vilchinsky
LLC "Institute Gipronikel", Saint-Petersburg, Russia

Valery Petrovich Marysiuk
*PJSC Mining and Metallurgical Company "Norilsk Nickel", Director of the Center
for Geodynamic Safety, Norilsk, Russia*

ABSTRACT: The design of stationary objects at deep mines is associated with the issue of finding the optimal balance between reducing the transportation distances for personnel, self-propelled equipment, possibility producing of technological processes to mining sites and ensuring long-term operation of mine workings located in the zone of influence of the front of the second workings.

To choose the optimal solution, it is necessary to take into account geomechanically safe functioning of a stationary object near zones with potential stress concentration (tectonically faults, weak rocks and etc.). To avoid negative consequences, the decision must be made on the basis of geotechnical calculations and modeling using specialized software—strength analysis systems.

Two different examples of the use of finite element modeling are described in the article. In the first example, it is possible based on the results of the simulation, to draw conclusions about the effectiveness of measures to ensure the stability of mine workings. Based on the results of the second example, it is possible to draw conclusions about the rationality of locating a capital construction of underground structures at great depths in the rocks with known physical-mechanical characteristics.

Keywords: deep mines, simulation, physical-mechanical characteristics, stress

1 INTRODUCTION

Modern high-performance mining and metallurgical production, includes complex, multi-stage processes mining of minerals, for which energy, material and labor resources are expended. The most important in this chain are processes of interaction with an massive of rocks, namely, ensuring the stability of excavations.

In connection with mining at greater depths the problems of the stability of mine workings is particular urgency. At greater depths the natural rock pressure is played a crucial role in the stability of underground structures. With depth for weak rocks, the largest principal stress can reached critical values, even at full observance technology of management the mountain pressure, it can cause processes of cracks formation in rocks - formation of new (induced) cracks under the influence of the redistribution of rock pressure near the workings. Induced fracturing of the rock massive is significantly change the structure of the rock massive near the workings, and changed its reaction to natural and technogenic effects [Odintsev V.N., 1997].

At the present stage of the development of computer modeling, the adoption of design solutions without carrying out simulation leads to a decrease in economic efficiency and safety of production, and sometimes inability to fully extract mineral resurces. This is especially true for deep sites of deposits, with a complex geological structure, which is expressed in the uncharacteristic stress-strain state of the rock massive and manifests itself in the form of critical deformations of the mines.

Numerical methods is widely used in the last few decades due to progress in computing power. In a broad sense, numerical methods can be classified as continuum and discontinuum (discrete) [Jing L., 2002]. There is a sufficiently large numbers of different methods for estimating in geomechanics. The most important or at least the most frequently used methods are: for continuum models; finite difference method (FDM), finite element method (FEM) and boundary element method (BEM); for discrete ones, discrete element method (DEM), discrete deformation analysis (DDA), and bonded particle model (BPM) [Bobet A., 2010].

The finite element method (FEM) is applied for analysis the stress-strain state of rocks in the article, because it has established itself as a reliable and affordable tool for modeling such problems.

The application of FEM for geomechanical tasks is widely used, for example, in the article [Pisetsky V.B., 2016], with the help of FEM, the stability of the rocks of the transport tunnel No. 6 in Sochi is estimated.

In the article [Elmo D., 2013] the use of the FEM in conjunction with DEM is used to calculate the stability of the fractured and blocky rocks, to evaluate the subsidence of the earth's surface and the stability of open pit mines.

In a work [Sepehri M., 2013] at an underground mine in Northern Canada using FEM is evaluated stresses in the rocks, and the predicted behavior of the solid rocks after ore is extracted.

In the thesis [Kardani M., 2012] the possibility of applying FEM in the case of large deformations is considered, various algorithms and methods for their implementation are presented.

In a work [Sarathchandran A., 2014] is showed the possibility of using FEM program Rocscience RS3 to assess the sustainability and development of underground mines in high horizontal stress, presented specific examples of calculations, for example, for North Selby coal mine in England.

2   EXAMPLE OF APPLICATION OF FINITE ELEMENT MODELING №1

In the presented study, a complex approach to the problem of simulation of critical convergences of excavations and methods of their reduction by studying the physico-mechanical properties of the rock mass by field and laboratory methods was considered. The analysis were carried out in the CAE Fidesys package [Vershinin A.V., 2015] is based on these results.

Survey excavations revealed substantial irreversible deformation of the supports (Figure 1). The nature of deformation of the supports is manifested in the periodic local predominance of horizontal deformations over vertical.

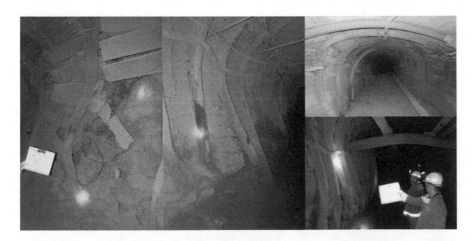

Figure 1.   Critical deformations of underground mines.

At this stage of the research is examined two ways to solve: first one is the choice of kind and parameters of the support with greater carrying capacity, which enhance the sustainability of the mine workings, the second is the change of physico-mechanical properties around the mines by the cement slurry. To simulate the hardening of the rocks around the mine, carried out laboratory and field tests, methods and results which is discussed in detail in [Trofimov A.V., 2016]. The main objective was to confirm the possibility of strengthening the destroyed rock mass and establish a relationship between the degree of consolidation (cohesion), the deformation modulus and the propagation velocity of transverse waves.

For calculations in CAE Fidesys was built three-dimensional model of the underground workings.

The model consists of several blocks differing in their physico-mechanical characteristics, which, with some assumptions, brings the model to natural.

The model presented in Figure 2 includes:

1. block No. 1 – massive block of rocks, which is limited by tectonic disturbances;
2. block No. 2 – block of rocks with intense fracturing;
3. block No. 3 – block of rocks, which fill the voids between the support and the walls of the mine;
   3.1. physical and mechanical characteristics correspond to the characteristics of block 2;
   3.2. strengthening of this block with the help of tamponage (physical and mechanical characteristics are taken on the basis of laboratory tests);
4. block No. 4 – block of steel supports—arched pliable support, in particular type SVP27 (existing support) and SVP33 (may application) with the specified step of installing 0.5 m between frames. On the one of the main frames rigidly secured to I-beam for installation of the pipeline, which corresponds to the actual situation in the mine.

Physical and mechanical properties to solve the model of buckling in the elastic formulation [Lalin V.V., 2008; Vershinin A.V., 2015] shown in Table 1.

Model of supports SVP 27 and SVP 33 were built in accordance with [GOST 18662-83].

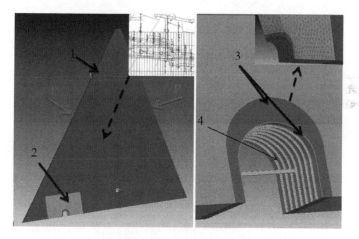

Figure 2.   3D model of excavation

Table 1.   Physico-mechanical characteristics of blocks in the model.

Number of block	The deformation modulus, GPA	Poisson's ratio
1	35	0,27
2	0,092	0,48
3.1	0,092	0,48
3.2	1	0,3
4	86	0,3

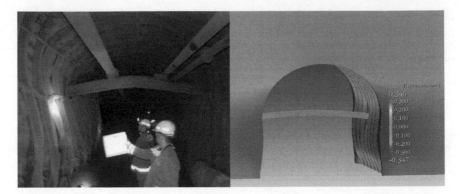

Figure 3.   Real deformations in excavation and result of modelling,

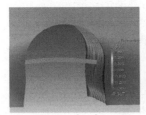

| Current situation horizontal convergence 70cm | Replacement of SVP 27 for SVP 33 horizontal convergence 64 cm | Reinforcement (tamponage) horizontal convergence 32 cm | Reinforcement (tamponage) + Replacement of SVP 27 at SVP 33 horizontal convergence 30 cm |

Figure 4.   The results of modeling technical solutions to strengthen workings.

To simulate deformations similar to those shown in Figure 3, the distributed pressure is applied to the lateral faces of model then it increased by the iteration method, with a loading step 10 MPa, in the range of 10 to 150 MPa. When deformations similar to the deformations in Figure 3 were achieved, the pressure obtained in the previous step was applied to the lateral faces of the model, in this case the physical and mechanical properties of the model blocks was replaced too, and a recalculation of model was performed.

Figure 4 shows the results of additional (re -) calculations.

Analyzing results it is seen that the horizontal convergence of mine in the calculation of the current situation was 70 cm, in the case of strengthening (tamponage) of rocks and making sites with improved mechanical characteristics horizontal convergence was 32 cm, which suggests that the consolidation of weak rock material is increased the resistance mine. The use of the support type SVP 33 are not able ceteris paribus ensure the best stability of mine, and only slightly reduces the deformation, the horizontal convergence of the sides of the excavation is 70 cm for profile SVP 27 and 64 cm for the SVP 33.

Based on the study, according to the simulation results in the CAE Fidesys, that to increase the stability of the excavation in conditions of increased horizontal stresses, one of the most effective methods for increase the stability of the excavation is strengthening of the rock mass behind of the support.

3   EXAMPLE OF APPLICATION OF FINITE ELEMENT MODELING №2

In the present study, the application of the strength analysis method for predicting the geomechanical situation in excavations at great depths in weak rocks with an obvious stratification and intensive fracturing is presented.

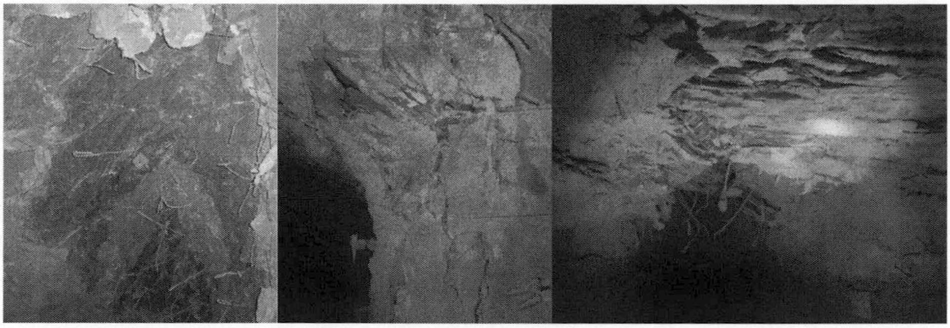

Figure 5. The destruction of workings, which were built in weak rocks at the depth more than 1 km.

When the excavations was examined it is seen everywhere significant deformations of the roof, the soil and sides of the workings, the destruction of the supports and cracking, some of the photos is presented in Figure 5.

A complex geomechanical situation is associated with many factors, and the mechanism of the behavior of rocks in the vicinity of underground excavation is based on the laws of continuum mechanics [Fadeev A., 1987]. Displacements of rocks and load on the supports are caused by the processes of formation around the workings of the zone of inelastic deformations. Since the stresses in the rocks are reduced in the zone of inelastic deformations, stresses are grown up at the boundary of this zone and under the influence of this stress, as well as through the development of cracks and dilatancy, the rocks are extruded into the workings.

The exact mechanism of destruction of the mine workings is almost impossible to describe, because the location of workings on the boundary of a large tectonic fault, causes the complex interactions of the block structures which comprise the rock mass. Also among the negative factors include the fact that all minings are located in weak rocks (mudstone and marl), the strength at uniaxial compression in which is comparable to the hydrostatic stresses at these depths.

Based on the above information the conclusion about the need for mathematical modeling (CAE Fidesys [Lalin V.V., 2008; Vershinin A.V., 2015]) to assess the possibilities of further existence of presented underground building.

The creation of the 3D model was carried out on the basis of plans and cuts, with the transfer of characteristic dimensions, in the Autocad3D. The result was a 3D model consisting of three-dimensional bodies (Figure 6), the model includes the ore unloading areas, the ore massive, the filling the worked out space with concrete and the rock massive with different types of rocks.

The calculation was carried out according to the model of plasticity of Drucker-Prager with hardening, i.e. the task was solved not only in the elastic, but also in plastic state.

Conditionally, the model was divided into 5 blocks:

1. block No. 1 is a barren rock, represents the main part of the rock massive up to the ore body, around the ore body and continues after the ore body to weak rocks (marl) in which is located the examine object;
2. block No. 2 at the lowermost block in the model is represented by the rocks with physical-mechanical properties of marl, in this rock is the main part of the structure is located;
3. block No. 3 is an ore body;
4. block No. 4 is the concrete filled space in the ore body;
5. block No. 5 part of the unloaded (using boreholes) areas in the ore body.

Table 2 shows the physico-mechanical properties of blocks of rock involved in the analysis model.

The model was calculated both under the existing conditions of mining works and for the completion of the mining works locating above of the examined excavations.

The calculation results are shown in Figures 7 and 8.

899

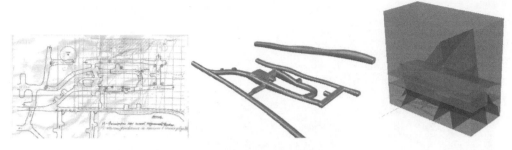

Figure 6.   Plan of workings, 3D model of workings, 3D model entirely.

Table 2.   Physical and mechanical properties of blocks in the model.

Physical and mechanical parameter	Barren rock block № 1	Marl block № 2	Ore body block № 3	Concrete filling block № 4	Unloaded area к block № 5
Deformation module, GPa	10	1	8	1	1
The coefficient of transverse deformation	0,22	0,49	0,3	0,4	0,45
Density, kg/m³	2700	2700	4300	2100	4000
Tensile yield strength, MPa	7,9	1,9	3,9	0,39	1,9
Ultimate tensile strength, MPa	8	2	4	0,4	2
Ultimate tensile strain	0,01	0,01	0,01	0,01	0,01
Compression yield strength, MPa	90	25	64	2,9	29
Compressive strength, MPa	100	27	65	3	30
Ultimate compression strain	0,02	0,03	0,02	0,02	0,02

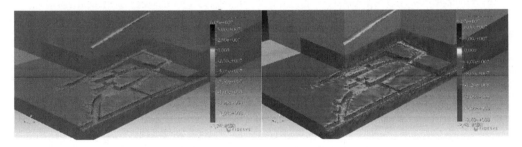

Figure 7.   The horizontal stress along the axis XX with unloaded areas: the left-at the moment, right - at the end of mining the ore body.

Figure 8.   Edler von Mises plastic deformations in the workings with unloading zones: the left - at the moment, right- at the end of mining the ore body.

900

The main evaluation criterion of the stability of workings, is the criterion of plasticity by Richard Edler von Mises, is one of the two criteria of plasticity used to date (the second important criterion belongs to Henri Édouard Tresca).

The simulation found that expected to improve geomechanical situation in the excavations (mining) at the end of mining the ore body will not happen, but on the contrary will worsen, as the excavations is built in heavily disturbed and weak physico-mechanical parameters of rocks.

It should be noted that the excavation locating over the examine object, built in strong rocks (block No. 1) does not experience plastic deformation.

As can be seen from the presented examples of application of FEM is a very effective tool for the evaluation of stresses and strains that can occur in the workings in deep mines.

REFERENCES

[1] Beck D., Arndt S., Thin I., Stone C., Butcher R. A conceptual sequence for a block cave in an extreme stress and deformation environment. Deep and High Stress Mining 2006.
[2] Bobet A. Numerical methods in geomechanics. School of Civil Engeneerin, Purdue University, West Lafayette, IN, USA 2010.
[3] Elmo D., Stead D., Eberthardt E., Vyazmensky A. Applications of Finite/Discrete Element Modeling to Rock Engeneering Problems 2013.
[4] Fadeev A.B. The finite element method in geomechanics. M.: Nedra, 1987. 221 p.
[5] GOST 18662-83 "Profiles hot-rolled SVP for support of mine workings".
[6] Jing L. and Hudson J.A., "Numerical Methods in Rock Mechanics", International Journal of Rock Mechanics and Mining Sciences, 39 (2002), pp. 409-427.
[7] Kardani M. Large deformation analysis in geomechanics using adaptive finite element methods. The University of Newcastle 2012.
[8] Lalin V.V., Kolosova G.S. The course of lectures on the theory of elasticity. SPb.: publishing house of Polytechnic University 2008.
[9] Odintsov V.N., Doctor of Technical Sciences Dissertation "Regularities of the formation of detached cracks in rocks near excavations at great depths" Moscow 1997.
[10] Pisetsky V.B., Lapin S.E., Levin V.A., Gorbunov V.A., Chevodar S.M. "On the choice of criterion for assessing the risk of loss of the state of stability of the mountain massif by seismic, airborne and geomechanical data." I International Scientific and Technical Conference "Occupational Safety and Efficiency of Mining Enterprises with Underground Development". Yekaterinburg, 2016, pp. 59-65.
[11] Sarathchandran A. Three Dimensional Numerical modelling Of Coal Mine Roadways Under High Horizontal Stress Fields. CSM Project Dissertation 2014.
[12] Sepehri M., Apel D.B., Szymanski J. Full Three-dimensional Finite Element Analysis of the Stress Redistribution in Mine Structural Pillar. Powder Metallurgy and Mining 2013.
[13] Trofimov A.V., Rumyantsev A.E., Andreev A.A. "The application of strength analysis methods to substantiate the technical means of ensuring the stability of mine workings when mining ore deposits by underground method." I International Scientific and Technical Conference "Occupational Safety and Efficiency of Mining Enterprises with Underground Development". Yekaterinburg, 2016 pp. 85-93.
[14] Vershinin A.V., Levin V.A., Morozov E.M. Strength analysis: Fidesis in the hands of an engineer. M.: LENAND, 2015. – 408 p.

Geomechanics and Geodynamics of Rock Masses – Litvinenko (Ed.)
© *2018 Taylor & Francis Group, London, ISBN 978-1-138-61645-5*

Numerical modelling of rock fracture with a Hoek-Brown viscoplastic-damage model implemented with polygonal finite elements

Timo Saksala

Laboratory of Civil Engineering, Tampere University of Technology, Tampere, Finland

ABSTRACT: In the present paper, the rock mineral structure is described as a Voronoi diagram where the Voronoi cells are polygonal finite elements. The minerals constituting the rock are represented by random clusters of polygonal finite elements. Rock fracture is described in the continuum sense by using a damage-viscoplasticity model based on the Hoek-Brown criterion. Due to the asymmetry of the tension and compression behavior of rocks, separate scalar damage variables, driven by viscoplastic strain, are employed in tension and compression. The equations of motion are solved by explicit time marching since the final aim of this research project is to model transient problems with contact loading (such as percussive rock drilling). In the numerical examples, the capabilities of the present numerical approach are demonstrated. Specifically, uniaxial tension and compression tests of a numerical rock sample are simulated under plane strain conditions. It is shown that the present method can capture the salient features, including the stress-strain response and the failure modes, of typical rock behavior in these constitutive tests.

1 INTRODUCTION

Polygonal finite elements have been drawing increasing attention during the last 15 years (Sukumar and Tabarraei 2004 & 2006; Talischi et al. 2012; Saksala 2017). In comparison to the standard finite elements, these elements offer greater flexibility in meshing arbitrary geometries, better accuracy in the numerical solution, better description of certain materials, and less locking-prone behavior under volume-preserving deformation (Sukumar 2004). Polygonal discrete elements based on Voronoi diagrams are a widely used method in geomechanics (see, for example, UDEC manual by Itasca, 2013). On the other hand, the disadvantages of the polygonal finite element method include less sparse system matrix and the need for a higher order numerical integration quadrature to achieve high-accuracy.

In the present paper, some further results on a project of modelling rock materials with polygonal finite elements based on the Voronoi tessellation are presented. Namely, the method to describe the rock microstructure based on the Voronoi tessellation and randomly mapped clusters of polygonal elements representing the mineral texture, presented in Saksala (2017), is equipped here with a rock constitutive model. This model is based on the Hoek-Brown viscoplasticity model and a bi-variable scalar damage model (Saksala 2015). The aim of the project is to model rock fracture under dynamic applications involving contact/impact. For this reason, the equations of motions are solved by explicit time marching. In the numerical examples, the performance of the present approach is demonstrated with uniaxial tension and compression test simulations.

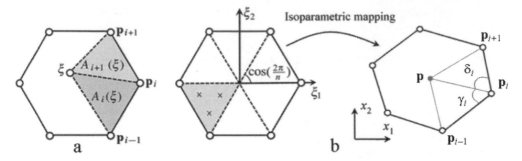

Figure 1. Illustration of the triangular areas used in the definition of Wachspress shape function (a), and the triangulation of the reference regular polygon, integration points of each triangle, and the isoparametric mapping to a physical element (b).

2 POLYGONAL FINITE ELEMENTS

Polygonal finite elements differ from the traditional triangle and rectangular elements in that they can have any number of angles (n-gon). Moreover, the resulting interpolation functions are not polynomials, as for the traditional elements, but usually rational functions. In the present paper, the finite element formulation based on Wachspress interpolation functions (Wachspress 1975) developed by Talischi et al. (2012) is adopted. The formulations relies on the standard isoparametric mapping from a reference element to the physical element, as illustrated in Figure 1.

For a reference n-gon, the barycentric Wachspress shape function at node i reads

$$N_i(\xi) = \frac{\alpha_i(\xi)}{\sum_{j=1}^{n} \alpha_j(\xi)}, \quad \alpha_i(\xi) = \frac{A(\mathbf{p}_{i-1}, \mathbf{p}_i, \mathbf{p}_{i+1})}{A(\mathbf{p}_{i-1}, \mathbf{p}_i, \xi) A(\mathbf{p}_i, \mathbf{p}_{i+1}, \xi)} = \frac{\cot \gamma_i + \cot \delta_i}{\| \mathbf{x} - \mathbf{x}_i \|^2} \tag{1}$$

where $A(a, b, c)$ denotes the signed area of triangle a, b, c (see Figure 1a). Moreover, the angles γ_i, δ_i are shown in Figure 1b. The numerical integration is carried out by sub-dividing the reference polygon into triangles and using a three integration points scheme for each triangle (resulting $3n$ integration points for each n-gon), as illustrated in Figure 1b. Finally, the variational (weak) formulation of the polygonal finite elements method is standard, see Sukumar and Tabarraei (2004) for details.

3 ROCK CONSTITUTIVE MODEL BASED ON VISCOPLASTICITY AND DAMAGE MECHANICS

The viscoplastic part of the model indicates the stress states leading to inelastic deformation and damage of the rock. It also accounts for the strain rate effects through viscosity. For present purposes, the Hoek-Brown (HB) failure criterion is written in form

$$f_{\mathrm{HB}}(\boldsymbol{\sigma}, \dot{\lambda}) = (\sigma_1 - \sigma_3)^2 + \sigma_{\mathrm{c}}(\dot{\lambda})^2 \left(\frac{\sigma_1}{\sigma_{\mathrm{t}}(\dot{\lambda})} - 1 \right) \quad \text{with}$$
$$\sigma_{\mathrm{c}}(\dot{\lambda}) = \sigma_{\mathrm{c}0} + s_{\mathrm{c}} \dot{\lambda}, \quad \sigma_{\mathrm{t}}(\dot{\lambda}) = \sigma_{\mathrm{t}0} + s_{\mathrm{t}} \dot{\lambda} \tag{2}$$

where σ_1 and σ_3 are the major and minor principal stresses of stress tensor $\boldsymbol{\sigma}$, σ_{c}, σ_{t} are the compressive and the tensile strengths, s_{c}, s_{t} are the constant viscosity moduli, and $\dot{\lambda}$ is the rate of the internal variable, respectively. Perfectly viscoplastic behavior is assumed here. The rate sensitivity in this modelling approach is provided by adding the viscosity term to the static values, $\sigma_{\mathrm{c}0}$, $\sigma_{\mathrm{t}0}$, of the compressive and tensile strength.

The damage part of the model is formulated with separate damage variables, both driven by the inelastic strain, in compression and tension. The damage functions, the equivalent viscoplastic strains, and the nominal-effective stress relation of the model are

$$\omega_t(\varepsilon_{eqvt}^{vp}) = A_t\left(1 - \exp\left(-\beta_t \varepsilon_{eqvt}^{vp}\right)\right), \quad \omega_c(\varepsilon_{eqvc}^{vp}) = A_c\left(1 - \exp\left(-\beta_c \varepsilon_{eqvc}^{vp}\right)\right) \quad \text{with}$$

$$\beta_t = \sigma_{t0} h_e / G_{Ic}, \quad \beta_c = \sigma_{c0} h_e / G_{IIc}$$

$$\dot{\varepsilon}_{eqvt}^{vp} = \sqrt{\sum_{i=1}^{3} \langle \dot{\varepsilon}_i^{vp} \rangle^2}, \quad \dot{\varepsilon}_{eqvc}^{vp} = \sqrt{\tfrac{2}{3} \dot{\varepsilon}^{vp} : \dot{\varepsilon}^{vp}} \quad \text{with} \quad \dot{\varepsilon}^{vp} = \lambda \frac{\partial f_{HB}}{\partial \sigma}$$

$$\sigma = (1 - \omega_t)\bar{\sigma}_+ + (1 - \omega_c)\bar{\sigma}_- \quad (\bar{\sigma} = \bar{\sigma}_+ + \bar{\sigma}_-)$$

(3)

where parameters A_t, A_c control the final value of the damage variables ω_t, ω_c in tension and in compression, respectively. Parameters β_t, β_c, which control the initial slope and the amount of damage dissipation, are defined by the fracture energies G_{Ic} and G_{IIc} and h_e is a characteristic length of a finite element. The equivalent viscoplastic strain in tension, ε_{eqvt}^{vp}, is defined by the ith principal value, $\dot{\varepsilon}_i^{vp}$, of the viscoplastic strain rate tensor, $\dot{\varepsilon}^{vp}$, using the Macauley brackets so that tensile damage evolution occurs only if the viscoplastic principal strains are positive. Finally, $\bar{\sigma}_+ = \max(\bar{\sigma}, 0)$ and $\bar{\sigma}_- = \min(\bar{\sigma}, 0)$ are the positive and negative parts, respectively, of the principal effective stress. This formulation naturally conveys the strain rate dependency to the damaging as well.

The coupling of the viscoplastic and damage parts of the model is formulated in the effective stress space, which enables the viscoplastic and damage computations to be separated so that first, the stress integration is performed independently of damage and then the damage variables are updated according to Equation (3). The stress integration (the return mapping of the trial stress onto the yield surface) for this viscoplastic consistency model is performed as in Saksala (2015).

4 ROCK MICROSTRUCTURE DESCRIPTION

Polycrystalline materials such as rocks can be naturally described with non-regular meshes of polygonal elements. Here, the minerals constituting the rock are represented by random clusters of Voronoi cells which themselves are the physical polygonal finite elements. This is

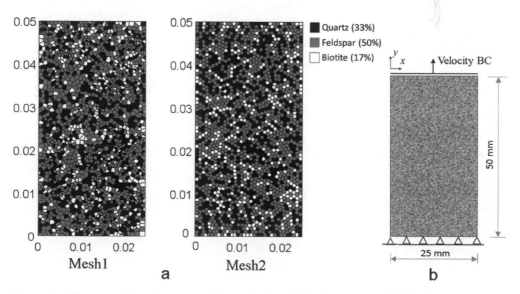

Figure 2.   The numerical rock samples with random seed (Mesh1) and centroidal Voronoi tessellation (Mesh2) (a), and boundary conditions (b).

905

accomplished by the PolyMesher Matlab code by Talischi et al. (2012). This code generates 2D Voronoi diagrams (tessellations) consisting of centroidal (or alternatively non-centroidal) Voronoi cells, which are then used as polygonal finite element meshes in the present work.

The rock material is assumed to consist of three main minerals (quartz, feldspar, biotite) which have their respective material properties. Thereby, the rock heterogeneity can be described. This method is illustrated in Figure 2a where two different realizations of the numerical rock sample with 2000 grains (Voronoi cells/polygon elements) are generated. In the first sample (Mesh1), the generating seeds are random yielding a realistic rock microstructure while in the second the polygons in the mesh are centroidal, i.e. the generating seeds are at the centroids of the polygons (Mesh2), leading to a more regular cell distribution.

5 SOLVING THE EQUATIONS OF MOTIONS

As the aim is to simulate transient dynamic problems involving impact and stress wave propagation, in addition to quasi-static tests, the equations of motion are discretized explicitly in time. The modified Euler method is chosen for this end. Accordingly, the response (velocity and displacement) of the system is predicted based on the acceleration solved from the Newton's second law of motion as follows

$$\ddot{\mathbf{u}}^t = \mathbf{M}^{-1}\left(\mathbf{f}_{ext}^t - \mathbf{f}_{int}^t\right), \dot{\mathbf{u}}^{t+\Delta t} = \dot{\mathbf{u}}^t + \Delta t \ddot{\mathbf{u}}^t, \mathbf{u}^{t+\Delta t} = \mathbf{u}^t + \Delta t \dot{\mathbf{u}}^{t+\Delta t} \tag{4}$$

where $\mathbf{M}$ is the lumped mass matrix, and $\mathbf{f}_{ext}^t, \mathbf{f}_{int}^t$ are the external and internal force vectors, respectively.

6 NUMERICAL EXAMPLES

Uniaxial laboratory scale compressive and tensile tests are simulated here in order to demonstrate the performance of the present numerical approach. The numerical rock samples (see Figure 2a and b) generated as the method explained in Section 4 above are simply supported at the bottom while constant velocity boundary conditions is applied at the top edge (see Figure 2b). The material properties as well as the model parameters are given in Table 1.

These values may not represent any single real rock but are chosen for demonstrative purposes only. These cohesion and internal friction values give the homogeneous compressive strength of 137 MPa.

Table 1. Material properties and model parameters used in simulations.

Parameter	Quartz	Feldspar	Biotite	Homogeneous
ρ (Density) [kg/m^3]	–	–	–	2630
E (Elastic modulus) [GPa]	80	60	20	–
ν (Poisson's ratio)	0.17	0.29	0.20	–
σ_t (Tensile strength) [MPa]	10	10	7	–
c_0 (Cohesion) [MPa]	–	–	–	25
φ (Internal friction angle)	–	–	–	50°
G_{Ic} (Mode I fracture energy) [J/m^2]	–	–	–	100
G_{IIc} (Mode II fracture energy) [J/m^2]	–	–	–	1000
A_t (Tensile damage final value)	–	–	–	0.98
A_c (Compr. damage final value)	–	–	–	0.95
s_c (Viscosity in compression) [MPa·s]	–	–	–	0.01
s_t (Viscosity in tension) [MPa·s]	–	–	–	0.01
Percentage in the sample	33%	50%	17%	100%

The uniaxial compression and tension tests are simulated with a constant velocity v_0 = ±0.05 m/s which results in a strain rate of 1 s⁻¹ with the present specimen dimensions. The simulation results for compression are shown in Figure 3 and tension in Figure 4.

According to the results in Figure 3, the compressive and tensile damage variables have virtually identical distributions at the end of the simulations. This means that tensile damaging occurs extensively in compression test as well. The final failure modes are quite similar, for Mesh1 even more so, to those observed in the experiments (see Figure 3f). The compressive strength is about 100 MPa for both numerical rock samples and the post-peak behavior is extremely brittle.

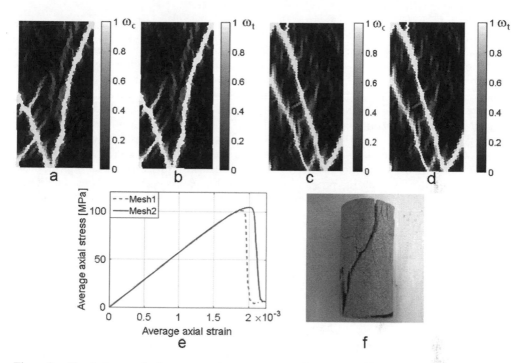

Figure 3.   Simulation results for compression test: compressive (a), tensile (b) damage distribution with Mesh1, compressive (c), tensile (d) damage distribution with Mesh2, corresponding average stress-strain curves (e), and an example of experimental failure mode (adapted from (Zhou and Zhao 2011)) (f).

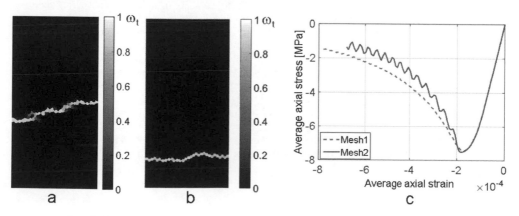

Figure 4.   Simulation results for tension test: tensile damage distribution with Mesh1 (a), Mesh2 (b), and corresponding average stress-strain curves (c).

The results for the tension test simulation in Figure 4 show that the tensile strength of both numerical rock specimen is 7.5 MPa. As for the failure modes, they exhibit the typical transverse non-smooth splitting crack with some unexpected inclination in case of Mesh1. Finally, the pre-peak nonlinearity is observed in both the compression and tension stress-strain responses. It results from the heterogeneity of the numerical rock through failure of individual elements upon reaching the compressive/tensile strength. This heterogeneity also resulted, through combined tensile and compressive stress states at the elements located to mineral boundaries, in a lower compressive strengths of the numerical specimens that was expected (the homogeneous compressive strength of the specimen was 137 MPa).

7 CONCLUSIONS

The 2D numerical approach based on polygonal finite elements representation of rock micro-structure and the Hoek-Brown viscoplastic-damage model for rock material description seems a promising avenue to rock fracture simulations. The numerical simulations demonstrated that the approach captures the salient features of rock observed in the uniaxial compression and tension tests. Especially, the description of rock material heterogeneity via random mapping of the mineral texture and the association of the corresponding mineral mechanical properties enabled the model to capture pre-peak non-linearity caused by isolated failure events of elements under mixed compressive-tensile stress states. This is a significant benefit, as no specific model component is needed in the constitutive description of the pre-peak nonlinearity. In conclusion, the present modelling technique seems to have some predictive capabilities and will thus be used in more advanced application oriented studies in future.

ACKNOWLEDGMENTS

This research was funded by Academy of Finland under grant number 298345.

REFERENCES

Itasca Consulting Group, Inc.: Universal Distinct Element Code: Theory and Background. Itasca Consulting Group, Inc., Minneapolis, 2013.
Saksala T. Damage-viscoplastic model based on the Hoek-Brown criterion for numerical modeling of rock fracture. *Rakenteiden Mekaniikka (Journal of Structural Mechanics)*. 2015; 48: 99–114.
Saksala T. Numerical modelling of rock materials with polygonal finite elements. *Rakenteiden Mekaniikka (Journal of Structural Mechanics)*. 2017; 50: 216–219.
Sukumar N, Tabarraei A. Application of polygonal finite elements in linear elasticity. *International Journal of Computational Methods*. 2006; 3: 503–520.
Sukumar N, Tabarraei A. Conforming polygonal finite elements. *International Journal for Numerical Methods in Engineering*. 2004: 61: 2045–2066.
Talischi C, Paulino GH, Pereira A, Menezes IFM. PolyMesher: a general-purpose mesh generator for polygonal elements written in Matlab. *Structural and Multidisciplinary Optimization*. 2012; 45: 309–328.
Talischi C, Paulino GH, Pereira A, Menezes IFM. PolyTop: a Matlab implementation of a general topology optimization framework using unstructured polygonal finite element meshes. *Structural and Multidisciplinary Optimization*, 45: 329–357, 2012.
Wachspress EL. A Rational Finite Element Basis. Academic Press: New York, NY, 1975.
Zhou Y, Zhao J. In: J. Zhao (Ed.), Advances in Rock Dynamics and Applications. *Chapter 1. Introduction*. CRC Press, 2011.

Geomechanics and Geodynamics of Rock Masses – Litvinenko (Ed.)
© 2018 Taylor & Francis Group, London, ISBN 978-1-138-61645-5

Numerical modeling and back analysis method for optimal design of Butterfly Valve Chamber (BVC) of Tehri HPP

D.V. Singh, R.K. Vishnoi, T.S. Rautela, U.D. Dangwal & A.K. Badoni
THDC India Limited, Rishikesh, Uttarakhand, India

ABSTRACT: Tehri Hydro Power Project (HPP; 4 × 250 MW) is an integral part of Tehri Hydro Power Complex (2400 MW) located in Uttarakhand State, India. The major HPP project components constructed are Machine Hall, HRT, Butterfly Valve Chamber (BVC), Penstock Assembly Chamber (PAC), Machine Hall, Downstream Surge shafts, a pair of Tailrace Tunnels (TRTs) and Outlet Structures. Initially the support design for the BVC & PAC caverns was based on a detailed evaluation of all available geological and geotechnical information. The convergence of different points located on the crowns and side walls of the chambers was estimated on the basis of results of the numerical analyses analysis carried out for the anticipated excavation sequence as well as specified timing of support installation. Convergence of the Chambers was monitored with Bi-reflex targets and special tape extensometers. Differences between predicted convergence by numerical modeling and the large deformations observed visually as well as revealed by monitoring instruments necessitated a review of the numerical models used for the detailed design. Review of numerical models was done to account for the changes in construction sequencing, changes in support system and changes in geotechnical parameters encountered during construction. Based on 2D plain strain analysis, cable anchors were installed for stabilization of BVC and PAC. However, the chamber walls continued to converge and finally it was decided to construct concrete buttresses (wall to wall) in both the chambers to improve the stability of these chambers.

This paper highlights the role of the buttresses for improving the long term stability of the BVC and PAC by carrying out Finite Element Analysis.

1 INTRODUCTION

THDCIL, a joint Venture of GOI & GOUP, had been entrusted with the construction of Tehri Hydro Power Complex. The complex envisaged as construction of Tehri Hydro Power Project (HPP) with installed capacity of 1000 MW in the stage I, construction of Koteshwer Hydro Electric Project (KHEP; 400 MW) in the downstream of Tehri Dam and an underground Tehri Pump Storage Plant (PSP; 1000 MW) at Tehri under stage II of the complex. Tehri HPP (1000 MW) and Koteshwar HEP (400 MW) have already been completed and commissioned and construction of Tehri PSP (1000 MW) is under progress.

Tehri HPP (4 × 250 MW) involved in construction of a 260.5 m high Earth and Rockfill Dam near old Tehri town. The major underground components of Tehri HPP are02 HRTs, Butterfly Valve Chamber (BVC), Penstock Assembly Chamber (PAC), Machine Hall, Transformer Hall, Downstream Surge shafts and a pair of Tailrace Tunnels (TRTs).

The BVC is 121.3 m in length, 23.8 m in height and 10 m in width. The PAC is 117.1 m in length, 19.3 m in height and 13.3 m in width. These chambers are parallel oriented along N303°. The rock pillar between these two chambers is only 18.5 m and the major horizontal stress is aligned nearly perpendicular to long walls of the chamber.

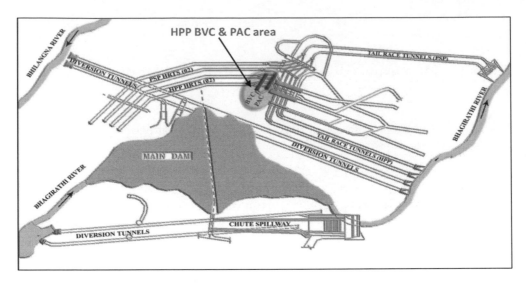

Figure 1. Layout of Tehri HPP.

2 THE PROJECT LAYOUT

The key structure of the project is the Earth and Rock-fill dam, 260.5 m high (above deepest rock foundation) across river Bhagirathi, 1.5 km downstream of its confluence with river Bhilangana, near old Tehri town in Uttarakhand. Other works completed under stage-I are:

- A chute spillway on the right Bank
- Four Shaft Spillways-two on the left bank and two on the Right bank with one intermediate Level out-let. These utilize horizontal parts of four diversion tunnels constructed for dam construction.
- A water conductor system comprising of four 8.5 m diameter head Race tunnel (of these two would be for pump storage plant in stage-II) and 2 nos., 9 m diameter Tail Race Tunnels.
- An underground power house with an installed capacity of 1000 MW, with four units of 250 MW each.

Layout of the Tehri Hydro Power Project is shown below in Figure 1.

3 GEOLOGICAL SETUP

Engineering geological conditions of rock mass enclosing the chambers are very complicated because of unfavorable orientation of discontinuities (schistosity, tectonic features and bedding planes) with respect to the axis of the chamber and the low rock pillar thickness between BVC & PAC. Presence of number of longitudinal tectonic faults further aggravated the excavation problems. The axis of BVC cavern is sub-parallel to the strike of steeply dipping bedding planes of the rockmass. The presence of this crushed zone was interpreted as being associated with the presence of series of tight folds.

Four major joint sets, encountered in this area, are shown below in the Table 1.The disposition of these joints sets revealed that the prominent joint was the bedding joint with strikes sub parallel to the chamber axis. Rock mass classification data along the length of cavern in roof of BVC & PAC and in u/s wall of PAC is given below in Table 2.

The rock mass classification reveals that the major part of the excavation is in the phylitic quartzite thinly bedded (PQT) and only a localized zone of sheared phylite and deformed PQT was identified.

910

Table 1.

Joint set	Average dip	Dip direction	Average strike continuity	Average spacing	Remarks
J1	52°	N 200°	30 m	0.20 m	Parallel to the cavity
J2	40°	N 170°	10 m	0.10 m	Sub-parallel to the cavity
J3	50°	N 325°	3 m	0.20 m	Joint oblique to cavern axis
J4	40°	N 055°	3 m	0.50 m	Joints dipping towards u/s wall.

Table 2.

Cavern	Rock type	Rock state	Q	GSI	UCS (MPa)
BVC Roof	Sequence of PQM/PQT Deformed QP/SP	Blocky/Very Blocky Seamy/Crushed.	13 2	55 25	40–60 –
PAC Roof	Sequence of PQM/PQT Deformed QP/SP	Blocky/Very Blocky Seamy/Crushed.	12 1.4–2	53 25	25–50 –

4 EXCAVATION METHODOLOGY & INITIAL SUPPORT SYSTEM

For excavation of both the caverns of BVC & PAC, initially a central gullet was excavated on the top of the chambers followed by widening to complete the roof arches. Thereafter, both the chambers were excavated in parallel by heading and benching method. Initially better rock conditions were encountered in the side walls and hence the roof and walls of both the chambers were initially planned to be supported by rock bolts and shotcreting with welded wire mesh. The walls were supported by 25 mm diameter, 5 m in length grouted rock bolts with spacing @ 2.5 m c/c (vertical) and @ 1.25 m c/c (horizontally staggered).

5 PROBLEMS ENCOUNTERED

Heavy rock falls were experienced due to structurally controlled wedge formations during excavation of the roof of both PAC and BVC and cracks were developed in the shotcrete on d/s wall of BVC in the zone where crushed rock was encountered. It was apprehended that further lowering of the benches may result in structurally controlled failures on the walls of both the chambers. Therefore further benching was suspended and numerical analysis was performed to check the changes in deformation and stress conditions in both chambers with simulated final excavation. Further, during benching down, at the lower elevation of these chambers, the instability of the Chamber walls was aggravated due to intersection of these walls with tunnels (04 Penstocks and 02 HRTs).

6 NUMERICAL ANALYSES

A detailed two dimensional analysis was performed treating the rock mass as continuum as well as discontinuum using 2-Dimensional software UDEC2.01 to optimize the rock support system. In discontinuum analysis the structural discontinuities were modeled explicitly. In this analysis only joint set J1 was incorporated in the model because the strike of joint J1 is almost parallel to the axis of the cavern where as strike of other joints is oblique to the axis of the cavern. The joint was considered continuous with dip = 52° and spacing of 2 m. Joint strength parameters considered in the analysis were ø = 27°, c = 4 KPa.

6.1 Rock mass properties

The rock mass properties for PQM/PQT and SP were estimated based on a) Intact Rock properties as UCS = 40 MPa & mi (Hoek & brown constant) = 10 and b) Geological Strength Index (GSI) as GSI = 50for PQM/PQT & GSI 25 for Sheared Phyllites. The Rock Mass Properties evaluated on the basis of the above data by THDC along with the values recommended by Dr. Evert Hoek, during his visit to the project in April 1998, are given in Table 3. Although the parameters recommended by Dr. Hoek were only based on his observations during site visit and not on the basis of any field/laboratory testing.

6.2 Analysis results and recommendations

The results of continuum analysis revealed that the entire rockpillar is under distress for the final excavation stage and results of discontinuum analysis showed that the bedding joints undergo slip and open up at the final stage of excavation. At some places in BVC wall, convergence about 50–80 mm was observed. Design Consultants, HPI Moscow conducted additional geophysical explorations (seismic sounding, seismic profiling, ultrasonic sounding of the holes) and observed that the rock pillar in between the two caverns is heavily distressed. In view of this, HPI suggested to strengthen the rock pillar with pre-stressed grouted cable anchors of 80 T design capacity and to be stressed to 50 T to counter the displacements in the walls. The prestressed cables were required to be finally grouted after pre-tensioning.

7 REVIEW OF THE NUMERICAL ANALYSIS

The Geotechnical issues, also got reviewed through Dr. Evert Hoek and after a quick preliminary analysis of PAC & BVC, he recommended cable anchors to strengthen the rock pillar. Based on the visual observations/field data and preliminary analysis, the recommended rock mass properties by Dr. Hoek are given in Table 3, above, along with the properties considered by THDC in the 2-D Numerical analyses. Dr. Evert Hoek also recommended carrying out the 2-D continuum analysis by incorporating the minor modification in dip continuity and taking into account the support provided by the tensioned and grouted cables planned to be installed at different excavation stages.

2-D analysis with the tensioned and grouted cables installed in different excavation stages, with new set of rock mass parameters, with the joint dip continuity to a maximum of 10 m and with random joint spacing (due to the presence of intense folding) was carried out to finalize the number of cables to be installed in the rock pillar between the two caverns and in the upstream wall of BVC. The analysis revealed that the installation of pre-stressed anchors

Table 3.

| Properties | Adopted In 2D Analysis By THDC | | Recommended By Dr. Hoek | |
	PQM/PQT	Sheared Phyllites (SP)	PQM/PQT	Sheared Phyllites (SP)
UCS of intact rock	40 MPa	40 MPa	40 MPa	40 MPa
Deformation modulus	3 GPa	3 GPa	6.3 GPa	1.5 GPa
Poisson's ratio	0.22	0.22	–	–
Cohesion	1.6 MPa	1.03 MPa	1.7 MPa	0.9 MPa
Angle of internal friction	31°	25°	31°	25°
Tensile strength	0.1 MPa	0	0.09 MPa	0
UCS of Rock Mass	5.72 MPa	3.26 MPa	6.1 MPa	2.8 MPa

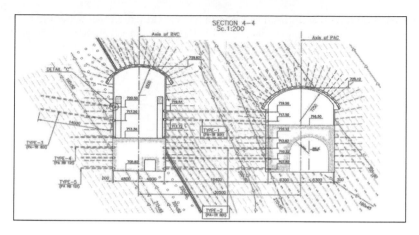

Figure 2.   Final Support System adopted in PAC & BVC.

of 80 T capacity have positive effect on stability of the upstream BVC wall. It was also shown
by computations that due to the pre-stressed anchors installed in the rock pillar between the
chambers the displacement of the inner walls of excavations is reduced by 10–15%.

8   FINAL SUPPORT ARRANGEMENTS

The final support system adopted in BVC & PAC is shown below in Figure 2.
   In addition to the initial support system following additional strengthening measures were
adopted:

- PAC wall adjoining the rock pillar was strengthened with a row of vertical steel (ISBM-
  300) supports and horizontal steel ribs (ISMB-250) spaced at 1 m.
- The rock pillar between the chambers has been strengthened with 80 through pre-stressed
  cable anchors of design working load of 80 T. The opposite wall in the BVC was strengthened with 94 blind pre-stressed anchors of design load of 80 T.
- Keeping in view the increased convergence of the chamber walls even after installation of
  Cable anchors during excavation of the chamber, four concrete buttresses in the BVC (from
  wall to wall) and four concrete buttresses (from wall to wall) in the PAC were erected.

9   MONITORING OF CHAMBER ROCKMASS

The behavior of the rockmass surrounding BVC & PAC chambers was monitored through
Visual observations, Instrumental observations, Seismic sounding of rock pillar between the
chambers & Monitoring of the tensioning in pre-stressed cable anchors. Convergence of
BVC & PAC walls was measured by special tape extensometer having an accuracy of 0.5 mm.
By the end of 2001 maximum deformation were observed about 100 mm and more than
130 mm respectively. After construction of the buttresses in BVC and PAC, the walls convergence rate at these sections reduced significantly.

10   DISCUSSIONS

The decision of installation of Cable anchors for stabilization of rock pillar between the
Chambers and u/s wall of BVC was based on the 2D plain strain analysis; however for properly taking into account the effect of intersections (chambers walls, HRTs & penstocks etc)

and effect of rockmass discontinuities on stability of the chambers a detailed 3D discontinuum stress analysis is an appropriate choice. Further, proper evaluation of geotechnical parameters is of utmost importance for an effective stress analysis and for designing of adequate support system.

The chamber walls continued to converge considerably even after installation of Cable anchors. Therefore, taking into consideration the distressed condition of the surrounding rockmass and keeping in view the forthcoming underground excavation for bifurcation chambers and penstock tunnels, it was decided to construct concrete buttresses (wall to wall) in both the chambers so that further convergence in chambers walls can be prevented.

A fresh finite element analysis (FEM) using rocscience software Phase2 v7.0 and with the parameters suggested by Dr. Evert Hoek (Ref: Table 2) has been carried out to illustrate the positive effect of buttresses in improving the stability of chambers. The Phase model has total 10 stages; in first 8 stages, complete excavation in both the chambers along with installation of cable anchors has been simulated.

The total displacement contours generated by the program for 8th stage corresponding to complete excavation of the chambers are shown in Figure 3 and maximum displacement computed is 135 mm on the right wall of BVC. In order to illustrate the positive effect of buttresses in improving the long term stability of these chambers, two cases have been analyzed in 9th and 10th stage of the Phase model. In 9th stage (Figure 4), the rockmass strength

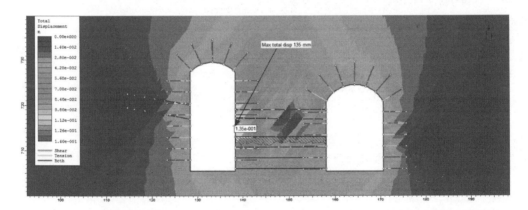

Figure 3.   Max displacement contours with complete excavation of chambers.

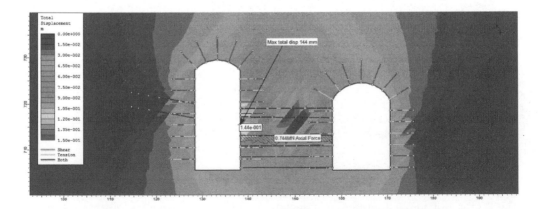

Figure 4.   Displacement contours with rockmass strength reduced to 80% (9th Stage).

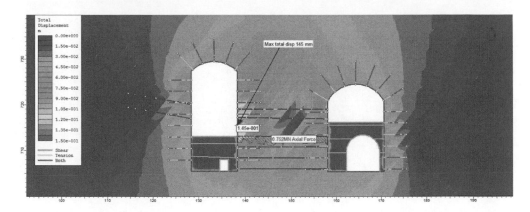

Figure 5.   Displacement contours with rockmass Strength reduced to 60% & with Buttresses (10th Stage).

parameters were reduced to 80% of the peak strength to account for the effect due to forth-coming nearby excavations of bifurcation chambers and penstock tunnels and it is found that the maximum displacement as computed by the program increases from 135 mm to 144 mm. However, in 10th stage (Figure 5) with buttresses erected in these chambers, the maximum displacement does not increase beyond 144 mm even by further reducing the rock-mass strength parameters to 60% of peak strength. With this it is evident that buttresses have positive effect in improving the stability of the chambers.

REFERENCES

[1] SHC, Institute Hydroproject, Moscow 2002, "Tehri HPP on Bhagirathi River, The Final Report".
[2] CBIP Volume 64, No. 1, Water and Energy International "Special issue on Tehri Dam Project".

Geomechanics and Geodynamics of Rock Masses – Litvinenko (Ed.)
© *2018 Taylor & Francis Group, London, ISBN 978-1-138-61645-5*

3D finite element modelling of chain-link drapery system

Soheil Tahmasbi & Anna Giacomini
University Drive, Callaghan NSW, Australia

Corinna Wendeler
Geohazard Solutions, Romanshorn, Switzerland

Olivier Buzzi
University Drive, Callaghan NSW, Australia

ABSTRACT: Rockfall can cause loss of lives and significant damage to infrastructure and requires adequately designed protection structures to reduce the risk to an acceptable level. In Australia, rockfall draperies are more and more used as a passive protection structure although some research is still required to fully characterize the performance of draperies. Full-scale experimentation on rockfall drapery systems is very expensive and time consuming so that numerical methods are a useful and more economical alternative in order to better understand the behaviour of such systems. This paper presents a realistic 3D model of a chain-link drapery system developed with the commercially available finite element package of ABAQUS. The dynamic response of the system was simulated by incorporating the elastoplastic constitutive model in the explicit procedure of the solution. The developed model was calibrated by comparing the numerical results with the results of the laboratory-scale experiments. Preliminary results showed that the calibrated model was capable of predicting the system response with reasonable accuracy. In future, the model will be used in order to conduct parametric studies on different aspects of rockfall drapery design including slope material properties, number and spacing of anchors, impact energy, slope inclination, height of drapery system and block size.

Keywords: ABAQUS, chain-link, drapery, numerical modelling, rockfall simulation

1 INTRODUCTION

Protection structures used to control the destructive effect of rockfall events are typically divided into active and passive measures. The former aims at dissipating the kinetic energy of falling blocks, controlling their trajectories or stopping them before critical zones. Rockfall draperies, as passive measures, are gaining more popularity due to the ease of installation and reduced need for maintenance. Such systems have been around for more than fifty years but only limited research has been conducted on their performance and dynamic response (e.g. Bertolo et al. 2009; Giacomini et al. 2012; Muhunthan et al. 2005). Physical testing of drapery systems typically incurs very high cost and significant technical constraints, all of which can be reduced by resorting to numerical simulations. Like experimental studies, only limited studies can be found on numerical simulations of rockfall draperies (Sasiharan et al. 2006; Thoeni et al. 2014). The objective of this study is to develop a three-dimensional finite element model of full-scale chain-link drapery system. This paper covers developing and calibrating the finite element model of laboratory-scale tests on chain-link drapery.

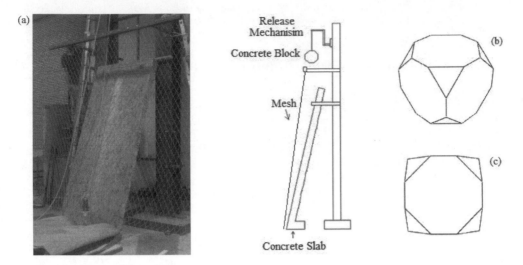

Figure 1. (a) Experimental set up. (b) 3D view of the Swiss-shaped block. (c) Side view of the Swiss-shaped block.

2 EXPERIMENTAL DATA

The experimental data used in this study was obtained from laboratory tests conducted at the University of Newcastle. Figure 1 shows the experimental set up.

A concrete slab of 3 m × 1.2 m was used as an impacted surface during the tests. The slab was leaning against a support structure with the possibility to adjust the dip angle. The drapery was hung from a steel bar with an adjustable horizontal distance from the surface. A Swiss-shaped concrete block (mass of 18.5 Kg) was released in between the slab and the wire net and its trajectory was captured by two high-speed cameras. Two cameras were positioned in a perpendicular direction. One of them was capturing the front view and the other one was capturing the side view of the tests. Two series of tests were conducted: one without drapery (in order to collect data for calibrating the interaction between the falling block and the surface) and one with the drapery in order to reproduce the interaction between the falling block, the surface and the drapery. Tests results were analysed using Tracker video analysis tool.

3 NUMERICAL SIMULATIONS

The 3D geometry of TECCO chain-link mesh was modelled in Abaqus (as per Figure 2). The size of the chain-link drapery is 3 m × 3.5 m, the aperture diameter of each rhomboid is 65 mm and the wire diameter is 4 mm. For the sake of simplicity, the nodes at the upper boundary of the drapery were fixed instead of modelling the steel bar. The concrete surface and the concrete block were modelled as rigid. A total number of 85492 2-node linear beam elements were used to discretise the chain-link drapery. The mesh material was assigned an elasto-plastic behaviour with a tensile strength of 1,770 MPa (as per high tensile steel wire) with a progressive damage requiring the definition of the plastic strain at the onset of damage (ε_0^{pl}), the equivalent plastic displacement (U_f) and the exponent (α_{damage}) defining the exponential relation between the damage factor and the equivalent plastic strain. These parameters were calibrated against experimental data (see Tahmasbi et al. 2017).

The interactions between the block, the concrete surface and the drapery were simulated using general contact algorithm of ABAQUS/Explicit. Tangential contact behaviour was modelled using Coulomb frictional model. Three friction coefficients were introduced into the model: $\mu_s = 0.25$ for steel/steel interaction (as suggested by Persson 2000), $\mu_c = 0.4$ for concrete/concrete interaction (as per BSI, 2008) and $\mu_w = 0.3$ for steel/concrete interaction which was

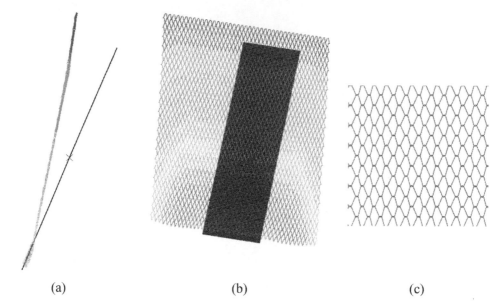

| (a) | (b) | (c) |

Figure 2. (a): Side view of the modelled drapery mesh and the impacted surface (b): 3D view of the drapery under the effect of gravity. (c): View of the 3D model of the chain-link mesh.

calibrated for the numerical model to reproduce the tests results. Furthermore, a small amount of damping factor was applied to the block/wire net interaction to reduce the solution noise. The damping factor for the block/surface interaction also requires calibration to improve the quality of prediction. The analysis was conducted in two steps: a gravity step was first applied in order for the loose chain-link wires to engage under the effect of gravity. The second step simulates the block free fall from a given height (governed by the tests) in between the surface and the drapery. Mass scaling was used to reduce the computational time.

4 RESULTS AND DISCUSSION

Several analyses were conducted to calibrate the damping factor for the impact of the concrete block onto the concrete surface. The impact test without drapery was used for this purpose. Block velocity and trajectory predicted by the model were compared to the experimental data (see Figure 3).

Figure 3a compares the evolution of block vertical velocity from the model and the test without wire net (negative value of velocity refers to the block downward movement). In this test, only one impact was observed, which corresponds to point A. Point B is the moment when the block bounces off the slab after impact. Before and after impact, the velocity increases under gravity.

A scattering of the experimental data can be explained by the inadequate lighting of the images that makes it difficult to track the block using Tracker software. Figure 3b and 3c depict the experimental and numerical block trajectory, respectively. A good agreement can be seen between the experimental and numerical evolution of velocities, in particular in the vicinity of impact. This suggests that the model can satisfactorily reproduce the interaction between the surface and the block. Note that accurate quantitative comparison of block trajectory was not possible because block rotation cannot be captured through two-dimensional analysis of the tests results. This is a limitation of the current set up.

Following due calibration of all model parameters, the trajectory of the block between the surface and the drapery mesh was modelled and compared to experimental data (see Figures 4a and 4b).

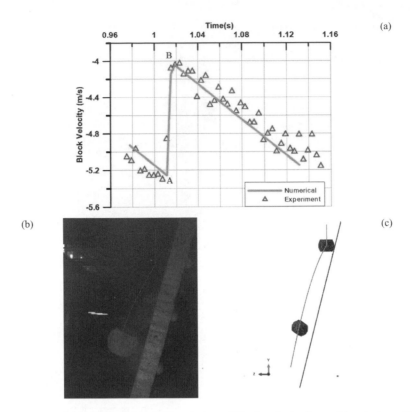

Figure 3. (a): Experimental and numerical evolution of block velocity (test without wire net). View of experimental (b) and numerical (c) block trajectory.

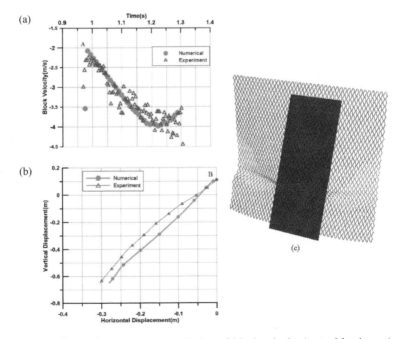

Figure 4. (a): Experimental and numerical evolution of block velocity (test with wire net). (b): Experimental and numerical block trajectory. (c): The concrete block travelling between the drapery and the concrete slab.

Point A in Figure 4a refers to the moment when the block bounced off the slab while point B in Figure 4b refers to the moment when the block touched the drapery. Block rotation decreases after it reaches the drapery. Therefore, the moment when the block touches the wire net was selected as a start point for trajectory comparison (point B in Figure 4b). The result shows that the model is capable of capturing the block velocity and trajectory with a reasonable accuracy.

5 CONCLUSIONS

A numerical model simulating the 3D chain-link drapery was developed in ABAQUS finite element code. The model was calibrated and validated against laboratory-scale tests. Two series of tests with and without draperies were used to calibrate and validate the block/surface and the drapery/surface interactions respectively. The model proved to be able to satisfactorily capture block velocity and trajectory. The results of this study will be used in order to develop a finite element model of a full-scale chain-link drapery that will be used for parametric study on different design factors of such systems.

REFERENCES

Bertolo P, Oggeri C, Peilab D (2009) Full-scale testing of draped nets for rock fall protection. Canadian Geotechnical Journal 46(3):306–317.
BSI (2008) Code of practice for temporary works procedures and the permissible stress design of falsework. In: BS 5975.
Giacomini A, Thoeni K, Lambert C, Booth S, Sloan SW (2012) Experimental study on rockfall drapery systems for open pit highwalls. International Journal of Rock Mechanics and Mining Sciences 56:171–181.
Muhunthan B, Shu S, Sasiharan N, Hattamleh OA (2005) Analysis and design of wire mesh/cable net slope protection. In: Washington State Department of Transportation, Washington.
Persson BNJ (2000) Sliding friction: physical principles and applications, nanoscience and technology. Springer.
Sasiharan N, Muhunthan B, Badger TC, Shu S, Carradine DM (2006) Numerical analysis of the performance of wire mesh and cable net rockfall protection systems. Engineering Geology 88(1–2):121–132.
Tahmasbi S, Giacomini A, Buzzi O, Wendeler C (2017) Preliminary 3d Modelling Of Chain-Link Tecco Mesh For Rockfall Protection. In: 6 t h Interdisciplinary Workshop on Rockfall Protection (Rocexs2017). Barcelona, Spain, pp 165–168.
Thoeni K, Giacomini A, Lambert C, Sloan SW, Carter JP (2014) A 3D discrete element modelling approach for rockfall analysis with drapery systems. International Journal of Rock Mechanics & Mining Sciences 68:107–109.

Geomechanics and Geodynamics of Rock Masses – Litvinenko (Ed.)
© 2018 Taylor & Francis Group, London, ISBN 978-1-138-61645-5

Analysis of instability mechanisms of a high rock prism standing on a cliff face

Luca Verrucci
Department of Structural and Geotechnical Engineering, Sapienza Università di Roma, Rome, Italy

Paolo Tommasi & Anita Di Giulio
CNR—Institute for Environmental Geology and Geo-Engineering, Rome, Italy

Paolo Campedel
Geological Survey, Provincia Autonoma di Trento, Trento, Italy

Tatiana Rotonda
Department of Structural and Geotechnical Engineering, Sapienza Università di Roma, Rome, Italy

ABSTRACT: The analysis of possible failure mechanisms of a 50-m-high rock prism is presented. The case well represents a situation typical of many rock cliffs. The reconstruction of the geometry through remote and direct surveys and the mechanical characterization through laboratory and in-situ tests allowed to build a geotechnical 3D model. Analyses were conducted with the finite difference code FLAC3D where discontinuities are reproduced by interface elements. Under different hypotheses, the severity of the state of stress was evaluated to investigate which mechanism, between the failure of the rock sustaining the prism at its base or shearing along discontinuities, is more critical. The results of the numerical analyses address a more realistic evaluation of safety margins and criteria for design of stabilization measures.

1 INTRODUCTION

Large rock prisms threatening infrastructures and inhabited areas are frequently described in literature (e.g. Adachi *et al.*, 1991; Barbero & Barla, 2010; Bonilla-Sierra *et al.*, 2014). Even though these prisms stand since historical times, the high potential risk for anthropic activities requires appraisal of safety margins against collapse and, possibly, monitoring of displacements.

Once detailed investigations are carried out many of these prisms appear to be either not completely isolated or isolated but not removable, so that stability conditions largely depend on strength and deformability of the surrounding rock. As a consequence, safety margins and failure mechanisms cannot be evaluated only with limit equilibrium methods but through analyses that explicitly model actual 3D geometry and consider stress-strain behaviour of both discontinuities and rock mass.

In the case here presented, a 50-m-high prism 5000 m^3 in volume, preliminary analyses suggests that discontinuity orientation and boundary conditions ensure prism locking. But since the prism stands 80 meters above the valley floor and threatens an inhabited area and a major highway, numerical analyses were conducted to evaluate the state of stress in the rock mass constraining the prism and to individuate possible collapse mechanisms, in case progressive plasticization of the rock mass occurs. Even though the prism has never shown tangible signs of movements, the present study aims to appraise its safety margins and is preparatory to the design of stabilization measures.

2   GEOLOGICAL, GEOMORPHOLOGICAL AND STRUCTURAL SETTING OF THE AREA

The prism stands on a 200-m-high sub-vertical cliff located on the left flank of the Valle dei Laghi, a NNE-SSW-oriented glacial valley (Fig. 1a) formed during the last Ice Age by a south-west branch of the large ice tongue of the Adige River Valley. The valley is carved into the Permian-Cenozoic sedimentary succession of the South Alpine structural domain. In particular the prism consists of a Jurassic carbonate peritidal sequence (Monte Zugna Formation) including fine-grained dolomitic limestones, stromatolithic limestones and micritic limestones.

The site is part of the hanging wall of the Paganella-Toblino Miocene thrust (buried by the quaternary valley sediments), a major transpressive structure verging toward SE and directed NE-SW, i.e. more or less parallel to the rock cliffs. Later tectonics (Upper Miocene-Pliocene) produced a set of sub-vertical left-lateral strike-slip NNW-SSE faults that are visible few hundreds of meters to the NE of the prism.

3   ROCK MASS STRUCTURE AND PRISM GEOMETRY

The morphology of the rock cliff was surveyed by the Geological Survey of Provincia Autonoma di Trento (PAT) through terrestrial laser scanner (TLS) and unmanned aerial vehicle (UAV). Data were locally integrated with the DTM of the PAT obtained from aerial LIDAR surveys. Rope access scanline surveys were conducted to measure spacing, orientation and surface conditions of joints (Andreis, 2016).

The prism has two free faces (Fig. 1c): the front F1 (cliff face) and the right side F2. Layers vary in thickness between few tens of centimetres to two metres; in the dolomitic facies they are often separated by thin pelitic horizons. Bedding joints, which dip on average at 16° towards NNE, delimit the prism both at top and bottom (ST and D3 in Fig. 1c). At the back the prism is delimited by the major discontinuity D1S1 and the D1-S2 joint, in the lower-left part, which together to the cliff face belong to the NNE-SSW set of faults. Finally, the right side of the prism is formed by the D2-S2 and D2-S1 joints (Fig. 1d). These two joints and the back discontinuity D1-S1 form a small dihedral angle which entails a deep embedding of the prism into the rock mass, thus giving a significant constraint. Discontinuities are generally gaping (opening is 0.1–0.2 m for D1-S1 and D2-S2) and persistent (trace lengths longer than 10 m).

At the very underneath of the prism, is visible the remnant of a pillar, being the left part collapsed thus causing overhanging of the prism. The pillar, with the shape of a slender column, constitutes a basal pedestal for the prism. At a closer observation the exposed ST joint surface forming the prism bottom is pervaded by closely-spaced irregular vertical joints parallel to the cliff, which seem an effect of a severe state of stress in the rock mass.

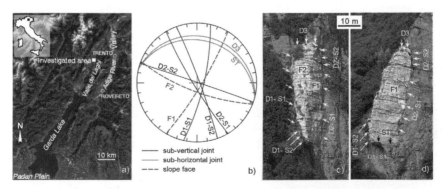

Figure 1.   Site location (a), stereo-plot with planes delimiting the prism (b), view of the prism from SE (c) and E (d). Joint traces are indicated with arrows.

4 MECHANICAL CHARACTERIZATION OF THE ROCK MASS

4.1 *The rock material*

Specimens, 42 mm in diameter, for laboratory tests were cored from ten blocks collected during the rope access surveys. According to texture two lithotypes were identified: limestone and cataclastic limestone. Nevertheless it was not possible to estimate two distinct sets of mechanical parameters. Their physical and mechanical properties are summarized in Table 1.

According to Deere-Miller classification, the intact material exhibits medium to high strength (classes C and B) and a high modulus ratio. Three triaxial compression tests were performed in the Hoek triaxial cell, at confining pressures up to 10 MPa. These results were compared with those obtained during a more extensive investigation campaign (Tommasi *et al.*, 2009) carried out by the authors on the same formation, south of Rovereto (Fig. 1a). Data were fitted with a non-linear Hoek-Brown curve, which provides values of the two indexes σ_{ci} and m_i of 85 MPa and 14 ± 2, respectively.

4.2 *Rock mass characterization*

The structural survey identified two homogeneous portions of the rock mass differing from each other in spacing and surface conditions of discontinuities: the prism and its pedestal. In particular the pedestal is more fractured and discontinuities are more weathered. For these two zones both Rock Mass Rating (Bieniawski, 1989) and *GSI* (Hoek *et al.*, 2002) were estimated and parameter of the Hoek-Brown strength criterion calculated assuming a disturbance factor D (Hoek *et al.*, 2002) of 0.7, i.e. accounting for the stress relief experienced by the slope. Values are reported in the next section. The Young modulus of the rock mass was estimated from GSI and the modulus measured on rock specimens through the expression proposed by Hoek and Diederichs (2006).

4.3 *Characterization of the discontinuities*

The shear strength of the discontinuities delimiting the prism, due to their high persistence, was considered to be dependent on a dilative and a frictional contribution of asperities. A friction angle of 32° was assumed along asperities, i.e. that obtained from tilt tests on saw-cut slabs from Rovereto. Dilative contribution at small scale was estimated from parameters JRC_0 and JCS_0 of Barton's criterion (Barton, 1973) measured during the scanline surveys and equal to 12 and 50 MPa, respectively. These values were scaled as function of the ratio L_0/L_n where L_0 was 0.1 m and L_n is the length of the sliding surface (Bandis, 1990), i.e. about 15 m. To the shear strength angle calculated with the Barton criterion, the angle of large-scale undulations (2.6°) calculated from dip measurements along joints, was added. The Barton curve was linearized in the range of normal stresses expected along the sliding surface (between 0 and 0.5 MPa) obtaining the equivalent friction angle assumed in the numerical analyses.

Table 1. Physical and mechanical properties of rock materials (Nr: number of specimens).

	Limestone		Cataclastic limestone	
	Nr	Mean value	Nr	Mean value
Bulk density (Mg/m³)	13	2.70	2	2.66
Porosity (%)	13	2.58	2	4.08
P-wave velocity (km/s)	14	6.2	4	5.7
Brazilian strength (MPa)	7	6.3	1	6.3
UCS (MPa)	7	87.1	2	78.2
Young modulus (GPa)	2	77.9	1	43.4

925

5 NUMERICAL MODEL

The geometrical model of the prism and neighbouring rock mass was based on the point cloud of TLS and UAV surveys. The surrounding slope surface was an interpolation of the PAT's DTM. The rock mass (Fig. 2) was discretized into 35 500 tetrahedral zones (1.5 m to 20 m in edge). Analyses were performed with the 3D finite difference code FLAC3D (Itasca, 2013). Vertical and bottom boundaries, where normal displacements are constrained, are some 100 m far from the prism. The prism, compounded by some 7000 zones, is separated from the rock mass by the six discontinuities (Fig. 2c), modelled as triangular interface elements.

Two models were analysed: in the LI model (limited interface), joints are to the prism faces; in the EI model (extended interface), the sub-vertical D1-S1 and D2-S1 joints, more realistically, extend into the lower portion of the rock mass thus delimiting a pillar on which the prism leans along the bottom joint ST. The constraint exerted on the prism by the small joint D1-S2 was investigated.

Parameters of the Hoek-Brown strength criterion, assigned with an associated flow rule (Cundall *et al.*, 2003), are listed in Table 2. Interfaces have a purely elastic-perfectly plastic behaviour with only frictional strength ($\varphi = 39.5°$).

Simulation of a realistic geological history would have been arduous and not essential for evaluating equilibrium condition. The geo-static equilibrium (phase-0) is reached assigning an elastic behaviour to both the continuum (a single homogeneous material with GSI = 55) and interface elements. Two phases follow: phase-1) interfaces are "unlocked" (actual shear strength is assigned) and the material below the ST joint is assigned deformability according to its quality (GSI = 36); phase-2) introduction of the plastic behaviour of the continua.

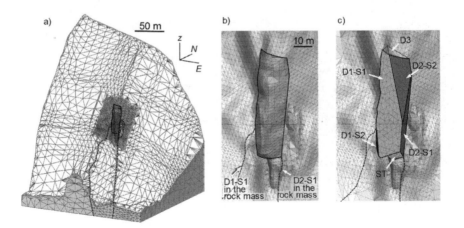

Figure 2.   Complete model grid (a), detail of the prism (b) and of the niche without the prism (c).

Table 2.   Mechanical properties of the zones forming the rock mass.

Model zone	Quality index GSI –	Bulk density (Mg/m³)	Young modulus (GPa)	Poisson ratio –	Parameters of HB criterion			
					σ_{ci} (MPa)	m_b –	s –	a –
Prism and rock mass above ST	55	2.6	9.7	0.3	85	1.18	0.0015	0.504
Rock mass below ST	36		3.1			0.42	0.0001	0.515

6 RESULTS OF NUMERICAL ANALYSES

The analyses on the LI model show that in the phase-2 displacements relative to the surrounding rock-mass are limited to 1–2 mm. The prism completely detaches along both the upper joint D3 and the right lateral joint D2-S2, while it loses contact only along the upper portions of the back joint D1-S1 and the lateral joint D2-S1. The higher normal reaction is given by the bottom joint ST (maximum normal stress $\sigma_{n,max} > 2.4$ MPa in Fig. 3b). Its outer portion (the edge of the basal pedestal) locally experiences small plastic strains (not exceeding 1%, Fig. 3a). In model LI the presence of the small D1-S2 joint does not influence stress and displacement distribution. Movements in the North-South plane, parallel to the cliff face, induce a slight rotation towards North (right) with a relative horizontal displacement of about 3–4 mm.

In the EI model, the displacement of the prism increases up to 4–5 mm. The pedestal is encompassed in the displacement field of the prism (Fig. 4a) and plastic strains reach 5%. Significant reaction stresses are spread also over the upper part of the joint D2-S1 on the right side of the pedestal and over the small surface of D1-S2 joint (normal and shear stresses reach 1 MPa and 0.7 MPa, respectively) which is activated by the larger movements of the prism (Fig. 4b). Nonetheless the strength assumed for the rock mass is sufficient to prevent plastic strains in the prism.

Once ascertained that friction along lateral joints and the support of the basal pedestal are the main contributions to the static equilibrium of the prism, safety margins were evaluated through a strength reduction method applied to two scenarios: a) both joints D1-S1 and D2-S1 extend into the lower rock mass, as in the EI model described above; b) only

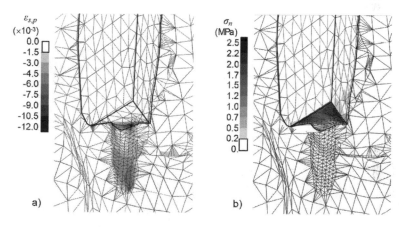

Figure 3. LI model. Octahedral shear plastic strains (a) and normal contact stress on the joints of the lower part of the prism (b). Views without the prism.

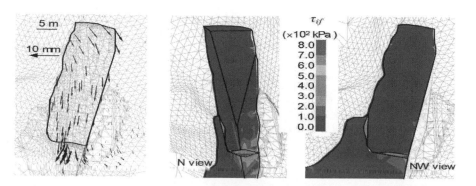

Figure 4. EI model: displacements (a); shear stresses along interfaces viewed from South (b) and SE (c).

D1-S1 extends downwards whilst D2-S1 terminates into the rock mass above the pedestal or is markedly discontinuous. The b) hypothesis is supported by the fading of the trace of joint D2-S1 on the cliff below the ST joint.

The prism maintains the equilibrium even when the shear strength angle is reduced to low values (10% of the initial value in the a-scenario) or nil (b-scenario). Conversely, collapse is attained if the original σ_{ci} value of the rock of the basal pedestal is reduced to the 95% or 75% in a- and b-scenario, respectively.

7 CONCLUSION

Evaluation of safety margins against failure of the limestone prism described in this study indicates that frequently, to account realistic collapse mechanisms, problems usually treated through limit equilibrium of rigid blocks requires modelling of deformability and irreversible deformations in the rock mass and along discontinuities. In the investigated case, kinematical analysis of the rigid prism based on the sole slip would not envisage vertical movements. Conversely if continuum deformability and plastic threshold of joints and rock mass are introduced, the prism detaches at the top (joint D3) and along the upper portions of lateral joints D1-S1 and D2-S1 due to the partial yielding of the rock extending below the prism base (pedestal). In this phase the vertical displacements are associated also to a slight lateral rotation and to an increase of normal stress on both the bottom joint ST and the inner portions of lateral joints. The prism is mainly sustained by the basal pedestal under the joint ST, with a low safety margin against its internal failure. Friction on lateral discontinuities provides however an important contribution, although their sub-vertical orientation.

The numerical model accounts for deformation and damage features observed on the cliff, though calculated displacements appear to be an order of magnitude lower than those observed. The study also leads to an increased preparedness in managing possible emergency conditions and is part of a transparent policy towards local institutions and citizens.

REFERENCES

Andreis F. (2016). Rilevamento geomeccanico e geostrutturale in sito di alcuni blocchi posti lungo la parete rocciosa a monte del campo sportivo della frazione Sarche nel comune di Calavino. *Technical Report,* Provincia Autonoma di Trento – Servizio Geologico.

Adachi T., Ohnishi Y., Arai K. (1991). Investigation of toppling slope failure at Route 305 in Japan. *International Congress of Rock Mechanics* Aachen, 2:843–846.

Barbero M., Barla G. (2010). Stability analysis of a rock column in seismic conditions. *Rock Mech. Rock Eng,*, 43:845–855.

Barton N. (1973). Review of a new shear strength criterion for rock joints. *Eng. Geol.*, 7, 287–332.

Bandis S. (1990). Scale effects in the strength and deformability of rocks and rock joints. *1st Int. Workshop on Scale effects in rock masses*, Loen, Norway, N. Barton & O. Stephansson (ed.), Balkema, Rotterdam, 59–76.

Bieniawski Z.T. (1989). Engineering Rock Mass Classifications. Wiley, Chichester.

Bonilla-Sierra V., Donzé F.V., Scholtès L., Elmouttie M. (2014). Coupling photogrammetric data with a discrete element model for rock slope stability assessment, *Eurock 2014: Rock Engineering and Rock Mechanics: Structures in and on Rock Masses*, Alejano, Perucho, Olalla, Jiménez Eds. 433–438.

Hoek E., Carranza-Torres C., Corkum B. (2002). Hoek-Brown failure criterion - 2002 Edition. *Proc. NARMS-TAC Conference*, Toronto, 1, 267–273.

Hoek E., Diederichs M. (2006). Empirical estimates of rock mass modulus. *Int. J Rock Mech. Min. Sci.*, 43, 203–215.

Itasca (2013). FLAC3D Fast Lagrangian Analysis of Continua in 3 Dimensions. Minneapolis.

Tommasi P., Verrucci L., Campedel P., Veronese L., Pettinelli E., Ribacchi R. (2009). Buckling of high natural slopes: The case of Lavini di Marco (Trento-Italy), *Eng. Geol.*, 109, 93–108.

The propagation of hydraulic fractures in coal seams based on discrete element method

Yanjun Lu

Department of Geology, Moscow Lomonosov State University, Moscow, Russian Federation

Zhaozhong Yang

State Key Laboratory of Oil and Gas Reservoir Geology and Exploitation, Southwest Petroleum University, Chengdu, China

V.V. Shelepov & Jinxuan Han

Department of Geology, Moscow Lomonosov State University, Moscow, Russian Federation

Xiaogang Li

State Key Laboratory of Oil and Gas Reservoir Geology and Exploitation, Southwest Petroleum University, Chengdu, China

Yongsheng Zhu

Itasca Consulting China Ltd., Wuhan, China

Junfeng Guo

Shanxi CBM Exploration and Development Subsidiary Company of SINOPEC, Jincheng, China

Zhongliang Ma

Hua Tugou Town, Haixi, Qinghai Province, Qinghai, China

ABSTRACT: The simulation of hydraulic fractures is an important research content that can guide the engineering practice to achieve the purpose of increasing production. Coal has approximately orthogonal face cleats and butt cleats resulting in the discontinuous performance. Discrete Element Method (DEM) has obvious advantages in studying mechanical properties of discontinuous materials, so based on the distribution characteristic of cleats in coal, the research on the propagation of hydraulic fractures is carried out via DEM. The simulated results show that: hydraulic fractures mainly propagate along the cleats towards the maximum principal stress. The fracture network can be formed due to the intersection of face cleats and butt cleats that can propagate at certain pressure. The general variation trends of 3DEC numerical simulated results are consistent with physical experimental results at the same condition. With the increase of injection rate and fracturing fluid viscosity, the maximum aperture of hydraulic fracture increases, while the length of principal hydraulic fracture shortens. Therefore, to achieve the purpose of forming hydraulic fracture network in coal seams with cleats, low viscosity fracturing fluid and low injection rate need be applied to the fracturing technology. As the cleat density increases, the number of branched fractures increases, but the length of principal hydraulic fracture becomes short.

Keywords: Coal, hydraulic fracturing, discrete element method, numerical simulation, cleat, hydraulic fracture

1 INTRODUCTION

Hydraulic fracturing is the key technology for realizing the industrial exploitation of CBM. It is estimated that hydraulic fracturing is applied in most of CBM wells in the United State, and the CBM wells with more than 1000 m³/d almost have been stimulated by hydraulic fracturing in China (Zhang et al., 2006). Artificial fractures induced by fracturing can effectively connect the wellbore with the reservoir, which can promote desorption and diffusion of CBM and increase the CBM production.

Geological parameters, fracturing fluid properties and construction parameters can have impact on the morphology of artificial fractures. Laboratory study on true triaxial fracturing of coal and fracture monitoring at the site show that hydraulic fractures in coal seams mainly initiate and propagate along the cleats, and complex and irregular multi-fractures are asymmetric distribution (Abass et al., 1990; Zhang et al., 2013; Yang et al., 2012). Conventional two-dimensional and three-dimensional fracture models can be applied in the specified coal seams to study the formation of symmetric fractures, but their applications are restricted in most of coal seams. To solve the problem of fracture propagation in the fractured reservoirs such as coal seams, a lot of work have been carried out in the numerical simulation.

At present, discrete model of fracture network (Meyer et al., 2010), wire-mesh model (Xu et al., 2010) and unconventional fracture model (Weng et al., 2011) are mainly applied in the fractured reservoirs. In recent years, discrete element method (DEM) has been applied to the simulation of hydraulic fractures Zangeneh et al. (2012) used two-dimensional DEM (UDEC software) to simulate the hydraulic fracture propagation. Nagel et al. (2011) researched the law of fracture initiation and propagation and their influence factors in the fractured reservoirs with three-dimensional DEM. Hamidi and Mortazavi (2014) developed and simulated fracture initiation and propagation in the fracturing process based on three-dimensional DEM. Savitski et al. (2013) studied the interaction between hydraulic fracture propagation and generated discrete fracture network via DEM.

The previous results show that complex fractures can be formed mainly due to the interference of natural fractures in the process of hydraulic fracturing, so the key of different simulated methods reveals the interaction of different fractures. The existing numerical simulations of fractures are mainly aimed at shale and tight sandstone, while the relevant simulated research on fracture propagation is less in the field of coal. Coal has approximately orthogonal distribution of face cleats and butt cleats that are not possessed in the other fractured reservoirs. DEM has a great advantage in dealing with the problems on discontinuous structural mechanics of natural fractured reservoirs. In this paper, the propagation laws of hydraulic fractures in coal seams are studied based on DEM, which can guide the engineering practice of fracturing in coal seams.

2 SIMULATED DETAILS

DEM firstly proposed by Cundall (1971) is a numerical simulated method to specially solve the problem of discontinuous medium. This method considers the rock composed of discrete blocks and cleats. Blocks can shift, rotate and deform, while cleats can be compressed,

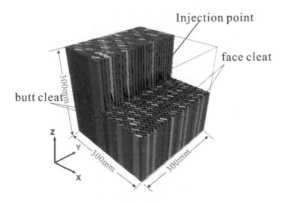

Figure 1. Physical model of coal.

Table 1. Simulated parameters of coal.

Type	Parameter	Value
Petrologic parameters	Density (g/cm^3)	1.4–1.6/1.5
	Cohesion (MPa)	6.02–8.06/7.04
	Internal friction angle (°)	20.38–24.48/22.43
	Tensile strength (MPa)	0.53–3.18/1.855
	Young's modulus (GPa)	4
	Poisson's ratio	0.2–0.4/0.3
	Residual hydraulic aperture (μm)	1
	Aperture at zero stress (μm)	10
	Maximum hydraulic aperture (μm)	1000
Boundary conditions &	Vertical stress (MPa)	10
Initial conditions	Maximum horizontal principal stress (MPa)	5
	Minimum horizontal principal stress (MPa)	2

separated and slid. Therefore, the rock can be considered as the discontinuous and discrete medium that more actually reflect the nonlinear deformation characteristics of the rock with cleats. In addition, this method can express any complicated constitutive relation as tiny increments in the calculation, and no solution equations do not exist. The program of 3DEC was developed by Cundall and ITASCA consulting group. The FISH language embedded in the 3DEC program can make users define new variables and functions, which can extend the computing program and add user-defined characteristics.

True triaxial fracturing experiment is a method of fracturing research that can be used to understand the mechanism of fracture initiation and propagation, and then experimental results can guide engineering practice of fracturing. However, true triaxial fracturing experiment has the problem of high cost and difficult sampling, so its widespread application is restricted. To solve the problem, numerical simulation and physical experiment are combined, which can not only improve the accuracy of numerical simulation but also widely research fracturing mechanism of rocks and its relevant influence due to the low cost. Numerical simulation model is established based on physical model. In this paper, the established numerical model is shown in Fig. 1, the parameters of model is listed in Table 1.

3 SIMULATED RESULTS

3.1 The comparison of physical experiment and numerical simulation

The physical experiment (Fan and Zhang, 2014) is simulated with 3DEC at the same condition. In the simulation, the injection rate is 10 ml/min, and viscosity and density of fracturing fluid are, respectively, 65 mPa.s and 1000 kg/m^3. Model establishment, parameter assignment and iterative simulation are carried out via FISH program language. The simulated results are shown in Fig. 2 and Fig. 3.

3DEC simulated results indicate that hydraulic fractures propagate along the maximum horizontal principal stress. Hydraulic fractures propagate along the cleat, and the asymmetrical branched fracture network is shown. When initial pressure is 4.1 MPa, with the continuous injection of fracturing fluid, the pressure spreads rapidly with the elliptic to the boundary in the fracture, and the pressure curve fluctuates during the stage of fracture propagation. When fluid pressure spreads to the cleat, fracture propagation is suppressed easily resulting in the pressure fluctuation. Finally, the fracture propagates along the cleat at high pressure, and the branched fracture is formed. With the continuous injection of fracturing fluid, the maximum aperture increases near the injection point, while the aperture gradually decreases along the flow surface. When hydraulic fractures encounter the cleats, the cleat at the maximum principal stress preferentially propagates.

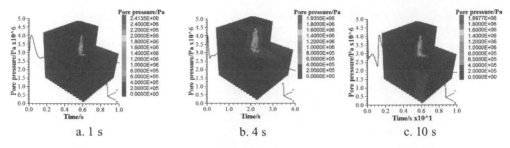

a. 1 s   b. 4 s   c. 10 s

Figure 2.   Pressure distribution in the hydraulic fracture.

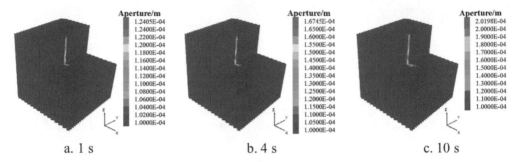

a. 1 s   b. 4 s   c. 10 s

Figure 3.   The distribution of hydraulic fracture aperture.

Table 2.   Simulated parameters at different influence factors.

Number	Injection rate (ml/min)	Fracturing fluid viscosity (mPa·s)	Cleat density (cleats per meter)	
			Butt cleat	Face cleat
01	10	65	50	100
02	20	65	50	100
03	30	65	50	100
11	20	10	50	100
12	20	30	50	100
13	20	65	50	100
21	20	65	12.5	25
22	20	65	25	50
23	20	65	50	100

Compared numerical simulation with physical experiment, the results show that the hydraulic fractures are the approximately vertical fractures propagating along the maximum principal stress. The injection pressure fluctuates during the propagation of fractures. When hydraulic fractures encounter the cleats, the fracture preferentially propagates along the cleat at the maximum principal stress. The final morphology of fractures is the multi-fractures, and the principal fracture locates at the center of model. In the physical experiment, hydraulic fractures in coal are mainly composed of propagated cleats, and the new fracture easily shifts due to its intersection at the cleat, which is consistent with the numerical simulated result. Therefore, the general trends of simulated results are in accordance with the experimental results, indicating that 3DEC simulated method has validity and feasibility.

3.2   *The analysis of influence factors*

To further study the influence factors of hydraulic fracture propagation, the simulations on fracture propagation are carried out at different parameters that are listed in Table 2. The same injection time or the same injection rate as the precondition is used to analyze the influence of different factors in the simulation.

3.2.1 Injection rate

Injection rate as an important parameter for fracturing operation can have impact on proppant carrying capacity of fracturing fluid, fracture height control, leak-off loss and so on. Therefore, simulated results on fracture parameters (Fig. 4) are analyzed at different injection rates.

As is shown in Fig. 4, with the increase of injection rate, the initial pressure of hydraulic fracture increases, and injection rates 10 ml/min, 20 ml/min and 30 ml/min correspond to initial pressures 4.1 MPa, 7.9 MPa and 11.8 MPa, respectively. The pressure fluctuates during the facture propagating, and when the propagation pressure tends to be stable, the greater the injection rate is, the higher the propagation pressure is. As the injection rate increases, the maximum hydraulic fracture aperture increases at the same injected fluid volume, while the length of principal hydraulic fracture becomes short, and branched fracture numbers are less.

3.2.2 Fracturing fluid viscosity

Fracturing fluid viscosity can affect the morphology of hydraulic fractures. The higher viscosity of fracturing fluid is not only beneficial to proppant carrying but also has the advantage in controlling fracture height. Simulated results at different viscosity of fracturing fluid are shown in Fig. 5.

With the increase of fracturing fluid viscosity, the initial pressure of hydraulic fracture increases, and when the propagation pressure tends to be stable, the higher the viscosity is, the

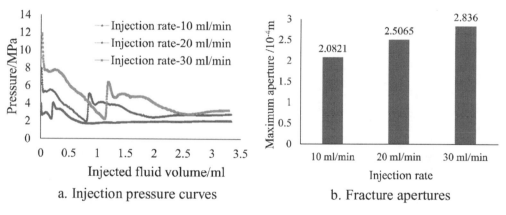

a. Injection pressure curves   b. Fracture apertures

Figure 4.   Simulated results at different injection rates.

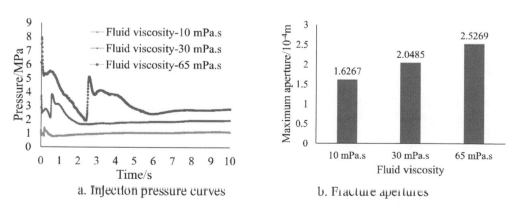

a. Injection pressure curves   b. Fracture apertures

Figure 5.   Simulated results at different viscosity of fracturing fluid.

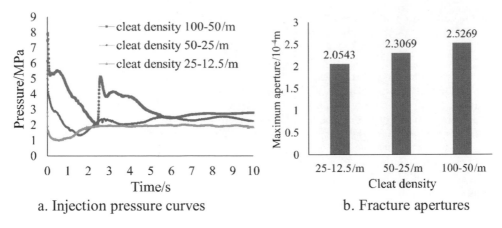

a. Injection pressure curves   b. Fracture apertures

Figure 6.   Simulated results at different cleat densities.

greater the propagation pressure is. As the viscosity of fracturing fluid increases, the maximum hydraulic fracture aperture increases at the same injection time, while the number of branched fractures lessens, and the principal hydraulic fracture with shorter length is formed.

3.2.3   *Cleat density*

Cleats are natural fractures formed in the process of coalification, and the number of cleats is the key geological factor for the fracture network formed. The principal hydraulic fracture usually propagates along the maximum horizontal principal stress. However, it will propagate along the cleats after encountering the cleats resulting in the formation of branched fractures and complicated fracture network. The influence of cleat density on fracture network is analyzed, and the simulated results at different cleat densities are shown in Fig. 6.

With the cleat density increasing, initial pressure and stable propagation pressure increase, and at the same injection time, the maximum hydraulic fracture aperture increases, but the propagated length of principal hydraulic fracture in the coal with high cleat density becomes short.

4   CONCLUSIONS

Relative to other fractured reservoirs, the cleats in coal seams have higher density and better regularity. However, the cleat existence makes coal have the discontinuous feature, so 3DEC has obvious advantages in mechanical study on discontinuous materials. In this paper, based on the simulated results from the 3DEC program, the following conclusions are obtained:

1. The numerical simulated results on initiation, propagation and morphology of hydraulic fractures are consistent with physical experimental results, indicating that 3DEC simulated method has validity and feasibility.
2. Hydraulic fractures propagate with the elliptic from the injection point to the boundary, and the spread of injection pressure has the priority, which means that fracturing fluid preferentially reaches the boundary with the leak-off, and yet the variation of fracture aperture is relative hysteresis. The pressure fluctuates when the fracture encounters the cleat during the propagation, and the higher fracture fluid pressure can break through the intersection of fractures. When the fracture reorients, the cleats have the advantage in propagating along the maximum principal stress.
3. Maximum fracture aperture near the injection point, initial pressure and propagation pressure are positively associated with injection rater, fracturing fluid viscosity and cleat density, but the length of principal hydraulic fracture is negatively correlated with these parameters. Therefore, to achieve the purpose of forming hydraulic fracture network

in coal seams with cleats, low viscosity fracturing fluid and low injection rate need be applied to the fracturing technology. At the same construction condition, long and narrow hydraulic fractures are easily formed in coal seams with lower cleat density.

REFERENCES

Abass H H, Van Domelen M L, El Rabaa W M. Experimental Observations of Hydraulic Fracture Propagation Through Coal Blocks [C]. SPE Eastern Regional Meeting, 31 October-2 November, Columbus. Ohio, USA, 1990.

Fan Tiegang, Zhang Guangqing. Influence of injection rate and fracturing fluid viscosity on hydraulic fracture geometry in coal [J]. Journal of China University of Petroleum (Edition of Natural Science), 2014, 38(4): 117–123.

Gundall P A. A computer model for simulating progressive large scale movement in block rock system [J].Symposium ISRM, 1971, Proc 2:129–136.

Hamidi F, Mortazavi A. A New Three Dimensional Approach to Numerically Model Hydraulic Fracturing Process [J]. Journal of Petroleum Science and Engineering, 2014, 124: 451–467.

Meyer B R, Bazan L W, R H Jacot, et al. Optimization of Multiple Transverse Hydraulic Fractures in Horizontal Wellbores [C]. SPE Unconventional Gas Conference, 23–25 February, Pittsburgh, Pennsylvania, USA, 2010.

Nagel N, Gil I, Sanchez-Nagel M. Simulating Hydraulic Fracturing in Real Fractured Rocks—Overcoming the Limits of Pseudo3D Models [C]. Paper SPE 140480 presented at the SPE Hydraulic Fracturing Technology Conference, The Woodlands, Texas, 24–26 January 2011.

Savitski A A, Lin M, Riahi A, Damjanac B, Nagel N B. Explicit modeling of hydraulic fracture propagation in fractured shales. IPTC-17073-MS, 2013.

Weng X, Kresse O, Cohen C, et al. Modeling of Hydraulic-Fracture-Network Propagation in a Naturally Fractured Formation [J]. SPE Production and Operations, 2011, 26(4): 362–368.

Wenyue Xu, Marc Thiercelin, Utpal Ganguly, et al. Wiremesh: A Novel Shale Fracturing Simulator [C]. Paper SPE 132218 presented at CPS/SPE International Oil & Gas Conference and Exhibition. Beijing China, 2010.

Yang Jiaosheng, Wang Yibing, Li Anqi, et al. Experimental study on propagation mechanism of complex hydraulic fracture in coal-bed [J]. Journal of China Coal Society, 2012, 37(01): 73–77.

Zangeneh N, Eberhardt E, Bustin R M. Application of the distinct-element method to investigate the influence of natural fractures and in situ stresses on hydrofrac propagation [C]. 46th US Rock Mechanics/Geomechanics Symposium, Chicago. 2012.

Zhang Ping, Wu Jianguang, Sun Hansen, et al. Analysis the results of the downhole microseismic monitoring technique in coalbed methane well fracturing [J]. Science Technology and Engineering, 2013, 23: 6681–6685.

Zhang Yapu, Yang Zhengming, Xian Baoan. Coal-bed gas stimulation technology [J]. Special Oil & Gas Reservoirs, 2006, 13(1): 95–98.

The influence of the interface of drilled socketed shafts and rock mass on their behavior

Mikhail G. Zertsalov & Valeriy E. Merkin
Department of Soil Mechanics and Geotechnics, Moscow State University of Civil Engineering (National Research University), Moscow, Russia

Ivan N. Khokhlov
LLC Scientific Engineering Center of Tunneling Association, USA

ABSTRACT: Socketed shafts are usually designed and constructed as foundations of high-rise buildings and bridge structures when layers of loose soil overlie bedrock. The behavior of socketed shafts in rock has a lot in common with the large diameter piles in soil. However, the structure of a rock mass and their highly variable mechanical characteristics considerably complicate the calculation of the socketed shaft bearing capacity and settlement in rocks.

The factor, significantly affecting the interaction of socketed shafts with rock mass—the mechanical properties of the sidewall interface, is analyzed in this article as well as some peculiarities of calculations of settlements of socketed shafts in rock under the action of vertical loading. The major factors, affecting the bearing capacity of socketed shafts under loads, are associated with composition, structure and mechanical properties of rock mass. The results of the study of the effect of sidewall interface of socketed shafts in conditions of elasto-plastic problem are presented to assess the tangential stiffness influence on the behavior of the socketed shaft, in particular, on its settlements.

1 INTRODUCTION

The socketed drilled shafts of large diameter are typically used to transfer loads from the structures through layers of soils to stronger underlying rock. The shafts can either be based on the rock, or socketed into it. As world practice shows, the diameter of sockets can usually be changed from 0.5 to 2.0 m, and the length of the socket to reach 10.0 m or more. The arrangement of drilled shafts, as indicated in (Zhang, 2004, NCHRP, 2006) has a number of advantages: the ability to transfer heavy loads and relatively low cost; relatively simple applied construction technology, including a fairly simple process of drilling of shafts; small noise and vibration arising during the process of drilling and construction. Taking this into account, during the construction of bridges and high-rise buildings on soil, underlying bedrock, socketed drilled shafts are considered as the most efficient and economical deep foundations.

Despite the fact that the behavior of the socketed shaft both in soil and in rock is similar, i.e., its bearing capacity is defined by the resistance at the side surface and the strength of the ground under the tip of the shaft, the nature of interaction between the shaft and surrounding rock mass is quite different and it is determined mainly by factors caused by the structural features and mechanical properties of rocks. The results of various studies (Osterberg, Gill, 1973; Pells, Turner, 1979; Donald et al. 1980; O'Neill, Reese, 1999; Zhang, 2004; NCHRP, 2006) show that these factors primarily include: the deformability and strength of rock mass, the ratio between the modulus of elasticity of the shaft material and the modulus of deformation of rock, mechanical characteristics of the interface between the shaft and the rock mass, the ratio between the length of the shaft socket in rock and its diameter.

In (Zertsalov, Nikishkin, 2015; Zertsalov et al., 2017) a method of calculation of bearing capacity and settlements of vertically loaded socketed shafts in rock is proposed. The method is based on the joint use of numerical modeling and the method of experimental design techniques. The experimental design techniques significantly expands the research opportunities, allowing to solve problems of interaction of engineering structures with rock mass, which cannot be solved analytically or by physical modeling. The results of carried out investigations allowed to obtain the *regression equations* – dependencies, linking the response functions (in this case, load applied to the top of the socketed shaft and, corresponding to these load, settlements in the key points *A, B* and *C* of the settlement curve) with the independent factors. As the key points the following ones had been chosen: the start and the end of the failure of the sidewall interface (points A and B), and the beginning of the failure of the shaft material or rock mass (point C). As the independent factors, which mostly affect the shafts behavior, three following factors were used: RQD, E_c/E_r, L/D, where RQD is the rock quality designation, E_c/E_r is the ratio of the modulus of elasticity of shaft material to the modulus of elasticity of the intact rock and L/D is the ratio of the length of the shaft to its diameter. All numerical calculations were performed in conditions of elasto-plastic problem, the Mohr-Coulomb model was used to analyze both the rock mass and interface behavior. The assignment of the limits of variation of the independent factors and values of mechanical characteristics of socketed shaft, rock mass and interface between shaft and rock mass was explicitly described in (Zertsalov, Nikishkin, 2015; Zertsalov et al., 2017). Curves of the shaft settlements based on the results of numerical calculations of the five-meter socketted shafts are presented in Fig. 1 as an example. The bearing capacity of the shaft was determined by summation of the resistance around its side surface and the strength of the rock under the tip of the shaft. The modulus of deformation of the rock mass E_m was determined using the empirical dependencies of ratio of rock mass deformation modulus to the modulus of elasticity of intact rock from RQD - rock quality designation factor (Zhang, 2004).

It is also worth saying, that in the above mentioned papers the factor, significantly affecting the performance of socketed shafts in rock mass—the mechanical properties of the sidewall interface, was not considered in the factor analysis. The influence of this factor was investigated separately. The results of these studies are presented in this paper.

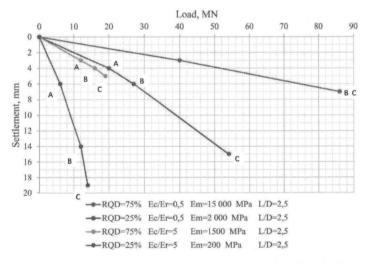

Figure 1.   Settlement curves of socketed shafts (A—the beginning of the failure of sidewall interface, B—the end of the failure of sidewall interface, C—the beginning of the failure of the rock mass or the shaft).

2 NUMERICAL SIMULATION AND ANALYSIS

Numerical simulations with FEM were performed using a special contact element, which allows simulation of behavior of the sidewall interface of the socketed shaft in rock—the process of elastic deformation of the interface and its failure. In Fig. 2a three-dimensional computational model, used in the research, generated using Z-Soil software package, is shown. The rock mass and the shaft body is modeled by 8-node continuum 3D elements. The bottom and outer boundaries of the model are fixed. The distance between side planes and shaft body B has been chosen after studying the influence of mesh boundaries on shaft-rock interaction. The initial stresses have to be unchanging at model boundaries. The depth beneath socketed shaft tip is equal to the embedment length of the shaft.

The analysis of the currently available studies, for example (Ooi, Carter, 1987; Hatami, Bathurst, 2006), shows that mechanical characteristics of the "concrete-rock" interface may vary in a wide range. This was confirmed by full-scale field shaft tests, which show that the value of sidewall interface stiffness can vary significantly depending on the properties of drilled shaft and rock mass. At the same time, the results of the in situ tests of socketed shafts (Williams, Pells, 1981; Horvath et al., 1983; Hoonil Seol et al, 2009) also showed that the state and, as a consequence, the mechanical characteristics of interface (the absence or presence of roughness of the borehole walls, the degree of roughness, the use of bentonite, etc.) have a significant impact on settlements of shaft (Fig. 2b).

Taking this into account, a series of calculations to study the influence of the tangential stiffness of the interface on the shaft settlements under the action of axial compressive loads were carried out. Two socketed shafts with length of 5 m and 20 m in rock mass were studied.

At first in each numerical experiment, the load on the shaft had been increasing to the value at which the failure of the interface between shaft and rock occurred (elastic portion of the settlements curve). The settlements of the shafts were determined for different values of tangential stiffness of the interface: $K_s = 50 \times 10^3$, 100×10^3, 200×10^3, 300×10^3, 500×10^3 and $1 \times 10^6 \, kN/m^3$.

Series of calculations, using the following relationship of the modulus of elasticity of the shaft to modulus of deformation of the rock mass: $E_c/E_m = 125$, 50, 25, 5, 2.5, 1.7, were conducted. Value of the normal stiffness K_n in the calculations was determined using the formula (Boresi, 1965):

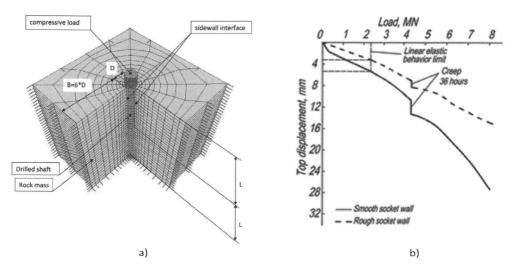

a)   b)

Figure 2. Finite element model of interaction of drilled shaft with rock mass (a), the socketed shaft settlement curves with smooth and rough borehole walls (Horvath et al, 1983) (b).

$$K_n = \frac{E_m}{R(1+\nu_m)} \tag{1}$$

where R is the drilled shaft radius, ν_m is the Poisson ratio of the rock mass and E_m is the rock mass deformation modulus.

In Fig. 3 shown the results of the studies, presented in the form of curves based on the relationships between shafts settlements and the change of the tangential stiffness of the interface (K_s) for different ratios of modulus of deformation of the shaft and rock mass (E_c/E_m), which correspond to rock mass of medium strength. Curves are drawn for values of the settlements at the point A (see Fig. 1) - the end of the elastic portion of the shaft settlement curve.

By analyzing the curves one can see, that with increasing of stiffness (K_s) in the range from 50×10^3 kN/m³ up to 500×10^3 kN/m³ the value of the shaft settlements are reduced in the range of from 2.8 to 3.1 times, depending on the changes of the tangential stiffness and the modulus of deformation of rock mass. At the same time, with increasing of the interface tangential stiffness in the range from 500×10^3 kN/m³ up to 1×10^6 KN/m^3, the value of the shaft settlements changes very little. In this range of the values of the interface stiffness (K_s) it becomes comparable with the stiffness of the rock mass, which primarily determines the magnitudes of the shaft settlements.

Since the limit of variation of tangential stiffness had some uncertainty before preceding studies (Zertsalov et al., 2017, Zertsalov, Nikishkin, 2015), shear stiffness values of "concrete-rock" interface K_s, basing on the analysis of available publications and studies, were taken averaged and equal to $K_s = 100\,000\ kN/m^3$. Therefore, the magnitude of the settlement, calculated using the obtained regression equations, is suitable only for the basic value of K_s. The curves presented on Fig. 3, allow, within the first linear part of shaft deformation curve, to determine shaft settlements in any combination of the values of the tangential stiffness of the interface and deformation modulus of rock mass in the range within the variation of values of these parameters ($K_s = 50 \times 10^3\ kN/m^3 - 1 \times 10^6\ kN/m^3$; $E_m = 0,2 \times 10^6\ kN/m^2 - 15 \times 10^6\ kN/m^2$). Using curves (Fig. 3) it is possible to determine the settlement conversion factor C_s, which is presented in Fig. 4.

With this chart, when the modulus of deformation of rock mass and the corresponding magnitude of the shaft settlement with $K_s = 100 \times 10^3\ kN/m^3$ are known, it is possible to determine the settlement for any value of the interface stiffness - K_s. This requires the settlement value of the socketed shaft with $K_s = 100 \times 10^3$ kN/m³ to be multiplied by the factor C_s.

Similar studies of the influence of the tangential stiffness of the "shaft-rock" interface and deformation modulus of the rock mass on the shafts settlements were made for the point B (Fig. 1) – the end of the interface failure. Analysis of the results showed that in this case the nature and intensity of change of the settlement curves are quite similar to the curves shown

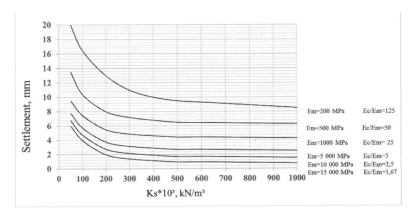

Figure 3. Curves of the dependence of the shaft settlements from the change of tangential stiffness of the interface under different ratios of modulus of deformation of the shaft and rock mass (E_c/E_m).

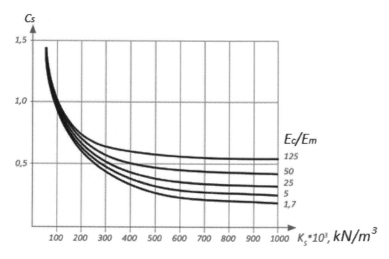

Figure 4. Conversion factor C_s depending from value K_s.

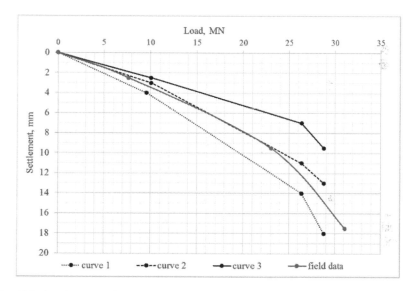

Figure 5. Calculated curves of the shaft settlements compared with field test data.

in Fig. 3, which can be explained by mentioned above, mild nonlinearity of the settlements curves within this part of deformation diagram.

The calculation results obtained using the proposed method were compared with the results of field test of the shaft. As example, the shaft behavior under compression with a diameter of 1.22 m and a length of socket into rock – 4.88 m (Zhang, 2010) is considered. Rock mass is formed by the weathered fractured limestone with the following characteristics: $RQD = 55\%$, the modulus of elasticity of the intact rock $E_r = 32 \times 10^3$ MPa, the modulus of deformation of rock mass $E_m = 4,8 \times 10^3$ MPa. In this case, bearing capacity of the shaft is provided by the resistance along the sidewall and the strength of the rock mass under the shaft. In Fig. 5 there are four curves of the shaft settlements: field test curve and three polylines, constructed using above mentioned regression equation (Zertsalov et al., 2017) and conversion factor C_s. For these three polylines different values of tangential stiffness of the "shaft-rock" interface were accepted: $K_s = 100 \times 10^3$ kN/m^3 (curve 1), 200×10^3 kN/m^3 (curve 2) and 300×10^3 kN/m^3 (curve 3). Comparison of the curves shows that the curves constructed on the basis of numerical simulations, correspond well with the experimental curve, however, the range of deviation (Fig. 5) is completely determined by the value of K_s. For example, values of shaft settlements of the curves 1 and 3 differ from the corresponding curve values of field tests of the shaft not more than 25%, and the difference of the curve 2 does not exceed 7%. The obtained results confirm the conclusion, made after analysis of results of experimental researches, by the necessity of considering the influence of the stiffness of "shaft-rock" interface and obligatory determination of its value by conducting a field full scale shaft test.

3 CONCLUSIONS

1. In the study of the interaction of the socketed shafts with rock mass it is necessary to determine the mechanical characteristics of "shaft-rock" interface, which has a significant influence on the behavior of the shaft not only on the portion of the settlement curve, where interface failures, but also on the portion of elastic deformation of interface.
2. Taking into account, the importance of the parameter of shear stiffness of interface K_s, when considering the behavior of drilled socketed shafts, it is necessary to conduct more detailed and precise analysis of the effect of the stiffness of the "shaft-rock" interface on shaft settlements and obligatory determination of value K_s, when conducting a field full scale shaft test.
3. Taking into account, that the source of information for assigning the values of mechanical properties of sidewall interface of the socketed shafts, used in the above researches, was taken from various sources or obtained empirically, the proposed method of calculation K_s can be used for the evaluation of the interaction between socketed shafts and rock mass only at the stage of preliminary design.

REFERENCES

Boresi, A.P. 1965. Elasticity in engineering mechanics. Prentice-Hall, Englewood Cliffs, N.J.
Donald, I.B., Sloan, S.W., Chiu, H.K. "Theoretical Analysis of Rock Socketed Piles", Proceedings, International Conference on Structural Foundations on Rock, Vol. 1, Sydney, Australia, 1980, pp. 303–316.
Hatami, K. and Bathurst, R.J. "Numerical Model for Reinforced Soil Segmental Walls under Surcharge Loading" Journal of Geotechnical and Geoenvironmental Engineering, 2006, pp. 673–684.
Horvath, R.G., Kenney, T.C., and Kozicki, P. "Methods of Improving the Performance of Drilled Piers in Weak Rock", Canadian Geotechnical Journal, Vol. 20, 1983, pp. 758–772.
NCHRP Synthesis 360, "Rock-Socketed Shafts for Highway Structure Foundations," Transportation Research Board, Washington D.C., 2006, pp. 136.
O'Neil, M.W. and Reese, L.C. "Drilled Shafts: Construction Procedures and Design Methods," Report FHWA-IF-99–025, Federal Highway Administration, Washington D.C., 1999, 758 pp.
Ooi, I.H. and J.P. Carter, "Direct Shear Behavior of Concrete-Sandstone Interface", Proceedings, 6th International Conference on Rock Mechanics, ON, Canada, 1987, pp. 467–470.
Osterberg, J.O. and. Gill, S.A. "Load Transfer Mechanism for Piers Socketed in Hard Soil or Rock," Proceedings, 9th Canadian Symposium on Rock Mechanics, Montreal, ON, Canada, 1973, pp. 235–262.
Pells, P.J.N. and. Turner, R.M. "Elastic Solution for the Design and Analysis of Rock-Socketed Piles," Canadian Geotechnical Journal, Vol. 16, 1979, pp. 481–487.
Seol. H., Jeong, S. and Cho, S., 2009. Analytical Method for Characteristics of Rock-Socketed Shafts. Journal of Geotechnical and Geoenvironmental Engineering, ASCE. June 2009: 778–790.
Vybornov K.A., Sainov M.P. "Influence of Seams Work on Spatial Deflected Mode of Concrete-Face Rock Fill Dam" Vestnik MGSU, Vol. 5, 2011, pp. 12–17.
Williams, A.F. and P.J.N. Pells, "Side Resistance Rock Sockets in Sandstone, Mudstone, and Shale," Canadian Geotechnical Journal, Vol. 18, 1981, pp. 502–513.
Zertsalov M.G., Nikishkin M.V., and Khokhlov, I.N. "On the calculation of bored piles under axial compressive loads in rocky soils" Soil Mechanics and Foundation Engineering, Vol.3, 2017, pp. 2–8.
Zertsalov M.G. and Nikishkin M.V. "Interaction of drilled shafts with rock masses" Procedia Engineering, Vol.111, 2015, pp. 877–881.
Zhang, L. "Drilled Shafts in Rock (Analysis and Design)," A.A. Balkema Publishers, 2004, 383 pp.
Zhang, L. Prediction of end-bearing capacity of rock-socketed shafts considering rock quality designation (RQD)" Can. Geotech. J. 47, 2010, pp. 1071–1084.

Geomechanics and Geodynamics of Rock Masses – Litvinenko (Ed.)
© 2018 Taylor & Francis Group, London, ISBN 978-1-138-61645-5

The determination of crack resistance of circular shaped fiber reinforced concrete tunnel lining by means of linear fracture mechanics

Mikhail Zertsalov
NRU MGSU, Moscow, Russia

Valery Merkin
Scientific Consultant LLC, "NIC TA", Moscow, Russia

Egor Khoteev
Head of Structural Analysis Department LLC, "Sigma Tau", Moscow, Russia

ABSTRACT: In the paper the results of research of fracture toughness of fiber reinforced concrete tunnel linings of circular shape are presented. Such lining is effectively used in rock in zones of considerable fracturing, occurrence of weak weathered rocks. On the basis of laboratory testing of linings and numerical simulation of their behavior (FEM) in the same conditions it is shown that the deformation of fiber concrete under load is linear and therefor this material can be modeled as solid, elastic, homogeneous medium.

In addition, the analysis of stress–strain state of these linings shows that their failure is caused by stable propagation of the single crack. At the same time, in accordance with the stress distribution along the contour of the lining, this single crack occurs, usually in its ceiling section, less in the floor section.

The described mechanism of failure of the circular shape lining allows for the calculations of fracture toughness (crack resistance) to use the linear fracture mechanics. Taking into consideration the symmetry of the stress distribution, the crack appears on the inner contour of the lining and spread with the growth of load under conditions of pure tension in its own plane. In this case, the criterion for the crack propagation is the critical stress intensity factor—K_{IC}, which is constant mechanical characteristics of the material.

The article presents the results of experimental determination on the samples values of K_{IC} and of the formula, allowing to calculate K_{IC} for fiber-reinforced concrete of various compositions. On the basis of the conducted experimental studies, numerical simulations of the experiments and laws of fracture mechanics developed a method of calculating the fracture toughness of fiber reinforced concrete lining of circular shape.

Keywords: fiber reinforced concrete tunnel lining, fracture mechanics, FEM

1 INTRODUCTION

Fiber-concrete is a modern and promising building material. It is a composition of concrete and fiber (steel or synthetic). Concrete has a high compressive strength. However, its tensile strength is negligible. Due to the use of fiber, an increase in the tensile strength of the material, an increase in its fracture toughness, and also its strength after cracking is achieved. This makes it possible to ensure the operational reliability and durability of tunnel lining made of fiber-reinforced concrete, using a minimum percentage of additional core reinforcement or completely abandoning it. Taking this into account, fiber-reinforced concrete can be

Figure 1.   An example of a tunnel lining made of fiber-reinforced concrete.

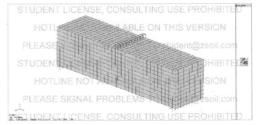

Figure 2.   Testing of samples of fiber-reinforced concrete for bending.

Figure 3.   Numerical simulation of tests of fiber-reinforced concrete samples for bending.

an effective substitute for reinforced concrete in the construction of tunnel lining in rocky grounds, including in highly fractured rocks or in fracture and weak ground areas.

Many works have been devoted to the research of fiber-reinforced concrete, as a result of which reliable methods of calculating its strength have been developed [5–8]. At the same time, methods for calculating the fracture toughness of structures made of fiber concrete, taking into account the features of its work, are currently lacking.

In laboratory tests of samples of fiber-reinforced concrete for bending (Figure 2) and their numerical modeling (Fig. 3) it was established that fibrous concrete has the properties of an elastic homogeneous isotropic material.

The tests also showed that the fracture in fibrous concrete can propagate steadily (Figure 4). The resulting crack propagates in the structure to a certain thickness and stops. Its further growth is associated only with an increase in the load.

Numerical simulation of the interaction of fiber-reinforced concrete tunnel lining with a surrounding solid, homogeneous, isotropic massif (Figure 5), performed in the ZSoil program [11], made it possible to reveal certain patterns of cracking in such structures. Particular attention should be paid to the fact that cracking in tunnel lining of circular contours operating under such conditions is characterized by crack formation, which is usually formed in a lining that spreads stably or unstable in its own plane. Initially, numerical studies were carried out for monolithic lining and for lining to be built by the method of spattering of concrete. Later, additional calculations were carried out, demonstrating that the obtained regularities are also valid for prefabricated tunnel lining of a circular outline.

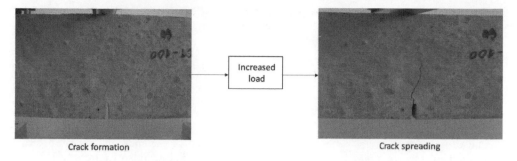

Crack formation   Crack spreading

Figure 4.   An example of a sustainable crack spread in fiber-reinforced concrete.

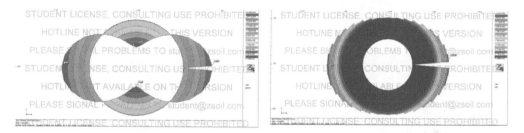

Figure 5.   Numerical simulation of stress state of tunnel lining. On the left is an example of the bending moment diagrams, on the right is an example of the longitudinal force diagram.

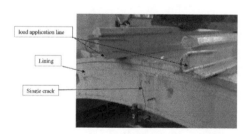

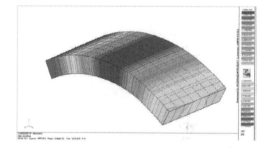

Figure 6.   Testing of blocks of tunnel lining made of fiber-reinforced concrete.

Figure 7.   Numerical simulation of tests of tunnel lining blocks made of fiber-reinforced concrete.

The performed laboratory studies of large-scale tunnel lining (Figures 6, 7) confirmed the conclusions drawn from the results of numerical simulation and testing of fiber-reinforced concrete samples. When testing large-scale fiber-reinforced concrete blocks of tunnel lining, deformations of the blocks were measured, the character of the cracking process was recorded, in particular, the width of the crack opening and the depth of its spread.

Proceeding from the above, to calculate the crack formation in tunnel lining circular outline of fiber concrete can be applied to linear fracture mechanics. The foundations of the mechanics of destruction were laid in the works of G. Kirsch, G.V. Kolosova [2, 3], K. Inglis [9]. Studies show that at the tip of a thin crack, for which the ratio of the lengths of the larger semiaxis to the smaller one tends to infinity, a singular region is formed in which the stresses increase infinitely. This means that the tensile stresses can not be used as a criterion for the propagation of a crack. J.R. Irvine proposed [10] to characterize the distribution of stresses in the region around the crack tip by the stress intensity factor (K_I, K_{II}, K_{III} – respectively for cases of normal separation, longitudinal and transverse shear), the critical values of which (K_{IC}, K_{IIC}, K_{IIIC}) allow to determine the moment of friction of the crack. Given that the tunnel lining is eccentrically-compressed, and the shear stresses in it are minimal, we can speak of the formation in it of only cracks of normal separation. Therefore, the condition of crack resistance of the tunnel lining will be:

$$K_I < K_{IC} \tag{1}$$

Once again, it must be emphasized that K_I is a function of the stress state of the lining, while K_{IC} is a constant characteristic of a particular material, in our case, a fiber concrete of a certain composition.

The value of the critical stress intensity factor K_{IC}, which depends on the composition of fiber-reinforced concrete, is determined experimentally. The results of laboratory studies on the determination of K_{IC} fiber-reinforced concrete, given in [1], give the following mathematical dependences of K_{IC} on the concentration of fiber and the class of concrete matrix for steel (2) and for polypropylene (3) fibers.

$$K_{IC\_S} = 0,04 \cdot FS + 0,08 \cdot B - 0,00072 \cdot FS \cdot B - 2,84 \tag{2}$$

$$K_{IC\_P} = -0,02 \cdot FP + 0,007 \cdot B - 0,00033 \cdot FS \cdot B + 0,83 \tag{3}$$

where FS, FP – is the content of steel and polypropylene fibers, respectively, kg/m³;
B – is the strength of concrete of the matrix for compression, MPa.

Calculation of the value of the stress intensity factor KI in the lining of a circular outline was carried out according to the procedure described in [4] for an eccentrically-compressed element, in accordance with the scheme shown in Fig. 8, according to the formula:

$$K_I = \sigma^0 \sqrt{l_0} f_1(\lambda_0) + \Delta\sigma \sqrt{l_0} f_2(\lambda_0) \tag{4}$$

where λ_0 – is the ratio of the length of the initial crack to the thickness of the structure;
σ^0 – stresses from compressive forces at the mouth of the crack;
$\Delta\sigma$ – is the difference in stresses at the mouth and at the apex of the initial crack;
l_0 – length of the initial crack, which is a set of microdefects in the lining;
$f_1(\lambda_0), f_2(\lambda_0)$ – are tabular functions given in [4].

For tunnel lining circular outlines of fiber-reinforced concrete, there are three design cases:

- Condition (1) is fulfilled and a crack is not formed (Fig. 9);
- Condition (1) does not hold. The resulting crack propagates steadily to a certain depth (Figure 10). Here, as the results of calculations show, two variants can be distinguished. In the first variant, the crack extends to 30% of the thickness of the lining. Such a crack is acceptable, since a sufficient thickness of undisturbed fiber-reinforced concrete remains. In

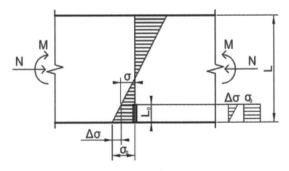

Figure 8. Scheme for calculating the stress intensity factor at the tip of a normal tear crack in an eccentrically-compressed element.

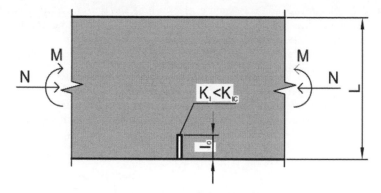

Figure 9.   The first calculation case. A crack is not formed.

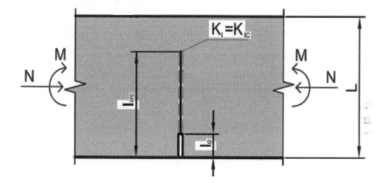

Figure 10.   The second calculation case. The crack forms and spreads steadily.

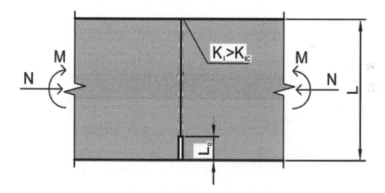

Figure 11.   The third calculation case. The crack is formed and spreads unsteadily.

the second variant, the crack passes through 90% of the thickness of the structure. Such a crack is inadmissible;

- Condition (1) does not hold. The resulting crack propagates unstably throughout the entire thickness of the lining and leads to its destruction (Figure 11). Such a crack is also inadmissible.

A joint analysis of the results of a laboratory experiment and numerical modeling allows us to develop a method for calculating the fracture toughness of tunnel linings made of fiber-reinforced concrete using the dependences of linear fracture mechanics given in [4].

The presented technique can be used also in the consideration of prefabricated tunnel lining of a circular outline. In this case, it is necessary to consider two cases:

- The joint of blocks in the ring is located in the arch of the lining. In this case, the KIC is assumed to be zero, since the joint is regarded as a crack in the lining and the amount of its opening determines the allowable stress-strain state of the lining;
- The joints are offset with respect to the arch of the lining. In this case, the calculation of crack resistance is carried out, both for monolithic lining.

Thus, a joint analysis of the results of a laboratory experiment and numerical simulation allows us to develop a method for calculating the fracture toughness of tunnel linings made of fiber-reinforced concrete using the dependences of linear fracture mechanics given in [4].

2   CONCLUSION

1. The theoretical and experimental justification of the possibility of applying linear fracture mechanics to the calculation of the fracture toughness of prefabricated and monolithic tunnel lining of circular outline from fiber-reinforced concrete;;
2. The possibility of developing a method for calculating the fracture toughness of tunnel lining made of fiber-reinforced concrete, which takes into account the characteristics of their interaction with the enclosing ground mass, is substantiated. As the basis of the method, an approach that was previously not used to solve this problem, combining the finite element method and linear fracture mechanics, is proposed. The method makes it possible to estimate the fracture toughness of the lining and to determine the depth spread of crack;
3. The main design cases of crack formation in tunnel lining of circular contours have been determined, which make it possible to draw a conclusion about the fracture toughness of the tunnel lining of circular outline from fiber-reinforced concrete. A method for estimating the admissibility of a crack in a fiber-reinforced concrete lining is shown.

REFERENCES

[1] Zertsalov MG, Khoteev EA Experimental determination of fracture toughness characteristics of fiber-reinforced concrete. // Bulletin of MGSU (VAK). 2014. №5. Pp. 91–99.
[2] Kolosov G.V. On an application of the theory of a function of a complex variable to a plane problem. Yuryev, 1909, 197 with.
[3] Kolosov G.V. Application of a complex variable to the theory of elasticity. ML, Gostekhizdat, 1935, 224 p.
[4] Orekhov V.G., Zertsalov M.G. Mechanics of destruction of engineering structures and mountain ranges. DIA Publishing House, 1999, 330 p.
[5] Rusanov V.E. Features of the calculation of prefabricated steel-fiber-concrete lining tunnels subway. // Problems of reliability and efficiency of tunnel structures. Collection of scientific papers, Issue No. 254 ed. V.E. Merkina - M.: OAO ZNIIS, 2009. - P. 44–82.
[6] Rusanov V.E. Designing tunnel structures from fiber-reinforced concrete (modern approaches). // Proceedings of the international scientific and technical conference "The main directions of the development of innovative technologies in the construction of tunnels and the development of the underground space of large megacities." - Moscow: "Timur", 2010. - P. 89–92.
[7] SP 52-104-2006. Stalefibrobetonnye constructions.
[8] ACI 544.4R-88 Design Consideration for Steel Fiber Reinforced Concrete.
[9] Inglis C. E. Stresses in a plate due to the presence of cracks and sharp corners, Trans. Institute of Naval Architects. 1913. V. 55. P. 219–241.
[10] Irwin, G. R., Analysis of Stress and Strains, Near the End of a Crack Traversing a Plate, Trans. ASME. J. Appl. Mech. 1957. V. 24. P. 361–364.
[11] Z_Soil User Manual by: Thomas Zimmermann, Andrzej Truty, Aleksander Urbanski, Stephane Commend, Krzysztof Podles.

Geomechanics and Geodynamics of Rock Masses – Litvinenko (Ed.)
© *2018 Taylor & Francis Group, London, ISBN 978-1-138-61645-5*

Kinematics and discreet modelling for ramp intersections

T. Zvarivadza & F. Sengani
School of Mining Engineering, University of the Witwatersrand, Johannesburg, South Africa

ABSTRACT: Geotechnical work was conducted in order to identify the required support length and other important parameters in ramp intersections at a deep level gold mine. The investigation included detailed scanline and joint mapping. Discreet models were constructed and eight different excavation orientations were analyzed in order to obtain a worst case scenario of mining direction. A sensitivity analysis was conducted and the effect of clamping stresses on the hanging wall was quantified. The probabilistic modeling program, JBlock, was utilized to quantify the probable maximum apex height of blocks forming in the ramp intersections. A geo-domain was constructed making use of all the data acquired and was analyzed. Eight excavations were constructed ranging in mining directions from 0 to 315 degrees. The excavations were assumed to have a dip of 8 degrees to ensure the largest blocks are created, subsequently catering for the worst case scenario. The output data created by the software was scrutinized for failed blocks, and cumulative distribution graphs created. In order to understand the effect of clamping stresses in the hanging wall, the worst case mining direction (180 degrees) was utilized and clamping stresses in the hanging wall were varied from 0 kPa to 30 kPa. A 32% reduction in the 50 percentile apex height was attained when the clamping stress increased to 5 kPa, a 43% reduction was attained when a clamping stress of 10 kPa was applied, and a 48% reduction in apex height was attained when the clamping stress was set to 30 kPa.

Keywords: Modelling, kinematics, geo-domain, ramps, probabilistic analysis, clamping stresses, JBlock, gold mining

1 INTRODUCTION

Block Theory, introduced by Goodman and Shi, 1985, presents a geometric approach to rock mechanics based on a "key block" concept. The overall stability of an excavation was assumed to depend, in the first analysis, on the orientations of the joint sets which cut the rock mass, with respect to the free surface(s). Blocks which are kinematically "free" to translate into space were termed removable. Removable blocks were then examined for stability under the applied forces. Blocks which were both removable and unstable were the "key blocks" of the excavation, and if these blocks were stabilized (e.g. with rock bolts) the entire excavation was assumed to be safe (Goodman, 1976 and 1979). The failure modes discussed in Block Theory (sliding and lifting) involve translation only. Further studies consider another possible failure mode: rotation. It is shown that a removable block, safe against sliding, might fail in rotation. Such a block would be deemed safe in the conventional block theory analysis. This paper is intended to supplement block theory and examines: (1) the geometric conditions necessary for rotation to be possible (kinematics); and (2) the rotational stability of blocks which satisfy the kinematic conditions for rotation (equilibrium). Rotations of rock wedges have been discussed by others, including [3–6] but not previously in the context of block theory.

2 DATA COLLECTION

The underground Scanline mapping was conducted along the sidewall of the ramp to quantify input parameters for use in the J-Block software package. Scanline mapping involves a process of measuring joint orientations, frequency, persistence, water and joint contact surface conditions present in the host rock. An accepted scanline has to at least have three joints traversing the line so as to be able to calculate joint spacing. Nine scanlines amounting to approximately 180 m in total length were mapped.

The following parameters were logged during the underground mapping process to be used as input parameters for the discreet modelling: Joint dip and direction, Joint persistence, the visibility of ends, Joint roughness coefficients, degree of moisture on joints, hardness of joint walls, degree of alteration of joints, separation of joints, joint infilling thickness and healing of joints and infilling material.

The DIPS (Stereonet analysis software) was used to analyses the joint data gathered. Schmidt rebound numbers were also obtained for the rock comprising the excavation sidewalls. These numbers were used to approximate the Joint wall Compressive Strength (JCS) for the different joint wall conditions observed. The JCS values obtained were used in the Barton-Bandis (1990) model in order to calculate the respective friction angles required by JBLOCK.

3 JOINT ORIENTATIONS

Based on the underground mapping, four joint sets were identified along the mining ramp. One joint set dipping approximately south represents the various layering of the strata. Two additional joint sets dip approximately east of south-east, one almost vertical, and the fourth set of joints were vertical, striking in a northerly direction. These joint sets were then represented using Dips software package (see Figure 1). Furthermore, analysis using in-house software to determine mean, min and max dip and dip direction has shown that relatively small standard deviations exist for the various joint sets, indicating a very good correlation between the joint sets at the various scanline locations. The large dip direction's standard deviation associated with joint set 1 (see Table 1) may be attributed to the very flat nature of its dip direction and was to be expected.

4 JOINT PERSISTENCE

The joint persistence is highly dependent on the number of ends that could be observed during the logging. For this reason, a set of general rules was constructed out of the experi-

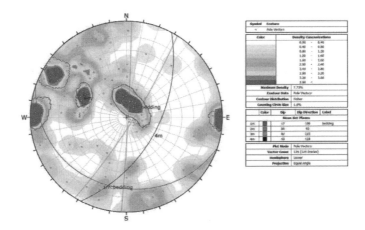

Figure 1.   Stereonet showing joint sets orientation.

Table 1. Joint orientations.

	Dip				Dip direction			
	Mean	Min	Max	Std.Dev	Mean	Min	Max	Std.Dev
Joint Set 1	20.96	9	34	6.2	187	131	278	40
Joint Set 2	86.4	83	89	2.3	75.12	71	82	4.45
Joint Set 3	81.9	76	89	4.8	104	97	113	5.7
Joint Set 4	52.9	48	56	2.6	100.3	93	111	5.7

ence. The rules include: When no ends are visible for a specific joint, the persistence of the joint for modeling purposes is obtained by multiplying the excavation width by a factor of two. When only one end is visible, the persistence for modeling purposes is obtained by multiplying the excavation width by a factor of 1.5. Based on that, joint persistence results have shown that majority of the joint sets have persistence which are relatively short, ranging from 0 m to 5 m.

5 JOINT ROUGHNESS COEFFICIENT

The JRC value was estimated from visual inspection of the various joints during the underground mapping phase and correlated to a chart proposed by Barton et al. (1977). The visual correlation between the chart and actual joints results in JRC values ranging from 0 to smooth, flat joints to 20 for stepped, rough joints.

6 JOINT SPACING

The spacing of joints has a profound impact on the size of the various key blocks formed in the ramp excavation's hanging wall. As a result of scanline bias, the joint spacing recorded during the in-situ mapping process required some correction to derive the actual/corrected joint spacing. This scanline bias was as a result of the orientation of the scanline relative to the various joint orientations. Therefore, the joint spacing representing the various joint sets in Table 2 were deemed to be very conservative (closer than the actual spacing), but adequate for this study.

7 JOINT WALL COMPRESSIVE STRENGTH

The joint wall compressive strength is required in order to moderate the base friction angle (assumed to be 32°) to a residual friction angle for use in the Barton and Bandis (1990) shear strength model. This model is then manipulated to obtain accurate friction angles per joint set. Figure 2 Depicts how the Schmidt Rebound number is converted to UCS. From Figure 2, it is evident that a JCS value of 175 MPa was attained. The average dispersion of strength can be assumed to be in the region of about 75 MPa. This dispersion value was taken as the standard deviation of JCS and incorporated in the friction angle calculations.

8 JOINT FRICTION ANGLE

Considering that the joint friction angle plays a major role in the stability of various blocks formed, accurate friction angles need to be used in the probabilistic analyses (Hencher and Richards, 1982). In order to obtain the various joint sets' friction angles, the Barton and Bandis (1990) shear strength formula was utilized, making use of the point estimate method.

Table 2. Joint sets spacing.

	Joint spacing		
	Mean	Min	Max
Joint Set 1	0.45	0.02	1.83
Joint Set 2	0.58	0.46	0.72
Joint Set 3	1.28	0.01	6.46
Joint Set 4	0.61	0.11	1.93
Random	1.1	0.02	7.12

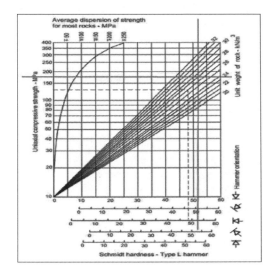

Figure 2. Determination of JCS using the Schmidt rebound number chart.

The results of the study have shown mean values of joint friction angle to range from 32.8° to 34.8°.

9   CLAMPING STRESS SENSITIVITY ANALYSIS

From the various collected data, it was evident that the 180° mining direction requires the greatest tendon length as well as tendon load bearing capability requirements. For this reason, and to limit modeling runtime, this mining direction was utilized in order to assess the effect of clamping stresses that might be present in the hanging wall. It was evident that the introduction of clamping stresses in the hanging wall had a significant impact on the height distribution of the formed blocks. A 95 percentile height where no clamping stresses were present equates to 1.03 m. With the introduction of a 30 kPa clamping stress, this height reduces to 0.83 m. This then shows that, with the introduction of clamping stresses, the larger blocks were reduced and subsequently the apex height. The assumption that no clamping stress was present in the hanging wall may thus be deemed to be the worst case scenario (see Figure 3).

10   DISCUSSION AND CONCLUSIONS

Based on the data collected, it was noted that the maximum 95% fall out height was attained when mining in a southerly or south-westerly direction. The 95% fall out height was approxi-

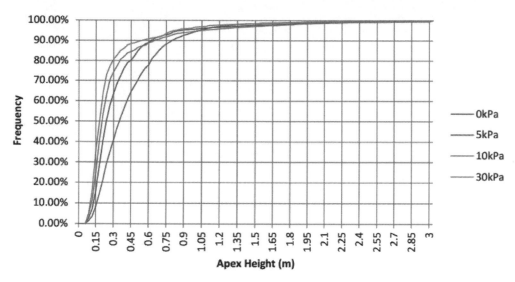

Figure 3.   Clamping sensitivity on the 180-degree mining direction.

Table 3.   Summary of the results.

Confidence level	Fall out height (m)	Fall out volume (m³)	Block volume (m³)	Density (kg/m³)	Block mass (kg)	Support resistance (kN/1.8 m²)	Support resistance (kN/m²)
95%	1.05	0.22	0.14	2750.00	396.77	3.89	2.16
99%	1.8	2.5	0.49	2750.00	1358.44	13.33	7.40
99.9%	3.5	14.69	1.51	2750.00	4148.05	40.69	22.61
100%	14.5	91.71	4.73	2750.00	13005.58	127.58	70.88

mately 1.05 m (see Table 3). Furthermore, analysis has indicated that the 99% fall out height for the same mining directions was approximately 1.8 m and the 99.9% fall out height for the data set was 3.5 m. The maximum fall out height recorded, taking cognizance of the fact that some 180,000 blocks were simulated was 14.5 m. However, it should be noted that the fall out the height of 14.5 m may be deemed as an outlier, based on the frequency plot. A more realistic value to consider was that of the 99 percentile (1.8 m). This parameter is needed to be considered in support design when analyzing the support load-bearing requirements. The maximum 95% volume for the various mining directions was found to be 0.22 m³. The maximum 99% volume for the various directions was 2.5 m³. The absolute maximum volume of the largest fall recorded was 91.71 m³. However, it should be noted that the fall out volume of 91.71 m³ may be deemed as an outlier. Based on the frequency plot, a more realistic value to consider was that of the 95 percentile, 0.22 m³.

Another important parameter which was found to assist in the determination of the spacing of support units was the block face area. For this parameter, a smaller area resulted in a denser support pattern requirement. However, smaller face areas were generally associated with smaller volume blocks. For this purpose, and as a starting point, the mine's standard 1.5 m × 1.2 m spacing representing an area of 1.8 m² was considered. A cumulative distribution of block volume for face areas between 0 m² and 1.8 m² was created in order to understand the size of probable blocks which might form between the support units.

REFERENCES

Goodman R.E. and Shi G.-H. (1985). Block Theory and Its Application to Rock Engineering. Prentice-Hall, Englewood Cliffs, New Jersey.

Goodman R.E. (1989). Introduction to Rock Mechanics, Second Ed. Wiley, New York.

Goodman R.E. (1976). Alethods of Geological Engineering in Discontinuous Rocks. West, St Paul, Minn.

Barton, N.R. and Bandis, S.C. 1990. Review of predictive capabilities of JRC-JCS model in engineering practice. In *Rock joints, proc. int. symp. on rock joints,* Loen, Norway, (eds N. Barton and O. Stephansson), 603–610. Rotterdam: Balkema.

Barton, N.R. and Choubey, V. 1977. The shear strength of rock joints in theory and practice. *Rock Mech.* 10(1–2), 1–54.

Hencher, S.R. & Richards, L.R. 1982. The basic frictional resistance of sheeting joints in Hong Kong granite *Hong Kong Engineer*, Feb., 21–25.

Author index

Estébanez, E. 617
Evseev, A. 985

Farinetti, A. 1287
Fedoseev, A.K. 979
Fedotova, I. 653, 1451
Fedotova, I.V. 1549
Feng, W. 791, 1173
Feoktistov, A.Ju. 1071
Fereidooni, D. 625
Fereshtenejad, S. 255
Fernández, C.C.G. 325
Fernández, M.I.Á. 325
Ferrero, A.M. 785, 1543
Ferrero, A.M. 1287
Fiorucci, M. 263
Florkowska, L. 499
Fomenko, I.K. 797, 1165
Franović, I. 539
Frid, V. 505, 513
Fujii, H. 1369
Fujii, Y. 479
Fukuda, D. 479

Gabova, A.V. 1351, 1357
Gabrieli, F. 877
Galperin, A.M. 1149
Garkov, I. 967
Garrido, C. 1259
Gautam, P.K. 269
Gawałkiewicz, R. 499
Gaziev, E.G. 31
Georgakiev, I. 967
Gessica, U. 1543
Ghamgosar, M. 991
Ghasempour, N. 999
Ghassemi, A. 1419
Gholami, M.A. 1209
Ghosh, C.N. 1005
Giacomini, A. 639
Giacomini, A. 917
Giot, R. 193
Giulio, A.D. 923
Gladyr, A.V. 1501
Glibota, A. 1585
Gojković, N. 1229
Gómez, C.L. 389
Gonçalves da Silva, B. 17
González-Gallego, J. 1569
Gorobtsov, D.N. 797,
  1165
Gorokhova, E.A. 557
Gottsbacher, L. 227
Gou, Y. 833, 1173
Gray, I. 41

Griffiths, V. 639
Grouset, C. 1101
Gubaidullin, V.M. 587,
  659
Gubaydullina, R. 1639
Guiheneuf, S. 275
Günther, C. 1179
Guo, J. 929
Guryev, D.V. 199
Gutierrez, M. 281, 1191
Guzev, M.A. 727

Hagan, P.C. 1249
Hahn, F. 287
Hamada, Y. 307
Han, J. 929
Han, J. 1507
Harrison, J.P. 599, 1543
Hartlieb, P. 1017
Hartzenberg, A.G. 293
Hasov, A.N. 587
Hassani, H. 1413
Hassanzadegan, A. 1363
He, M. 63
Hedtmann, N. 1185
Hirose, T. 307, 563
Hosoda, K. 1317
Hotchenkov, Eu.I. 593
Hou, M.Z. 833, 1173
Hou, Z. 791
Houshmand, N. 1197
Howald, E.P. 1529
Huang, S. 1217

Iannucci, R. 485
Ikegami, S. 479
Ilin, M.M. 557
Ilinov, M.D. 669
Ilyasov, B. 805
Invernici, M. 599
Ishida, T. 1369
Iusupov, G.A. 1071
Iwano, K. 1341
Iwasaki, H. 479

Jabs, T. 287
Jacobsson, L. 633
Jalili Kashtiban, Y. 1025
Janiszewski, M. 821
Jeffery, M. 639
Jeon, S. 451, 465
Jha, M.K. 395
Jimenez, R. 1555
Jovanovski, M. 519
Justo, J. 1203

Kagan, M.M. 551
Kajzar, V. 711
Kalamaras, G. 747
Kalender, A. 345
Kalinin, E.V. 1521
Kallam, N.R. 889
Kamali, A. 1197, 1209
Kamiya, N. 563
Kang, K. 1217
Kantia, P. 647, 1223
Karakus, M. 525, 581
Karasev, M.A. 739, 1645
Karev, V.I. 1375, 1381
Karpov, I.A. 1357
Kashnikov, Yu. 533
Kashnikov, Yu.A. 1425
Kasparian, E. 1451
Katayama, S. 457
Kauther, R. 1179
Kazakov, A. 1059
Kelfoun, K. 785
Kharisov, T.F. 1597
Kharisova, O.D. 1597
Khatibi, S. 1387, 1395
Khokhlov, I.N. 937
Khoteev, E. 943
Khvostantcev, D. 533
Kikumoto, M. 1317
Kishimoto, Y. 1369
Kiuru, R. 647, 1223
Kızıltaş, Z. 353
Klimov, D.M. 1375, 1381
Klykov, P.I. 1407
Kodama, J.-i. 479
Kolikov, K.S. 593
Kong, L. 1401
Konicek, P. 711
Konovalov, O.L. 781
Konstantinov, K.N. 551
Kornilkov, S.V. 131
Korolev, V.M. 189
Korshunov, V.A. 299, 1053
Kossovich, E. 1603
Kostić, S. 539, 1229
Kostina, A. 1273
Kotiukov, P.V. 241
Kotlov, S. 827
Kovács, L. 439, 1235
Kovalenko, M. 1011
Kovalenko, Y.F. 1375, 1381
Kovaleva, Y. 1387, 1395
Kozlova, E.V. 1357
Kozyrev, A.A. 139, 1031,
  1457
Krasnov, S.A. 575

Vagnon, F. 785, 1123, 1543
Valero, J.D.L. 389
Vallejos, J. 1259
Vasilieva, A.D. 1645
Vasović, N. 539
Vavilova, V.K. 1279
Velkov, T. 967
Verbilo, P.E. 1659
Vergara, M.R. 401
Verma, A.K. 269, 395
Verrucci, L. 923
Vilchinsky, V. 695
Vilchinsky, V.B. 895
Vishnoi, R.K. 909
Vlastelica, G. 1585
Vostrikov, V.I. 705, 1477

Waclawik, P. 711
Wagner, H. 1047
Wang, K. 1515
Wang, S. 465
Wang, S. 859
Weekes, G. 761
Welideniya, S. 761

Wendeler, C. 917, 1333
Were, P. 791
Wicaksana, Y. 451
Wilfing, L. 567
Wu, S. 1115

Xhao, X. 41

Yamagami, M. 457
Yamamoto, Y. 563
Yamamoto, Y. 563
Yang, Z. 929
Yogo, J.-C. 1101
Yokota, Y. 1341
Yu, B. 1011
Yu, G. 1011
Yu, H. 1469
Yuan, S. 1011
Yurchenko, G. 1311

Zakharov, V.N. 167
Zamahaev, A.M. 1279
Zare, M. 1555
Zarei, H. 1197

Zemtsovskii, A.V. 1031
Zeng, X. 1011
Zerkal, O.V. 1217, 1521
Zertsalov, M. 943
Zertsalov, M.G. 937
Zhang, H. 1507
Zhang, N. 465
Zhao, Y. 525
Zhao, Z. 1341
Zhelnin, M. 1273
Zhou, L. 833
Zhou, X. 1507
Zhu, Y. 929
Zhuravkov, M.A. 781
Zhuravleva, O.G. 1457
Zinchenko, A. 1311
Znamenskiy, E.A. 659
Zoback, M.D. 313
Zubkov, V.V. 1075, 1081
Zubkova, I.A. 1075, 1081
Zuev, B.Yu. 423
Zuzin, R.S. 1279
Zvarivadza, T. 411, 417,
    471, 949, 1085